Finite Elements for Wave Electromagnetics

IEEE Press
445 Hoes Lane, PO Box 1331
Piscataway, NJ 08855-1331

IEEE Antennas and Propagation Society, *Sponsor*

AP-S Liaison to IEEE Press
Robert J. Mailloux

Technical Reviewers

Daniel T. McGrath
USAF Phillips Laboratory

Andrew F. Peterson
Georgia Institute of Technology

John L. Volakis
The University of Michigan

Finite Elements for Wave Electromagnetics

Methods and Techniques

Edited by
Peter P. Silvester
McGill University
Giuseppe Pelosi
University of Florence

A Selected Reprint Volume
IEEE Antennas and Propagation Society, *Sponsor*

The Institute of Electrical and Electronics Engineers, Inc., New York

This book may be purchased at a discount from the publisher when ordered in bulk quantities. For more information contact:

IEEE Press Marketing
Attn: Special Sales
P.O. Box 1331
445 Hoes Lane
Piscataway, NJ 08855-1331
Fax: (908) 981-8062

Printed in the United States of America

10 9 8 7 6 5 4 3 2 1

ISBN 0–7803–1040–3

IEEE Order Number: PC0390-5

Library of Congress Cataloging-in-Publication Data

Finite elements for wave electromagnetics : methods and techniques /
edited by Peter P. Silvester, Giuseppe Pelosi.
p. cm.
"A selected reprint volume."
"IEEE Antennas and Propagation Society, sponsor."
Includes bibliographical references and index.
ISBN 0–7803–1040–3
1. Electromagnetic waves—Transmission. 2. Finite element method.
I. Silvester, Peter P. II. Pelosi, Giuseppe. III. IEEE Antennas and Propagation Society."
QC665.T7F56 1994
621.382'4—dc20 94-3161
CIP

Contents

Preface

The finite element method has recently celebrated its fiftieth birthday. Electrical engineers have made use of this group of mathematical techniques for just half that long; in turn, most members of the community interested in antenna and scattering problems "discovered" finite elements only in the latter half of that. Despite the relative youth of finite element analysis as applied to this problem class, its published literature has enjoyed vigorous growth. It has now reached the limits of what a single individual can conveniently read, so the time seems ripe for an anthology of the key papers to be collected into a single volume.

Just which of the many hundreds of published articles are truly central is of course a matter of the anthologist's judgement, a matter of striking a balance between the conflicting demands of historical accuracy, lucidity in presentation and completeness. The authors have chosen to remain with the English-language literature, and within that restriction have laid their main emphasis on the major professional journals.

The authors are deeply indebted to many colleagues for their advice in both technical and editorial matters, and for their forbearance and patience with this work. Particular thanks are due to Dr. R. Coccioli and Dr. D. Omeragić, two ever-helpful and indefatigable collaborators.

P. P. Silvester and G. Pelosi
Westmount and Florence

Introduction

The reader of a book bearing the title *Finite Elements Wave Electromagnetics* may well expect it to begin by defining its terms. What do its authors consider to be a finite element method? What sorts of wave propagation problems do they propose to discuss? Indeed, why should wave problems be treated differently from any other physical problem?

This anthology does indeed begin by defining its terms, then seeks to summarize the state of the art, and occasionally ventures to guess at its future course. Like most anthologies, it reviews and selects existing methods, it does not seek to present new ones — despite the substantial, perhaps unusually large, amount of evaluative comment and description added by the editors. It is intended mainly for engineers and software developers well acquainted with electromagnetics, but not necessarily experts in finite elements. Its approach is methodological rather than applications-oriented. Consequently, many specific technical areas have been treated only lightly, even if their practical importance might seem to call for more attention. Occasionally, this may have happened because the problem appears excessively specialized, but more often because the finite element methods currently used in that field are not distinctive, so there is not very much literature to review.

Finite Elements

For the purposes of this book, the term *finite element method* will mean a numerical approximation method for solving field problems that possesses the following characteristics:

a) The geometric domain of the problem is subdivided into a set of subdomains or *elements*, which cover the entire domain but do not overlap.

b) The quantity to be approximated has a uniquely defined value everywhere within each element.

c) The approximate solution is correct (i.e., has minimal error) in a global sense that takes into account the solution values at all points.

d) Only some continuity requirements of the physical problem are strictly enforced at element interfaces; the remainder are implicit in the element formulation.

A few clarifying notes may be of value, particularly with regard to what this three-part definition does *not* include. First, it does not say that the geometric domain of the problem is of finite size, nor that it is subdivided into geometric subregions of finite size. The term "finite element" is viewed as referring not to finite geometric size but to finiteness of the fundamental electromagnetic quantities involved: energy or power, reaction or content, Lagrangians or Hamiltonians of the field. Second, the definition does not specify how the original problem is to be formulated mathematically: as a Fredholm integral equation, as a boundary value problem of differential equations, or as any mixture of integrodifferential operator equations. Finite elements may be quite legitimately defined for any of these. Third, there is no requirement that the approximation must be locally polynomial, nor that it must be interpolative; although most finite element methods embody these properties, some others do not. What is considered essential is the existence of a uniquely defined approximation everywhere in the problem space. This requirement sets the finite element method apart from finite difference methods and collocation methods, which produce approximate solution values at a finite set of points but say nothing about the behavior of the solution elsewhere.

It is sometimes said that "the finite element method is a special case of the moment method." The latter term has unfortunately lost its descriptive value through overuse. It has become fashionable to call any projective or collocative approximation "the moment method," even though many of these have at best a tenuous mathematical relationship. Asking whether the finite element method is a moment method no longer helps to define or classify numerical methods, so this book will neither put the question nor attempt to give an answer.

Electromagnetic Waves

This anthology addresses the sorts of electromagnetic wave problems likely to interest the IEEE Antennas and Propagation Society as well as (probably in descending order) such other IEEE societies as Magnetics, Microwave Theory and Techniques, or Electromagnetic Compatibility: propagating fields adequately described by the Maxwell equations in comparatively simple media (usually linear though inhomogeneous) that extend very far in at least one direction. Such waves are characterized by three interesting peculiarities: great propagation distances, wide wavefronts, and the essentially vectorial nature of the **E** and **H** fields.

Modeling Problems

Most electromagnetic wave propagation problems encompass very large geometric domains. Problem domains often extend over many wavelengths — so many wavelengths that many electromagnetic devices are best described with reference to waves that propagate to infinitely distant points, or arrive from infinitely far away. For example, such basic concepts as the radar cross-section of an object or the characteristic impedance of a transmission line are defined in terms of infinite propagation distance. The analyst is thus faced with the "wavelength problem": the need to construct analytic models able to represent extremely great propagation distances or even to approach the asymptotic case of a semi-infinite space.

Many electromagnetic waves possess wavefronts of great breadth. For instance, the fields that surround a classical two-wire transmission line or an optic fiber extend infinitely far laterally. Even from the point of view of practical approximation, they are significant over a considerably greater distance than the dimensions of the wires or fibers themselves. These large distances give rise to a mathematical modeling problem different from the wavelength problem, one sometimes called the "infinite region" problem. Here the broad geometric extent to be modeled is the breadth of the wavefront rather than the length of the propagation path; the problem is infinite in directions orthogonal to the direction of propagation rather than along it. The difficulties in modeling great distances along the propagation direction are not quite the same as those encountered in the crosswise direction, and they yield to different computational methods.

The third major problem encountered in analyzing wave problems arises from the vector nature of electromagnetic fields. Scalar representations are sometimes possible, but they often fail where waves propagate in mixed media. This difficulty is traditionally known as the "spurious mode problem," because a typical symptom in resonator or waveguide problems is the appearance of several wrong solutions in addition to the correct ones. Spurious modes are common, for most really interesting wave problems arise when propagating waves interact with material objects — antennas, scatterers, or guiding structures. Wherever the physical objects are made up of multiple dielectric materials, for example an optic fiber or the polar ice-cap, simultaneous solution for the electric and magnetic fields is required. For this reason, the spurious mode problem is sometimes called the "coupled field problem," and more rarely, the "inhomogeneous material problem."

Modeling Approaches

Each of the three key modeling problems of electromagnetic wave propagation can yield to a variety of approaches.

The wavelength problem and the infinite region problem are geometrically related, even though they differ electromagnetically. In virtually all cases that have appeared in the literature to date, the electromagnetic field problem in question consists of a finite region of interest, surrounded by a large space which cannot be neglected but where full field detail is not required. Figure 1 indicates this situation schematically. Here Ω_i is a finite region of major interest, typically containing the device — antenna, scatterer, waveguide junction, or obstacle — to be analyzed. It is surrounded by a region Ω_e, which is homogeneous and source-free, extending infinitely far. The dividing boundary $\partial\Omega$ that separates Ω_i from Ω_e is chosen at will; it need not coincide with any material boundary in the original problem. The general line of attack is to model the interior region Ω_i as a collection of classical finite elements, and the exterior region Ω_e as one or more "exterior elements" or "infinite elements."

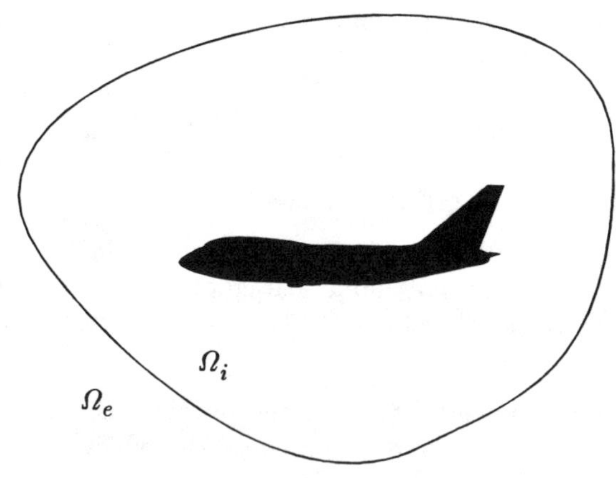

Fig. 1. The interior-exterior modeling technique: the object of interest is encased in a fictitious boundary.

The interior region is usually modeled with finite elements based on a differential equation model of the electromagnetic fields. The exterior elements may be derived either from an integral equation representation or a differential equation representation. In either case, field behavior at infinite distances must be taken into account correctly. Here the wavelength problem differs from the infinite wavefront problem. In the direction of propagation, boundary conditions at infinity must be so arranged that energy transfer is possible, e.g., by imposing a Sommerfeld radiation condition asymptotically. On the other hand, in the lateral directions (i.e., across the wavefront) fields are evanescent at a great distance, so simpler terminations can serve, easing the modeling task.

A Taxonomy of Wave Problems

Almost all problems to be considered in this book relate to traveling waves. For present purposes, a traveling wave is a field that depends on time t only through a quantity of the form $w \pm v_p t$, where w is a spatial coordinate unbounded in one or both directions, and v_p is a propagation velocity. Thus, a major feature that distinguishes wave problems from other field problems is the existence of a privileged coordinate w, the direction of propagation. The geometric region of the problem must be open (semi-infinite) at least in this direction — possibly also in other directions — even though the field may be evanescent (attenuated) in all directions. Given that all wave problems possess such a privileged direction, the following subdivision of problems suggests itself:

1. *Rectilinear propagation.* Propagation takes place in a single coordinate, say z, with wavefronts preserving their size and shape. A familiar example is the propagation of a TEM_z wave along an open-wire transmission line. Most often, rectilinear propagation is guided by a structure of some sort. Field problems of this class divide naturally into two subclasses:

Transverse problems require solving a field problem in the plane transverse to the propagation direction. Typical examples include determining the modal structure of classical closed waveguides, optic fibers, and dielectric or topographic guides. Many such problems have been solved by finite element methods over the past quarter-century; in fact, finite elements are probably the most common way of tackling them.

Longitudinal problems require solving a field problem in a longitudinal plane. Examples include finding the mode structure of surface waves, or computing the scattering parameters of guide discontinuities or fiber junctions. The literature in this area is at present very thin indeed; this may represent a productive growth area for the future.

2. *Radial propagation.* Propagation takes place in a radial direction r. Wavefronts change amplitude and energy density in the direction of propagation as the wave progresses. A common example is the radiated wave emanating from an antenna. Such waves often propagate in free space, without any local guiding structure. Two problem subclasses again arise:

Cylindrical waves, in which the propagation is outward from a rectilinear axis. The classical example is the scattering of a normally incident plane wave from an infinitely long cylindrical obstacle.

Spherical waves. Propagation takes place in a spherically radial direction r. Finding the polar pattern of an antenna, or the radar cross-section calculation of a finite obstacle illuminated by an incident plane wave, are classical examples.

Organization of This Book

The main material in this book follows the above problem hierarchy, with occasional minor deviations where common solution methods make this appropriate, or where the uneven distribution of available literature suggests otherwise. It is organized into four chapters, plus an extensive annotated bibliography. This first (and shortest) chapter deals with broad issues of methodology and technique. The second concentrates on rectilinearly propagating waves, in which the relative energy distribution remains substantially unaltered as the wave propagates. The third and fourth chapters, which form by far the largest part of the book, deal with radially propagating waves: waves in which the wavefront area expands or contracts as the wave propagates, so that the energy propagated diminishes in intensity and may also undergo substantial changes of relative temporal and spatial distribution. In a rough and loose manner, the second chapter might be described as dealing with "guided wave problems," the third and fourth as dealing with "radiated wave problems," though these terms are obviously too restrictive to be accurate. The final chapter contains an annotated bibliography of the finite element art as applied to wave problems.

The organization of material mainly follows methodological lines, taking into account the mathematical properties of the problems at hand. In other words, the questions addressed include "how are finite elements used for analyzing anisotropic media?" but not "how do I calculate the radar cross-section of this object?" Fortunately, methods often correlate well with application areas. Organization by methodology rather than applications is preferable in a book such as this, simply because its central topic is a computational technique, not a technological area. Where a choice exists, selections have generally been chosen from the antenna engineering literature, with survey or overview papers to illustrate the more peripheral application areas and to provide bibliographic depth.

Bibliographic References and Reprints

The major portion of this book, indeed its *raison d'être*, is a collection of papers selected mainly from recent periodical literature. Its object is to provide an overview of the present state of the art, as well as to sketch out its historical and logical development, by selected readings. These are supported by an extensive annotated bibliography and by the introductory text that opens each chapter. The introductory overviews are not intended to form a complete book by themselves — if they did, there would be little point in the reprints — but rather to relate the various selected reprints to each other and to fill in the gaps between where such may occur.

There are many bibliographic references in a work such as this, so the practice often followed in scientific articles of numbering all references sequentially becomes too cumbersome to use. In both oral and and informal written communication, most engineers speak of "the paper Bloggs published in 1989," not of "Reference [217]" — so when the quantity of references exceeds a modest handful, it makes good sense to refer to them by name and date rather than to use a code number with no intrinsic meaning or mnemonic value. For this reason, references are given here in a slightly modified version of the Royal Society bibliographic style, rather than in the style commonly employed for brief scientific papers. Most in-text references simply mention the authors' names in the natural fashion, so they might, for example, refer to the work of Bloggs and Doe (1989). Where explicit mention of names is not appropriate in the context, references are placed in parentheses (Bloggs and Doe 1989); either way, they always include the authors' names and the year of publication. Where the same authors have published several papers in the same year, italic letters are suffixed to the year of publication, so that reference may be made to Bloggs and Doe (1990*a*). Recourse to this quite arbitrary scheme is fortunately not often required, and the letters rarely need go beyond *b* or *c*.

The bibliographic references for each chapter are grouped at the end of the chapter, sorted alphabetically by first authors' names, so references are easily found. Reprints included in this volume are identified by the use of square brackets instead of parentheses. For example, the paper by Bloggs [1991] would be found among the reprints as well as in the bibliography, but another cited in parentheses (Bloggs 1988) appears only in the bibliography.

Finite Elements and Waves

The Beginnings

The finite element method in its presently accepted forms can be credited to no lesser a person than Courant. When he prepared the published version of his 1942 address to the American Mathematical Society, he added a two-page Appendix to show by example how the variational methods first described by Lord Rayleigh could be put to wider use in potential theory (Courant 1943). Using piecewise-linear approximants on a set of triangles which he called "elements," he dashed off a few two-dimensional examples — and the finite element method was born. It remained dormant for half a generation, probably waiting for computers to be invented. Courant's approach, amplified by the mathematical ideas of Synge (1957), next appeared in the work of Duffin (1959).

Finite element activity in electrical engineering began in earnest another decade later, about 1968–1969. A paper giving the finite element formulation of the classical hollow waveguide problem was presented at the 1968 URSI symposium and subsequently published in *Alta Frequenza* [Silvester 1969*a*]. A concurrent but independent paper by Arlett, Bahrani, and Zienkiewicz (1968) addressed waveguides and cavities, but unfortunately based the development on an incorrect formulation of the electromagnetic field problem. There followed a rapid succession of papers on magnetostatics, dielectric loaded waveguides (Csendes and Silvester 1971), and other well-known boundary value problems of electromagnetics. The method was quickly applied to integral operators as well, both in microwave devices (Benedek and Silvester 1972) and wire antenna problems (Silvester and Chan 1972).

The approach to finite elements throughout the early period was variational. This made it relatively hard to follow for classically trained engineers, few of whom had any knowledge of variational methods. In spite of this difficulty, finite element methods spread quickly in the 1980s and assumed a dominant role in several technical areas. One way of assessing the growth of finite element applications is to examine the INSPEC bibliographic data base over the 1968–1993 period. In 1968, finite element papers with electrical engineering content amounted to a small handful. By 1993 the total had reached about 6000, with some 600 articles being added each year. Figure 1 shows the growth of the finite element literature of electrical engineering, as recorded in the INSPEC data base. As is evident from this graph, the rate of publication (i.e., the annual output of articles) has doubled about every four years since 1969, though it has slowed somewhat since the turn of the last decade.

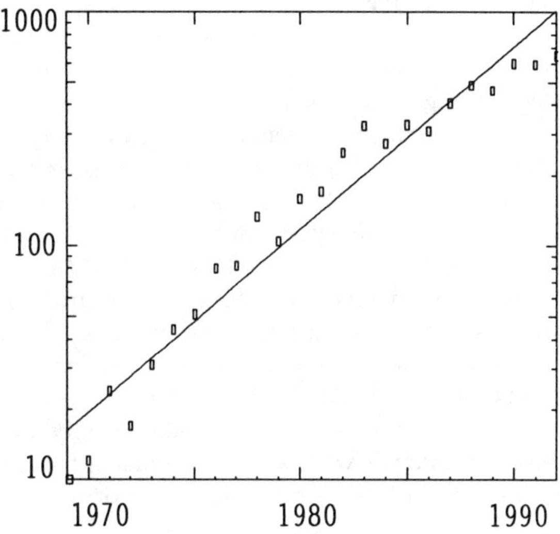

Fig. 1. Annual production of finite element papers in electrical engineering, 1969–1992.

Of the professional journals, the premier publisher of finite element research and applications is undoubtedly the *IEEE Transactions on Magnetics*, whose subject coverage far transcends the traditional magnetic devices and materials implied by its title. In 1990–1992, *IEEE Transactions on Antennas and Propagation* averaged about one article per issue on finite element methods. Other major sources for this material include *IEE Proceedings–H*, as well as *Electromagnetics* and numerous other journals.

Textbook material in this area is still a bit thin. The first book in the area is now passing into its third edition (Silvester and Ferrari 1983, 1990), growing thicker as the field matures. A second one has recently appeared (Jin 1993); it is particularly welcome because it stresses high-frequency applications. There are quite a few general texts on computational electromagnetics, and several include a chapter or two on the finite element method; but these present the method mainly in the context of static or quasi-static potential fields (Sabonnadière and Coulomb 1987; Steele 1987; Hoole

1989; Binns, Lawrenson, and Trowbridge 1992). The monograph literature of finite element mathematics, on the other hand, is now extensive. Unfortunately, it often employs mathematical techniques and notations difficult for electrical engineers, and much of it is divorced from applications. Of the books now available, Oden and Reddy (1976) is highly readable (like the curate's egg) in parts — fortunately, in the important parts. Strang and Fix (1973) describe the basic theory well, and can be recommended for a first reading; but the book is neither new enough, nor extensive enough, to suffice for the serious reader.

The creation of finite element methods for wave problems, like for most other problems, involves at least two distinct areas of research. One concerns finite element technology proper: families of suitable approximating functions, modeling of regions by aggregates of finite elements, and exploitation of any peculiar properties the matrices may possess. The other research area lies in electromagnetics: finding variationally stationary quantities that describe fields, choosing how the fields are to be described (field vectors *vs.* potentials), determining what the natural boundary conditions are, and eliminating superfluous variables. Despite appearances, the two areas are in large measure interdependent. This book emphasizes the areas of interdependence; material that touches only upon one or the other is dealt with lightly. Details of these may be found in textbooks on electromagnetics or finite element theory.

Basics: Elements for Scalar Fields

The essence of the finite element method may perhaps best be stated by example, as indeed it was in the early papers [Silvester 1969*a*]. The fields in a hollow waveguide are governed by the Helmholtz equation

$$(\nabla^2 + k^2)\Pi = 0 \tag{1}$$

where Π is the magnitude of a longitudinally directed Hertz vector, proportional to E_z or H_z in the guide for the TM_z and TE_z modes, respectively. The solutions of this equation are known to render stationary the functional

$$\mathcal{F}(\varphi) = \int_\Omega (\nabla\varphi \cdot \nabla\varphi + k^2\varphi^2)d\Omega \tag{2}$$

so that $\mathcal{F}(\varphi)$ reaches a maximum, minimum, or saddle point for $\varphi = \Pi$. To perform the extremization approximately, the guide cross-section Ω is subdivided into finite patches or *elements* and a piecewise polynomial approximation for φ is written on the assemblage of all elements,

$$\varphi(x,y) \simeq \sum_{i=1}^{N} \varphi_i \alpha_i(x,y) \tag{3}$$

where the φ_i are numeric coefficients, as yet unknown, and the $\alpha_i(x,y)$ are N piecewise polynomial functions. Substitution of (3) into $\mathcal{F}(\varphi)$ then produces an approximate expression $F(\varphi)$, which is no longer a functional but an ordinary function of the N unknown coefficients,

$$\mathcal{F}(\varphi) \simeq F(\varphi_1, \varphi_2, \ldots, \varphi_N). \tag{4}$$

It merely remains to find its maxima and minima,

$$\frac{\partial F}{\partial \varphi_i} = 0, \qquad i = 1, 2, \ldots, N. \tag{5}$$

This approximation to $\mathcal{F}(\varphi)$ is a quadratic function of the coefficients φ_i. Differentiation therefore turns (5) into a linear matrix equation of order N,

$$\mathbf{S\Phi} = k^2\mathbf{T\Phi} \tag{6}$$

where the matrices $\mathbf{S}$ and $\mathbf{T}$ are defined by

$$S_{jk} = \int_\Omega \nabla\alpha_j \cdot \nabla\alpha_k d\Omega, \tag{7}$$

$$T_{jk} = \int_\Omega \alpha_j \alpha_k d\Omega, \tag{8}$$

and $\mathbf{\Phi}$ is the column vector of coefficients φ_i. In the more general case of the inhomogeneous Helmholtz equation containing a source function g,

$$(\nabla^2 + k^2)\Pi = g, \tag{9}$$

a similar procedure is followed; the result is the matrix equation

$$\mathbf{S\Phi} - k^2\mathbf{T\Phi} = \mathbf{G} \tag{10}$$

where

$$G_j = \int_\Omega \alpha_j g d\Omega. \tag{11}$$

A few points should be observed about this procedure. First, the matrix equations (6) and (10) involve only first derivatives of the approximating functions $\alpha_j(x,y)$. This permits use of piecewise polynomials, for although Π is a function differentiable many times, the approximation $\varphi \simeq \Pi$ need only be differentiable once. Second, the approximating functions $\alpha_j(x,y)$ must satisfy all boundary conditions that fix the *value* of Π; in the TM_z case, where $\Pi = 0$ along the guide wall, this constraint must be built into the function set itself. Third, the approximating functions need not satisfy any homogeneous boundary conditions that fix the value of the *normal derivative* of Π. For example, in the TE_z case $\partial\Pi/\partial n = 0$ is required on the guide wall, but this requirement automatically imposes itself as part of the minimization process.

The minimization process of the finite element method effectively partitions all boundary conditions into two classes: the *natural* boundary conditions and the *essential* boundary conditions (sometimes also termed *principal*). The latter must be imposed on the approximating functions themselves; the former can simply be ignored because the minimization process enforces them naturally. The natural boundary conditions often correspond to the physical requirement of no energy transport. Their exact nature depends on the way the electromagnetic field problem is formulated; typical cases are symmetry planes in static problems and perfect reflection in propagation problems. This partitioning of boundary conditions is a major strength of the method, for it simplifies the problem formulation and the computational mathematics a great deal.

An alternative approach to the formulation of field problems derives from *projective approximation* and *weighted residual methods.* The essence of these techniques lies in rewriting the Helmholtz equation (9) in the transposed form

$$(\nabla^2 + k^2)\Pi - g = 0, \tag{12}$$

then projecting both sides onto the same space of functions $\mathbb{W}$. If $\mathcal{P}$ is a projector (projection operator) onto $\mathbb{W}$, then

$$\mathcal{P}\big((\nabla^2 + k^2)\Pi - g\big) = 0. \tag{13}$$

Projective approximation methods, of which several variants of the Galerkin approach are most popular, achieve this goal by applying inner product projections to specific families of approximating functions $\alpha_i \in \mathbb{W}$, so that $\Pi \simeq \phi = \sum \phi_i\alpha_i$ as in the variational approach. Typically, a Bubnov-Galerkin method for (13) will then be derived in the form

$$\int_\Omega \Big((\nabla^2 + k^2)\sum \phi_i\alpha_i - g\Big)\alpha_j d\Omega = 0, \qquad j = 1, \ldots, N. \tag{14}$$

To get rid of the second derivatives, which would require approximating functions at least twice differentiable, an integration by parts is again performed by applying Green's second identity. The resulting equations are very similar — for self-adjoint operators, identical — to those obtained by the variational method, so the computational task is no different either way. The variational approach has the theoretical strength of relying on physical minimum principles, thereby affording the analyst better understanding. The projective approach has the complementary strength that no variational principle need be known — indeed, one need not even exist — so it can be applied in a broader range of situations.

Vector Fields

In waveguide problems and various other special cases, the electromagnetic field can generally be represented by a pair of Hertz vectors or a pair of scalar variables. However, this reduction to two variables is often not convenient at all. In general, an electromagnetic field is naturally described by the vectors **E** and **H**. The Helmholtz equation in a vector variable assumes the form

$$\nabla \times \nabla \times \mathbf{\Pi} + k^2\mathbf{\Pi} = 0 \tag{15}$$

where $\mathbf{\Pi}$ is the vector quantity sought, typically **E** or **H**, sometimes a vector potential. The corresponding functional to be minimized is then

$$\mathcal{F}(\boldsymbol{\phi}) = \int_\Omega (\nabla \times \boldsymbol{\phi} \cdot \nabla \times \boldsymbol{\phi} + k^2\boldsymbol{\phi} \cdot \boldsymbol{\phi})d\Omega. \tag{16}$$

The natural boundary condition here is

$$\mathbf{1}_n \times \nabla \times \boldsymbol{\phi} = 0 \tag{17}$$

and the essential condition is

$$\boldsymbol{\phi} \times \mathbf{1}_n = \mathbf{\Pi} \times \mathbf{1}_n \tag{18}$$

at all boundary points where the tangential component of $\mathbf{\Pi}$ is prescribed as part of the problem. Here, and throughout this book, $\mathbf{1}_n$ denotes the unit normal vector. Other unit vectors will be written similarly, with the subscript identifying direction, as $\mathbf{1}_x$, $\mathbf{1}_y$, and so on.

To discretize, that is, to approximate the functional $\mathcal{F}$ by a function F, two approaches are possible. One is to use the scalar approximating functions $\alpha_i(x, y, z)$ and to multiply them by coefficients whose values are spatial vectors,

$$\boldsymbol{\phi}(x, y, z) \simeq \sum_{i=1}^{N} \boldsymbol{\phi}_i\alpha_i(x, y, z). \tag{19}$$

The other approach is to create vector-valued approximating functions $\boldsymbol{\alpha}_i(x, y, z)$ and to combine them with scalar multiplying coefficients,

$$\boldsymbol{\phi}(x, y, z) \simeq \sum_{i=1}^{N} \phi_i \boldsymbol{\alpha}_i(x, y, z). \tag{20}$$

The first approach may be called the structural mechanics approach, for it has an extensive history of use in that field. At least some theory and much practical experience should be transferable to electromagnetics. Further, it has the appeal of simplicity and familiarity. The same approximating functions can be called upon to serve for both scalar and vector cases, and the vectorial coefficients have clear meaning as component representations of **E**, **H**, or whatever vector variable may be modeled. In fact, this approach was initially used and remained the main tool for vector fields until some time in the 1980s.

The structural mechanics approach has one major flaw for electromagnetic field analysis: it doesn't work very well. The reason is simply that the fields that occur in structural mechanics and those encountered in electromagnetics are fundamentally different. The electromagnetic field vectors not only obey the Maxwell curl equations, but they are also constrained by the divergence equations

$$\nabla \cdot \mathbf{B} = 0, \qquad \nabla \cdot \mathbf{D} = -\frac{\rho}{\epsilon}. \tag{21}$$

Solenoidality is only required of a limited range of structural materials or fluids, but of practically all fields in electromagnetics. If the scalar finite element functions are used, and no particular care is taken to enforce solenoidality, solutions are obtained that satisfy the Maxwell curl equations but not the divergence equations (21). An analogous difficulty can arise in applied mechanics, but it seems to be much more rare.

Four approaches to the divergence problem are commonly encountered in the literature. The first is to reformulate the field representation itself, so that

$$\nabla \cdot \mathbf{\Pi} = 0 \tag{22}$$

does not need to be satisfied. This is impossible if $\mathbf{\Pi}$ is either **E** or **H**, but it is feasible if $\mathbf{\Pi}$ is a potential function, e.g., the magnetic vector potential **A**. The second approach is to reformulate the problem, reducing the number of independent components of $\mathbf{\Pi}$ by one so that the missing component is implicitly supplied by equation (22). This technique works well where there is a natural privileged direction, for example the direction of propagation in a waveguide, but it is not convenient in general three-dimensional problems. The third method is sometimes called the "penalty function" approach. It settles for making the divergence small rather than zero, by adding an extra term to the functional $\mathcal{F}$:

$$\mathcal{F}(\boldsymbol{\phi}) = \int_{\Omega} \left(\nabla \times \boldsymbol{\phi} \cdot \nabla \times \boldsymbol{\phi} + k^2 \boldsymbol{\phi} \cdot \boldsymbol{\phi} + \lambda (\nabla \cdot \boldsymbol{\phi})^2\right) d\Omega. \tag{23}$$

The coefficient λ may be viewed as a Lagrange multiplier and included in the problem solution itself, or it may be given a value *a priori*. Either way, the physically impossible solutions (those with nonzero divergence) still appear, but a good choice of λ can shift them into a remote part of the eigenvalue spectrum of the Helmholtz operator, where they are likely to do no great harm. The fourth, and probably the only truly sound, technique is not to use scalar approximating functions but to adopt the vector interpolation approach of equation (20) instead.

Even where the scalar approximation approach does work reasonably well, it still has disadvantages. Equation (19) must be set up in some coordinate system, so the three numbers $\boldsymbol{\phi}_i$ that form the vector-valued summation coefficients describe coordinate-directed vector components (e.g., the x, y, and z components in Cartesians). However, the boundary conditions that arise from physical conditions are invariably framed in terms of normal and tangential components to a bounding surface. The coordinate-directed components represent some linear combination of normal and tangential components, thus mixing essential boundary conditions and natural ones. The result is extensive algebraic complexity, for the essential boundary values are known *a priori*, the natural ones flow from the solution itself. Scalar approximants have the compensating advantage, on the other hand, that their theory and computational technique have been developed quite extensively, much more extensively than their vector counterparts.

Finite Element Construction

The methods available for constructing finite element approximating functions will now be taken up. The scalar functions are well known and well documented in textbooks. The vector functions are well known in some respects, but their theoretical basis is mathematically much more complex than their scalar counterparts' and their practical development is still incomplete. Both will be described briefly in the following, but with somewhat greater stress laid on the vector method.

Mesh Generation

To be useful, a finite element ideally should satisfy several requirements. It should be mathematically sound, in the sense that all continuity and boundary conditions implied by the appropriate functionals must be satisfied. It should be convergent, meaning that subdivision of a region into smaller and smaller finite elements should provide increasingly accurate results. It should be computationally tractable, particularly with respect to allowing automatic mesh generation. These are not easy requirements to meet simultaneously, but there are several families of finite elements that do so well enough.

The subject of automatic mesh generation is an important part of finite element software technology, for the industrial success of any such method will depend heavily on its ease of use and on its reliability in applications. The review paper by Ho-Le (1988) and the more recent book by George (1992) deal well with this rather specialized area of the finite element art, whose mathematical basis lies in computational geometry. Success in software construction also depends on the right choice of methods for solving simultaneous algebraic equations, a specialist area within linear algebra. These topics are important, but they have little to do with waves and less with electromagnetics, so they will not be dealt with in this book. They will occasionally be noted, however, in connection with individual element types.

Scalar Simplex Elements

The basic scalar element shapes are simplexes: line segments, triangles, and tetrahedra. These have several attractive properties, not the least of which is that any polygon or polyhedron can always be expressed as a union of simplexes. While this statement may appear almost tautologically self-evident, it has considerable importance, for this simple observation leads to proofs that an automatic mesh generator can be built that *always* produces a valid simplicial finite element mesh. This point is particularly important in three-dimensional analysis. Errors and anomalies are fairly easy to spot in two dimensions, by visual inspection of a graphical picture. Three dimensions are much more difficult; even experienced analysts have great difficulty comprehending the shape of a three-dimensional object from a wire-frame picture of its decomposition into tetrahedra. They certainly cannot be relied upon to spot errors, even if they did have the time and will to do so. The feasibility of totally reliable mesh generation must be counted as one of the most important points in favor of simplex elements.

A typical triangular finite element mesh, consisting of just two elements, is shown in Fig. 2. Each element is endowed with a regular array of ten nodes, and has four nodes on each edge. These numbers stem from the nature of polynomial approximation. To form a *complete* cubic approximation in two dimensions, ten independent terms are required, as can be seen immediately from the classical Pascal triangle,

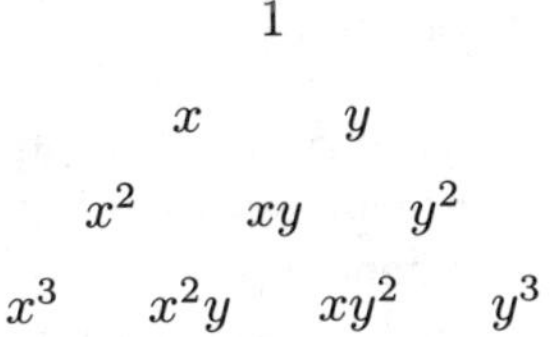

so specifying ten nodal values of the function $\Pi(x, y)$ fully specifies its cubic approximation. Along any triangle edge the two-dimensional cubic expression degenerates to a cubic in edge distance (a one-dimensional cubic) which is fully specified by four nodal values — exactly the number along one edge of an element. Consequently, insisting that the local approximations in the two adjacent elements match exactly at the four edge nodes guarantees continuity of the function everywhere along the edge. These simple observations lead directly to the use of interpolative approximating polynomials $\alpha_i(x, y)$ with the following properties (Silvester 1969*b*) :

a) The functions are interpolative on a set of node points P_i, that is, $\alpha_j(P_i) = \delta_{ij}$.
b) The functions are interpolative on the set of points along each triangle edge.
c) They form a *complete* polynomial family of degree N, that is, they contain all possible terms of degree N or lower in x and y jointly.

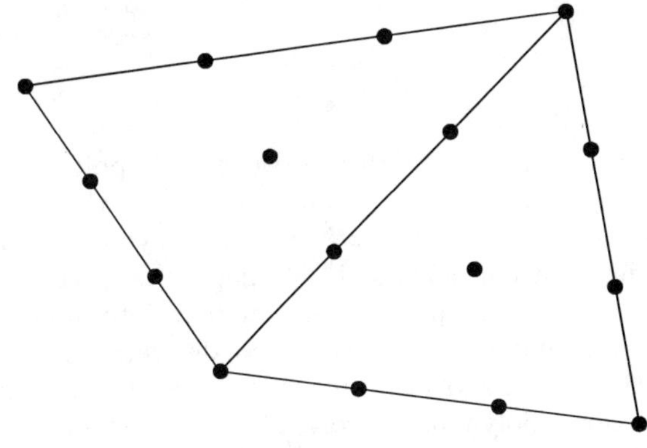

Fig. 2. Mesh of two triangular finite elements, N= 3.

The property of completeness is significant because it guarantees that the placement of a triangle with respect to a global x–y system is unimportant; translations and rotations with respect to the global coordinates merely transform terms of degree N into differently expressed terms still of degree N, so the approximation is unaltered. The clear conclusion is that simplex elements are *geometrically isotropic*; they possess no privileged directions or axes.

Exactly the same arguments apply with respect to tetrahedra. There are more nodes, of course, but the same continuity arguments apply to tetrahedron faces as to triangle edges. In fact, comparable arguments can be carried into higher dimensionalities, and at least a five-dimensional simplex element has been generated and described.

The formation of element matrices **S** and **T** for simplex elements is particularly easy, for the procedure is fortunately independent of the polynomial order. The essence of the method is given by Silvester [1982*a*], (1982*b*). On a given set of polynomials, complete to degree N, the differentiation operator $\partial/\partial x$ can be represented exactly by a numeric matrix, because differentiation of polynomials always produces polynomials of lower order. Any polynomial $p(x, y)$ of order no greater than N can be written in terms of the complete set $\alpha_i(x, y)$, and so can its derivative $\partial p/\partial x$

$$\frac{\partial p}{\partial x} = \sum_{i=1}^{N} p_i \frac{\partial \alpha_i}{\partial x} = \sum_{i=1}^{N} \sum_{j=1}^{N} D_{ji}^{(x)} p_j \alpha_i \tag{24}$$

where $\mathbf{D}^{(x)}$ is the matrix of derivative values at element nodal points,

$$D_{ji}^{(x)} = \left. \frac{\partial \alpha_i}{\partial x} \right|_{P_j} . \tag{25}$$

To put the matter in different words: A polynomial function can always be represented by giving the set of coefficients p_i that multiply the basis polynomials α_i. The differentiation matrix $\mathbf{D}^{(x)}$ maps the coefficient string that represents a given function into another coefficient string that represents its derivative. Such differentiation matrices can be worked out for interpolation polynomials on simplex elements in an element-independent manner, so that all the differentiations and integrations required to form element matrices can be carried out once and for all for a typical tetrahedron [Silvester 1982*a*] or triangle (Silvester 1969*b*). A similar argument applies to the integration of polynomials, hence to the integrations needed to form element matrices. Tables of universal element matrices have been computed in this way and published as integer fractions (i.e., with infinite precision). These are universal in the sense of being independent of element shape and size. The algebraic burden of computing their values is actually quite large, but that is largely irrelevant because that work has already been done and need never be done again. Because they are known exactly, recalculation of the finite element matrices makes even less sense than recalculation of the value of π.

Scalar Cartesian Product Elements

Engineers generally mature professionally in a universe of Cartesian coordinates. In spite of the geometric flexibility of simplexes, the tendency persists to favor Cartesian constructions, so families of rectangular elements (and rectangular parallelepipeds) have been developed. To guarantee that all requisite continuity requirements are met, these must have exactly the correct number of boundary nodes (e.g., $N + 1$ nodes along each edge of a square with polynomial approximants of order N). For elements of practical interest, say $N \leq 4$, the number of edge nodes required exceeds the minimum number of terms prescribed by the Pascal triangle for the whole element, so additional higher-order terms are introduced. Although these complicate the mathematics, here too the element matrices are tabulated so the underlying complexity is no longer of concern. The shortcoming of Cartesian products is that they are geometrically inflexible: there is no way to decompose a general polyhedron into parallelepipeds, and any approximation to curved boundaries is crude indeed. Their major strength lies elsewhere: Cartesian product elements can be constructed to have derivative continuity, not just function continuity, across element boundaries. This property endears them to the applied mechanics community, which commonly must address the biharmonic equation $\nabla^4 \phi = 0$. It is of minor importance in scalar electromagnetics, though it reappears in vector field modeling, as described further below.

Isoparametric Elements

The real importance of Cartesian products in scalar field problems lies in the construction of curvilinear elements. In general, element matrices are computed and expressed most conveniently in terms of local coordinates u, v, w attached to each element, while the entire problem to be solved is framed in a coordinate system x, y, z global to all elements. On each element, the relationship between coordinates may be expressed by a coordinate transformation such as

$$x = k_{xu}u + k_{xv}v + k_{xw}w \tag{26}$$

(and similarly for y, z), which maps rectangles in $u - v - w$ into quadrilaterals in $x - y - z$. By making the transformation piecewise polynomial instead of linear, curvilinear elements can be generated. The procedure to follow emerges from the simple but profound observation that the continuity and differentiability requirements of such a coordinate transformation are exactly the same as the corresponding requirements of potential or wave functions — that modeling the coordinate quantities is no different from modeling a potential, for both are single-valued continuous functions. Consequently, the same finite element functions may be used for approximating the potential function and the coordinate transformation! It is only necessary to know where in the global coordinate system each finite element node must be mapped — that is, to know their global coordinate values $x_i, y_i, z_i,$ — and to use

$$x(u, v, w) = \sum_{i=1}^{N} x_i \alpha_i(u, v, w). \tag{27}$$

Note that this statement is *not* an approximation; it defines the coordinate transformation and does not approximate it. Such elements are generally called *isoparametric*; Fig. 3 illustrates the transformation. The method was first developed by Ergatoudis, Irons, and Zienkiewicz (1968) in virtually its final form, and has been widely used in mechanics.

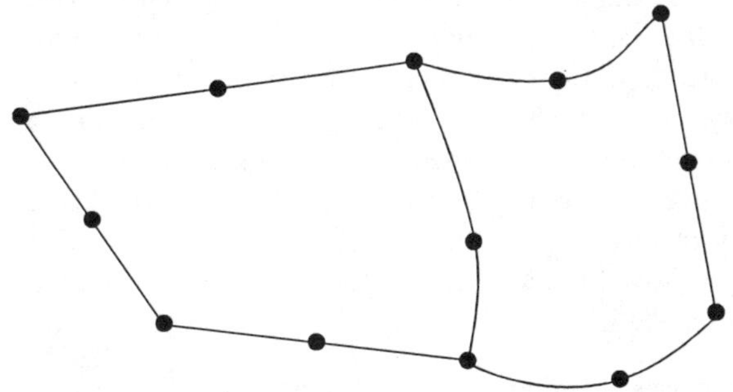

Fig. 3. Isoparametric coordinate transformations on two elements combine to yield a continuous interface.

Mesh generation is the major problem with isoparametric elements. In effect, transformation (27) is too flexible; it is possible to write singular transformations, or transformations that make elements overlap with themselves. A secondary though minor difficulty lies in forming the element matrices. Because the coordinate mappings enter into all integrations and differentiations, once-and-for-all generation of element matrices is no longer possible and numerical integration is commonly used. Although it is relatively costly, its cost is only proportional to the number of elements N_E and is usually overwhelmed by the cost of solving simultaneous equations, which is always considerably greater than $O(N_E)$ and can easily rise to $O(N_E^3)$ for eigenvalue problems.

The use of local coordinates enormously simplifies the creation of finite elements and their matrices; indeed, it it hard to imagine how tabulated universal matrices could be arrived at in any other way. For rectangular elements Cartesians are the obvious choice, but simplex elements are better dealt with using *simplex coordinates*, also known as *barycentric* coordinates. Figure 4 shows a triangle, with a point P identified in its interior. Clearly, this point partitions the triangle 1–2–3 into three subtriangles P–2–3, 1–P–3, and 1–2–P. If the relative sizes of the subtriangles are given, the point P is uniquely determined. Setting

$$\zeta_1 = \frac{\text{area}(P - 2 - 3)}{\text{area}(1 - 2 - 3)} \tag{28}$$

defines one of these coordinate quantities; the other two follow analogously. An alternative geometric interpretation of equation (28) is obtained by noting that the area of any triangle is half the product of base and height. Triangles P–2–3 and 1–2–3 obviously share the same base 2–3, so the ratio (28) may be viewed as measuring relative distance toward vertex 1 from the side opposite. A similar argument holds for tetrahedra, save that volumes are used instead of areas. Practically all the element development work on simplex elements has relied on the use on such coordinates, for the scalar as well as the vector elements.

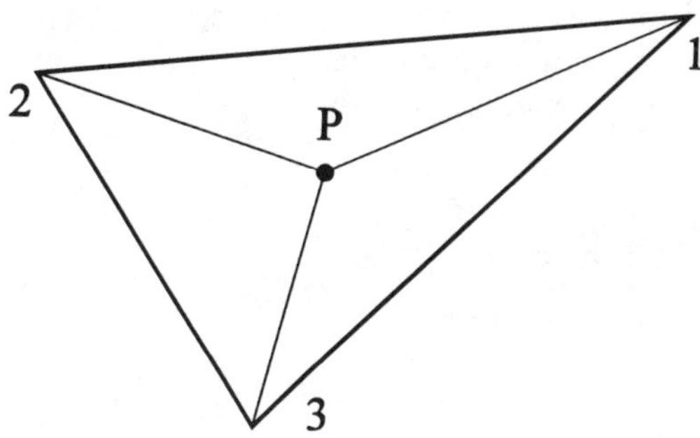

Fig. 4. Point P partitions 1-2-3 into three triangles.

The placement of a tetrahedron or a triangle into a global Cartesian framework may be viewed as a mapping, in exactly the same manner as the local-global coordinate identification in Cartesians. Thus, the global coordinate value x is associated with the local coordinate values $\zeta_1, \ldots, \zeta_4$ by

$$x(\zeta_1, \zeta_2, \zeta_3, \zeta_4) = \sum_{i=1}^{N} x_i \alpha_i(\zeta_1, \zeta_2, \zeta_3, \zeta_4) \tag{29}$$

and similarly for $y(\zeta_1, \zeta_2, \zeta_3, \zeta_4)$ and $z(\zeta_1, \zeta_2, \zeta_3, \zeta_4)$. By assigning node coordinate values x_i, y_i, z_i so as to place the nodes where they are wanted, isoparametric tetrahedra and isoparametric triangles may be created, exactly like isoparametric quadrilaterals and hexahedra. Recent examples of their use in microwave field analysis have been given by Webb and McFee (1991).

An interesting and theoretically sound alternative to polynomials is to use rational functions, which often provide better performance in one-dimensional function modeling than polynomials. Wachspress (1975) worked out the theoretical details of rational basis functions, but it does not seem from the engineering literature that they have enjoyed wide application.

Vector Simplex Elements

Solenoidal-field vector elements are of fairly recent origin. In contrast to their scalar counterparts, a complete theory has not yet been presented, so any treatment will appear piecemeal by comparison. Nevertheless, sufficient classes of solenoidal-field elements have now been derived for much useful work to be done.

Although the vector-based finite element methods may have originated with Bossavit and Verite (1982), the derivation now commonly given for tetrahedra is that of Barton and Cendes [1987]. They noted that any arbitrary vector $\mathbf{A}$,

$$\mathbf{A} = \mathbf{f} + \mathbf{g} \times \mathbf{r}, \tag{30}$$

is solenoidal, provided that $\mathbf{f}$ and $\mathbf{g}$ are constant vectors and $\mathbf{r}$ is the position (radius) vector of a point. Let the specific choices be made

$$\mathbf{f} = \frac{1}{6V} \sum_{i=1}^{6} (\mathbf{r}_{i+1} \times \mathbf{r}_{i+2}) A_{7-i} \tag{31}$$

$$\mathbf{g} = \frac{1}{6V} \sum_{i=1}^{6} \mathbf{e}_i A_{7-i} \tag{32}$$

where $\mathbf{r}_i$ is the position vector of vertex i and $\mathbf{e}_i$ is an edge of the tetrahedron, $\mathbf{e}_i = \mathbf{r}_j - \mathbf{r}_k$ (though the rules for numbering are a little complicated). V is the tetrahedron volume, and A_i is the projection of $\mathbf{A}$ onto edge $\mathbf{e}_i$, $A_i = \mathbf{A} \cdot \mathbf{e}_i$. There are six edges in a tetrahedron, hence six projections of a vector onto the edges. Taking only $A_1 = 1$, $A_2 = A_3 = \ldots = A_6 = 0$, then taking $A_2 = 1$ and setting the remainder zero, and so on, the above three equations produce a family of six vector-valued functions $\boldsymbol{\alpha}_i$ with the following properties:

a) Their three components vary linearly over the tetrahedral element.

b) They are edge-interpolative, that is, each possesses a nonzero edge-parallel component along only one edge.

c) They are solenoidal everywhere in the element.

These are exactly the properties required for a vector-valued element, for modeling $\mathbf{H}$, or $\mathbf{E}$ in empty (charge-free) space. Elements built in this fashion are known as "edge elements." Both simplicial and Cartesian product edge elements have been described in the literature. They fall short of perfection in not being solenoidal *everywhere*; their fields are free from sources within each element, but may possess surface sources at element interfaces. Such elements cannot guarantee absence of spurious modes, but they do endow them with the correct zero eigenvalue so they do not appear in solutions.

Tetrahedral and hexahedral elements have been viewed as important for some time, because the three-dimensional field problem is very difficult to master without them. Edge-based triangular elements are almost equally useful, but (as will be seen in connection with guided waves) many two-dimensional problems permit alternative reformulations that manage to avoid solenoidal fields. Tanner and Peterson (1989) gave a description of triangular elements which resembles the Barton and Cendes [1987] work but differs in notation. This development, however, is still only valid only for linear interpolation between edges, that is, for elements of first order. Given the spectacularly higher accuracy they provide, higher-order methods would be more than welcome.

For triangles, a major step ahead came with the work of Cendes [1991]. He showed, in considerable algebraic detail, how high-order approximating functions with tangential-component continuity can be constructed, and gave as an example the approximating functions for an element with quadratic interpolation of the edge components. This theory is capable of extension to tetrahedral elements, and a higher-order tetrahedral element was described by Lee, Sun, and

Cendes [1991]. However, the work is so recent that experimental verification is as yet incomplete, and tables of elements are not yet available. Webb and Forghani [1993] described another grouping of tangentially-continuous functions of second and third order on tetrahedra, and gave some experimental results; this work is even more recent and remains to be evaluated by other researchers. A good summary of the problems of this field and an assessment of the state of the art will be found in the review paper by Webb [1993].

Vector-Based Cartesian Products

Cartesian product edge elements have been used about as long as edge-based simplexes. The elements used by Mur and de Hoop (1985) or van Welij [1985] are hexahedral and satisfy similar requirements as the tetrahedral elements already described. Kameari [1990] described how they could be generalized to higher orders and gave details of a second-order family of vector interpolants. Because the coordinate-directed components of vectors coincide with their tangential and normal components, the scalar and vector approximation schemes are sometimes difficult to tell apart on rectangular parallelepipeds or rectangles. This observation led Crowley, Silvester, and Hurwitz [1988] to use covariant (curvilinear) vector components on isoparametrically transformed rectangles or parallelepipeds — but with a difference. Crowley (1988) proved an inclusion theorem that (like many good theorems) seems obvious in retrospect: if the linear manifold spanned by the finite element approximating functions properly includes its own projection onto the space of all possible irrotational functions that satisfy the Helmholtz equation boundary-value problem, then all eigensolutions of the boundary-value operator will approximate true eigensolutions of the physical problem. Close examination of the interpolation polynomials for rectangles then suggests a method for constructing suitable functions. The resulting elements have been termed "spectrally correct" or "orthospectral" elements, or simply "mixed-order" elements. They have been used in a variety of three-dimensional problems (Miniowitz and Webb 1991; Foo and Silvester 1992) and some two-dimensional ones.

Hierarchal Elements

The error theory of finite elements shows that, under hypotheses satisfied in many practical problems, the error ε is bounded by

$$\varepsilon \leq K h^{-2p} \qquad (33)$$

where h is the longest element edge in the finite element mesh, p is the polynomial order to which the approximating functions are complete, and K is some constant that depends on the problem being solved, but not on the finite element approximation. It is evident that the error can be reduced in two ways: by using a finer mesh (i.e., making h smaller) or by using more sophisticated elements (i.e., making p larger). Of the two ways, the latter is usually the more effective, as can be seen from equation (33): increases in the exponent p bring rewards much more quickly than decreases in the base h. Much of the work described above has in fact been directed to building elements of increasing sophistication so as to benefit from the very considerable accuracy improvement that results from increased p.

In most physical problems, error is largest where field gradients are largest, and it is there that improvements should be sought. Rather than to raise p globally, with the attendant large cost in computing effort, it would be attractive to mix elements of varying orders. High-order approximations could then be used in sensitive areas only. This approach is not feasible with the standard element types, for interelement continuity of function values can only be guaranteed if compatible approximations are used on both sides of each element interface. Mixed-order approximation is made possible by "hierarchal elements," whose approximating functions are constructed in families so that the functions of degree p, say $\mathbb{A}^p : \{\alpha_i^p, i = 1, \ldots, N_p\}$ form a proper subset of the functions of next higher degree $p+1$, $\mathbb{A}^{p+1} : \left\{\alpha_i^{p+1}, i = 1, \ldots, N_{p+1}\right\}$. (Note that the inclusion refers to a specific set of basis functions, *not* merely the spaces spanned by these functions.) This can be done by matching function values on interfaces between elements of order 1, but function values and derivatives at interfaces involving higher-order elements. Where a low-order element abuts one of higher order, some of the derivatives are automatically zero, but continuity is maintained. The idea was neatly implemented by Rossow and Katz (1978), who gave a detailed formulation as well as test results and tables of element matrices. Unfortunately, the latter were presented in floating-point form, so recomputation is necessary if arithmetic of higher precision is to be used. This technique as well as others is covered in the review paper by Zienkiewicz, Gago, and Kelly (1983).

The use of hierarchal elements is particularly attractive in combination with adaptive mesh generation, where both the element layout and the order of elements can be varied. McFee and Webb (1992) have presented experimental results on triangular elements, and Sears (1988) discusses isoparametric quadrilaterals.

Variational Principles

Finite element solutions to most electromagnetic field problems have traditionally been framed in variational terms, so it is appropriate to close this chapter by examining some of the key formulations.

The key early work, deservedly often cited, is the article by Berk (1956). Although engineers of that time were generally not well trained in variational methods, the specific community concerned with scattering theory was exceptionally well versed in such techniques. The object most often in that precomputer era was to obtain reliable estimates of stationary global quantities, typically resonant frequencies or scattering cross-sections, from rough approximations to the electromagnetic field distribution. Berk generalized the work of earlier researchers to cover quite general (anisotropic and inhomogeneous) linear materials and showed that the resonant frequency of a finite volume of material, e.g., a resonant cavity, could be framed in terms of the electric field as

$$\omega^2 = \frac{1}{\int \mathbf{E}^* \cdot \epsilon \mathbf{E} d\Omega} \left\{ \int \mathbf{E}^* \cdot \nabla \times (\mu^{-1} \cdot \nabla \times \mathbf{E}) d\Omega - \oint (\mathbf{E} \times (\mu^{-1*} \cdot \nabla \times \mathbf{E}^*) \cdot \mathbf{dS} \right\}. \tag{34}$$

This quantity is not of direct interest for the finite element analyst, for it does not furnish estimates of the field **E** itself. However, a corollary of Berk's argument is that the functional

$$\mathcal{F}(\mathbf{E}) = \int \mathbf{E}^* \cdot \nabla \times (\mu^{-1} \cdot \nabla \times \mathbf{E}) d\Omega - \oint (\mathbf{E} \times (\mu^{-1*} \cdot \nabla \times \mathbf{E}^*) \cdot \mathbf{dS} - \omega^2 \int \mathbf{E}^* \cdot \epsilon \mathbf{E} d\Omega \tag{35}$$

is stationary at the correct solution **E**. Here the asterisk denotes complex conjugation (if the fields are complex), while ϵ and μ may be tensors or scalars. There is also a corresponding dual expression in **H**. Berk also gives stationary expressions, on which functionals may be based, for the propagation constant of guided waves, which will be considered in more detail in the next chapter of this anthology. The expressions derived by Berk have served as a basis for numerous finite element solutions, usually in somewhat simplified forms (e.g., scalar material properties). For example, Ferrari and Maile (1978) formulated and solved three-dimensional problems in this manner, though without the benefit of the extensive edge-element work that has been accomplished since.

Berk's work and its modifications harbor two shortcomings. First, the variationally stationary quantities are defined only for self-adjoint boundary value problems; in more general cases, recourse must be had to adjoint operators and adjoint solutions, whose physical significance is sometimes quite obscure. A second difficulty lies in the apparently *ad hoc* nature of the argument. Variationally stationary expressions seem to drop from the sky, followed by proofs that they do indeed work; but it would be more satisfying to deduce them from some broad first principle. A major effort to overcome the second lacuna was made by Morishita and Kumagai [1977], who strove to derive classical electromagnetism, and by extension all of its variational principles, from the unifying principle of least action. Their work included rederivation of Berk's formulae (one of which turned out to contain an error of sign) as well as some new results. Their main contribution, however, is that of unification and clarification. Chen and Lien [1980] extended this work further to cover non-self-adjoint problems as well. Although these formulations have been available for some time, it seems that relatively few workers have actually used them to derive new finite element formulations. Those few, on the other hand, have contributed work of considerable value. Much of this work has concentrated on guided waves and will be described in the next chapter. Throughout the 1980s, most finite element developers were largely preoccupied with such more immediate issues as vector-valued approximating functions and the suppression of spurious solutions. It is highly probable that further applications of variational methods will yield valuable new formulations in the future.

References

Arlett, P. L., A. K. Bahrani, and O. C. Zienkiewicz (1968): "Application of finite elements to the solution of Helmholtz's equation." *Proceedings of the Institution of Electrical Engineers*, vol. **115**, pp. 1762–66, 1968.

Barton, M. L. and Z. J. Cendes [1987]: "New vector finite elements for three-dimensional magnetic field computation." *Journal of Applied Physics*, vol. **61**, pp. 3919–21, 1987.

Benedek, P. and P. Silvester (1972): "Capacitance of parallel rectangular plates separated by a dielectric sheet." *IEEE Transactions on Microwave Theory and Techniques* vol. **MTT-20**, pp. 504–10, 1972.

Berk, A. D. (1956): "Variational principles for electromagnetic resonators and waveguides." *IRE*

Transactions on Antennas and Propagation, vol. **AP-4**, pp. 104–11, 1956.

Binns, K. J., P. J. Lawrenson, and C. W. Trowbridge (1992): *The Analytical and Numerical Solution of Electric and Magnetic Fields.* Chichester, UK: John Wiley, 1992.

Bossavit, A. and J. C. Verite (1982): "A mixed FEM-BIEM method to solve 3-D eddy current problems." *IEEE Transactions on Magnetics*, vol. **MAG-18**, pp. 431–35, 1982.

Chen, C. H. and C.-D. Lien [1980]: "The variational principle for non-self-adjoint electromagnetic problems." *IEEE Transactions on Microwave Theory and Techniques*, vol. **MTT-28**, pp. 878–86, 1980.

Cendes, Z. J. [1991]: "Vector finite elements for electromagnetic field computation." *IEEE Transactions on Magnetics*, vol. **27**, pp. 3958–66, 1991.

Courant, R. L. (1943): "Variational methods for the solution of problems of equilibrium and vibration." *Bulletin of the American Mathematical Society*, vol. **49**, pp. 1–23, 1943.

Crowley, C. W. (1988): *Mixed Order Covariant Projection Vector Fields.* Ph. D. dissertation. Montreal: McGill University, 1988.

Crowley, C. W., P. P. Silvester, and H. Hurwitz, Jr. [1988]: "Covariant projection elements for 3D vector field problems." *IEEE Transactions on Magnetics*, vol. **24**, pp. 397–400, 1988.

Csendes, Z. J. and P. Silvester (1971): "Numerical solution of dielectric loaded waveguides. I — Finite-element analysis." *IEEE Transactions on Microwave Theory and Techniques*, vol. **MTT-19**, pp. 504–9, 1971.

Duffin, R. J. (1959): "Distributed and lumped networks." *Journal of Mathematics and Mechanics*, vol. **8**, pp. 793–826, 1959.

Ergatoudis, I., B. M. Irons, and O. C. Zienkiewicz (1968): "Curved isoparametric 'quadrilateral' elements for finite element analysis." *International Journal for Solids and Structures*, vol. **4**, pp. 31–42, 1968.

Ferrari, R. L. and G. L. Maile (1978): "Three-dimensional finite element method for solving electromagnetic problems." *Electronics Letters*, vol. **14**, pp. 467–68, 1978.

Foo, S. L. and P. P. Silvester (1992): "Boundary-marching method for discontinuity analysis in waveguides of arbitrary cross section." *IEEE Transactions on Microwave Theory and Techniques*, vol. **40**, pp. 1889–93, 1992.

George, P. L. (1992): *Automatic Mesh Generation: Applications to Finite Element Methods.* New York: John Wiley, 1992.

Ho-Le, K. (1988): "Finite element mesh generation methods: A review and classification." *Computer Aided Design*, vol. **20**, pp. 27–38, 1988.

Hoole, S. R. H (1989): *Computer-Aided Analysis and Design of Electromagnetic Devices.* New York: Elsevier, 1989.

Jin, J. (1993): *The Finite Element Method in Electromagnetics.* New York: John Wiley, 1993.

Kameari, A. [1990]: "Calculation of transient 3-D eddy current using edge-elements." *IEEE Transactions on Magnetics*, vol. **26**, pp. 466–69, 1990.

Lee, J.-F., D.-K. Sun, and Z. J. Cendes (1991): "Full-wave analysis of dielectric waveguides using tangential vector finite elements." *IEEE Transactions on Microwave Theory and Techniques*, vol. **39**, pp. 1262–71, 1991.

McFee, S. and J. P. Webb (1992): "Adaptive finite elements analysis of microwave and optical devices using hierarchal triangles." *IEEE Transactions on Magnetics*, vol. **28**, pp. 1708–11, 1992.

Miniowitz, R. and J. P. Webb (1991): "Analysis of 3-D microwave resonators using covariant-projection elements." *IEEE Transactions on Microwave Theory and Techniques*, vol. **39**, pp. 1895–99, 1991.

Morishita, K. and N. Kumagai [1977]: "Unified approach to the derivation of variational expression for electromagnetic fields." *IEEE Transactions on Microwave Theory and Techniques*, vol. **MTT-25**, pp. 34–40, 1977.

Mur, G. and A. T. de Hoop (1985): "A finite-element method for computing three-dimensional electromagnetic fields in inhomogeneous media." *IEEE Transactions on Magnetics*, vol. **MAG-21**, pp. 2188–91, 1985.

Oden, J. T. and J. N. Reddy (1976): *An Introduction to the Mathematical Theory of Finite Elements.* New York: John Wiley, 1976.

Rossow, M. P. and I. N. Katz (1978): "Hierarchal finite elements and precomputed arrays." *International Journal for Numerical Methods in Engineering*, vol. **12**, pp. 977–99, 1978.

Sabonnadière, J.-C. and J.-L. Coulomb (1987): *Finite Element Methods in CAD: Electrical and Magnetic Fields.* S. Salon, translator. New York: Springer-Verlag, 1987.

Sears, G. J. (1988): "Hierarchical quadrilateral elements in field analysis." *IEEE Transactions on Magnetics*, vol. **24**, pp. 404–7, 1988.

Silvester, P. [1969*a*] : "Finite-element solution of homogeneous waveguide problems." *Alta Frequenza*, vol. **38**, pp. 313–17, 1969.

Silvester, P. (1969*b*) : "High-order polynomial triagular finite elements for potential problems." *International Journal for Engineering Science*, vol. **7**, pp. 849–61, 1969.

Silvester, P. P. [1982*a*] : "Universal finite element matrices for tetrahedra." *International Journal for Numerical Methods in Engineering*, vol. **18**, pp. 1055–61, 1982.

Silvester, P. P. (1982*b*) : "Permutation rules for simplex finite elements." *International Journal for Numerical Methods in Engineering*, vol. **18**, pp. 1245–48, 1982.

Silvester, P. and K. K. Chan (1972): "Bubnov-Galerkin solutions to wire-antenna problems." *Proceedings IEE*, vol. **119**, pp. 1095–99, 1972.

Silvester, P. P. and R. L. Ferrari (1983): *Finite Elements for Electrical Engineers.* Cambridge: Cambridge University Press, 1983; second edition, 1990.

Steele, C. W. (1987): *Numerical Computation of Electric and Magnetic Fields.* New York: Van Nostrand Reinhold, 1987.

Strang, G. and G. J. Fix (1973): *An Analysis of the Finite Element Method.* Englewood Cliffs: Prentice-Hall, 1973.

Synge, J. L. (1957): *The Hypercircle in Mathematical Physics.* Cambridge: Cambridge University Press, 1957.

Tanner, D. R. and A. F. Peterson (1989): "Vector expansion functions for the numerical solution of Maxwell's equations." *Microwave and Optical Technology Letters*, vol. **2**, pp. 331–34, 1989.

van Welij, J. S. [1985]: "Calculation of eddy currents in terms of H on hexahedra." *IEEE Transactions on Magnetics*, vol. **MAG-21**, pp. 2239–41, 1985.

Wachspress, E. L. (1975): *A Rational Finite Element Basis.* New York: Academic Press, 1975.

Webb, J. P. [1993]: "Edge elements and what they can do for you." *IEEE Transactions on Magnetics*, vol. **29**, pp. 1460–65, 1993.

Webb, J. P. and Forghani, B. [1993]: "Hierarchal scalar and vector tetrahedra." *IEEE Transactions on Magnetics*, vol. **29**, pp. 1495–98, 1993.

Webb, J. P. and S. McFee (1991): "The use of hierarchical triangles an finite-element analysis of microwave and optical devices." *IEEE Transactions on Magnetics*, vol. **27**, pp. 4040–43, 1991.

Zienkiewicz, O. C., J. P. de S. R. Gago, and D. W. Kelly (1983): "The hierarchical concept in finite element analysis." *Computers and Structures*, vol. **16**, pp. 53–65, 1983.

6-9 FINITE-ELEMENT SOLUTION OF HOMOGENEOUS WAVEGUIDE PROBLEMS

P. SILVESTER (*)

This paper presents a new method of waveguide analysis, which permits finding not only the dominant propagating mode, but also the higher modes, of guides of arbitrary cross-sectional shape. While certain methods have been developed to determine some higher modes in guides of comparatively simple shape, e. g., the rectangular coaxial guide [1] *or the lunar guide* [2], *difficulty is encountered in attempts to extend them to very general cross-sections. On the other hand, finite-difference methods have proved very powerful for finding the dominant modes of complicated guides, but lend themselves ill to the determination of any further propagating modes* [3]. *The new method therefore fills a significant gap in the range of available analytic techniques.*

FINITE-ELEMENT FORMULATION OF THE WAVEGUIDE PROBLEM.

As is well known, the propagating modes of a hollow, uniform, waveguide may be determined by solving the two-dimensional scalar Helmholtz equation

$$(\nabla^2 + \lambda^2)\, \Pi = 0\,. \tag{1}$$

Here Π represents the magnitude of an axially directed electric or magnetic Hertz vector. If the transverse electric (TE) modes of the guide are desired, Π must be the electric Hertz vector, and satisfy the homogeneous Neumann boundary condition

$$\frac{\partial \Pi}{\partial n} = 0 \tag{2}$$

while the transverse magnetic (TM) modes are obtained if Π is taken to be the magnetic Hertz vector, and subject to the boundary condition

$$\Pi = 0 \tag{3}$$

everywhere along the guide walls.

Rather than attempting to solve this eigenvalue problem directly, the new method reframes it in variational terms. It is shown in books on mathematical physics [4] that the solution of the Helmholtz equation with homogeneous boundary conditions is equivalent to minimising the functional $\mathscr{F}$ defined by

$$\mathscr{F} = \iint_R (|\,\mathrm{grad}\, \varphi\,|^2 + k^2 \varphi^2)\, dx\, dy \tag{4}$$

The region of integration is, of course, the waveguide cross-section. If a trial solution $\varphi(x, y)$ is represented geometrically as a surface spanning the region R over the $x - y$ plane, the correct surface $\Pi(x, y)$ is that which yields the smallest possible value of $\mathscr{F}$. One may view $\mathscr{F}$ as a measure of error; among any class of trial surfaces $\varphi(x, y)$, the one that best matches $\Pi(x, y)$ yields the smallest $\mathscr{F}$. Instead of attempting to find the exact solution surface $\Pi(x, y)$, a particular restricted class of surfaces will now be considered, and the smallest possible $\mathscr{F}$ among these determined.

In the finite-element method, the correct solution will be approximated to by a surface $\varphi(x, y)$ made up of finite surface elements of a simple kind. Let the region R be subdivided into a number of elementary regions of finite size. For simplicity, all the elementary regions are taken to be polygonal. Corresponding to each elementary region, a surface element will be defined by requiring $\varphi(x, y)$ to be a linear function of its values at the polygon vertices (or, more generally, of a finite number of parameters associated with the elementary region). For a general polygonal elementary region with n vertices, φ is then defined by

$$\varphi(x, y) = \sum_{i=1}^{n} \alpha_i(x, y)\, \varphi_i \tag{5}$$

If desired, the finite-element method may be defined in terms of the method of moments [5]; equation (5) is then the definition of a family of multiparameter basis functions.

$\mathscr{F}$ is now no longer a true functional, but an ordinary function of the parameters that define φ, in this case the vertex values φ_i:

$$\mathscr{F}(\varphi) = F(\varphi_1, \varphi_2, \ldots, \varphi_N) \tag{6}$$

The best possible set of parameter values, and hence the best possible φ within the class defined by (5), is found by minimising F with respect to all the parameters. That is, it will be required that

$$\frac{\partial F}{\partial \varphi_m} = 0 \tag{7}$$

for all m. If the region R contains altogether N vertices, this minimisation requirement changes (4) into a matrix eigenvalue problem of order N, as will be shown next.

THE MATRIX EIGENVALUE PROBLEM.

In view of equation (6), the minimisation requirement (7) is equivalent to

$$\iint_R \frac{\partial}{\partial \varphi_m} |\,\mathrm{grad}\, \varphi\,|^2\, dx\, dy = k^2 \iint_R \frac{\partial}{\partial \varphi_m} \varphi^2\, dx\, dy \tag{8}$$

Differentiating repeatedly, one finds from (5) that

$$\frac{\partial}{\partial \varphi_m} |\,\mathrm{grad}\, \varphi\,|^2 = = 2 \sum_{k=1}^{N} \left(\frac{\partial \alpha_m}{\partial x} \frac{\partial \alpha_k}{\partial x} + \frac{\partial \alpha_m}{\partial y} \frac{\partial \alpha_k}{\partial y} \right) \varphi_k \tag{9}$$

everywhere within an elementary region that abuts on vertex m. Elsewhere, α_m is zero so that the right-hand side

(*) P. SILVESTER - Dr., McGill University, Montreal, P. Q. (Canada).

Reprinted with permission from *Alta Frequenza*, vol. 38, pp. 313–317, 1969.

vanishes. By carrying out a similar differentiation, one next finds that

$$(10) \qquad \frac{\partial}{\partial \varphi_m} (\varphi^2) = 2 \sum_{k=1}^{N} \alpha_m \, \alpha_k \, \varphi_k$$

where again nonzero terms arise on the right-hand side only if both vertices m and k are vertices of the elementary polygon in question.

It will be convenient to define the purely geometric quantities

$$(11) \qquad S_{mk} = \iint_R \left(\frac{\partial \alpha_m}{\partial x} \frac{\partial \alpha_k}{\partial x} + \frac{\partial \alpha_m}{\partial y} \frac{\partial \alpha_k}{\partial y} \right) dx\, dy$$

and

$$(12) \qquad T_{mk} = \iint_R \alpha_m \, \alpha_k \, dx\, dy$$

which express the nature of the region R, and the manner of its subdivision into elementary regions. In terms of these integral quantities, equation (8) reads

$$(13) \qquad S\,\Phi = k^2\, T\,\Phi$$

where Φ is the column matrix of vertex values φ_i ; S and T are square matrices of order N, whose elements are S_{mk} and T_{mk} as defined above. Minimisation of $\mathcal{F}$ is thus equivalent to the eigenvalue problem (13), and solution of the latter will provide approximate eigenvalues and eigenfunctions for the boundary-value problem. It will be noted that S and T are always symmetric, and the eigenvalues therefore always real.

MATRICES S AND T FOR TRIANGULAR ELEMENTS.

While it is possible in principle to consider any shape of polygonal elementary regions, it is most convenient to keep the algebraic details simple by restricting the elementary regions, and therewith the solution surface elements, to be triangular. The logical choice of linear functions in equation (5) is then such as to make the value of φ anywhere within an elementary region simply the linear interpolate of its vertex values [6]. Geometrically, the true solution surface is now to be approximated by a surface $\varphi\ (x, y)$ consisting of triangular flats. It should be noted that this approximation forms a continuous surface throughout the region.

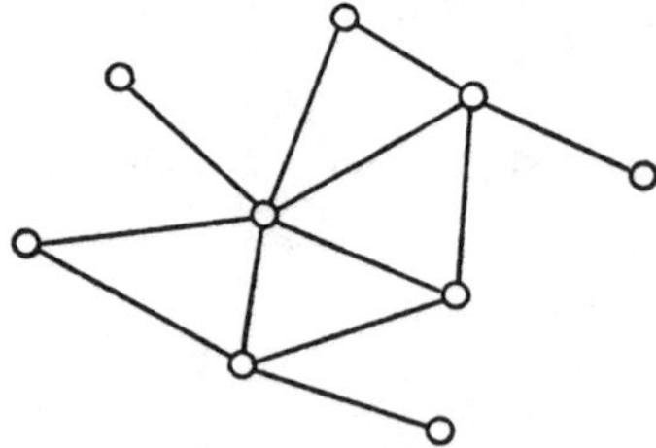

Fig. 1. — Subdivision of region R into triangular elementary regions.

Consider now a triangular elementary region with vertices numbered 1, 2, 3, as in fig. 1. The value of φ for all x, y within the triangle is given by

$$(14) \qquad \varphi = \frac{1}{2A} \sum_{j=1}^{3} (a_j + b_j\, x + c_j\, y)\, \varphi_j$$

where A represents the triangle area. It can be shown [6] that the three coefficients that result in linear interpolation over the triangle are

$$(15) \qquad a_j = \begin{vmatrix} x_{j+1} & y_{j+1} \\ x_{j+2} & y_{j+2} \end{vmatrix}$$

$$(16) \qquad b_j = y_{j+2} - y_{j+1}$$

$$(17) \qquad c_j = x_{j+1} - x_{j+2}$$

the subscripts being cyclic modulo 3.

For ease in both analysis and computer programming, it is useful to regard the matrices S and T as being composed of sums of sparse matrices of order N, typically $s^{(i)}$ and $t^{(i)}$, made up of the contributions to the S and T matrices that are attributable to only one elementary region, say the ith. That is to say, one may write

$$(18) \qquad S_{mk} = \sum_n s_{mk}^{(n)}$$

$$(19) \qquad T_{mk} = \sum_n t_{mk}^{(n)} .$$

Clearly $s_{mk}^{(n)} = t_{mk}^{(n)} = 0$ whenever n equals neither m nor k, i. e., for all elements which touch neither vertex m nor k. Explicit expressions for the element matrix components are

$$(20) \qquad s_{mk}^{(i)} = \iint_i \left(\frac{\partial \alpha_m}{\partial x} \frac{\partial \alpha_k}{\partial x} + \frac{\partial \alpha_m}{\partial y} \frac{\partial \alpha_k}{\partial y} \right) dx\, dy$$

$$(21) \qquad t_{mk}^{(i)} = \iint_i \alpha_m \, \alpha_k \, dx\, dy$$

the integrations being carried out over the ith elementary region. For triangular elementary regions, $s^{(i)}$ and $t^{(i)}$ in general possess nine nonzero components, and are symmetric. These component values are found by substituting (14) into (20):

$$(22) \qquad s_{mk}^{(i)} = \frac{1}{4A^2} \iint_i (b_m\, b_k + c_m\, c_k)\, dx\, dy .$$

On carrying out the necessary integration,

$$(23) \qquad s_{mk}^{(i)} = \frac{-1}{4A} [(x_m - x_i)(x_k - x_i) + + (y_m - y_i)(y_k - y_i)]$$

Similarly, substitution into (21) yields

$$(24) \qquad t_{mk}^{(i)} = \frac{1}{4A^2} \iint_i (a_m + b_m\, x + c_m\, y) \cdot \cdot (a_k + b_k\, x + c_k\, y)\, dx\, dy .$$

After a lengthy integration, one obtains the surprisingly simple result

$$(25)\qquad t_{mk}^{(i)} = \begin{cases} \dfrac{1}{12} A\,, & m \neq k \\[2ex] \dfrac{1}{6} A\,, & m = k \end{cases}$$

Leaving out all rows and columns composed of zeros only, a single triangular element is thus described by the matrix contributions

$$(26)\qquad \begin{bmatrix} s_{11} & s_{12} & s_{13} \\ s_{21} & s_{22} & s_{23} \\ s_{31} & s_{32} & s_{33} \end{bmatrix} \begin{bmatrix} \varphi_1 \\ \varphi_2 \\ \varphi_3 \end{bmatrix} = k^2 \frac{A}{12} \begin{bmatrix} 2 & 1 & 1 \\ 1 & 2 & 1 \\ 1 & 1 & 2 \end{bmatrix} \begin{bmatrix} \varphi_1 \\ \varphi_2 \\ \varphi_3 \end{bmatrix}$$

where the s_{mk} are calculated from (23), and the area A is given by

$$(27)\qquad A = \frac{1}{2} \left| (y_m - y_i)(x_k - x_i) - (y_k - y_i)(x_m - x_i) \right| .$$

Comparison of Finite-Element and Finite-Difference Methods.

Were the waveguide problem to be solved by a finite-difference instead of finite-element method, one would subdivide the region R by a suitable mesh, most commonly regular and square. Further subdivision of the mesh into triangles by means of added lines but whithout adding new mesh-points, is always possible, and permits alternative treatment by the above finite-element method. A direct comparison of the two techniques can then be made.

In the finite-difference method, the Helmholtz equation is replaced by the matrix equation

$$(28)\qquad S\,\Phi = k^2\,\Phi\,.$$

The matrix S arises from discretisation of the Laplacian operator, and closely resembles the matrix S defined above. In fact, if the boundary conditions are of Dirichlet type, the matrices S in (13) and (28) are always identical; in the case of Neumann boundaries, it depends on which of the several possible methods of boundary treatment is employed in the finite-difference approach. However, the matrix T in equation (13) is replaced by the unit matrix in equation (28). It can be shown that this replacement is physically equivalent to replacing distributed sources by point sources located at the mesh-points. Therefore equation (13) seeks to establish a least-squares fit to the solution surface, while (28) requires a solution surface consistent with its own point-source replacement. One is thus led to the unexpected conclusion that the weakness of the usual finite-difference method lies not in its relatively crude modelling of the Laplacian operator, as is sometimes thought, but rather in its inaccurate discretisation of the identity operator. Considerable support is lent to this view by the numerical results reported below, which show that the finite-element method provides by far the more accurate results for any given number of points.

A General Computer Programme.

The procedure described above has been implemented in a computer programme coded in Fortran IV language. Typical running times of this programme are about 1-2 minutes, using an IBM 7090 computer (32K word memory).

Assembly of the matrices S and T constitutes the first, and most important, section of the programme. The input consists of a description of the waveguide cross-section, in the form of a listing of triangle vertex coordinates followed by an incidence listing which identifies the connections to be made between triangle vertices. Ordinarily, 50-100 vertices suffice to provide useful results for most guide cross-sections. Consequently both S and T can be stored in immediate-access memory concurrently, and subsequent matrix operations can be performed by explicit rather than iterative methods. The assembly of S and T typically occupies about 10 seconds of machine time.

From (26), S is always nonnegative definite and T always positive definite. It is therefore possible to perform a Choleski decomposition on T,

$$(29)\qquad L\,L' = T$$

where L is a positive definite lower triangular matrix. Inversion of L, and subsequent multiplication, then convert (13) into the form

$$(30)\qquad (L^{-1}\,S\,L^{-1'})\,(L'\,\Phi) = k^2\,(L'\,\Phi)\,.$$

This is an eigenvalue problem in its standard form. It is worth noting that Choleski decomposition, instead of inversion, of T is necessary to guarantee symmetry of the final matrix.

The actual solution of the eigenvalue problem uses techniques which may be regarded as standard. Tridiagonalisation is first performed, using the Householder transformation algorithm, and eigenvalues sought by the bisection method, employing Wilkinson's modified Sturm sequences [7]. Eigenvectors are determined by Wielandt iteration, and transformed back to find the desired Φ. No computational difficulty is encountered in locating as many eigenvalues as are thought desirable or physically meaningful. Neumann boundaries do not give rise to any special problems; the bisection method produces the redundant zero eigenvalue, of course, but it is only necessary to ignore it.

Modes of a Circular Waveguide.

In order to assess the efficacy of the new method, some rectangular and circular waveguides were investigated. The circular cross-section was subjected to a lengthy series of numerical experiments whose most important

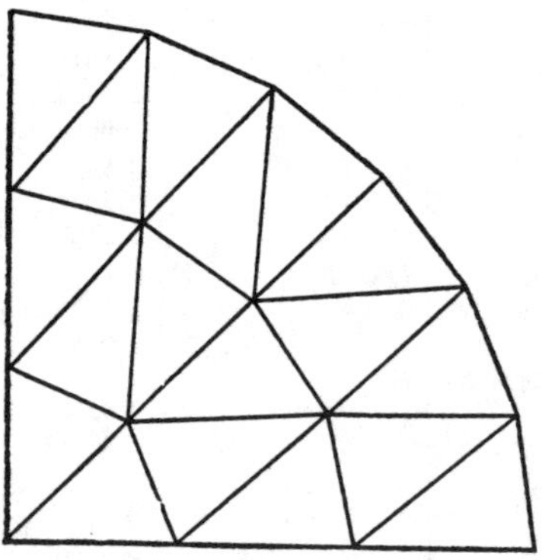

Fig. 2. — 18-element model of one quadrant of a circular waveguide

results are summarised here. These experiments sought to determine the accuracy to which the first sixteen modes (eighteen, counting degeneracies) of a circular guide could be determined. The E_{31} mode (the eleventh most important) was omitted from consideration, in order to permit modelling the guide cross-section by only one quadrant. Meshes of the type shown in fig. 2 were employed. These were generated by computer, and fited to the circular boundary so as to produce the correct enclosed area. Although the elementary regions are all roughly equilateral, their precise shapes and sizes vary so as to permit accurate fitting to the given boundaries.

The table below exhibits the cutoff wavelengths, normalised to unit guide radius, for several levels of approximation, and compares them with the known exact solution. The eigenvalues are found, in effect, as Rayleigh quotients in appropriate subspaces, so one would expect the computed cutoff wavelengths to represent lower bounds; examination of the table shows that this indeed appears to be so. It is also seen that the error varies essentially inversely with the number of elementary regions used. As a consequence, extrapolation can be used to improve the cut-off wavelengths so as to achieve very high accuracies indeed.

Tabulated cutoff wavelengths below should be compared with the results given by Davies and Huilwyk [3] for their solution of the same problem (for the dominant modes only) by finite differences. The maximum number of vertex points used in the present method was 100, as compared with about 10,000 for finite-difference work reported by the latter authors. However, the inherent accuracy of the finite-element technique is obviously higher; the error in cutoff wavelength of the H_{11} and E_{01} modes achieved by finite differences, using roughly the same amount of immediate-access storage and about the same length of computer time, is higher than that resulting from finite-element analysis. One is therefore inclined to conclude that the finite-element method is to be preferred. It is highly competitive for the dominant modes, and produces in addition a number of higher modes, with no extra programming effort and virtually no extra computing time.

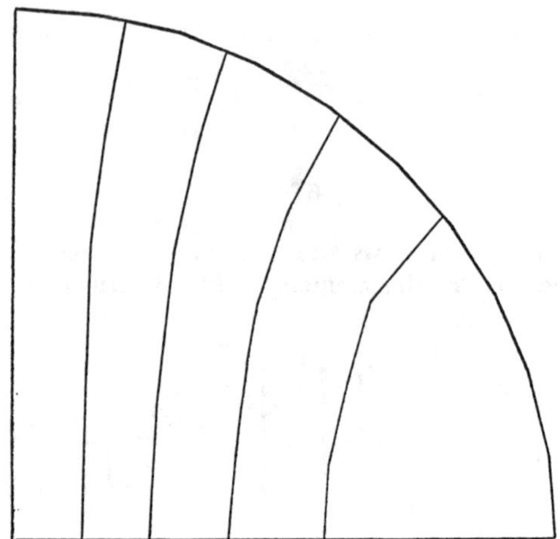

Fig. 3. — Round guide modelled by 50 elements, H_{11} mode.

Figures 3-4 show two modal patterns generated for the circular guide, subdivided into 50 triangular elements per quadrant. This comparatively crude model illustrates clearly, if somewhat exaggeratedly, the essential characteristics of finite-element solutions. The contour lines of each solution surface are composed of straight-line segments, as would be expected from a surface consisting of flat facets.

The mode plots shown here are of course much more crude than those obtainable with a finer subdivision. In fact, subdivision into 162 elementary regions per quadrant produces contour lines almost indistinguishable from smooth curves, especially for the lower-order modes. With increasing mode number, the complexity of the solution surface grows,

Calculated Cutoff Wavelengths of a Circular Waveguide.

Mode	Number of elements in one waveguide quadrant							Exact
	18	32	50	72	98	128	162	
H_{11}	3.3564	3.3915	3.3991	3.4032	3.4056	3.4073	3.4083	3.4129
E_{01}	2.5773	2.6002	2.6044	2.6069	2.6084	2.6094	2.6100	2.6125
H_{21}	1.9977	2.0286	2.0388	2.0439	2.0473	2.0496	2.0511	2.0574
H_{01}, E_{11}	1.5452	1.5899	1.6097	1.6183	1.6230	1.6269	1.6296	1.6397
H_{31}	1.4251	1.4580	1.4707	1.4780	1.4825	1.4855	1.4875	1.4956
E_{21}	1.1102	1.1598	1.1813	1.1938	1.2015	1.2065	1.2100	1.2234
H_{41}	1.1156	1.1428	1.1473	1.1574	1.1674	1.1706	1.1728	1.1817
H_{12}	1.0394	1.0988	1.1265	1.1417	1.1513	1.1575	1.1619	1.1786
E_{02}	1.0285	1.0755	1.0968	1.1086	1.1162	1.1212	1.1246	1.1383
H_{51}	0.8891	0.9264	0.9436	0.9540	0.9604	0.9647	0.9677	0.9793
H_{22}	0.7881	0.8470	0.8761	0.8941	0.9050	0.9122	0.9173	0.9369
H_{02}, E_{12}	0.7538	0.8023	0.8350	0.8540	0.8618	0.8695	0.8748	0.8955
H_{61}	0.7397	0.7783	0.7943	0.8088	0.8160	0.8208	0.8242	0.8376
E_{41}	0.7003	0.7499	0.7715	0.7907	0.8002	0.8065	0.8109	0.8280
H_{32}	0.6373	0.6849	0.7163	0.7353	0.7475	0.7556	0.7614	0.7839
E_{22}	0.5843	0.6437	0.6760	0.6771	0.7077	0.7163	0.7224	0.7465

and plots in all cases present an increasingly ragged appearance.

Solution of Difficult Waveguide Problems.

The general programme described above is immediately usable for analysis of waveguides of nearly any shape, and requires only the preparation of an appropriate data deck for each cross-section. To verify the correct functioning of the programme, a rectangular coaxial guide was analysed, for which there exist solutions based on orthonormal block analysis [1, 8]. Fig. 5 shows one of the two fundamental TE modes of such a guide, and again exhibits the characteristic piecewise-linear behaviour of contour lines. This pattern was calculated using a partly machine-generated subdivision mesh consisting of equilateral triangles over most of the guide cross-section, and odd-shaped triangles near the edges where the boundary lines are represented exactly. The cutoff frequencies for this guide agree with those of Bräckelmann [1] to the accuracy to which the given curves can be read.

Other guide shapes have been investigated by means of the technique described, for example the T-septate lunar cross-section. Not surprisingly, the accuracies obtained vary with the nature of the cross-section, being generally lower where re-entrant corners and sharp edges are encountered, and higher where the cross-sectional shape is essentially convex. In all cases, accuracy has been comparable to what can be achieved by finite-difference methods with the same computational effort.

Conclusions.

A new method for analysing hollow uniform waveguides has been developed. This method permits determination of the propagating modes with sufficient accuracy for most practical purposes; in contrast to finite-difference techniques, to which it bears a certain resemblance, it produces the higher-order as well as the dominant propagating modes.

Numerical experiments have shown that the new method is capable of accuracy similar to that obtained by finite differences, assuming comparable computational facilities. The accuracy deteriorates with increasing mode order, limiting the number of useful modes to about 10-20 where no symmetries exist, or 20-50 where computational advantage can be taken of symmetry. It is believed that this number is adequately large to permit solution of various waveguide problems involving the higher modes. A distinctive feature of the new method is that the mode patterns are uniquely defined for all points of the guide cross-section, and not only for a discrete set of "sampling" points.

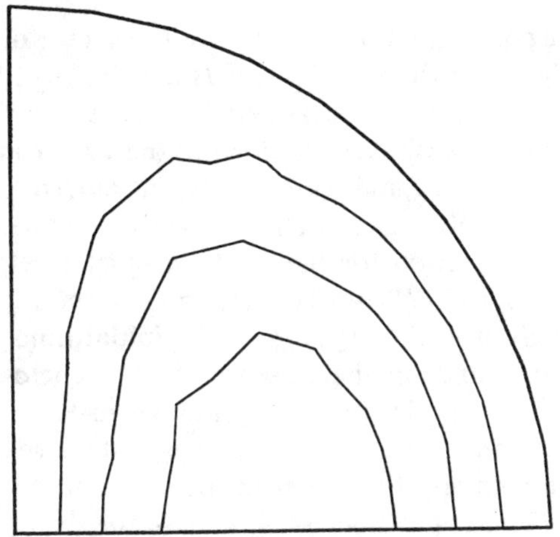

Fig. 4. — Round guide modelled by 50 elements, E_{11} mode.

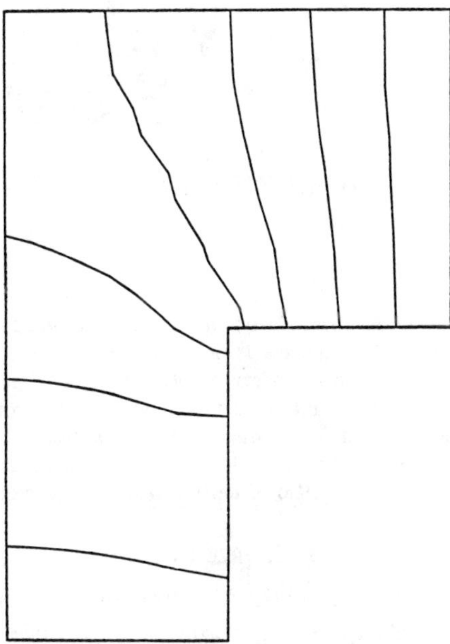

Fig. 5. — One quadrant of a rectangular coaxial guide, ${}_sH_{11}$ mode in Bräckelmann's nomenclature.

Acknowledgments.

The author wishes to thank Professor J. Brown, Department of Electrical Engineering, Imperial College of Science and Technology, for provision of research facilities. The work was supported financially in part by the National Research Council of Canada.

REFERENCES

[1] W. Bräckelmann: *Wellentypen auf der Streifenleitung mit rechteckigem Schirm.* - « Archiv der elektr. Übertragung », 21, 1967, pp. 641÷648.

[2] H. H. Meinke, W. Baier: *Die Eigenschaften von Hohlleitern allgemeineren Querschnitts.* - « Nachrichtentechn. Zeitschr. », 19, 1966, pp. 662÷670.

[3] J. B. Davies, C. A. Muilwyk: *Numerical solution of uniform hollow waveguides with boundaries of arbitrary shape.* - « Proceedings I.E.E. », 113, 1966, pp. 277÷284.

[4] P. M. Morse, H. Feshbach: *Methods of theoretical physics* (Part I). New York: McGraw-Hill 1953.

[5] R. F. Harrington: *Field computation by moment methods.* - New York: Macmillan 1968.

[6] O. C. Zienkiewicz, Y. K. Cheung: *Finite elements in the solution of field problems.* - « The Engineer », 1965, pp. 507÷510.

[7] W. Barth, R. S. Martin, J. H. Wilkinson: *Calculation of the eigenvalues of a symmetric tridiagonal matrix by the method of bisection.* - « Numerische Mathematik », 9, 1967, pp. 386÷393.

[8] L. Gruner: *Higher-order modes in rectangular coaxial waveguides.* - « I.E.E.E. Transact. on Microwave Theory and Tech. », MTT-15, 1967, pp. 483÷485.

Unified Approach to the Derivation of Variational Expression for Electromagnetic Fields

KATSUMI MORISHITA, STUDENT MEMBER, IEEE, AND NOBUAKI KUMAGAI, SENIOR MEMBER, IEEE

***Abstract*—A unified procedure to derive the variational expressions for electromagnetic field problems is presented. It is shown that the variational expressions for a variety of electromagnetic parameters such as, for instance, a resonant frequency, a propagation constant, and an impedance matrix, can be formulated systematically all from "the principle of least action" point of view. It is pointed out that the Maxwell's equations themselves can also be derived from the least action principle.**

I. Introduction

THE variational method is one of the very powerful techniques capable of treating a large variety of electromagnetic fields problems. So far, however, the appropriate variational expressions have been derived for each specific problem by a trial-and-error approach which is attended with laborious and time-consuming difficulties.

To avoid such a tricky approach, the procedures which are systematic to some extent have been proposed previously by several authors. Rumsey [1], Berk [2], and Harrington [3] have described the method to derive the variational expressions by using the "reaction concept," while Cairo and Kahan [4] have shown that the variational expressions can be obtained by means of "transpose operator and field." However, the method of "reaction concept" lacks uniformity in the strict sense because the definitions of reaction are different for the three-dimensional problems and the two-dimensional problems. In Cairo and Kahan's method, the physical meaning of the "transpose field" is hard to understand and also the technical skills are still required.

In the present paper, a unified procedure to derive the variational expressions for the electromagnetic field problems is given from "the principle of least action" point of view. We shall show first that the Maxwell's equations themselves can be yielded from the principle of least action, and then derive the variational expressions for a resonant frequency, a propagation constant, and an impedance matrix, systematically, all from the least action principle.

II. A Review of the Principle of Least Action

It is well known that a general feature of the classical physical phenomenon can be explained by the principle of least action, or, in other words, the stationarity of the actions which are suitably defined fundamental physical quantities such as energy, time, and others. The Fermat's principle in optics and the Thomson's theorem in the electrostatic fields are the typical examples. In dynamics, the system satisfies the Lagrange equation when the action, that is, the integration of Lagrange function with respect to time, is stationary. It follows that the principle of least action in dynamics states that "the motion of a matter passes through the path for which the action has the stationary value."

Fig. 1. Region V enclosed by closed surface S containing the electromagnetic fields, charge, and current. $\boldsymbol{n}$ is a unit vector normal to S and outwardly directed.

In electromagnetism, the action J is expressed in terms of Lagrange function L as follows [5]:

$$J = \int_{t_0}^{t_1} L \, dt \tag{1}$$

where

$$L = \int_V \left(\tfrac{1}{2}\{\boldsymbol{E}(\boldsymbol{r},t) \cdot \boldsymbol{D}(\boldsymbol{r},t) - \boldsymbol{H}(\boldsymbol{r},t) \cdot \boldsymbol{B}(\boldsymbol{r},t)\} + \boldsymbol{A}(\boldsymbol{r},t) \cdot \boldsymbol{J}(\boldsymbol{r},t) - \rho(\boldsymbol{r},t)\phi(\boldsymbol{r},t)\right) dv \tag{2}$$

$$\boldsymbol{B}(\boldsymbol{r},t) = \nabla \times \boldsymbol{A}(\boldsymbol{r},t)$$

$$\boldsymbol{E}(\boldsymbol{r},t) = -\frac{\partial \boldsymbol{A}(\boldsymbol{r},t)}{\partial t} - \nabla\phi(\boldsymbol{r},t). \tag{3}$$

In the foregoing equation, $\boldsymbol{r}$ and t are the vector distance from the origin and the time, $\boldsymbol{E}$ and $\boldsymbol{H}$ are the electric and the magnetic field intensities, $\boldsymbol{D}$ and $\boldsymbol{B}$ are the electric and the magnetic flux densities, $\boldsymbol{A}$ and ϕ are the vector and the scalar potentials, and $\boldsymbol{J}$ and ρ are the electric current density and the electric charge density, respectively. Then, the principle of least action for the electromagnetic fields can be stated as follows: "Provided that the correct values of $\boldsymbol{A}$ and ϕ are given in the region V at the initial time t_0 and the final time t_1 and on the closed surface S enclosing V during the time interval $[t_0,t_1]$, then $\boldsymbol{A}$ and ϕ in V for which J becomes stationary can be determined, and those $\boldsymbol{A}$ and ϕ thus determined give the true fields throughout the region V" (Fig. 1). The true values of $\boldsymbol{E}$ and $\boldsymbol{B}$ are then yielded from (3) using these $\boldsymbol{A}$ and ϕ, and further, the true expres-

Reprinted from *IEEE Trans. Microwave Theory Tech.*, vol. MTT-25. no. 1, pp. 34–40, Jan. 1977.

sions for $\boldsymbol{D}$ and $\boldsymbol{H}$ can be obtained with the aid of the constitutive relations.

Let us consider next another expression of the principle of least action for the electromagnetic fields in the frequency region. Extending the time interval $[t_0,t_1]$ to $(-\infty,\infty)$, the time domain problem of the electromagnetic fields can be transformed into the frequency domain problem by means of Fourier transformation. Thus the integration with respect to the time t in (1) becomes the integration with respect to the frequency f as follows:

$$J = \int_{-\infty}^{\infty} df \int_V (\tfrac{1}{2}\{\boldsymbol{E}(\boldsymbol{r},\omega)\cdot\boldsymbol{D}(\boldsymbol{r},\omega)^* - \boldsymbol{H}(\boldsymbol{r},\omega)\cdot\boldsymbol{B}(\boldsymbol{r},\omega)^*\} + \boldsymbol{A}(\boldsymbol{r},\omega)\cdot\boldsymbol{J}(\boldsymbol{r},\omega)^* - \rho(\boldsymbol{r},\omega)\phi(\boldsymbol{r},\omega)^*)\, dv \quad (4)$$

where * stands for the complex conjugate, $\boldsymbol{A}(\boldsymbol{r},\omega)$ and $\phi(\boldsymbol{r},\omega)$ are the Fourier transforms of $\boldsymbol{A}(\boldsymbol{r},t)$ and $\phi(\boldsymbol{r},t)$, respectively, and ω is an angular frequency. The frequency domain expressions of (3) are

$$\boldsymbol{B}(\boldsymbol{r},\omega) = \nabla\times\boldsymbol{A}(\boldsymbol{r},\omega)$$
$$\boldsymbol{E}(\boldsymbol{r},\omega) = -j\omega\boldsymbol{A}(\boldsymbol{r},\omega) - \nabla\phi(\boldsymbol{r},\omega). \quad (5)$$

Let us assume that the materials involved are linear and dissipationless, for simplicity, but are inhomogeneous and anisotropic, in general. The constitutive relations can then be expressed in the form

$$\boldsymbol{D}(\boldsymbol{r},\omega) = \hat{\varepsilon}(\boldsymbol{r},\omega)\boldsymbol{E}(\boldsymbol{r},\omega)$$
$$\boldsymbol{B}(\boldsymbol{r},\omega) = \hat{\mu}(\boldsymbol{r},\omega)\boldsymbol{H}(\boldsymbol{r},\omega). \quad (6)$$

Since the materials involved are assumed to be loss free, both the tensor permittivity $\hat{\varepsilon}$ and the tensor permeability $\hat{\mu}$ are Hermitian.

Since $\boldsymbol{E}(\boldsymbol{r},t)$, $\boldsymbol{H}(\boldsymbol{r},t)$, $\boldsymbol{D}(\boldsymbol{r},t)$, $\boldsymbol{B}(\boldsymbol{r},t)$, $\boldsymbol{J}(\boldsymbol{r},t)$, and $\rho(\boldsymbol{r},t)$ are real quantities, both $\boldsymbol{A}(\boldsymbol{r},t)$ and $\phi(\boldsymbol{r},t)$ are also the real functions. Therefore, the Fourier transformations of these real functions, $\boldsymbol{E}(\boldsymbol{r},\omega)$, $\boldsymbol{H}(\boldsymbol{r},\omega)$, $\boldsymbol{D}(\boldsymbol{r},\omega)$, $\boldsymbol{B}(\boldsymbol{r},\omega)$, $\boldsymbol{A}(\boldsymbol{r},\omega)$, $\phi(\boldsymbol{r},\omega)$, $\boldsymbol{J}(\boldsymbol{r},\omega)$, and $\rho(\boldsymbol{r},\omega)$, possess the following properties:

$$\boldsymbol{E}(\boldsymbol{r},\omega) = \boldsymbol{E}(\boldsymbol{r},-\omega)^*$$
$$\boldsymbol{H}(\boldsymbol{r},\omega) = \boldsymbol{H}(\boldsymbol{r},-\omega)^* \quad (7)$$

etc. Substituting the relations given by (6) and (7) into (4), we obtain

$$J = \int_0^{\infty} df \int_V (\boldsymbol{E}(\boldsymbol{r},\omega)^*\cdot\hat{\varepsilon}(\boldsymbol{r},\omega)\boldsymbol{E}(\boldsymbol{r},\omega) - \boldsymbol{H}(\boldsymbol{r},\omega)^* \cdot\hat{\mu}(\boldsymbol{r},\omega)\boldsymbol{H}(\boldsymbol{r},\omega) + \boldsymbol{A}(\boldsymbol{r},\omega)^*\cdot\boldsymbol{J}(\boldsymbol{r},\omega) + \boldsymbol{A}(\boldsymbol{r},\omega)\cdot\boldsymbol{J}(\boldsymbol{r},\omega)^* - \rho(\boldsymbol{r},\omega)\phi(\boldsymbol{r},\omega)^* - \rho(\boldsymbol{r},\omega)^*\phi(\boldsymbol{r},\omega))\, dv. \quad (8)$$

The first-order variation of the action J due to the small variation in ϕ is given as

$$\delta J_\phi = \int_0^{\infty} df \int_V \delta\phi^*(\nabla\cdot\hat{\varepsilon}\boldsymbol{E} - \rho)\, dv - \int_0^{\infty} df \int_{S_d} \delta\phi^*\hat{\varepsilon}\boldsymbol{E}\cdot\boldsymbol{n}\, ds + \text{c.c.} \quad (9)$$

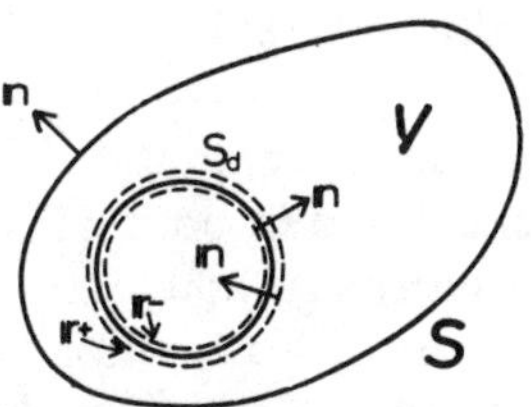

Fig. 2. Volume V enclosed by closed surface S. S_d indicates the surface across which $\hat{\varepsilon}$ and $\hat{\mu}$ change discontinuously. $\boldsymbol{r}^+$ and $\boldsymbol{r}^-$ designate the vector distances indicating outer and inner sides of discontinuous boundary S_d.

where the term designated as c.c. represents the complex conjugate of the first two terms on the right-hand side of the preceding equation. The surface integral must be carried out over both sides of the surface S_d across which the material constants $\hat{\varepsilon}$ and $\hat{\mu}$ change discontinuously. $\boldsymbol{n}$ is a unit vector normal to the boundary surface and directed as shown in Fig. 2.

In order that the action J be stationary for a variation in ϕ, the following equation

$$\nabla\cdot\hat{\varepsilon}(\boldsymbol{r},\omega)\boldsymbol{E}(\boldsymbol{r},\omega) = \rho(\boldsymbol{r},\omega) \quad (10)$$

must be satisfied in the region where $\hat{\varepsilon}$ and $\hat{\mu}$ are continuous, and at the same time, the following condition

$$\boldsymbol{n}\cdot\{\hat{\varepsilon}(\boldsymbol{r}^+,\omega)\boldsymbol{E}(\boldsymbol{r}^+,\omega) - \hat{\varepsilon}(\boldsymbol{r}^-,\omega)\boldsymbol{E}(\boldsymbol{r}^-,\omega)\} = 0 \quad (11)$$

must also be satisfied on both sides of discontinuity surface S_d. In the foregoing equation, $\boldsymbol{r}^+$ and $\boldsymbol{r}^-$ are the vector distance indicating outer and inner sides of the discontinuity surface S_d as shown in Fig. 2.

Similarly, the first-order variation of the action J with respect to $\boldsymbol{A}$ is given by

$$\delta J_A = \int_0^{\infty} df \int_V \delta\boldsymbol{A}^*\cdot(j\omega\hat{\varepsilon}\boldsymbol{E} - \nabla\times\boldsymbol{H} + \boldsymbol{J})\, dv + \int_0^{\infty} df \int_{S_d} (\boldsymbol{H}\times\delta\boldsymbol{A}^*)\cdot\boldsymbol{n}\, ds + \text{c.c.} \quad (12)$$

The notations in the preceding equation are the same as those used in (9). To make δJ_A zero for a variation in $\boldsymbol{A}$, we get

$$\nabla\times\boldsymbol{H}(\boldsymbol{r},\omega) = j\omega\hat{\varepsilon}(\boldsymbol{r},\omega)\boldsymbol{E}(\boldsymbol{r},\omega) + \boldsymbol{J}(\boldsymbol{r},\omega) \quad (13)$$

in the region where $\hat{\varepsilon}$ and $\hat{\mu}$ are continuous, and also

$$\boldsymbol{n}\times\{\boldsymbol{H}(\boldsymbol{r}^+,\omega) - \boldsymbol{H}(\boldsymbol{r}^-,\omega)\} = 0 \quad (14)$$

on both sides of discontinuity surface S_d. By choosing the trial functions for $\boldsymbol{A}$ and ϕ in such a way that $\boldsymbol{n}\times\boldsymbol{A}$, $\boldsymbol{n}\times\nabla\phi$, and ϕ are continuous across the discontinuity surface S_d, the following relation

$$\boldsymbol{n}\times\{\boldsymbol{E}(\boldsymbol{r}^+,\omega) - \boldsymbol{E}(\boldsymbol{r}^-,\omega)\} = 0 \quad (15)$$

is satisfied on both sides of S_d. Therefore, if $\boldsymbol{A}$ and ϕ are determined in such a manner that for those $\boldsymbol{A}$ and ϕ the action J becomes stationary and also $\boldsymbol{n}\times\boldsymbol{A}$, $\boldsymbol{n}\times\nabla\phi$, and ϕ are continuous across S_d, (5), (10), and (13), i.e., the Maxwell's equations, are satisfied in the region where $\hat{\varepsilon}$ and $\hat{\mu}$ are continuous, and at the same time, (14) and (15), i.e., the boundary conditions, are also satisfied on the discontinuity surface S_d.

We can conclude, therefore, that the physically realizable electromagnetic fields $\boldsymbol{E}$ and $\boldsymbol{H}$, both satisfying the Maxwell's equations and the necessary boundary conditions, can be derived uniquely from the least action principle. It should be noted that the additional relation, that is, the so-called gauge condition, is required to determine the vector and scalar potentials $\boldsymbol{A}$ and ϕ uniquely.

III. Variational Expressions for Resonant Frequency

In this section, we shall derive the variational expressions for the resonant frequency of the cavity resonator using the results obtained in the preceding section.

We will assume that the cavity resonator is formed by a perfect conducting wall, and the materials involved in the resonator are inhomogeneous and anisotropic, in general. But the materials involved are assumed to be linear, nondispersive, and dissipation free, for simplicity. As a gauge condition, let ϕ be specified, for later convenience, as

$$\phi(\boldsymbol{r},\omega) = 0. \tag{16}$$

Under this condition, the relation between $\boldsymbol{A}$ and $\boldsymbol{E}$ given by (5) is reduced to

$$\boldsymbol{E}(\boldsymbol{r},\omega) = -j\omega\boldsymbol{A}(\boldsymbol{r},\omega). \tag{17}$$

We take for V in (8) the charge and current free region within the cavity resonator. Making use of (17), together with the aforementioned assumptions, reduces (8) to

$$J = \int_0^\infty df \int_V \Big(\boldsymbol{E}(\boldsymbol{r},\omega)^* \cdot \hat{\varepsilon}(\boldsymbol{r})\boldsymbol{E}(\boldsymbol{r},\omega) - \frac{1}{\omega^2}\{\nabla \times \boldsymbol{E}(\boldsymbol{r},\omega)\}^* \cdot \hat{\mu}^{-1}(\boldsymbol{r})\{\nabla \times \boldsymbol{E}(\boldsymbol{r},\omega)\}\Big)\, dv. \tag{18}$$

Thus the stationary problem being expressed in terms of $\boldsymbol{A}$ was transformed to the stationary problem expressed in terms of $\boldsymbol{E}$.

Since the electric field in a cavity resonator can be expressed by a linear combination of the electric fields of the individual resonant modes, $\boldsymbol{E}(\boldsymbol{r},\omega)$ can be written in the form

$$\boldsymbol{E}(\boldsymbol{r},\omega) = \sum_i \delta(\omega - \omega_i)\boldsymbol{E}_i(\boldsymbol{r}) \tag{19}$$

where $\boldsymbol{E}_i(\boldsymbol{r})$ and ω_i are the electric field and the resonant frequency of the ith mode, respectively, and $\delta(\omega - \omega_i)$ signifies the delta function. Substituting (19) into (18), and carrying out the integration with respect to the frequency f, we get

$$J = \sum_i L_i\delta(0) \tag{20}$$

where

$$L_i = \int_V \Big(\boldsymbol{E}_i(\boldsymbol{r})^* \cdot \hat{\varepsilon}(\boldsymbol{r})\boldsymbol{E}_i(\boldsymbol{r}) - \frac{1}{\omega_i^2}\{\nabla \times \boldsymbol{E}_i(\boldsymbol{r})\}^* \cdot \hat{\mu}^{-1}(\boldsymbol{r})\{\nabla \times \boldsymbol{E}_i(\boldsymbol{r})\}\Big)\, dv. \tag{21}$$

The amplitude of $\boldsymbol{E}_i(\boldsymbol{r})$ is determined by the electric field distributions at $t = -\infty$. If we assume that there exists only ith mode alone at $t = -\infty$, (20) becomes

$$J = L_i\delta(0). \tag{22}$$

Therefore, the stationary problem for J is reduced to that for L_i. Hence we can determine the correct values of the electric field $\boldsymbol{E}(\boldsymbol{r})$ and the resonant frequency ω in such a manner that for those $\boldsymbol{E}(\boldsymbol{r})$ and ω,

$$L = \int_V \Big(\boldsymbol{E}(\boldsymbol{r})^* \cdot \hat{\varepsilon}(\boldsymbol{r})\boldsymbol{E}(\boldsymbol{r}) - \frac{1}{\omega^2}\{\nabla \times \boldsymbol{E}(\boldsymbol{r})\}^* \cdot \hat{\mu}^{-1}(\boldsymbol{r})\{\nabla \times \boldsymbol{E}(\boldsymbol{r})\}\Big)\, dv \tag{23}$$

becomes stationary. The subscript i will be omitted, for simplicity, hereafter.

The first integral on the right-hand side of (21) gives the electric energy, while the second represents the magnetic energy, both stored in the cavity resonator. Since the stored electric and the magnetic energies in the resonator are equal at the resonant frequency, L given by (23) must be zero for the correct value of the resonant frequency ω, which in turn leads to the expression [3]

$$\omega^2 = \frac{\int_V (\{\nabla \times \boldsymbol{E}(\boldsymbol{r})\}^* \cdot \hat{\mu}^{-1}(\boldsymbol{r})\{\nabla \times \boldsymbol{E}(\boldsymbol{r})\})\, dv}{\int_V (\boldsymbol{E}(\boldsymbol{r})^* \cdot \hat{\varepsilon}(\boldsymbol{r})\boldsymbol{E}(\boldsymbol{r}))\, dv}. \tag{24}$$

In other words, the stationary problem for L given by (23) is equivalent to the stationary problem for ω^2 given by (24). We can conclude, therefore, that (24) represents the variational expression for the resonant frequency of the cavity resonator. The variational expression given by (24) coincides with that obtained previously by Berk [2].

The foregoing variational expression is expressed in terms of the electric field alone. Alternative to the preceding formulation, we may derive a similar expression which contains both electric and magnetic fields. By steps similar to those leading to (23) from (8), we get

$$L = \int_V (\{j\omega\boldsymbol{A}(\boldsymbol{r})\}^* \cdot \hat{\varepsilon}(\boldsymbol{r})\{j\omega\boldsymbol{A}(\boldsymbol{r})\} - \{\nabla \times \boldsymbol{A}(\boldsymbol{r})\}^* \cdot \hat{\mu}^{-1}(\boldsymbol{r})\{\nabla \times \boldsymbol{A}(\boldsymbol{r})\})\, dv \tag{25}$$

where

$$\boldsymbol{H}(\boldsymbol{r}) = \hat{\mu}^{-1}(\boldsymbol{r})\nabla \times \boldsymbol{A}(\boldsymbol{r})$$
$$\boldsymbol{E}(\boldsymbol{r}) = -j\omega\boldsymbol{A}(\boldsymbol{r}). \tag{26}$$

Substituting the vector identity

$$(\nabla \times \boldsymbol{A})^* \cdot \hat{\mu}^{-1}(\nabla \times \boldsymbol{A}) = \boldsymbol{A}^* \cdot \nabla \times \hat{\mu}^{-1}\nabla \times \boldsymbol{A} - \nabla \cdot \{(\hat{\mu}^{-1}\nabla \times \boldsymbol{A}) \times \boldsymbol{A}^*\} \tag{27}$$

into (25) and applying Gauss' theorem, and further, in view of (26), we get the expression for L in the form

$$L = \int_V \Big(\boldsymbol{E}(\boldsymbol{r})^* \cdot \hat{\varepsilon}(\boldsymbol{r})\boldsymbol{E}(\boldsymbol{r}) + \boldsymbol{H}(\boldsymbol{r})^* \cdot \hat{\mu}(\boldsymbol{r})\boldsymbol{H}(\boldsymbol{r}) - \frac{j}{\omega}\{\boldsymbol{H}(\boldsymbol{r})^* \cdot \nabla \times \boldsymbol{E}(\boldsymbol{r}) - \boldsymbol{E}(\boldsymbol{r})^* \cdot \nabla \times \boldsymbol{H}(\boldsymbol{r})\}\Big)\, dv - \frac{j}{\omega}\int_{S+S_d} \{\boldsymbol{H}(\boldsymbol{r}) \times \boldsymbol{E}(\boldsymbol{r})^*\} \cdot \boldsymbol{n}\, ds \tag{28}$$

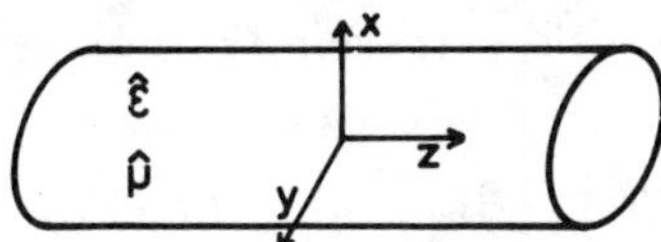

Fig. 3. Transmission line containing inhomogeneous and anisotropic materials. Direction of wave propagation is in z axis along which the line is uniform.

TABLE I
SYMMETRICAL TRANSFORMATIONS BETWEEN ELECTROMAGNETIC QUANTITIES

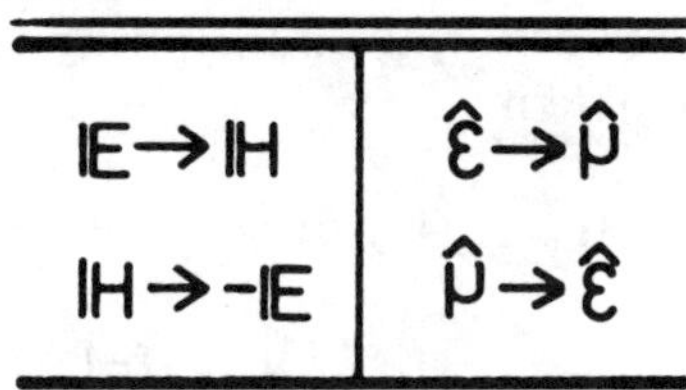

$\mathbf{E} \to \mathbf{H}$	$\hat{\varepsilon} \to \hat{\mu}$
$\mathbf{H} \to -\mathbf{E}$	$\hat{\mu} \to \hat{\varepsilon}$

where S indicates the surface of a perfect conducting wall enclosing a cavity resonator while S_d is the surface across which $\hat{\varepsilon}$ and $\hat{\mu}$ change discontinuously. Using the same procedure as that used to obtain (24), we found that

$$\omega = \frac{j\int_V \{\boldsymbol{H}(\boldsymbol{r})^* \cdot \nabla \times \boldsymbol{E}(\boldsymbol{r}) - \boldsymbol{E}(\boldsymbol{r})^* \cdot \nabla \times \boldsymbol{H}(\boldsymbol{r})\}\, dv + j\int_{S+S_d} \{\boldsymbol{H}(\boldsymbol{r}) \times \boldsymbol{E}(\boldsymbol{r})^*\} \cdot \boldsymbol{n}\, ds}{\int_V \{\boldsymbol{E}(\boldsymbol{r})^* \cdot \hat{\varepsilon}(\boldsymbol{r})\boldsymbol{E}(\boldsymbol{r}) + \boldsymbol{H}(\boldsymbol{r})^* \cdot \hat{\mu}(\boldsymbol{r})\boldsymbol{H}(\boldsymbol{r})\}\, dv}. \tag{29}$$

This is another variational expression for the resonant frequency of the cavity resonator which involves both electric and magnetic fields. It can be shown that the trial function for $\boldsymbol{E}$ in (29) must satisfy the conditions such that $\boldsymbol{n} \times \delta\boldsymbol{E} = 0$ on the conductor surface S and also $\boldsymbol{n} \times \delta\boldsymbol{E}$ is continuous across the discontinuity boundary S_d, where δ signifies a small variation in $\boldsymbol{E}$.

Making use of the symmetry of the Maxwell's equations shown in Table I, the variational expression (29) can be rewritten in slightly different form

$$\omega = \frac{j\int_V \{\boldsymbol{H}(\boldsymbol{r})^* \cdot \nabla \times \boldsymbol{E}(\boldsymbol{r}) - \boldsymbol{E}(\boldsymbol{r})^* \cdot \nabla \times \boldsymbol{H}(\boldsymbol{r})\}\, dv - j\int_{S+S_d} \{\boldsymbol{E}(\boldsymbol{r}) \times \boldsymbol{H}(\boldsymbol{r})^*\} \cdot \boldsymbol{n}\, ds}{\int_V \{\boldsymbol{E}(\boldsymbol{r})^* \cdot \hat{\varepsilon}(\boldsymbol{r})\boldsymbol{E}(\boldsymbol{r}) + \boldsymbol{H}(\boldsymbol{r})^* \cdot \hat{\mu}(\boldsymbol{r})\boldsymbol{H}(\boldsymbol{r})\}\, dv} \tag{30}$$

which coincides with that obtained by Berk [2]. The trial function for $\boldsymbol{H}$ in (30) must satisfy the condition that $\boldsymbol{n} \times \delta\boldsymbol{H}$ is continuous across the discontinuity boundary S_d.

IV. VARIATIONAL EXPRESSIONS FOR PROPAGATION CONSTANT

Let us derive next the variational expressions for the propagation constant of the guided waves traveling along a uniform transmission line. It is assumed that the materials involved are inhomogeneous and anisotropic, in general, but are linear, nondispersive, and dissipation free. Fig. 3 illustrates the transmission line under consideration which is uniform in a direction of wave propagation z.

To derive the variational expressions for the propagation constant, divide the volume integral in (28) into the surface integral over the transverse (xy) plane S and the integral along the propagation axis z. Further, transforming the integration with respect to z into the integration with respect to the propagation constant β by performing the

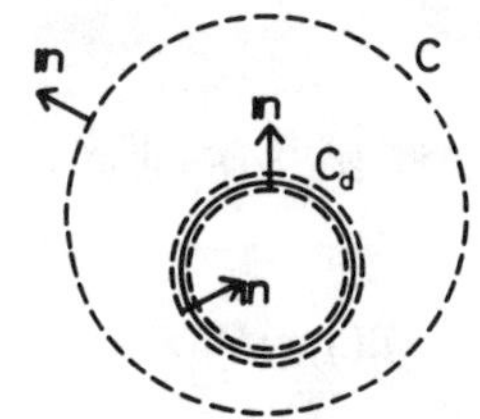

Fig. 4. Contours used in the evaluation of integration.

Fourier transformation, we get

$$\begin{aligned} L = \frac{1}{2\pi\omega}\int_{-\infty}^{\infty} d\beta \Bigg[\iint_S & (\{\omega \boldsymbol{E}(x,y,\beta)^* \cdot \hat{\varepsilon}(x,y)\boldsymbol{E}(x,y,\beta) \\ & + \omega \boldsymbol{H}(x,y,\beta)^* \cdot \hat{\mu}(x,y)\boldsymbol{H}(x,y,\beta) \\ & - j\boldsymbol{H}(x,y,\beta)^* \cdot \nabla_t \times \boldsymbol{E}(x,y,\beta) \\ & + j\boldsymbol{E}(x,y,\beta)^* \cdot \nabla_t \times \boldsymbol{H}(x,y,\beta)\} \\ & + \beta\{\boldsymbol{H}(x,y,\beta)^* \cdot \boldsymbol{i}_z \times \boldsymbol{E}(x,y,\beta) \\ & - \boldsymbol{E}(x,y,\beta)^* \cdot \boldsymbol{i}_z \times \boldsymbol{H}(x,y,\beta)\})\, ds \\ & - j\int_{C+C_d} \{\boldsymbol{H}(x,y,\beta) \times \boldsymbol{E}(x,y,\beta)^*\} \cdot \boldsymbol{n}\, dl \Bigg]. \end{aligned} \tag{31}$$

In the foregoing equation, C represents a closed contour in the transverse plane as shown in Fig. 4. For the open-type transmission lines, C must be a closed contour enclosing the transmission line at infinity, while in the case of the metallic waveguide, C is a cross-sectional boundary of the guide wall. C_d indicates the line in the transverse plane across which $\hat{\varepsilon}$ and $\hat{\mu}$ change discontinuously. $\boldsymbol{E}(x,y,\beta)$ and $\boldsymbol{H}(x,y,\beta)$ are the Fourier transforms of $\boldsymbol{E}(\boldsymbol{r})$ and $\boldsymbol{H}(\boldsymbol{r})$, respectively, and $\nabla_t = \boldsymbol{i}_x(\partial/\partial x) + \boldsymbol{i}_y(\partial/\partial y)$ where $\boldsymbol{i}_x$ and $\boldsymbol{i}_y$ are the unit vectors directed in x and y directions, respectively.

The electromagnetic fields of the wave propagation in the z direction can be expressed in terms of the linear combinations of the electromagnetic fields of each propagation mode. Hence

$$\boldsymbol{E}(x,y,\beta) = \sum_i \delta(\beta + \beta_i)\boldsymbol{E}_i(x,y)$$

$$\boldsymbol{H}(x,y,\beta) = \sum_i \delta(\beta + \beta_i)\boldsymbol{H}_i(x,y) \tag{32}$$

where $\boldsymbol{E}_i(x,y)$ and $\boldsymbol{H}_i(x,y)$ represent the electric and the magnetic fields of the ith mode, respectively, and β_i is its propagation constant. Substituting (32) into (31), and carrying out the integration with respect to β, we get

$$L = \frac{1}{2\pi}\sum_i M_i \frac{\delta(0)}{\omega} \tag{33}$$

where

$$M_i = \int_S (\{\omega E_i(x,y)^* \cdot \hat{\varepsilon}(x,y)E_i(x,y) + \omega H_i(x,y)^* \cdot \hat{\mu}(x,y)H_i(x,y) - jH_i(x,y)^* \cdot \nabla_t \times E_i(x,y) + jE_i(x,y)^* \cdot \nabla_t \times H_i(x,y)\} - \beta_i\{H_i(x,y)^* \cdot i_z \times E_i(x,y) - E_i(x,y)^* \cdot i_z \times H_i(x,y)\})\, ds - j\int_{C+C_d} \{H_i(x,y) \times E_i(x,y)^*\} \cdot n\, dl. \tag{34}$$

The amplitude of the ith mode is determined by the electromagnetic fields at $t = -\infty$. If we assume that there exists only the ith mode alone at $t = -\infty$, the stationary problem for L is reduced to that for M_i, and hence we can determine the correct values of E_i, H_i, and β_i in such a way as M_i becomes stationary for those correct values. We shall neglect the subscript i for brevity hereafter. According to the similar reasoning used in the preceding section, M given by (34) must be zero for the correct values of E, H, and β (see Appendix). Therefore, the stationary problem for M is equivalent to the stationary problem for the propagation constant β given by

$$\beta = \frac{\int_S (\omega E(x,y)^* \cdot \hat{\varepsilon}(x,y)E(x,y) + \omega H(x,y)^* \cdot \hat{\mu}(x,y)H(x,y) - jH(x,y)^* \cdot \nabla_t \times E(x,y) + jE(x,y)^* \cdot \nabla_t \times H(x,y))\, ds - j\int_{C+C_d} \{H(x,y) \times E(x,y)^*\} \cdot n\, dl}{\int_S (H(x,y)^* \cdot i_z \times E(x,y) - E(x,y)^* \cdot i_z \times H(x,y))\, ds}. \tag{35}$$

The conditions to the trial function for E in (35) are $n \times \delta E = 0$ on C and $n \times \delta E$ changes continuously across C_d, where δ signifies a small variation in E. In view of the transformation given by Table I, (35) can be rewritten in the form

$$\beta = \frac{\int_S (\omega E(x,y)^* \cdot \hat{\varepsilon}(x,y)E(x,y) + \omega H(x,y)^* \cdot \hat{\mu}(x,y)H(x,y) - jH(x,y)^* \cdot \nabla_t \times E(x,y) + jE(x,y)^* \cdot \nabla_t \times H(x,y))\, ds + j\int_{C+C_d} \{E(x,y) \times H(x,y)^*\} \cdot n\, dl}{\int_S (H(x,y)^* \cdot i_z \times E(x,y) - E(x,y)^* \cdot i_z \times H(x,y))\, ds} \tag{36}$$

which coincides with that derived by Berk [2].[1] The trial function for H in (36) must satisfy the conditions $n \times \delta H = 0$ on C and $n \times \delta H$ changes continuously across C_d.

Provided that $\hat{\varepsilon}$ and $\hat{\mu}$ are in the form

$$\hat{\varepsilon}(x,y) = \begin{pmatrix} \varepsilon_{xx} & \varepsilon_{xy} & 0 \\ \varepsilon_{xy}^* & \varepsilon_{yy} & 0 \\ 0 & 0 & \varepsilon_{zz} \end{pmatrix} \qquad \hat{\mu}(x,y) = \begin{pmatrix} \mu_{xx} & \mu_{xy} & 0 \\ \mu_{xy}^* & \mu_{yy} & 0 \\ 0 & 0 & \mu_{zz} \end{pmatrix} \tag{37}$$

the variational expression for the propagation constant can be expressed in terms of only the transverse component of the electric field alone, or only the transverse component of the magnetic field alone. According to the same procedure used to yield (31) from (28), we get from (21)

$$L = \frac{1}{2\pi}\int_{-\infty}^{\infty} d\beta \int_S \left(E(x,y,\beta)^* \cdot \hat{\varepsilon}(x,y)E(x,y,\beta) - \frac{1}{\omega^2}\{\nabla \times E(x,y,\beta)^* \cdot \hat{\mu}^{-1}(x,y)\nabla \times E(x,y,\beta)\}\right) ds. \tag{38}$$

Dividing the electric field E into the transverse component e_t and the z component $i_z e_z$, and applying Gauss' law (10) with $\rho = 0$, we obtain

$$E(x,y,\beta) = e_t(x,y,\beta) + i_z e_z(x,y,\beta) \tag{39}$$

$$e_z(x,y,\beta) = -\frac{i_z \cdot \hat{\varepsilon}^{-1} i_z \nabla_t \cdot \hat{\varepsilon} e_t(x,y,\beta)}{j\beta}. \tag{40}$$

For the correct value of e_t the following equation, which is derived from the Maxwell's equations, must be satisfied:

$$-i_z \times \{-\beta^2 e_t(x,y,\beta) + \nabla_t i_z \cdot (\hat{\varepsilon}^{-1} i_z \nabla_t \cdot \hat{\varepsilon} e_t(x,y,\beta))\} = \omega^2 \hat{\mu}\{i_z \times \hat{\varepsilon} e_t(x,y,\beta)\} - \hat{\mu}\{i_z \times \nabla_t \times \hat{\mu}^{-1}\nabla_t \times e_t(x,y,\beta)\}. \tag{41}$$

Substituting (39)–(41) into (38), we get

$$L = \frac{1}{2\pi\omega^2}\int_{-\infty}^{\infty} d\beta \int_S \Big(\{\omega^2 e_t(x,y,\beta)^* \cdot \hat{\varepsilon} e_t(x,y,\beta) - (\nabla_t \times e_t(x,y,\beta)^*) \cdot \hat{\mu}^{-1}(\nabla_t \times e_t(x,y,\beta))\} - \frac{1}{\beta^2}[\{i_z \times (\omega^2 \hat{\varepsilon}^* e_t(x,y,\beta)^* - \nabla_t \times \hat{\mu}^{-1*}\nabla_t \times e_t(x,y,\beta)^*)\} \cdot \hat{\mu}\{i_z \times (\omega^2 \hat{\varepsilon} e_t(x,y,\beta) - \nabla_t \times \hat{\mu}^{-1}\nabla_t \times e_t(x,y,\beta))\} - \omega^2(i_z \nabla_t \cdot \hat{\varepsilon}^* e_t(x,y,\beta)^*) \cdot \hat{\varepsilon}^{-1}(i_z \nabla_t \cdot \hat{\varepsilon} e_t(x,y,\beta))]\Big)\, ds. \tag{42}$$

[1] Note that the last term's sign in the numerator of (14) in Berk's paper [2] is in error.

In the same way as that used to derive (35), we obtain from (42) the variational expression for β which is expressed in terms of the transverse electric field component $\boldsymbol{e}_t$ as follows:

$$\beta^2 = -\frac{\int_S (\{\boldsymbol{i}_z \times (\omega^2\hat{\varepsilon}^*\boldsymbol{e}_t^* - \nabla_t \times \hat{\mu}^{-1*}\nabla_t \times \boldsymbol{e}_t^*)\} \cdot \hat{\mu}\{\boldsymbol{i}_z \times (\omega^2\hat{\varepsilon}\boldsymbol{e}_t - \nabla_t \times \hat{\mu}^{-1}\nabla_t \times \boldsymbol{e}_t)\} - \omega^2(\boldsymbol{i}_z\nabla_t \cdot \hat{\varepsilon}^*\boldsymbol{e}_t^*) \cdot \hat{\varepsilon}^{-1}(\boldsymbol{i}_z\nabla_t \cdot \hat{\varepsilon}\boldsymbol{e}_t))\, ds}{\int_S ((\nabla_t \times \boldsymbol{e}_t^*) \cdot \hat{\mu}^{-1}(\nabla_t \times \boldsymbol{e}_t) - \omega^2\boldsymbol{e}_t^* \cdot \hat{\varepsilon}\boldsymbol{e}_t)\, ds}. \tag{43}$$

In the preceding equation, $\boldsymbol{e}_t = \boldsymbol{e}_t(x,y)$, $\hat{\varepsilon} = \hat{\varepsilon}(x,y)$, and $\hat{\mu} = \hat{\mu}(x,y)$ are the functions of x and y (independent of z). In particular, if materials involved are isotropic, $\hat{\varepsilon}$ and $\hat{\mu}$ in (43) become scalar constants, and (43) reduces to Kurokawa's variational expression [6] for the waveguide consists of perfectly conducting walls.

The first variation of (43) is given by

$$\begin{aligned}
\delta\beta^2 \int_S &\{(\nabla_t \times \boldsymbol{e}_t^*) \cdot \hat{\mu}^{-1}(\nabla_t \times \boldsymbol{e}_t) - \omega^2\boldsymbol{e}_t^* \cdot \hat{\varepsilon}\boldsymbol{e}_t\}\, ds \\
= -\int_S &(\{\boldsymbol{i}_z \times (\nabla_t \times \hat{\mu}^{-1*}\nabla_t \times \delta\boldsymbol{e}_t^* - \omega^2\hat{\varepsilon}^*\delta\boldsymbol{e}_t^*)\} \\
&\cdot \{\boldsymbol{i}_z \times (\beta^2\boldsymbol{e}_t - \nabla_t\boldsymbol{i}_z \cdot \hat{\varepsilon}^{-1}(\boldsymbol{i}_z\nabla_t \cdot \hat{\varepsilon}\boldsymbol{e}_t)) - \hat{\mu}(\boldsymbol{i}_z \times (\omega^2\hat{\varepsilon}\boldsymbol{e}_t \\
&- \nabla_t \times \hat{\mu}^{-1}\nabla_t \times \boldsymbol{e}_t))\})\, ds \\
&- \beta^2 \int_{C+C_d} \{\boldsymbol{e}_t \times \hat{\mu}^{-1*}(\nabla_t \times \delta\boldsymbol{e}_t^*)\} \cdot \boldsymbol{n}\, dl \\
&- \int_{C+C_d} ((\nabla_t \times \hat{\mu}^{-1*}\nabla_t \times \delta\boldsymbol{e}_t^* - \omega^2\hat{\varepsilon}^*\delta\boldsymbol{e}_t^*) \\
&\cdot \{\boldsymbol{i}_z \cdot \hat{\varepsilon}^{-1}(\boldsymbol{i}_z\nabla_t \cdot \hat{\varepsilon}\boldsymbol{e}_t)\}) \cdot \boldsymbol{n}\, dl + \text{c.c.}
\end{aligned} \tag{44}$$

where C is a closed conductor in the transverse plane as in the case of (31), while C_d indicates the line in the transverse plane across which $\hat{\varepsilon}$, $\hat{\mu}$, and their first derivatives change discontinuously. The term designated as c.c. represents the complex conjugate of the terms before it. The correct value of $\boldsymbol{e}_t$ in (44) must satisfy the equation given by (41), and also both $\boldsymbol{n} \times \boldsymbol{e}_t$ and $\boldsymbol{i}_z \cdot \hat{\varepsilon}^{-1}(\boldsymbol{i}_z\nabla_t \cdot \hat{\varepsilon}\boldsymbol{e}_t)$ must be continuous across C_d. Therefore, the conditions to the trial function for $\boldsymbol{e}_t$ in (43) can be stated as follows: 1) $\boldsymbol{n} \times \hat{\mu}^{-1}\nabla_t \times \delta\boldsymbol{e}_t$ and $\boldsymbol{n} \cdot (\nabla_t \times \hat{\mu}^{-1}\nabla_t \times \delta\boldsymbol{e}_t - \omega^2\hat{\varepsilon}\delta\boldsymbol{e}_t)$ must vanish on C; 2) $\boldsymbol{n} \times \hat{\mu}^{-1}\nabla_t \times \delta\boldsymbol{e}_t$ and $\boldsymbol{n} \cdot (\nabla_t \times \hat{\mu}^{-1}\nabla_t \times \delta\boldsymbol{e}_t - \omega^2\hat{\varepsilon}\delta\boldsymbol{e}_t)$ must change continuously across C_d. In order to simplify the conditions to the trial function for $\boldsymbol{e}_t$, let us add the term of line integral to both the denominator and the numerator in (43) as follows:

$$\beta^2 = -\frac{\begin{aligned}&\int_S (\{\boldsymbol{i}_z \times (\omega^2\hat{\varepsilon}^*\boldsymbol{e}_t^* - \nabla_t \times \hat{\mu}^{-1*}\nabla_t \times \boldsymbol{e}_t^*)\} \cdot \hat{\mu}\{\boldsymbol{i}_z \times (\omega^2\hat{\varepsilon}\boldsymbol{e}_t - \nabla_t \times \hat{\mu}^{-1}\nabla_t \times \boldsymbol{e}_t)\} \\ &- \omega^2(\boldsymbol{i}_z\nabla_t \cdot \hat{\varepsilon}^*\boldsymbol{e}_t^*) \cdot \hat{\varepsilon}^{-1}(\boldsymbol{i}_z\nabla_t \cdot \hat{\varepsilon}\boldsymbol{e}_t))\, ds - \int_{C+C_d} ((\nabla_t \times \hat{\mu}^{-1*}\nabla_t \times \boldsymbol{e}_t^* - \omega^2\hat{\varepsilon}^*\boldsymbol{e}_t^*)\{\boldsymbol{i}_z \cdot \hat{\varepsilon}^{-1}(\boldsymbol{i}_z\nabla_t \cdot \hat{\varepsilon}\boldsymbol{e}_t)\} \\ &+ (\nabla_t \times \hat{\mu}^{-1}\nabla_t \times \boldsymbol{e}_t - \omega^2\hat{\varepsilon}\boldsymbol{e}_t)\{\boldsymbol{i}_z \cdot \hat{\varepsilon}^{-1*}(\boldsymbol{i}_z\nabla_t \cdot \hat{\varepsilon}^*\boldsymbol{e}_t^*)\}) \cdot \boldsymbol{n}\, dl\end{aligned}}{\int_S ((\nabla_t \times \boldsymbol{e}_t^*) \cdot \hat{\mu}^{-1}(\nabla_t \times \boldsymbol{e}_t) - \omega^2\boldsymbol{e}_t^* \cdot \hat{\varepsilon}\boldsymbol{e}_t)\, ds - \int_{C+C_d} (\boldsymbol{e}_t \times \hat{\mu}^{-1*}(\nabla_t \times \boldsymbol{e}_t^*) + \boldsymbol{e}_t^* \times \hat{\mu}^{-1}(\nabla_t \times \boldsymbol{e}_t)) \cdot \boldsymbol{n}\, dl}. \tag{45}$$

Then the conditions to the trial function for $\boldsymbol{e}_t$ derived from the first variation of (45) become in simpler form as follows: 1) $\boldsymbol{n} \times \delta\boldsymbol{e}_t$ and $\boldsymbol{i}_z \cdot \hat{\varepsilon}^{-1}(\boldsymbol{i}_z\nabla_t \cdot \hat{\varepsilon}\delta\boldsymbol{e}_t)$ must vanish on C; 2) $\boldsymbol{n} \times \delta\boldsymbol{e}_t$ and $\boldsymbol{i}_z \cdot \hat{\varepsilon}^{-1}(\boldsymbol{i}_z\nabla_t \cdot \hat{\varepsilon}\delta\boldsymbol{e}_t)$ must change continuously across C_d.

The variational expression for the propagation constant expressed in terms of transverse magnetic field component $\boldsymbol{h}_t$ can easily be obtained from (45) with the aid of the transformation given by Table I. The result is

$$\beta^2 = -\frac{\begin{aligned}&\int_S (\{\boldsymbol{i}_z \times (\omega^2\hat{\mu}^*\boldsymbol{h}_t^* - \nabla_t \times \hat{\varepsilon}^{-1*}\nabla_t \times \boldsymbol{h}_t^*)\} \cdot \hat{\varepsilon}\{\boldsymbol{i}_z \times (\omega^2\hat{\mu}\boldsymbol{h}_t - \nabla_t \times \hat{\varepsilon}^{-1}\nabla_t \times \boldsymbol{h}_t)\} \\ &- \omega^2(\boldsymbol{i}_z\nabla_t \cdot \hat{\mu}^*\boldsymbol{h}_t^*) \cdot \hat{\mu}^{-1}(\boldsymbol{i}_z\nabla_t \cdot \hat{\mu}\boldsymbol{h}_t))\, ds - \int_{C+C_d} ((\nabla_t \times \hat{\varepsilon}^{-1*}\nabla_t \times \boldsymbol{h}_t^* - \omega^2\hat{\mu}^*\boldsymbol{h}_t^*)\{\boldsymbol{i}_z \cdot \hat{\mu}^{-1}(\boldsymbol{i}_z\nabla_t \cdot \hat{\mu}\boldsymbol{h}_t)\} \\ &+ (\nabla_t \times \hat{\varepsilon}^{-1}\nabla_t \times \boldsymbol{h}_t - \omega^2\hat{\mu}\boldsymbol{h}_t)\{\boldsymbol{i}_z \cdot \hat{\mu}^{-1*}(\boldsymbol{i}_z\nabla_t \cdot \hat{\mu}^*\boldsymbol{h}_t^*)\}) \cdot \boldsymbol{n}\, dl\end{aligned}}{\int_S ((\nabla_t \times \boldsymbol{h}_t^*) \cdot \hat{\varepsilon}^{-1}(\nabla_t \times \boldsymbol{h}_t) - \omega^2\boldsymbol{h}_t^* \cdot \hat{\mu}\boldsymbol{h}_t)\, ds - \int_{C+C_d} (\boldsymbol{h}_t \times \hat{\varepsilon}^{-1*}(\nabla_t \times \boldsymbol{h}_t^*) + \boldsymbol{h}_t^* \times \hat{\varepsilon}^{-1}(\nabla_t \times \boldsymbol{h}_t)) \cdot \boldsymbol{n}\, dl}. \tag{46}$$

The conditions to the trial function for $\boldsymbol{h}_t$ in (46) can be found from the first variation of (46). Those conditions are as follows: 1) $\boldsymbol{n} \times \delta\boldsymbol{h}_t$ and $\boldsymbol{i}_z \cdot \hat{\mu}^{-1}(\boldsymbol{i}_z\nabla_t \cdot \hat{\mu}\delta\boldsymbol{h}_t)$ must vanish on C; 2) $\boldsymbol{n} \times \delta\boldsymbol{h}_t$ and $\boldsymbol{i}_z \cdot \hat{\mu}^{-1}(\boldsymbol{i}_z\nabla_t \cdot \hat{\mu}\delta\boldsymbol{h}_t)$ must change continuously across C_d.

V. Variational Expression for Impedance Matrix

Let us derive in this section the variational expression for the impedance matrix of waveguide junctions from the least action principle. The n-port waveguide junction under consideration is shown schematically in Fig. 5. Suppose that the materials involved in the junction are linear, nondispersive, and dissipation free. In Fig. 5, S_i $(i = 1,2,\cdots,n)$ are the reference planes chosen in the connected waveguides far enough from the junction so that all the higher modes above cutoff are extinguished. Let the electric and magnetic fields within the junction be $\boldsymbol{E}_i$ and $\boldsymbol{H}_i$, respectively, when all ports except the ith port are open circuited and a unit current is fed into the ith port. Similarly, $\boldsymbol{E}_j$ and $\boldsymbol{H}_j$ may be regarded as the electric and magnetic fields within the junction caused by an input unit current at port j when all ports except the jth port are open circuited.

Because of the linearity of the system under consideration, the electromagnetic fields within the junction corresponding to the unit input currents at ports i and j can then be expressed as

$$\begin{aligned}\boldsymbol{E} &= \boldsymbol{E}_i + \boldsymbol{E}_j \\ \boldsymbol{H} &= \boldsymbol{H}_i + \boldsymbol{H}_j.\end{aligned} \tag{47}$$

We take for V in (8) the charge and current free region within the junction, and assume that the junction is operated

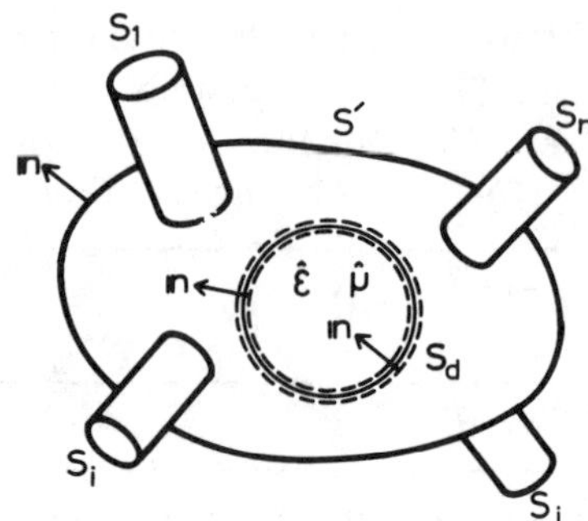

Fig. 5. Schematic diagram of n-port waveguide junction.

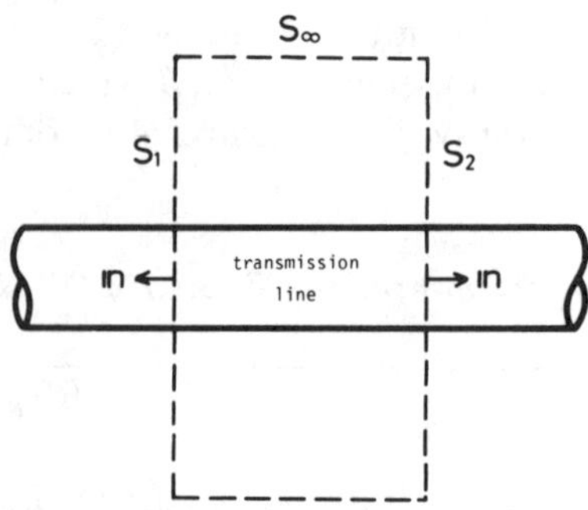

Fig. 6. Surface $S(S_1 + S_2 + S_\infty)$ used in evaluation of integration in (A3). S_∞ is a cylindrical side surface with infinite radius, and S_1 and S_2 are parallel surfaces transverse to the propagation direction of the transmission line.

with single angular frequency ω. Then (8) becomes as

$$J = \delta(0) \int_V (\boldsymbol{E}(\boldsymbol{r})^* \cdot \hat{\varepsilon}(\boldsymbol{r})\boldsymbol{E}(\boldsymbol{r}) - \boldsymbol{H}(\boldsymbol{r})^* \cdot \hat{\mu}(\boldsymbol{r})\boldsymbol{H}(\boldsymbol{r}))\, dv$$

$$= -\frac{\delta(0)}{j\omega} \sum_{i,j} Z_{ij} \tag{48}$$

where Z_{ij} is a component of an impedance matrix given by

$$Z_{ij} = j\omega \int_V (\boldsymbol{H}_i^* \cdot \hat{\mu}\boldsymbol{H}_j - \boldsymbol{E}_i^* \cdot \hat{\varepsilon}\boldsymbol{E}_j)\, dv. \tag{49}$$

Since both $\boldsymbol{E}$ and $\boldsymbol{H}$ in the foregoing equation are the functions of $\boldsymbol{A}$ and ϕ, we can rewrite (49), with the aid of (5) and (16), in the form

$$Z_{ij} = j\omega \int_V \left\{ \boldsymbol{H}_i^* \cdot \hat{\mu}\boldsymbol{H}_j - \frac{1}{\omega^2} (\nabla \times \boldsymbol{H}_i^*) \cdot \hat{\varepsilon}^{-1} (\nabla \times \boldsymbol{H}_j) \right\} dv. \tag{50}$$

Thus the stationary problem for J given by (48) has been reduced to the stationary problem for Z_{ij} given by (50) expressed in terms of $\boldsymbol{H}$.

The first-order variation of Z_{ij} is given by

$$\delta Z_{ij} = \int_V \{\delta \boldsymbol{H}_i^* \cdot (\nabla \times \boldsymbol{E}_j + j\omega\hat{\mu}\boldsymbol{H}_j) + \delta \boldsymbol{H}_j \cdot (-\nabla \times \boldsymbol{E}_i^* + j\omega\hat{\mu}^*\boldsymbol{H}_i^*)\}\, dv + \int_{S'+S_d+S_r} \{\delta \boldsymbol{H}_i^* \times \boldsymbol{E}_j - \delta \boldsymbol{H}_j \times \boldsymbol{E}_i^*\} \cdot \boldsymbol{n}\, ds \tag{51}$$

where S' is a conductor wall forming a junction, S_d is a surface across which $\hat{\varepsilon}$ and $\hat{\mu}$ change discontinuously, and S_r are the reference planes in the connected waveguides, as shown in Fig. 5. The conditions to the trial function for $\boldsymbol{H}$ in (50) can then be found from (51) as follows: 1) $\boldsymbol{n} \times \delta\boldsymbol{H} = 0$ on S_r; 2) $\boldsymbol{n} \times \delta\boldsymbol{H}$ changes continuously across S_d.

VI. Conclusion

A novel approach to derive the variational expressions for electromagnetic field problems has been proposed. It has been shown that the variational expressions for a resonant frequency of a cavity resonator, a propagation constant of a uniform transmission line, and an impedance matrix of a waveguide junction can be derived systematically all from the least action principle. We have shown that the Maxwell's equations themselves can also be yielded by solving the stationary problem of the action J.

Though we have assumed the materials contained in the system under consideration are linear and loss free, the variational expressions derived in the present paper are applicable to the system which involves anisotropic and/or inhomogeneous materials.

Appendix

For the charge and the current free region V, (8) is reduced to

$$J = \int_0^\infty df \int_V (\boldsymbol{E}(\boldsymbol{r},\omega)^* \cdot \hat{\varepsilon}(\boldsymbol{r},\omega)\boldsymbol{E}(\boldsymbol{r},\omega) - \boldsymbol{H}(\boldsymbol{r},\omega)^* \cdot \hat{\mu}(\boldsymbol{r},\omega)\boldsymbol{H}(\boldsymbol{r},\omega))\, dv. \tag{A1}$$

Provided that $\boldsymbol{E}$ and $\boldsymbol{H}$ are the correct fields satisfying the Maxwell's equations for the loss-free region, the following equation

$$\nabla \cdot (\boldsymbol{E} \times \boldsymbol{H}^*) = j\omega(\boldsymbol{E}^* \cdot \hat{\varepsilon}\boldsymbol{E} - \boldsymbol{H}^* \cdot \hat{\mu}\boldsymbol{H}) \tag{A2}$$

is derived. Substituting (A2) into (A1) and applying Gauss' theorem, reduce (A1) to

$$J = \int_0^\infty df \int_S \left(\frac{1}{j\omega} \boldsymbol{E} \times \boldsymbol{H}^*\right) \cdot \boldsymbol{n}\, ds. \tag{A3}$$

As shown in Fig. 6, the surface S consists of S_1, S_2, and S_∞. J given by (A3) vanishes for a propagation mode because $(\boldsymbol{E} \times \boldsymbol{H}^*) \cdot \boldsymbol{n}$ is zero on S_∞ and also the unit normal vectors $\boldsymbol{n}$ on S_1 and S_2 direct to opposite directions. Therefore, the action J, and hence L, becomes zero for a propagation mode. We can conclude, therefore, that M must be zero for a correct propagation mode.

References

[1] V. H. Rumsey, "Reaction concept in electromagnetic theory," *Phys. Rev.*, vol. 94, pp. 1483–1491, June 1954.

[2] A. D. Berk, "Variational principles for electromagnetic resonator and waveguides," *IRE Trans. Antennas Propagat.*, vol. AP-4, pp. 104–111, Apr. 1956.

[3] R. F. Harrington, *Time-Harmonic Electromagnetic Fields.* New York: McGraw-Hill, 1961, ch. 7.

[4] L. Cairo and T. Kahan, *Variational Techniques in Electromagnetism.* New York: Gordon and Breach, 1965.

[5] L. D. Landau and E. Lifshitz, *Classical Theory of Fields.* Cambridge: Addison-Wesley, 1951, ch. 2. and ch. 4.

[6] K. Kurokawa, "Electromagnetic waves in waveguides with wall impedance," *IRE Trans. Microwave Theory Tech.*, vol. MTT-10, pp. 314–320, Sept. 1962.

The Variational Principle for Non-Self-Adjoint Electromagnetic Problems

CHUN HSIUNG CHEN AND CHUEN-DER LIEN

Abstract—**A systematic and intuitive procedure is proposed to derive the variational (or stationary) principle for non-self-adjoint electromagnetic problems with various boundary conditions. Several physical interpretations of this principle in terms of generalized reactions, time-average stored energy, and reactive powers, respectively, are discussed in detail. This general variational principle which makes the generalized reactions a stationary value is actually an extension of the least action principle in physics. The applications of the principle to establish the variational expressions for a waveguide, a cavity resonator, and a lossy one-dimensional inhomogeneous slab are presented.**

I. Introduction

ALMOST ALL physical problems can be formulated mathematically in terms of differential equations, integral equations, or variational equations [1], [2]. Before the advent of computer, the applicability of the variational formulation is somewhat limited, because its solution eventually has to go back to that of solving the differential equations. But in recent years, the variational formulation has received much attention for three reasons. The primary reason is that we have computers capable of solving the variational equations directly and practically, using the direct methods [2], [3], such as Ritz method and finite-element method. The second reason is that the variational formulation itself also permits a physical interpretation, thus, it may supply another picture to a physical problem. The third reason is that the variational formulation can be used not only for computing the field but also for establishing the stationary formula of a quantity such as the eigenvalue. Of course the variational formulation is not suitable in discussing the problem for which the functional does not exist [4]. Actually, only when the functional exists can the variational formulation have physical and mathematical significance.

In physics, both differential formulation (Newton's law) and variational formulation (the principle of least action) have been utilized extensively in studying a mechanical system. But the variational formulation of an electromagnetic system is still not well-established in comparison with the differential formulation (Maxwell's equations) of the same problem. With a few exceptions [4]–[7], most investigations on the variational formulation are mainly concerned with the self-adjoint electromagnetic problems. The purpose of this paper is to establish a general variational principle (formulation) for dealing with a non-self-adjoint electromagnetic system and then to present a physical interpretation of this novel principle.

Manuscript received October 19, 1979; revised March 19, 1980.

The authors are with the Department of Electrical Engineering, National Taiwan University, Taipei, Taiwan, Republic of China.

In the field of electromagnetism, some derivations and applications of the variational formulation have been reported recently [8]–[23]. Rumsey [8] and Harrington [9] have created the reaction concept for deriving the stationary expressions in an isotropic medium. Konrad [10] has investigated the variational expressions in an anisotropic medium, while Morishita and Kumagai [11], [12] have studied the same problem, using the principle of least action. There are many authors who have derived the variational expressions for each specific problem [13]–[23]. But all these investigations are mainly concerned with the self-adjoint problems (with respect to either real- or complex-type inner product). Although a few authors [4], [5], [24] have proposed a scalar theory for handling some specific non-self-adjoint problems, the theory is still not powerful enough in dealing with the general problems for a vector field. While the variational expressions for a non-self-adjoint vector field have been obtained by means of "transpose operator and field" [7], it is still worthwhile to have an intuitive derivation and physical insight of these expressions on the other hand. Moreover, only boundary conditions of Dirichlet and/or Neumann types are considered by the previous investigations. There is still little information concerning more general types of boundary conditions.

In this paper, a systematic and intuitive procedure is suggested to derive the variational expressions for a non-self-adjoint electromagnetic system. The generalized reaction concept is introduced to interpret the general variational principle for this electromagnetic system. The principle is then applied to the problems of establishing the variational expressions for a waveguide, a cavity resonator, and a lossy one-dimensional inhomogeneous slab.

II. Variational Formulation of a Non-Self-Adjoint Problem

The variational solution of a non-self-adjoint problem

$$Lf = s \tag{1}$$

will be summarized in this section. Although only the electromagnetic field problem is considered in this study, the theory is, in general, applicable to any non-self-adjoint problem in physics and mathematics. In (1), L is a non-self-adjoint linear operator for describing a physical or

Reprinted from *IEEE Trans. Microwave Theory Tech.*, vol. MTT-28, no. 8, pp. 878–886, Aug. 1980.

mathematical problem, f is the field (unknown function) to be determined, and s is a given (known) source function.

A method of solving the original non-self-adjoint problem (1) is to introduce an auxiliary problem, the adjoint problem [4]–[6], [24], [25] as follows:

$$L^a f^a = s^a \tag{2}$$

where L^a is the adjoint operator of L, f^a is another unknown function (the adjoint field) to be determined, and s^a is another known source function.

Both real- and complex-type inner products [25] will be considered in this study. The latter is accepted conventionally by many authors and is easily interpreted as the complex power physically, however, the former is more convenient in mathematical manipulation [26]. Either real- or complex-type inner product of two vectors u and w is defined as a scalar $\langle u,w\rangle$ such that

$$\langle u,w\rangle = \langle w,u\rangle^p$$
$$\langle \sigma_1 u_1 + \sigma_2 u_2, w\rangle = \sigma_1^p\langle u_1, w\rangle + \sigma_2^p\langle u_2, w\rangle \tag{3a}$$

where u_1,u_2 are vectors and σ_1,σ_2 are scalars. For the complex-type inner product the superscript p should be interpreted as "complex conjugate*" so that

$$\sigma_i^p = \sigma_i^*, \qquad i=1,2$$
$$\langle u,w\rangle = \langle w,u\rangle^p = \langle w,u\rangle^* \tag{3b}$$

while for the real-type inner product p should represent "no operation" and should be removed and interpreted as

$$\sigma_i^p = \sigma_i, \qquad i=1,2$$
$$\langle u,w\rangle = \langle w,u\rangle^p = \langle w,u\rangle. \tag{3c}$$

The symbol p will always have the above interpretation throughout this study.

We now conduct a systematic and intuitive derivation of the variational formulation for solving both original and adjoint problems, equations (1) and (2), simultaneously. The idea is to express the left-hand side of the equation

$$\langle \delta f^a, Lf - s\rangle + \langle L^a f^a - s^a, \delta f\rangle = 0 \tag{4}$$

as the first variation δI of some functional I [2], where δf and δf^a are the variations of f and f^a, respectively. Then it can be shown that the problem of solving f and f^a simultaneously from (1) and (2) is completely equivalent to that of determining the stationary functions (both f and f^a) from the following variational equation:

$$\delta I(f,f^a) = 0$$
$$I(f,f^a) = \langle f^a, Lf\rangle - \langle s^a, f\rangle - \langle f^a, s\rangle. \tag{5}$$

Note that for the problems defined by differential operators L and L^a with their boundary conditions $B(f)=0$ and $B^a(f^a)=0$ regarded as essential ones, the stationary functions f and f^a of (5) should also be subject to the constraints $B(f)=0$ and $B^a(f^a)=0$, respectively. However, if these boundary conditions are regarded as natural ones, then some modifications should be made on (4) and (5), as demonstrated in Section III, with the stationary functions f and f^a subject to no constraints on the boundary.

Note also that with the symbol $\langle , \rangle$ defined by (3), the variational formulations for real- and complex-type inner products can both be written in the same form as indicated by (5). The expression in (5) is identical to that of the stationary principle [4], [24] and that of another derivation [5]. The expression (5), of course, includes that adopted in the previous investigation of the self-adjoint problem [27].

It seems that both f (the desired field) and f^a (the adjoint or auxiliary field) have to be solved simultaneously in the variational problem (5). However, the process of determining both f and f^a can actually be decoupled, as explained later, when Rayleigh–Ritz or finite element method is employed in the solution.

Let us express the solution in the form

$$f = \sum_{n=1}^{N} C_n \phi_n$$
$$f^a = \sum_{n=1}^{N} C_n^a \phi_n^a \tag{6}$$

where ϕ_n and ϕ_n^a are known functions, and C_n and C_n^a are constants to be determined. Note that both ϕ_n and ϕ_n^a may or may not form the bases of the domains of L and L^a, respectively, however they should be linearly independent and should form the complete sets as N approaches infinity. By substituting (6) into (5) or (4) and adjusting C_n and C_n^a such that $\delta I(f,f^a)=0$, one obtains two decoupled systems as follows:

$$\sum_{n=1}^{N} \langle \phi_m^a, L\phi_n\rangle C_n = \langle \phi_m^a, s\rangle$$
$$\sum_{n=1}^{N} \langle \phi_m, L^a\phi_n^a\rangle C_n^a = \langle \phi_m, s^a\rangle, \qquad m=1,2,\cdots,N. \tag{7}$$

The positive integer N in (6) and (7) may be finite or infinite if an approximate or exact solution is to be determined. The fact that C_n and C_n^a are decoupled in (7) has greatly simplified the process of determining the stationary functions f and f^a from (5).

The discrete systems (7) from the Ritz method are identical, in form, to those from the moment (or Petrov–Galerkin's) method [26] of simultaneously solving the original and adjoint problems (1) and (2). Since there is a basic difference or mathematical distinction between these two methods [4], the above statement is meaningful only when both methods can make sense. If the functions ϕ_n and ϕ_n^a are selected from the eigenfunctions of L and L^a, respectively, then the systems (7) will reduce to the conventional ones derived from the method of eigenfunction expansion [25]. If f^a in (6) is expanded into a series of ϕ_n instead of ϕ_n^a, then the resultant systems obtained will be identical to those from the Galerkin's (or Bubnov–Galerkin's) method.

The introduction of the auxiliary problem (2) for supplementing the original problem (1) has an interesting physical interpretation as follows:

$$\langle f^a, s\rangle = \langle f^a, Lf\rangle = \langle L^a f^a, f\rangle = \langle s^a, f\rangle. \tag{8}$$

This is the generalized reciprocity theorem which states that the generalized reaction of the adjoint field f^a on the source s of the original problem is identical to that of the original field f on the source s^a of the adjoint problem. The term $\langle f^a, s\rangle$, for example, may be interpreted as a generalized reaction since it can be reduced to the conventional reaction [8] if the real-type inner product is employed.

The general reciprocity theorem (8), in the case of a self-adjoint problem using the real-type inner product, has an important result as follows. By setting $f=f_1$, $s=s_1$, $f^a=f_2$, $s^a=s_2$, one then has the conventional reciprocity theorem [28] of relating the reactions between two different problems: (f_1, s_1) and (f_2, s_2), i.e., $\langle f_2, s_1\rangle = \langle f_1, s_2\rangle$.

III. Variational Principle for Electromagnetic System

The variational principle for dealing with electromagnetic field problems will be investigated in this section. This study is mainly concerned with the solution of the non-self-adjoint electromagnetic problem (the original problem) described by the equations

$$\begin{aligned} -\nabla\times\bar{E} &= j\omega\,\bar{\bar{\mu}}\cdot\bar{H}+\bar{M} \\ \nabla\times\bar{H} &= j\omega\,\bar{\bar{\epsilon}}\cdot\bar{E}+\bar{J}, \qquad \text{in } V \\ \bar{B}(\bar{E},\bar{H}) &= 0, \qquad \text{on } S \end{aligned} \tag{9}$$

where V is a closed region bounded by the closed surface S, Fig. 1. In (9), $\bar{E}$ and $\bar{H}$ are the electric and magnetic fields to be determined, $\bar{J}$ and $\bar{M}$ are the known source distributions (electric and magnetic current densities) within V, $\bar{\bar{\epsilon}}$ and $\bar{\bar{\mu}}$ are the permittivity and permeability tensors (dyadics) of the medium contained in V, and $\bar{B}(\bar{E},\bar{H})=0$ is the required boundary condition on the surface S. The anisotropic medium in V may be lossy and inhomogeneous, thus, the tensors $\bar{\bar{\epsilon}}$ and $\bar{\bar{\mu}}$ are in general complex quantities and functions of the position. Only time-harmonic variation of the form $e^{j\omega t}$ will be discussed in this study.

As before, an auxiliary adjoint problem

$$\begin{aligned} -\nabla\times\bar{E}^a &= j\omega\,\bar{\bar{\mu}}^a\cdot\bar{H}^a+\bar{M}^a \\ \nabla\times\bar{H}^a &= j\omega\,\bar{\bar{\epsilon}}^a\cdot\bar{E}^a+\bar{J}^a, \qquad \text{in } V \\ \bar{B}^a(\bar{E}^a,\bar{H}^a) &= 0, \qquad \text{on } S \end{aligned} \tag{10}$$

is introduced to supplement the original problem (9). In (10), $\bar{E}^a$ and $\bar{H}^a$ are the unknown adjoint fields, $\bar{J}^a$ and $\bar{M}^a$ are known source distributions, and $\bar{B}^a(\bar{E}^a,\bar{H}^a)=0$ is the adjoint boundary condition. Note that the condition $\bar{B}^a(\bar{E}^a,\bar{H}^a)=0$ in (10) should be chosen according to $\bar{B}(\bar{E},\bar{H})=0$ in (9), and the dyadics $\bar{\bar{\epsilon}}^a$ and $\bar{\bar{\mu}}^a$ should be equal to $(\bar{\bar{\epsilon}}^T)^p$ and $(\bar{\bar{\mu}}^T)^p$, respectively, in order to make (10) the adjoint of (9). The symbol T denotes "transpose" and p again denotes no operation or complex conjugate* as discussed in (3).

The inner product, for studying an electromagnetic system, of vectors $\bar{U}$ and $\bar{W}$ is defined as

$$\langle\bar{U},\bar{W}\rangle = \int_V \bar{U}^p\cdot\bar{W}\,dV. \tag{11}$$

Two kinds of boundary conditions are discussed, i.e.,

$$\begin{aligned} \bar{B}_1(\bar{E}) &= \hat{n}\times\bar{E}-\bar{\gamma}=0 \\ \bar{B}_1^a(\bar{E}^a) &= \hat{n}\times\bar{E}^a-\bar{\gamma}^a=0, \qquad \text{on } S \end{aligned} \tag{12}$$

and

$$\begin{aligned} \bar{B}_2(\bar{E},\bar{H}) &= \bar{H}\times\hat{n}-j\omega\,\bar{\bar{\alpha}}\cdot\bar{E}-\bar{\beta}=0 \\ \bar{B}_2^a(\bar{E}^a,\bar{H}^a) &= \bar{H}^a\times\hat{n}-j\omega\,\bar{\bar{\alpha}}^a\cdot\bar{E}^a-\bar{\beta}^a=0, \qquad \text{on } S. \end{aligned} \tag{13}$$

In (12) and (13), $\bar{B}_i=0$ and $\bar{B}_i^a=0$, $i=1, 2$, are the boundary conditions associated with the original and adjoint problems (9) and (10), respectively, $\hat{n}$ is the unit outward normal vector to the surface S (Fig. 1), $\bar{\gamma}$, $\bar{\gamma}^a$, $\bar{\beta}$, $\bar{\beta}^a$ are given constant vectors, and $\bar{\bar{\alpha}}$, $\bar{\bar{\alpha}}^a$ are given constant dyadics. To make the problem (10) the adjoint of (9), $\bar{\bar{\alpha}}^a$ should also be chosen such that $\bar{\bar{\alpha}}^a=(\bar{\bar{\alpha}}^T)^p$.

Physically, the vectors $\bar{\beta}$, $\bar{\beta}^a$ and $\bar{\gamma}$, $\bar{\gamma}^a$ may be interpreted as electric- and magnetic-surface currents, and the dyadic $\bar{\bar{\alpha}}$ as the permittivity of an infinitesimally thin layer over the surface S. In the expressions for the variational principles given by (19)–(21), (22)–(24), and those for the generalized reactions given by (25), these constants $\bar{\beta}$, $\bar{\beta}^a$, $\bar{\gamma}$, $\bar{\gamma}^a$, and $\bar{\bar{\alpha}}$ have actually served as sheets of additional sources and sheet of extra anisotropic material at the boundary S.

The variational equivalent of (9) and (10) may be written in terms of $(\bar{E},\bar{E}^a)$, $(\bar{H},\bar{H}^a)$, or $(\bar{E},\bar{H},\bar{E}^a,\bar{H}^a)$ and is named E-, H-, or E,H- formulation, respectively.

A. E-Formulation

First consider the variational formulation in terms of $\bar{E}$ and $\bar{E}^a$. To this end, the unknowns $\bar{H}$ and $\bar{H}^a$ in the original and adjoint problems (9) and (10) are considered as functions of $\bar{E}$ and $\bar{E}^a$, respectively, i.e.,

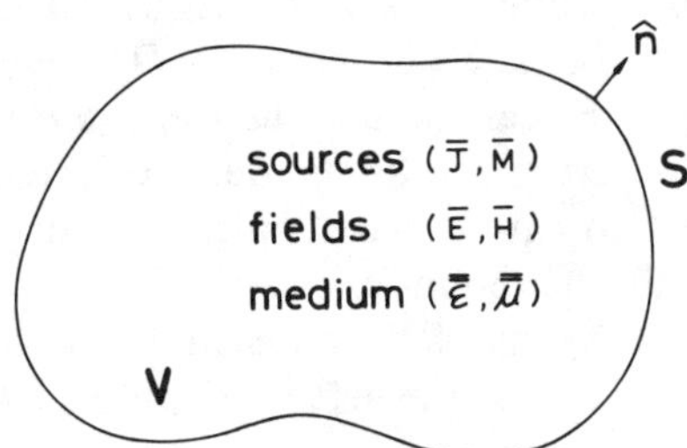

Fig. 1. Geometry of non-self-adjoint electromagnetic problem (original problem).

$$\bar{H}(\bar{E}) = -\bar{\bar{\mu}}^{-1}\cdot(\nabla\times\bar{E}+\bar{M})/j\omega$$
$$\bar{H}^a(\bar{E}^a) = -\bar{\bar{\mu}}^{a^{-1}}\cdot(\nabla\times\bar{E}^a+\bar{M}^a)/j\omega. \tag{14}$$

Then $\bar{H}$ and $\bar{H}^a$ in (9) and (10) may be eliminated to obtain the wave equations as follows:

$$L\bar{E} \equiv -\nabla\times\left(\bar{\bar{\mu}}^{-1}\cdot\nabla\times\bar{E}\right)+\omega^2\bar{\bar{\epsilon}}\cdot\bar{E}$$
$$=j\omega\bar{J}+\nabla\times\left(\bar{\bar{\mu}}^{-1}\cdot\bar{M}\right), \quad \text{in } V$$
$$\bar{B}(\bar{E},\bar{H}(\bar{E}))=0, \quad \text{on } S \tag{15}$$

$$L^a\bar{E}^a \equiv -\nabla\times\left(\bar{\bar{\mu}}^{a^{-1}}\cdot\nabla\times\bar{E}^a\right)+\omega^2\bar{\bar{\epsilon}}^a\cdot\bar{E}^a$$
$$=j\omega\bar{J}^a+\nabla\times\left(\bar{\bar{\mu}}^{a^{-1}}\cdot\bar{M}^a\right), \quad \text{in } V$$
$$\bar{B}^a(\bar{E}^a,\bar{H}^a(\bar{E}^a))=0, \quad \text{on } S. \tag{16}$$

Here we have the electric field $\bar{E}$ to be determined from the original problem (15) and the adjoint electric field $\bar{E}^a$ to be determined from the adjoint problem (16).

To absorb the boundary conditions into the variational formulation as well as to simplify the derivation, a symbol $\Delta(S)$ is introduced for converting volume integral into surface integral and vice versa,

$$\int_V F(\bar{r})\Delta(S)\,dV=\int_S F(\bar{r})\,dS$$
$$\Delta(S)^*=\Delta(S). \tag{17}$$

In (17), F is a scalar function of the position $\bar{r}$ so it may be replaced by any component of a vector or a tensor.

To seek for the variational formulation of (15) and (16) with boundary conditions (13), we now start from the equation of the form

$$\langle\delta\bar{E}^a,\, L\bar{E}-j\omega\bar{J}-\nabla\times\left(\bar{\bar{\mu}}^{-1}\cdot\bar{M}\right)\rangle$$
$$+\langle L^a\bar{E}^a-j\omega\bar{J}^a-\nabla\times\left(\bar{\bar{\mu}}^{a^{-1}}\cdot\bar{M}^a\right),\,\delta\bar{E}\rangle$$
$$+\langle\delta\bar{E}^a,\, D\bar{B}(\bar{E},\bar{H}(\bar{E}))\Delta(S)\rangle$$
$$+\langle D^a\bar{B}^a(\bar{E}^a,\bar{H}^a(\bar{E}^a))\Delta(S),\,\delta\bar{E}\rangle=0 \tag{18}$$

and try to reduce the left-hand side of (18) into the first variation δI of some functional I. Note that we have also included the boundary conditions $\bar{B}=\bar{B}_2=0$ and $\bar{B}^a=\bar{B}_2^a=0$, (13), as natural ones into the derivation. The constants D and D^a are chosen ($D=D^a=j\omega$) so that the dimension in (18) can be matched and the left-hand side of (18) can be expressed as the desired form δI.

Note that (18) can also be applied to the problems (15) and (16) with the boundary conditions (12), $\bar{B}=\bar{B}_1=0$ and $\bar{B}^a=\bar{B}_1^a=0$, regarded as essential ones. In this case, the last two terms in the left-hand side of (18) are then dropped automatically.

For the problems (15) and (16) with the same boundary conditions (12), $\bar{B}=\bar{B}_1=0$ and $\bar{B}^a=\bar{B}_1^a=0$, regarded as natural ones, one should start from (18) with $\delta\bar{E}$ and $\delta\bar{E}^a$ in the surface integral terms replaced by $\delta(j\bar{H})$ and $\delta(j\bar{H}^a)$, and $D=D^a=-\omega$.

By a straightforward manipulation it can be shown, from (18), that the mathematical equivalent of solving (15) and (16) is the variational problem as follows:

$$\delta I^e(\bar{E},\bar{E}^a)=0 \tag{19}$$

where the functional $I^e(\bar{E},\bar{E}^a)$ takes the form

$$I_{B1}^e(\bar{E},\bar{E}^a)=I_{B1E}^e(\bar{E},\bar{E}^a)$$
$$-\langle j\omega\bar{H}^a(\bar{E}^a),\,(\hat{n}\times\bar{E}-\bar{\gamma})\Delta(S)\rangle$$
$$-\langle(\hat{n}\times\bar{E}^a-\bar{\gamma}^a)\Delta(S),\,j\omega\bar{H}(\bar{E})\rangle \tag{20a}$$

$$I_{B1E}^e(\bar{E},\bar{E}^a)=-\omega^2\langle j\bar{H}^a(\bar{E}^a),\,\bar{\bar{\mu}}\cdot j\bar{H}(\bar{E})\rangle$$
$$+\omega^2\langle\bar{E}^a,\,\bar{\bar{\epsilon}}\cdot\bar{E}\rangle-\langle\bar{E}^a,j\omega\bar{J}\rangle-\langle j\omega\bar{J}^a,\,\bar{E}\rangle \tag{20b}$$

for the conditions (12), or the form

$$I_{B2}^e(\bar{E},\bar{E}^a)=-\omega^2\langle j\bar{H}^a(\bar{E}^a),\,\bar{\bar{\mu}}\cdot j\bar{H}(\bar{E})\rangle$$
$$+\omega^2\langle\bar{E}^a,\,(\bar{\bar{\epsilon}}+\bar{\bar{\alpha}}\Delta(S))\cdot\bar{E}\rangle$$
$$-\langle\bar{E}^a,j\omega(\bar{J}+\bar{\beta}\Delta(S))\rangle$$
$$-\langle j\omega(\bar{J}^a+\bar{\beta}^a\Delta(S)),\,\bar{E}\rangle \tag{21}$$

for the conditions (13). In (20) and (21), $\bar{H}(\bar{E})$ is actually a function of $\bar{E}$, $\bar{H}^a(\bar{E}^a)$ a function of $\bar{E}^a$, as given by (14), and $\Delta(S)$ is the symbol defined by (17).

Precisely, the variational principle states that the problem of solving $\bar{E}$ and $\bar{E}^a$ simultaneously from (15) and (16) with (12) (or (13)) is completely equivalent to that of determining the stationary functions $\bar{E}$ and $\bar{E}^a$ from the variational problem: $\delta I^e=0$, $I^e=I_{B1}^e$ (or I_{B2}^e), with both $\bar{E}$ and $\bar{E}^a$ unrestricted on the boundary surface S. In this case the conditions (12) (or (13)) are just the natural boundary conditions of the variational problem (19). Alternatively, we may solve the variational problem (19) with the trial functions $\bar{E}$ and $\bar{E}^a$ subject to the same constraints (12) $\bar{B}_1=\bar{B}_1^a=0$. With these constraints, the functional I_{B1}^e then reduces to I_{B1E}^e and the same conditions (12) are now the essential boundary conditions of the variational problem.

B. H-Formulation

The formulas for H-formulation, which may be obtained from those for E-formulation by means of duality transformation: $\bar{E}\to j\bar{H}$, $j\bar{H}\to\bar{E}$, $\bar{\bar{\epsilon}}\to\bar{\bar{\mu}}$, $\bar{\bar{\mu}}\to\bar{\bar{\epsilon}}$, $-j\bar{J}\to\bar{M}$, $\bar{M}\to -j\bar{J}$, etc., (or that given by Harrington [9]) are omitted in this investigation.

C. E, H-Formulation

Instead of treating $\bar{H}$ as a function of $\bar{E}$ and $\bar{H}^a$ as a function of $\bar{E}^a$ as did in Section III-A, we may also regard $\bar{E},\bar{H},\bar{E}^a,\bar{H}^a$ as the unrelated unknown fields to be determined from (9) and (10). In this manner, we then have the equivalent variational formulation of (9) and (10) in terms of $\overline{E},\overline{H},\overline{E}^a,\overline{H}^a$

$$\delta I^{e,h}(\overline{E},\overline{H},\overline{E}^a,\overline{H}^a)=0 \tag{22}$$

where the functional $I^{e,h}(\overline{E},\overline{H},\overline{E}^a,\overline{H}^a)$ is

$$I^{e,h}_{B1}(\overline{E},\overline{H},\overline{E}^a,\overline{H}^a)=I^{e,h}_{B1E}(\overline{E},\overline{H},\overline{E}^a,\overline{H}^a) - \langle j\omega\overline{H}^a, (\hat{n}\times\overline{E}-\overline{\gamma})\Delta(S)\rangle - \langle(\hat{n}\times\overline{E}^a-\overline{\gamma}^a)\Delta(S), j\omega\overline{H}\rangle$$

$$I^{e,h}_{B1E}(\overline{E},\overline{H},\overline{E}^a,\overline{H}^a)=\omega^2\langle j\overline{H}^a, \overline{\overline{\mu}}\cdot j\overline{H}\rangle+\omega^2\langle\overline{E}^a, \overline{\overline{\epsilon}}\cdot\overline{E}\rangle + \langle j\omega\overline{H}^a, \nabla\times\overline{E}+\overline{M}\rangle + \langle\nabla\times\overline{E}^a+\overline{M}^a, j\omega\overline{H}\rangle - \langle\overline{E}^a, j\omega\overline{J}\rangle-\langle j\omega\overline{J}^a, \overline{E}\rangle \tag{23}$$

for the conditions (12), or

$$I^{e,h}_{B2}(\overline{E},\overline{H},\overline{E}^a,\overline{H}^a)=\omega^2\langle j\overline{H}^a, \overline{\overline{\mu}}\cdot j\overline{H}\rangle + \omega^2\langle\overline{E}^a,(\overline{\overline{\epsilon}}+\overline{\overline{\alpha}}\Delta(S))\cdot E\rangle + \langle j\omega\overline{H}^a,\nabla\times\overline{E}+\overline{M}\rangle + \langle\nabla\times\overline{E}^a+\overline{M}^a, j\omega\overline{H}\rangle - \langle\overline{E}^a, j\omega(\overline{J}+\overline{\beta}\Delta(S))\rangle - \langle j\omega(\overline{J}^a+\overline{\beta}^a\Delta(S)),\overline{E}\rangle \tag{24}$$

for the conditions (13).

The variational principle again states that the solutions $(\overline{E},\overline{H})$ of (9) and $(\overline{E}^a,\overline{H}^a)$ of (10) with (12) (or (13)) are also the solutions $(\overline{E},\overline{H},\overline{E}^a,\overline{H}^a)$ of the variational problem: $\delta I^{e,h}=0$, $I^{e,h}=I^{e,h}_{B1}$ (or $I^{e,h}_{B2}$) with the trial functions $(\overline{E},\overline{H},\overline{E}^a,\overline{H}^a)$ completely arbitrary on the boundary S, and vice versa. This is the case of regarding (12) (or (13)) as the natural boundary conditions. For the case of regarding the same conditions (12) as the essential ones, the variational problem should be solved from: $\delta I^{e,h}=0, I^{e,h}=I^{e,h}_{B1E}$ with the trial functions $(\overline{E},\overline{H},\overline{E}^a,\overline{H}^a)$ subject to the constraints (12).

Note that by regarding $\overline{H}$ and $\overline{H}^a$ in (23) and (24) as functions of $\overline{E}$ and $\overline{E}^a$, respectively, as given by (14), one may show that the functionals in (23) and (24) for E,H-formulation can be reduced to those in (20) and (21) for E-formulation, respectively. The reduction of the unknowns, from $(\overline{E},\overline{H},\overline{E}^a,\overline{H}^a)$ to $(\overline{E},\overline{E}^a)$, does not raise the order of the derivatives in the resultant expressions, since (20), (21) and (23), (24) all contain the derivatives of at most order one. In this sense, the formulas of E-formulation are more attractive than those of E,H-formulation.

Note also that with $(j\overline{H})$ as well as $(j\overline{H}^a)$ treated as a single quantity in the derivation, the variational formulations for real- and complex-type inner products can both be expressed in the same form as given by (19)–(21) and (22)–(24).

IV. The Generalized Reaction Concept

We now present physical interpretation of the general variational principle described by (19)–(21) and (22)–(24). For this purpose, we define the "generalized reactions" as

$$\langle f^a, s\rangle=j\{\langle j\overline{H}^a, \overline{M}+\overline{M}_s\Delta(S)\rangle-\langle\overline{E}^a,j[\overline{J}+\overline{\beta}\Delta(S)]\rangle\}$$

$$\langle s^a, f\rangle=j\{\langle\overline{M}^a+\overline{M}^a_s\Delta(S),j\overline{H}\rangle-\langle j[\overline{J}^a+\overline{\beta}^a\Delta(S)],\overline{E}\rangle\}$$

$$\langle f^a, s(f)\rangle=j\{\langle j\overline{H}^a, \overline{M}(\overline{E},\overline{H})+\overline{M}_s(\overline{E},\overline{H})\Delta(S)\rangle - \langle\overline{E}^a,j[\overline{J}(\overline{E},\overline{H})+\overline{\beta}(\overline{E},\overline{H})\Delta(S)]\rangle\}$$

$$\langle s^a(f^a), f\rangle=j\{\langle\overline{M}^a(\overline{E}^a, \overline{H}^a)+\overline{M}^a_s(\overline{E}^a, \overline{H}^a)\Delta(S), j\overline{H}\rangle - \langle j[\overline{J}^a(\overline{E}^a,\overline{H}^a)+\overline{\beta}^a(\overline{E}^a,\overline{H}^a)\Delta(S)],\overline{E}\rangle\}. \tag{25}$$

Then it can be shown that the functionals $I^e(\overline{E},\overline{E}^a)$ in (20), (21) and $I^{e,h}(\overline{E},\overline{H},\overline{E}^a,\overline{H}^a)$ in (23), (24) can be rewritten in terms of these generalized reactions as follows:

$$I(f,f^a)=j\omega[\langle f^a, s(f)\rangle-\langle s^a,f\rangle-\langle f^a,s\rangle]. \tag{26}$$

Note that (26) is actually an extension of (5) if one defines a field dependent source such that $s(f)=Lf$ in (5).

In (25), $(\overline{J},\overline{M})$ and $(\overline{J}^a,\overline{M}^a)$ are the true sources of the original and adjoint problems, (9) and (10), respectively. The fields $(\overline{E},\overline{H})$ and $(\overline{E}^a,\overline{H}^a)$ are the trial fields used in the original and adjoint systems (9) and (10). The field-dependent sources

$$\overline{J}(\overline{E},\overline{H})=\nabla\times\overline{H}-j\omega\overline{\overline{\epsilon}}\cdot\overline{E}$$

$$\overline{M}(\overline{E},\overline{H})=-\nabla\times\overline{E}-j\omega\overline{\overline{\mu}}\cdot\overline{H} \tag{27}$$

are the ones to support the trial fields $(\overline{E},\overline{H})$, and

$$\overline{J}^a(\overline{E}^a,\overline{H}^a)=\nabla\times\overline{H}^a-j\omega\overline{\overline{\epsilon}}^a\cdot\overline{E}^a$$

$$\overline{M}^a(\overline{E}^a,\overline{H}^a)=-\nabla\times\overline{E}^a-j\omega\overline{\overline{\mu}}^a\cdot\overline{H}^a \tag{28}$$

are those to support the trial fields $(\overline{E}^a,\overline{H}^a)$.

The electric surface currents $\overline{\beta}$, $\overline{\beta}^a$, $\overline{\beta}(\overline{E},\overline{H})$, $\overline{\beta}^a(\overline{E}^a,\overline{H}^a)$ and magnetic surface currents $\overline{M}_s$, $\overline{M}^a_s$, $\overline{M}_s(\overline{E},\overline{H})$, $\overline{M}^a_s(\overline{E}^a,\overline{H}^a)$ in (25) are chosen according to the problems. For the problem with the boundary conditions $\overline{B}_1=\overline{B}^a_1=0$, (12), the currents are chosen such that

$$\overline{\beta}=\overline{\beta}^a=\overline{\beta}(\overline{E},\overline{H})=\overline{\beta}^a(\overline{E}^a,\overline{H}^a)=0$$

$$\overline{M}_s(\overline{E},\overline{H})=\hat{n}\times\overline{E}$$

$$\overline{M}^a_s(\overline{E}^a,\overline{H}^a)=\hat{n}\times\overline{E}^a \tag{29}$$

and $\overline{M}_s=\hat{n}\times\overline{E}$, $\overline{M}^a_s=\hat{n}\times\overline{E}^a$ or $\overline{M}_s=\overline{\gamma}$, $\overline{M}^a_s=\overline{\gamma}^a$ depending on whether the conditions (12) are regarded as essential or natural ones. For the problem with the natural boundary conditions $\overline{B}_2=\overline{B}^a_2=0$, (13), the currents should be chosen such that

$$\overline{M}_s=\overline{M}^a_s=\overline{M}_s(\overline{E},\overline{H})=\overline{M}^a_s(\overline{E}^a,\overline{H}^a)=0$$

$$\overline{\beta}(\overline{E},\overline{H})=-j\omega\overline{\overline{\alpha}}\cdot\overline{E}+\overline{H}\times\hat{n}$$

$$\bar{\beta}^a(\bar{E}^a,\bar{H}^a) = -j\omega\,\bar{\bar{\alpha}}^a\cdot\bar{E}^a + \bar{H}^a\times\hat{n} \tag{30}$$

and $\bar{\beta},\bar{\beta}^a$ are defined by (13).

The generalized reactions in (26) should be calculated from (25), (27)–(30) with $\bar{J},\bar{M},\bar{J}^a,\bar{M}^a$ denoting the true sources and $\bar{E},\bar{H},\bar{E}^a,\bar{H}^a$ denoting the trial fields.

The expressions in (25)–(30) are mainly written for the functionals $I^{e,h}(\bar{E},\bar{H},\bar{E}^a,\bar{H}^a)$ of E,H-formulation. The same expressions may be applied to the functionals $I^e(\bar{E},\bar{E}^a)$ of E-formulation if one regards $\bar{H}$ as a function of $\bar{E}$ and $\bar{H}^a$ as a function of $\bar{E}^a$ as given by (14). Then $\bar{M}(\bar{E},\bar{H})$ and $\bar{M}^a(\bar{E}^a,\bar{H}^a)$ automatically reduce to the true sources $\bar{M}$ and $\bar{M}^a$, respectively. Note that (25) and (26) are invariant under duality transformation, thus, they are also applicable to the expressions for H-formulation.

The term $\langle f^a, s\rangle$ may be interpreted as the "generalized reaction" of $(\bar{E}^a,\bar{H}^a)$, the trial fields in the adjoint system (10), on $(\bar{J}+\bar{\beta}\Delta(S),\ \bar{M}+\bar{M}_s\Delta(S))$, the sources of the original problem (9). The term $\langle f^a, s(f)\rangle$ is the reaction of the trial fields $(\bar{E}^a,\bar{H}^a)$ on the sources $(\bar{J}(\bar{E},\bar{H})+\bar{\beta}(\bar{E},\bar{H})\Delta(S),\ \bar{M}(\bar{E},\bar{H})+\bar{M}_s(\bar{E},\bar{H})\Delta(S))$ to support the trial fields $(\bar{E},\bar{H})$. Similar interpretations may be applied to $\langle s^a, f\rangle$ and $\langle s^a(f^a), f\rangle$.

Note that we have included surface sources (currents) $\bar{\beta}, \bar{\beta}^a,\cdots,\bar{M}_s, \bar{M}_s^a,\cdots$ and surface materials $\bar{\bar{\alpha}}, \bar{\bar{\alpha}}^a$ in the definition of the reactions. The generalized reactions (25) reduce to the conventional ones proposed by Rumsey [8] when the real-type inner product is adopted in formulation and the surface quantities $\bar{\bar{\alpha}}, \bar{\bar{\alpha}}^a, \bar{\beta}, \bar{\beta}^a,\cdots,\bar{M}_s, \bar{M}_s^a,\cdots$ are removed from the above equations.

The general variational principle described by (22)–(24) (or (19)–(21)) now has a physical interpretation as follows: the true solutions (fields) of the original and adjoint problems (9) and (10) (or (15) and (16)) are just the ones that give the sum of the generalized reactions in (26) a stationary value.

The trial fields $(\bar{E},\bar{H})$ and $(\bar{E}^a,\bar{H}^a)$ used in the original and adjoint systems possess the following symmetric property:

$$\langle f^a, s(f)\rangle = \langle s^a(f^a), f\rangle. \tag{31}$$

In words, the reaction of the trial fields $(\bar{E}^a, \bar{H}^a)$ (in the adjoint system) on the equivalent sources to produce the trial fields $(\bar{E}, \bar{H})$ (in the original system) is always equal to that of $(\bar{E}, \bar{H})$ on the equivalent sources to produce $(\bar{E}^a, \bar{H}^a)$.

Note that for the problems (9) and (10) with the boundary conditions (12) (or (13)), the reaction $\langle f^a, s(f)\rangle_t$ is equal to $\langle f^a, s\rangle_t$ and $\langle s^a(f^a), f\rangle_t$ is equal to $\langle s^a, f\rangle_t$ whenever the trial fields are equal to the true fields. The subscript t in $\langle\ ,\ \rangle_t$ indicates that the true fields and sources of (9) and (10) are utilized in the expressions for the reactions. One consequence is that the true fields and sources of the original and adjoint problems, (9) and (10), should be connected by the relation

$$\langle f^a, s\rangle_t = \langle s^a, f\rangle_t. \tag{32}$$

This is the generalized reciprocity theorem [29] which states that the reaction of the true fields of the adjoint system (10) on the true sources of the original system (9) should be equal to that of the true fields of (9) on the true sources of (10). Another consequence is that the functional I in (26) has a stationary value equal to $-j\omega\langle f^a, s\rangle_t$ $(=-j\omega\langle s^a, f\rangle_t)$.

It remains to consider the important special case of a self-adjoint problem such that

$$(\bar{\bar{\epsilon}}^a, \bar{\bar{\mu}}^a, \bar{\bar{\alpha}}^a) = (\bar{\bar{\epsilon}}, \bar{\bar{\mu}}, \bar{\bar{\alpha}}). \tag{33}$$

The dyadics in (33) are then symmetric or Hermitian depending on whether real- or complex-type inner product is adopted. For this case, there are many choices for the adjoint quantities. A convenient one is that

$$(\bar{J}^a,\bar{M}^a,\bar{E}^a,\bar{H}^a,\bar{\beta}^a,\bar{\gamma}^a) = (\bar{J},\bar{M},\bar{E},\bar{H},\bar{\beta},\bar{\gamma}). \tag{34}$$

Then the functional I in (26) can be written in terms of self-reactions as follows:

$$I(f) = j\omega[\langle f, s(f)\rangle - \langle s, f\rangle - \langle f, s\rangle]. \tag{35}$$

For the self-adjoint problem discussed by the complex-type inner product, the expressions in (20) and (21) then automatically reduce to the ones studied by the previous investigator [10]. In particular, the variational expression I^e_{B1E} in (20) has an interesting physical interpretation when it is rewritten in the following form:

$$\frac{1}{2\omega}I^e_{B1E}(\bar{E}) = \omega[1/2\langle\bar{E},\bar{\bar{\epsilon}}\cdot\bar{E}\rangle - 1/2\langle\bar{H}(\bar{E}),\bar{\bar{\mu}}\cdot\bar{H}(\bar{E})\rangle] + (-\mathrm{Im}\langle\bar{J},\bar{E}\rangle) \tag{36}$$

where Im means the imaginary part of a complex quantity. The first and second terms in the right-hand side of (36) may be interpreted as the time-average electric and magnetic energy stored within the region V, and the third term as the reactive power supplied by the electric source $\bar{J}$. The variational principle for an electromagnetic system then states that the true solution $\bar{E}$ of the variational problem $\delta I^e_{B1E}=0$ should be the one that makes the difference in the energy, electric minus magnetic, of the system plus the (electric) source reactive power a stationary value. The same interpretation may be applied to the expression in (21) if the surface quantities $\bar{\bar{\alpha}}$ and $\bar{\beta}$ are properly included in (36) (i.e., $\bar{\bar{\epsilon}}\to\bar{\bar{\epsilon}}+\bar{\bar{\alpha}}\Delta(S)$ and $\bar{J}\to\bar{J}+\bar{\beta}\Delta(S)$).

Using the complex-type inner product to describe a self-adjoint problem it can also be shown that the self-reactions $\langle f, s(f)\rangle$ and $\langle s(f), f\rangle$ are all imaginary, since

$$\langle f, s(f)\rangle = \langle s(f), f\rangle = -\langle f, s(f)\rangle^*. \tag{37}$$

This implies that the self-reaction such as $\langle f,s\rangle_t$ has no real part. Physically, the imaginary part of $\langle f,s\rangle_t$ is just the reactive power delivered by the true magnetic source $\bar{M}+\bar{M}_s\Delta(S)$ minus that delivered by the true electric source $\bar{J}+\bar{\beta}\Delta(S)$. These reactive powers are exactly the stationary value of the functional $I/(-j\omega)$.

Note that the expression in (35) is actually an alternative version of that given by (36). The sum of the reactive powers in (35) are, therefore, related to the difference of the average stored energy plus the (electric) source reactive power given by (36).

The electromagnetic variational (or stationary) principle described by $\delta I=0$ (where I is defined by (20), (21), (23), (24), (26), (35), and (36)) may be considered as an extension of the principle of least action in physics. The principle now states that the true solution of a self-adjoint (or non-self-adjoint) electromagnetic system is just the one that makes the sum of the reactive powers in (35) or (36) (or the sum of the generalized reactions in (26)) a stationary value.

Another choice of the adjoint quantities for a self-adjoint problem is worthy of further investigation. If the self-adjoint problem is discussed by the real-type inner product, then one may use the alternative choice of the form

$$(\bar{J},\bar{M},\bar{E},\bar{H},\bar{\beta},\bar{\gamma})=(\bar{J}_1,\bar{M}_1,\bar{E}_1,\bar{H}_1,\bar{\beta}_1,\bar{\gamma}_1)$$
$$(\bar{J}^a,\bar{M}^a,\bar{E}^a,\bar{H}^a,\bar{\beta}^a,\bar{\gamma}^a)=(\bar{J}_2,\bar{M}_2,\bar{E}_2,\bar{H}_2,\bar{\beta}_2,\bar{\gamma}_2). \quad (38)$$

The generalized reactions defined by (25) then automatically reduce to the ones originally proposed by Rumsey [8], [9]. The generalized reciprocity theorem (32) also reduces to the conventional one [9]

$$\langle f_2, s_1\rangle_t=\langle f_1, s_2\rangle_t \quad (39)$$

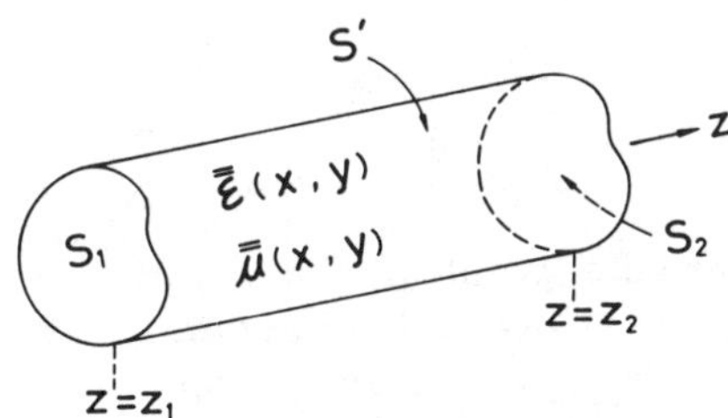

Fig. 2. A section of infinite uniform guide for establishing two-dimensional formulation.

for relating two different problems $(\bar{J}_1,\bar{M}_1,\bar{E}_1,\bar{H}_1)$ and $(\bar{J}_2,\bar{M}_2,\bar{E}_2,\bar{H}_2)$ in the same environment.

V. Reduction to Two-Dimensional Problems

We now apply the three-dimensional variational expressions in Sections III and IV to the derivation of the two-dimensional ones for an infinite uniform guide (Fig. 2). Although these expressions can also be obtained directly from the two-dimensional governing differential equations, it is interesting to establish the same results with a different approach.

Consider a section of uniform guide, from $z=z_1$ to $z=z_2$, to form a three-dimensional region (V of Fig. 1) with the boundary S consisting of the surface S' plus two cross-sectional planes S_1 and S_2 at $z=z_1$ and $z=z_2$, respectively, as shown in Fig. 2. Note that the two-dimensional problem in Fig. 2 should only include S' as its boundary, consequently, the starting equation (18) for this purpose should not contain the surface integral terms over S_1 and S_2.

Let us consider the variational expressions for the cases

$$[\bar{E}(x,y,z),\bar{H}(x,y,z),\bar{\beta}(x,y,z),\bar{\gamma}(x,y,z)]$$
$$=[\bar{E}(x,y),\bar{H}(x,y),\bar{\beta}(x,y),\bar{\gamma}(x,y)]e^{-jkz} \quad (40)$$
$$[\bar{E}^a(x,y,z),\bar{H}^a(x,y,z),\bar{\beta}^a(x,y,z),\bar{\gamma}^a(x,y,z)]$$
$$=[\bar{E}^a(x,y),\bar{H}^a(x,y),\bar{\beta}^a(x,y),\bar{\gamma}^a(x,y)]e^{-jk^az}. \quad (41)$$

Note that only under the condition

$$(jk^a)^p=-jk \quad (42)$$

it is possible to get a useful expression such as (43). By substituting (40) and (41) into (18) with the condition (42) it is discovered that the left-hand side of (18) (i.e., the volume integral plus the surface integral over S') may be written as

$$(z_2-z_1)\delta K=0 \quad (43)$$

where K is a functional $K(\bar{E}(x,y),\ \bar{H}(x,y),\ \bar{E}^a(x,y),\ \bar{H}^a(x,y))$ independent of z.

The two-dimensional functional K is related to the three-dimensional one I defined by (20), (21), (23), (24), (26), (35), or (36) with the fields given by (40)–(42), the boundary conditions on S' specified by (12) or (13), and some suitable boundary conditions, for example (12) or (13), specified on S_1 and S_2. The relation is

$$K=dI/dz_2 \quad (44)$$

i.e., the two-dimensional functional K can be obtained from the three-dimensional one I by performing the integration with respect to x, y, and z to get I first and then taking the derivative of I with respect to z_2. Actually, this functional K can directly be obtained from I by simply removing the integration with respect to z.

In conclusion, we have the two-dimensional formulation for a guide

$$\delta K=0 \quad (45)$$

since (z_2-z_1) in (43) is arbitrary.

By using the complex-type inner product to discuss a self-adjoint problem and choosing the adjoint quantities according to (34), it is found that the condition (42) is fulfilled automatically. Then one can yield, from (44), the same results investigated by the previous authors [7], [11]–[15].

VI. Applications of the Variational Principle

Examples of deriving the variational expressions for fields computation and establishing the stationary formulas for certain physical quantities, such as the eigenvalues, will be included to demonstrate the applications of the variational principle.

Consider first the normal incidence propagation problem for an isotropic inhomogeneous dielectric slab [16]. From the region 1 ($x<0$) of homogeneous medium ($\mu_0, \epsilon_0\epsilon_1$), a uniform plane wave is normally incident upon the region 2 ($0<x<a$) of inhomogeneous medium ($\mu_0, \epsilon_0\epsilon_2(x)$), and is then transmitted to the region 3 ($x>a$) of homogeneous medium ($\mu_0, \epsilon_0\epsilon_3$). By means of the continuity conditions for tangential electric and magnetic fields over the boundaries at $x=0$ and $x=a$, one can obtain the constants associated with the problem as follows:

$$\bar{\bar{\alpha}}(0)=\frac{k_1}{j\omega^2\mu_0}\hat{y}\hat{y} \qquad \bar{\beta}(0)=\frac{-2k_1}{\omega\mu_0}\hat{y}$$

$$\bar{\bar{\alpha}}(a)=\frac{k_3}{j\omega^2\mu_0}\hat{y}\hat{y} \qquad \bar{\beta}(a)=0. \tag{46}$$

For computational simplicity by adopting the real-type inner product and choosing (34) one has, from (21), the variational expression for computing the electric field $\bar{E}(=\hat{y}\psi(x))$

$$-\mu_0 I(\psi)=\int_0^a\left[\psi'(x)^2-k_0^2\epsilon_2(x)\psi^2(x)\right]dx$$
$$+jk_3\psi^2(a)+jk_1\psi^2(0)-j4k_1\psi(0). \tag{47}$$

The constants ϵ_0 and μ_0 are the permittivity and permeability of free space; ϵ_1, $\epsilon_2(x)$, and ϵ_3 are the relative permittivities of region 1, 2, and 3; and k_0, k_1, and k_3 are the propagation constants of free space, region 1 and region 3, respectively.

For the functional $I(f,f^a)$ in (26) (or (5)) which contains an unknown physical parameter such as resonance frequency, cutoff frequency, or propagation constant, the stationary formula for that parameter may be obtained from the equation

$$I(f,f^a)=I(f_t,f_t^a) \tag{48}$$

where f,f^a are the trial fields and f_t,f_t^a are the true fields of the original and adjoint problems. Note that the functional (26) for the true fields is zero, i.e.,

$$I(f_t,f_t^a)=0 \tag{49}$$

whenever the true sources $\bar{J}$, $\bar{M}$, $\bar{J}^a$, $\bar{M}^a$ and the surface sources $\bar{\beta}$, $\bar{\beta}^a$, $\bar{M}_s$, $\bar{M}_s^a$ are all zero.

It remains to apply (48) and (49) to the problems of establishing the stationary formulas for a cavity resonator and a waveguide.

Consider the resonator which has a perfectly conducting wall S to enclose a region V of Hermitian medium. The complex type inner product is adopted in formulation. Then the problem is self-adjoint. The adjoint quantities are chosen according to (34). Note that the true sources $(\bar{J},\bar{M})$ are zero within V so that the relation in (49) can be satisfied. By considering the boundary condition, $\hat{n}\times\bar{E}(S)=\bar{\gamma}=0$, as a natural one and using (14), (20), (34), (48), and (49), one then has the stationary formula for resonance frequency ω [15] of the form

$$\omega^2=\left\{\int_V\nabla\times\bar{E}^*\cdot\bar{\bar{\mu}}^{-1}\cdot\nabla\times\bar{E}\,dV\right.$$
$$-\int_S\hat{n}\cdot\left[\bar{E}\times\left(\bar{\bar{\mu}}^{-1}\cdot\nabla\times\bar{E}\right)^*\right]dS$$
$$\left.-\int_S\hat{n}\cdot\left[\bar{E}^*\times\left(\bar{\bar{\mu}}^{-1}\cdot\nabla\times\bar{E}\right)\right]dS\right\}\Big/\int_V\bar{E}^*\cdot\bar{\bar{\epsilon}}\cdot\bar{E}\,dV. \tag{50}$$

Finally consider the waveguide with a perfectly conducting wall S' to enclose the anisotropic medium

$$\bar{\bar{\epsilon}}=\bar{\bar{\epsilon}}(x,y)=\epsilon_{xx}\hat{x}\hat{x}+\epsilon_{xy}\hat{x}\hat{y}+\epsilon_{xz}\hat{x}\hat{z}$$
$$+\epsilon_{yx}\hat{y}\hat{x}+\epsilon_{yy}\hat{y}\hat{y}+\epsilon_{yz}\hat{y}\hat{z}$$
$$-\epsilon_{xz}\hat{z}\hat{x}-\epsilon_{yz}\hat{z}\hat{y}+\epsilon_{zz}\hat{z}\hat{z}$$
$$\bar{\bar{\mu}}=\bar{\bar{\mu}}(x,y)=\mu_{xx}\hat{x}\hat{x}+\mu_{xy}\hat{x}\hat{y}+\mu_{xz}\hat{x}\hat{z}$$
$$+\mu_{yx}\hat{y}\hat{x}+\mu_{yy}\hat{y}\hat{y}+\mu_{yz}\hat{y}\hat{z}$$
$$-\mu_{xz}\hat{z}\hat{x}-\mu_{yz}\hat{z}\hat{y}+\mu_{zz}\hat{z}\hat{z}. \tag{51}$$

With the fields expressed as

$$\bar{E}(x,y,z)=\bar{E}(x,y)e^{-jkz}=\left(\bar{E}_t+\hat{z}E_z\right)e^{-jkz}$$

$$\bar{H}(x,y,z)=\bar{H}(x,y)e^{-jkz}=\left(\bar{H}_t+\hat{z}H_z\right)e^{-jkz} \tag{52}$$

and adopting real-type inner product it can be shown, from Maxwell's equations and (42), that the adjoint fields may be written as

$$\bar{E}^a(x,y,z)=\bar{E}^a(x,y)e^{jkz}=\left(\bar{E}_t-\hat{z}E_z\right)e^{jkz}$$

$$\bar{H}^a(x,y,z)=\bar{H}^a(x,y)e^{jkz}=\left(-\bar{H}_t+\hat{z}H_z\right)e^{jkz}. \tag{53}$$

For this specific problem, the relation in (49) is again satisfied. By considering the boundary condition, $\hat{n}\times\bar{E}=\bar{\gamma}=0$, as an essential one and using (23), (44), (45), (48), and (49) one then has the stationary formula for the propagation constant k:

$$k=\left\{\int_{S'}\left[\omega\bar{E}^a(x,y)\cdot\bar{\bar{\epsilon}}\cdot\bar{E}(x,y)-\omega\bar{H}^a(x,y)\cdot\bar{\bar{\mu}}\cdot\bar{H}(x,y)\right.\right.$$
$$\left.\left.+j\bar{H}^a(x,y)\cdot\nabla_t\times\bar{E}(x,y)+j\bar{H}(x,y)\cdot\nabla_t\times\bar{E}^a(x,y)\right]dS\right\}$$
$$\Big/2\int_{S'}\bar{E}_t\times\bar{H}_t\cdot\hat{z}\,dS. \tag{54}$$

The symbol ∇_t in (54) represents the "transverse" part of the del operator ∇, i.e., $\nabla_t=\nabla-\hat{z}\ \partial/\partial z$. Note that (54) reduces to the special one in [9] whenever $\bar{\bar{\epsilon}}$ and $\bar{\bar{\mu}}$ are scalar.

Stationary formulas for other quantities, such as the impedance of an antenna and the echo area of a scatterer, etc., [7], [9], can also be derived from this variational principle.

VII. Conclusions

The general variational (or stationary) principle has been established for dealing with the non-self-adjoint problem in an electromagnetic system with various boundary conditions. This principle has been interpreted physically in terms of generalized reactions, time-average stored energy, and reactive powers. It has been shown that the true field of a non-self-adjoint system (or a self-adjoint system) is the one that makes the sum of the generalized reactions (or the sum of the reactive powers) a stationary value. In this sense, this general variational principle is indeed an extension of the least action principle in physics. The non-self-adjoint variational expressions have been found automatically leading to the self-adjoint ones discussed by the previous investigators. The general formulation is proved very useful in deriving the variational expressions for fields and eigenvalues computation.

Acknowledgment

The authors are indebted to Prof. C. M. Huang, Y. W. Kiang, and C. T. Liu for helpful discussions, and Miss S. L. Chang for various assistance during the preparation of this manuscript. Valuable suggestions and reference materials from one of the reviewers are much appreciated.

References

[1] P. M. Morse and H. Feshbach, *Methods of Theoretical Physics.* New York: McGraw-Hill, 1953.

[2] F. B. Hildebrand, *Methods of Applied Mathematics*, 2nd. ed. Englewood Cliffs, NJ: Prentice-Hall, 1956.

[3] G. Strang and G. J. Fix, *An Analysis of the Finite Element Method.* Englewood Cliffs, NJ: Prentice-Hall, 1973.

[4] T. K. Sarker, "Mathematical distinctions between variational method (Rayleigh–Ritz) and projection methods (Galerkin and method of moments)," Tech. Rep. 1, Dep. Elec. Eng., Rochester Inst. Technol., Rochester, NY, Jan. 1979.

[5] G. Jeng and A. Wexler, "Self-adjoint variational formulation of problems having non-self-adjoint operators," *IEEE Trans. Microwave Theory Tech.*, vol. MTT-26, pp. 91–94, Feb. 1978.

[6] A. D. McAulay, "Variational finite-element solution for dissipative waveguides and transportation application," *IEEE Trans. Microwave Theory Tech.*, vol. MTT-25, pp. 382–392, May 1977.

[7] L. Cairo and T. Kahan, *Variational Techniques in Electromagnetism.* New York: Gordon and Breach, 1965.

[8] V. H. Rumsey, "Reaction concept in electromagnetic theory," *Phys. Rev.*, vol. 94, pp. 1483–1491, June 1954.

[9] R. F. Harrington, *Time-Harmonic Electromagnetic Fields.* New York: McGraw-Hill, 1961.

[10] A. Konrad, "Vector variational formulation of electromagnetic fields in anisotropic media," *IEEE Trans. Microwave Theory Tech.*, vol. MTT-24, pp. 553–559, Sept. 1976.

[11] K. Morishita and N. Kumagai, "Unified approach to the derivation of variational expression for electromagnetic fields," *IEEE Trans. Microwave Theory Tech.*, vol. MTT-25, pp. 34–40, Jan. 1977.

[12] K. Morishita and N. Kumagai, "Systematic derivation of variational expressions for electromagnetic and/or acoustic waves," *IEEE Trans. Microwave Theory Tech.*, vol. MTT-26, pp. 684–689, Sept. 1978.

[13] W. J. English and F. J. Young, "An E vector variational formulation of the Maxwell equations for cylindrical waveguide problems," *IEEE Trans. Microwave Theory Tech.*, vol. MTT-19, pp. 40–46, Jan. 1971.

[14] W. J. English, "Vector variational solutions of inhomogeneously loaded cylindrical waveguide structures," *IEEE Trans. Microwave Theory Tech.*, vol. MTT-19, pp. 9–18, Jan. 1971.

[15] A. D. Berk, "Variational principle for electromagnetic resonators and waveguides," *IRE Trans. Antennas Propagat.*, vol. AP-4, pp. 104–110, Apr. 1956.

[16] C. H. Chen and C. D. Lien, "A finite element solution of the wave propagation problem for an inhomogeneous dielectric slab," *IEEE Trans. Antennas Propagat.*, vol. AP-27, pp. 877–880, Nov. 1979.

[17] M. A. Morgan, S. K. Chang, and K. K. Mei, "Coupled azimuthal potentials for electromagnetic field problems in inhomogeneous axially symmetric media," *IEEE Trans. Antennas Propagat.*, vol. AP-25, pp. 413–417, May 1977.

[18] K. Okamoto and T. Okoshi, "Vectorial wave analysis of inhomogeneous optical fibers using finite element method," *IEEE Trans. Microwave Theory Tech.*, vol. MTT-26, pp. 109–114, Feb. 1978.

[19] S. K. Chang and K. K. Mei, "Application of the unimoment method to electromagnetic scattering of dielectric cylinders," *IEEE Trans. Antennas Propagat.*, vol. AP-24, pp. 35–42, Jan. 1976.

[20] M. A. Morgan and K. K. Mei, "Finite-element computation of scattering by inhomogeneous penetrable bodies of revolution," *IEEE Trans. Antennas Propagat.*, vol. AP-27, pp. 202–214, Mar. 1979.

[21] D. T. Thomas, "Functional approximations for solving boundary value problems by computer," *IEEE Trans. Microwave Theory Tech.*, vol. MTT-17, pp. 447–454, Aug. 1969.

[22] B. H. McDonald and A. Wexler, "Finite-element solution of unbounded field problems," *IEEE Trans. Microwave Theory Tech.*, vol. MTT-20, pp. 841–847, Dec. 1972.

[23] N. G. Alexopoulos and N. K. Uzunoglu, "A simple analysis of thick microstrip on anisotropic subtrates," *IEEE Trans. Microwave Theory Tech.*, vol. MTT-26, pp. 455–456, June 1978.

[24] I. Stakgold, *Boundary Value Problems of Mathematical Physics*, vol. II. New York: Macmillan, 1968.

[25] B. Friedman, *Principles and Techniques of Applied Mathematics.* New York: Wiley, 1956.

[26] R. F. Harrington, *Field Computation by Moment Methods.* New York: Macmillan, 1968.

[27] T. G. Hazel and A. Wexler, "Variational formulation of the Dirichlet boundary condition," *IEEE Trans. Microwave Theory Tech.*, vol. MTT-20, pp. 385–390, June 1972.

[28] D. S. Jones, "A critique of the variational method in scattering problems," *IRE Trans. Antennas Propagat.*, vol. AP-4, pp. 297–301, 1956.

[29] R. E. Collin and F. J. Zucker, *Antenna Theory*, Part I. New York: McGraw-Hill, 1969.

UNIVERSAL FINITE ELEMENT MATRICES FOR TETRAHEDRA

PETER P. SILVESTER

Computational Analysis and Design Laboratory, Department of Electrical Engineering, McGill University, Montreal, Quebec, Canada

SUMMARY

Methods are described for forming element matrices for a wide variety of operators on tetrahedral finite elements, in a manner similar to that previously employed for line segments and triangles. This technique models the differentiation and product-embedding operators as rectangular matrices, and produces finite element matrices by replacing all required analytic operations by their finite matrix analogues. The method is illustrated by deriving the conventional matrix representation for Laplace's equation. Computer programs are available, which generate universal finite element matrices for use in various applications.

INTRODUCTION

The finite element analyst has traditionally had two choices for evaluating the matrix elements required for any given finite element model. In one approach, advocated by Zienkiewicz,[1] Irons and others, the matrix elements are evaluated numerically as and when required, using quadrature formulae to compute the necessary integrals. The second approach, first employed by Silvester,[2,3] is to evaluate the matrix elements analytically in terms of parametric factors for a representative element. The precomputed matrix values are then combined in weighted sums to the overall finite element matrix.

Both of the accepted procedures have advantages and disadvantages. The numerical integration approach is simple, and easy to implement; but it gives rise to high computing costs and sometimes to poor accuracy. Analytic integration is much less costly, but it requires precomputing and storing many different numeric matrices for the various differential operators encountered in applications, and their associated functionals.

In recent years a third approach, variously called an 'elementary matrix' or 'universal matrix' approach, has been developed.[4–8] In this approach, exact numeric representations are developed for certain elementary operators, such as the differentiation operator. Finite element matrices are then generated in specific cases as weighted combinations of the universal matrices. This third approach shares the precision advantages of the precomputed matrix technique, since all necessary differentiations and integrations are performed analytically, not numerically. Yet it shares the simplicity of the numerical integration approach, because the elementary matrices are few and are combined in simple ways. Not surprisingly, the computing time demands of the new method lie between those of the two classical techniques.

In the majority of applications, it is found that at most four elementary matrices suffice to model problems involving arbitrary linear differential operators. For practical use, one has the choice of either tabulating these matrices, or of giving programs capable of generating them as needed. The usual course in the past has been to tabulate the matrices, preferably in the form of integer quotients; for only in that form is full precision preserved. In the present work, the alternative approach is taken: short computer programs are described which generate the elementary matrices in floating-point form. The disadvantage of finite machine-dependent precision is avoided by employing the same computer, or a computer of at least the same precision, for both the elementary matrix generation and subsequent finite element problem solving.

INTERPOLATION POLYNOMIALS ON TETRAHEDRA

Interpolation polynomials of the closed Newton–Cotes type are commonly used on triangular and tetrahedral elements in field analysis. To set up these polynomials in a convenient form, let

Reprinted with permission from *International Journal for Numerical Methods in Engineering*, vol. 18, pp. 1055–1061, 1982.

$$\zeta_1 = \begin{vmatrix} x & y & z & 1 \\ x_2 & y_2 & z_2 & 1 \\ x_3 & y_3 & z_3 & 1 \\ x_4 & y_4 & z_4 & 1 \end{vmatrix} \Bigg/ \begin{vmatrix} x_1 & y_1 & z_1 & 1 \\ x_2 & y_2 & z_2 & 1 \\ x_3 & y_3 & z_3 & 1 \\ x_4 & y_4 & z_4 & 1 \end{vmatrix} \tag{1}$$

denote one of the homogeneous (volume) co-ordinates[9] on a tetrahedron; the remaining three are defined similarly by cyclic interchange of subscripts. Silvester[2] has defined a family of semi-interpolative (one-sidedly interpolative) polynomials by

$$\begin{aligned} P_m(z) &= \prod_{i=1}^{m} \frac{Nz-i+1}{i}, \qquad m \geqslant 1 \\ &= 1 \qquad\qquad\qquad\quad m = 0 \end{aligned} \tag{2}$$

These are semi-interpolative because they possess zeros at $z=(i-1)/N$, for $i=1,\ldots,m$. They are very convenient for defining the set of Lagrangian interpolation polynomials on a tetrahedron, with interpolation nodes of the closed Newton–Cotes pattern. The latter are given by

$$\alpha_{ijkl} = P_i(\zeta_1)P_j(\zeta_2)P_k(\zeta_3)P_l(\zeta_4) \tag{3}$$

subject to the requirement that $i+j+k+l=N$, where N is the degree of the desired polynomial. On a tetrahedron there are $M(N)=(N+1)(N+2)(N+3)/6$ such nodes and corresponding polynomials. The quadruple index $ijkl$ identifies the polynomial associated with each interpolation node clearly. However, in most applications it is preferable to use single indices to identify the polynomials, so as to avoid cluttering expressions with long subscript strings. In principle, the quadruple indices may be mapped on to single indices in any consistent fashion. In practice, the mapping is usually accomplished by regarding each quadruple index as a four-digit integer, and taking these in descending order.

THE DIFFERENTIATION OPERATOR

The directional derivative of a polynomial finite element approximation is best expressed in a tetrahedral element by writing the derivative in terms of interpolation polynomials. Consider for example a potential function u, given in a tetrahedron as a polynomial of degree N in the space co-ordinates,

$$u = \sum_{i=1}^{M(N)} u_i \alpha_i^{(N)}(x, y, z) \tag{4}$$

and suppose that its directional derivative is desired in some direction, say s. If the interpolation polynomials used for approximating are of degree N, this derivative is clearly a polynomial of degree at most $N-1$. Thus, one may write

$$\frac{\partial u}{\partial s} = \sum_{i=1}^{M(N)} u_i \sum_{j=1}^{4} \frac{\partial \alpha_i^{(N)}}{\partial \zeta_j} \frac{\partial \zeta_j}{\partial s} \tag{5}$$

where the chain rule of differentiation has been used to move the operation of differentiation from the space direction s to the tetrahedron co-ordinates. But since the derivative is a polynomial of degree $N-1$, it may be expressed exactly in terms of the interpolation polynomials of degree $N-1$:

$$\frac{\partial u}{\partial s} = \sum_{k=1}^{M(N-1)} d_k \alpha_k^{(N-1)} \tag{6}$$

The coefficients in equation (6) are most easily determined by equating right-hand sides of equations (5) and (6), and then observing that the summation of equation (6) collapses to a single term if evaluated at an interpolation node, say node k, of the family of interpolation polynomials of degree $N-1$:

$$d_k = \sum_{i=1}^{M(N)} u_i \sum_{j=1}^{4} \frac{\partial \zeta_j}{\partial s} \left[\frac{\partial \alpha_i^{(N)}}{\partial \zeta_j} \right]_{P_k^{(N-1)}} \tag{7}$$

Let four purely numeric matrices $D^{(j)}$ be defined by

$$D_{ki}^{(j)} = \left.\frac{\partial \alpha_i^{(N)}}{\partial \zeta_j}\right|_{P_k^{(N-1)}} \tag{8}$$

These matrices are pure numerics, independent of the size and shape of the tetrahedron. In terms of these matrices, equation (7) may be written in the form

$$d = \left(\sum_{j=1}^{4} \frac{\partial \zeta_j}{\partial s} D^{(j)}\right) u \tag{9}$$

It should be observed that although there are in principle four distinct coefficient matrices D, the very nature of homogeneous co-ordinates dictates that they must be row and column permutations of each other. Thus, tabulation and calculation of only one matrix suffices.

The directional differentiation operator may be regarded as a mapping between the space spanned by the interpolation polynomials of degree N, and the space spanned by those of degree $N-1$. These spaces are of dimensionality $M(N) = (N+1)(N+2)(N+3)/6$ and $M(N-1) = (N)(N+1)(N+2)/6$, respectively. One possible representation of the finite directional differentiation operator is therefore a rectangular matrix with $M(N)$ columns but only $M(N-1)$ rows. This representation is advantageous in many applications because of its compactness, as well as because the matrices are guaranteed to have full row rank. However, if directional derivative values are desired, this representation suffers from the shortcoming that the values are obtained on an interpolation node set different from that used for the function values. In this circumstance, it is more convenient to express the derivatives in terms of polynomials of degree N. Thus, one may replace equation (6) by

$$\frac{\partial u}{\partial s} = \sum_{k=1}^{M(N)} \bar{d}_k \alpha_k^{(N)} \tag{10}$$

This equation is exact, since the directional derivative is a polynomial of degree $N-1$, and may therefore be expressed in terms of the polynomials of degree N. In this case, the equation corresponding to (9) becomes

$$\bar{d} = \left(\sum_{j=1}^{4} \frac{\partial \zeta_j}{\partial s} \bar{D}^{(j)}\right) u \tag{11}$$

where

$$\bar{D}_{ki}^{(j)} = \left.\frac{\partial \alpha_i^{(N)}}{\partial \zeta_j}\right|_{P^{(N)}} \tag{12}$$

the derivatives being evaluated at the interpolation nodes of the set of degree N, not $N-1$.

Again, the four numeric matrices D are row and column permutations of each other, so that only one needs to be calculated and stored. However, this matrix is square, having $M(N)$ rows and columns. Of course, it has a row nullity of $M(N) - M(N-1)$, and rank $M(N-1)$.

THE METRIC MATRICES

A matrix frequently required in finite element analysis is the metric of the interpolation polynomials in each element. This matrix is occasionally also termed the 'mass matrix' by analysts whose background is rooted in elasticity theory or structural analysis. Given the set of interpolation polynomials of degree N, the metric T is defined as the matrix whose elements are given by

$$T_{ij}^{(N)} = \int \alpha_i^{(N)} \alpha_j^{(N)} \, d\Omega \tag{13}$$

Here and in the following, it is assumed that the tetrahedral element has unit volume; for any other element, T must be multiplied by the element volume. There will of course be a distinct metric, of order $M(N)$, for each order of tetrahedral element; orders will be distinguished by superscripts in parentheses, as above. Metrics for the first three orders of tetrahedra have been published[3] in the form of integer quotients, so that the first three are known exactly.

An interesting point to observe is that the sequence of metrics T for the various orders of tetrahedron is not independent. Since the interpolation polynomials (3) of the various orders are all complete in the sense of Dunne,[11] the family of polynomials of any given order must embed all polynomial families of all lower orders. Consequently, the metric of any given order must also embed, in some sense, the metrics of all lower orders. Just exactly how, will become evident on brief examination of the manner in which the embeddings of the polynomials themselves can be represented.

EMBEDDING OPERATORS

Suppose that a certain polynomial p has an exact representation in terms of the interpolation polynomials of degree N, say

$$p = \sum_{i=1}^{M(N)} p_i^{(N)} \alpha_i^{(N)} \tag{14}$$

Then it must also have an exact representation in terms of the interpolation polynomials of degree $N+1$,

$$p = \sum_{j=1}^{M(N+1)} p_j^{(N+1)} \alpha_j^{(N+1)} \tag{15}$$

and it is interesting to enquire how the coefficients in equation (15) can be derived from those in equation (14). To determine the necessary mapping, it suffices to equate the right sides of these two equations,

$$\sum_{j=1}^{M(N+1)} p_j^{(N+1)} \alpha_j^{(N+1)} = \sum_{i=1}^{M(N)} p_i^{(N)} \alpha_i^{(N)} \tag{16}$$

and to evaluate both sides at interpolation node k of order $N+1$, for $k = 1, \ldots, M(N+1)$. Since the polynomials are interpolative, the left-hand summation collapses, leaving only a single surviving term:

$$p_k^{(N+1)} = \sum_{i=1}^{M(N)} p_i^{(N)} \alpha_i^{(N)} |_{P_k^{(N+1)}} \tag{17}$$

Let a rectangular matrix, with $M(N+1)$ rows and $M(N)$ columns, be defined by

$$B_{ki}^{(N)} = \alpha_i^{(N)} |_{P_k^{(N+1)}} \tag{18}$$

The mapping of coefficient vectors between equations (14) and (15) is then clearly given, in matrix form, by

$$p^{(N+1)} = Bp^{(N)} \tag{19}$$

The matrix B may be termed a finite embedding operator, or an embedding matrix, for it embeds the coefficients related to degree N in the next higher-order set.

While the matrix B could easily be computed and tabulated for various orders, it may be useful to consider another matrix, which is more general than B, but allows B to be derived easily. Consider again the polynomial of equation (14); but this time let it be multiplied by some quantity which varies linearly with one of the tetrahedron co-ordinates. This time,

$$p^{\zeta_l} = \sum_{i=1}^{M(N)} p_i^{(N)} \zeta_l \alpha_i^{(N)} \tag{20}$$

is of interest, instead of equation (14). Equating and evaluating it at node k of the next higher order node set, as above, one is quickly led to define a matrix C by

$$C_{ki}^{(l)} = [\zeta_l \alpha_i^{(N)}]_{P_k^{(N+1)}} \tag{21}$$

Once again there exist four matrices C, one corresponding to weighting p with respect to each tetrahedron co-ordinate; the appropriate co-ordinate is identified by the bracketed subscript. The four matrices C are again row and column permutations of each other, so that there is no need to compute more than one of them.

Since the tetrahedron co-ordinates must add to exactly unity in any tetrahedron, the matrix B must be given by the sum of the matrices C:

$$B_{ki} = \sum_{l=1}^{4} C_{ki}^{(l)} \tag{22}$$

The matrices C provide a more general product embedding operation than does the matrix B. Yet the cost of computing them is virtually the same. Hence, it is well worth while to use computer programs that calculate and tabulate the matrices C, rather than B.

METRICS AND PROJECTORS

An interesting special case of embeddings arises when the polynomial p of equation (14) is in fact one of the interpolation polynomials of degree N. In this case the right-hand coefficient vector in equation (19) becomes one column of the unit matrix, and

$$\alpha_k^{(N)} = \sum_{i=1}^{M(N+1)} B_{ki}\alpha_i^{(N+1)} \tag{23}$$

This property is very useful in evaluating projection matrices. Csendes[4] shows that the best approximation to a polynomial of degree N in a subspace spanned by polynomials of degree $N-1$ is obtained by application of the projector

$$p^{(N)} = [T^{(N-1)}]^{-1}A^{(N)} \tag{24}$$

where

$$A_{ij}^{(N)} = \int \alpha_i^{(N-1)}\alpha_j^{(N)}\, \mathrm{d}\Omega \tag{25}$$

These matrices are easily evaluated. Substituting equation (19) into (25), there immediately results

$$A^{(N)} = [B^{(N-1)}]'T^{(N)} \tag{26}$$

where the prime denotes transposition. A separate evaluation of equation (25) from first principles, by actual integration, is never required. In a comparable fashion, one easily derives

$$T^{(N-1)} = [B^{(N-1)}]'T^{(N)}B^{(N-1)} \tag{27}$$

This equation indicates that, at least in principle, there is no need for programs to calculate metrics of all orders. If the metric of the highest order element to be employed is known, then the metrics of all lower orders can be derived by successive applications of the embedding operator. The projector of equation (24) may thus be written in the alternative form

$$P^{(N)} = [(B^{(N-1)})'T^{(N)}B^{(N-1)}]^{-1}B^{(N-1)}T^{(N)} \tag{28}$$

It might be observed in passing that the two forms of differentiation operators, rectangular and square, are also related to each other by an embedding operation:

$$\bar{D}^{(N)(j)} = B^{(N-1)}D^{(N)(j)} \tag{29}$$

Thus there is no fundamental need to possess both types of differentiation matrices, although it may at times be convenient to do so.

THE DIRICHLET MATRIX

The Dirichlet matrix is very commonly encountered in finite element analysis of potential field problems, and will be employed to illustrate the use of the universal matrices described here. On a tetrahedral element of unit volume, the Dirichlet matrix is given by

$$S_{ij} = \int \nabla\alpha_i^{(N)} \,.\, \nabla\alpha_j^{(N)}\, \mathrm{d}\Omega \tag{30}$$

Written out in detail, this equation reads

$$\int \nabla\alpha_i \,.\, \nabla\alpha_j\, \mathrm{d}\Omega = \sum_{m=1}^{4}\sum_{n=1}^{4}\left(\frac{dx}{\partial\zeta_m}\frac{\partial x}{\partial\zeta_n} + \frac{\partial y}{\partial\zeta_m}\frac{\partial y}{\partial\zeta_n} + \frac{\partial z}{\partial\zeta_m}\frac{\partial z}{\partial\zeta_n}\right)\int \frac{\partial\alpha_i}{\partial\zeta_m}\frac{\partial\alpha_j}{\partial\zeta_n}\mathrm{d}\Omega \tag{31}$$

The crucial quantity is obviously the integrand on the right-hand side; the term in parentheses is simply a geometric constant that expresses the relationship of the four homogeneous co-ordinate directions to the three Cartesian axes. Using the relationships above, however, this integrand is readily written as

$$\frac{\partial \alpha_i}{\partial \zeta_m}\frac{\partial \alpha_j}{\partial \zeta_n}=\sum_{k=1}^{M(N-1)}\sum_{l=1}^{M(N-1)}[D_{ik}^{(j)}]'T^{(N-1)}D_{jl}^{(j)} \tag{32}$$

in terms of the rectangular differentiation matrices; or as an analogous expression in terms of the square differentiation matrices.

CONCLUSIONS

To derive finite element matrices for tetrahedral elements, using the conventional tetrahedron interpolation polynomials, it suffices to possess the following primitive matrices: (1) a finite differentiation operator; (2) the metric of the interpolation polynomial basis; (3) an embedding operator that maps low-order polynomials to a representation one order higher; (4) a projection operator that projects polynomials on to a space one order lower. The differentiation operator may be expressed in two different ways, each of which has advantages in certain applications.

The projection operators, and the two forms of the differentiation operator, may be derived easily from the first three primitive matrices above by simple matrix manipulations. Further, and more importantly, finite element matrix representations of many linear operators may be constructed from three primitives: (1) the rectangular differentiation matrix of order N, (2) the metric of order N, and (3) the embedding between orders $N-1$ and N. Computation of these matrices is relatively straightforward. Programs for doing so have been written, and are publicly available from the CPC Program Library, under catalogue number AARB.[12]

REFERENCES

1. O. C. Zienkiewicz, *The Finite Element Method in Engineering Science*, McGraw-Hill, New York, 1971.
2. P. Silvester, 'High-order polynomial triangular finite elements for potential problems', *Int. J. Engng Sci.* **7**, 849–861 (1969).
3. P. Silvester, 'Tetrahedral polynomial finite elements for the Helmholtz equation', *Int. J. num. Meth. Engng.* **4**, 405–413 (1972).
4. Z. J. Csendes, 'A finite element method for the general solution of ordinary differential equations', *Int. J. num. Meth. Engng.* **9**, 551–561 (1975).
5. P. Silvester and C. R. S. Haslam, 'Magnetotelluric modelling by the finite element method', *Geophys. Prospect.* **20**, 872–891 (1972).
6. Z. J. Csendes, 'A Fortran program to generate finite difference formulas', *Int. J. num. Meth. Engng*, **9**, 579–597 (1975).
7. E. Kisak, P. Silvester and W. M. Telford, 'A recursive method in the *E*-polarization of magnetotelluric modelling by high-order finite elements', *Acta Geodaet., Geophys. Montanist,* Acad. Sci., Hung. **12**, 255–266 (1977).
8. P. Silvester, 'Construction of triangular finite element universal matrices', *Int. J. num. Meth. Engng*, **12**, 237–244 (1978).
9. E. A. Maxwell, *General Homogeneous Coordinates*, Cambridge University Press, 1960.
10. P. Silvester, 'Symmetric quadrature formulae for simplexes', *Maths. Comp.* **24**, 95–100 (1970).
11. P. C. Dunne, 'Complete polynomial displacement fields for finite element method', *J. Roy. Aeronaut. Soc.* **72**, 245–246 (1968).
12. Z. J. Csendes, F. U. Minhas and P. P. Silvester, 'Tetrahedral finite element matrix primitives', *Comput. Phys. Commun.* **15**, 1981 (in press).

Calculation of Eddy Currents in Terms of H on Hexahedra

J.S. VAN WELIJ

Abstract. The eddy current equations are formulated here in terms of H [1] instead of the more usual vector potential A [2,3]. In both cases a combination with a scalar potential is necessary in the non conducting regions. In the case of A this leads to either a non symmetric or not positive definite system of equations to be solved. This is avoided by the formulation in H, but special elements have to be chosen in order to satisfy the interface conditions between the two regions. The construction of such elements on tetrahedra or rectangular blocks can be found in [4]. The construction for hexahedra, or more specifically, isoparametric bricks, is shown here. Compared to filling a hexahedron with tetrahedra, this method reduces the number of unknowns by a half. The computed approximations for the eddy currents are exactly non divergent

THE EQUATIONS TO BE SOLVED

The equations that describe the eddy currents are:

$$\text{curl } H = J \qquad \text{with } J = \sigma E$$
$$\text{div } B = 0 \qquad \text{with } B = \mu H + Br$$
$$\text{curl } E = -B'$$

In a part of space the current is a known function of time. This is the driving current. Its effect in the formulae can be split off by defining H = Hm + Hc, where Hc is the field resulting from the driving current in absence of any magnetic or conducting material. This field can be calculated explicitly.

For ease of statement the place of the driving current in the variational formulation will not be discussed. Also the term Br, the remanence of permanent magnets and the conditions at the far boundaries will be omitted. However no assumptions are made that restrict introduction of these terms.

Although no explicit attention is paid to non linearities, these are fully included.

THE VARIATIONAL FORMULATIONS

Formulation in A

In the formulation in terms of a vector potential A, the equation div B = 0 ensures that A exists such that B = curl A. Then set $H = 1/\mu\, B$, $E = -A'$, $J = \sigma E$ and solve J = curl H [2,3]. To obtain the variational formulation, multiply by a vector test function W and integrate over space. Partial integration delivers:

$$-\int \sigma A'.W = \int 1/\mu \text{ curl } A \,.\, \text{curl } W - \oint W \,.\, Hxn$$

After expansion of A in the functions W, this leads to a singular matrix. Therefore in the non conducting region one uses J = 0 to write H = -grad u. The preceding variational form is then only applied in the conducting region. By the continuity of Hxn on the interface, the surface term in the expression above becomes:

$$+\oint (W \times \text{grad } u).n$$

In the non conducting region one defines $B = \mu H$ and solves for 0 = div B. Again project on a test function and use continuity of B.n across the interface to express the surface integral in terms of data from the other side:

$$0 = \int \mu \text{ grad } u \,.\, \text{grad } w + \oint w(\text{curl } A).n$$

For the surface integral one can use the differentiation rule:

$$\text{curl}(wA) = \text{grad } w \times A + w \text{ curl } A$$

and the fact that the integral of the normal component of a curl over a closed suface vanishes, which leaves as surface term:

$$+\oint (A \times \text{grad } w).n$$

Realising that on the interface the normal has different signs in the two regions, we may choose one and combine the surface integrals. This has as total result:

$$-\int \sigma A'.W = \int 1/\mu \text{ curl } A \,.\, \text{curl } W + \int \mu \text{ grad } u \,.\, \text{grad } w + \oint (W \times \text{grad } u - A \times \text{grad } w).n$$

Clearly after discretisation the minus sign in the surface integral disturbs the symmetry. This can be avoided by solving for 0 = -div B instead, but then the second volume integral in the right hand side spoils the positive definiteness.

Formulation in H

In the formulation directly in the field H, it is assumed that the initial solution satisfies div B = 0. In that case div B will automatically stay zero because B' is equated to a curl.

In the conducting region take as unknown the field H. Then define J = curl H, $E = (1/\sigma)J$, $B = \mu H$ and solve for -B' = curl E. As above, multiply by a vector test function W and integrate over the whole region. After partial integration this leads to:

$$-\int (\mu H)' \,.\, W = \int 1/\sigma \text{ curl } H \,.\, \text{curl } W - \oint W \,.\, Exn$$

In the non conducting region J = 0, so there exists a u such that H = -grad u. Starting from a non divergent B, solve for 0 = div B'. Multiplying by a scalar test function w and integrating partially gives:

$$0 = \int (\mu \text{ grad } u)' \,.\, \text{grad } w - \oint w(\text{curl } E).n$$

The surface integral is rewritten by the same rules as above, resulting in

$$0 = \int (\mu \text{ grad } u)' \,.\, \text{grad } w + \oint \text{grad } w \,.\, Exn$$

Noting again the sign reversal in the normal direction on the interface, the total result is:

$$-\int (\mu H)'.W = \int 1/\sigma \text{ curl } H \,.\, \text{curl } W + \int (\mu \text{ grad } u)'.\, \text{grad } w - \oint (W + \text{grad } w) \,.\, Exn$$

The author is with Nederlandse Philips Bedrijven BV, Eindhoven, Netherlands.

Reprinted from *IEEE Trans. Magn.*, vol. MAG-21, no. 6, pp. 2239–2241, Nov. 1985.

If it is possible to define test functions such that W = - grad w on the interface, this integral vanishes. Closer inspection shows that it is sufficient if only the tangential components are equal.

For these functions to be used also as basis functions in the space of H, clearly the tangential components must be continuous across element boundaries [4]. At the same time this condition guarantees that the discretised eddy currents calculated are exactly non divergent.

When meeting these two requirements, we obtain, after discretisation in time and space, a set of symmetric and positive definite equations to solve.

When the u depends on the field, it is important to discretise time in terms of B. If the time derivation is carried out first and the discretisation is done in H and u, symmetry may be lost.

THE ELEMENTS

We consider isoparametric bricks, i.e. each element is seen as the image of the unit cube under a trilinear mapping x(s), where x and s are three dimensional vectors (Fig. 1.)

Restricted to the element also the inverse mapping s(x) exists. The surface defined by the equation s3(x) = 1 for example is the top of the element. So evidently grad s3 is normal to the top face.

In the non conducting region, the standard trilinear functions on the unit cube are used for the scalar potential.

The new elements have nodes on the edges instead of on the vertices. The degree of freedom belonging to an edge is the tangential component of H along it. In practice it is easier to multiply this by the length of the edge, so the dimension of the unknown becomes Ampere. Now the test function for the edge where s2 = 1 and s3 = 1, so running along s1 (see Fig. 1.) is defined as:

W = s2 * s3 * grad s1

On both faces s1 = 0 and s1 = 1 this vector function is normal to it, so it does no contribute to the tangential components there; this includes all edges not running in direction s1. On the edges in direction s1 to which this function does not belong, the product s2*s3 is always zero, so the whole function vanishes. On its own edge the tangential component appears to be a constant (the inverse of length of the edge). This can be seen with help of the following definition:

Vi = dx/dsi

Since the functions x are trilinear polynomials in s, the Vi are bilinear and readily calculated. In every point on an edge of the element in direction si, Vi is the vector describing that edge. The Jacobian matrix ds/dx, which has grad si as rows, is the inverse of dx/ds, which has Vi as columns, so:

Vi.(grad sj) = delta(i,j)

This also gives an explicit expression for grad si. For example:

grad s1 = (V2 x V3)/((V2 x V3).V1)

Cyclical permutation gives the others. The validity of these expressions is easily checked. Note that the denominator is actually the Jacobian of the transformation and has the same value for all three gradients.

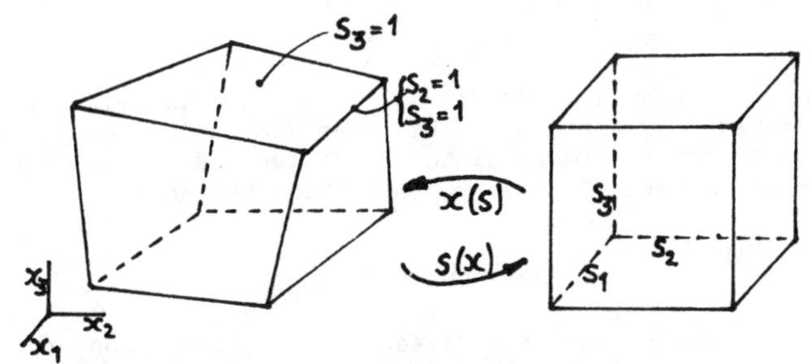

Fig. 1. The mapping x(s) transforms the unit cube into the hexahedron. Its inverse s(x) is a vector function on the original space. The equation s3(x)=1 describes the top face.

THE INTER ELEMENT CONTINUITY

With the help of the above, the proof of inter element continuity follows easily. On all faces not containing the edge in question, the tangential components have been shown to vanish. From the remaining two faces, take for example the top face and the same W as before. The normal is parallel to grad s3:

n = (V1 x V2)/||V1 x V2||

Strictly speaking, the decomposition of a vector into normal and tangential components is given by

W = n(W.n) + nx(Wxn)

As usual we use here just Wxn as tangential component. This is justified by the observation that continuity across the surface of Wxn is equivalent to continuity of nx(Wxn). Substituting the above stated form we obtain:

$$Wxn = s2 * s3 * \frac{(V2 \times V3)}{(V2 \times V3).V1} \times \frac{(V1 \times V2)}{||V1 \times V2||}$$

$$= s2 * s3 * \frac{((V2 \times V3).V2)V1-((V2 \times V3).V1)V2}{((V2 \times V3).V1) * ||V1 \times V2||}$$

$$= - s2 * s3 * V2 / ||V1 \times V2||$$

All quantities appearing in this expression are continuous going through the top face to the next element. By simple symmetry all the other cases follow immediately.

THE INTERFACE CONDITION

The ordinary scalar test function on the point s1 = s2 = s3 = 1 is given by w = s1*s2*s3. By the chain rule of differentiation and the implicit summation convention the gradient becomes:

grad w = dw/dsi grad si

On the face s3 = 1 the normal n has the direction of grad s3, so only two terms remain:

(grad w)xn = (s2*s3 grad s1 + s3*s1 grad s2)xn
= (W1 + W2)xn

where W1 and W2 belong to the two edges in the face ending in the point considered. The definition of the W orientates the edges. Edges beginning in a point

1163

appear with a minus sign. So indeed, the edge type degrees of freedom in the interface can be condensed by setting them equal to the difference of the vertex type degrees of freedom, that bound the edge.

SOME FURTHER REMARKS

It might seem that differentiating the functions W is rather tedious. This however is not really necessary, because in the equations only curl W turns up, which can be reduced by the product rule and the fact that a gradient is irrotational. For the same example:

curl W = grad(s2 * s3) x grad s1

By developing the right hand side as in the continuity proof, this reduces further to:

curl W = (s2 * V2 - s3 * V3) / ((V2 x V3).V1)

in which all factors are explicit expressions in the coordinates of the vertices of the element and the parameters s. The expressions for the other edges are similarly simple and can be found by symmetry considerations.

Per hexahedron, there are 12 unknowns, which are all shared by four elements, giving 3 unknowns per unit cell.

A hexahedron with planar faces, can be divided into 5 tetrahedra by adding one diagonal in each face. This adds 6 unknowns, each shared by two elements, giving a total of 6 unknowns per unit cell.

When calculating a periodic linear problem, the resulting complex linear system, written as a twice bigger real system, is not symmetric. This is more fundamental and is not remedied by the proposed elements.

Making some vertices of a hexahedron coincide, other shapes, such as prisms and pyramids can be created. In this process the mapping x(s) becomes singular at those vertices. However, the integrals in the variational formulation are still defined and by using internal integration points, such as the Gauss points, practical problems are circumvented.

Following a comparable reasoning as used in the above for edges and tangential components, it is possible to define elements which have nodes on the faces, and exhibit continuity in their normal components. For the face s3 = 1 the basis function would be:

W = s3 * (grad s1 x grad s2)
  = s3 * V3 / ((V1 x V2).V3)

It is conceivable to use this type of element for B or J.

REFERENCES

[1] A. Bossavit, J.C. Verite, "A Mixed FEM-BIEM Method to Solve 3-D Eddy Current Problems" in IEEE Transactions on Magnetics vol18 nr 2 March 1982, p 431.

[2] C.R.I Emson, J. Simkin "An Optimal Method for 3-D Eddy Currents" in IEEE Transactions on Magnetics 1983, vol19 nr 6.

[3] S.J. Polak, A.J.H. Wachters, J.S. van Welij, "A New 3-D Eddy Current Model" in IEEE Transactions on Magnetics 1983, vol19 nr 6.

[4] J.C. Nedelec, "Mixed Finite Elements in R3" in Numerische Matematik 1980.

New vector finite elements for three-dimensional magnetic field computation

M. L. Barton
Westinghouse Research & Development Center, Pittsburgh, Pennsylvania 15235
Z. J. Cendes
Department of Electrical and Computer Engineering, Carnegie Mellon University, Pittsburgh, Pennsylvania 15213

Finite-element vector potential solutions of three-dimensional magnetic field problems are usually obtained by approximating each component of the vector potential by a separate set of scalar finite-element approximation functions and by imposing continuity conditions between elements on all three components. This procedure is equivalent to imposing continuity of both the normal and the tangential components of the vector potential. We show in this paper that this procedure is too restrictive: While continuity of the tangential component of the vector potential is required, continuity of the normal components is not essential in the variational formulation. We introduce a new type of vector finite-element approximation function that has the property that it interpolates not to point values of each component of vector potential, but rather to the tangential projection of the vector potential on each edge of tetrahedral finite elements. With the new basis functions, continuity of the normal component of the vector potential is provided only approximately by means of the natural interface conditions inherent in the variational procedure. This results in a more efficient procedure for the solution of three-dimensional magnetostatic field problems than is obtained by enforcing normal component continuity exactly.

INTRODUCTION

The finite-element solution of three-dimensional magnetostatic problems via the vector potential has generated considerable interest in recent years. In the published procedures,[1-5] each component of the vector potential is approximated separately, and continuity of all three components between elements is imposed by setting nodal values in adjacent elements to be identical. We show in this paper that a more efficient procedure for solving magnetostatics problems is to employ vector finite-elements basis functions in which only the tangential component of the field is made continuous.

The use of mixed-order vector-valued finite elements was first suggested by Raviart and Thomas[6] for two-dimensional fluid flow. These elements were subsequently extended to three dimensions by Nedelec.[7] Recently, Bossavit and Verite[8] have made extensive use of tangential continuity elements to solve eddy-current problems in terms of the magnetic field variable **H**. In this paper, we show that tangential vector finite elements can be used to solve magnetostatic field problems in terms of the vector potential **A**. Indeed, we show that the new tangential elements exactly match the continuity requirements of the underlying variational procedure and that they provide a natural fit for approximating magnetostatic fields.

CONTINUITY CONDITIONS

Consider the interface between two different materials of permeabilities μ_1 and μ_2. We know that the continuity requirements of the magnetic field at the interface are expressed mathematically as

$$\mathbf{1}_n \cdot \mathbf{B}_1 = \mathbf{1}_n \cdot \mathbf{B}_2 , \qquad (1)$$

$$\mathbf{1}_n \times \mathbf{H}_1 = \mathbf{1}_n \times \mathbf{H}_2 , \qquad (2)$$

where **B** is the magnetic flux, **H** is the magnetic field, and $\mathbf{1}_n$ is the unit normal to the interface. Since $\mathbf{B} = \nabla \times \mathbf{A}$, where **A** is the vector potential, we may rewrite Eq. (1) and (2) as

$$\mathbf{1}_n \cdot (\nabla \times \mathbf{A})_1 = \mathbf{1}_n \cdot (\nabla \times \mathbf{A})_2 , \qquad (3)$$

$$\mathbf{1}_n \times (1/\mu_1)(\nabla \times \mathbf{A})_1 = \mathbf{1}_n \times (1/\mu_2)(\nabla \times \mathbf{A})_2 . \qquad (4)$$

Let (n,τ,λ) provide a local orthogonal coordinate system with τ and λ tangential to the interface. Expanding Eq. (3) in components then gives

$$\left(\frac{\partial A_\lambda}{\partial \tau} - \frac{\partial A_\tau}{\partial \lambda}\right)_1 = \left(\frac{\partial A_\lambda}{\partial \tau} - \frac{\partial A_\tau}{\partial \lambda}\right)_2 . \qquad (5)$$

Since Eq. (5) involves only the tangential variables τ and λ, it follows that *continuity of the normal component of* **B** *is ensured by setting the tangential components of* **A** *to be continous.*

From the tangential continuity requirements, Eq. (2) or (4), we obtain

$$\frac{1}{\mu_1}\left(\frac{\partial A_\tau}{\partial n} - \frac{\partial A_n}{\partial \tau}\right)_1 = \frac{1}{\mu_2}\left(\frac{\partial A_t}{\partial n} - \frac{\partial A_n}{\partial \tau}\right)_2 , \qquad (6a)$$

$$\frac{1}{\mu_1}\left(\frac{\partial A_n}{\partial \lambda} - \frac{\partial A_\lambda}{\partial n}\right)_1 = \frac{1}{\mu_2}\left(\frac{\partial A_n}{\partial \lambda} - \frac{\partial A_\lambda}{\partial n}\right)_2 . \qquad (6b)$$

We shall show that these two equations are obtained as natu-

Reprinted with permission from *Journal of Applied Physics*, vol. 61, no. 8, pp. 3919–3921, April 1987.

ral boundary conditions in the finite element process provided that the tangential component of **A** is made continuous. Note that it is not necessary to set $A_{n1} = A_{n2}$ to satisfy Eqs. (6a) and (6b).

VARIATIONAL PRINCIPLES

It is well known[1-5] that, for linear materials, a variational functional for magnetostatics is given by

$$F(\mathbf{A}) = \frac{1}{2}\int \frac{1}{\mu} |\nabla\times\mathbf{A}|^2 d\Omega - \int \mathbf{A}\cdot\mathbf{J}\, d\Omega . \tag{7}$$

To derive the natural boundary conditions for $F(\mathbf{A})$, we need to set its first variation about the exact solution $\mathbf{A}_{ex}$ to zero,

$$\delta F(\mathbf{A}) = \frac{\partial F}{\partial \epsilon}(\mathbf{A}_{ex} + \epsilon\boldsymbol{\xi})|_{\epsilon=0} = 0 , \tag{8}$$

where ϵ is a small real number and $\boldsymbol{\xi}$ is an arbitrary function that satisfies all essential boundary conditions. Substituting Eq. (7) into (8) gives

$$0 = \int \frac{1}{\mu}(\nabla\times\boldsymbol{\xi})\cdot(\nabla\times\mathbf{A}_{ex})d\Omega - \int \boldsymbol{\xi}\cdot\mathbf{J}\, d\Omega . \tag{9}$$

By means of standard vector identities, Eq. (9) may be written as

$$0 = \int \boldsymbol{\xi}\cdot\left(\nabla\times\frac{1}{\mu}\nabla\times\mathbf{A}_{ex} - \mathbf{J}\right)d\Omega + \oint\left(\boldsymbol{\xi}\times\frac{1}{\mu}\nabla\times\mathbf{A}_{ex}\right)\cdot d\mathbf{s} , \tag{10}$$

where the second integral must be evaluated on all surface—interior and exterior—of the finite-element mesh.

As is well known,[1-5] Eq. (10) shows that the Euler equation of $F(\mathbf{A})$ is

$$\nabla\times\frac{1}{\mu}\nabla\times\mathbf{A}_{ex} = \mathbf{J} . \tag{11}$$

However, let us examine the surface integral in (10) at interfaces in detail. Expanding by components gives

$$\oint\left[\xi_\tau \frac{1}{\mu}\left(\frac{\partial A_\tau}{\partial n} - \frac{\partial A_n}{\partial \tau}\right) - \xi_\lambda \frac{1}{\mu}\left(\frac{\partial A_n}{\partial \lambda} - \frac{\partial A_\lambda}{\partial n}\right)\right] ds = 0 . \tag{12}$$

If we impose continuity of the tangential component of **A**, then

$$\xi_\tau|_1 = \xi_\tau|_2 , \quad \xi_\lambda|_1 = \xi_\lambda|_2 , \tag{13}$$

and (12) yields on an interior surface

$$\int \xi_\tau \left[\frac{1}{\mu_1}\left(\frac{\partial A_\tau}{\partial n} - \frac{\partial A_n}{\partial \tau}\right)_1 - \frac{1}{\mu_2}\left(\frac{\partial A_\tau}{\partial n} - \frac{\partial A_n}{\partial \tau}\right)_2\right] ds , - \int \xi_\lambda \left[\frac{1}{\mu_1}\left(\frac{\partial A_n}{\partial \lambda} - \frac{\partial A_\lambda}{\partial n}\right)_1 - \frac{1}{\mu_2}\left(\frac{\partial A_n}{\partial \tau} - \frac{\partial A_\lambda}{\partial n}\right)_2\right] ds = 0 . \tag{14}$$

Since ξ_τ and ξ_λ are arbitrary, we see that the *natural interface conditions of the functional F(A) are exactly the required interface conditions in Eq. (6).*

VECTOR ELEMENTS WITH TANGENTIAL CONTINUITY

Since continuity of the normal component of **A** is nowhere required in the above analysis, we wish to develop finite-element trial functions that possess only continuity of the tangential components. For this purpose, consider the tetrahedron for which each of the six edges lies between the following two vertices:

Edge number i	Vertex number i_1	Vertex number i_2
1	1	2
2	1	3
3	1	4
4	2	3
5	4	2
6	3	4

This provides a natural ordering of the edges in the element in the sense that all other ordering result in minus signs in the upcoming expressions. Now define the 18 quantities

$$b_i = x_{i2} - x_{i1} , \quad c_i = y_{i2} - y_{i1} , \quad d_i = z_{i2} - z_{i1} . \tag{15}$$

The six vectors along the six edges can be represented as

$$\mathbf{e}_i = b_i \mathbf{1}_x + c_i \mathbf{1}_y + d_i \mathbf{1}_z . \tag{16}$$

The length of e_i is the length of the ith edge. Also define the quantity

$$\delta = \begin{vmatrix} b_1 & b_2 & b_3 \\ c_1 & c_2 & c_3 \\ d_1 & d_2 & d_3 \end{vmatrix} , \tag{17}$$

where δ gives six times the signed volume of the tetrahedron.

Now define the following vector in the tetrahedron,

$$\mathbf{A} = \mathbf{f} + \mathbf{g}\times\mathbf{r}, \tag{18}$$

where **f** and **g** are three-component constant vectors and **r** is the vector

$$\mathbf{r} = x\mathbf{1}_x + y\mathbf{1}_y + z\mathbf{1}_z . \tag{19}$$

By direct evaluation, we see that **A** possesses no divergence and has a constant curl

$$\nabla\cdot\mathbf{A} = 0 ,$$

$$\nabla\times\mathbf{A} = 2\mathbf{g} . \tag{20}$$

The vector **A** is linear but does not allow for all possible linear variations in the tetrahedron. The expression for **A** has six degrees of freedom (the six values of the components of **f** and **g**) while a full three-component linear vector on a tetrahedron would possess twelve.

Since there are six degrees of freedom in **A** and six edges in a tetrahedron, let us convert the variables **f** and **g** to variables defined by the tangential projections of **A** onto the six edges:

$$A_i = \mathbf{A}\cdot\mathbf{e}_i\big|_{\mathrm{edge}\,i}\,. \tag{21}$$

After some algebra, Eq. (18) and (21) yield

$$\mathbf{f} = \frac{1}{\delta}\sum_{i=1}^{6}(\mathbf{p}_{i1}\times\mathbf{p}_{i2})A_{7-i}\,,$$

$$\mathbf{g} = \frac{1}{\delta}\sum_{i=1}^{6}\mathbf{e}_i A_{7-i}\,, \tag{22}$$

where the vector $\mathbf{p}_i$ is defined as

$$\mathbf{p}_i = x_i\mathbf{1}_x + y_i\mathbf{1}_y + z_i\mathbf{1}_z\,. \tag{23}$$

The vector $\mathbf{p}_i$ is the location vector for the point (x_i,y_i,z_i), i.e., the vector from the origin to the ith vertex of the element. Note that the function defined in Eq. (21)–(23) interpolates to the tangential components of the vector field on the tetrahedron edges. Thus, the tangential components of the vector provide the unknowns in the formulation, and while the normal component in general exists, it is not directly represented.

It must be noted that while the above formulation appears to be complicated at first glance, it is actually simpler than the conventional approach. Only elements sharing edges are connected in the tangential formulation, while all elements sharing nodes are connected in the traditional approach. The net result is that there are fewer connections in the new formulation, and thus the finite-element coefficient matrix generated by tangential elements is more sparse than that produced by traditional elements.

RESULTS AND CONCLUSIONS

The formulation outlined above has been tested by solving for the magnetic field in a simple cube, one meter a simple cube, one meter on each side, in which a uniform current flows perpendicular to one face. The boundary conditions are that no flux crosses any face. This results in a problem whose statement contains no variation in the direction of current flow and whose solution is thus two-dimensional in character.

It is difficult to compare the results in terms of point values. The vector potential in either procedure is not unique since its gauge is not specified. The only meaningful comparison is between values of flux obtained by taking the curls of the vector potentials. Table I presents some of the results from a convergence study for the cube using $J = 10^7$ A/m^2. For the arbitrary point (0.27, 0.34, 0.18), the flux converges to the exact solution $B = 2.063$ T obtained by Fourier series. However, convergence is not monotonic since the point listed is sometimes in the center of an element, sometimes near its edge, as the mesh is refined. For this reason, we also evaluated the energy in the solution; the lower the energy the better the solution. The true energy, obtained from the accurate two-dimensional solution, is -2.204 J.

When merit of the solution (the energy) is plotted on logarithmic scales versus the cost of the solution which is taken to be proportional to the size of the required matrix, the plot reveals that the new tangential finite elements require, on the average, only 60% of the work to achieve an approximation of a given accuracy compared with the traditional approach. However, when the energy is plotted against the number of degrees of freedom in the problem, this plot reveals that in terms of the degrees of freedom the two methods are equally effective in approximating the vector potential. The advantage of the tangential formulation is that for a given number of degrees of freedom, it results in a much sparser matrix and thus costs less to set up and solve.

TABLE I. Convergence results for finite-element solutions. The upper set of numbers was obtained by using the tangential vector elements; the lower set by the conventional elements.

	Subdivision				
	1	2	4	8	16
Number of parameters	19 24	98 81	604 375	4 184 2 187	31 024 14 739
Degrees of freedom	1 0	26 9	316 135	3 032 1 323	26 416 11 475
Number of matrix entries	19 24	190 108	2392 1914	24 484 23 742	221 404 232 998
Flux at (0.27,0.34,0.18)			3.177 2.725	2.260 2.103	2.186 2.384
Energy	− 0.436 0.0	− 1.30 − 0.982	− 1.93 − 1.81	− 2.14 − 2.10	− 2.19 − 2.18

[1]M. V. K. Chari, Z. J. Cendes, P. P. Silvester, A. Konrad and M. A. Palmo, IEEE Trans. Power Apparatus Syst. **PAS-100**, 4007 (1981).

[2]J. L. Coulomb, IEEE Trans. Magn. **MAG-17**, 3241 (1981).

[3]N. A. Demerdash, T. W. Nehl, F. A. Fouad, and O. A. Mohammed, IEEE Trans. Power Apparatus Syst. **PAS-100**, 4104 (1981).

[4]Z. J. Cendes, J. Weiss, and S. R. H. Hoole, IEEE Trans. Magn. **MAG-18**, 362 (1982).

[5]M. V. K. Chari, A. Konrad, J. D'Angelo, and M. A. Palmo, IEEE Trans. Magn. **MAG-19**, 2321 (1983).

[6]P. A. Raviart and J. M. Thomas, Math. Comp. **31**, 292 (1977).

[7]J. C. Nedelec, Numer. Math. **35**, 315 (1980).

[8]A. Bossavit, and J. C. Verite, IEEE Trans. Magn. **MAG-18**, 431 (1982).

Covariant Projection Elements for 3D Vector Field Problems

C. W. CROWLEY, P. P. SILVESTER, and H. HURWITZ, Jr.

DEPARTMENT OF ELECTRICAL ENGINEERING, MCGILL UNIVERSITY, MONTREAL, CANADA

HERBITS ASSOC., SCHENECTADY, N.Y., U.S.A.

ABSTRACT

Covariant projection elements for vector finite element problems are discussed. These elements facilitate imposing boundary and interface conditions on tangential components of the vector field. The elements' vector trial functions satisfy the inclusion condition, which guarantees that no spurious modes are computed.

INTRODUCTION

Finite element methods provide robust and accurate numerical solutions to scalar formulations of many electromagnetic field problems [1]. For general three-dimensional electromagnetic field problems, it is desirable to extend finite elements to vector field problems. However, vector field problems are less tractable than scalar field problems because of complications such as the necessity of imposing boundary and interface conditions on the individual vector field components. Additionally, spurious divergent vector fields are often computed when scalar finite element methods are extended directly to vector field problems. In this paper, covariant projection elements are introduced for vector field problems. With these elements, vector finite element methods can be implemented as directly as scalar finite element methods: boundary and interface conditions are easily imposed and spurious modes are removed without resorting to global penalty terms or constraints.

COVARIANT PROJECTIONS AND VECTOR TRIAL FUNCTIONS

Finite element analysis of vector field problems requires a set of vector trial functions defined over finite elements. Unlike trial functions for scalar problems, vector trial functions have both magnitude and direction. To solve a variety of problems, it is desirable to model the geometric domain of the problem with parametrically defined curvilinear finite elements (Fig. (1)).

The parameters (ξ,η,ν) defining the curvilinear element constitute a local coordinate system of the element. It is convenient to define a vector field over a curvilinear element in terms of its covariant projections. The covariant projections of a

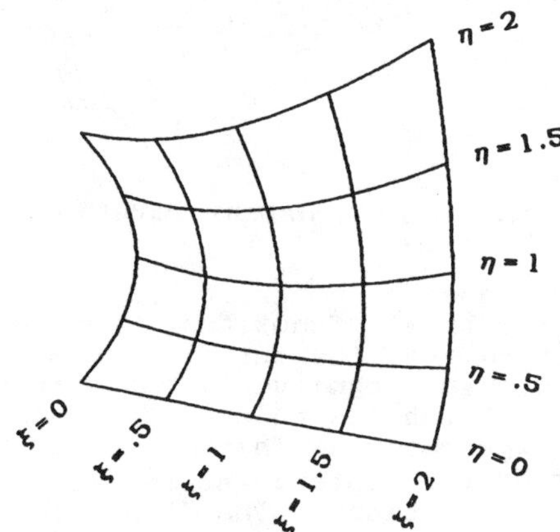

Figure 1: Parametrically Defined Finite Element

vector field **E** are simply its components in a curvilinear coordinate system [2]:

$$E_\xi = \mathbf{E}\cdot\vec{\xi} \qquad E_\eta = \mathbf{E}\cdot\vec{\eta} \qquad E_\nu = \mathbf{E}\cdot\vec{\nu} \tag{1}$$

The covariant components of **E** are related to its cartesian components by

$$\begin{bmatrix} E_\xi \\ E_\eta \\ E_\nu \end{bmatrix} = \begin{bmatrix} \frac{\partial\xi}{\partial x} & \frac{\partial\xi}{\partial y} & \frac{\partial\xi}{\partial z} \\ \frac{\partial\eta}{\partial x} & \frac{\partial\eta}{\partial y} & \frac{\partial\eta}{\partial z} \\ \frac{\partial\nu}{\partial x} & \frac{\partial\nu}{\partial y} & \frac{\partial\nu}{\partial z} \end{bmatrix}^{-1} \begin{bmatrix} E_x \\ E_y \\ E_z \end{bmatrix} \tag{2}$$

For example, the field defined by

$$\begin{pmatrix} E_\xi \\ E_\eta \\ E_\nu \end{pmatrix} = \begin{pmatrix} const \\ 0 \\ 0 \end{pmatrix} \tag{3}$$

Reprinted from *IEEE Trans. Magn.*, vol. 24, no. 1, pp. 397–400, Jan. 1988.

is shown in Fig. (2). Note that the constant covariant projection is curved in cartesian space.

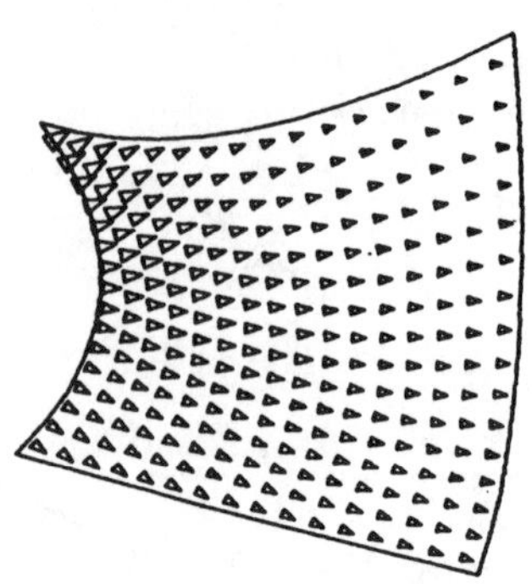

Figure 2: Constant Covariant Projection

Based on covariant projections, a set of vector trial functions, denoted by T_k, is defined over the curvilinear element as follows. The magnitude of the trial functions is unity at one node of the element and zero at other nodes. This is akin to the usual polynomial interpolation used for scalar trial functions. The direction of the vector trial functions is defined by the covariant projections in the curvilinear coordinate system. Two such covariant trial functions are illustrated in Fig. (3).

(a)

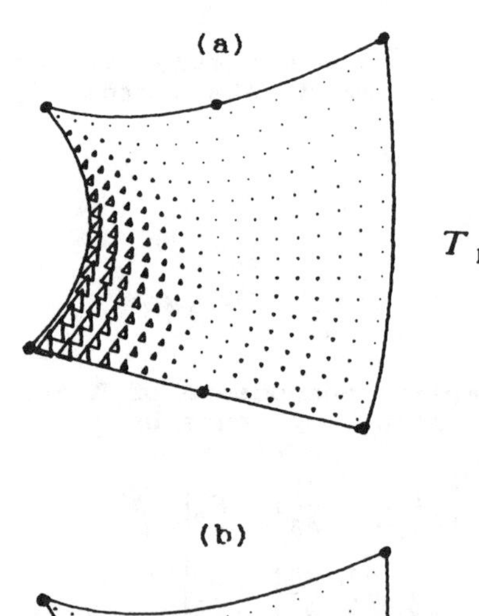

(b)

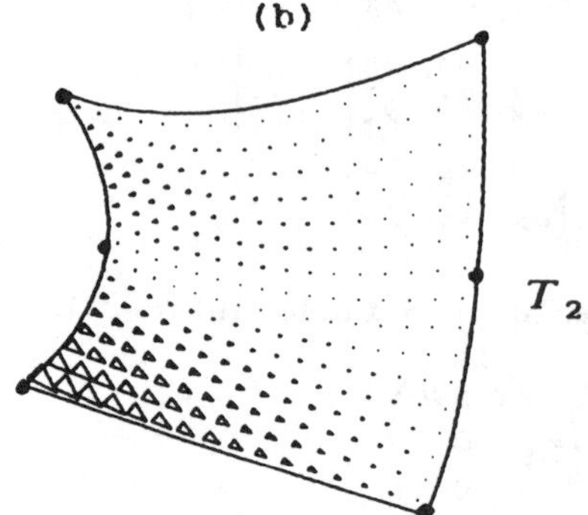

Figure 3: Covariant Projection Trial Functions

As in scalar finite element methods, the vector field over an element is restricted to a linear combinations of the trial functions:

$$\tilde{E} = \sum_{k=1}^{m} e_k T_k \tag{4}$$

Substituting Eq. (4) into the following variational expression gives a matrix quadratic form whose stationary points are easily found [3]:

$$F = \int \frac{1}{\mu} |\nabla \times \mathrm{E}|^2 - \frac{k^2}{\mu} |\mathrm{E}|^2 dv \tag{5}$$

Boundary conditions on E arising from boundary excitations or perfect conductors require the tangential components to be constrained on curvilinear surfaces. With covariant projection elements, tangential components are constrained simply by fixing the appropriate nodal values in Eq. (4). Similarly, interface continuity of tangential components is enforced over curvilinear surfaces by matching nodal values on the interface. The correct interface conditions on the curl of E follow as natural conditions from stationarity of the functional [4,5].

SPURIOUS CORRUPTIONS AND THE INCLUSION CONDITION

Unlike the corresponding functional for scalar electromagnetic field problems, the vector functional in Eq. (5) has a nontrivial set of stationary points with zero eigenvalues [6]. The set of eigenfunctions corresponding to these zero eigenvalues spans the irrotational space, I:

$$\mathrm{I} = \{C:\ \nabla \times C = 0\} \tag{6}$$

With the possible exception of the static field, all of the zero stationary points have divergent eigenfunctions that do not correspond to physical oscillations or modes.

Now it can be shown that the smallest nonzero eigenvalue in Eq. (5) has the following characterization:

$$\lambda_1^+ = \inf \frac{\int |\nabla \times \mathrm{E}|^2 dv}{\int |\mathrm{E}|^2 dv} \tag{7}$$

subject to

$$\mathrm{E} \in \mathrm{I}(perp) \tag{8}$$

The restriction on E in Eq. (8) simply makes the eigenfunction corresponding to λ_1^+ orthogonal to all the zero eigenfunctions which, in this case, span I.

When E is restricted to a linear combination of vector trial functions as in Eq. (4), the first nonzero eigenvalue has the characterization

$$\tilde{\lambda}_1^* = \inf \frac{\int |\nabla \times \tilde{E}|^2 dv}{\int |\tilde{E}|^2 dv} \tag{9}$$

subject to

$$\tilde{E} \in M(perp) \tag{10}$$

The tilde distinguishes the finite element eigenpairs from the true eigenpairs. Again, the constraint in Eq. (10) simply says that the eigenfunctions are mutually orthogonal. In this case, however, M is finite dimensional and is spanned only by the irrotational vector fields in the finite element trial function space.

Now the first nonzero eigenvalue defined by Eq. (9) in the finite element spectrum may correspond to a nonphysical spurious mode for the following reason. The space M(perp) which is orthogonal to the irrotational trial functions may not be contained in the space I(perp), which is orthogonal to all irrotational vector functions. Nothing prevents the first nonzero eigenvector in the finite element space from having a component in the space of irrotational vector functions. Because of this, Eq. (7) shows that $\tilde{\lambda}_1^*$ may be much smaller than the true λ_1^*. Additionally, because the corresponding finite element eigenfunction may have a component in the irrotational space I, it can have substantial nonzero divergence. The same argument applies for the second, third, and subsequent nonzero eigenvalues. In short, the finite element spectrum of nonzero eigenvalues can be corrupted by nonphysical eigenpairs.

It is possible to shift or remove spurious modes in vector eigenvalue problems by imposing penalty terms [6], or by globally imposing irrotationality on the assembled finite element matrices [7]. It is more direct, however, to avoid these global adjustments to the finite element method. With covariant projection elements, it is possible to sidestep the occurrence of spurious modes in vector eigenvalue problems.

Let T be the space of vector trial functions and let $P_I T$ be the projection of the vector trial functions on the irrotational space, I. Then it is easy to show that if

$$P_I T \subset T \tag{11}$$

then

$$M(perp) \subset I(perp) \tag{12}$$

This condition in turn guarantees that

$$\tilde{\lambda}_1^* \geq \lambda_1^* \tag{13}$$

Eq. (11) also guarantees that the eigenfunction corresponding to $\tilde{\lambda}_1^*$ has zero divergence. The condition in Eq. (11) is called the inclusion condition. It is easy to show that the inclusion condition guarantees the same results apply for all the nonzero eigenvalues. Thus, generating vector trial functions satisfying the inclusion condition generates a correct finite element spectrum without spurious modes.

How easy is it to satisfy the inclusion condition with vector finite element trial functions? Hano [8] and Csendes [9] note that no spurious modes are computed when mixed order interpolation functions are used for the different field components in vector trial functions. The reason for this is the inclusion condition: it can be shown that mixed orders of polynomial interpolation for the different field components creates a trial function space that satisfies the condition in Eq. (11) [10]. Using covariant projection elements, curvilinear vector trial functions satisfying the inclusion condition are generated by using mixed orders of interpolation in the various covariant components of the vector field. For example, the vector trial function shown in Fig. (3b) is a mixed order function, since the ξ directed component varies linearly with ξ and quadratically with η. The η directed component in Fig. (3a) varies linearly with η and quadratically with ξ.

EXAMPLES

As a simple computational example, a quarter section of a circular waveguide was modeled with 3 of the proposed curvilinear covariant projection elements. Fig. (4) displays one of the computed TM modes. In Fig. (4), the solid lines indicate the boundary of the curvilinear finite elements and the triangles indicate the relative field strength and direction. The computed θ directed components along the bottom edge of Fig. (4) are plotted in Fig (5). The analytical values for the TM21 mode are also plotted in Fig. (5) for comparison. Both the eigenvalue and eigenfunction agree with the analytical values. No spurious modes were found. The software for the vector formulation was adapted directly from software for a scalar finite element formulation: boundary and interface conditions were imposed by fixing nodal values, and no penalty functions or global constraints were imposed.

Because covariant projection elements correctly model the true spectrum for vector problems, they are also necessary for driven problems such as eddy current problems [11]. As an example, a dielectric slab was modeled with unmixed order finite elements. The slab was driven at the top edge with a sinusoidal source; the remaining boundaries were assumed to be perfect conductors. As shown in Fig. (6), the driven solution is nonphysical because it diverges in a source-free region. Essentially, this driven problem corresponds to a spurious mode of the unmixed element model which is driven by the boundary source. The existence of the spurious corruption in Fig. (6) is due to vector trial functions that violate the inclusion condition For contrast, Fig. (7) shows the same problem modeled with the covariant projection elements. The field has zero divergence and has the correct exponential decay into the slab.

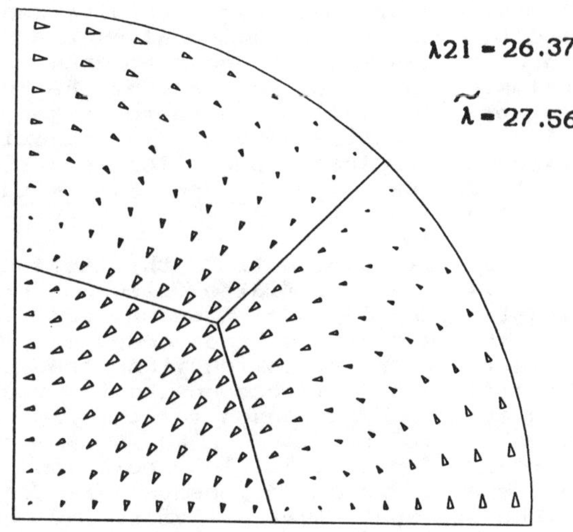

Figure (4): Computed Mode For Circular Waveguide

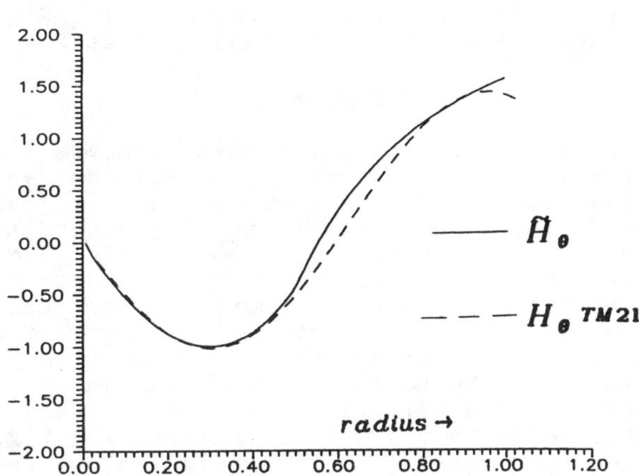

Figure (5): Comparison of Exact and Computed Azimuthal Field Components

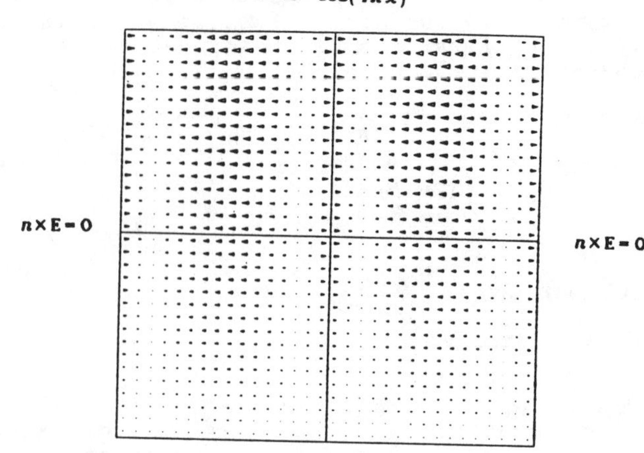

Figure (6): Driven Spurious Mode

n×E=cos(4πx)

n×E=0 n×E=0

n×E=0

Figure (7): Driven Problem Correctly Modeled

CONCLUSION

This paper has introduced covariant projection elements for vector field problems. As with finite elements for scalar problems, the covariant projection elements allow boundary and interface conditions to be directly imposed by constraining nodal field values. The mixed order covariant projection trial functions avoid spurious modes, so existing scalar methods can be extended directly to vector methods without resorting to ad hoc adjustments to remove divergent corruptions. The covariant projection elements are useful for both driven and eigenvalue problems in vector electromagnetic fields.

Acknowledgments

The authors wish to thank J. P. Webb for many useful conversations. They also acknowledge the financial support provided by The Natural Sciences and Engineering Research Council of Canada, and Centre de recherche informatique de Montreal.

REFERENCES

[1] Silvester, P. P. and Ferrari, R. L., **Finite Elements for Electrical Engineers.** Cambridge: Cambridge Univ. Press, 1983.

[2] Stratton, J. A., **Electromagnetic Theory.** New York: McGraw-Hill, 1941.

[3] Webb, J. P., Maile, G. L., Ferrari, R. L., "Finite-element solution of three-dimensional electromagnetic problems," **I.E.E. Proc.**, Vol. 130, Pt. H, No. 2, pp. 153-159, 1983.

[4] Ferrari, R. L., "Complementary variational formulation for eddy current problems using the variables E and H directly," **I.E.E. Proc.**, Pt. A, Vol. 132, No. 4, pp. 157-164, 1985.

[5] Barton, M. L. and Csendes, Z. L. "New vector finite elements for three-dimensional magnetic field computation," **Journal of Applied Physics**, 61 (8), pp. 3919-3921, 15 April, 1987

[6] Webb, J. P., "The finite-element method for finding modes of dielectric-loaded cavities," **IEEE trans. on Microwave Theory and Techniques**, MTT-33, No. 7, pp.635-639, July 1985.

[7] Kobelansky, A.J., Webb, J.P. " Eliminating spurious modes in finite element wavequide problems by using divergence-free fields," **Electronic letters**, Vol. 22, No. 11, pp. 569-570, May 1986.

[8] Hano, M., "Finite element analysis of dielectric-loaded waveguides," **IEEE trans. on Microwave Theory and Techniques**, MTT-32, No. 10, pp. 1275-1279,October, 1984

[9] Csendes, Z. L., Oral Presentation, **IEEE Conference on Electromagnetic Field Computation**, Schenectady, N.Y., November, 1986

[10] Crowley, C. W. Ph.D. Dissertation (to be published) McGill University

[11] Pinchuk, A. R., Crowley, C. C., and Silvester, P. P., "Spurious solutions to vector diffusion and wave field problems", **IEEE trans. on Magnetics**, to be published this issue.

Calculation of Transient 3-D Eddy Current Using Edge-Elements

A. KAMEARI
MITSUBISHI ATOMIC POWER INDUSTRIES, INC.
4-1, SHIBAKOUEN, 2-CHOME, MINATO-KU, TOKYO 105, JAPAN

Abstract - Calculation of transient 3D eddy current is presented using edge-elements in the modified A-method by the finite elements method. Quadratic 36-edge elements on the 20-node isoparametric elements are newly proposed. Some numerical results by the method are presented and the linear and quadratic elements are compared. The results show the applicability of the elements.

Introduction

In the formulations of eddy current analysis, the magnetic vector potential **A** in the A-V method and the current vector potential **T** in the T-Ω method are naturally represented by the shape functions of edge-elements, which are in the vector functional space of H(curl)[1,2]. With the vector functional space of edge-elements and the corresponding C^0 scalar space, it has been shown[3] that the gauge transformation is possible in the finite element space just in the same way as in the continuous space and, to fix the gauge, the vector space is decomposed to subspaces of a tree and a co-tree of the graph of edges[4]. The tree space is a rotationless space and can be removed from the functional space of analysis and the uniqueness of the solution is ensured. Or, we can adopt a gauge that the electric scalar potential V is zero in the A-V method or a gauge that the magnetic scalar potential Ω is zero in the T-Ω method. These gauges correspond to the modified A-method[5] and the H-method[6]. The difficulty at discrete conductivity in the modified A-method using nodal C^0 space is solved by the edge-elements.

The construction of edge-elements on tetrahedra or rectangular blocks was shown by J.C. Nedelec[1] and the extension to hexahedral isoparametric elements was shown by J.S. van Welij[7]. In this paper, further extension to quadratic edge-elements on 20-node isoparametric elements are newly proposed and examined in the modified A-method. The usage of the quadratic elements is familiar in field calculations using nodal elements and is known to result in higher accuracy by less degree of freedom. The possibility of usage of higher order edge-elements is interesting and significant in the development of methods for eddy current calculation.

Modified A-Method

Consider a conductive region R_1 of conductivity σ and permeability μ surrounded by a non-conductive region R_2 of permeability μ which is bounded by a surface S_2 as shown in Fig.1. Assume that forcing current $\mathbf{J}_s$ is in the region R_2.

The modified A-method is corresponding to a formulation of the A-V method with a gauge of V=0. The both methods are completely equivalent when edge-elements are used[3]. In the modified A-method, the magnetic and electric fields are given as

$$\mathbf{B} = \text{rot}\ \mathbf{A} \tag{1}$$
$$\mathbf{E} = -\mathbf{A}' \tag{2}$$

where (') denotes a derivative in time. The governing equation is

$$\text{rot}\ (\frac{1}{\mu}\ \text{rot}\ \mathbf{A}\ ) = -\sigma\ \mathbf{A}' + \mathbf{J}_s \tag{3}$$

The equation is solved using the Galerkin form of weighted residual given by

$$\int_{R_1} \sigma\ \mathbf{N}\ \mathbf{A}'\ dV + \int_{R_1+R_2} \frac{1}{\mu}\ \text{rot}\ \mathbf{N}\ \text{rot}\ \mathbf{A}\ dV - \int_{R_2} \mathbf{N}\ \mathbf{J}_s\ dV = \int_{S_2} \frac{1}{\mu}\ (\mathbf{N} \times \text{rot}\ \mathbf{A})\ \mathbf{n}\ ds \tag{4}$$

Here, **N** is the vector test function, which is in the same functional space as **A**. The tangential components of **A** or **B** (=rot **A**) are given at S_2 as the boundary condition.

Edge-Elements

Using edge-elements, variables are assigned to edges on finite elements. The tangential components of shape functions are continuous and the normal components are discontinuous at inter-element boundaries. It is noted that the Coulomb gauge is not implied on the shape functions.

To construct base shape functions of edge-elements, the following criteria are necessary[3].

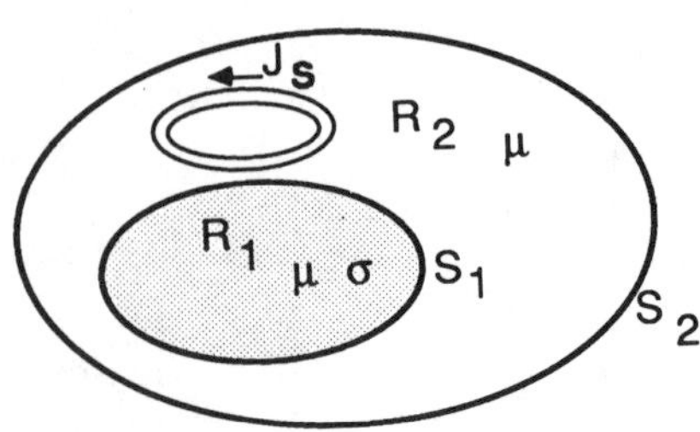

Fig.1. Model configuration of eddy current.

Reprinted from *IEEE Trans. Magn.*, vol. 26, no. 2, pp. 466–469, March 1990.

(i) The rotationless (rot **A**=0) subspace in the functional space of the base shape functions coincides with the space of the gradients of the functions in the corresponding C^0 scalar space. This results that the gauge transformation between **A** and V is possible and we can adopt the gauge of V=0, i.e. the modified A-method.

(ii) The line integration of the tangential component of the shape function $\mathbf{N}_i$ along the i-th edge is one and integrations along other edges are zero. It makes possible to eliminate the extra degree of freedom using the graph of the tree and co-tree.

Linear 12-edge Elements

Let R be a bounded domain and discretized into brick elements. Each brick has a local coordinate **r**(r,s,t) and the coordinates r, s and t range from -1 to 1 in each element. A mapping **x**(**r**) and the inverse mapping **r**(**x**) exist, where **x**(x,y,z) is the global Cartesian coordinate.

The scalar shape functions on linear 8-node isoparametric elements are represented by tri-linear functions of r, s and t. The functions are linear combinations of {1, r, s, t, rs, st, tr, rst} and the gradients are represented by linear combinations of {1, s, t, st} grad r, {1, t, r, tr} grad s and {1, r, s, rs} grad t. The independent 12 functions are set on 12 edges so that the above criterion (ii) is satisfied as

$$\mathbf{N}_i = \frac{1}{8}(1+s_i s)(1+t_i t)\ \mathrm{grad}\ r \qquad (5)$$

for edges along r, where s_i and t_i are equal to ±1 and the local coordinates of the i-th edge. Cyclic permutations give the shape functions for edges running along s and t. The linear 12-edge elements are shown in Fig.2(a). The shape functions coincide with the ones shown in references[1,7].

Quadratic 36-edge Elements

By the same consideration as the 12-edge elements, we can give quadratic 36-edge elements on the usual 20-node isoparametric elements as shown in Fig.2(b). The 12 edges are in r direction and 8 edges out of these are along the element's edges and 4 edges are on the element's surfaces. The shape functions are given as

$$\mathbf{N}_i = \frac{1}{8}(1+s_i s)(1+t_i t)(r_i r+s_i s+t_i t-1)\ \mathrm{grad}\ r \qquad (6)$$

for edges on the element's edges and

$$\mathbf{N}_i = \frac{1}{4}(1+t_i t)(1-s^2)\ \mathrm{grad}\ r \qquad (7)$$

for edges on the element's surfaces of t=±1. Here r_i=±1/2, s_i=±1 and t_i=±1 and are the local coordinate of the center of the i-th edge. The other functions are given by exchanges and cyclic permutations of r, s and t.

Numerical Results

Square Plate

A test problem of a square plate is solved using the 36-edge elements and the result is compared with the ones solved using 12-edge elements and the analytical solution. Consider a square plate (0.2 m x 0.2 m) with thickness 1 cm and resistivity 2 μΩcm and uniform magnetic field in z direction with a constant rate B'=1 T/sec is applied. One eighth of the plate is discretized by 36-edge elements as shown as Fig. 3. The number of the elements is 9 in the plate. The problem is solved at t=1 sec, when the eddy current is steady and analytical solution is given.

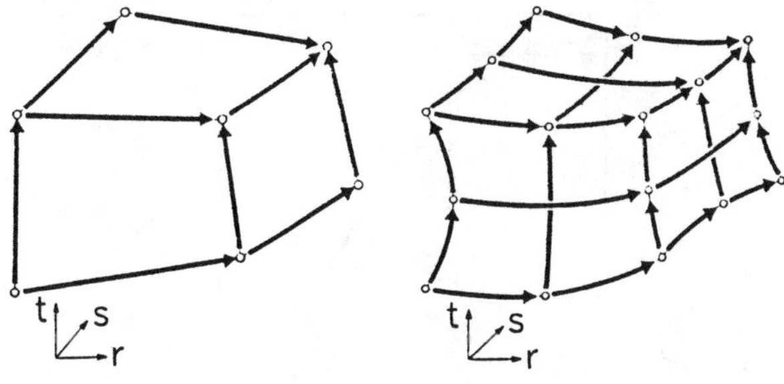

Fig.2. Linear 12-edge element (a) and quadratic 36-edge element (b).

Figure 4 shows the distribution of the eddy current in y direction at x=0.1 m and z=0.0 m. The analytical solution and the results by 12-edge elements with 9 and 36 elements in the plate are also shown in Fig. 4. The distribution in elements are interpolated using the edge shape functions. The shape functions in the direction normal to inter-element interfaces are discontinuous and the distributions in the direction are constant in 12-edge elements and linear in 36-edge elements. It can be seen that all results agree well with analytical solution on an average in elements. The distribution in elements resulted by 36-edge elements agrees with analytical solution better than the one by 12-edge elements using the finer mesh.

Hollow Sphere in a Step Field

The benchmark problem of a hollow sphere in a step field[8] is solved by linear and quadratic elements and results are compared with analytical solution. The sphere consists of a non-permeable conducting material, with conductivity $\sigma=5.0\times10^8$ S/m. The inner and outer radii of the sphere are 5 cm and 5.5 cm, respectively. The incident field is spatially uniform, with zero value for time t<0. At t=0, the field is instantaneously switched to 1 Tesla in the +z direction, where it remains for time t≥0.

One eighth of the region is discretized as shown in Fig.5 by linear elements and Fig.6 by quadratic elements. The total number of elements in the linear case is 928 with 192 elements on the sphere. The number of elements in the quadratic case is one eighth

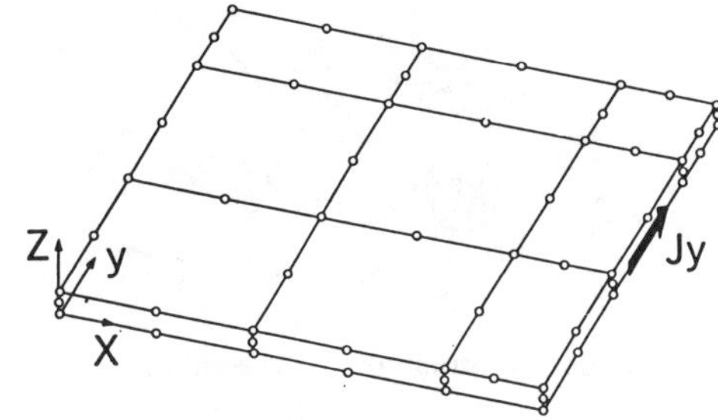

Fig. 3. Mesh division of a square plate using 36-edge elements.

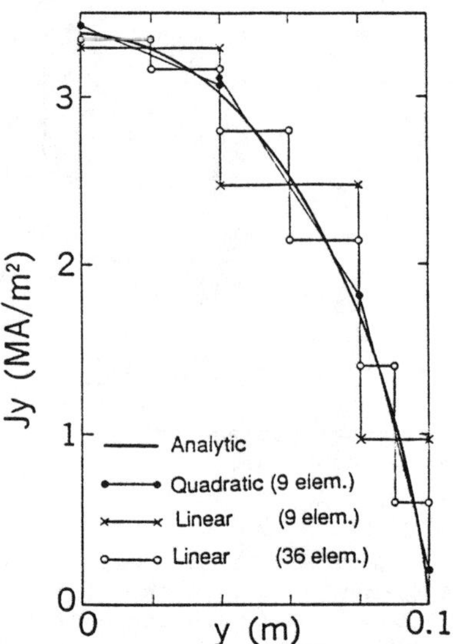

Fig.4. Eddy current distributions in a square plate.

of the one in the linear case. The initial state at t=0 is solved with the condition that the normal components of B is zero, i.e. tangential components of **A** is zero, on the outer surface of the sphere. At far boundary, the tangential components of B by induced field are assumed to be zero. The time steps of the calculation are 0.1 msec (0 to 1 msec), 0.5 msec (1 to 5 msec) and 5 msec (5 to 50 msec). The time difference method is Crank-Nicholson method with 'θ'=2/3 and the solution method is the ICCG method.

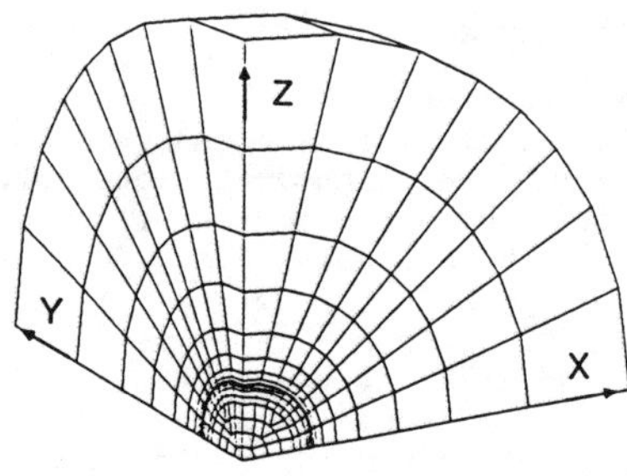

Fig.5. Mesh by linear elements.

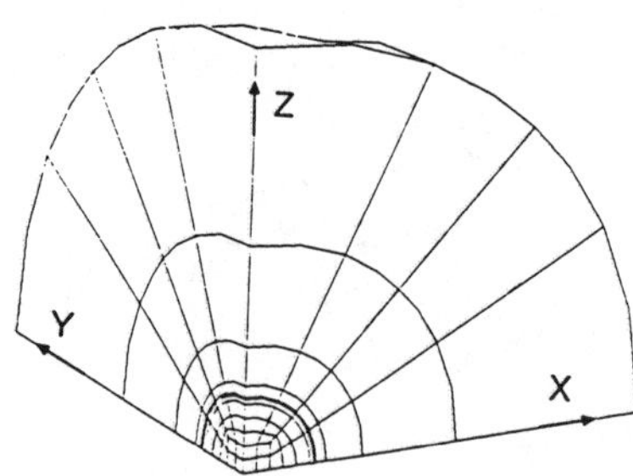

Fig.6. Mesh by quadratic elements.

The calculated current distributions along x-axis are shown in Fig.7 with the analytical solution at 0.1, 0.2, 1 and 5 msec. The distributions in elements are interpolated by shape functions. The current distributions are well approximated in the both cases after 0.2 msec. The discrepancy is large at 0.1 msec because the skin depth is comparable with the mesh size, but the agreement is good on an average and the total currents calculated agree well with analytical solution as shown later.

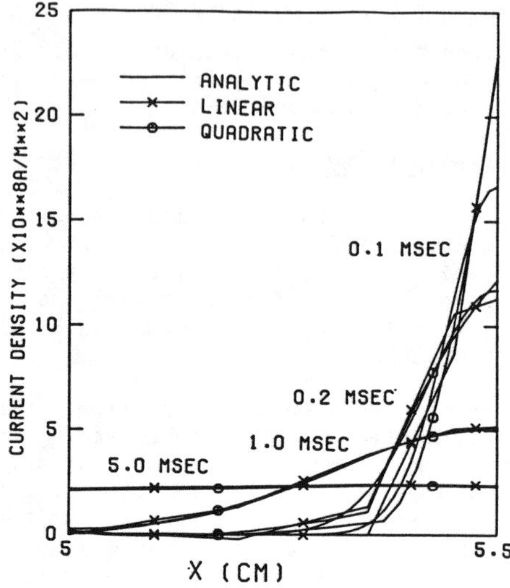

Fig.7. Eddy current distributions along x-axis.

The distributions of magnetic flux density along x- and z-axes at 10 msec are shown in Fig.8 in the linear case and in Fig.9 in the quadratic case. The distribution in elements are calculated by the rotations of shape functions. The distributions are continuous along z-axis but discontinuous along x-axis because of the nature of the shape functions. The calculated distribution agrees well on average and the values at the centers of elements agree very well with analytical solution. The distribution inside the sphere should be uniform but the calculated results along x-axis show sawteeth-like distributions, especially in the linear case. This is caused by the usage of non-parallelepiped hexahedral elements.

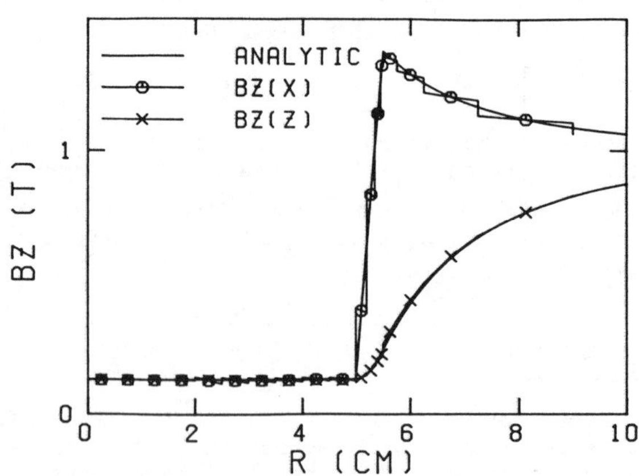

Fig.8. Distribution of magnetic flux density at 10 msec in linear case.

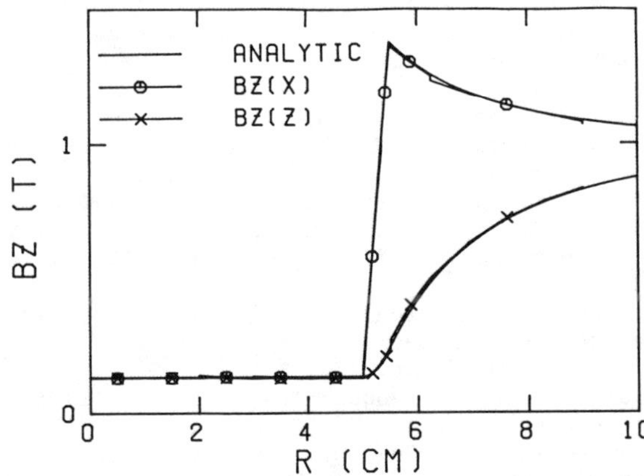

Fig.9. Distribution of magnetic flux density at 10 msec in quadratic case.

The total eddy current crossing the y=0 half plane, the power loss in the complete sphere and the magnetic flux density at the origin are shown in Table 1, 2 and 3, respectively. The calculated values by linear and quadratic elements reasonably agree with analytical solutions. The accuracy by quadratic elements is nearly on the same level with the one by linear elements. At 0.1 msec, the total currents agree well with the analytical value but the power losses differ a little from the analytical value, because the mesh is too coarse to simulate the skin current distribution.

Totally, the accuracy of the results by quadratic elements is nearly same with the one by linear elements, although the current and magnetic distributions are approximated better by quadratic elements. The total number of unknowns in the quadratic case (1002) is about half of one of the linear case (2020), but the number of the non-zero elements in the system matrix are nearly same (22755 and 24904 in the linear and quadratic cases, respectively) and the calculation time in the quadratic case is about doubled.

The same problem is solved with the time step 10 msec from the initial state and the result agrees also well with the analytical solution as shown in Table 1 and 2. It is interesting that reasonable results are calculated with the time step although the current and magnetic field distributions change drastically at initial phase in shorter time than the time step. This probably shows that the fast decaying mode of the eddy current does not affect the numerical results at later time.

Table 1. Total eddy current (kA) crossing the y=0 half plane. The fifth column shows results by linear elements with time step 10 msec

Time (msec)	Analytic	Linear	Quadratic	Linear*
0.1	130.222	131.470	128.851	
0.2	129.770	130.672	128.000	
0.4	129.126	130.024	127.345	
1.0	127.805	128.781	126.098	
2.0	125.902	126.771	124.134	
5.0	119.465	120.185	117.788	
10.0	108.687	107.978	106.019	107.257
20.0	89.826	89.210	87.880	89.005
50.0	50.703	50.296	50.047	50.550

Table 2. Power loss (kW) in the complete sphere.

Time (msec)	Analytic	Linear	Quadratic	Linear*
0.1	70.904	75.413	79.351	
0.2	49.752	49.701	49.149	
0.4	34.795	34.383	34.123	
1.0	21.540	21.234	21.567	
2.0	15.389	15.158	15.309	
5.0	12.016	11.954	11.926	
10.0	9.900	9.617	9.622	9.652
20.0	6.762	6.563	6.611	6.537
50.0	2.154	2.086	2.145	2.107

Table 3. Magnetic Flux density B_z (Tesla) at the origin.

Time (msec)	Analytic	Linear	Quadratic
0.1	0.0000	0.0000	0.0003
0.2	0.0000	0.0000	0.0002
0.4	0.0000	0.0001	0.0008
1.0	0.0004	-0.0008	0.0012
2.0	0.0057	0.0037	0.0060
5.0	0.0479	0.0468	0.0478
10.0	0.1326	0.1317	0.1307
20.0	0.2831	0.2827	0.2793
50.0	0.5953	0.5954	0.5890

Conclusion

Quadratic 36-edge elements on the usual 20-node isoparametric elements are proposed in 3D eddy current calculations with the modified A-method. The test problems of a square plate and a hollow sphere are solved with the linear 12-edge and quadratic 36-edge elements and the results are compared with the analytical solutions. The results by both types of elements agree well mutually and with the analytical solutions, and applicability of these elements in the eddy current calculation are demonstrated. The spatial distributions of the current and magnetic field in elements interpolated by shape functions is approximated better by the quadratic elements with less degree of freedom than the linear elements. The accuracy of the both types of the elements is nearly same in the test problem of a hollow sphere if the size of quadratic elements is double of the one of linear elements.

The procedure to construct edge-elements shown in this paper is in general. It is possible to construct edge-elements on other types of elements, for example 54-edge elements on quadratic Lagrange C^0 elements with 27 nodes. And also this procedure is applicable to tetrahedral and triangular prism elements.

One problem of the edge-elements in the modified A-method is that the distribution of magnetic field in elements calculated by the rotations of the shape functions is deformed when non-parallelepiped hexahedral elements are used, although values at element centers and averages in elements are relatively accurate. Some post-processing to smooth the distribution is necessary by the method like shown in reference[9]. Or the T-Ω or H methods are possibly more effective to obtain directly the magnetic field distribution with less calculative effort but the same problem would occur in the current distribution.

Acknowledgment

The author would like to appreciate Mr. M. Yamada and Dr. M. Nishikawa of Mitsubishi Atomic Power Ind., Inc. for their continuous encouragement.

References

[1] J.C. Nedelec, 'Mixed finite elements in R^3', Numer. Math, Vol.35, pp. 315-341, 1980.

[2] A. Bossavit, 'Two dual formulations of the 3-D eddy-currents problem', COMPEL, Vol.4, No.2, pp. 103-116, 1985.

[3] A. Kameari, 'Three dimensional eddy current calculation using edge elements for magnetic vector potential' in Applied Electromagnetics in Materials, Pergamon press, pp. 225-236, 1986.

[4] R. Albanese and G. Rubinacci, 'Integral formulation for 3D eddy-current computation using edge elements', IEE Proc., Vol. 135, pt. A, No. 7, p457, 1988.

[5] C.R.I. Emson and J. Simkin, 'An optimal method for 3-D eddy currents', IEEE Trans. on Magnetics, Vol. MAG-19, No.6, pp. 2450-2452, 1983.

[6] A. Bossavit and J.C. Verite, 'A mixed FEM-BEM method to solve 3-D eddy-current problems', IEEE Trans. on Magnetics, Vol. MAG-18, pp. 431-435, 1982.

[7] J.S. van Welij, 'Calculation of eddy current in terms of H on hexahedra', IEEE Trans. on Magnetics, Vol. MAG-21, No.6, pp. 2239-2241, 1985.

[8] L. Turner, TEAM Workshops: Test Problems, 1988, private circulation.

[9] J.T. Oden and H.J. Brauchli, 'On the calculation of consistent stress distribution in finite element approximation', Int. J. Numer. Method in Engrg., Vol. 3, pp. 317-325, 1971.

Vector Finite Elements for Electromagnetic Field Computation

Z. J. CENDES

ELECTRICAL AND COMPUTER ENGINEERING
CARNEGIE MELLON UNIVERSITY
PITTSBURGH, PA 15213

Abstract - A new structure is presented for the finite element analysis of vector fields. This structure employs the affine transformation to represent vectors and vector operations over triangular domains. Two-dimensional high order vector elements are derived that are consistent with Whitney forms. 1-form elements preserve the continuity of the tangential components of a vector field across element boundaries while 2-form elements preserve the continuity of the normal components. The 1-form elements are supplemented with additional variables to achieve p'th order completeness in the range space of the curl operator. The resulting elements are called tangential vector finite elements and provide consistent, reliable and accurate methods for solving electromagnetic field problems.

1. INTRODUCTION

The finite element method has proved to be successful in solving electromagnetic field problems involving scalar variables in both two and three dimensions [1]. However, the finite element analysis of field problems involving vector variables has been an enigma. The analysis of both two or three dimensional vector eddy current and microwave problems has generated numerical instabilities called spurious modes [2-6] while the application of gauge conditions for vector magnetostatic problems has been problematic [7-12]. It has been demonstrated that even simple variations in the finite element mesh can produce large errors in the vector potential solution of both eddy current [21] and magnetostatic field problems [13].

In this paper, we examine the representation of vectors over triangular finite elements. We show that the structure of vector finite elements is fundamentally different from that of scalar elements. A consistent relationship between scalar and vector elements is derived that matches the continuity conditions provided by variational calculus. This procedure eliminates the numerical instabilities that plague traditional solutions of vector electromagnetic field problems.

The structure presented in this paper is an extension of recent work in edge elements [14-16,32,33], in tangential vector elements [17-23], and in normal vector elements [24,25]. It is related to a mathematical construction known as Whitney forms [26,27,34]. In these forms, scalar functions that provide function but not derivative continuity are called 0-forms; vector functions that possess tangential but not normal continuity are called 1-forms; vector functions that possess normal but not derivative continuity are called 2-forms; and discontinuous scalar functions are called 3-forms.

The structure of electromagnetics parallels that of Whitney forms. The continuity requirement of a scalar variable such as the electric potential is that the potential is continuous but that its derivative may jump across inter-element boundaries. The electric field, which is the negative gradient of the potential, satisfies the continuity requirement that its tangential component is continuous but that its normal component may jump across inter-element boundaries. The electric flux, which is obtained by multiplying the electric field by the permittivity, satisfies the continuity requirement that its normal component is continuous but that its tangential component may jump across inter-element boundaries. Finally, the divergence of the electric flux is a charge distribution that is a discontinuous scalar.

In the numerical realm, if the potential is approximated by piecewise continuous polynomials - that is, separate polynomials defined over each finite elements possessing function but not derivative continuity across inter-element boundaries - then to be consistent and to avoid numerical instabilities, the electric field must be approximated by tangential vector finite elements in which the tangential component of the field across the element faces is continuous but the normal component may jump. Further, in this numerical structure, permittivity is an operator that converts tangential elements into normal elements in which the normal component of the field is continuous but the tangential component may jump across the element boundaries.

High-order vector elements in two-dimensions are presented that interpolate to the appropriate Whitney form. These elements mimic the inherent continuity of electromagnetic fields. We show that the resulting vector finite elements are consistent with the requirements of variational calculus and that they are complete to p'th order in the range space of the curl operator.

2. TRIANGLES

2.1 The Affine Transformation - As is well known, homogeneous or barycentric coordinates $\{\zeta_i, i = 1,2,3\}$ are defined over a triangle as follows [1,30]

$$X = \Delta\, F\, Z \tag{2.1}$$

where

$$X = \begin{bmatrix} 1 \\ x \\ y \end{bmatrix} \qquad F = \frac{1}{\Delta}\begin{bmatrix} 1 & 1 & 1 \\ x_1 & x_2 & x_3 \\ y_1 & y_2 & y_3 \end{bmatrix} \qquad Z = \begin{bmatrix} \zeta_1 \\ \zeta_2 \\ \zeta_3 \end{bmatrix} \tag{2.2}$$

Here (x_i, y_i), $i = 1,2,3$ are the three sets of triangle vertex coordinates and

$$\Delta = \sum_{i=1}^{3} (x_i y_{i1} - x_i y_{i2}) \tag{2.3}$$

with $(i, i1, i2) = \{(1,2,3), (2,3,1), (3,1,2)\}$. The transformation in (2.1) is due to Moebius and is known as the affine transformation.

The affine transformation is nonsingular and may be inverted to yield

$$Z = \frac{1}{\Delta} F^{-1} X \tag{2.4}$$

where the elements of the inverse matrix F^{-1} are

$$F^{-1} = \begin{bmatrix} a_1 & b_1 & c_1 \\ a_2 & b_2 & c_2 \\ a_3 & b_3 & c_3 \end{bmatrix} \tag{2.5}$$

Reprinted from *IEEE Trans. Magn.*, vol. 27, no. 5, pp. 3958–3966, September 1991.

Here

$$a_i = x_{i1}y_{i2} - x_{i2}y_{i1}$$
$$b_i = y_{i1} - y_{i2} \quad (2.6)$$
$$c_i = x_{i2} - x_{i1}$$

We note that

$$\Delta = \sum_{i=1}^{3} x_i b_i = \sum_{i=1}^{3} y_i c_i$$

Unit vectors tangent to the element edges are given by

$$\hat{t}_i = \frac{1}{l_i}(c_i\,\hat{x} - b_i\,\hat{y}) \quad (2.7)$$

where side i is opposite vertex i, $\hat{x}$ and $\hat{y}$ are the unit vectors in the x and y directions, respectively, and $l_i^2 = b_i^2 + c_i^2$ is the length of side i. See Figure 1. Unit normal vectors are given by

$$\hat{n}_i = \hat{t}_i \times \hat{z} = -\frac{1}{l_i}(b_i\hat{x} + c_i\hat{y}) \quad (2.8)$$

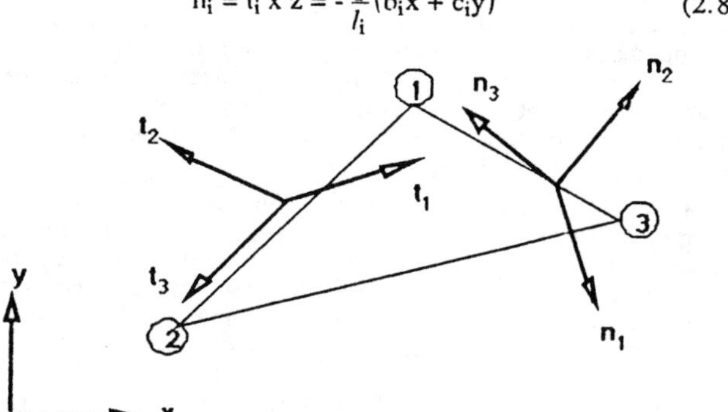

Figure 1. The unit tangent and unit normal vectors on a triangle where side i is opposite vertex i.

Writing (2.8) in matrix form gives

$$\hat{N} = -L^{-1}F^{-1}\hat{X} \quad (2.9)$$

where

$$\hat{N} = \begin{bmatrix} \hat{n}_1 \\ \hat{n}_2 \\ \hat{n}_3 \end{bmatrix} \quad L = \begin{bmatrix} l_1 & 0 & 0 \\ 0 & l_2 & 0 \\ 0 & 0 & l_3 \end{bmatrix} \quad \hat{X} = \begin{bmatrix} \hat{0} \\ \hat{x} \\ \hat{y} \end{bmatrix} \quad (2.10)$$

Even though the parameters a_i are multiplied by zero in equation (2.9), they are included in this equation to preserve all of the columns of the affine matrix F^{-1}. This allows (2.9) to be inverted

$$\hat{X} = -F\,L\,\hat{N} \quad (2.11)$$

In a similar way, the unit tangent vectors are written as

$$\hat{T} = L^{-1}G^{-1}\hat{X} \quad (2.12)$$

where G^{-1} is the matrix

$$G^{-1} = \begin{bmatrix} a_1 & c_1 & -b_1 \\ a_2 & c_2 & -b_2 \\ a_3 & c_3 & -b_3 \end{bmatrix} \quad (2.13)$$

Inverting (2.12) gives

$$\hat{X} = G\,L\,\hat{T} \quad (2.14)$$

where

$$G = \frac{1}{\Delta}\begin{bmatrix} 1 & 1 & 1 \\ y_1 & y_2 & y_3 \\ -x_1 & -x_2 & -x_3 \end{bmatrix} \quad (2.15)$$

From (2.11) and 2.14), the zero "unit vector" in $\hat{X}$ equals

$$\hat{0} = \sum_i l_i\,\hat{n}_i = \sum_i l_i\,\hat{t}_i \quad (2.16)$$

Also, combining the above equations provides $\hat{N} = -L^{-1}F^{-1}G\,L\,\hat{T}$ and $\hat{T} = -L^{-1}G^{-1}F\,L\,\hat{N}$ where the elements of the product matrices $F^{-1}G$ and $G^{-1}F$ are

$$\left(F^{-1}G\right)_{ij} = (a_i + b_i y_j - c_i x_j)/\Delta$$
$$\left(G^{-1}F\right)_{ij} = (a_i + c_i x_j - b_i y_j)/\Delta \quad (2.17)$$

2.2 Angles Between Unit Vectors - The angles between the unit tangent and unit normal vectors are obtained by using the dot product

$$\hat{N}\cdot\hat{N}^T = \hat{T}\cdot\hat{T}^T = C \quad (2.18)$$

where the matrix C is given by

$$C = \begin{bmatrix} 1 & -\cos\theta_3 & -\cos\theta_2 \\ -\cos\theta_3 & 1 & -\cos\theta_1 \\ -\cos\theta_2 & -\cos\theta_1 & 1 \end{bmatrix} \quad (2.19)$$

Here θ_i represents the included angle at vertex i. We also find that

$$\hat{N}\cdot\hat{T}^T = \hat{T}\cdot\hat{N}^T = \begin{bmatrix} 0 & -\sin\theta_3 & \sin\theta_2 \\ \sin\theta_3 & 0 & -\sin\theta_1 \\ -\sin\theta_2 & \sin\theta_1 & 0 \end{bmatrix} \quad (2.20)$$

Equation (2.20) is cast into a simpler form by noting that

$$\sin\theta_i = \frac{h_{i1}}{l_{i2}} = \frac{h_{i2}}{l_{i1}} \quad (2.21)$$

where h_i is the altitude to vertex i. Thus

$$\hat{N}\cdot\hat{T}^T = \hat{T}\cdot\hat{N}^T = H\,S\,L^{-1} = L^{-1}S\,H \quad (2.22)$$

where

$$H = \begin{bmatrix} h_1 & 0 & 0 \\ 0 & h_2 & 0 \\ 0 & 0 & h_3 \end{bmatrix} \quad S = \begin{bmatrix} 0 & -1 & 1 \\ 1 & 0 & -1 \\ -1 & 1 & 0 \end{bmatrix} \quad (2.23)$$

3. ANALYSIS

3.1 Vectors - A vector v is expressed in terms of $\hat{X}$ as

$$v = v_x\hat{x} + v_y\hat{y} = \hat{X}^T V^{(X)} = V^{(X)T}\hat{X} \quad (3.1)$$

where

$$V^{(X)} = [\,q \quad v_x \quad v_y\,]^T \quad (3.2)$$

and q is to be determined. To convert (3.1) to normal and tangential forms, use (2.11) and (2.14). The result is

$$v = \widehat{N}^T V^{(N)} = \widehat{T}^T V^{(T)} \tag{3.3}$$

where

$$\begin{aligned} V^{(N)} &= -L\,F^T\,V^{(X)} \\ V^{(T)} &= L\,G^T\,V^{(X)} \end{aligned} \tag{3.4}$$

Solving for $V^{(X)}$ from (3.4) gives

$$\begin{aligned} V^{(X)} &= -\,F^{-T}\,L^{-1}\,V^{(N)} \\ V^{(X)} &= \ G^{-T}\,L^{-1}\,V^{(T)} \end{aligned} \tag{3.5}$$

The top row of (3.5) yields

$$q = -\sum_{i=1}^{3} \frac{a_i}{l_i} V_i^{(N)} = \sum_{i=1}^{3} \frac{a_i}{l_i} V_i^{(T)} \tag{3.6}$$

The components of v in the three directions tangent to the element edges are

$$\begin{aligned} v_t &= \widehat{T}\cdot\widehat{N}^T\,V^{(N)} = H\,S\,L^{-1}V^{(N)} = L^{-1}\,S\,H\,V^{(N)} \\ &= \widehat{T}\cdot\widehat{T}^T\,V^{(T)} = C\,V^{(T)} \end{aligned} \tag{3.7}$$

where v_t is a vector containing the total components of v in the three tangential directions. Notice that the components of v_t are in general not the same as the components of $V^{(T)}$. Referring to Figure 1, we see that the representation of a vector v in terms of the unit vectors $\widehat{N}$ and $\widehat{T}$ is not unique. The unique components of the vector v in the three directions $\hat{t}_i$ are provided by the three components of v_t.

Substituting (3.4) into (3.7) yields

$$v_t = -\,H\,S\,F^T\,V^{(X)} = C\,L\,G^T\,V^{(X)} \tag{3.8}$$

It follows that

$$C = -\,H\,S\,F^T\,G^{-T}\,L^{-1} \tag{3.9}$$

In a similar way, the components of v in the directions normal to the element edges are

$$v_n = \widehat{N}\cdot\widehat{T}^T\,V^{(T)} = H\,S\,L^{-1}V^{(T)} = H\,S\,G^T\,V^{(X)} \tag{3.10}$$

or

$$v_n = \widehat{N}\cdot\widehat{N}^T\,V^{(N)} = C\,V^{(N)} = -\,C\,L\,F^T\,V^{(X)} \tag{3.11}$$

Let $D^{-1} = \Delta\,S\,F^T$ and $E^{-1} = -\,\Delta\,S\,G^T$. Direct evaluation yields

$$D^{-1} = \begin{bmatrix} 0 & c_1 & -b_1 \\ 0 & c_2 & -b_2 \\ 0 & c_3 & -b_3 \end{bmatrix} \qquad E^{-1} = \begin{bmatrix} 0 & b_1 & c_1 \\ 0 & b_2 & c_2 \\ 0 & b_3 & c_3 \end{bmatrix} \tag{3.12}$$

Thus we may write

$$\begin{aligned} v_t &= -\ L^{-1}\,D^{-1}\,V^{(X)} \\ v_n &= -\,L^{-1}\,E^{-1}\,V^{(X)} \end{aligned} \tag{3.13}$$

An example of the above is provided by decomposing the vector $v = v_y\,\hat{y}$ in a triangle with vertices located at (0, 0), (x_2, 0) and (x_3,y_3). In this case, F^T is lower triangular and

$$V^{(N)} = -\frac{L}{\Delta}\begin{bmatrix} 1 & 0 & 0 \\ 1 & x_2 & 0 \\ 1 & x_3 & y_3 \end{bmatrix}\begin{bmatrix} q \\ 0 \\ v_y \end{bmatrix} = -\begin{bmatrix} h_1^{-1}\,q \\ h_2^{-1}\,q \\ h_3^{-1}\,q + v_y \end{bmatrix} \tag{3.14}$$

On the other hand, the decomposition in (3.13) yields

$$v_n = -\,L^{-1}\begin{bmatrix} 0 & -y_3 & x_3-x_2 \\ 0 & y_3 & -x_3 \\ 0 & 0 & x_2 \end{bmatrix}\begin{bmatrix} q \\ 0 \\ v_y \end{bmatrix} = -\begin{bmatrix} \cos\theta_2\,v_y \\ \cos\theta_1\,v_y \\ v_y \end{bmatrix} \tag{3.15}$$

From (3.7) we note that in general the vector $V^{(N)} = [\,0 \quad 0 \quad V_3\,]^T$ has a zero tangential component in the direction $\hat{t}_3$

$$v_t = H\begin{bmatrix} 0 & -1 & 1 \\ 1 & 0 & -1 \\ -1 & 1 & 0 \end{bmatrix}\begin{bmatrix} 0 \\ 0 \\ V_3/l_3 \end{bmatrix} = \begin{bmatrix} \sin\theta_2\,V_3 \\ -\sin\theta_1\,V_3 \\ 0 \end{bmatrix} \tag{3.16}$$

Also, a vector $V^{(N)} = [\,V_1 \quad V_2 \quad 0\,]^T$ satisfying the ratio $V_1 / V_2 = \sin\theta_1 / \sin\theta_2$ has a zero tangential component in the direction $\hat{t}_3$

$$v_t = \begin{bmatrix} -\sin\theta_3\,V_2 \\ \sin\theta_3\,V_1 \\ \sin\theta_1\,V_2 - \sin\theta_2\,V_1 \end{bmatrix} = \begin{bmatrix} -\sin\theta_3\,V_2 \\ \sin\theta_3\,V_1 \\ 0 \end{bmatrix} \tag{3.17}$$

3.2 The Gradient - In two dimensions, the gradient operator is written in terms of the vector $\widehat{X}$ as

$$\nabla\phi = \widehat{X}^T\,\partial W\,\phi = \partial W^T\,\phi\,\widehat{X} \tag{3.18}$$

where

$$\partial W = [\ w \quad \partial_x \quad \partial_y\ ]^T \tag{3.19}$$

Here we adopt the notation $\partial x \equiv \partial/\partial x$, $\partial y \equiv \partial/\partial y$ et cetera and w is to be determined.

Applying the chain rule gives

$$\partial x = \frac{1}{\Delta}\sum_{i=1}^{3} b_i\,\partial\zeta_i \qquad \partial y = \frac{1}{\Delta}\sum_{i=1}^{3} c_i\,\partial\zeta_i \tag{3.20}$$

Thus we may write

$$\partial W = \frac{1}{\Delta}\ F^{-T}\ \partial Z \tag{3.21}$$

where

$$\partial Z = [\ \partial\zeta_1 \quad \partial\zeta_2 \quad \partial\zeta_3\ ]^T \tag{3.22}$$

Equation (3.21) defines w. Substituting (3.21) into (3.18) gives

$$\nabla\phi = \ \widehat{X}^T\,\Delta^{-1}\,F^{-T}\,\partial Z\,\phi \tag{3.23}$$

To express $\nabla\phi$ in terms of the unit normal vectors, use (2.11)

$$\nabla\phi = -\ \widehat{N}^T\,H^{-1}\,\partial Z\,\phi \tag{3.24}$$

where we have employed $h_i\,l_i = \Delta$. Thus, the gradient is expressed "naturally" (i.e. as partials of homogeneous coordinates) in terms of the unit normals to the sides of the element.

The gradient is given in terms of the unit tangent vectors by substituting (2.14) into (3.23). The result is

$$\nabla\phi = \widehat{T}^T\,H^{-1}\,G^T\,F^{-T}\,\partial Z\,\phi \tag{3.25}$$

The components of $\nabla\phi$ in the directions tangent to the element edges are obtained by taking the dot product of (3.24) with $\widehat{T}$

$$\nabla\phi_t = -\,\widehat{T}\cdot\widehat{N}^T\,H^{-1}\,\partial Z\,\phi = -\,L^{-1}\,S\,\partial Z\,\phi \tag{3.26}$$

Similarly, the components of ϕ in the normal directions are

$$\nabla \phi_n = - C\ H^{-1}\ \partial Z\ \phi \tag{3.27}$$

3.3 The Divergence - In two dimensions, we may write the divergence of the vector **v** as

$$\nabla \cdot \mathbf{v} = \partial X^T\ V^{(X)} \tag{3.28}$$

where

$$\partial X = \begin{bmatrix} 0 & \partial x & \partial y \end{bmatrix}^T \tag{3.29}$$

Using (3.20) gives

$$\partial X^T = \partial Z^T\ E^{-1}\ /\ \Delta \tag{3.30}$$

Thus

$$\nabla \cdot \mathbf{v} = \partial Z^T\ \Delta^{-1}\ E^{-1}\ \ V^{(X)} \tag{3.31}$$

In terms of $V^{(T)}$, this yields from (3.5)

$$\nabla \cdot \mathbf{v} = \partial Z^T\ \Delta^{-1}\ L^{-1}\ E^{-1}\ G^{-T}\ V^{(T)} \tag{3.32}$$

Using (3.12), we find that

$$E^{-1}\ G^{-T} = -\ \Delta\ S \tag{3.33}$$

Thus (3.32) becomes

$$\nabla \cdot \mathbf{v} = -\ \partial Z^T\ S\ \ L^{-1}\ V^{(T)} \tag{3.34}$$

It is interesting to note that the divergence operator expressed in terms of unit tangent vectors is

$$\mathrm{div}^{(T)} = -\ \partial Z^T\ S\ \ L^{-1} \tag{3.35}$$

while the transpose of the tangential components of the gradient operator is from (3.26)

$$\mathrm{grad}^{(T)^T} = -\ \partial Z^T\ S\ \ L^{-1} \tag{3.36}$$

Thus, as expected, the adjoint of the gradient is equal to the divergence.

To write the divergence in terms of $V^{(N)}$, substitute (3.5) into (3.31). This gives

$$\nabla \cdot \mathbf{v} = -\partial Z^T\ K\ \ L^{-1}\ V^{(N)} \tag{3.37}$$

where $K = E^{-1}F^{-T}/\Delta$. The elements of K are

$$K_{ij} = (\ b_i b_j + c_i c_j\ )\ /\ \Delta \tag{3.38}$$

Using the cotangent identity [1], K may be written as

$$K = \begin{bmatrix} \cot\theta_2 + \cot\theta_3 & \cot\theta_3 & \cot\theta_2 \\ \cot\theta_3 & \cot\theta_1 + \cot\theta_3 & \cot\theta_1 \\ \cot\theta_2 & \cot\theta_1 & \cot\theta_1 + \cot\theta_2 \end{bmatrix} \tag{3.39}$$

3.4 The Curl - The curl operator is similar to the divergence operator in two dimensions

$$\nabla \times \mathbf{v} = \ \hat{z}\ \ \partial Y^T\ V^{(X)} \tag{3.40}$$

where

$$\partial Y = \begin{bmatrix} 0 & -\partial y & \partial x \end{bmatrix}^T \tag{3.41}$$

We may write this as

$$\nabla \times \mathbf{v} = -\ \hat{z}\ \ \partial Z^T\ D^{-1}\ V^{(X)} \tag{3.42}$$

Using (3.5) and recognizing that $D^{-1}F^{-T} = \Delta S$ and that $D^{-1}G^{-T} = \Delta K$ allows us to write (3.42) in terms of unit normal and tangent vectors

$$\begin{aligned} \nabla \times \mathbf{v} &= \ \ \hat{z}\ \partial Z^T S\ H\ V^{(N)} \\ \nabla \times \mathbf{v} &= -\ \hat{z}\ \partial Z^T K\ H\ V^{(T)} \end{aligned} \tag{3.43}$$

3.5 The Laplacean - Using (3.34) and (3.25) we find that

$$\begin{aligned} \nabla \cdot \nabla\phi &= -\ \Delta^{-1}\ \partial Z^T\ \ S\ \ G^T\ \ F^{-T}\ \partial Z\ \ \phi \\ &= \ \ \Delta^{-1}\ \partial Z^T\ K\ \partial Z\ \ \phi \end{aligned} \tag{3.44}$$

Thus the relationship of the Laplacean operator to the triangle cotangents [1] is derived directly from the affine transformation.

4. VECTOR ELEMENTS

4.1 Compatibility - The finite element solution of electromagnetic field problems requires that the approximation functions satisfy the proper continuity conditions across element boundaries. Finite elements satisfying the proper continuity conditions are said to be *compatible*. For 1-form elements, compatibility requires that the tangential components of the vector field be continuous while for 2-form elements the normal components must be continuous.

Consider 1-form elements first. To set the tangential component of the field to be continuous, we need to evaluate the tangential components of the field along the element edges. From (3.7), we see that the simplest expression for the tangential component of the field is given in terms of the unit normal vectors

$$\mathbf{v}_t = L^{-1}\ S\ H\ V^{(N)} \tag{4.1}$$

A two dimensional vector function $\mathbf{v}(x,y)$ is expressed in terms of normal unit vectors as

$$\mathbf{v}(x,y) = \hat{N}^T\ V^{(N)}\left(\zeta_1,\ \zeta_2,\ \zeta_3\right) \tag{4.2}$$

We now approximate this function as follows

$$V^{(N)}\left(\zeta_1,\ \zeta_2,\ \zeta_3\right) = H^{-1}\ \ \Gamma\left(\zeta_1,\ \zeta_2,\ \zeta_3\right)\ \ \underset{\sim}{\tau} \tag{4.3}$$

Here the columns of the matrix Γ provide the desired vector basis functions and $\underset{\sim}{\tau}$ is a column vector containing coefficients that interpolate to the tangential components of the vector field, as indicated in Figure 2. The matrix H^{-1} is included in (4.3) to simplify (4.1). Combining (4.2) with (4.3) and (2.9) gives

$$\mathbf{v}(x,y) = \ \hat{N}^T\ H^{-1}\ \Gamma\ \underset{\sim}{\tau} = -\frac{1}{\Delta}\ \hat{X}^T\ F^{-T}\ \Gamma\ \underset{\sim}{\tau} \tag{4.4}$$

Written in block form, (4.3) is

$$V^{(N)} = \begin{bmatrix} h_1^{-1} & 0 & 0 \\ 0 & h_2^{-1} & 0 \\ 0 & 0 & h_3^{-1} \end{bmatrix} \begin{bmatrix} \tilde{\gamma}_{11} & \tilde{\gamma}_{12} & \tilde{\gamma}_{13} \\ \tilde{\gamma}_{21} & \tilde{\gamma}_{22} & \tilde{\gamma}_{23} \\ \tilde{\gamma}_{31} & \tilde{\gamma}_{32} & \tilde{\gamma}_{33} \end{bmatrix} \begin{bmatrix} \underset{\sim}{\tau}^{1} \\ \underset{\sim}{\tau}^{2} \\ \underset{\sim}{\tau}^{3} \end{bmatrix} \tag{4.5}$$

where the contribution to $V^{(N)}$ in direction i along side j is given by

$$\begin{aligned} \tilde{\gamma}_{ij} &= \begin{bmatrix} \gamma_0^{ij} & \gamma_1^{ij} & \cdots & \gamma_p^{ij} \end{bmatrix} \\ \underset{\sim}{\tau}^{i} &= \begin{bmatrix} \tau_0^{i} & \tau_1^{i} & \cdots & \tau_p^{i} \end{bmatrix}^T \end{aligned} \tag{4.6}$$

The product matrix $S\Gamma$ arising from (4.1) and (4.3) equals

$$S\Gamma = \begin{bmatrix} \tilde{\gamma}_{31}-\tilde{\gamma}_{21} & \tilde{\gamma}_{32}-\tilde{\gamma}_{22} & \tilde{\gamma}_{33}-\tilde{\gamma}_{23} \\ \tilde{\gamma}_{11}-\tilde{\gamma}_{31} & \tilde{\gamma}_{12}-\tilde{\gamma}_{32} & \tilde{\gamma}_{13}-\tilde{\gamma}_{33} \\ \tilde{\gamma}_{21}-\tilde{\gamma}_{11} & \tilde{\gamma}_{22}-\tilde{\gamma}_{12} & \tilde{\gamma}_{23}-\tilde{\gamma}_{13} \end{bmatrix} \tag{4.7}$$

Consider the variables associated with side 1 first. We want the basis functions in the top row of the matrix $S\Gamma$ are to interpolate to the components indicated in Figure 2. Since $\mathbf{v}_t = L^{-1} S\, \Gamma\, \underset{\sim}{\tau}$ we need to impose the following conditions

$$\tilde{\gamma}_{31} - \tilde{\gamma}_{21} = l_1\, \tilde{\beta}(\zeta_2,\zeta_3) \quad \text{along } \zeta_1=0 \tag{4.8a}$$
$$\tilde{\gamma}_{32} - \tilde{\gamma}_{22} = 0 \quad \text{along } \zeta_1=0 \tag{4.8b}$$
$$\tilde{\gamma}_{33} - \tilde{\gamma}_{23} = 0 \quad \text{along } \zeta_1=0 \tag{4.8c}$$

where $\tilde{\beta}(\zeta_2,\zeta_3)$ is the row vector containing the interpolation polynomials along side 1. Similar equations are obtained for the other two sides

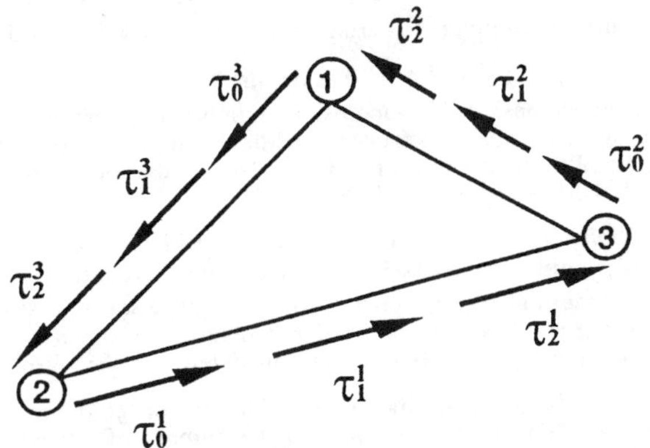

Figure 2. The tangential component interpolation coefficients τ_i^j where j is the triangle side and i is the node number.

$$\tilde{\gamma}_{11} - \tilde{\gamma}_{31} = 0 \quad \text{along } \zeta_2=0 \tag{4.8d}$$
$$\tilde{\gamma}_{12} - \tilde{\gamma}_{32} = l_2\, \tilde{\beta}(\zeta_1,\zeta_3) \quad \text{along } \zeta_2=0 \tag{4.8e}$$
$$\tilde{\gamma}_{13} - \tilde{\gamma}_{33} = 0 \quad \text{along } \zeta_2=0 \tag{4.8f}$$
$$\tilde{\gamma}_{21} - \tilde{\gamma}_{11} = 0 \quad \text{along } \zeta_3=0 \tag{4.8g}$$
$$\tilde{\gamma}_{22} - \tilde{\gamma}_{12} = 0 \quad \text{along } \zeta_3=0 \tag{4.8h}$$
$$\tilde{\gamma}_{23} - \tilde{\gamma}_{13} = l_3\, \tilde{\beta}(\zeta_1,\zeta_2) \quad \text{along } \zeta_3=0 \tag{4.8i}$$

Along side i, the desired interpolation polynomials are given by

$$\beta_k(\zeta_{i1}, \zeta_{i2}) = P_{p-k}(\zeta_{i1})\, P_k(\zeta_{i2}) \tag{4.9}$$

where p is the polynomial order and $P_m(\zeta)$ are the Silvester polynomials [1]

$$P_o(\zeta) = 1$$
$$P_m(\zeta) = \prod_{j=1}^{m} \left(\frac{p\zeta - j + 1}{j}\right), \quad m \neq 1 \tag{4.10}$$

The solution of equations (4.8) is not unique. For the first column of (4.7), consider the solution

$$\tilde{\gamma}_{11} = \tilde{0} \qquad \tilde{\gamma}_{21} = \tilde{0} \qquad \tilde{\gamma}_{31} = l_1 \tilde{\beta}(\zeta_2,\zeta_3) \tag{4.11}$$

Equations (4.11) satisfy equations (4.8a) and (4.8g) everywhere but violate equation (4.8d) at the one point $\zeta_3 = 1$. On the other hand, the solution

$$\tilde{\gamma}_{11} = 0 \qquad \tilde{\gamma}_{21} = -l_1 \tilde{\beta}(\zeta_2,\zeta_3) \qquad \tilde{\gamma}_{31} = 0 \tag{4.12}$$

satisfies (4.8a) and (4.8d) everywhere but violates (4.8g) at the one point $\zeta_2 = 1$. As a result, provided that we use r + s = 1, we may write

$$\Gamma_1 = l_1\, B_1 \tag{4.13}$$

where Γ_1 is the side 1 part of Γ and

$$B_1 = \begin{bmatrix} 0 & 0 & \cdots & 0 \\ 0 & -r\beta_1(\zeta_2,\zeta_3)\cdots & & -\beta_p(\zeta_2,\zeta_3) \\ \beta_0(\zeta_2,\zeta_3) & s\,\beta_1(\zeta_2,\zeta_3)\cdots & & 0 \end{bmatrix} \tag{4.14}$$

Similar arguments with the variables on sides 2 and 3 lead to

$$\Gamma = [\; l_1 B_1 \quad l_2 B_2 \quad l_3 B_3 \;] \tag{4.15}$$

where

$$B_2 = \begin{bmatrix} \beta_0(\zeta_3,\zeta_1) & s\beta_1(\zeta_3,\zeta_1)\cdots & 0 \\ 0 & 0 \quad \cdots & 0 \\ 0 & -r\beta_1(\zeta_3,\zeta_1)\cdots & -\beta_p(\zeta_3,\zeta_1) \end{bmatrix} \tag{4.16}$$

$$B_3 = \begin{bmatrix} 0 & -r\beta_1(\zeta_1,\zeta_2)\cdots & -\beta_p(\zeta_1,\zeta_2) \\ \beta_0(\zeta_1,\zeta_2) & s\beta_1(\zeta_1,\zeta_2)\cdots & 0 \\ 0 & 0 \quad \cdots & 0 \end{bmatrix} \tag{4.17}$$

2-form elements possessing continuity of the normal components are derived in a similarly with the letters referring to normal and tangential directions interchanged. The final result is

$$\mathbf{v}(x,y) = \hat{\underset{\sim}{T}}^T H^{-1}\, \Gamma\, \underset{\sim}{\eta} = -\frac{1}{\Delta}\, \hat{\underset{\sim}{N}}^T F^{-T}\, \Gamma\, \underset{\sim}{\eta} \tag{4.18}$$

where $\underset{\sim}{\eta}$ is a column vector containing the coefficients that interpolate to the normal components of the vector field.

4.2 Low Order Elements - For first order, (4.15) becomes

$$\Gamma = \begin{bmatrix} 0 & 0 & l_2\zeta_3 & 0 & 0 & -l_3\zeta_2 \\ 0 & -l_1\zeta_3 & 0 & 0 & l_3\zeta_1 & 0 \\ l_1\zeta_2 & 0 & 0 & -l_2\zeta_1 & 0 & 0 \end{bmatrix} \tag{4.19}$$

The six functions in (4.19) satisfy the compatibility requirements of 1 and 2-form finite elements. However, these functions do not provide finite elements that are complete in the range space of the curl operator. As we shall see below, we need to add two extra functions to this set to make curl v complete to first order.

Zeroth order elements are derived from first order by setting $\tau_0^1 = \tau_1^1$, $\tau_0^2 = \tau_1^2$ and $\tau_0^3 = \tau_1^3$ in the p = 1 approximation for v(x,y). The result is

$$\Gamma = \begin{bmatrix} 0 & l_2\zeta_3 & -l_3\zeta_2 \\ -l_1\zeta_3 & 0 & l_3\zeta_1 \\ l_1\zeta_2 & -l_2\zeta_1 & 0 \end{bmatrix} \tag{4.20}$$

These elements are equivalent to the edge elements of [28,14] and to the vector elements of [24,25]. Interpolation polynomials of the type (4.20) were first given in [16].

4.3 Unisolvence - As already noted, the above analysis is useful but incomplete. Nedelec [28,29] makes the point that vector elements must satisfy both compatibility and unisolvence conditions.

Compatibility requires that the proper interface conditions be satisfied along element boundaries; this is achieved in a 1-form sense by using the above tangential elements and in a 2-form sense by using the above normal elements. Unisolvence means that the finite element basis functions must be linearly independent to provide a unique, particular solution of the operator equation.

The range space of the curl operator is

$$R(\text{curl}) = \{ \mathbf{r} : \mathbf{r} = \nabla \times \mathbf{v} \;\; \forall \, \mathbf{v} \in C^T \} \tag{4.21}$$

where C^T is the space of tangentially continuous 1-form vectors. The nullspace of the curl operator is

$$N(\text{curl}) = \{ \mathbf{v} : \mathbf{v} = -\nabla\phi \;\; \forall \; \phi \in C^o \} \tag{4.22}$$

where C^o is the space of 0-form continuous scalars. Obviously, the solution v of the curl equation $\mathbf{r} = \nabla \times \mathbf{v}$ is not unique since $\mathbf{v} = \mathbf{v}' - \nabla\phi$ will satisfy the curl equation for any ϕ.

There is a second issue however. Even if one takes care of model $\nabla\phi$ correctly, it is possible to satisfy $\nabla \times \mathbf{v} = 0$ with $\mathbf{v} \neq 0$ and $\mathbf{v} \neq -\nabla\phi$ if the functions approximating v are not linearly independent. Thus, to get correct solutions, we must ensure the linear independence of the approximating functions. Unisolvence requires that the only non-trivial solution of $\nabla \times \mathbf{v} = 0$ is $\mathbf{v} = -\nabla\phi$ for some ϕ. Thus, if $\nabla \times \mathbf{v} = 0$ and $\mathbf{v} \neq -\nabla\phi$ then $\mathbf{v} = 0$ may be the only solution.

4.4 Bezier Forms - The unisolvence of vector finite elements is most easily understood by representing the finite element approximation functions in terms of Bezier polynomials. Bezier polynomials have a number of advantages over ordinary interpolation polynomials: they are simpler, easier to evaluate, and have a clearer geometric interpretation. Their disadvantage is that only the Bezier coefficients corresponding to the element vertices equal field values.

In general, p'th order Bezier polynomials in two-dimensions are given by [30]

$$\beta_{ijk}(\zeta_1,\zeta_2,\zeta_3) = \frac{p!}{i!\,j!\,k!}\, \zeta_1^i \, \zeta_2^j \, \zeta_3^k \tag{4.23}$$

where $i + j + k = n$. We may approximate a scalar $\phi(x,y)$ as

$$\phi(x,y) = \sum_{\substack{i,j,k \\ i+j+k=p}} \phi_{ijk} \; \beta_{ijk}(\zeta_1,\zeta_2,\zeta_3) \tag{4.24}$$

As noted, the parameters ϕ_{ijk} in (4.24) are not equal to the values of the function $\phi(x,y)$ except at the vertices. Instead, the ϕ_{ijk} form a faceted surface called the "control net" for $\phi(x,y)$. This control net may be loosely thought of as a first-order finite element approximation of the p'th order polynomial where $\phi(x,y)$ fits inside the convex hull of the control net.

As is well known [1], the number of terms m_s in a complete polynomial of order p is

$$m_s = (p+1)\,(p+2)\,/\,2 \tag{4.25}$$

We now ask the following questions:

1. How many independent variables are there in $\mathbf{v} = \mathbf{v}' - \nabla\phi$ if ϕ is a p'th order polynomial?
2. Of these variables, how many are required to make $\mathbf{r} = \text{curl } \mathbf{v}$ complete to order (p-1) ?

Consider p=1 first. In this case, we may write

$$\phi = \sum_{i=1}^{3} \phi_i \; \zeta_i \tag{4.26}$$

Equation (3.24) implies that $\nabla\phi$ is a constant. We find that

$$\mathbf{v} = \nabla\phi = -\hat{N}^T L \begin{bmatrix} \phi_1 \\ \phi_2 \\ \phi_3 \end{bmatrix} \tag{4.27}$$

Thus there are three parameters in a single element in $\nabla\phi$ and U parameters in a connected mesh, where U is the number of vertices.

Thinking geometrically, we know that ϕ has a different, constant slope along each edge and can therefore be represented by zeroth order tangential elements. The number of variables in this representation is equal to the number of edges in the mesh. The number of edges E in a triangular mesh is given by $E = 3(U - 1) - B$ where U is the total number of vertices and B is the number of vertices on the boundary.

The curl of a zeroth order edge element is equal to a constant in each element. Since the number of independent variables in a single zeroth order edge element is three, we are led to the conclusion that three independent variables are required to represent $\mathbf{v} = \mathbf{v}' - \nabla\phi$ such that $\mathbf{r} = \text{curl } \mathbf{v}$ is complete to zeroth order. Note that ϕ is linear in this expression and that $\nabla \times \nabla \phi = 0$ as required.
In a connected mesh of T triangles, the number of parameters in $\nabla\phi$ is equal to the number of vertices U while the number of parameters in v' is equal to the number of edges E. The number of independent curls is $T = 2(U - 1) - B = E - U + 1$.

4.5 Higher-Order - The number of variables in $\nabla\phi$ for higher order polynomials is derived by noticing that the control net of a Bezier polynomial $\phi(x,y)$ consists of a triangular set of first order finite elements. Since the control net determines the associated Bezier polynomial completely, we can determine the number of variables in $\nabla\phi$ by counting the variables in the gradient of the control net. This number is equal to the number of edges in the Bezier control net and is given by $m_b = 3p(p+1)\,/\,2$.

To answer question two, note that in high order elements the number of independent curls is equal to the number of facets in the control net and is given by $m_f = p^2$. Yet $\mathbf{r} = \nabla \times \mathbf{v}$ should be a complete polynomial of order (p-1) in which there are $m_r = p(p+1)/2$ variables. It follows that $m_x = m_f - m_r = p(p-1)/2$ of the variables are redundant. Thus, the number of independent variables in v complete to order (p-1) is

$$m_c = m_b - m_x = p\,(p+2) \tag{4.28}$$

Equation (4.28) was derived previously by Nedelec using a different approach [28]. We note that the first four terms in the series for m_c are 3, 8, 15, and 24.

4.6 Tangential Elements - Equation (4.28) shows that the polynomials derived in Section 4.1 are incomplete. The number of variables in these edge elements is $m_e = 3p$. We must add $m_c - m_e = p(p-1)$ additional variables to achieve completeness. The first four terms in this series are 0, 2, 6, 12. Note that only the second of the first four terms in this sequence is not divisible by three.

No additional terms beyond the three in (4.20) are needed to generate zeroth order complete tangential elements. However, two additional terms must be added to the first order elements in (4.19) to make it complete. These additional basis functions may not contribute to the tangential components of the field along the edges of the element. We therefore use (3.16) to introduce functions that have a zero tangential component along the appropriate side.

Consider the matrix formed by augmenting the matrix Γ with two additional columns such that

$$\Gamma = [\; l_1 B_1 \quad l_2 B_2 \quad l_3 B_3 \quad \Lambda \;] \tag{4.29}$$

where Λ is the matrix

$$\Lambda = \begin{bmatrix} 4\,\zeta_2\,\zeta_3 & 0 \\ 0 & 4\,\zeta_1\,\zeta_3 \\ 0 & 0 \end{bmatrix} \tag{4.30}$$

The two extra functions in (4.29) do not contribute to the tangential components of v(x,y): the function in column 1 of Λ is zero along edges 2 and 3 of the triangle and has a zero tangential component along edge 1 while the function in column 1 of Λ is zero along edges 2 and 3 of the triangle and has a zero tangential component along edge 2. Thus the element formed by (4.29) retains the linear tangential components on the three sides. However, its normal components are quadratic on two of the three sides with the addition of Λ. Since column 1 of Λ equals 1 at $\zeta_2 = \zeta_3 = 1/2$ while column 2 of Λ equals 1 at $\zeta_1 = \zeta_3 = 1/2$, these functions interpolate the quadratic part of the normal at the mid sides. This is depicted pictorially in Figure 3. Note, however, that the total function defined by Γ does **NOT** interpolate to the normal components of v(x,y) at the mid sides since the first six functions in Γ contribute non-zero normal components there.

Elements of the type given in (4.29) were first derived by Din Sun and the author [17]. A different derivation is provided in a companion paper [31]. An obvious disadvantage of first order tangential elements is that 2 variables cannot be located in a triangle in a symmetric way. This disadvantage does not occur with second order tangential elements. Using arguments similar to the above yields the element in Figure 4. The basis functions for this element are given in Table 1.

5. VARIATIONAL PRINCIPLES

5.1 The Vector Wave Equation - At this point, we need to determine how the elements derived above are used in finite element analysis. To make the analysis concrete, we examine the variational solution of the vector wave equation. The following analysis expands on that given in [3] and first appeared in this form in [17].

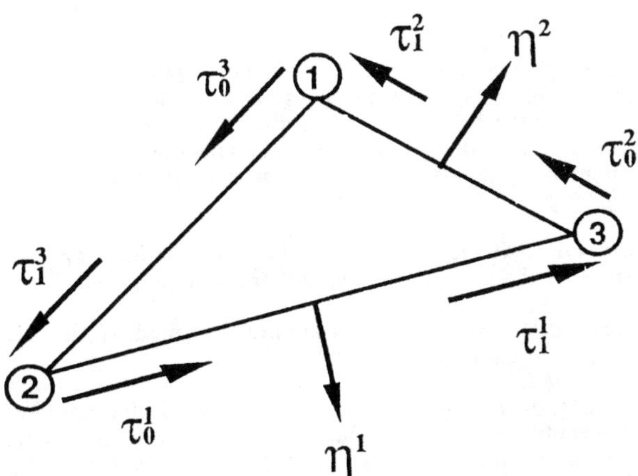

Figure 3. A first order complete tangential vector finite element. Note that only two of the three sides have a parameter in the normal direction and that these normal parameters are not interpolatory.

The vector wave equation is

$$\nabla \times \frac{1}{\mu} \nabla \times \mathbf{E} = \frac{k^2}{\mu} \mathbf{E} \tag{5.1}$$

where $k^2 = \omega^2 \mu \varepsilon$. We shall show that the functional

$$F(\mathbf{E}) = \int \frac{1}{\mu} (\nabla \times \mathbf{E})^2 d\Omega - \int \frac{k^2}{\mu} \mathbf{E}^2 d\Omega \tag{5.2}$$

may be used to solve equation (5.1) provided that the tangential component of **E** is made continuous. To minimize F(**E**), let $\mathbf{E} = \mathbf{E}_{ex} + \beta\xi$ where $\mathbf{E}_{ex}$ is the exact solution of the wave equation (5.1), β is a number, and ξ is an arbitrary vector function. The first variation of F(**E**) is

$$\delta F(\mathbf{E}) = \frac{\partial F(\mathbf{E}_{ex} + \beta\xi)}{\partial \beta} \Big|_{\beta=0}$$
$$\int \frac{1}{\mu} (\nabla \times \xi) \cdot (\nabla \times \mathbf{E}_{ex}) d\Omega - \int \frac{k^2}{\mu} \xi \cdot \mathbf{E}_{ex} d\Omega. \tag{5.3}$$

By integrating the vector identity

$$\nabla \cdot (\mathbf{a} \times \mathbf{b}) = (\nabla \times \mathbf{a}) \cdot \mathbf{b} - \mathbf{a} \cdot (\nabla \times \mathbf{b}) \tag{5.4}$$

we find that

$$\oint (\mathbf{a} \times \mathbf{b})) \cdot d\mathbf{S} = \int \left[(\nabla \times \mathbf{a}) \cdot \mathbf{b} - \mathbf{a} \cdot (\nabla \times \mathbf{b}) \right] d\Omega \tag{5.5}$$

Thus (5.3) is converted into

$$\delta F(A) = \int \xi \cdot \left[\nabla \times \frac{1}{\mu} \nabla \times \mathbf{E}_{ex} - \frac{k^2}{\mu} \mathbf{E}_{ex} \right] d\Omega + \oint \left(\xi \times \frac{1}{\mu} \nabla \times \mathbf{E}_{ex} \right) \cdot d\mathbf{S}. \tag{5.6}$$

Since $\mathbf{E}_{ex}$ satisfies (5.1) exactly, the volume integral in (5.6) drops out, leaving

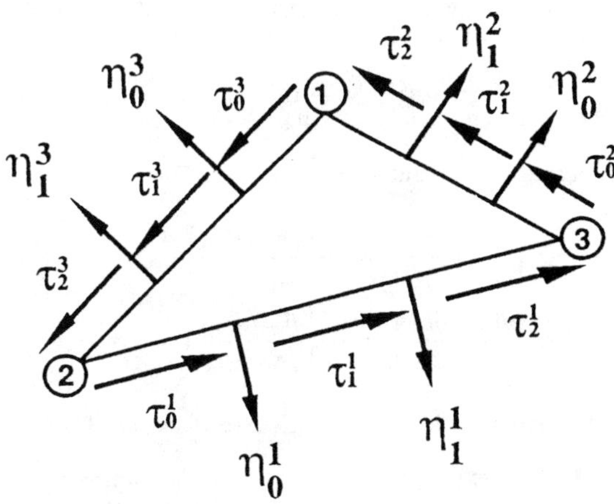

Figure 4. A second order complete tangential vector finite element. The tangential components of this element are quadratic and interpolate to 3 points on each side while the normal components are cubic and are not interpolated.

Table 1. The basis functions for the element in Figure 4.

$$\Gamma = \begin{bmatrix} 0 & 0 & 0 & l_2\zeta_3(2\zeta_3-1) & 4sl_2\zeta_1\zeta_3 & 0 & 0 & -4rl_3\zeta_1\zeta_2 & -l_3\zeta_2(2\zeta_2-1) & 9\zeta_2\zeta_1(2\zeta_1-1)/2 & 9\zeta_2\zeta_3(2\zeta_2-1)/2 & 0 & 0 & 0 & 0 \\ 0 & -4rl_1\zeta_2\zeta_3 & -l_1\zeta_3(2\zeta_3-1) & 0 & 0 & 0 & l_3\zeta_1(2\zeta_1-1) & 4sl_1\zeta_1\zeta_2 & 0 & 0 & 0 & 9\zeta_1\zeta_1(2\zeta_1-1)/2 & 9\zeta_1\zeta_3(2\zeta_3-1)/2 & 0 & 0 \\ l_1\zeta_2(2\zeta_2-1) & 4sl_1\zeta_2\zeta_3 & 0 & 0 & -4rl_2\zeta_1\zeta_3 & -l_2\zeta_1(2\zeta_1-1) & 0 & 0 & 0 & 0 & 0 & 0 & 0 & 9\zeta_2\zeta_3(2\zeta_2-1)/2 & 9\zeta_2\zeta_3(2\zeta_2-1)/2 \end{bmatrix}$$

$$\delta F(\mathbf{E}) = \oint \left(\xi \times \frac{1}{\mu} \nabla \times \mathbf{E}_{ex}\right) \cdot \mathbf{dS} = -jw \oint (\xi \times \mathbf{H}) \cdot \mathbf{dS} \quad (5.7)$$

Finally, setting the first variation of F(E) equal to zero yields

$$\oint (\xi \times \mathbf{H}) \cdot \mathbf{dS} = 0 \quad (5.8)$$

(5.8) provides the natural boundary conditions for the functional.

5.2 Interface Conditions - At interfaces, the electric field **E** satisfies the condition that its tangential component is continuous

$$\hat{n} \times \mathbf{E}_1 = \hat{n} \times \mathbf{E}_2 \quad (5.9)$$

while the flux $\mathbf{D} = \varepsilon\mathbf{E}$ satisfies the condition that its normal component of is continuous

$$\hat{n} \cdot \varepsilon_1 \mathbf{E}_1 = \hat{n} \cdot \varepsilon_2 \mathbf{E}_2 \quad (5.10)$$

For the magnetic field **H**, the corresponding interface conditions are

$$\hat{n} \times \mathbf{H}_1 = \hat{n} \times \mathbf{H}_2 \quad (5.11)$$

$$\hat{n} \cdot \mu_1 \mathbf{H} = \hat{n} \cdot \mu_2 \mathbf{H}_2 \quad (5.12)$$

(5.12) is expressed in terms of **E** by using Faraday's law. This yields

$$\hat{n} \cdot \nabla \times \mathbf{E}_1 = \hat{n} \cdot \nabla \times \mathbf{E}_2 \quad (5.13)$$

Adopting an (n,τ,λ) coordinate system, where n is normal to the interface and τ and λ are parallel to it, (5.13) becomes

$$\left(\frac{\partial E_\lambda}{\partial \tau} - \frac{\partial E_\tau}{\partial \lambda}\right)_1 = \left(\frac{\partial E_\lambda}{\partial \tau} - \frac{\partial E_\tau}{\partial \lambda}\right)_2 \quad (5.14)$$

Notice that (5.14) involves only the tangential components and the tangential derivatives of **E**. We therefore conclude that continuity of the normal component of magnetic flux $\mathbf{B} = \mu\mathbf{H}$ is ensured by setting the tangential component of the electric field **E** to be continuous. A corollary of this is that continuity of the normal component of electric flux $\mathbf{D} = e\mathbf{E}$ is ensured by setting the tangential component of the magnetic field **H** to be continuous. We express the tangential component of the magnetic field in terms of the electric field by using Faraday's law. The result is

$$\hat{n} \times \frac{1}{\mu_1}(\nabla \times \mathbf{E}_1) = \hat{n} \times \frac{1}{\mu_2}(\nabla \times \mathbf{E}_2) \quad (5.15)$$

Writing this out in components gives

$$\frac{1}{\mu_1}\left(\frac{\partial E_\tau}{\partial n} - \frac{\partial E_n}{\partial \tau}\right)_1 = \frac{1}{\mu_2}\left(\frac{\partial E_\tau}{\partial n} - \frac{\partial E_n}{\partial \tau}\right)_2 \quad (5.16)$$

$$\frac{1}{\mu_1}\left(\frac{\partial E_n}{\partial \lambda} - \frac{\partial E_\lambda}{\partial n}\right)_1 = \frac{1}{\mu_1}\left(\frac{\partial E_n}{\partial \lambda} - \frac{\partial E_\lambda}{\partial n}\right)_2 \quad (5.17)$$

Provided that (5.16) and (5.17) are satisfied, (5.10) is satisfied.

5.3 Natural Interface Conditions - Evaluating (5.8) by components yields

$$\oint (\xi_\tau H_\lambda - \xi_y H_\tau)\, dS = 0 \quad (5.18)$$

In terms of the electric field **E** this is

$$\oint \left[\xi_\tau \frac{1}{\mu}\left(\frac{\partial E_\tau}{\partial n} - \frac{\partial E_n}{\partial \tau}\right) - \xi_\lambda \frac{1}{\mu}\left(\frac{\partial E_n}{\partial \lambda} - \frac{\partial E_\lambda}{\partial n}\right)\right] dS = 0 \quad (5.19)$$

On exterior boundaries, if we set the tangential components of **E** equal to a given value, then ξ_τ and ξ_λ are zero and (5.19) is satisfied.

On the other hand, along boundaries where we leave ξ_τ and ξ_λ to be arbitrary, we will automatically obtain a solution for **E** that satisfies

$$\hat{n} \times \mathbf{H} = 0 \quad (5.20)$$

On interior boundaries, if we impose continuity of the tangential component of **E** then

$$\xi_\tau|_1 = \xi_\tau|_2 \qquad \xi_\lambda|_1 = \xi_\lambda|_2 \quad (5.21)$$

and (5.19) yields

$$\int \xi_\tau \left[\frac{1}{\mu_1}\left(\frac{\partial E_\tau}{\partial n} - \frac{\partial E_n}{\partial \tau}\right)_1 - \frac{1}{\mu_2}\left(\frac{\partial E_\tau}{\partial n} - \frac{\partial E_n}{\partial \tau}\right)_2\right] dS - \int \xi_\lambda \left[\frac{1}{\mu_1}\left(\frac{\partial E_n}{\partial \lambda} - \frac{\partial E_\lambda}{\partial n}\right)_1 - \frac{1}{\mu_2}\left(\frac{\partial E_n}{\partial \lambda} - \frac{\partial E_\lambda}{\partial n}\right)_2\right] dS = 0 \quad (5.22)$$

Since ξ_τ and ξ_λ are arbitrary on interior boundaries, (5.22) yields exactly the required interface conditions (5.16) and (5.17).

6. CONCLUSIONS

In view of the above analysis, specifying the continuity of the normal component of electric or magnetic field is *not* required in the finite element solution of the vector wave equation. As discussed in [16] and [23], this is also true in the magnetostatic and eddy current cases. In fact, imposing continuity on the normal component of the field is wrong with finite elements of polynomial order less than 5 since it is then impossible to have C^1 or derivative continuous finite elements with an arbitrary mesh. Tangential vector finite elements provide correct solutions because the continuity of the normal component of the electric field is *not* specified, so that the discontinuity in the normal component is correctly produced by the natural boundary conditions in the variational principle.

The zeroth and first order elements presented in this paper are used in references [14-25] to solve electromagnetic field problems. In all cases, the solutions obtained are reported to be accurate and reliable.

The analysis presented in Section 4 shows that spurious solutions in finite element methods arise from two sources: (1) approximations that do not satisfy the compatibility conditions between elements, and (2) approximations that do not satisfy the unisolvence condition within an element. This allows us to define two types of spurious modes:

Type 1 Spurious Modes - Incorrect solutions generated by poor approximations of the nullspace of the curl operator.

Type 2 Spurious Modes - Incorrect solutions formed by linearly dependent approximation functions.

References [20-21] show that tangential vector finite elements eliminate the problem of Type 1 spurious modes. These modes are formed with conventional Cartesian finite elements because polynomials of order less than five do not in general satisfy the proper compatibility conditions between elements. A characteristic of Type 1 spurious modes is that the number of such modes equals the dimension of the nullspace of the differential operator.

Type 2 spurious modes arise when one or more of the approximation functions in a finite element is a linear combination of the others. An example of an element generating this type of spurious mode is provided by adding the third normal on the vacant side in the element in Figure 3. While such an element is symmetric, the analysis in Section 4 shows that the resulting set of nine functions is linearly dependent. Computations confirm that adding this ninth function introduces Type 2 spurious modes. A characteristic of Type 2 spurious modes is that their number equals the product of the number of elements times the number of dependent approximation functions per element.

In summary, we find that the affine transformation provides a powerful tool for defining vector finite elements. This representation allows us to simplify the derivation of finite elements that interpolate to either the tangential or the normal components of vector fields. The methods presented in this paper can be extended to three dimensions; work is in progress on a paper describing this extension.

7. ACKNOWLEDGEMENTS

The author would like to thank Tze-Ting Fang and Xingchao Yuan for proof reading the manuscript and for improving the notation. He would also like to thank Din Sun, Jin-Fa Lee and the students in his Finite Elements Class for stimulating discussions regarding vector finite elements.

8. REFERENCES

[1] P. P. Silvester and R. L. Ferrari, *Finite Elements for Electrical Engineers*, Second Edition, New York: Cambridge University Press, 1990.

[2] Z. J. Cendes and P. P. Silvester, "Numerical Solution of Dielectric Loaded Waveguides", *IEEE Trans. Microwave Theory Tech.*, Vol. MTT-18, pp. 1124-1131, 1970.

[3] A. Konrad, "Vector Variational Formulation of Electromagnetic Fields in Anisotropic Media", *IEEE Trans. Microwave Theory Tech.*, Vol. MTT-24, pp. 553-559, 1976.

[4] J. R. Winkler and J. B. Davies, "Elimination of Spurious Modes in Finite Element Analysis", *J. Computational Physics*, Vol. 56, pp. 1-14, 1984.

[5] M. Hara, T. Wada, T. Fukasawa and F. Kikuchi, "Three-Dimensional Analysis of RF Electromagnetic Field by the Finite Element Method", *IEEE Trans. Magnetics*, Vol. MAG-19, pp. 2417-2420, 1983.

[6] C. W. Crowley, *Mixed Order Covariant Projection Finite Elements for Vector Fields,* PhD dissertation, McGill University, 1988.

[7] M. V. K. Chari, Z. J. Cendes, P. P. Silvester, A. Konrad and M. A. Palmo, "Three-Dimensional Magnetostatic Field Analysis of Electric Machinery by the Finite Element Method", *IEEE Trans. Power App. Syst.*, Vol. PAS-100, pp. 4007-4015, 1981.

[8] N. A. Demerdash, T. W. Nehl, F. A. Fouad and O. A. Mohammed, "Three Dimensional Finite Element Vector Potential Formulation of Magnetic Fields in Electrical Apparatus", *IEEE Trans. Power App. Syst.*, Vol. PAS-100, pp. 4104-4111, 1981.

[9] J. L. Coulomb, "Finite Element Three Dimensional Field Computation", *IEEE Trans. Magnetics*, Vol. MAG-17, pp. 3241-3246, 1981.

[10] Z. J. Cendes, J. Wiess and S. R. H. Hoole, "Alternative Vector Potential Formulations of 3-D Magnetostatic Field Problems", *IEEE Trans. Magnetics*, Vol. MAG-18, pp. 367-372, 1982.

[11] P. R. Kotiuga and P. P. Silvester, "Vector Potential Formulation for Three-Dimensional Magnetostatics", *J. Appl. Phys.*, Vol. 53, pp. 8399-8401, 1983.

[12] J. Rikabi, C. F. Bryant and E. M. Freeman, "On the Solvability of Magnetostatic Vector Potential Formulations", *IEEE Trans. Magnetics*, Vol. MAG-26, pp. 2866-2874, 1990.

[13] R. Wang and N. A. Demerdash, "On the Effects of Grid Ill-Conditioning in Three Dimensional Finite Element Vector Potential Magnetostatic Field Computations", *IEEE Trans. Magnetics*, Vol. MAG-26, pp. 2190-2192, 1990.

[14] A. Bossavit and J. C. Verite, "A Mixed FEM-BIEM Method to Solve 3-D Eddy Current Problem", *IEEE Trans. Magnetics*, Vol. MAG-18, pp. 431-435, 1982.

[15] A. Bossavit and I. Mayergoyz, "Edge-Elements for Scattering Problems", *IEEE Trans. Magnetics*, Vol. MAG-25, pp. 2816-2821, 1989.

[16] M. L. Barton and Z. J. Cendes, "New Vector Finite Elements for Three-Dimensional Magnetic Field Computation", *J. Appl. Phys.*, Vol. 61, pp. 3919-3921, 1987.

[17] Z. J. Cendes, D. Hudak, J. F. Lee and D. K. Sun, *Development of New Methods for Predicting the Bistatic Electromagnetic Scattering from Absorbing Shapes*, RADC Final Report, Hanscome Air Force Base, MA, April, 1986.

[18] D. K. Sun, D. N. Shenton and Z. J. Cendes, "High-Order Tangential Vector Finite Elements for Three-Dimensional Magnetic Field Computation", Joint MMM/Intermag Conference, Vancouver, July, 1988.

[20] S. H. Wong and Z. J. Cendes, "Combined Finite Element-Modal Solution of Three-Dimensional Eddy Current Problems", *IEEE Trans. Magnetics*, Vol. MAG-24, pp. 2685-2687, 1988.

[21] S. H. Wong and Z. J. Cendes, "Numerically Stable Finite Element Methods for the Galerkin Solution of Eddy Current Problems", *IEEE Trans. Magnetics*, Vol. MAG-25, pp. 3019-3021, 1989.

[22] J. F. Lee, D. K. Sun and Z. J. Cendes, "Full-Wave Analysis of Dielectric Waveguides using Tangnetial Vector Finite Elements", *IEEE Trans. Microwave Theory Tech.*, accepted for publication.

[23] Z. J. Cendes and D. N. Shenton, "Three-Dimensional Eddy Current Analysis using Tangnetial Finite Elements", Intermag Conference, Tokyo, Japan, April, 1987.

[24] A. W. Glisson and D. R. Wilton, "Simple and Efficient Numerical Methods for Problems of Electromagnetic Radiation and Scattering from Surfaces", *IEEE Trans. Antennas Propagat.*, Vol. AP-28, pp. 593-603, 1980.

[25] S. M. Rao, D. R. Wilton and A. W. Glisson, "Electromagnetic Scattering by Surfaces of Arbitrary Shape", *IEEE Trans. Antennas Propagat.*, Vol. AP-30, pp. 409-418, 1982.

[26] H. Whitney, *Geometric Intergation Theory*, Princton University Press, 1957.

[27] A. Bossavit, "Whitney Forms: A Class of Finite Elements for Three-Dimensional Computations in Electromagnetism", *IEE Proceedings*, Vol. 135, Pt. A, pp. 493-500, 1988.

[28] J. C. Nedelec, "Mixed Finite Elements in R^3", *Numer. Math.*, Vol. 35, pp. 315-341, 1980.

[29] J. C. Nedelec, "A New Family of Mixed Finite Elements in R^3", *Numer. Math.*, Vol. 50, pp. 57-81, 1986.

[30] G. Farin, *Curves and Surfaces for Computer Aided Geometric Design*", Second Edition, New York: Academic Press, 1990.

[31] J. F. Lee, D. K. Sun and Z. J. Cendes, "Tangential Vector Finite Elements for Electromagnetic Field Computation", CEFC, Toronto, Oct. 1990, *IEEE Trans. Magn.*, this issue.

[32] M. Hano, "Finite-Element Analysis of Dielectric-Loaded Waveguides", *IEEE Trans. Microwave Theory Tech.*, Vol. MTT-32, pp. 1275-1279, 1984.

[33] M. Hano, *Finite Element Vector Potential Solution of Toroidal Magnet Forminga Magnetic Flux Closure*, CEFC, Toronto, Oct. 1990, *IEEE Trans. Magn.*, this issue.

[34] G. A. DesChamps, "Electromagnetics and Differential Forms", Proceedings IEEE, Vol. 69, pp. 676-696, 1981.

Edge Elements and What They can do for You

J. P. Webb
Computational Analysis and Design Laboratory
Electrical Engineering Department, McGill University
3480 University Street, Montreal, PQ, H3A 2A7, Canada

Abstract—**Edge elements are finite elements whose degrees of freedom are associated with edges rather than with nodes. They have been used to compute 3D vector electromagnetic fields in both power and microwave engineering. They have several useful properties: they allow the normal component of the vector to be discontinuous from one element to the next; they are better at handling field singularities; they eliminate spurious modes; they permit a unique separation of a vector field into the gradient of a scalar and a remaining part. Various tetrahedral and hexahedral edge elements exist, and triangular and quadrilateral edge elements have been developed for 2D problems.**

I. Introduction

The earliest finite elements were scalar, and had as unknown parameters the values of the scalar field at the nodes of the element. For example, a first-order scalar tetrahedron interpolates the values at the four vertices:

$$\phi = \sum_{i=1}^{4} \phi_i \zeta_i \tag{1}$$

ζ_i is the simplex coordinate associated with vertex i. When attention was turned to vector problems, the natural approach was to treat the vector field $\boldsymbol{H}$ simply as three coupled scalar fields, H_x, H_y and H_z. Scalar elements were used again, only this time each node had three unknowns instead of one (Fig. 1):

$$\boldsymbol{H} = \sum_{i=1}^{4} (H_{xi}\hat{x} + H_{yi}\hat{y} + H_{zi}\hat{z})\, \zeta_i \tag{2}$$

Elements of this kind are *node-based*. They have been, and continue to be, widely used.

However, vectors are not simply triplets of numbers. They have a physical and mathematical identity that goes beyond their representation in any particular coordinate frame. By dividing the vector into three Cartesian parts, node-based elements fail to take this into account. For example, boundary conditions in electromagnetics often take the form of a specification of only the part of the vector field that is tangent to the boundary. With node-based elements, this physical and simple unary constraint must be transformed into linear relationships between the Cartesian components; in addition, at nodes where the boundary changes direction an average tangential direction must first be determined. None of this is impossible, but it is awkward and unnatural.

An alternative to the node-based vector tetrahedron shown in Fig. 1 is the *edge element* [1] [2] shown in Fig. 2. This paper presents a review of edge elements: what they are and how they have been used in electromagnetics. Note that many of the useful properties of edge elements can be explained in somewhat more rigorous, mathematical terms (e.g. as in [1]); such explanations are outside the scope of this review.

II. Edge elements

The field in the edge element is represented by six basis functions associated with edges, rather than nodes:

$$\boldsymbol{H} = \sum_{i=1}^{6} H_i \boldsymbol{e}_i \tag{3}$$

where:

$$\begin{aligned} \boldsymbol{e}_1 &= \zeta_1\nabla\zeta_2 - \zeta_2\nabla\zeta_1 & \boldsymbol{e}_2 &= \zeta_1\nabla\zeta_3 - \zeta_3\nabla\zeta_1 \\ \boldsymbol{e}_3 &= \zeta_1\nabla\zeta_4 - \zeta_4\nabla\zeta_1 & \boldsymbol{e}_4 &= \zeta_2\nabla\zeta_3 - \zeta_3\nabla\zeta_2 \\ \boldsymbol{e}_5 &= \zeta_2\nabla\zeta_4 - \zeta_4\nabla\zeta_2 & \boldsymbol{e}_6 &= \zeta_3\nabla\zeta_4 - \zeta_4\nabla\zeta_3 \end{aligned} \tag{4}$$

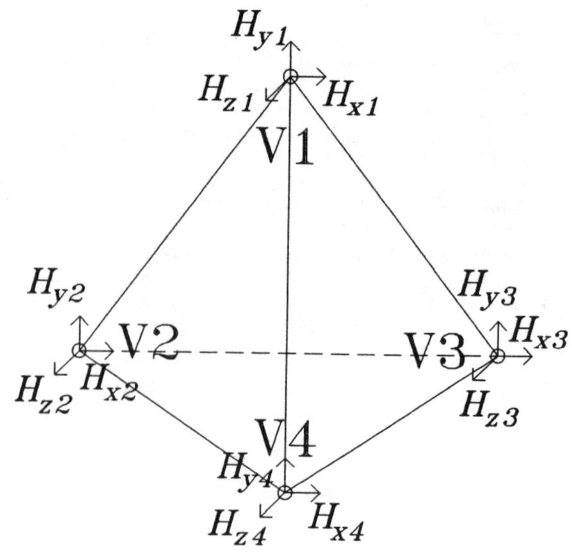

Fig. 1. A node-based tetrahedral element.

Reprinted from *IEEE Trans. Magn.*, vol. 29, no. 2, pp. 1460–1465, March 1993.

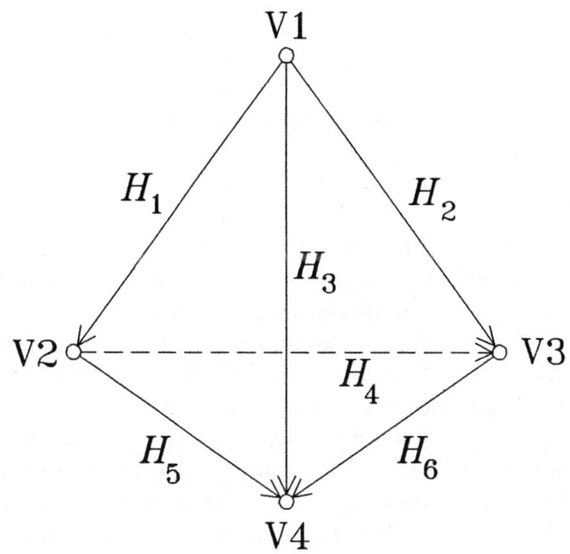

Fig. 2. An edge element.

The six unknown parameters in this element are H_1, H_2,...,H_6. They are *not* the values of any components of $\boldsymbol{H}$. However, it may be shown that on edge i:

$$\boldsymbol{t}_i \cdot \boldsymbol{H} = \frac{H_i}{L_i} \tag{5}$$

where $\boldsymbol{t}_i$ is a unit vector tangent to edge i, in the direction of the edge (Fig. 2), and L_i is the length of edge i. In other words, H_i controls the tangential field on edge i. Equation (5) makes it simple to constrain the tangential field to a prescribed boundary value.

Further, it makes it simple to impose continuity. Suppose two tetrahedra share an edge; let it be edge number i in tetrahedron A, and edge number j in tetrahedron B. The component of the field in A tangent to the edge can be made equal to the component of the field in B tangent to the edge by imposing:

$$H_i^A = \pm H_j^B \tag{6}$$

The ± arises because the direction of the edge in A may be opposite the its direction in B. Applying equation (6) throughout a mesh of tetrahedra gives one unknown per edge, and makes the field tangentially continuous across all edges. In fact, it also makes the field tangentially continuous across all faces too.

Note that the field so constructed is not also normally continuous. The part of the field perpendicular to an element interface is discontinuous, in general. This is in contrast to a node-based element, where, since each Cartesian component is continuous, the entire vector is continuous. The lack of normal continuity is often needed in electromagnetics. For example, the normal part of the electric field $\boldsymbol{E}$ changes abruptly across a dielectric interface.

A variety of finite elements with characteristics similar to those of the element shown in Fig. 2 has been invented, and in this paper "edge element" is used as a general term for all of them. Specifically, an edge element is here defined as a finite element which:

- has unknown parameters (degrees of freedom) which are associated with edges; and
- enforces tangential continuity, but not normal continuity, on the vector field.

Edge elements have been used in a wide range of applications, from DC to microwave frequencies. They have many useful properties, mostly related to the reduced continuity they impose.

III. Continuity

From Maxwell's equations in time-harmonic form:

$$\begin{aligned} \nabla \times \boldsymbol{E} &= -j\omega\mu\boldsymbol{H} \\ \nabla \times \boldsymbol{H} &= \alpha\boldsymbol{E} \end{aligned} \tag{7}$$

where $\alpha = j\omega\epsilon$ for microwave problems, and $\alpha = \sigma$ for quasi-static, induced-current problems. In 3D, unless $\alpha = 0$ (7) cannot be solved with a single scalar potential. On the other hand, it can be solved with a vector potential such as $\boldsymbol{A}$; further, if a vector potential and scalar potential are used, both may be continuous, even in the presence of discontinuities in the materials. Node-based elements have been widely used with such formulations.

An alternative to the use of potentials is to solve directly for $\boldsymbol{E}$ or $\boldsymbol{H}$. The normal component of $\boldsymbol{E}$ is discontinuous at an abrupt change in α; similarly, the normal component of $\boldsymbol{H}$ is discontinuous at an abrupt change in μ. For problems involving changes in both α and μ, direct solution for either $\boldsymbol{E}$ or $\boldsymbol{H}$ is therefore not straightforward with node-based elements, which impose full continuity – though the discontinuities can be built into a node-based mesh [3].

However, direct solution is handled naturally by edge elements. Eliminating $\boldsymbol{E}$ from (7):

$$\nabla \times \frac{1}{\alpha} \nabla \times \boldsymbol{H} + j\omega\mu\boldsymbol{H} = 0 \tag{8}$$

This can be solved numerically with a weighted residual approach, or via a variational principle. The functional required for the variational method is:

$$F(\boldsymbol{H}) = \int_V [\frac{1}{\alpha}(\nabla \times \boldsymbol{H})^2 + j\omega\mu\boldsymbol{H}^2] dV \tag{9}$$

It can be shown that the stationary point of this functional solves the differential equation (8), and *has the correct normal continuity* across material interfaces, i.e. μH is normally continuous everywhere, even though it is not imposed explicitly on the basis functions. In other words, normal continuity of the flux density is a natural condition of the variational principle. It is this that allows us to use edge elements for the magnetic field.

Edge elements have also been used to solve the quasi-static problem by means of a modified magnetic vector potential, which is normally discontinuous [4].

IV. SINGULARITIES

Relaxation of normal continuity is also important in dealing with sharp points and edges.

There are two difficulties in modeling, say, the electric field at a sharp, perfectly conducting edge (Fig. 3):

(a) it is infinite at the edge; and

(b) its direction changes infinitely rapidly at the edge.

Without the use of singular trial functions, neither node-based nor edge-based elements can help with (a). However, edge-based elements, since they only impose tangential continuity, allow the field to change direction abruptly at the sharp edge. This is illustrated in Fig. 4. The field in element A has a quite different direction to the field in B, even though their components tangent to the interface between them are the same. Such a discontinuity is not possible with node-based elements. Consequently, good results can only be obtained with node-based elements by introducing singular trial functions [5]. With edge elements, good accuracy in global quantities - such as resonant frequencies and scattering parameters - can be obtained without any need for special trial functions [6] [7] [8].

V. SPURIOUS MODES

Consider equation (8) with $\alpha = j\omega\epsilon$:

$$\nabla\times\frac{1}{\epsilon}\nabla\times \boldsymbol{H} - \omega^2\mu \boldsymbol{H} = 0 \qquad (10)$$

This is the equation governing the field in a 3D, time-harmonic, microwave device. Even if the device is driven by known, non-zero boundary conditions, the quality of the solution will be determined in part by the eigensolutions of this equation [9], i.e. the solutions $(\boldsymbol{H},\omega^2)$ that are possible when all the boundary conditions are zero. A long-standing difficulty with solving (10) numerically is the existence of spurious modes - eigensolutions that are nonphysical.

The reason these occur is now fairly well understood [10] [9] [2] [6]. The eigenproblem has infinitely many solutions of the form $(\boldsymbol{H},\omega^2)=(\nabla\phi, 0)$. No numerical scheme can model accurately all of these static fields; it will get some of them right, and produce approximations to the rest. These approximations will have *non-zero* frequencies (since they are approximate), which show up in the spectrum as spurious resonances.

One way of removing spurious modes is to impose $\nabla\cdot\boldsymbol{H}=0$ on the solution, thereby reducing the infinite number of static eigensolutions to a few or possibly none. Various methods have been proposed for doing this, but none is entirely satisfactory. However, there is an alternative approach. It appears that if the finite elements used are sufficiently good at modeling $\nabla\phi$ fields, no approximations to them show up in the computed spectrum, i.e. there are no spurious modes. A key factor in improving the modeling of $\nabla\phi$ fields is the reduced continuity provided by edge elements.

Consider a single tetrahedron of the type shown in Fig. 1. It has 4x3=12 degrees of freedom, or to put it another

Fig. 3. The singular field at a sharp metal edge.

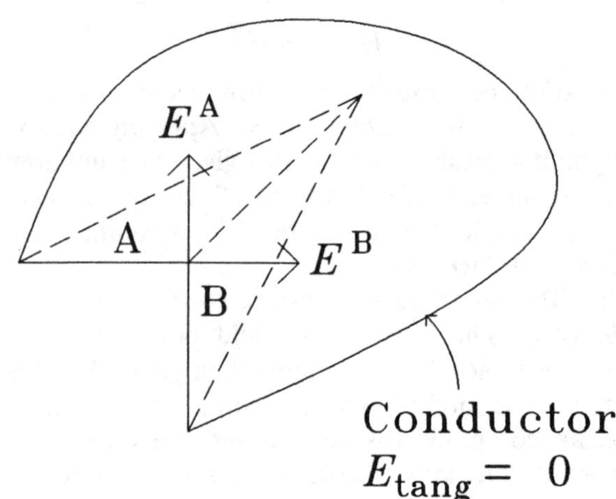

Fig. 4. Field singularity modeled by edge elements.

way, its basis functions span a space V of dimension 12. V has a subspace I of fields of the form $\nabla\phi$. Because the vectors must be of polynomial order no greater than one, ϕ must be of polynomial order no greater than two, i.e. I consists of all vectors $\nabla\phi$ where ϕ is a second-order scalar field. Now a second-order scalar field in 3D has 10 degrees of freedom, but one of these is the uniform field whose gradient is zero, so I has dimension 9.

Now consider a mesh of these tetrahedra, with N nodes. The space V spanned by all the basis functions has dimension $3N$. Its subspace I is formed from the 9 basis functions per tetrahedron, by connecting them together so that $\nabla\phi$ is fully continuous from one element to the next. Tangential continuity can be ensured by making ϕ continuous; normal continuity requires, in addition, the continuity of the normal derivative of ϕ across element faces. For a general mesh, there are relatively few such C^1-continuous scalar fields [10]. The $\nabla\phi$ subspace is therefore very small.

For a single edge element (Fig. 2), V has dimension 6. I consists of all $\nabla\phi$ where ϕ is first-order only, so I has dimension $4-1=3$. On a mesh of these tetrahedra with E edges, V has dimension E because there is one degree of freedom per edge. I is formed from the 3 basis functions per tetrahedron, by connecting them together so that $\nabla\phi$ is *tangentially* continuous, which can be achieved simply by making ϕ continuous - there is no need for continuity of the normal derivative of ϕ in this case. Representing ϕ with 4 first-order basis functions in each tetrahedron (equation (1)), continuity of ϕ results when the vertex values of ϕ are shared by adjacent tetrahedra in the usual way. In other words, I in this case is just $\nabla\phi$ where ϕ is any continuous, first-order scalar field on the same tetrahedra. Therefore, the dimension of I is the same as the number of vertices, minus one for the uniform ϕ. The $\nabla\phi$ subspace is considerably larger than it is with node-based tetrahedra. Consequently, edge elements do a much better job at representing the static eigensolutions, and spurious modes do not appear.

It should be mentioned that tangential continuity is not sufficient to eliminate spurious modes with all edge element types. For example, with brick-shaped edge elements it has also been found necessary to use mixed-order basis functions - vector functions which are of a lower polynomial order in the direction of the vector than in the other two directions [11]. However, the effect of this additional requirement seems to be the same: to increase the relative size of the $\nabla\phi$ subspace.

VI. SEPARATION

Each of the E edges of a mesh of tetrahedral edge elements corresponds to a basis function, and the space spanned by all the basis functions is the function space, V, with dimension E.

The edges and vertices of the mesh form a connected graph, G. Let T be a spanning tree of the graph, i.e. let T be a subgraph containing every vertex of G, but no cycles (closed loops). The edges *not* in T form the *co-tree*. Each edge of the co-tree corresponds to a basis function, and the space spanned by the co-tree basis functions is a subspace, C, of V.

The interesting property of C is that it contains no functions of the form $\nabla\phi$ (apart, that is, from the zero function). For, suppose it did; suppose such a function is f. Pick any edge of the co-tree. Let its end vertices be A and B. From the definition of the tree, there must be a path in the tree form A to B. Consider the closed cycle L consisting of this path, together with edge AB. Since f is $\nabla\phi$, its path integral around L must be zero. But only AB contributes to the path integral, because all of the other edges of the path are not in the co-tree, and therefore the degrees of freedom associated with them are zero. So the path integral of f along edge AB is zero. Therefore, from (5), the degree of freedom associated with AB is also zero. This argument can be repeated for every edge of the co-tree, establishing that f must be identically zero.

It follows that the intersection of the co-tree subspace and the subspace I (see "Spurious Modes", above) is just the zero vector. But the dimension of the co-tree space is the total number of edges, E, minus the number of tree edges, $N-1$ [12]. The dimension of I is $N-1$. Consequently, the union of C and I is a space of dimension E, and must therefore be the whole space V. In other words, V *separates* into a subspace of $\nabla\phi$ functions, and the co-tree space [13]. This property has found applications in static and quasi-static formulations.

For example, it has been used to gauge the magnetic vector potential, A [14]. A is defined by:

$$\nabla\times A = B \tag{11}$$

which leaves it arbitrary to within the gradient of a scalar field, i.e. if A is a valid vector potential, then $(A+\nabla\phi)$ is also valid, for any ϕ. The usual way to fix A is to specify its divergence, e.g. $\nabla\cdot A=0$. An alternative [14] is to represent A with edge elements, and separate it into two parts:

$$A = -\nabla\phi + A_c \tag{12}$$

ϕ is any first-order scalar field, and A_c is any co-tree field. Since the decomposition (12) is unique, A is now gauged by setting ϕ to zero.

Another example is in the use of scalar potentials in magnetostatics. It was recognized early in the development of 3D methods for magnetostatics that since

$$\nabla\times H = J \tag{13}$$

the magnetic field H can be written in terms of a scalar

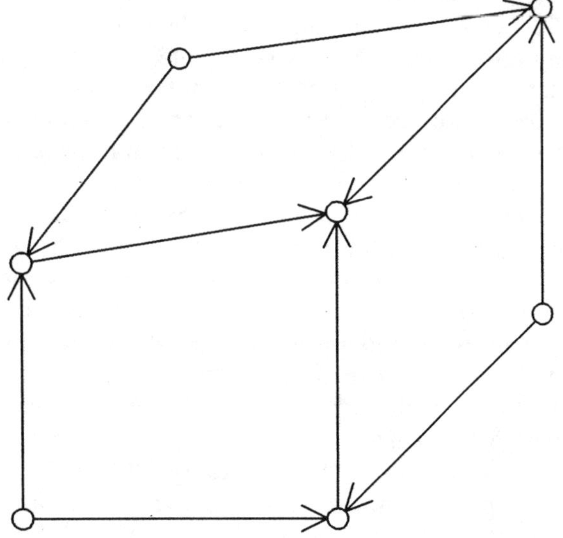

Fig. 5. An 8-node-brick edge element.

potential ϕ:

$$H = -\nabla\phi + H_s \tag{14}$$

where H_s is a source field that can be determined from the known currents J. However, if the basis functions used to represent ϕ and H_s are not carefully chosen, the two terms on the right in (14) will not be able to cancel one another in regions of very high permeability, where H should be vanishingly small. The result is that large errors can occur in ferromagnetic materials. An alternative is to use edge elements to represent H [15]. This overcomes the cancellation difficulty - there is nothing to prevent H becoming arbitrarily small. The edge element field H is then separated:

$$H = -\nabla\phi + H_c \tag{15}$$

The co-tree field is uniquely fixed by the requirement:

$$\nabla\times H_c = J \tag{16}$$

It can be determined by a relatively inexpensive iterative solution of Ampère's Law. The scalar field ϕ is then determined variationally, or by the method of weighted residuals.

VII. Other Edge Elements

The edge element shown in Fig. 2 is complete to order zero, i.e. it can represent exactly any vector that is a zero-order polynomial in x, y and z. Unlike the node-based version (Fig. 1), it is not quite complete to first order. Higher-order edge elements have also been developed.

In [16], a tetrahedron with 20 degrees of freedom was introduced. This element is complete to first order, but not quite to second. In [17], a hierarchal set of tetrahedral edge elements is developed, each member of the set incorporating the basis functions of the lower members, and adding some more basis functions of higher polynomial order. The highest element is complete to second order. The advantage of hierarchal elements is that they allow different orders to be incorporated into the same mesh without any loss of tangential continuity. Note that for the higher order elements, it is necessary to include basis functions that are associated with the faces of the tetrahedron, as well as several edge functions on each edge.

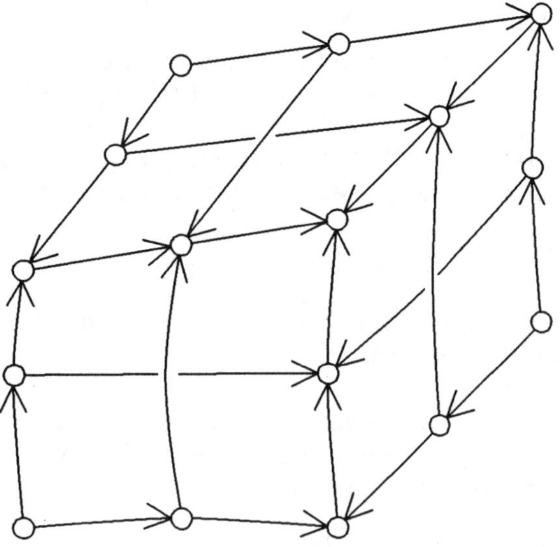

Fig. 6. A 20-node-brick edge element.

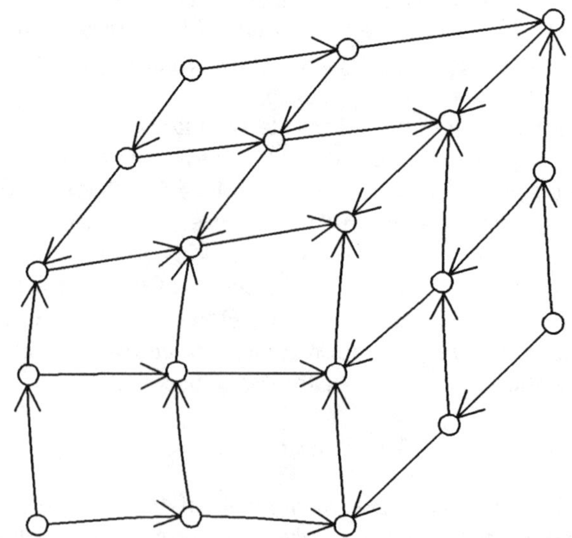

Fig. 7. A 27-node-brick edge element (covariant-projection element).

Hexahedral, or brick-shaped, edge elements have also been developed. The simplest is shown in Fig. 5 [18] [4] [7]: 12 degrees of freedom on an 8-node brick. Each arrow corresponds to an edge basis function - though note that the arrow does not give the direction of the basis function, which is not in general tangent to its own edge. The vector basis functions are uniform (order zero) in the direction of the vector, and first order in the other two directions.

Figs. 6 and 7 show the degrees of freedom on higher-order, curvilinear bricks: 36 degrees of freedom on a 20-node brick [4], and 54 degrees of freedom on a 27-node brick [11], which is also known as a covariant-projection element.

VIII. 2D PROBLEMS

Translationally-symmetric and axisymmetric problems, in which there is no field variation in one of three space coordinates, can generally be solved with a scalar variable on a two dimensional slice. However, there are also cases in which the variation with the third coordinate is sinusoidal. These can still be solved on a two-dimensional slice, but a vector unknown is generally needed. Vector problems in two dimensions exhibit many of the difficulties of three dimensions: discontinuities, spurious modes, singularities, etc. To overcome these, two-dimensional edge elements have been developed. They interpolate a two or three-component vector over a plane surface. For example, both triangular [19] [20] [21] and quadrilateral [6] edge elements have been used to find the propagating modes of uniform waveguides of arbitrary cross-section.

IX. CONCLUSIONS

Are edge elements, then, always preferable to node-based elements in vector electromagnetics? Given the large number of possible formulations and problem types, it is impossible to give a definitive answer. There may be some problems, and some regions of some problems, for which greater accuracy can be achieved by explicitly imposing normal continuity. In such cases, edge elements may not be the best answer. However, as this paper has tried to show, there are many circumstances in which edge elements have clear advantages over the node-based alternatives.

REFERENCES

[1] J. C. Nedelec, "Mixed finite elements in R^{3}", *Numer. Math.*, vol. 35, pp.315-341, 1980.

[2] A. Bossavit, "A rationale for 'edge-elements' in 3-D fields computations", *IEEE Trans. Magn.*, vol. 24, no. 1, pp.74-79, January 1988.

[3] X. Yuan, D. R. Lynch, K. Paulsen, "Importance of normal field continuity in inhomogeneous scattering calculations", *IEEE Trans. on Microwave Theory and Techniques*, vol. 39, no. 4, pp.638-641, April 1991.

[4] A. Kameari, "Calculation of transient 3D eddy current using edge-elements", *IEEE Trans. Magn.*, vol. 26, no. 2, pp.466-469, March 1990.

[5] J. P. Webb, "Finite element analysis of dispersion in waveguides with sharp metal edges", *IEEE Trans. on Microwave Theory and Techniques*, vol 36, pp.1819-1824, December 1988.

[6] R. Miniowitz, J. P. Webb, "Covariant-projection quadrilateral elements for the analysis of waveguides with sharp edges", *IEEE Trans. on Microwave Theory and Techniques*, MTT-39, no. 3, pp.501-505, March 1991.

[7] K. Ise, K. Inoue, M. Koshiba, 'Three-dimensional finite-element method with edge elements for electromagnetic waveguide discontinuities', IEEE Trans., MTT-39, no. 8, pp.1289-1295, August 1991.

[8] J. P. Webb, R. Miniowitz, "Analysis of 3D microwave resonators using covariant-projection elements", *IEEE Trans. on Microwave Theory and Techniques*, MTT-39, no. 11, pp.1895-1899, November 1991.

[9] A. R. Pinchuk, C. W. Crowley, P. P. Silvester, "Spurious solutions to vector diffusion and wave field problems", *IEEE Trans. Magn.*, vol. 24, no. 1, pp.158-161, January 1988.

[10] S. H. Wong, Z. J. Cendes, "Combined finite element - modal solution of three-dimensional eddy current problems", *IEEE Trans. Magn.*, vol. 24, no. 6, pp.2685-2687, November 1988.

[11] C. W. Crowley, P. P. Silvester, H. Hurwitz, "Covariant projection elements for 3D vector field problems", *IEEE Trans. Magn.*, vol. 24, no. 1, pp.397-400, January 1988.

[12] A. Gibbons, *Algorithmic graph theory*, Cambridge University Press, 1985.

[13] R. Albanese, C. Rubinacci, "Solution of three dimensional eddy current problems by integral and differential methods", *IEEE Trans. Magn.*, vol. 24, no. 1, pp.98-101, January 1988.

[14] K. Preis, I. Bardi, O. Biro, C. Magele, G. Vrisk, K. R. Richter, 'Different finite element formulations of 3D magnetostatic fields', *IEEE Trans. Magn.*, vol. 28, no. 2, pp.1056-1059, March 1992.

[15] J. P. Webb, B. Forghani, "A single scalar potential method for 3D magnetostatics using edge elements", *IEEE Trans. Magn.*, vol. 25, no. 5, pp.4126-4128, September 1989.

[16] J. F. Lee, "Analysis of passive microwave devices by using three-dimensional tangential vector finite elements", *International Journal of Numerical Modelling: Electronic Networks, Devices and Fields*, vol. 3, no. 4, pp.235-246, December 1990.

[17] J. P. Webb, B. Forghani, "Hierarchal scalar and vector tetrahedra", Paper TOF2, *Fifth Biennial IEEE Conference on Electromagnetic Field Computation*, Claremont, August 1992; to appear in *IEEE Trans. Magn.*, March 1993.

[18] S. J. van Welij, "Calculation of eddy currents in terms of H on hexahedra", *IEEE Trans. Magn.*, vol. 21, no. 6, pp.2239-2241, November, 1985.

[19] M. Hano, "Vector finite-element solution of anisotropic waveguides using novel triangular elements", *Electronics and Communications in Japan*, part 2, vol. 71, no. 8, pp.71-80, 1988.

[20] Z. J. Cendes, 'Vector finite elements for electromagnetic field computation', *IEEE Trans. Magn.*, vol. 27, no. 5, pp.3958-3966, September 1991.

[21] J.F. Lee, D. K. Sun, Z. J. Cendes, "Full-wave analysis of dielectric waveguides using tangential vector finite elements", *IEEE Trans. on Microwave Theory and Techniques*, vol. 39, no. 8, pp.1262-1271, August 1991.

Hierarchal Scalar and Vector Tetrahedra

J. P. Webb
Computational Analysis and Design Laboratory
Electrical Engineering Department, McGill University
3480 University Street, Montreal, PQ, H3A 2A7, Canada

B. Forghani
Infolytica Corporation
Suite 1160, 1140 Boul. de Maisonneuve W.
Montreal, PQ, H3A 1M8, Canada

Abstract—**A new set of scalar and vector tetrahedral finite elements are presented. The elements are hierarchal, allowing mixing of polynomial orders; scalar orders up to 3 and vector orders up to 2 are defined. The vector elements impose tangential continuity on the field but not normal continuity, making them suitable for representing the vector electric or magnetic field. Further, the scalar and vector elements are such that they can easily be used in the same mesh, a requirement of many quasi-static formulations. Results are presented for two 50 Hz problems: the Bath Cube, and TEAM Problem 7.**

I. Introduction

Tetrahedral finite elements are widely used in 3D electromagnetics. They are the simplest shape into which a 3D region can be broken, and are well-suited to automatic mesh generation. In the simplest tetrahedral elements, the potential or field is interpolated with functions that are first-order polynomials in x, y and z. Examples are the first-order scalar element, with one function per vertex, and the Whitney edge element, with one vector function per edge [1] [2]. By going to higher polynomial orders, greater accuracy can be achieved with fewer tetrahedra, and generally with fewer degrees of freedom. For this reason, various higher-order elements have been invented.

This paper describes a new set of high-order tetrahedral elements with these properties:

- They are **hierarchal**; that is, elements of different orders can be used together in the same mesh.
- Both **scalar and vector** elements are defined, and can be used together in the same mesh. This is important for magneto-quasistatic problems, for which a vector is needed in the conducting regions, but a scalar is sufficient elsewhere.
- The elements are **tangentially-continuous**. Like Whitney edge elements [2] and Tangential Elements [3], they impose only the continuity of the tangential part of field at interfaces. The normal field is allowed to be discontinuous. This is the correct continuity for solving the curl-curl equation for E or H; it allows abrupt material interfaces to be handled easily; and it is important in the elimination of spurious modes or spurious corruptions.

Manuscript received June 1, 1992; revised October 28, 1992.

II. The New Elements

Table I defines the new elements, which are shown schematically in Fig. 1. ζ_1, ζ_2, ζ_3, ζ_4 are simplex coordinates within a tetrahedron. Each basis function is associated with a vertex (V), an edge (E) or a face (F). Vertex functions come in sets of four, one per vertex; the table gives the function for vertex i. Edge functions come in sets of six, one per edge; the table gives the functions for the edge connecting vertices i and j. Face functions come in sets of four, one per face; the table gives the functions for the face connecting vertices i, j and k. For an element that is complete to a given polynomial order m, all the functions down to the corresponding shaded row are required; n is the number of functions. For the vector functions, certain intermediate sets of functions are also of interest. For convenience in referring to them, they are

TABLE I
BASIS FUNCTIONS FOR THE NEW ELEMENTS
m = polynomial order; n = number of functions

SCALAR				VECTOR		
m	n	Basis Function		m	n	Basis Function
		ζ_i	V			
			E			$\zeta_i \nabla \zeta_j - \zeta_j \nabla \zeta_i$
1	4			0.5	6	
		$\zeta_i \zeta_j$	E			$\nabla [\zeta_i \zeta_j]$
2	10			1	12	
			F			$\zeta_k \zeta_j \nabla \zeta_i$
			F			$\zeta_k \zeta_i \nabla \zeta_j$
				1.5	20	
		$\zeta_i \zeta_j (\zeta_i - \zeta_j)$	E			$\nabla [\zeta_i \zeta_j (\zeta_i - \zeta_j)]$
		$\zeta_i \zeta_j \zeta_k$	F			$\nabla [\zeta_i \zeta_j \zeta_k]$
3	20			2	30	

Reprinted from *IEEE Trans. Mag.*, vol. 29, no. 2, pp. 1495–1498, 1993.

given intermediate orders: 0.5 is the Whitney edge element [2]; 1.5 spans the same space as the Tangential Element [3], though the table gives a different, hierarchal basis for the space.

III. Hierarchal Property

Finite elements are said to be hierarchal when the basis functions for an element are a subset of the basis functions for any element of higher order [4]. For example, in Table I, the vector element m=0.5 has n=6 basis functions. The vector element m=1 is formed from it by retaining these 6, and adding a further 6 (one per edge); the vector element m=1.5 is obtained by adding 8 more (two per face).

The great advantage of hierarchal basis functions is that elements of different orders may be used in the same mesh. For example, suppose a scalar element A of order 1 shares an edge with a scalar element B of order 2. If the elements were non-hierarchal, the common edge would present a problem: element A's basis functions and element B's basis functions would not match on the edge, and continuity would be lost. However, when A and B are hierarchal, A's basis functions are a subset of B's, and to maintain continuity it is sufficient to set to zero the coefficient of the edge function $\zeta_i\zeta_j$ of B, since this is not shared by A.

IV. Tangential Continuity

When several vector elements share an edge, the field tangent to the edge can be made the same in the tetrahedra by setting the coefficients of the edge functions

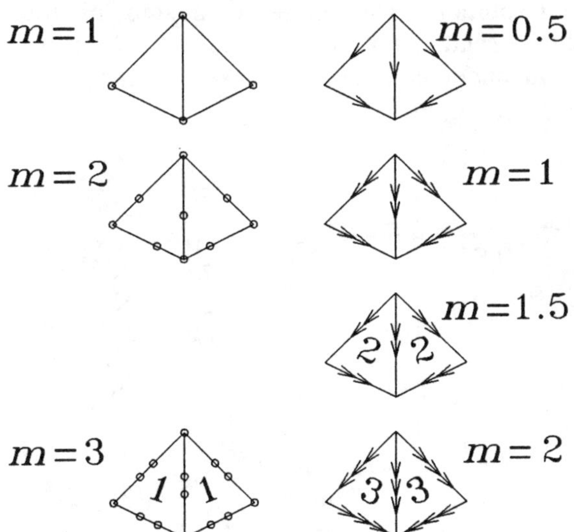

Fig. 1. Scalar (left) and vector (right) elements. Degrees of freedom are shown as circles (for vertices and edges), arrows (for edges) and numbers (for faces).

in each tetrahedron equal to the coefficients of the corresponding edge functions in the other tetrahedra. For tangential continuity across a face, the same equality must be imposed between the coefficients of all the edge and face functions associated with the face.

This process of ensuring continuity by locally matching coefficients of basis functions at edges and faces is characteristic of finite elements, and an important property of the vector elements. It follows from a few basic properties of $\nabla\zeta_i$:

1. $\nabla\zeta_i$ is a constant vector in a tetrahedron. It depends on the Cartesian coordinates of all four vertices of the tetrahedron.
2. $\nabla\zeta_i$ is perpendicular to the face ζ_i=0, i.e. the face opposite vertex i.
3. The part of $\nabla\zeta_i$ tangent to an edge depends only on the coordinates of the two vertices of the edge, and not on the coordinates of the other two vertices of the tetrahedron.
4. The part of $\nabla\zeta_i$ tangent to a face depends only on the coordinates of the three vertices of the face.

It can then be shown that for each edge function in Table I, associated with edge e:

- The part of the function tangent to an edge is zero on all edges except e, and on edge e it depends only on the coordinates of the two vertices of e.
- The part of the function tangent to a face is zero on a face not sharing e, and on a face sharing e it depends only on the coordinates of the three vertices of the face.

Similarly, for each face function in Table I, associated with face f:

- The part of the function tangent to an edge is zero on all edges.
- The part of the function tangent to a face is zero on all faces except f, and on face f it depends only on the coordinates of the three vertices of f.

In other words, on edge e the tangential field is controlled solely by the coefficients of the edge functions associated with the edge; and the magnitude, direction and variation of these functions along e are determined by the geometry of e, not by the geometry of the rest of the tetrahedron. Similarly, on face f the tangential field is controlled solely by the coefficients of the edge and face functions associated with the face; and the magnitude, direction and variation of these functions over f are determined by the geometry of f, not by the geometry of the rest of the tetrahedron. Given these properties, tangential continuity is guaranteed by locally matching coefficients of edge and face functions.

V. Continuity between Scalar and Vector

The vector elements may be used to solve fully vector problems, such as finding the time-harmonic fields in microwave components. They may also be used with scalar elements in the same mesh.

In magneto-quasistatics, the neglect of the displacement current means that the magnetic field is governed by:

$$\begin{aligned} \nabla \times \boldsymbol{H} &= \boldsymbol{J} \\ \nabla \cdot \boldsymbol{B} &= 0 \end{aligned} \tag{1}$$

In regions where the current density $\boldsymbol{J}$ is known - either because it is zero or because it is flowing in a finely-stranded coil - the first of these two equations allows $\boldsymbol{H}$ to be written in terms of a scalar potential ϕ:

$$\boldsymbol{H} = -\nabla\phi + \boldsymbol{H}_s \tag{2}$$

where $\boldsymbol{H}_s$ is a source field that can be calculated directly from the current density [5]. In conducting regions, however, Ohm's Law relates the current density to the electric field, and thereby couples the magnetic and electric fields. It is necessary in these regions to solve for a vector field.

Vector elements are used for conductors, therefore, and scalar elements everywhere else. On the interface between, the tangential magnetic field must be made continuous. If the source field $\boldsymbol{H}_s$ is extended to the conductors, the continuity takes the form:

$$(-\nabla\phi)_{tang} = \boldsymbol{H}_{tang} \tag{3}$$

where $\boldsymbol{H}$ here is the part of the magnetic field in the conductors *not* represented by the source field.

For the scalar element m=1 and the vector element m=0.5, this is done as follows [6]: for each edge ij on the interface between the scalar and vector regions, the coefficient ϕ_i of ζ_i and the coefficient ϕ_j of ζ_j are related to the coefficient h_{ij} of $(\zeta_i\nabla\zeta_j - \zeta_j\nabla\zeta_i)$ by

$$\phi_j - \phi_i = h_{ij} \tag{4}$$

This equation is used to eliminate h_{ij} in favour of the two variables ϕ_i and ϕ_j.

For higher-order basis functions, the procedure is simpler. For each higher-order scalar basis function, there is a corresponding vector basis function which is its gradient (see Table I). Therefore, to impose equation (3) it is only necessary to set the coefficient of the scalar basis function equal to the negative of the corresponding vector basis function. On a face shared by scalar and vector elements, the two vector face functions which do not have scalar equivalents are eliminated by setting their coefficients to zero.

VI. Results

Inset in Fig. 2 is the cross-section of a three-dimensional, 50 Hz, quasi-static problem, the well-known Bath cube, details of which may be found elsewhere [7]. The solution method used involves both scalar and vector fields [6]. To simplify the discussion of results, the terms "order 1", "order 2" and "order 3" are used to denote the following scalar and vector element types (see Table I):

Order 1: scalar m=1, vector m=0.5
Order 2: scalar m=2, vector m=1
Order 3: scalar m=3, vector m=2

The problem was first solved with all the elements set to the same order. The results are shown in Fig. 2. Because a coarse mesh of 1824 tetrahedra was used, the order 1 results are poor. However, orders 2 and 3 are increasingly close to the measured results [7], as expected. The number of degrees of freedom increases rapidly with order, as shown in Table II.

Next the problem was solved with most of the elements at order 1, and with just a few layers of elements at order 2. In Fig. 3, those elements in regions marked A were set to order 2, while those elements in regions marked B were set to order 1. The accuracy obtained is, as expected, intermediate between that of full order 1 and full order 2 - though it is substantially more accurate than the full order 1 solution. The number of degrees of freedom is somewhat smaller than that of full order 2. It was computed in under half the time taken by the full order 2

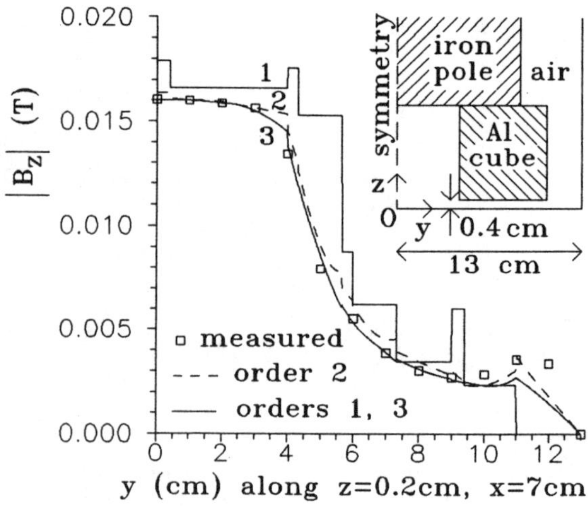

Fig. 2. Results for the Bath Cube - orders 1, 2 and 3.

TABLE II
NUMBER OF DEGREES OF FREEDOM FOR THE BATH CUBE

	Degrees of Freedom
Full order 1	684
Full order 2	3233
Full order 3	10748
Orders 1 and 2	1878

solution.

A second example is the TEAM Workshop Problem number 7, "Asymmetrical Conductor with Hole" [7], Fig. 4. Again, this is a 50 Hz, quasi-static problem. It consists of a current-carrying coil that induces currents in an aluminum plate with a hole in it. The quantity plotted in Fig. 4 is the magnitude of B_z, multiplied by -1 when the real part of B_z is negative. The measured results are from [7]. A mesh of 25,952 tetrahedra was constructed. At order 1, there were 6,056 degrees of freedom. When the elements in the plate and in the region above the plate upwards into the coil were made second order, the number of degrees of freedom increased to 18,737, and the results improved considerably.

VII. Conclusions

The scalar and vector elements described in this paper may be applied to a wide range of electromagnetic problems. They can be used at one fixed order, or different orders can be mixed in the same mesh, to allow for greater control over the distribution of degrees of freedom. So far this control has taken the form of manually setting the orders in various problem regions; in the future the orders may be set automatically, by an adaptive scheme.

References

[1] J. C. Nedelec, "Mixed finite elements in R^3", *Numer. Math.*, vol. 35, pp.315-341, 1980.

[2] A. Bossavit, "A rationale for 'edge-elements' in 3-D fields computations", *IEEE Trans. Magn.*, vol. 24, no. 1, pp.74-79, January 1988.

[3] J. F. Lee, "Analysis of passive microwave devices by using three-dimensional tangential vector finite elements", *International Journal of Numerical Modelling: Electronic Networks, Devices and Fields*, vol. 3, no. 4, pp.235-246, December 1990.

[4] O. C. Zienkiewicz, R. L. Taylor, *The Finite Element Method*, fourth edition, vol. 1, McGraw-Hill, London 1988, pp.139-149.

[5] J. P. Webb, B. Forghani, "A single scalar potential method for 3D magnetostatics using edge elements", *IEEE Trans. Magn.*, vol. 25, no. 5, pp.4126-4128, September 1989.

[6] J. P. Webb, B. Forghani, "A scalar-vector method for 3D eddy current problems using edge elements", *IEEE Trans. Magn.*, vol. 26, no. 5, pp.2367-2369, September 1990.

[7] L. Turner (editor), *Proceedings of the Vancouver TEAM Workshop*, University of B.C., 18-19 July 1988.

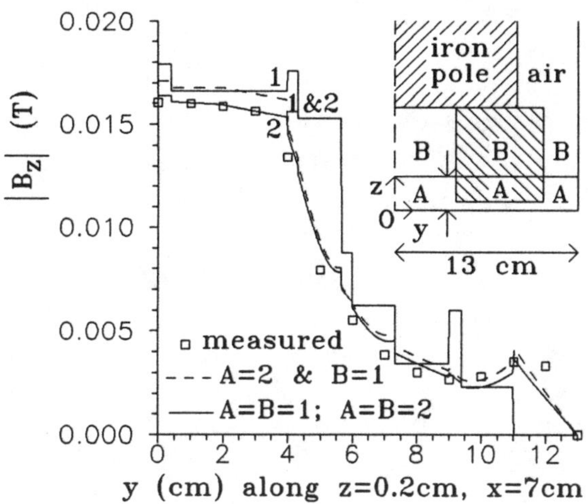

Fig. 3. Results for the Bath Cube - orders 1 & 2 together.

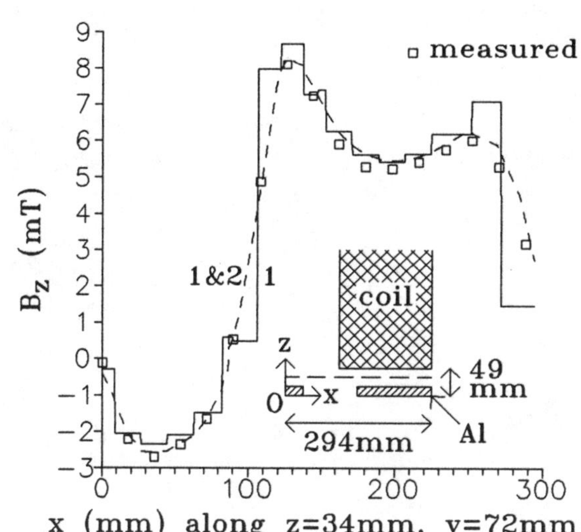

Fig. 4. TEAM Problem 7.

Rectilinearly Propagating Waves

Introduction

The finite element method was first introduced into electrical engineering practice in connection with waveguide and resonator problems [Silvester 1969*a* — included with the previous chapter]. It has remained an important tool for solving problems of guided wave propagation, while scoring great successes in other areas spanning the breadth of electrical engineering, as reviewed by Konrad (1985) and more recently but narrowly by Tsukerman, Konrad, Bedrosian, and Chari (1992). The term "guided waves" as used in this chapter should really be understood as referring to rectilinear propagation: waves propagating in a single privileged direction, the energy carried by the propagating wave in unvarying packets. With rare exceptions (e.g., the physically unrealizable infinitely-extending plane wave), such waves are guided by some physical structure. In other words, the rectilinearly propagating wave can be described by

$$\begin{bmatrix} \mathcal{E}(x,y,z,\omega) \\ \mathcal{H}(x,y,z,\omega) \end{bmatrix} = \begin{bmatrix} \mathbf{E}(x,y) \\ \mathbf{H}(x,y) \end{bmatrix} \exp j(\omega t \pm \beta z) \qquad (1)$$

in the frequency domain, or by an analogous expression in the time domain; z denotes the direction of wave propagation. The frequency and time domain representations of a wave are in principle interchangeable, but the interchange is not always easy, for the time-domain description is well suited to nondispersive waves, the frequency-domain representation to waves with dispersion. Either way, the great majority of interesting problems solved to date have been characterized by the separation of variables implied by these two representations.

Rectilinearly propagating waves present several challenges to the finite element analyst. The major difficulties that arise are subsequently echoed in three-dimensional propagation problems, where waves propagate radially outward from a central point or region. The initial analytic challenge, that of correctly modeling a two-dimensional field, was successfully dealt with in the early days (Silvester 1969*b*). The second, often called the "spurious mode problem" in the literature, arose very quickly thereafter (Csendes and Silvester 1970). It concerns the fact that not all solutions of the two vector Helmholtz equations (in **E** and **H**) also satisfy the Maxwell equations. Consequently, solutions to the Helmholtz equations that cannot physically exist are often found. These most often occur in mixed or anisotropic media. A third difficulty concerns infinite regions: **E** and **H** in many rectilinearly propagating wave problems are not bounded by any physical barrier, so the transverse ($x - y$ plane) region they occupy is infinite, at least in principle. Finally, the infinite region problem may recur in the direction of propagation, for it is often necessary to describe the absorption of a wave propagating indefinitely, quite likely without reflection.

Waves in Closed Guides

Rectilinearly propagating waves can generally be formulated in terms of two scalar quantities. The choice of formulations is extensive; the two quantities may be magnitudes of Hertzian potential vectors, classical potential functions, or simply two of the six field components, typically but not necessarily E_z and H_z. Corresponding to the various field representations, numerous distinctive finite element formulations of the traveling wave problem are possible.

Longitudinal Fields

The simplest description of guided wave propagation applies to TE_z or TM_z mode fields in homogeneously filled waveguides, where the propagation problem reduces to solving the scalar Helmholtz equation

$$(\nabla^2 + k^2)\Pi = 0, \qquad (2)$$

where $\Pi = E_z$ or $\Pi = H_z$ (for TM_z and TE_z modes, respectively), subject to boundary conditions appropriate to the guide at hand. The problem is stated in variational form as requiring that the functional

$$\mathcal{F}(\Pi) = \int_\Omega (\nabla\Pi \cdot \nabla\Pi - (\omega^2\mu\epsilon - \beta^2)\Pi^2) d\Omega \qquad (3)$$

assume a stationary value. This formulation is conventionally used to determine the modal spectra of lossless hollow guides. However, it can provide good approximations to other problems as well, and has served variously for the analysis of somewhat lossy guides (Hayata, Koshiba, and Suzuki 1986) and mildly anisotropic guides (Chiang 1985; Mustacich 1988). Reasonably good results have been obtained in these, even though the formulation is not strictly valid.

Two-Component Representations

The single scalar formulation does not correctly describe the modal structure, consisting of hybrid modes with neither **E** nor **H** purely transverse, of a guide containing inhomogeneous or anisotropic material, as shown in Fig. 1. A description in terms of two components is then necessary, and at times it is convenient to include even more. In terms of the axial (z–directed) field components, the functional to be rendered stationary is now (Csendes and Silvester 1970; Daly 1971)

$$\begin{aligned}\mathcal{F}(E_z, H_z) = &\int_{\Omega} \left(\frac{\nabla E_z \cdot \nabla E_z}{\omega^2 \mu\epsilon - \beta^2 - E_z^2} \right) d\Omega \\ &+ \int_{\Omega} \left(\frac{\nabla H_z \cdot \nabla H_z}{\omega^2 \mu\epsilon - \beta^2} - H_z^2 \right) d\Omega \\ &+ \int_{\Gamma} \left(\frac{\epsilon_1 E_1}{\omega^2 \mu_1 \epsilon_1 - \beta^2} \frac{\partial E_1}{\partial n} - \frac{\epsilon_2 E_2}{\omega^2 \mu_2 \epsilon_2 - \beta^2} \frac{\partial E_2}{\partial n} \right) d\Gamma \\ &+ \int_{\Gamma} \left(\frac{\mu_1 H_1}{\omega^2 \mu_1 \epsilon_1 - \beta^2} \frac{\partial H_1}{\partial n} - \frac{\mu_2 H_2}{\omega^2 \mu_2 \epsilon_2 - \beta^2} \frac{\partial H_2}{\partial n} \right) d\Gamma\end{aligned} \tag{4}$$

where the last two integrals are contour integrals taken over the boundary Γ that separates the two distinct materials. The key point here is that the boundary integrals make **E** dependent on the permeability μ of the materials and **H** dependent on their permittivities; it is not possible to solve for the **E** and **H** fields independently. While this coupling (and the hybrid modes it implies) may seem obvious, it was the point missed by Arlett, Bahrani, and Zienkiewicz (1968) in their early but unsuccessful attempt to employ finite elements in wave propagation problems. This formulation has been employed for inhomogeneous guiding structures loaded with linear but otherwise complicated materials (Tzuang and Itoh 1986) [Gibson and Helszajn 1989]. It has been applied to the problem of propagation in optical fibers (Yeh, Dong, and Brown 1979; Oyamada and Okoshi 1982). However, Rahman, Fernandez, and Davies [1991] have pointed out that in optical fiber analysis the accuracy of this formulation may be questionable, since the longitudinal field components are the feeblest and the least interesting in most optical propagation problems. The early work in this area was thoroughly reviewed by Rahman and Davies [1984*a*]. The formulation in terms of two longitudinal components has also proved useful in analyzing guides with some anisotropic materials (Vandenbulcke and Lagasse 1976). With only the slight generalization of allowing complex propagation constants, it has also been applied to lossy guides, with good results provided the loss is not very great (McAulay 1977; Aubourg, Villotte, Godon, and Garault 1983).

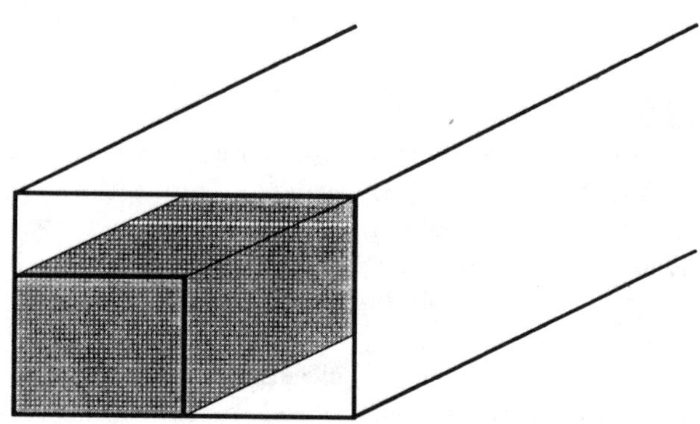

Fig. 1. Inhomogeneously filled (loaded) waveguide.

Spurious Modes

Throughout the 1970s and most of the 1980s, the E_z-H_z formulation was implemented in finite element methods by choosing approximating functions $\alpha_i(x,y)$ on each element and combining them linearly,

$$E_z(x,y) \simeq \sum_i E_{zi}\alpha_i(x,y), \tag{5}$$

and similarly for H_z. Difficulties with so-called *spurious modes*, solutions with no physical existence, were encountered throughout this period. Fields whose z–directed components extremize (or at least render stationary) the functional $\mathcal{F}(E_z, H_z)$ satisfy the Maxwell curl equations, but not necessarily the divergence equations. Consequently, the number of solutions obtained by extremizing $\mathcal{F}(E_z, H_z)$ exceeds the number of physically possible solutions, with the superfluous solutions violating one or both Maxwell divergence equations. They can be identified by this characteristic, but the identification is neither unequivocal (how close to zero is close enough?) nor convenient to implement in computer programs.

Spurious solutions in two-dimensional problems have a mathematical form that differs a little from the general three-dimensional case discussed in the preceding chapter. The techniques used to avoid them are partly comparable but not the same. The direct and fundamentally sound approach is to use approximating functions guaranteed to be globally solenoidal (Kobelansky and Webb, 1986). This is unfortunately a costly approach, so three others have gained popularity: penalty functions, field reformulations, and vector interpolation. These will be briefly reviewed in the following. A more extensive review of these three and various other methods, somewhat dated now but still excellent in its systematic coverage, was published by Koshiba, Hayata, and Suzuki [1987].

Penalty Functions

One straightforward approach to the spurious mode problem is to constrain the minimization of $\mathcal{F}(E_z, H_z)$ by adding *penalty terms* which shift the minimum value of the functional away from the nonsolenoidal and therefore wrong solutions. This converts the already familiar functional $\mathcal{F}$ into a related functional $\mathcal{F}_p$,

$$\mathcal{F}_p(E_z, H_z) = \mathcal{F}(E_z, H_z) + \int_\Omega \lambda_B^2 |\nabla \cdot \mathbf{B}|^2 d\Omega + \int_\Omega \lambda_D^2 |\nabla \cdot \mathbf{D}|^2 d\Omega. \tag{6}$$

Here λ_B^2 and λ_D^2 are arbitrarily chosen weighting constants. In problems with only a single permeability value but several dielectrics, it suffices to set $\lambda_D^2 = 0$ so only the appearance of isolated magnetic poles is penalized; conversely, where all permittivities are equal but all permeabilities are not, $\lambda_B^2 = 0$ is permissible. Adding one or both penalty terms forces the minimization process to avoid candidate solutions with substantial divergence values, as discussed by Rahman and Davies (1984*b*) or Koshiba, Hayata, and Suzuki (1985). This and related methods are more broadly reviewed by Young (1988). Inclusion of the penalty term actually does not eliminate the spurious modes; it merely forces their apparent cutoff frequencies to much higher values. The cutoff frequency spectrum of the true modes, however, remains unaffected. With some skill, the spectral range of the spurious modes can be shifted so as to be well separated from the true modes.

Solenoidality of Flux Densities

To obtain a satisfactory general solution of the spurious mode problem, the magnetic and electric flux density vectors **B** and **D** may be constrained to be truly solenoidal. Two elegant and fully satisfactory methods for doing so have been proposed: one relies on reformulation of the functional to be minimized, the other on a clever choice of finite elements. Two principal reformulations have been documented extensively. One uses a full three-component field **E** or **H**, depending on the analyst's preference and the nature of the problem); it appears to work satisfactorily, but at a high computing cost. The more elegant approach, mathematically sound and computationally stable, retains only the minimally necessary two variables, but chooses for this purpose the transverse components of **E** or **H**, rather than the longitudinal field components E_z and H_z used traditionally.

By using the transverse field components, the functional $\mathcal{F}(E_z, H_z)$ is recast in another closely related functional form, say $\mathcal{F}'(H_x, H_y)$. This recasting permits the solenoidality requirement to be introduced into the functional itself, so that H_x and H_y can only take on values that lead to a solenoidal field **H**. (The analogous formulation in **E** is equally valid, but less convenient where the permittivity varies spatially while the permeability remains constant.) This method was described by Hayata, Koshiba, Eguchi, and Suzuki (1986) and subsequently elaborated upon by Hayata, Eguchi, and Koshiba [1989]. Chew and Nasir [1989] showed that the **E** and **H** formulations are equivalent, lead to equivalent variational principles, and may be chosen on a basis of convenience in any given problem. Fernandez and Lu (1991) used a similar formulation, restricting it to the **H** vector. While this technique is strictly applicable only to rectilinearly propagating waves and does not easily generalize to three dimensions, it works superbly well for rectilinearly propagating wave problems. It has the very useful property of allowing the functional to be formulated in terms of the propagation constant β rather than the frequency ω, thus allowing dispersion curves to be computed in the "natural" way, by explicitly finding the propagation constant that corresponds to a given frequency. The formulation of Angkaew, Matsuhara, and Kumagai [1987] shares this desirable property, in addition to separating the true modes from the spurious.

Vector Interpolation

The requirement for solenoidality of flux density vectors can be approached in two fundamentally distinct ways. One is to incorporate this vector property into the functional to be extremized, as in the approaches described above. The second is to ensure that the extremization only employs solenoidal vector functions. The latter approach is useful with the two- component transverse-field formulations of the preceding section. It is perhaps most naturally implemented by approximating the fields themselves, rather than individual components. Taking only solenoidal vector approximating functions $\boldsymbol{\alpha}_i$ from a set $\{\boldsymbol{\alpha}_i \mid \nabla \cdot \boldsymbol{\alpha}_i = 0,\ i = 1, \ldots, N\}$, strictly solenoidal approximations result:

$$\mathbf{E}(x, y) \simeq \sum_i E_i \boldsymbol{\alpha}_i(x, y). \tag{7}$$

Such approximations require construction of edge-interpolative or face-interpolative vector functions, as discussed in the preceding and succeeding chapters in this book. Their use provides what may be the best approach to the field approximation problem. In two dimensions, often appropriate to rectilinearly propagating guided waves, the vector interpolation functions most often used are those already discussed in the preceding (and following) chapters, even though they guarantee solenoidality within each element but not at element interfaces — not unless continuity of normal components is strictly enforced as a side constraint. A very recent addition to the available techniques is the method of Sachdev and Cendes (1993), who propose two-dimensional elements that possess tangential field continuity and also provide normal flux continuity on an averaged basis. Such elements and (one hopes) their eventual three-dimensional counterparts may prove very useful in scattering problems also, even though at present there are no suitable methods for matching them to exterior regions with a similar degree of continuity.

Because appropriate vector functions have only become known in the past decade, much of the literature of guided-wave solutions of this kind is quite recent. This class of techniques allows ready generalization to fully three-dimensional problems and is therefore dealt with in detail in the following chapter.

Open-Sided Guides

Many rectilinear wave propagation problems are defined on geometric domains that extend infinitely far both in the direction of propagation z and laterally in the $x-y$ plane. Typical examples are the classical Lecher wire pair or an optic fiber. Pure TEM_z waves can propagate along the open wires, while hybrid modes are supported by the mixed-dielectric structures. In either case, the guided fields stick close to the guiding structure; they are evanescent in the lateral directions where the fields decay roughly exponentially. In open guides with two metallic conductors, a TEM_z-like propagation mode is possible down to and including zero frequency. With one conductor or none at all, propagation can take place only above a certain cutoff frequency, near which the fields surrounding the guide decay slowly with distance. The space region containing fields of significant value can then be very large.

Techniques Available

The simplest approach to open guides is to choose some arbitrary lateral distance and to declare the fields to vanish there. This approach was employed by various workers even after better techniques had become available (Yeh, Dong, and Oliver 1975; Mabaya, Lagasse, and Vandenbulcke 1981). There is no firm guideline for choosing the truncation distance, especially in programs where human estimation and manual data entry are to be avoided, so it is unsatisfactory outside the research laboratory. Better methods exist, and the major ones will be briefly reviewed in the following.

The major techniques for representing infinite regions in finite element analysis have been reviewed by Emson [1988]. They can be loosely grouped in five families: (1) *ad hoc* methods, (2) infinite elements, (3) recursive condensation or "ballooning," (4) transformation methods, (5) absorbing boundary conditions. In all methods, the infinite region is divided into two parts: an *interior* region, that is, the finite part of space where the field is computed in detail, using classical finite elements; and an *exterior* or infinite region in which the detailed solution is not wanted. This situation is sketched in Fig. 2, where Γ represents the boundary between interior and exterior regions.

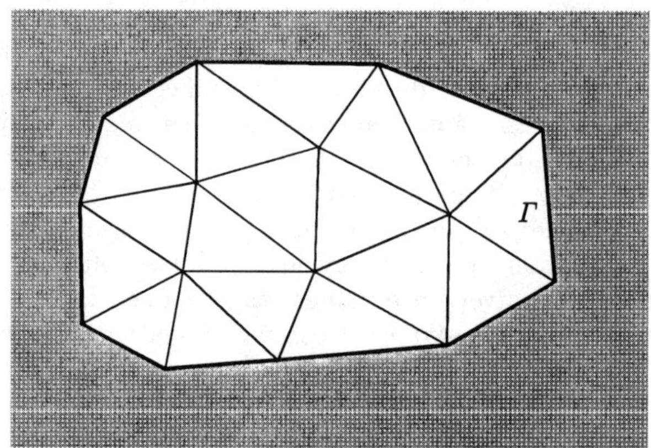

Fig. 2. Interior finite element region surrounded by hypothetical boundary Γ.

The nomenclature of this area is somewhat uncertain, with various authors adopting terms not only in conflict with each other but also contrary to the conventions of neighboring fields of engineering. Historical precedence has been used as the main guide here, with popularity a second important criterion. Perhaps this terminological confusion underscores the need for an anthology of the existing literature!

Assumed Decay Methods

The *ad hoc* techniques include the simple truncation already discussed — taking Γ far enough away from all regions of interest to permit setting all fields to zero on Γ — and assumed decay rate methods, in which it is assumed that fields exist outside Γ but decay according to a predetermined law such as an exponential function. The assumed decay law is imposed in the radial outward direction from the boundary Γ, thereby creating approximating functions for the field in the exterior so that an exterior finite element (finite in energy though not in geometric extent) can be constructed. Such elements have the advantage of being computationally very cheap. Despite their rough-and-ready nature, the assumed decay law methods have been found useful by various workers in the field (Pantic and Mittra 1986; McDougall and Webb 1989).

Infinite Elements

Infinite elements can be incorporated in the usual variational finite element method by representing fields in the exterior region in terms of an integral expression. This technique is perhaps easiest explained by an electrostatic example. Suppose the exterior field is represented by a scalar potential φ governed by Laplace's equation

$$\nabla^2\varphi = 0. \tag{8}$$

The value of φ at some point P on the boundary Γ is given by

$$\varphi(P) = \oint_{\Gamma} \frac{\partial \mathfrak{G}(P;Q)}{\partial n_Q}\varphi(Q)dQ. \tag{9}$$

Here $\mathfrak{G}(P;Q)$ is the appropriate Green's function (for unbounded space, the fundamental solution of the differential equation), and $\partial/\partial n_Q$ is the derivative taken normal to the boundary at the point Q. This point ranges over the entire boundary Γ to perform the integration. Suppose now the potential φ is to be approximated by some functions β_i in the exterior region. The Dirichlet matrix for the exterior region may be written in the usual form

$$S_{ij} = \int_{\Omega} \nabla\beta_i \cdot \nabla\beta_j d\Omega \tag{10}$$

where the region of integration is the space exterior to Γ. Equation (10) may be rewritten, using Green's second identity, as

$$S_{ij} = -\int_{\Omega} \beta_i \nabla^2 \beta_j d\Omega + \oint_{\Gamma} \beta_i \frac{\partial \beta_j}{\partial n_P} dP \tag{11}$$

where the integration point P ranges over the boundary Γ. To simplify matters, suppose now that the approximating functions β_i are chosen to have two properties: they satisfy Laplace's equation in the exterior region, and their values on the boundary Γ are interpolation polynomials. The first integral in equation (11) vanishes because $\nabla^2\beta_j = 0$. By virtue of the same property, β_j may be expressed in the form (9). Substituting into equation (11),

$$S_{ij} = \oint_{\Gamma}\oint_{\Gamma} \frac{\partial^2 \mathfrak{G}(P;Q)}{\partial n_P \partial n_Q}\beta_j(Q)\beta_i(P)dPdQ. \tag{12}$$

This form of the Dirichlet matrix involves no integration over the infinite region; only integration over the boundary is required. That, however, implies that a knowledge of the functions β_i is unnecessary, other than on the boundary Γ. So long as the β_i are interpolation polynomials along Γ, they fit perfectly with classical polynomials in the interior. Despite its elegance, this method is unfortunately not usable as its stands, for the integrations implied by equation (12) involve strong singularities (second derivatives of a singular function!) so approximation schemes are conventionally used. The most common scheme solves directly for a numerical approximation to the exterior Green's function. It is ordinarily called the "hybrid method" and was initially presented by Silvester and Hsieh [1971, included in the next chapter]. It provides continuity of φ, but solutions can be obtained even if strict continuity is not observed. To what extent it can be relaxed, and what constraints must be preserved, was established in detail by Kagawa, Muratai, and Kitagami (1982). Extensions to vector problems have been contributed, among others, by Onuki (1990).

The method described above can be generalized directly to wave functions, with the Helmholtz operator in place of the Laplacian, and to vector variables that require a dyadic rather than a scalar Green's function. The work of Ramahi and Mittra (1991) or Hirayama and Koshiba (1990) may serve as a representative examples. Computational algorithms based on this approach have been used widely for the past two decades or so. They are usually referred to as *hybrid methods* because they may be viewed as involving simultaneous solution of a differential-equation boundary value problem in the interior and an integral equation in the exterior. They are of considerable interest for three-dimensional problems, and will therefore be discussed at much greater length in the following chapter.

Recursive Condensation

Recursive condensation methods in essence build a truly infinite mesh, or at any rate a large one, diameters of miles or light-years being almost equally easy to achieve. The large mesh is constructed by adding an annular strip (in three dimensions, a shell) of elements outside Γ so as to enlarge the region. The added elements are so chosen that the enlarged outer boundary is geometrically similar to Γ and centered on the same point as Γ, so that the enlargement can be repeated recursively. At each recursion step the matrix equations describing the finite element mesh are rewritten so as to eliminate all the newly added nodal variables in the mesh. Each newly added layer of elements is geometrically similar to the last, so the element matrices can simply be scaled without recomputing. The recursion can therefore be carried out quickly. Although the sparsity of the global matrix description is reduced somewhat by this process, the total size of the matrix equation is the same as for the interior region alone, so the field computation is relatively cheap. The method was developed by Silvester, Lowther, Carpenter, and Wyatt (1977) for static fields and further refined by Dasgupta [1984] as well as by Lowther, Rajanathan, and Silvester (1978) in a series of papers. It was later applied to wave problems by Chiang (1985). This method was subsequently refined to its limiting case of an infinite sequence of infinitesimal steps, thereby creating a new method called "infinitesimal scaling" [Hurwitz 1984]. Infinitesimal scaling was then successfully applied to radially propagating wave problems by Hurwitz (1988).

In its conventional finite form, the recursive condensation method moves the exterior boundary far away but does not eliminate it. It can be applied to radially evanescent problems (such as the rectilinear guided wave problem) easily; but for radially propagating waves, an approximate radiation condition or other absorbing condition must be placed at the distant boundary. Furthermore, the rate of element growth must be controlled so as to avoid instabilities that amount to internal resonances within the successively added space regions. The theoretical requirements of doing so were worked out in detail by Goldstein (1982) in a sequence of papers, and successfully applied in several scalar wave programs. Unfortunately, Goldstein's papers do not make easy reading for those not well versed in the mathematical theory of finite elements.

Transformation Methods

Transformation methods have an ancient lineage, with the major techniques known to Maxwell and Lord Kelvin. They are based on the observation that the Laplacian operator ∇^2 is invariant under an *inversion*: if the coordinate system (r, θ) is transformed with respect to some constant R^2 by setting

$$\begin{bmatrix} r' \\ \theta' \end{bmatrix} = \begin{bmatrix} \frac{R^2}{r} \\ \theta \end{bmatrix}, \tag{13}$$

harmonic functions remain harmonic in the inverted coordinates. However, equation (13) transforms the exterior of the circle $r = R$ into its interior. This suggests that an infinite element could be constructed for Laplace's equation by taking Γ to be the circle $r = R$. The exterior infinite element matrix is then identical to the straightforward finite element matrix for the circular region, which can be constructed by wholly standard methods. More generally, conformal mappings can be used (Wu and Chen 1985; Wong and Ciric 1985).

The inversion transformation has been used, in both two and three dimensions, by various workers for more than a century. In recent years, the geometric inflexibility of the circular bounding contour has been alleviated by generalizations to other shapes by Imhoff, Meunier, Brunotte, and Sabonnadière (1990). Further work by Stochniol [1992] has indicated that the peculiarities of the resulting element matrices are similar to those arising from material anisotropies and inhomogeneities. The Stochniol transformations appear promising and will no doubt see much use in future years.

Geometric transformation methods are primarily applicable to static field problems and to the classical rectilinear propagation problem where fields decay laterally. They therefore show considerable potential for further development in this class of problem, less so for the radially propagated waves commonly associated with antennas and scatterers.

Absorbing Boundary Conditions

Absorbing boundary conditions are boundary conditions so contrived that normally incident traveling waves are absorbed at the boundary without reflection. They are of great value in radiation and scattering problems — in other words, in problems involving radially rather than rectilinearly propagating waves. In the latter area, their use to date has not been very extensive. Absorbing boundary conditions will therefore be considered in detail in the following chapter.

Models for Semi-Infinite Guides

The semi-infinite straight guide occurs as an idealization in many problems of guided wave propagation; indeed, the majority of guiding structures are conventionally described by stating how they would behave if the were of nominally infinite (actually *semi*-infinite) length. For example, the characterization of various waveguide discontinuities and junctions relies heavily on the notion of infinitely-extending guides attached to all ports. This problem in analysis has been surprisingly neglected, with a variety of ingenious but fundamentally unsatisfactory *ad hoc* schemes used for specific problems.

In principle, the semi-infinite guide represents nothing more than a one-dimensional version of the infinite-field problem already discussed. Thus it is amenable to much the same techniques, even if these are surprisingly ill-documented in the literature. A technique that bears perhaps the greatest promise is recursive condensation, termed "marching" by Foo and Silvester [1992] because the guide is initially terminated in an artificial boundary close by, and the boundary is then "marched" out a long distance, typically thousands of wavelengths. An alternative, perhaps better known and established, method is that of expanding the fields in the eigenmodes of the guide; the work of Baumeister (1987) furnishes an example of this approach.

The infinitely long guide problem would seem at first glance a perfect application of absorbing boundary conditions: it requires most of the strengths of the absorbing boundary condition technique and avoids most of its weaknesses. The use of absorbing boundary conditions has been reported in the literature (e.g., Wilkins, Lee, and Mittra 1991); but so far, surprisingly little use seems to have been made of this very promising technique.

References

Angkaew, T., M. Matsuhara, and N. Kumagai [1987]: "Finite-element analysis of waveguide modes: A novel approach that eliminates spurious modes." *IEEE Transactions on Microwave Theory and Techniques*, vol. **MTT-35**, pp. 117–23, 1987.

Arlett, P. L., A. K. Bahrani, and O. C. Zienkiewicz (1968): "Application of finite elements to the solution of Helmholtz's equation." *Proceedings of the Institution of Electrical Engineers*, vol. **115**, pp. 1762–66, 1968.

Aubourg, M., J.-P. Villotte, F. Godon, and Y. Garault (1983): "Finite element analysis of lossy waveguides —application to microstrip lines on semiconductor substrate." *IEEE Transactions on Microwave Theory and Techniques*, vol. **MTT-31**, pp. 326–31, 1983.

Baumeister, K. J. (1987): "Finite element modeling of electromagnetic propagation in composite structures." *1987 A-P International Symposium Digest*, vol. **1**, pp. 272–75, 1987.

Chew, W. C. and M. A. Nasir [1989]: "A variational analysis of anisotropic, inhomogeneous dielectric waveguides." *IEEE Transactions on Microwave Theory and Techniques*, vol. **MTT-37**, pp. 661–68, 1989.

Chiang, K. S. (1985): "Finite-element analysis of optical fibres with iterative treatment of the infinite 2-D space." *Optical and Quantum Electronics*, vol. **17**, pp. 381–91, 1985.

Csendes, Z. J. and P. Silvester (1970): "Numerical solution of dielectric loaded waveguides. I — Finite element analysis." *IEEE Transactions on Microwave Theory and Techniques*, vol. **MTT-18**, pp. 1124–31, 1970.

Daly, P. (1971): "Hybrid-mode analysis of microstrip by finite-element methods." *IEEE Transactions on Microwave Theory and Techniques*, vol. **MTT-19**, pp. 19–25, 1971.

Dasgupta, G. [1984]: "Computation of exterior potential fields by infinite substructuring." *Computer Methods in Applied Mechanics and Engineering*, vol. **46**, pp. 295–305, 1984.

Emson, C. R. I. [1988]: "Methods for the solution of open-boundary electromagnetic-field problems." *IEE Proceedings*, Part A, vol. **135**, pp. 151–58, 1988.

Fernandez, F. A. and Y. Lu (1991): "A variational finite element formulation for dielectric waveguides in terms of transverse magnetic fields." *IEEE Transactions on Magnetics*, vol. **27**, pp. 3864–67, 1991.

Foo, S. L. and P. P. Silvester [1992]: "Boundary-marching method for discontinuity analysis in waveguides of arbitrary cross section." *IEEE Transactions on Microwave Theory and Techniques*, vol. **40**, pp. 1889–93, 1992.

Gibson, A. A. P. and J. Helszajn [1989]: "Finite element solution of longitudinally magnetized elliptical gyromagnetic waveguides." *IEEE Transactions on Microwave Theory and Techniques*, vol. **37**, pt. 1, pp. 999–1005, 1989.

Goldstein, C. I. (1982): "A finite element method for solving Helmholtz type equations in waveguides and other unbounded domains." *Mathematics of Computation*, vol. **39**, pp. 309–24, 1982.

Hayata, K., M. Koshiba, M. Eguchi, and M. Suzuki (1986): "Vectorial finite-element method without any spurious solutions for dielectric waveguiding problems using transverse magnetic-field component." *IEEE Transactions on Microwave Theory and Techniques*, vol. **MTT-34**, pp. 1120–24, 1986.

Hayata, K., M. Koshiba, and M. Suzuki (1986): "Lateral mode analysis of buried heterostructure diode lasers by the finite-element method." *IEEE Journal of Quantum Electronics*, vol. **QE-22**, pp. 781–88, 1986.

Hayata, K., M. Eguchi, and M. Koshiba [1989]: "Finite element formulation for guided-wave problems using transverse electric field component." *IEEE Transactions on Microwave Theory and Techniques*, vol. **37**, pp. 256–58, 1989.

Hirayama, K. and M. Koshiba (1990): "Numerical analysis of arbitrarily shaped discontinuities between planar dielectric waveguides with different thicknesses." *IEEE Transactions on Microwave Theory and Techniques*, vol. **38**, pp. 260–64, 1990.

Hurwitz, H., Jr. [1984]: "Infinitesimal scaling — A new procedure for modeling exterior field problems." *IEEE Transactions on Magnetics*, vol. **MAG-20**, pp. 1918–23, 1984.

Hurwitz, H., Jr. (1988): "The surface projection approach to vector field calculations." *Mathematical and Computer Modelling*, vol. **11**, pp. 313–16, 1988.

Imhoff, J. F., G. Meunier, X. Brunotte, and J. C. Sabonnadière (1990): "An original solution for unbounded electromagnetic 2D- and 3D-problems throughout the finite element method." *IEEE Transactions on Magnetics*, vol. **26**, pp. 1659–61, 1990.

Kagawa, Y., T. Murai, and S. Kitagami (1982): "On the compatibility of finite element-boundary element coupling in field problems." *COMPEL*, vol. **1**, pp. 197–217, 1982.

Kobelansky, A. J. and J. P. Webb (1986): "Eliminating spurious modes in finite-element waveguide problems by using divergence-free fields." *Electronics Letters*, vol. **22**, pp. 569–70, May 1986.

Konrad, A. (1985): "Electromagnetic devices and the application of computational techniques in their design." *IEEE Transactions on Magnetics*, vol. **MAG-21**, pp. 2382–87, 1985.

Koshiba, M., K. Hayata, and M. Suzuki [1987]: "Finite-element method analysis of microwave and optical waveguides — Trends in countermeasures to spurious solutions." *Electronics and Communications in Japan*, Part 2, vol. **70**, no. 9, pp. 96–107, 1987. (Translated from *Densi Tsushin Gakkai Ronbunshi*, vol. **69-C**, pp. 1477–86, 1986.)

Lowther, D. A., C. B. Rajanathan, and P. P. Silvester (1978): "A finite element technique for solving 2-D open boundary problems." *IEEE Transactions on Magnetics*, vol. **MAG-14**, pp. 467–69, 1978.

Mabaya, N., P. E. Lagasse, and P. Vandenbulcke (1981): "Finite element analysis of optical waveguides." *IEEE Transactions on Microwave Theory and Techniques*, vol. **MTT-29**, pp. 600–605, June 1981.

McAulay, A. D. (1977): "Variational finite-element solution for dissipative waveguides and transportation application." *IEEE Transactions on Microwave Theory and Techniques*, vol. **MTT-25**, pp. 382–92, 1977.

McDougall, M. J. and J. P. Webb (1989): "Infinite elements for the analysis of open dielectric waveguides." *IEEE Transactions on Microwave Theory and Techniques*, vol. **MTT-37**, pp. 1724–31, 1989.

Mustacich, R. V. (1988): "Scalar finite element analysis of electro-optic modulation in diffused channel waveguides and poled waveguides in polymer thin films." *Applied Optics*, vol. **27**, pp. 3732–37, 1988.

Onuki, T. (1990): "Hybrid finite element and boundary element method applied to electromagnetic problems." *IEEE Transactions on Magnetics*, vol. **26**, pp. 582–87, 1990.

Oyamada, K. and T. Okoshi (1982): "Two-dimensional finite-element calculation of propagation characteristics of axially nonsymmetrical optical fibres." *Radio Science*, vol. **17**, pp. 109–16, 1982.

Pantic, Z. and R. Mittra (1986): "Quasi-TEM analysis of microwave transmission lines by the finite-element method." *IEEE Transactions on Microwave Theory and Techniques*, vol. **MTT-34**, pp. 1096–1103, 1986.

Rahman, B. M. A. and J. B. Davies [1984*a*] : "Finite-element analysis of optical and microwave waveguide problems." *IEEE Transactions on Microwave Theory and Techniques*, vol. **MTT-32**, pp. 20–28, 1984.

Rahman, B. M. A. and J. B. Davies (1984*b*) : "Penalty function improvement of waveguide solution by finite elements." *IEEE Transactions on Microwave Theory and Techniques*, vol. **MTT-32**, pp. 922–28, 1984.

Rahman, B. M. A., F. A. Fernandez, and J. B. Davies [1991]: "Review of finite element methods for microwave and optical waveguides." *Proceedings of the IEEE*, vol. **79**, pp. 1442–48, 1991.

Ramahi, O. M. and R. Mittra (1991): "A surface integral equation method for the finite element solution of waveguide discontinuity problems." *IEEE Transactions on Microwave Theory and Techniques,* vol. **39**, pp. 604–8, 1991.

Sachdev, S. and Z. J. Cendes (1993): "Combined tangential-normal vector elements for computing electric and magnetic fields." *IEEE Transactions on Magnetics*, vol. **29**, pp. 1456–59, 1993.

Silvester, P. (1969*a*): "Finite-element solution of homogeneous waveguide problems." *Alta Frequenza*, vol. **38**, pp. 313–17, 1969.

Silvester, P. (1969*b*): "A general high-order finite -element waveguide analysis program." *IEEE Transactions on Microwave Theory and Techniques*, vol. **MTT-17**, pp. 204–10, 1969.

Silvester, P. and M. S. Hsieh (1971): "Finite-element solution of 2-dimensional exterior-field problems." *Proceedings of the Institution of Electrical Engineers*, vol. **118**, pp. 1743–47, 1971.

Silvester, P. P., D. A. Lowther, C. J. Carpenter, and E. A. Wyatt (1977): "Exterior finite elements for 2-dimensional field problems with open boundaries." *Proceedings IEE*, vol. **124**, pp. 1267–70, 1977.

Stochniol, A. [1992]: "A general transformation for open boundary finite element method for electromagnetic problems." *IEEE Transactions on Magnetics*, vol. **28**, pp. 1679–81, 1992.

Tsukerman, I. A., A. Konrad, G. Bedrosian, and M. V. K. Chari (1993): "A survey of numerical methods for transient eddy current problems." *IEEE Transactions on Magnetics*, vol. **29**, pp. 1711–16, 1993.

Tzuang, C.-K. and T. Itoh (1986): "Finite-element analysis of slow-wave Schottky contact printed lines." *IEEE Transactions on Microwave Theory and Techniques*, vol. **MTT-34**, pp. 1483–89, 1986.

Vandenbulcke, P. and P. E. Lagasse (1976): "Eigenmode analysis of anisotropic optical fibres or integrated optical waveguides." *Electronics Letters*, vol. **12**, pp. 120–22, March 1976.

Yeh, C., S. B. Dong, and W. Oliver (1975): "Arbitrarily shaped inhomogeneous optical fiber or integrated optical waveguides." *Journal of Applied Physics*, vol. **46**, pp. 2125–29, 1975.

Yeh, C., K. Ha, S. B. Dong, and W. P. Brown (1979): "Single-mode optical waveguides." *Applied Optics*, vol. **18**, pp. 1490–1504, 1979.

Young, T. P. (1988): "Design of integrated optical circuits using finite elements." *IEE Proceedings, Pt. A*, vol. **135**, pp. 135–44, 1988.

Computation of Exterior Potential Fields by Infinite Substructuring

G. DASGUPTA
DEPARTMENT OF CIVIL ENGINEERING AND ENGINEERING MECHANICS
COLUMBIA UNIVERSITY
NEW YORK, NY 10027, U.S.A.

A numerical method is presented to evaluate approximate solutions of elliptic partial differential equations outside an arbitrary convex region. The exterior region is conceived as an infinite assembly of geometrically similar finite element cells. A complete set of spatially discretized shape functions is derived for the entire unbounded exterior within which Laplace's equation is required to be satisfied. These functions with the characteristic of outward decay are evaluated by solving a quadratic eigenproblem. The coefficient matrices therein are furnished by the finite element system matrix (which relates the resulting nodal fluxes to the corresponding field strengths) of a typical cell. Elements of that matrix are viewed in the light of the virtual work principle. The energy terms associated with any pair of those basic functions form a convergent geometrical series over a sequence of similarly shaped elements with increasing characteristic dimensions. The exact infinite sum of the virtual work quantities yields the elements of the boundary matrix for the unbounded region. This enables one to carry out the proposed infinite substructuring scheme over the entire infinite collection of those cells. Finally, an expression to estimate the exterior field at an arbitrary point is also presented. The present study has applications to elastostatic, hydrodynamic, electro- and magneto-static protentials in two and three dimensions.

1. Introduction

The finite element method [1] has been extensively employed to generate numerical solutions for field problems whenever complicated geometrical boundaries are encountered. The conventional formulations are, of course, used entirely to model areas and volumes which have closed boundaries. However, in a number of important applications, it is desirable to obtain field variables within the infinite and semi-infinite domains which surround the bounded regions (Fig. 1). This class of evaluations is identified as the exterior field problem. Some examples of such problems in engineering mechanics are the evaluation of elastostatic and hydrodynamic potentials; in electrical engineering representative computations are busbar impedance, microwave transmissions and flux leakage and electromagnetic prospecting for mineral deposits in geophysical engineering.

The study of potential fields constitutes a fundamental discipline in computational mechanics. The discrete analog of the system operator is called the Dirichlet matrix in the literature [2]. This relates the total fluxes ('forces') and field intensities ('responses') at the nodes of a spatial mesh. With the aid of digital computers, the numerical inversions and eigensolutions of the associated algebraic problems, *for finite regions*, have been reduced to a straightforward task. An unbounded region, on the other hand, when modeled in an analogous fashion, demands an infinite number of elements. This difficulty has been alleviated by the introduction of infinite elements [3] which are themselves unbounded geometrically. Therein the approximate interpolations have been constructed from the analytical solutions of infinite regions of various standard geometrical forms. The exact shape functions are indeed higher-order and transcendental in nature [4]. Consequently, their intrinsic singularities pose enormous computational difficulties during the evaluation of the spatial derivatives and subsequent quadrature operations. Simplified polynomial forms of shape functions, as suggested in [3], significantly reduce these complexities. In another approach [2], a sequential substructuring has been proposed in which the exterior boundary nodes are repeatedly condensed out in

Reprinted with permission from *Computer Methods in Applied Mechanics and Engineering*, vol. 46, pp. 295–305, 1984.

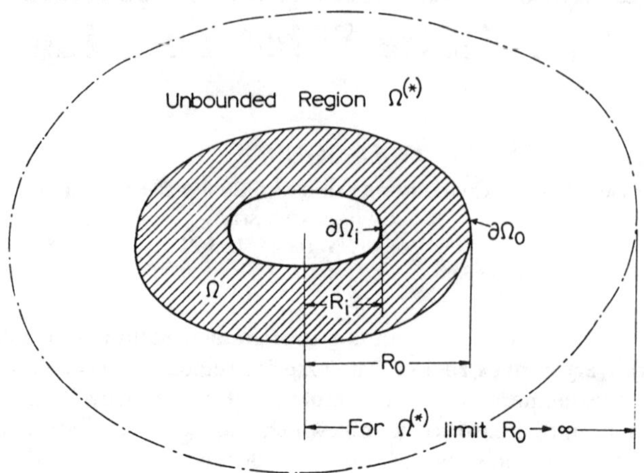

Fig. 1. Exterior field problem.

stages. This enables one to obtain the reduced Dirichlet matrix for a significantly large domain. This region could be accepted as a workable approximation for an unbounded one.

In this paper the proposed method furnishes an algorithm to evaluate the infinite element shape functions for elliptic partial differential equations. It subsequently yields a procedure to substructure over that infinite collection of finite element cells which completely cover the infinite region in question. The computation is also accelerated by accounting exactly for the energy contributions of infinite number of cells. The method does not incur numerical inaccuracies which could originate from the basic simplifications inherent in [2, 3].

2. Problem statement

The canonical problem to be addressed in this paper is

$$\nabla^2 u = 0 \quad \text{in } \mathbb{R}^k, \tag{1}$$

in which the field variable u satisfies the Laplacian ∇^2 in the k-dimensional Eucledian space $\mathbb{R}^k$. In the interest of engineering application, $k \geq 2$. However, the most commonly occurring cases are concerned with two- and three-dimensional spaces, i.e., $k = 2$ and 3.

Consider an annular region Ω (refer to Fig. 1). The interior and exterior boundaries are denoted by $\partial\Omega_i$ and $\partial\Omega_o$, respectively. Let R_i and R_o be some characteristic dimensions for $\partial\Omega_i$ and $\partial\Omega_o$, respectively. The conventional boundary value problem (for bounded regions where R_o is finite) is to evaluate the field variable u of (1) in Ω. An appropriate (Dirichlet, Neumann or Robin type) boundary condition then needs to be prescribed on $\partial\Omega_i$ and $\partial\Omega_o$. The infinite region $\Omega^{(*)}$ with an interior boundary $\partial\Omega_i$ can now be conceived as a limiting case when the outer dimension R_o tends to be indefinitely large. It is important to note that unlike the problem of finite regions, a qualitative statement of some regularity condition is to be imposed 'on the infinite boundary'. It is then necessary to state that u should be bounded on the entire infinite region $\Omega^{(*)}$, instead of a quantitative description involving u and its normal derivative $\partial u/\partial n$ as in the case of finite regions. For some simple geometry (circular, elliptic, etc.) of the interior boundary $\partial\Omega_i$, analytical solutions in the form of infinite series of general harmonics are possible. Practical engineering applications, however, demand $\partial\Omega_i$ to be of any arbitrary geometry. The analytical solutions are then intractable. The resolution of this difficulty suggests the implementation of an appropriate spatial discretization. The finite element method has proved to be a convenient choice.

The finite element analog of the continuum field equation $\nabla^2 u = 0$ in (1) along with an acceptable boundary condition is

$$S\phi = \psi . \tag{2}$$

Therein, ϕ and ψ are the vectors representing the field intensity and the total nodal flux, respectively. The convergence criterion for this approximate scheme is based on the energy norm E,

$$E(u) \equiv \int_{\Omega} (\nabla u)^2 \, \mathrm{d}\Omega , \quad \Omega \text{ volume in } \mathbb{R}^k ,$$

where u is an approximate solution of (1). Refer to [5, p. 349] for details. An analogous definition of the virtual work quantity W for a pair of such test functions u and v will then be

$$W(u, v) \equiv \int_{\Omega} (\nabla u) \cdot (\nabla v) \, \mathrm{d}\Omega . \tag{3}$$

For finite element applications it is intended that the distributions u and v satisfy the lower-order (i.e., the field intensity) boundary condition. In the case of an unbounded region Ω^* (refer to Fig. 1) the Dirichlet matrix is S^*. An algebraic relationship is required in the form

$$S^* \phi_{\mathrm{i}} = \psi_{\mathrm{i}} . \tag{4}$$

Therein, ϕ_{i} is the vector which consists of the nodal point field intensities, and the vector ψ_{i} contains the corresponding total flux quantities integrated over the spatially discretized boundary pieces on $\partial\Omega_{\mathrm{i}}$.

For conventional problems of bounded regions, such as Ω in Fig. 1, the finite element Dirichlet Matrix S can be evaluated following the standard procedure described in [1] even for the kth dimensional space. A number of computer programs is currently available to readily evaluate S especially for two- and three-dimensional finite bodies. For the purpose of this presentation, it will be assumed that the Dirichlet matrix S is available for any finite domain Ω with arbitrary boundaries $\partial\Omega_{\mathrm{i}}$ and $\partial\Omega_{\mathrm{o}}$. In the following sections we shall discuss two methods that yield S^* for a geometrically unbounded region from S, which pertains to a convenient bounded element, as the outward characteristic dimension R_{o} tends to infinity.

3. Repeated substructuring

This method has been described in [2] specifically for two-dimensional geometrical regions, i.e., $k = 2$ in (1). The unbounded region $\Omega^{(*)}$ in Fig. 2 is covered with an infinite assembly of annular regions $\Omega^{(i)}$, $i = 1, 2, \ldots$, to ∞. All these cells are selected to be geometrically similar. It is therefore imperative that basically convex shapes (along with a limited number of *concave* regions) are permissible for $\partial\Omega_i$. Throughout this paper this restriction is enforced. The advantage in such a choice is that the numerical values of the Dirichlet matrices for each cell $\Omega^{(i)}$ are identical in this particular two-dimensional case. The Dirichlet matrix for a given shape of $\partial\Omega_i$ depends only on the enlargement factor γ defined by

$$\gamma = \frac{R_2}{R_1} = \frac{R_3}{R_2} = \cdots = \frac{R_{i+1}}{R_i} , \quad i = 1, 2, 3, \ldots . \tag{5}$$

The flux-potential transformation matrix $S^{(i)}$ of a typical cell $\Omega^{(i)}$ in Fig. 2 can be represented in the following partitioned form:

$$\begin{bmatrix} \boldsymbol{S}^{(i)}_{i,i} & \boldsymbol{S}^{(i)}_{i,i+1} \\ \boldsymbol{S}^{(i)}_{i+1,i} & \boldsymbol{S}^{(i)}_{i+1,i+1} \end{bmatrix} \begin{Bmatrix} \boldsymbol{\phi}_i \\ \boldsymbol{\phi}_{i+1} \end{Bmatrix} = \begin{Bmatrix} \boldsymbol{\psi}^{(i)}_i \\ \boldsymbol{\psi}^{(i)}_{i+1} \end{Bmatrix}. \tag{6}$$

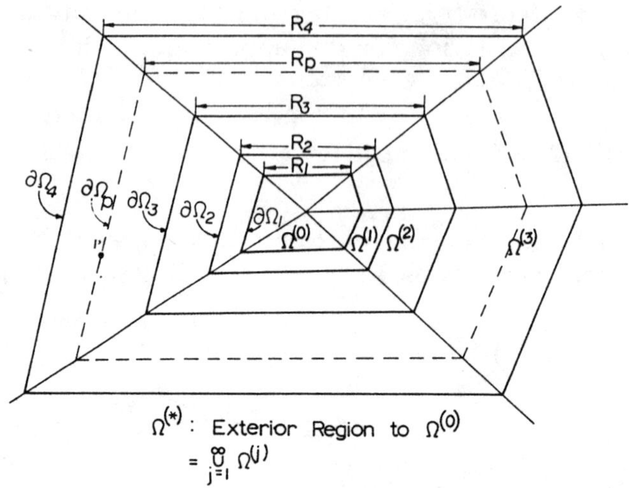

Fig. 2. Infinite assembly of elements.

The variables are associated with, say, n nodes on $\partial\Omega_i$ and an equal number of nodes on $\partial\Omega_{i+1}$. The numerical values of these $(n \times n)$ square submatrices are expected to be known in this derivation. Note, the convention of indicating the element numbers as superscripts and the boundary contour designations as subscripts is maintained throughout this paper.

Consider two adjacent cells $\Omega^{(1)}$ and $\Omega^{(2)}$. Their assembly forms a region $\mathcal{B}^{(1)}$,

$$\mathcal{B}^{(1)} = \Omega^{(1)} \cup \Omega^{(2)}; \tag{7}$$

this leads to

$$\begin{bmatrix} \boldsymbol{S}^{(1)}_{1,1} & \boldsymbol{S}^{(1)}_{1,2} & \boldsymbol{O} \\ \boldsymbol{S}^{(1)}_{2,1} & \boldsymbol{S}^{(1)}_{2,2} + \boldsymbol{S}^{(2)}_{2,2} & \boldsymbol{S}^{(2)}_{2,3} \\ \boldsymbol{O} & \boldsymbol{S}^{(2)}_{3,2} & \boldsymbol{S}^{(2)}_{3,3} \end{bmatrix} \begin{Bmatrix} \boldsymbol{\phi}_1 \\ \boldsymbol{\phi}_2 \\ \boldsymbol{\phi}_3 \end{Bmatrix} = \begin{Bmatrix} \boldsymbol{\psi}_1 \\ \boldsymbol{\psi}_2 \\ \boldsymbol{\psi}_3 \end{Bmatrix}. \tag{8}$$

An externally applied flux along the interface $\partial\Omega_2$ is zero, hence $\boldsymbol{\psi}_2$ is zero. Thus a relationship between the boundary quantities of this 'super element' $\mathcal{B}^{(1)}$ is obtained in the form

$$\begin{bmatrix} \boldsymbol{S}^{(1)}_{1,1} + \boldsymbol{S}_{1,1} & \boldsymbol{S}_{1,2} \\ \boldsymbol{S}^{t}_{1,2} & \boldsymbol{S}^{(2)}_{3,3} + \boldsymbol{S}_{3,3} \end{bmatrix} \begin{Bmatrix} \boldsymbol{\phi}_1 \\ \boldsymbol{\phi}_3 \end{Bmatrix} = \begin{Bmatrix} \boldsymbol{\psi}_1 \\ \boldsymbol{\psi}_3 \end{Bmatrix}, \tag{9}$$

in which

$$\begin{aligned} \boldsymbol{S}_{1,1} &= -\boldsymbol{S}^{(1)}_{1,2}[\boldsymbol{S}^{(1)}_{2,2} + \boldsymbol{S}^{(1)}_{1,1}]^{-1}\boldsymbol{S}^{(1)}_{2,1}, \qquad \boldsymbol{S}_{1,2} = -\boldsymbol{S}^{(1)}_{1,2}[\boldsymbol{S}^{(1)}_{2,2} + \boldsymbol{S}^{(1)}_{2,1}]^{-1}\boldsymbol{S}^{(1)}_{1,2}, \\ \boldsymbol{S}_{3,3} &= -\boldsymbol{S}^{(1)}_{2,1}[\boldsymbol{S}^{(1)}_{2,2} + \boldsymbol{S}^{(1)}_{1,1}]^{-1}\boldsymbol{S}^{(1)}_{1,2}, \end{aligned} \tag{10}$$

since $\boldsymbol{S}^{(2)}_{2,2} = \boldsymbol{S}^{(1)}_{1,1}$ and $\boldsymbol{S}^{(2)}_{2,3} = \boldsymbol{S}^{(1)}_{1,2}$ (for this two-dimensional case). Thus a larger element $\mathcal{B}^{(1)}$ could be constructed with the enlargement factor

$$R_3/R_1 = \gamma^2. \tag{11}$$

Now, by considering $\mathcal{B}^{(1)}$ to be the basic cell such as $\Omega^{(1)}$ in (7), an even larger element $\mathcal{B}^{(2)}$, i.e., of an enlargement factor γ^4, can be constructed. The authors of [2] initiated their calculations with $\gamma = 1.5$ and in seven repeated substructuring operations obtained the outer characteristic dimension to be $3.46 \times 10^{22} R_1$. This approaches the floating point capability of standard digital computers. Thus $\mathcal{B}^{(7)}$ was accepted to be a reasonable representation of the *infinite* $\Omega^{(*)}$. The authors therein commented that the above formulation was significantly more economical as compared with the boundary integral method and series expansion techniques.

The innovative computational scheme of repeated substructuring [2] suffers from the following deficiencies. It is often essential, as in geological prospecting, to estimate the field strength at some intermediate points on the exterior region $\Omega^{(*)}$. This method [2] fails to yield a procedure for such calculations. Secondly, after m cycles of intermediate node condensations, the outer field could best be approximated by m number of piecewise polynomials of the order p, where p is the order of interpolations employed to construct the initial matrix $\boldsymbol{S}$ in (6). Thus, with constant gradient triangular elements, for which $p = 1$, the order of accuracy would substantially differ from those of the analytical results which decay according to the inverse power law. However, the selection of an initial enlargement factor γ in (5) close to unity could improve the accuracy. This in turn would demand a large number of condensation cycles. Consequently, substantial computer time would be needed.

The aforementioned shortcomings are alleviated in the proposed computational scheme of infinite substructuring. The outer region $\Omega^{(*)}$, refer to Fig. 2, is indeed represented as the infinite union

$$\Omega^{(*)} = \bigcup_{j=1}^{\infty} \Omega^{(j)} . \tag{12}$$

Since all the subregions $\Omega^{(j)}$ are geometrically similar, the order of accuracies, which is obtained from a nondimensional frame, for the element flux-field formulation (refer to (6)) are identical. Thus the infinite assembly procedure adheres to the preassigned error bound which is the same as that at the element level. In particular, the proposed method amounts to setting the condensation cycle in [2] to *infinity*. Furthermore, it has been demonstrated that computational efficiency need not be sacrificed in order to account for the entire infinite collection of finite element regions. As an intermediate step, the decay coefficients of a complete set of characteristic fields are evaluated. It then becomes possible to estimate the potential field at an arbitrary exterior location.

4. Infinite substructuring

The proposed algorithm is developed for Laplace's equation $\nabla^2 u = 0$ in $\mathbb{R}^k$, $k \geqslant 2$. The interior region $\Omega^{(0)}$, in Fig. 2, is taken to be convex. Each intercell boundary is symmetrically discretized with n nodal degrees-of-freedom. The Dirichlet matrix $\boldsymbol{S}$ is required to be supplied only for one annular region $\Omega^{(1)}$. Geometrical similarity between the cells $\Omega^{(j)}$, $j = 1, 2, \ldots$ indicates the following scalar relationship between the system matrices of the two adjacent cells $\Omega^{(j)}$ and $\Omega^{(j+1)}$:

$$\boldsymbol{S}^{(j+1)}_{\alpha+1,\beta+1} = \gamma^{k-2} \boldsymbol{S}^{(j)}_{\alpha,\beta} . \tag{13}$$

This statement can be easily verified from the dimensional analysis. The spatial dimension k, enlargement ratio γ and submatrices $\boldsymbol{S}^{(j)}_{\alpha,\beta}$ are defined in (1), (5) and (6), respectively.

4.1. Exterior fields

A complete set of interpolations for the 'infinite element' $\Omega^{(*)}$ is derived first. These functions are the finite element solutions of Laplace's equation in the spatially discretized unbounded region shown in Fig. 2. These characteristic shapes should then be derived by considering the balance of flux and compatibility of field strength at each interface $\partial\Omega_j$.

Let us hypothesize the existence of a nonzero field profile $\boldsymbol{q}$ on $\partial\Omega_1$ such that the corresponding potential distributions on three consecutive cell boundaries $\partial\Omega_j$, $\partial\Omega_{j+1}$ and $\partial\Omega_{j+2}$ are $\lambda^{j-1}\boldsymbol{q}$, $\lambda^j\boldsymbol{q}$ and $\lambda^{j+1}\boldsymbol{q}$, respectively. The existence of a corresponding nonzero scalar λ (the field decay coefficient, as dictated by the physics of the problem) is also assumed at this stage. The flux-field relationship for the enclosed cells $\Omega^{(j)}$ and $\Omega^{(j+1)}$, from (6), are, respectively,

$$\begin{bmatrix} \boldsymbol{S}^{(j)}_{j,j} & \boldsymbol{S}^{(j)}_{j,j+1} \\ \boldsymbol{S}^{(j)}_{j+1,j} & \boldsymbol{S}^{(j)}_{j+1,j+1} \end{bmatrix} \begin{Bmatrix} \lambda^{j-1}\boldsymbol{q} \\ \lambda^j\boldsymbol{q} \end{Bmatrix} = \begin{Bmatrix} \boldsymbol{\psi}^{(j)}_j \\ \boldsymbol{\psi}^{(j)}_{j+1} \end{Bmatrix} \tag{14a}$$

and

$$\begin{bmatrix} \boldsymbol{S}^{(j+1)}_{j+1,j+1} & \boldsymbol{S}^{(j+1)}_{j+1,j+2} \\ \boldsymbol{S}^{(j+1)}_{j+2,j+1} & \boldsymbol{S}^{(j+1)}_{j+2,j+2} \end{bmatrix} \begin{Bmatrix} \lambda^{j}\boldsymbol{q} \\ \lambda^{j+1}\boldsymbol{q} \end{Bmatrix} = \begin{Bmatrix} \boldsymbol{\psi}^{(j+1)}_{j+1} \\ \boldsymbol{\psi}^{(j+1)}_{j+2} \end{Bmatrix} \tag{14b}$$

From (13),

$$\lambda\gamma^{k-2}\begin{bmatrix} \boldsymbol{S}^{(j)}_{j,j} & \boldsymbol{S}^{(j)}_{j,j+1} \\ \boldsymbol{S}^{(j)}_{j+1,j} & \boldsymbol{S}^{(j)}_{j+1,j+1} \end{bmatrix} \begin{Bmatrix} \lambda^{(j-1)}\boldsymbol{q} \\ \lambda^j\boldsymbol{q} \end{Bmatrix} = \begin{Bmatrix} \boldsymbol{\psi}^{(j+1)}_{j+1} \\ \boldsymbol{\psi}^{(j+1)}_{j+2} \end{Bmatrix}. \tag{14c}$$

Since there is no external flux input on the boundary $\partial\Omega_{j+1}$,

$$\boldsymbol{\psi}^{(j)}_{j+1} + \boldsymbol{\psi}^{(j+1)}_{j+1} = \boldsymbol{o},$$

which leads to

$$[\boldsymbol{S}^{(j)}_{j+1,j} + \lambda(\boldsymbol{S}^{(j)}_{j+1,j+1} + \gamma^{k-2}\boldsymbol{S}^{(j)}_{j,j}) + \lambda^2\gamma^{k-2}\boldsymbol{S}^{(j)}_{j,j+1}]\boldsymbol{q} = \boldsymbol{o}.$$

By repeated application of the similarity result of (13), the above expression can be reduced to the following form which includes the submatrices pertaining only to cell element $\boldsymbol{\Omega}^{(1)}$:

$$[\boldsymbol{S}^{(1)}_{2,1} + \lambda(\boldsymbol{S}^{(1)}_{2,2} + \gamma^{k-2}\boldsymbol{S}^{(1)}_{1,1}) + \lambda^2\gamma^{k-2}\boldsymbol{S}^{(1)}_{1,2}]\boldsymbol{q} = \boldsymbol{o}. \tag{15}$$

It is convenient to introduce the variables

$$\begin{aligned} &\boldsymbol{H} = \boldsymbol{S}^{(1)}_{2,1}, \qquad \eta = \lambda\sqrt{\gamma^{k-2}}, \\ &\boldsymbol{G} = \sqrt{\gamma^{k-2}}\boldsymbol{S}^{(1)}_{1,1} + \boldsymbol{S}^{(1)}_{2,2}/\sqrt{\gamma^{k-2}}; \quad \boldsymbol{G} = \boldsymbol{G}^{\mathrm{t}}, \end{aligned} \tag{16}$$

in order to reduce (15) to the following form of the quadratic eigenvalue problem:

$$[\boldsymbol{H} + \eta\boldsymbol{G} + \eta^2\boldsymbol{H}^{\mathrm{t}}]\boldsymbol{q} = \boldsymbol{o}. \tag{17}$$

The eigenvalue η and the eigenvector $\boldsymbol{q}$ can be computed from the linear eigenvalue problem

$$\begin{bmatrix} \boldsymbol{O} & \boldsymbol{I} \\ -\boldsymbol{H} & \boldsymbol{O} \end{bmatrix} \begin{Bmatrix} \boldsymbol{q} \\ \eta\boldsymbol{q} \end{Bmatrix} = \eta \begin{bmatrix} \boldsymbol{I} & \boldsymbol{O} \\ \boldsymbol{G} & \boldsymbol{H}^{\mathrm{t}} \end{bmatrix} \begin{Bmatrix} \boldsymbol{q} \\ \eta\boldsymbol{q} \end{Bmatrix}, \tag{18}$$

in which $\boldsymbol{I}$ and $\boldsymbol{O}$ are the $n \times n$ identity and null matrices, respectively.

It is now essential to demonstrate that a complete set of shape function $\boldsymbol{q}$ with appropriate decay coefficients λ (or its invariant counterpart η) can be obtained from the quadratic eigenproblem (17). The eigenvalues η can be computed, in principle, as the roots of the determinant equation

$$\det[\boldsymbol{H} + \eta \boldsymbol{G} + \eta^2 \boldsymbol{H}^{\mathrm{t}}] = 0 .$$

The zero value of the determinant is not altered by transposition or by scaling with a factor $(1/\eta^2)$; hence,

$$\det\left[\boldsymbol{H} + \frac{1}{\eta}\boldsymbol{G} + \frac{1}{\eta^2}\boldsymbol{H}^{\mathrm{t}}\right] = 0 .$$

Comparison of the above two equations indicates that for each root η, its reciprocal $(1/\eta)$ is also another root. Furthermore, for n degrees-of-freedom on $\partial\Omega_j$, there are $2n$ number of η and corresponding $\boldsymbol{q}$, provided $\boldsymbol{H}$ is nonsingular. The discrete analog of Almansi's theorem for potentials guarantees $\det[\boldsymbol{S}^{(1)}_{1,2}] \neq 0$. The general quadratic eigenvalue problem associated with the cloning algorithm [6] addresses this question. Consequently, $\boldsymbol{H}$ is nonsingular according to (16). Hence there could be at best n values of η such that $|\eta| < 1$. In particular, if there is no η such that $|\eta| = 1$, then there will be exactly n number of those roots which would yield decaying values of n number of λ's in (14).

The degenerate case of $|\eta| = 1$ occurs only for two-dimensional problems. The expression of energy in (3) involves gradients of fields. The elements of a Dirichlet matrix can be interpreted as a 'virtual work term' as in [1]. Hence the constant field, with $|\eta| = 1$ *does not* contribute nontrivially to the Dirichlet matrix for each of $\Omega^{(j)}$. Consequently, this distribution should not be accounted for in generating $\boldsymbol{S}^*$ for $\Omega^{(*)} \equiv \bigcup_{j=1}^{\infty} \Omega^{(j)}$.

4.2. *Infinite assembly*

After we have chosen the set of n potential distributions as shape functions, $\boldsymbol{q}_s$, $s = 1, 2, \ldots, n$, from the solutions of the quadratic eigenproblem (17), with the decay condition on the corresponding eigenvalues $|\eta_s| < 1$, a transformation of coordinates to the modal one which is constituted by $\{\boldsymbol{q}_s\}$ is proposed. The transformed Dirichlet matrix for an element $\Omega^{(j)}$ is indicated by $\boldsymbol{E}^{(j)}$ and for $\Omega^{(*)}$ by $\boldsymbol{E}^{(*)}$. The elements $\boldsymbol{E}^{(j)}_{s,t}$ (or $\boldsymbol{E}^{(*)}_{s,t}$) are then the virtual work quantities, in regions $\Omega^{(j)}$ (or $\Omega^{(*)}$), due to the conjugation of the modes labeled s and t. The discrete analog of the energy term, (3), for an element $\Omega^{(j)}$ associated with a pair of vectors $\boldsymbol{q}_s$ and $\boldsymbol{q}_t$ is $\boldsymbol{E}^{(j)}_{s,t}$ and is given by

$$\boldsymbol{E}^{(j)}_{s,t} = \langle \lambda_s^{j-1} \boldsymbol{q}_s^{\mathrm{t}}, \lambda_s^j \boldsymbol{q}_s^{\mathrm{t}} \rangle \begin{bmatrix} \boldsymbol{S}^{(j)}_{j,j} & \boldsymbol{S}^{(j)}_{j,j+1} \\ \boldsymbol{S}^{(j)}_{j+1,j} & \boldsymbol{S}^{(j)}_{j+1,j+1} \end{bmatrix} \begin{Bmatrix} \lambda_t^{j-1} & \boldsymbol{q}_t \\ \lambda_t^j & \boldsymbol{q}_t \end{Bmatrix} .$$

Now, by applying the similarity result (13) repeatedly on $\boldsymbol{S}^{(1)}$, $\boldsymbol{E}^{(j)}_{s,t}$ becomes

$$\boldsymbol{E}^{(j)}_{s,t} = (\lambda_s \lambda_t \gamma^{k-2})^{j-1} \boldsymbol{E}^{(1)}_{s,t} ,$$

or from (16),

$$\boldsymbol{E}^{(j)}_{s,t} = (\eta_s \eta_t)^{j-1} \boldsymbol{E}^{(1)}_{s,t} .$$

Hence the total energy associated with the s and t modes for the entire infinite region $\Omega^{(*)}$ is

$$E_{s,t}^{(*)} = \sum_{j=1}^{\infty} (\eta_s\eta_t)^{j-1} E_{s,t}^{(1)} .$$

Since we have selected only the decaying fields with $|\eta_s$ and $\eta_t| < 1$, consequently $|\eta_s\eta_t| < 1$. The above infinite geometric progression is summed *exactly* leading to

$$E_{s,t}^{(*)} = \frac{E_{s,t}^{(1)}}{1 - \eta_s\eta_t} . \tag{19}$$

Finally, S^* is recovered according to the transformation

$$S^{(*)} = [\{q\}^t]^{-1} E^* [\{q\}]^{-1} . \tag{20}$$

This concludes the derivation of the scheme of infinite substructuring.

5. Estimation of outer field

The potential u at a point P inside $\Omega(*)$ due to an arbitrary distribution of a field ϕ_1 on $\partial\Omega_1$ can be approximated in terms of the shape functions $\{q_s\}$ and the decay coefficients η_s (refer to Fig. 2). Let x_s denote the participation factor for the mode q_s. Then

$$[\{q_s\}]\{x_s\} = \phi_1 . \tag{21}$$

In order to compute the decay of the mode q_s, we identify the contour $\partial\Omega$, which is geometrically similar to those in the family $\{\partial\Omega^{(j)}\}$, such that $\partial\Omega$ contains the observation point p. The real number p indexes the family $\{\partial\Omega_j\}$ such that $\partial\Omega$ can be labeled as $\partial\Omega_p$. Let R be the characteristic dimension for $\partial\Omega_p$. Then, from (5),

$$\gamma^p = R/R_1$$

or

$$p = \frac{\mathrm{Log}_\alpha R - \mathrm{Log}_\alpha R_1}{\mathrm{Log}_\alpha R_2 - \mathrm{Log}_\alpha R_1} \quad (\alpha \text{ constant}) .$$

In general p is a fraction nested by the integers l and $l+1$. The potential distributions on $\partial\Omega_{(l)}$ and $\partial\Omega_{(l+1)}$ are then

$$\phi_l = [q_s]\{\lambda_s^{l-1} x_s\}$$

and

$$\phi_{l+1} = [q_s]\{\lambda_s^l x_s\} ,$$

respectively. Finally, these nodal values are interpolated according to the geometrical distributions which were originally prescribed for the element $\Omega^{(1)}$ leading to the required approximated field.

6. Numerical example

The proposed numerical technique is illustrated for a sample problem whereby the field is evaluated within a two-dimensional region having a circular deletion (Fig. 3). The polar coordinates (r, θ) represent the geometry. The analytical solution for Laplace's equation can be easily obtained in terms of a Fourier series. The two-dimensional problem is most critical when evaluating the effectiveness of any solution procedure for unbounded domains since the response decay is comparatively slower than those in the higher-dimensional exterior spaces. It is also essential to isolate the case of the constant field, i.e., $|\eta| = 1$ in (17).

The radius of the circular deletion $\Omega^{(0)}$ is taken to be unity. (This facilitates presentation of the nondimensional results.) The boundary contours $\{\partial\Omega_i\}$ are discretized with eighteen elements. Quadrilateral constant gradient elements are used to construct the cell matrix $\boldsymbol{S}$ in (6). In order to procure the best possible aspect ratio for the elements, the mesh magnification factor γ in (5) is set to 1.33.

An important aspect regarding the nature of two-dimensional potentials was satisfactorily captured in the proposed finite element calculation of infinite substructuring. The existence of a uniform field was revealed by one of the pairs of eigenvalue η of (17) to be very close to unity. The corresponding eigenvector $\boldsymbol{q}$ indeed had all entries differing from unity in the fifth decimal place. This suggests that in a discrete computation, the average field on $\partial\Omega_1$ can be confidently accepted as the datum. Consequently, this particular potential distribution does not contribute to the Dirichlet matrix for the unbounded region. Hence $\boldsymbol{S}^{(*)}$ is of rank one.

The exterior field due to a sinusoidal distribution on $\partial\Omega_1$ is examined numerically. The analytical solution is $\sin\theta/r$ as illustrated in Fig. 3. Along a contour $\partial\Omega_4$, the computed distribution is plotted. At the node $\theta = \frac{1}{2}\pi$, the potential was calculated to be 0.4221 whereas the analytical result yields $\frac{27}{64}$ (i.e., 0.421875). The overall comparison with the analytical result is excellent as evidenced in Fig. 3.

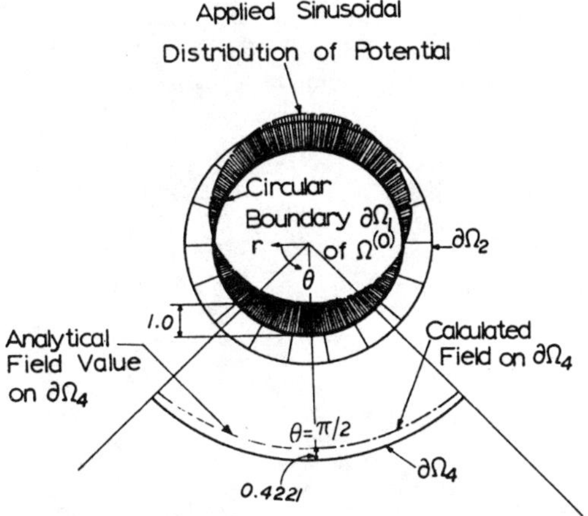

Fig. 3. Illustrated example.

7. Conclusion

Applications of the conventional substructuring procedure have traditionally been confined to bounded domains. The nodes on the common boundary are then condensed out to construct the system matrix for a larger geometrical region. Such a method is not computationally possible for unbounded domains. In this paper, a technique is described which in essence amounts to carrying out the substructuring scheme over an infinite collection of finite elements.

This stand-alone finite element method as opposed to the commonly used hybrid procedures, does not require any knowledge of analytical fields in unbounded domains.

It is demonstrated herein that a conventional finite element package with eigensolution routines should suffice for the construction of an infinite element. This method of infinite substructuring has been observed to be numerically accurate and computationaly efficient in benchmark problems.

References

[1] O.C. Zienkiewicz, The Finite Element Method (McGraw-Hill, New York, 1977).

[2] P.P. Silvester, D.A. Lowther, C.J. Carpenter and E.A. Wyatt, Exterior finite elements for two-dimensional field problems with open boundaries, Proc. IEE 124 (12) (1977) 1267–1270.

[3] O.C. Zienkiewicz, C. Emson and P. Bettess, A novel boundary infinite element, Internat. J. Numer. Meths. Engrg. 19 (1983) 393–404.

[4] G. Dasgupta, Numerical solution for viscoelastic half planes, J. Engrg. Mech. Div., ASCE 102 (EM4) (1976) 601–612.

[5] I. Stakgold, Boundary Value Problems of Mathematical Physics, Vol. II (Macmillan, New York, 1971).

[6] G. Dasgupta, A finite element formulation for unbounded homogeneous continua, J. Appl. Mech., ASME 49 (1982) 136–140.

[7] E. Almansi, Un teorema culle deformazioni elastiche dei solidi isotropi, Atti Reale Accad. Nazionale Lincei 16 (1907) 865–868.

Infinitesimal Scaling—A New Procedure for Modeling Exterior Field Problems

H. HURWITZ, JR.

Abstract — The ability to treat extended uniform linear regions (particularly unbounded external regions) as single elements is a desirable addition to electromagnetic finite element techniques. The use of extended elements requires basis functions that are better approximations to the solutions of the underlying partial differential equations than the conventional polynomial basis functions. This problem can be circumvented by means of the infinitesimal scaling technique, which enables the coefficient matrix required for assembling the contribution of an extended element into the global functional to be determined without explicit knowledge of internal basis functions.

INTRODUCTION

Applied problems in magnetics often have the characteristic that extended regions (particularly unbounded external regions) in which the material properties are linear and uniform must be considered. When finite element techniques are utilized, these regions may substantially increase the number of elements and nodes that must be considered. It is thus desirable to be able to treat extended uniform and linear regions as single elements while retaining the option of treating the more complicated (but frequently less extended) regions by utilizing physically smaller elements. (The size of these latter elements may in practice be limited by nonlinearities in material properties which, in effect, cause material properties to be strongly position-dependent, even in a region of fixed material composition.) In order for the technique utilized in treating the extended elements to be easily integrated with the conventional technique utilized in treating the small elements, it is desirable that the former technique be cast in terms of the same formulation utilized in dealing with the small elements. The variational approach fortunately provides the needed common ground because the global functional that describes the full system is a volume integral, and thus is just the sum of the contributions from the individual elements, irrespective of their topological extent. The only requirement is that C0 continuity of the appropriate field variable be assured by utilizing the same nodal values and boundary basis functions on the two sides of an interface between adjacent elements. (Boundary conditions on the normal derivatives of the field quantities are satisfied in an integral sense by finding the extremum of the global functional with respect to variations in the nodal field values. Errors that are orthogonal to the boundary basis functions cannot be excluded.)

Standard methods for expressing the contribution of a small element to the global functional in terms of the trial function values on its boundary nodes are based on utilizing explicit internal basis functions of some simple form (e.g., polynomials in isoparametric coordinates). The fact that these basis functions do not satisfy the underlying field differential equation is of little consequence if the element is of sufficiently small size. In large or unbounded elements, however, the use of such simple trial functions may introduce unacceptably large errors. Ideally, one should instead utilize basis functions that rigorously satisfy the underlying differential equations within the element, while having a specified variation on the boundary [1]. Such internal basis functions would have the property of providing an extremum to the contribution of the element to the global functional subject to fixed trial function behavior on the element boundary. (The latter constraint is essential for the purpose of assuring C0 continuity of the global trial function.)

The infinitesimal scaling method [4,5] seeks to determine the coefficient matrix which expresses the contribution of an element to the global functional in terms of the trial function values on the boundary nodes. Instead of attempting to explicitly determine the basis functions which have the desired boundary variation and which satisfy the underlying differential equation, the method makes use of the extremum property mentioned above, together with a scaling concept analogous to that used in the "ballooning" technique [2,3], but in the limit that the increment in scaling parameter approaches zero. By going to this limit it is possible to derive a nonlinear first-order differential equation for the coefficient matrix (which we refer to as **C**). In the Laplacian case, this matrix equation reduces to an algebraic equation which can be solved by quadratically convergent iteration techniques.

It may appear surprising that the coefficient matrix can be determined by making use only of the extremum property of the contribution of an element to the global functional and the comparatively simple relationship that exists between the coefficient matrices of regions that are of nearly the same size. The price that is paid is the need to solve the resulting nonlinear matrix equation. In practice, this is less of a penalty than might be feared because of the fact that the nonlinear matrix equations can be solved fairly quickly on modern high-speed computers. Once the coefficient matrix has been calculated for a particular geometry, it may be retained in a library and used in a number of different calculations. (Matching internal finite element grids to a standard boundary for which the external region coefficient matrix is known is not difficult and may, in fact, be good practice for other reasons.) The infinitesimal scaling matrix equation is of interest to numerical analysts so that progress may be anticipated in the development of robust and efficient algorithms for its solution. While "on the fly" calculation of coefficient matrices by infinitesimal scaling is probably not appropriate in most finite element packages, progress in developing improved solvers for the nonlinear matrix equation should facilitate use of the method in special cases in which the library approach is feasible and the accuracy that can be achieved by infinitesimal scaling is needed.

An important consideration is that the coefficient matrix is specific not only to the geometric properties of the extended element being analyzed, but also to the particular boundary basis functions that are utilized to couple this element to adjacent elements. Therefore, while the problem of determining the coefficient matrix is intrinsically of a linear nature, the use of a nonlinear method is warranted because of the fact that it then becomes possible to incorporate the specification of boundary basis functions into the analysis in a natural way at an early stage in the calculation (specifically, in the determination of the matrices **M**, **N**, and **Q** defined below). Furthermore, the complex interactions between pairs of distant boundary basis functions on the same extended element do not have to be dealt with explicitly since these interactions are automatically treated accurately in the process of solving the nonlinear matrix equation.

The boundary of an extended element will usually consist of a number of segments corresponding to interfaces with adjacent elements in the finite element grid. In the infinitesimal scaling method, each of these segments will correspond to a boundary element which has the same number of nodes as the adjacent smaller element. The boundary basis functions must coincide with the values of the basis functions utilized in the adjacent elements evaluated on the interface surface. Assembling the contributions

Manuscript received March 2, 1984
The author is with General Electric Company, Corporate Research and Development, Schenectady, NY 12345

Reprinted from *IEEE Trans. Magn.*, vol. MAG-20, no. 5, pp. 1918–1923, Sept. 1984.

of individual boundary elements into the auxiliary matrices that appear in the fundamental infinitesimal scaling matrix equation is a relatively simple process. Not only do these boundary elements have one less spatial dimension than the adjacent elements, but also the contributions of the boundary elements can be assembled into the auxiliary matrices one by one. There is no need to consider pairs of boundary elements and to deal with Green's function singularities, as is the case with boundary integral methods. (In the latter methods, the singularities can be removed by basis function integrations, but in infinitesimal scaling the singularities do not appear explicitly.) Since the assembly process tends to be a major component of finite element calculations, the fact that infinitesimal scaling does not greatly increase the difficulty of this part of the overall task is an advantage which tends to compensate for the necessity of dealing with a nonlinear matrix equation. The avoidance of the need to consider pairs of boundary elements in the assembly process is also a feature of the ballooning method, but ballooning requires the selection of a specific scaling parameter and does not have the advantage of reducing the dimensionality of the boundary elements.

The long range interactions encountered in electromagnetic problems cause the coefficient matrix for an extended element to be full rather than banded. This may increase the bandwidth of the global system of nodal equations that must be solved. Fortunately, modern finite element packages often use preconditioned conjugate gradient solvers in which bandwidth is not a primary problem so long as the number of non-zero matrix elements is not greatly increased. (It is, of course, always possible to neglect small contributions to the coefficient matrix from nodes that are far apart. Once the coefficient matrix has been calculated, this may be done in a systematic manner.)

As has been indicated, infinitesimal scaling and ballooning are closely related. Indeed, the process of eliminating (or "condensing") exterior nodes in the ballooning technique is implicitly equivalent to solving a nonlinear matrix difference equation by a linearly convergent iteration technique. While ballooning does not possess the advantage of reducing the dimensionality of the boundary elements and eliminating possible errors resulting from a difference equation approach, it does retain a tie to a specific finite element grid in the extended or unbounded region. Hence the advantages of infinitesimal scaling may, in practice, be limited to situations in which an unusually high degree of accuracy is required.

In the following sections, the fundamental equation of infinitesimal scaling will be derived and discussed. Applications to simple test examples in *x-y* and *r-z* geometry which demonstrate the feasibility of coupling infinitesimal scaling to conventional finite element algorithms will be summarized.

MATHEMATICAL FORMULATION

While the infinitesimal scaling technique is based on the same variational principle utilized in conventional finite element methods, there is the important distinction that infinitesimal scaling in effect makes two successive applications of the extremum property of the functional. The first application, which leads to the fundamental equation of infintesimal scaling (Eq. 7 below), is carried out with respect to behavior of the basis functions in the interior of the extended element. This first extremum is obtained subject to the constraint that the trial function on the boundary of the element is fixed by the choice of trial function values on the boundary nodes together with the boundary basis functions that interpolate between these nodes. While the boundary nodal values are considered to be arbitrary, they are not allowed to vary in the process of obtaining the first extrememum. As a result of constructing and solving the fundamental equation of infinitesimal scaling, the coefficient matrix, **C**, is obtained. This enables the contribution of the extended element to the global functional to be completely determined by the values of the trial function on the boundary nodes.

The interior extremum process is absent in conventional finite element methods; it being assumed that the coefficient matrix can be determined with acceptable accuracy by specifing the interior behavior of the basis functions in a reasonable manner for the purpose of interpolation. While the coefficient matrices obtained in this manner have large errors (including errors in sign of some of the matrix elements) these errors are of limited consequence if the size of the finite element is sufficiently small. In any case, it is allowable to combine the infinitesimal scaling technique with the conventional technique for evaluating the coefficient matrix for different elements in the same calculation.

The second application of the extremum principle, which has the effect of satisfying normal derivative boundary conditions in an integral sense, is with respect to the nodal trial function values that determine the amplitude of the basis functions. This second extremum process is not affected by the method used in determining the coefficient matrices for the various finite elements that appear in the global problem.

We now give a general derivation of the fundamental equation of infinitesimal scaling. The basic assumption is that the contribution of the element to the global functional is a volume integral over the element of a quadratic form in the trial function and its first-order spatial partial derivatives. (Other variables which depend on position in a known way may also appear in the quadratic form, e.g. the radius.) It is also implicitly assumed that, for trial functions which exhibit C0 continuity and have first derivatives with respect to position almost everywhere, the extremum of the functional is uniquely determined by the behavior of the trial function on the element boundary (i.e., by the nodal values and the boundary basis functions). For unbounded elements, this requires that the appropriate boundary conditions at infinity be satisfied. In the case of multicomponent field quantities (i.e., vector fields) certain questions of overdetermination of the interior solution must be considered, but such problems do not arise when only one component of the vector field need be considered (e.g., the azimuthal component of the vector potential in a problem with circular cylindrical symmetry).

The above assumptions are adequate to assure that the contribution of the element to the global functional will be a quadrative form of the trial function values on the boundary nodes (ϕ_i) with uniquely determined coefficients (C_{ij}). Thus if the contribution of the region R to the global functional is designated by F_R, then

$$F_R = \bar{\phi}\, \mathbf{C}\, \phi = \sum_{i,j} C_{ij}\, \phi_i \phi_j \tag{1}$$

Here ϕ is the column vector made up of the trial function values on the boundary nodes and $\bar{\phi}$ is the associated row vector. Equation 1 does not include the effect of sources in the element under consideration. (Methods for making this extension are discussed elsewhere.[5])

The value of F_R given by (1) is actually the extremum of the functional subject to the boundary constraint on the trial function. The extremum occurs when the behavior of the trial function within the element has the form

$$\phi(\mathbf{r}) = \sum \phi_i\, f_i\,(\mathbf{r}) \tag{2}$$

where the functions $f_i(\mathbf{r})$ satisfy the partial differential equation associated with the functional within the element, and match the boundary basis functions on the boundary. (As usual, we have adopted the convention that the ith basis function is equal to unity on boundary node i, and zero on all the other boundary nodes.) A comparison of (1) and (2) shows that the coefficient C_{ij} is just the cross term between f_i and f_j in the volume integral of the functional integrand over the element R. As we will see from the following analysis, it is not necessary to determine how the functions f_i behave in the interior of the element in the calculation of the coefficients C_{ij}. Instead, we utilize the extremum property of

F_R to derive a matrix equation for **C** which can be solved to simultaneously determine all the elements C_{ij}.

In order to utilize the extremum property of the function to derive a matrix differential equation for **C** we consider a region R_δ which is slightly smaller than the original region. The boundaries of R and R_δ will be nested closed surfaces which we consider to belong to a family of nested surfaces characterized by a parameter s. (If R is an external unbounded element, the boundary of the region R_δ will lie outside that of the region R.) Let the difference in s between the two surfaces be equal to a small quantity δ which will be allowed to approach zero. When δ is small, the region between the two surfaces will degenerate to a thin shell. For any trial function, the functional for region R is just the sum of the functional evaluated for region R_δ, and the functional evaluated for the shell. We furthermore assume that the trial function on the surface of region R_δ can be expressed in terms of boundary basis functions for this smaller region, and that the nodal values of the trial function on the boundary of region R_δ will differ from those on region R by amounts of order δ. Then the vector of nodal values for region R_δ will be given by an expression of the form

$$\phi = \phi + \delta \cdot \psi \tag{3}$$

Here ψ is an unknown vector. If we now assume that the trial function within region R_δ satisfies the underlying partial differential equation, the contribution of region R_δ to the functional for region R can be written in a form analogous to (1). Thus

$$F_R = F_\delta + (\bar{\phi} + \delta \cdot \bar{\psi})\, \mathbf{C}_{R_\delta}\, (\phi + \delta \cdot \psi) + O(\delta^2) \tag{4}$$

Here $\mathbf{C}_{R_\delta}$ represents the coefficient matrix for the region R_δ.

Since the region between the boundaries of regions R and R_δ is a thin shell when δ is small, it should be possible to determine F_δ by knowing the vectors ϕ and ψ, the boundary basis functions, and the behavior of the nested surfaces as a function of the parameter s. The most general form of the limiting expression for F_δ is

$$F_\delta = \delta \cdot (\bar{\phi}\mathbf{M}_R\phi + \bar{\psi}\mathbf{N}_R\psi + 2 \cdot \bar{\phi}\mathbf{Q}_R\psi) + O(\delta^2) \tag{5}$$

The matrices $\mathbf{M}_R$, $\mathbf{N}_R$, and $\mathbf{Q}_R$, will in general depend on the parameter s. $\mathbf{M}_R$ and $\mathbf{N}_R$ can be assumed to be symmetrical, but $\mathbf{Q}_R$ is in general not symmetrical.

The fundamental equation of infinitesimal scaling can be derived from (4) and (5) without further assumptions concerning the characteristics of the nested surfaces. It is simply necessary to combine these two equations into a single expression for F_R. The only free parameters are the components of the vector ψ which describe the variation of the trial function in the thin shell between the boundaries of regions R and R_δ. The expression for F_R contains both linear and quadratic terms in these unknown vector components. Hence it is possible to eliminate the components of ψ by simply requiring that F_R be an extremum with respect to variations in the vector ψ. (This is the interior extremum process mentioned above.)

By straightforward algebra, we find that the extremum of (5) with respect to ψ can be expressed in matrix notation as follows:

$$\bar{\phi}\mathbf{C}_R\phi = \bar{\phi}\mathbf{C}_{R_\delta}\phi + \delta(\bar{\phi}\mathbf{M}_R\phi - \bar{\phi}(\mathbf{C}_R + \mathbf{Q}_R)\mathbf{N}_R^{-1}(\mathbf{C}_R + \mathbf{Q}_R^T)\phi) \tag{6}$$

Since δ represents the small change in the parameter s, and since the vector ϕ is completely arbitrary, (6) reduces to a first-order nonlinear matrix differential equation in the limit of zero δ. We find

$$+\frac{d\mathbf{C}}{ds} = \mathbf{M} - (\mathbf{C} + \mathbf{Q})\, \mathbf{N}^{-1}(\mathbf{C} + \mathbf{Q}^T) \tag{7}$$

(We have dropped the subsript R for notational convenience, and assume that the parameter s increases as the region R expands. For external regions, this means that s increases as the boundary surface decreases in geometric size.)

Equation 7 is the fundamental equation of infinitesimal scaling, although it has been derived without any specific assumptions concerning the properties of the family of nested surfaces described by the parameter s. Before discussing the simplifications which arise when the nested surfaces differ only in size and not in shape, we call attention to the fact that this equation is rigorously correct only in the limit of infinitely closely spaced boundary nodes. When only a finite number of nodes (and hence boundary basis functions) are used, there is no assurance that the function which satisfies the underlying partial differential equations will in fact be expandable in these basis functions throughout the region between R and R_δ. The error can persist to order δ because the normal derivative of the correct solution is not necessarily expandable in a finite set of basis functions, and may, in fact, not even be finite in the neighborhood of sharp corners. Fortunately, the fact that (7) has been derived from a variational principle tends to reduce the magnitude of the errors which arise. Thus accurate results have been obtained by using (7) even for external regions which do have corners. It is possible to further reduce the magnitude of residual errors in the application of (7) by utilizing a larger number of surface nodes (and hence boundary basis functions) than are required in the final problem. This yields a coefficient matrix with more components than are required, but the larger coefficient matrix can be reduced by a synthesis process to the smaller desired size.

The infinitesimal scaling method derives its name from the fact that the matrices **M**, **N**, and **Q** can be most easily calculated if it is assumed that the family of nested surfaces are simply scaled versions of the specific boundary surface that is of interest. By this we mean that a central point can be defined such that corresponding points on the nested surfaces will lie on a ray extending from the central point, and that the distances of pairs of corresponding points will have the same ratio for all members of the family of nested surfaces. (This is the same concept utilized in the ballooning technique.) A scaled family of nested surfaces can be constructed from any closed surface providing that a central point can be found such that all points of the surface can be "seen" from this point. (Surfaces of toroidal topology cannot be scaled in this simple manner.) For a particular closed surface, it is generally possible to select many alternative central points on which the scaling can be based, and one of the first tests of the infinitesimal scaling technique was to check that the resulting coefficient matrices do not depend sensitively on the choice of the central point.

When scaled surfaces are used, it is convenient to define the parameter s which specifies a particular member of the family of nested surfaces to be the logarithm of a characteristic length of the corresponding surface. (For external regions, s can be the negative of this logarithm to preserve the convention that s increases as the extent of the region increases.) For Laplacian cases, the matrices which appear in (7) will all be proportional to the $(n - 2)$ power of scale where n is the dimension of the space. Hence the derivative of **C** which appears on the left-hand side of (7) is simply $(n - 2)$ ***C**. Furthermore, both sides of (7) will vary in the same way with s, so that a unique algebraic matrix equation is obtained for **C**. In Helmholtz and diffusion applications, the matrix **M** will contain an additional term proportional to physical scale to the power n. The matrix **C** no longer varies in a trivial way with scale, and it is necessary to retain the full differential equation form of (7).

CALCULATIONAL TECHNIQUES

In infinitesimal scaling applications, the first step is to decide on the specific geometry and boundary element structure of the extended or unbounded region for which the coefficient matrix is desired. Boundary basis functions must be associated with each boundary node in accordance with the requirement of matching the basis functions to be utilized in adjacent elements. In general the boundary of interest will be divided into a number of small boundary elements which will ultimately be matched to individual elements adjacent to the region treated by infinitesimal scaling.

The matrices **M**, **N**, and **Q** which appear in (7) must then be assembled from the contributions of the individual segments in much the same manner as the conventional S-matrix of finite element analysis is assembled from the contributions of individual elements.

The technique of determining the contribution of a single boundary element to the **M**, **N**, and **Q** matrices is similar to conventional isoparametric methods, with the exception that the dependence of the trial function on the local variable associated with the scaling direction is always assumed to be linear. This simplification, which is legitimate because of the fact that only the leading term in the scaling increment is required, has the effect of reducing the dimensionality of the calculation by one. A typical calculation in $r-z$ geometry is described in detail in the appendix of Ref. 4.

It is important to note that the matrices **M**, **N**, and **Q** are sparse since they incorporate interactions only between boundary nodes in the same boundary element. (The long range interactions automatically appear in the final coefficient matrix as a result of solving (7) Therefore the number of non-zero matrix elements that must be assembled is the product of the number of boundary elements times the square of the number of nodes per element. Thus the assembly process in an infinitesimal scaling calculation is, in principle, much easier than the corresponding process in boundary integral methods in which the number of matrix elements to be computed is of order of the square of the number of boundary nodes.

As an illustration of the ease of determining the matrices **M**, **N**, and **Q** for a boundary element, we note that the special case of a general straight line boundary element in the two-dimensional Laplacian situation has been treated in closed form by the Macsyma algebraic computer program [7]. It was possible to treat elements up to sixth order with an acceptable amount of computer time. This closed form solution was useful in checking the results of a more general fully isoparametric subroutine based on the usual Gauss point integration approach.

After the matrices **M**, **N**, and **Q** have been assembled, the next task is to solve the fundamental equation of infinitesimal scaling, (7), for the coefficient matrix **C**. In situations in which the full differential equation form of this equation must be retained, it is necessary to start with the known (or easily determinable) form of the coefficient matrix in the limit of zero or infinite scale depending on whether an internal or external region is being considerd. Skeleton calculations performed for spherical boundaries utilizing a resolution of the matricies into spherical harmonic components have indicated that the integration of (7) from zero or infinity to an arbitrary finite scale can be carried out in a small number of steps, provided that a logarithmic scale variable is employed and an implicit integration scheme is used.

The applications of infinitesimal scaling that have been investigated thus far have all been based on Laplacian situations (i.e., scalar or vector potentials in magnetostatic problems) for which the (7) reduces to a nonlinear algebraic equation. The initial technique used in solving this equation involved a simple linearly convergent iteration scheme. While no fundamental convergence problems were encountered, the computer time required for convergence was undesirably long (e.g., 9.5 minutes on a CRAY computer for a 96 node square). Therefore attention has been focused on developing quadratically convergent algorithms analogous to a Newton-Raphson iteration.

As may be seen from inspection of (7), standard quadratically convergent iteration techniques for nonlinear systems lead to a Lyapunov equation for the correction to the coefficient matrix in one iteration. ACM algorithm 432 [6] can be used to solve this equation, and an updated version of the algorithm has been provided by R. Beyers. S. Lamson has made modifications to facilitate vectorization, and has incorporatated the resulting algorithm into a utility subroutine for solving (7) in magnetostatic situations. Reductions in computer time, as compared to the original linearly convergent algorithm of about a factor 30, were obtained. The factor of improvement was found to be somewhat greater for scalar computers than for vector computers because of the fact that the quadratically convergent algorithm is less easily vectorizable than the linearly convergent algorithm. E. Wachspress has pointed out the possibility of developing algorithms which are not only more efficient than the standard Lyapunov algorithm when applied to (7), but are also more easily vectorizable. Meanwhile, it has been found satisfactory to adopt the "library" approach in which standard exterior region coefficient matrices are determined in advance and then utilized in a variety of applied problems which can be easily matched to one of these standard exterior regions.

ADVANCED TECHNIQUES

In this section we outline certain advanced techniques which can improve the accuracy or extend the applicability of infinitesimal scaling. As mentioned previously, (7) is not rigorously correct when only a finite number of boundary basis functions are utilized. The solutions to the Laplace equation in the region external to a two-dimensional rectangle or a three-dimensional finite cylinder have singularities at the corners which may cause errors in the coefficient matrix obtained by solving this equation. In order to reduce such errors, it is possible to utilize an expanded set of boundary basis funtions which spans a larger region of Hilbert space than the set of boundary basis functions for which the coefficient matrix is to be calculated. A rectangluar matrix **T** can be defined which gives the expansion coefficients of the latter boundary basis functions in terms of the former basis functions. Then the desired coefficient matrix can be obtained by multiplying the coefficient matrix for the expanded basis function set by **T** on the left and the transpose of **T** on the right. In some interior region cases, this synthesis process yields a rigorously correct result for the desired coefficient matrix. A useful example is the synthesis of second-order (quadratic) elements from fourth- or sixth-order elements. (The analytic results for the coefficient matrix for the interior of an equilateral triangle with quadratic boundary basis functions on each side have been successfully reproduced by a synthesized infinitesimal scaling calculation.)

Since the computer time required to solve (7) increases rapidly with the order of the matrices, it is very important to exploit the symmetry properties of the region being treated by infinitesimal scaling whenever possible. Group theory makes it possible to express the coefficient matrix of a symmetrical region in terms of the much smaller partial coefficient matrices associated with the different irreducible representations of the symmetry group of the region under consideration. Once the full coefficient matrix of a symmetrical region has been determined in this manner, this matrix can be used in global problems which, due to asymmetries in other regions, do not possess the symmetry properties of the infinitesimal scaling region. The group theoretical approach is useful in the case of rectangular regions in which reflection symmetry exists with respect to the coordinate planes. Interior or exterior cylindrical regions which are symmetrical with respect to axial rotations through an angle $2\pi/m$ where m is an integer can also benefit greatly from group theory since the order of the matrices associated with the irreducible representations is reduced by a factor m. Therefore, such regions should be particularly useful in three-dimensional applications.

A potentially useful approach for applying infinitesimal scaling results to complex regions that are not amenable to scaling is to combine simpler regions by means of a node condensation technique. In some problems of practical interest, the main complexities are restricted to a toroidal annular region lying between two bounding surfaces which partially coincide. The exterior region defined by the outer surface may then be combined with the interior region defined by the inner surface to form a single "exterior" region. Furthermore, if the nodes and basis functions on the common regions of the two surfaces also coincide, they can be eliminated by a condensation process. The result is a coefficient matrix for the entire exterior of the toroidal region that requires more detailed analysis. Hence the number of finite elements that must be considered is greatly reduced.

TEST EXAMPLES

Numerical tests of the infinitesimal scaling method have been reported elsewhwere in considerable detail [5], and will only be summarized here. Tests of various different types have been carried out. These include (a) the comparison of results obtained for the functional in two dimensional Laplacian regions with results obtained for harmonic trial functions by direct numerical quadrature or by closed form analysis, (b) the use of infinitesimal scaling coefficient matrices to treat "two element" problems in which one element is an extended interior region, the second element is the exterior region with the same boundary, and source terms are assumed to be concentrated on boundary nodes, (c) treatment of global problems in which one exterior region and a plurality of interior regions are all treated by infinitesimal scaling, and (d) tests of cases in which the infinitesimal scaling coefficient matrix for an exterior region is combined with coefficient matrices obtained with standard interpolation trial functions for small interior elements in one global problem. (The latter tests are most representative of intended production applications.)

The type (a) tests were for squares in two dimensions and spheres in $r-z$ geometry. Harmonic functions of the form $\cos(n\theta)/r^n$ and $\sin(n\theta)/r^n$ were utilized in the former case, and $P_n(\cos\theta)/r^{n+1}$ in the latter case. In general, the functional provided by infinitesimal scaling was accurate to better than one part per million for small n, and one part per thousand for values of n such that $n\theta$ changes by less than 45 degrees within one boundary element. For larger values of n, the behavior of the trial function on the boundary could not be accurately represented by the boundary basis functions that were utilized, so that disagreements were to be expected. For example, for a square with four sixth-order boundary elements per side, the agreement was to order one part in ten thousand for values of n up to 9. It must be emphasized that the existence of disagreements for the larger values of n does not necessarily imply that there are errors in the coefficient matrix since, by definition, the coefficient matrix is only valid for harmonic functions that are rigorously expandable in the boundary basis functions everywhere on the boundary.

The tests of type (b) and (c) were possible because, by assuming that source terms are concentrated on boundary nodes, it becomes possible to evaluate the source-dependent part of the global functional in a trivial manner. The solution of the global problem gives only the nodal values of the potential which, in type (b) examples, are restricted to the boundary between the interior and exterior regions. In addition to comparing these boundary node values with theoretical expressions in simple cases (e.g., the potential due to a pair of line charges in the two-dimensional Laplacian tests) it is possible to calculate interior and exterior fields at other points in space by the method of Appendix A in Ref. 5. By treating the problem of a pair of current loops on the surface of a finite cylinder, utilizing the infinitesimal scaling interior and exterior region coefficient matrices for the azimuthal component of the vector potential, it was possible to accurately reproduce the elliptic integral expressions for the vector potential due to circular current loops.

An interesting set of tests was made with the interior and exterior region coefficient matrices computed for the two-dimensional Laplacian case with a square boundary having sixteen fourth-order elements per side (256 boundary nodes in all). Coefficient matrices for 128, 32, and 8 node squares (all with quadratic boundary elements) were then synthesized from the original coefficient matrices. The external region of the 32 node square was coupled to a four by four array of 8 node interior squares, and the global problem was solved for pairs of positive and negative line charges (or line currents) located on various pairs of nodes. Good agreement was obtained with the theoretical potentials for nodes not in the same element as one of the localized sources. (It was, of course, not possible to match the logarithmic singularity at closer distances to the localized sources.) When the pair of source points was moved around within the interior grid, the potential contours were found to retain their shape regardless of the distance of the source pair from the boundary of the interior grid. This gave assurance that the exterior region coefficient matrix was correctly incorporating the long-range interactions that are present in this problem.

Similar tests were made utilizing the conventional "serendipity" coefficient matrix for the interior eight-node squares. Despite the substantial errors in the individual elements of this coefficient matrix, the global accuracy was not seriously impaired. This confirms the expectation that it is not essential to utilize accurate trial functions in small interior elements.

Another test of combining an infinitesimal scaling exterior region coefficient matrix was made for a finite circular cylinder in $r-z$ geometry. Symmetry was assumed around the plane $z=0$, and sixteen quadratic boundary elements were used on the outer radius and on the end face of the cylinder. A 16 × 16 grid of quadratic finite elements was used in the interior of the positive z half of the cylinder. The field lines resulting from a pair of symmetrically placed current loops interior to the cylinder are shown in Fig. 1. It is apparent from this figure that the proximity of the inner boundary of the exterior region to the current loops does not significantly distort the field lines.

When the coefficient matrices are known for the interior and exterior regions associated with a particular boundary, the expansion coeffieients of the magnetic field at the center of the interior region can be determined from the vector potential values on the boundary nodes [5]. This provides a useful check of the consistency of the coefficient matrices, and, for simple field distributions within an axially symmetric finite cylindrical boundary, the error is typically of order 1 part in 10^8.

CONCLUSION

As demonstrated by the tests described in the previous section, infinitesimal scaling enables the coefficient matrices for extended

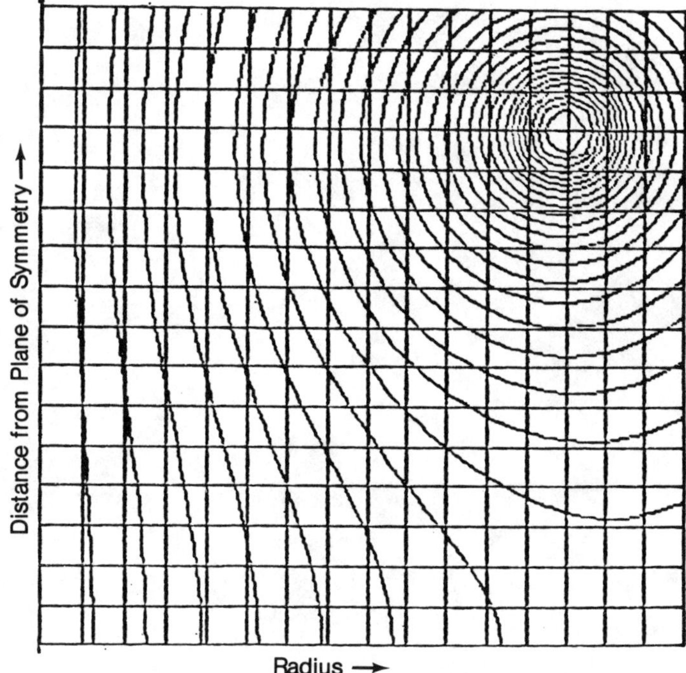

Fig. 1. Magnetic field lines (contours of γA_ϕ) due to two symmetrically placed filamentary circular current loops. A standard finite element grid was used within the finite circular cylindrical region shown, and infinitesimal scaling was used for the exterior region.

interior elements and unbounded external elements to be calculated with high degree of accuracy. These coefficient matrices can be conveniently combined with coefficient matrices computed by conventional techniques for smaller interior elements in the same finite element problem. The coupling of infinitesimal scaling with conventional finite element techniques may considerably reduce the number of nodes that must be considered in a global problem.

It should be possible to apply the infinitesimal scaling technique to full three-dimensional problems, particularly when symmetry properties of the boundary of the exterior region can be utilized to reduce the order of the matrices appearing in the fundamental infinitesimal scaling equation.

The algorithm that has been developed for solving the nonlinear infinitesimal scaling matrix equation for the coefficient matrix in the magnetostatic case provides accurate results with an acceptable amount of computer time. Nevertheless, rather than calculating a new coefficient matrix for each individual applied problem, it appears desirable to build up libraries of coefficient matrices for a selected number of standard boundary surfaces. These standard coefficient matrices can then be used for a number of different problems of practical interest by matching the interior grids to the appropriate standard boundary.

Additional investigations appear warranted to further improve the algorithms for solving the infinitesimal scaling matrix equation and to demonstrate the applicability of the method to full three-dimensional problems and to Helmholtz and diffusion problems. Finally, the extension of infinitesimal scaling to the treatment of full vector field problems, though more complicated than the scalar case, may afford unique advantages that are not achievable by other methods.

It is of particular interest and importance to make detailed quantitative comparisons of the coefficient matrices obtained by infinitesimal scaling with the results of other methods such as boundary integral techniques and ballooning. The ease of developing flexible programs for dealing with general boundaries and choices of boundary element basis functions will be as important as comparative computing requirements in determining the overall attractiveness of the alternative techniques.

ACKNOWLEDGEMENTS

The author expresses appreciation to the large number of individuals who have provided encouragement and help in developing the method of infinitesimal scaling. Dr. M.V.K. Chari provided initial impetus by suggesting the investigation of unbounded regions, P. Moore authored inititial programs for applying infinitesimal scaling to two-dimensional circles and squares, E. Wachspress provided invaluable advice in regard to quadratically convergent algorithms for solving the fundamental infinitesimal scaling equation, S. Lamson authored the Lyapunov solver, M. Palmo carried out the coupling of infinitesimal scaling to standard methods in $r-z$ geometry, and C. Crowley has extended this work to more complex situations.

REFERENCES

[1] P.P. Silvester, M. Hsieh, "Finite element solution of 2 dimensional exterior field problems," in *Proc. IEE*, Vol. 118, No. 12, Dec. 1971.

[2] P. Silvester et al., "Exterior finite elements for 2 dimensional field problems with open boundaries," in *Proc. IEE*, Vol. 124, No. 12, Dec. 1977, p. 1267.

[3] C.F.R.L. Antunes, E.M. Freeman, D. Lowther, and P. Silvester, "A static ballooning technique for 2-D open boundary problems," in *J. Appl. Phys. 53 (11)*, Nov. 1982, p. 8360.

[4] H. Hurwitz, Jr. "Tests of a nonlinear matrix equation formulation of infinitesimal scaling," General Electric Corporate Research and Development Report 83CRD084, May 1983.

[5] H. Hurwitz, Jr. "Application of infinitesimal scaling to finite element analysis," General Electric Corporate Research and Development Report 83CRD085, May 1983.

[6] R.H. Bartels and G.W. Stewart "Algorithm 432 Solution of the matrix equation AX + XB = C [F4]," in *Comm. ACM*, Vol. 15, No. 9, Sept. 1972, p. 820.

[7] Mathlab Group, "MACSYMA Reference Manual," Laboratory for Computer Science, Massachusetts Institute of Technology, Cambridge, MA, December 1977.

Finite-Element Analysis of Optical and Microwave Waveguide Problems

B. M. AZIZUR RAHMAN, MEMBER, IEEE, AND J. BRIAN DAVIES, MEMBER, IEEE

***Abstract*—A vector H-field formulation is developed for electromagnetic wave propagation for a wide range of guided-wave problems. It is capable of solving microwave or optical waveguide problems with arbitrarily anisotropic materials. We have introduced infinite elements to extend the region of explicit field representation to infinity, to consider open-type waveguides more accurately. Computed results are given for a variety of optical planar guides, image lines, and waveguides containing skew anisotropic dielectric.**

I. Introduction

As THE RANGE of guiding structures becomes more intricate, so the need for computer analysis becomes greater and more demanding. Some guides are so important as to justify specialized techniques adapted to their needs, such as microstrip or the various optical fibers. But, for its flexibility or versatility, the finite-element method has become a powerful tool throughout engineering.

In this work, we are concerned with guiding structures that are uniform in the direction of wave propagation, but where the "guiding" arises from nonuniformity of material in the transverse dimensions. Any guide cross section can be divided into a patchwork of triangular elements, and the appropriate field components represented by polynomials over these elements. For open-type guides, beyond the cross section represented by these orthodox triangular elements, we add "infinite elements"—rectangular strips which give explicit field representations to infinity in all of the necessary transverse directions. This is shown in Fig. 1 for a channel waveguide, where, because of the mirror symmetry, the problem is reduced to a half.

The finite-element variational formulation is made via a full $\boldsymbol{H}$ vector field [1], where each field component H_x, H_y, and H_z is separately represented by a function continuous over the whole transverse plane. This is particularly convenient for guides with permittivity discontinuities (e.g., image guide) where continuity of $\boldsymbol{H}$-field components is automatically satisfied.

Each element may have a different dielectric constant, and the constant may be arbitrarily tensor (loss-free). Applying stationariness of the variational form reduces the problem to a standard eigenvalue matrix equation. For a large-order matrix, most of the terms are zero, and so we have used a truly sparse method [2] to solve the equation.

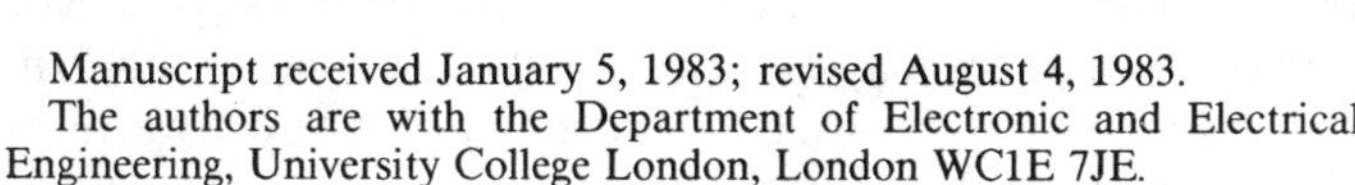

Manuscript received January 5, 1983; revised August 4, 1983.
The authors are with the Department of Electronic and Electrical Engineering, University College London, London WC1E 7JE.

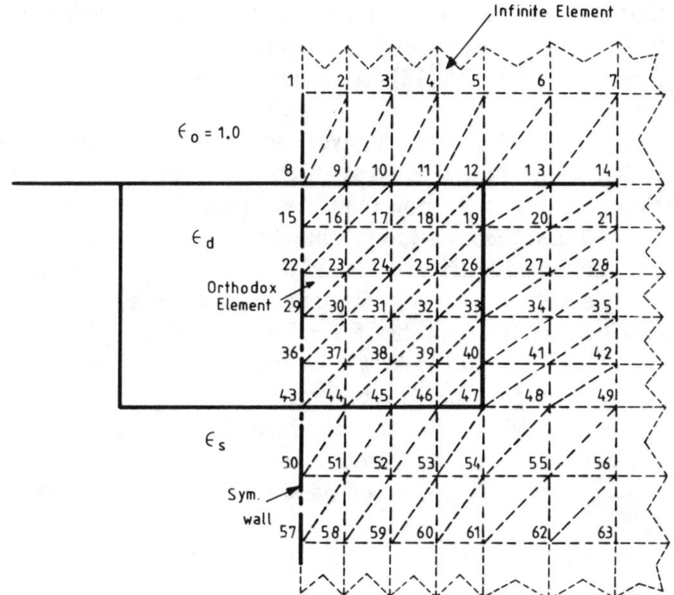

Fig. 1. Node representation of a channel guide, with orthodox and infinite elements.

II. Variational Formulation for Finite Elements

There are different forms of variational formulation used for finite-element methods. The scalar form of finite-element method has been used for solving homogeneous waveguide problems [3], [4]. This single scalar formulation is inadequate for the inherently hybrid mode situation of inhomogeneous or anisotropic problems. A finite-element formulation using an axial component of the fields (E_z and H_z) has been used to solve many different types of guiding structures problems [5]–[9]. This formulation cannot, without destroying the canonical form of the eigenvalue matrix, treat general anisotropic problems. For a waveguide with arbitrary dielectric distribution, the satisfying of boundary conditions in this method can be quite difficult. Another fundamental disadvantage of this method for optical dielectric waveguide problems is that it considers axial components of the fields, which are the least important components of the six-vector field. Berk [1] derived vector variational formulations in the form of a Rayleigh quotient for loss-free anisotropic waveguides and resonators. Later, Morishita and Kumagai [10] derived similar vector variational formulations. By contrast, a vector $\boldsymbol{E}$ field was applied by English and Young [11]. This formula-

Reprinted from *IEEE Trans. Microwave Theory Tech.*, vol. MTT-32, no. 1, pp. 20–28, Jan. 1984.

tion can solve general anisotropic problems. They considered a vector $\boldsymbol{E}$ formulation as it involved more Dirichlet (rather than Neumann) boundary conditions associated with the fields. But for this formulation, the natural boundary condition is that of a magnetic wall, which cannot be left free for an electric wall boundary. As the necessary boundary condition $\boldsymbol{n} \times \boldsymbol{E} = 0$ must be imposed on any conducting boundaries, it is an added difficulty to implement that boundary condition on arbitrarily shaped guide walls. Another disadvantage of this formulation is that at the dielectric interface it needs special consideration to ensure continuity of the tangential components of the fields. Ohtaka *et al.* [12] used a variational form in terms of the transverse component of $\boldsymbol{H}$; compared with the full three-vector $\boldsymbol{H}$, this involves additional differentiation which would be particularly disadvantageous with a finite element approach. For many purposes, a vector $\boldsymbol{H}$-field formulation [13] has the advantage over all others. In this formulation, the natural boundary condition is that of an electric wall, so that for arbitrary conducting guide walls it can be left free. In this formulation, the chosen field is continuous at the dielectric interfaces, and so it is very convenient for the finite-element solution of dielectric waveguide problems. In this formulation, we can also consider general anisotropic problems, which are important for many active and passive integrated optics structures.

This vector $\boldsymbol{H}$-field formulation can be written as [1]

$$\omega^2 = \frac{\int (\text{Curl}\, \boldsymbol{H})^* \cdot \hat{\epsilon}^{-1} \cdot \text{Curl}\, \boldsymbol{H}\, dV}{\int \boldsymbol{H}^* \cdot \hat{\mu} \cdot \boldsymbol{H}\, dV} \tag{1}$$

where $\hat{\epsilon}$ and $\hat{\mu}$ can be general anisotropic permittivity and permeability of the loss-free medium, respectively. As the natural boundary condition is that of an electric wall, we need not force any boundary condition on conducting guide walls. But for regularly shaped waveguides, and at the symmetric walls (if applicable), we can enforce the boundary conditions to reduce the problem size.

III. Finite-Element Formulation

In the finite-element method [14], we first discretize the entire problem domain into a finite number of triangular subregions, called elements. In general, using many elements, we can approximate any continuum with a complex boundary and with an arbitrary index distribution to such a degree that an accurate analysis can be carried out. The field functions (each of H_x, H_y, and H_z) are defined by a set of algebraic polynomials over each element in the transverse plane, and longitudinal dependence $\exp(-j\beta z)$ is assumed, for given β. By expressing these fields in terms of "nodal values," most or all of which occur on the element boundaries, the resulting field components can be continuous over the whole domain. The extremum functional from (1) is then minimized with respect to the nodal values of the field components. In this way, the minimization generates a set of linear algebraic eigenvalue equations. This eigenvalue equation can be written as

$$Ax - \lambda Bx = 0 \tag{2}$$

where A is a complex hermitian, B is a real symmetric matrix, and λ is proportional to ω^2. We can solve this eigenvalue problem by one of various standard subroutines to get different eigenvectors and eigenvalues.

In general, the matrices A and B are quite sparse and have the same sparsity pattern. In the finite-element method, we discretize the whole problem domain in many smaller triangles, so that the continuous field problem is reduced to finding fields at discrete nodes, where the unknown field value is only coupled to the field values of the neighboring nodes. When we divide the problem domain into many regularly shaped triangles, then the number of nonzero terms per row and/or per column in A or B matrices will be a maximum of seven, irrespective of the order of the matrices (for scalar field formulation with first-order "shape functions," viz. first-degree polynomials). The nonzero coefficients can be even lower than seven per row or column when we consider nodal points on, or adjacent to, the boundary. Their coupling increases when we use higher order shape functions or more field components per nodal points, but always a very high proportion of the A and B matrix elements are zero. As an example, if we take a typical problem, using vector $\boldsymbol{H}$ field and a first-order shape function, when the order of the matrices is 397, then the percentage of nonzero terms is only 4.69 percent. We have exploited geometric symmetry and applied required boundary conditions at the time of assembling the global matrices, so reducing the storage required, as only the nonzero elements of the reduced matrices have been stored. Using a "search-limiting algorithm," we have greatly reduced the searching time required to decide whether a particular contribution to the global matrix has been made earlier or not.

In contrast to the usual dense and band-matrix algorithms [15], we have used a specially developed [16] "arbitrary sparse matrix" algorithm. Like the "subspace iteration" and "Sturm count" algorithms described by Bathe and Wilson [15], any band of adjacent eigenvalues can be safely computed, together with their associated eigenvectors. The advantage of the sparse matrix version is in taking into account all of the many zeros of (2), and so allowing a larger matrix order for given computer store. Its storage requirement is roughly proportional to the order of the matrices, rather than the square of the order of matrices, as in the dense version. This subroutine can find one or any number of eigenvalues (along with their eigenvectors) close to a specified point in the spectrum.

One question of tradeoffs in finite elements is (for given computer store and/or computer time) whether to use low-degree polynomials in many elements or high-degree polynomials in few elements. The "correct" choice, for most accurate results, depends on the application and on the matrix algorithm used to solve (2). For any arbitrarily shaped waveguide with arbitrary index distribution, our experience is that better accuracy is achieved (for given computer time and storage) by using many small-order triangles rather than fewer higher order triangles. The lower order shape function can represent the problem domain more accurately for a given order of matrix eigenvalue problem and gives higher sparsity of the matrices, and so it is advantageous when (as in our case) a completely sparse algorithm is used to solve the matrix eigenvalue problems. For all our modeling of 1) optical stripe guides with smooth refractive profiles, 2) image guides with a rather irregularly shaped cross section, and 3) guides with skew anisotropic permittivity (all illustrated later in this paper), the first-degree polynomial has been used and has proved to be effective.

IV. Infinite Elements

One property of the open-type waveguide is that finite fields exist in the region outside the guide. Outside the guide core, the field decays and the region of interest extends to infinity. The modeling of the infinite transverse extent of the waveguide presents a problem, and in orthodox finite-element discretization [3]–[9] one does not extend the region of consideration up to infinity. This extension of the problem domain is particularly important for the solutions close to cutoff, as (by definition of cutoff) the fields decay slowly and the region of significant field value can be arbitrarily large. To date, the customary approach is the simple truncation, which sets artificial boundary walls enclosing the guide [9]. In this case, care has to be taken so that it is sufficiently remote for the influence of the position of this artificial wall to be negligible, and at the same time it is sufficiently close to limit the necessary number of finite elements. One technique [8] involves shifting the virtual boundary wall to satisfy a given criterion for the maximum field strength at that wall.

Another approach is to use a recursion technique [17] to generate the matrix representing the region outside the main domain. It is possible to find the internal and external solutions and to match them on an imaginary boundary, some sort of integral solution being possible for the outside region. McDonald and Wexler [18] have used the finite-element method for a bounded region along with an integral equation for an unbounded region. In field matching techniques, these "two region solutions" are quite common [19].

Boundary elements have been used by Yeh *et al.* [20], considering an exponential decay outside the guide core. But in their method, because of nonconformity of the two coordinate systems, the fields used were not continuous. Bettes [21] has used infinite elements with a Cartesian coordinate system, using an exponential decay and Lagrange multipliers, for fluid flow problems.

An infinite element is a finite element that indeed extends to infinity, and shape functions for such an element should be realistic to represent the fields and should be square integrable over an infinite-element area, to satisfy the radiation condition [22]. For an infinite element extending towards infinity in the x-direction, we can assume exponential decay in x and conventional shape function

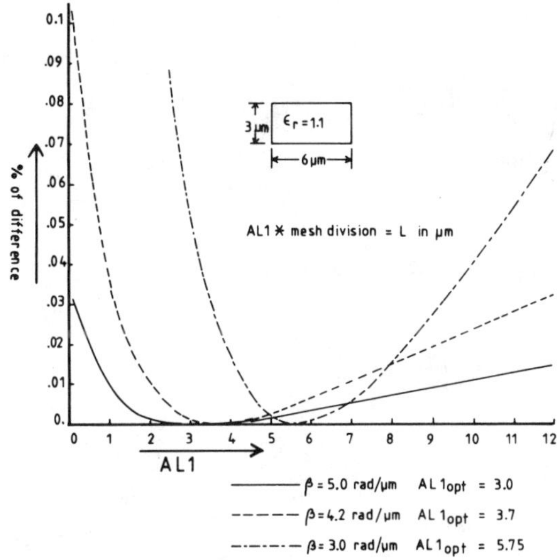

Fig. 2. Optimization of a decay parameter at three different frequencies.

dependence in the y-direction. It can be written as

$$N_i(x, y) = f_i(y)\exp(-x/L) \tag{3}$$

where L is a decay length. Similarly, we can have infinite elements extending towards the y-direction by considering exponential decay in y. Similarly, by assuming exponential decay in both x and y, we can consider a rectangular or quadrant element extending towards infinity in both x- and y-directions. Integration of these shape functions or their derivatives over the infinite elements are finite, and can be simply carried out. Combining all of these elements along with the conventional finite element, we can represent any open-type waveguide cross-sectional domain very conveniently, with each field component being continuous over the whole infinite domain.

Fig. 1 illustrates the use of infinite elements for the analysis of a "channel" guide [23]. Orthodox triangular elements are used in the central region. The infinite elements can be seen extending in all Cartesian directions. In infinite elements we need to assume exponential decay factors, which are unknown at the beginning. The best value of decay parameter depends not only on the specific problem, but also on the operating frequency, and can also be different for different elements. Nevertheless, we would contend that any reasonable parameter will be better than considering a perfect reflector at the same boundary location. Instead of assuming many decay parameters, we restrict ourselves to a few, and subsequently optimize them. Varying any one of them, we can find a stationary solution, and the value of the parameter which makes the solution stationary is taken as the optimized value. In Fig. 2, we show the stationary nature of the solution for ω^2 as ordinate with variation of a particular decaying factor $AL1$ for three different propagation constants (i.e., for three frequencies). In Fig. 3, we show the variation of optimized $AL1$ with a propagation constant. We observe that as the frequency increases, the $AL1_{\text{opt}}$ is reduced, which corresponds to the fact that fields are more confined inside the guide at the

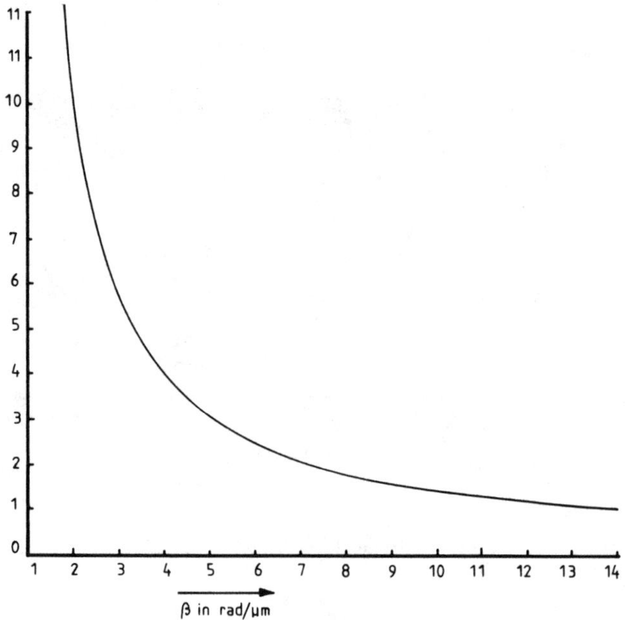

Fig. 3. Variation of an optimized decay parameter with propagation constants.

higher frequencies. Figs. 2 and 3 are for a rectangular dielectric waveguide with dimension 6.0 μm $\times$ 3.0 μm having $\epsilon_r = 1.1$.

V. Spurious Modes

Like other workers [2], [6], [8], [9], and [13], we have observed the presence of spurious solutions along with the physical solutions. These spurious solutions apparently exist for all types of vector formulations. Their exact cause of appearance still is not resolved. It could be due to various factors such as enforcement of boundary conditions, the positive definiteness of the operator, possibly due to the nonzero divergence of the trial fields, or maybe because the system is too flexible [24]. They can be identified from the true solutions by different ways, one of which is by inspecting corresponding eigenvectors. We have developed a procedure which gives a very easy identification of the probability of being a physical or spurious solution. The logic behind the scheme is the fact that a physical eigenvector should obey $\operatorname{div} \boldsymbol{B} = 0$. Thus we calculate $\operatorname{div} \boldsymbol{H}$ for each eigenvector of interest and compare their values. We have observed that, for all problems, a true solution has lower $\operatorname{div} \boldsymbol{H}$ than the spurious solutions. Comparing the $\operatorname{div} \boldsymbol{H}$ value of a particular mode with others helps in identifying their validity. In all these cases, we have calculated $\operatorname{div} \boldsymbol{H}$ from the discrete nodal field values obtained after the solution of the eigenvalue equation (2).

VI. Rectangular Dielectric Waveguides

A dielectric fiber with a refractive index higher than its surrounding region is a simple form of an optical waveguide. These guides support various hybrid modes, which can be grouped into two families H_{mn}^x and H_{mn}^y, and we have used the mode designation as used by Marcatili [23]. We show propagation characteristics for H_{11}^x and H_{31}^x modes, for guide dimension 6.0 μm $\times$ 3.0 μm and $\epsilon_r = 1.1$ in Fig. 4, including a comparison with Yeh [20], Marcatili [23], and Goell [25]. Fig. 5(a) and (b) shows the eigenvec-

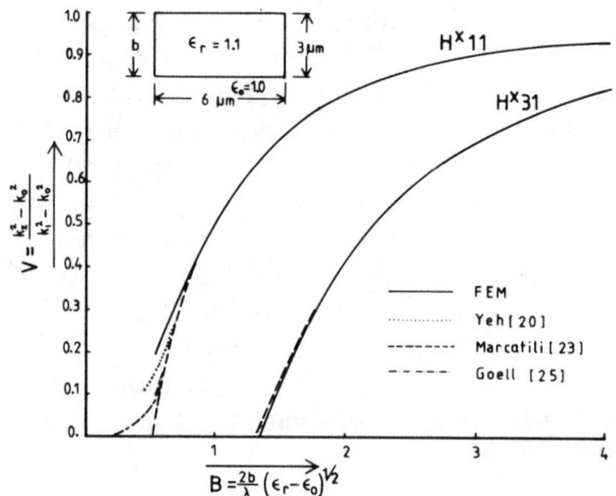

Fig. 4. Propagation characteristics for a rectangular dielectric waveguide, finite element divisions 8×8 (128 triangles).

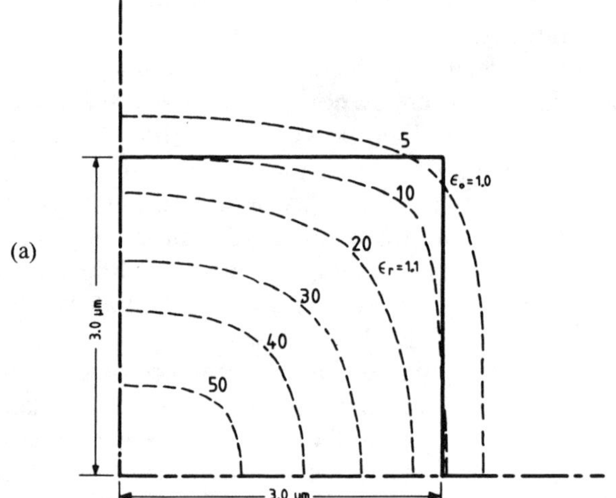

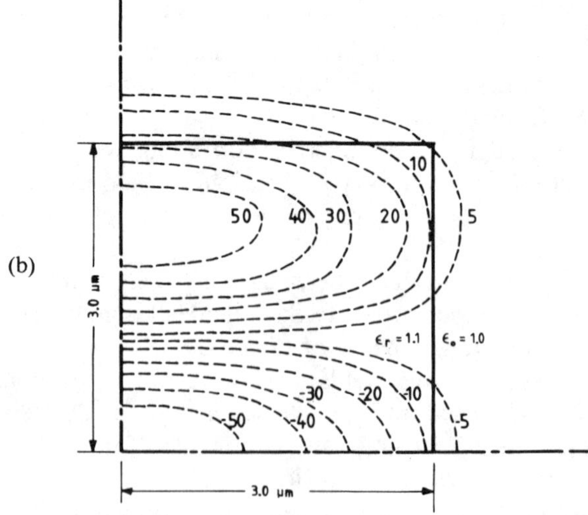

Fig. 5. H_x field contour in a rectangular dielectric waveguide. (a) H_{11}^x mode. (b) H_{13}^x mode.

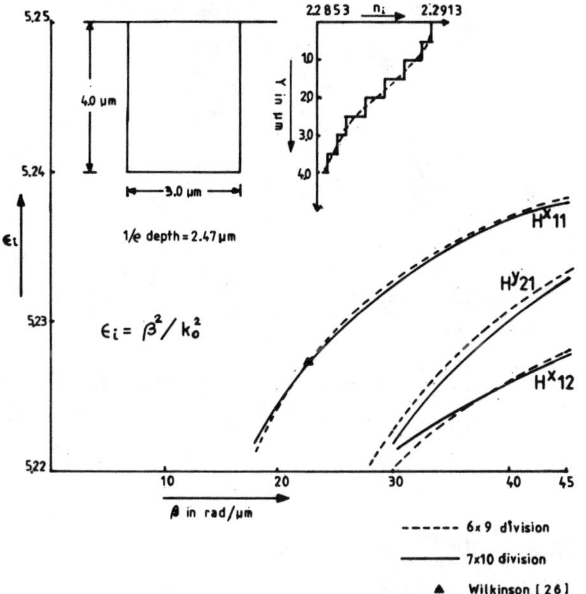

Fig. 6. Propagation characteristics of a titanium-diffused channel waveguide with a refractive index varying as a Gaussian function.

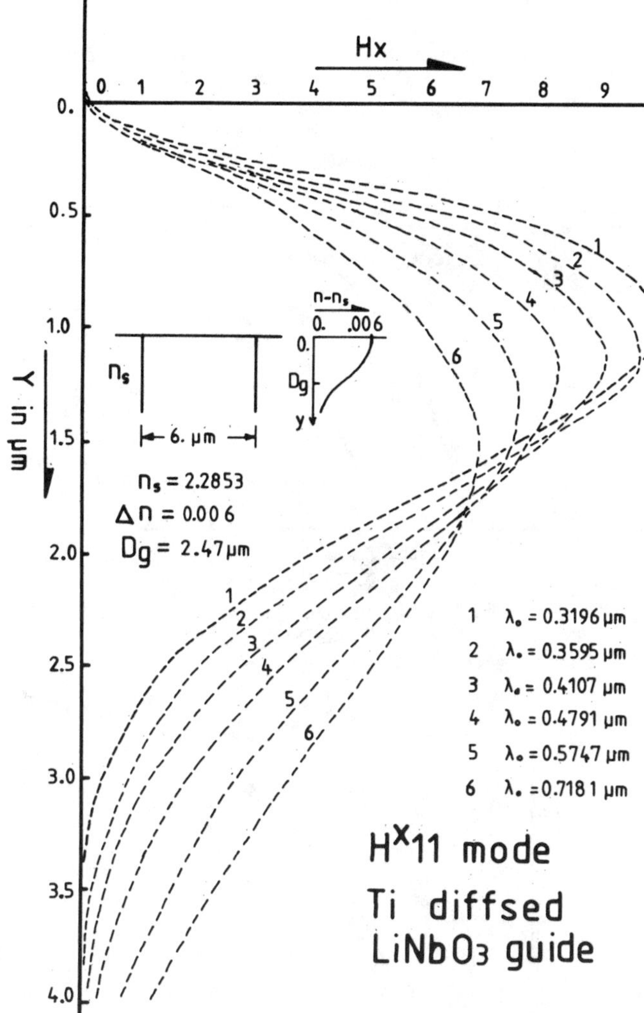

Fig. 7. Field distribution with depth y for titanium-diffused channel waveguide at different wavelengths.

tors for the H_{11}^x mode and the H_{13}^x mode, respectively, for a guide dimension of 6.0 μm$\times$6.0 μm with $\epsilon_r = 1.1$.

VII. Channel Waveguide

A channel, or embedded-strip, waveguide is a special waveguide of practical interest. The guide can be fabricated with more relaxed requirements for the resolution and edge roughness than for the rectangular dielectric waveguide. The refractive index of the guiding section is higher than that of the substrate, and this difference can be achieved in just one step, or smoothly, such as a linear or exponential variation. For comparison with the experiment, we have solved a titanium-diffused channel waveguide, with a strip width of 3.0 μm and with a refractive index following a Gaussian function of guide depth. At the top of the guide, the refractive index is 2.2913 and that of the substrate is 2.2853. At the depth of 2.47 μm, it attains the value $2.2853+(2.2913-2.2853)/e$. Propagation characteristics were computed for the H_{11}^x, H_{21}^y, and H_{12}^x modes, and the results are shown in Fig. 6. It shows good agreement with the point obtained experimentally by Wilkinson [26]. The field variation with guide depths for different frequencies is shown in Fig. 7.

VIII. Rib Waveguide

The rib waveguide is another type of dielectric waveguide having considerable practical importance. This is a special type of slab-coupled waveguide, where the loading strip is made out of the same material as the slab. We have analyzed a GaAs/GaAlAs rib waveguide, where the dielectric constant of the rib and the slab is 11.8336 and that of the substrate is 11.792356, as shown in Fig. 8. For all of these rib waveguide problems, the dimension XDL of Fig. 8 is taken as 5.0 μm, and $YDL + YARM$ is taken to be constant and equal to 3.0 μm. Fig. 8 shows the propagation characteristics for the H_{11}^x mode for five different rib heights, including a point from Austin [27]. Fig. 9 shows propagation characteristics for four modes when the rib height is 0.5 μm. Fig. 10 shows the field variation in the symmetry plane normal to the surface, for different frequencies. Fig. 11(a) and (b) shows eigenvectors for H_{11}^x modes for two different rib heights.

IX. Image Lines

At millimeter and submillimeter wavelengths, the dielectric image guide is a convenient form of waveguide. The geometry of an image line, as shown in Fig. 12, consists of a rectangular dielectric slab with relative permittivity ϵ_r, backed by a perfectly conducting metal plate, and surrounded by a semi-infinite medium of lower dielectric constant (usually air). We have solved a typical structure (10.0 mm$\times$5.0 mm guide) with dielectric constant 1.05, and its propagation characteristics are shown in Fig. 12. In Fig. 13, we compare the solution of image lines having different dielectric constants. Fig. 14(a) and (b) shows eigenvectors for two modes of an image line. We have also solved an image line with a deformed cross section due to the fabrication process used. Its dielectric constant was unknown, but its experimental dispersion curve was available [28]. We concluded that a dielectric constant value about 7.5 is a reasonable approximation for the unknown value. Associated dispersion curves are shown in Fig. 15.

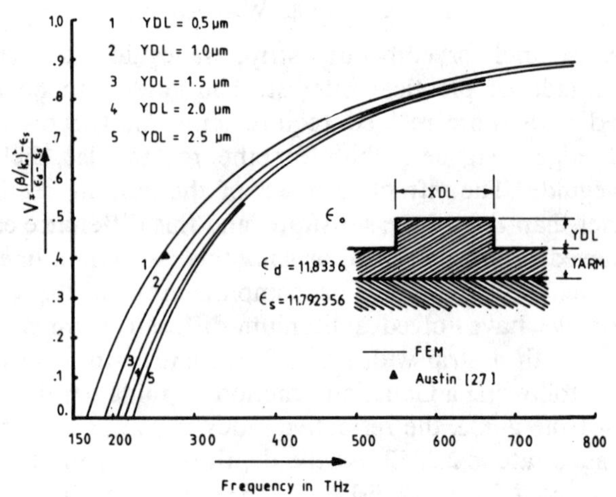

Fig. 8. Propagation characteristics for H_{11}^x modes in rib waveguides for different rib heights.

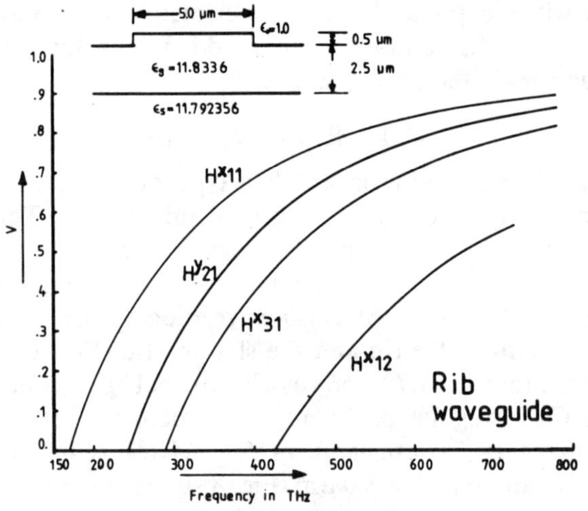

Fig. 9. Propagation characteristics for four modes in rib waveguide, $YDL = 0.5$ μm, and $YARM = 2.5$ μm.

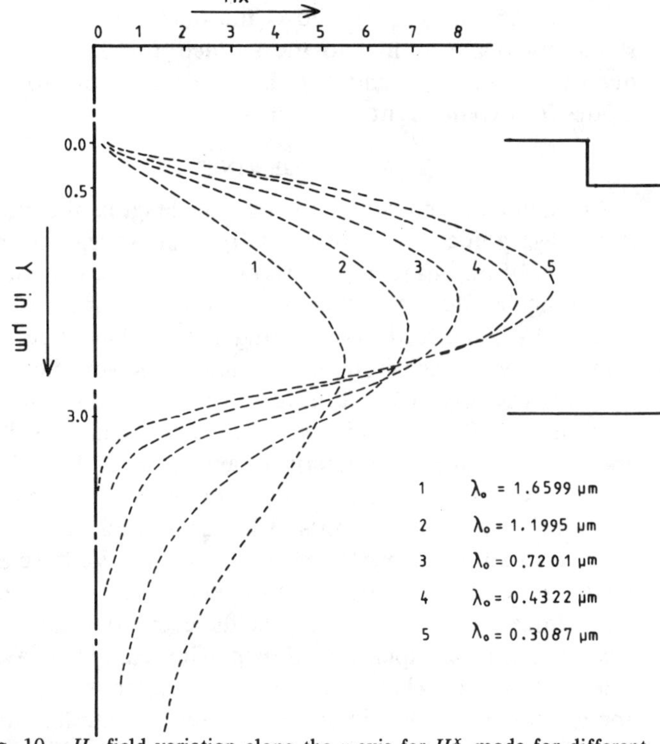

Fig. 10. H_x field variation along the y-axis for H_{11}^x mode for different frequencies $YDL = 0.5$ μm.

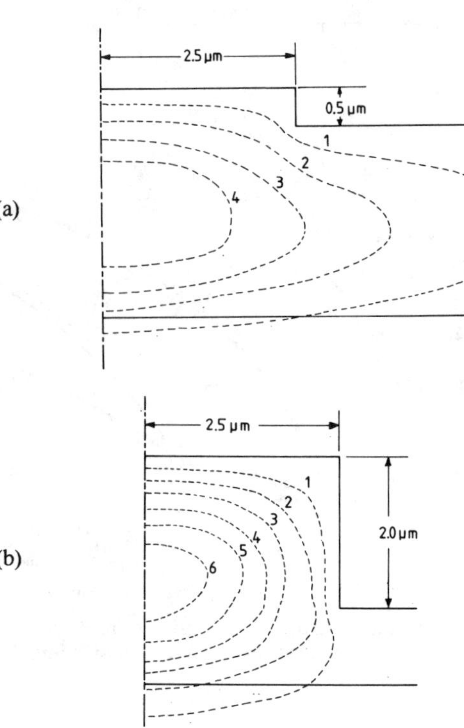

Fig. 11. H_x field contour at 0.4502-μm wavelength. (a) $YDL = 0.5$ μm. (b) $YDL = 2.0$ μm.

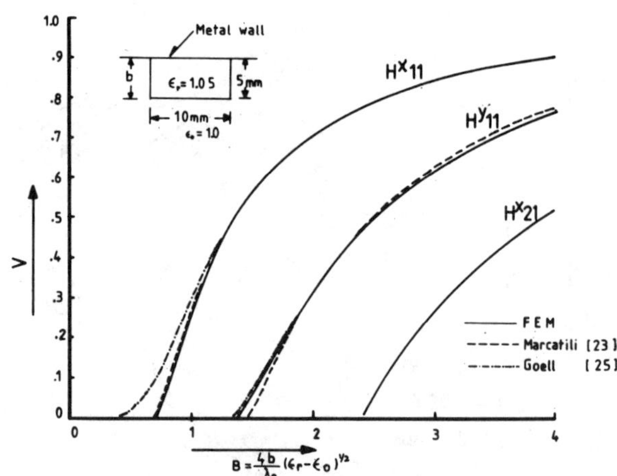

Fig. 12. Normalized propagation characteristics for an image line, mesh division 9×9.

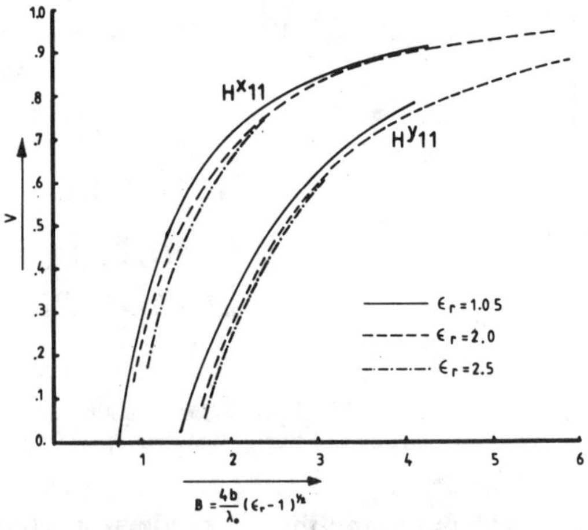

Fig. 13. Comparison of normalized propagation characteristics for image lines with different dielectric constants, mesh division 9×9.

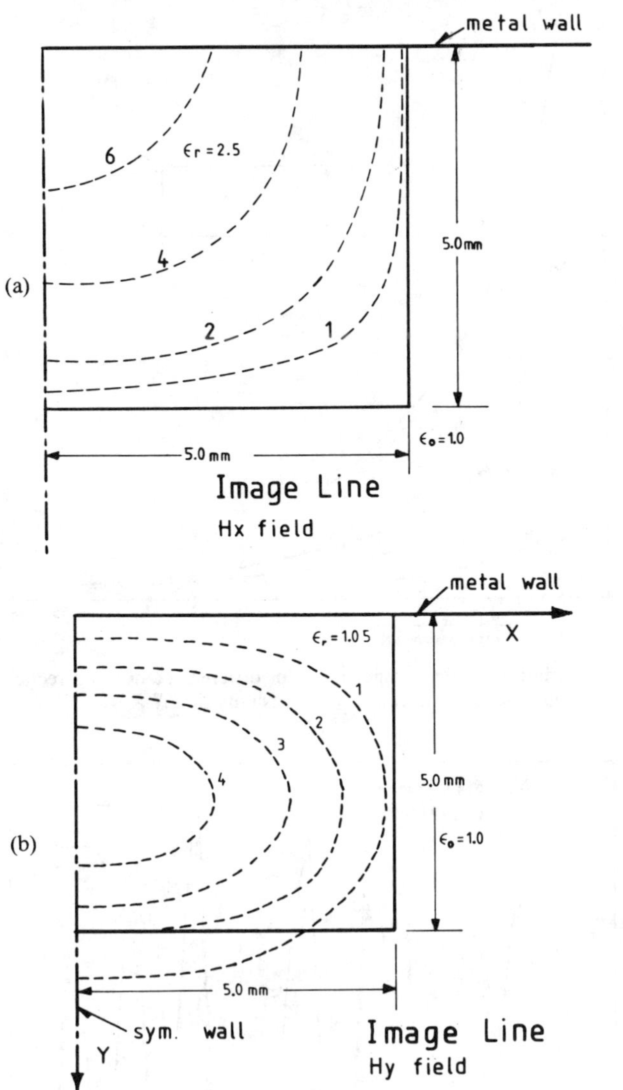

Fig. 14. Field contour for image line. (a) H_x field, $\epsilon_r = 2.50$, H_{11}^x mode. (b) H_y field, $\epsilon_r = 1.05$, H_{11}^y mode.

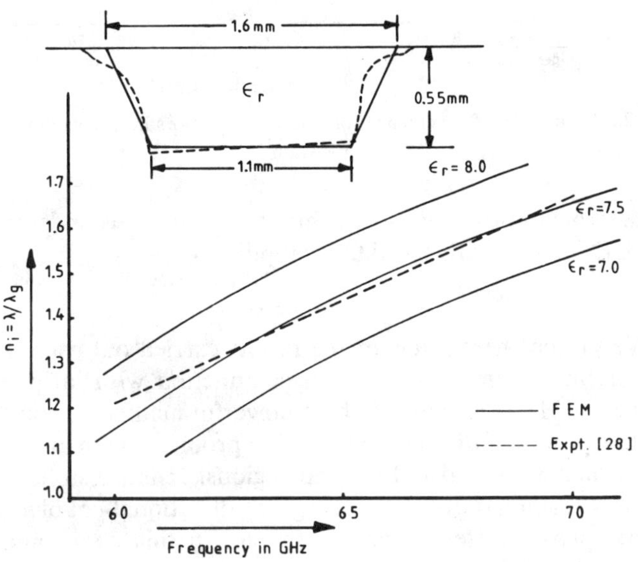

Fig. 15. Propagation characteristics for H_{11}^x mode in the trapezoidal image line, approximating the indicated experimental cross section.

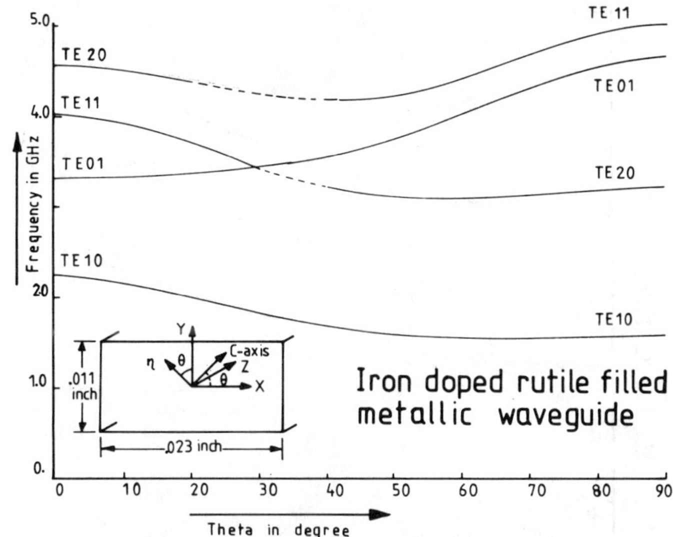

Fig. 16. Variation of cutoff frequency, with theta for four TE modes in an anisotropic waveguide, mesh division 14×14.

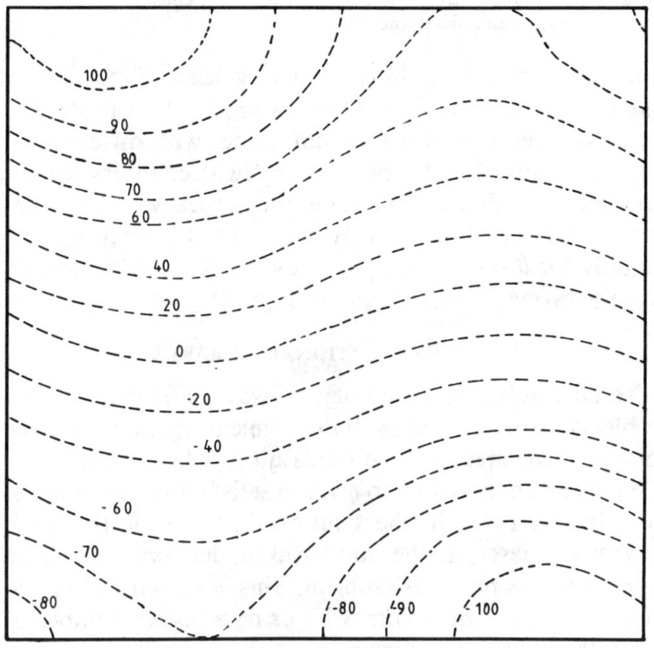

Fig. 17. H_z field contour for TE_{10} mode in anisotropic waveguide at theta = 30°.

X. Anisotropic Waveguide

A number of important applications of anisotropic media have come to the forefront in the field of electromagnetic waves. It is quite important for crystalline materials, such as $LiNbO_3$ or GaAs, as used for optical waveguides. Their special tensor properties can be exploited to design phase shifters, modulators, etc. [29]. To properly analyze these structures, we need a method which is able to handle structures with general anisotropy, and the ***H*** field FEM is capable of doing so. Here we show the results of a metallic rectangular waveguide loaded with iron-doped rutile. Its crystal *c*-axis lies in the transverse plane at any arbitrary angle θ with the larger side. The dielectric constants of this material are 260.0 along the *c*-axis, and 130.0 along the

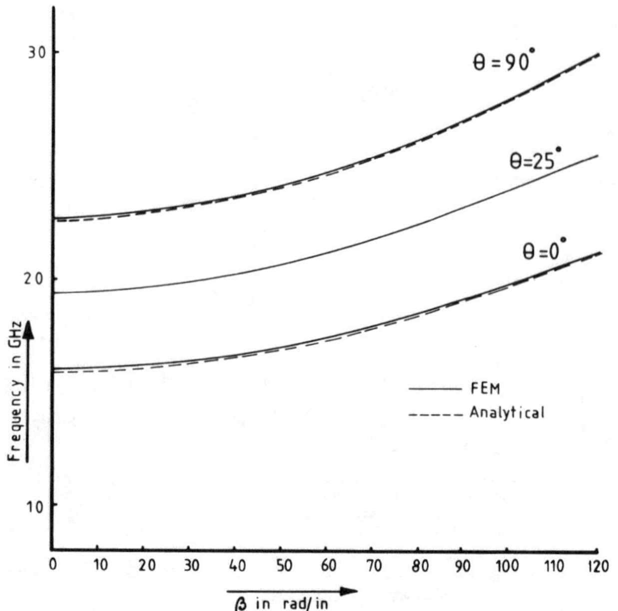

Fig. 18. Variation of frequency, with propagation constant, for a TE_{10} mode in an anisotropic waveguide, mesh division 6×6.

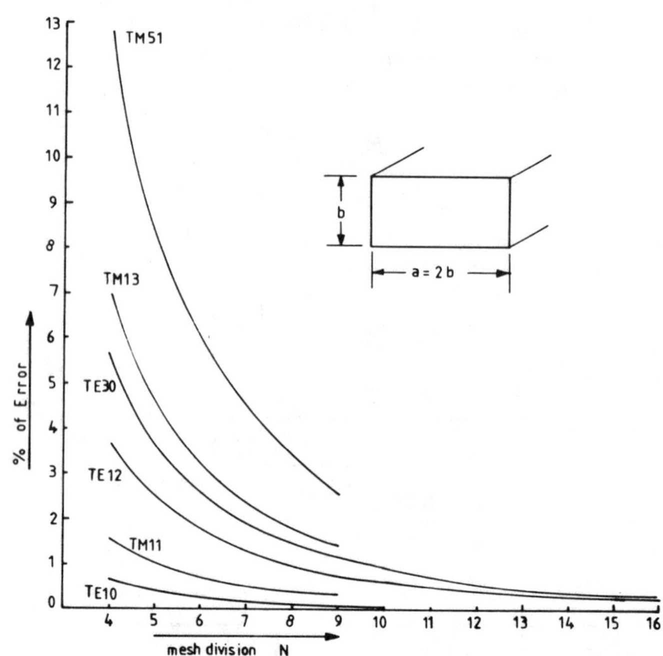

Fig. 19. Variation of percentage errors for different modes in a rectangular metallic waveguide with mesh division $N \times N$.

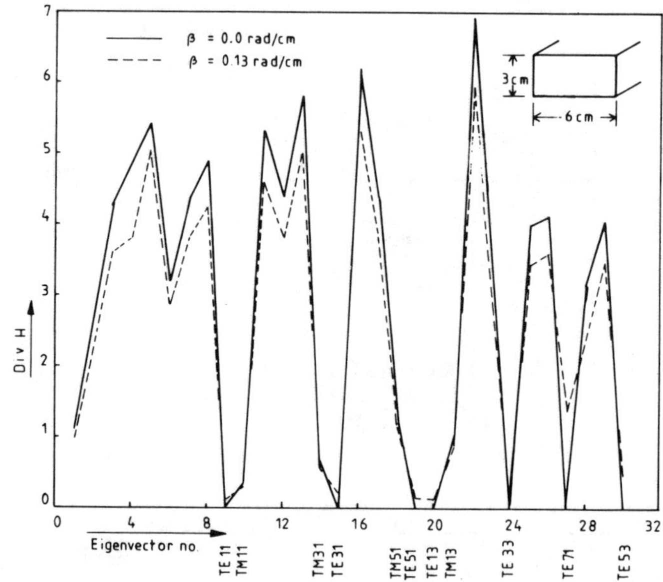

Fig. 20. Variation of divergence for a few eigenvectors in a rectangular metallic waveguide.

transverse axes. The fully loaded guide's dimension is taken to be 0.023 in × 0.011 in, as in [30]. Its variation of the cutoff frequency for different modes with different θ's is given in Fig. 16. We plot the H_z field contours for the TE_{10} mode in Fig. 17. We show the accuracy of the FEM method compared to an analytical method (which is possible only for $\theta = 0°$ or 90°), and the variation of frequency with the propagation constant in Fig. 18.

XI. Computational Remarks

We have solved a wide range of waveguide propagation problems using a vector finite-element method. In the finite-element method, one needs quite a large number of elements or nodal points to achieve satisfactory accuracy of the solution. Error in the solution decreases as the mesh number increases, as the numerical model comes closer to the continuous physical problem. This is shown in Fig. 19. We can increase the accuracy by using a higher number of elements, or by using higher order shape functions, or by using extra-precision word length.

Mention was made in Section V of the appearance of "spurious" modes in addition to the true physical modes. We have identified the spurious solutions from the true solutions by using the "divergence test" [2], where we calculate $|\text{div}\boldsymbol{H}|$ of each possible mode from the computed nodal field values, and then integrate and compare them. A true eigenvector invariably has a smaller value of $|\text{div}\boldsymbol{H}|$ than a spurious one. Fig. 20 shows $|\text{div}\boldsymbol{H}|$ for a few modes for an empty rectangular metallic waveguide.

For most of our problems, we have divided the structure domain into 100 to 200 triangles, and corresponding matrices' orders were around 150 to 350. We store *only* the nonzero elements of the highly sparse matrices, in linear arrays, and solve by using the "subspace iteration method" [2], using single-precision arithmetic. The time required by our GEC minicomputer 4082 or 4090 is around 150 to 600 i.e.u., which corresponds, roughly, to 1–5 cpu seconds on the CDC 7600 (with the FTN compiler, opt = 2).

XII. Conclusions

We present here a few of the results carried out using a vector finite-element method in conjunction with infinite elements [31]. It is found to be a powerful method to solve a wide variety of dielectric waveguide problems with anisotropic materials and unbounded regions. It is possible to use this formulation for the design of directional couplers, filters, phase shifters, modulators, etc. In this paper and earlier work [31], we have ascertained the accuracy, generality, and diversity of this formulation.

Acknowledgment

One of the authors, B. M. A. Rahman, would like to thank the Commonwealth Scholarship Commission for supporting his tenure during the research work.

References

[1] A. D. Berk, "Variational principles for electromagnetic resonators and waveguides," *IRE Trans. Antennas Propagat.*, vol. AP-4, pp. 104–111, Apr. 1956.

[2] J. B. Davies, F. A. Fernandez, and G. Y. Philippou, "Finite element analysis of all modes in cavities with circular symmetry," *IEEE Trans. Microwave Theory Tech.*, vol. MTT-30, pp. 1975–1980, Nov. 1982.

[3] S. Ahmed, "Finite element method for waveguide problems," *Electron Lett.*, vol. 4, pp. 387–389, Sept. 6, 1968.

[4] P. Silvester, "A general high-order finite-element waveguide analysis program," *IEEE Trans. Microwave Theory Tech.*, vol. MTT-17, pp. 204–210, Apr. 1969.

[5] S. Ahmed and P. Daly, "Finite element methods for inhomogeneous waveguides," *Proc. IEE*, vol. 116, pp. 1661–1664, Oct. 1969.

[6] Z. J. Csendes and P. Silvester, "Numerical solution of dielectric loaded waveguide: I-finite-element analysis," *IEEE Trans. Microwave Theory Tech.*, vol. MTT-18, pp. 1124–1131, Dec. 1970.

[7] P. Daly, "Hybrid-mode analysis of microstrip by finite element methods," *IEEE Trans. Microwave Theory Tech.*, vol. MTT-19, pp. 19–25, Jan. 1971.

[8] M. Ikeuchi, H. Sawami, and H. Niki, "Analysis of open-type dielectric waveguides by the finite-element iterative method," *IEEE Trans. Microwave Theory Tech.*, vol. MTT-29, pp. 234–239, Mar. 1981.

[9] N. Mabaya, P. E. Lagasse, and P. Vandenbuleke, "Finite element analysis of optical waveguides," *IEEE Trans. Microwave Theory Tech.*, vol. MTT-29, pp. 600–605, June 1981.

[10] K. Morishita and N. Kumagai, "Unified approach to the derivation of variational expression for electromagnetic fields," *IEEE Trans. Microwave Theory Tech.*, vol. MTT-25, pp. 34–40, Jan. 1977.

[11] W. J. English and F. J. Young, "An *E* vector variational formulation of the Maxwell equations for cylindrical waveguide problems," *IEEE Trans. Microwave Theory Tech.*, vol. MTT-19, pp. 40–46, Jan. 1971.

[12] M. Ohtaka, M. Matsuhara, and N. Kumagai, "Analysis of the guided modes in slab-coupled waveguides using a variational method," *IEEE J. Quantum Electron*, vol. QE-12, pp. 378–382, July 1976.

[13] A. Konrad, "Vector variational formulation of electromagnetic fields in anisotropic media," *IEEE Trans. Microwave Theory Tech.*, vol. MTT-24, pp. 553–559, Sept. 1976.

[14] O. C. Zienkiewicz, *The Finite Element Method*, 3rd ed. London: McGraw-Hill, 1977.

[15] K. J. Bathe and E. L. Wilson, *Numerical Methods in Finite Element Analysis*. Englewood Cliffs, NJ: Prentice Hall, 1976.

[16] F. A. F. Fernandez, "Finite element analysis of rotationally symmetric electromagnetic cavities," Ph.D. thesis, Dept. of Elec. and Elec. Eng., University College, London, 1981.

[17] P. Silvester, D. A. Lowther, C. J. Carpenter, and E. A. Wyatt, "Exterior finite elements for 2-dimensional field problems with open boundaries," *Proc. IEE*, vol. 124, pp. 1267–1270, Dec. 1977.

[18] B. H. McDonald and A. Wexler, "Finite element solution of unbounded field problems," *IEEE Trans. Microwave Theory Tech.*, vol. MTT-20, pp. 841–847, Dec. 1972.

[19] G. Mur and P. J. Fondse, "Cylindrical inhomogeneity in a homogeneous medium of infinite extent," *Microwave Opt. Acoustics*, vol. 3, pp. 224–230, Nov. 1979.

[20] C. Yeh, K. Ha, S. B. Dong, and W. P. Brown, "Single-mode optical waveguides," *Appl. Opt.*, vol. 18, pp. 1490–1504, May 1979.

[21] P. Bettes, "Infinite elements," *Int. J. Num. Meth. Eng.*, vol. 11, no. 1, pp. 53–64, 1977.

[22] D. S. Jones, *The Theory of Electromagnetism*. Oxford: Pergamon Press, 1964.

[23] E. A. J. Marcatili, "Dielectric rectangular waveguide and directional coupler for integrated optics," *Bell Syst. Tech. J.*, vol. 48, pp. 2079–2102, Sept. 1969.

[24] J. R. Winkler and J. B. Davies, "Elimination of spurious modes in finite element analysis," to be published.

[25] J. E. Goell, "A circular-harmonic computer analysis of rectangular dielectric waveguides," *Bell Sys. Tech. J.*, vol. 48, pp. 2133–2160, Sept. 1969.

[26] J. Wilkinson, Dept. of Elec. and Elec. Eng., University College, London, private communication.

[27] M. W. Austin, British Telecom Research Lab., Ipswich, England, private communication and "GaAs/GaAlAs curved rib waveguide," *IEEE Trans. Microwave Theory Tech.*, vol. MTT-30, pp. 641–646, Apr. 1982.

[28] R. V. Gelthrope, Electrical Research Association Ltd., Leatherhead, Surrey, England, private communication.

[29] R. C. Alferness, "Waveguide electrooptic modulators," *IEEE Trans. Microwave Theory Tech.*, vol. MTT-30, pp. 1121–1137, Aug. 1982.

[30] J. B. Davies, "Propagation in rectangular waveguide filled with skew uniaxial dielectric," *IEEE Trans. Microwave Theory Tech.*, vol. MTT-15, pp. 372–376, June 1967.

[31] B. M. A. Rahman, "Finite element solution of optical and microwave guides," Ph.D. thesis, Dept. of Elec. and Elec. Eng., University College, London, Oct. 1982.

Finite-Element Analysis of Waveguide Modes: A Novel Approach That Eliminates Spurious Modes

TUPTIM ANGKAEW, MASANORI MATSUHARA, AND NOBUAKI KUMAGAI, FELLOW, IEEE

Abstract —**An efficient finite-element method for analyzing the propagation characteristics of a wide variety of waveguides is presented. A variational expression suited for the finite-element method is formulated in terms of the transverse electric and magnetic field components. In this approach, all guided-mode solutions are real, while the spurious-mode solutions are not real. Therefore, discrimination of the spurious-mode solutions can be achieved merely by imposing the simple condition that guided-mode solutions be real. Three numerical examples, two for the isotropic case and the other for the magnetic anisotropic case, are carried out.**

I. Introduction

RECENTLY, a method employing finite-element analysis to investigate the propagation characteristics of any arbitrarily shaped waveguide has attracted the attention of many researchers. The finite-element method is a powerful means which enables one to analyze a wide variety of waveguide problems. Several variational formulations for use with the finite-element method have been proposed [1]–[7]. Most of the variational expressions previously used are a functional of frequency and can be classified into the following three types [1]–[6]:

1) variational expressions which are formulated in terms of the longitudinal components of the electric field ($\boldsymbol{e}_z$) and the magnetic field ($\boldsymbol{h}_z$) and can be written as $\omega =$ functional $(\beta, \boldsymbol{e}_z, \boldsymbol{h}_z)$ [1], [2];
2) variational expressions employing the longitudinal component ($\boldsymbol{e}_z$) and the transverse component ($\boldsymbol{e}_t$) of the electric field, which can be written as $\omega =$ functional $(\beta, \boldsymbol{e}_z, \boldsymbol{e}_t)$ [3]; and
3) variational expressions employing all three components of the magnetic field, namely $\boldsymbol{h}_z$ and $\boldsymbol{h}_t$, which can be written as $\omega =$ functional $(\beta, \boldsymbol{h}_z, \boldsymbol{h}_t)$ [4]–[6].

Here, ω is the angular frequency and β is the propagation constant.

The most serious drawback associated with the finite-element method is the appearance of spurious-mode solutions. Up to this time, much effort has been devoted to finding a criterion to eliminate the spurious-mode solutions. A second drawback occurs in the application of variational expressions which are a functional of frequency. Let us consider, for example, the most common requirement in a waveguide problem, namely which mode can exist for a given value of frequency ω and what is the value for the propagation constant β? In order to fulfill this requirement, the calculation of the $\omega - \beta$ diagram must be performed. Further, if the permeability or permittivity of the medium is a function of frequency, the calculation becomes almost impossible.

The approaches proposed by Hano [3] and Koshiba *et al.* [6] are typical methods whereby the mixing of guided-mode solutions with the spurious-mode solutions is prevented. Hano has used a unique method that employs the variational expression of type 2, where the longitudinal and the transverse components of electric field are expressed as quadratic and linear functions of transverse coordinates, respectively. As a consequence of this approach, the values for ω of the spurious-mode solutions do not lie in the region of lower order guided modes. However, an overlap between the existence region of the values for ω^2 of the spurious-mode solutions and guided-mode solutions still occurs. Koshiba *et al.* have modified the variational expression of type 3 given in [4].[1] Besides the continuity condition for $\boldsymbol{i}_z \cdot (\boldsymbol{n} \times \boldsymbol{h}_t)$ and $\boldsymbol{i}_z \cdot \boldsymbol{h}_z$ at the boundary between element and element, a supplementary boundary condition, which requires the continuity of $\boldsymbol{n} \cdot \boldsymbol{h}_t$, has been introduced. Consequently, the values for ω of the spurious-mode solutions do not lie in the existence region of slow guided modes. However, an overlap between the existence region of ω^2 of the spurious-mode solutions and guided-mode solutions cannot be avoided. The supplementary boundary condition also has a constraint which restricts the scope of applications to cases where the permeability of the medium is uniform.

In this paper, a finite-element formulation intended to overcome the two drawbacks previously mentioned is proposed. A novel variational expression is established in terms of the transverse electric and magnetic field components. In our finite-element program, we assign the parameters according to the necessary boundary conditions $\boldsymbol{i}_z \cdot (\boldsymbol{n} \times \boldsymbol{e}_t)$ and $\boldsymbol{i}_z \cdot (\boldsymbol{n} \times \boldsymbol{h}_t)$ required in the variational expres-

Manuscript received July 22, 1986; revised October 10, 1986.
The authors are with the Department of Electrical Communication Engineering, Osaka University, Yamada Oka, Suita, Osaka 565 Japan.
IEEE Log Number 8611939.

[1]Recently, Hayata *et al.* [7] have demonstrated a method more advanced than that outlined in [4].

Reprinted from *IEEE Trans. Microwave Theory Tech.*, vol. MTT-35, no. 2, pp. 117–123, Feb. 1987.

sion. In our approach, the spurious-mode solutions are not real while the guided-mode solutions are real. Thus, the discrimination between spurious-mode solutions and guided-mode solutions can be achieved merely by imposing the simple condition that the latter be real. In addition to the advantage that the spurious-mode solutions can be eliminated, our method can also be applied to the case where the permeability or permittivity of the medium is a function of frequency. Therefore, our finite-element formulation is very useful for the analysis of arbitrarily shaped waveguides.

The application of our finite-element method to isotropic and anisotropic waveguides is discussed. In particular, the case where the permeability of the medium is a function of frequency is considered.

II. Variational Formulation in Terms of the Transverse Electromagnetic Field Components

Consider the anisotropic waveguide with an arbitrary cross section in the $x-y$ plane. The waveguide is assumed to be uniform along its longitudinal z-axis. Then the cross section of the waveguide is subdivided into a finite number of elements according to the finite-element method. The permittivity and permeability tensors of the anisotropic medium are assumed to be Hermite tensors and are defined in matrix form as

$$\epsilon = \begin{bmatrix} \epsilon_{tt} & \epsilon_{tz} \\ \epsilon_{zt} & \epsilon_{zz} \end{bmatrix} \quad \mu = \begin{bmatrix} \mu_{tt} & \mu_{tz} \\ \mu_{zt} & \mu_{zz} \end{bmatrix}$$

where the subscripts tt, tz, zt, and zz refer to 2×2, 2×1, 1×2, and 1×1 submatrices, respectively. From Maxwell's equations, the equations that govern the electromagnetic fields in each element are

$$\begin{aligned} &\omega\epsilon_{tt}\cdot \boldsymbol{e}_t + \omega\epsilon_{tz}\cdot \boldsymbol{e}_z + j\nabla\times \boldsymbol{h}_z + \beta \boldsymbol{i}_z \times \boldsymbol{h}_t = 0 \\ &\omega\mu_{tt}\cdot \boldsymbol{h}_t + \omega\mu_{tz}\cdot \boldsymbol{h}_z - j\nabla\times \boldsymbol{e}_z - \beta \boldsymbol{i}_z \times \boldsymbol{e}_t = 0 \end{aligned} \tag{1}$$

where $\boldsymbol{e}_t$ and $\boldsymbol{h}_t$ are the transverse components of the electric and magnetic fields, respectively. The longitudinal components, $\boldsymbol{e}_z$ and $\boldsymbol{h}_z$, can be expressed in terms of the transverse electromagnetic field components as

$$\begin{aligned} \boldsymbol{e}_z &= \frac{1}{j\omega\epsilon_{zz}}(\nabla\times \boldsymbol{h}_t - j\omega\epsilon_{zt}\cdot \boldsymbol{e}_t) \\ \boldsymbol{h}_z &= -\frac{1}{j\omega\mu_{zz}}(\nabla\times \boldsymbol{e}_t + j\omega\mu_{zt}\cdot \boldsymbol{h}_t). \end{aligned} \tag{2}$$

At the boundary of each element, the boundary conditions for the electromagnetic fields in (1) require the following.

For the boundary between element and element

$$\left.\begin{aligned} \boldsymbol{i}_z\cdot(\boldsymbol{n}\times \boldsymbol{e}_t) &= \text{continuous function} \\ \boldsymbol{i}_z\cdot \boldsymbol{e}_z &= \text{continuous function} \\ \boldsymbol{i}_z\cdot(\boldsymbol{n}\times \boldsymbol{h}_t) &= \text{continuous function} \\ \boldsymbol{i}_z\cdot \boldsymbol{h}_z &= \text{continuous function.} \end{aligned}\right\} \tag{3a}$$

For the boundary between element and electric wall

$$\boldsymbol{i}_z\cdot(\boldsymbol{n}\times \boldsymbol{e}_t) = 0, \quad \boldsymbol{i}_z\cdot \boldsymbol{e}_z = 0. \tag{3b}$$

For the boundary between element and magnetic wall

$$\boldsymbol{i}_z\cdot(\boldsymbol{n}\times \boldsymbol{h}_t) = 0, \quad \boldsymbol{i}_z\cdot \boldsymbol{h}_z = 0. \tag{3c}$$

Here, $\boldsymbol{n}$ is a normal unit vector to the boundary of each element.

Next, the guided-mode solutions that satisfy (1) and the boundary conditions in (3) make the following functional of propagation constant stationary. In other words, we can prove that this functional is a variational expression

$$\beta(\boldsymbol{e}_t, \boldsymbol{h}_t) = \sum A(\boldsymbol{e}_t, \boldsymbol{h}_t) \Big/ \sum B(\boldsymbol{e}_t, \boldsymbol{h}_t) \tag{4}$$

where

$$\begin{aligned} A(\boldsymbol{e}_t, \boldsymbol{h}_t) = \int \Big[&\boldsymbol{e}_t^*\cdot\omega\epsilon_{tt}\cdot \boldsymbol{e}_t + \boldsymbol{h}_t^*\cdot\omega\mu_{tt}\cdot \boldsymbol{h}_t \\ &- \frac{1}{\omega\mu_{zz}}(\nabla\times \boldsymbol{e}_t + j\omega\mu_{zt}\cdot \boldsymbol{h}_t)^* \\ &\cdot(\nabla\times \boldsymbol{e}_t + j\omega\mu_{zt}\cdot \boldsymbol{h}_t) \\ &- \frac{1}{\omega\epsilon_{zz}}(\nabla\times \boldsymbol{h}_t - j\omega\epsilon_{zt}\cdot \boldsymbol{e}_t)^* \\ &\cdot(\nabla\times \boldsymbol{h}_t - j\omega\epsilon_{zt}\cdot \boldsymbol{e}_t)\Big]\, ds \end{aligned}$$

$$B(\boldsymbol{e}_t, \boldsymbol{h}_t) = \int [\boldsymbol{i}_z\cdot(\boldsymbol{e}_t^*\times \boldsymbol{h}_t + \boldsymbol{e}_t\times \boldsymbol{h}_t^*)]\, ds.$$

Here, Σ denotes a summation taken over all elements and $\int ds$ denotes a surface integral taken in each element.

The trial functions of the transverse electric and magnetic field components used in (4) must necessarily satisfy the following boundary conditions at the boundary of each element.

For the boundary between element and element

$$\left.\begin{aligned} \boldsymbol{i}_z\cdot(\boldsymbol{n}\times \boldsymbol{e}_t) &= \text{continuous function} \\ \boldsymbol{i}_z\cdot(\boldsymbol{n}\times \boldsymbol{h}_t) &= \text{continuous function.} \end{aligned}\right\} \tag{5a}$$

For the boundary between element and electric wall

$$\boldsymbol{i}_z\cdot(\boldsymbol{n}\times \boldsymbol{e}_t) = 0. \tag{5b}$$

For the boundary between element and magnetic wall

$$\boldsymbol{i}_z\cdot(\boldsymbol{n}\times \boldsymbol{h}_t) = 0. \tag{5c}$$

In the following, we shall prove the validity of the variational expression in (4). First, we assume that the trial functions of the transverse electric and magnetic field components used in (4) differ from the true guided-mode solutions satisfying (1) and (3) by small admissible changes $\delta\boldsymbol{e}_t$ and $\delta\boldsymbol{h}_t$, respectively. Let $\delta\beta$ denote a variation of the propagation constant corresponding to $\delta\boldsymbol{e}_t$ and $\delta\boldsymbol{h}_t$. Under the condition that β and $\delta\beta$ be real, substituting the trial functions written in variation notations into (4) results in the following equations:

$$\delta\beta \sum B = \sum (A - \beta B) + \sum (\delta A - \beta\delta B) \tag{6}$$

where

$$A-\beta B=\int[\boldsymbol{e}_t^*\cdot(\omega\epsilon_{tt}\cdot\boldsymbol{e}_t+\omega\epsilon_{tz}\cdot\boldsymbol{e}_z+j\nabla\times\boldsymbol{h}_z+\beta\boldsymbol{i}_z\times\boldsymbol{h}_t)+\boldsymbol{h}_t^*\cdot(\omega\mu_{tt}\cdot\boldsymbol{h}_t+\omega\mu_{tz}\cdot\boldsymbol{h}_z-j\nabla\times\boldsymbol{e}_z-\beta\boldsymbol{i}_z\times\boldsymbol{e}_t)]\,ds+j\oint[\boldsymbol{h}_z\cdot(\boldsymbol{n}_0\times\boldsymbol{e}_t)^*-\boldsymbol{e}_z\cdot(\boldsymbol{n}_0\times\boldsymbol{h}_t)^*]\,dl \tag{7a}$$

and

$$\delta A-\beta\delta B=\int[\delta\boldsymbol{e}_t^*\cdot(\omega\epsilon_{tt}\cdot\boldsymbol{e}_t+\omega\epsilon_{tz}\cdot\boldsymbol{e}_z+j\nabla\times\boldsymbol{h}_z+\beta\boldsymbol{i}_z\times\boldsymbol{h}_t)+\delta\boldsymbol{h}_t^*\cdot(\omega\mu_{tt}\cdot\boldsymbol{h}_t+\omega\mu_{tz}\cdot\boldsymbol{h}_z-j\nabla\times\boldsymbol{e}_z-\beta\boldsymbol{i}_z\times\boldsymbol{e}_t)+\delta\boldsymbol{e}_t\cdot(\omega\epsilon_{tt}\cdot\boldsymbol{e}_t+\omega\epsilon_{tz}\cdot\boldsymbol{e}_z+j\nabla\times\boldsymbol{h}_z+\beta\boldsymbol{i}_z\times\boldsymbol{h}_t)^*+\delta\boldsymbol{h}_t\cdot(\omega\mu_{tt}\cdot\boldsymbol{h}_t+\omega\mu_{tz}\cdot\boldsymbol{h}_z-j\nabla\times\boldsymbol{e}_z-\beta\boldsymbol{i}_z\times\boldsymbol{e}_t)^*]\,ds+j\oint[\boldsymbol{h}_z\cdot(\boldsymbol{n}_0\times\delta\boldsymbol{e}_t)^*-\boldsymbol{e}_z\cdot(\boldsymbol{n}_0\times\delta\boldsymbol{h}_t)^*-\boldsymbol{h}_z^*\cdot(\boldsymbol{n}_0\times\delta\boldsymbol{e}_t)+\boldsymbol{e}_z^*\cdot(\boldsymbol{n}_0\times\delta\boldsymbol{h}_t)]\,dl. \tag{7b}$$

Here $\oint dl$ denotes a line integral taken along the entire boundary line in each element and $\boldsymbol{n}_0$ denotes an outward normal unit vector at the boundary of each element. The $\boldsymbol{e}_t$, $\boldsymbol{h}_t$, $\boldsymbol{e}_z$, $\boldsymbol{h}_z$, and β in (7) satisfy (1), (2), and (3). Thus, by substituting (1), (2), and (3) into (7) we get

$$\sum(A-\beta B)=0$$

$$\sum(\delta A-\beta\delta B)=\sum j\oint[\boldsymbol{h}_z\cdot(\boldsymbol{n}_0\times\delta\boldsymbol{e}_t)^*-\boldsymbol{e}_z\cdot(\boldsymbol{n}_0\times\delta\boldsymbol{h}_t)^*-\boldsymbol{h}_z^*\cdot(\boldsymbol{n}_0\times\delta\boldsymbol{e}_t)+\boldsymbol{e}_z^*\cdot(\boldsymbol{n}_0\times\delta\boldsymbol{h}_t)]\,dl. \tag{8}$$

Since $\delta\boldsymbol{e}_t$ and $\delta\boldsymbol{h}_t$ also satisfy the boundary conditions in (5) according to the trial functions $\boldsymbol{e}_t$ and $\boldsymbol{h}_t$, the term $\Sigma(\delta A-\beta\delta B)$ in the above equation becomes zero. Hence, (6) can be written as

$$\delta\beta=0. \tag{9}$$

From (9), we can prove that the functional in (4) is the variational expression accounting for (1) and (3). This variational expression has the advantage that for a given value of angular frequency, the values of the propagation constant can be obtained directly. The variational expression is also suited for a finite-element formulation in the sense that all guided-mode solutions are real while the spurious-mode solutions are not real. This is confirmed in the numerical examples described in Section IV.

III. Finite-Element Method

According to the standard finite-element method, the cross section of the waveguide is subdivided into a finite number of triangular elements. Fig. 1 shows an arbitrary triangular element. The node number and the corresponding node coordinate at each vertex are assigned as 1, (x_1, y_1), 2, (x_2, y_2), and 3, (x_3, y_3), respectively. The normal unit vector at each side is assigned as $\boldsymbol{n}_{12}$, $\boldsymbol{n}_{23}$, and $\boldsymbol{n}_{31}$ normal to, respectively, the sides (1–2), (2–3), and (3–1). The trial functions of the transverse electric and magnetic

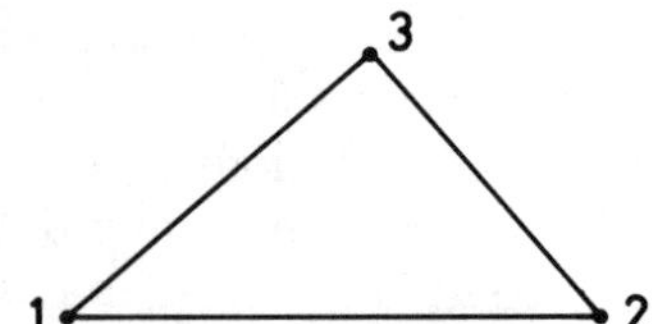

Fig. 1. An arbitrary triangular element.

field components in each element are expressed by using 12 unknown parameters as

$$\boldsymbol{e}_t=\sum_{m=1}^{6}N_m(x,y)\phi_m$$
$$\boldsymbol{h}_t=\sum_{m=7}^{12}N_{m-6}(x,y)\phi_m \tag{10}$$

where ϕ_m (for $m=1\sim12$) denotes unknown parameters and $N_m(x, y)$ (for $m=1\sim6$) represents linear vector shape functions. The vector shape functions are determined by the scalar shape functions and the normal unit vectors in the element. These vector shape functions can be expressed as

$$N_{2i-1}=\frac{\boldsymbol{n}_{ij}}{\boldsymbol{i}_z\cdot(\boldsymbol{n}_{ki}\times\boldsymbol{n}_{ij})}N_i$$
$$N_{2i}=\frac{\boldsymbol{n}_{ki}}{\boldsymbol{i}_z\cdot(\boldsymbol{n}_{ij}\times\boldsymbol{n}_{ki})}N_i$$
$$N_i=\frac{(x_jy_k-x_ky_j)+(y_j-y_k)x+(x_k-x_j)y}{x_1y_2+x_2y_3+x_3y_1-x_2y_1-x_3y_2-x_1y_3} \tag{11}$$

where (i, j, k) permutes in a natural order. The equations for the unknown parameters ϕ_m (for $m=1\sim12$) can be recognized from (10) and (11) as follows

$$\phi_{2i-1}=\boldsymbol{i}_z\cdot(\boldsymbol{n}_{ki}\times\boldsymbol{e}_{ti})$$
$$\phi_{2i}=\boldsymbol{i}_z\cdot(\boldsymbol{n}_{ij}\times\boldsymbol{e}_{ti})$$
$$\phi_{2i+5}=\boldsymbol{i}_z\cdot(\boldsymbol{n}_{ki}\times\boldsymbol{h}_{ti})$$
$$\phi_{2i+6}=\boldsymbol{i}_z\cdot(\boldsymbol{n}_{ij}\times\boldsymbol{h}_{ti}) \tag{12}$$

where $\boldsymbol{e}_{ti}$ and $\boldsymbol{h}_{ti}$ (for $i=1\sim3$) denote the transverse electric and magnetic field components at node i. Equations (12) mean that the parameters ϕ_m (for $m=1\sim12$) are $\boldsymbol{i}_z\cdot(\boldsymbol{n}\times\boldsymbol{e}_t)$ and $\boldsymbol{i}_z\cdot(\boldsymbol{n}\times\boldsymbol{h}_t)$ taken at each side of the triangular element. By means of (12), the trial functions can be expressed in terms of parameters ϕ_m instead of explicitly using the transverse electric and magnetic field components. In this way, the trial functions can be forced to satisfy the necessary boundary conditions in (5) in a simple manner.

After substituting (10) into (4) and integrating, the following equations for A and B are obtained:

$$A=\phi^{*t}p\phi$$
$$B=\phi^{*t}q\phi. \tag{13}$$

Here, ϕ denotes a column vector composed of 12 unknown parameters used in the triangular element; p and q are

12×12 matrices composed of the values of permittivity, permeability, and vector shape functions in the element; and t denotes a transposed column vector.

Imposing the boundary conditions in (5) to (13) and summing over all elements, ΣA and ΣB are obtained. Then, substituting ΣA and ΣB into (4), the variational expression can be written as

$$\beta = \frac{\Phi^{*t} P \Phi}{\Phi^{*t} Q \Phi} \tag{14}$$

where Φ is a column vector composed of all unknown parameters used in the waveguide. P is a regular Hermitian matrix[2] and Q is a singular real symmetric matrix. The dimensions of P, Q, and Φ are exactly equal to six times the total number of elements in the case of linear shape functions.

The stationary condition of (4) requires the derivative of (14) with respect to Φ to be equal to zero. From this condition, we can obtain a generalized eigenvalue problem as

$$Q\Phi = \frac{1}{\beta} P \Phi \tag{15}$$

or, in detailed matrix form, as

$$\begin{bmatrix} 0 & Q_2 \\ Q_1 & 0 \end{bmatrix} \begin{bmatrix} \Phi_1 \\ \Phi_2 \end{bmatrix} = \frac{1}{\beta} \begin{bmatrix} P_1 & P_4 \\ P_3 & P_2 \end{bmatrix} \begin{bmatrix} \Phi_1 \\ \Phi_2 \end{bmatrix} \tag{16}$$

where Φ_1 and Φ_2 are subcolumn vectors of Φ composed of the unknown parameters used to express $\boldsymbol{e}_t$ and $\boldsymbol{h}_t$, respectively, in the waveguide. P_1, P_2, P_3, and P_4 are the submatrices of matrix P. Q_1 and Q_2 are the submatrices of matrix Q. For the common case where $\epsilon_{zt} = \mu_{zt} = 0$, the submatrices P_3 and P_4 become equal to zero and the following two equations can be obtained from (16):

$$Q_1 \Phi_1 = \frac{1}{\beta} P_2 \Phi_2 \tag{17}$$

$$Q_2 \Phi_2 = \frac{1}{\beta} P_1 \Phi_1. \tag{18}$$

From (17) and (18), we have

$$\left(P_1^{-1} Q_2 P_2^{-1} Q_1\right) \Phi_1 = (1/\beta)^2 \Phi_1 \tag{19}$$

$$\left(P_2^{-1} Q_1 P_1^{-1} Q_2\right) \Phi_2 = (1/\beta)^2 \Phi_2. \tag{20}$$

It is clear that one of the above two equations can be used to solve for β and that the dimension of both eigenvalue problems is approximately one half the dimension of (15). Thus, the eigenvalue problem obtained by using $\boldsymbol{e}_t$ and $\boldsymbol{h}_t$ can be reduced to an eigenvalue problem in which the unknown parameters are expressed only in terms of $\boldsymbol{e}_t$ or $\boldsymbol{h}_t$. It should also be mentioned that for the case where $\epsilon_{zt} \neq 0$ or $\mu_{zt} \neq 0$, (15) can be transformed into the standard eigenvalue problem: $P^{-1} Q \Phi = 1/\beta \Phi$.

By solving the eigenvalue problem, the eigenvectors Φ and the eigenvalues β are obtained. Substituting the eigenvectors into (10), we obtain the eigenfunctions of the transverse electric and magnetic field components. Thus, by following this method, we can calculate the electromagnetic field distribution in the waveguide.

[2] In the case of isotropic and some anisotropic waveguides, P can be reduced to a real symmetric matrix.

IV. Numerical Examples and Considerations

A. Dielectric-Loaded Waveguide

As a first numerical example of the method described in Sections II and III, we investigate a dielectric-loaded waveguide. Fig. 2 shows the cross section of the dielectric-loaded waveguide of size $a \times 2a$ bounded by a perfect conductor. Half of the waveguide is filled with dielectric material whose relative permittivity and permeability are equal to 2.25 and 1, respectively. The other half of the waveguide is assumed to be vacuum. The dielectric-loaded waveguide is a test case for the finite-element method, which is widely used in many papers, such as [3] and [6].

Now let us consider our result at, for example, $k_0 a = \omega\sqrt{\epsilon_0 \mu_0}\, a = 3$, where the cross section of the dielectric-loaded waveguide is subdivided into 64 triangular elements, as shown in Fig. 3. Here, k_0, ϵ_0, and μ_0 are the wavenumber, permittivity, and permeability in free space, respectively.

The total number of eigenvalues obtained with our method is equal to six times the total number of elements used in the finite-element mesh. Thus, for the present case, the total number of eigenvalues is equal to $64 \times 6 = 384$. These eigenvalues can be expressed in the form

$$k_0/\beta = u + jv \qquad (u,\ v \text{ real}).$$

All 384 eigenvalues can be classified into four types, as shown in Table I. From the eigenvalues in Table I, we may conclude that the eigenvalues of types 1, 2, and 3 are definitely nonphysical spurious-mode solutions, because these eigenvalues are not bounded real numbers. All eigenvalues of type 4 are real, as shown in the following:

$$\frac{\beta}{k_0} = \begin{cases} \pm 1.27102 \text{ (the corresponding exact solution is } \pm 1.27576) \\ \pm 0.94546 \text{ (the corresponding exact solution is } \pm 0.97154) \\ \pm 0.67683 \text{ (the corresponding exact solution is } \pm 0.72865) \\ \pm 0.55718 \text{ (the corresponding exact solution is } \pm 0.59390). \end{cases}$$

The eigenvalues of type 4 clearly represent the approximate values of the exact guided-mode solutions, and we also observe the one-to-one correspondence between the values of type 4 and the exact solutions. The positive and negative values of β in type 4 can be regarded as the

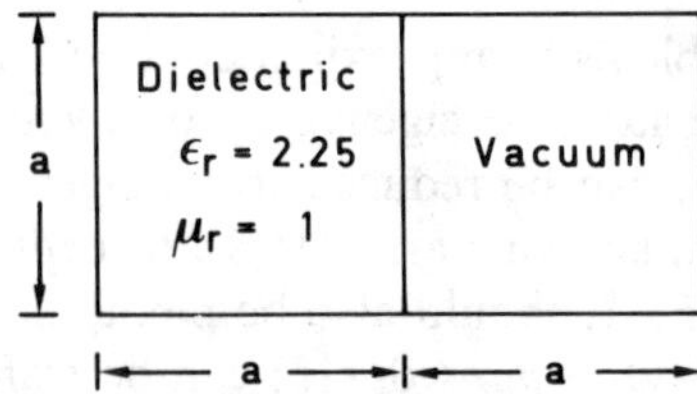

Fig. 2. The cross section of a dielectric-loaded waveguide.

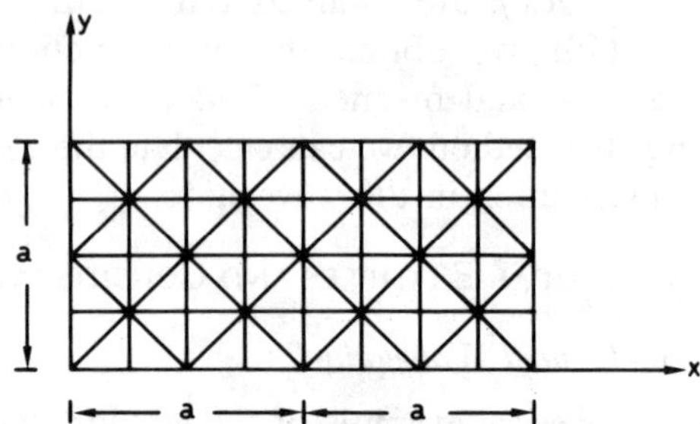

Fig. 3. Illustration of the finite-element division of a dielectric-loaded waveguide.

TABLE I
CLASSIFICATION OF 384 EIGENVALUES IN THE FINITE-ELEMENT ANALYSIS OF THE DIELECTRIC-LOADED WAVEGUIDE IN FIG. 3

Type 1	u = 0	v = 0	132 values
Type 2	u = 0	v ≠ 0	240 values
Type 3	u ≃ 0	v ≠ 0	4 values
Type 4	u ≠ 0	v = 0	8 values

values corresponding to the waves propagating in $+z$ and $-z$ directions, respectively. Thus, if the condition that guided-mode solutions be real is imposed, all the spurious-mode solutions can be completely discriminated.

The finite-element analysis for the dielectric-loaded waveguide in Fig. 3 has been carried out by using a mesh with 64 triangular elements. In Fig. 4, the solid lines represent the finite-element analysis, while dots show the exact solutions. In case of the fundamental mode, our finite-element analysis agrees almost exactly with the exact solutions. The higher order modes, obviously, cannot be reproduced so well because of an insufficient number of elements in the finite-element mesh. However the accuracy in the higher order modes can be improved by increasing the number of elements or using quadratic shape functions instead of linear shape functions. Note that the cutoff frequency of each mode in Fig. 4 is the extrapolated value from the values near the cutoff frequency.

B. Ferrite-Loaded Waveguide

As a numerical example for the magnetic anisotropic case, we investigate a ferrite-loaded waveguide, shown in Fig. 5. The ferrite-loaded waveguide is a version of a nonreciprocal microwave components in which a ferrite

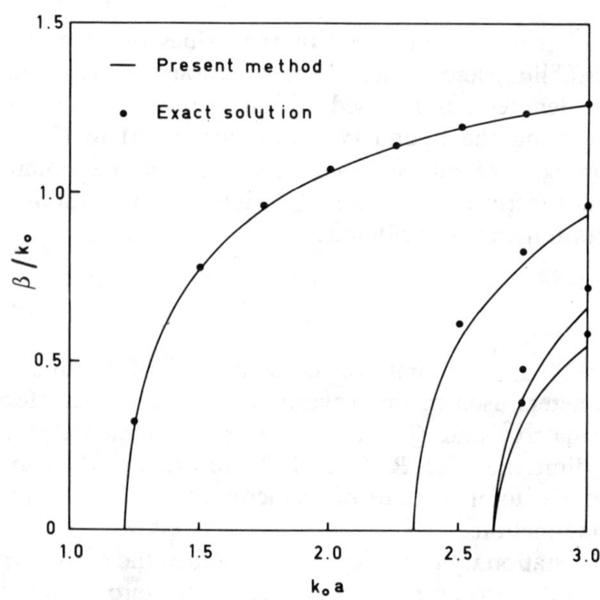

Fig. 4. Dispersion characteristics for a dielectric-loaded waveguide.

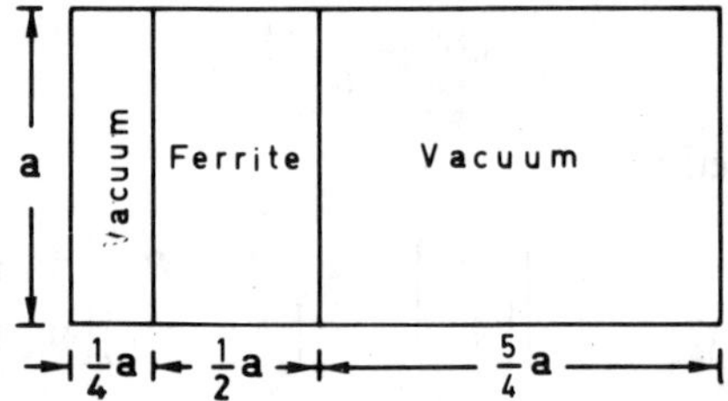

Fig. 5. The cross section of a ferrite-loaded waveguide.

slab is placed asymmetrically into the waveguide, and the static magnetic field is imposed transverse to the direction of propagation. The ferrite slab in the waveguide can be characterized by the tensor permeability $\bar{\mu}$ and the scalar permittivity ϵ as

$$\bar{\mu} = \begin{bmatrix} \mu & 0 & j\kappa \\ 0 & \mu_0 & 0 \\ -j\kappa & 0 & \mu \end{bmatrix}$$

$$\mu = \mu_0\left(1 + \frac{\omega_m \omega_0}{\omega_0^2 - \omega^2}\right)$$

$$\kappa = \mu_0 \frac{\omega_m \omega}{\omega_0^2 - \omega^2}$$

$$\omega_0\sqrt{\epsilon_0\mu_0}\, a = \omega_m\sqrt{\epsilon_0\mu_0}\, a = 0.5$$

$$\epsilon = 10\epsilon_0.$$

Here, μ_0 and ϵ_0 are the permeability and permittivity in free space, respectively. The tensor permeability of the ferrite is frequency-dependent. Thus, the ferrite-loaded waveguide can be regarded as a proof for the validity of the finite-element method for the general case where the permeability of the medium is a function of frequency. For the case of the variational expression in (4), the finite-element analysis of the ferrite-loaded waveguide can be

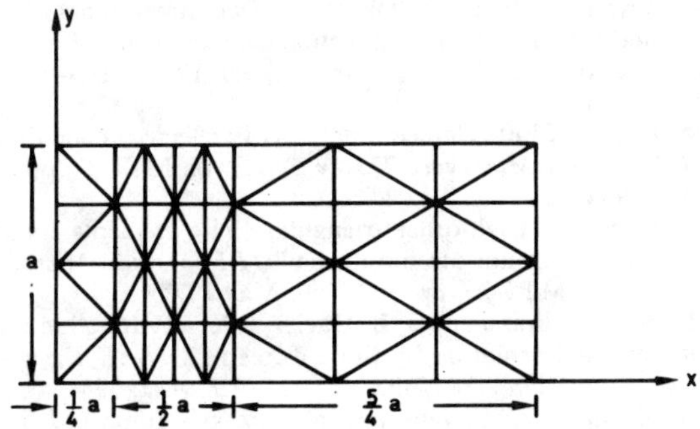

Fig. 6. Illustration of the finite-element division of a ferrite-loaded waveguide.

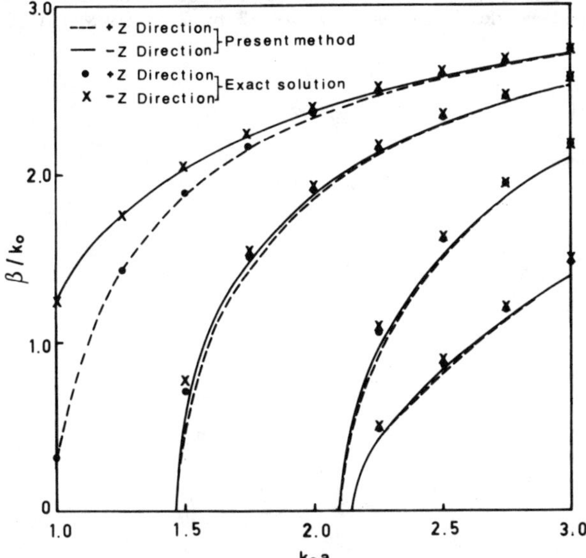

Fig. 7. Dispersion characteristics for a ferrite-loaded waveguide.

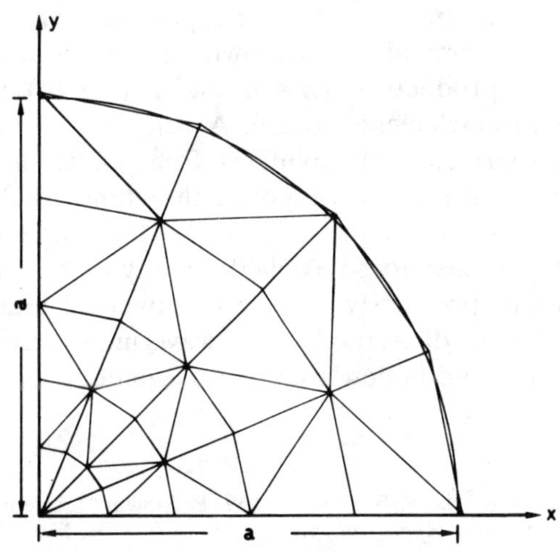

Fig. 8. Illustration of the finite-element division of one quarter of a hollow circular waveguide.

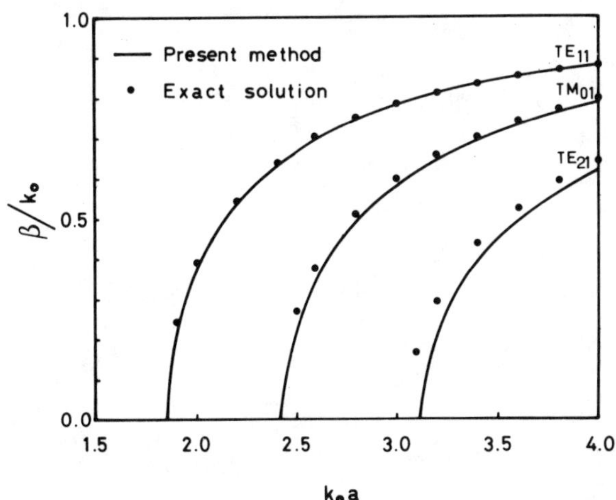

Fig. 9. Dispersion characteristics for a hollow circular waveguide.

accomplished because this variational expression is not a functional of frequency.

In Fig. 6, the cross section of the ferrite-loaded waveguide is subdivided into 64 triangular elements. The calculations are carried out by employing the finite-element mesh shown in Fig. 6. In Fig. 7, the dashed lines and solid lines represent the finite-element analysis corresponding to the waves propagating in the $+z$ and $-z$ directions, respectively, while the exact solutions are represented by dots and crosses. Again, our finite-element analysis is in good agreement with the exact solutions for the fundamental mode, but as in the numerical example of a dielectric-loaded waveguide previously discussed, deviations in the higher order modes still occur. The cause of the deviations in the higher order modes lies in the insufficient number of elements and in the circumstance that the shape functions being used do not yield the correct approximate functions fitted to the electromagnetic field distribution.

Note also that for the case of a ferrite-loaded waveguide, all spurious-mode solutions are not real while all guided-mode solutions are real. Thus, again, all spurious-mode solutions can be discriminated by imposing the condition that the guided-mode solutions be real.

C. *Hollow Circular Waveguide*

An additional numerical example has been carried out with a hollow circular waveguide. This example can be regarded as a test case with curved boundary. The computation is carried out with the mesh in Fig. 8, where a quarter of the waveguide is considered. A one-to-one correspondence between the real eigenvalues in our method and exact solutions has been observed, and no spurious-mode solutions appear to be real numbers. Fig. 9 shows the dispersion characteristics for the hollow circular waveguide. The solid lines and dots show the finite-element solutions and the exact solutions, respectively. Thus, the validity of the present method is also confirmed by this example.

V. Conclusions

A variational expression of the propagation constant, employing transverse electric and magnetic field components has been formulated and used in a finite-element method in order to investigate the propagation characteristics of waveguides. The trial functions of the transverse electric and magnetic field components, which are expressed in terms of the unknown parameters according to the vector products. $i_z \cdot (n \times e_t)$ and $i_z \cdot (n \times h_t)$, were used in this finite-element method. A complete discrimination of the spurious-mode solutions from guided-mode solutions was confirmed by imposing the simple condition that the guided-mode solutions be real. This simple condition does not appear to be available in any of the variational expressions previously proposed. Numerical examples are given for a dielectric-loaded waveguide, ferrite-loaded waveguide, and hollow circular waveguide.

References

[1] C. Yeh, K. Ha, S. B. Dong, and W. P. Brown, "Single-mode optical waveguides," *Appl. Opt.*, vol. 18, pp. 1490–1504, May 1979.

[2] K. Oyamada and T. Okoshi, "Two-dimensional finite-element method calculation of propagation characteristics of axially nonsymmetrical optical fibers," *Radio Sci.*, vol. 17, pp. 109–116, Jan.–Feb. 1982.

[3] M. Hano, "Finite-element analysis of dielectric-loaded waveguides," *IEEE Trans. Microwave Theory Tech.*, vol. MTT-32, pp. 1275–1279, Oct. 1984.

[4] A. Konrad, "High-order triangular finite elements for electromagnetic waves in anisotropic media," *IEEE Trans. Microwave Theory Tech.*, vol. MTT-25, pp. 353–360, May 1977.

[5] B. M. A. Rahman and J. B. Davies, "Penalty function improvement of waveguide solution by finite elements," *IEEE Trans. Microwave Theory Tech.*, vol. MTT-32, pp. 922–928, Aug. 1984.

[6] M. Koshiba, K. Hayata, and M. Suzuki, "Improved finite-element formulation in terms of the magnetic field vector for dielectric waveguides," *IEEE Trans. Microwave Theory Tech.*, vol. MTT-33, pp. 227–233, Mar. 1985.

[7] K. Hayata, M. Koshiba, M. Eguchi, and M. Suzuki, "Novel finite-element formulation without any spurious solutions for dielectric waveguides," *Electron. Lett.*, vol. 22, pp. 295–296, Mar. 1986.

Finite-Element Method Analysis of Microwave and Optical Waveguides—Trends in Countermeasures to Spurious Solutions

Masanori Koshiba, Kazuya Hayata and Michio Suzuki, Members

Faculty of Engineering, Hokkaido University, Sapporo, Japan 060

SUMMARY

This paper presents a review of finite-element methods for the analysis of microwave and optical waveguides, particularly focusing on the recent results for suppressing or eliminating techniques of the spurious solutions that have become a subject of discussion to date. The finite-element method analysis of electromagnetic waveguides can be divided into four classes: (1) scalar wave analysis by the one-dimensional finite-element method; (2) vectorial wave analysis by the one-dimensional finite-element method; (3) scalar wave analysis by the two-dimensional finite-element method; and (4) vectorial wave analysis by the two-dimensional finite-element method. Spurious solutions appear in (4), and, depending on the variational expression used, they also appear in (2). Placing emphasis on (2) and (4), particularly on (4), various finite-element formulations are presented such as axial fields of E and H, full vector H field, full vector E field, transverse fields of E and H, and transverse fields of E or H formulations, and various approaches capable of suppressing or eliminating the spurious solutions are reviewed in detail.

1. Introduction

The finite-element method (FEM) has been applied recently to a wide range of problems in electrical and electronic engineering, and many successful results have also been obtained in the field of electromagnetic-wave engineering dealing with a wave field [1 - 15]. In particular, as computers become larger, faster, and more manifold (from personal and minicomputers to supercomputer), a new interdisciplinary field called Computational Mechanics, in which methods such as finite-element method, finite-difference method (FDM), and boundary-element method (BEM) play an important role, is now being developed in cooperation with various fields of study [16]. Consequently, more interest is presently being paid to the finite-element method.

Research on the application of the finite-element method to electromagnetic-wave engineering has been made in earnest from the latter half of the 1960's, particularly from the 1970's, and is now extensive. Among many applications of the finite-element method, research on application to electromagnetic waveguide problems is continuous. The finite-element method is being established as one of the powerful numerical methods for electromagnetic waveguides through application to various waveguiding structure from microwave to optical wavelength regions. However, the most serious difficulty in applying the finite-element method to such electromagnetic waveguide problems is the appearance of so-called spurious, nonphysical solutions. Consequently, the development of a method to suppress or eliminate such spurious solutions is a pressing need and research on this topic has been extensive in recent years.

This paper reviews recent results in the development of methods to suppress or eliminate spurious solutions encountered in the finite-element method analysis of electromagnetic waveguides. Here, the dielectric waveguide, widely used from microwave through optical wavelength regions, is considered. No magnetic material is considered unless it is particularly mentioned.

The electromagnetic waveguide can be classified into two categories from its cross-sectional shape. One is the layered waveguide such as a layered film waveguide (planar waveguide) or a layered circular (axially symmetric waveguide), which can be treated as an equivalent one-dimensional problem; the other is the more general arbitrarily shaped waveguide that should be treated as a two-dimensional problem. Furthermore, depending on the eigenmode property in a guide, there are two cases: one

Reprinted with permission from *Electronics and Communications in Japan*, part 2, vol. 70, no. 9, pp. 96–108, 1987.

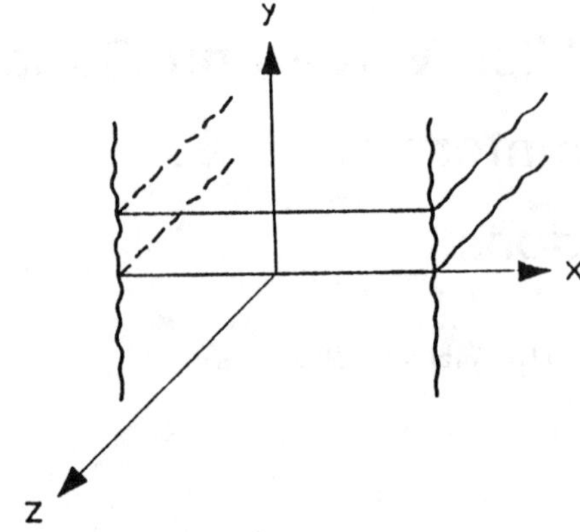

Fig. 1. Planar waveguide ($\partial/\partial x \equiv 0$).

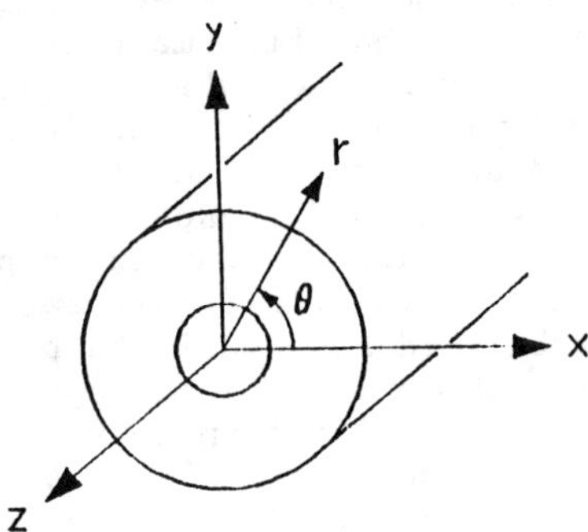

Fig. 2. Axially symmetrical waveguide.

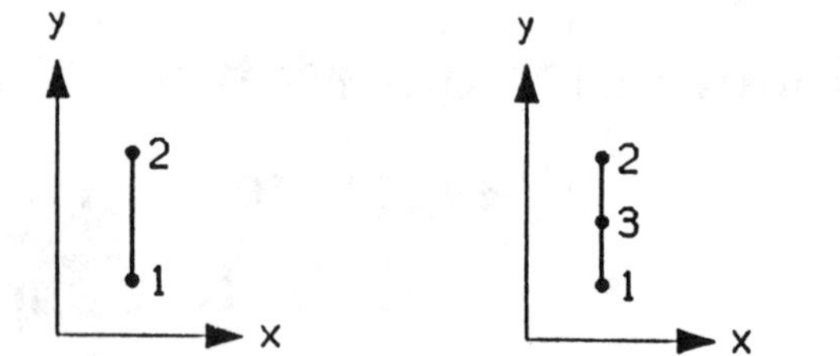

Fig. 3. Line element.

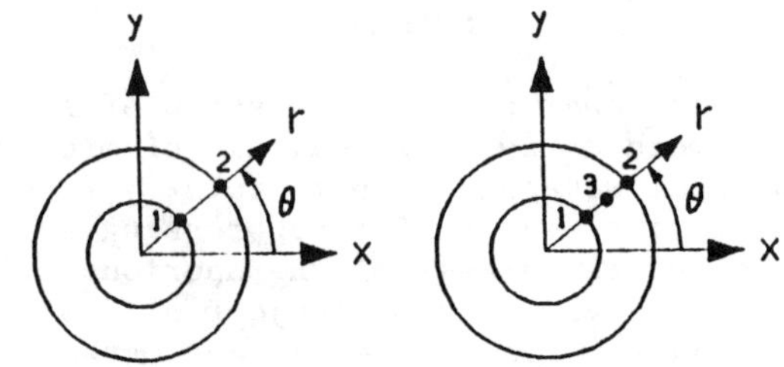

(a) Linear element. (b) Quadratic element.

Fig. 4. Ring element.

is the case in which scalar wave analysis is possible, the other is the case in which vectorial wave analysis is required. From this point of view, the following classification is adopted here*:

(1) scalar wave analysis by the one-dimensional finite-element method;

(2) vectorial wave analysis by the one-dimensional finite-element method;

(3) scalar wave analysis by the two-dimensional finite-element method; and

(4) vectorial wave analysis by the two-dimensional finite-element method.

*Here, the one-dimensional and two-dimensional finite-element methods analyze one-dimensional and two-dimensional problems, respectively. Furthermore, the scalar wave and vectorial wave analysis indicate guided-mode analyses for electromagnetic waveguides via the scalar wave and vectorial wave equations, respectively. In addition, the approximate scalar wave analysis corresponds to the case in which scalar wave approximation is applied to the vectorial wave analysis.

Since the variational principle is used in ordinary finite-element formulations, a variational expression for the system under consideration should be available in advance. Various variational expressions have already been available for the preceding four approaches (it is usually possible to discretize the problem directly via the Galerkin procedure without using the variational principle).

Spurious solutions do appear in (4) and, depending on the variational expression used, they also appear in (2). Here, placing emphasis on (2) and (4), particularly on (4), various approaches are reviewed; for the case in which spurious solutions do not appear, related references are presented. To save space, the references already presented in the previous reviews or books [1 - 15] may be omitted from the present reference list. On the other hand, since the dielectric waveguide, particularly the optical waveguide, is unbounded, it is of importance how to treat these unbounded domains. Although various approaches have been developed to deal with such a problem, we would like to describe this topic on another occasion.

2. One-Dimensional Finite-Element Method

The guided-wave problem in a layered waveguide as shown in Figs. 1 and 2 can be treated as an equivalent one-dimensonal problem. The finite-element method using the elements as shown in Figs. 3 and 4 (both belong to the one-dimensional element) has been developed for the waveguides of Figs. 1 and 2.

2.1 Scalar wave analysis

The scalar wave analysis [17] of the planar inhomogeneous waveguide and the approximate scalar wave analysis [18, 19] of the inhomogeneous-core optical fiber have been made using the one-dimensional finite-element method. The spurious solutions do not appear in this case [17 - 19].

2.2 Vectorial wave analysis

For the finite-element analysis of the hybrid modes in the waveguide of Fig. 1, the variational expression coupled between x-component of electric field and that of magnetic field has been used, and spurious solutions do not appear [20 - 23].

By contrast, for the finite-element analysis of the hybrid modes in the waveguide of Fig. 2, the variational expression in terms of the z-components (axial components) of electromagnetic field has been used, and the axially symmetrical waveguide with complicated refractive-index profile like the inhomogeneous-core optical fiber has been analyzed [10, 18, 24]. In applying this approach, one should pay much attention to how to assign nodal points. We write the phase constant along the propagation axis 'β', angular frequency ω, vacuum permittivity ε_0, vacuum permeability μ_0, free-space wavenumber $k_0 = \omega\sqrt{\varepsilon_0\mu_0}$, and the permittivities at the neighboring nodal points i, j 'ε_i, ε_j', respectively. When the effective permittivity $\varepsilon_0(\beta/k_0)^2$ approaches the average of these permittivities, $(\varepsilon_i + \varepsilon_j)/2$, a numerically ill-condition is found for the matrix derived from discretization, and consequently spurious solutions can be involved [24]. To avoid such solutions, it is necessary to assign the nodal point i to the position $r = r_i$ where the permittivity is equal to the effective permittivity, i.e., $\varepsilon(r_i) = \varepsilon_0 (\beta/k_0)^2$ [24]. However, since the term $1/[\varepsilon(r)/\varepsilon_0 - (\beta/k_0)^2]$ is involved in the variational expression in terms of axial electromagnetic field, an integrand included in the variational expression diverges at $r = r_i$. Fortunately, a skillful technique capable of avoiding divergence at such singular points has been developed [10, 18]. In this technique, a special function satisfying a nondivergent condition is used for the element with the singular point whereas the electromagnetic field is expanded using a piecewise polynomial for the element without it as is done in the ordinary finite-element analysis. The spurious solutions do not appear there [10, 18, 24].

Additionally, it is difficult to avoid divergence at the singular point in the Rayleigh-Ritz method, which is one of the main approaches for variational problems.

As described, when the variational expression in terms of axial electromagnetic-field components is used, assigning nodal points and dealing with the singular point are cumbersome. Recently, to get rid of such a difficulty, a one-dimensional finite-element method has been developed using the variational expression in terms of full vector E field or full vector H field, and the vectorial wave analysis of the inhomogeneous-core optical fiber [25] has been made. However, if one uses the conventional variational expression in terms of full vector E or H field, many spurious solutions do appear which do not satisfy the following condition:

$$\nabla \cdot \varepsilon_r \boldsymbol{E} = 0 \tag{1}$$

or

$$\nabla \cdot \boldsymbol{H} = 0 \tag{2}$$

where $\boldsymbol{E}$ and $\boldsymbol{H}$ are the electric and magnetic field vectors, respectively, and ε_r is the relative permittivity. The full details of this topic are given in Sect. 3.3.2 because one can suppress or eliminate these spurious solutions in the same way as in the two-dimensional finite-element method [25].

3. Two-Dimensional Finite-Element Method

Wave propagation in a waveguide having an arbitrarily shaped cross section as shown in Fig. 5 can be treated as an equivalent two-dimensional problem, and the finite-element methods have been developed using triangular elements as shown in Fig. 6 or rectangular elements, both of which are two-dimensional elements.

3.1 Scalar wave analysis

Two-dimensional finite-element method for scalar wave analysis has been developed in the latter half of the 1960's [4, 12, 26, 27]. Since then, research for improving its use or accuracy has been extensively made [4, 12, 28, 29]. Spurious solutions are not involved in this method, and it is now one of the powerful approaches for analyzing waveguides with arbitrarily shaped cross section. Recently, it has been applied to the analysis of radio wave propagation in tunnels and the effect of vehicles on the

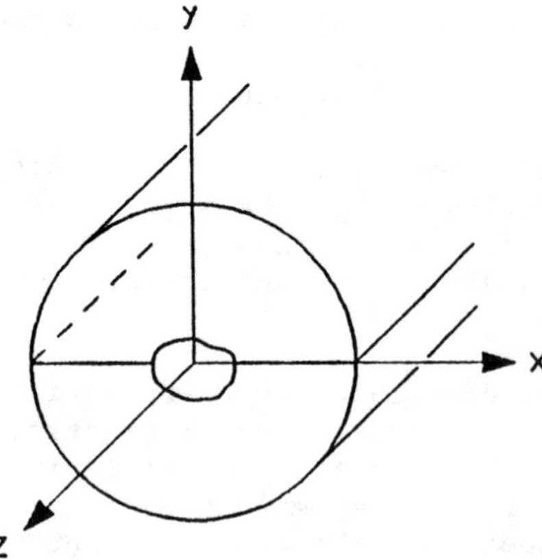

Fig. 5. Waveguide with arbitrarily shaped cross section.

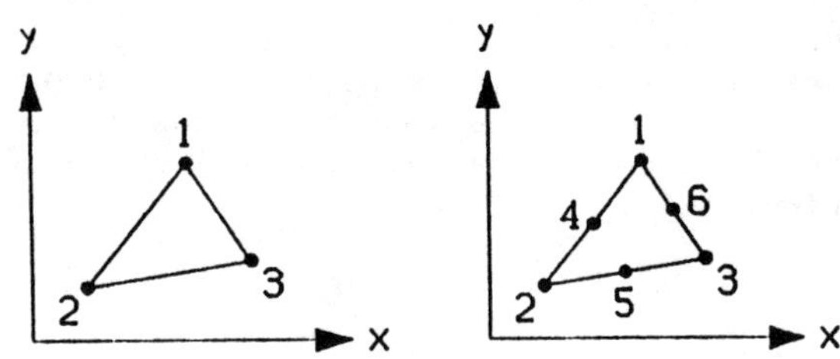

(a) Linear element (b) Quadratic element

Fig. 6. Triangular element

electromagnetic field in them has been investigated [30]. Furthermore, it has been applied to the search of optimum cross-sectional shape of a waveguide for minimizing propagation loss [31, 32] and it is now being used for an optimum design procedure such as the CAD sysetm.

3.2 Approximate scalar wave analysis

Since the modes propagating in dielectric waveguides are, in general, hybrid modes, vectorial wave analysis is required. However, depending on waveguiding structures or propagating modes, the quasi-TEM, quasi-TE, or quasi-TM mode approximation is practically available. From this point of view, approximate scalar wave analyses of dielectric waveguides have been performed using two-dimensional finite-element method. Under this approximation, spurious solutions are not involved [33 - 42].

3.3 Vectorial wave analysis

To evaluate rigorously the propagation characteristics of an inhomogeneous waveguide, vectorial wave analysis is necessary. Although various two-dimensional finite-element formulations have been developed to date, spurious solutions do appear in the vectorial wave analysis in terms of this two-dimensional finite-element method.

3.3.1 Finite-element method using axial electromagnetic-field components

Two-dimensional finite-element method using the following variational expression in terms of axial electromagnetic-field component:

$$F(E_z, H_z) = \iint \frac{1}{\varepsilon/\varepsilon_0 - (\beta/k_0)^2} \Big[\frac{\varepsilon}{\varepsilon_0} |\nabla_t E_z|^2 + \frac{\mu_0}{\varepsilon_0} |\nabla_t H_z|^2 + 2\frac{\beta}{k_0}\sqrt{\frac{\mu_0}{\varepsilon_0}}\, \hat{z} \cdot (\nabla_t E_z \times \nabla_t H_z) \Big] dx\, dy - k_0^2 \iint \Big[\frac{\varepsilon}{\varepsilon_0} |E_z|^2 + \frac{\mu_0}{\varepsilon_0} |H_z|^2 \Big] dx\, dy$$

$\hat{z}$: unit vector for z-direction (3)

is known as a standard finite-element method for vectorial wave analysis, and has been utilized for analyzing various waveguides for use in microwave and millimeter wave regions [4, 23, 43 - 48]. Recently, it has also been utilized for analyzing the waveguide for use in optical wavelength region [33, 49 - 54]. While this approach has been applied to the anisotropic waveguide without off-diagonal elements in a permittivity tensor [33], nor formulation for the arbitrarily anisotropic waveguide with non-zero off-diagonal elements in it has been reported to date.

Application of the finite-element method to Eq. (3) gives:

$$[A]\{X\} - k_0^2 [B]\{X\} = \{0\} \quad (4)$$

where $[A]$ and $[B]$ are matrices obtained from the first and second terms of the right-hand side of Eq. (3), respectively, $\{0\}$ is a null vector, $\{X\}$ is:

$$\{X\} = \begin{bmatrix} \{E_z\} \\ \sqrt{\mu_0/\varepsilon_0}\,\{H_z\} \end{bmatrix} \quad (5)$$

and $\{E_z\}$ and $\{H_z\}$ are the nodal axial electric- and magnetic-field vectors, respectively. While the matrix $[B]$ is positive definite, the matrix $[A]$ is non-negative (semi-positive definite) or indefinite, depending on the fast-wave ($0 < \beta/k_0 < 1$) or slow-wave ($1 < \beta/k_0$) region, resepctively [4, 12, 46, 47]. Consequently, the spurious solutions appear in the slow-wave region, and what is worse, these are coupled with physical solutions. This causes great difficulty in the analysis of the dielectric waveguides whose guided modes exist in the slow-wave region. Furthermore, although the natural boundary condition for Eq. (3) must correspond to the continuity of the tangential electromagnetic-field components at the interface of neighboring media, the spurious solutions, in fact, do not satisfy this condition. Recently, a new variational expression imposing explicitly the continuity of the tangential electromagnetic-field components on Eq. (3) has been devised by means of Lagrange multipliers, and it has been reported that, although not perfect, most of the spurious solutions can be removed [33]. However, this approach has not yet been applied to various waveguides because of the difficulty in computing time and accuracy of solutions. Under these circumstances, in the finite-element method in terms of axial electromagnetic-field components, no effective countermeasure has yet been found. It is therefore necessary to discriminate the physical solutions from the spurious solutions by their field shapes.

3.3.2 Finite-element method using full vector H field

As has already been stated, the finite-element method in terms of axial electromagnetic-field components is not applicable to arbitrarily anisotropic waveguides. Thus, the finite-element method using the variational expression in terms of full vector H or E field is attracting attention because it can analyze arbitrarily anisotropic waveguides.

Since the magnetic field is continuous anywhere in dielectric waveguides, it is convenient to use the following variational expression [55] in terms of full vector H field in performing the finite-element analysis:

$$F(\boldsymbol{H})=\iint (\nabla\times\boldsymbol{H})^{*}\cdot([\varepsilon_r]^{-1}\nabla\times\boldsymbol{H})\,dx\,dy - k_0^2\iint \boldsymbol{H}^{*}\cdot\boldsymbol{H}\,dx\,dy \tag{6}$$

*: complex conjugate

where $[\varepsilon_r]$ is the relative permittivity tensor.

Application of the standard finite-element method yields:

$$[S]\{H\}-k_0^2[T]\{H\}=\{0\} \tag{7}$$

where $[S]$ and $[T]$ are matrices obtained from the first and second terms of the right-hand side of Eq. (6), respectively, and $\{H\}$ is the nodal magnetic-field vector.

Although the finite-element method [56] in terms of the variational expression Eq. (6) has been applied later than that in terms of Eq. (3), in recent years attempts have been made to apply it to various waveguiding structures for use in microwave, millimeter-wave, and optical wavelength regions [57, 58] because of its ability to analyze arbitrarily anisotropic waveguide systems. However, unfortunately, the spurious solutions also appear in this finite-element analysis using full vector H field. Since the matrix $[S]$ in Eq. (7) is semi-positive definite, it seems that these spurious solutions appear via a different mechanism from those encountered in Eq. (4). Actually, it was confirmed in the 1980's that the spurious solutions encountered in Eq. (6) do not satisfy the divergence-free condition $\nabla\cdot\boldsymbol{H}=0$ (see Eq. (2)) [57, 58]. Following this confirmation, it was expected that the spurious solutions can be eliminated if the condition $\nabla\cdot\boldsymbol{H}=0$ is considered. To that end, several approaches, in which the condition $\nabla\cdot\boldsymbol{H}=0$ is explicitly imposed, have been imposed since 1984. The approaches now available are classified into the following.

1. a method choosing a basis function satisfying the condition $\nabla\cdot\boldsymbol{H}=0$;

2. a method using Lagrange multipliers to enforce the condition $\nabla\cdot\boldsymbol{H}=0$ [59];

3. a method enforcing the condition $\nabla\cdot\boldsymbol{H}=0$ in a least squared way [59 - 72];

4. a method transforming the finite-element equation in terms of full vector H field (see Eq. (7)) into that in terms of only transverse magnetic-field component, using the condition $\nabla\cdot\boldsymbol{H}=0$ [73 - 76]; and

5. a method expressing the nodal magnetic-field vectors with a linear combination of basis vectors for the subspace satisfying the condition $\nabla\cdot\boldsymbol{H}=0$ [77].

In 1, it is not necessary to modify the variational expression. However, if one wants to consider the condition $\nabla\cdot\boldsymbol{H}=0$, it is impossible to use standard finite elements and it is therefore necessary to use a nonstandard basis function or nonstandard finite elements. Hence, to date no attempt to suppress or eliminate the spurious solutions has been made using this approach.

Although method 2 is a usual approach to imposing a constraint condition, it has been confirmed to be of no effect in suppressing or eliminating the spurious solutions in the finite-element analysis using full vector H field [59].

Method 3, called the penalty method, has been studied most extensively and has been applied to various waveguiding problems in recent years because it can suppress or eliminate the spurious solutions [59, 61 - 72]. In this approach, the variational expression Eq. (6) is modified into:

$$\tilde{F}(\boldsymbol{H})=\iint (\nabla\times\boldsymbol{H})^{*}\cdot([\varepsilon_r]^{-1}\nabla\times\boldsymbol{H})\,dx\,dy - k_0^2\iint \boldsymbol{H}^{*}\cdot\boldsymbol{H}\,dx\,dy$$

$$+ s \iint (\nabla \cdot \boldsymbol{H})^* (\nabla \cdot \boldsymbol{H})\, dx\, dy \quad (8)$$

Equation (8) is a variational expression satisfying the condition $\nabla \cdot \boldsymbol{H} = 0$ in a least squared way, where s is a positive constant called penalty coefficient.

Application of the standard finite-element method to Eq. (8) yields:

$$([S]+s[U])\{H\} - k_0^2 [T]\{H\} = \{0\} \quad (9)$$

where $[U]$ is a matrix obtained via discretization of the third term in the right-hand side of Eq. (8).

Although the spurious solutions involved in Eq. (7) are completely eliminated in Eq. (9), the solutions:

$$\nabla \times \boldsymbol{H} = 0, \quad \nabla \cdot \boldsymbol{H} \neq 0 \text{ for } k_0^2 > 0 \quad (10)$$

are involved because the variational expression Eq. (6) has been modified into Eq. (8). These solutions correspond to a new type of spurious solutions [65, 66]. However, fortunately, this type of spurious solutions can be predicted as a solution of the following equation:

$$\boldsymbol{H} = \nabla \phi \quad (11a)$$

$$(s\nabla^2 + k_0^2)\phi = 0 \quad \text{in waveguide cross section} \quad (11b)$$

$$\partial \phi / \partial n = 0 \quad \text{on electric wall} \quad (11c)$$

$$\phi = 0 \quad \text{on magnetic wall} \quad (11d)$$

$\partial / \partial n$: normal derivative

where ϕ is a scalar function. Distinct from other spurious solutions encountered in the finite-element method, it is not necessary to worry about these solutions because, as is found from Eq. (11), they can be suppressed into the following region [65, 66]:

$$\beta / k_0 < 1 / \sqrt{s} \quad (12)$$

Equation (12) indicates that the larger s, the narrower the region for the spurious solutions. For example, if one wants to suppress the spurious solutions from the region $\beta/k_0 \geq n_{min}$, one has only to choose the value as $s \geq 1/n_{min}^2$. In this connection, for $s = 1$, the spurious solutions can be completely eliminated from the slow-wave region. Therefore, this approach is useful particularly for the analysis of dielectric waveguides. Recently, it has begun to be successfully applied to various waveguiding structures for use in millimeter or optical integrated circuits [61 - 64, 66, 67, 69] and single-polarization optical fibers [59, 68, 70 - 72] such as stress-applied and side-tunnel optical fibers.

Method 4, developed only recently, can completely eliminate the spurious solutions [73 - 76]. The key point of this approach is that the condition $\nabla \cdot \boldsymbol{H} = 0$, i.e.:

$$H_z = \frac{1}{j\beta}\left(\frac{\partial H_x}{\partial x} + \frac{\partial H_y}{\partial y}\right) \quad (13)$$

is discretized into the following form:

$$[D_z]\{H_z\} = [D_t]\{H_t\} \quad (14)$$

via a Galerkin procedure. Here, $\{H_z\}$ and $\{H_t\}$ are, respectively, the nodal magnetic-field vectors for axial and transverse directions, and the matrices $[D_z]$ and $[D_t]$ are given by:

$$[D_z] = \sum_e \iint_e \{N\}\{N\}^T dx\, dy \quad (15)$$

$$[D_t] = \sum_e \frac{1}{j\beta} \iint_e \left[\{N\} \frac{\partial \{N\}^T}{\partial x} \quad \{N\} \frac{\partial \{N\}^T}{\partial y} \right] dx\, dy \quad (16)$$

$\sum_e$: summation over all elements

T: transposition

where $\{N\}$ is the shape function vector [1 - 12].

Using Eq. (14), the nodal magnetic-field vector $\{H\}$ is expressed as

$$\{H\} = [D]\{H_t\} \quad (17)$$

$$[D] = \begin{bmatrix} [1] \\ [D_z]^{-1}[D_t] \end{bmatrix} \quad (18)$$

where [1] is a unit matrix.

Considering Eq. (17) into the finite-element Eq. (7) in terms of full vector H field, the following finite-element equation in terms of transverse magnetic-field component is obtainable:

$$[S_{tt}]\{H_t\} - k_0^2 [T_{tt}]\{H_t\} = \{0\} \quad (19)$$

where

$$[S_{tt}] = [D]^\dagger [S][D] \quad (20)$$

$$[T_{tt}] = [D]^\dagger [T][D] \quad (21)$$

$\dagger$: complex conjugate and transposition.

In Eq. (19), the spurious solutions are completely eliminated, and the matrix size is two-thirds that of Eqs. (7) and (9). Moreover, an artificial coefficient such as included in the penalty method (the accuracy of solutions depends on the penalty coefficient s in the penalty method) is not included at all. For these reasons, method 4 is very useful for analyzing various electromagnetic waveguides, and more work is expected in the future.

Method 5, also developed recently, succeeds in the complete elimination of the spurious solutions [77]. In this approach, the following variational expression:

$$G(\boldsymbol{H})=\iint (\nabla\cdot\boldsymbol{H})^*(\nabla\cdot\boldsymbol{H})\,dx\,dy - k_0^2\iint \boldsymbol{H}^*\cdot\boldsymbol{H}\,dx\,dy \tag{22}$$

is first considered in order to obtain basis vectors for the subspace satisfying the condition $\nabla\cdot\boldsymbol{H}=0$, and subsequently m eigenvectors $\{D_i\}$ ($i = 1, 2, \cdots, m$) corresponding to zero eigenvalues ($k_0^2 = 0$) are computed by applying the finite-element method. The nodal magnetic-field vector $\{H\}$ is expressed as a linear combination of these basis vectors:

$$\{H\}=[\widetilde{D}]\{\widetilde{H}\} \tag{23}$$

$$[\widetilde{D}]=[\{D_1\}\{D_2\}\cdots\{D_m\}] \tag{24}$$

Considering Eq. (23) into the finite-element Eq. (7) in terms of full vector H field, the final finite-element equation is derivable:

$$[\widetilde{S}]\{\widetilde{H}\}-k_0^2[\widetilde{T}]\{\widetilde{H}\}=\{0\} \tag{25}$$

where

$$[\widetilde{S}]=[\widetilde{D}]^{\dagger}[S][\widetilde{D}] \tag{26}$$

$$[\widetilde{T}]=[\widetilde{D}]^{\dagger}[T][\widetilde{D}] \tag{27}$$

The spurious solutions are completely eliminated in Eq. (25). The problem is how accurately zero eigenvalues of Eq. (22) and the corresponding eigenvectors can be computed. In particular, as the value of phase constant β is larger, discrimination of zero eigenvalues from non-zero eigenvalues may be difficult. In addition, it is necessary to solve the eigenvalue problem twice to compute k_0 corresponding frequency. For these reasons, future efforts to improve the accuracy of solutions and for enhancing efficiency in numerical computations will be necessary.

Since the eigenvalues of the finite-element method using full vector H field correspond to k_0^2, in practical calculations, β is first given as an input datum, and subsequently k_0 is obtained as a solution. Therefore, iterations are required if one wants to deal with the medium whose material constant varies with k_0 (dispersive medium) or the medium with loss or gain (complex medium). To avoid such a difficulty, a method which can directly obtain β by giving k_0 as an input datum has recently been developed in connection with the above case 4 [74]. In this approach, iterations are not necessary. While this approach is applicable in principle to case 3, it is not applicable to case 5.

Moreover, from another point of view from the methods 1 ~ 5, a method has been developed for eliminating the spurious solutions, in which only the continuity of the tangential components of magnetic field (or electric field for the case using full vector E field) is imposed on a basis function at the interface between neighboring elements [78]. In this approach, the variational expression Eq. (6) can be utilized without modification. However, to consider the continuity of the tangential magnetic field (or electric field) components at the interface between elements, it is necessary to devise how to construct a trial function. For example, it is necessary to express x-, y-, and z-components of magnetic field (or electric field) with a linear polynomial of y, a linear polynomial of x, and a quadratic polynomial of x and y, respectively [78].

3.3.3 Finite-element method using full vector E field

Although the finite-element method in terms of full vector H field is applicable to arbitrarily anisotropic waveguides with tensor permittivity, it is difficult to apply it to anisotropic waveguides with tensor permeability. In this case, the variational expression in terms of full vector E field [55] is useful. However, spurious solutions are involved also in this approach, and these spurious solutions do not satisfy the condition $\nabla\cdot\varepsilon_r\boldsymbol{E}=0$ (see Eq. (1)). To suppress or eliminate such unwanted solutions, method 3 described in Sect. 3.3.2 has been used.

Using the same procedure as in Eq. (8), the variational expression in terms of full vector E field [55] is modified into [79, 80]:

$$\widetilde{F}(\boldsymbol{E})=\iint (\nabla\times\boldsymbol{E})^*\cdot([\mu_r]^{-1}\nabla\times\boldsymbol{E})\,dx\,dy - k_0^2\iint \varepsilon_r\boldsymbol{E}^*\cdot\boldsymbol{E}\,dx\,dy + s\iint \varepsilon_r^{-1}(\nabla\cdot\varepsilon_r\boldsymbol{E})^*(\nabla\cdot\varepsilon_r\boldsymbol{E})\,dx\,dy \tag{28}$$

where $[\mu_r]$ is the relative permeability tensor.

In the finite-element analysis using Eq. (28), the spurious solutions encountered in the conventional variational expression [55] ($s = 0$ in Eq. (28)) are completely eliminated. In Eq. (28), however, the solutions:

$$\nabla \times \boldsymbol{E}=0, \quad \nabla \cdot \varepsilon_r \boldsymbol{E} \neq 0 \quad \text{for } k_0^2 > 0 \tag{29}$$

involved are encountered as a new type of spurious solutions. Fortunately, this type of spurious solutions is predictable as a solution of the following equation:

$$\varepsilon_r \boldsymbol{E}=\nabla \phi \tag{30a}$$

$$(s\nabla^2 + k_0^2)\phi = 0 \text{ in waveguide cross section} \tag{30b}$$

$$\phi = 0 \text{ on electric wall} \tag{30c}$$

$$\partial \phi / \partial n = 0 \text{ on magnetic wall} \tag{30d}$$

where ψ is a scalar function. In practical computations, it is not necessary to worry about them because, as is found from Eq. (30), they can be restricted within the region given by Eq. (12). Furthermore, if one considers the continuity of the tangential electric-field component and the normal component of electric-flux density at the interface between the media with different permittivity, one can discretize Eq. (28) using standard finite elements [78, 80].

3.3.4 Finite-element method using transverse electromagnetic-field components

Recently, a new variational expression has been proposed using transverse electromagnetic-field components, and a finite-element procedure has been developed using this variational expression [81]. Different from the method using conventional variational expressions, this method, in principle can obtain β from k_0 because the eigenvalue corresponds to β. If the eigenvalues are written $k_0/\beta = u + jv$, they are grouped into four types:

$$\left.\begin{array}{ll} \text{Mode 1} : u=0, & v=0 \\ \text{Mode 2} : u=0, & v\neq 0 \\ \text{Mode 3} : u\simeq 0, & v\neq 0 \\ \text{Mode 4} : u\neq 0, & v=0 \end{array}\right\} \tag{31}$$

Since the eigenvalues for types 1, 2, and 3 are not bounded real numbers, there are spurious solutions, and the eigenvalue for type 4 corresponds to a physical solution. Namely, only imposing the condition that is a real number, all spurious solutions can be avoided. However, this criterion may cause difficulty in applying it to the waveguides below cutoff or those composed of complex media. Furthermore, the size of the final matrix eigenvalue problem is comparatively larger, i.e., six times the number of elements. More effort is needed in the future for enhancing numerical efficiency and for improving the accuracy of solutions [81].

3.3.5 Finite-element method using transverse electric- or magnetic-field component

The finite-element method in terms of transverse electric- or magnetic-field component has been developed for use in the case where the permittivity varies continuously over the waveguide cross section [82]. In this approach, β is obtained from k_0 in the same way as described in Sect. 3.3.4, and no spurious solution has been reported [82]. However, this approach cannot be applied straightforwardly if the permittivity varies discontinuously in the cross section. For example, assuming the permittivities $\varepsilon = \varepsilon_-$, ε_+ for the regions $x < 0$ and $x > 0$, respectively, one should approximate the permittivity as

$$\left.\begin{array}{l} \varepsilon = \begin{cases} \varepsilon_- & x<-\delta \\ (\varepsilon_- + \varepsilon_+)/2 + [(\varepsilon_+ - \varepsilon_-)/2](x/\delta) & |x|<\delta \\ \varepsilon_+ & x>\delta \end{cases} \\ \delta \to 0 \end{array}\right\} \tag{32}$$

This procedure may be laborious in practical computation.

4. Conclusion

Keeping in mind the electromagnetic waveguides, particularly the dielectric waveguides for use in microwave to optical wavelength regions, we have presented a review of recent progress in the development of methods to suppress or eliminate the spurious solutions encountered in their finite-element analysis as well as for several finite-element procedures to analyze their propagation characteristics. As has been described in the text, research on the spurious solutions, a long-pending question, has advanced in recent years, and several countermeasures have been found. In the near future, the time will come when these are ensured and applied to a wide range of problems.

Acknowledgement. We would like to thank the members of the Editorial Board and Electronics Group of the Transactions Editorial Committee, the Institute of Electronics and Communication Engineers of Japan, for affording the opportunity to write this review. We would also like to thank many Japanese and foreign authors of the books and papers to which we have referred in completing the work.

REFERENCES

1. T. Okoshi and T. Miyoshi. Planar Circuit, Ohm Pub. Co., Tokyo, Japan (1975).
2. Y. Kagawa. An Introduction to Finite-Element Method for Electrical and Electronic Engineers, Ohm Pub. Co., Tokyo, Japan (1977).
3. M.V.K. Chari and P.P. Silvester, Ed. Finite Elements in Electrical and Magnetic Field Problems, John Wiley & Sons (1980).
4. Japan Society for Simulation Technology. Ed. Simulation Technology (III), The Finite-Element Method, Corona Pub. Co., Tokyo, Japan (1981).
5. Y. Kagawa. Vibration and Acoustic Engineering by Finite-Element Method/ Fundamentals and Applications, Baifukan Pub. Co., Tokyo, Japan (1981).
6. Y. Kagawa. Practical Finite-Element Method for Electrical and Electronic Engineers, Ohm Pub. Co., Tokyo, Japan (1982).
7. T. Nakata and N. Takahashi. Finite-Element Method for Electrical Engineers, Morikita Pub. Co., Tokyo, Japan (1982).
8. P.P. Silvester and R.L. Ferrari. Finite Elements for Electrical Engineers, Cambridge University Press (1983).
9. T. Onuki, T. Nakata and S. Nemoto. Modern Finite Elements for Electrical Engineering System, Ohm Pub. Co., Tokyo, Japan (1983).
10. T. Okoshi, K. Okamoto and K. Hotate. Optical Fibers, Ohm Pub. Co., Tokyo, Japan (1983).
11. Y. Kagawa. Finite/Boundary Elements for Unbounded Field Problems, Science Pub. Co., Tokyo, Japan (1983).
12. Y. Kagawa, M. Koshiba, M. Ikeuchi and S. Kagami. Finite/Boundary Element Method for Electrical and Electronic Engineers--Application to Wave Problems, Ohm Pub. Co., Tokyo, Japan (1984).
13. M. Koshiba. Analysis of electromagnetic field problems, Jour. ITE Japan, 38, 12, pp. 1045-1052 (Dec. 1984).
14. M. Koshiba and M. Suzuki. On the finite-element analysis of electromagnetic and optical waveguides, Tech. Res. Rep. I.E.C.E., Japan, MW85-39 (July 1985).
15. M. Koshiba and M. Suzuki. On the finite-element analysis of electromagnetic eigenvalue problems, Tech. Res. Rep. I.E.E. Japan, EMT-86-3 (Jan. 1986).
16. T. Kawai, Ed. Special Issue on Computational Mechanics, Computrol, 8, Corona Pub. Co., Tokyo, Japan (Oct. 1984).
17. H. Kumagami, M. Koshiba and M. Suzuki. Numerical analysis of silicon-clad planar optical waveguides, Trans. I.E.C.E., Japan, J69-C, 7, pp. 856-864 (July 1986).
18. K. Okamoto and T. Okoshi. Vectorial wave analysis of inhomogeneous optical fibers using finite element method, I.E.E.E. Trans. Microwave Theory & Tech., MTT-26, 2, pp. 109-114 (Feb. 1978).
19. T.A. Lenahan. Calculation of modes in an optical fiber using the finite element method and EISPACK, Bell Syst. Tech. J., 62, 9, pp. 2663-2694 (Nov. 1983).
20. M. Hano and H. Kayano. An analysis of anisotropic and inhomogeneous optical waveguide using finite-element method, Trans. I.E.E. Japan, 101-C, 9, pp. 213-219 (Sept. 1981).
21. M. Koshiba and M. Suzuki. Numerical analysis of planar arbitrarily anisotropic diffused optical waveguides using finite-element method, Electron. Lett., 18, 13, pp. 579-581 (June 1982).
22. H. Kumagami, M. Koshiba and M. Suzuki. Finite-element analysis of planar anisotropic inhomogeneous optical waveguides, Trans. I.E.C.E. Japan, J67-C, 12, pp. 949-956 (Dec. 1984).
23. M. Koshiba, H. Kumagami and M. Suzuki. Finite-element solution of planar arbitrarily anisotropic diffused optical waveguides, I.E.E.E./OSA J. Lightwave Technol., LT-3, 4, pp. 773-778 (Aug. 1985).
24. C.C. Su. Origin of spurious modes in the analysis of optical filter using the finite-element or finite-difference technique, Electron Lett., 21, 19, pp. 858-860 (Sept. 1985).
25. H. Kumagami, M. Koshiba, K. Hayata and M. Suzuki. Finite-element solution of graded-core optical fibers, Nat. Conf. Rec., 1986, Opt. Radio Wave Electron., I.E.C.E. Japan, 22 (Sept. 1986).
26. P.L. Arlett, A.K. Bahrani and O.C. Zienkiewicz. Applications of finite elements to the solution of Helmholtz's equation, Proc. I.E.E., 115, 12, pp. 1762-1766 (Dec. 1968).
27. P. Silvester. A general high-order finite-element waveguide analysis program, I.E.E.E. Trans. Microwave Theory & Tech., MTT-17, 4, pp. 204-210 (April 1969).
28. M. Ikeuchi, K. Inoue, H. Sawami and H. Niki. Analysis of hollow waveguide by the α-method, Trans. I.E.C.E., Japan, J62-A, 4, pp. 224-231 (April 1979).
29. P. Daly. Finite element approach to propagation in elliptical and parabolic waveguides, Int. J. Numer. Methods Eng., 20, pp. 681-688 (1984).
30. J. Chiba and K. Sugiyama. Effects of trains on cutoff frequency and field in rectangular tunnel as waveguide, I.E.E.E. Trans. Microwave Theory & Tech., MTT-30, 5, pp. 757-759 (May 1982).
31. M. Suzuki and T. Hosono. Numerical

analysis of waveguide cross-sections giving minimum attenuation constants, Trans. I.E.C.E. Japan, J65-B, 7, pp. 921-927 (July 1982).

32. M. Suzuki and T. Hosono. Optimum sectional shape of dominant mode waveguide, I.E.E.E. Trans. Microwave Theory & Tech., MTT-31, 10, pp. 836-841 (Oct. 1983).

33. N. Mabaya, P.E. Lagasse and P. Vandenbulcke. Finite element analysis of optical waveguides, I.E.E.E. Trans. Microwave Theory & Tech., MTT-29, 6, pp. 600-605 (June 1981).

34. M. Koshiba, K. Hayata and M. Suzuki. Approximate scalar finite-element analysis of anisotropic optical waveguides, Electron. Lett., 18, 10, pp. 411-413 (May 1982).

35. K. Hayata, M. Koshiba and M. Suzuki. Approximate scalar finite-element analysis of anisotropic dielectric optical waveguides, Trans. I.E.C.E. Japan, J66-C, 10, pp. 752-759 (Oct. 1983).

36. K. Hayata, M. Koshiba and M. Suzuki. Analysis of partially metal-clad and dielectric overlay-loaded diffused optical waveguide, Trans. I.E.C.E. Japan, J67-C, 1, pp. 39-46 (Jan. 1984).

37. M. Koshiba, K. Hayata and M. Suzuki. Approximate scalar finite-element analysis of anisotropic optical waveguides with off-diagonal elements in a permittivity tensor, I.E.E.E. Trans. Microwave Theory & Tech., MTT-32, 6, pp. 587-593 (June 1984).

38. K.S. Chiang. Finite element method for cutoff frequencies of weakly guiding fibers of arbitrary cross-section, Opt. Quantum Electron., 16, pp. 487-493 (1984).

39. M. Matsuhara and N. Kumagai. Vector analysis of the arbitrary polarization holding fibers, Trans. I.E.C.E. Japan, J68-C, 10, pp. 796-802 (Oct. 1985).

40. R.B. Wu and C.H. Chen. A scalar variational conformal mapping technique for weakly guiding dielectric waveguides, I.E.E.E. J. Quantum Electron., QE-22, 5, pp. 603-609 (May 1986).

41. K. Hayata, M. Koshiba and M. Suzuki. Lateral mode analysis of buried heterostructure diode lasers by the finite-element method, I.E.E.E. J. Quantum Electron., QE-22, 6, pp. 781-788 (June 1986).

42. K. Hayata, M. Koshiba and M. Suzuki. Stress-applied optical fiber having inhomogeneous core, Trans. I.E.C.E. Japan, J69-C, 11, pp. 1404-1410 (Nov. 1986).

43. S. Ahmed and P. Daly. Finite-element methods for inhomogeneous waveguides, Proc. I.E.E., 116, 10, pp. 1661-1664 (Oct. 1969).

44. Z.J. Csendes and P. Silvester. Numerical solution of dielectric loaded waveguides: I-finite-element analysis, I.E.E.E. Trans. Microwave Theory & Tech., MTT-18, 12, pp. 1124-1131 (Dec. 1970).

45. P. Daly. Hybrid-mode analysis of microstrip by finite-element methods, I.E.E.E. Trans. Microwave Theory & Tech., MTT-19, 1, pp. 19-25 (Jan. 1971).

46. D.G. Corr and J.B. Davies. Computer analysis of the fundamental and higher order modes in single and coupled microstrip, I.E.E.E. Trans. Microwave Theory & Tech., MTT-20, 10, pp. 669-678 (Oct. 1972).

47. M. Ikeuchi, K. Inoue, H. Sawami and H. Niki. Spurious solutions in the finite element analysis of microstrip lines, Trans. I.E.E. Japan, 98-A, 8, pp. 415-422 (Aug. 1978).

48. M. Ikeuchi, H. Sawami and N. Niki. Analysis of open-type dielectric waveguides by the finite-element iterative method, I.E.E.E. Trans. Microwave Theory & Tech., MTT-29, 3, pp. 234-239 (March 1981).

49. P.M. Pelosi, P.V. Vandenbulcke, C.D.W. Wilkinson and R.M. De La Rue. Propagation characteristics of trapezoidal cross-section ridge optical waveguides: an experimental and theoretical investigation, Appl. Opt., 17, 8, pp. 1187-1193 (April 1978).

50. C. Yeh, K. Ha, S.B. Dong and W.P. Brown. Single-mode optical waveguides, Appl. Opt., 18, 10, pp. 1490-1504 (May 1979).

51. K. Oyamada and T. Okoshi. Two-dimensional finite-element method calculation of propagation characteristics of axially nonsymmetrical optical fibers, Radio Sci., 17, 1, pp. 109-116 (Jan.-Feb. 1982).

52. T. Okoshi, T. Aihara and K. Kikuchi. Finite-element method (FEM) analysis of side-tunnel type single-polarization fibers having circular tunnels, Tech. Res. Rep. I.E.C.E., Japan, MW83-10 (May 1983).

53. D. Welt and J. Webb. Finite-element analysis of dielectric waveguides with curved boundaries, I.E.E.E. Trans. Microwave Theory & Tech., MTT-33, 7, pp. 576-585 (July 1985).

54. R.B. Wu and C.H. Chen. A variational analysis of dielectric waveguides by the conformal mapping technique, I.E.E.E. Trans. Microwave Theory & Tech., MTT-33, 8, pp. 681-685 (Aug. 1985).

55. A.D. Berk. Variational principles for electromagnetic resonators and waveguides, IRE Trans. Antenna & Prop., AP-4, 2, pp. 104-111 (April 1956).

56. A. Konrad. High-order triangular finite elements for electromagnetic waves in anisotropic media, I.E.E.E. Trans. Microwave Theory & Tech., MTT-25, 5, pp. 353-360 (May 1977).

57. J.B. Davies, F.A. Fernandez and G.Y. Philippou. Finite element analysis of

all modes in cavities with circular symmetry, I.E.E.E. Trans. Microwave Theory & Tech., MTT-30, 11, pp. 1975-1980 (Nov. 1982).

58. B.M.A. Rahman and J.B. Davies. Finite-element analysis of optical and micro-waveguide problems, I.E.E.E. Trans. Microwave Theory & Tech., MTT-32, 1, pp. 20-28 (Jan. 1984).

59. T. Okoshi, Y. Shindo, K. Kikuchi and M. Fukawa. Finite-element method (FEM) analysis of side-tunnel type single-polarization fibers, Tech. Res. Rep. I.E.C.E. Japan, OQE85-28 (June 1985).

60. M. Hara, T. Wada, T. Fukasawa and F. Kikuchi. A three-dimensional analysis of RF electromagnetic fields by the finite element method, I.E.E.E. Trans. Magn., MAG-19, 6, pp. 2417-2420 (Nov. 1983).

61. M. Koshiba, K. Hayata and M. Suzuki. Vectorial finite-element formulation without spurious modes for dielectric waveguides, Trans. I.E.C.E. Japan, E67, 4, pp. 191-196 (April 1984).

62. M. Koshiba, K. Hayata and M. Suzuki. Vectorial finite-element method without spurious solutions for dielectric waveguide problems, Electron Lett.. 20, 10, pp. 409-410 (May 1984).

63. B.M.A. Rahman and J.B. Davies. Penalty function improvement of waveguide solution by finite elements, I.E.E.E. Trans. Microwave Theory & Tech., MTT-32, 8, pp.922-928 (Aug. 1984).

64. B.M.A. Rahman and J.B. Davies. Finite element solution of integrated optical waveguides, I.E.E.E./OSA J. Lightwave Technol., LT-2, 5, pp. 682-688 (Oct. 1984).

65. M. Koshiba, K. Hayata and M. Suzuki. Study of spurious solutions of finite-element methods in the three-component magnetic-field formulation for dielectric waveguide problems, Trans. I.E.C.E. Japan, J67-B, 12, pp. 1333-1338 (Dec. 1984).

66. M. Koshiba, K. Hayata and M. Suzuki. Improved finite-element formulation in terms of the magnetic field vector for dielectric waveguides, I.E.E.E. Trans. Microwave Theory & Tech., MTT-33, 3, pp. 227-233 (March 1985).

67. M. Koshiba, K. Kayata and M. Suzuki. Finite-element analysis of anisotropic waveguides with tensor permittivity, Trans. I.E.C.E. Japan, J68-B. 8, pp. 888-895 (Aug. 1985).

68. K. Hayata, M. Koshiba and M. Suzuki. Vectorial wave analysis of stress-applied polarization-maintaining optical fibers, Trans. I.E.C.E. Japan, J69-C, 1, pp. 17-24 (Jan. 1986).

69. M. Koshiba, K. Hayata and M. Suzuki. Finite-element solution of anisotropic waveguides with arbitrary tensor permittivity, I.E.E.E./OSA J. Lightwave Technol., LT-4, 1, pp. 121-126 (Feb. 1986).

70. K. Hayata, M. Koshiba and M. Suzuki. Vectorial wave analysis of stress-applied polarization-maintaining optical fibers by the finite-element method, I.E.E.E./OSA J. Lightwave Technol., LT-4, 2, pp. 133-139 (Feb. 1986).

71. T. Okoshi, O. Ishida and K. Kikuchi. Finite element method (FEM) analysis of side-tunnel optical fibers, Tech. Res. Rep. I.E.C.E. Japan, MW86-19 (June 1986).

72. K. Hayata, M. Eguchi, M. Koshiba and M. Suzuki. Vectorial wave analysis of side-tunnel type polarization-maintaining optical fibers by variational finite-elements, I.E.E.E./OSA J. Lightwave Technol., LT-4, 8, pp. 1090-1096 (Aug. 1986).

73. K. Hayata, M. Koshiba, M. Eguchi and M. Suzuki. Novel finite-element formulation without any spurious solutions for dielectric waveguides, Electron. Lett., 22, 6, pp. 295-296 (March 1986).

74. K. Hayata, M. Koshiba, M. Eguchi, K. Miura and M. Suzuki. Elimination of spurious solutions in finite-element analysis of dielectric waveguides--A formulation in terms of tranverse magnetic-field component, Tech. Res. Rep. I.E.E. Japan, EMT-86-28 (July 1986).

75. K. Hayata, M. Koshiba, M. Eguchi and M. Suzuki. Elimination of spurious solutions in vectorial finite-element analysis of dielectric waveguides--A method using transverse magnetic-field component, Trans. I.E.C.E. Japan, J69-C, 12, pp. 1487-1493 (Dec. 1986).

76. K. Hayata, M. Koshiba, M. Eguchi and M. Suzuki. Vectorial finite-element method without any spurious solutions for dielectric waveguiding problems using transverse magnetic-field component, I.E.E.E. Trans. Microwave Theory & Tech., MTT-34, (1986).

77. A.J. Kobelansky and J.P. Webb. Eliminating spurious modes in finite-element waveguide problems by using divergence-free fields, Electron. Lett., 22, 11, pp. 569-570 (May 1986).

78. M. Hano. Finite-element analysis of dielectric-loaded waveguides, I.E.E.E. Trans. Microwave Theory & Tech., MTT-32, 10, pp. 1275-1279 (Oct. 1984).

79. M. Koshiba, K. Hayata and M. Suzuki. Finite-element formualtion in terms of the electric-field vector for electromagnetic waveguide problems, I.E.E.E. Trans. Microwave Theory & Tech., MTT-33, 10, pp. 900-905 (Oct. 1985).

80. M. Koshiba, K. Hayata and M. Suzuki. Vector E-field finite-element analysis of dielectric optical waveguides, Appl. Opt., 25, 1, pp. 10-11 (Jan. 1986).

81. M. Matsuhara, T. Angkaew and N. Kumagai. Finite element analysis of the waveguide eigenmodes--A novel method to employ the transverse electromagnetic

fields, Trans. I.E.C.E. Japan, J69-C, 5, pp. 548-553 (May 1986).
82. J. Katz. Novel solution of 2-D waveguides using the finite element method, Appl. Opt., 21, 15, pp. 2747-2750 (Aug. 1982).

Methods for the solution of open-boundary electromagnetic-field problems

C.R.I. Emson, BSc, DipEng, MSc, PhD

Indexing terms: Electromagnetic theory, Mathematical techniques, Algorithms, Magnetic fields

Abstract: A review of methods suitable for the solution of open-boundary field problems is given. Emphasis is placed on algorithms that can be used in conjunction with the finite-element method. The requirements for any algorithm to model an exterior region is outlined, as well as a description of some of the implications of the possible methods. It is hoped that users of such models will then appreciate the assumptions made in any method and thus its limitations, and so be able to choose the most appropriate one for a particular application.

1 Introduction

In the analysis of electromagnetic-field problems, the technique of finite elements has become increasingly popular, and is probably the most commonly used method for solving all but the most trivial problems. However, the technique does suffer the disadvantage that, in general, only a finite number of elements can be used to discretise a given problem, owing to the finite amount of computer storage available.

Unfortunately, it is a feature of electromagnetic fields that the extent of the problem being solved can be infinite. There is then a contradiction in requirements when solving such problems — the need for an infinite number of finite elements to model an infinite space, but only the resources to allow a finite (although these days, on modern computers, a very large) number of finite elements.

It has therefore become common practice, when using the finite-element method, to truncate the mesh at some finite distance from the centre of the problem. As this artificial boundary is made more remote, its effect diminishes, until the error thus created can be regarded as being part of the 'noise' in the solution. Although this requires a large number of elements, it does mean that a solution can be obtained, theoretically to within any desired accuracy.

The aim of this paper is to describe alternative procedures currently available for dealing with exterior regions, all of which could prove more efficient than simple truncation. A description will also be given of some of the implications of using such procedures.

2 Types of problem to be modelled

Before attempting to deal with models for the infinite exterior domain, it is first necessary to examine the types of problem that can be solved. The two cases of static fields and travelling waves are quite different, and will be discussed separately. Transient problems will also be briefly mentioned.

2.1 Static fields

For any well posed problem, the fields associated with any sources in the problem should decay to zero with distance from those sources. This implies that the energy of the system is finite, in other words there are no sources positioned at infinity. If there were to be sources at infinity, the field would not decay to zero, the value being unknown, and the energy of the system would be infinite. The problem would then not be well posed, i.e. not a physically possible problem.

For 3-dimensional problems, there is usually no difficulty in identifying well posed problems — all sources should be within the finite-element mesh, leaving the exterior source free (including values at infinity).

In two dimensions, the situation is less straightforward. Considering the field due to a single current-carrying conductor, it would at first sight be thought reasonable to assume the field decays to zero away from the conductor in the x- and y-directions in Fig. 1. However, by stating that the problem is 2-dimensional (in the xy-plane), there is the assumption of no variation in the z-direction. There is therefore a current existing at $z = \pm\infty$ (the current return path). From the above argument, this problem is not well posed, and as such the exterior cannot be modelled correctly.

Consider next the situation in Fig. 2, with a pair of conductors with currents in opposing directions. This time the currents are balanced, with no return path required at infinity. With no sources at infinity, the problem is well posed, and the exterior can be modelled by one of the techniques to be described.

In a 2-dimensional analysis of an infinite problem therefore, there must only appear balanced sources (no

Reprinted with permission from *IEE Proceedings-A*, vol. 135, no. 3, pp. 151–158, March 1988.

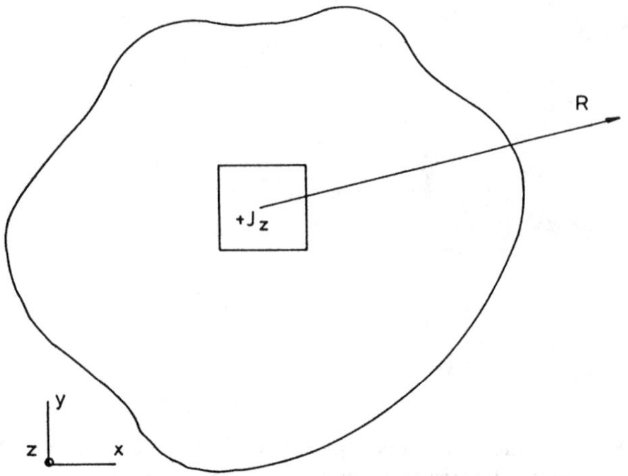

Fig. 1 *Current carrying conductor (current in z-direction)*

net charge or current), in order that the exterior be modelled correctly by some special technique.

It should be mentioned however that a uniform incident field is an exceptuon to this statement, as there is an infinite energy implied in the system (there must be sources at infinity to create the incident field). In this case, since the field at infinity is known, it can be imposed as a boundary condition; although a nonphysical problem it can be solved.

2.2 Travelling waves

The situation is somewhat different with travelling-wave problems, as the problem to be solved can often be separated into two parts. First, it is assumed that there is some form of wave incident on the region of interest (within the finite-element mesh). This implies sources at infinity to generate the incoming wave. Secondly there is a scattered wave travelling outwards, and it is this wave that can be modelled by some special technique, because its amplitude must decay to zero at infinity.

2.3 Transient fields

The same arguments as above apply equally well to transient problems. For example, it is common practice to move the outer boundary far enough away, such that any waves reflected back into the interior will not reach the region of interest until after the time scale being examined.

3 Algorithms for exterior regions

There are many methods currently available for exterior problems, each having its own advantages and disadvantages. The following is a list of many of these methods, and some of the implications of using them. The list is an extension of that presented in a previous review paper [72], including the many techniques that have been developed since that time.

The list is divided into global and elemental methods: global methods deal with the exterior as a whole, whereas elemental methods subdivide the exterior into a finite number of subregions (or elements).

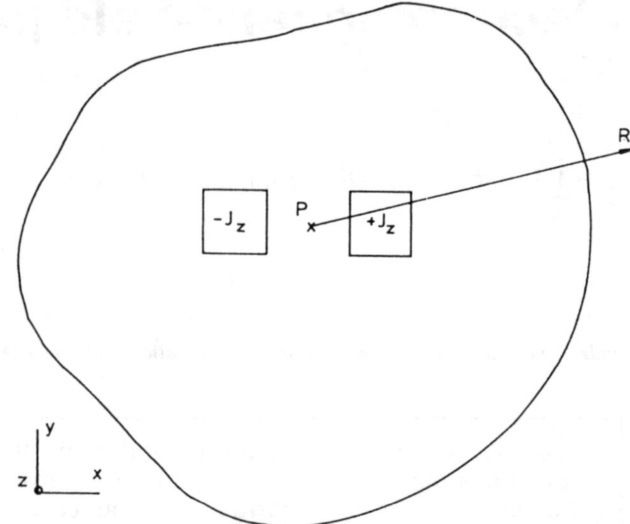

Fig. 2 *Pair of current carrying conductors (currents in z-direction)*

3.1 Global methods

3.1.1 Truncation: The most commonly used method to date has been that of truncation, being simple to apply and not needing any modifications to an existing computer program. By regarding the application of boundary conditions as equivalent to adding extra sources, some useful consequences arise.

If, in a magnetostatic problem a boundary condition $\phi = 0$ is applied, $H_t = 0$ is imposed on the outer boundary. This can also be regarded as equivalent to placing a source of equal strength, but opposite sign, an equal distance beyond the boundary. In Fig. 3, the added source

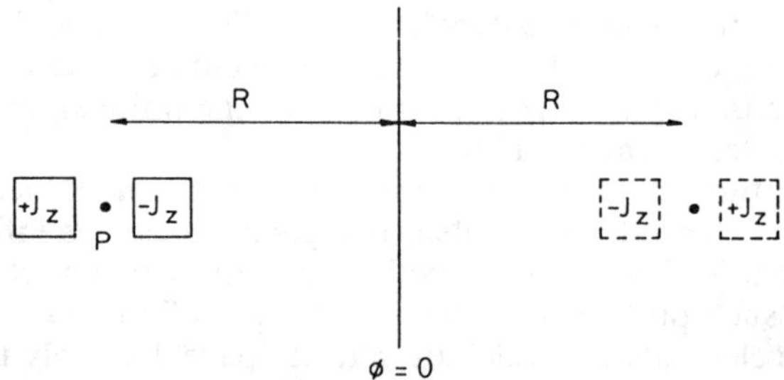

Fig. 3 *Implied sources due to Dirichlet boundary conditions*

will generate a field at point P of strength $1/4R^2$ times that of the original source. Similarly, if $d\phi/dn = 0$, $B_n = 0$ is imposed, which can be regarded as a second source of equal strength and same sign placed an equal distance beyond the boundary.

It is therefore possible to predict the likely effect of any truncated boundary on the results by considering how significant the implied secondary source field is, compared with that due to the original source. As a further check, the boundary can be moved further away and a new solution obtained. If this is repeated a few times, it is possible to extrapolate the solution to give a better estimate of the solution when the boundary is lying at infinity [63].

Difficulties arise when simple truncation is applied to wave-type problems however, as any outgoing wave will be reflected by the artificial boundary. For a Neumann boundary condition (normal derivative constrained) the reflected wave is in phase with the incident (outgoing) wave, and, converely, for a Dirichlet (constrained potential) condition, the reflected wave is out of phase. Unless there is damping present in the system, this technique is likely to lead to totally erroneous results.

3.1.2 Mapping: Simple truncation described above will generally lead to a large number of elements to model a given problem, many of which will be used to define the 'exterior' region. Mapping can therefore be used as a means of moving the outer boundary farther away, without requiring such a large number of elements.

As well as using a finite co-ordinate mapping, it is also possible to use a singular mapping function to map the outside of a unit circle back onto its interior using $1/R$ [48] (in fact the exterior of any polygon can be mapped onto the exterior of a unit circle, and hence onto the interior of the circle). This works well when the region outside the polygon is source-free and contains linear materials, but leads to complications otherwise.

Care must be taken when using singular mappings to ensure that the mapped region is adequately discretised, especially towards the centre point (which maps to infinity) where the solution can be changing very rapidly. When applying singular mappings to wave problems [49], an outgoing wave of a given wavelength will map onto a wave whose wavelength is decreasing to zero towards the centre point. This could also cause problems when performing the necessary integrals.

3.1.3 Analytic solutions: For many problems, suitable boundary conditions can be applied to a finite-element mesh based on the far-field analytic solution to the governing equations. This technique has been applied in both static (originally with the finite-difference method [54]) and wave-type problems [38].

As the outer boundary of the finite-element mesh is moved further away from the region containing the sources in the problem, the solution over the outer boundary will resemble that due to point sources at the centre of the problem [58, 60]. The exterior can therefore be approximated by applying, as boundary conditions, the analytic solution due to appropriate point sources at the centre of the problem, either in terms of Dirichlet or Neumann conditions [63]. If, for example, the field due to a pair of current-carrying bars is required (as in Fig. 2), an approximate boundary condition could involve the analytic solution due to a dipole positioned at the point P. If more bars were added, higher order multipoles could then be added to generate the correct boundary condition.

Alternatively, the analytic solution on the boundary can be used to generate an equivalent 'stiffness' matrix for the exterior region, as if it were a single element. This involves all the boundary nodes being connected, although if the magnitude of the coupling were sufficiently small for a pair of nodes, that term could be removed from the matrix, slightly decreasing the storage required [51].

The advantage of these techniques over truncation is that the outer boundary can be placed much closer to the region of interest, thus requiring fewer elements for a given accuracy of solution. Of course, the far-field solution for the given problem needs to be known beforehand.

3.1.4 Approximate solutions: It is sometimes not possible to use an analytic solution to approximate the boundary conditions, for example when the analytic solution is not known to sufficient accuracy on the proposed boundary. In these circumstances, the exterior field can be expressed in terms of eigenvectors over the boundary [30, 34]. For example, in terms of cylindrical [19, 44, 64] or spherical [14, 45] harmonics, or Fourier series [27, 59]. If the boundaries are parallel to the co-ordinate system axis, separation of variables may be applied, and appropriate boundary conditions thus imposed [55]. The constants in all of these approximations would then appear as unknowns in the final equations to be solved, usually corresponding in number to the nodes on the exterior boundary.

In the case of wave problems, it is also required that any outgoing wave is not reflected, in other words they are completely damped by the boundary. A technique known as 'boundary dampers' is an attempt to do just that [37, 65]. The method is an approximate one because an arbitrary wave cannot be completely damped. Instead, the outgoing wave is assumed to be a plane wave, and appropriate boundary conditions can then be imposed to completely damp out plane waves. An improvement on this was to assume the outgoing wave was cylindrical [68], and this has then led to a series of higher order dampers being developed [4, 3]. The higher the order the more the harmonics present are damped, but inevitably at a higher cost computationally.

Iterative methods for approximating the boundary conditions have also been applied [18]. If the boundary condition can be written as a function of the solution derivative, the boundary conditions can be continuously updated as a new estimate for the solution derivative is found. For example, in electrostatics the exterior can be represented by an equivalent charge density on the outer boundary. The charge strengths can be written in terms of the solution derivatives on the boundary, and hence this technique can be applied.

3.1.5 Global exterior elements: There are also a series of methods that rely on treating the exterior region as a single 'super' element, with the obvious result that all nodes on the boundary are coupled to each other, thus increasing the bandwidth of the final matrices. However, some of the methods to be described have been found to give very good results, and for that reason are still very popular.

One of the simplest of these global methods is probably that known as ballooning [57, 35, 15]. The technique involves generating a single 'super' element outside the finite-element mesh (Fig. 4). This super element is

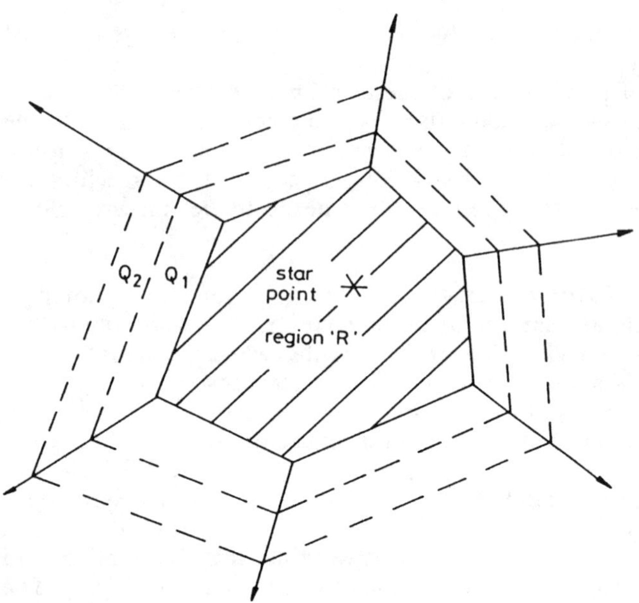

Fig. 4 *Concentric rings of elements around a finite-element mesh, used in the method of 'ballooning'*

constructed by merging successive concentric rings of elements (Q_1, Q_2 etc. in Fig. 4) in an iterative manner, until the outer boundary has a sufficiently large radius. All nodes between adjacent rings are removed so that there are only nodes on the boundary with the finite-element mesh, and at infinity. Boundary conditions at infinity can then be imposed.

The radius of the outer boundary created increases in a geometric series (hence, therefore, it never becomes infinite), but, after some seven iterations, the radius of this boundary is measured in light years, and so it can be regarded as an extremely effective method of finite mapping.

In the same way as other forms of mapping, ballooning models the exterior region using progressively larger elements away from a predefined 'star point'. Care must be taken therefore that the solution is not changing too rapidly for these constructed elements to model satisfactorily.

Because the method of ballooning effectively deletes nodes in the exterior region, it becomes quite difficult to impose symmetry conditions on a plane extending to infinity, or to recover solutions in the exterior region itself. It must also be remembered that, as with other methods of mapping, the technique cannot be applied to travelling-wave-type problems if Dirichlet or Neumann boundary conditions are imposed at infinity (but could be used to great advantage if, for example, analytic or approximate boundary conditions from the preceding Sections were employed).

There are a few other methods also making use of iterative techniques to construct a model of an infinite exterior region. It is possible, for example, to generate an infinite number of elements in the exterior, created in a systematic manner, and then to reduce this to a finite set of equations that give a best minimum for the energy functional in the exterior [61].

Another technique involves applying a similar method to ballooning an infinite number of times (infinite substructuring [23]). It is then possible to generate the 'stiffness' matrix for the exterior (as well as provide a means for recovering the solution at points outside the finite-element mesh) by solving the resulting quadratic eigenvalue matrix equation. It should be mentioned that because an infinite number of iterations are made, the sizes of the elements generated in the exterior are vanishingly small, so a rapidly varying solution can be modelled with no difficulty (cf. ballooning). Also, an infinite number of elements implies the outer boundary lies at infinity, and this technique is therefore also suitable for wave-type applications [22].

A related method is 'infinitesimal scaling' [28], which also generates an expression for the energy functional for the exterior. In this method, a solution to a nonlinear ordinary differential matrix equation is required.

In fact, identical results to infinitesimal scaling can be obtained by employing another well known method in the exterior, namely the boundary element method. Boundary elements are often used on their own to solve linear problems, both finite and infinite in extent, and the literature on this subject is vast (see Reference 16 for a general description of the technique). More recently, boundary elements have been used in conjunction with finite elements to solve infinite-domain problems [8, 56, 69, 70, 17]. In this way, the finite elements can model nonlinear interiors, and the boundary elements used to give a very accurate model of the exterior region.

The boundary element method itself can be visualised as being a method of replacing a region by a set of equivalent sources on its boundary (for example, charge density or currents) (Fig. 5*A*). In general, there will be an equivalent source at each node on the boundary (corresponding to the nodes of the finite-element mesh on the same boundary). The field anywhere due to these

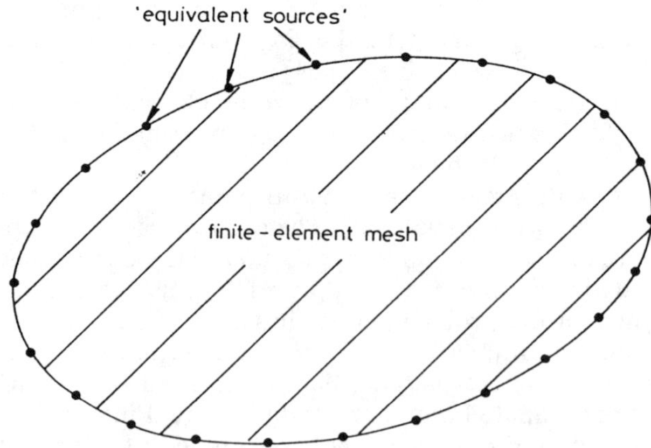

Fig. 5A *Distribution of equivalent sources on exterior boundary of finite-element mesh, to model the exterior space (boundary element method)*

sources can then be expressed in terms of appropriate Green's functions (the solution to the governing equation due to a point source). The strengths of these equivalent sources are then solved, either directly or indirectly [47]. The technique does, however, involve the evaluation of some singular integrals, requiring special consideration. This has been overcome to some extent by regarding the equivalent sources as not lying on the boundary itself, but slightly inside the boundary (Fig. 5B) [39]. It has also

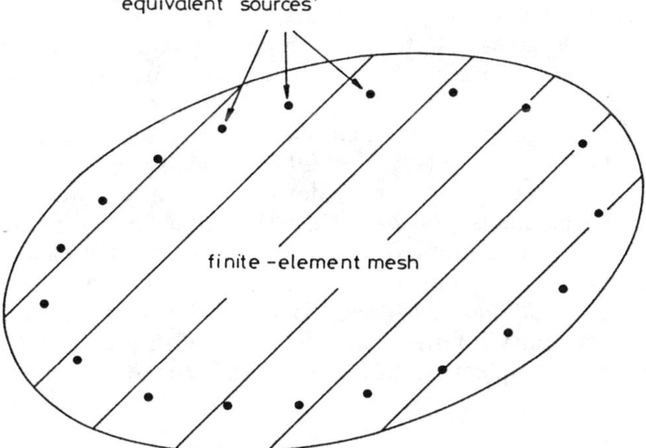

Fig. 5B *Equivalent sources placed within the finite-element mesh rather than on the boundary itself, to avoid singular integrals (modified boundary element method)*

been proposed that it is not even necessary to have as many sources as nodes on the boundary; and, by using less equivalent sources, a best set of source strengths can be determined [46].

In the boundary element method the outer boundary is implied at infinity, and is therefore ideally suited to solving propagating-wave problems [43, 46].

3.2 Elemental methods (infinite elements)

The idea of dividing the exterior into a series of 'infinite elements' (Fig. 6) was first proposed in 1973 [62], and, since then, many variations have been published. Most of the methods can, in fact, be separated into two broad categories, one that modifies the functions representing the solution in the exterior (using decay functions), and the second that modifies the way in which the infinite

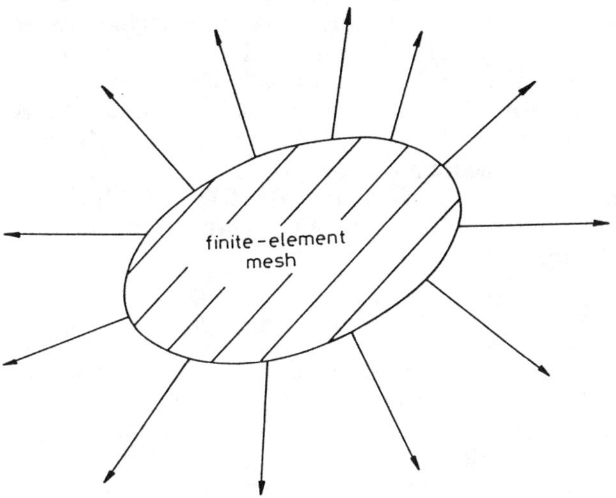

Fig. 6 *Exterior region subdivided into semi-infinite regions or elements (often called 'infinite elements')*

element is mapped onto a finite dimensionless region (over which the necessary integrals can be performed more readily).

3.2.1 Decay functions: In the first application of infinite elements to static problems, the solution variation in the infinite direction was modified by the introduction of some form of decay function (Fig. 7). In this way, the

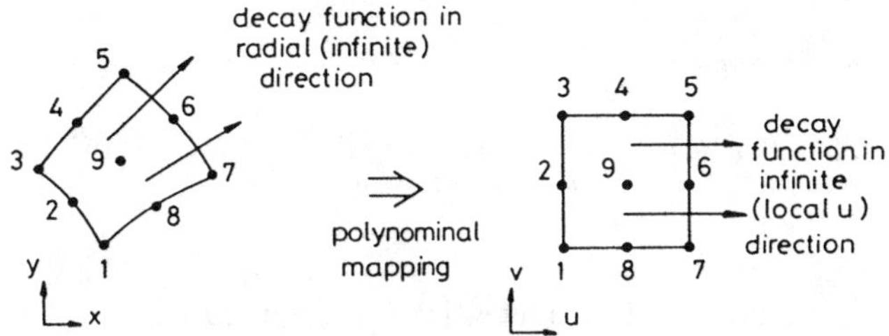

Fig. 7 *Decay function included in infinite (radial) direction, maps onto decay in local u-direction (with this orientation of element).*

solution was forced to decay either reciprocally [62, 1, 11, 41] or exponentially [66, 9, 24, 20] with distance.

The choice of reciprocal decay seems obvious if the far-field solution is also known to be reciprocal (in three-dimensional statics problems), but the use of exponential decay functions may not appear quite so obvious and will need some elaboration.

It is, in fact, not necessary for an infinite element to try to impose an analytic solution in the exterior, rather it is sufficient to ensure that the solution decays to zero with distance to ensure a reasonable solution inside the region of interest, the finite-element mesh (see Section 4 for further details). It is then to be expected that the solution in the exterior itself will not be particularly good.

In some cases, however, it may be important to be able to generate good approximations to the solution, even at large distances from the finite part of the problem. In these cases, the use of more specialised decay functions may be necessary.

Reciprocal decay functions have already been mentioned, and at large distances they may be an adequate representation of the solution. They can also be extended to include higher orders by the introduction of more nodes in the element (in the same way as higher order finite elements). As more nodes are added, more terms in the series $1/R^n$ are introduced, thus improving the accuracy of the solution. Another form of higher order decay function has been suggested [36] using $(a/R)^n$, but, in fact, as $R \Rightarrow 0$ the solution becomes $1/R$ again, and so it is not truly a higher order method.

Of a more specialised nature, it is possible to use functions very much problem-dependent [2], or indeed almost any function that may be wanted [53].

3.2.2 Co-ordinate mapping: It is usual in the finite-element method to employ some form of co-ordinate mapping between the global system and a local one. Typically, Lagrange polynomial mappings could be used, leading to the usual set of parametric finite elements [29]. However, it is also feasible to use some form of singular mapping function (e.g. based on $1/R$) [26, 5, 6], where one of the nodes in the local element is mapped onto a point lying at infinity (Fig. 8).

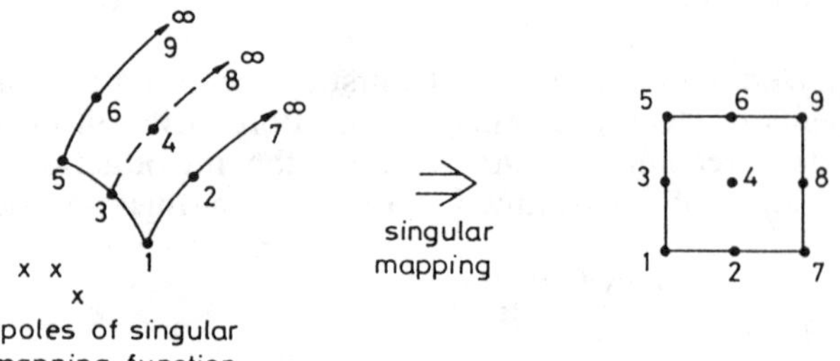

Fig. 8 *Semi-infinite region mapped onto finite local element using singular mapping function (including implied poles of mapping), (nodes 7, 8 and 9 map onto infinity)*

A variation of this was later developed [72, 73, 50] such that normal Lagrange polynomials could be used in the local element to represent the solution variation. This was shown to be equivalent (after performing the mapping) to a global variation of the form $1/R^n$. As ordinary polynomial functions are used for the solution variation in the local element, standard numerical integration methods could be used without modification (for static problems). By varying the order of the singular mapping function used, different powers of decay of the global solution can be obtained [21].

A form of reciprocal decay element has also been devised [52] using techniques from both of the categories listed above. The solution variation over the local element is modified, and a nonpolynomial mapping is performed (based on $1/R$ again). The advantage of this technique is that a numerical integration rule can be developed to integrate the necessary term exactly.

3.2.3 Extension to wave problems: Many of the methods mentioned so far have been extended to include periodic time variation. The first method to implement this [10, 67] included the term $\exp(i\omega t) \cdot \exp(ikR)$ in the decay functions representing the solution in the exterior. The first part, $\exp(i\omega t)$, can be neglected (being present in all terms in the equations it can be cancelled out). The second part is used to indicate that, in the infinite region, the wave is propagating in a radial direction, with wave number k. Inevitably, this requires a new set of integration routines to be developed, to deal with the new terms arising.

As before, reciprocal and exponential decay functions can be used in the infinite direction, with the amount of decay being determined by comparing the decay function with the expected far-field solution. (In three dimensions this is $1/R$, and in two dimensions this is typically a zeroth-order Hankel function, decaying as approximately $R^{-1/2}$ for large R.)

The same technique has been applied to the method of co-ordinate mapping [72, 13], again with great success. Whereas, with the static version of the method, standard numerical integration routines could be used, on introducing the wave term new integration rules had to be developed. A modified form of mapping was also used to allow for the fact that, in two dimensions, the far field does not decay as $1/R$, but as $R^{-1/2}$, giving improved accuracy for such problems [74, 25].

As the wave number of the outgoing wave is included in the description of the solution, these methods can only be applied to problems where a single wave of known frequency is propagating. Problems involving more than a single wave have been tackled [20, 40], but these methods have not, to date, been applied to transient problems, where, in principle, an infinite number of frequencies are propagating.

One further method that should be mentioned is one in which the boundary element method is applied in the exterior, but over the boundaries of a finite number of infinite elements [31, 32]. This technique does not suffer the same disadvantage of the normal boundary element method of coupling all boundary nodes to every other. The method has also been found to be more accurate than ordinary infinite elements, although not as accurate as the full implementation of boundary elements.

4 Requirements of a valid algorithm

A list of requirements for a valid infinite element has been previously published [42], and a few of the more important features that apply to all exterior domain algorithms are expanded here.

In Section 2, the types of problem that could be solved using special techniques for the exterior region were discussed. It was stated that only well posed problems could be solved, and these were described as problems where the solution decays to zero with distance away from the finite part of the problem. This is a rather loose statement, and will need elaboration.

When applying the finite-element method, it is common practice to reduce the order of the Laplacian operator using Green's theorem (integration by parts) [67]. This results in first-order differential terms over the volume and the boundaries. Within the finite part of the problem, the volume terms are modelled with finite elements. In the exterior, the volume term can be modelled using some form of infinite element, or, alternatively, they can be transformed into a boundary term, which is then modelled using boundary elements, analytic or approximate methods, or some form of iterative technique.

Most of the methods mentioned therefore rely, in some way, on the application of Green's theorem in the exterior. However, care must be taken as there are certain restrictions on the use of Green's theorem. In infinite regions, the theorem can only be applied if the solution is decaying to zero at a certain rate [33] (the solution must be regular at infinity). In other words, for a solution ϕ and radial distance R,

$$\phi R, \quad \frac{\partial \phi}{\partial x} R^2, \quad \frac{\partial \phi}{\partial y} R^2, \quad \frac{\partial \phi}{\partial z} R^2$$

must be bounded for all sufficiently large R. Therefore, the solution must decay to zero at least as fast as $1/R$ (in three dimensions, and $R^{-1/2}$ in two dimensions).

This is inherent in all the methods mentioned that in some way approximate the solution in the exterior (rather than extend the finite-element mesh in some way as in the mapping techniques). In particular, infinite elements choose decay functions that decay at least as rapidly as this, or employ mappings such that this is implied [12]. Boundary element methods also satisfy this constraint, because, although the boundary is represented by equivalent sources (e.g. charges), the interior has zero net charge; all surface charges will therefore balance, and, in the far field, the effect is equivalent to that of a distribution of half the number of dipoles.

If wave problems are being solved, there is a further constraint to be imposed, the so called 'Sommerfeld radiation condition' [68]. This requires that, for an n-dimension problem:

$$\lim_{r \Rightarrow \infty} R^{\left(\frac{n-1}{2}\right)} \left[\frac{\partial \phi}{\partial R} + ik\phi \right] = 0$$

that is, the magnitude of the solution decays as for the static problem, and the wave is always propagating outwards.

If finite elements are being used, the usual constraint of compatibility applies. The solution in adjacent elements must be continuous, both between adjacent infinite elements (if being used), and between whatever technique is employed and the finite-element mesh. (In fact, only one of the methods mentioned above [62, 1] does not satisfy this constraint.)

5 Summary

In Section 3 of this paper, many algorithms suitable for infinite domains have been mentioned. For convenience, they were divided into two main groups: global and elemental methods.

Many of the algorithms presented under the heading 'global' methods treat the exterior region in terms of a single element, with the natural consequence that all boundary nodes become coupled to every other. If some form of banded matrix solver is to be employed, this will greatly increase the cost of solution. Even when using sparse solvers, some increase in cost is to be expected because the matrix associated with the exterior will be fully populated.

The advantage, then, of using elemental methods (infinite elements) is that there is not the drastic increase in bandwidth of the matrix equation, and, indeed, few extra degrees of freedom are usually needed.

What must be remembered is that, for all the methods described (with the exception of the methods based on mappings), some knowledge of the far-field solution is required. For analytic and approximate methods this will appear reasonable, but maybe less so for some of the other techniques. Infnite elements, for example, require the evaluation of some form of decay parameter so that the chosen decay function best matches the far-field solution (albeit over part of the element only). Boundary element methods require a suitably chosen Green's function, an analytic solution to the governing equation.

The infinite elements that make use of singular coordinate mappings will also require the choice of some star point to define the mapping. The position of this must be chosen with some care, as the far field will appear to originate from a point source at this position (the same situation arises when applying analytic boundary conditions).

As far as relative accuracy is concerned, it is generally accepted [69, 71, 31, 32, 7] that boundary elements are more accurate in modelling an exterior region than infinite-element methods. The reason for this is that the boundary element method is employing an analytic solution to the governing equation, to represent the solution in the exterior, whereas with infinite elements it is more usual to simply require that the solution deays to zero with distance. However, there is the added complexity with boundary elements, not only due to the more fully populated matrices mentioned here, but also with regard to the evaluation of singular integrals inherent in the formulation.

It is interesting to note that, for all those global methods which require the evaluation of coefficients, generally corresponding to the nodes on the boundary, it is not possible to increase the accuracy of the exterior solution without increasing the number of these boundary nodes. The infinite-element methods, however, allow an increase in the order of the solution in the infinite direction, quite independently of the number of nodes or elements on the boundary itself. In principle then, the accuracy of the infinite-element methods can be increased to any desired level, at the expense of more nodes being required in the exterior.

In general, it is not possible to say which of the methods mentioned is the best, as the choice of method depends on many factors, many of which may be problem-specific. For example, if accuracy is more important than speed of solution, or the size of the model, then boundary element methods are probably the best choice. On the other hand, if size and speed are to be kept to a minimum, some form of infinite element may be more suitable. The type of approach to be used may then depend on whether the solution is required in the exterior region or not.

6 Conclusions

In the preceding Sections, the types of physical problems that can be solved have been described as well as a few guidelines concerning the requirements for such methods. In these Sections, it has been emphasised that, to solve an exterior problem, the solution must decay to zero with distance; physically interpreted as requiring all sources to balance, and necessary numerically in order to apply Green's theorem in the infinite region.

A list of many of the algorithms currently available for exterior problems has been given. It is not particularly easy to draw any conclusions concerning the many techniques, and it is probably not appropriate to try. Instead, a few points of comparison have been given, particularly concerning the relative accuracy of some of the techniques.

The use of the methods for wave-type problems has been discussed, although some have been found to be completely inappropriate. Most of the methods mentioned have been frequency-dependent and, as such, are not suitable for transient types of problem (where an infinite number of frequencies can be expected). Although infinite-element methods have been applied to problems with more than one frequency, it is still only the boundary element method that has been applied to transient problems, using a time dependence in the Green's function.

7 Acknowledgments

I would like to thank SERC for financial support during my period of research at University College Swansea, during which time most of this work was carried out. I would also like to thank Professor P. Bettess for his help and guidance during that time.

8 References

1 ANDERSON, D.L., and UNGLESS, R.L.: 'Infinite finite elements'. *Int. Symp. Innovative Numer. Analysis in Appl. Eng. Sci.*, Versailles, 1977

2 ASKAR, H.G., and LYNN, P.P.: 'Infinite elements for ground freezing problems', *J. Eng. Mech. (ASCE)*, 1984, **110**, (2), pp. 157–172

3 BANDO, K., BETTESS, P., and EMSON, C.: 'The effectiveness of dampers for the analysis of exterior scalar wave diffraction by cylinders and ellipsoids', *Int. J. Numer. Method Fluids*, 1984, **4**, pp. 599–617

4 BAYLISS, A., GUNZBERGER, M., and TURKEL, E.: 'Boundary conditions for the numerical solution of elliptic equations in exterior regions'. ICASE report 80-1, NASA, Langley, 1980

5 BEER, G., and MEEK, J.L.: 'A boundary finite element for underground mining applications', *in* BREBBIA, C.A. (Ed.): 'New developments in boundary element methods' (CML Publications, Southampton, 1980), pp. 281–294

6 BEER, G., and MEEK, J.L.: 'Infinite domain elements', *Int. J. Numer. Methods Eng.*, 1981, **17**, pp. 43–52

7 BEER, G.: 'Finite element, boundary element and coupled analysis of unbounded problems in elastostatics', *ibid.*, 1983, **19**, pp. 567–580

8 BERKHOFF, J.C.W.: 'Linear wave propagation problems and the finite element method', *in* GALLAGHER, R.H. (Ed.): 'Finite elements in fluids—vol. 1' (Wiley, 1975), Chap. 13

9 BETTESS, P.: 'Infinite elements', *Int. J. Numer. Methods Eng.*, 1977, **11**, pp. 53–64

10 BETTESS, P., and ZIENKIEWICZ, O.C.: 'Diffraction and refraction of surface waves using finite and infinite elements', *ibid.*, 1977, **11**, pp. 1271–1290

11 BETTESS, P.: 'More on infinite elements', *ibid.*, 1980, **15**, pp. 1613–1626

12 BETTESS, P., EMSON, C., and BANDO, K.: 'Some useful techniques for the testing of infinite elements', *Appl. Math. Modelling*, 1982, **6**, pp. 436–440

13 BETTESS, P., EMSON, C., and CHIAM, T.C.: 'A new mapped infinite element for exterior wave problems', *in* LEWIS, R.W. (Ed.): 'Numerical methods in coupled systems' (Wiley, 1984), p. 489

14 BRAMBLE, J.H., and PASCIAK, J.E.: 'A new computational approach for the linearised scalar potential formulation of the magnetostatic field problem', *IEEE Trans.*, 1982, **MAG-18**, (2), pp. 357–361

15 BRAUER, J.R.: 'Open boundary finite elements for axisymmetric magnetic and skin effect problems', *J. Appl. Phys.*, 1982, **53**, (11), pp. 8366–8368

16 BREBBIA, C.A.: 'The boundary element method for engineers' (Pentech Press, 1980)

17 BRYANT, C.: 'A mixed integral differential method for low frequency electromagnetic fields'. PhD thesis, Dept. Electrical Engineering, Imperial College of Science and Technology, London, 1981

18 CERMAK, I.A., and SILVESTER, P.: 'Solution of 2-dimensional field problems by boundary relaxation', *Proc. IEE*, 1968, **115**, (9), p. 1341

19 CHEN, H.S., and MEI, C.C.: 'Oscillations and wave forces in a man made harbor in the open sea'. *Proc. 10th Symp. Naval Hydrodyn.*, Cambridge Mass., USA, 1974

20 CHOW, Y.K., and SMITH, I.M.: 'Static and periodic infinite solid elements', *Int. J. Numer. Methods Eng.*, 1981, **17**, pp. 503–526

21 CURNIER, A.: 'A static infinite element', *ibid.*, 1983, **19**, pp. 1479–1488

22 DASGUPTA, G.: 'A finite element formulation for unbounded homogeneous continua', *Trans. ASME. J. Appl. Mech.*, 1982, **49**, pp. 132–140

23 DASGUPTA, G.: 'Computation of exterior potential fields by infinite substructuring', *Comput. Methods Appl. Mech. & Eng.*, 1984, **46**, pp. 295–305

24 EMSON, C., and BETTESS, P.: 'Application of infinite elements to external electromagnetic field problems'. Paper presented at *Int. Conf. Coupled Problems*, Swansea, UK, 1980

25 EMSON, C.R.I.: 'A new infinite element for static and dynamic electromagnetic field problems'. PhD thesis, Dept. Civil Engineering, University College Swansea, University of Wales, 1985, C/Ph/88/85 (Also Rutherford Appleton Laboratory report RAL-T-019)

26 GARTLING, D.K., and BECKER, E.B.: 'Finite element analysis of viscous, incompressible fluid flow, Part I: Basic methodology', *Comput. Methods Appl. Mech. & Eng.*, 1976, **8**, pp. 51–60

27 GUPTA, S., PENZIEN, J. LIN, T.W., and YEH, C.S.: 'Three dimensional hybrid modelling of soil-structure interaction', *Earthquake Eng. & Struct. Dyn.*, 1982, **10**, pp. 69–87

28 HURWITZ, H. (Jr.): 'Infinitesimal scaling, a new procedure for modelling exterior field problems'. Invited paper, INTERMAG'84 Conference, Hamburg, 1984

29 IRONS, B., and AHMAD, S.: 'Techniques of finite elements' (Ellis Horwood, 1980)

30 JUNG, H.K., LEE, G.S., and HAHN, S.Y.: '3-D magnetic field computations by infinite elements', *J. Appl. Phys.*, 1984, **55**, (6), Part IIB, pp. 2201–2203

31 KAGAWA, Y., YAMABUCHI, T., and KITAGAMI, S.: 'The infinite boundary element and its application to a combined finite boundary element technique for unbounded field problems', *COMPEL*, 1983, **2**, (4), pp. 179–193

32 KAGAWA, Y., YAMABUCHI, T., and ARAKI, Y.: 'The infinite boundary element and its application to the unbounded Helholtz problem', *ibid.*, 1985, **4**, (1), pp. 29–41

33 KELLOG, O.D.: 'Foundations of potential theory' (Dover Publications, 1953)

34 KIM, I.H., JUNG, H.K., LEE, G.S., and HAHN, S.Y.: 'Magnetic field computations by infinite elements', *J. Appl. Phys.*, 1982, **53**, (11), Part II, pp. 8372–8374

35 LOWTHER, D.A., and SILVESTER, P.P.: 'Finite element models of exterior regions containing moving conductors'. COMPUMAG Conference Paper 3.4, Grenoble, 4–6 September 1978

36 LYNN, P.P., and HADID, H.A.: 'Infinite elements with $1/r^n$ type decay', *Int. J. Numer. Methods Eng.*, 1981, **17**, pp. 347–355

37 LYSMER, J., and KUHLMEYER, R.L.: 'Finite dynamic model for infinite media', *J. Eng. Mech. Div., Proc. ASCE*, 1969, **EM-4**, pp. 859–877

38 LYSMER, J., and WAAS, G.: 'Shear waves in plane infinite structures', *ibid.*, 1972, **EM-1**, pp. 85–105

39 McDONALD, B.H., and WEXLER, A.: 'Finite element solution of unbounded field problems', *IEEE Trans.*, 1972, **MTT-20**, (12), pp. 841–847

40 MEDINA-MELO, F.J.: 'Modelling of soil structure interaction by finite and infinite elements'. PhD thesis, University California, Berkeley, 1981

41 MEDINA, F.: 'An axisymmetric infinite element', *Int. J. Numer. Methods Eng.*, 1981, **17**, pp. 1177–1185

42 MEDINA, F., and Taylor, R.L.: 'Problems of unbounded domains', *Int. J. Numer. Methods Eng.*, 1983, **19**, pp. 1209–1226

43 MEI, C.C.: 'Numerical methods in water wave diffraction and radiation', *Ann. Rev. Fluid Mech.*, 1978, **10**, pp. 393–416

44 MEI, K.K.: 'Unimoment method of solving antenna and scattering problems', *IEEE Trans.*, 1974, **AP-22**, pp. 760–766

45 MIYOSHI, T., and MAEDA, G.: 'Finite element analysis of leakage

magnetic flux from an induction heating system', *ibid.*, 1982, **MAG-18,** pp. 917–920

46 MURAKAMI, H., SHIOYA, S., and YAMADA, R.: 'Transmitting boundaries for time harmonic elastodynamics on infinite domains', *Int. J. Numer. Methods Eng.*, 1981, **17,** pp. 1697–1716

47 MUSTOE, G.G.W.: 'Symmetric variational boundary integral and finite element procedure in continuum mechanics'. PhD thesis, Dept. Civil Engineering, University College Swansea, 1981

48 NATH, B., and JAMSHADI, J.: 'The w-plane finite element method for the solution of scalar field problems in two dimensions', *Int. J. Numer. Methods Eng.*, 1980, **15,** pp. 361–379

49 NATH, B., and JAMSHADI, J.: 'The w-plane finite element method for the solution of the steady state scalar wave equation in 2-D', *Earthquake Eng. & Struct. Dyn.*, 1980, **8,** pp. 181–191

50 OKABE, M.: 'Generalized Lagrange family for the cube with special reference to infinite-domain interpolations', *Comput. Methods Appl. Mech. & Eng.*, 1983, **38,** pp. 153–168

51 PIRCHER, H., and BEER, G.: 'On the treatment of infinite boundaries in the finite element method', *Int. J. Numer. Methods Eng.*, 1977, **11,** pp. 1194–1197

52 PISSANETZKY, S.: 'An infinite element and a formula for numerical quadrature over an infinite interval', *ibid.*, 1983, **19,** pp. 913–927

53 PISSANETZKY, S.: 'A simple infinite element', *COMPEL*, 1984, **3,** (2), pp. 107–114

54 RICHARDSON, L.F.: 'The approximate arithmetical solution by finite differences of physical problems involving differential equations, with an application to the stresses in a masonry dam', *Trans. Roy. Soc. London Ser. A*, 1911, **210,** pp. 307–357

55 ROLICZ, P.: 'Use of Bubnov-Galerkin method to skin effect problems for the case of unknown boundary conditions', *IEEE Trans.*, 1982, **MAG-18,** pp. 527–530

56 SILVESTER, P., and HSIEH, M.S.: 'Finite element solution of 2-dimensional exterior field problems', *Proc. IEE*, 1971, **118,** pp. 1743–1747

57 SILVESTER, P.P., LOWTHER, D.A., CARPENTER, C.J., and WYATT, E.A.: 'Exterior finite elements for 2-dimensional field problems with open boundaries', *Proc. IEE*, 1977, **124,** p. 1267

58 SMYTHE, W.R.: 'Static and dynamic electricity' (McGraw-Hill, 1968, 3rd edn.)

59 STEELE, C.W.: 'Iterative algorithm for magnetostatic problems with saturable media', *IEEE Trans.*, 1982, **MAG-18,** pp. 393–396

60 STRATTON, J.A.: 'Electromagnetic theory' (McGraw-Hill, 1941)

61 THATCHER, R.W.: 'On the finite element method for unbounded regions', *SIAM J. Numer. Anal.*, 1978, **15,** pp. 466–477

62 UNGLESS, R.F.: 'An infinite finite element'. MASc thesis, University of British Columbia, 1973

63 WOOD, W.L.: 'On the finite element solution of an exterior boundary value problem', *Int. J. Numer. Methods Eng.*, 1976, **10,** pp. 885–891

64 YUE, D.K.P., CHEN, H.S., and MEI, C.C.: 'A hybrid method for diffraction of water waves by three dimensional bodies', *Int. J. Numer. Methods Eng.*, 1978, **12,** pp. 245–266

65 ZIENKIEWICZ, O.C., and NEWTON, R.E.: 'Coupled vibration of a structure immersed in a compressible fluid'. *Proc. Symp. Finite Element Techniques*, Stuttgart, 1969

66 ZIENKIEWICZ, O.C., and BETTESS, P.: 'Infinite elements in the study of fluid structure interaction problems'. *Proc. 2nd Int. Symp. Comput. Methods Appl. Sci.*, Versailles, 1975

67 ZIENKIEWICZ, O.C.: 'The finite element method' (McGraw-Hill, 1977, 3rd edn.)

68 ZIENKIEWICZ, O.C., KELLY, D.W., and BETTESS, P.: 'The Sommerfeld (radiation) condition on infinite domains and its modelling in numerical procedures'. *Proc. 3rd Int. Symp. Comput. Methods Appl. Sci.*, IRIA Laboria, 1977

69 ZIENKIEWICZ, O.C., KELLY, D.W., and BETTESS, P.: 'Marriage *a la mode* — the best of both worlds (finite elements and boundary integrals)'. *Int. Symp. Innovative Numer. Anal. Appl. Eng. Sci.*, Versailles, 1977

70 ZIENKIEWICZ, O.C., KELLY, D.W., and BETTESS, P.: 'The coupling of the finite element method and boundary solution procedure', *Int. J. Numer. Methods Eng.*, 1977, **11,** pp. 355–375

71 ZIENKIEWICZ, O.C., KELLY, D.W., and BETTESS, P.: 'The finite element method for determining fluid loadings on rigid structures', *in* LEWIS, R.W. (Ed.): 'Numerical methods in offshore engineering' (Wiley, 1978), Chap. 4

72 ZIENKIEWICZ, O.C., BETTESS, P., CHIAM, T.C., and EMSON, C.: 'Numerical methods for unbounded field problems and a new infinite element formulation'. Winter Annual Meeting, ASME, Washington, **AMD-46,** pp. 115–148 (& University College Swansea report C/R/384/81)

73 ZIENKIEWICZ, O.C., EMSON, C., and BETTESS, P.: 'A novel boundary infinite element', *Int. J. Numer. Methods Eng.*, 1983, **19,** pp. 393–404

74 ZIENKIEWICZ, O.C., BANDO, K., BETTESS, P., EMSON, C., and CHIAM, T.C.: 'Mapped infinite elements for exterior wave problems', *ibid.*, 1985, **21,** pp. 1229–1251

A Variational Analysis of Anisotropic, Inhomogeneous Dielectric Waveguides

WENG CHO CHEW, SENIOR MEMBER, IEEE, AND MUHAMMAD A. NASIR

Abstract —**We derive a variational formulation for anisotropic, dielectric waveguides using only the (E_x, E_y) or only the (H_x, H_y) components of the electromagnetic field. We show that the (E_x, E_y) formulation is completely equivalent to the (H_x, H_y) formulation. In fact, they are the transpose problems of each other. Given the variational formulation, one can derive the finite element solution quite easily. We also show how to derive a variational expression where the natural boundary conditions are incorporated as an optimal solution of the variational expression. We illustrate our theory with a simple implementation of a finite element solution. Our solutions agree with previous results, and there is no occurrence of spurious modes. Furthermore, our formulation allows the easy inclusion of loss and frequency dispersion in the $\bar{\bar{\mu}}$ and $\bar{\bar{\epsilon}}$.**

I. Introduction

THERE HAVE been numerous papers published on the analysis of dielectric waveguides due to their growing importance in integrated optics and millimeter waves. Because of the complexity of the structure, numerical methods [1]–[15], such as the finite element method, are popular for solving this class of problems. In the finite element method, there is always an associated variational expression. Many different variational formulations have been introduced to derive the guided modes of a dielectric waveguide. Some of these formulations produce spurious modes [1]–[6], [8], which are not the physical modes of the dielectric waveguide. The spurious modes are usually observed with the (E_z, H_z) formulation where only the z components of the electromagnetic field are kept as unknowns [1], [2], [6], [8], or they are observed in a formulation where the $\nabla \cdot \boldsymbol{H} = 0$ condition is not imposed [4], [5].

Recently, several papers have been published in which the (H_x, H_y) formulation is advocated [11]–[13]. The authors showed that with the use of such a formulation, the spurious modes disappear. Even more recently, some workers have suggested using both (E_x, E_y) and (H_x, H_y) formulations to remove the spurious modes [14].

In all of the above formulations, the finite element method was used to solve the pertinent equations. Hence, one important consideration is the minimization of required computer memory. Therefore, it is most expedient to formulate the problem with a minimum number of the components of the electromagnetic fields. Some formulations use three components of the $\boldsymbol{H}$ field, and the unknowns for all three components of the $\boldsymbol{H}$ field have to be stored [4], [5], [12]. In other formulations, the unknowns for all four components of the fields, (H_x, H_y) and (E_x, E_y), have to be stored [14].

In this paper, we will use either the (E_x, E_y) only or the (H_x, H_y) only for the formulation of anisotropic, inhomogeneous waveguides. Contrary to popular belief, we show that the (E_x, E_y) formulation is as good as the (H_x, H_y) formulation. By properly defining the differential equations and the transpose problems [16], we show that the (E_x, E_y) and (H_x, H_y) formulations are completely equivalent to each other. Su [13] suggested that the (E_x, E_y) formulation was inherently less accurate than the (H_x, H_y) formulation. However, it can be shown that for the isotropic case, if the same basis set is used for the expansion functions and the testing functions, our (H_x, H_y) formulation is similar to Su's; hence, there is no reason to prefer the (H_x, H_y) formulation over the (E_x, E_y) formulation.

In all the formulations in which the spurious modes do not occur, either the $\nabla \cdot \boldsymbol{D} = 0$ or the $\nabla \cdot \boldsymbol{B} = 0$ condition is imposed. We believe that these conditions remove the possible accumulation of fictitious charges in the finite element approximation. These fictitious charges, which can store energy, can give rise to extraneous guided modes.

We will first derive the pertinent equations involving either the (E_x, E_y) or the (H_x, H_y) for an anisotropic waveguide with reflection symmetry [17]. Assuming that the field has $e^{ik_z z}$ dependence, our k_z^2 appears explicitly as the desired eigenvalue in the eigenequation. As opposed to formulations in which k^2 or the frequency appears explicitly as the desired eigenvalue [1], [4], [6], [8], [12], our formulation makes it possible to include loss and frequency dispersion.

After having derived the differential equations, we define their transpose problems [16] appropriately. Then, we derive the variational expression corresponding to the differential equations. We also discuss how the natural boundary conditions can be imposed in such a formulation. Finally, we show the numerical implementation of such a theory.

Manuscript received July 26, 1988; revised October 24, 1988. This work was supported in part by the National Science Foundation under Grant NSF ECS-85-25891 and by a grant from TRW.

The authors are with the Department of Electrical and Computer Engineering, University of Illinois, Urbana, IL 61801.

IEEE Log Number 8826054.

Reprinted from *IEEE Trans. Microwave Theory Tech.*, vol. 37, no. 4, pp. 661–668, April 1989.

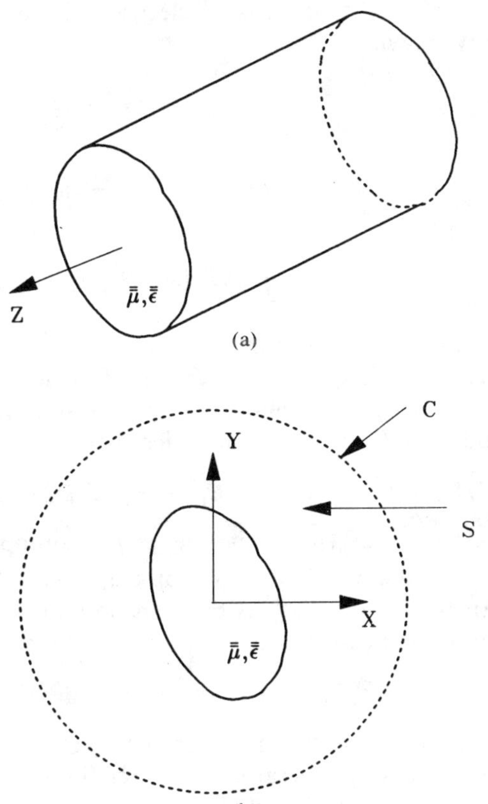

Fig. 1. Anisotropic, dielectric waveguide to be analyzed.

II. Derivation of the Differential Equations

In applying the finite element method, we have to first derive a variational expression. Before we do that, we need to derive the differential equation governing the field. Once the differential equation is derived, we can derive the variational expression quite easily once we identify the transpose problem (e.g., see [16] and [18]). In order to reduce the number of unknowns required, we reduce the differential equation to an equation involving (E_x, E_y) alone for the $\boldsymbol{E}$ field. The differential equation governing (H_x, H_y) can be derived via the duality principle.

The electric field in an inhomogeneous, anisotropic waveguide satisfies the vector wave equation

$$\nabla \times \bar{\bar{\mu}}^{-1}\cdot\nabla \times \boldsymbol{E}(\boldsymbol{r}) - \omega^2\bar{\bar{\epsilon}}\cdot\boldsymbol{E}(\boldsymbol{r}) = 0. \tag{1}$$

We will assume that

$$\bar{\bar{\mu}} = \begin{pmatrix} \bar{\bar{\mu}}_s & 0 \\ 0 & \mu_{zz} \end{pmatrix} \qquad \bar{\bar{\epsilon}} = \begin{pmatrix} \bar{\bar{\epsilon}}_s & 0 \\ 0 & \epsilon_{zz} \end{pmatrix} \tag{2}$$

are functions of position and that $\bar{\bar{\mu}}_s$ and $\bar{\bar{\epsilon}}_s$ are 2×2 tensors. In this manner, the geometry in Fig. 1 has reflection symmetry about the z axis; i.e., a mode propagating in the $+z$ direction is degenerate with a mode propagating in the $-z$ direction [17]. Equation (1) has three components of the electric field. To reduce the number of electric field components, we next proceed to separate the transverse and longitudinal components of (1). We assume that the field has $e^{ik_z z}$ dependence so that $\partial/\partial_z = ik_z$. Defining

$$\nabla = \nabla_s + \hat{z}ik_z \qquad \boldsymbol{E} = \boldsymbol{E}_s + \hat{z}E_z \tag{3}$$

where the subscript s denotes a vector transverse to z, the transverse components of (1) become

$$\nabla_s \times \nu_{zz}\nabla_s \times \boldsymbol{E}_s + ik_z\hat{z}\times\bar{\bar{\nu}}_s\cdot\nabla_s\times\hat{z}E_z - \omega^2\bar{\bar{\epsilon}}_s\cdot\boldsymbol{E}_s - k_z^2\hat{z}\times\bar{\bar{\nu}}_s\cdot\bar{\bar{z}}_s\cdot\hat{z}\times\boldsymbol{E}_s = 0 \tag{4}$$

where we have defined $\bar{\bar{\nu}}_s = \bar{\bar{\mu}}^{-1}$. We can remove E_z in (4) using the $\nabla\cdot\boldsymbol{D}=0$ condition, from which we deduce that

$$E_z = -\frac{\nabla_s\cdot\bar{\bar{\epsilon}}_s\cdot\boldsymbol{E}_s}{ik_z\epsilon_{zz}} = \frac{i\kappa_{zz}}{k_z}\nabla_s\cdot\bar{\bar{\epsilon}}_s\cdot\boldsymbol{E}_s. \tag{5}$$

We have defined $\bar{\bar{\kappa}} = \bar{\bar{\epsilon}}^{-1}$ in the above. Consequently, (4) can be reduced to an equation involving only $\boldsymbol{E}_s$, viz.,

$$\nabla_s \times \nu_{zz}\nabla_s \times \boldsymbol{E}_s + \hat{z}\times\bar{\bar{\nu}}_s\cdot\hat{z}\times\nabla_s\left(\kappa_{zz}\nabla_s\cdot\bar{\bar{\epsilon}}_s\cdot\boldsymbol{E}_s\right) - \omega^2\bar{\bar{\epsilon}}_s\cdot\boldsymbol{E}_s - k_z^2\hat{z}\times\bar{\bar{\nu}}_s\cdot\hat{z}\times\boldsymbol{E}_s = 0. \tag{6}$$

The above is an eigenequation for k_z^2. The dependence on k_z^2 implies that $e^{\pm ik_z z}$ modes are degenerate. This would not have been possible without assuming the form of $\bar{\bar{\mu}}$ and $\bar{\bar{\epsilon}}$ in (2). In (6), k_z^2 multiplies a rather complicated expression. We can premultiply (6) by $\bar{\bar{\mu}}_s\cdot\hat{z}\times$ to arrive at

$$\bar{\bar{\mu}}_s\cdot\hat{z}\times\nabla_s\times\nu_{zz}\nabla_s\times\boldsymbol{E}_s - \hat{z}\times\nabla_s\kappa_{zz}\nabla_s\cdot\bar{\bar{\epsilon}}_s\cdot\boldsymbol{E}_s - \omega^2\bar{\bar{\mu}}_s\cdot\hat{z}\times\bar{\bar{\epsilon}}_s\cdot\boldsymbol{E}_s + k_z^2\hat{z}\times\boldsymbol{E}_s = 0. \tag{7}$$

Invoking the duality principle [18], we arrive at the equation for the transverse magnetic field:

$$\bar{\bar{\epsilon}}_s\cdot\hat{z}\times\nabla_s\times\kappa_{zz}\nabla_s\times\boldsymbol{H}_s - \hat{z}\times\nabla_s\nu_{zz}\nabla_s\cdot\bar{\bar{\mu}}_s\cdot\boldsymbol{H}_s - \omega^2\bar{\bar{\epsilon}}_s\cdot\hat{z}\times\bar{\bar{\mu}}_s\cdot\boldsymbol{H}_s + k_z^2\hat{z}\times\boldsymbol{H}_s = 0. \tag{8}$$

Equations (7) and (8) are very dissimilar in appearance. However, we shall show next that they are related via the definition of transpose operators [16].

III. The Transpose Problems

Equation (7) is a differential equation of the form

$$\mathscr{L}_e\cdot\boldsymbol{E}_s + k_z^2\mathscr{B}\cdot\boldsymbol{E}_s = 0 \tag{9}$$

where $\mathscr{L}_e$ and $\mathscr{B}$ are linear operators. Defining the inner product $\langle \boldsymbol{A}_s, \boldsymbol{B}_s\rangle = \int_S dS\boldsymbol{A}_s\cdot\boldsymbol{B}_s$, where S is the whole cross-sectional area of the waveguide (see Fig. 1(b)), the transpose operator[1] of $\mathscr{L}$ is an operator $\mathscr{L}^t$ such that $\langle \boldsymbol{A}_s, \mathscr{L}\cdot\boldsymbol{E}_s\rangle = \langle \boldsymbol{E}_s, \mathscr{L}^t\cdot\boldsymbol{A}_s\rangle$. An operator is symmetric if $\mathscr{L}^t = \mathscr{L}$. The operator $\mathscr{L}_e$ in this case is not symmetric because

$$\langle \boldsymbol{A}_s, \mathscr{L}_e\cdot\boldsymbol{E}_s\rangle \neq \langle \boldsymbol{E}_s, \mathscr{L}_e\cdot\boldsymbol{A}_s\rangle. \tag{10}$$

[1]We use the term *transpose* rather than *adjoint* because, to be strictly correct, the term *adjoint* is used only when we define the inner product as $\langle \boldsymbol{A}_s^*, \boldsymbol{B}_s\rangle$ [16], [20].

We can prove this readily because

$$\langle \boldsymbol{A}_s, \mathscr{L}_e \cdot \boldsymbol{E}_s \rangle = \int_S dS \boldsymbol{A}_s \cdot \bar{\bar{\mu}}_s \cdot \hat{z} \times \nabla_s \times \nu_{zz} \nabla_s \times \boldsymbol{E}_s - \int_S dS \boldsymbol{A}_s \cdot \hat{z} \times \nabla_s \kappa_{zz} \nabla_s \cdot \bar{\bar{\epsilon}}_s \cdot \boldsymbol{E}_s - \omega^2 \int_S dS \boldsymbol{A}_s \cdot \bar{\bar{\mu}}_s \cdot \hat{z} \times \bar{\bar{\epsilon}}_s \cdot \boldsymbol{E}_s. \quad (11)$$

Using integration by parts, or the appropriate vector identities, the above becomes

$$\langle \boldsymbol{A}_s, \mathscr{L}_e \cdot \boldsymbol{E}_s \rangle = -\int_S dS \hat{z} \cdot \nabla_s \times \boldsymbol{E}_s \nu_{zz} \nabla_s \cdot \bar{\bar{\mu}}_s^t \cdot \boldsymbol{A}_s + \int_S dS \hat{z} \cdot \nabla_s \times \boldsymbol{A} \kappa_{zz} \nabla_s \cdot \bar{\bar{\epsilon}}_s \cdot \boldsymbol{E}_s - \omega^2 \int_S dS \hat{z} \cdot (\bar{\bar{\epsilon}}_s \cdot \boldsymbol{E}_s) \times (\bar{\bar{\mu}}_s^t \cdot \boldsymbol{A}_s) + \int_C dl \hat{n} \cdot \bar{\bar{\mu}}_s^t \cdot \boldsymbol{A}_s \nu_{zz} \hat{z} \cdot \nabla_s \times \boldsymbol{E}_s - \int_C dl \hat{z} \cdot \hat{n} \times \boldsymbol{A}_s \kappa_{zz} \nabla_s \cdot \bar{\bar{\epsilon}}_s \cdot \boldsymbol{E}_s. \quad (12)$$

If we think of $\boldsymbol{A}_s$ as a magnetic field, the last terms involving line integrals vanish if we have a PEC (perfect electric conductor) or PMC (perfect magnetic conductor) at C or if we assume C is at infinity. Then,

$$\langle \boldsymbol{A}_s, \mathscr{L}_e \cdot \boldsymbol{E}_s \rangle = -\int_S dS \hat{z} \cdot \nabla_s \times \boldsymbol{E}_s \nu_{zz} \nabla_s \cdot \bar{\bar{\mu}}_s^t \cdot \boldsymbol{A}_s + \int_S dS \hat{z} \cdot \nabla_s \times \boldsymbol{A} \kappa_{zz} \nabla_s \cdot \bar{\bar{\epsilon}}_s \cdot \boldsymbol{E}_s - \omega^2 \int_S dS \hat{z} \cdot (\bar{\bar{\epsilon}}_s \cdot \boldsymbol{E}_s) \times (\bar{\bar{\mu}}_s^t \cdot \boldsymbol{A}_s). \quad (12a)$$

Clearly, (10) holds true from the above expression.

Similarly, (8) is of the form

$$\mathscr{L}_h \cdot \boldsymbol{H}_s + k_z^2 \mathscr{B} \cdot \boldsymbol{H}_s = 0. \quad (13)$$

By the same token,

$$\langle \boldsymbol{C}_s, \mathscr{L}_h \cdot \boldsymbol{H}_s \rangle \neq \langle \boldsymbol{H}_s, \mathscr{L}_h \cdot \boldsymbol{C}_s \rangle. \quad (14)$$

In other words, $\mathscr{L}_h$ is not a symmetric operator. More specifically,

$$\langle \boldsymbol{C}_s, \mathscr{L}_h \cdot \boldsymbol{H}_s \rangle = -\int_{S_0} dS \hat{z} \cdot \nabla_s \times \boldsymbol{H}_s \kappa_{zz} \nabla_s \cdot \bar{\bar{\epsilon}}_z^t \cdot \boldsymbol{C}_s + \int_S dS \hat{z} \cdot \nabla_s \times \boldsymbol{C}_s \nu_{zz} \nabla_s \cdot \bar{\bar{\mu}}_s \cdot \boldsymbol{H}_s - \omega^2 \int_S dS \hat{z} \cdot (\bar{\bar{\mu}}_s \cdot \boldsymbol{H}_s) \times (\bar{\bar{\epsilon}}_s^t \cdot \boldsymbol{C}_s) + \int_C dl \hat{n} \cdot \bar{\bar{\epsilon}}_s^t \cdot \boldsymbol{C}_s \kappa_{zz} \hat{z} \cdot \nabla_s \times \boldsymbol{H}_s - \int_C dl \hat{z} \cdot \hat{n} \times \boldsymbol{C}_s \nu_{zz} \nabla_s \cdot \bar{\bar{\mu}}_s \cdot \boldsymbol{H}_s. \quad (15)$$

Again the line integrals vanish if we assume that C is a PEC or PMC and that $\boldsymbol{C}_s$ is an electric field, or that C is at infinity. Then,

$$\langle \boldsymbol{C}_s, \mathscr{L}_h \cdot \boldsymbol{H}_s \rangle = -\int_S dS \hat{z} \cdot \nabla_s \times \boldsymbol{H}_s \kappa_{zz} \nabla_s \cdot \bar{\bar{\epsilon}}_z^t \cdot \boldsymbol{C}_s + \int_S dS \hat{z} \cdot \nabla_s \times \boldsymbol{C}_s \nu_{zz} \nabla_s \cdot \bar{\bar{\mu}}_s \cdot \boldsymbol{H}_s - \omega^2 \int_S dS \hat{z} \cdot (\bar{\bar{\mu}}_s \cdot \boldsymbol{H}_s) \times (\bar{\bar{\epsilon}}_s^t \cdot \boldsymbol{C}_s). \quad (15a)$$

Clearly, (14) holds true from the above expression.

In (12a) and (15a), $\mathscr{L}_e$ and $\mathscr{L}_h$ are functions of $\bar{\bar{\mu}}$ and $\bar{\bar{\epsilon}}$. By comparing (12a) and (15a) and letting $\boldsymbol{A} = \boldsymbol{H}_s$ in (12a) and $\boldsymbol{C}_s = \boldsymbol{E}_s$ in (15a), we notice that

$$\langle \boldsymbol{H}_s, \mathscr{L}_e(\bar{\bar{\mu}}, \bar{\bar{\epsilon}}) \cdot \boldsymbol{E}_s \rangle = \langle \boldsymbol{E}_s, -\mathscr{L}_h(\bar{\bar{\mu}}^t, \bar{\bar{\epsilon}}^t) \cdot \boldsymbol{H}_s \rangle. \quad (16)$$

Therefore, $-\mathscr{L}_h(\bar{\bar{\mu}}^t, \bar{\bar{\epsilon}}^t)$ is the transpose operator of $\mathscr{L}_e(\bar{\bar{\mu}}, \bar{\bar{\epsilon}})$. If the medium is reciprocal, then $\bar{\bar{\mu}}^t = \bar{\bar{\mu}}$ and $\bar{\bar{\epsilon}}^t = \bar{\bar{\epsilon}}$. In this case, $-\mathscr{L}_h$ is the transpose operator of $\mathscr{L}_e$ from the same medium. However, in general, we have

$$\mathscr{L}_e^t(\bar{\bar{\mu}}, \bar{\bar{\epsilon}}) = -\mathscr{L}_h(\bar{\bar{\mu}}^t, \bar{\bar{\epsilon}}^t). \quad (17)$$

In other words, the transpose of the operator $-\mathscr{L}_e$ in medium $(\bar{\bar{\mu}}, \bar{\bar{\epsilon}})$ is the operator $\mathscr{L}_h$ in medium $(\bar{\bar{\mu}}^t, \bar{\bar{\epsilon}}^t)$.

It is easy to show that the operator $\mathscr{B}$ in (9) and (13) satisfies

$$\langle \boldsymbol{A}, \mathscr{B} \cdot \boldsymbol{E}_s \rangle = \langle \boldsymbol{E}_s, -\mathscr{B} \cdot \boldsymbol{A} \rangle. \quad (18)$$

Therefore,

$$\mathscr{B}^t = -\mathscr{B}. \quad (19)$$

The above derivation establishes the transpose problems of (9) and (13), respectively. Given (9), its transpose problem exists, and can be associated with a physical problem. Hence, given

$$\mathscr{L}_e(\bar{\bar{\mu}}, \bar{\bar{\epsilon}}) \cdot \boldsymbol{E}_s + k_z^2 \mathscr{B} \cdot \boldsymbol{E}_s = 0 \quad (20)$$

the transpose problem is

$$\mathscr{L}_e^t(\bar{\bar{\mu}}, \bar{\bar{\epsilon}}) \cdot \boldsymbol{E}_s^t + k_z^2 \mathscr{B}^a \cdot \boldsymbol{E}_s^t = 0. \quad (21)$$

Using (17) and (19), the transpose problem is the same as

$$\mathscr{L}_h(\bar{\bar{\mu}}^t, \bar{\bar{\epsilon}}^t) \cdot \tilde{\boldsymbol{H}}_s + k_z^2 \mathscr{B} \cdot \tilde{\boldsymbol{H}}_s = 0. \quad (22)$$

Hence, the transpose solution, $\boldsymbol{E}_s^t$, is $\tilde{\boldsymbol{H}}_s$, which is the transverse magnetic field solution of a medium with the electromagnetic property of $(\bar{\bar{\mu}}^t, \bar{\bar{\epsilon}}^t)$. From this point onward, we use the tilde (~) to denote that $\tilde{\boldsymbol{H}}_s$ is the magnetic field solution with the medium $\bar{\bar{\mu}}^t, \bar{\bar{\epsilon}}^t$. If the medium is reciprocal, $\bar{\bar{\mu}}^t = \bar{\bar{\mu}}$ and $\bar{\bar{\epsilon}}^t = \bar{\bar{\epsilon}}$, the transpose solution is just the transverse magnetic field solution of the same medium, that is, $\tilde{\boldsymbol{H}}_s = \boldsymbol{H}_s$. Similarly, we can establish the transpose problem and the solution associated with (13). A similar result could be obtained if we use $\langle \boldsymbol{A}^*, \boldsymbol{B} \rangle$ as our inner products [16], [18], [20]. In this case, the term *transpose operators* will be replaced by *adjoint operators*. Equations (9) and (13) can be proved to be non-self-adjoint.

IV. Variational Principle

Given a nonsymmetric eigenequation (20) and its transpose equation (21), we can derive a variational expression for the eigenvalues k_z^2. The variational expression is

$$k_z^2 = -\frac{\langle \boldsymbol{E}_s^t, \mathscr{L}_e \cdot \boldsymbol{E}_s \rangle}{\langle \boldsymbol{E}_s^t, \mathscr{B} \cdot \boldsymbol{E}_s \rangle} = -\frac{\langle \tilde{\boldsymbol{H}}_s, \mathscr{L}_e \cdot \boldsymbol{E}_s \rangle}{\langle \tilde{\boldsymbol{H}}_s, \mathscr{B} \cdot \boldsymbol{E}_s \rangle}. \tag{23}$$

We can establish the variational property of the above expression about the exact solutions $\tilde{\boldsymbol{H}}_{se}$ and $\boldsymbol{E}_{se}$ by letting

$$\tilde{\boldsymbol{H}}_s = \tilde{\boldsymbol{H}}_{se} + \delta \boldsymbol{H} \qquad \boldsymbol{E}_s = \boldsymbol{E}_{se} + \delta \boldsymbol{E} \tag{24}$$

and showing that the first variation of k_z^2 is zero.[2]

Since (23) is a variational expression, we can apply the Rayleigh–Ritz procedure to find the optimal solution. To do this, we let

$$\boldsymbol{E}_s = \sum_{n=1}^{N} a_n \boldsymbol{E}_{ns} \qquad \tilde{\boldsymbol{H}}_s = \sum_{m=1}^{N} b_m \boldsymbol{H}_{ms}. \tag{25}$$

We next substitute (25) into (23) and require that the first variations of (23) with respect to the a_n's and b_m's vanish. We find that the optimal solution is given by the solution of the following matrix equations:

$$\left[\bar{\bar{L}}_e + k_z^2 \bar{\bar{B}}\right] \cdot \bar{a} = 0 \tag{26a}$$

$$\left[\bar{\bar{L}}_e^t + k_z^2 \bar{\bar{B}}^t\right] \cdot \bar{b} = 0 \tag{26b}$$

where

$$\left[\bar{\bar{L}}_e\right]_{mn} = \langle \boldsymbol{H}_{ms}, \mathscr{L}_e \cdot \boldsymbol{E}_{ns} \rangle \tag{27a}$$

$$\left[\bar{\bar{B}}\right]_{mn} = \langle \boldsymbol{H}_{ms}, \mathscr{B} \cdot \boldsymbol{E}_{ns} \rangle. \tag{27b}$$

In the above, $\bar{\bar{L}}_e$ and $\bar{\bar{B}}$ are also known as the matrix representations [21][3] of the operators $\mathscr{L}_e$ and $\mathscr{B}$. From the definition of transpose operators,

$$\langle \boldsymbol{H}_{ms}, \mathscr{L}_e \cdot \boldsymbol{E}_{ns} \rangle = \langle \boldsymbol{E}_{ns}, \mathscr{L}_e^a \cdot \boldsymbol{H}_{ms} \rangle = \langle \boldsymbol{E}_{ns}, -\mathscr{L}_h(\bar{\bar{\mu}}^t, \bar{\bar{\epsilon}}^t) \cdot \boldsymbol{H}_{ms} \rangle \tag{28}$$

or

$$\left[\bar{\bar{L}}_e\right]_{mn} = -\left[\bar{\bar{L}}_h\right]_{nm}$$

we see that

$$\bar{\bar{L}}_e^t = -\bar{\bar{L}}_h \tag{29}$$

where $\bar{\bar{L}}_h$ is the matrix representation of the operator $\mathscr{L}_h$. Since $\bar{\bar{B}}^t = -\bar{\bar{B}}$, (26b) can be rewritten as[4]

$$\left[\bar{\bar{L}}_h + k_z^2 \bar{\bar{B}}\right] \cdot \bar{b} = 0. \tag{30}$$

Equations (26a) and (30) can also be directly obtained from (9) and (13) by the application of the method of weighted residuals or the Petrov–Galerkin method [22] (also known as the method of moments [23]). In solving (9) using this method, for example, we will use $\boldsymbol{E}_{ns}$'s as expansion functions and $\boldsymbol{H}_{ms}$'s as weighting functions. Intuitively, for fast convergence, we would want the $\boldsymbol{E}_{ns}$'s to approximate $\boldsymbol{E}_s$ well, while we would want the $\boldsymbol{H}_{ms}$'s to approximate the transpose solution, $\tilde{\boldsymbol{H}}_s$, well. When we choose $\boldsymbol{H}_{ms}$ and $\boldsymbol{E}_{ns}$ to be the same basis set, the method is commonly known as Galerkin's method. When the basis functions are finite domain basis functions, the method is also known as the finite element method.

When the same basis set is used for $\boldsymbol{H}_{ms}$ and $\boldsymbol{E}_{ns}$, viz., using the Galerkin's method, our formulation here is equivalent to the $\boldsymbol{E}_s$ method where the testing field is also $\boldsymbol{E}_s$ or to the $\boldsymbol{H}_s$ method where the testing field is also $\boldsymbol{H}_s$.

Since the eigenequation (30) is the transpose of the eigenequation (26a), we need only solve one of them for the eigenvalues, since they share the same eigenvalues.[5] Furthermore, we can prove that for $\bar{a}_i$ and $\bar{b}_j$ corresponding to the ith and jth eigenvalues, respectively, $\bar{b}_j \cdot \bar{\bar{B}} \cdot \bar{a}_i = D_j \delta_{ij}$. Hence, once all the $\bar{a}_i$'s are found, we can find the $\bar{b}_j$'s easily. In addition, since $\bar{\bar{L}}_e$ is not a symmetric matrix and is non-Hermitian in general (a consequence of nonsymmetric and non-self-adjoint operator), k_z^2 can be complex. This corresponds to the complex modes of the structure which are not due to losses. The existence of these complex modes is a direct consequence of the non-self-adjoint natures of (9) and (13). A similar variational principle can be derived using $\langle \boldsymbol{A}^*, \boldsymbol{B} \rangle$ type inner products.

V. Natural Boundary Conditions

In the previous derivation, we have assumed C to be a PEC or a PMC. Under such an assumption, the line integral terms in (12) and (15) vanish. In choosing the basis set for $\boldsymbol{E}_s$ and $\tilde{\boldsymbol{H}}_s$, we have to ensure that the PEC or PMC boundary conditions are satisfied or that the fields vanish on C. Boundary conditions that are imposed by choosing the basis set are known as the essential boundary conditions (also known as forced or principal boundary conditions). However, it is sometimes desirable that the boundary conditions be a natural result of the optimal solution of a variational expression. These boundary conditions are known as the natural boundary conditions [18], [22]. We show next how such a variational expression can be derived, in which the boundary conditions are the natural result of the variational expression.

We can rewrite (12) with $\boldsymbol{A}_s = \tilde{\boldsymbol{H}}_s$ as

$$\langle \tilde{\boldsymbol{H}}_s, \mathscr{L}_e \cdot \boldsymbol{E}_s \rangle = \langle \tilde{\boldsymbol{H}}_s, \mathscr{L}_e^o \cdot \boldsymbol{E}_s \rangle + \int_C dl \hat{n} \cdot \bar{\bar{\mu}}_s^t \cdot \tilde{\boldsymbol{H}}_s \nu_{zz} \hat{z} \cdot \nabla_s \times \boldsymbol{E}_s - \int_C dl \hat{z} \cdot \hat{n} \times \tilde{\boldsymbol{H}}_s \kappa_{zz} \nabla_s \cdot \bar{\bar{\epsilon}}_s \cdot \boldsymbol{E}_s \tag{31}$$

where $\langle \tilde{\boldsymbol{H}}_s, \mathscr{L}_e^o \cdot \boldsymbol{E}_s \rangle$ is used to denote terms in (12) involving only integrals over S. Similarly, we can rewrite (15)

[2] To show that the first variation in k_z^2 vanishes, it is easiest to cross-multiply out (23) first.

[3] In quantum mechanics, a matrix representation is defined with the inner product $\langle \boldsymbol{A}^*, \boldsymbol{B} \rangle$.

[4] Note that the definitions of transpose operators and symmetric operators are very similar to the definitions of transpose matrices and symmetric matrices.

[5] This follows from $\det(\bar{\bar{L}}_e + k_z^2 \bar{\bar{B}}) = \det(\bar{\bar{L}}_e^t + k_z^2 \bar{\bar{B}}^t) = 0$.

with $\boldsymbol{C}_s = \boldsymbol{E}_s$ as

$$\langle \boldsymbol{E}_s, \mathscr{L}_h \cdot \tilde{\boldsymbol{H}}_s \rangle = \langle \boldsymbol{E}_s, \mathscr{L}_h^o \cdot \tilde{\boldsymbol{H}}_s \rangle + \int_C dl \hat{n} \cdot \bar{\bar{\epsilon}}_s \cdot \boldsymbol{E}_s \kappa_{zz} \hat{z} \cdot \nabla_s \times \tilde{\boldsymbol{H}}_s - \int_C dl \hat{z} \cdot \hat{n} \times \boldsymbol{E}_s \nu_{zz} \nabla_s \cdot \bar{\bar{\mu}}_s^t \cdot \tilde{\boldsymbol{H}}_s \quad (32)$$

where $\langle \boldsymbol{E}_s, \mathscr{L}_h^o \cdot \tilde{\boldsymbol{H}}_s \rangle$ is used to denote terms in (15) involving only integrals over S. If we have an impedance boundary condition on C, the line integral terms in (31) and (32) need not necessarily vanish. To impose the impedance boundary conditions, we can assume that

$$H_z = -Y_s \hat{z} \cdot \hat{n} \times \boldsymbol{E}_s \quad (33a)$$

$$\tilde{E}_z = \tilde{Z}_s \hat{z} \cdot \hat{n} \times \tilde{\boldsymbol{H}}_s \quad (33b)$$

for the original and transpose problems, respectively. After some manipulations, it is clear that the expression

$$k_z^2 = -\frac{\langle \tilde{\boldsymbol{H}}_s, \mathscr{L}_e^o \cdot \boldsymbol{E}_s \rangle - \langle \tilde{\boldsymbol{H}}_s, \partial \mathscr{L}_e \cdot \boldsymbol{E}_s \rangle}{\langle \tilde{\boldsymbol{H}}_s, \mathscr{B} \cdot \boldsymbol{E}_s \rangle} \quad (34)$$

is the required variational expression [18]. In the above, we define

$$\langle \tilde{\boldsymbol{H}}_s, \partial \mathscr{L}_e \cdot \boldsymbol{E}_s \rangle = i\omega \int_C dl \hat{n} \cdot \bar{\bar{\mu}}_s^t \cdot \tilde{\boldsymbol{H}}_s Y_s \hat{z} \cdot \hat{n} \times \boldsymbol{E}_s - i\omega \int_C dl \hat{n} \cdot \bar{\bar{\epsilon}}_s \cdot \boldsymbol{E}_s \tilde{Z}_s \hat{z} \cdot \hat{n} \times \tilde{\boldsymbol{H}}_s. \quad (35)$$

The functional is chosen judiciously so that symmetric boundary conditions will ensue for the actual and transpose problems, as we shall see later.

Taking the first variation of (34) with respect to the exact solution, we have

$$\langle \delta \boldsymbol{H}_s, \mathscr{L}_e^o \cdot \boldsymbol{E}_{se} \rangle + \langle \tilde{\boldsymbol{H}}_{se}, \mathscr{L}_e \cdot \delta \boldsymbol{E}_s \rangle - \langle \delta \boldsymbol{H}_s, \partial \mathscr{L}_e \cdot \boldsymbol{E}_{se} \rangle - \langle \tilde{\boldsymbol{H}}_{se}, \partial \mathscr{L}_e \cdot \delta \boldsymbol{E}_s \rangle + k_{ze}^2 \langle \delta \boldsymbol{H}_s, \mathscr{B} \cdot \boldsymbol{E}_{se} \rangle + k_{ze}^2 \langle \tilde{\boldsymbol{H}}_{se}, \mathscr{B} \cdot \delta \boldsymbol{E}_s \rangle + \delta k_z^2 \langle \tilde{\boldsymbol{H}}_{se}, \mathscr{B} \cdot \boldsymbol{E}_{se} \rangle = 0. \quad (36)$$

For arbitrary variations of $\delta \boldsymbol{E}_s$ and $\delta \boldsymbol{H}_s$, we find that the condition for δk_z^2 to vanish is that

$$\langle \delta \boldsymbol{H}_s, \mathscr{L}_e^o \cdot \boldsymbol{E}_{se} \rangle - \langle \delta \boldsymbol{H}_s, \partial \mathscr{L}_e \cdot \boldsymbol{E}_{se} \rangle + k_{ze}^2 \langle \delta \boldsymbol{H}_s, \mathscr{B} \cdot \boldsymbol{E}_{se} \rangle = 0 \quad (37a)$$

$$\langle \tilde{\boldsymbol{H}}_{se}, \mathscr{L}_e^o \cdot \delta \boldsymbol{E}_s \rangle - \langle \tilde{\boldsymbol{H}}_{se}, \partial \mathscr{L}_e \cdot \delta \boldsymbol{E}_s \rangle + k_{ze}^2 \langle \tilde{\boldsymbol{H}}_{se}, \mathscr{B} \cdot \delta \boldsymbol{E}_s \rangle = 0. \quad (37b)$$

Using (31) for the definition of $\langle \delta \boldsymbol{H}_s, \mathscr{L}_e^o \cdot \boldsymbol{E}_{se} \rangle$, and (35) for the definition of $\langle \delta \boldsymbol{H}_s, \partial \mathscr{L}_e \cdot \boldsymbol{E}_{se} \rangle$, we see that (37a) is satisfied only if

$$\langle \delta \boldsymbol{H}_s, \mathscr{L}_e \cdot \boldsymbol{E}_{se} \rangle + k_{ze}^2 \langle \delta \boldsymbol{H}_s, \mathscr{B} \cdot \boldsymbol{E}_{se} \rangle = 0 \quad (38)$$

and

$$\int_C dl \hat{n} \cdot \bar{\bar{\mu}}_s^t \cdot \delta \boldsymbol{H}_s \nu_{zz} \hat{z} \cdot \nabla_s \times \boldsymbol{E}_s - \int_C dl \hat{z} \cdot \hat{n} \times \delta \boldsymbol{H}_s \kappa_{zz} \nabla_s \cdot \bar{\bar{\epsilon}}_s \cdot \boldsymbol{E}_s = i\omega \int_C dl \hat{n} \cdot \bar{\bar{\mu}}_s^t \cdot \delta \boldsymbol{H}_s Y_s \hat{z} \cdot \hat{n} \times \boldsymbol{E}_s - i\omega \int_C dl \hat{n} \cdot \bar{\bar{\epsilon}}_s \cdot \boldsymbol{E}_s \tilde{Z}_s \hat{z} \cdot \hat{n} \times \delta \boldsymbol{H}_s. \quad (39)$$

For arbitrary $\delta \boldsymbol{H}_s$, (38) and (39) imply that

$$\mathscr{L}_e \cdot \boldsymbol{E}_{se} + k_{ze}^2 \mathscr{B} \cdot \boldsymbol{E}_{se} = 0 \quad (40a)$$

and

$$\nu_{zz} \hat{z} \cdot \nabla_s \times \boldsymbol{E}_s = -i\omega Y_s \hat{z} \cdot \hat{n} \times \boldsymbol{E}_s$$

$$\kappa_{zz} \nabla_s \cdot \bar{\bar{\epsilon}}_s \cdot \boldsymbol{E}_s = -i\omega \tilde{Z}_s \hat{n} \cdot \bar{\bar{\epsilon}}_s \cdot \boldsymbol{E}_s \quad (40b)$$

on C. Equation (40b) is equivalent to

$$H_z = -Y_s \hat{z} \cdot \hat{n} \times \boldsymbol{E}_s \qquad E_z = \frac{\omega}{k_z} \tilde{Z}_s \hat{n} \cdot \boldsymbol{D}_s. \quad (41)$$

Similarly, from (37b), by noting that $\langle \tilde{\boldsymbol{H}}_{se}, \mathscr{L}_e^o \cdot \delta \boldsymbol{E}_s \rangle = \langle \delta \boldsymbol{E}_s, -\mathscr{L}_h^o \cdot \tilde{\boldsymbol{H}}_{se} \rangle$, and using its definition from (32), we conclude that

$$\mathscr{L}_h \cdot \boldsymbol{H}_{se} + k_{ze}^2 \mathscr{B} \cdot \boldsymbol{H}_{se} = 0 \quad (42a)$$

and

$$\kappa_{zz} \hat{z} \cdot \nabla_s \times \tilde{\boldsymbol{H}}_s = -i\omega \tilde{Z}_s \hat{z} \cdot \hat{n} \times \tilde{\boldsymbol{H}}_s$$

$$\nu_{zz} \nabla_s \cdot \bar{\bar{\mu}}_s^t \cdot \tilde{\boldsymbol{H}}_s = -i\omega Y_s \hat{n} \cdot \bar{\bar{\mu}}_s^t \cdot \tilde{\boldsymbol{H}}_s \quad (42b)$$

on C. Equation (42b) is equivalent to

$$\tilde{E}_z = \tilde{Z}_s \hat{z} \cdot \hat{n} \times \tilde{\boldsymbol{H}}_s \qquad \tilde{H}_z = \frac{\omega}{k_z} Y_s \hat{n} \cdot \tilde{\boldsymbol{B}}_s. \quad (43)$$

Note that in (41) and (43), auxiliary boundary conditions for E_z and $\tilde{H}_z$ have surfaced in addition to those assumed in (33a) and (33b). With these natural boundary conditions, to obtain a PEC at C, we let $Z_s = 0$ and $\tilde{Z}_s = 0$. Furthermore, if we were to assume an essential boundary condition for PEC where $\hat{n} \times \boldsymbol{E}_s$ or $\hat{n} \cdot \bar{\bar{\mu}}_s^t \cdot \boldsymbol{H}_s$ is zero, then the first term in (35) would vanish. For a PEC, the natural boundary condition dictates that $\tilde{Z}_s = 0$, and the second term in (35) vanishes. In such a case, we can remove the functional (35) from (34) completely. Similar statements can be made for a PMC at C.

With the variational expression (34) available, we can apply the Rayleigh–Ritz procedure to find the optimal solution as in Section IV. The resultant matrix equations are

$$\left[\bar{\bar{L}}_e - \partial \bar{\bar{L}}_e + k_z^2 \bar{\bar{B}}\right] \cdot \bar{a} = 0 \quad (44a)$$

$$\left[\bar{\bar{L}}_e^t - \partial \bar{\bar{L}}_e^t + k_z^2 \bar{\bar{B}}^t\right] \cdot \bar{b} = 0. \quad (44b)$$

The resultant equations are similar to (26a) and (26b) except for the $\partial \bar{\bar{L}}_e$ term, which is defined as

$$\left[\partial \bar{\bar{L}}_e\right]_{mn} = \langle \boldsymbol{H}_{ms}, \partial \mathscr{L}_e \cdot \boldsymbol{E}_{ns} \rangle. \quad (45)$$

If finite domain basis functions are used in (25), since (45) involves line integrals on C, only the elements touching the boundary C are involved in calculating $\partial \bar{\bar{L}}_e$. As before, we need only solve one of the two equations (44a) and (44b). Furthermore, once $\bar{a}_i$'s are found, the $\bar{b}_j$'s can be found easily.

VI. Implementation and Numerical Results

We have implemented the above theory for an inhomogeneous anisotropic dielectric waveguide. Without loss of generality, we let

$$\boldsymbol{E}_s = \begin{pmatrix} E_x \\ E_y \end{pmatrix} = \sum_{n=1}^{N} \bar{\bar{f}}_n(x, y) \cdot \bar{\alpha}_n$$

$$\boldsymbol{H}_s = \begin{pmatrix} H_x \\ H_y \end{pmatrix} = \sum_{m=1}^{N} \bar{\bar{f}}_m(x, y) \cdot \bar{\beta}_m \tag{46}$$

where $\bar{\bar{f}}_n(x, y)$ is a 2×2 matrix, and $\bar{\alpha}_n$ and $\bar{\beta}_n$ are 2×1 column vectors. In general, $\bar{\bar{f}}_n$ need not be diagonal. For simplicity, we have assumed

$$\bar{\bar{f}}_n(x, y) = \begin{pmatrix} f_n(x, y) & 0 \\ 0 & f_n(x, y) \end{pmatrix}. \tag{47}$$

The choice of $f_n(x, y)$ is such that it should be good enough to approximate the x and y components of the electric and magnetic fields. With the above choice of basis functions,

$$\left[\bar{\bar{L}}_e\right]_{mn} = \langle \bar{\bar{f}}_m, \mathscr{L}_e \cdot \bar{\bar{f}}_n \rangle \tag{48a}$$

$$\left[\bar{\bar{B}}\right]_{mn} = \langle \bar{\bar{f}}_m, \mathscr{B} \cdot \bar{\bar{f}}_n \rangle \tag{48b}$$

are 2×2 submatrices. More specifically,

$$\left[\bar{\bar{L}}_e\right]_{mn} = -\int_S dS\,\hat{z}\cdot\nabla_s \times \bar{\bar{f}}_m \nu_{zz} \nabla_s \cdot \bar{\bar{\mu}}_s^t \cdot \bar{\bar{f}}_n$$
$$+ \int_S dS\,\hat{z}\cdot\nabla_s \times \bar{\bar{f}}_n \kappa_{zz} \nabla_s \cdot \bar{\bar{\epsilon}}_s \cdot \bar{\bar{f}}_m$$
$$- \omega^2 \int_S dS\,\hat{z}\cdot\left(\bar{\bar{\epsilon}}_s \cdot \bar{\bar{f}}_n\right) \times \left(\bar{\bar{\mu}}_s^t \cdot \bar{\bar{f}}_m\right) \tag{49a}$$

$$\left[\bar{\bar{B}}\right]_{mn} = \int_S dS \bar{\bar{f}}_m \cdot \hat{z} \times \bar{\bar{f}}_n. \tag{49b}$$

A detailed analysis of (49) is given in [24].

We have chosen $f_n(x, y)$ to be first-order elements. They correspond to interpolating the three points of a triangle with a plane. They can also be thought of as pyramidal functions with a hexagonal base and six plane surfaces. The general mesh we use for the computation is shown in Fig. 2. This mesh is good for an arbitrarily shaped waveguide. If the waveguide is symmetric with respect to the vertical and horizontal midplanes through the waveguide, the mesh can be reduced to a quarter of its original size by imposing the necessary magnetic and electric wall boundary conditions at the midplanes. We found that results similar to those before are obtained by reducing the mesh to one quarter of its original size. The reduction in mesh size is done by a number of workers to reduce storage requirement, but this makes the program more restrictive. For a guided mode, we found that the boundary conditions on the outermost wall were not important because the field was, in general, exponentially small there.

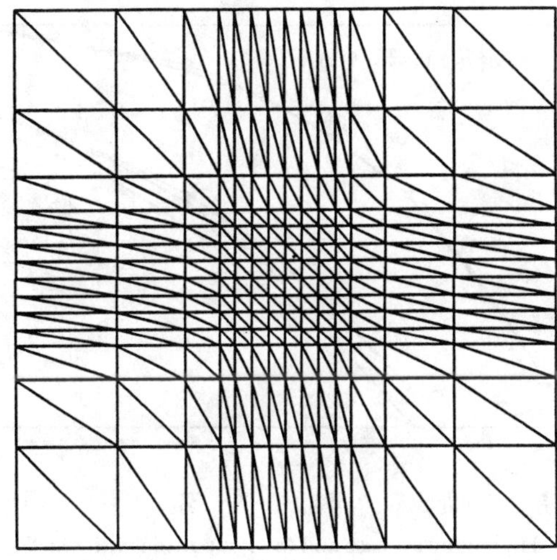

Fig. 2. A sample finite element mesh.

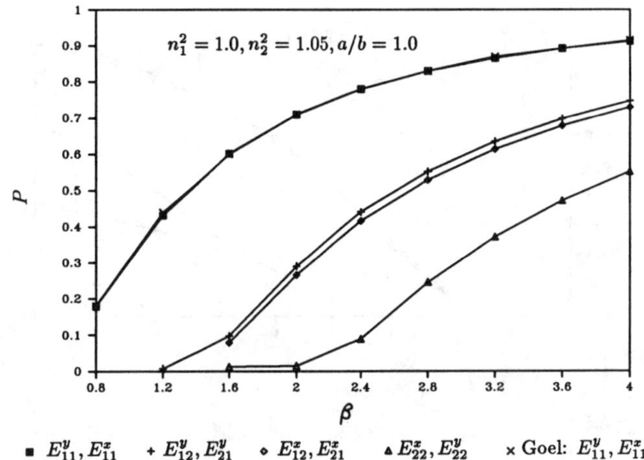

Fig. 3. Low-contrast case and the comparison with Goell's result.

We have solved the eigenequation (26a) with a standard eigenvalue solver from IMSL (International Mathematical Subroutine Library). We have compared the results generated by our code with previous results and found good agreement for both the isotropic and anisotropic cases (Figs. 3–5). We did not observe the occurrence of spurious modes.

In the following plots

$$\beta = \frac{2b}{\lambda_0}\left[\left(\frac{k_1}{k_0}\right)^2 - 1\right]^{\frac{1}{2}} \quad \text{and} \quad P = \frac{(k_z/k_0)^2 - 1}{(k_1/k_0)^2 - 1}.$$

The case considered is a rectangular dielectric waveguide with aspect ratio a/b. The simulations are done using 17×17 pyramidal functions for each E_x and E_y, corresponding to 578 unknowns. It took about 30 seconds on a CRAY-XMP/48 to solve once and for all the eigenvalues of the matrix. The first six modes are plotted, and the lowest two modes are compared with previous results.

In Fig. 3, we compare our fundamental mode with the solution of Goell [25], showing excellent agreement. Goell's solution is derived from cylindrical harmonic analysis, and

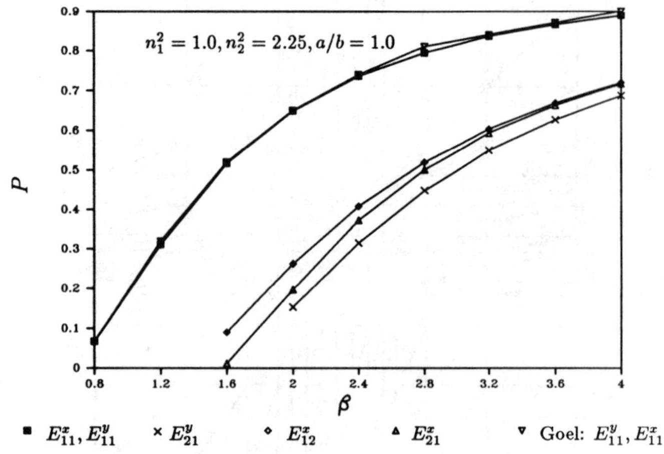

Fig. 4. High-contrast case and the comparison with Goell's result.

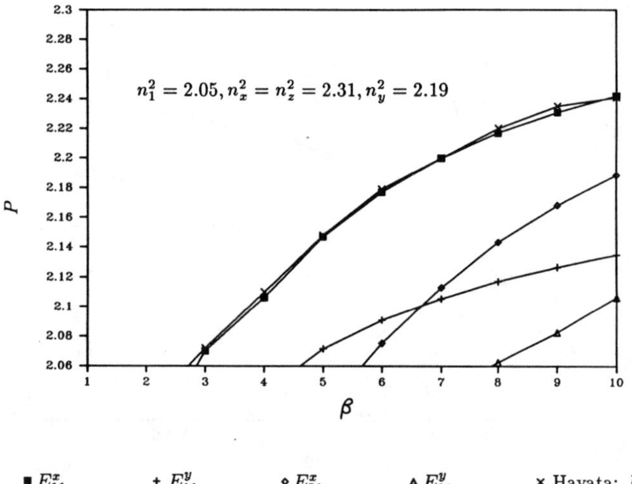

Fig. 5. Anisotropic case which is uniaxial, and the comparison with Hayata's result.

has been standard for comparison in the literature. However, a finite element solution is more versatile than a cylindrical harmonic analysis.

In Fig. 4, we display the dispersion diagram for the high-contrast case. The fundamental mode again agrees very well with Goell's result.

In Fig. 5, we display the dispersion curve for the anisotropic case, and compare the result with that of Hayata *et al.* [12]. We have rescaled β and P to agree with those defined by Hayata. The agreement is very good.

VII. Conclusions

We have developed a variational formulation for anisotropic dielectric waveguides with reflection symmetry using $\boldsymbol{E}_s$ or $\boldsymbol{H}_s$ components only. We cast our variational formulation via the use of transpose operators and shed a different light on the formulation. We show that the $\boldsymbol{E}_s$ and $\boldsymbol{H}_s$ formulations are completely equivalent to each other; hence, there is no reason to prefer one over the other. Our formulation also allows us to easily study the effect of loss and frequency dispersion. We show how the natural boundary conditions can be included in the formulation. The results of our numerical implementation also compare favorably with these results of previous formulations.

Acknowledgment

The authors would like to thank the National Center for Supercomputing Applications at the University of Illinois for the use of the CRAY computer, and they are grateful to Dr. C. H. Chan for useful discussions.

References

[1] M. Ikeuchi, H. Sawami, and H. Niki, "Analysis of open-type dielectric waveguides by the finite-element iterative method," *IEEE Trans. Microwave Theory Tech.*, vol. MTT-29, pp. 234–239, Mar. 1981.

[2] N. Mabaya, P. E. Lagasse, and P. Vandenbulcke, "Finite element analysis of optical waveguides," *IEEE Trans. Microwave Theory Tech.*, vol. MTT-29, pp. 600–605, June 1981.

[3] J. Katz, "Novel solution of 2-D waveguide using the finite element method," *Appl. Opt.*, vol. 21, no. 15, p. 2747, 1982.

[4] B. M. A. Rahman and J. B. Davies, "Finite-element analysis of optical and microwave waveguide problems," *IEEE Trans. Microwave Theory Tech.*, vol. MTT-32, pp. 20–28, Jan. 1984.

[5] M. Koshiba, K. Hayata, and M. Suzuki, "Vectorial finite-element formulation without spurious modes for dielectric waveguides,' *Trans. IECE Japan*, vol. E 67, no. 4, pp. 191–196, 1984.

[6] E. Schweig and W. B. Bridges, "Computer analysis of dielectric waveguides: A finite-difference method," *IEEE Trans. Microwave Theory Tech.*, vol. MTT-32, pp. 531–541, May 1984.

[7] M. Koshiba, K. Hayata, and M. Suzuki, "Approximate scalar finite element analysis of anisotropic optical waveguides with off-diagonal elements in a permittivity tensor," *IEEE Trans. Microwave Theory Tech.*, vol. MTT-32, pp. 587–593, June 1984.

[8] R. B. Wu and C. H. Chen, "A variational analysis of dielectric waveguides by the conformal mapping technique," *IEEE Trans. Microwave Theory Tech.*, vol. MTT-33, no. 8, pp. 681–685, Aug. 1985.

[9] C. C. Su, "A surface integral equation method for homogeneous optical fibers and coupled image lines of arbitrary cross sections," *IEEE Trans. Microwave Theory Tech.*, vol. MTT-33, pp. 1114–1119, Nov. 1985.

[10] R. B. Wu and C. H. Chen, "A scalar variational conformal mapping technique by weakly guiding dielectric waveguides," *IEEE J. Quantum Electron.*, vol. QE-22, pp. 595–608, May 1986.

[11] K. Bierwirth, N. Schulz, and F. Arndt, "Finite-difference analysis of rectangular dielectric waveguide structures," *IEEE Trans. Microwave Theory Tech.*, vol. MTT-34, pp. 1104–1114, Nov. 1986.

[12] K. Hayata, M. Koshiba, M. Eguchi, and M. Suzuki, "Vectorial finite-element method without any spurious solutions for dielectric waveguiding problems using transverse magnetic field component," *IEEE Trans. Microwave Theory Tech.*, vol. MTT-34, pp. 1120–1124, Nov. 1986.

[13] C. C. Su, "A combined method for dielectric waveguides using finite-element techniques and the surface integral equations," *IEEE Trans. Microwave Theory Tech.*, vol. MTT-34, pp. 1140–1146, Nov. 1986.

[14] T. Angkaew, M. Matsuhara, and N. Kumagai, "Finite-element analysis of waveguide modes: A novel approach that eliminates spurious modes," *IEEE Trans. Microwave Theory Tech.*, vol. MTT-35, pp. 117–123, 1987.

[15] W. C. Chew, "Analysis of optical and millimeter wave dielectric waveguide," *J. Electromagnetic Waves and Appl.*, to be published.

[16] L. Cairo and T. Kahan, *Variational Techniques in Electromagnetism*. New York: Gordon and Breach, 1965.

[17] R. E. Collin, *Field Theory of Guided Waves*. New York: McGraw-Hill, 1960. p. 231.

[18] C. H. Chen and C. D. Lien, "The variational formulation for non-self-adjoint electromagnetic problems," *IEEE Trans. Microwave Theory Tech.*, vol. MTT-28, pp. 878–886, 1980.

[19] J. A. Kong, *Electromagnetic Wave Theory*. New York: Wiley, 1986.

[20] I. Stakgold, *Boundary Value Problems of Mathematical Physics*, vol. I. New York: Macmillian, 1968.

[21] E. Merzbacher, *Quantum Mechanics*. New York: Wiley, 1970.

[22] R. Wait and A. R. Mitchell, *Finite Element Analysis and Applications*. New York: Wiley, 1985.

[23] R. Harrington, *Field Computation by Moment Method*. Malabar, FL: Krieger, 1983.

[24] M. Nasir, "Finite element analysis of an inhomogeneous, anisotropic dielectric waveguide," M.S. thesis, Dept. of Elect. and Comp. Eng., Univ. of Illinois, Urbana, 1988.

[25] J. E. Goell, "A circular-harmonic computer analysis of rectangular dielectric waveguides," *Bell Syst. Tech. J.*, vol. 48, pp. 2133–2160, 1969.

Finite Element Solution of Longitudinally Magnetized Elliptical Gyromagnetic Waveguides

ANDREW A. P. GIBSON AND JOSEPH HELSZAJN, SENIOR MEMBER, IEEE

Abstract — **One of the most common methods employed in the study of waveguides with irregular electric or magnetic walls is the finite element approach. This method is based on the minimization of a functional, the solution of which satisfies the boundary value problem. It is utilized in this paper to study the elliptical gyromagnetic waveguide with either an electric or a magnetic wall. The functionals for the four possible planar solutions are separately summarized.**

I. Introduction

THE MODERN approach to the derivation of propagation in inhomogeneous waveguides or in waveguides with irregular cross sections is often based on the derivation of a functional and the use of the finite element method [1], [2]. The functional encountered in gyromagnetic waveguides with the direct field along the direction of propagation has been developed in terms of the coupled longitudinal electric (E_z) and magnetic (H_z) fields entering into the description of the coupled wave equations of this type of problem [3]. Since the transverse fields may be specified from a knowledge of the longitudinal ones, this approach gives the complete solution to the problem. This is in fact a common procedure in the study of waveguide problems which support hybrid modes [4]–[8]. The functional in the situation for which the direction of the direct field makes an arbitrary angle with that of propagation has been historically posed for the gyrotropic situation using a three-component electric or magnetic field vector and in terms of the transverse electric and magnetic fields [9], [10]. If the direct field is parallel to that of the direction of propagation, then E_z and H_z are coupled. If the direct field and the direction of propagation are perpendicular to each other, then E_z and H_z are decoupled provided the alternating fields do not vary along the direction of the direct field. The functional for the first arrangement has been tackled using the finite difference method [3] and that for the second one using the finite element method [10], [11]. The purpose of this paper is to investigate propagation along an elliptical gyromagnetic waveguide with electric or magnetic walls with the direct field parallel to that of propagation using the finite element method. Fig. 1 depicts the case of a longitudinally magnetized elliptical waveguide with an electric wall.

The required functional is a scalar quantity which may be constructed by starting with a Rayleigh quotient [3], [12] and expressing it as a quadratic form [4], [5], [13] or by constructing the quadratic form directly from a knowledge of the coupled or uncoupled wave equations [9], [14]–[16]. It is reproduced here by directly constructing it in terms of the properties of a quadratic form as a preamble to carrying out some calculations. If the minimum of the functional corresponds to the solution of the boundary value problem then the final result is the energy functional of the system. In such cases the quadratic form is understood to be an energy function. While this approach gives the classical result [3], the language is more akin to that met in network rather than in electromagnetic theory [17]. Therefore it may be more comprehensible to the nonspecialist engineer. The solutions also correctly reduce to those met frequently in the description of the planar gyromagnetic problems with electric and magnetic walls in all combinations [18]–[20]. The cutoff space for this type of waveguide has been dealt with separately in [20] and [21]. The finite element method is currently being employed to analyze lossy three-dimensional problems through a variational approach [22]–[24] and by using alternative weighted residual schemes [25], [26].

II. Derivation of a Functional

The solution to irregular isotropic or gyromagnetic waveguides is often solved by constructing a quadratic form from a knowledge of the wave equation in either the z-directed scalar [3]–[8] or the three-dimensional vector [11]–[13], [16] form. It may be readily demonstrated that all energy functions have the nature of a quadratic form [17]. A property of any waveguide is that its field distribution in the transverse plane may always be formed from a knowledge of E_z and H_z. In a gyromagnetic waveguide magnetized along the direction of propagation, E_z and H_z are coupled. The classic derivation of the required functional starts with the coupled wave equations

$$L_{ee}E_z + L_{eh}H_z = 0 \quad (1a)$$

$$L_{he}E_z + L_{hh}H_z = 0. \quad (1b)$$

Manuscript received May 3, 1988; revised December 20, 1988.

The authors are with the Department of Electrical and Electronic Engineering, Heriot-Watt University, 31-35 Grassmarket, Edinburgh, Scotland, U.K.

IEEE Log Number 8927133.

Reprinted from *IEEE Trans. Microwave Theory Tech.*, vol. 37, no. 6, pp. 999–1005, June 1989.

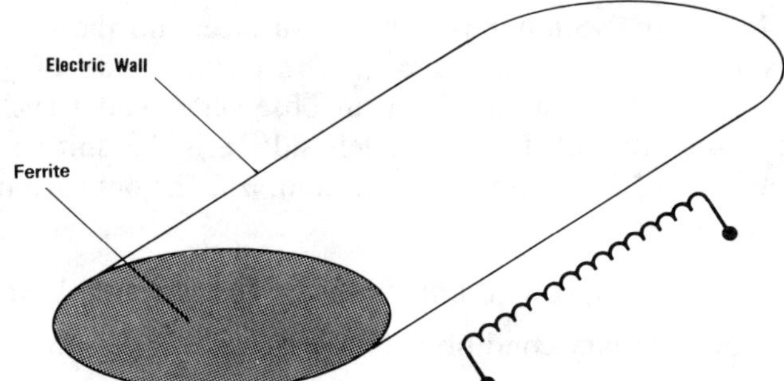

Fig. 1. Schematic diagram of elliptical gyromagnetic waveguide with an electric wall.

L_{ee}, L_{eh}, L_{he}, and L_{hh} are dependent upon the propagation constant, the frequency, the constitutive parameters, and second-order linear derivatives. This system of equations may be written in matrix form as

$$\begin{bmatrix} L_{ee} & L_{eh} \\ L_{he} & L_{hh} \end{bmatrix} \begin{bmatrix} E_z \\ H_z \end{bmatrix} = [0]. \tag{2}$$

The square matrix in this relationship is known as the operator and is denoted by

$$\bar{L} = \begin{bmatrix} L_{ee} & L_{eh} \\ L_{he} & L_{hh} \end{bmatrix}. \tag{3}$$

An energy function can be associated with the square matrix $\bar{L}$ and column vector

$$\begin{bmatrix} E_z \\ H_z \end{bmatrix}$$

by forming the related quadratic form

$$[E_z^* \quad H_z^*] \begin{bmatrix} L_{ee} & L_{eh} \\ L_{he} & L_{hh} \end{bmatrix} \begin{bmatrix} E_z \\ H_z \end{bmatrix}$$

and by recognizing that all energy functions have this type of form. If the quadratic form is greater than but not equal to zero it is said to be positive definite (P.D.). If it is equal to or greater than zero it is said to be positive semidefinite (P.S.D.). In the case of a lossless homogeneous gyromagnetic waveguide the tensor permeability is Hermitian, and a property of the matrix $\bar{L}$ of the quadratic form is that it is self-adjoint. A necessary condition for the resulting matrix eigenvalue problem to produce real eigenvalue frequencies is that the quadratic form be P.S.D. This condition requires that the elements of the matrix $\bar{L}$ must satisfy

$$L_{ee} \geqslant 0 \tag{4a}$$

$$L_{hh} \geqslant 0 \tag{4b}$$

$$(L_{ee}L_{hh} - L_{eh}L_{he}) \geqslant 0. \tag{4c}$$

An energy function of this type, once integrated over the cross-sectional area of the waveguide, is referred to as the functional of the system. This operation is usually written as

$$F(E_z, H_z) = \left\langle \begin{bmatrix} E_z \\ H_z \end{bmatrix}, \begin{bmatrix} L_{ee} & L_{eh} \\ L_{he} & L_{hh} \end{bmatrix} \begin{bmatrix} E_z \\ H_z \end{bmatrix} \right\rangle. \tag{5}$$

This is given in integral form by

$$F(E_z, H_z) = \iint_s E_z^* L_{ee} E_z \, ds + \iint_s E_z^* L_{eh} H_z \, ds + \iint_s H_z^* L_{he} E_z \, ds + \iint_s H_z^* L_{hh} H_z \, ds. \tag{6}$$

In order to obtain a complete solution of the boundary value problem it is necessary to minimize the quadratic functional in conjunction with the boundary conditions of the waveguide problem.

III. The Matrix of the Form in Gyromagnetic Waveguides

The derivation starts by decomposing the electric and magnetic fields ($\boldsymbol{E}$ and $\boldsymbol{H}$) into transverse ($\boldsymbol{E}_t$, $\boldsymbol{H}_t$) and longitudinal components ($\boldsymbol{a}_z E_z$, $\boldsymbol{a}_z H_z$). Adopting this notation and assuming the time and z variation to be

$$\exp(j\omega t)$$

$$\exp(-j\beta z)$$

gives

$$\boldsymbol{E} = (\boldsymbol{E}_t + \boldsymbol{a}_z E_z) \exp(-j\beta z) \exp(j\omega t) \tag{7}$$

$$\boldsymbol{H} = (\boldsymbol{H}_t + \boldsymbol{a}_z H_z) \exp(-j\beta z) \exp(j\omega t). \tag{8}$$

The constitutive parameters in a gyromagnetic waveguide with the direction of the direct magnetic field along that of propagation are given in terms of a tensor permeability ($[\mu]$) and a scalar permittivity (ϵ) by

$$[\mu] = \mu_0 \begin{bmatrix} \mu & -j\kappa & 0 \\ j\kappa & \mu & 0 \\ 0 & 0 & \mu_z \end{bmatrix} \tag{9}$$

$$\epsilon = \epsilon_0 \epsilon_f \tag{10}$$

respectively, where ϵ_0 and μ_0 denote the permittivity and permeability of free space.

The wave equations for E_z and H_z are in this instance coupled. The solution to this problem is a standard result in the literature [27]. This gives the entries of the matrix of the form as

$$L_{ee} = \nabla_t^2 - \beta^2 + k_0^2 \mu_{\text{eff}} \epsilon_f \tag{11a}$$

$$L_{eh} = j\sqrt{\left(\frac{\mu_0}{\epsilon_0}\right)} \frac{\beta k_0 \kappa \mu_z}{\mu} \tag{11b}$$

$$L_{he} = L_{eh}^* \tag{11c}$$

$$L_{hh} = \frac{\mu_0 \mu_z}{\epsilon_0 \epsilon_f} \left(\nabla_t^2 + \frac{k_1^2 \mu_z}{\mu} \right) \tag{11d}$$

where

$$k_1^2 = -\beta^2 + k_0^2 \mu \epsilon_f \tag{12}$$

and

$$\mu_{\text{eff}} = \frac{\mu^2 - \kappa^2}{\mu} \tag{13a}$$

$$k_0 = \frac{2\pi}{\lambda_0}. \tag{13b}$$

In this description of the matrix of the form each element is a function of both β and k_0 to first and second order. It is usual to recast it, for computational purposes, in terms of the normalized propagation constant ($\bar{\beta} = \beta/k_0$) and k_0^2 [3]. Using (11c) and (11d) the coupled wave equation in (1b) may be written as

$$H_z = j\sqrt{\left(\frac{\epsilon_0}{\mu_0}\right)} \frac{\beta \kappa k_0 \epsilon_f}{\mu_z k_1^2} E_z - \frac{\mu}{\mu_z k_1^2} \nabla_t^2 H_z \tag{14}$$

and substituting it together with (11a) and (11b) into the other coupled wave equation defined by (1a) gives

$$L'_{ee} = a\nabla_t^2 + k_0^2 \sqrt{\left(\frac{\epsilon_0}{\mu_0}\right)} \epsilon_f \tag{15a}$$

$$L'_{eh} = -jb\nabla_t^2. \tag{15b}$$

Starting with the coupled wave equation (1a) and making use of the relationships in (11a) and (11b) gives

$$L'_{he} = L'^{*}_{eh} \tag{15c}$$

$$L'_{hh} = c\nabla_t^2 + k_0^2 \sqrt{\left(\frac{\mu_0}{\epsilon_0}\right)} \mu_z. \tag{15d}$$

E_z and H_z are coupled through the imaginary term L'_{eh}, which is nonzero for β and κ nonzero. The quantities a, b, c, d, e, and f are given in the Appendix.

IV. Quadratic Functional for a Gyromagnetic Region

The quadratic functional applicable to a finite space region is readily deduced by combining (6) and (15). The result is in terms of second-order linear derivatives. Integrating this equation by parts and applying Green's identity in a plane reduces it to one involving first-order linear derivatives only. The products appearing in the quadratic form then become

$$\iint_s E_z^* L'_{ee} E_z \, ds = -a \iint_s |\nabla_t E_z|^2 \, ds + a \int_\xi E_z^* \frac{\partial E_z}{\partial n} \, dl + k_0^2 \sqrt{\left(\frac{\epsilon_0}{\mu_0}\right)} \epsilon_f \iint_s |E_z|^2 \, ds \tag{16a}$$

$$\iint_s E_z^* L'_{eh} H_z \, ds = jb \iint_s \nabla_t E_z^* \cdot \nabla_t H_z \, ds - jb \int_\xi E_z^* \frac{\partial H_z}{\partial n} \, dl \tag{16b}$$

$$\iint_s H_z^* L'_{he} E_z \, ds = -jb \iint_s \nabla_t H_z^* \cdot \nabla_t E_z \, ds + jb \int_\xi H_z^* \frac{\partial E_z}{\partial n} \, dl \tag{16c}$$

$$\iint_s H_z^* L'_{hh} H_z \, ds = -c \iint_s |\nabla_t H_z|^2 \, ds + c \int_\xi H_z^* \frac{\partial H_z}{\partial n} \, dl + k_0^2 \sqrt{\left(\frac{\mu_0}{\epsilon_0}\right)} \mu_z \iint_s |H_z|^2 \, ds. \tag{16d}$$

In this derivation s is a closed surface, and the boundary of s is a closed curve ξ. The positive side of ξ is defined to be that in which an observer would travel to have the interior of s on the left side. $\bar{\xi}$ is the unit vector tangent to ξ in the positive direction. $\boldsymbol{n}$ is the outward unit vector normal to s.

V. Elliptical Waveguide with an Electric Wall

The boundary conditions in the electric wall problem

$$E_\xi = 0 \tag{17a}$$

$$E_z = 0 \tag{17b}$$

are applied to the tangential electric field E_ξ:

$$E_\xi = -jf \frac{\partial E_z}{\partial \xi} + b \frac{\partial E_z}{\partial n} - d \frac{\partial H_z}{\partial \xi} - jc \frac{\partial H_z}{\partial n} \tag{17c}$$

to give the required boundary equation as [28]

$$b \frac{\partial E_z}{\partial n} - d \frac{\partial H_z}{\partial \xi} - jc \frac{\partial H_z}{\partial n} = 0. \tag{17d}$$

Introducing this boundary equation into the functional for the finite gyromagnetic region and making use of the following relationship between the coupled variables,

$$j \iint_s \nabla_t E_z^* \cdot \nabla_t H_z - (\nabla_t E_z^* \cdot \nabla_t H_z)^* \, ds = -2 \iint_s \nabla_t E_z \cdot \nabla_t H_z \, ds \tag{18}$$

gives the required form of the functional for a longitudinally gyromagnetic waveguide with an electric wall:

$$\begin{aligned} F(E_z, H_z) = &-a \iint_s |\nabla_t E_z|^2 \, ds + k_0^2 \sqrt{\left(\frac{\epsilon_0}{\mu_0}\right)} \epsilon_f \iint_s |E_z|^2 \, ds \\ &-2b \iint_s \nabla_t E_z \cdot \nabla_t H_z \, ds \\ &-c \iint_s |\nabla_t H_z|^2 \, ds + jd \int_\xi H_z^* \frac{\partial H_z}{\partial \xi} \, dl \\ &+ k_0^2 \sqrt{\left(\frac{\mu_0}{\epsilon_0}\right)} \mu_z \iint_s |H_z|^2 \, ds. \end{aligned} \tag{19}$$

The electromagnetic field solution may be obtained from the stationary condition of the functional. Adopting the Rayleigh–Ritz minimization method yields the following matrix eigenvalue equation:

$$\begin{bmatrix} a[S] & b[S] \\ b[S] & c[S] \end{bmatrix} \begin{bmatrix} e_z \\ h_z \end{bmatrix} - jd \begin{bmatrix} [0] & [0] \\ [0] & [C] \end{bmatrix} \begin{bmatrix} e_z \\ h_z \end{bmatrix} = k_0^2 \begin{bmatrix} \sqrt{\left(\frac{\epsilon_0}{\mu_0}\right)} \epsilon_f [T] & [0] \\ [0] & \sqrt{\left(\frac{\mu_0}{\epsilon_0}\right)} \mu_z [T] \end{bmatrix} \begin{bmatrix} e_z \\ h_z \end{bmatrix} \tag{20}$$

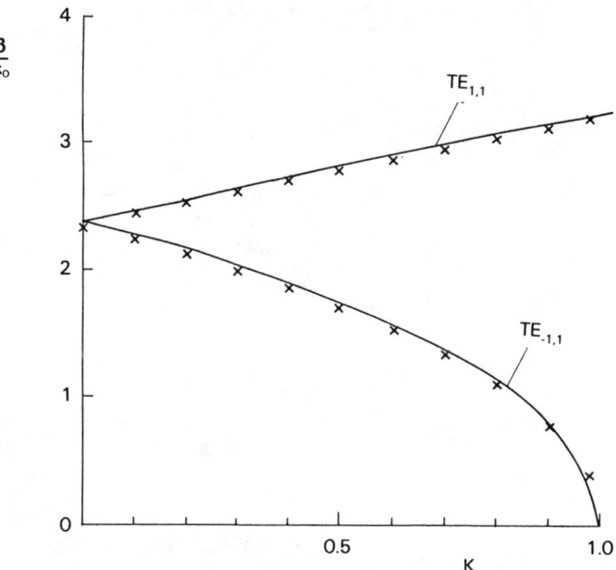

Fig. 2. Comparison between exact and finite element method of the propagation constants of the $\mathrm{TE}_{\pm 1,1}$ mode in a gyromagnetic circular waveguide with an electric wall ($k_0 R = 0.6$, $\mu = 1$, $\mu_z = 1$, $\epsilon_f = 15$).

Fig. 3. Comparison between exact and finite element method of the propagation constants of the ${}_e\mathrm{TE}_{1,1}$ and ${}_o\mathrm{TE}_{1,1}$ modes in an elliptical gyromagnetic waveguide with an electric wall ($k_0 R = 0.9696$, $\mu = 1$, $\mu_z = 1$, $\epsilon_f = 10$, $e = 0.648$).

for a waveguide of general cross section bounded by an electric wall. This quantity differs from the isotropic formulation in the coupling terms [4]–[8]. It is of note that the imaginary skew-symmetric part of the Hermitian matrix only makes a contribution to the real part of the Hermitian form, and the resultant form is therefore real and will produce real eigenfrequencies. $[S]$ and $[T]$ are real, symmetric square matrices [2]:

$$S_{i,j} = \iint_s \nabla_t \alpha_i \cdot \nabla_t \alpha_j \, ds \tag{21a}$$

$$T_{i,j} = \iint_s \alpha_i \alpha_j \, ds \tag{21b}$$

and the imaginary skew-symmetric square matrix $[C]$ is given by [18]

$$C_{i,j} = \int_\xi \alpha_i \frac{\partial \alpha_j}{\partial \xi} \, dl. \tag{21c}$$

Here $\alpha_{i,j}$ are a set of linearly independent real basis functions. The constitutive parameters and $\bar{\beta}$ are inputs to the finite element program; k_0^2 is the calculated eigenvalue. From a knowledge of the relationship between $\bar{\beta}$ and k_0 over a range of κ values, it is possible to plot $\bar{\beta}$ versus κ diagrams.

Fig. 2 illustrates the agreement between the exact and finite element solutions in the case of a round waveguide. Fig. 3 depicts the phase constants for an elliptical gyromagnetic waveguide. The segmentation employed here consists of 15 third-order elements. This is indicated in Fig. 4.

VI. Elliptical Waveguide with Magnetic Walls

The boundary conditions relating to a waveguide with a magnetic wall surface,

$$H_\xi = 0 \tag{22a}$$

$$H_z = 0 \tag{22b}$$

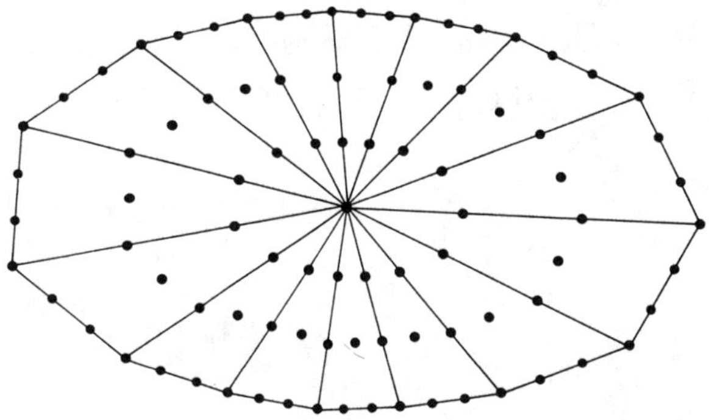

Fig. 4. Segmentation of an elliptical cross section into 15 third-order triangles.

reduce the expression for the tangential magnetic field,

$$H_\xi = e\frac{\partial E_z}{\partial \xi} + ja\frac{\partial E_z}{\partial n} - jf\frac{\partial H_z}{\partial \xi} + b\frac{\partial H_z}{\partial n} \tag{22c}$$

to the following boundary equation:

$$b\frac{\partial H_z}{\partial n} + e\frac{\partial E_z}{\partial \xi} + ja\frac{\partial E_z}{\partial n} = 0. \tag{22d}$$

Once again making use of the relationship between the coupled variables (18), the functional for the magnetic wall waveguide is obtained as

$$\begin{aligned} F(E_z, H_z) = &-a\iint_s |\nabla_t E_z|^2 \, ds + je\int_\xi E_z^* \frac{\partial E_z}{\partial \xi} \, dl \\ &+ k_0^2 \sqrt{\left(\frac{\epsilon_0}{\mu_0}\right)} \epsilon_f \iint_s |E_z|^2 \, ds \\ &- 2b\iint_s \nabla_t E_z \cdot \nabla_t H_z \, ds - c\iint_s |\nabla_t H_z|^2 \, ds \\ &+ k_0^2 \sqrt{\left(\frac{\mu_0}{\epsilon_0}\right)} \mu_z \iint_s |H_z|^2 \, ds. \end{aligned} \tag{23}$$

The Rayleigh–Ritz minimization of this functional gives

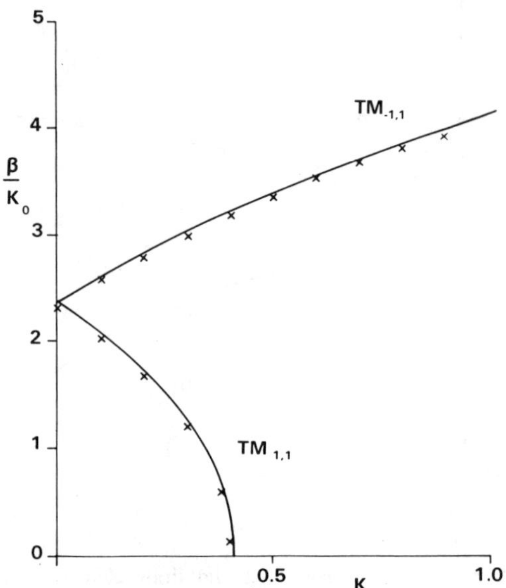

Fig. 5. Comparison between exact and finite element method of the propagation constants of the TM$_{\pm 1,1}$ mode in a circular gyromagnetic waveguide with a magnetic wall ($k_0R = 0.6$, $\mu = 1$, $\mu_z = 1$, $\epsilon_f = 15$).

the matrix eigenvalue equation as

$$\begin{bmatrix} a[S] & b[S] \\ b[S] & c[S] \end{bmatrix}\begin{bmatrix} e_z \\ h_z \end{bmatrix} - je\begin{bmatrix} [C] & [0] \\ [0] & [0] \end{bmatrix}\begin{bmatrix} e_z \\ h_z \end{bmatrix} = k_0^2 \begin{bmatrix} \sqrt{\left(\frac{\epsilon_0}{\mu_0}\right)}\epsilon_f[T] & [0] \\ [0] & \sqrt{\left(\frac{\mu_0}{\epsilon_0}\right)}\mu_z[T] \end{bmatrix}\begin{bmatrix} e_z \\ h_z \end{bmatrix}. \quad (24)$$

Fig. 5 compares the finite element solution and the exact one for a circular gyromagnetic waveguide with a magnetic wall. Fig. 6 indicates one result in the case of an elliptical waveguide. This solution is compatible with the approximate phase constants deduced from the split frequencies of a planar circuit with magnetic sidewalls and top and bottom electric walls [20].

VII. Functionals for Planar Gyromagnetic Resonators

The cutoff plane of the gyromagnetic waveguide problem with an electric sidewall may be constructed by minimizing the functionals of the related planar geometries with either magnetic or electric top and bottom walls together with a knowledge of the rules governing mode nomenclature [29]–[31]. The functional for the transverse electric (TE) modes of a planar circuit with top and bottom magnetic walls is obtained by letting

$$\beta = 0 \quad (25a)$$

$$E_z = 0 \quad (25b)$$

in (19). This gives

$$F(H_z) = -\iint_s |\nabla_t H_z|^2 \, ds + k_0^2 \epsilon_f \mu_z \iint_s |H_z|^2 \, ds. \quad (26)$$

The functional associated with the transverse magnetic (TM) modes of the planar resonator bounded by an electric sidewall and top and bottom electric walls is derived by letting

$$\beta = 0 \quad (27a)$$

$$H_z = 0 \quad (27b)$$

in (19). The result is

$$F(E_z) = -\iint_s |\nabla_t E_z|^2 \, ds + k_0^2 \epsilon_f \mu_{\text{eff}} \iint_s |E_z|^2 \, ds. \quad (28)$$

These functionals, (26) and (28), have been used in [20] to construct the cutoff space of elliptical gyromagnetic waveguides with electric walls.

The functional for the planar circuit with a magnetic sidewall may also be obtained from the related functional for the magnetic wall waveguide. The conditions for electric top and bottom walls, stated in (27), reduce the functional in (23) to the following:

$$F(E_z) = -\iint_s |\nabla_t E_z|^2 \, ds + j\frac{\kappa}{\mu} \int_\xi E_z^* \frac{\partial E_z}{\partial \xi} \, dl + k_0^2 \epsilon_f \mu_{\text{eff}} \iint_s |E_z|^2 \, ds. \quad (29)$$

This is exactly the result given in [18] and [19] and may be used to deal with the planar gyromagnetic problem. The imaginary part of (29) results in split cutoff numbers for the TM modes of this waveguide.

Similarly the conditions for magnetic top and bottom walls, stated in (25), applied to (23) yield the functional for the planar geometry with magnetic side, and top and

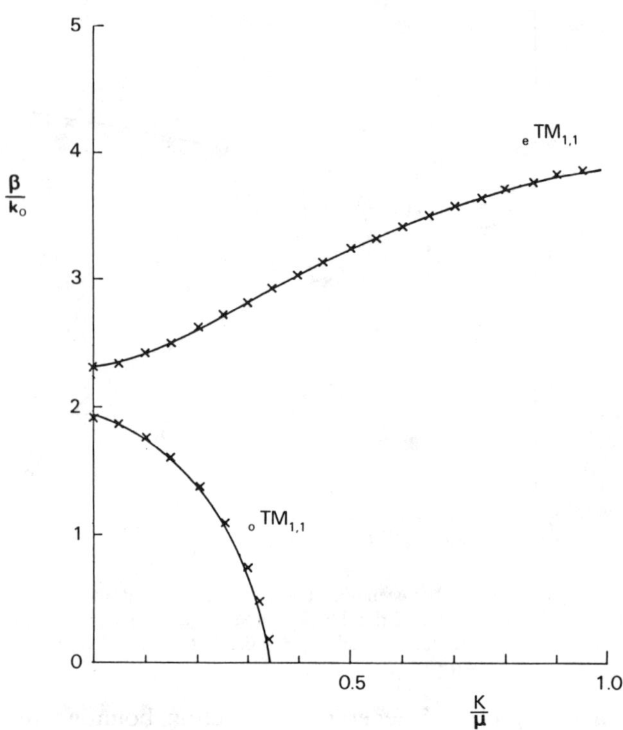

Fig. 6. Comparison between perturbation theory and the finite element method of the propagation constants of the $_e$TM$_{1,1}$ and $_o$TM$_{1,1}$ modes in an elliptical gyromagnetic waveguide with a magnetic wall ($k_0R = 0.6$, $\mu = 1$, $\mu_z = 1$, $\epsilon_f = 15$, $e = 0.4$).

bottom walls as

$$F(H_z) = -\iint_s |\nabla_t H_z|^2\, ds + k_0^2 \epsilon_f \mu_z \iint_s |H_z|^2\, ds. \quad (30)$$

This functional is identical to that in (26) but the solution to this problem requires that the functional be minimized in conjunction with the boundary condition that H_z is zero on the magnetic sidewall.

The functionals given by (29) and (30) may be used to construct the TM and TE modes respectively in the cutoff plane of the related gyromagnetic waveguide problem with a magnetic wall [20].

VIII. Conclusion

The classic method employed to derive a functional is to construct a quadratic form, by premultiplying the wave equation by the conjugate field, and then to integrate this quantity over the cross section of the problem region. This procedure has been utilized in this paper to construct a solution of the longitudinally magnetized gyromagnetic problem using the z-directed coupled wave equations. It has been used in conjunction with the finite element method to evaluate the propagation constants of an elliptical gyromagnetic waveguide with either an electric or a magnetic wall. The well-known functionals associated with the four possible planar circuits made up of electric and magnetic side and top and bottom walls are directly obtained from those of the related waveguide problems.

Appendix

The coefficients met in the text are defined below:

$$a = \epsilon_f \sqrt{\left(\frac{\epsilon_0}{\mu_0}\right)} \frac{\left(\epsilon_f \mu - \bar{\beta}^2\right)}{\left[\bar{\beta}^2 - \epsilon_f(\mu + \kappa)\right]\left[\bar{\beta}^2 - \epsilon_f(\mu - \kappa)\right]}$$

$$b = \frac{\epsilon_f \kappa \bar{\beta}}{\left[\bar{\beta}^2 - \epsilon_f(\mu + \kappa)\right]\left[\bar{\beta}^2 - \epsilon_f(\mu - \kappa)\right]}$$

$$c = \mu \sqrt{\left(\frac{\mu_0}{\epsilon_0}\right)} \frac{\left(\epsilon_f \mu_{\text{eff}} - \bar{\beta}^2\right)}{\left[\bar{\beta}^2 - \epsilon_f(\mu + \kappa)\right]\left[\bar{\beta}^2 - \epsilon_f(\mu - \kappa)\right]}$$

$$d = \sqrt{\left(\frac{\mu_0}{\epsilon_0}\right)} \frac{\bar{\beta}^2 \kappa}{\left[\bar{\beta}^2 - \epsilon_f(\mu + \kappa)\right]\left[\bar{\beta}^2 - \epsilon_f(\mu - \kappa)\right]}$$

$$e = \sqrt{\left(\frac{\epsilon_0}{\mu_0}\right)} \frac{\epsilon_f^2 \kappa}{\left[\bar{\beta}^2 - \epsilon_f(\mu + \kappa)\right]\left[\bar{\beta}^2 - \epsilon_f(\mu - \kappa)\right]}$$

$$f = \frac{\bar{\beta}\left(\epsilon_f \mu - \bar{\beta}^2\right)}{\left[\bar{\beta}^2 - \epsilon_f(\mu + \kappa)\right]\left[\bar{\beta}^2 - \epsilon_f(\mu - \kappa)\right]}$$

$$\bar{\beta} = \frac{\beta}{k_0}.$$

References

[1] S. Ahmed, "Finite element method for waveguide problems," *Electron Lett.*, vol. 4, pp. 387–389, Sept. 1968.

[2] P. Silvester, "A general high order finite element waveguide analysis program," *IEEE Trans. Microwave Theory Tech.*, vol. MTT-17, pp. 204–210, Apr. 1969.

[3] C. D. Hannaford, "Finite difference/variational method for inhomogeneous, isotropic or anisotropic waveguides," Ph.D. dissertation, University of Leeds, Leeds, England, 1967.

[4] S. Ahmed and P. Daly, "Finite-element methods for inhomogeneous waveguides," *Proc. Inst. Elec. Eng.*, vol. 116, pp. 1661–1664, Oct. 1969.

[5] P. Daly, "Finite-element coupling matrices," *Electron. Lett.*, vol. 5, pp. 613–615, Nov. 1969.

[6] Z. J. Csendes and P. Silvester, "Numerical solution of dielectric loaded waveguides—Part 1: Finite element analysis," *IEEE Trans. Microwave Theory Tech.*, vol. MTT-18, pp. 1124–1131, Dec. 1970.

[7] N. Mabaya, P. E. Lagasse, and P. Vandenbulcke, "Finite element analysis of optical waveguides," *IEEE Trans. Microwave Theory Tech.*, vol. MTT-29, pp. 600–605, 1981.

[8] P. Daly, "Hybrid-mode analysis of microstrip by finite-element methods," *IEEE Trans. Microwave Theory Tech.*, vol. MTT-19, pp. 19–25, Jan. 1971.

[9] A. Konrad, "Vector variational formulation of electromagnetic fields in anisotropic media," *IEEE Trans. Microwave Theory Tech.*, vol. MTT-24, pp. 553–559, 1976.

[10] T. Angkaew, M. Matsuhan, N. Kumagai, "Finite element analysis of waveguide modes: A novel approach that eliminates spurious modes," *IEEE Trans. Microwave Theory Tech.*, vol. MTT-35, pp. 117–123, Feb. 1987.

[11] A. Konrad, "High order triangular finite elements for electromagnetic waves in anisotropic media," *IEEE Trans. Microwave Theory Tech.*, vol. MTT-25, pp. 353–360, May 1977.

[12] A. D. Berk, "Variational principles for electromagnetic resonators and waveguides," *IRE Trans. Antennas Propagat.*, vol. AP-4, pp. 104–111, Apr. 1956.

[13] B. M. A. Rahman and J. B. Davies, "Finite-element analysis of optical and microwave waveguide problems," *IEEE Trans. Microwave Theory Tech.*, vol. MTT-32, pp. 20–28, 1984.

[14] I. Stakgold, *Boundary Value Problems of Mathematical Physics*, vols. I and II. New York: Macmillan, 1968.

[15] A. Wexler, "Computation of electromagnetic fields," *IEEE Trans. Microwave Theory Tech.*, vol. MTT-17, Aug. 1969.

[16] W. J. English and F. J. Young, "An E vector variational formulation of Maxwell equations for cylindrical waveguide problems," *IEEE Trans. Microwave Theory Tech.*, vol. MTT-19, pp. 40–46, Jan. 1971.

[17] N. Balabanian and T. A. Bickart, *Electrical Network Theory*. New York: Wiley, 1969.

[18] T. Miyoshi, S. Yamaguchi, and S. Goto, "Ferrite planar circuits in microwave integrated circuits," *IEEE Trans. Microwave Theory Tech.*, vol. MTT-25, July 1977.

[19] R. W. Lyon and J. Helszajn, "A finite element analysis of planar circulators using arbitrarily shaped resonators," *IEEE Trans. Microwave Theory Tech.*, vol. MTT-30, pp. 1967–1974, Nov. 1982.

[20] J. Helszajn and A. Gibson, "Cut-off spaces of elliptical gyromagnetic planar circuits and waveguides using finite elements," *IEEE Trans. Microwave Theory Tech.*, vol. 37, pp. 71–80, Jan. 1989.

[21] W. E. Salmond and C. Yeh, "Ferrite filled elliptical waveguides," *J. Appl. Phys.*, vol. 41, no. 8, pp. 3210–3226, July 1970.

[22] A. D. McAuley, "Variational finite element solution for dissipative waveguides and transportation systems," *IEEE Trans. Microwave Theory Tech.*, vol. MTT-25, pp. 382–392, May 1977.

[23] C. H. Chen and D. L. Chuen, "The variational principle for non-self adjoint-problems," *IEEE Trans. Microwave Theory Tech.*, vol. MTT-28, pp. 878–886, Aug. 1980.

[24] J. P. Webb, G. L. Maile, R. L. Ferrari, "Finite element solution of three dimensional electromagnetic problems," *Proc. Inst. Elec. Eng.*, vol. 130, pt. H, no. 2, pp. 153–159, Mar. 1983.

[25] J. Simkin and C. W. Trowbridge, "Three-dimensional nonlinear electromagnetic field computations using scalar potentials," *Proc. Inst. Elec. Eng.*, vol. 127, pt. B, pp. 368–374, 1980.

[26] C. S. Biddlecombe, E. A. Heighway, J. Simkin, and C. W. Trowbridge, "Methods for eddy current computations in three dimensions," *IEEE Trans. Magn.*, vol. MAG-18, pp. 492–497, 1982.

[27] M. L. Kales, "Modes in waveguides containing ferrites," *J. Appl. Phys.*, vol. 24, pp. 604–608, 1953.

[28] L. G. Chambers, "Propagation in a ferrite filled waveguide," *J. Mech. Appl. Maths.*, pp. 435–447, 1955.

[29] R. A. Waldron, "Properties of ferrite-loaded cylindrical waveguides in the neighbourhood of cut-off," *Proc. Inst. Elec. Eng.*, vol. 109B, suppl. 21, pp. 90–94, 1962.

[30] R. A. Waldron, *Theory of Waveguides and Cavities*. McLaren, 1967.

[31] J. Helszajn and A. Gibson, "Mode nomenclature of circular gyromagnetic and anisotropic waveguides with magnetic and open walls," *Proc. Inst. Elec. Eng.*, pt. H, Dec. 1987.

Finite Element Formulation for Guided-Wave Problems Using Transverse Electric Field Component

K. HAYATA, M. EGUCHI, and M. KOSHIBA, SENIOR MEMBER, IEEE

Abstract —**A finite-element formulation for electromagnetic waveguide problems is described using the transverse electric field component. In this approach, the divergence relation $\nabla \cdot \boldsymbol{D} = 0$ is satisfied and spurious solutions can be eliminated in the entire region of a propagation diagram. The validity of the formulation is examined via applications to a few canonical guided-wave problems.**

I. Introduction

The most serious difficulty in applying the finite element method to waveguide problems has been the appearance of so-called spurious, nonphysical solutions. To overcome this difficulty, various approaches have recently been developed; these are reviewed in [1]. More recently, a new finite element formulation for the analysis of dielectric waveguide modes has been developed by the authors in terms of the transverse magnetic field component ($\boldsymbol{H}_t$) [2]. The key point of this method, which is distinctly different from other transverse field methods [3]–[7], is that it transforms the finite element equation in terms of a full vector $\boldsymbol{H}$ field [8], [9] into one in terms of only the transverse magnetic field component, using the condition $\nabla \cdot \boldsymbol{H} = 0$. In this approach, the spurious solutions can be completely eliminated in the entire region of a propagation diagram, and the final matrix dimension is reduced to two thirds that of the conventional three-component approach using the penalty function method [10]–[14]. However, in the finite element analysis based on this approach, the magnetic field components are first obtained as an eigenvector, and the electric field components are later derived from them via Maxwell's equations. This additional operation based on spatial differentiations of the original data may cause an unnatural field profile when one uses lower order Lagrange elements.

In this paper, as an electric field version of the method described in [2], a finite element method for electromagnetic waveguide problems is formulated using the transverse electric field component ($\boldsymbol{E}_t$). In this approach, the spurious solutions can be eliminated in the entire region of a propagation diagram, and the dimension of the final matrix equation is reduced to two thirds that of the full vector $\boldsymbol{E}$ field approach based on a penalty function [15], [16]. The validity of the present method is confirmed via applications to a few representative waveguiding problems.

II. Formulation

We consider a waveguide with a tensor permeability and a scalar permittivity. With a time dependence of the form $\exp(j\omega t)$ being implied, from Maxwell's equations the following wave equation is derived:

$$\nabla \times \left([\mu]^{-1} \nabla \times \boldsymbol{E} \right) - k_0^2 \epsilon \boldsymbol{E} = 0 \tag{1}$$

where ω is the angular frequency, k_0 is the free-space wavenumber, $[\mu]$ is the relative permeability tensor, and ϵ is the relative permittivity, which is assumed to be constant in each material.

The divergence relation for source-free media, $\nabla \cdot \boldsymbol{D} = 0$, can be written

$$\epsilon E_z = (j\beta)^{-1} \left(\epsilon \partial E_x / \partial x + \epsilon \partial E_y / \partial y \right) \tag{2}$$

where β is the phase constant in the propagation direction (z direction).

Application of the standard finite element technique [2] to (1) and (2) gives the following matrix equations:

$$[S]\{E\} - (k_0/\beta)^2 [T]\{E\} = \{0\} \tag{3}$$

$$[D_z]\{E_z\} = [D_t]\{E_t\} \tag{4}$$

where

$$[S] = \sum_e \iint_e [B]^* [\mu]_e^{-1} [B]^T \, d\bar{x}\, d\bar{y} \tag{5}$$

$$[T] = \sum_e \iint_e \epsilon_e [N]^* [N]^T \, d\bar{x}\, d\bar{y} \tag{6}$$

$$[D_z] = \sum_e \iint_e \epsilon_e \{N\}\{N\}^T \, d\bar{x}\, d\bar{y} \tag{7}$$

$$[D_t] = -\sum_e \iint_e \left[\epsilon_e \{N\}\{N\}_x^T \quad \epsilon_e \{N\}\{N\}_y^T \right] d\bar{x}\, d\bar{y} \tag{8}$$

$$\{E_t\} = \begin{bmatrix} \{E_x\} \\ \{E_y\} \end{bmatrix}. \tag{9}$$

Here, $\{N\}$ is the shape function vector; $\{0\}$ is a null vector; T, $\{\cdot\}$, and $\{\cdot\}^T$ denote a transpose, a column vector, and a row vector, respectively; the components of vectors $\{E_x\}$, $\{E_y\}$, and $\{E_z\}$ are the values of E_x, E_y, and E_z at nodal points in the cross section, respectively; * denotes complex conjugate; $\bar{x} = \beta x$ and $\bar{y} = \beta y$; and $[N]$ and $[B]$ are given in [14].

Using (4), we can express the nodal electric field vector $\{E\}$ in terms of $\{E_t\}$:

$$\{E\} = [D]\{E_t\} \tag{10}$$

where

$$[D] = \begin{bmatrix} [U] \\ [D_z]^{-1}[D_t] \end{bmatrix}. \tag{11}$$

Here $[U]$ is a unit matrix.

Substituting (10) into (3) and multiplying (3) by $[D]^T$ from the left, we obtain the following final matrix equation with respect to the transverse electric field component $\{E_t\}$:

$$[\tilde{S}_{tt}]\{E_t\} - (k_0/\beta)^2 [\tilde{T}_{tt}]\{E_t\} = \{0\} \tag{12}$$

Reprinted from *IEEE Trans. Microwave Theory Tech.*, vol. 37, no. 1, pp. 256–258, Jan. 1989.

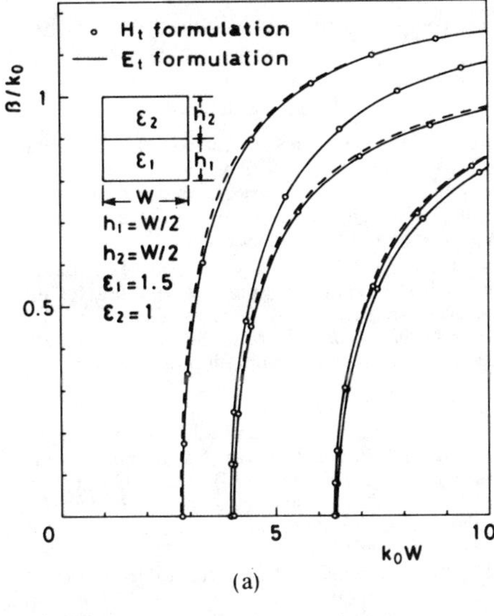

(a)

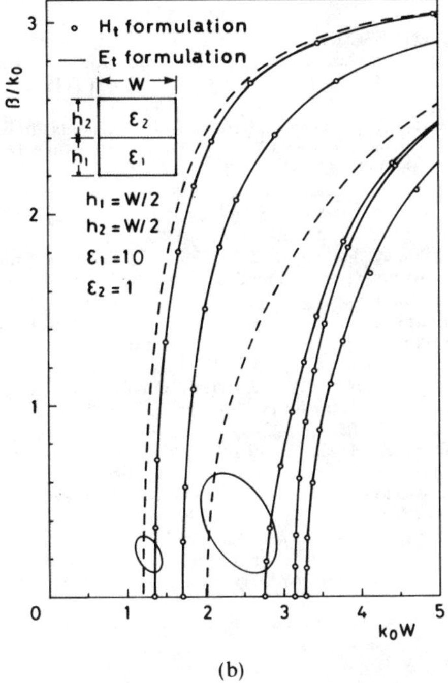

(b)

Fig. 1. Dispersion characteristics of half-filled dielectric waveguides. (a) $\epsilon_1 = 1.5$. (b) $\epsilon_1 = 10$.

where

$$[\tilde{S}_{tt}] = [D]^T [S][D] \tag{13}$$

$$[\tilde{T}_{tt}] = [D]^T [T][D]. \tag{14}$$

In (12), the nodal electric field vector should be forced to satisfy the boundary conditions at the interface between two media with different permittivities [15], [16]. We consider an interface Γ with an abrupt discontinuity in the permittivity as shown in [15, fig. 1]. The tangential components of $\boldsymbol{E}$ and the normal component of $\epsilon \boldsymbol{E}$ should be continuous at the interface Γ. These boundary conditions can be written as

$$\{E_x\}_2 = q_{xx}\{E_x\}_1 + q_{xy}\{E_y\}_1 \tag{15}$$

$$\{E_y\}_2 = q_{xy}\{E_x\}_1 + q_{yy}\{E_y\}_1 \tag{16}$$

$$\{E_z\}_2 = \{E_z\}_1 \tag{17}$$

where the components of $\{E_i\}_j$ are the values of E_i at the nodal points on Γ included in the element e_j. The quantities q_{xx}, q_{xy},

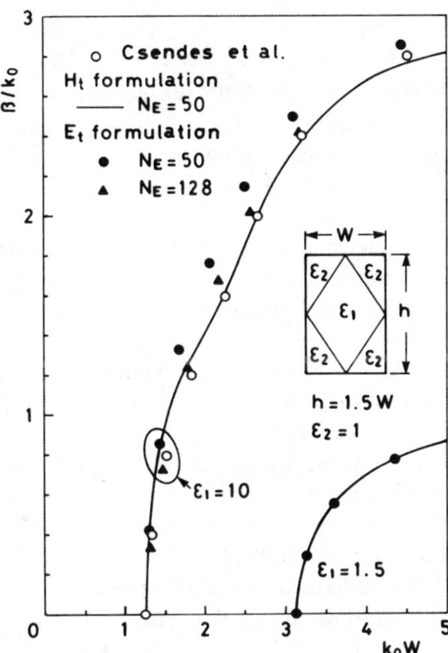

Fig. 2. Dispersion characteristics of rectangular waveguide with diamond-shaped insert.

and q_{yy} are given by

$$q_{xx} = \sin^2\theta + (\epsilon_1/\epsilon_2)\cos^2\theta \tag{18}$$

$$q_{xy} = (\epsilon_1/\epsilon_2 - 1)\sin\theta\cos\theta \tag{19}$$

$$q_{yy} = \cos^2\theta + (\epsilon_1/\epsilon_2)\sin^2\theta \tag{20}$$

where ϵ_1 and ϵ_2 are the relative permittivities of regions 1 and 2, respectively, and θ is the angle between the unit vector normal to Γ and the x axis.

By using the original functional for e_1 and the modified functional for e_2, which can be obtained by considering (15)–(17) in the original one, the boundary conditions of the electric field $\boldsymbol{E}$ at the interface with an abrupt discontinuity in the permittivity are satisfied [15].

Equation (12) is an ordinary matrix eigenvalue problem whose eigenvalue and eigenvector are $(k_0/\beta)^2$ and $\{E_t\}$, respectively. It should be noted that the divergence condition $\nabla \cdot \boldsymbol{D} = 0$ is implicitly included in (12) and the matrix dimension is reduced to two thirds that of the penalty function method [15], [16]. The electric field can be obtained first as an eigenvector in (12), and the tensor permeability is considered. (Note that a constant scalar permeability is assumed in [2].)

III. Numerical Examples

In this section, we present computed results obtained by the present formulation. In numerical computations, double precision is used to avoid roundoff errors and Householder's method is used as an eigenvalue solution method. Since the matrices (13) and (14) are dense, unlike those generated by the penalty approach, there is no possibility of taking advantage of sparsity with a sophisticated eigenvalue solver. An arbitrary small value of the input datum, e.g., $\beta W = 10^{-3}$, 10^{-2}, virtually gives a cutoff value of $k_0 W$.

Fig. 1(a) and (b) shows the dispersion characteristics for the first five modes of half-filled dielectric waveguides, where the plane of symmetry is assumed to be a perfect magnetic conductor and one half of the cross section is divided into second-order triangular elements (number of elements $(N_E) = 36$, number of nodal points $(N_p) = 91$). In Fig. 1, the broken lines correspond to the case in which no additional boundary conditions are explicitly imposed at the interface between media with different permittivities. It is found from Fig. 1 that the agreement is very good between the results obtained by the present formulation (solid lines) and those obtained by the formulation using the transverse magnetic field component (hollow circles) [2]. The spurious solutions do not appear in the entire region of a propagation diagram.

As a more complicated waveguiding configuration, we next consider the rectangular waveguide with a diamond-shaped insert studied by Csendes and Silvester [17], as illustrated in Fig. 2. In this waveguide, there are abrupt changes in the permittivity at the interface, the normal direction of which does not coincide with the direction of a coordinate axis. Fig. 2 shows the dispersion characteristics for the fundamental mode, where two planes of symmetry are assumed to be perfect magnetic conductors and one quarter of the cross section is divided into second-order triangular elements. In Fig. 2, the results of the $\boldsymbol{H}_t$ field formulation [2] with $N_E = 50$ and $N_p = 121$ and those of the modal approximation technique [17] are also presented. For $\epsilon_1 = 1.5$, the results of the present $\boldsymbol{E}_t$ field formulation with $N_E = 50$ and $N_p = 121$ agree well with those of the $\boldsymbol{H}_t$ counterpart. On the other hand, for a larger value of relative permittivity, $\epsilon_1 = 10$, the results of the $\boldsymbol{E}_t$ field formulation with $N_E = 50$ deviate from those of the $\boldsymbol{H}_t$ counterpart at higher frequencies. This deviation at higher frequencies reflects the singularity of the normal electric field component at the tips of wedges of the dielectric insert [18]–[20]. Such a singularity near the tip of a dielectric corner may cause electrical breakdown in high-power applications. It is evident from Fig. 2 that the $\boldsymbol{E}_t$ field finite element solutions can be improved by increasing the number of elements. Indeed, the results of the $\boldsymbol{E}_t$ field formulation with $N_E = 128$ and $N_p = 289$ are closer to those of the $\boldsymbol{H}_t$ field counterpart. No spurious solutions are involved in this case as well.

IV. Conclusions

We have formulated a vectorial finite element scheme for solving guided-wave problems using the transverse electric field component. Considering the duality between electric and magnetic field vectors in Maxwell's equations, we have used the same procedure as the transverse magnetic field counterpart except that additional conditions are enforced on the boundary between different dielectric materials. In this approach, the electric field components can be directly obtained as an eigenvector of a matrix eigenvalue problem. Furthermore, no spurious solutions are involved in the entire region of a propagation diagram, and the dimension of the final matrix equation is reduced to two thirds that of the penalty function method. We have confirmed the validity of the present formulation via applications to some canonical waveguide problems.

The approach described in this paper is also applicable to waveguides containing anisotropic media such as ferrites because the tensor permeability may vary from material to material. Furthermore, extension to waveguides containing lossy and/or active media is straightforward if one uses the procedure that has recently been proposed by the authors [21] for the magnetic field case.

References

[1] M. Koshiba, K. Hayata, and M. Suzuki, "Finite-element method analysis of microwave and optical waveguides—Trends in countermeasures to spurious solutions," *Trans. Inst. Electron. Commun. Eng. Japan*, vol. J69-C, pp. 1477–1486, Dec. 1986 (in Japanese); to be published in English in *Electron. Commun. Japan*.

[2] K. Hayata, M. Koshiba, M. Eguchi, and M. Suzuki, "Vectorial finite-element method without any spurious solutions for dielectric waveguiding problems using transverse magnetic-field component," *IEEE Trans. Microwave Theory Tech.*, vol. MTT-34, pp. 1120–1124, Nov. 1986.

[3] D. Gelder, "Numerical determination of microstrip properties using the transverse field components," *Proc. Inst. Elec. Eng.*, vol. 117, pp. 699–703, Apr. 1970.

[4] C. G. Williams and G. K. Cambrell, "Numerical solution of surface waveguide modes using transverse field components," *IEEE Trans. Microwave Theory Tech.*, vol. MTT-22, pp. 329–330, Mar. 1974.

[5] K. Bierwirth, N. Schulz, and F. Arndt, "Finite-difference analysis of rectangular dielectric waveguide structures," *IEEE Trans. Microwave Theory Tech.*, vol. MTT-34, pp. 1104–1114, Nov. 1986.

[6] C.-C. Su, "A combined method for dielectric waveguides using the finite-element technique and the surface integral equations method," *IEEE Trans. Microwave Theory Tech.*, vol. MTT-34, pp. 1140–1146, Nov. 1986.

[7] T. Angkaew, M. Matsuhara, and N. Kumagai, "Finite-element analysis of waveguide modes: A novel approach that eliminates spurious modes," *IEEE Trans. Microwave Theory Tech.*, vol. MTT-35, pp. 117–123, Feb. 1987.

[8] A. Konrad, "High-order triangular finite elements for electromagnetic waves in anisotropic media," *IEEE Trans. Microwave Theory Tech.*, vol. MTT-25, pp. 353–360, May 1977.

[9] B. M. A. Rahman and J. B. Davies, "Finite-element analysis of optical and microwave waveguide problems," *IEEE Trans. Microwave Theory Tech.*, vol. MTT-32, pp. 20–28, Jan. 1984.

[10] M. Koshiba, K. Hayata, and M. Suzuki, "Vectorial finite-element formulation without spurious modes for dielectric waveguides," *Trans. Inst. Electron. Commun. Eng. Japan*, vol. E67, pp. 191–196, Apr. 1984.

[11] M. Koshiba, K. Hayata, and M. Suzuki, "Vectorial finite-element method without spurious solutions for dielectric waveguide problems," *Electron. Lett.*, vol. 20, pp. 409–410, May 1984.

[12] B. M. A. Rahman and J. B. Davies, "Penalty function improvement of waveguide solution by finite elements," *IEEE Trans. Microwave Theory Tech.*, vol. MTT-32, pp. 922–928, Aug. 1984.

[13] M. Koshiba, K. Hayata, and M. Suzuki, "Study of spurious solutions of finite-element method in the three-component magnetic-field formulation for dielectric waveguide problems," *Electron. Commun. Japan*, pt. 1, vol. 68, pp. 114–119, Aug. 1985.

[14] M. Koshiba, K. Hayata, and M. Suzuki, "Improved finite-element formulation in terms of the magnetic-field vector for dielectric waveguides," *IEEE Trans. Microwave Theory Tech.*, vol. MTT-33, pp. 227–233, Mar. 1985.

[15] M. Koshiba, K. Hayata, and M. Suzuki, "Finite-element formulation in terms of the electric-field vector for electromagnetic waveguide problems," *IEEE Trans. Microwave Theory Tech.*, vol. MTT-33, pp. 900–906, Oct. 1985.

[16] M. Koshiba, K. Hayata, and M. Suzuki, "Vector *E*-field finite-element analysis of dielectric optical waveguides," *Appl. Opt.*, vol. 25, pp. 10–11, Jan. 1986.

[17] Z. J. Csendes and P. Silvester, "Numerical solution of dielectric loaded waveguides: II—Modal approximation technique," *IEEE Trans. Microwave Theory Tech.*, vol. MTT-19, pp. 504–509, June 1971.

[18] J. Meixner, "The behavior of electromagnetic fields at edges," *IEEE Trans. Antennas Propagat.*, vol. AP-20, pp. 442–446, July 1972.

[19] J. B. Andersen and V. V. Solodukhov, "Field behavior near a dielectric wedge," *IEEE Trans. Antennas Propagat.*, vol. AP-26, pp. 598–602, July 1978.

[20] J. V. Bladel, "Field singularities at metal-dielectric wedges," *IEEE Trans. Antennas Propagat.*, vol. AP-33, pp. 450–455, Apr. 1985.

[21] K. Hayata, K. Miura, and M. Koshiba, "Finite-element formulation for lossy waveguides," *IEEE Trans. Microwave Theory Tech.*, vol. 36, pp. 268–276, Feb. 1988.

Review of Finite Element Methods for Microwave and Optical Waveguides

B. M. AZIZUR RAHMAN, MEMBER, IEEE, F. ANIBAL FERNANDEZ, MEMBER, IEEE, AND J. BRIAN DAVIES, MEMBER, IEEE

This paper reviews the application of the finite element method to analysis of microwave and optical waveguide problems. It also discusses the significance of different variational formulations, the modeling of the infinite domain of open-boundary waveguides, techniques to avoid spurious solutions, and matrix solution techniques. We will also briefly refer to the application of these techniques to waveguides containing nonlinear materials and to three dimensional problems.

I. Introduction

The optimization of the performance of microwave and optical waveguide devices requires the knowledge of the propagation characteristics and field distributions and their dependence on the fabrication parameters. Therefore, there is a great interest in theoretical methods of waveguide analysis. It is appropriate to refer to two early review papers, by Davies [1] and Ng [2] on microwave waveguides. More recent review papers by Saad [3] and Sorrentino [4] discuss some of the techniques for microwave and optical waveguides.

This paper discusses two-dimensional finite element techniques applied to microwave and optical waveguide problems. The waveguide is assumed to be uniform along its longitudinal z axis. The electromagnetic fields at the frequency ω have the form:

$$\begin{aligned}\mathcal{H}(x,y,z,t) &= \boldsymbol{H}(x,y)\exp j(\omega t-\beta z)\\ \mathcal{E}(x,y,z,t) &= \boldsymbol{E}(x,y)\exp j(\omega t-\beta z)\end{aligned} \qquad (1)$$

where β is the propagation constant in the positive z-direction. The general geometry of the guide can be quite complicated, with an arbitrary permittivity profile $\varepsilon(x,y)$ in the transverse directions.

As the range of guiding structures becomes more intricate, so the need for computer analysis becomes greater and more demanding. The finite element method has become a powerful tool throughout engineering for its flexibility and versatility, being used in complicated structural, thermal, fluid flow, semiconductor, and electromagnetic problems. This method is particularly advantageous for electromagnetic field problems, because of its applicability to waveguides with arbitrary shape, arbitrary refractive index profile, and even anisotropy.

II. Finite Element Formulation

Finite element formulations are usually established via a variational or a Galerkin (method of moments [5] or weighted residuals) approach. The latter is more flexible but when it is possible, it is advantageous to take a variational approach, especially when one global parameter (like propagation constant) is needed. We will only refer to this form of derivation.

A. Scalar Approximation

Several different variational formulations have been proposed for use with the finite element method. The simplest is a scalar one in terms of the longitudinal components of TE or TM modes. This can be used in situations where the fields can be described as predominantly TE or TM. It has been used for solving homogeneous waveguide problems [6]–[9], for approximate analysis of lossy guides [10], [11], for open-boundary problems [12], [13], and for analysis of anisotropic waveguides [11], [14]–[16]. This formulation can be written as [3], [14], [16]:

$$F(\phi) = \int_S |\nabla_t \phi|^2\, dS - \int_S (k^2-\beta^2)\phi^2\, dS \qquad (2)$$

where, ϕ is either H_z or E_z.

Reprinted from *Proc. IEEE*, vol. 79, no. 10, pp. 1442–1448, Oct. 1991.

In the finite element approximation ([17]–[20]; see footnote[1]), the primary dependent variables are replaced by a system of discretized variables over the domain of consideration. One first divides the entire waveguide cross section into a patchwork of subregions or elements, usually triangles or quadrilaterals. Triangles are commonly used because they are easy to adapt to complex shapes. Using many elements, any cross section with a complex boundary and with arbitrary permittivity distribution can be accurately approximated. The simplest triangular element assumes a linear interpolation between the field values at the vertices of the triangle. Higher order interpolation polynomials are also used, with a larger number of nodal values (unknowns of the problem) in each element. By expressing the fields in terms of nodal values and with the assumed shape functions, the resulting field components are continuous over the whole domain.

To find these nodal fields, the usual (Rayleigh-Ritz) procedure is to force stationarity of the functional with respect to each nodal variable. This yields a matrix eigenvalue equation, with vector $\boldsymbol{x}$ of nodal variables:

$$\boldsymbol{A}\boldsymbol{x} - \lambda \boldsymbol{B}\boldsymbol{x} = 0 \tag{3}$$

where $\boldsymbol{A}$ and $\boldsymbol{B}$ are real symmetric sparse matrices, and $\boldsymbol{B}$ is also positive definite. The eigenvalue λ may be chosen as k^2 or β^2 from (2). We emphasize that it is most desirable for the resulting matrix equation to be of this canonical form, to allow efficient and robust solution. This equation can be solved by one of various standard subroutines to get the different modal eigenvalues and fields.

B. Vector Formulations

The single scalar formulation is inadequate (except as an approximation) for the inherently hybrid mode situation of anisotropic or genuinely two-dimensional inhomogeneous cross section waveguide problems. To represent accurately general waveguide fields, a vector formulation is necessary with at least two field components. However, as we will see later, some vector formulations are affected by spurious or nonphysical solutions which appear mixed with the correct ones in the computations.

Finite element methods in terms of the axial electric and magnetic field components, as the two-parameter basis for a variational formulation, have been used for analyzing various microwave [21]–[27] and optical waveguides [28]–[33]. They have also been applied to anisotropic waveguides with diagonal permittivity tensor [34], [16], and to lossy waveguides [24]–[26], [35].

This $E_z - H_z$ formulation cannot treat general anisotropic problems without destroying the canonical form of (3). Also, for a waveguide with arbitrary dielectric distribution, enforcing boundary conditions in this method can be quite difficult. Another fundamental disadvantage for optical waveguide problems is that it is based on axial components which are the least important of the six components of the vector fields. Additionally, this formulation is also affected by spurious solutions, and techniques to reduce them [16] do so at the expense of greatly increasing the computing cost.

In 1956, Berk [36] presented a number of vector variational formulations in the form of Rayleigh quotients for loss-free anisotropic waveguides and resonators in terms of the $\boldsymbol{H}$ field, the $\boldsymbol{E}$ field or a combination of both. Later, Morishita and Kumagai [37] and Chen and Lien [38] established general procedures to derive variational formulations for self-adjoint and nonself-adjoint operators.

A vector $\boldsymbol{E}$ formulation has been applied to analyze cylindrical waveguides [39], microstrip [40], optical fibers [41], magnetically anisotropic waveguides [42], and three-dimensional cavities [43]. The $\boldsymbol{H}$-field formulation has been extensively used for the solution of a variety of optical and microwave waveguides [44]–[55]. These formulations attracted attention because they can analyze general anisotropic problems. The $\boldsymbol{H}$-field formulation is more suitable to dielectric waveguide problems where the magnetic field is continuous everywhere; furthermore, as the natural boundary condition [17]–[20] corresponds to that of electric walls, one need not force the trial fields at conducting boundaries. This formulation can be written as [36], [45]:

$$\omega^2 = \frac{\int (\nabla \times \boldsymbol{H})^* \cdot \bar{\bar{\varepsilon}}^{-1} \cdot (\nabla \times \boldsymbol{H})\, d\Omega}{\int \boldsymbol{H}^* \cdot \bar{\bar{\mu}}^{-1} \cdot \boldsymbol{H}\, d\Omega}. \tag{4}$$

$\bar{\bar{\varepsilon}}$ and $\bar{\bar{\mu}}$ are respectively the permittivity and permeability (which may be of arbitrary anisotropy) of the loss-free medium. Application of the Rayleigh-Ritz procedure to (4) leads to a similar eigenvalue problem as in (3), where now $\boldsymbol{A}$ is a complex Hermitian matrix (which can be reduced to real symmetric), $\boldsymbol{B}$ is real symmetric and positive definite, and the eigenvectors $\boldsymbol{x}$ represent the unknown field components at the nodal points for different modes with ω^2 as their corresponding eigenvalues. Unfortunately, spurious solutions also appear in this formulation (as in the $\boldsymbol{E}$-field formulation), although we will see later that these nonphysical modes can be avoided.

Variational formulations in terms of the transverse $\boldsymbol{H}$-field or $\boldsymbol{E}$-field components have also been recently considered, including an implicit satisfaction of the divergence-free condition [56]–[61]. These formulations use the minimum number of variables required and can completely eliminate spurious solutions but at the expense of the complexity of the matrix problem or the sparsity of the matrices [57]. In this case, the matrix order can be reduced to two thirds of what is needed in the full vector formulation but more work would be needed in the future, both in their development and in matrix techniques before this formulation becomes fully usable.

Othaka *et al.* [62] used a variational form and Williams and Cambrell [63] used the method of moments with the

[1] These references, [17]–[20] are ones that we suggest for tutorial, general treatment of the finite element method. References [17] and [20] are classic, comprehensive accounts of the method, though [17] unfortunately gives all its applications to structural mechanics. References [18] and [19] are written for application to electromagnetics, and include computer programs. Reference [18] is the more comprehensive.

transverse component of $\boldsymbol{H}$. Compared with the full vector $\boldsymbol{H}$, this form can involve additional differentiation which is particularly disadvantageous with a finite element approach. Although there is recently a trend towards the employment of vector formulations based on only two components (of $\boldsymbol{E}$ or $\boldsymbol{H}$) [59], [61], a full vector form still has many advantages. It is easiest to implement, most versatile with respect to tensor $\bar{\bar{\varepsilon}}$, and leads to standard matrix form for which robust and efficient solvers exist.

The full 6-component $\boldsymbol{E}$ and $\boldsymbol{H}$ formulation [64], [65] does not appear to have much advantage over the $\boldsymbol{H}$-field or $\boldsymbol{E}$-field formulation, since in this case the degrees of freedom per node double. Apparent advantages reside in the simpler modeling of the basic equations, which allows elimination of spurious modes and can be important when considering problems with special material properties.

III. Elaborating the Finite Element Method

A. Spurious Modes

Perhaps the most serious difficulty in using some vector formulations is the appearance of extraneous nonphysical or spurious modes [16], [22], [29], [32], [44], [45], [66]–[68]. In the conventional vector finite element formulations such as (4), the associated Euler equation is consistent with the two Maxwell curl equations but does not imply $\nabla \cdot \boldsymbol{B} = 0$. This causes the system to be "under-determined" or excessively flexible, which in turn is believed [44], [46] to be responsible for the spurious modes. A useful review of the problem appeared four years ago [69].

Computing a set of eigenmodes, it is difficult and quite cumbersome to distinguish between spurious and physical modes. Spurious modes can sometimes be identified by observing their dispersion curve [29], [32], [66]. Another simplistic way to identify them is to inspect their eigenvectors [22], [16], [44], [45], [68], since the nonphysical field distributions usually vary in an unreasonable, sometimes random fashion over the guide cross section. For a full vector formulation where the divergence-free condition is neither implied nor forced, it has been found that spurious modes occur and that they are distinctive in giving particularly high values of $\nabla \cdot \boldsymbol{B}$ [44], [45]. Threshold criteria have been successfully used [44], [45] that automatically filter out most of the spurious solutions from a set of calculated eigenvectors. Following this, one would expect that the spurious solutions can be eliminated if the condition $\nabla \cdot \boldsymbol{H} = 0$ is somehow forced. To that end, several approaches have been used. The penalty function method is a useful way of imposing certain constraints on solution variables. As in the standard penalty function method [17], [20], one adds to the original full vector $\boldsymbol{H}$-field formulation (4) an integral so that the resulting Euler equation is the Helmholtz equation plus the vanishing of $\nabla \cdot \boldsymbol{B}$ [46].

One advantage of the penalty method is that it does not increase the matrix order or the storage and the additional computational time is negligible. It is found that this removes many of the spurious solutions from the spectrum and also improves the quality of the field eigenfunctions [46], [49]. This method has been applied extensively to various waveguiding problems in recent years [42], [46], [47], [49]–[53], [55], [70]. A similar method has also been used to eliminate spurious modes from acoustic waveguides [71].

Another scheme for removing spurious modes has been proposed by Kobelansky and Webb [54], who suggest generating divergence free trial functions by satisfying the condition $\nabla \cdot \boldsymbol{H} = 0$ through a variational expression, but at the expense of the sparsity of the matrix. In an alternative approach, Angkaew *et al.* [72] have arranged a formulation where real eigenvalues are the distinctive physical solutions of a complex eigenvalue problem. One might expect difficulty in choosing the threshold figure to distinguish "real" and "complex" eigenvalues. Again, the scheme suffers considerable expense of the increasing computing effort, as four complex unknowns per node are needed.

A recent scheme for the elimination of spurious modes involves so-called "edge elements" [73] and their generalization "tangential elements" [74]. In this approach, the tangential field components between elements are forced to be continuous.

Another promising approach [59], [61] consists in using a variational formulation in terms of only the transverse components of $\boldsymbol{H}$, by implicitly forcing the divergence-free condition. This eliminates spurious modes, has the scope to include significant loss (see Section III-G), and works (to advantage) for given ω rather than given β.

B. Open-Boundary Problems

Practical electromagnetic problems often involve finding fields in an infinitely extended region. In optical waveguides, the region of interest extends to infinity outside some guide "core," where the field decay is roughly exponential. This is particularly important close to cutoff, as in this situation fields decay slowly and the region of significant field value can be arbitrarily large. The crudest approach is the simple truncation at a certain distance [16], [28], which sets artificial boundary walls enclosing the guide. But this approach either introduces a significant error when the boundary is too close, or needs to consider an excessively large domain. One adaptive technique involves shifting the virtual boundary wall recursively [32] to satisfy a criterion for maximum field strength at that wall.

Open-boundary problems may be solved more accurately by several techniques, all having in common the idea that the open infinitely extending region is divided into interior and exterior regions [33], [75]–[77]. It is possible to find the internal and external solutions and to match them on an imaginary boundary, some choice of integral solution being possible for the outside region [56], [78]. But it is quite complicated to consider an integral equation method for this unbounded region without losing the canonical form of the matrix problem. Another approach is to use a recursion technique [12], [79] to represent the region outside the main domain. Wu and Chen [13], [80] used conformal mapping to condense the exterior region, but this technique can only be applied to open guides having a plane of symmetry. Yeh *et al.* [31] used sectorial infinite elements with radial exponential decay outside the guide core, but because of nonconformity between the two coordinate systems, the fields are not continuous over the problem domain.

Infinite elements have also been used with a Cartesian coordinate system using suitable exponentially decaying functions [45], [47], [55], [81], [82]. These infinite elements extend the domain of explicit field representation to infinity without increasing the matrix order, so that computational time is virtually unchanged. The shape functions for such an element should be realistic to represent the fields and should be square integrable over the infinite area to satisfy the radiation condition. Integration of these shape functions—or their derivatives, as necessary for (2) or (4)—are finite over the infinite elements and can be simply carried out as in conventional elements. Adding these infinite elements along the outer boundary of orthodox finite elements, any open-type optical waveguide cross sectional domain can be represented very conveniently.

C. Matrix Solution Techniques

Solving waveguide problems by finite elements, the key factor affecting storage requirements and computational effort is the choice of algorithm to solve the matrix equation. In (3), the matrices $\boldsymbol{A}$ and $\boldsymbol{B}$ are highly sparse, the sparsity increasing with the order of the matrices and decreasing with the polynomial order of the shape functions [18]. The advantage of higher order basis functions for the fields is that it gives a more accurate solution for a given matrix order, but it involves an increased programming effort, particularly when considering anisotropic materials, infinite elements, and penalty functions. Another disadvantage of higher order polynomials (again for given matrix order) is that it increases the density of the matrix. It must be emphasized here that the trade-off and the optimum choice between low- and high-order polynomials depends crucially on the matrix algorithm used. For instance, to take the extreme case of a fully dense matrix algorithm being used to solve (3), the optimum choice of polynomial order would be high—but of course this would be in the knowledge of the gross inefficiency of using a dense matrix algorithm. By contrast, at the other extreme one can use a matrix routine which considers only the nonzero elements of the upper or lower triangle of the symmetric real matrices. This has been used [44], [45] with a specially developed arbitrary sparse matrix algorithm which solves the eigenvalue problem by an iteration process applied simultaneously to a subspace of eigenvectors; the method of subspace iteration [83].

Accuracy of results can be considerably increased by careful use of extrapolation techniques [7], [32], [44], [50]. The technique is based on solving the physical problem with say two or three coarse meshes and extrapolating systematically from these results with considerable savings in computer time.

D. Special Elements

Ordinary first-order elements are widely used for their simplicity but cannot represent curved boundaries correctly. The waveguide with curved boundaries can be computed efficiently by shaping the element to fit the real geometry with curved-sided or isoparametric elements [8], [18], [52], [68]. In this case, the polynomial representing the geometry of an element is coincident with that interpolating fields in the element. Formulation of the element matrices follows the usual principle but involves more computation than with ordinary elements. When there are field singularities caused by edges in the cross section of a guide, a very fine mesh is required locally to obtain accurate results. Alternatively, for such regions special element functions may be used to represent unbounded local fields [81], [84], [85].

E. Three-Dimensional Problems

The finite element method has established itself as one of the most popular techniques for two-dimensional problems, but unfortunately in a general three-dimensional problem, the computer manipulation requires a great deal of memory and time. The penalty incurred may in fact be greater than that associated with changing from n^2 to n^3 variables. Despite this, the finite element method is very powerful for arbitrarily shaped three-dimensional regions and has been applied to model some three-dimensional problems [9], [86], [87].

Sometimes, useful subclasses of three dimensional problems are essentially two-dimensional and can be amenable to two-dimensional finite element analysis in conjunction with other techniques as in the case of the analysis of waveguide discontinuities.

F. Nonlinear Optical Waveguides

With the guides described in earlier sections, we sought modal solutions $\mathcal{E}(x, y, z, t) = \boldsymbol{E}(x, y) \exp j(\omega t - \beta z)$ consistent with a prescribed permittivity profile $\varepsilon(x, y)$. With nonlinear optical waveguides [88], the profile is now $\varepsilon(x, y, |\boldsymbol{E}|)$ where indeed the local permittivity is field dependent. Any resulting modal analysis must be used and interpreted with care; physically the modes can exist (as can solitons) but the absence of linearity or superposition means that "expansion in terms of modes" is no longer appropriate. Modes can even bifurcate with changing power level and can also be unstable.

Nonlinear optical waveguides can be analyzed by studying stationary states of the modes as an eigenvalue problem. Most theoretical studies over the 1980's used a "semianalytic" approach, where carefully chosen laws of nonlinearity (e.g., with $\boldsymbol{D} = (n_0 + n_2 S)^2 \boldsymbol{E}$ and S being the local energy density) allowed special closed-form solutions. However, in the last three years, finite elements have been used with great effect [88]–[93]. Just as with the earlier consideration of linear material but rather arbitrary profiles, the finite element method has a supreme advantage of versatility.

As the refractive index profile of the guide depends on the field-intensity profile, the finite element approach [88]–[93] has to use an iterative scheme to seek a consistent solution. In this way, a solution is approached via the satisfaction of alternately a) Maxwell's equations and b) the specified $n(\boldsymbol{E})$ law. This method appears to be much more powerful (because of its versatility) than other of the semianalytical methods.

G. Lossy Waveguides

The date, most of the applications of the finite element method has been focused on a loss-free system. For the case of low loss, a perturbation method can be applied to the loss-free finite element results [10], [26], [81]. For many cases this is sufficient as most of the practical dielectric materials are low-loss media.

For nonsmall loss, McAulay [35] used an $E_z - H_z$ formulation, where a complex eigenvalue problem is solved iteratively until a real eigenvalue (frequency) is obtained. A similar technique has been used for lossy microstrip [24], and MMIC structures [25]. Hayata *et al.* [10] analyzed lateral modes of laser diodes by using a scalar formulation.

A more satisfactory approach is a solution that, for given ω, yields complex γ using a formulation in terms of transverse magnetic field components. This allows one to consider significant losses (or gain) but the matrices will become complex [58]–[61].

IV. Concluding Remarks

Finite element methods have undoubtedly become a powerful tool for the analysis of electromagnetic problems. Several ways to formulate this method using variational approaches have been discussed.

The full vector ($\boldsymbol{E}$ or $\boldsymbol{H}$) formulation appears to be still the most convenient for practical large scale problems since, although it uses three rather than the minimum of two components to describe the fields, the resultant matrix problem can be solved using standard software library packages. This formulation is fundamentally more accurate than scalar forms since it can represent true hybrid modes. Its main drawback, the existence of spurious modes, can be partially cured using a penalty method.

One special and redeeming feature of a scalar approach is that it should not give rise to spurious modes. Although the scalar formulation is approximate, it takes significantly lower computational time, and hence is attractive for design procedures in CAD systems.

A desirable formulation that has yet to be found should be free of any spurious modes, should use a robust matrix algorithm, taking full advantage of sparsity, and should be defined in terms of only two field components, with facility of changing to one where a scalar approximation is adequate. It would also be convenient to be able to deal with anisotropic lossless or lossy materials. A formulation based only on the transverse field components satisfies most of the previous requirements but no efficient matrix solver is currently available, preventing its use in practical problems.

Recent advances in computer technology have brought a rapid change in the form of traditional finite element analysis. The advent of fast processing and computer graphics has allowed the incorporation of pre- and post-processing facilities to such an extent that nowadays it is hard to conceive a finite element program without some of these.

The enormous increase of available memory in present computers with fast processors, together with the use of parallel techniques now available even in modest workstations, has caused a trend towards even more massive computing, allowing solution of problems that only a few years ago were too big even for the largest machine. Undoubtedly this trend will mean that in the near future, more and more electromagnetic field problems will be solved routinely using these numerical techniques.

References

[1] J. B. Davies, "Review of methods for numerical solution of the hollow-waveguide problem," *Proc. IEE,* vol. 119, pp. 33–37, Jan. 1972.

[2] F. L. Ng, "Tabulation of methods for the numerical solution of the hollow waveguide problem," *IEEE Trans. Microwave Theory Tech.,* vol. MTT-22, pp. 322–329, Mar. 1974.

[3] S. M. Saad, "Review of numerical methods for the analysis of arbitrarily-shaped microwave and optical dielectric waveguides," *IEEE Trans. Microwave Theory Tech.,* vol. MTT-33, pp. 894–899, Oct. 1985.

[4] R. Sorrentino, "Numerical methods for passive components," in *1988 IEEE MTT-S Int. Microwave Symp. Dig.,* New York, May 1988, vol. 2, pp. 619–622.

[5] R. F. Harrington, *Field Computation by Moment Methods.* Malabar, Florida: R. E. Krieger, 1968.

[6] P. Silvester, "A general high-order finite-element waveguide analysis program," *IEEE Trans. Microwave Theory Tech.,* vol. MTT-17, pp. 204–210, Apr. 1969.

[7] P. Daly, "Polar geometry waveguides by finite-element methods," *IEEE Trans. Microwave Theory Tech.,* vol. MTT-22, pp. 202–209, Mar. 1974.

[8] P. Daly, "Finite element approach to propagation in elliptical and parabolic waveguides," *Int. J. Num. Meth. Eng.,* vol. 20, pp. 681–688, 1984.

[9] P. Silvester, "Tetrahedral polynomial finite elements for the Helmholtz equation," *Int. J. Num. Meth. Eng.,* vol. 4, pp. 405–413, 1972.

[10] K. Hayata, M. Koshiba and M. Suzuki, "Lateral mode analysis of buried heterostructure diode lasers by the finite-element method," *IEEE J. Quantum Electron.,* vol. QE-22, pp. 781–788, June 1986.

[11] M. Koshiba, K. Hayata and M. Suzuki, "Approximate scalar finite-element analysis of anisotropic optical waveguides with off-diagonal elements in a permittivity tensor," *IEEE Trans. Microwave Theory Tech.,* vol. MTT-32, pp. 587–593, June 1984.

[12] K. S. Chiang, "Finite-element analysis of optical fibres with iterative treatment of the infinite 2-D space," *Opt. Quantum Electron.,* vol. 17, pp. 381–391, 1985.

[13] R. B. Wu and C. H. Chen, "A scalar variational conformal mapping technique for weakly guiding dielectric waveguides," *IEEE J. Quantum Electron.,* vol. QE-22, pp. 603–609, May 1986.

[14] M. Koshiba, K. Hayata and M. Suzuki, "Approximate scalar finite-element analysis of anisotropic optical waveguides," *Electron. Lett.,* vol. 18, pp. 411–413, May 1982.

[15] R. V. Mustacich, "Scalar finite element analysis of electro-optic modulation in diffused channel waveguides and poled waveguides in polymer thin films," *Appl. Opt.,* vol. 27, pp. 3732–3737, Sept. 1988.

[16] N. Mabaya, P. E. Lagasse and P. Vandenbulcke, "Finite element analysis of optical waveguides," *IEEE Trans. Microwave Theory Tech.,* vol. MTT-29, pp. 600–605, June 1981.

[17] O. C. Zienkiewicz, *The Finite Element Method, 3rd. ed.* London, U.K.: McGraw-Hill, 1977.

[18] P. P. Silvester and R. L. Ferrari, *Finite Elements for Electrical Engineers, 2nd ed.* Cambridge, U.K.: Cambridge Press, 1990.

[19] J. B. Davies, "The finite element method," in *Numerical Techniques for Microwave and Millimeter-wave Passive Structures*, T. Itoh, Ed. New York: Wiley, 1989, pp. 33–132.
[20] J. N. Reddy, *An Introduction to the Finite Element Method.* New York: McGraw-Hill, 1984.
[21] S. Ahmed and P. Daly, "Finite-element methods for inhomogeneous waveguides," *Proc. IEE*, vol. 116, pp. 1661–1664, Oct. 1969.
[22] Z. J. Csendes and P. Silvester, "Numerical solution of dielectric loaded waveguides: I-Finite-element analysis," *IEEE Trans. Microwave Theory Tech.*, vol. MTT-18, pp. 1124–1131, Dec. 1970.
[23] P. Daly, "Hybrid-mode analysis of microstrip by finite-element methods," *IEEE Trans. Microwave Theory Tech.*, vol. MTT-19, pp. 19–25, Jan. 1971.
[24] M. Aubourg, J. P. Villotte, F. Godon, and Y. Garault, "Finite element analysis of lossy waveguides—Application to microstrip lines on semiconductor substrate," *IEEE Trans. Microwave Theory Tech.*, vol. MTT-31, pp. 326–331, Apr. 1983.
[25] C.-K. Tzuang and T. Itoh, "Finite-element analysis of slow-wave Schottky contact printed lines," *IEEE Trans. Microwave Theory Tech.*, vol. MTT-34, pp. 1483–1489, Dec. 1986.
[26] Eswarappa, G. I. Costache, and W. J. R. Hoefer, "Finlines in rectangular and circular waveguide housings including substrate mounting and bending effects—Finite element analysis," *IEEE Trans. Microwave Theory Tech.*, vol. MTT-37, pp. 299–306, Feb. 1989.
[27] A. A. P. Gibson and J. Helszajn, "Finite element solution of longitudinally magnetized elliptical gyromagnetic waveguides," *IEEE Trans. Microwave Theory Tech.*, vol. MTT-37, pp. 999–1005, June 1989.
[28] C. Yeh, S. B. Dong, and W. Oliver, "Arbitrarily shaped inhomogeneous optical fiber or integrated optical waveguides," *J. Appl. Phys.*, vol. 46, pp. 2125–2129, May 1975.
[29] T. S. Bird, "Propagation and radiation characteristics of rib waveguide," *Electron. Lett.*, vol. 13, pp. 401–403, July 1977.
[30] K. Okamoto and T. Okoshi, "Vectorial wave analysis of inhomogeneous optical fibres using finite element method," *IEEE Trans. Microwave Theory Tech.*, vol. MTT-26, pp. 109–114, 1978.
[31] C. Yeh, K. Ha, S. B. Dong and W. P. Brown, "Single-mode optical waveguides," *Appl. Opt.*, vol. 18, pp. 1490–1504, May 1979.
[32] M. Ikeuchi, H. Sawami and H. Niki, "Analysis of open-type dielectric waveguides by the finite-element iterative method," *IEEE Trans. Microwave Theory Tech.*, vol. MTT-29, pp. 234–239, Mar. 1981.
[33] K. Oyamada and T. Okoshi, "Two-dimensional finite-element calculation of propagation characteristics of axially nonsymmetrical optical fibres," *Radio Sci.*, vol. 17, pp. 109–116, Jan.-Feb. 1982.
[34] P. Vandenbulcke and P. E. Lagasse, "Eigenmode analysis of anisotropic optical fibres or integrated optical waveguides," *Electron. Lett.*, vol. 12, pp. 120–122, Mar. 1976.
[35] A. D. McAulay, "Variational finite-element solution for dissipative waveguides and transportation application," *IEEE Trans. Microwave Theory Tech.*, vol. MTT-25, pp. 382–392, May 1977.
[36] A. D. Berk, "Variational principles for electromagnetic resonators and waveguides," *IRE Trans. Antennas Propagat.*, vol. AP-4, pp. 104–111, Apr. 1956.
[37] K. Morishita and N. Kumagai, "Unified approach to the derivation of variational expression for electromagnetic fields," *IEEE Trans. Microwave Theory Tech.*, vol. MTT-25, pp. 34–40, Jan. 1977.
[38] C. H. Chen and C. Lien, "The variational principle for non-self-adjoint electromagnetic problems," *IEEE Trans. Microwave Theory Tech.*, vol. MTT-28, pp. 878–886, Aug. 1980.
[39] W. J. English and F. J. Young, "An E vector variational formulation of the Maxwell equations for cylindrical waveguide problems," *IEEE Trans. Microwave Theory Tech.*, vol. MTT-19, pp. 40–46, Jan. 1971.
[40] M. Hano, "Finite-element analysis of dielectric-loaded waveguides," *IEEE Trans. Microwave Theory Tech.*, vol. MTT-32, pp. 1275–1279, Oct. 1984.
[41] J. Katz, "Novel solution of 2-D waveguides using the finite element method," *Appl. Opt.*, vol. 21, pp. 2747–2750, Aug. 1982.
[42] M. Koshiba, K. Hayata and M. Suzuki, "Finite-element formulation in terms of the electric-field vector for electromagnetic waveguide problems," *IEEE Trans. Microwave Theory Tech.*, vol. MTT-33, pp. 900–905, Oct. 1985.
[43] M. Hara, T. Wada, T. Fukasawa and F. Kikuchi, "A three-dimensional analysis of RF electromagnetic fields by the finite element method," *IEEE Trans. Magnetics*, vol. MAG-19, pp. 2417–2420, Nov. 1983.
[44] J. B. Davies, F. A. Fernandez and G. Y. Philippou, "Finite element analysis of all modes in cavities with circular symmetry," *IEEE Trans. Microwave Theory Tech.*, vol. MTT-30, pp. 1975–1980, Nov. 1982.
[45] B. M. A. Rahman and J. B. Davies, "Finite-element analysis of optical and microwave waveguides problems," *IEEE Trans. Microwave Theory Tech.*, vol. MTT-32, pp. 20–28, Jan. 1984.
[46] B. M. A. Rahman and J. B. Davies, "Penalty function improvement of waveguide solution by finite elements," *IEEE Trans. Microwave Theory Tech.*, vol. MTT-32, pp. 922–928, Aug. 1984.
[47] B. M. A. Rahman and J. B. Davies, "Finite-element solution of integrated optical waveguides," *J. Lightwave Technol.*, vol. LT-2, pp. 682–688, Oct. 1984.
[48] A. Konrad, "Vector variational formulation of electromagnetic fields in anisotropic media," *IEEE Trans. Microwave Theory Tech.*, vol. MTT-24, pp. 553–559, Sept. 1976.
[49] M. Koshiba, K. Hayata, and M. Suzuki, "Improved finite-element formulation in terms of the magnetic field vector for dielectric waveguides," *IEEE Trans. Microwave Theory Tech.*, vol. MTT-33, pp. 227–233, Mar. 1985.
[50] B. M. A. Rahman and J. B. Davies, "Vector-H finite element solution of GaAs/GaAlAs rib waveguides," *IEE Proc. Pt. J, Optoelectron.*, vol. 132, pp. 349–353, Dec. 1985.
[51] M. Koshiba, K. Hayata, and M. Suzuki, "Finite-element solution of anisotropic waveguides with arbitrary tensor permittivity," *J. Lightwave Technol.*, vol. LT-4, pp. 121–126, Feb. 1986.
[52] K. Hayata, M. Eguchi, M. Koshiba and M. Suzuki, "Vectorial wave analysis of side-tunnel type polarization-maintaining optical fibers by variational finite elements," *J. Lightwave Technol.*, vol. LT-4, pp. 1090–1096, Aug. 1986.
[53] T. P. Young, "Design of integrated optical circuits using finite elements," *IEE Proc. Pt. A*, vol. 135, pp. 135–144, Mar. 1988.
[54] A. J. Kobelansky and J. P. Webb, "Eliminating spurious modes in finite-element waveguide problems by using divergence-free fields," *Electron. Lett.*, vol. 22, pp. 569–570, May 1986.
[55] M. J. McDougall and J. P. Webb, "Infinite elements for the analysis of open dielectric waveguides," *IEEE Trans. Microwave Theory Tech.*, vol. MTT-37, pp. 1724–1731, Nov. 1989.
[56] C.-C. Su, "A combined method for dielectric waveguides using the finite-element technique and the surface integral equations method," *IEEE Trans. Microwave Theory Tech.*, vol. MTT-34, pp. 1140–1146, Nov. 1986.
[57] K. Hayata, M. Koshiba, M. Eguchi, and M. Suzuki, "Vectorial finite-element method without any spurious solutions for dielectric waveguiding problems using transverse magnetic-field component," *IEEE Trans. Microwave Theory Tech.*, vol. MTT-34, pp. 1120–1124, Nov. 1986.
[58] K. Hayata, K. Miura, and M. Koshiba, "Finite-element formulation for lossy waveguides," *IEEE Trans. Microwave Theory Tech.*, vol. MTT-36, pp. 268–276, Feb. 1988.
[59] W. C. Chew and M. A. Nasir, "A variational analysis of anisotropic, inhomogeneous dielectric waveguides," *IEEE Trans. Microwave Theory Tech.*, vol. MTT-37, pp. 661–668, Apr. 1989.
[60] S. R. Cvetkovic, F. A. Fernandez, and J. B. Davies, "Finite element analysis of wave propagation in lossy inhomogeneous anisotropic dielectric media based on variational principles," in *1988 IEEE AP-S Int. Antenna Propagat. Symp. Dig.*, Syracuse, New York, June 1988, pp. 1130–1133.
[61] F. A. Fernandez and Y. Lu, "Variational finite element analysis of dielectric waveguides with no spurious solutions," *Electron. Lett.*, vol. 26, pp. 2125–2126, Dec. 1990.
[62] M. Ohtaka, M. Matsuhara, and N. Kamagai, "Analysis of the guided modes in slab-coupled waveguides using a variational method," *IEEE J. Quantum Electron.*, vol. QE-12, pp. 378–382, July 1976.
[63] C. G. Williams and G. K. Cambrell, "Numerical solution of surface waveguide modes using transverse field components," *IEEE Trans. Microwave Theory Tech.*, vol. MTT-22, pp. 329–330, Mar. 1974.
[64] W. J. English, "Vector variational solutions of inhomogeneously loaded cylindrical waveguide structures," *IEEE Trans. Microwave Theory Tech.*, vol. 19, pp. 9–18, Jan. 1971.
[65] J. A. M. Svedin, "A numerically efficient finite-element formulation for the general waveguide problem without spurious modes," *IEEE Trans. Microwave Theory Tech.*, vol. MTT-37, pp. 1708–1715, Nov. 1989.
[66] D. G. Corr and J. B. Davies, "Computer analysis of the fundamental and higher order modes in single and coupled microstrip," *IEEE Trans. Microwave Theory Tech.*, vol. MTT-20, pp. 669–678, Oct. 1972.

[67] C.-C. Su, "Origin of spurious modes in the analysis of optical fibre using the finite-element or finite-difference technique," *Electron. Lett.*, vol. 21, pp. 858–860, Sept. 1985.

[68] D. Welt and J. Webb, "Finite-element analysis of dielectric waveguides with curved boundaries," *IEEE Trans. Microwave Theory Tech.*, vol. MTT-33, pp. 576–585, July 1985.

[69] M. Koshiba, K. Hayata and M. Suzuki, "Finite-element method analysis of microwave and optical waveguides—Trends in countermeasures to spurious solutions," *Electron. and Comm. in Japan, Pt. 2*, vol. 70, pp. 96–108, 1987.

[70] M. S. Stern, "Semivectorial polarised H-field solutions for dielectric waveguides with arbitrary index profiles," *IEE Proc. Pt. J, Optoelectron.*, vol. 135, pp. 333–338, Oct. 1988.

[71] J. R. Winkler and J. B. Davies, "Elimination of spurious modes in finite element analysis," *J. Comp. Phys.*, vol. 56, Oct. 1984.

[72] T. Angkaew, M. Matsuhara, and N. Kumagai, "Finite-element analysis of waveguide modes: A novel approach that eliminates spurious modes," *IEEE Trans. Microwave Theory Tech.*, vol. MTT-35, pp. 117–123, Feb. 1987.

[73] A. Bossavit and I. Mayergoyz, "Edge-elements for scattering problems," *IEEE Trans. Magnetics*, vol. 25, pp. 2816–2821, July 1989.

[74] J. F. Lee, D. K. Sun, and Z. J. Cendes, "Tangential vector finite elements for electromagnetic field computation," presented at IEEE Conf. Electromagnetic Field Computation, paper DB-04, Toronto, Oct. 22–24, 1990.

[75] P. Silvester and M. S. Hsieh, "Finite-element solution of 2-dimensional exterior-field problems," *Proc. IEE*, vol. 118, pp. 1743–1747, Dec. 1971.

[76] Z. J. Csendes, "A note on the finite-element solution of exterior-field problems," *IEEE Trans. Microwave Theory Tech.*, vol. MTT-16, pp. 468–473, July 1976.

[77] C. R. I. Emson, "Methods for the solution of open-boundary electromagnetic field problems," *Proc. IEE Pt. A.*, vol. 135, pp. 151–158, 1988.

[78] B. H. McDonald and A. Wexler, "Finite-element solution of unbounded field problems," *IEEE Trans. Microwave Theory Tech.*, vol. MTT-20, pp. 841–847, Dec. 1972.

[79] P. Silvester, D. A. Lowther, C. J. Carpenter, and E. A. Wyatt, "Exterior finite elements for 2-dimensional field problems with open boundaries," *Proc. IEE*, vol. 124, pp. 1267–1270, Dec. 1977.

[80] R. B. Wu and C. H. Chen, "A variational analysis of dielectric waveguides by the conformal mapping technique," *IEEE Trans. Microwave Theory Tech.*, vol. MTT-33, pp. 681–685, Aug. 1985.

[81] Z. Pantic and R. Mittra, "Quasi-TEM analysis of Microwave transmission lines by the finite-element method," *IEEE Trans. Microwave Theory Tech.*, vol. MTT-34, pp. 1096–1103, Nov. 1986.

[82] K. Hayata, M. Eguchi and M. Koshiba, "Self-consistent finite/infinite element scheme for unbounded guided wave problems," *IEEE Trans. Microwave Theory Tech.*, vol. 36, pp. 614–616, Mar. 1988.

[83] K. J. Bathe and E. L. Wilson, *Numerical Methods in Finite Element Analysis.* Englewood Cliffs, NJ: Prentice Hall, 1976.

[84] Z. J. Csendes and J. F. Lee, "Transfinite element method for modeling MMIC devices," *IEEE Trans. Microwave Theory Tech.*, vol. MTT-36, pp. 1639–1649, Dec. 1988.

[85] J. P. Webb, "Finite element analysis of dispersion in waveguides with sharp metal edges," *IEEE Trans. Microwave Theory Tech.*, vol. MTT-36, pp. 1819–1824, Dec. 1988.

[86] R. L. Ferrari and G. L. Maile, "Three-dimensional finite-element method for solving electromagnetic problems," *Electron. Lett.*, vol. 14, pp. 467–468, July 1978.

[87] J. P. Webb, G. L. Maile and R. L. Ferrari, "Finite-element solution of three-dimensional electromagnetic problems," *IEE Proc. Pt. H*, vol. 130, pp. 153–159, Mar. 1983.

[88] B. M. A. Rahman and J. B. Davies, "Finite element solution of nonlinear bistable optical waveguides," *Int. J. Optoelectronics*, vol. 4, pp. 153–161, Mar. 1989.

[89] K. Hayata, M. Nagai and M. Koshiba, "Finite-element formalism for nonlinear slab-guided waves," *IEEE Trans. Microwave Theory Tech.*, vol. MTT-36, pp. 1207–1215, July 1988.

[90] K. Hayata and M. Koshiba, "Full vectorial analysis of nonlinear-optical waveguides," *J. Opt. Soc. Amer. B*, vol. 5, pp. 2494–2501, Dec. 1988.

[91] B. M. A. Rahman, J. R. Souza, and J. B. Davies, "Numerical analysis of nonlinear bistable optical waveguides," *IEEE Photonics Technol. Lett.*, vol. 2, Apr. 1990.

[92] B. M. A. Rahman, "Finite element solution of nonlinear optical waveguides," *Proc. SPIE*, vol. 1177, pp. 228–235, Sept. 1989.

[93] J. B. Davies, R. Ettinger, F. A. Fernandez, R. Souza, and A. Rahman, "Variational and finite element solutions of TE modes in waveguides with nonlinear material," presented at the U. R. S. I. Nat. Radio Science Meeting, Boulder, CO, Jan. 1989.

Boundary-Marching Method for Discontinuity Analysis in Waveguides of Arbitrary Cross Section

S. L. Foo and P. P. Silvester, *Fellow, IEEE*

Abstract—**A recursive algorithm previously used in diffusion problems of geophysics and in electrostatics, is extended to wave phenomena. It is used to construct a matrix representation for an infinitely long waveguide of arbitrary cross-sectional shape. This representation is used in finite element analysis of waveguide discontinuities. In numerical tests, scattering matrices for the long guides converge to nearly full word-length in 6–7 recursion steps, and discontinuity characteristics are within 1%–2% of known results where they exist.**

I. Introduction

FINITE elements formulated in terms of vector field components have been widely used in characterizing arbitrarily shaped waveguides. Recently, the finite element method has also been successfully applied in analyzing some subclasses of waveguide discontinuity problems which are essentially two-dimensional, such as E-Plane and H-plane discontinuities. However, a general waveguide discontinuity is three-dimensional. It joins two or more waveguides, possibly dissimilar, it has an arbitrary shape in all directions, and may contain inhomogeneous materials. In such cases, a full three dimensional analysis is required. When using the finite element method, one also needs to model the two infinite waveguides directly attached to the discontinuity section. The most common way of dealing with such infinite guides is to truncate the guides at a distance sufficiently far from the discontinuity, with a large number of mesh nodes. Proper boundary conditions are then applied at the truncated surfaces, assuming that the field decays significantly before reaching the truncations. This approach results in an undesirably large mesh and therefore is not practical in many three-dimensional problems. For some special cases, where the waveguide geometry is rectangular, circular or elliptical, an equivalent boundary condition, derived from analytically known guided modes, may be used to truncate the mesh at a smaller distance from the discontinuity. However, if the geometry of the waveguide structure is more general, there is no analytic solution for the waveguide and an equivalent boundary condition for the infinite guide section cannot be found easily. This is especially true in the case of inhomogeneously dielectric-loaded waveguides.

This paper presents a very general finite element scheme which can be used to model an arbitrarily-shaped guide that may be inhomogeneous in the transverse direction. The algorithm uses a simple recursive method to generate a submatrix which relates the field characteristics on the near-field surface to the field conditions on the far-field surface. As a result, it can be used to truncate the finite element mesh at a distance very close to the discontinuity without losing any generality, for any arbitrarily-shaped guide. This procedure resembles the "roof-raising" process used in static and diffusion fields by Kisak, Silvester and Telford [1], and the related but more general two-dimensional "ballooning" algorithm applied to electrostatics problems by Silvester, Lowther, Carpenter and Wyatt [2].

Waveguide analysis with the finite element method has long been troubled by the appearance of spurious modes. Although these are commonly encountered in eigenvalue problems, it has been shown that spurious modes can affect solutions to deterministic problems also [3], [4]. The orthospectral ("spectrally correct") elements obtained by using mixed-order approximating functions on hexahedra [5], [6] have been shown to produce solutions free of spurious modes. All the results reported in this paper are based on elements of this type and, as would be expected, no spurious-mode corruption of solutions has been encountered.

II. Variational Formulation

The general configuration of the class of problems considered here is illustrated in Fig. 1. The waveguide discontinuity or junction region is viewed as being composed of three subregions: (i) a uniform guide Ω_1, (ii) the discontinuity region proper, Ω_d, and (iii) the second uniform guide Ω_2, not necessarily similar to Ω_1. In the uniform guides and in the discontinuity region Ω_d the electric field must everywhere satisfy the vector Helmholtz equation:

$$\frac{1}{\mu_r} \nabla \times \nabla \times \boldsymbol{E} - k_0^2 \epsilon_r \boldsymbol{E} = 0, \tag{1}$$

Manuscript received September 17, 1991; revised March 10, 1992.
The authors are with McGill University, Department of Electrical Engineering, 3480 University Street, Montreal, PQ, Canada H3A 2A7.
IEEE Log Number 9202131.

Reprinted from *IEEE Trans. Microwave Theory Tech.*, vol. 40, no. 10, pp. 1889–1893, Oct. 1992.

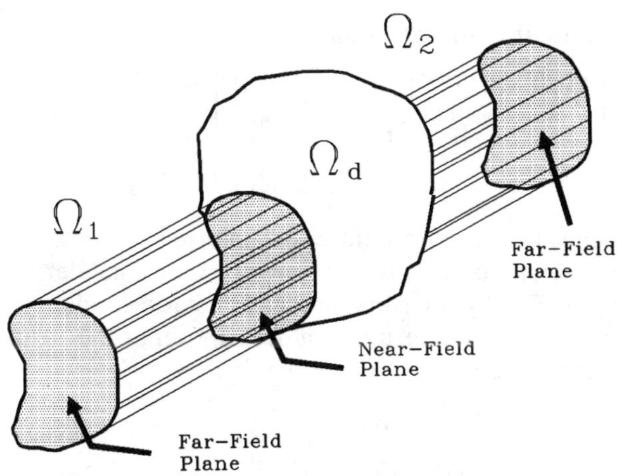

Fig. 1. The three subregions of a waveguide junction or discontinuity problem: the discontinuity Ω_d proper, flanked by two infinite waveguides Ω_1 and Ω_2.

subject to boundary conditions of the following types:

$$(\nabla \times \boldsymbol{E}) \times \mathbf{1}_n = 0 \quad \text{(homogeneous Neumann on magnetic wall)}, \tag{2}$$

$$\boldsymbol{E} \times \mathbf{1}_n = 0 \quad \text{(homogeneous Dirichlet on perfect electric conductors)}, \tag{3}$$

$$\boldsymbol{E} \times \mathbf{1}_n = \boldsymbol{E}_0 \quad \text{(inhomogeneous Dirichlet on excitation planes)}. \tag{4}$$

As is well known [7], solving the Helmholtz equation for the electric field vector in the lossless case is equivalent to extremizing the variational functional

$$\mathfrak{F}(\boldsymbol{E}) = \frac{1}{2}\int_\Omega \left\{ -\frac{1}{\mu_r}(\nabla \times \boldsymbol{E}) \cdot (\nabla \times \boldsymbol{E}) + k_0^2 \epsilon_r \boldsymbol{E} \cdot \boldsymbol{E} \right\} d\Omega. \tag{5}$$

The finite element methods contemplated here rely on this formulation, and the boundary-marching method fits into the general framework of variational methods very well. The results given by Marcuvitz [8] are quite extensive enough to serve for independent verification of all calculations.

III. Discretization

In order to avoid the appearance of spurious modes, orthospectral (mixed-order hexahedral) finite elements cast in terms of the projection components [5] are utilized. That is, the electric field vector $\boldsymbol{E}$ is written

$$\boldsymbol{E} = 1_\xi E_\xi + 1_\eta E_\eta + 1_\nu E_\nu, \tag{6}$$

where 1_ξ, 1_η, 1_ν are reciprocal unitaries of the local coordinates, and E_ξ, E_η, E_ν are the covariant projection components of $\boldsymbol{E}$. Each component of the electric field $\boldsymbol{E}$ in each element is approximated by element functions $\alpha^\xi_m(\xi, \eta, \nu)$, $\alpha^\eta_m(\xi, \eta, \nu)$, $\alpha^\nu_m(\xi, \eta, \nu)$:

$$E_\xi = \sum_{m=1}^{18} E^\xi_m \alpha^\xi_m(\xi, \eta, \nu), \tag{7}$$

$$E\eta = \sum_{m=19}^{36} E^\eta_m \alpha^\eta_m(\xi, \eta, \nu), \tag{8}$$

$$E_\nu = \sum_{m=37}^{54} E^\nu_m \alpha^\nu_m(\xi, \eta, \nu). \tag{9}$$

Discretizing the entire region into finite elements and applying the standard finite element minimization procedure to the functional [7], [9] gives the following system of equations:

$$\left\{ -\frac{1}{\mu_r}[S] + k_0^2 \epsilon_r [T] \right\} [E] = [0]. \tag{10}$$

Here $[E]$ is a column matrix representing the nodal electric fields, $[S]$ is the square matrix that results from the curl-curl term in the functional, and $[T]$ corresponds to the dot-product term in the functional.

IV. Boundary-Marching Algorithm

Four steps of the boundary-marching process are illustrated in Fig. 2. As indicated, the far-field and near-field planes are initially placed at the same location; then the far-field plane is moved away step by step, with the distance at each step growing larger as the far-field plane recedes. To develop the computational algorithm, let Ω^0 be the volume of the initial segment of a uniform guide; let Γ_1 and Γ_2 be the two surfaces enclosing this segment of the guide. Because the guide is uniform, Γ_1 and Γ_2 are congruent. The guide segment is discretized into finite elements, with the following restriction on the manner of subdivision: the placement of finite element nodes and edges must leave Γ_1 and Γ_2 congruent, i.e., the element and node placement on Γ_1 must correspond exactly to that on Γ_2. On discretizing the segment into a number of finite elements and minimizing the corresponding electric-field functional, the following set of simultaneous equations is obtained:

$$\begin{bmatrix} [W]_{11} & [W]_{1i} & [W]_{12} \\ [W]_{i1} & [W]_{ii} & [W]_{i2} \\ [W]_{21} & [W]_{2i} & [W]_{22} \end{bmatrix} \begin{bmatrix} [E]_1 \\ [E]_i^0 \\ [E]_2 \end{bmatrix} = 0. \tag{11}$$

The matrix equation had been partitioned so that the submatrices identified by subscript 1 correspond to finite element nodes located on plane 1 (the near-field plane), those identified by subscript 2 correspond to finite element nodes located on plane 2, and those identified by i are in the interior of the guide segment. (The superscripts are of no significance for the moment; they are introduced only for notational consistency with further development.) Thus $[E]_1$ represents the nodal electric fields on boundary surface 1, $[E]_2$ the nodal electric fields on boundary plane 2,

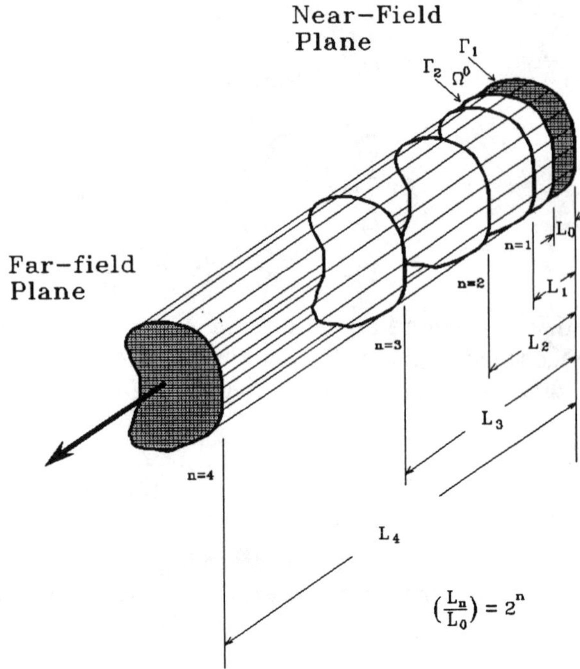

Fig. 2. The boundary marching process. The finite guide length L_n doubles at every step, reaching full floating-point precision after 5 or 10 steps.

and $[E]_i^0$ is the set of all nodal electric fields inside the region. The internal field nodes $[E]_i^0$ are not of interest, they are eliminated by a static condensation scheme [9], [10]. The system of equations for waveguide segment Ω^0 is thereby reduced to the more compact form

$$\begin{bmatrix} [S]^0_{11} & [S]^0_{12} \\ [S]^0_{21} & [S]^0_{22} \end{bmatrix} \begin{bmatrix} [E]^0_1 \\ [E]^0_2 \end{bmatrix} = 0, \tag{12}$$

where the number of matrix rows and columns equals the number of nodal variables on the bounding surfaces Γ_1 and Γ_2 only; the subscript i has disappeared altogether. The new submatrices $[S]^0_{mn}$ are obtained from the nine submatrices $[W]_{mn}$ of the finite element functional by

$$[S]^0_{11} = [W]_{11} - [W]_{1i}([W]_{ii})^{-1}[W]_{i1} \tag{13}$$

$$[S]^0_{12} = [W]_{12} - [W]_{1i}([W]_{ii})^{-1}[W]_{i2} \tag{14}$$

$$[S]^0_{21} = [W]_{21} - [W]_{2i}([W]_{ii})^{-1}[W]_{i1} \tag{15}$$

$$[S]^0_{22} = [W]_{22} - [W]_{2i}([W]_{ii})^{-1}[W]_{i2}. \tag{16}$$

The notation follows that for the full matrix representation: submatrix $[S]^0_{mn}$ interrelates field components associated with nodes on surface Γ_m of the condensed element with those on surface Γ_n.

Now the coefficient matrix of (9) describes the interrelationship of electric field components at the two ends of a fixed length of uniform guide, with no assumptions as to the length (it need not be smaller than a wavelength). A section of guide twice that length can therefore be modeled without loss of accuracy by cascading two such matrices. The description of a double-length segment of guide is thus obtained by combining two identical segments as described by (9) and enforcing the field continuity on their common boundary:

$$\begin{bmatrix} [S]^0_{11} & [S]^0_{12} & 0 \\ [S]^0_{21} & [S]^0_{11} + [S]^0_{22} & [S]^0_{12} \\ 0 & [S]^0_{21} & [S]^0_{22} \end{bmatrix} \begin{bmatrix} [E]^0_1 \\ [E]^0_i \\ [E]^0_2 \end{bmatrix} = 0. \tag{17}$$

As before, the electric field at the internal nodes is of no interest. It can be eliminated from further consideration, removing $[E]^0_i$ by the same process of static condensation as previously. What results is a description of the "super-element" twice as long as the original one:

$$\begin{bmatrix} [S]^1_{11} & [S]^1_{12} \\ [S]^1_{21} & [S]^1_{22} \end{bmatrix} \begin{bmatrix} [E]^1_1 \\ [E]^1_i \end{bmatrix} = 0, \tag{18}$$

where the superscript 1 indicates that one doubling of the guide length has taken place. The four new submatrices are obtained much as before:

$$[S]^1_{11} = [S]^0_{11} - [S]^0_{1i}([S]^0_{ii})^{-1}[S]^0_{i1}, \tag{19}$$

$$[S]^1_{12} = -[S]^0_{1i}([S]^0_{ii})^{-1}[S]^0_{i2}, \tag{20}$$

$$[S]^1_{21} = -[S]^0_{2i}([S]^0_{ii})^{-1}[S]^0_{i1}, \tag{21}$$

$$[S]^1_{22} = [S]^0_{22} - [S]^0_{2i}([S]^0_{ii})^{-1}[S]^0_{i2}. \tag{22}$$

The new super-element representing the guide section is twice as long, and therefore has twice the volume of the original section: $\Omega^1 = 2\Omega^0$.

Further lengthening of the guide segment is achieved by applying the same procedure recursively. After each of k recursions, the matrix relating the excitation field on boundary surface 1 to the field distribution on the far-field plane can be expressed in terms of the submatrices of the previous recursion as follows:

$$[S]^{k+1}_{11} = [S]^k_{11} - [S]^k_{1i}([S]^k_{ii})^{-1}[S]^k_{i1} \tag{22}$$

$$[S]^{k+1}_{12} = -[S]^k_{1i}([S]^k_{ii})^{-1}[S]^k_{i2} \tag{23}$$

$$[S]^{k+1}_{21} = -[S]^k_{2i}([S]^k_{ii})^{-1}[S]^k_{i1} \tag{24}$$

$$[S]^{k+1}_{22} = [S]^k_{22} - [S]^k_{2i}([S]^k_{ii})^{-1}[S]^k_{i2}. \tag{25}$$

At each recursion step, the length of the super-element (i.e., of the guide segment) is augmented by a factor of 2. Consequently, in the course of N recursions, the length of the uniform guide in the propagation direction grows by a factor of 2^N. In effect the procedure is equivalent to marching out the boundary of the uniform guide from the excitation plane to the far-field plane, sufficiently far for all evanescent modes to decay to a negligible level. The method therefore provides a simple way for simulation and investigation of wave propagation in any arbitrarily-shaped guide. The advantage of the algorithm becomes even more pronounced in the case of inhomogeneous dielectric-loaded guides, where any competitive method currently available (e.g., any integral equation method) becomes too complicated, if not impossible.

To summarize the full computational algorithm: The finite element matrix for a finite, generally quite short, length of guide is constructed using standard techniques. It is partitioned into submatrices according to whether the node numbers refer to bounding plane 1 (the near-field plane), plane 2 (the farther plane), or the interior i. The nine submatrices are then manipulated in an N-step recursion as follows.

Step 1: Initialize.

$$[S]^0_{11} = [W]_{11} - [W]_{1i}([W]_{ii})^{-1}[W]_{i1}$$

$$[S]^0_{12} = [W]_{12} - [W]_{1i}([W]_{ii})^{-1}[W]_{i2}$$

$$[S]^0_{21} = [W]_{21} - [W]_{2i}([W]_{ii})^{-1}[W]_{i1}$$

$$[S]^0_{22} = [W]_{22} - [W]_{2i}([W]_{ii})^{-1}[W]_{i2}.$$

Step 2: March out, for $k = 0, \cdots, N - 1$.

$$[S]^{k+1}_{11} = [S]^k_{11} - [S]^k_{1i}([S]^k_{ii})^{-1}[S]^k_{i1}$$

$$[S]^{k+1}_{12} = -[S]^k_{1i}([S]^k_{ii})^{-1}[S]^k_{i2}$$

$$[S]^{k+1}_{21} = -[S]^k_{2i}([S]^k_{ii})^{-1}[S]^k_{i1}$$

$$[S]^{k+1}_{22} = [S]^k_{22} - [S]^k_{2i}([S]^k_{ii})^{-1}[S]^k_{i2}$$

Step 3: Record result.

$$\begin{bmatrix} [S]^N_{11} & [S]^N_{12} \\ [S]^N_{21} & [S]^N_{22} \end{bmatrix} \begin{bmatrix} [E]^N_1 \\ [E]^N_i \end{bmatrix} = 0. \tag{27}$$

The guide length grows as 2^N, so only 20 recursion steps will turn an initial explicit guide model millimeters long into the same number of kilometers.

V. Numerical Results

The boundary-marching technique has been extensively tested on guides of several different cross-sectional shapes. All test programs used Crowley-type orthospectral elements [5] and were written in the Ada language. Two computers were used: an 80386-based (MS-DOS) machine for program development and debugging, followed by a Cray X-MP supercomputer for subsequent production runs. The algorithm is first tested here by investigating the decay behavior of the evanescent modes, as the boundary of the far-field plane is marched out from the excitation plane. Fig. 3 shows how the reflected waves of the $TE_{0,1}$, $TE_{0,2}$ and $TE_{2,0}$ modes decay with guide length in a typical rectangular waveguide. All computations were carried out in 64-bit arithmetic, so that noticeable roundoff error accumulation in the fourteenth or fifteenth digit is to be expected. In fact, the roundoff error falls substantially below that level in two of the modes. The magnitude of the forward transmission of the scattering coefficient of the propagating mode ($TE_{1,0}$ mode), as a function of the length of the waveguide is also shown in Fig. 3. Clearly, about 6 or 7 recursions are more than adequate in this case to make the guide "infinite" for all practical purposes. Even quite near cutoff, about 20 or 30 recursion steps suffice, yielding a waveguide length of 10^5–10^8 times guide width.

To illustrate how the boundary-marching procedure can

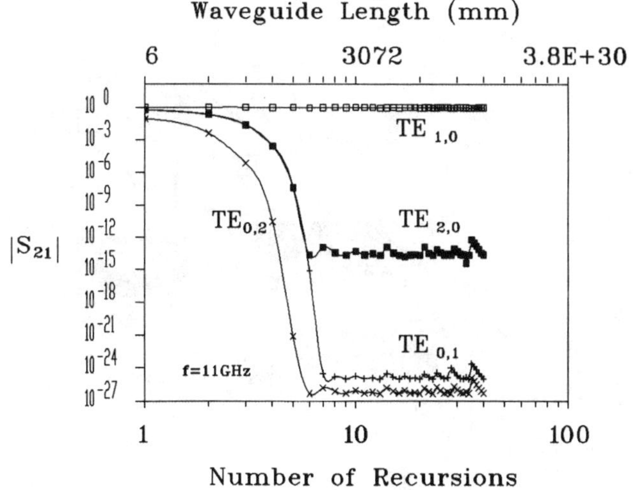

Fig. 3. Scattering of dominant and evanescent modes in a rectangular guide of 20.32 mm × 10.16 mm cross-section, as a function of the number of recursions or equivalent guide length.

be incorporated into waveguide analysis, Fig. 4 shows the finite element scheme for the solution of a zero-thickness capacitive window. This problem is well established, and results may be found in Marcuvitz [8]. The reference plane of the transmission and reflection parameters is calibrated by using a matched "through" segment. The calibration also eliminates any error that may arise as a result of signal loss or phase distortions in the waveguide segments. The scattering parameters of the structure are first computed with the discontinuity section replaced by a segment of empty waveguide of known finite length. The calibration factors are determined by setting the magnitude of the forward transmission of the scattering coefficients to unity and the phase to 90 degrees. Then the scattering parameters of the structure including the discontinuity are computed by replacing the "calibration segment" with the discontinuity. Dividing the transmission and reflection parameters by the proper calibration factor, and subtracting the phase offset from the phase angles of the parameters, one obtains the final scattering parameters at the desired reference plane. By using the same segment length for the calibration segment as for the discontinuity segment, the transmission and reflection parameters are calibrated to the plane where the zero-thickness window resides. If desired, the parameters could also be calibrated to some other reference plane, by choosing different lengths for the calibration segment and the discontinuity segment.

Phase and amplitude of the forward transmission coefficient of the scattering parameters of the capacitive window are shown in Fig. 5, for different frequencies. The results agree with the analytical approximation given by Marcuvitz [8] to within 1–2 percent. Since the Marcuvitz solution is not exact, it cannot be used to establish firm error bounds. However, the finite element solution and Marcuvitz's approximation are thought to incur errors of roughly similar magnitude, so their agreement is held to confirm the validity of the boundary-marching technique.

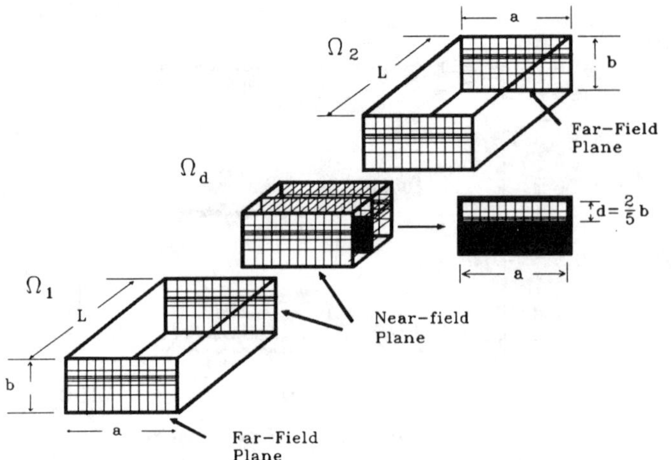

Fig. 4. The classical zero-thickness capacitive window problem, modeled as the three regions of Fig. 1.

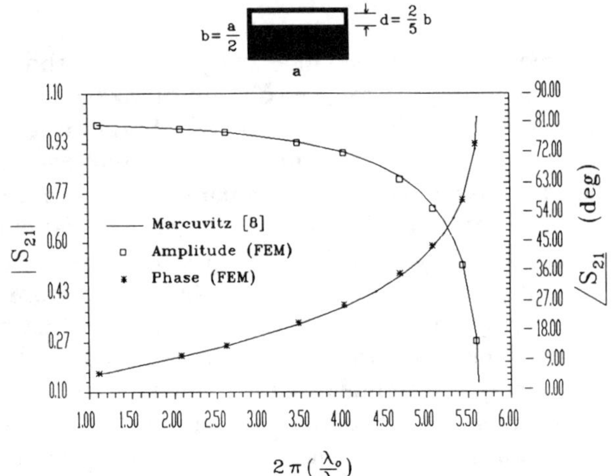

Fig. 5. Forward transfer scattering parameter for the capacitive window problem: comparison of curves given by Marcuvitz (solid line) with finite element computations.

VI. Conclusions

A general recursive method has been proposed and validated for creating finite element models of very great lengths (thousands of free-space wavelengths) of arbitrarily-shaped waveguide. The method is valid for any guide, so long as a technique is available for constructing a finite element model of a finite length of the guide. It is particularly useful for analysis of waveguide components and discontinuity regions, where it permits truncation of the finite element mesh very close to the discontinuity region without compromising result accuracy. It does not introduce any error beyond the discretization error inherited from the finite element meshing; it is unconditionally stable, except possibly at frequencies very close to the cutoff frequency of the lowest eigenmode, where the ratio of free-space wavelength to guided wavelength approaches the floating-point precision available. This algorithm appears to be particularly useful for discontinuity analyses involving inhomogeneous dielectric-loaded guides, such as finline and shielded microstrips, but further verifying work is needed to establish what limits there may be to its use.

Acknowledgment

The authors wish to thank Mr. Nicholas Chepurnyi of Environment Canada, and Mr. Phil Garnatz of Cray Research, for their extensive and generous assistance.

References

[1] E. Kisak, P. Silvester, W. M. Telford, "A recursive method in the E-polarization of magnetotelluric modeling by high-order finite elements," *Acta Geodaetica, Geophysica et Montanistica Acad. Sci. Hungaricae*, vol. 12, pp. 255-266, 1977.

[2] P. P. Silvester, D. A. Lowther, C. J. Carpenter, E. A. Wyatt, "Exterior finite elements for 2-dimensional field problems with open boundaries," *Proc. Inst. Elec. Eng.*, vol. 124, pp. 1267-1270, Dec. 1977.

[3] C. W. Crowley, A. R. Pinchuk, and P. P. Silvester, "Spurious solutions to vector diffusion and wave field problems," *IEEE Transactions on Magn.*, vol. MAG-24, no. 1, pp. 158-61, Jan. 1988.

[4] K. Ise, K. Inoue, and M. Koshiba, "Three-dimensional finite-element solution of dielectric scattering obstacles in a rectangular waveguide," *IEEE Trans. Microwave Theory Tech.*, vol. 38, no. 9, pp. 1352-1359, Sept. 1990.

[5] C. W. Crowley, P. P. Silvester, H. Hurwitz, Jr., "Covariant projection elements for 3-D vector field problems," *IEEE Trans. Magn.*, vol. MAG-24, no. 1, pp. 397-400, Jan. 1988.

[6] A. R. Pinchuk, C. W. Crowley, and P. P. Silvester, "Spectrally correct finite element operators for electromagnetic field problems," *J. Appl. Phys.*, vol. 63, no. 8, pt. 2A, pp. 3025-3027, Apr. 1987.

[7] J. P. Webb, G. L. Maile, R. L. Ferrari, "Finite-element solution of three-dimensional electromagnetic problems," *Proc. Inst. Elec. Eng.*, vol. 130, pt. H, no. 2, pp. 153-159, Mar. 1983.

[8] N. Marcuvitz, *Waveguide Handbook*. London: Peregrinus, 1986.

[9] P. P. Silvester and R. L. Ferrari, *Finite Elements for Electrical Engineers*, 2nd ed. Cambridge: Cambridge University Press, 1990.

[10] T. J. R. Hughes, *The Finite Element Method: Linear Static and Dynamic Finite Element Analysis*. Englewood Cliffs, NJ: Prentice-Hall, 1987.

A General Transformation for Open Boundary Finite Element Method for Electromagnetic Problems

A. Stochniol
Mechanical Engineering Department
Imperial College, London SW7 2BX, UK

Abstract **- A general transformation applicable to the solution of open boundary field problems using the finite element method is presented and discussed. The principle of the solution method lies in the spatial mapping of the infinite unbounded region, exterior to the main region of interest, onto a finite region, followed by the solution of the closed boundary problem for two connected regions.**

I. INTRODUCTION

Considerable efforts have been made to devise approaches which can overcome the apparent deficiencies of differential methods used to solve unbounded problems in electromagnetics. A good survey of them is given in [1]. Unfortunately, it is not always easy to implement them in existing finite element software, sometimes they require some judgement by experience, and moreover they might be quite demanding in terms of computing resources, particularly for 3-D applications. It has become apparent in the last few years that an approach of spatial transformations seems to be a good alternative [2], [3], [4]. This technique, in the form which is presented in this paper, can be used in a standard finite element package without the need for any software modification, can be fully automated and can be used in the presence of material inhomogeneities as well as anisotropies.

II. FORMULATION

Let the n-dimensional unbounded region Ω_{ex} be the exterior of a bounded region Ω_{in}. The region Ω_{in} contains all windings, iron cores and permanent magnets. The boundary between Ω_{in} and Ω_{ex} is $\Gamma = \Omega_{in} \cap \Omega_{ex}$ as shown in Fig. 1. To demonstrate the theoretical formulation of the method, consider the following boundary value problem for a scalar field Φ in Ω_{ex}:

Manuscript received July 7, 1991.

$$\nabla\cdot(K\cdot\nabla\Phi) = g \qquad \text{in } \Omega_{ex} \tag{1}$$

$$\Phi(p) = \Phi_0(p)\,, \qquad p \in S_1 \tag{2}$$

$$(K\cdot\nabla\Phi)\cdot\mathbf{n}\,|_{S_2} = 0 \qquad \text{on } S_2 \tag{3}$$

where g , $\Phi_0(p)$ are given functions of point, $S_1 \cup S_2$ is the boundary of Ω_{ex}. The symmetric tensor K represents the material properties of the anisotropic and inhomogeneous medium in Ω_{ex}.

The finite element solution of the above boundary value problem can be obtained through the minimization of the functional:

$$F = \int_{\Omega_{ex}} [(\nabla\Phi)\cdot(K\cdot\nabla\Phi) + 2g\Phi]\, d\Omega \tag{4}$$

with trial functions which satisfy the Dirichlet boundary condition (2). Let $\mathcal{T}$ be defined as a real, single-valued, reversible transformation which transforms the region Ω_{ex} onto the region Ω_{tr} :

$$\mathcal{T}:\ x'_i = x'_i(x_1, x_2, \dots, x_n) = x'_i(x)\,, \quad (i = 1, 2, \dots, n) \tag{5}$$

where $n = 2$ or 3 is the space dimension. The original cartesian coordinates are denoted $x = (x_1, x_2, \dots, x_n)$ and the transformed ones: $x' = (x'_1, x'_2, \dots, x'_n)$.

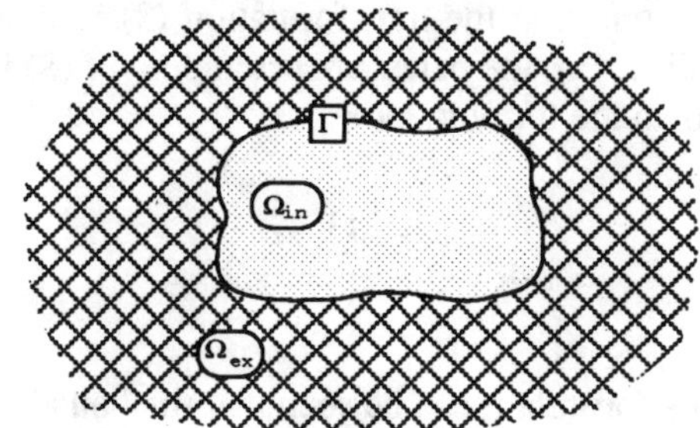

Fig. 1. Open boundary problem.

Reprinted from *IEEE Trans. Magn.*, vol. 28, no. 2, pp. 1679–1681, March 1992.

After such a transformation the state function Φ is not modified but the functional in (4) is transformed into the form:

$$F = \int_{\Omega_{tr}} [(\nabla'\Phi')\cdot(K'\cdot\nabla'\Phi') + 2g'\Phi']\, d\Omega' \quad (6)$$

where Φ' is the Φ but dependent on x', ∇' is the del operator in transformed (primed) coordinates. The tensor K' is related to K by [2]:

$$K' = T\cdot K\cdot T^{t}\cdot |\det(T^{-1})| = T\cdot K(x')\cdot T^{t}\cdot |\det(T^{-1})| \quad (7)$$

where $T = (\partial x'_i/\partial x_j)$ is the Jacobian of the $\mathcal{T}$ transformation, T^{t} - transposition of T, $|\det(T^{-1})|$ - the absolute value of the determinant of T^{-1}; g' is related to g by: $g' = g(x')\cdot|\det(T^{-1})|$.

III. THE PROPOSED TRANSFORMATION $\mathcal{T}$

Let us introduce transformation $\mathcal{T}$ of the form:

$$x'_i = \left(\frac{a\cdot R_\Gamma}{r} - \frac{(a-1)\cdot R_\Gamma^2}{r^2}\right)\cdot x_i\ , \quad (i = 1, 2, \dots, n) \quad (8)$$

where $0 \le a \ne 1$, $r^2 = \sum_{i=1}^{n} x_i^2$, $R_\Gamma^2 = \sum_{i=1}^{n} \left(x_i^\Gamma\right)^2$

and a point $x^\Gamma \in \Gamma$ and it lies on the radial line which connects the origin of the transformation (point $\{0\}$) with the point x (so R_Γ depends on x) (Fig. 2). For the sake of simplification of formulas, the coordinate system X was chosen in such a way that all coordinates of the origin of this transformation are equal to 0. It should be noted that the origin of the transformation (8) must be positioned inside the bounded inner region.

We do not need to specify the shape of the inner region now, but we assume that it has such a shape that any radial line from the origin intersects the boundary Γ only once. On such a condition the transformation (8) is single-valued and reversible. For the better understanding of (8) it is useful to write down the formula for r':

$$r' = a\cdot R_\Gamma - \frac{(a-1)\cdot R_\Gamma^2}{r}\ . \quad (9)$$

From this formula it can be seen that the boundary Γ is an invariant for the transformation. From (9) it follows that infinity ($r = \infty$) is transformed into a new boundary Γ_∞ which is described by: $R_{\Gamma_\infty} = a\cdot R_\Gamma$.

So the Γ_∞ is in fact created by scaling Γ taking each point of Γ radially outward (or inward) by a scaling factor $a > 1$ (or $a < 1$). For a special case when $a = 0$ the infinity is transformed into one point, the origin.

Some authors have already used spatial transformation to solve open boundary problems in electromagnetics, which are in fact special cases of the general transformation (8). When n-dimensional sphere is chosen as the Ω_{in} region, (8) simplifies to:

$$x'_i = \left(\frac{a\cdot R}{r} - \frac{(a-1)\cdot R^2}{r^2}\right)\cdot x_i\ , \quad (i = 1, 2, \dots, n) \quad (10)$$

where R is the radius of that n-dimensional sphere. For $a=0$ the transformation (10) simplifies to the well known Kelvin transformation which was investigated, for example, in [2], [3]. A physical interpretation of such an approach based on construction of an alternative set of impedances for the exterior region was also described [3].

For $a > 1$ a transformation equivalent to (10) was originally proposed and successfully used in [4]. The method there was slightly different in comparison with the one proposed here. Instead of introducing the material transformation the process of integration for elements inside the transformed region was

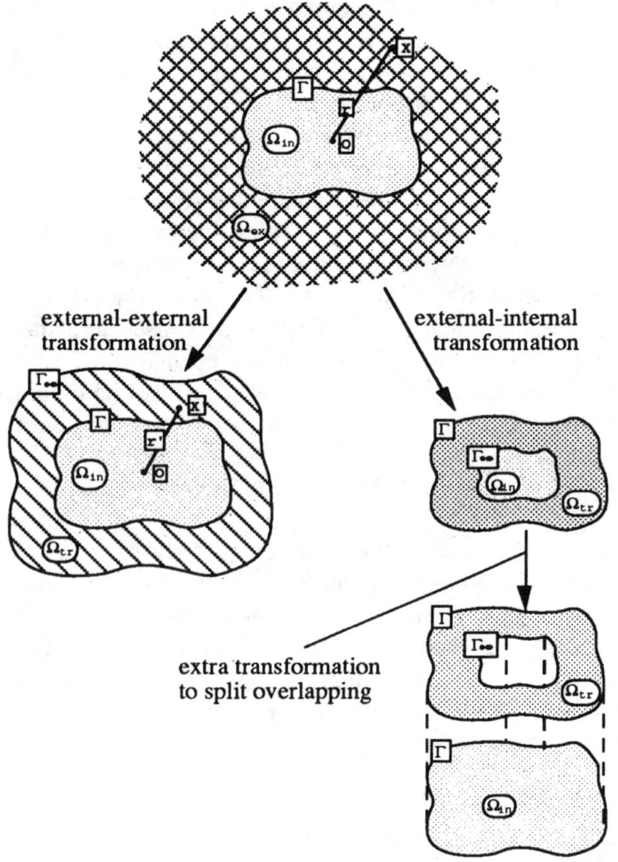

Fig. 2. Spatial transformations of the unbounded region Ω_{ex}.

modified [4]. In the approach proposed here the finite element code does not need any changes at all.

IV. PRACTICAL APPLICATIONS

For finite element packages which use the tensor approach for specifying the value of K, the application of the proposed method is straightforward for any transformation (8). One of the most interesting, especially for the study of systems with one dimension much bigger than the others, is that which uses n-dimensional ellipsoids. But very often electromagnetic finite element packages give the possibility to specify the isotropic materials only. For such packages the application of the proposed method is still possible, but only with some transformations, for example with (10).

For the transformation (10) K' is equal to (for isotropic material $K=k$):

$$K' = k \cdot T \cdot T^{t} \cdot |\det(T^{-1})| = k \cdot Q \cdot |\beta \cdot \alpha^{n}| \qquad (11)$$

where $Q = T \cdot T^{t}$, $\beta = \dfrac{r'}{aR - r'}$, $\alpha = \dfrac{R^2(a-1)}{r'(aR-r')}$.

The elements of Q are equal to:

$$Q_{ii} = \frac{1}{\alpha^2} + \frac{1}{\alpha^2}\frac{aR}{r'}\left(\frac{1}{\beta} - 1\right)\frac{x_i'^2}{r'^2} \qquad (12)$$

$$Q_{ij} = \frac{1}{\alpha^2}\frac{aR}{r'}\left(\frac{1}{\beta} - 1\right)\frac{x_i' x_j'}{r'^2}, \quad \text{for } i \neq j. \qquad (13)$$

As can be seen from (13), the elements Q_{ij}, $i \neq j$ are equal to 0 only when $a = 0$, i.e. when the exterior region is mapped onto the interior one, and infinity is mapped onto the origin point. For $a = 0$ the diagonal elements $Q_{11} = Q_{22} = \dots = Q_{nn}$, so instead of using tensor K' we can use the variable $k' = k \cdot \alpha^{n-2}$. For 2-dimensional problems we obtain that $k' = k$, i.e. no material transformation is needed, for 3-dimensional problems: $k' = k \cdot \alpha$. Unfortunately for $a \neq 0$ the K' is a tensor, but there is a way to avoid this difficulty. Taking into account the form of minimized functional F (6), it can be shown after some transformations that instead of minimizing (6) with K' defined by (11), it is possible to minimize this functional replacing the tensor K' by a variable k' defined by:

$$k' = k \cdot \frac{1}{\alpha^2\beta^2} \cdot |\beta \cdot \alpha^{n}| = \begin{cases} k \cdot \dfrac{\alpha}{\beta} = k \cdot \dfrac{R^2|a-1|}{r'^2} & \text{for 3D} \\ k \cdot \dfrac{1}{\beta} = k \cdot \dfrac{|aR - r'|}{r'} & \text{for 2D} \end{cases} \qquad (14)$$

(for: $a > 1$, $R \leq r' \leq a \cdot R$; $0 \leq a < 1$, $a \cdot R \leq r' \leq R$).

The formula (14) covers the previous result for a=0, as well.

In fact we have two classes of transformation (10) : external-internal ($0 \leq a < 1$) and external-external ($a > 1$) as shown in Fig. 2. In the second case, after obtaining the transformed region the mesh must be set there and join into the mesh inside the main region. The only thing to remember is to use a new version of k' inside the transformed region. Then the closed boundary problem should be solved. For the external-internal transformation, the transformed region overlaps with the main, inner region, and therefore an additional translation is needed to split this overlapping (Fig. 2). Moreover, additional boundary conditions must be set to enforce Φ equality on such split boundaries. It should be done by binary constraints. There is only one hidden limitation for this transformation in the case of axisymmetric problems. Common practice in axisymmetric solvers is to introduce an internally modified potential to avoid singularity at $r = 0$ (such as Φ/r or $\Phi \cdot r$). Because of that the transformed region should be coaxial with that of the interior one.

In general it is not possible to say which of the two classes of transformation (8), exterior-exterior or exterior-interior, is better. Numerical calculations carried out for axisymmetric 3-D problems using transformation (10), and the modified vector potential obtained by radial scaling, show that accuracy and speed of solution for both methods depends on many factors, many of which are problem specific.

ACKNOWLEDGMENTS

The results of this study were obtained during the residency of the author as a Leverhulme Trust (UK) Fellow at the Imperial College, Electrical Engineering Department, London, UK. The author is most grateful to the Leverhulme Trust for the award.

REFERENCES

[1] C.R.I. Emson, "Methods for the solution of open-boundary electromagnetic-field problems", *IEE Proc.*, Vol. 135, Pt. A, No. 3, March 1988, pp. 4135-4137.

[2] I.R. Ciric and S.H. Wong, "Inversion transformation for the finite-element solution of three-dimensional exterior-field problems", *COMPEL*, Vol. 5, No. 2, 1986, pp. 109-119.

[3] E.M. Freeman and D.A. Lowther, "An open boundary technique for axisymmetric and three dimensional magnetic and electric field problems", *IEEE Trans. Magn.*, Vol. 25, No. 5, September 1989, pp. 4135-4137.

[4] J.F. Imhoff, G. Meunier, X. Brunotte and J.C. Sabonnadiere, "An original solution for unbounded electromagnetic 2D- and 3D-problems throughout the finite element method", *IEEE Trans. Magn.*, Vol. 26, No. 5, September 1990, pp. 1659-1661.

Radially Propagating Waves I-Hybrid Methods

Introduction

This chapter and the following one deal with the application of finite element methods to two- and three-dimensional scattering and antenna problems, also referred to as "open problems" since they are inherently defined on unbounded domains. Because the electromagnetic waves propagate in radial directions in such problems, it seems appropriate to speak of "radially propagating waves."

Open problems have traditionally been tackled by using the moment method (Harrington 1968) to solve surface integral equations, which appropriately describe metallic or impedance surfaces and stratified media, or by volume integral equations, which are more convenient in case of heterogeneous structures. Both formulations have been exploited to solve two- and three-dimensional problems, and they constitute well established and thoroughly refined techniques to solve electromagnetic problems. An exhaustive treatment of the subject can be found in the collection of papers edited by Miller, Medgyesi-Mitschang, and Newman (1992).

Integral formulations have numerous attractive strengths, but they also possess some disadvantages which suggest that methods based on differential equations may be preferable in many problems. First of all, integral equations can become quite cumbersome when dealing with electromagnetically complex structures. Second, they give rise to systems of linear equations with dense coefficient matrices. Consequently, solving such systems requires a large amount of memory and long computing times, especially when treating three-dimensional problems.

Early Work

The first work concerning the application of the finite element method to electromagnetic wave problems on unbounded domains was published in 1972 by McDonald and Wexler [included with this chapter]. The number of procedures developed to analyze open problems by means of the finite element method has grown markedly since then, so their classification is in many respects arbitrary. However, a first clear division can be established on the basis of the approach used to formulate the problem: integral equations or differential equations with boundary conditions. In the first case, the finite element method is employed to solve a surface or contour integral equation (in three or two dimensions, respectively), much like those solved by the moment method (for example, the electric field integral equation or the magnetic field integral equation). From this point of view, the finite element method can be seen as a particular application of the Galerkin projection method, where the basis functions are polynomials defined on a subsection of the domain of integration. The technique that flows from this approach is sometimes referred to as the "boundary element method." It is obviously applicable when the problem can be formulated via an integral equation. This method is most appropriate to homogeneous, stratified, or otherwise electromagnetically fairly simple structures. Although it is basically an application of the finite element method to integral equations, many workers consider it a different method because the mathematical challenges it poses — in both theory and computation — are distinctive. A large specialist literature now exists on the boundary element family of methods, particularly in applied mechanics. For this reason, boundary element methods are only included in this survey where they are directly relevant to classical finite element methods as applied to boundary value problems of differential equations or to hybrid formulations.

When the structure of a scatterer or radiating system is electromagnetically complex in either material composition or geometry, the integral formulation may become impractical because it requires a knowledge of the Green's function, which is seldom known, for the situation to be analyzed. In contrast to the integral equation approach, the differential equation formulation is simple to set up regardless of the complexity of the structures analyzed. It amounts to describing the whole structure to be analyzed by the basic equations of electromagnetics, which are usually simple enough to state, accompanied by suitable boundary conditions.

The solution of open problems by means of finite element methods based on a differential equation approach can be cast in two different ways.

(a) The first way consists of discretizing the entire unbounded domain either by using special elements of infinite extension — for example, infinite elements (Bettess and Bettess 1977), recursive condensation

(Silvester, Lowther, Carpenter, and Wyatt 1977), or infinitesimal scaling [Hurwitz 1984, included with the preceding chapter] — or by means of suitable geometric transformations such as a conformal mapping (Chang, Wong, and Chen 1990). This group of methods has proven very popular in static and quasi-static field analysis.

(b) The second way, in contrast to the first one, discretizes only a finite portion of the original infinite domain. The field representation in the finite portion of space must then be joined to a suitable representation of the field in the unbounded portion of the domain.

The present survey includes only works dealing with the analysis of radially propagating wave problems by the latter way. However, some leads to the low-frequency literature and to the mathematical theory underlying mapping methods may be found in the survey papers of Emson [1988, with preceding chapter] or Bettess and Bettess (1991).

The following section presents the different schemes of classification of finite element methods for radially propagating waves which have been proposed in the literature.

A Taxonomy of Available Methods

Solving an open problem without discretizing the whole infinite domain requires bounding the geometric region where the finite element method is to be applied. This is accomplished by introducing a fictitious finite surface, henceforth referred to as surface S (Γ will denote its two-dimensional counterpart), arbitrarily placed but so chosen as to enclose all sources and discontinuities constituting the electromagnetically complex region of interest.

The basic point now is that of imposing appropriate boundary conditions on the surface S, which account for the infinite extent of the original geometric domain. In other words, the solution of Maxwell's equations inside S with such boundary conditions — the equivalent internal problem — must yield a good approximation of the solution of the original unbounded problem. Silvester and Hsieh [1971] and McDonald and Wexler [1972] have referred to such techniques as "picture-frame methods," for, like a frame, the fictitious boundary S (or Γ) limits what can be seen of a finite region, without altering what is visible within the frame. Several different procedures have been reported in the literature for imposing suitable boundary conditions, and they give rise to methods to solve unbounded domain problems, which have been classified by various authors in three different ways.

Peterson and Castillo (1989) grouped all picture-frame methods into two general classes. The first class comprises all techniques employing non-local boundary conditions on S. They are also called "hybrid methods" because they were originally devised as ways to combine the finite element method with other methods better suited for representing the electromagnetic field on unbounded domains — such as integral equation methods or eigenfunctions expansion methods. Their second class comprises all methods that use local boundary conditions on the fictitious surface S.

Pearson, Peterson, Bahrmasel, and Whitaker [1992] based an alternative taxonomy of picture-frame methods on the numerical procedures used to solve the resulting system of equations. They denote as "inward-looking formulations" those formulations in which the equivalent surface current densities on the surface S are chosen as the primary unknowns to be determined. In their terminology, "outward-looking formulations" are those using as primary unknowns the coefficients of the finite element expansion of the field in the region enclosed by S. This kind of classification is significant because it allows the presence (if any) of spurious solutions due to equivalent internal problem non-physical resonances to be predicted. In particular, while outward-looking formulations can always avoid resonant solutions, provided they are built up with appropriate precautions in mind, the inward-looking approaches always remain subject to non-physical solutions. Consequently, this classification will often be alluded to as individual papers included in this survey are examined more closely.

A third classification scheme is due to Volakis (1994). Here, the methods for solving unbounded problems by means of finite elements are categorized on the basis of the mesh truncation scheme used. The name "exact mesh truncation schemes" is given to procedures that give rise to non-local boundary conditions while destroying the sparsity of the finite element matrices. Correspondingly, "approximate mesh truncation schemes" are those that lead to local boundary conditions and maintain matrix sparsity.

The taxonomy followed here — which does not run counter to previous ones, but indeed can be seen as complementary to them — is based on the kind of computational approximation used to represent the exact boundary condition for the equivalent internal problem. Such a boundary condition is deduced by using the equivalence principle and by introducing suitable surface current distributions on both sides of the fictitious surface S. The electric and magnetic fields $[\mathbf{E}(\mathbf{r}), \mathbf{H}(\mathbf{r})]$ satisfy

$$\mathbf{E}(\mathbf{r}) = \mathbf{E}^I(\mathbf{r}) + \mathbf{E}^S(\mathbf{r}) = \mathbf{E}^I(\mathbf{r}) - jk_0\zeta_0\mathcal{J}(\mathbf{J}) - \mathcal{M}(\mathbf{M}),$$
$$\mathbf{H}(\mathbf{r}) = \mathbf{H}^I(\mathbf{r}) + \mathbf{H}^S(\mathbf{r}) = \mathbf{H}^I(\mathbf{r}) + \mathcal{M}(\mathbf{J}) - j\frac{k_0}{\zeta_0}\mathcal{J}(\mathbf{M}). \quad (1)$$

Here, $\mathcal{J}$ and $\mathcal{M}$ are the radiation integrals involving the free-space Green's function $\mathcal{G}(\mathbf{r}, \mathbf{r}')$,

$$\mathcal{J}(\mathbf{J}) = \oint_S \mathcal{G}(\mathbf{r}, \mathbf{r}') \cdot \mathbf{J}(\mathbf{r}')\, dS',$$
$$\mathcal{J}(\mathbf{M}) = \oint_S \mathcal{G}(\mathbf{r}, \mathbf{r}') \cdot \mathbf{M}(\mathbf{r}')\, dS',$$
$$\mathcal{M}(\mathbf{J}) = \oint_S \nabla \times \mathcal{G}(\mathbf{r}, \mathbf{r}') \cdot \mathbf{J}(\mathbf{r}')\, dS',$$
$$\mathcal{M}(\mathbf{M}) = \oint_S \nabla \times \mathcal{G}(\mathbf{r}, \mathbf{r}') \cdot \mathbf{M}(\mathbf{r}')\, dS'. \quad (2)$$

In the above equations, $[\mathbf{E}(\mathbf{r}), \mathbf{H}(\mathbf{r})]$, $[\mathbf{E}^I(\mathbf{r}), \mathbf{H}^I(\mathbf{r})]$ and $[\mathbf{E}^S(\mathbf{r}), \mathbf{H}^S(\mathbf{r})]$ denote the total, incident, and scattered electromagnetic fields, respectively, at the generic point $\mathbf{r}$ on the fictitious surface S, which is usually a closed surface; $\mathbf{J}$ and $\mathbf{M}$ are the equivalent surface current densities on S, which decouple the internal and external problems. Note that, in equations (2) as well as in the following discussion, a harmonic time dependence $e^{j\omega t}$ has been assumed and suppressed. Furthermore, in equations (2), the free-space Green function $\mathcal{G}(\mathbf{r}, \mathbf{r}')$ is a dyadic,

$$\mathcal{G}(\mathbf{r}, \mathbf{r}') = \left(\mathcal{I} + \frac{1}{k_0^2}\nabla\nabla\right) \frac{\exp(-jk_0|\mathbf{r} - \mathbf{r}'|)}{|\mathbf{r} - \mathbf{r}'|} \quad (3)$$

where $\mathcal{I}$ is the unitary dyad, and ζ_0 and k_0 are the characteristic impedance and the propagation constant of free space, respectively. In equations (2) and (3), the operator ∇ acts on the variable $\mathbf{r}$.

The boundary condition expressed by equations (1) is non-local, that is, the value of the electromagnetic field $[\mathbf{E}, \mathbf{H}]$ at the generic point $\mathbf{r}$ on S depends on the values of the field itself at all the other points of the surface S. Its use in conjunction with the finite element method to solve the equivalent internal problem has been implemented in several ways. These can be classified on the basis of the approximations used for the equivalent surface current densities $\mathbf{J}$ and $\mathbf{M}$ in equations (2), in the following fashion:

(a) Approximations that retain validity in both the near and far fields, and maintain the non-local character of the exact boundary condition (1).

(b) Approximations valid only in the far field zone, which make local the exact boundary condition (1).

This scheme leads to the same subdivision as that proposed by Peterson and Castillo (1989) or Volakis (1994), but it stems from different considerations.

In the case (a), the unknown equivalent surface current densities $\mathbf{J}$ and $\mathbf{M}$ on S — or, equivalently, the components of the electromagnetic field tangential to S — are expressed as a linear combination of suitable vector basis functions $\mathbf{\Psi}_m(\mathbf{r})$, with unknown coefficients a_m and b_m, respectively. The techniques that use this kind of approximation coincide with the exact mesh truncation schemes of Volakis (1994). In particular, the basis functions usually employed are polynomials, preferably chosen to match the shape functions of the finite element method used to discretize the region enclosed by S. Another useful type of exterior basis function consists of the eigenfunctions of the Helmholtz equation. Their use is particularly convenient when the surface S is a constant coordinate surface in some coordinate system where the Helmholtz operator is separable. This occurs, for example, in the unimoment method. Let us consider a two-dimensional problem (TE$_z$ case, where the transverse plane is taken as the x–y coordinate plane); equation (1) becomes [Pearson *et al.* 1992]

$$H_z(\boldsymbol{\rho}) = H_z^I(\boldsymbol{\rho}) + H_z^S(\boldsymbol{\rho}) = H_z^I(\boldsymbol{\rho}) + \mathcal{M}(\mathbf{J}) - j\frac{k_0}{\zeta_0}\mathcal{J}(\mathbf{M}). \quad (4)$$

Since in this simple case the eigenfunctions $\Psi_m(\boldsymbol{\rho})$ of the scalar Helmholtz equation are also eigenfunctions of the integral operators $\mathcal{J}$ and $\mathcal{M}$ [Pearson *et al.* 1992], their use in discretizing equation (4) is equivalent to expressing the scattered field $H_z^S(\boldsymbol{\rho})$ as a linear combination of the same basis functions $\Psi_m(\boldsymbol{\rho})$:

$$H_z(\boldsymbol{\rho}) = H_z^I(\boldsymbol{\rho}) + H_z^S(\boldsymbol{\rho}) \simeq H_z^I(\boldsymbol{\rho}) + \sum_{m=1}^{N} a_m \Lambda_m^{\mathcal{M}} \Psi_m(\boldsymbol{\rho}) - j\frac{k_0}{\zeta_0}\sum_{m=1}^{N} b_m \Lambda_m^{\mathcal{J}} \Psi_m(\boldsymbol{\rho}), \quad (5)$$

where $\Lambda_m^{\mathcal{J}}$ and $\Lambda_m^{\mathcal{M}}$ are the eigenvalues of the operators $\mathcal{J}$ and $\mathcal{M}$, respectively.

Both sets of basis functions previously mentioned — polynomial or eigenfunctions — lead to a non-local boundary condition. While this is obvious in the case of the polynomial basis functions, it is not so apparent when the basis functions are the eigenfunctions of the Helmholtz equation. As a matter of fact, equation (5) appears at a first glance to be a local boundary condition. However, the determination of the coefficients a_m and b_m in (5) requires the scattered magnetic field $H_z(\boldsymbol{\rho})$ to be known at least at N different points. Actually, it follows that the boundary condition in equation (5) is still non-local.

Exact mesh truncation schemes are used in a variety of ways in conjunction with finite elements, and they give rise to a class of methods usually referred to as hybrid methods in the literature, which will be reviewed in this chapter. All the methods belonging to this category are characterized by a partially full coefficient matrix of the global system (interior region plus fictitious boundary). By partially full is meant that the submatrix associated with boundary points is dense, the submatrices associated with interior region elements are sparse.

An alternative technique for approximating the exact boundary condition (1) consists in evaluating equations (2) by retaining just their dominant contributions from the points close to $\mathbf{r}' = \mathbf{r}$. Apparently, with this approximation, the boundary condition (1) becomes local. Various factors affect the validity of the approximation adopted. Among them, the most important are the frequency of the field, the shape of the surface S, and the field distribution on the same surface. It is important to observe that such an approximation leads to an open problem formulation that employs the finite element method in conjunction with the well known "absorbing boundary conditions." These boundary conditions are of the local type, since they express the electromagnetic field at any point $\mathbf{r}$ on the surface S as a function of the electromagnetic field in the neighborhood of $\mathbf{r}$. Consequently, the finite element method with absorbing boundary conditions has the advantage of leading to a coefficient matrix of the global system that maintains the sparsity characteristic of finite element matrix equations.

A valuable feature of the finite element method with absorbing boundary conditions is that it yields a unique approximation to the solution. Perhaps surprisingly, uniqueness is not always guaranteed by non-local boundary conditions. In fact, as noted first by Maue (1949), the surface integral equation formulation of an open problem — whether by the electric field integral equation or the magnetic field integral equation — may give rise to non-physical solutions related to the resonances in the region enclosed by the surface S. This computational artifact is sometimes referred to as the "interior resonance problem." Non-physical solutions can also arise when the surface integral equation acts as boundary condition on the surface S for an equivalent internal problem solved by a finite method (Cwik 1992). To avoid interior resonances, non-local boundary conditions should be imposed following specific formulations given in the literature (Peterson 1990) or by introducing small losses in the medium surrounding the structures analyzed [Collins, Jin, and Volakis 1992].

The major drawback of the finite element–absorbing boundary condition formulation is that its use often requires a larger volume to be discretized than would be necessary using hybrid methods. Because the absorbing boundary conditions are derived from asymptotic approximations, they should be imposed on a surface S placed at a suitable distance from all sources and scatterers. On the other hand, the fictitious surface S can be chosen arbitrarily when employing hybrid methods, thereby minimizing the geometric domain for computation. Some of the relative advantage in preserving matrix sparsity is thus paid for in a larger matrix size.

The methods belonging to this second category will be more thoroughly considered in the next chapter.

Hybrid Methods

The term "hybrid methods" denotes all those methods for solving open-domain problems that combine the finite element method with a boundary condition derived from a non-local, near-field approximation of the operators $\mathcal{J}$ and $\mathcal{M}$ to the exact boundary condition (1). Such an approximation expresses the unknown equivalent surface current densities $\mathbf{J}$ and $\mathbf{M}$ on S — or equivalently, the components of the electromagnetic field tangential to S — as a finite linear combination of suitable basis functions.

Many hybrid methods have been proposed and applied during the past 20 years, and their taxonomy can flow from different considerations. The criteria used here focus on the kind of basis functions employed. Generally, two kinds of basis functions are common: polynomials or eigenfunctions of the Helmholtz equation. To the latter category belong the "unimoment method," the "finite element with extended boundary conditions method," the "transfinite element method," and the "bymoment method."

Hybrid methods that use polynomial basis functions to approximate the exact boundary condition allow an arbitrary choice of the fictitious boundary (contour Γ or surface S) separating interior and exterior regions. This is a noteworthy strength of such methods, for it allows Γ or S to be placed near the scatterer or antenna to be analyzed, thus minimizing the geometric domain for computation. The main disadvantage of these hybrid methods is due to the required numerical evaluation of the Green's integral appearing in equations (2). Indeed, when the integrand contains singularities, such evaluation requires specialized analytical and/or numerical techniques depending on both the order of the singularity and the domain of integration. Hybrid methods that approximate the equivalent surface current densities **J** and **M** by eigenfunctions of the Helmholtz equation, on the other hand, have the obvious advantage that they avoid the numerical evaluation of the Green's integral in equations (2) when they coincide with the eigenfunctions of the operators $\mathcal{J}$ and $\mathcal{M}$. However, their use requires the fictitious boundary S to be a constant coordinate surface in some coordinate system where the Helmholtz operator is separable. This constraint causes a major loss of efficiency when the structures in the interior region are thin and slender, for in this case the region discretized via the finite element method becomes much wider than necessary for the analysis of the region of interest. The large number of superfluous unknowns thus introduced increases both the memory needed and computing time.

The order of presentation of the works included in this part of the present survey follows the above classification by type of basis functions used to express the equivalent surface current densities **J** and **M** on S. Those methods that employ the eigenfunctions of the Helmholtz equation will be discussed first, the polynomial techniques thereafter. However, the various different methods that have been devised since the 1970s are often closely related, for in most cases new methods have been proposed in an attempt to overcome the restrictions and difficulties of already existing ones. Thus, it is nearly impossible to discuss any one method in isolation without reference to those developed earlier. For this reason, forward as well as backward references occur in the discussion to follow.

Each technique considered in the ensuing exposition will be placed within the framework of the taxonomy related to the basis functions used, as well as of the classification based on solution algorithms as suggested by Pearson, Peterson, Bahrmasel, and Whitaker [*op. cit.* 1992] — the inward-looking and outward-looking formulations. This latter classification has the important advantage of bringing into evidence the cases in which interior problem resonances may arise.

Eigenfunctions as Basis Functions

Hybrid methods that employ eigenfunctions of the Helmholtz equation to approximate the equivalent surface current densities **J** and **M** in equations (2) naturally divide into four classes: the unimoment method, the finite element with extended boundary conditions method, the transfinite element method, and the bymoment method. These will be taken up individually in the following.

Unimoment method

The unimoment method was devised by Mei (1974) as an alternative method to that previously proposed by McDonald and Wexler (see next section). In his original work, Mei applied the unimoment method to solve two-dimensional scattering and antenna problems. Since it allows a simple numerical implementation, the unimoment method has been employed widely to solve antenna problems and to analyze the scattering from arbitrarily complex objects. Among the various applications, the following are particularly worth mentioning:

(a) Axially symmetric antenna and scattering problems involving heterogeneous dielectric materials were dealt with by Stovall and Mei (1975) and Chang and Mei [1976].

(b) Scattering from two bodies of revolution and from buried objects were treated by Hunka and Mei (1981) and Chang and Mei (1981), respectively.

(c) Scattering by raindrops was dealt with by Morgan (1980), who also computed microwave absorption by the cranial structure (Morgan 1981).

Works that describe or apply the unimoment method are numerous. The theory is probably best presented in more recent papers as, for instance, in the work by Mei [1987]. This statement of the technique is probably more complete and clearer than any of the many others.

The application of the unimoment method for the solution of an unbounded domain problem requires the use of a fictitious contour Γ, characterized by a fixed value of the radial coordinate ρ, to separate the interior and exterior regions. The scattered electromagnetic field outside Γ is expressed in terms of a series of eigenfunctions Ψ_m ($m = 1, \ldots, N$) of the Helmholtz equation. As pointed out by Pearson *et al.* [1992], this representation of the scattered field is tantamount to expressing the equivalent surface current densities **J** and **M** on Γ as a linear combination of the eigenfunctions Ψ_m. In fact, in this particular case, the eigenfunctions Ψ_m of the Helmholtz equation are also eigenfunctions of the integral operators $\mathcal{J}$ and $\mathcal{M}$. The field in the region interior to Γ is expressed as a linear combination of numerically generated basis functions u_m ($m = 1, \ldots, N$). These basis functions are derived by using the finite element method to solve N interior Dirichlet problems, with values of the eigenfunction Ψ_m along Γ serving as Dirichlet-type boundary values. The unknown coefficients of the linear combinations representing the interior and exterior fields are subsequently determined by enforcing the continuity of the field and its normal derivative on Γ. Solution of a problem by the unimoment method thus requires multiple solutions of a large but sparse matrix equation ($N_T \times N_T$, where N_T is the total number of nodes of the finite element mesh) and a smaller full matrix equation to effect matching of the interior and exterior fields.

Clearly, the unimoment method consists of two separate and successive phases. The first phase deals with the construction of the basis functions for the interior problem; the second phase consists of matching the expansions of the interior and exterior fields and their normal derivatives. This matching can be accomplished in several ways. The easiest to implement is probably that suggested by Mei [1987]. It consists in requiring continuity of the internal and external fields along two adjacent circles, spaced by only one finite element of the mesh, so that the normal derivative is implicitly replaced by a normal first difference.

Because the unimoment method leads to a final linear system of equations whose unknowns are the expansion coefficients of the internal and external fields along Γ (or the equivalent surface current densities on the same contour), it can be considered an inward-looking formulations in the classification framework of Pearson *et al.* [1992]. Consequently, the unimoment method is subject to potential difficulties due to the interior resonance problem, which may occur when solving the interior problem with Dirichlet-type boundary conditions. This may be shown by all the inward-looking formulations. As compared with the earlier McDonald and Wexler [1972] technique, it enjoys the following advantages:

(a) The large matrix arising from the solution of the interior problems is sparse and possibly banded. Sparsity can be exploited to good effect by choosing appropriate matrix algorithms.

(b) Most of the calculations, for example the finite element solution of the N interior problems, do not involve the incident field. Multiple scattering problems with different incident fields therefore require comparatively minor numerical effort.

The major drawback of the unimoment method, most evident in three-dimensional applications, arises from the large number of superfluous unknowns introduced when the scattering object has a long and slender shape — disks, thin wire antennas, elongated ellipsoids, and so on. In such cases, most of the nodes of the finite element mesh may belong to free space or to the homogeneous space surrounding the region of interest.

Finite elements with extended boundary conditions

In 1984, Morgan, Chen, Hill, and Barber [included with this chapter] proposed a hybrid technique for solving three-dimensional scattering problems, which combines finite elements with the extended boundary condition method (Waterman 1965). This method overcomes the inefficiency of the unimoment method in treating elongated objects, which arises from restrictions on the shape of the fictitious surface S by choosing an arbitrarily-shaped surface to enclose the region of interest. The electromagnetic field outside S is written as the sum of the incident and scattered fields, and both these fields are expanded in vector spherical harmonic functions. Each of these harmonic functions is employed in turn as a Dirichlet boundary condition for an internal problem, and the several internal problems are solved by the finite element method. These solutions are subsequently employed as basis functions for expanding the field in the interior region, similarly to the unimoment method. The matching between internal and external solutions is accomplished by using the same surface integral equation as in the extended boundary condition method (Waterman 1965), first developed for the analysis of electromagnetic scattering from dielectric bodies.

This method has not been extensively used because it expresses the fields in the exterior region in terms of special functions, which are cumbersome to evaluate. Moreover, although the choice of the surface S is more flexible than in the unimoment method, thus reducing the geometric domain for computation, the finite element method with extended boundary conditions still presents convergence problems when applied to long and slender structures.

Transfinite element method

We denote with this term the set of techniques that introduce the analytical expression of the outward normal derivative of the scattered field, as derived from the eigenfunction expansion of the scattered field itself, into either the variational expression or the weighted residual statement of the equivalent internal problem. Consequently, the transfinite element method has the advantage, with respect to the unimoment method, of avoiding the numerical approximation of the internal field normal derivative. On the other hand, the introduction of a non-local boundary condition in the formulation of the equivalent internal problem makes the coefficient matrix of the finite element method lose its characteristic sparsity, even though it still contains a large number of zero entries.

Although the term "transfinite element method" was coined by Lee and Cendes [1987*a*], other researchers had previously developed similar ideas. In particular, Jeng and Chen (1984), in solving a two-dimensional scattering problem to validate their fundamental variational principle, employed a technique that could be classified as a transfinite element method. Much like the unimoment procedure, they used eigenfunctions of the Helmholtz equation to expand the electromagnetic field in the unbounded region exterior to a fictitious contour Γ. This boundary contour (or surface) has to be chosen of the constant-coordinate kind, so this method suffers from the same efficiency defects as the unimoment method when applied to the study of elongated antennas or scatterers. The main difference between the Jeng-Chen formulation of the transfinite element method and the unimoment method lies in the procedure employed to match the exterior and the interior fields. The unimoment method matches the expansions of the interior and exterior fields after the solution of N internal problems, which determine the N internal region basis functions u_m $(m = 1, \ldots, N)$. On the other hand, Jeng and Chen directly introduced the eigenfunction expansion of the external field normal derivative in the expression of the functional associated with the interior region. Moreover, by exploiting the orthogonality of the eigenfunctions of the Helmholtz equation, they were able to analytically express the unknown coefficients in the expansion of the scattered field as a function of the same scattered field on the contour Γ. The final system of equations built in this way involves only the values of the field at the finite element mesh nodes, and implicitly establishes the continuity of the field and its normal derivative on the contour Γ. Subsequently, the same technique was applied by Wu and Chen [1986] to solve problems concerning the electromagnetic scattering by anisotropic dielectric cylinders in two dimensions. Their formulation, as well as that by Jeng and Chen, is of the outward-looking type, and its accuracy has later on been compared with that achievable by using the second-order absorbing boundary conditions by Pearson, Whitaker, and Bahrmasel (1989). A noteworthy feature of the Jeng-Chen formulation of the transfinite element method is that, even though it employs just one contour Γ to separate the interior and the exterior regions, it is not affected by the interior resonance problem [Pearson *et al.* 1992].

In 1987, Lee and Cendes [1987*a*] slightly modified this formulation and initially applied it to solve the Laplace's equation on unbounded domains, employing a weighted residual formulation of the problem. Probably, Lee and Cendes arrived at their formulation independently of the previous works, since they mentioned neither the work of Jeng and Chen nor of Wu and Chen in their paper. Subsequently, the same authors (Lee and Cendes 1987*b*) extended its field of applicability to two-dimensional scattering problems and to the determination of scattering parameters in microwave monolithic integrated circuit devices (Lee and Cendes 1988). In their implementation of the technique, Lee and Cendes (1987*b*) did not analytically eliminate the coefficients of the scattered field expansion in the weighted residual statement of the internal problem. Consequently, only the continuity of the normal derivative of the field on the contour Γ is implicitly established. The continuity of the field on Γ is separately imposed by requiring the interior and exterior field approximations to be the same at the finite element nodes. Apparently, the Lee-Cendes transfinite element solution algorithm leads to a system of linear equations whose unknowns include both the values of the field at the nodes of the finite element mesh and the coefficients of the field expansion in the unbounded region outside Γ. Thus, their implementation fits ill into the inward-looking and outward-looking classification of Pearson *et al.* [1992], since the final system of linear equations contains all the variables of the problems, both internal and external. Although the Lee-Cendes procedure leads to a system of equations that involves a greater number of unknowns than the Jeng-Chen technique, it allows the scattering field to be immediately evaluated after the solution of the system of equation. In contrast to the former procedure, the Jeng-Chen technique requires further calculation for evaluating the scattered field.

The Lee-Cendes formulation of the transfinite element method has also been successfully applied to the analysis of electromagnetic scattering from periodic structures by Pelosi, Freni, and Coccioli [1993] and from rotating cylinders by Graglia, Freni, and Pelosi [1993]. However, in these papers continuity of the field along the fictitious contour Γ is enforced by means of a weighted residuals technique. This procedure allows the external field expansion to be truncated at a number of eigenfunctions different from the number of the finite element mesh nodes on the contour Γ. This approach is particularly advantageous if the number of finite element mesh nodes on Γ is much larger than the number of eigenfunctions required to provide a good approximation of the exterior field. Both these papers solve the global system of linear equations by using a technique similar to that suggested by Jin and Liepa [1988*b*]. They therefore make use of an inward-looking formulation.

Bymoment method

This technique was initially presented by Cangellaris and Lee [1990] as a way to solve simple two-dimensional scattering problems. Subsequently, Cangellaris and Lee [1991] themselves applied the same method to the analysis of the electromagnetic scattering from one or more inhomogeneous dielectric cylinders of arbitrary cross-section at oblique incidence.

The bymoment method tries to combine the unimoment method and the "field feedback formulation" (discussed in detail in the next section) to obtain the best of both worlds. In particular, it permits use of a non-constant-coordinate mathematical boundary so the geometric domain for computation can be kept fairly minimal. This is generally characteristic of the field feedback formulation, while maintaining the straightforward matching procedure for the interior and exterior fields of the unimoment method, without involving the Green's function.

Here, the bymoment method will be described in a slightly different way than that of Cangellaris and Lee [1990]. The presentation is intended to emphasize the similarities shared by the bymoment and unimoment methods, and also to justify placing the bymoment method in the class of methods that use the eigenfunctions of the Helmholtz equation to approximate the equivalent surface current densities **J** and **M** on the fictitious contour Γ. The bymoment method actually employs two contours — Γ_o and Γ_i, the outer and inner, respectively — of arbitrary shape to separate the interior and the exterior problems. Both the fictitious contours Γ_o and Γ_i completely enclose the scatterer following its natural shape. The use of non-constant-coordinate contours allows the minimization of the number of unknowns relevant to the interior problem. This latter is solved N times with different appropriate Dirichlet-type boundary conditions, much like the unimoment method and the field feedback formulation. This procedure has the advantage that it leads to a very efficient solution of the interior problem. By solving the N different interior problems, the N basis functions u_m $(m = 1, \ldots, N)$ are obtained that subsequently serve to approximate the total field Φ in the region enclosed by the boundary Γ_o,

$$\Phi = \sum_{m=1}^{N} a_m u_m. \tag{6}$$

If the transverse plane is taken as the x–y coordinate plane, then in equation (6) $\Phi = E_z$ (in the TM$_z$ case) or $\Phi = H_z$ (in the TE$_z$ case).

The first step of the matching procedure between the interior region and exterior region fields consists of expanding the scattered field Φ^S outside the fictitious boundary Γ_i in terms of the eigenfunctions Ψ_m of the Helmholtz equation

$$\Phi^S = \sum_{m=1}^{N} b_m \Psi_m. \tag{7}$$

The continuity conditions for the field and its normal derivative along the contour Γ_i are imposed through the weighted residual method, much as they are in the unimoment method. Entire-domain weighting functions may be chosen as:

(a) The normal derivatives of the eigenfunctions $\partial\Psi_j/\partial n$ to impose the continuity of the field.

(b) The eigenfunctions Ψ_j to impose the continuity of normal derivative of the field.

The matching procedure described above then leads to the equations

$$\oint_{\Gamma_i} \Phi \frac{\partial \Psi_j}{\partial n} d\Gamma_i = \oint_{\Gamma_i} \Phi^S \frac{\partial \Psi_j}{\partial n} d\Gamma_i + \oint_{\Gamma_i} \Phi^I \frac{\partial \Psi_j}{\partial n} d\Gamma_i \qquad j = 1, \ldots, N, \tag{8}$$

$$\oint_{\Gamma_i} \Psi_j \frac{\partial \Phi}{\partial n} d\Gamma_i = \oint_{\Gamma_i} \Psi_j \frac{\partial \Phi^S}{\partial n} d\Gamma_i + \oint_{\Gamma_i} \Psi_j \frac{\partial \Phi^I}{\partial n} d\Gamma_i \qquad j = 1, \ldots, N, \tag{9}$$

where Φ^I is the incident field. If equations (6) and (7) are substituted into equations (8) and (9), and then Green's theorem is applied to the difference obtained on subtracting the resulting equations, the following system of equations is obtained:

$$\sum_{m=1}^{N} a_m \oint_{\Gamma_i} \left[\Psi_j \frac{\partial u_m}{\partial n} - u_m \frac{\partial \Psi_j}{\partial n} \right] d\Gamma_i = \oint_{\Gamma_i} \left[\Psi_j \frac{\partial \Phi^I}{\partial n} - \Phi^I \frac{\partial \Psi_j}{\partial n} \right] d\Gamma_i \qquad j = 1, \ldots, N. \tag{10}$$

This coincides with the equations obtained by Cangellaris and Lee [1990] following a somewhat different procedure. Solution of the system of linear equations (10) finally yields the unknown coefficients a_m of expression (6) for interior field.

Viewed in this light, the bymoment method appears as an extension of the unimoment method. The difference between the two techniques consists of using a non-constant-coordinate contour to separate the interior and exterior problems in the bymoment method. This is made possible by slightly changing the matching procedure by introducing equation (10). Although the use of a non-constant-coordinate contour minimizes the number of unknowns in the interior problems, it makes the implementation of the method more involved, as already noted by Morgan and Welch (1986). In fact, *appropriate* Dirichlet boundary conditions, to be imposed in the solution of the N

interior problems, are no longer unequivocally defined; their choice is left to the user's experience and judgment. This is a drawback of the bymoment method, for the efficiency of the method and the accuracy of the results strongly depend on these boundary conditions — as clearly stated by Cangellaris and Lee [1990] themselves.

Except for the previously mentioned characteristic of using an arbitrary contour to separate the interior and exterior problems, the bymoment method has the same advantages and disadvantages as the unimoment method. In particular, it should be remembered that it is subject to the problem of interior resonances.

Polynomial or Other Basis Functions

The methods that use polynomials or other basis functions to approximate the equivalent surface current densities **J** and **M** in equations (2) have the advantage that they allow an arbitrary choice of the contour Γ (or the surface S) that encloses the region to be treated by the finite element method. Because it is possible to place such a fictitious boundary arbitrarily close to the scatterer or to the radiating system of interest, it is apparent that these methods allow the geometric domain of computation to be minimal. This advantage is traded against the requirement of a numerical evaluation of the Green's integral appearing in equations (2).

Hybrid finite element–boundary integral method

The first published works on this topic date to the early 1970s when Silvester and Hsieh [1971] and then McDonald and Wexler [1972] suggested a technique to solve unbounded domain problems by means of the finite element method. Initially, Silvester and Hsieh devised a technique to solve Laplace's equation on unbounded domains. McDonald and Wexler modified this formulation to apply to the Helmholtz equation as well. Subsequently, a wide class of procedures, denoted also as "hybrid finite element methods" (HFEM) or "finite element–boundary integral methods" (FE–BI), have been derived by different authors from this technique.

McDonald and Wexler [1972] dealt with two-dimensional problems, where equation (1) is replaced by its two-dimensional scalar counterpart

$$\Phi(\boldsymbol{\rho}) = \Phi^I(\boldsymbol{\rho}) + \int_\Gamma (G(\boldsymbol{\rho},\boldsymbol{\rho}')\nabla\Phi(\boldsymbol{\rho}') \cdot \mathbf{1}_{n'} \\ -\Phi(\boldsymbol{\rho}')\nabla G(\boldsymbol{\rho},\boldsymbol{\rho}') \cdot \mathbf{1}_{n'})\, d\Gamma' . \tag{11}$$

If the transverse plane is taken as the x–y coordinate plane, then in equation (11) $\Phi = E_z$ (in the TM$_z$ case) or $\Phi = H_z$ (in the TE$_z$ case). Moreover, $\Phi^I(\boldsymbol{\rho})$ is the incident field, and $G(\boldsymbol{\rho},\boldsymbol{\rho}')$ is the two-dimensional free-space Green's function

$$G(\boldsymbol{\rho},\boldsymbol{\rho}') = -\frac{j}{4} H_0^{(2)}(k_0|\boldsymbol{\rho}-\boldsymbol{\rho}'|)\,, \tag{12}$$

where $H_0^{(2)}(k_0|\boldsymbol{\rho}-\boldsymbol{\rho}'|)$ is the zeroth-order Hankel function of the second kind. The formulation of the method as described by McDonald and Wexler makes use of two arbitrary contours, both circumscribing all the sources and discontinuities of the problem. The outer contour, here referred to as Γ_o, encloses the region discretized by the finite element method. The inner contour Γ_i defines the domain of integration in equation (11); it is placed just inside Γ_o and parallels Γ_o topologically. It is worth noting that the two boundaries Γ_o and Γ_i may be multiply connected if the structure of interest is made up of disconnected parts, once again allowing the geometric domain for computation to be minimized.

The solution procedure proposed by McDonald and Wexler consists of two steps. In the first step, equation (11) is discretized by expressing the unknown integrand function $\Phi(\boldsymbol{\rho}')$ on the contour Γ_i in terms of the polynomial shape functions used in the finite element method. This leads to the matrix equation

$$\mathbf{P}_{BB}\mathbf{\Phi}_B + \mathbf{P}_{BI}\mathbf{\Phi}_I = \mathbf{\Phi}_B^I\,, \tag{13}$$

where $\mathbf{\Phi}_B$ and $\mathbf{\Phi}_I$ are two column vectors containing the unknown values of the function Φ at the N_B nodes of the finite element mesh that lie on the contour Γ_o and at the N_I nodes interior to the region delimited by Γ_o, respectively. $\mathbf{\Phi}_B^I$ is the column vector whose entries are the values of the incident field Φ^I at the N_B nodes on the contour Γ_o, and finally, $\mathbf{P}_{BB}$ and $\mathbf{P}_{BI}$ are two complex $N_B \times N_B$ and $N_B \times N_I$ matrices, respectively. The use of a double contour to separate the interior and exterior fields avoids the difficulties due to the Green's function singularities otherwise encountered in calculating the entries of the matrix $\mathbf{P}_{BB}$. In fact, in equation (11) the domain of integration is the inner contour Γ_i, while the observation point $\boldsymbol{\rho}$ lies on the outer boundary Γ_o. The second step of the McDonald and Wexler procedure consists of deriving an expression for the vector $\mathbf{\Phi}_B$ from equation (13) and substituting it into the matrix equation

$$\mathbf{K}_{IB}\mathbf{\Phi}_B + \mathbf{K}_{II}\mathbf{\Phi}_I = \mathbf{0}\,. \tag{14}$$

This latter is obtained by minimizing the functional relevant to the interior region problem, as usual in the finite element method. The final result is the system of linear equations

$$\mathbf{K}'_{II}\mathbf{\Phi}_I = \mathbf{0}\,, \tag{15}$$

whose solution yields the values of the field at the N_I nodes interior to the region delimited by Γ_o.

The technique employed by McDonald and Wexler to impose the non-local boundary condition in equation (11) by means of two mathematical contours is similar to the so called "extended boundary condition method" of Waterman (1965). It has the advantage of eliminating non-physical resonances of the equivalent internal problem (Peterson 1990). It belongs to the category denoted by Pearson *et al.* [1992] as outward-looking formulations. In fact, it leads to the final system of linear equations (15) whose unknowns are the values of the field at the finite element mesh nodes that lie inside the outer contour Γ_o. The solution of the original unbounded domain problem requires the inversion of two matrices. The first is the complex full matrix $\mathbf{P}_{BB}$, with dimensions $N_B \times N_B$. The second is the complex $N_I \times N_I$ matrix $\mathbf{K}'_{II}$. This latter matrix has large dimensions and is partly full. Hence, the solution of the system of linear equation (15) may be numerically inefficient using direct matrix methods.

The method of McDonald and Wexler has been applied — sometimes with more or less effective modifications — by several authors to the analysis of both two- and three-dimensional problems. For instance, the method employed by Marin (1982) to analyze the scattering from cylinders of arbitrary cross-section and composition is basically that suggested by McDonald and Wexler with the minor modification that only one fictitious contour Γ is introduced. Consequently, such formulation does not provide accurate results when the quantity $-k^2$ (the propagation constant of the homogeneous medium in the region outside Γ) is close to an eigenvalue of the Laplacian in the interior region delimited by Γ.

Another example of the McDonald-Wexler method — without any modification — was published by Jin and Liepa [1988*a*], who solved the two-dimensional problem of electromagnetic scattering from coated cylinders of arbitrary cross-section. In a further work, the same authors [Jin and Liepa 1988*b*] improved the numerical efficiency of the method by using a better solution scheme for equations (13) and (14). In particular, Jin and Liepa suggested substituting equation (14) into equation (13) instead of the other way around, as had been usual till then. This procedure yields a final system of linear equations of the form

$$\mathbf{P}'_{BB}\mathbf{\Phi}_B = \mathbf{\Phi}^I_B. \tag{16}$$

Although this new solution algorithm is equivalent to the previous one from a mathematical point of view, it is more efficient from a numerical point of view, even though it still requires inversion of two matrices: the $N_I \times N_I$ matrix $\mathbf{K}_{II}$ and subsequently the $N_B \times N_B$ matrix $\mathbf{P}'_{BB}$. However, while $\mathbf{P}'_{BB}$ is still a complex full matrix, the matrix $\mathbf{K}_{II}$, which is of larger dimensions than $\mathbf{P}'_{BB}$, is sparse and symmetric, since it is a standard finite element matrix. Moreover, its entries are real if the materials involved are lossless. Consequently, $\mathbf{K}_{II}$ can be very efficiently inverted by use of sparsity-exploiting algorithms. A further advantage is realized if solutions are required for several directions of incidence. For different directions of propagation of the incident wave, the McDonald and Wexler formulation requires construction and inversion of the matrix $\mathbf{K}'_{II}$ for each different incidence direction. On the other hand, the Jin and Liepa [1988*b*] formulation requires only a new vector $\mathbf{\Phi}^I_B$ for each incidence direction. Both the matrices $\mathbf{K}_{II}$ and $\mathbf{P}'_{BB}$, relevant to the interior and exterior problem, respectively, need to be computed and factorized just once.

Although the formulation given by Jin and Liepa [1988*b*] presents several numerical advantages, it affects the solution uniqueness adversely. Indeed, the changes introduced by Jin and Liepa [1988*b*] convert the McDonald-Wexler method into a formulation of the inward-looking type [Pearson *et al.* 1992], so that non-physical internal resonances may occur. Inversion of the matrix $\mathbf{K}_{II}$, which constitutes the first step of the solution procedure by Jin and Liepa, is equivalent to solving an interior problem with boundary conditions of the Neumann type. Only subsequently is the interior problem solution joined to the exterior one. Consequently, non-physical solutions may arise from interior-region resonances.

A further interesting modification of the original method by McDonald and Wexler was presented by Collins, Volakis, and Jin [1990] for a two-dimensional problem. In this case, the two contours that bound the electromagnetically complex interior region are chosen to be rectangular. This choice permits the boundary condition in equation (11) to be expressed by convolution integrals, which can be efficiently evaluated by the Fast Fourier Transform. In particular, this is accomplished by using pulse basis functions to discretize the unknown field in the convolution integrals. A technique is obtained that has been called "FE–CGFFT," because the global system of equations is subsequently solved by applying the conjugate gradient method. Later, Collins, Jin, and Volakis (1990) extended their technique to the case of circular and oval contours, but without using a double boundary. In this case, non-physical solutions related to the resonances of the interior region may arise. Recently, the same authors [Collins, Jin, and Volakis 1992] proposed a technique to avoid interior-region resonances based on the "complexification" (Murphy, Rokhlin, and Vassiliou 1990) of the propagation constant, that is, by introducing small losses in the medium that surrounds the scatterer or the antenna of interest. The formulation suggested by Collins, Jin, and Volakis [1992] makes the solution relatively insensitive to the attenuation factor α introduced in the expression of the propagation constant k.

The method of McDonald and Wexler, even with the successive improvements made by other authors, does not allow the analysis of infinite structures such as metallic planes or wedges with localized inhomogeneities or anisotropies. Indeed, it is apparent that such structures cannot be completely enclosed inside a finite-sized fictitious surface. This problem has been considered in a general form by Ramahi and Mittra [1991], while Jin and Volakis (1991*a*) solved it for the particular case of an open cavity in an infinite metallic plane. The first step of their procedure is to separate the interior and exterior problems. This is done by applying the equivalence principle and by introducing suitable surface current distributions on both sides of the surface that coincides with the aperture of the cavity. Hence, the original problem is reduced to the solution of two independent simpler problems:

(a) The exterior problem of an electromagnetic field impinging on an infinite conducting plane on which flows a surface magnetic current distribution $\mathbf{M}^{ext} = +\mathbf{M}$.

(b) The interior problem of radiation from a surface magnetic current distribution $\mathbf{M}^{int} = -\mathbf{M}$, which lies on the internal side of a metallic cavity.

The continuity of the tangential component of the electric field at the cavity aperture is automatically ensured, as a consequence of the definitions of the surface magnetic current distributions $\mathbf{M}^{ext}$ and $\mathbf{M}^{int}$. However, continuity of the tangential component of the magnetic field must be imposed independently at the same aperture,

$$(\mathbf{H}^{ext}(\mathbf{M}) + \mathbf{H}^{inc}) \times \mathbf{1}_n = \mathbf{H}^{int}(-\mathbf{M}) \times \mathbf{1}_n, \qquad (17)$$

where $\mathbf{1}_n$ is the unit outward normal vector on the cavity aperture surface S_{ap}. This condition is directly imposed into the functional relevant to the internal problem (Silvester and Ferrari 1990)

$$\mathcal{F}(\mathbf{E}^{int}) = \frac{1}{2}\int_V \left(\frac{1}{\mu_r}(\nabla \times \mathbf{E}^{int}) \cdot (\nabla \times \mathbf{E}^{int}) + \frac{s}{\mu_r \epsilon_r^2}(\nabla \cdot \epsilon_r \mathbf{E}^{int})^2 - k_0^2 \epsilon_r \mathbf{E}^{int} \cdot \mathbf{E}^{int}\right) dV + jk_0\zeta_0 \int_{S_{ap}} (\mathbf{E}^{int} \times \mathbf{H}^{int}) \cdot \mathbf{dS}. \qquad (18)$$

This latter functional is minimized by applying a standard finite element procedure over the cavity volume V. Since Jin and Volakis used nodal finite elements, their functional equivalent to equation (18) also contained the penalty term $s(\nabla \cdot \epsilon_r \mathbf{E}^{int})^2/\mu_r\epsilon_r^2$, where s is the penalty factor, so that the problem of spurious solutions is partially eliminated (Rahman and Davies 1984). Magnetic field continuity is taken into account by substituting equation (17) into equation (18). It is worth noting that the surface magnetic currents are expressed by the same basis functions as are employed in the finite element method.

The Jin-Volakis technique is apparently an outward-looking formulation, and it is not subject to non-physical solutions due to interior resonances, since the boundary integral equation is not defined on a closed surface. Since Jin and Volakis used just one fictitious surface to separate the interior and exterior regions, they had to cope with the singularities of the Green's function when calculating the magnetic field at the aperture surface S_{ap} generated by the equivalent magnetic current placed on S_{ap}.

Subsequently, Jin and Volakis (1991*b*) applied the same technique to analyze the electromagnetic transmission through a three-dimensional slot in a conducting plane, and in a subsequent paper determined the radiating performance of microstrip patch antennas and arrays residing in a cavity [Jin and Volakis 1991*c*]. Both solutions were carried out using solenoidal vector basis functions in the finite element method.

The extended approach suggested by Ramahi and Mittra [1991] to solve unbounded problems deals with two-dimensional scatterers whose exterior surface defines a separable geometry, as is the case of a wedge, and uses a double contour scheme. Ramahi and Mittra suggested using the available (Harrington 1961) Green's function of the particular geometry, denoted as Green's function of the first kind $G^{(1)}$, rather than the free space Green's function. The very narrow support of the term $\partial G^{(1)}/\partial n$ allows the boundary integral to be satisfactorily approximated by integrating over the small interval spanned by the support of $\partial G^{(1)}/\partial n$. This approximation leads to a reduction of the non-local behavior of the boundary condition, thus largely preserving the sparsity of the finite element matrix. Although Ramahi and Mittra carried out the theoretical part of the paper for a general wedge geometry, the numerical results given pertain only to a plane geometry.

Hybrid finite element and moment method

Yuan, Lynch, and Strohbehn [1990] presented a hybrid technique that combines the finite element method with the method of moments (MoM). A fictitious surface S is introduced that completely encloses the scattering body. S is unrestricted; indeed, it may be chosen to coincide with the exterior surface of the scatterer. The equivalence principle is then applied to transform the original problem into separate interior and exterior problems. A suitable set of equivalent surface currents is defined for each problem. The two problems are coupled by imposing continuity requirements on both the electric and magnetic fields produced by the two sets of currents on opposite sides of the boundary surface S. The resulting integral equation is then solved by applying the standard MoM procedure, which involves expressing the unknown currents by a linear combination of suitable basis functions. The basis functions used for the interior problem must be congruent with the shape functions of the finite elements, while the basis functions for the exterior can be arbitrarily chosen. Yuan, Lynch, and Strohbehn [*op. cit.* 1990] in fact found it convenient to use the same functions for both interior and exterior current densities. The finite element method is applied to evaluating the elements of the moment matrix that express the coupling between elements of the surface current distributions through the region interior to S. This latter region can be arbitrarily complex.

The hybrid finite element–moment method has also been applied to the scattering from three-dimensional bodies [Yuan 1990] and from two-dimensional periodic structures (Gedney, Lee, and Mittra 1992).

As stated by Yuan, Lynch, and Strohbehn themselves, this method has the typical advantages and shares the inconveniences of every finite element–boundary integral formulation. In particular, they noted the occurrence of non-physical resonant solutions of the interior problem. This is the most important defect of their hybrid method, and it is due to the inward-looking character of the formulation adopted. Although this method coincides in practice with previous formulations, it may appear simpler because it makes use of the finite element method in the context of the well known moment method and extends the latter to electromagnetically complex problems. In other words, the finite element method is used to determine the Green's function — otherwise unknown and possibly very complicated — of the interior problem numerically.

Field feedback formulation

The "field feedback formulation," or "F^3," was first introduced by Morgan and Welch (1986) in an effort to avoid limitations on the choice of the boundary surface S separating the interior and exterior problems as they arise in the unimoment method, transfinite elements, and the finite element method with extended boundary conditions. However, it has been applied only to solve two-dimensional scattering problems from finite metallic circular cylinders (Morgan and Welch 1986) and from penetrable objects [Morgan and Welch 1992].

The original Morgan and Welch formulation employs two fictitious surfaces, an external surface S_o and an internal surface S_i, to impose the exact boundary condition (1). These surfaces are conformal to the complex region to be analyzed. In order to avoid the convergence difficulties of the eigenfunctions of the Helmholtz equation over non-spherical surfaces (non-circular contours in two dimensions) typical of the finite element method with extended boundary conditions, the tangential component of the scattered field $\mathbf{\Phi}^S(\mathbf{r}_o)$ on S_o is expressed as a linear combination of suitable vector basis functions $\mathbf{\Psi}_m(\mathbf{r}_o)$ with unknown coefficients. The tangential component $\mathbf{\Phi}(\mathbf{r}_o)$ of the total electromagnetic field on the same surface S_o can be expressed as the sum of the tangential components of the incident field $\mathbf{\Phi}^I(\mathbf{r}_o)$ and of the scattered field $\mathbf{\Phi}^S(\mathbf{r}_o)$,

$$\mathbf{\Phi}(\mathbf{r}_o) = \mathbf{\Phi}^I(\mathbf{r}_o) + \mathbf{\Phi}^S(\mathbf{r}_o) = \\ = \mathbf{\Phi}^I(\mathbf{r}_o) + \sum_{m=1}^{N} c_m \mathbf{\Psi}_m(\mathbf{r}_o)\,. \tag{19}$$

The finite element method is used to determine the $(N+1)$ solutions of the interior problem that result from imposing on S_o the inhomogeneous Dirichlet boundary values $\mathbf{\Phi}^I(\mathbf{r}_o)\,\mathbf{\Psi}_1(\mathbf{r}_o),\ \mathbf{\Psi}_2(\mathbf{r}_o), \ldots, \mathbf{\Psi}_N(\mathbf{r}_o)$. Once the $(N+1)$ solutions of the interior problems are known, it is possible to evaluate both the electric and magnetic equivalent current surface distributions ($\mathbf{J}_m$, $\mathbf{M}_m$, $m = 1, \ldots, N+1$) that are induced on the internal fictitious surface S_i by each solution.

It should be apparent that the equivalent surface current densities $\mathbf{J}_m$ and $\mathbf{M}_m$ are thus approximated in terms of the polynomials of the finite element method used in the interior region bounded by S_o. The electromagnetic field generated by these current densities on the external surface S_o can be calculated by applying the operators $\mathcal{J}$ and $\mathcal{M}$ in equation (2) to the equivalent currents; the result must be equal to expression (19). Consequently, an integral equation is obtained which, when solved (e.g., by the method of weighted residuals), yields the values of the unknown coefficients c_m.

The field feedback formulation is to a certain extent similar to the unimoment method. Just like the unimoment method, it allows numerically efficient solution, in two successive and separate steps, of the interior and exterior problems. Consequently, the large but sparse matrix of the interior problem may be treated separately from the smaller but full exterior matrix. Computations on the large sparse matrix need only be executed once, regardless of the direction of propagation of the incident wave. This latter feature is common also to the hybrid finite element method of Jin and Liepa [1988*b*]. The main difference between the field feedback formulation and the Jin-Liepa hybrid method is that the former requires solving the interior problem $(N+1)$ times with the Dirichlet boundary conditions as given, while in the latter method the interior problem is solved just once, with a Neumann boundary condition that is imposed implicitly.

Since it is apparent that the F^3 method uses an inward-looking formulation, it presents all the difficulties related to resonant solutions for the interior problem. Such non-physical solutions may occur when solving the interior problem with Dirichlet-type boundary conditions.

The major drawback of the field feedback formulation is connected with the choice of the vector basis functions $\mathbf{\Psi}_m(\mathbf{r}_o)$, that are needed to start the solution procedure. This is the same disadvantage as attaches to the bymoment method, which was in fact derived from the field feedback formulation. The choice of the basis functions $\mathbf{\Psi}_m(\mathbf{r}_o)$ is not easy, particularly in three-dimensional problems and in the presence of electromagnetically complex structures composed of heterogeneous materials. Yet this is exactly the kind of situation where the finite element method is at its best!

Comparison of Hybrid Methods

To make a reasonable and useful comparison among the different hybrid methods discussed in this survey, suppose each method to be applied to a general two-dimensional scattering problem involving a two-dimensional scatterer of finite cross-section. The transverse plane is taken as the x–y coordinate plane, and an incident plane wave with $\mathrm{T}M_z$ polarization is assumed. Consequently, the unknown of the problem is the longitudinal component of the electric field E_z. The features of each technique in solving the considered scattering problem are summarized in Table I. Some of them are evaluated in a qualitative "smile" – "serious" – "sad" scale where the "serious" qualitative level is assigned to those characteristics that are not definitely good or bad in themselves but whose goodness depends on other factors also.

The first four columns of Table I describe the characteristics of the fictitious contour Γ that separates the interior and exterior problems. Γ can have an arbitrary shape (column A) or may have to be a constant-coordinate contour (CC). Moreover, some methods use a single fictitious boundary (S) while others employ a double fictitious contour (D) to avoid singularities in the integral equation used to furnish boundary conditions for the equivalent internal problem. It is worth noting that the quality judgment applies to a general scattering problem in which the scatterer has an arbitrary cross-sectional shape (for instance, it could be of the worst kind: thin and long). Indeed, a method employing a constant-coordinate contour Γ — such as the unimoment method [Mei 1987] or transfinite elements [Pelosi, Freni, and Coccioli 1993] — is not inherently inefficient: it might work quite efficiently in case of scatterers with certain cross-sectional shapes not too different from a circle.

The fifth to tenth columns of the table should be self-explanatory. The only comments that could be added relate to the work of Lee and Cendes [1987] and Silvester and Hsieh [1971]. The former fits ill into the inward-looking and outward-looking classification of Pearson, Peterson, Bahrmasel, and Whitaker [1992]. The second is not subject to non-physical resonant solutions because the original article deals with Laplace's equation.

The three columns that deal with the matching technique between interior and exterior solutions (columns 11 to 13) deserve a few explanatory words. All the methods that use eigenfunctions of the Helmholtz equation as basis functions to express the equivalent surface current densities **J** and **M** — or equivalently, the tangential components of the exterior electromagnetic field — except the transfinite elements of Lee and Cendes [1987], match both the electric and magnetic fields in an integral manner. The transfinite elements of Lee and Cendes [1987] match the magnetic field in an integral manner, while the interior and exterior electric fields are joined via a point-matching technique. On the other hand, many methods use polynomials to approximate **J** and **M**. When the polynomials used match the shape functions of the finite element method of the interior problem, the matching of the tangential component of the electric field is global. In fact, imposing continuity of the interior and exterior electric fields at a suitable number of points on Γ guarantees that the interior and exterior

Table 1: The Features of Techniques in Solving the Considered Scattering Problem

	Fictitious Contour Γ				Equivalent Current Basis Functions		Formulation		Nonphysical Resonance		Inner and Outer Solutions Matching			Number of Matrices to Be Solved	Number of Inner Problems to Be Solved	Structure of the Coefficient Matrix/ces to be Solved		
							Inward	Outward										
Method	A	CC	S	D	Eigenfunctions	Polynomials	Looking	Looking	Yes	No	Punctual	Integral	Global			Full	Partially Full	Sparse
Mei 1987		☹	😐		😐		😐		☹			😐 E,H		2	N	$N \times N$		$N_t \times N_t$
Morgan, Chen, Hill, and Barber 1984	☺			😐	😐		😐		☹			😐 E,H		2	N	$N \times N$		$N_t \times N_t$
Lee and Cendes 1987		☹	😐		😐		—	—		☺	☹ E	😐 H		1	1		$(N_t + N) \times (N_t + N)$	
Pelosi, Freni, and Coccioli 1993		☹	😐		😐		😐		☹			😐 E,H		2	1	$N \times N$		$N_t \times N_t$
Cangellaris and Lee 1990	☺			😐	😐		😐		☹			😐 E,H		2	N	$N \times N$		$N_t \times N_t$
Silvester and Hsieh 1971	☺		😐			😐		😐	—	—			☺ E	2	1	$N_b \times N_b$	$N_t \times N_t$	
Mc Donald and Wexler 1972	☺			😐		😐		😐		☺		😐 H	☺ E	2	1	$N_b \times N_b$	$N_i \times N_i$	
Marin 1982	☺		😐			😐		😐	☹			😐 H	☺ E	2	1	$N_b \times N_b$	$N_t \times N_t$	
Jin and Liepa 1988 a	☺			😐		😐		😐		☺		😐 H	☺ E	2	1	$N_b \times N_b$	$N_i \times N_i$	
Jin and Liepa 1988 b	☺			😐		😐	😐		☹			😐 H	☺ E	2	1	$N_b \times N_b$		$N_i \times N_i$
Collins, Volakis, and Jin 1990		☹		😐		😐		😐		☺	☹ E	😐 H		1	1		$N_t \times N_t$	
Collins, Jin, and Volakis 1992		☹	😐			😐		😐		☺	☹ E	😐 H		1	1		$N_t \times N_t$	
Yuan, Lynch, and Strohbehn 1990	☺		😐			😐	😐		☹			😐 E,H		2	N_b	$N_b \times N_b$		$N_t \times N_t$
Morgan and Welch 1992	☺			😐		😐	😐		☹			😐 H	☺ E	2	N	$N \times N$		$N_t \times N_t$

fields coincide at every point on Γ. The continuity of the tangential component of the magnetic field is imposed in an integral manner using the Green's integral. The papers by Collins, Volakis, and Jin [1990] and by Collins, Jin, and Volakis [1992] express the the tangential component of the exterior electric field by means of pulse basis functions, and then match the interior and exterior electric fields via a point-matching technique. The tangential component of the magnetic field is still imposed in an integral manner using the Green's integral. The hybrid finite element and moment method of Yuan, Lynch, and Strohbehn [1990] matches both the electric and magnetic fields in an integral manner.

The last five tabular columns summarize the computational effort required by each method. They show the number of coefficient matrices that must be decomposed, the number of interior problems with distinct boundary conditions that must be solved, and the dimension and the structure of the coefficient matrix to be decomposed or inverted. In these last four columns:

(a) N_b is the number of the nodes of the finite element mesh that lie on the fictitious contour Γ.

(b) N_i is the number of the nodes of the finite element mesh that lie inside the region enclosed by Γ.

(c) N_t is the total number of the nodes of the finite element mesh ($N_t = N_b + N_i$).

(d) N is the total number of basis functions employed to express the equivalent surface current densities **J** and **M** or, equivalently, the tangential component of the exterior electric field.

Note that this analysis does not apply to procedures explicitly devised to treat unbounded geometries. Hence, the works by Jin and Volakis [1991*c*] and Ramahi and Mittra [1991] are not covered by the comparison discussed here.

References

Bettess, P. and J. A. Bettess (1991): "Infinite elements for dynamic problems: Part 2." *Engineering Computations*, vol. **8**, pp. 125–51, 1991.

Cangellaris, A. C. and R. Lee [1990]: "The bymoment method for two-dimensional electromagnetic scattering." *IEEE Transactions on Antennas and Propagation*, vol. **38**, pp. 1429–37, 1990.

Cangellaris, A. C. and R. Lee [1991]: "Finite element analysis of electromagnetic scattering from inhomogeneous cylinders at oblique incidence." *IEEE Transactions on Antennas and Propagation*, vol. **39**, pp. 645–50, 1991.

Chang, C. N., Y. C. Wong, and C. H. Chen (1990): "Full-wave analysis of coplanar waveguides by variational conformal mapping technique." *IEEE Transactions on Microwave Theory and Techniques*, vol. **38**, pp. 1339–44, 1990.

Chang, S. K. and K. K. Mei [1976]: "Application of the unimoment method to electromagnetic scattering of dielectric cylinders." *IEEE Transactions on Antennas and Propagation*, vol. **AP-24**, pp. 35–42, 1976.

Chang, S. K. and K. K. Mei (1981): "Multipole expansion technique for electromagnetic scattering by buried objects." *Electromagnetics*, vol. **1**, pp. 73–89, 1981.

Collins, J. D., J. M. Jin, and J. L. Volakis (1990): "A combined finite element–boundary element formulation for solution of two-dimensional scattering problems via CGFFT." *Electromagnetics*, vol. **10**, pp. 423–37, 1990.

Collins, J. D., J. M. Jin, and J. L. Volakis [1992]: "Eliminating interior resonances in finite element–boundary integral methods for scattering." *IEEE Transactions on Antennas and Propagation*, vol. **40**, pp. 1583–85, 1992.

Collins, J. D., J. L. Volakis, and J. M. Jin [1990]: "A combined finite element–boundary integral formulation for solution of two-dimensional scattering problems via CGFFT." *IEEE Transactions on Antennas and Propagation*, vol. **38**, pp. 1852–58, 1990.

Cwik, T. (1992): "Coupling finite element and integral equation solutions using decoupled boundary meshes." *IEEE Transactions on Antennas and Propagation*, vol. **40**, pp. 1496–504, 1992.

Emson, C. R. I. (1988): "Methods for the solution of open-boundary electromagnetic-field problems." *Proceedings IEE*, Part A, vol. **135**, pp. 151–58, 1988 [with preceding chapter].

Gedney, S. D., J. F. Lee, and R. Mittra (1992): "A combined FEM/MoM approach to analyze the plane wave diffraction by arbitrary gratings." *IEEE Transactions on Microwave Theory and Techniques*, vol. **40**, pp. 363–70, 1992.

Graglia, R. D., A. Freni, and G. Pelosi [1993]: "A finite element approach to the electromagnetic interaction with rotating penetrable cylinder of arbitrary cross section." *IEEE Transactions on Antennas and Propagation*, vol. **41**, pp. 635–50, 1993.

Harrington, R. F. (1961): *Time-Harmonic Electromagnetic Fields.* New York: McGraw-Hill, 1961.

Harrington, R. F. (1968): *Field Computation by Moment Methods.* New York: Macmillan, 1968.

Hunka, J. F. and K. K. Mei (1981): "Electromagnetic scattering by two bodies of revolution."

Electromagnetics, vol. **1**, pp. 329–47, 1981.

Hurwitz, H., Jr. (1984): "Infinitesimal scaling — A new procedure for modeling exterior field problems." *IEEE Transactions on Magnetics*, vol. **MAG-20**, pp. 1918–23, 1991.

Jeng, S. K. and C. H. Chen (1984): "On variational electromagnetics: theory and application." *IEEE Transactions on Antennas and Propagation*, vol. **AP-32**, pp. 902–7, 1984.

Jin, J. M. and V. V. Liepa [1988*a*] : "Application of hybrid finite element method to electromagnetic scattering from coated cylinders." *IEEE Transactions on Antennas and Propagation*, vol. **36**, pp. 50–54, 1988.

Jin, J. M. and V. V. Liepa [1988*b*] : "A note on hybrid finite element method for solving scattering problems." *IEEE Transactions on Antennas and Propagation*, vol. **36**, pp. 1486–90, 1988.

Jin, J. M. and J. L. Volakis (1991*a*) : "A finite element–boundary integral formulation for scattering by three-dimensional cavity-backed apertures." *IEEE Transactions on Antennas and Propagation*, vol. **39**, pp. 97–104, 1991.

Jin, J. M. and J. L. Volakis (1991*b*) : "Electromagnetic scattering by and transmission through a three-dimensional slot in a thick conducting plane." *IEEE Transactions on Antennas and Propagation*, vol. **39**, pp. 543–50, 1991.

Jin, J. M. and J. L. Volakis [1991*c*] : "A hybrid finite element method for scattering and radiation by microstrip patch antennas and arrays residing in a cavity." *IEEE Transactions on Antennas and Propagation*, vol. **39**, pp. 1598–604, 1991.

Lee, J. F. and Z. J. Cendes [1987*a*] : "Transfinite elements: a highly efficient procedure for modeling open field problems." *Journal of Applied Physics*, vol. **61**, pp. 3913–15, 1987.

Lee, J. F. and Z. J. Cendes (1987*b*) : "The transfinite element method for computing electromagnetic scattering from arbitrary lossy cylinders." *IEEE AP-S International Symposium Digest*, AP03-5, pp. 99–102, 1987.

Lee, J. F. and Z. J. Cendes (1988): "The transfinite element method for modeling MMIC devices." *IEEE Transactions on Microwave Theory and Techniques*, vol. **36**, pp. 1639–49, 1988.

Marin, S. P. (1982): "Computing scattering amplitudes for arbitrary cylinders under incident plane waves." *IEEE Transactions on Antennas and Propagation*, vol. **AP-30**, pp. 1045–49, 1982.

Maue, A. W. (1949): "Toward formulation of a general diffraction problem via an integral equation." *Zeitschrift für Physik*, vol. **126**, pp. 601–18, 1949.

McDonald, B. H. and A. Wexler [1972]: "Finite-element solution of unbounded field problems." *IEEE Transactions on Microwave Theory and Techniques*, vol. **MTT-20**, pp. 841–47, 1972.

Mei, K. K. (1974): "Unimoment method of solving antenna and scattering problems." *IEEE Transactions on Antennas and Propagation*, vol. **AP-22**, pp. 760–66, 1974.

Mei, K. K. [1987]: "Unimoment method for electromagnetic wave scattering." *Journal of Electromagnetic Waves and Applications*, vol. **1**, pp. 201–22, 1987.

Miller, E. K., L. Medgyesi-Mitschang, and E. H. Newman (1992): *Computational Electromagnetics*. New York: IEEE Press, 1992.

Morgan, M. A. (1980): "Finite element computation of microwave scattering by raindrops." *Radio Science*, vol. **15**, pp. 1109–19, 1980.

Morgan, M. A. (1981): "Finite element calculation of microwave absorption by the cranial structure." *IEEE Transactions on Biomedical Engineering*, vol. **BME-28**, pp. 687–95, 1981.

Morgan, M. A., C. H. Chen, S. C. Hill, and P. W. Barber [1984]: "Finite element–boundary integral formulation for electromagnetic scattering." *Wave Motion*, vol. **6**, pp. 91–103, 1984.

Morgan, M. A. and B. E. Welch (1986): "The field feedback formulation for electromagnetic scattering computations." *IEEE Transactions on Antennas and Propagation*, vol. **AP-34**, pp. 1377–82, 1986.

Morgan, M. A. and T. B. Welch [1992]: "Field feedback computation of scattering by 2-D penetrable objects." *IEEE Transactions on Antennas and Propagation*, vol. **40**, pp. 445–50, 1992.

Murphy, W. D., V. Rokhlin, and M. S. Vassiliou (1990): "Solving electromagnetic scattering problems at resonance frequencies." *Journal of Applied Physics*, vol. **67**, pp. 6061–65, 1990.

Pearson, L. W., A. F. Peterson, L. J. Bahrmasel, and R. A. Whitaker [1992]: "Inward-looking and outward-looking formulations for scattering from penetrable objects." *IEEE Transactions on Antennas and Propagation*, vol. **40**, pp. 714–20, 1992.

Pearson, L. W., R. A. Whitaker, and L. J. Bahrmasel (1989): "An exact radiation boundary condition for the finite-element solution of electromagnetic scattering on an open domain." *IEEE Transactions on Magnetics*, vol. **25**, pp. 3046–48, 1989.

Pelosi, G., A. Freni, and R. Coccioli [1993]: "A hybrid technique for analyzing the scattering from periodic structures." *IEE Proceedings-H*, vol. **140**, pp. 65–70, 1993.

Peterson, A. F. (1990): "The 'interior resonance' problem associated with surface integral equations of electromagnetics: Numerical consequences and a survey of remedies." *Electromagnetics*, vol. **10**, pp. 293–312, 1990.

Peterson, A. F. and S. P. Castillo (1989): "A frequency-domain differential equation formulation for electromagnetic scattering from inhomogeneous cylinders." *IEEE Transactions on Antennas and Propagation*, vol. **37**, pp. 601–7, 1989.

Rahman, B. M. A. and J. B. Davies (1984): "Penalty function improvement of waveguide solution by finite elements." *IEEE Transactions on Microwave Theory and Techniques*, vol. **MTT-32**, pp. 922–28, 1984.

Ramahi, O. M. and R. Mittra [1991]: "Finite element solution for a class of unbounded geometries." *IEEE Transactions on Antennas and Propagation*, vol. **39**, pp. 245–50, 1991.

Silvester, P. P. and R. L. Ferrari (1990): *Finite Elements for Electrical Engineers*, 2nd edition. Cambridge: Cambridge University Press, 1990.

Silvester, P. P. and M. S. Hsieh [1971]: "Finite-element solution of 2-dimensional exterior-field problems." *IEE Proceedings–H*, vol. **118**, pp. 1743–47, 1971.

Silvester, P. P., D. A. Lowther, C. J. Carpenter, and E. A. Wyatt (1977): "Exterior finite elements for 2-dimensional field problems with open boundaries." *IEE Proceedings*, vol. **124**, pp. 1267–70, 1977.

Stovall, R. E. and K. K. Mei (1975): "Application of a unimoment technique to a biconical antenna with inhomogeneous dielectric loading." *IEEE Transactions on Antennas and Propagation*, vol. **AP-23**, pp. 335–41, 1975.

Volakis, J. L. (1994): "Progress on the applications of the finite element method for 3-D electromagnetic scattering and radiation." *COMPEL*, vol. **13** (to be published), 1994.

Waterman, P. C. (1965): "Matrix formulation of electromagnetic scattering." *Proceedings IEEE*, vol. **53**, pp. 805–12, 1965.

Wu, R. B. and C. H. Chen [1986]: "Variational reaction formulation of scattering problem for anisotropic dielectric cylinders." *IEEE Transactions on Antennas and Propagation*, vol. **AP-34**, pp. 640–45, 1986.

Yuan, X. [1990]: "Three dimensional electromagnetic scattering from inhomogeneous objects by the hybrid moment and finite element method." *IEEE Transactions on Microwave Theory and Techniques*, vol. **38**, pp. 1053–58, 1990.

Yuan, X., D. R. Lynch, and J. W. Strohbehn [1990]: "Coupling of finite element and moment methods for electromagnetics scattering from inhomogeneous objects." *IEEE Transactions on Antennas and Propagation*, vol. **38**, pp. 386–93, 1990.

Finite-Element Solution of 2-Dimensional Exterior Field Problems

P. P. SILVESTER and M. S. HSIEH

Indexing terms: *Difference equations, Partial differential equations*

Abstract

A new method, based on the finite-element method of discretisation and compatible with existing finite-element techniques, is described for the solution of field problems in which the region of prime interest is embedded in an infinitely extending region where Laplace's equation holds. The essence of this method lies in representing the infinitely extending region by a single finite element, which may be included in an element assembly descriptive of the finite region of major interest. No iterative computation is necessary, and the computing effort required for solution is essentially the same as that required for solving the interior-field problem alone.

1 Introduction

Numerous 2-dimensional problems arising in the analysis of electric and magnetic fields lead to a requirement for solving Laplace's equation in an unbounded region. Very often, such problems involve a bounded region, in which some more complicated equations must be satisfied, embedded in an infinitely extending Laplacian region, the relevant potential function and its derivatives being required to be continuous. For example, the determination of magnetic fields in the neighbourhood of an infinitely long, straight, ferromagnetic conductor carrying time-varying electric currents, involves solving a nonlinear diffusion equation within the conductive material, and Laplace's equation everywhere outside. Similarly, the electrostatic problems associated with a variety of strip transmission lines for microwave use involve the solution of Laplace's equation in a region containing conductors and inhomogeneous dielectrics, embedded in an infinitely extending region where Laplace's equation holds. In the context of this paper, such problems will be termed *exterior* problems or *unbounded* problems.

Numerical solution of unbounded problems poses certain difficulties. The well known finite-difference methods are successful in treating finite regions, and can, of course, be extended by the simple, if brutal, artifice of truncating the unbounded region by means of a wholly artifical boundary; examples of this approach abound in the literature. Two principal difficulties are involved. First, the resulting solution cannot give any indication of the error involved in the geometric truncation (except perhaps by repeated solution for regions of varying size); secondly, the number of mesh points tends to be large if the truncating boundary is at a reasonable distance, so that fine modelling of the region of prime interest is not possible. The second difficulty may be partially circumvented by mesh grading, an approach which unfortunately makes programming considerably more complicated.

A second, more rigorous, approach is that of boundary relaxation, in which a purely artificial boundary is imposed, but the potential values on this boundary are refined iteratively until they assume values equal to those that would be obtained on an infinitely extending mesh.[1–4] There is then no longer any need to place the artificial boundary far from the region of prime interest, so that very considerable computational economies result. The method does have the defect, however minor, that the iterative correction of values on the artificial boundary requires one relatively large auxiliary matrix to be available. Although this matrix depends only on the mesh size, and is therefore applicable to many problems, it does present a programming nuisance.

In some special circumstances, particularly where the potentials in the interior region as well as in the exterior satisfy Laplace's equation, the problem may be stated in terms of an integral equation, which can then be solved approximately by a projection method (the 'equivalent-source method'). Numerous important problems of engineering interest have been solved by this technique, and there exists a large literature on the subject. However, this approach fails when the interior region is not purely Poissonian or Laplacian.

Recently, the finite-element method has found increasing application to electric, magnetic, and electromagnetic-field problems. It has been found to be far superior to finite differences in a number of bounded problems. Its use in solving unbounded problems, on the other hand, has suffered restrictions similar to those that limit the finite-difference methods; the only solutions published to date seem to rely once again on truncation of the unbounded region into a bounded one. In the present paper it will be shown that such truncation is not necessary. Indeed, all that is required is the development of an *exterior element*; such a development is given below. The resulting formulation of the unbounded problem not only makes more economical use of computer memory than either of the older, approximate, approaches, but also it requires no iterative calculation whatever. In contrast to all past methods, therefore, there are no problem-dependent acceleration parameters to be chosen; furthermore, computing time for a given problem can be forecast in advance from the known size of all matrices generated.

2 Finite-element formulation of unbounded problem

Development of the exterior element is intimately related to the methods used for interior elements in the standard finite-element approach. The latter has been described extensively elsewhere, and will be stated below only in those aspects necessary to show its relationship to the exterior formulation; for details, the reader is referred to the various published papers.

Reprinted with permission from *IEE Proceedings*, vol., 118, no. 12, pp. 1743–1747, Dec. 1971.

2.1 Division into interior and exterior problems

Let the structure shown in Fig. 1*a*, a twin-strip transmission line with dielectric separator, serve as an example of an unbounded problem typically encountered. One very simple method for approximate solution would be to substitute for this structure another, similar except for an enclosing metal box; if the box is comparatively large, it might be assumed that the results obtained for the two problems would be similar. Insertion of such a zero-potential boundary, shown in Fig. 1*b*, converts the unbounded problem into an ordinary interior problem of Laplace's equation. A solution is then found by seeking the potential function ψ in the interior which will minimise the functional

$$\mathscr{F}(\psi) = \int_{\Omega} (\nabla\psi \cdot \nabla\psi - 2\psi g)d\Omega \quad . \quad . \quad . \quad (1)$$

integration being extended over the region enclosed by the broken line. It is well known[5] that the potential function which satisfies the prescribed Dirichlet boundary conditions, and minimises this energy-related functional, is precisely the solution $\psi = \phi$ of

$$\nabla^2\phi = g \quad . \quad . \quad . \quad (2)$$

In the problem of Fig. 1, of course, $g = 0$; however, the term will be retained in this paper to preserve generality.

To achieve a practical, computable formulation of this minimum problem, the interior region to the dotted boundary of Fig. 1*c* is subdivided into triangular (or other convenient)

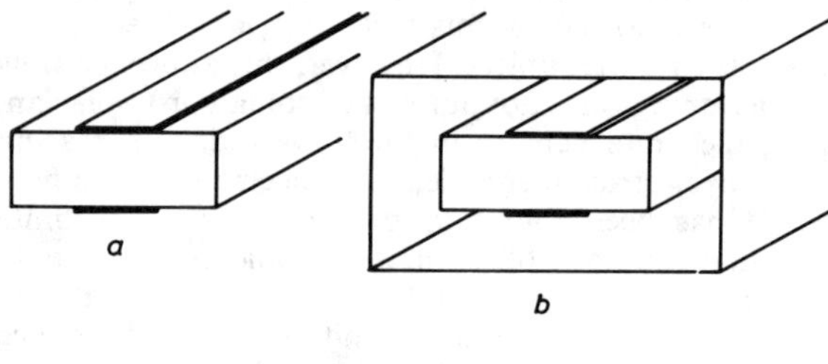

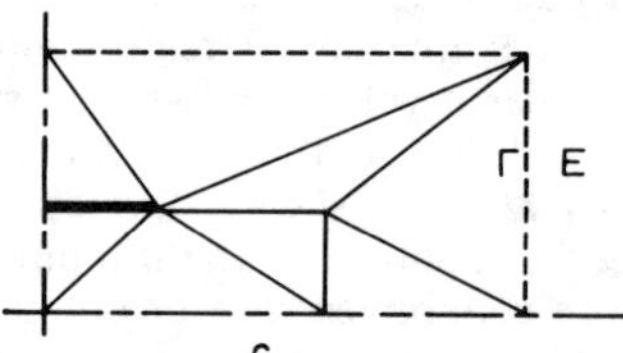

Fig. 1
Double stripline, illustrating the nature of 2-dimensional unbounded static-field problems

a Double stripline on dielectric separator
b Stripline enclosed in metal box, as analytic convenience
c Subdivision of plane into triangular elements in the interior, and the exterior region E, by boundary Γ

elements. On each element, a set of approximation functions $\{\alpha_i\}$ is defined. These may in principle be any linearly independent set. However, it has been normal practice to use interpolation polynomials which possess numerous convenient and simple (though not necessary) properties.[6–8] Using these functions, the potential and source functions ψ and g may be approximated over each element by

$$\psi \sim \sum_i a_i\alpha_i(x, y) \quad . \quad . \quad . \quad (3)$$

and $$g \sim \sum_i f_i\alpha_i(x, y) \quad . \quad . \quad . \quad (4)$$

Details of this procedure will be found elsewhere.[9] Substitution of these approximations into eqn. 1 gives

$$\mathscr{F}(\psi) = \sum_k \sum_i \sum_j (a_i a_j \int_{\Omega} \nabla\alpha_i \cdot \nabla\alpha_j d\Omega - a_i f_j \int_{\Omega} \alpha_i \, \alpha_j d\Omega) \quad . \quad . \quad . \quad (5)$$

where the summation index k refers to the triangular elements, and the indices i and j run over the interpolation nodes in the kth triangle. In matrix notation, eqn. 5 may be written as

$$\mathscr{F}(\psi) = \sum_k (\boldsymbol{a}^T\boldsymbol{S}\boldsymbol{a} - 2\boldsymbol{a}^T\boldsymbol{T}\boldsymbol{f}) \quad . \quad . \quad . \quad (6)$$

where $\boldsymbol{a}$ and $\boldsymbol{f}$ are the column vectors of a_i and f_i, and $\boldsymbol{S}$ and $\boldsymbol{T}$ are matrices, formed element by element, in accordance with

$$S_{ij} = \int \nabla\alpha_i \cdot \nabla\alpha_j d\Omega \quad . \quad . \quad . \quad (7)$$

$$T_{ij} = \int \alpha_i\alpha_j d\Omega \quad . \quad . \quad . \quad (8)$$

If interpolation polynomials are employed as the approximating functions, it is possible to formulate the numerical values of the elements of $\boldsymbol{S}$ and $\boldsymbol{T}$ so as to avoid all integration and differentiation in problem solving.[9]

2.2 Unbounded problem formulation

To solve the problem of Fig. 1*a* correctly, no boundary such as that shown dotted in Fig. 1*c*, with artificial restrictions on the boundary values, can be permitted. If, however, Fig. 1*c* is regarded as defining a finite-element subdivision of an unbounded problem, with all the plane exterior to the dotted boundary Γ considered to constitute one elment, the minimisation problem of eqn. 1 for the unbounded problem becomes merely

$$\mathscr{F}(\psi) = \int_{\Omega} (\nabla\psi \cdot \nabla\psi - 2\psi g)d\Omega + \int_E \nabla\psi \cdot \nabla\psi dE \quad . \quad . \quad (9)$$

where the region of integration E is the whole exterior of the dotted boundary Γ. Clearly, for a solution to exist, the latter integral must exist. This requirement, however, implies that the stored energy associated with the exterior region must be finite, a requirement known to be met by any exterior solution of Laplace's equation.

The practical difficulty with the formulation of eqn. 9 is that no very convenient approximating functions, which would meet the following simultaneous requirements, immediately suggest themselves:

(*a*) the exterior stored energy must be finite
(*b*) the potential must be continuous across the boundary Γ
(*c*) the functions should preferably not introduce additional degrees of freedom
(*d*) if possible, they should be computationally convenient.

Fortunately, a conversion, by Green's theorem, suggests a method for meeting these needs. It is clearly possible to write

$$\int_E \nabla\psi \cdot \nabla\psi dE = -\int_E \psi\nabla^2\psi dE - \oint_{\Gamma} \psi \frac{\partial\psi_E}{\partial n} d\Gamma = \mathscr{F}_E \quad . \quad (10)$$

The negative sign on the contour-integral term is occasioned by the fact that an exterior region is under consideration, with the conventional normal along Γ pointing *into E*.

Recognising that the exterior region E is Laplacian throughout,

$$\nabla^2 \psi_E = 0 \quad . \quad . \quad . \quad . \quad . \quad . \quad . \quad . \quad . \quad . \quad . \quad . \quad (11)$$

and the contribution to $\mathscr{F}$ arising from the exterior region, $\mathscr{F}_E$ of eqn. 10, becomes simply

$$\mathscr{F}_E = -\oint_\Gamma \psi_E \frac{\partial \psi_E}{\partial n} d\Gamma \quad . \quad . \quad . \quad . \quad . \quad . \quad . \quad . \quad (12)$$

The unbounded problem may be regarded as solved if this contribution can be stated in terms of the values of ψ_E along the boundary Γ, for these boundary values must be such as to satisfy the minimality requirement of $\mathscr{F}$ for the interior and exterior regions taken together.

As has been extensively documented elsewhere,[5] the formulation given in eqn. 9 has the particularly attractive property that the functional $\mathscr{F}$ is variationally stationary about the correct solution ψ. This property is not altered in the present development; consequently, the known smooth convergence of the finite-element method may be expected to apply to the method presented here.

2.3 Relation between exterior potentials and derivatives

Evaluation of the contour integral in eqn. 12 is possible if an expression can be found for the normal derivatives in terms of function values along the boundary Γ. Such an expression can be determined by solving eqn. 11 in the exterior region, and differentiating the solution; however, it is easier and more direct to restate the problem in terms of an integral equation and solve for the normal derivative directly. In general, ψ_E is known to be given by

$$\psi_E(x) = -\int_\Gamma \frac{\partial \psi_E}{\partial n} G(x;\xi) d\xi \quad . \quad . \quad . \quad . \quad . \quad . \quad (13)$$

where $G(x;\xi)$ is the Green's function of Laplace's equation

$$G(x;\xi) = \frac{1}{4\pi\epsilon} \ln |x - \xi| \quad . \quad . \quad . \quad . \quad . \quad . \quad . \quad (14)$$

and x represents any point on Γ or in the exterior region E. To solve approximately for the normal derivative, let β_k be a a convenient set of functions defined on Γ. It is prudent to choose this set to be compatible with functions used in the interior region; i.e. sectionally polynomial so as to match the finite-element approximations in the interior. Setting

$$\left.\frac{\partial \psi_E}{\partial n}\right|_\Gamma = \sum_i b_i \beta_i \quad . \quad . \quad . \quad . \quad . \quad . \quad . \quad . \quad . \quad . \quad (15)$$

eqn. 13 becomes

$$\psi = -\sum_i b_i \oint \beta_i(\xi) G d\xi \quad . \quad . \quad . \quad . \quad . \quad . \quad . \quad (16)$$

To solve projectively,[10] multiply both sides by $\beta_j(x)$ and integrate over the boundary Γ so as to obtain

$$\oint_\Gamma \psi(x)\beta_j(x)dx = \sum_i b_i \oint_\Gamma \oint_\Gamma \beta_j(x)\beta_i(\xi) G d\xi dx \quad . \quad (17)$$

This may be regarded as the matrix equation

$$\boldsymbol{\Psi} = \boldsymbol{Q}\boldsymbol{b} \quad . \quad . \quad . \quad . \quad . \quad . \quad . \quad . \quad . \quad . \quad . \quad . \quad (18)$$

if matrices $\boldsymbol{\Psi}$ and $\boldsymbol{Q}$ are defined by

$$\boldsymbol{\Psi}_j = \oint_\Gamma \psi \beta dx \quad . \quad . \quad . \quad . \quad . \quad . \quad . \quad . \quad (19)$$

$$Q_{ji} = \oint \oint \beta_j \beta_i G dx d\xi \quad . \quad . \quad . \quad . \quad . \quad . \quad . \quad (20)$$

Finally, set

$$\psi|_\Gamma = \sum_i a_i \beta_i \quad . \quad . \quad . \quad . \quad . \quad . \quad . \quad . \quad (21)$$

It should be noted that continuity of exterior and interior potentials can be guaranteed, provided the functions β_k are chosen to match their interior counterparts:

$$\beta_k = \alpha_k|_\Gamma \quad . \quad . \quad . \quad . \quad . \quad . \quad . \quad . \quad . \quad . \quad . \quad (22)$$

Since triangular finite elements with polynomial approximation functions are employed in the interior region, eqn. 22 states in effect that the β_k should be piecewise polynomial, the degree of polynomial to be chosen everywhere along Γ to accord with that of the neighbouring element in the interior.

Define the matrix $\boldsymbol{R}$ by

$$R_{ij} = \oint_\Gamma \beta_i(x)\beta_j(x)dx \quad . \quad . \quad . \quad . \quad . \quad . \quad . \quad (23)$$

Then eqns. 19 and 21 may be combined into the matrix equation

$$\boldsymbol{\Psi} = \boldsymbol{R}\boldsymbol{a} \quad . \quad . \quad . \quad . \quad . \quad . \quad . \quad . \quad . \quad . \quad . \quad . \quad (24)$$

or, substituting in eqn. 18,

$$\boldsymbol{R}\boldsymbol{a} = \boldsymbol{Q}\boldsymbol{b} \quad . \quad . \quad . \quad . \quad . \quad . \quad . \quad . \quad . \quad . \quad . \quad (25)$$

which may be solved for the coefficient vector $\boldsymbol{b}$, since $\boldsymbol{Q}$ cannot be singular:

$$\boldsymbol{b} = \boldsymbol{Q}^{-1}\boldsymbol{R}\boldsymbol{a} \quad . \quad . \quad . \quad . \quad . \quad . \quad . \quad . \quad . \quad (26)$$

For any given set of boundary-potential values, eqn. 26 prescribes the coefficients $\boldsymbol{b}$, and hence the value of normal derivative everywhere on Γ. The contour integral of eqn. 12 can therefore be evaluated directly.

2.4 Exterior element matrix

To obtain the element matrix corresponding to the exterior contribution $\mathscr{F}_E$ to the functional $\mathscr{F}$, substitute eqns. 15 and 21 into eqn. 12, which becomes

$$\mathscr{F}_E = -\sum_i \sum_j b_i b_j \oint \oint \beta_i \beta_i G d\xi dx \quad . \quad . \quad . \quad (27)$$

which may be recognised as equivalent to the matrix quadratic form

$$\mathscr{F}_E = \boldsymbol{b}^T \boldsymbol{Q} \boldsymbol{b} \quad . \quad . \quad . \quad . \quad . \quad . \quad . \quad . \quad . \quad . \quad (28)$$

By eqn. 26, and using the fact, evident from eqn. 20, that $\boldsymbol{Q}$ is symmetric, one has

$$\mathscr{F}_E = \boldsymbol{a}^T(\boldsymbol{R}^T\boldsymbol{Q}^{-1}\boldsymbol{R})\boldsymbol{a} \quad . \quad . \quad . \quad . \quad . \quad . \quad . \quad (29)$$

where, now, the coefficients $\boldsymbol{b}$ of the normal derivatives are no longer involved at all. Assuming that the functions β_k have been chosen according to the prescription of eqn. 22, however, the coefficients a_k have the same significance in eqn. 29 as in eqn. 6. Hence the functional for the unbounded problem, eqn. 9, may be written as

$$\mathscr{F} = \sum (\boldsymbol{a}^T\boldsymbol{S}\boldsymbol{a} - 2\boldsymbol{a}^T\boldsymbol{T}\boldsymbol{f}) \quad . \quad . \quad . \quad . \quad . \quad . \quad (30)$$

the summation being carried out over all triangles in the interior subdivision, and the exterior element. Indeed, it is evident that the exterior region, treated as a finite element (finite in energy, though not in geometric extent), fits precisely into the formalism of the interior problem, provided only that it is understood to contribute the matrix

$$\boldsymbol{S}_E = \boldsymbol{R}^T\boldsymbol{Q}^{-1}\boldsymbol{R} \quad . \quad . \quad . \quad . \quad . \quad . \quad . \quad . \quad . \quad (31)$$

to the formation of the overall functional.

2.5 Solution of unbounded problem

For either the bounded, or the unbounded, problem, the energy functional $\mathscr{F}$ has now been approximated by a matrix quadratic form. Minimisation of this quadratic form with respect to each and every coefficient a_k immediately leads to the matrix equation

$$\boldsymbol{S}\boldsymbol{a} = \boldsymbol{T}\boldsymbol{f} \quad \quad (32)$$

whose solution provides the coefficient vector $\boldsymbol{a}$, and hence the approximate solution in the interior, eqn. 3.

A point of some importance is that, although the solution thus obtained is valid throughout the entire plane, explicit pontential values are easily available only in the interior region. Should potential values outside the boundary Γ be desired, eqn. 13 may of course be used to continue the solution outward. In this respect, the new method is similar to the boundary-relaxation process: the boundary Γ may be thought of as a mathematical window-frame, which permits only a small portion of the infinite plane to be examined, but does not in any way distort that which is seen. As a corollary, the placement of Γ is perfectly arbitrary, provided only that the exterior is always governed by Laplace's equation.

3 Computer implementation

The steps outlined above for forming the exterior element are relatively easily carried out in a computer program. Such a program, allowing up to 100 interpolation nodes in the interior and along the boundary, has been written to verify the effectiveness of the method. Its most important details will be described briefly below.

3.1 Solution of integral equation

Formation of the matrices $\boldsymbol{Q}$ and $\boldsymbol{R}$ is relatively straightforward, except for the singularity in the Green's function, which requires special treatment in order to preserve numerical accuracy. Elements of $\boldsymbol{R}$ are easily formed by using Gaussian quadrature formulas in two-dimensions, of sufficiently high order to guarantee exact integration. To avoid difficulties with the elements of $\boldsymbol{Q}$, the artifice used earlier[10] for projective solutions is convenient. Define coordinates u and v by

$$u = x + \xi \quad \quad (33)$$

$$v = x - \xi \quad \quad (34)$$

so that eqn. 20 becomes

$$Q_{ji} = \frac{1}{4\pi\epsilon}\oint\oint \beta_i\left(\frac{u-v}{2}\right)\beta_j\left(\frac{u-v}{2}\right)\ln|v|\,dudv \quad \quad (35)$$

The limits of integration in u and v are complicated and render detailed evaluation rather more tedious than in the 'natural' coordinates x and ξ. However, Gaussian integration formulas with the weight function $\ln|v|$ are readily available,[11] so that no difficulty attaches to exact evaluation of the matrix elements Q_{ij}, provided the functions β_i are polynomials.

3.2 Time and memory requirements of program

As indicated above, the exterior element formulated in Section 2.4 requires no interpolation nodes outside the boundary Γ, so that the total size of the finite-element matrices needed is exactly the same whether solution of an unbounded problem, or a bounded problem of similar interior structure, is contemplated. The additional memory requirement imposed by the exterior matrix therefore comprises only the additional space needed to store the program segment for calculating the exterior-element matrix, a few hundred words. The time taken to solve the matrix equation (eqn. 32) is also the same as for an interior problem of similar complexity, since solution time depends on matrix size. However, some extra time is needed to construct the exterior matrix; in typical problems, this extra time amounts to perhaps 5–10% of total problem running-time, the major portion of time being taken by the solution of eqn. 32.

Compared with the traditional approach, in which the region to be examined is simply truncated by an equipotential boundary at a considerable distance, the new method is therefore much to be preferred on both counts. The storage requirement is lowered by storing only the matrix portion which actually corresponds to the problem region to be investigated. There is no longer any reason to place the boundary Γ far away, indeed there is every reason to bring it as close to the region of prime interest as is practicable. If desired, the storage and running-time economies gained may be traded for enhanced accuracy of solution.

4 Examples of solved problems

To illustrate the flexibility of the new approach, a few of the problems solved by this method are described.

4.1 Single flat charged strip

This problem, first solved by Maxwell and shown in numerous modern textbooks,[12] is of interest largely because an analytic solution is available for comparison. An equipotential plot near the strip, automatically plotted by computer,

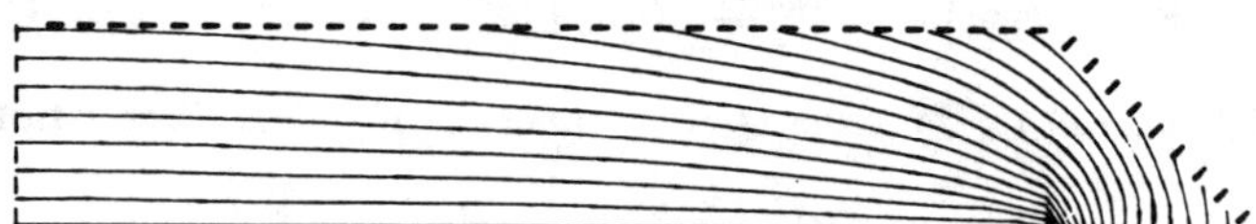

Fig. 2
Equipotential lines near a long, flat, thin charged ribbon conductor

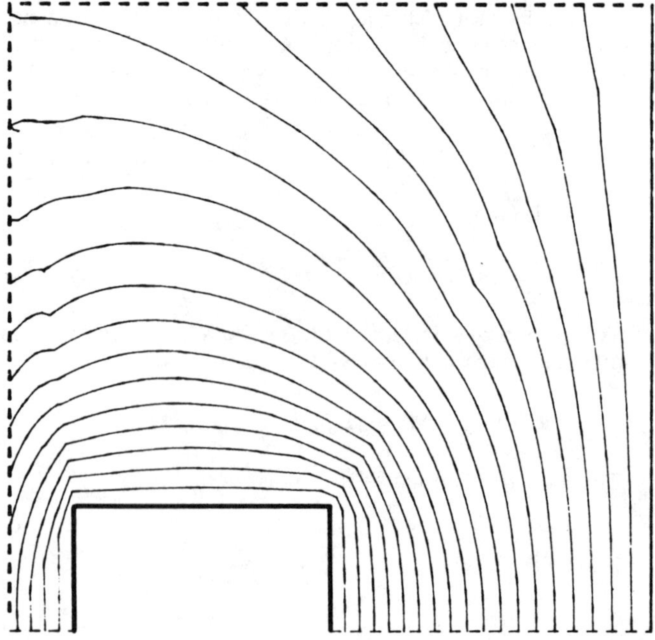

Fig. 3
Equipotential map near one member of open-wire pair made of square wires

appears in Fig. 2. The exterior bounding contour Γ was in this case taken far enough away to permit the equipotential plot to be examined, and the interior region was subdivided into three triangular elements. Using cubic approximating functions (ten functions per element), the matrix size, after constraining the potential on the strip itself, is 16 × 16, a trivial problem for modern computers.

4.2 Two parallel square conductors

This problem, equivalent to that of a single square conductor near a ground plane, appears in Fig. 3, again plotted automatically. It is worth noting that the equipotential lines meet the artificial bounding contour at a variety of angles, ranging from near-orthogonal to grazing incidence —clear indication that the exterior element does not terminate the solution in the same fashion as an imposed equipotential or flux boundary, which would require either parallel or orthogonal equipotentials. The interior region in this case was subdivided into six elements.

A certain amount of raggedness in the contour plot may be discerned near element interfaces. The prime source of this inaccuracy lies in the interpolation routine used by the plotting program, which uses quadratic interpolation, even though the approximating surfaces are cubic or quartic.

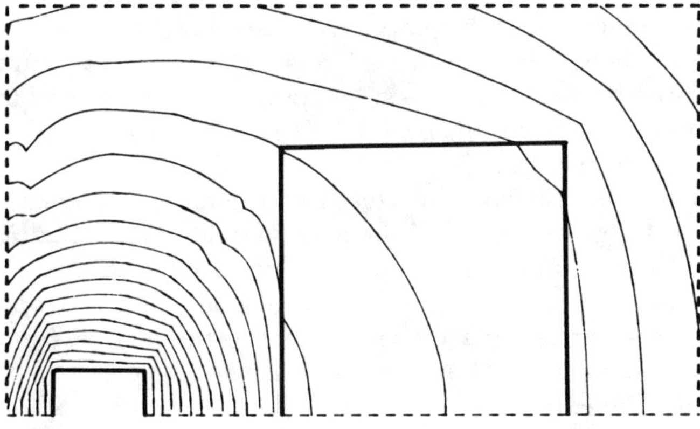

Fig. 4

Equipotential lines near square conductor placed parallel to dielectric slab of relative permittivity 6

4.3 Square conductor flanked by dielectric slab

Finally, a more complicated interior region is shown in Fig. 4, where a square, charged, metallic conductor is placed parallel to, and near, a rectangular dielectric slab of relative permittivity 6. In this case, as in the previous ones, symmetry is taken into account so as to reduce the computation required. Again, the equipotential lines are seen to proceed to the bounding countour in the manner expected; they are not distorted by the introduction of the contour.

5 Conclusions

The finite-element method shown in this paper for the solution of unbounded field problems requires no iterative computation, and is computationally very cheap. It is therefore held that it supersedes the boundary-relaxation method for the solution of 2-dimensional problems with translational symmetry. It is probable that an extension of this technique to the rotationally symmetric case, and to the Helmholtz operator, will prove possible; such problems are now under investigation.

6 References

1 CERMAK, I. A., and SILVESTER, P.: 'Solution of 2-dimensional field problems by boundary relaxation', *Proc. Inst. Elec. Electron. Eng.*, 1968, **115**, pp. 1341–1348

2 SILVESTER, P., and CERMAK, I. A.: 'Analysis of coaxial line discontinuities by boundary relaxation', *IEEE Trans.*, 1969, **MTT–17**, pp. 489–495

3 CERMAK, I. A., and SILVESTER, P.: 'Boundary-relaxation analysis of rotationally symmetric electric field problems', *ibid.*, 1970, **PAS–89**, pp. 925–932

4 RICHTER, S. L.: 'Extension of boundary relaxation for finite-difference solution of field problems', *Proc. Inst. Elec. Electron. Eng.*, 1970, **68**, pp. 1146–1148

5 ZIENKIEWICZ, O. C., with CHEUNG, Y. K.: 'The finite element method in structural and continuum mechanics' (McGraw-Hill, 1967)

6 SILVESTER, P.: 'A general high-order finite-element waveguide analysis program', *IEEE Trans.*, 1969, **MTT–17**, pp. 204–210

7 DALY, P.: 'Hybrid-mode analysis of microstrip by finite-element methods', *ibid.*, 1971, **MTT–19**, pp. 19–25

8 CSENDES, Z. J., and SILVESTER, P.: 'Numerical solution of dielectric loaded waveguides. I—Finite-element analysis', *ibid.*, 1970, **MTT–18**, pp. 1124–1131

9 SILVESTER, P.: 'High-order polynomial triangular finite elements for potential problems', *Int. J. Eng. Sci.*, 1969, **7**, pp. 849–861

10 SILVESTER, P., and HSIEH, M.-S.: 'Projective solution of integral equations arising in electric and magnetic field problems', *J. Comput. Phys.*, 1971, **8** (to be published)

11 STROUD, A. H., and SECREST, D.: 'Gaussian quadrature formulas' (Prentice-Hall, 1966)

12 MAXWELL, J. C.: 'A treatise on electricity and magnetism—Vol. II' (Dover, 1954), art. 191

Finite-Element Solution of Unbounded Field Problems

BRUCE H. MCDONALD AND ALVIN WEXLER

Abstract—**An unbounded region is divided into local picture-frame regions where a partial differential-equation solution is obtained, with the remaining unbounded region represented by an integral equation. (The method permits the use of free-space Green's functions, and thus special problem-dependent Green's functions need not be found.) The integral equation is formulated as a constraint upon the local picture-frame solutions, whence these local solutions are solved directly by a variational method, using finite elements, in a manner such that the problem of the Green's-function singularity is side-stepped. The technique is applicable where sources and media inhomogeneities and anisotropies are local, and can all be placed within one or several picture frames. It is in these cases that the integral-equation approach is at a particular disadvantage, and the use of a partial differential-equation technique is advisable if not necessary. Examples presented include the static and harmonic fields of a parallel-plate capacitor, a microstrip line on a dielectric substratum, and a radiating antenna with dielectric obstacles.**

I. Introduction

FIELD PROBLEMS, where the region of interest is bounded, are customarily solved by formulating a partial differential equation and using knowledge of the boundary conditions to obtain the solution. When the region of interest is unbounded it is customary to use an integral-equation method to obtain a local charge distribution, and from this the fields as required. Integral equations often possess singular kernels that can produce "algorithmic nightmares" [1], and such methods depend upon the existence of an "action-at-a-distance" Green's function that may be very difficult to find, especially when inhomogeneities or anisotropies occur. Moreover, certain algorithmic niceties about a partial differential-equation solution, along with demonstrable capabilities for handling inhomogeneities and anisotropies, invite the analyst to seek ways of solving unbounded field problems using differential techniques.

If sources, media inhomogeneities, and anisotropies are local, a surface may be defined to enclose them. The internal region so defined is bounded, and differential techniques may be used there. In the remaining unbounded region an integral equation may be formulated using only a simple free-space Green's function. The field at the surface may be interpreted two ways: as a boundary condition for the differential-equation formulation of a boundary-value problem; and as an equivalent source distribution for the exterior integral equation. If continuity of field and first derivatives are enforced between these two formulations at the surface, the problem is readily solved.

When iterative procedures are used to solve this problem, the technique belongs to the well-known class of boundary-relaxation methods. This paper presents a method whereby the bounded region is solved directly by a variational technique using finite elements, with the integral equation as a constraint.

II. General Description of the Method

A two-dimensional, or plane, problem is discussed in detail, although extension to three-dimensional problems can readily be made.

Fig. 1(a) shows a portion of the infinite x–y plane in Cartesian coordinates (for definiteness, $z=0$). A bounded region called a picture frame R_1 is placed so as to contain all sources, inhomogeneities, and anisotropies. The boundary of the picture frame is designated B_1, and the unbounded exterior region is R_E. The picture frame represents a boundary-value problem with boundary conditions as yet unspecified. It is solved by the finite-element method.

We now introduce a closed contour C_1, as shown in Fig. 1(b). The region of the plane outside C_1 is called R_c, and it is stipulated that the region of overlap between R_c and R_1 be source free, homogeneous, and isotropic. This is not really a restriction, since the picture frame and contour can be placed as desired.

In the exterior region R_c an integral equation is used to relate the field at any point in R_c to the field (or a function of the field that can be considered as an equivalent charge distribution) on the contour C_1. In particular, we evaluate the integral equation at points on B_1, the picture-frame boundary. This produces a set of constraints upon the picture-frame field that forces it to assume, at the boundary B_1, the correct value for the unbounded problem. The overlap region precludes evaluation of the integral equation at a source point, thus side-stepping problems with Green's-function singularities.

Extension to several picture frames is straightforward —a contour C may be defined consisting of a closed contour C_1 in each frame with cuts defined between these C_1, which disappear in integration and which link the picture frames together. See Fig. 1(c).

The method is applicable to problems having unbounded regions provided that:

1) sources, inhomogeneities, and anisotropies are local, so that they may be completely enclosed within picture frames:

Manuscript received May 30, 1972; revised August 21, 1972. This work was supported by the National Research Council of Canada. This paper was presented at the 1972 IEEE International Microwave Symposium, Chicago, Ill., May 22–24, 1972.

The authors are with the Department of Electrical Engineering, University of Manitoba, Winnipeg, Man., Canada.

Reprinted from *IEEE Trans. Microwave Theory Tech.*, vol. MTT-20, no. 12, pp. 841–847, Dec. 1972.

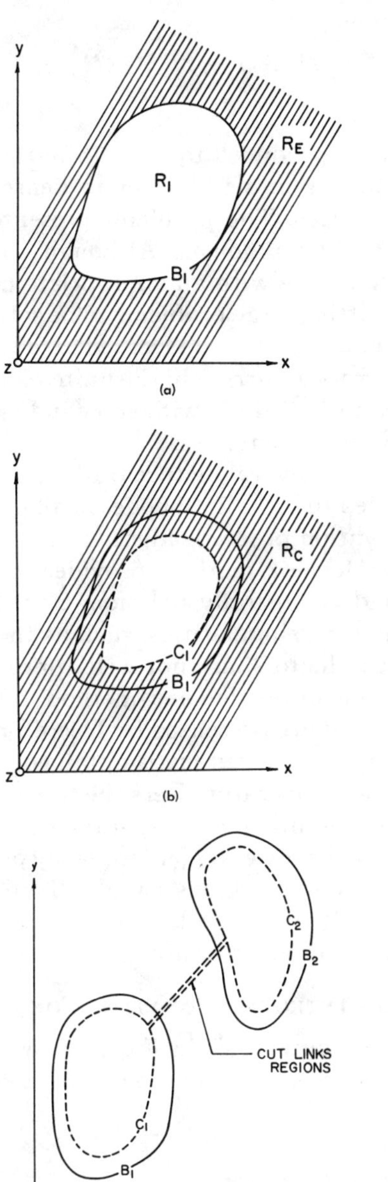

Fig. 1. The infinite plane, the picture frame, and associate contours. (a) The picture frame R_1, its boundary B_1, and the external region R_E. (b) The contour C_1 and the region external to it R_C. (c) The linking of two picture frames.

2) a free-space Green's function exists;
3) local picture-frame solutions are constructed so that field values at the boundary can be linearly expressed in terms of field values on the contour C_1.

Physically, it is possible to interpret the exterior integral-equation formulation as the replacement of the picture-frame solution by an equivalent-source or charge distribution along the contour C_1. Alternatively, the picture-frame solution involves the replacement of the free-space problem by an equivalent-boundary condition on B_1.

Static and harmonic electromagnetic-field problems are easily solved, and the finite-element method readily admits the linear constraint equations.

III. Previous Work

Several related schemes have recently been reported, which, in the main:

1) have used finite-difference techniques for the picture-frame solutions;
2) have been used for only one picture frame, with no indication otherwise;
3) have been restricted to Laplace's equation;
4) have had to cope with singularities of Green's functions.

Silvester and Cermak [2], [3] in effect chose C_1 very close to B_1 within one grid point in their finite-difference representation. This choice allows a direct linking of potentials on C_1 and B_1 with finite differences, but introduces problems with Green's-function singularities. The scheme does not preclude choosing C_1 further away from B_1, although such a choice would require additional equations to relate potentials, and could possibly affect convergence of the iterative process.

More recently Silvester and Hsieh [4] have applied finite elements, using a functional which effectively minimizes energy in all space, including R_E. This approach appears to be limited to the Laplace operator (where total energy is finite), and the problem of singularity of the Green's function remains.

Sandy and Sage [5] in effect choose C_1 to lie along the surface of a conductor in the picture frame (a TEM microstrip problem). Problems with Green's-function singularities are avoided, since potentials are not evaluated at source points. They use an iterative procedure to obtain their solution, beginning with an estimate of potential on B_1, evaluating the field by SOR with finite differences to determine sources on C_1, and then computing a potential for B_1 from the integral equation. This potential is used to make a better estimate, and the procedure recycles. They point out that convergence is not guaranteed, and, in fact, use underrelaxation to improve their estimate for potential on B_1. Computation time is undoubtedly high, and media inhomogeneities must be very carefully provided for. Their technique permits sources outside the picture frame when these can be added as a lumped contribution.

Greenspan and Werner [6] discuss a finite-difference approach for the Helmholtz operator, and show that solutions do exist, are unique, and can be numerically obtained. The contour C_1 is taken coincident with B_1, thus producing singularity problems.

IV. The Finite-Element Method

The finite-element method, used to produce a field solution for a bounded region, such as R_1 in Fig. 1, is described in detail by Silvester [16], and Richards and Wexler [7], with extension to provide for anisotropic, continuously inhomogeneous regions given by Wexler [8]. The book by Zienkiewicz [18] is also a good reference.

Using the example of Poisson's equation in a continuous inhomogeneous region, we have

$$-\nabla\cdot(\epsilon\nabla\phi) = f \qquad \text{within } R_1. \tag{1}$$

The quadratic functional is discussed fully in [14, pp. 291–318], and using the forms given in [17, eqs. (96) and (112)], we have, in general

$$F = \int_{R_1} \epsilon\nabla\phi^*\cdot\nabla\phi\, dxdy - \int_{R_1} \phi f^*\, dxdy - \int_{R_1} \phi^* f\, dxdy \tag{2a}$$

$$F = \int_{R_1} \epsilon\,|\nabla\phi|^2\, dxdy - 2\,\mathrm{Re}\left\{\int_{R_1} \phi f\, dxdy\right\} \tag{2b}$$

where clearly $\nabla\phi^*\cdot\nabla\phi = |\nabla\phi|^2$. The superscript $*$ denotes the complex conjugate. For positive ϵ, F possesses the desirable characteristic of positive definiteness.

The field within R_1 is represented by a combination of appropriate polynomial interpolatory functions α_j, defined on an element basis, and node potentials ϕ_j, which are field values at n specific points in R_1

$$\phi = \sum_{j=1}^{n} \alpha_j\phi_j \tag{3a}$$

where by definition α_j takes on the values

$$\alpha_j = \begin{cases} 1, & \text{at node } j \\ 0, & \text{at other nodes.} \end{cases} \tag{3b}$$

The subscript j refers to node positions, some internal (subscripted I) and some on the boundary (subscripted B). Then, in vector notation, (3a) may be written

$$\phi = \boldsymbol{\alpha}_B{}^T\boldsymbol{\phi}_B + \boldsymbol{\alpha}_I{}^T\boldsymbol{\phi}_I = \boldsymbol{\phi}_B{}^T\boldsymbol{\alpha}_B + \boldsymbol{\phi}_I{}^T\boldsymbol{\alpha}_I. \tag{4}$$

We may define four matrices, A_{II}, A_{IB}, A_{BI}, and A_{BB}, for example

$$A_{BI} = \int_{R_1} \epsilon\nabla\boldsymbol{\alpha}_B\nabla\boldsymbol{\alpha}_I{}^T\, dxdy \tag{5}$$

and vectors $\boldsymbol{b}_I$ and $\boldsymbol{b}_B$, for example

$$\boldsymbol{b}_I = \int_{R_1} \boldsymbol{\alpha}_I f\, dxdy. \tag{6}$$

The integral-equation constraint produces a linear relationship between $\boldsymbol{\phi}_B$ and $\boldsymbol{\phi}_I$, which is written, as shown in Section V, (17)

$$\boldsymbol{\phi}_B = M\boldsymbol{\phi}_I. \tag{7}$$

Now we make use of (4)–(7) to rewrite (2)

$$F = \boldsymbol{\phi}_I{}^T\{A_{II} + A_{IB}M + M^TA_{BI} + M^TA_{BB}M\}\boldsymbol{\phi}_I - 2\boldsymbol{\phi}_I{}^T\{\boldsymbol{b}_I + M^T\boldsymbol{b}_B\}. \tag{8}$$

Positive definiteness of the functional is not altered by the substitutions. At the solution point, F is a minimum and its derivative with respect to the vector $\boldsymbol{\phi}_I$ is a zero vector. Performing this differentiation, and equating this derivative to a zero vector, we obtain the positive-definite system

$$[A_{II} + A_{IB}M + M^TA_{BI} + M^TA_{BB}M]\boldsymbol{\phi}_I = \boldsymbol{b}_I + M^T\boldsymbol{b}_B. \tag{9}$$

The matrices produced by this finite-element formulation are smaller in size but more dense than finite-difference matrices. The problems presented here consist of about 60 unknowns. Although these matrices may be block sparse with an appropriate node ordering system, permitting large problems to be solved by block-iterative techniques, they are treated here as being dense.

Types of errors occurring in the finite-element method are discussed in [7], along with some indication of computer-time requirements.

For the case of several picture frames, the representation is extended to include all nodes in all picture frames, and the derivation proceeds identically.

The finite-element method provides good solutions when the field is naturally polynomic or close to polynomic. Derivative singularities, such as those occurring at edges of conductors, are not well approximated. Extensions of the finite-element method to provide for such singularities are presented by Décreton and Wexler [9]; much improved accuracy is obtained.

The Helmholtz operator alters the functional slightly, and positive definiteness is guaranteed only for frequencies below the lowest eigenfrequency of any picture-frame region. These regions should therefore be kept sufficiently small to ensure that this is so. This will be discussed in a subsequent publication.

V. The Integral-Equation Constraint

In the region outside C_1 in Fig. 1(b), which we have called R_e, Laplace's equation holds

$$\nabla^2\phi(x, y) = 0. \tag{10}$$

Also, the Green's relation

$$-\nabla^2G(x, y\,|\,x_0, y_0) = \delta(x - x_0, y - y_0) \tag{11}$$

is applicable, and since R_e is assumed to be source free, homogeneous, and isotropic, the free-space Green's function may be used

$$G(x, y\,|\,x_0, y_0) = -\frac{1}{2\pi\epsilon_0}\ln\sqrt{(x - x_0)^2 + (y - y_0)^2}. \tag{12}$$

Multiplying (10) by G, (11) by ϕ, adding the resulting equations, and integrating over R_e, we obtain

$$\phi(x, y) = \int_{R_e} \{G(x, y\,|\,\xi, \rho)\nabla^2\phi(\xi, \rho) - \phi(\xi, \rho)\nabla^2G(x, y\,|\,\xi, \rho)\}\, d\xi d\rho \tag{13}$$

for any point (x, y) in R_e. Applying Green's theorem, we obtain:

$$\phi(x, y) = \oint_{C_1} \left(G\frac{\partial\phi}{\partial n} - \phi\frac{\partial G}{\partial n}\right) ds. \tag{14}$$

Note that C_1 is only part of a larger closed-contour C, the outer parts of which are cuts which disappear in integration, and a part at infinity whose contribution to the integration vanishes. See Stakgold [10, pp. 142–143].

Field quantities on C_1 in (14) may be represented by the node functions and potentials given by (4)

$$\phi = \left\{\oint_{C_1}\left(G\frac{\partial \alpha_I^T}{\partial n} - \alpha_I^T\frac{\partial G}{\partial n}\right) ds\right\}\phi_I + \left\{\oint_{C_1}\left(G\frac{\partial \alpha_B^T}{\partial n} - \alpha_B^T\frac{\partial G}{\partial n}\right) ds\right\}\phi_B. \quad (15)$$

Once all node potentials are established, (15) is used to determine the field at any point in space outside the picture frame.

To produce the constraint equation, (15) is evaluated at all node points on the boundary B_1. Thus

$$\phi_B = P_I\phi_I + P_B\phi_B \quad (16)$$

where rows of the matrices P_I and P_B are determined from (15) and evaluated for each of the boundary node points.

An adjustment of (16) puts it into the form of (7)

$$\phi_B = (I - P_B)^{-1}P_I\phi_I = M\phi_I. \quad (17)$$

Again there is no restriction to just one picture frame. If more than one picture frame is employed, each has its own boundary nodes and contour C_1. The derivation proceeds similarly for the Helmholtz Green's function, excepting that complex arithmetic must be used.

Contour integration along C_1 is performed numerically using the trapezoidal rule at present; three-figure accuracy is easily obtained. The computer time required for these integrations has been found not to exceed 30 percent of that required to build the finite-element matrices.

VI. Examples

Several examples are presented using the method described. The square parallel-plate capacitor shows the accuracy of the scheme for an electrostatic problem, while other examples show placement of dielectric material, harmonic-field problems, and the use of several picture frames in one problem.

A. The Square Parallel-Plate Capacitor

A cross section of the geometry is shown in Fig. 2(a), representing two infinitely long, thin, conducting plates of width equal to their separation. A single picture-frame boundary and contour are labelled B_1 and C_1, respectively. Considering the symmetries involved, the solution can be obtained in the positive quadrant, and, accordingly, finite elements are placed there. This reduced picture frame is bounded by one-quarter of B_1, where the integral constraint is applied, and by parts of the x and y axes, where homogeneous Dirichlet and

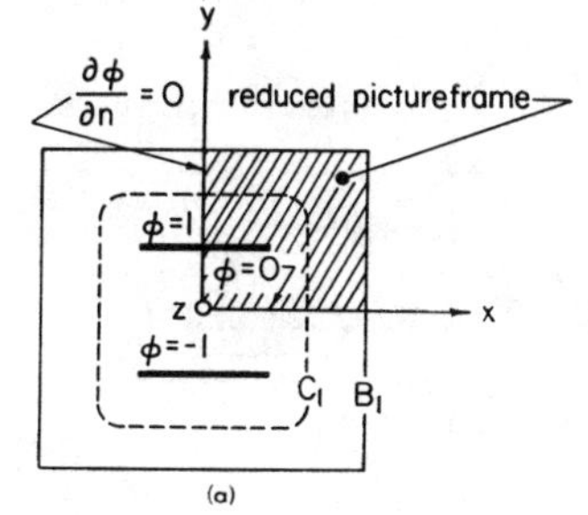

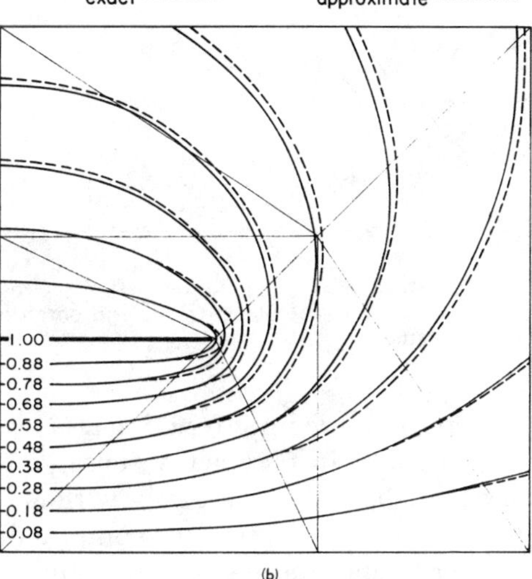

Fig. 2. The parallel-plate capacitor. (a) Cross section showing the picture frame. (b) Electrostatic equipotentials in the reduced picture frame.

Neumann boundary conditions are imposed, respectively, to provide for the symmetries. Integration along the contour C_1 is facilitated by choosing it symmetrically, in order that appropriately signed images of the field values in the positive quadrant may be used in the other three.

The edge of the capacitor plate within the reduced picture frame causes a derivative singularity. Triangular finite elements are judiciously placed in order to minimize this effect, as shown in Fig. 2(b). A Dirichlet boundary condition of 1 V is specified on the plate, and the electrostatic solution is obtained using fourth-order polynomials of the form $x^m y^n$ [7], which results in 65 interior node potentials, the unknowns in (9).

Capacitance is computed from the electrostatic energy of the field obtained by an integration of $(\nabla\phi)^2$ over the picture frame plus an integration of $\phi\, \partial\phi/\partial n$ along B_1, the picture frame boundary [4]. With free-space permittivity assumed throughout, the capacitance is computed to be 18.9 pF/m. An equipotential plot of the solution within the picture frame is shown as the dashed lines in Fig. 2(b).

Several methods are available to determine the exact capacitance [11]–[13], which is given as 18.7 pF/m, indicating our value to be high by about 1 percent. The derivative singularity no doubt contributes to this error, while the nature of the convergence of the finite-element method predicts an energy (and hence capacitance) higher than the correct value. As Mikhlin [14, ch. II]

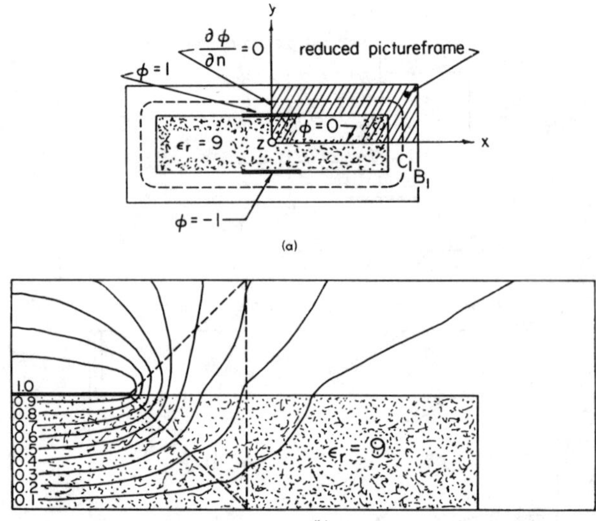

Fig. 3. Parallel plates with dielectric slab $\epsilon_r=9$. (a) Cross section showing the picture frame. (b) Electrostatic equipotentials in the reduced picture frame.

points out, the positive-definite functional (2) produces a minimizing sequence in the energy sense, with exact convergence occurring in the limit only with an energy-complete function set. Therefore, our computed capacitance is an upper bound, but is not a least upper bound since our set of polynomials is not energy complete.

An "exact" equipotential plot, shown as solid lines in Fig. 2(b), is obtained by solving, with care, the relevant integral equation [1], and is reliable to three significant figures. Discrepancies near the edge of the plate are observed, while the equipotential plot tends to exaggerate errors in the relatively flat field near the picture-frame boundary. The constrained finite-element solution of the field within the picture frame is clearly quite good, and does represent a snapshot of the entire field.

B. Square Capacitor with a Dielectric Slab

Fig. 3(a) shows, in cross section, an infinitely long slab of dielectric ($\epsilon_r=9$), filling the space between the plates and extending outward from the edges a small distance. This configuration resembles some microstrip problems discussed in the literature [11], [12], although normally the slab is taken as being infinitely wide as well as long. Following identically the steps of the previous example, the reduced picture frame is solved producing the equipotential plot shown in Fig. 3(b). Electrostatic energies are used ([4] and [12]) to compute an effective dielectric constant of 6.38. Since most of the energy in the dielectric is concentrated near the plates, an infinitely wide dielectric slab should produce a dielectric constant just slightly higher; Bryant and Weiss [12] published 6.47, which is about 1 percent higher than our value.

The effect of the derivative singularity is clearly observed [Fig. 3(b)] in the wobble of the equipotentials in the two triangular elements (indicated by dashed lines) abutting the edge of the plate. (The other elements are not indicated.)

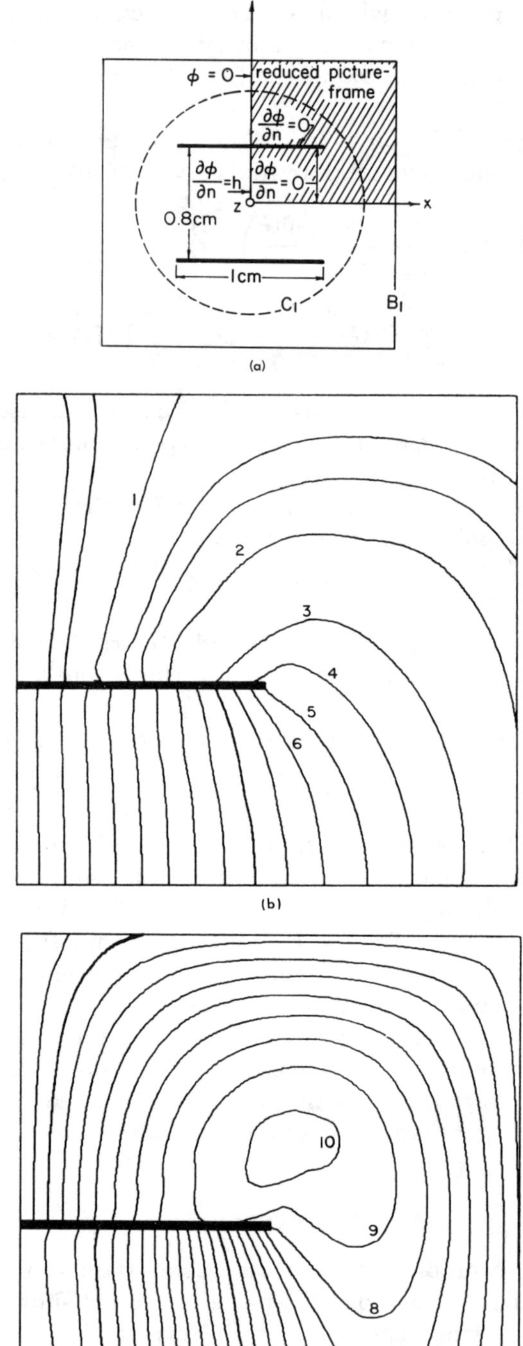

Fig. 4. Parallel plates, using electric vector potential, with static and harmonic solutions. (a) Cross section showing the picture frame. (b) Static solution, with contours proportional to H_z. (c) Harmonic solution, with contours proportional to H_z, at 19 GHz at phase zero, in the reduced picture frame.

C. Parallel Plates; the Harmonic Problem

For this example, the z-directed electric-vector potential is used. Equipotentials, then, are proportional to contours of constant H_z and prescribe the direction of the electric field (see [15, pp. 129–131]).

Fig. 4(a) shows a familiar geometry—infinitely long parallel plates with specified dimensions. Symmetry

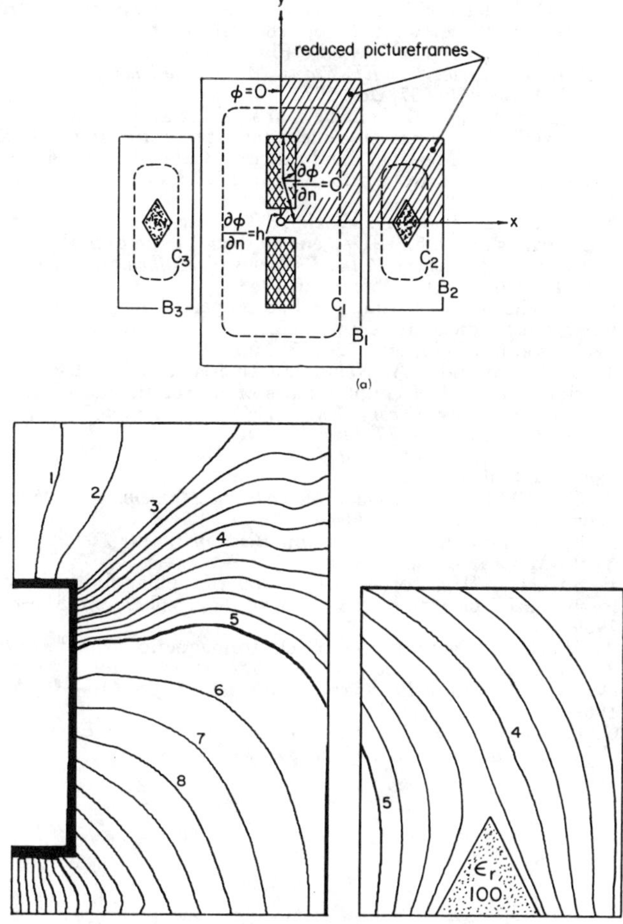

Fig. 5. Antenna-type problem with dielectric obstacles using 3 picture frames. (a) Cross section showing the picture frames. (b) Harmonic solution, contours proportional to H_z, at 5 GHz, at phase zero, in the reduced picture frames.

conditions are used to reduce the picture frame, with appropriate axis boundary conditions. A line dipole source is connected across the plates at the origin, producing the inhomogeneous Neumann boundary condition shown. Free-space media constants are assumed throughout, and a homogeneous Neumann boundary condition is specified on the plate.

Fig. 4(b) shows the static solution. Comparison with Fig. 2(b) shows the effect of the change in potential function, the contours of one being normal to the contours of the other.

Operating the dipole harmonically presents us with the Helmholtz operator. The Green's function is a Hankel function—see [10, p. 266–268] and [15, pp. 198–204] for outward travelling waves, resulting in a complex-valued field. A frequency of 19 GHz produces a marginally positive-definite system, and Fig. 4(c) shows equipotentials of the real part of the field at phase zero. Harrington [15, p. 277] in discussing free-space modes in spherical coordinates, presents the TM_{01} mode, which shows a marked similarity to the contours of Fig. 4(c).

D. A Radiating Antenna with Dielectric Obstacles

Fig. 5(a) shows a cross section of two conducting slabs and two dielectric bars, all infinite in length and parallel. The conducting slabs, about the $x=0$ plane, represent an antenna with a thick-line dipole source connected across the gap between them. The two dielectric bars ($\epsilon_r=100$) represent obstacles, and are placed symmetrically on the x axis. Three picture frames and contours are placed as shown. The symmetry of the problem again permits solution in one positive quadrant, resulting in two reduced picture frames to consider. We proceed to a harmonic solution, using the z-directed electric-vector potential, and show equipotentials of H_z [Fig. 5(b)] for the source at 5 GHz at phase zero.

A rather small number of node poetntials was specified with large triangular elements, resulting in some field errors observed in part of the larger picture frame R_1. (Finite-difference users might note that 60 nodes represent a very coarse grid. With finite elements much more field information is usually provided by each node potential.) Nevertheless, the field is continuous from one picture frame to the other, as evidenced by the contour numbered 5. Also behavior of the field near the dielectric bar is clearly correct.

Theoretically, the boundaries of the picture frames should be transparent to the final solution. In practice errors do occur, but the error at a picture-frame boundary is no worse than the error at any element interface.

VII. Conclusions

The use of free-space integral equations to constrain the local finite-element solution in picture frames appears to be entirely satisfactory for the solution of static and time-harmonic problems in unbounded regions. Time is spent solving problems in local regions of interest; none is wasted elsewhere. However, the field at intermediate locations can be computed at small marginal cost. Field values in all space are available from the picture-frame solutions. The finite-element software at hand permits analysis with sources, inhomogeneities, and anisotropies, when these can all be placed within picture frames, and its power will be greatly enhanced when derivative singularities are provided for. Thus two recurring problems of integral methods are alleviated as follows.

1) Particular problem-dependent Green's functions need not be found.

2) Difficulties associated with singularity of Green's functions are side-stepped.

The advantages of the partial differential-equation approach are as follows.

1) Inhomogeneities and anisotropies are fairly easily provided for.

2) The problem is formulated as the solution of a positive-definite system.

In summary, this paper has attempted to show how

to use the partial differential and integral techniques, each in its appropriate region, to solve problems that could be intractable by either one individually.

Acknowledgment

The authors wish to thank their colleagues for the assistance given them. They wish to thank D. J. Richards, for his excellent programming, and M. C. Décreton for many stimulating discussions about integral equations, Green's functions, and singularities.

References

[1] F. S. Acton, *Numerical Methods That Work.* New York: Harper & Row, 1970, pp. 410–430.

[2] I. A. Cermak and P. Silvester, "Solution of two-dimensional field problems by boundary relaxation," *Proc. Inst. Elec. Eng.*, vol. 115, pp. 1341–1348, Sept. 1968.

[3] ——, "Analysis of coaxial line discontinuities by boundary relaxation," *IEEE Trans. Microwave Theory Tech.* (*Special Issue on Computer-Oriented Microwave Practices*), vol. MTT-17, pp. 489–495, Aug. 1969.

[4] P. Silvester and M. S. Hsieh, "Finite-element solution of 2-dimensional exterior field problems," *Proc. Inst. Elec. Eng.*, vol. 118, no. 12, pp. 1743–1747, Dec. 1971.

[5] F. Sandy and J. Sage, "Use of finite difference approximations to partial differential equations for problems having boundaries at infinity," *IEEE Trans. Microwave Theory Tech.* (Corresp.), vol. MTT-19, pp. 484–486, May 1971.

[6] D. Greenspan and P. Werner, "A numerical method for the exterior Dirichlet problem for the reduced wave equation," *Arch. Ration. Mech. Anal.*, vol. 23, pp. 288–316, 1966.

[7] D. J. Richards and A. Wexler, "Finite-element solutions within curved boundaries," *IEEE Trans. Microwave Theory Tech.*, vol. MTT-20, pp. 650–657, Oct. 1972.

[8] A. Wexler, "Finite-element field analysis of an inhomogeneous, anisotropic, reluctance machine rotor," presented at the IEEE Summer Power Meeting, San Francisco, Calif., July 9–14, 1972, Paper T 72 422-4, to be published in *IEEE Trans. Power App. Syst.*

[9] M. C. Décreton and A. Wexler, "The care and treatment of singularities in the finite-element method," to be published.

[10] I. Stakgold, *Boundary Value Problems of Mathematical Physics*, vol. II. New York: Macmillan, 1968.

[11] H. A. Wheeler, "Transmission-line properties of parallel strips separated by dielectric sheet," *IEEE Trans. Microwave Theory Tech.*, vol. MTT-13, pp. 172–185, Mar. 1965.

[12] T. G. Bryant and J. A. Weiss, "Parameters of microstrip transmission lines and of coupled pairs of microstrip lines," *IEEE Trans. Microwave Theory Tech.* (*1968 Symposium Issue*), vol. MTT-16, pp. 1021–1027, Dec. 1968.

[13] W. R. Smythe, *Static and Dynamic Electricity.* New York: McGraw-Hill, 1968, p. 115.

[14] S. G. Mikhlin, *Variational Methods in Mathematical Physics.* New York: Macmillan, 1964.

[15] R. F. Harrington, *Time Harmonic Electromagnetic Fields.* New York: McGraw-Hill, 1961.

[16] P. Silvester, "High order polynomial triangular finite-element for potential problems," *Int. J. Eng. Sci.*, vol. 7, pp. 849–861, 1969.

[17] A. Wexler, "Computation of electromagnetic fields," *IEEE Trans. Microwave Theory Tech.* (*Special Issue on Computer-Oriented Microwave Practices*), vol. MTT-17, pp. 416–439, Aug. 1969.

[18] O. C. Zienkiewicz, *The Finite Element Method in Engineering Science.* London, England: McGraw-Hill, 1971.

Application of the Unimoment Method to Electromagnetic Scattering of Dielectric Cylinders

SHU-KONG CHANG, STUDENT MEMBER, IEEE, AND KENNETH K. MEI, MEMBER, IEEE

Abstract—**The unimoment method is applied to calculate the scattered fields of dielectric cylinders of arbitrary cross section or of inhomogeneous material. The basic technique of the method is the use of the finite element methods inside a mathematical circle, which encloses the inhomogeneous body. The fields outside are expanded in the usual cylindrical harmonics. The interior and exterior problems are then coupled at the circle. The versatility is increased greatly by introducing the method of "inhomogeneous element." The advantage of the proposed method is the simplicity and efficiency in programming. The validity of the computer program has been verified by comparing results with calculations from other methods for i) an off-centered circular cylinder, ii) two circular cylinders, and iii) a circular cylindrical shell.**

I. Introduction

THE PROBLEMS of electromagnetic scattering by conducting obstacles have been investigated mostly by solving integral equations for the induced surface current densities. The scattered fields are then calculated from the induced surface currents [2]–[6]. Waterman [7] and Okamoto [8] extended the method to the dielectric and anisotropic scattering problems by representing the scattering fields with infinite series of Hankel's functions. However, their method is limited to the cases in which the scatterers are not gross perturbations to circular shapes, because the expansion may otherwise diverge on the dielectric surface. The analytic continuation method applied by Wilton and Mittra [11] is very complicated for the dielectric case, where the complexity in constructing the circles in the method discourages the systemmatic solutions to general problems.

The above mentioned techniques in the literatures available are not suitable for inhomogeneous dielectric cylinders. The only known method which can solve inhomogeneous dielectric cylinder problems is that given by Richmond [9], [10]. However, his method, which represents the fields inside the dielectric by the equivalent current, is limited to thin cylinders. When the cylinders are of large cross section area, Richmond's method requires the inversion of a large full matrix, which is numerically impractical.

The object of this paper is to demonstrate the unimoment method [1] in solving the scattering fields of dielectric cylinders of arbitrary cross section and/or of inhomogeneous dielectric material. The finite element methods are applied to solve the partial differential equations inside a mathematical circle, which encloses the inhomogeneous body. The fields outside are expanded in Hankel's functions. The interior and exterior problems are first solved separately and later coupled on the circle by the continuity conditions of the E- and H-fields. The versatility is increased greatly by introducing the method of "inhomogeneous elements." The advantage of the proposed method is the simplicity and efficiency in computer programming. Various problems can be solved by the same program. And the input control for the shape and material constants can be made very simple. The finite element methods end up with a sparse banded linear system of equations which can be solved very efficiently by various methods [1].

The principles of the techniques to be discussed apply to both E-wave ($H_z = 0$) and H-wave ($E_z = 0$) scattering from dielectric cylinders with two different approaches of the unimoment method, i.e., the Dirichlet type and the Neumann type. The incident wave remains unrestricted; it may be a plane wave, waves generated by line sources, or waves generated by other kind of sources. The fields concerned in the following discussions are all uniform in the z direction and are all assumed to have the conventional exp $(j\omega t)$ harmonic time dependence.

II. The Unimoment Method

Assume that the incident waves propagate in the direction normal to the z axis, i.e., the axis of the dielectric cylinder. Since all quantities are uniform in z, the differential form of Maxwell's equations imply that E-wave and H-wave are separable. The basic idea of the method is to draw a circle C' with radius $r = a$ to enclose the entire scatterer (Fig. 1). The boundary conditions of E- and H-fields are considered at C' instead of the actual dielectric boundary C. The total

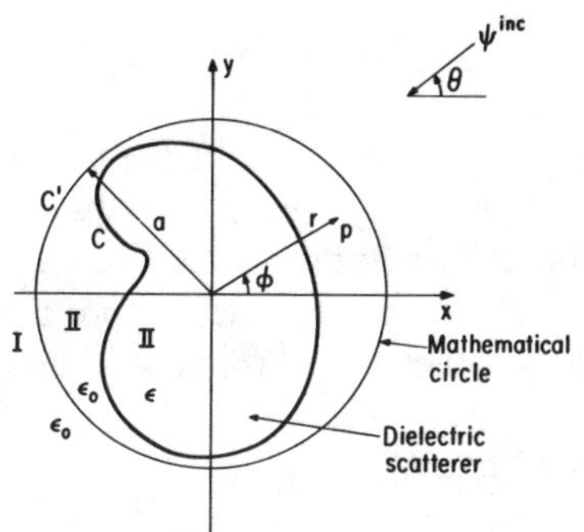

Fig. 1. Dielectric cylinder and artificial circle.

Manuscript received May 31, 1975. This work was sponsored by the U.S. Army Mobility Equipment Research and Development Center under Contract DAAK02-71-C-0206.

The authors are with the Department of Electrical Engineering and Computer Sciences and the Electronics Research Laboratory, University of California, Berkeley, CA 94720.

Reprinted from *IEEE Trans. Antennas Propagat.*, vol. AP-24, no. 1, pp. 35–42, Jan. 1976.

field ψ^{I} in region I ($r > a$) is the sum of the incident and scattered fields, that is,

$$\psi^{\mathrm{I}}(r,\phi) = \psi^{\mathrm{inc}}(r,\phi) + \psi^{\mathrm{sca}}(r,\phi) \tag{1}$$

where

$$\psi(r,\phi) = \begin{cases} E_z(r,\phi), & \text{for } E\text{-wave} \\ H_z(r,\phi), & \text{for } H\text{-wave.} \end{cases} \tag{2}$$

The incident field ψ^{inc} is a know function, e.g.,

$$\exp(jk_0 r \cos(\theta - \phi))$$

for a plane wave and $-\frac{1}{4}jH_0^{(2)}(k_0|\bar{r} - \bar{r}'|)$ for a unit line source. The scattered field ψ^{sca} is an outgoing wave which can be approximated in terms of a finite series of Hankel's functions as

$$\psi^{\mathrm{sca}}(r,\phi) = A_0 H_0^{(2)}(k_0 r) + \sum_{n=1}^{N} H_n^{(2)}(k_0 r) \cdot (A_n \cos n\phi + B_n \sin n\phi) \tag{3}$$

where $k_0 = \omega(\mu_0\varepsilon_0)^{1/2}$, and A_n and B_n are arbitrary constants to be determined. According to the analysis of Hankel's functions [11] and the numerical data available, N should be chosen slightly larger than $k_0 a$ for best results. The continuity conditions on C' are

$$\psi^{\mathrm{II}}(a,\phi) = \psi^{\mathrm{inc}}(a,\phi) + \psi^{\mathrm{sca}}(a,\phi) \tag{4}$$

and

$$\left.\frac{\partial\psi^{\mathrm{II}}}{\partial r}(r,\phi)\right|_{r=a} = \left.\frac{\partial\psi^{\mathrm{inc}}}{\partial r}(r,\phi)\right|_{r=a} + \left.\frac{\partial\psi^{\mathrm{sca}}}{\partial r}(r,\phi)\right|_{r=a}. \tag{5}$$

The preceding two equations hold in the entire interval $0 \leq \phi \leq 2\pi$. Since the normal derivative value of H_z at a dielectric boundary is discontinuous, it is convenient for H-wave to have the artificial circle a little larger to ensure both (4) and (5) valid on C'.

The fields in region II will be assumed to be linear combinations of a set of functions $\psi_n^{\mathrm{II}}(r,\phi)$ each of which is a solution of the differential equation

$$\nabla^2\psi_n^{\mathrm{II}}(\mathbf{r}) + k_0^2\varepsilon(\mathbf{r})\psi_n^{\mathrm{II}}(\mathbf{r}) = 0, \qquad \text{for } E\text{-wave} \tag{6}$$

or

$$\nabla\cdot\left[\frac{1}{\varepsilon_r(\mathbf{r})}\nabla\psi_n^{\mathrm{II}}(\mathbf{r})\right] + k_0^2\psi_n^{\mathrm{II}}(\mathbf{r}) = 0, \qquad \text{for } H\text{-wave.} \tag{7}$$

Thus,

$$\psi^{\mathrm{II}}(\mathbf{r}) \cong \sum_{n=0}^{2N} C_n\psi_n^{\mathrm{II}}(\mathbf{r}). \tag{8}$$

Equation (8) may be enforced if the set of boundary values $\psi_n^{\mathrm{II}}(a,\phi)$ or $(\partial\psi_n^{\mathrm{II}}/\partial r)(a,\phi)$ are capable of representing those of the true solution on C'. In the case of a two-dimensional scattering problem, the obvious choice for the boundary values of $\psi_n^{\mathrm{II}}(\mathbf{r})$ should be the Fourier series

$$\psi_n^{\mathrm{II}}(a,\phi)\left[\text{or } \frac{\partial\psi_n^{\mathrm{II}}}{\partial r}(a,\phi)\right] = \begin{cases} \cos n\phi, & \text{for } n \leq N \\ \sin(n-N)\phi, & \text{for } n > N \end{cases} \tag{9}$$

where $n = 0,1,2,\cdots,2N$.

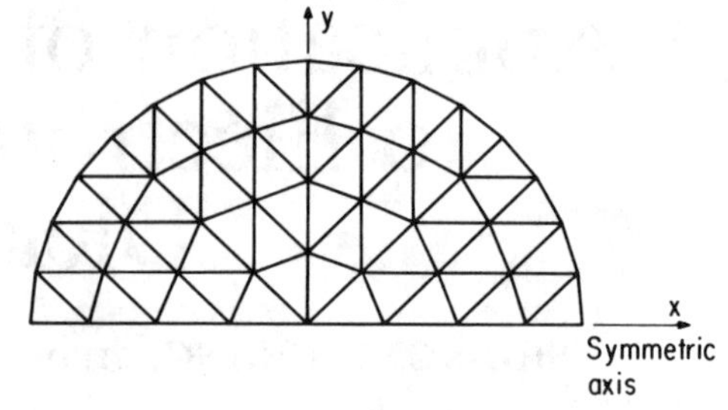

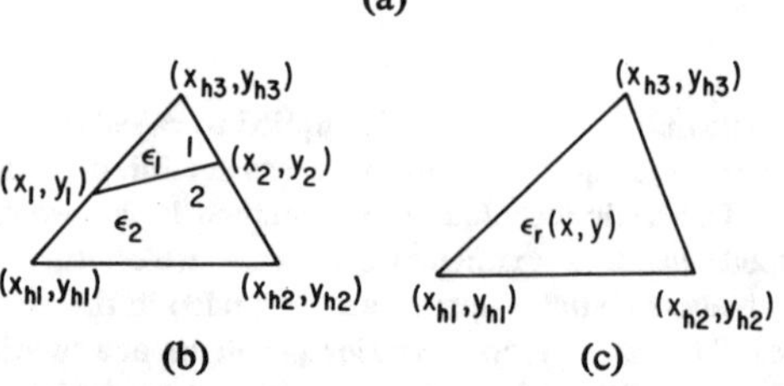

Fig. 2. (a) Construction of triangular elements. (b) Triangular element with two materials. (c) Triangular element with inhomogeneous dielectric material.

The mathematical foundation of the finite element methods is the variational representation of the differential equations. For (6) the variational equivalence is the minimization of the functional,

$$I = \iint_S \{|\nabla\psi_n^{\mathrm{II}}|^2 - k_0^2\varepsilon_r(\mathbf{r})(\psi_n^{\mathrm{II}})^2\}\, ds \tag{10}$$

and that for (7) is

$$I = \iint_S \left\{\frac{1}{\varepsilon_r(\mathbf{r})}|\nabla\psi_n^{\mathrm{II}}|^2 - k_0^2(\psi_n^{\mathrm{II}})^2\right\} ds \tag{11}$$

where S is the area bounded by the circle C'. The preceding functionals are sufficient for differential equations with Dirichlet type boundary conditions. For Neumann type boundary conditions, the boundary integral

$$I' = -2a\int_0^{2\pi} \psi_n^{\mathrm{II}}(a,\phi)\frac{\partial\psi_n^{\mathrm{II}}}{\partial r}(a,\phi)\, d\phi \tag{12}$$

should be added to I. It is noticed that for H-wave, the integrand of (12) should be divided by $\varepsilon_r(a,\phi)$ but we have assumed that the circumscribing circle C' is larger than the obstacle thus, $\varepsilon_r(a,\phi) = 1$ on C'.

In the following, we shall describe the nomenclature of the finite element technique which will facilitate the ensuing discussions.

A. Elements and Element Integrals

The finite element methods as developed by structure engineers traditionally use local matrix formulation for each element. In this section we shall rederive the finite element formulas in a direct form. The discretization in the finite element method in general results in a set of triangular subdivisions of the region S bounded by C'. Fig. 2(a) represents one of many possible subdivisions. We shall assign a number to each element (subdivision) and a number to each nodal point. Let L be the total number of elements and $K + K'$ be the total number of nodes, where

K represents the total number of nodes in the interior of S and K' on the boundary C'.

The value of ψ_n^{II} within the hth element is represented by a linear function

$$U_{(h)}(\xi) = a_{(h)1}\xi_1 + a_{(h)2}\xi_2 + a_{(h)3}\xi_3 = \bar{a}_{(h)}^T \cdot \xi \quad (13)$$

where $\xi_1 = 1$, $\xi_2 = x$, and $\xi_3 = y$. We shall use the rectangular coordinates in this section for convenience. $U_{(h)}(\xi)$ is bounded by three nodal values of ψ_n^{II} which are denoted by U_{hi} ($i = 1,2,3$) and hi may be one of the values $1,2,\cdots,(K+K')$. Let $\bar{U} = [U_1,U_2,\cdots,U_K]$ be a vector representing all the interior nodal points and $\bar{U}' = [U_{K+1},U_{K+2},\cdots,U_{K+K'}]$ all the boundary nodal points of $\psi_n^{\text{II}}(r)$, respectively. Therefore,

$$\bar{a}_{(h)} = N_{(h)}^{-1}\hat{U}_{(h)} = M_{(h)}\hat{U}_{(h)} \quad (14)$$

where

$$\hat{U}_{(h)} = [U_{h1},U_{h2},U_{h3}]^T$$

and

$$N_{(h)} = \begin{pmatrix} 1 & x_{h1} & y_{h1} \\ 1 & x_{h2} & y_{h2} \\ 1 & x_{h3} & y_{h3} \end{pmatrix}. \quad (15)$$

Equation (13) can now be represented by

$$U_{(h)} = [M_{(h)}\hat{U}_{(h)}]^T\xi \quad (16)$$

and

$$\nabla U_{(h)} = \sum_{i=1}^{3} m_{(h)i2}U_{hi}\xi_2 + \sum_{i=1}^{3} m_{(h)i3}U_{hi}\xi_3$$

$$= \sum_{k=2}^{3}\sum_{i=1}^{3} m_{(h)ik}U_{hi}\xi_k \quad (17)$$

where $m_{(h)ij}$ is the ijth element of $M_{(h)}$.

Thus

$$|\nabla U_{(h)}|^2 = \sum_{k=2}^{3}\left(\sum_{i=1}^{3} m_{(h)ik}U_{hi}\right)^2$$

and

$$\frac{\partial}{\partial U_n}|\nabla U_{(h)}|^2 = \sum_{k=2}^{3}\left(2\sum_{i=1}^{3} m_{(h)ik}U_{hi}\sum_{j=1}^{3} m_{(h)jk}\delta_{n,hj}\right)$$

$$= 2\sum_{k=2}^{3}\sum_{i=1}^{3}\sum_{j=1}^{3} m_{(h)ik}m_{(h)jk}U_{hi}\delta_{hj,n} \quad (18)$$

where $\delta_{i,j}$ is the well known Kronecker delta. Similarly, we may represent

$$U_{(h)}^2 = \{[M_{(h)}\hat{U}_{(h)}]^T\xi\}^2 = \left[\sum_{i=1}^{3}\sum_{j=1}^{3} m_{(h)ij}U_{hj}\xi_i\right]^2$$

and

$$\frac{\partial U_{(h)}^2}{\partial U_n} = 2\left(\sum_{i=1}^{3}\sum_{j=1}^{3} m_{(h)ij}U_{hj}\xi_i\right)\left(\sum_{k=1}^{3}\sum_{l=1}^{3} m_{(h)kl}\xi_k\delta_{hl,n}\right)$$

$$= 2\sum_{i=1}^{3}\sum_{j=1}^{3}\sum_{k=1}^{3}\sum_{l=1}^{3} m_{(h)ij}m_{(h)kl}U_{hj}\delta_{hl,n}\xi_i\xi_k. \quad (19)$$

Thus,

$$\frac{\partial I}{\partial U_n} = 0, \qquad n = 1,2,\cdots,K \quad (20)$$

results in the matrix equation of $\bar{U}$, i.e.,

$$Q\bar{U} = T\bar{U}' \quad (21)$$

where, for I of (10),

$$q_{n,m} = \left(2\sum_{h=1}^{L}\sum_{k=2}^{3}\sum_{i=1}^{3}\sum_{j=1}^{3} m_{(h)ik}m_{(h)jk}\delta_{hi,m}\delta_{hj,n}\int_{(h)} d\xi_2\, d\xi_3\right)$$

$$- \left(2k_0^2\sum_{h=1}^{L}\sum_{i=1}^{3}\sum_{j=1}^{3}\sum_{k=1}^{3}\sum_{l=1}^{3} m_{(h)ij}m_{(h)kl}\delta_{hj,m}\delta_{hl,n}\right.$$

$$\left.\cdot\int_{(h)} \varepsilon_r\xi_i\xi_k\, d\xi_2\, d\xi_3\right) \quad (22)$$

for $m = 1,2,\cdots,K$. Equation (22) may be used to find $t_{n,m-K}$, where $m = K+1, K+2,\cdots, K+K'$. Equation (22) appears to be most convenient to use for programming purposes.

While the preceding derivations are for linear trial functions only, their extensions to polynomials of higher degrees are straightforward.

B. Inhomogeneous Elements

Traditionally the geometry of the elements is determined by the boundary shape of the problem. It is frequently more convenient to let the geometry of the element be fixed while the material within each element is allowed to have step discontinuity. Let the boundary of the discontinuity within an element be approximated by a straight line such as shown in Fig. 2(b). For an E-wave, the linear approximation of $U_{(h)}$ is still valid even with material discontinuity, owing to the continuity of E_z and its normal derivative across a material boundary. Thus, the element integral of an inhomogeneous element need only slight modification, i.e.,

$$I_{(h)} = \iint_{(h)} |\nabla U_{(h)}|^2\, dx\, dy - k_0^2\varepsilon_2 \iint_{(h)} U_{(h)}^2\, dx\, dy$$

$$+ k_0^2(\varepsilon_2 - \varepsilon_1)\iint_{(h')} U_{(h)}^2\, dx\, dy \quad (23)$$

where (h') means the triangle bounded by the points 1, 2, and $h3$ (Fig. 2(b)).

For an H-wave, however, a linear approximation of $U_{(h)}$ within an element would fail, because the normal derivative of H_z is now discontinuous across a material boundary. In this case, we shall use two piecewise linear functions to represent $U_{(h)}$. Referring to Fig. 2(b), we have,

$$U_{(h)} = \bar{a}_{(h)}^T \cdot \xi, \qquad \text{in } \Delta_{1,2,h_3}$$

$$U_{(h)} = \bar{a}'^T \cdot \xi, \qquad \text{in } \Delta_{h_1,h_2,h_3} - \Delta_{1,2,h_3} \quad (24)$$

The trial function given by (24) should satisfy the following conditions.

a) They must pass through the nodal values $\hat{U}_{(h)}$ at the nodal points, which results in

$$U_{h1} = a_{(h)1}' + a_{(h)2}'x_{h1} + a_{(h)3}'y_{h1}$$

$$U_{h2} = a_{(h)1}' + a_{(h)2}'x_{h2} + a_{(h)3}'y_{h2}$$

$$U_{h3} = a_{(h)1} + a_{(h)2}x_{h3} + a_{(h)3}y_{h3}. \quad (25)$$

b) They must be continuous across the material boundary, which results in

$$(a_{(h)1} + a_{(h)2}x_1 + a_{(h)3}y_1) - (a_{(h)1}' + a_{(h)2}'x_1 + a_{(h)3}'y_1) = 0$$

$$(a_{(h)1} + a_{(h)2}x_2 + a_{(h)3}y_2) - (a_{(h)1}' + a_{(h)2}'x_2 + a_{(h)3}'y_2) = 0 \quad (26)$$

c) They must satisfy the continuity condition of the electric field on line (1,2), which results in

$$\varepsilon_2 \nabla U_{(h)} \cdot \hat{n}|_{U \text{ in } 1} = \varepsilon_1 \nabla U_{(h)} \cdot \hat{n}|_{U \text{ in } 2} \quad (27)$$

where $\hat{n}$ is the unit normal vector of line (1,2). Replacing $U_{(h)}$ by those of (24), (27) becomes

$$\varepsilon_2[a_{(h)3}(x_1 - x_2) - a_{(h)2}(y_1 - y_2)] = \varepsilon_1[a_{(h)3}'(x_1 - x_2) - a_{(h)2}'(y_1 - y_2)]. \quad (28)$$

Equations (25), (26), and (28) are six equations from which the coefficient vectors $\bar{a}_{(h)}$ and $\bar{a}_{(h)}'$ may be determined. The element integral can then be obtained accordingly.

It should be pointed out that the inhomogeneous elements of H-waves are not conforming, i.e., the representing polynomials of two neighboring triangles are not equal at their common edge. Our computations show however, such nonconformity has no ill effect on numerical solutions.

For continuous varying dielectric material within an element we need only represent the relative dielectric value or its reciprocal by polynomials of x and y, i.e.,

$$\varepsilon_r(x,y) \left[\text{or } \frac{1}{\varepsilon_r(x,y)}\right] = \sum_{p=0}^{n} \sum_{\substack{q=0 \\ p+q \le n}}^{n} a_{pq} x^p y^q \quad (29)$$

for E-wave or H-wave, respectively. Where n is the degree of the polynomial. The purpose of the approximation of (29) is to keep the integrals in (22) of

$$\iint_{(h)} x^i y^j \, dx \, dy$$

type, which can be evaluated in various analytical forms [12], [15].

C. Neumann Problems

The preceding discussions on the finite element method are in general valid only for Dirichlet problems. Its extension to the Neumann problem is immediate. The minimization of the functional should now include the boundary nodes, i.e.,

$$\frac{\partial(I + I')}{\partial U_n} = 0, \qquad n = 1, 2, 3, \cdots, K, K + 1, \cdots, K + K'. \quad (30)$$

Equation (22) is still valid and the resulting Q matrix is now of dimension $K + K'$, and

$$Q \begin{bmatrix} \bar{U} \\ \bar{U}' \end{bmatrix} = \begin{bmatrix} \bar{0} \\ \bar{b} \end{bmatrix}. \quad (31)$$

The vector $\bar{b}$ is obtained from the boundary integral of (12). Specifically,

$$b_i = -\frac{\partial I'}{\partial U_{K+i}}, \qquad i = 1,2,\cdots,K'. \quad (32)$$

The accuracy of b_i depends solely on the representing polynomials of $\psi_n^{\text{II}}(a,\phi)$. It is found that the following approximation is better than any continuous interpolating spline. That is, let

$$\psi_n^{\text{II}}(a,\phi) = \psi_n^{\text{II}}(a,\theta + \phi_{i+K}) = U_{i+k}\left(1 - \frac{\theta^2}{\delta^2}\right) + U_{i-1+K}\left(\frac{\theta^2}{2\delta^2} - \frac{\theta}{2\delta}\right) + U_{i+1+K}\left(\frac{\theta^2}{2\delta^2} + \frac{\theta}{2\delta}\right),$$

$$-\frac{\delta}{2} < \theta < \frac{\delta}{2} \text{ and } i = 1,\cdots,K' \quad (33)$$

where $\delta = \phi_{i+1} - \phi_i$, $i = K + 1,\cdots, K + K' - 1$.

It is important to note that the computation of I' with the representation (33) of $\psi_n^{\text{II}}(a,\phi)$ is numerically unstable when δ is small, because the integration results in small values divided by small numbers. Frequently it is necessary to expand the known function $(\partial\psi_n^{\text{II}}/\partial r)(a, \phi_{i+K} + \theta)$ in a Taylor's series of θ before integration. Substituting the truncated series into the integrand of I' and the integration will become polynomials of δ which have no numerical instabilities.

III. Numerical Results

The techniques discussed in this paper have been implemented by a Fortran program on CDC 6400 and 7600 computers. The program is able to solve the scattered waves of dielectric cylinders of arbitrary shape or of arbitrary inhomogeneity illuminated by plane waves. Both E-wave and H-wave, Dirichlet type and Neumann type unimoment methods are considered in the program. It uses the Fourier components $\cos n\phi$ and $\sin n\phi$, for $\psi_n^{\text{II}}(a,\phi)$ or $(\partial\psi_n^{\text{II}}/\partial r)(a,\phi)$. And, scatterer symmetries are considered in each case to reduce computational efforts.

Many cases have been tested on the program. They all show that both the Dirichlet and Neumann type unimoment methods have the same order of efficiency, accuracy, and rate of convergence. The accuracy of the Neumann problems is very sensitive to the interpolating polynomials of $\psi_n(a,\phi)$ chosen. Although satisfactory results are obtained by (33), the optimal ways of interpolation still have to be investigated.

At large distances ($r \gg a$, $r \gg \lambda$), the scattered fields in terms of (3) can be written as

$$\psi^{\text{sca}} = \left(\frac{2j}{\pi k_0 r}\right)^{1/2} \exp(-jk_0 r) \cdot \left[A_0 + \sum_{n=1}^{\infty} j^n (A_n \cos n\phi + B_n \sin n\phi)\right].$$

The far-field scattering pattern is defined by

$$g(\phi) = A_0 + \sum_{n=1}^{\infty} j^n (A_n \cos n\phi + B_n \sin n\phi). \quad (34)$$

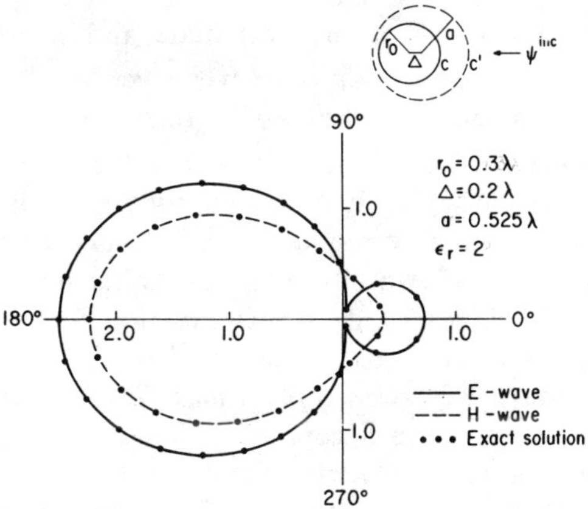

Fig. 3. Amplitude of E-wave and H-wave scattering far-field pattern for off-centered circular cylinder.

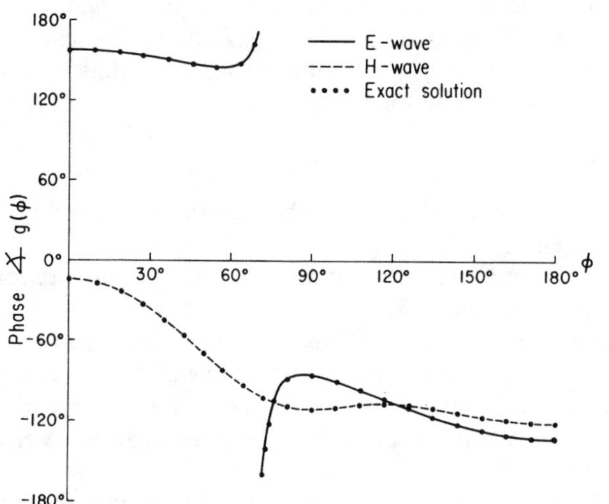

Fig. 4. Phase of E-wave and H-wave scattering far-field pattern for off-centered circular cylinder.

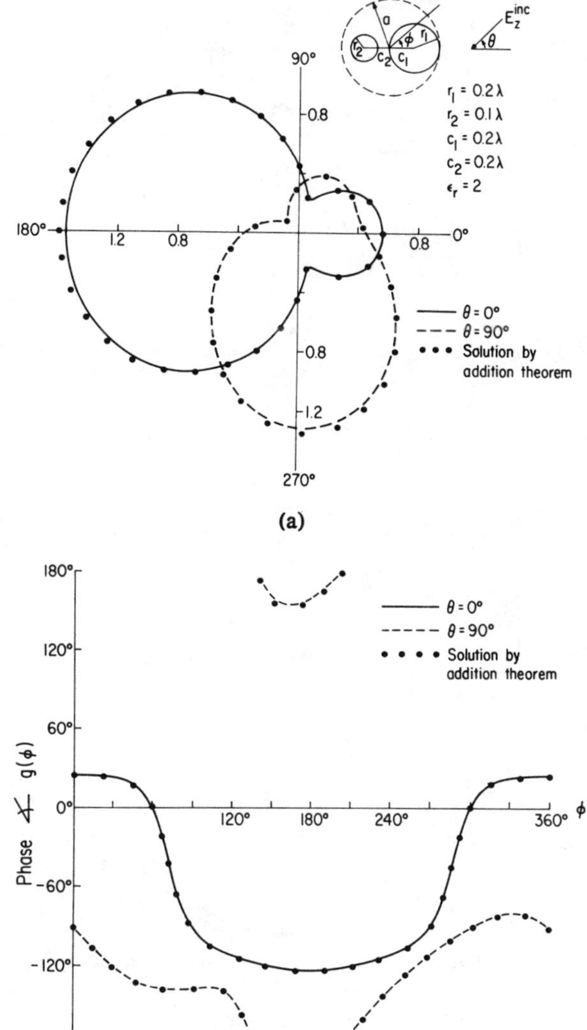

Fig. 5. E-wave scattering far-field pattern for two circular cylinders. (a) Amplitude. (b) Phase.

The scattering cross section, or echo width [9], is defined as follows:

$$W(\phi) = \lim_{r\to\infty} 2\pi r \left|\frac{\psi^{\text{sca}}(r,\phi)}{\psi^{\text{inc}}}\right|^2 = \frac{4}{k_0}\frac{|g(\phi)|^2}{|\psi^{\text{inc}}|^2}$$

$$= \frac{2\lambda}{\pi}|g(\phi)|^2 \tag{35}$$

where we have assumed $|\psi^{\text{inc}}| = 1$.

The validity and versatility of the techniques presented here are demonstrated by the following cases.

i) *A Circular Cylinder with Its Center Offset from the Origin:* Exact solutions are obtained from the method of separation of variables. Numerical solutions are generated by the unimoment methods with the center of the artificial circle at the origin, while the center of the scatterer, which has radius r_0, is offset by Δ from the origin. Figs. 3 and 4 show excellent agreement between numerical and exact solutions for both E-wave and H-wave. Fig. 3 shows the amplitude of $g(\phi)$, and Fig. 4 shows the phase. In both figures, $\varepsilon_r = 2$, $r_0 = 0.3\lambda$, $\Delta = 0.2\lambda$, $a = 0.525\lambda$. Since the off-centered circular scatterer disrupts the symmetry of the problem, the finite element method is no longer an integration of the Bessel's differential equation. We believe our accurate results here do give good evidence of the general validity of the method.

ii) *Two Parallel Circular Cylinders with Different Radii:* The solutions from another approach are obtained by expanding the scattered waves into two series of Hankel's functions and then by the application of the addition theorem. The artificial circle, which encloses both cylinders, has radius $a = \max[c_1 + r_1, c_2 + r_2]$, where c_i is the position of the center, and r_i is the radius of the ith cylinder. The amplitude and phase of the scattering pattern $g(\phi)$, produced by two cylinders of $\varepsilon_r = 2$, $c_1 = 0.2\lambda$, $r_1 = 0.2\lambda$, $c_2 = 0.2\lambda$, $r_2 = 0.1\lambda$, $a = 0.433\lambda$, are shown in Figs. 5 and 6, for E-wave and H-wave incidence, respectively. In both figures two different directions of incidence are presented. The excellent match of the unimoment solutions with the solutions by the addition theorem is another solid evidence of the validity of the technique.

iii) *A Circular Cylindrical Shell:* The unimoment solution is compared with the data provided by Richmond [10] for H-wave scattering of a circular shell cylinder of $\varepsilon_r = 4$,

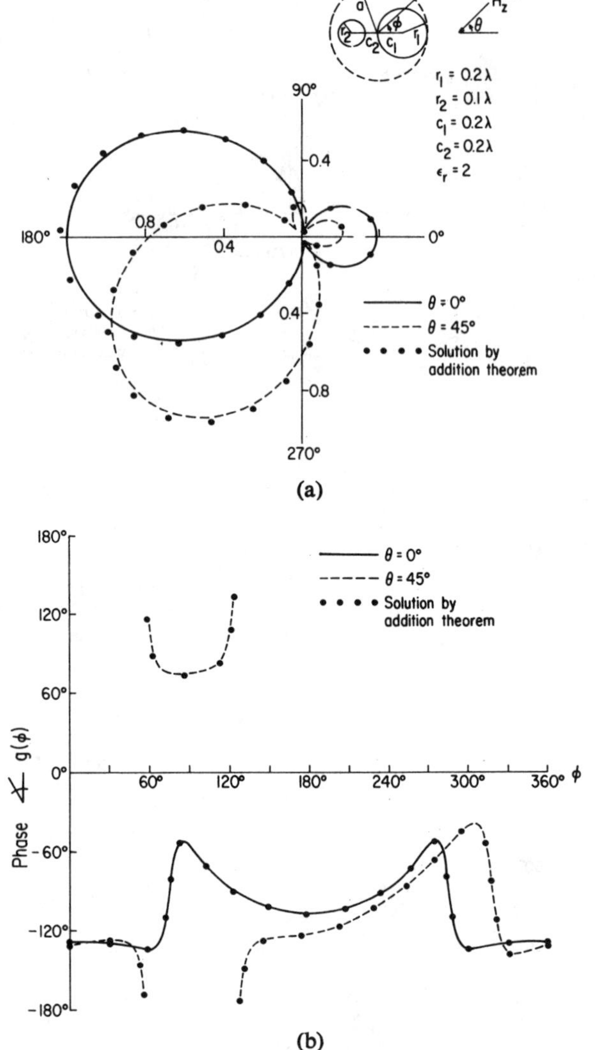

Fig. 6. *H*-wave scattering far-field pattern for two circular cylinders. (a) Amplitude. (b) Phase.

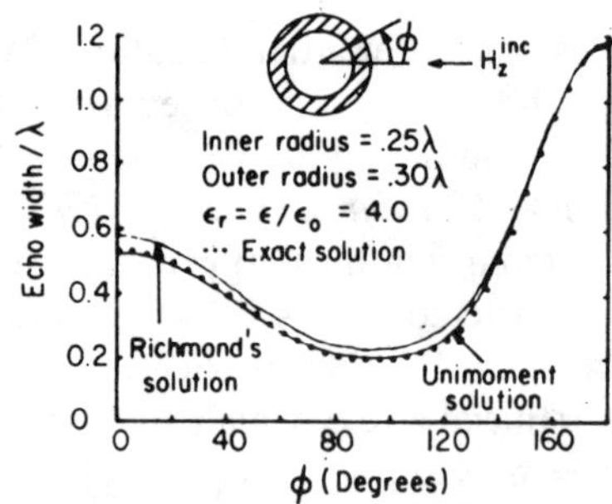

Fig. 7. Echo width pattern for circular cylindrical shell.

inner radius $= 0.25\lambda$, and outer radius $= 0.3\lambda$. In Fig. 7 the unimoment method is shown to give better agreement with the exact solution than Richmond's solution.

iv) *A Circular Cylinder to Test Convergence of the Method:* A circular cylinder with $\varepsilon_r = 4$, radius $r_0 = 0.4\lambda$ is applied to test the relation of the accuracy to the size of the meshes used in the finite element method. Let h be the length of the edge of the largest element. As shown in Figs. 8 and 9, the error of the amplitude and phase of $g(\phi)$ converge to a very small value whenever $kh = \sqrt{\varepsilon_r}\, k_0 h$ is sufficiently small. It is shown in these figures, for both E-wave and H-wave, that $\sqrt{\varepsilon_r}\, k_0 h = 0.2$ is required to reach the accuracy such that amplitude error is less than 1 percent and phase error less than 2°. Only three different angles of ϕ are shown in the figures. Other numerical data for various values of ϕ, which are not shown here, all have the same convergence characteristics.

v) *Dielectric Cylinders of Various Shapes and Inhomogeneity:* The problem is capable of handling the scattering properties of any dielectric cylinder with arbitrary shape and/or arbitrary inhomogeneity. Only the shape function or the function of $\varepsilon_r(x,y)$ is necessary to be defined in a subroutine for a new case. Fig. 10 is the scattering of an inhomogeneous dielectric cylinder. The rectangular cylinder problem is shown in Fig. 11 for a rectangle of width $= 0.1\lambda$, height $= 0.3\lambda$, and $\varepsilon_r = 4$. Fig. 12 shows the scattering of a dielectric elliptical cylinder with semimajor axis $= 0.56752526\lambda$, semiminor axis $= 0.26226316\lambda$, and $\varepsilon_r = 2$.

IV. Conclusion

Several advantages of the unimoment methods are apparent.

i) The artificial circle enables us to solve the interior and exterior problems separately. Only the interior problem involves inhomogeneous materials of various scatterers.

ii) In case the boundary conditions of the interior problem are expanded into Fourier series, the number of terms in the infinite summations to be truncated can be determined *a priori.*

iii) Since the interior and exterior problems are coupled at the circle, the Hankel's function and its derivatives are computed only once at $r = a$, the radius of the artificial circle.

iv) Most of the computations do not involve the incident fields. It is very economical of computer time to calculate the scattering for various incident waves at the same time.

v) The versatility is developed in the finite element methods which can solve the irregular problems relatively easily and systematically. The method of "inhomogeneous elements" enables us to compute various problems by a single program and to simplify the input control.

The validity of the method was confirmed by three cases: an off-centered circular cylinder, two parallel circular cylinders, and a circular cylindrical shell. The convergence of the method has also been shown to be very good. Virtually there is no restriction on the shape or inhomogeneity of the scatterer to be solved. Since the techniques we have discussed are suitable for scatterers with localized inhomogeneity or bounded cross section, the program works very efficiently for the scatterers with $\sqrt{\varepsilon_r}\, k_0 d \leq 10$, where d is the largest dimension of the scatterer. Like all other existing techniques (except the geometrical theory of diffraction approximations), the size of the cylinder is a restriction of the unimoment method. The largest size we can solve on the CDC 7600 computer is about $\sqrt{\varepsilon_r}\, k_0 d \doteq 40$.

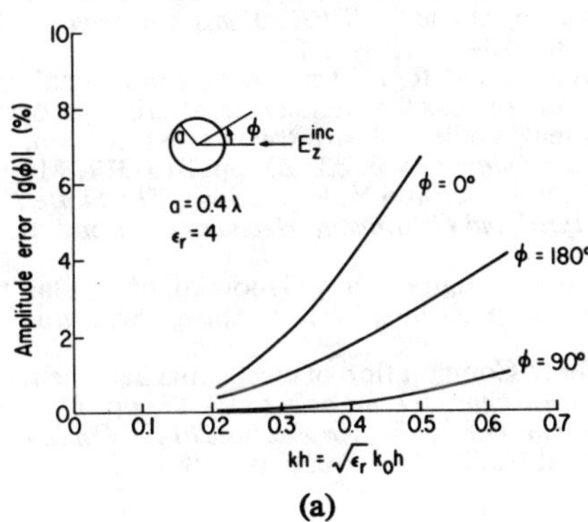

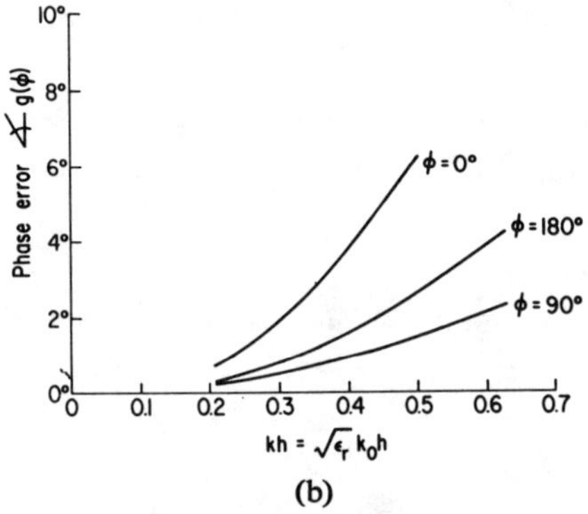

Fig. 8. Convergence of unimoment method for *E*-wave scattering. (a) Far-field amplitude. (b) Phase.

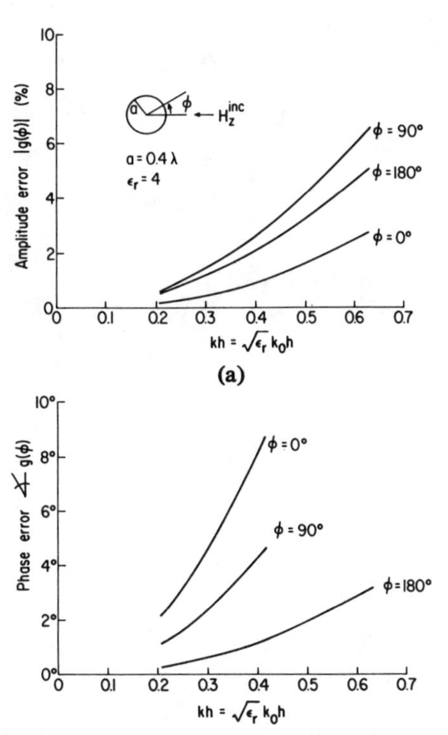

Fig. 9. Convergence of unimoment method for *H*-wave scattering. (a) Far-field amplitude. (b) Phase.

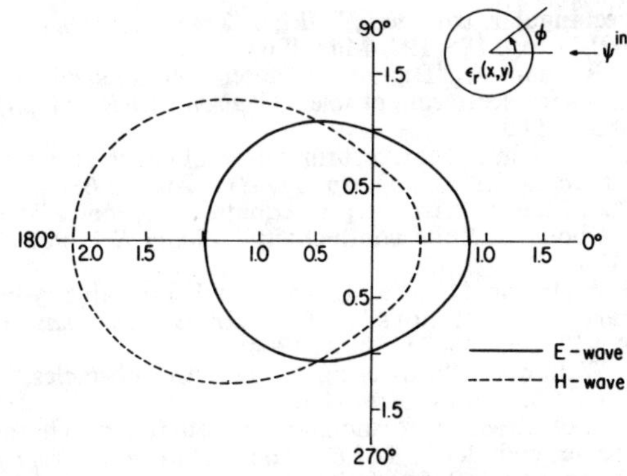

Fig. 10. Scattering far-field pattern $g(\phi)$ of inhomogeneous dielectric cylinder with $\varepsilon_r = [5 - 4(x/a)^2]/[1 + 4(y/a)^2]$.

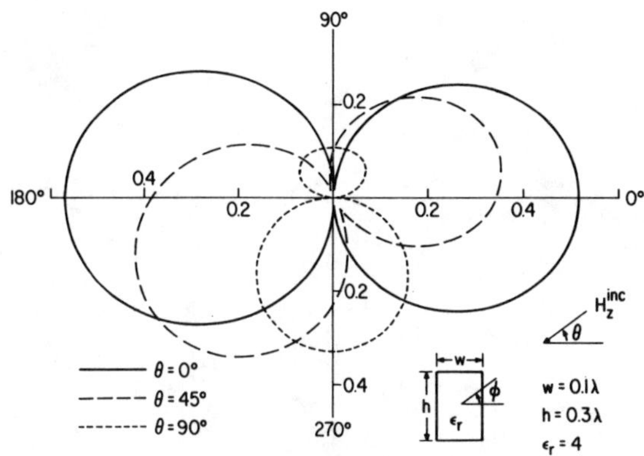

Fig. 11. *H*-wave scattering far-field pattern $|g(\phi)|$ of rectangular cylinder.

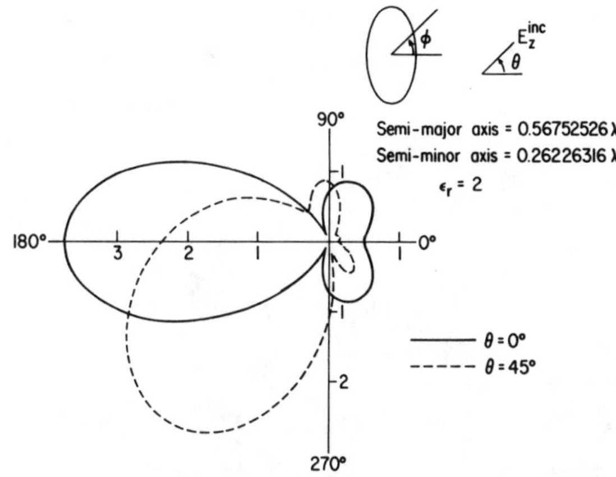

Fig. 12. *E*-wave scattering far-field pattern $|g(\phi)|$ of elliptical cylinder.

The same technique can be applied to scattering by conductors, dielectric coated conductors, and anisotropic scatterers. For conducting cylinders, it is expected to be even more efficient because fields are zero inside the conductor so that no finite element solution is required there.

References

[1] K. K. Mei, "Unimoment method of solving antenna and scattering problems," *IEEE Trans. Antennas Propagat.*, vol. AP-22, pp. 760–766, Nov. 1974.

[2] K. K. Mei and J. G. Van Bladel, "Scattering by perfectly-conduct-

ing rectangular cylinders," *IEEE Trans. Antennas Propagat.*, vol. AP-11, pp. 185–192, Mar. 1963.
[3] J. H. Richmond, "Digital computer solutions of the rigorous equations for scattering problems," *Proc. IEEE*, vol. 53, pp. 796–804, Aug. 1965.
[4] P. C. Waterman, "Matrix formulation of electromagnetic scattering," *Proc. IEEE*, vol. 53, pp. 805–812, Aug. 1965.
[5] K. M. Mitzner, "An integral equation approach to scattering from a body of finite conductivity," *Radio Sci.*, vol. 2, no. 12, Dec. 1967.
[6] M. G. Andreson, "Scattering from parallel metallic cylinders with arbitrary cross sections," *IEEE Trans. Antennas Propagat.*, vol. AP-12, pp. 746–754, Nov. 1964.
[7] P. C. Waterman, "Scattering by dielectric obstacles," in *URSI Symp. on EM Waves*, Stress-Italy, June 1968.
[8] N. Okamoto, "Matrix formulation of scattering by a homogeneous gyrotropic cylinder," *IEEE Trans. Antennas Propagat.*, vol. AP-18, pp. 642–649, Sept. 1970.
[9] J. H. Richmond, "Scattering by a dielectric cylinder of arbitrary cross section shape," *IEEE Trans. Antennas Propagat.*, vol. AP-13, pp. 334–341, May 1965.
[10] ——, "TE-wave scattering by a dielectric cylinder of arbitrary cross-section shape," *IEEE Trans. Antennas Propagat.*, vol. AP-14, pp. 460–464, July 1966.
[11] D. R. Wilton and R. Mittra, "A new numerical approach to the calculation of electromagnetic scattering properties of two-dimensional bodies of arbitrary cross section," *IEEE Trans. Antennas Propagat.*, vol. AP-20, pp. 310–317, May 1972.
[12] O. C. ZienKiewicz and Y. K. Cheung, *The Finite Element Method in Structural and Continuum Mechanics*. London: McGraw-Hill, 1967.
[13] P. Silvester, "High-order polynomial triangular finite elements for potential problems," *Int. J. Engng. Sci.*, vol. 7, pp. 849–861, 1969.
[14] A. Wexler, "Computation of electromagnetic fields," *IEEE Trans. Microwave Theory Tech.*, vol. MTT-17, pp. 416–439, Aug. 1969.
[15] G. Strang and G. J. Fix, *An Analysis of the Finite Element Method.* Englewood Cliffs, N.J.: Prentice-Hall, 1973.

FINITE ELEMENT–BOUNDARY INTEGRAL FORMULATION FOR ELECTROMAGNETIC SCATTERING*

M.A. MORGAN

Department of Electrical Engineering, Naval Postgraduate School, Monterey, CA 93940, USA

C.H. CHEN and S.C. HILL

Department of Bioengineering, University of Utah, Salt Lake City, UT 84112, USA

P.W. BARBER

Department of Electrical Engineering, Clarkson College of Technology, Potsdam, NY 13676, USA

Received 15 October 1982

A new formulation is described which combines the most robust attributes of the volume finite element and surface integral equation approaches to electromagnetic boundary value solutions. The result is a numerical technique which may be applied to scattering problems involving configurations having metallic surfaces and inhomogeneous penetrable material situated in open spatial regions. This is accomplished by way of coupling internal region finite element modal field solutions to equivalent currents on the surrounding boundary surface through an appropriate surface integral equation. The method is demonstrated for the special case of scattering by axisymmetric inhomogeneous penetrable objects. Example numerical calculations are presented for validation of the procedure and potential problem areas are discussed.

1. Introduction

The development of 'hybrid' techniques for analysis and computation is often motivated by the limitations of standard available methods. Such is the case here, where a new formulation is described which combines the finite element method (FEM) and the surface integral equation embodied in the extended boundary condition method (EBCM). This merging of the FEM and EBCM techniques results in a method that retains the most efficient characteristics of each for the computation of electromagnetic scattering.

The finite element method has recently been used in conjunction with the unimoment technique to obtain scattering solutions for inhomogeneous axisymmetric penetrable bodies, [1, 2]. This approach uses the FEM to obtain expressions for the fields within a spherical mesh region which encloses the inhomogeneous body. The incident and scattered fields outside the enclosing sphere are expanded in spherical harmonics and the boundary conditions at the surface of the sphere are used to couple the interior and exterior field expansions and thus obtain a solution.

The combined method is similar to the unimoment approach except that the finite element mesh region needs only to be used inside a boundary conforming to the surface of the object, rather than inside a circumscribed separable boundary, i.e., the mesh is only required in the inhomogeneous region. This minimization of the extent of the mesh region can result in a considerable savings in computation vis-à-vis a unimoment solution using a spherical mesh for the cases of elongated or flattened scatterers.

The finite element solution in the closed inhomogeneous region is coupled to the unbounded outside volume by way of the surface integral equation used in the EBCM associated with penetrable scatterers. The extended boundary condition method was originally developed for calculating the electromagnetic scattering by perfectly conducting objects of arbitrary shape [3, 4]. The method has since been applied to homogeneous [5, 6] and layered [7, 8] dielectric scatterers, to a wide variety of elastic wave and acoustic scattering problems [9], and most recently to electromagnetic problems involving buried inhomogeneities [10]. The extension of the EBCM to inhomogenous material scatterers uses the interior FEM generated solutions to replace the interior region spherical harmonic expansion incorporated in the homogenous penetrable scattering body case.

The combined method can thus be viewed as either a modification to the FEM-Unimoment approach or as an extension of the EBCM concept.

* This paper was presented at the Workshop/Symposium on Research Techniques in Wave Propagation and Scattering, organized by V.K. Varadan and V.V. Varadan, October 18–21, 1982, The Ohio State University, Columbus, Ohio.

Reprinted with permission from *Wave Motion*, vol. 6, pp. 91–103, 1984.

The presentation in this paper tends to follow the latter outlook (viewing the method as an extension of EBCM) only because this appears to result in a more concise description than that of the former idea. In Section 2 the combined method is described via the standard equations of EBCM. Then, in Section 3, the basis functions incorporated in the interior region and their generation by way of the FEM are discussed both from the aspects of philosophy and practical computation. Example scattering calculations are presented in Section 4 for the cases of spheres and spheroids with these results being compared to Mie series and direct EBCM computations. Finally, in Section 5 conclusions and suggestions for future efforts are discussed.

2. Boundary integral formulation

The theory of the extended boundary condition method is well known and only the equations which are required for the extension of the technique to the inhomogeneous problem will be given here. The interested reader should consult the literature [3, 4, 5], for further details. A new approach [11] which employs the concept of conserved flux is particularly suggested as an alternative to the original formulation which employed the concepts of analytical continuation and null-fields.

Fig. 1 shows a scatterer illuminated by a plane wave. The incident field $\bar{E}_i$ and the scattered field $\bar{E}_s$ can be expanded in the spherical vector harmonics $\bar{M}$ and $\bar{N}$ as in Stratton [12].

$$\bar{E}_i = \sum_{\nu} D_\nu [a_\nu \bar{M}^1_\nu (k_0\bar{r}) + b_\nu \bar{N}^1_\nu (k_0\bar{r})], \quad (1)$$

$$\bar{E}_s = \sum_{\nu} D_\nu [f_\nu \bar{M}^4_\nu (k_0\bar{r}) + g_\nu \bar{N}^4_\nu (k_0\bar{r})]. \quad (2)$$

The a_ν and b_ν coefficients are known, while f_ν and q_ν are initially unknown. D_ν is a normalization constant, ν is a combined index incorporating the three spherical harmonic indices, and k_0 is the free space wave number. The spherical vector harmonics of the first kind are used in (1) to assure that $\bar{E}_i$ is finite at the origin, while those of the fourth kind are used in (2) to denote an outgoing scattered wave. An $e^{j\omega t}$ time variation is assumed.

Using the expansions of (1) and (2), the EBCM procedure gives the following four basic boundary integral equations:

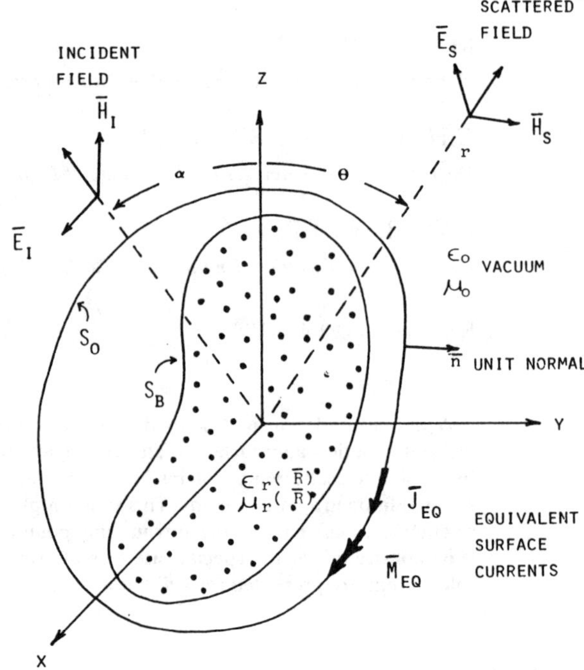

Fig. 1. Inhomogeneous scattering body.

$$\oint_{k_0^2 S_0} [\bar{N}^4_\nu \cdot (\bar{n} \times \bar{E}_+) - j\bar{M}^4_\nu \cdot (\bar{n} \times \eta_0 \bar{H}_+)] d(k_0^2 S) = -j\pi a_\nu, \quad (3a)$$

$$\oint_{k_0^2 S_0} [\bar{M}^4_\nu \cdot (\bar{n} \times \bar{E}_+) - j\bar{N}^4_\nu \cdot (\bar{n} \times \eta_0 \bar{H}_+)] d(k_0^2 S) = j\pi b_\nu, \quad (3b)$$

and

$$j\pi f_\nu = \oint_{k_0^2 S_0} [\bar{N}^1_\nu \cdot (\bar{n} \times \bar{E}_+) - j\bar{M}^1_\nu \cdot (\bar{n} \times \eta_0 \bar{H}_+)] d(k_0^2 S), \quad (4a)$$

$$j\pi g_\nu = \oint_{k_0^2 S_0} [\bar{M}^1_\nu \cdot (\bar{n} \times \bar{E}_+) - j\bar{N}^1_\nu \cdot (\bar{n} \times \eta_0 \bar{H}_+)] d(k_0^2 S), \quad (4b)$$

where the integrations are over any closed surface S_0 (not necessarily S_B of Fig. 1) and $(\bar{n} \times \bar{E}_+)$ and $(\bar{n} \times \bar{H}_+)$ are equivalent surface currents on S_0. The above four equations are always true without regard to the shape of S_0 as long as S_0 encloses the scatterer.

An important feature of (3) and (4) which needs to be emphasized is that the scatterer's characteristics were not used in the derivation. In fact, (3) and (4) merely relate the equivalent surface currents on S_0 to the coefficients of the incident and scattered field expansions by surface integral relations. This independence of the basic equations from the scatterer's characteristics distinguishes this from other surface integral formulations, which have the scatterer's characteristics embeded in the equations [13, 14].

While the scatterer-dependent surface integral equations will become very complicated or may not even exist if the scatterer is a general inhomogeneous object, the formulation presented thus far remains unchanged. The independence of these basic boundary integral equations from the geometry and composition of the scatterer indeed makes this procedure very suitable for being extended to inhomogeneous objects.

The usual EBCM approach to solving (3) and (4) for homogeneous or layered objects involves expanding the surface currents on the body surface S_B in terms of proper interior region spherical vector harmonics and then applying the surface integration on S_B. Such a technique is not applicable for a general inhomogeneous scatterer because analytical expressions for the required surface current expansions are not available. A more general solution procedure would be to initially assume

$$\begin{bmatrix} \bar{n}\times\bar{E}_+ \\ \bar{n}\times\bar{H}_+ \end{bmatrix} = \sum_\mu c_\mu \begin{bmatrix} \bar{n}\times\bar{E}^c_\mu \\ \bar{n}\times\bar{H}^c_\mu \end{bmatrix} + \sum_\mu d_\mu \begin{bmatrix} \bar{n}\times\bar{E}^d_\mu \\ \bar{n}\times\bar{H}^d_\mu \end{bmatrix}, \tag{5}$$

where $(\bar{n}\times\bar{E}_\mu, \bar{n}\times\bar{H}_\mu)$ are coupled pairs of basis functions which are left to be selected later, being subject to certain requirements, such as solving Maxwell's equations in the inhomogeneous region as well as satisfying boundary conditions at S_B. The summation in (5) is a single series of $2N$ terms. For later mathematical convenience, it is separated into two N-term sections. The superscripts c and d relate the basis functions to the coefficients.

Substituting (5) into (3) and (4) yields two $2N\times 2N$ linear systems whose functional forms are identical to those found in the extended boundary condition method:

$$\frac{1}{\mathrm{j}\pi}\begin{bmatrix} I^4_{\nu\mu} & J^4_{\nu\mu} \\ K^4_{\nu\mu} & L^4_{\nu\mu} \end{bmatrix} \cdot \begin{bmatrix} c_\mu \\ d_\mu \end{bmatrix} = \begin{bmatrix} -a_\nu \\ -b_\nu \end{bmatrix}, \tag{6}$$

and

$$\begin{bmatrix} f_\nu \\ g_\nu \end{bmatrix} = \frac{1}{\mathrm{j}\pi}\begin{bmatrix} I^1_{\nu\mu} & J^1_{\nu\mu} \\ K^1_{\nu\mu} & L^1_{\nu\mu} \end{bmatrix} \cdot \begin{bmatrix} c_\mu \\ d_\mu \end{bmatrix}, \tag{7}$$

where

$$I^t_{\nu\mu} = \oint_{k_0^2 S_0} [\bar{N}^t_\nu(\bar{n}\times\bar{E}^c_\mu) - \mathrm{j}\bar{M}^t_\nu(\bar{n}\times\eta_0\bar{H}^c_\mu)]\,\mathrm{d}(k_0^2 S),$$

$$J^t_{\nu\mu} = \oint_{k_0^2 S_0} [\bar{N}^t_\nu(\bar{n}\times\bar{E}^d_\mu) - \mathrm{j}\bar{M}^t_\nu(\bar{n}\times\eta_0\bar{H}^d_\mu)]\,\mathrm{d}(k_0^2 S),$$

$$K^t_{\nu\mu} = \oint_{k_0^2 S_0} [\bar{M}^t_\nu(\bar{n}\times\bar{E}^c_\mu) - \mathrm{j}\bar{N}^t_\nu(\bar{n}\times\eta_0\bar{H}^c_\mu)]\,\mathrm{d}(k_0^2 S),$$

$$L^t_{\nu\mu} = \oint_{k_0^2 S_0} [\bar{M}^t_\nu(\bar{n}\times\bar{E}^d_\mu) - \mathrm{j}\bar{N}^t_\nu(\bar{n}\times\eta_0\bar{H}^d_\mu)]\,\mathrm{d}(k_0^2 S),$$

with $t = 4$ in (6) and $t = 1$ in (7).

Since the a and b coefficients are known, the c_μ and d_μ coefficients, and thus the equivalent surface currents, can be found by solving (6). The c_μ and d_μ coefficients are then used in (7) to compute f and g, the scattered field coefficients. Equations (6) and (7) can be combined to give

$$\begin{bmatrix} f_\omega \\ g_\nu \end{bmatrix} = [T]\cdot\begin{bmatrix} a_\nu \\ b_\nu \end{bmatrix}, \tag{8}$$

where

$$[T] = \begin{bmatrix} I^1_{\nu\mu} & J^1_{\nu\mu} \\ K^1_{\nu\mu} & L^1_{\nu\mu} \end{bmatrix} \cdot \begin{bmatrix} I^4_{\nu\mu} & J^4_{\nu\mu} \\ K^4_{\nu\mu} & L^4_{\nu\mu} \end{bmatrix}^{-1}. \tag{9}$$

The 'transition matrix' or T-matrix in (9) is a characteristic matrix that transform the incident field coefficients into the scattered field coefficients.

The procedure just described will yield a non-unique solution unless the scatterer's characteristics are incorporated in the selection of the basis functions of (5) and subsequently embodied in the T-matrix. Therefore, it is necessary to find appropriate basis functions for each particular class of problems.

For example, for a perfect conductor, the basis functions can be generated through spherical harmonics.

$$\bar{n}\times\bar{E}^c_\mu = 0,$$
$$\bar{n}\times\bar{H}^c_\mu = \bar{n}\times\bar{M}^1_\mu(k_0\bar{r}). \tag{10a}$$
$$\bar{n}\times\bar{E}^d_\mu = 0,$$
$$\bar{n}\times\bar{H}^d_\mu = \bar{n}\times\bar{N}^1_u(k_0\bar{r}), \tag{10b}$$

where $k_0 = 2\pi/\lambda_0$. Also, the T-matrix is evaluated on the body surface, i.e., $S_0 = S_B$ in (6) and (7). The fact that $\bar{n} \times \bar{E}_\mu = 0$ on the body surface identifies the scatterer as a perfect conductor, while the evaluation of the T-matrix directly on the body surface guarantees that the T-matrix contains information about the scatterer's shape. Consequently, the solution so obtained is that for a perfect conductor with the given shape.

The application to a homogeneous dielectric object is similar in that the T-matrix is also evaluated on the body surface. However, in this case, the basis functions are

$$\bar{n} \times \bar{E}^c_\mu = \bar{n} \times \bar{M}^1_\mu(k\bar{r}),$$

$$\bar{n} \times \bar{H}^c_\mu = \bar{n} \times \left[\frac{\mathrm{j}}{\eta_0}\left(\frac{\varepsilon_r}{\mu_r}\right)^{1/2} \bar{N}^1_\mu(k\bar{r})\right]. \tag{11a}$$

$$\bar{n} \times \bar{E}^d_\mu = \bar{n} \times \bar{N}^1_\mu(k\bar{r}),$$

$$\bar{n} \times \bar{H}^d_\mu = \bar{n} \times \left[\frac{\mathrm{j}}{\eta_0}\left(\frac{\varepsilon_r}{\mu_r}\right)^{1/2} \bar{M}^1_\mu(k\bar{r})\right]. \tag{11b}$$

Notice here that the argument is $kr = \sqrt{\varepsilon_r \mu_r} k_0 r$. Therefore, the constitutive parameters of the scatterer, ε_r and μ_r, are included in the basis functions and therefore in the T-matrix. As in the case of the perfect conductor, the shape information is included by evaluating the surface integrals on S_B. Note that for the case of spherical harmonic basis functions, the basis function index μ symbolizes the three spherical harmonic indices.

3. Finite element basis function generation

The application of the boundary integral method to a general inhomogeneous object requires the production of a set of basis functions which contain adequate characteristic information about the scatterer. These basis functions will be called the 'informed bases'. In fact, for a general inhomogeneous object, the informed bases cannot be found without 'internally' solving the field equations. Since the internally obtained informed bases will also contain information about the scatterer's shape, it is no longer necessary to evaluate the T-matrix exactly on the body surface as in the two cases of the previous section. This is especially advantageous when the body shape has sharp angles which may give problems in the numerical evaluation of the surface integrals in the T-matrix.

Before showing how the informed bases can be generated (using the finite element method), it may be interesting to compare the boundary integral method with the method of moments solution [13, 14]. The two methods follow somewhat different concepts in their origination, but possess similar mathematics in many respects. To those who are familiar with the method of moments, (3) and (4) may be regarded as one type of inner product relation using spherical vector wave functions as the testing functions. However, unlike the testing functions in the method of moments, the testing functions in the boundary integral method are vector functions and are used in pairs, i.e., when the testing function for $\bar{n} \times \bar{E}_+$ is $\bar{M}$, that for $\bar{n} \times \bar{H}_+$ is $\bar{N}$ and vice versa. Because of this particular choice of testing functions, the basis functions in the boundary integral method are also required to be vectors and in pairs (see (5)), for compatibility. Thus (5), which is the trial solution in the boundary integral method, consists of four sets of scalar bases (one set for each component of the two surface currents) sharing the same set of coefficients as opposed to four sets of coefficients sharing the same single set of scalar bases in the trial solution of the method of moments. As will be seen shortly, the trial solution resulting from the finite element method is of the form of (5).

The generation of the informed basis functions will be performed by way of the finite element implementation of the coupled azimuthal potential (CAP) formulation [15]. The CAP formulation is a solution of Maxwell's equations originally implemented in conjunction with the unimoment method to solve scattering problems for axisymmetric inhomogeneous objects, [1]. The geometry employed in describing the CAP formulation is illustrated in Fig. 2 which shows the meridian cross-sectional plane of an inhomogeneous scattering body. A wave-number normalized system of circular cylindrical coordinates $(R, Z, \phi) = (k_0\rho, k_0 z, \phi)$ is being used to conform to the CAP notation. The scatterer as well as the closed surface, S_G, are rotationally symmetric about the z-axis. Because of this symmetry, the fields everywhere are separable into azimuthal modes and can be expressed in terms of an exponential Fourier series.

$$\bar{E}(R, Z, \phi) = \sum_{m=-\infty}^{\infty} \bar{e}_m(R, Z) \exp(\mathrm{j}m\phi), \tag{12a}$$

$$\eta_0 \bar{H}(R, Z, \phi) = \sum_{m=-\infty}^{\infty} \bar{h}_m(R, Z) \exp(\mathrm{j}m\phi), \tag{12b}$$

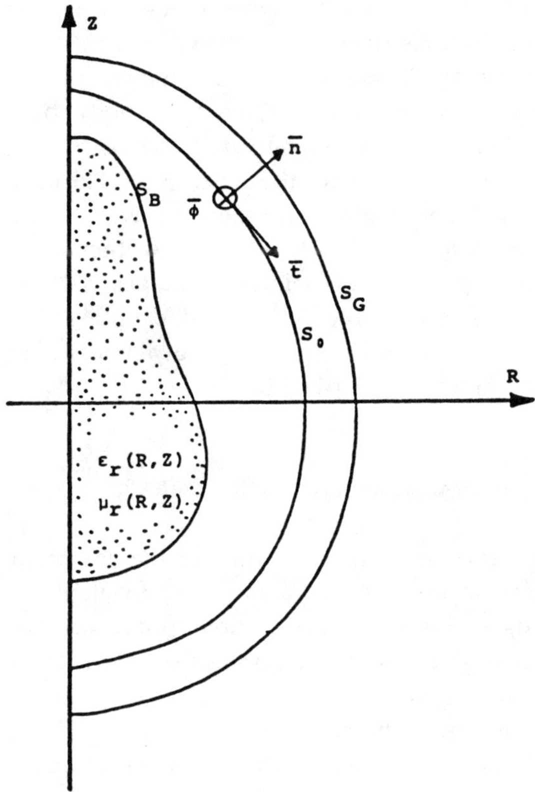

Fig. 2. Axisymmetric body meridian cross section.

where $\eta_0 = 120\pi\Omega$.

As has been shown in the CAP formulation [1, 15], the modal fields can be represented in terms of two uniformly continuous coupled azimuthal potentials which are, themselves, proportional to the azimuthal components of the electric and magnetic fields.

$$\bar{e}_m(R,Z) = \mathrm{j}f_m(m\nabla\psi_1 + R\mu_r\bar{U}_\varphi \times \nabla\psi_2) + \bar{U}_\varphi\psi_1/R, \tag{13a}$$

$$\bar{h}_m(R,Z) = \mathrm{j}f_m(m\nabla\psi_2 - R\varepsilon_r\bar{U}_\varphi \times \nabla\psi_1) + \bar{U}_\varphi\psi_2/R, \tag{13b}$$

where ∇ is the 2-D gradient operator in (R, Z), with

$$f_m(R,Z) = [\mu_r(R,Z)\varepsilon_r(R,Z)R^2 - m^2]^{-1} \tag{14}$$

and

$$\psi_1(R,Z,m) = R\bar{U}_\varphi \cdot \bar{e}_m(R,Z),$$

$$\psi_2(R,Z,m) = R\bar{U}_\varphi \cdot \bar{h}_m(R,Z).$$

The coupled azimuthal potentials satisfy a system of second-order coupled partial differential equations (PDE)

$$[f_m(R\varepsilon_r\nabla\psi_1 + m\bar{U}_\varphi \times \nabla\psi_2)] + \varepsilon_r\psi_1/R = 0, \tag{15a}$$

$$[f_m(R\mu_r\nabla\psi_2 - m\bar{U}_\varphi \times \nabla\psi_1)] + \mu_r\psi_2/R = 0. \tag{15b}$$

This PDE system may be solved inside of a closed region, such as that represented by S_G, for specified Dirichlet boundary conditions by use of either finite element or finite difference methods. The details of the finite element solution, which has been used in conjunction with the unimoment method for scattering calculations, are given in [1]. Many of the same computer algorithms employed in this previous effort were adopted for the work being described here. The major change in the finite element calculation has been in the development of a new mesh structure whose outer boundary provides a conformal enclosure for the body surface, as shown in Fig. 3.

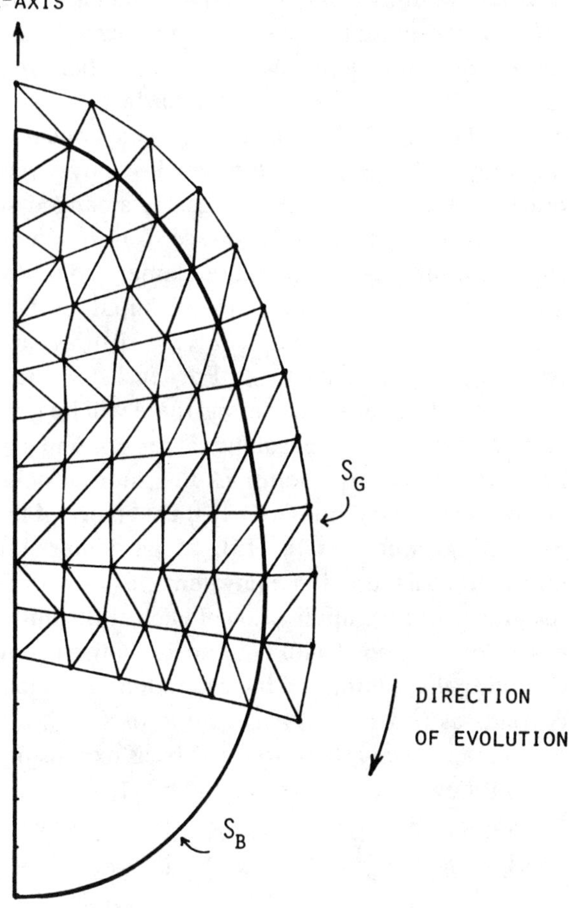

Fig. 3. Conformal finite element mesh structure.

The generation of the informed bases, to be used in (5), begins by expanding ψ_1 and ψ_2 on S_G in terms of individual asymptotically complete basis function expansions

$$\psi_1(l, m)|_{S_G} = \sum_{\mu=1}^{N} c_{m,\mu} b^c_{m,\mu}(l), \quad (16a)$$

$$\psi_2(l, m)|_{S_G} = \sum_{\mu=1}^{N} d_{m,\mu} b^d_{m,\mu}(l), \quad (16b)$$

where l is the surface path length on S_G starting from the positive z-axis, c and d are unknown coefficients, and the superscripts c and d represent an association with the corresponding coefficients. This same notation was employed in (5). The $b(l)$ basis functions are not the informed bases and need not result from a priori information concerning the scatterer composition. We shall hereafter refer to the $b(l)$'s as the 'initial bases'. These initial bases can be either subdomain or entire-domain functions [16] with the basis sets being complete in the Hilbert space of continuous functions, C^0, as $N \to \infty$.

It is the task of the finite element solution to provide the transformation of the initial basis sets, $b(l)$, for the potentials over S_G to informed basis sets for the equivalent current tangential field vectors $\bar{n} \times \bar{e}_m$ and $\bar{n} \times \bar{h}_m$ over the surface S_0, as is depicted in Fig. 2. The surface S_0, which encloses the body surface S_B is, in turn, enclosed by S_G. In practice, the finite element mesh is arranged so that S_0 is equal to the body surface so as to minimize the mesh extent. The finite element solutions are performed for two sets of initial surface bases on S_G: $(\psi_1, \psi_2)|_{S_G} = \{(b^c_{m,\mu}(l), 0)\}^N_{\mu=1}$ and $(\psi_1, \psi_2)|_{S_G} = \{(0, b^d_{m,\mu}(l)\}^N_{\mu=1}$. For each of these two sets of boundary conditions there will be generated interior region solutions for ψ_1 and ψ_2. From these potential solutions the internal fields can be constructed numerically through use of the generating equations in (13). In particular, the informed basis function equivalent currents on S_0 resulting from the application of each of the initial bases for ψ_1 and ψ_2 on S_G can be constructed. Denoting these informed basis function equivalent currents as $(\bar{n} \times \bar{e}^c_{m,\mu}, \bar{n} \times \bar{h}^c_{m,\mu})$ and $(\bar{n} \times \bar{e}^d_{m,\mu}, \bar{n} \times \bar{h}^d_{m,\mu})$ respectively, the informed basis expansions in (5) follow from superposition and (16),

$$\begin{bmatrix} \bar{n} \times \bar{e}^+_m \\ \bar{n} \times \bar{h}^+_m \end{bmatrix} = \sum_{\mu=1}^{N} c_{m,\mu} \begin{bmatrix} \bar{n} \times \bar{e}^c_{m,\mu} \\ \bar{n} \times \bar{h}^C_{m,\mu} \end{bmatrix} + \sum_{\mu=1}^{N} d_{m,\mu} \begin{bmatrix} \bar{n} \times \bar{e}^d_{m,\mu} \\ \bar{n} \times \bar{h}^d_{m,\mu} \end{bmatrix}. \quad (17)$$

The informed bases in (17) are obtained by solving the field equations everywhere inside the inhomogeneous body and thus incorporate the characteristics of the scatterer. These bases are then used to evaluate the elements of the T-matrix via integrations over S_0, as shown in eq. (7). A minor change has occurred in the indexing notation in comparing (5) to (17). In (5) the body could have been a non-axisymmetric 3-dimensional structure in which case the index μ would have embodied both azimuthal index, m, as well as a meridian surface index. In (17), on the other hand, the orthogonality of azimuthal modes for the axisymmetric case allows explicit decoupling of m from μ which now represents only a meridian surface index as described for (16).

4. Numerical computations

The development and testing of the computer program which incorporates the finite element-boundary integral formulation proceeded in definitive stages. A first step was the construction of a finite element program for generation of the 'informed basis functions' on the surface of the body. The mesh design is illustrated in Fig. 3 where the outer geometric surface, S_G, conforms to the contour of the scatterer being considered. Generation of this mesh proceeds in azimuthal steps, incrementing in θ from the positive z-axis. In parallel with this sequential construction of the mesh is the computation of the block matrices of the system finite element matrix and the block-by-block inversion of this matrix via the Riccati transform technique as described in [1].

To validate the accuracy of the finite element portion of the code, a z-axis offset sphere lossy dielectric body surface was used so as to produce a somewhat general conformal mesh structure. Initial bases for this test were chosen to be total field spherical harmonic modes composed of both incident modes and corresponding scattered field modes which combined to solve the boundary conditions (BC) at the spherical interface. The finite element computations resulting from each initial basis total field mode applied as a BC on S_G could then be compared to the same mode evaluated on S_B. Excellent accuracy was observed in these initial texts with the exception of relatively large errors in $\bar{u}_\varphi \times \bar{e}_m$ and $\bar{u}_\varphi \times \bar{h}_m$ at isolated points on S_B. Further investigation revealed that the generation of these field components through (15) provides large amplification of errors in the numerical evaluation of the indicated gradients near points where the multiplicative f_m-function becomes very large. As can be seen, for the case of lossless media with real $\mu_r\varepsilon_r$, there exists spatial contours defined by $R = m/\sqrt{\mu_r\varepsilon_r}$ where f_m is infinitely singular. Since the fields are not singular at these points, the gradient terms in parentheses in (15) must combine to be zero at these same spatial points. Thus, even small errors in $\nabla\psi_1$ and $\nabla\psi_2$ which result in incomplete cancellation of terms near the singular points of f_m will be amplified to produce much enhanced errors in $\bar{u}_\varphi \times \bar{e}_m$ and $\bar{u}_\varphi \times \bar{h}_m$. To circumvent this singularity induced problem, a restricted averaging scheme was incorporated, whereby the nodal values of $\bar{u}_\varphi \times \bar{e}_m$ and $\bar{u}_\varphi \times \bar{h}_m$ are obtained by evaluating these at the node in question in each of the elements surrounding the node and then averaging these calculations using only the elements where f_m is not singular. At most nodes six values are averaged while at some nodes either the free-space element values will be discarded (if f_m is singular therein) or those evaluated in elements just inside S_B for a lossless body may be ignored at surface nodes near singular points. This technique, although quite simple, does indeed provide an accurate conversion of initial bases into the desired informed bases.

A second test employed prior to assembling the complete program was to evaluate the accuracy of the boundary-integral computation and T-matrix evaluation. This was performed by substituting interior region spherical harmonics for the informed bases. Such a procedure produces an extended boundary condition program which is valid only for homogeneous scatterers. Results of scattering computations for spheres and spheroids were excellent thus verifying the operation of this portion of the code.

The complete Finite Element-Boundary Integral (FEBI) code was assembled using the previously tested sub-programs. Several scattering calculations were performed to verify the accuracy of the FEBI formulation. Some representative results of these computations are shown in Figs. 4 through 9 for various cases of spheres and prolate spheroids for which readily obtainable comparison calculations were available via Mie series and EBCM codes, respectively. Computations for inhomogeneous scatterers having bioelectromagnetic interest and further verifications will follow in a future publication.

The quantity being displayed in Figs. 4 through 9 is the magnitude of the 'complex scattering amplitude', $|F(\theta, \phi)|$, which is defined in terms of the far-zone scattered electric field and the intensity of the incident plane wave electric field.

$$\bar{E}_s(r, \theta, \phi) = |\bar{E}_i| \bar{F}(\theta, \phi) \frac{e^{-jk_0 r}}{k_0 r}. \tag{18}$$

The parameters associated with the various cases shown are summarized in Table 1 where "P.S." indicates prolated spheroid, "a" is the z-axis dimension, "ε_r" is complex dielectric constant, "α" is the incidence angle relative to the positive z-axis and "DENS" is the mesh density defined in terms of the number of elements spaced along the z-axis per unit *interior wavelength* $\lambda_0/\mathrm{Re}(\sqrt{\varepsilon_r})$.

Some comments regarding these results are necessary. The closest comparison appears in Case 1 (Fig. 4) with RMS variation of less then 5%. This case was computed on a CDC 7600 machine having a 60 bit single precision word length and very large core RAM which permitted using a relatively high 'DENS' of 22. This case, which is compared against a highly accurate Mie series, demonstrates the achievable accuracy of the FEBI method under somewhat idealized computational conditions. All other cases displayed had the FEBI computations performed on a UNIVAC machine having 48 bit single precision word length and smaller RAM capacity than the CDC computer, thus restricting the DENS value that could be employed on computations for larger scatterers (e.g. Cases 2 through 4). The EBCM computations were performed on a PDP-11 mini-computer having a 32 bit word length. Another restriction which limited the number and quality of the computations that were performed was a rather austere budget at the end of the fiscal year to fund computer time. Even with these computational limitations, the results are relatively good with RMS variations between FEBI and EBCM computations ranging in the 10 to 15% region. Cases 2 and 3 for a respective sphere and spheroid ($a/b = 2$) have similar RMS variations and also have identical DENS factors. Case 4, which is a 2-wavelength long prolate spheroid, has the highest RMS variation of all cases. This resulted from the course mesh (DENS = 8.75) that was necessary due to RAM limitations on the UNIVAC computer. Case 5, which is a highly elongated ($a/b = 4$) prolate spheroid provided a good comparison using the same DENS factor as Case 1. Case 5

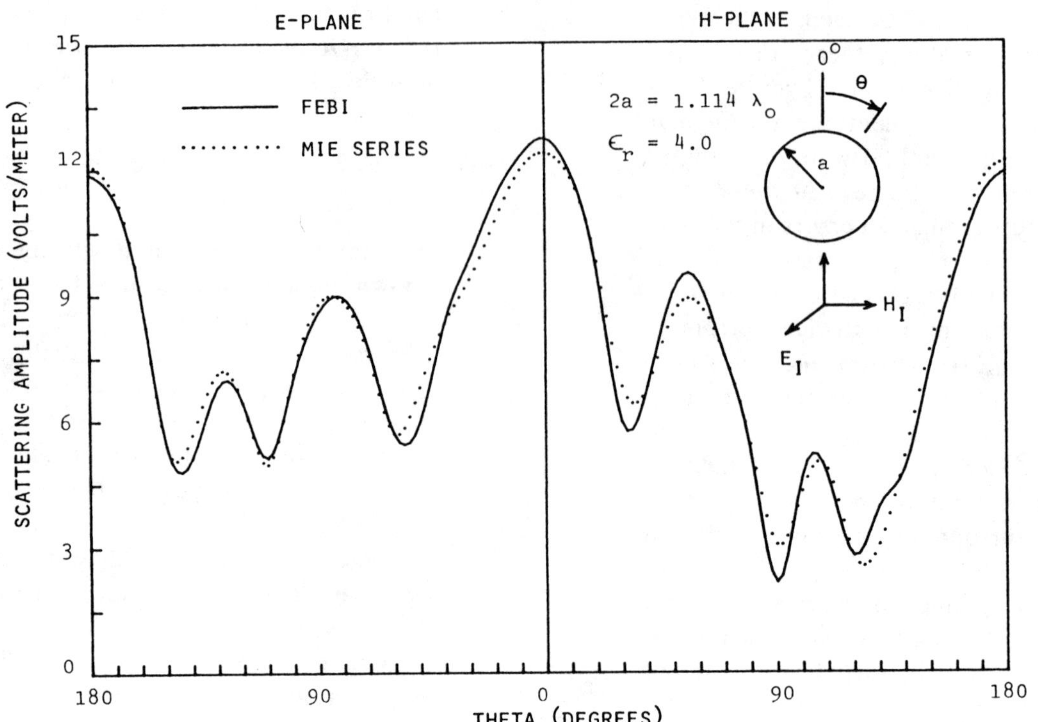

Fig. 4. Case 1: Sphere scattering amplitude comparison.

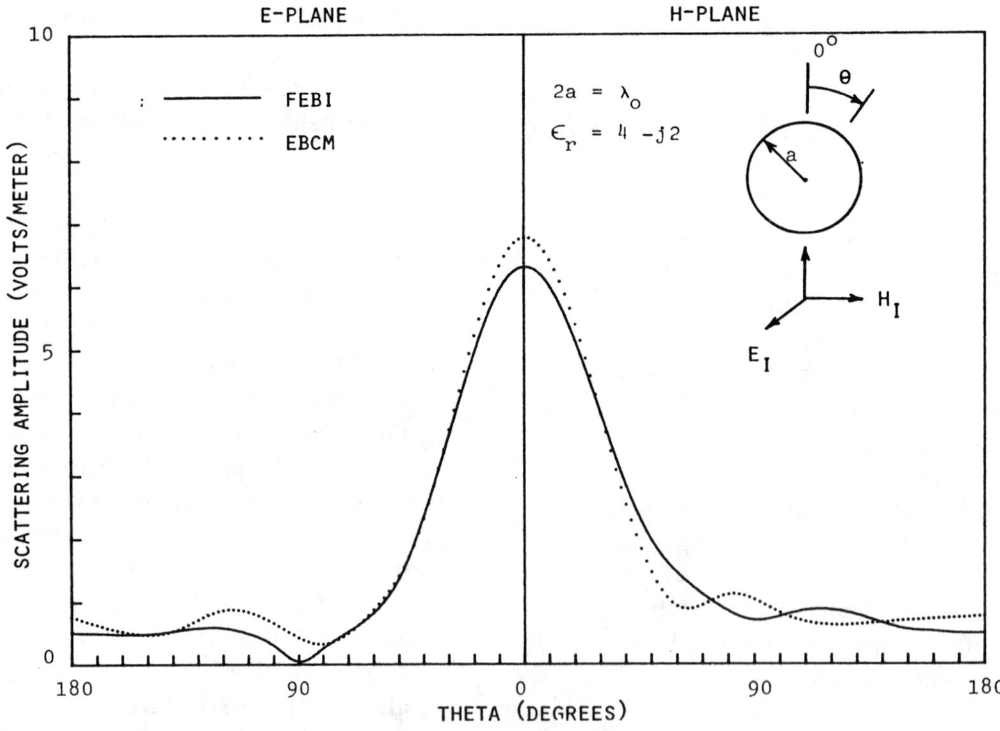

Fig. 5. Case 2: Sphere scattering amplitude comparison.

did not, however, attain the same level of agreement as Case 1 due possibly to higher roundoff errors in a shorter wordlength machine as well as possible errors in the EBCM calculation for such eccentric spheroids. In all cases but Case 6 only axial incident ($\alpha = 0°$) was considered in order to conserve financial resources (only the $m = 1$ azimuthal mode is required). In Case 6 broadside

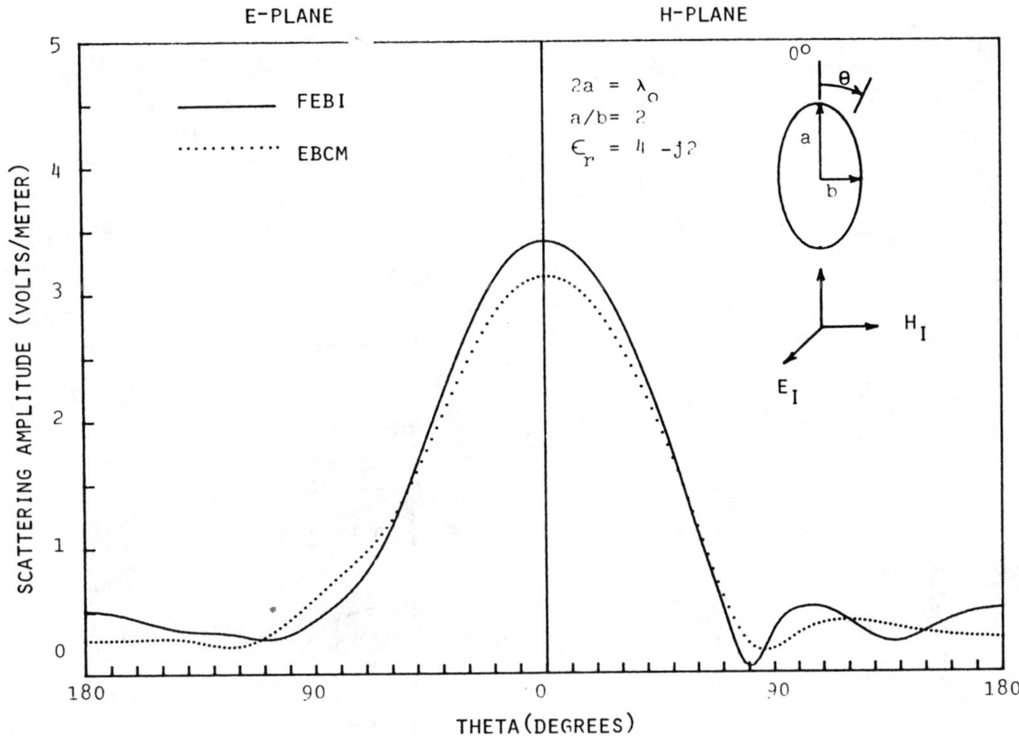

Fig. 6. Case 3: Prolate spheroid scattering amplitude comparison.

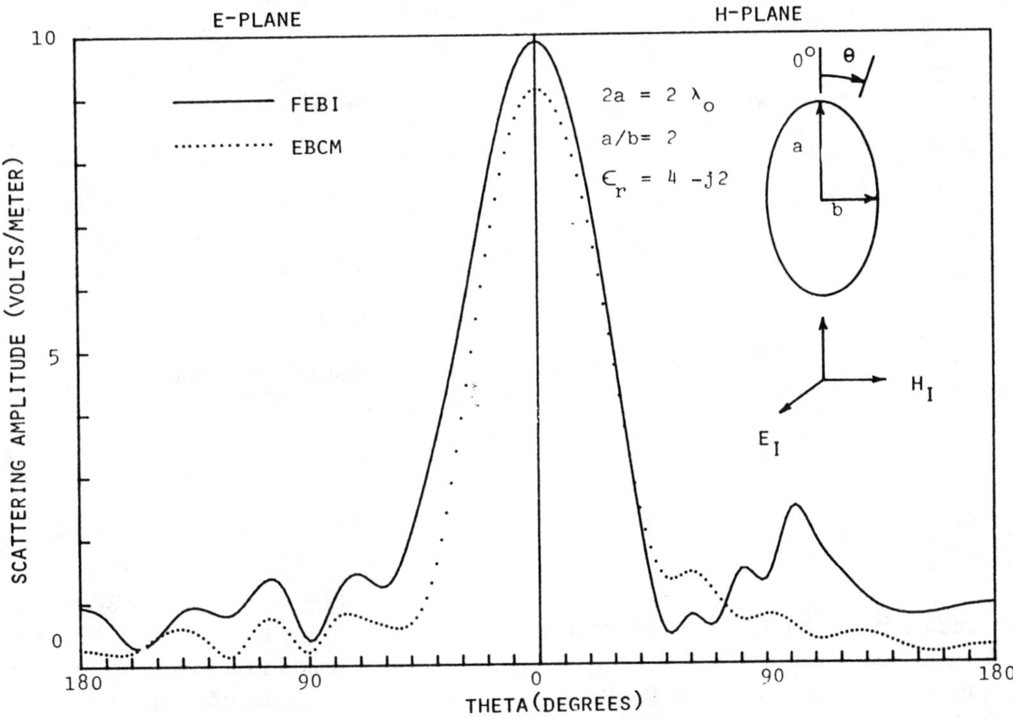

Fig. 7. Case 4: Prolate spheroid scattering amplitude comparison.

incidence ($\alpha = 90°$) was considered using modes $m = 0$, 1 and 2 since $k_0 b = 1{\cdot}57$ in this case. The comparison is relatively good as shown. It should be noted that the left- and right-hand plots for E- and H-plane data do not align at $\theta = 0°$ in this case because they are computed for two separate incident polarizations.

Table 1
FEBI scattering computations

Case	Fig.	Body	$2a/\lambda_0$	ε_r	α	DENS	Comparison
1	4	Sphere	1.114	4.0	0°	22	MIE
2	5	Sphere	1.0	4−j2	0°	17.5	EBCM
3	6	P.S. $a/b=2$	1.0	4−j2	0°	17.5	EBCM
4	7	P.S. $a/b=2$	2.0	4−j2	0°	8.75	EBCM
5	8	P.S. $a/b=4$	0.8	4−j1	0°	22	EBCM
6	9	P.S. $a/b=2$	0.5	4−j1	90°	36	EBCM

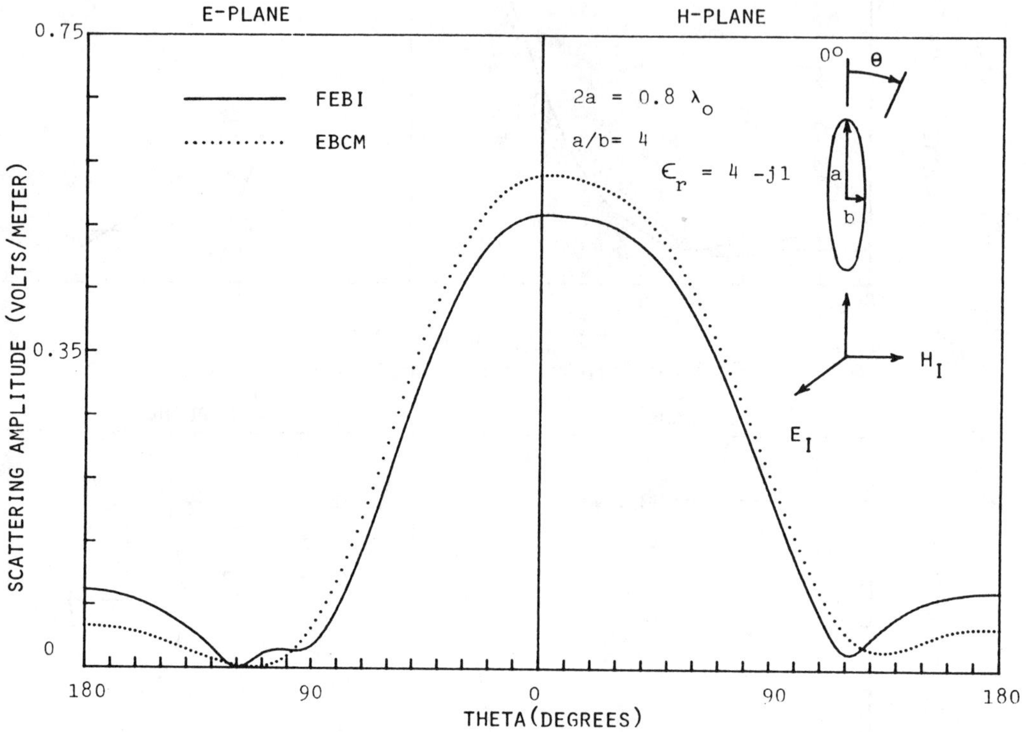

Fig. 8. Case 5: Prolate spheroid scattering amplitude comparison.

5. Conclusions

A new procedure has been described for combining the finite-element method and the boundary integral equation contained in the extended boundary condition method thus expanding the realm of T-matrix scattering calculations to include inhomogeneous objects. The implementation of this technique requires the use of a computational method for solving Dirichlet boundary value problems in the region containing the scatterer. These computations are used to generate a set of informed bases for the expansion of surface equivalent currents. The exterior and interior region fields are coupled by use of the informed bases to compute the surface integrals which are used in the calculation of the T-matrix.

To illustrate the proposed method scattering by axisymmetric penetrable objects was considered. The interior computation for construction of the informed bases were conducted through use of finite element solutions based on the CAP formulation. Excellent results were obtained in test calculations of scattering by homogeneous spheres and prolate spheroids. Future computations are planned for complex inhomogeneous objects of interest in bioelectromagnetic applications.

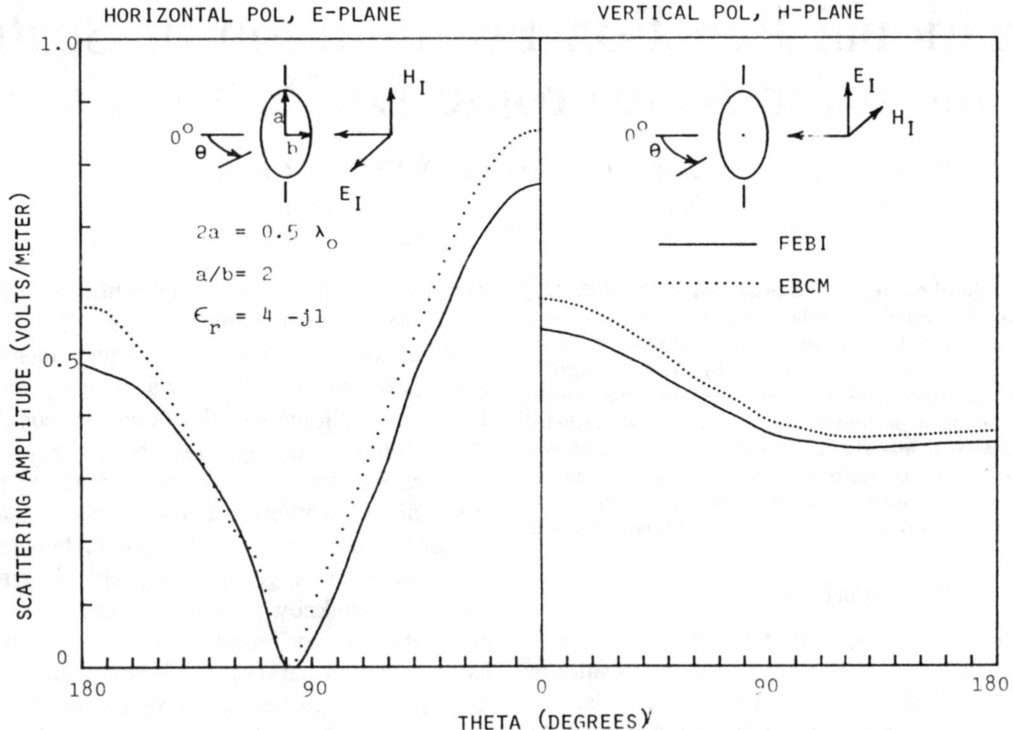

Fig. 9. Case 6: Prolate spheroid scattering amplitude comparison.

Acknowledgement

This work was partially supported by funding from the U.S. Army Research Office and employed the facilities of the University of Utah and the Naval Postgraduate School.

References

[1] M.A. Morgan and K.K. Mei, "Finite-element computation of scattering by inhomogeneous penetrable bodies of revolution", *IEEE Trans. Antennas Propag. 27,* 202–214 (1979).

[2] K.K. Mei, "Unimoment method of solving antenna and scattering problems", *IEEE Trans. Antennas Propag. 22,* 760–766 (1974).

[3] P.C. Waterman, "Matrix formulation of electromagnetic scattering", *Proc. IEEE 53,* 805–812 (1965).

[4] P.C. Waterman, "Symmetry, unitarity, and geometry in electromagnetic scattering", *Phys. Rev. D 3,* 825–839 (1971).

[5] P.C. Waterman, "Scattering by dielectric obstacles", *Alta Frequenza 38 (Speciale),* 348–352 (1969).

[6] P.W. Barber and C. Yeh, "Scattering of electromagnetic waves by arbitrarily shaped dielectric bodies", *Appl. Opt. 14,* 2864–2872 (1975).

[7] B. Peterson and S. Strom, "*T*-matrix formulation of electromagnetic scattering from multilayered scatterers", *Phys. Rev. D 10,* 2670–2684 (1974).

[8] D.S. Wang and P.W. Barber, "Scattering by inhomogeneous nonspherical objects", *Appl. Opt. 18,* 1190–1197 (1979).

[9] V.K. Varadan and V.V. Varadan, *Acoustic, Electromagnetic, and Elastic Wave Scattering-Focus on the T-Matrix Approach,* Pergamon Press, New York (1980).

[10] G. Kristensson, "Electromagnetic scattering from buried inhomogeneities—a general three-dimensional formalism", *J. Appl. Phys. 51,* 3486–3500 (1980).

[11] P.C. Waterman, "Matrix methods in potential theory and electromagnetic scattering", *J. Appl. Phys. 50,* 4550–4566 (1979).

[12] J.A. Stratton, *Electromagnetic Theory,* McGraw-Hill, New York (1941), Chapter 7.

[13] T.K. Wu and L.L. Tsai, "Scattering from arbitrarily-shaped lossy dielectric bodies of revolution", *Radio Science 12,* 709–718 (1977).

[14] L.N. Medgyesi-Mitschang and C. Eftimiu, "Scattering from axisymmetric obstacles embedded in axisymmetric dielectrics: the method of moment solution", *Appl. Phys. 19,* 275–285 (1979).

[15] M.A. Morgan, S.K. Chang, and K.K. Mei, "Coupled azimuthal potentials for electromagnetic field problems in inhomogeneous axially symmetric media", *IEEE Trans. Antennas Propag. 5,* 413–417 (1977).

[16] E.K. Miller and F.J. Deadrick, "Some Computational Aspects of Thin-Wire Modeling", in: R. Mittra, Ed., *Numerical and Asymptotic Techniques in Electromagnetics,* Springer, New York (1975) 89–127.

[17] M. Kerker, *The Scattering of Light and Other Electromagnetic Radiation,* Academic Press, New York (1969) Ch. 3.

Variational Reaction Formulation of Scattering Problem for Anisotropic Dielectric Cylinders

RUEY-BEEI WU AND CHUN HSIUNG CHEN

***Abstract*—The variational reaction theory is applied to obtain the (E_z, H_z) formulation for the scattering problem when a plane wave is obliquely incident upon an inhomogeneous and anisotropic dielectric cylinder. The variational equation is then solved by the finite element method together with the frontal solution technique. Numerical results for scattering cross sections of anisotropic $LiNbO_3$ cylinders are included to discuss the influence due to material anisotropy and oblique incidence. Also depicted is the guiding characteristics of the cylinder when the inhomogeneous wave is considered. In addition, the formulas of scattering cross sections at low frequencies are derived and compared with the numerical results.**

I. INTRODUCTION

EXCEPT FOR SOME special structures for which closed-form solutions exist [1], general scattering problems have to be dealt with by numerical methods. The extended boundary condition method, originally developed by Waterman [2], is effective in handling homogeneous and isotropic dielectric objects [3], [4] or several specific anisotropic cylinders [5]. The integral equation formulation for equivalent polarization current is frequently used to solve scattering problems [6]–[8]. This formulation can be applied to tackle inhomogeneous cylinders but requires the inversion of a large full matrix [6] or a presumed surface impedance [7]. The analytic continuation method tailored by Wilton and Mittra [9] is very complicated and is difficult to apply in general.

In recent years, the finite element method has received much attention in handling scattering problems. The general approach is to enclose the scatterer by an artificial boundary. The field inside the boundary, the interior field, is tackled by a variational formula which is obtained from the operator equation. The field outside the boundary, the exterior field, is tackled by the eigenfunction expansion method or the Green's function technique which may model the outgoing characteristics of the scattered field. Finally the problem is solved by imposing continuity conditions along the artificial boundary.

For the scattering problem of an isotropic cylinder illuminated by a normally incident wave, the operator equation is self-adjoint and the boundary conditions are of the Dirichlet or Neumann type. Chang and Mei [10] make use of the unimoment method in which the exterior field is expanded in terms of Hankel functions. Marin [11] adopts a nonlocal boundary condition in which the exterior field is characterized by an integral equation. Jeng and Chen [12] apply the variational electromagnetics to give an E-field formulation with natural boundary conditions such that the radiation condition at infinity and the continuity conditions over the boundary can be satisfied automatically.

When the incident wave is obliquely incident or the cylinder consists of anisotropic materials, the operator equation is no longer self-adjoint and the boundary conditions are mixed. The solution of this general problem is of great interest and finds application in plane-wave scattering and in computing the radiation patterns of surface-wave antennas. It is thus worthy of separate consideration. Although the E-principle [12] can be employed in this problem, it requires, for each node, three unknowns which correspond to the three components of the vector electric field. In a previous study [13], we have proposed an (E_z, H_z)-formulation to deal with the dielectric waveguides, which requires only two unknowns, E_z and H_z, for each node. Here, the formulation is employed to tackle an anisotropic cylinder and to investigate the effects of material anisotropy and oblique incidence on the scattering characteristics.

II. THEORY

Consider an arbitrarily shaped, inhomogeneous, and anisotropic dielectric cylinder of infinite length which is uniform along the z-axis (Fig. 1). The permittivity tensor $\bar{\bar{\epsilon}}$ and permeability tensor $\bar{\bar{\mu}}$ are written as

$$\bar{\bar{\epsilon}} = \begin{bmatrix} \epsilon_{xx} & \epsilon_{xy} & \epsilon_{xz} \\ \epsilon_{yx} & \epsilon_{yy} & \epsilon_{yz} \\ \epsilon_{zx} & \epsilon_{zy} & \epsilon_{zz} \end{bmatrix} = \begin{bmatrix} \bar{\bar{\epsilon}}_{tt} & \bar{\epsilon}_{tz} \\ \bar{\epsilon}_{zt}^{T} & \epsilon_{zz} \end{bmatrix}$$

$$\bar{\bar{\mu}} = \begin{bmatrix} \mu_{xx} & \mu_{xy} & \mu_{xz} \\ \mu_{yx} & \mu_{yy} & \mu_{yz} \\ \mu_{zx} & \mu_{zy} & \mu_{zz} \end{bmatrix} = \begin{bmatrix} \bar{\bar{\mu}}_{tt} & \bar{\mu}_{tz} \\ \bar{\mu}_{zt}^{T} & \mu_{zz} \end{bmatrix} \qquad (1)$$

where the subscript t means the component transverse to the z direction and the superscript T means the transpose of a matrix. The tensors $\bar{\bar{\epsilon}}$ and $\bar{\bar{\mu}}$ may be not only complex quantities but also complicated functions of x and y.

For the time harmonic problem, we assume that the cylinder is illuminated by an obliquely incident plane wave ($\bar{E}^i$, $\bar{H}^i$) at an angle θ with respect to the z-axis. The incident wave and hence the scattered wave must have the same space harmonic variation $e^{-j\beta z}$ in the z direction, where $\beta = k \cos \theta$ is the propagation constant in z direction and k is the free space propagation constant. The problem now is reduced to a two-dimensional one. The common time and space harmonic

Reprinted from *IEEE Trans. Antennas Propagat.*, vol. AP-34, no. 5, pp. 640–645, May 1986.

dependence $e^{j(\omega t-\beta z)}$ is thus assumed and omitted throughout this paper.

The variational formula for the unknown longitudinal field (E_z, H_z) can be obtained by variational reaction theory [13], i.e.,

$$\delta I=0,$$

$$I=\int_\Omega d\Omega \begin{bmatrix} \bar{E}_t^a \\ \bar{H}_t^a \end{bmatrix}^T \cdot \begin{bmatrix} j\omega\bar{\bar{\epsilon}}_{tt} & j\beta\hat{z}\times \\ j\beta\hat{z}\times & -j\omega\bar{\bar{\mu}}_{tt} \end{bmatrix} \cdot \begin{bmatrix} \bar{E}_t \\ \bar{H}_t \end{bmatrix} - \int_\Omega d\Omega \begin{bmatrix} E_z^a \\ H_z^a \end{bmatrix}^T \cdot \begin{bmatrix} j\omega\epsilon_{zz} & 0 \\ 0 & -j\omega\mu_{zz} \end{bmatrix} \cdot \begin{bmatrix} E_z \\ H_z \end{bmatrix} + \oint_\Gamma d\Gamma \begin{bmatrix} E_z^a \\ H_z^a \end{bmatrix}^T_{r=r_0} \cdot \begin{bmatrix} H_\phi^s + H_\phi^i \\ E_\phi^s + E_\phi^i \end{bmatrix}_{r=r_0^+}. \tag{2}$$

Here, the relation between $(\bar{E}_t, \bar{H}_t)$ and (E_z, H_z) inside Ω $(r < r_0)$ is

$$\begin{bmatrix} \bar{E}_t \\ \bar{H}_t \end{bmatrix} = \begin{bmatrix} j\omega\bar{\bar{\epsilon}}_{tt} & j\beta\hat{z}\times \\ j\beta\hat{z}\times & -j\omega\bar{\bar{\mu}}_{tt} \end{bmatrix}^{-1} \cdot \begin{bmatrix} -j\omega\bar{\epsilon}_{tz} & -\hat{z}\times\nabla_t \\ -\hat{z}\times\nabla_t & j\omega\bar{\mu}_{tz} \end{bmatrix} \cdot \begin{bmatrix} E_z \\ H_z \end{bmatrix}. \tag{3}$$

While $(\bar{E}_t^a, \bar{H}_t^a)$ is related to (E_z^a, H_z^a) by replacing $(\bar{\bar{\epsilon}}, \bar{\bar{\mu}}, \beta)$ by $(\bar{\bar{\epsilon}}^T, \bar{\bar{\mu}}^T, -\beta)$ in (3). The tangential field components over the boundary Γ $(r = r_0)$ are related to the longitudinal ones by

$$\begin{bmatrix} H_\phi^s \\ E_\phi^s \end{bmatrix} = -\frac{1}{k_t^2} \sum_{m=-\infty}^{\infty} \cdot \begin{bmatrix} j\omega\epsilon_0 k_t H_m^{(2)\prime} & -m\beta H_m^{(2)} \\ -m\beta H_m^{(2)} & -j\omega\mu_0 k_t H_m^{(2)\prime} \end{bmatrix} \cdot \begin{bmatrix} e_m^s \\ h_m^s \end{bmatrix} e^{jm\phi},$$

where

$$k_t = \sqrt{k^2-\beta^2},$$

$$\begin{bmatrix} e_m^s \\ h_m^s \end{bmatrix} = \frac{1}{2\pi} \int_0^{2\pi} \begin{bmatrix} E_z - E_z^i \\ \eta_0(H_z - H_z^i) \end{bmatrix}_{r=r_0^-} \frac{e^{-jm\phi}}{H_m^{(2)}(k_t r_0)}\, d\phi, \tag{4}$$

and $\eta_0 = \sqrt{\mu_0/\epsilon_0}$. The argument of the functions $H_m^{(2)}, H_m^{(2)\prime}$ is $k_t r_0$ and the prime on $H_m^{(2)}$ means the derivative.

For the normally incident case $(\beta = 0)$, the original and the adjoint variable systems are identical when the cylinder is isotropic. The formula in (2) is thus reduced to the E-wave formulation or the H-wave formulation by merely enforcing $H_z = 0$ or $E_z = 0$, respectively. The E-wave formula is the same as that adopted in the previous study [12]. It is also noticed that E-wave and H-wave may be uncoupled for an anisotropic cylinder which is isotropic in the z direction, i.e., $\bar{\epsilon}_{zt}^T, \bar{\epsilon}_{tz}, \bar{\mu}_{zt}^T$, and $\bar{\mu}_{tz}$, are zero. However, for obliquely incident waves $(\beta \neq 0)$, both the E-wave and H-wave components are coupled and the full (E_z, H_z) formulation is called for.

To solve the variational equation (2), the finite element method [14] is employed to give a matrix equation where the associated matrix is of a banded type. In case that the permittivity and permeability tensors are Hermitian, the matrix can be made Hermitian if the adjoint field is chosen as $(E_z^a, H_z^a) = (E_z^*, -H_z^*)$. Then, only the upper triangular part of the matrix is required in numerical solutions. Also, the frontal solution technique [15] can be employed directly to find the unknown field on Γ, and hence the scattered field everywhere. For more general material, this matrix is not Hermitian. The whole banded matrix needs to be handled and a slight modification for the frontal solution technique should be involved to solve the matrix equation.

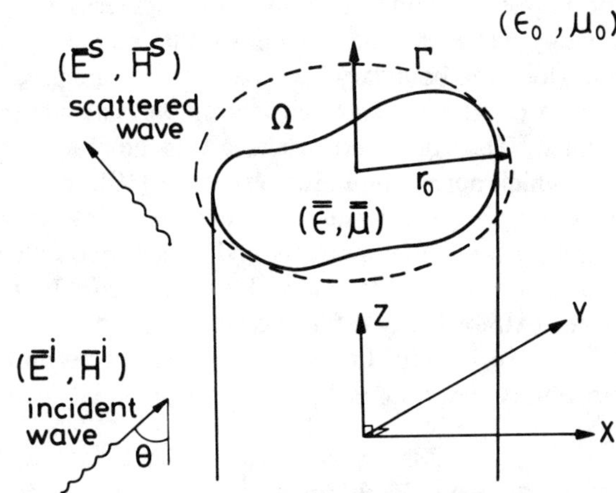

Fig. 1. Geometry of an arbitrarily shaped, inhomogeneous, and anisotropic dielectric cylinder illuminated by an obliquely incident plane wave.

At large distances $(r \gg r_0, r \gg \lambda)$, the scattered field in (4) can be written as

$$\begin{bmatrix} E_z^s \\ \eta_0 H_z^s \end{bmatrix} = \sqrt{\frac{2j}{k_t r}}\, e^{-jkr} \begin{bmatrix} g_e(\phi) \\ g_h(\phi) \end{bmatrix}, \tag{5}$$

where

$$\begin{bmatrix} g_e(\phi) \\ g_h(\phi) \end{bmatrix} = \sum_m j^m \begin{bmatrix} e_m^s \\ h_m^s \end{bmatrix} e^{jm\phi}.$$

The scattering cross section, or the echo width, corresponding to scattered E-wave and H-wave components is thus defined by [6], [10]

$$W_e(\phi) = \frac{2\lambda}{\pi} |g_e(\phi)|^2$$

$$W_h(\phi) = \frac{2\lambda}{\pi} |g_h(\phi)|^2, \tag{6}$$

where we have assumed $|E_z^i| = 1$ for a plane E-wave incident and $|\eta_0 H_z^i| = 1$ for a plane H-wave incident, respectively.

III. Numerical Results

A Fortran program has been implemented to solve the scattering problem for an arbitrarily shaped, inhomogeneous, and anisotropic cylinder. The computing time is nearly proportional to the square of the number of the unknowns inside the region Ω, while the required computer memory is approximately equal to the square of the number of the unknowns along the boundary Γ. As a check of the program, the scattering cross sections of an isotropic circular cylinder are computed. The computed results are in good agreement with those which appear in a previous study [10].

In this paper, we concentrate on the problem of electromagnetic scattering by a homogeneous and anisotropic cylinder. The cylinder is made of a uniaxial crystal $LiNbO_3$ whose ordinary and extraordinary refractive indices are $n_0 = 2.3$ and $n_e = 2.208$, respectively. In general, the relative permittivity and permeability tensors $\bar{\bar{\epsilon}}$ and $\bar{\bar{\mu}}$ can be written as

$$\frac{\bar{\bar{\epsilon}}}{\epsilon_0} = n_0^2 \bar{\bar{I}} + (n_e^2 - n_0^2) \begin{bmatrix} \sin\theta_c \cos\phi_c \\ \sin\theta_c \sin\phi_c \\ \cos\theta_c \end{bmatrix} \cdot \begin{bmatrix} \sin\theta_c \cos\phi_c \\ \sin\theta_c \sin\phi_c \\ \cos\theta_c \end{bmatrix},$$

$$\bar{\bar{\mu}} = \mu_0 \bar{\bar{I}}. \tag{7}$$

Here $\bar{\bar{I}}$ is the unit tensor, while θ_c and ϕ_c are the polar and azimuthal angles of the c-axis of the uniaxial crystal.

First, we consider scattering by a normally incident wave ($\beta = 0$). Fig. 2 shows the backscattering cross sections of a square cylinder (side length $= a$, and $(\theta_c, \phi_c) = (90°, 90°)$, i.e., c-axis $= y$-axis) illuminated by a plane wave with incident angles $\phi = 0°, 45°$, and $90°$ as parameters. Since the electric fields are polarized in the z direction for E-wave incident, the results for $\phi = 0°$ and $90°$ are identical in the W_e/λ curves. For H-wave incident, the electric fields are polarized in the transverse plane (x, y-plane). Hence, due to anisotropy in the x and y directions, there is a difference in the results for $\phi = 0°$ and $90°$. However, the results for $\phi = 0°$ and $90°$ are close to each other when compared with the results for $\phi = 45°$. This implies that the shape of the cylinder with respect to the incident angle is a stronger factor in determining the backscattering cross section.

Another dominant factor is the frequency of the incident wave. At low frequencies, the scattering cross sections per wavelength vary according to the factor λ^{-4} and are almost independent of the incident angles. In the Appendix, approximate formulas are given and used to check with the results in Fig. 2. Good agreement is found in the range $a/\lambda \leq 0.1$. At higher frequencies, the backscattering cross sections are unstable and rely much on the incident angles.

Fig. 3 shows the bistatic scattering cross sections of the same square cylinder which is illuminated by a plane wave with incident angle $\phi = 0°$. Both cases with E-wave incident (solid curves) and H-wave incident (dashed curves) are depicted in this figure. At low frequencies, e.g., $a/\lambda = 0.1$, the cylinder scatters like a small dipole polarized in the z direction (E-wave incident) or in the y direction (H-wave incident). At higher frequencies, e.g., $a/\lambda = 1.0$, interference between fundamental term and higher order terms in k [16] takes place and ripple appears in the bistatic scattering cross section.

Figs. 2 and 3 show that the polarization of the scattered field is the same as that of the incident wave. When the c-axis of the cylinder is oriented such that $\bar{\epsilon}_{zt}^T$ and $\bar{\mu}_{tz}$ are not zero, the E-wave and H-wave components are coupled and appear in the scattered field no matter whether the incident wave is of E-wave or H-wave. Fig. 4 shows the backscattering cross sections of circular cylinders (diameter $= a$, $\phi_c = 0°$, $\theta_c = 0°, 30°$, and $90°$) which are normally illuminated by a plane wave with incident angle $\phi = 90°$. It is noted that the cross-polarized field component appears in the scattered field for $\theta_c = 30°$, while it vanishes for $\theta_c = 0°$ and $90°$. Here, the incident wave field is taken as the reference of polarization. At

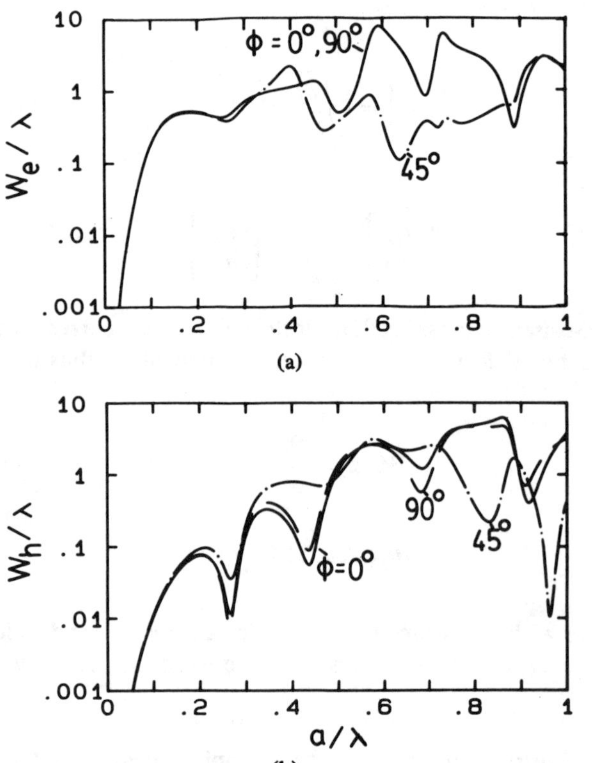

Fig. 2. Backscattering cross sections per wavelength of square cylinder (side length $= a$, c-axis $= y$-axis). Normal incidence case. (a) E-wave incident. (b) H-wave incident. Result for E-wave incident with incident angle $\phi = 90°$ is identical to that with $\phi = 0°$.

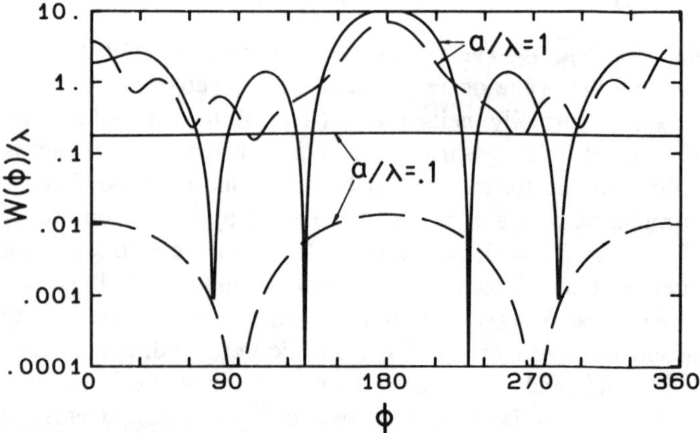

Fig. 3. Bistatic scattering cross sections per wavelength of square cylinders with side length a/λ as parameters (c-axis $= y$-axis). Normal incidence case with $\phi = 0°$. Solid and dashed curves correspond to cases with E-wave incident and H-wave incident, respectively.

low frequencies, the cross-polarized component is much smaller than that of the same-polarized one. As frequencies increase, however, these two components become comparable and the effects of varying θ_c's are significant. Since the material is symmetric, i.e., $\bar{\bar{\epsilon}} = \bar{\bar{\epsilon}}^T$, it is also found that the W_h/λ curve in Fig. 4(a) is identical to the W_e/λ curve in Fig. 4(b).

Next, we consider the case of oblique incidence such that $0<|\beta|<k$. The circular cylinders (diameter $a=\lambda/2$, $\phi_c=$ with incident angle $\phi = 90°$. Fig. 5 shows the backscattering cross sections versus the incident angle θ with respect to the z-axis. The backscattering cross sections of the cross-polarized component are shown for $\theta_c = 30°$ only, while those for $\theta_c = 0°$ and $90°$ are zero since the cylinders are symmetric with respect to the incident plane (y, z-plane). Again, the W_h/λ curve in Fig. 5(a) is identical to the W_e/λ curve in Fig. 5(b).

Fig. 6 shows the bistatic scattering cross sections of the same circular cylinder under plane waves with incident angle $\phi = 90°$ and $\theta = 60°$ ($\beta = k/2$). In Fig. 6(a), θ_c of the c-axis is $90°$ and the results are symmetric with respect to the incident plane. Meanwhile, the cross-polarized components vanish in the incident plane. When θ_c of the c-axis is $30°$ (Fig. 6(b)), the cylinder is no more symmetric with respect to the incident plane. Hence, the results in Fig. 6(b) are not symmetric with respect to the $\phi = 90°$ plane.

Finally, we consider the case with an inhomogeneous wave incident ($|\beta| > k$) [18]. The inhomogeneous wave associated with the sources near the scatterer is significant in the near zone. In this case, the Hankel functions in (4) have the pure imaginary argument $k_t r_0$ and are replaced by modified Bessel functions. The scattering coefficients are found by (4). Fig. 7(a) shows $|e_0^s|$, $|e_1^s|$ (for E-wave incident) and $|h_0^s|$, $|h_1^s|$ (for H-wave incident) of the circular cylinder with $a = \lambda/2$, $\phi_c = 0°$, $\theta_c = 0°$ and incident angle $\phi = 90°$. It is noted that the scattering coefficients tend to infinity at values of β which correspond to the guided propagation constants of the cylinder. These modes are identified as the HE_{11} ($\beta/k = 1.858$), TE_{01} ($\beta/k = 1.448$), and TM_{01} ($\beta/k = 1.118$) modes, respectively. Thus, sources which have the proper space variation and polarization can excite guided modes propagating in the cylinder. It is also noted that the present method has successfully avoided the appearance of the spurious modes. This may be attributed to the proper handling of the exterior field by the cylindrical harmonics expansion.

A similar case but with $\theta_c = 30°$ is considered in Fig. 7(b). It is found that the HE_{11} mode now splits into two guided modes, HE_{11}^y and HE_{11}^x, where the superscripts indicate the polarization of the electric field. Guided wave modes are observed at $\beta/k = 1.864$ (HE_{11}^y), 1.843 (HE_{11}^x), 1.434 (TE_{01}), and 1.132 (TM_{01}), respectively.

The HE_{11} mode splits further when $\theta_c = 90°$. Guided wave modes now appear at $\beta/k = 1.879$ (HE_{11}^y), 1.799 (HE_{11}^x), 1.398 (TE_{01}), and 1.145 (TM_{01}). The birefringence of the HE_{11} mode, $\Delta(\beta/k)/(n_0 - n_e)$, is 0, 0.23, and 0.87 for $\theta_c = 0°$, $30°$, and $90°$, respectively. Hence, the birefringent effect

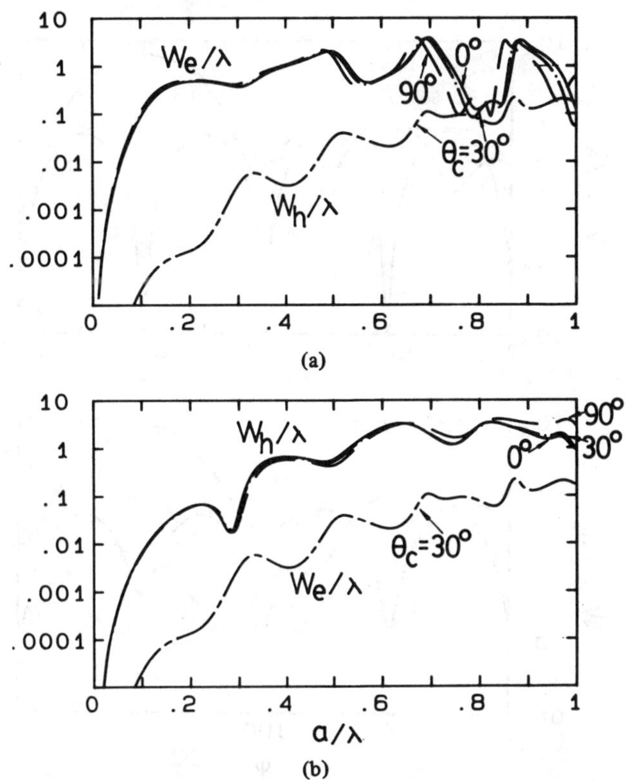

Fig. 4. Backscattering cross sections per wavelength of circular cylinders (diameter = a, $\phi_c = 0°$) with θ_c as parameters. Normal incidence case with $\phi = 90°$. (a) E-wave incident. (b) H-wave incident.

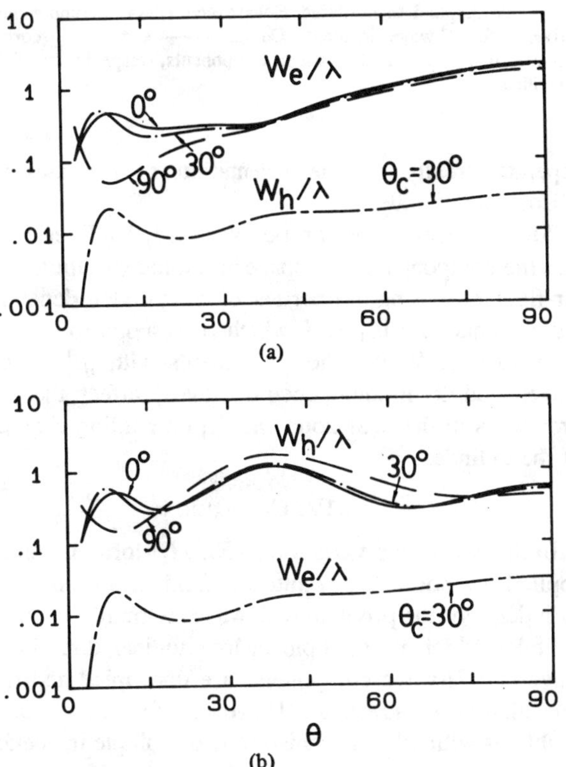

Fig. 5. Backscattering cross sections per wavelength of circular cylinders (diameter $a = \lambda/2$, $\phi_c = 0°$) with θ_c as parameters. Oblique incidence case with $\phi = 90°$ and θ is incident angle with resepct to z-axis. (a) E-wave incident. (b) H-wave incident.

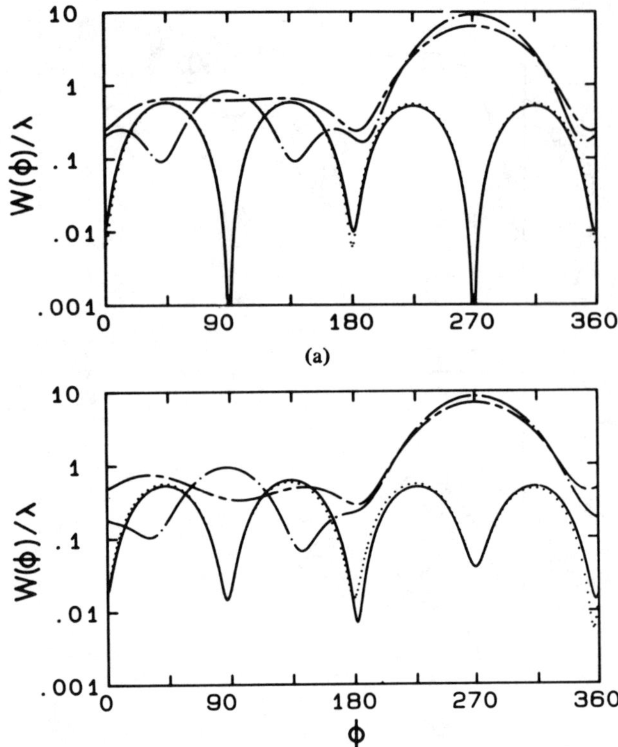

Fig. 6. Bistatic scattering cross sections per wavelength of circular cylinders (diameter $a = \lambda/2$, $\phi_c = 0°$). Oblique incidence case with $\phi = 90°$ and $\theta = 60°$. (a) $\theta_c = 90°$ (c-axis = x-axis). (b) $\theta_c = 30°$. Curves $- \cdot -$ and —— correspond to scattered E-wave and H-wave components, respectively, for E-wave incident. Curves ---- and $\cdots\cdots$ correspond to scattered H-wave and E-wave components, respectively, for H-wave incident.

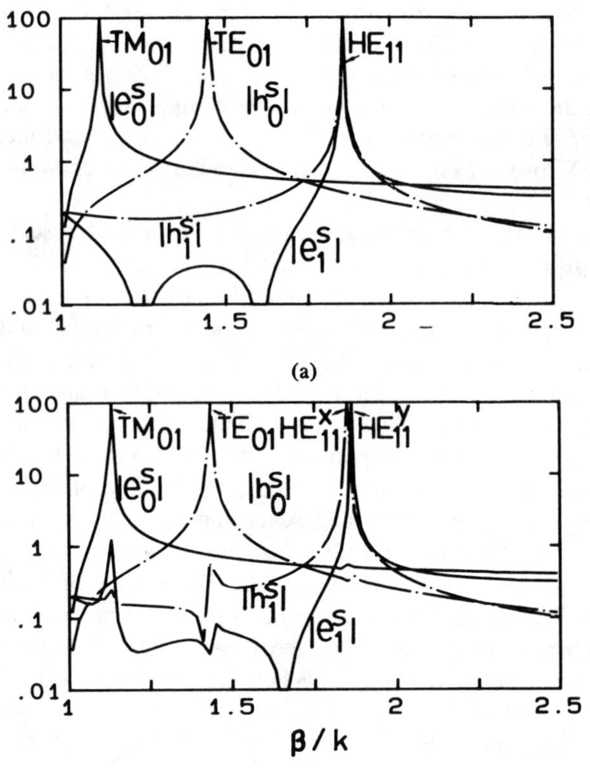

Fig. 7. Scattering coefficients of circular cylinders (diameter $a = \lambda/2$, $\phi_c = 0°$) when cases with inhomogeneous incident wave of incident angle $\phi = 90°$ are considered. β is propagation constant in z direction. (a) $\theta_c = 0°$ (c-axis = z-axis). (b) $\theta_c = 30°$.

depends greatly on the orientation of the c-axis of the anisotropic cylinder.

Any incident wave can be expanded, by Fourier integral, into the components with space harmonic variation $e^{-j\beta z}$. The far-field scattering properties of a cylinder depend on the components with $|\beta| < k$ which correspond to the real angle of incidence. While, the components with $|\beta| > k$, which correspond to inhomogeneous waves, affect the scattering properties in the near zone and depict guiding characteristics of the cylinder.

IV. Conclusion

In this work, the variational (E_z, H_z)-formulation has been applied to solve the scattered field from an anisotropic cylinder. For the problem with waves normally incident on the cylinder which is isotropic in longitudinal direction, the E-wave and H-wave components are uncoupled and the scalar formulation is adequate. However, for more complicated problems with arbitrary anisotropy or oblique incidence, these two components are coupled and a general (E_z, H_z)-formulation is called for.

At low frequencies, the cylinder scatters like a simple dipole. The scattering properties can be obtained directly by the approximate formulas in the appendix. At higher frequencies, other methods are required to tackle scattering problems. The present method has been found to be satisfactory for the scatterer with medium size and to be very efficient in the computer memory storage.

Appendix

Low Frequency Scattering Formulas for an Anisotropic Cylinder

When the dimensions of the cross section A of the cylinder are small with respect to the wavelength, the scattered field, in its first approximation, takes the form [16], [17]

$$\bar{E}^s(\phi) = \frac{k^2 e^{-jkr}}{\sqrt{8j\pi kr}} A \cdot (P_z\hat{z} + \bar{P}_t \cdot \hat{\phi}\hat{\phi}), \qquad \text{as } r \to \infty. \tag{8}$$

Here, the vector $\bar{P}$ is just the polarization induced in the dielectric cylinder by a static incident field $\bar{E}^i = E^i\hat{u}_i$. The induced polarization is related to the total field $\bar{E}$ by

$$\bar{P} = (\bar{\bar{\epsilon}} - \bar{\bar{I}}) \cdot \bar{E}. \tag{9}$$

Here, $\bar{\bar{\epsilon}}$ means the relative permittivity tensor and

$$\bar{E} = \bar{E}^s + \bar{E}^i. \tag{10}$$

Inside the cylinder, the scattered field $\bar{E}^s$ can be written as

$$E_z^s = 0,$$

$$\bar{E}_t^s = -\bar{P}_t/2. \tag{11}$$

The polarization can thus be found by (9)–(11).

When the incident wave is a plane E-wave, i.e., $\hat{u}_i = \hat{z}$, the scattering cross sections for the E-wave and H-wave components are, respectively,

$$W_e(\phi) = \frac{k^3}{4} \cdot A^2 \cdot |(\epsilon_{zz} - 1) - \bar{\epsilon}_{zt}^T \cdot (\bar{\bar{\epsilon}}_{tt} + \bar{\bar{I}})^{-1} \cdot \bar{\epsilon}_{tz}|^2,$$

$$W_h(\phi) = \frac{k^3}{4} \cdot A^2 \cdot |2\hat{\phi} \cdot (\bar{\bar{\epsilon}}_{tt} + \bar{\bar{I}})^{-1} \cdot \bar{\epsilon}_{tz}|^2, \tag{12}$$

where we have assumed $|E^i| = 1$. Similarly, when the incident wave is a plane H-wave, the scattering cross sections are

$$W_e(\phi) = \frac{k^3}{4} \cdot A^2 \cdot |2\bar{\epsilon}_{zt}^T \cdot (\bar{\bar{\epsilon}}_{tt} + \bar{\bar{I}})^{-1} \cdot \hat{u}_i|^2,$$

$$W_h(\phi) = \frac{k^3}{4} \cdot A^2 \cdot |2\hat{\phi} \cdot (\bar{\bar{\epsilon}}_{tt} - \bar{\bar{I}}) \cdot (\bar{\bar{\epsilon}}_{tt} + \bar{\bar{I}})^{-1} \cdot \hat{u}_i|^2. \tag{13}$$

References

[1] J.H. Richmond, "Digital computer solution of the rigorous equations for scattering problems," *Proc. IEEE*, vol. 53, pp. 796–804, Aug. 1965.

[2] P. C. Waterman, "Matrix formulation of electromagnetic scattering," *Proc. IEEE*, vol. 53, pp. 805–812, Aug. 1965.

[3] P. W. Barber, "Resonance electromagnetic absorption by nonspherical dielectric objects," *IEEE Trans. Microwave Theory Tech.*, vol. MTT-25, pp. 373–381, May 1977.

[4] N. Morita, "Another method of extending the boundary condition for the problem of scattering by dielectric cylinders," *IEEE Trans. Antennas Propagat.*, vol. AP-27, pp. 97–98, Jan. 1979.

[5] N. Okamoto, "Matrix formulation of scattering by a homogeneous gyrotropic cylinder," *IEEE Trans. Antennas Propagat.*, vol. AP-18, pp. 642–649, Sept. 1970.

[6] J. H. Richmond, "Scattering by a dielectric cylinder of arbitrary cross section shape," *IEEE Trans. Antennas Propagat.*, vol. AP-13, pp. 334–341, May 1965.

[7] M. G. Andreasen, "Scattering from cylinders with arbitrary surface impedance," *Proc. IEEE*, vol. 53, pp. 812–817, Aug. 1965.

[8] N. Morita, "Surface integral representations for electromagnetic scattering from dielectric cylinders," *IEEE Trans. Antennas Propagat.*, vol. AP-26, pp. 261–266, March 1978.

[9] D. R. Wilton and R. Mittra, "A new numerical approach to the calculation of electromagnetic scattering properties of two-dimensional bodies of arbitrary cross section," *IEEE Trans. Antennas Propagat.*, vol. AP-20, pp. 310–317, May 1972.

[10] S. C. Chang and K. K. Mei, "Application of the unimoment method to electromagnetic scattering of dielectric cylinders," *IEEE Trans. Antennas Propagat.*, vol. AP-24, pp. 35–42, Jan. 1976.

[11] S. P. Marin, "Computing scattering amplitudes for arbitrary cylinders under incident plane waves," *IEEE Trans. Antennas Propagat.*, vol. AP-30, pp. 1045–1049, Nov. 1982.

[12] S. K. Jeng and C. H. Chen, "On variational electromagnetics: theory and applications," *IEEE Trans. Antennas Propagat.*, vol. AP-32, pp. 902–907, Sept. 1984.

[13] R. B. Wu and C. H. Chen, "On the variational reaction theory for dielectric waveguides," *IEEE Trans. Microwave Theory Tech.*, vol. MTT-33, pp. 477–483, June 1985.

[14] O. C. Zienkiewicz, *The Finite Element Method*. New York: McGraw-Hill, 1977.

[15] E. Hinton and D. R. J. Owen, *Finite Element Programming*. New York: Academic, 1977.

[16] J. Van Bladel, "Scattering of low-frequency E-waves by dielectric cylinders," *IEEE Trans. Antennas Propagat.*, vol. AP-24, pp. 255–258, Mar. 1976.

[17] ——, *Electromagnetic Fields*. New York: McGraw-Hill, 1964.

[18] L. M. Brekhovskikh, *Waves in Layered Media*. New York: Academic, 1960.

Transfinite Elements: A Highly Efficient Procedure for Modeling Open Field Problems

J.-F. LEE and Z. J. CENDES

This paper presents a new approach for modeling open boundary conditions in electric and magnetic field problems that is considerably simpler and more efficient than the existing alternatives. The procedure is called the transfinite element method because it employs analytic basis functions to transfer the exterior boundary condition over an infinite domain to a finite-sized numerical solution region. In the transfinite element method, the problem region is divided into an interior region Ω_i and an exterior region Ω_e by a circular boundary Γ. Since the exterior region involves only a circular boundary, it may be solved analytically using the separation of variables procedure. Substituting this solution into the variational principle for the electric or magnetic field, minimizing it, and imposing continuity between the finite element solution in the interior region and the analytical solution in the exterior region results in a real, symmetric matrix equation that is solved for the field.

INTRODUCTION

In the numerical modeling of open boundary problems, a common feature is to subdivide the problem into a finite closed region, often called the interior or "picture frame" region, and the surrounding open space called the exterior region. A finite element or finite difference approximation is then made for the fields within the picture frame and some numerical device used to couple the exterior field boundary condition to the picture frame boundary.

One approach, often called the hybrid method, was proposed by Silvester and Hsieh.[1] In this procedure, a single picture frame is defined and the energy functional of the entire region exterior to the picture frame evaluated and added to the interior region energy functional. The solution involves a judicious choice of exterior field approximation functions and reduces the computation to a single boundary integral along the picture frame. In a related approach, proposed by McDonald and Wexler,[2] two concentric picture frames are defined and the integral equation relating the potentials between the two picture frames is used to specify the boundary conditions. Subsequently, Salon[3] extended this approach to a number of important cases by evaluating a boundary integral along the picture frame boundary. Unfortunately, however, in all cases, the integral is difficult and expensive to evaluate.

Bettess[4] in 1977 proposed a procedure he dubbed the "infinite element" approach. In this method, a series of shape functions analogous to Lagrange polynomials but including a decay term are generated. Unfortunately the accuracy of this method depends on the ability of the user to choose the right decay factor.

A newer group of methods are "ballooning" and its variants. Ballooning was invented by Silvester[5] for modeling Laplacian problems of the open type. In this method, the exterior boundary is modeled by using uniform layers of finite elements on the outside. Dasgupta[6] has modified this approach into a procedure called cloning. In cloning, a single layer of finite elements is used to produce a quadratic matrix equation that represents the behavior of the field in the infinite region. More recently, a method called "infinitesimal scaling" has been proposed by Hurwitz.[7] In this method, the exterior field contribution is determined by taking the limit as the thickness of the ballooning boundary layer approaches zero. The price to be paid in this approximation procedure is the need to solve a nonlinear matrix differential equation.[8]

FORMULATION

Consider the solution of the boundary value problem

$$-\nabla \epsilon_r \nabla \psi = g \quad \text{in } \Omega_\infty,$$
$$\psi = 0 \quad \text{at } \infty. \tag{1}$$

According to the Rayleigh–Ritz method, the solution of (1) is obtained by minimizing the functional

$$F(\psi) = \int_{\Omega_\infty} \epsilon_r \nabla\psi \, \nabla\psi d\Omega - 2\int_{\Omega_\infty} g\psi \, d\Omega, \tag{2}$$

where the solution region Ω_∞ is unbounded. By using a circle Γ with radius r_1 to enclose all inhomogenities and sources, we may separate Ω_∞ into two regions: an interior region Ω_i and the exterior region Ω_e, as shown in Fig. (1).

The integral in (2) can be integrated separately over Ω_i and Ω_e as follows:

$$F(\psi) = \int_{\Omega_i} \epsilon_r \nabla\psi \, \nabla\psi \, d\Omega - 2\int_{\Omega_i} g\psi \, d\Omega + \int_{\Omega_e} \nabla\psi \, \nabla\psi \, d\Omega. \tag{3}$$

By Green's theorem we way rewrite this as

FIG. 1. Separation of the problem region into Ω_i and Ω_e by a circular boundary Γ. All inhomogeneities and sources are contained within Γ.

Reprinted with permission from *Journal of Applied Physics*, vol. 61, no. 8, pp. 3913–3915, April 1987.

$$F(\psi) = \int_{\Omega_i} \epsilon_r \nabla\psi \nabla\psi \, d\Omega - 2\int_{\Omega_i} g\psi \, d\Omega - \int_\Gamma \psi \frac{\partial\psi}{\partial r} d\Gamma - \int_{\Omega_e} \psi \nabla^2 \psi \, d\Omega. \quad (4)$$

The reason for converting the functional from (3) to (4) will become clear in the following section.

THE SOLUTION SPACE

Note that the exterior region Ω_e is homogeneous and source free. If the net charge contained in the interior region Ω_i is zero, the function space Λ_e obtained by using the method of separation of variables in the exterior region is

$$\Lambda_e = \left(\psi^e | \psi^e = \sum_{i=1}^{M} \frac{1}{r^i} [a_i \cos(i\theta) + b_i \sin(i\theta)]\right), \quad r \geqslant r_1, \quad (5)$$

where M is the number of terms used to approximate the solution in the region Ω_e, and a_i and b_i are scalar numbers. If the net charge in Ω_i is not zero, a logarithmic term that vanishes at a reference point must be included in the expansion.

We subdivide the interior region Ω_i into second-order triangular finite elements. The function space Λ_i for the solution in the Ω_i then is

$$\Lambda_i = \left(\psi^i | \psi^i = \sum_{i=1}^{N} \Psi_i^I \alpha_i^I + \sum_{i=1}^{P} \Psi_i^\Gamma \alpha_i^\Gamma\right), \quad r \leqslant r_1, \quad (6)$$

where we have separated the interpolation polynomials into α_i^I for the nodes in Ω_i but not on the boundary Γ and α_i^Γ for the nodes on the boundary Γ. Corresponding to α_i^I and α_i^Γ are the solution values Ψ_i^I and Ψ_i^Γ, respectively.

The combined function space $\Lambda \equiv \Lambda_i \cup \Lambda_e$ provides a set of trial functions for ψ

$$\Lambda = \left\{\psi | \psi = \begin{cases} \psi^i \text{ in } \Omega_i \\ \psi^e \text{ in } \Omega_e \end{cases}, \quad \psi \in C^0\right\}. \quad (7)$$

Continuity of the field is imposed by setting the analytic solution on Γ equal to the numerical solution on Γ at the finite element nodes

$$\tilde{\Psi}^\Gamma = \theta R \tilde{A}, \quad (8)$$

where (r_1, θ_i) are the polar coordinates of the boundary point Ψ_i^Γ, and

$$\theta = \begin{bmatrix} \cos(\theta_1) & \sin(\theta_1)\ldots & \cos(M\theta_1) & \sin(M\theta_1) \\ \cos(\theta_2) & \sin(\theta_2) & \cos(M\theta_2) & \sin(M\theta_2) \\ \cdot & \cdot & \cdot & \cdot \\ \cdot & \cdot & \cdot & \cdot \\ \cdot & \vdots & \cdot & \cdot \\ \cdot & \cdot & \cdot & \cdot \\ \cos(\theta_p) & \sin(\theta_p)\ldots & \cos(M\theta_p) & \sin(M\theta_p) \end{bmatrix},$$

$$R = \begin{bmatrix} 1/r_1 & & & & 0 \\ & 1/r_1 & & & \\ & & \ddots & & \\ & & & 1/r_1^M & \\ 0 & & & & 1/r_1^M \end{bmatrix},$$

$\tilde{\Psi}^\Gamma = [\Psi_1^\Gamma \Psi_2^\Gamma \ldots \Psi_p^\Gamma]^T, \quad \tilde{A} = [a_1 b_1 \ldots a_M b_M]^T.$

In Eq. (8) and elsewhere, a bold-faced variable indicates a row vector, a tilde(~) above a column vector, and T represents the transpose.

THE BOUNDARY INTEGRAL

Since the functions in Λ_e are analytical solutions of the Laplace equation, the integral over Ω_e in (4) is zero for any choice of the coefficients a_i and b_i. Thus the functional reduces to the following form:

$$F(\psi) = \int_{\Omega_i} \epsilon_r \nabla\psi^i \nabla\psi^i \, d\Omega - 2\int_{\Omega_i} g\psi^i \, d\Omega - \int_\Gamma \psi^e \frac{\partial\psi^e}{\partial r} d\Gamma. \quad (9)$$

Thus by the particular choice of the function space Λ_e, the functional $F(\psi)$ is reduced from integrals over the entire unbounded region into integrals over the interior region plus a boundary integral term.[1]

The boundary integral term

$$B = \int_\Gamma \psi^e \frac{\partial\psi^e}{\partial r} d\Gamma \quad (10)$$

in (9) is evaluated simply by analytical integration. From (5), we have that

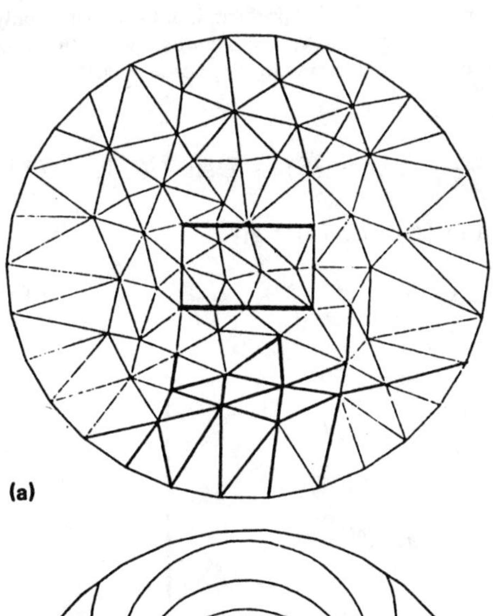

(a)

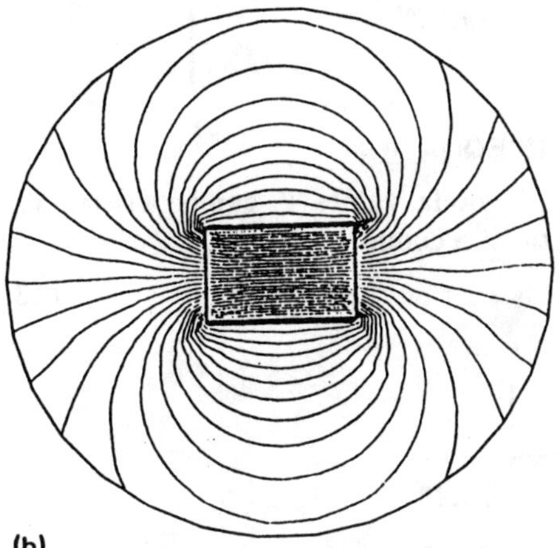

(b)

FIG. 2. (a) Cross section of a parallel-plate capacitor (shown in heavy lines) together with the circular picture frame boundary and the finite element mesh used in the transfinite element solution. (b) An equipotential plot of the solution.

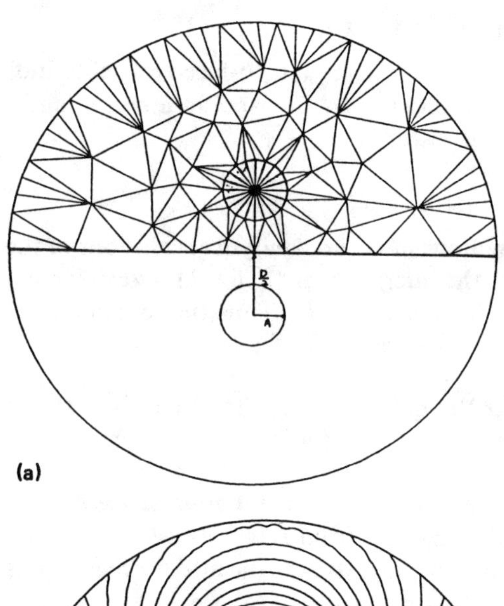

(a)

(b)

FIG. 3. (a) A pair of circular conductors in free space surrounded by a circular picture frame boundary. Since this problem is antisymmetric only one-half was triangulated as shown. (b) An equipotential plot of the solution when the potentials on the conductors are of opposite polarity.

$$\frac{\partial \psi^e}{\partial r} = -\sum_{i=1}^{M} \frac{i}{r^{i+1}} [a_i \cos(i\theta) + b_i \sin(i\theta)]. \tag{11}$$

Substituting (11) and (5) into (10) gives

$$\int_\Gamma \psi^e \frac{\partial \psi^e}{\partial r} d\Gamma = \mathbf{A}D\tilde{\mathbf{A}}, \tag{12}$$

where

$$D = \begin{bmatrix} \pi/r_1^2 & & & & 0 \\ & \pi/r_1^2 & & & \\ & & \ddots & & \\ & & & M\pi/r_1^{2M} & \\ 0 & & & & M\pi/r_1^{2M} \end{bmatrix}.$$

THE FINAL MATRIX EQUATION

From Eq. (8), the nodal values Ψ_i can be written as a connection matrix times a column vector

$$\tilde{\Psi} = C\tilde{X}, \tag{13}$$

where

$$C = \begin{bmatrix} {}_N\mathscr{I}^N & 0 \\ 0 & \theta R \end{bmatrix},$$

$$\tilde{\Psi} = [\Psi_1^I \ldots \Psi_N^I \quad \Psi_1^\Gamma \ldots \Psi_p^\Gamma]^T,$$

$$\tilde{X} = [\Psi_1^I \ldots \Psi_N^I \quad a_1 b_1 \ldots a_M b_M]^T,$$

and ${}_N\mathscr{I}^N$ is an N by N identity matrix.

Substituting (12) and (13) into the functional (9) results in

$$F(X) = \mathbf{X}C^T SC\tilde{X} - 2\mathbf{g}C\tilde{X} + \mathbf{A}D\tilde{A}, \tag{14}$$

where

$$S = \int_{\Omega_i} \nabla\tilde{\alpha}\cdot\nabla\alpha\, d\Omega, \quad \mathbf{g} = \int_{\Omega_i} g\alpha d\Omega.$$

Finally minimizing the functional $F(X)$ with respect to the variables X gives the matrix equation

$$\begin{bmatrix} S_{II} & S_{I\Gamma}\theta R \\ R\theta^T S_{\Gamma I} & R\theta^T S_{\Gamma\Gamma}\theta R \\ & +D \end{bmatrix} \tilde{X} = \mathbf{g}C. \tag{15}$$

Notice that the matrix on the left-hand side is real, symmetric, and positive definite and can therefore be solved very efficiently using the incomplete Choleski preconditioned conjugate gradient method.

RESULTS AND CONCLUSIONS

Figures 2 and 3 provide examples of solutions obtained by using the transfinite element procedure with some test problems. As is evident from the equipotential plots, the behavior of the solution inside the picture frame is the same as if all space were modeled.

In Fig. 3, the two infinitely long circular conductors both have radius A and are separated center to center by distance D, with the ratio A/D equal to 0.25. The exact value of the capacitance per unit length of this system is given by the formula

$$C = \pi\epsilon_0/[\cosh^{-1}(D/2A)].$$

In this case, the value of C is $2.3855\epsilon_0$. The capacitance obtained by using the transfinite element method with the mesh shown is $2.3783\epsilon_0$, an error of only 0.3%.

Since R is just a diagonal matrix of inverse powers of r, and since θ is just a matrix of sines and cosines of the boundary nodes, the programming time and matrix assembly time required to set up Eq. (15) is minimal.

[1]P. Silvester and M. S. Hsieh, Proc. IEE **118**, 1743 (1971).

[2]B. H. Mcdonald and A. Wexler, IEEE Trans. Microwave Theory Tech. **MTT-20**, 841 (1972).

[3]S. J. Salon, IEE Trans. Magn. **MAG-21**, 1829 (1985).

[4]Peter Bettess, Int. J. Numer. Methods Eng. **11**, 53 (1977).

[5]P. P. Silvester, D. A. Lowther, and C. J. Carpenter, Proc. IEE **124**, 1267 (1977).

[6]G. Dasgupta, Comput. Methods Appl. Mech. Eng. **46**, 295 (1984).

[7]H. Hurwitz, Jr., IEEE Trans. Magn. **MAG-20**, 1918 (1984).

[8]M. V. K. Chari, J. D'Angelo, C. Crowley, J. Roeth, and H. Hurwitz, IEEE Trans. Magn. **MAG-22**, 1037 (1986).

Unimoment Method for Electromagnetic Wave Scattering

K. K. MEI
DEPARTMENT OF ELECTRICAL ENGINEERING
UNIVERSITY OF CALIFORNIA
BERKELEY, CA 94720, USA

Abstract–A simple problem in 2-dimensional space is used as an example for the application of the unimoment method. The discussion includes fundamentals, mesh generation, loading of equations, sparse matrix inversion and boundary matching. The objective of this paper is to help uninitiated readers understand and apply this method.

INTRODUCTION

The unimoment method, since its inception, has been applied to numerous scattering problems [1–7]. The purpose of this paper is to give sufficient details of the application of the unimoment method so that an uninitiated reader can apply the method to problems of special interest.

The steps leading to a successful application of the unimoment method include (1) Fundamentals, (2) Mesh generation and node numbering scheme, (3) Loading of equations, (4) Sparse matrix inversion, and (5) Boundary matching. The following sections of this paper shall present the details of the steps just mentioned.

Consistent with the stated purpose of this paper, clarity is the first priority of the discussion and it can only be made at the expense of sophistication. A simple scattering problem in 2-dimensional space is used as a vehicle for this task. Readers who are interested in more advanced applications of this method may wish to consult references [4], [5], [6], [7], [8], and [9].

FUNDAMENTALS

The motivation of the unimoment method is to solve scattering problems consisting of penetrable inhomogeneous targets. Its principal idea is to approach the solution via differential equations directly rather than the integral equations. The most appealing advantage of differential equations is that the equations are basically unchanged whether the target is homogeneous or inhomogeneous, whereas inhomogeneity puts great strain on the integral equation method. The elliptical partial differential equation resulting from the associated Maxwell's equations can be solved numerically with conventional finite difference or finite element methods. However, concomitant with those methods are the problems of numerical stability and that of radiation condition. The mesh of the finite methods must be terminated at a finite surface, while the radiation condition is to be enforced in the far field zone. Numerous ways of enforcing the radiation condition have been suggested, e.g. Morgan et al. [8], McDonald, et al. [10] and Sylvester, et al. [11] used integral equations and Mei, et al. [12] used harmonic expansions, in which the finite difference or finite element equations are directly coupled to the integral equation or the harmonic expansion. Those approaches, however, add complications to numerical solutions in that the resulting matrix is partially full (from integral equations) and partially sparse (from finite difference). The unimoment method provides a simple numerical strategy, which separates the sparse matrix from the full matrix, thus making the problem more manageable.

The unimoment method is most conveniently viewed as a modification of the classical method of the separation of variables, wherein the modal solutions of

Reprinted with permission from *Journal of Electromagnetic Waves and Applications*, vol. 1, no. 3, p. 201–222, 1987.

the differential equations inside a separable surface is replaced by a sequence of computer generated solutions. A typical example of the essence of unimoment method is that of two-dimensional scattering by an arbitrary dielectric cylinder.

The equation to be solved is:

$$[\nabla^2 + k^2(\bar{r})]\phi(\bar{r}) = 0$$

where

$$\begin{aligned} k(\bar{r}) &= k_0\sqrt{\epsilon} && \text{inside the scatterer and} \\ k(\bar{r}) &= k_0 && \text{outside the scatterer} \end{aligned}$$

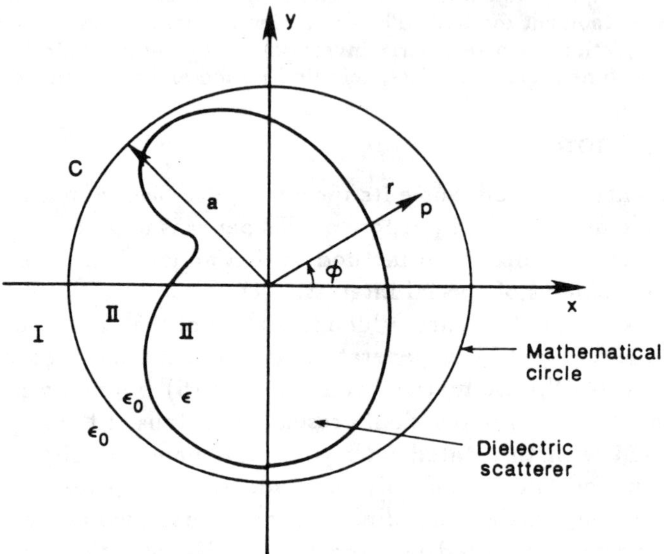

Figure 1. Scattering geometry of a dielectric cylinder.

If the medium inside the circle C of Fig. 1, were uniform, one would use the separation of variables to expand the interior fields by cylindrical harmonics,

$$\phi(\bar{r}) = \sum_{n=0}^{\infty} a_n^{e,o}\, J_n(k\rho) \begin{cases} \cos n\phi \\ \sin n\phi \end{cases} \qquad \rho \leq a \tag{1}$$

$$\phi(\bar{r}) = \phi^{\text{inc}}(\bar{r}) + \sum_{n=0}^{\infty} b_n^{e,o}\, H_n^{(2)}(k\rho) \begin{cases} \cos n\phi \\ \sin n\phi \end{cases} \qquad \rho \geq a \tag{2}$$

When $\phi^{\text{inc}}(\bar{r})$ is expanded in Fourier series, and the continuity of $\phi(\bar{r})$ and $\frac{\partial \phi(\bar{r})}{\partial r}$ are enforced on C, the coefficients $a_n^{e,o}$ and $b_n^{e,o}$ can be determined. If the medium inside C is not uniform, the expansion in (1) will not be available and the method of separation of variables fails. The unimoment method of solving the same problem proceeds essentially in parallel with the method of separation of variables, except the interior solution is expanded by a set of numerically generated solutions $\phi_n(\bar{r})$. That is,

$$\phi(\bar{r}) = \sum_{n=0}^{N} A_n\, \phi_n(\bar{r}) \tag{3}$$

where $\phi_n(\bar{r})$ can be generated by finite difference equations [2], finite element equations [3], or Monte Carlo methods [11]. To generate $\phi_n(\bar{r})$, one needs to solve inhomogeneous wave equations within C.

For the convenience of doing so numerically, we assign a set of linearly independent functions on C and solve the fields inside for each case. From the numerical solutions, we may obtain the normal derivatives $\frac{\partial \phi_n}{\partial n}$ on C. One need not solve the interior problem N times for N number of functions on C. The special sparse matrix method described in this paper shall show that it is possible to obtain a matrix operator, which operates on the boundary value $\phi_n(\bar{r})$ to produce the value of $\phi_n(\bar{r})$ on the nodes next to the boundary. Using the finite difference technique, we may then obtain $\frac{\partial \phi_n}{\partial n}$ on the boundary.

The interior solutions of Maxwell's equations may be obtained by the finite difference or finite element methods. But, it is not limited to these techniques. For example, the Monte Carlo technique [13] has been successfully used. If the scatterer is a uniform dielectric, modal solutions may sometimes be used. But, caution must be exercised when using analytic solutions within an irregular geometry. For example, the expansion of (1) may be used inside a subscribing circle within the scatterer of Fig. 1, but may not be used everywhere in the scatterer. This is because the expansion is only complete in a uniform medium inside a circle. However, if the boundary of the scatterer does not deviate very much from a circle, the expansion of (1) may be tolerated and offers reasonable results.

Since finite difference and finite element methods are by far the most versatile in generating solutions to wave equations, we shall briefly discuss their applications with regard to the unimoment method in the following.

If the boundary of the scatterer is described by simple coordinate surfaces it is convenient to use the finite difference technique, using orthogonal meshes [12]. Otherwise, finite difference would have undesirable features which discourage many uses [14]. The standard form of finite difference equations of the wave equation,

$$\phi_{i,j-1} + \phi_{i-1,j} - 4\phi_{i,j} + \phi_{i+1,j} + \phi_{i,j+1} + h^2k^2\phi_{i,j} = 0 \tag{4}$$

of a typical 5-point mesh cannot easily conform to the shape of an arbitrary scatterer. For a finite difference mesh to be non-orthogonal and conformal to an arbitrary geometry, a six-point mesh will be needed and it would be more difficult to draw such a mesh [12]. This is one important reason the finite element method is often preferred.

To use the finite element method, the differential equation is recast to the problem of mimimizing the integral

$$I = \int_S \left(|\nabla \phi|^2 - k^2(\bar{r})\phi^2 \right) ds \tag{5}$$

Numerical minimization of (5) is obtained by subdividing the circumscribing circle into triangular patches such as those shown in Fig. 2. The values of $\phi(\bar{r})$ between nodal points are linearly interpolated within each triangular element. When a value of $\phi(\bar{r})$ on C is assigned to be, say $\phi_n(C)$, $\phi_n(\bar{r})$ inside C can be uniquely determined, by minimizing the integral (5) with the assumption of linear $\phi_n(C)$ within each element.

The linear equations of the finite element method, like the finite difference method, are sparse. Each equation will contain only nodal values of ϕ_n directly connected by the mesh. Therefore, a simple finite difference equation shall contain only five unknowns and a linear triangular shaped finite element equation will typically contain seven unknowns. The ideal condition to solve the scattering problem is to find the values of $\phi_n(\bar{r})$ on nodes next to C, without calculating $\phi_n(\bar{r})$ at the other nodal points. This condition can be reached by proper inversion techniques of the sparse matrix.

In Fig. 2 it is noticed that the center region is void of a mesh. This area is deliberately left open to be filled with analytical solutions, if the center area is of a homogeneous medium. The advantage of this arrangement is a reduced mesh area, which will result in reduced computational cost.

Once a set of solutions $\phi_n(r)$ are generated inside C, the coefficients A_n in (3) and b_n in (2) may be obtained by solving the sets of linear equations resulting from the weighting integrals on C,

$$
\begin{aligned}
\sum_{n=0}^{2N-1} A_n \int_C \phi_n(\bar{r}) W_j(\bar{r})\, dc = \int_C \phi^{\text{inc}}(\bar{r}) W_j(\bar{r})\, dc \\
+ \sum_{n=0}^{N} b_n^e H_n^{(2)}(ka) \int_C \cos n\phi \, W_j(r)\, dc \\
+ \sum_{n=1}^{N} b_n^o H_n^{(2)}(ka) \int_C \sin n\phi \, W_j(r)\, dc
\end{aligned}
\tag{6}
$$

and

$$
\begin{aligned}
\sum_{n=0}^{2n-1} A_n \int_C \frac{\partial \phi}{\partial r}(\bar{r}) W_j(\bar{r})\, dc = \int_C \frac{\partial \phi^{\text{inc}}}{\partial r}(\bar{r}) W_j(\bar{r})\, dc \\
+ \sum_{n=0}^{N} b_n^e \frac{\partial H_n^{(2)}}{\partial r}(ka) \int_C \cos n\phi \, W_j(\bar{r})\, dc \\
+ \sum_{n=1}^{N} b_n^o \frac{\partial H_n^{(2)}}{\partial r}(ka) \int_C \sin n\phi \, W_j(\bar{r})\, dc
\end{aligned}
\tag{7}
$$

$W_j(\bar{r})$, $j = 1, 2, \ldots, 2N-1$, may be any set of linear independent functions on C. In this particular case for a two dimensional scattering problem the Fourier series, $\sin n\phi$, $\cos n\phi$, will be the obvious choices of both $W_j(\bar{r})$ and $\phi_n(\bar{r})$ on C.

As soon as the coefficients $b^{e,o}$ are determined, the scattered fields may be obtained by the summation in (2).

The sequence of steps involved in obtaining the scattered fields can be summarized as, mesh formation, matrix formation, sparse matrix inversion and weighting integrals, which will be discussed in the ensuing sections.

MESH FORMATION

One of the necessary tasks in applying the finite element method is to discretize the area where the differential equation is to be numerically solved. When linear base functions are to be used, it is convenient to subdivide the area into triangular patches. Each patch is called an element. In the unimoment method, the numerical area is either the space between the scatterer and a circumscribing circle (for a conducting scatterer) or between two concentric circles (for a dielectric scatterer). In the following we shall present several mesh formation schemes and node numbering schemes which have been successfully implemented. These schemes are by no means the best, but they are easy to use, although they may be limited to smooth scatterers.

The simplest mesh to form for the unimoment method is one between two concentric circles, such as the one shown in Fig. 2. In this mesh, the nodes are located on equally distributed radial lines. There are equal numbers of nodes on each radial line. This simple mesh scheme is applicable to smooth dielectric scatterers, but is not as good when applied to a metal scatterer. The region inside the inner circle, assumed to be uniform, will be filled by analytical solutions using Bessel functions.

To form the mesh, we first find the points where the scatterer surface crosses the radial lines. Let the radius of the outer circle be b and that of the inner circle be a, and the radius of the crossing point be c. Let the total number of nodes on each radial line be L. We shall assign M node points on $r > c$, and N

node points on $r < c$, where $M + N = L - 1$, and $\frac{M}{L} \simeq \frac{b-c}{b-a}$ and $\frac{N}{L} \simeq \frac{c-a}{b-a}$. After all the radial lines are divided, we have one M and one N for each radial line. Let M_i, N_i be those for the ith radial line counting from the top Z-axis. We have to check that the adjacent M_i's do not differ by more than one, i.e. $|M_i - M_{i-1}| \leq 1$. If $|M_j - M_{j-1}| > 1$ M_i and N_i should be adjusted.

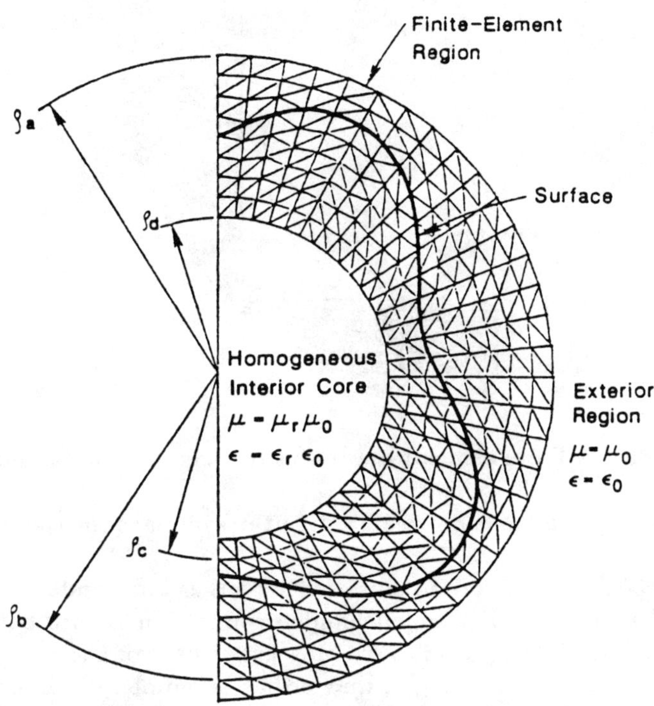

Figure 2. A typical finite element mesh.

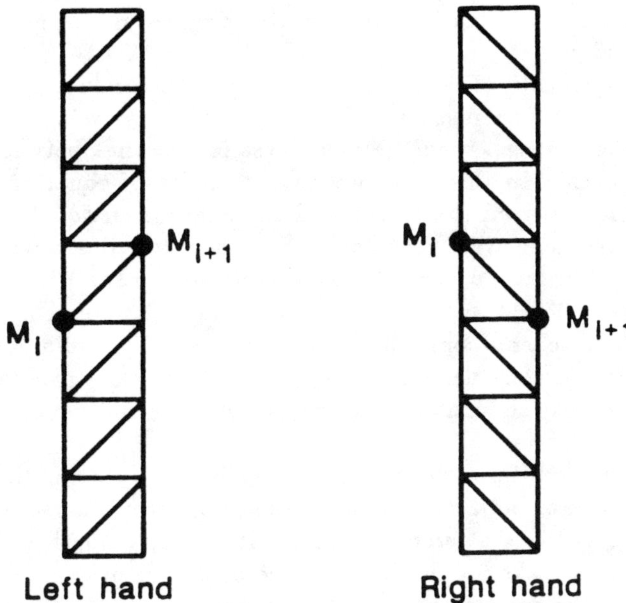

Figure 3. Left- and right-handed connections.

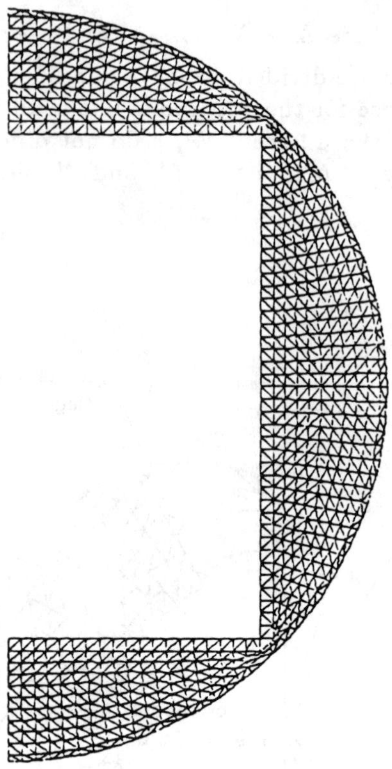

Figure 4. A finite element mesh of a rectangular conducting system.

This restriction in general is not required, but it will make the connection of nodes much easier.

The connection of the nodal points can be either left-handed as in Fig. 3a, or right-handed as in Fig. 3b. The left-handed connection is used between i th and $i+1$ th radial line if $M_{i+1} > M_i$. We may use either connection if $M_{i+1} = M_i$.

The nodes in the mesh of Fig. 2 may either be numbered circumferentially or radially. The dimension of the matrices to be inverted will be proportional to the number of nodes in each row (either radial row or circumferential row). In general radial rows should have fewer number of nodes. However, the inversion process of the finite element equations is sometimes easier if the nodes are numbered circumferentially.

For some applications, such as scattering by a conducting object, the mesh formation strategy just discussed does not work very well. Consider the target of Fig. 2 to be a metal, then the mesh area is just that between the target and the exterior cycle. In this case, we have a choice of equal or unequal number of nodal points between the target and the exterior circle. An equal number of nodes will make some areas very dense and others too dilute, which is undesirable intuitively, and in fact, a cause for instability in the resulting matrix. To make them unequal will destroy the circumferential numbering scheme and make the matrix inversion scheme somewhat more involved. The radial numbering scheme is almost always preferable, even though it requires more programming to invert the matrix, because the matrix sizes will be smaller and hence the target can be larger.

The procedure to form the mesh radially is greatly simplified if we limit the number of nodes on adjacent radial lines to differ by no more than one. This constraint limits the scatterers to those with fairly smooth sufaces, but it is a small sacrifice to make for the convenience of programming. The mesh between the object and the exterior circle of Fig. 2 is indeed such a mesh.

The mesh of a rectangular scatterer shown in Fig. 4 uses a similar scheme, but the connecting lines between the target and the exterior circle are not radial lines. They are instead formed by dividing the exterior circle and the target curve into an equal number of nodes and then connecting them.

The same ideas discussed above can be extended to objects with multilayer dielectric coating. Figure 5 shows such a mesh formation. The restriction is that the number of nodes on adjacent "radial" lines can differ by only one within each layer.

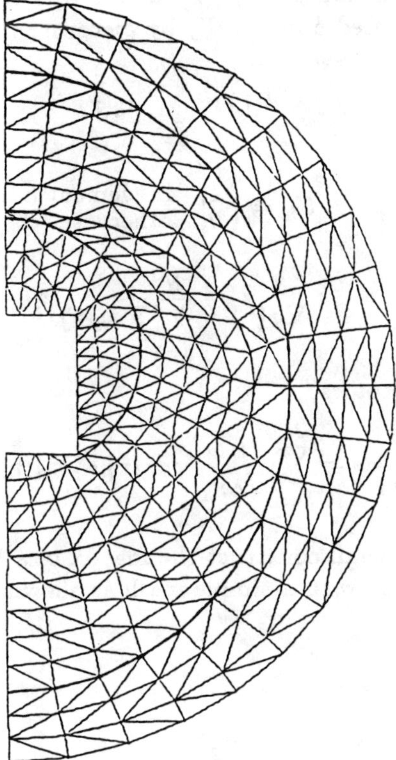

Figure 5. A mesh of a multilayer scatterer.

LOADING OF EQUATIONS

We shall use a two dimensional wave equation as a typical example for the unimoment method. The matrix equation representing the finite element formulation in the numerical region must be computed and the matrix element stored in proper locations.

The linearization of the field within each element makes $\phi(\overline{r})$ in element h a linear function

$$\phi_h(x,y) = a_{(h)1} + a_{(h)2}x + a_{(h)3}y \tag{8}$$

Let $\phi_{(h)i}$, $(i = 1,2,3)$ be the nodal values of element h, and x_i, y_i be the coordinates of the nodes. Then

$$\phi_{(h)i} = a_{(h)1} + a_{(h)2}x_i + a_{(h)3}y_i \qquad (i = 1,2,3) \tag{9}$$

and the vector $\underline{a}_h = \left[a_{(h)1}, a_{(h)2}, a_{(h)3}\right]^T$ may be obtained as

$$\underline{a}_h = \begin{bmatrix} 1 & x_1 & y_1 \\ 1 & x_2 & y_2 \\ 1 & x_3 & y_3 \end{bmatrix}^{-1} \begin{bmatrix} \phi_{(h)1} \\ \phi_{(h)2} \\ \phi_{(h)3} \end{bmatrix} = N_h^{-1}\underline{\phi}_h \tag{10}$$

Therefore,

$$\phi_h(\bar{r}) = (N_h^{-1}\underline{\phi_h})^T(1, x, y)^T \tag{11}$$

$$\nabla\phi_h(\bar{r}) = (N_h^{-1}\underline{\phi_h})^T \begin{bmatrix} 0 & 0 & 0 \\ 0 & 1 & 0 \\ 0 & 0 & 1 \end{bmatrix} \tag{12}$$

It is noticed that $\phi_h(\bar{r})$ is a linear function in the h th element while $\underline{\phi_h}$ is a vector containing three nodal values of $\phi_h(\bar{r})$ at the tips of the element. The function $\phi_h(\bar{r})$ may be represented numerically by piecewise linear functions $\phi_h(\bar{r})$ as

$$\phi(\bar{r}) = \sum_{h=1}^{N} \phi_h(\bar{r})U_h \tag{13}$$

where

$$U_h = \begin{cases} 1, & \text{in element } h \\ 0, & \text{otherwise} \end{cases} \tag{14}$$

and

$$\nabla\phi(\bar{r}) = \sum_{h=1}^{N} \nabla\phi_h(\bar{r})\, U_h \tag{15}$$

the integration (5) becomes,

$$\int_s \left[|\nabla\phi|^2 - k^2(\bar{r})\phi^2\right] ds = \sum_{h=1}^{N} \int_{\Delta S_h} \nabla\phi_h(\bar{r}) \cdot \nabla\phi_h(\bar{r}) ds - \sum_{h=1}^{N} \int_{\Delta S_h} k^2\phi_h^2(\bar{r})\, ds \tag{16}$$

As an illustration, we shall construct the matrix of a Laplace equation using the finite element technique with a simple mesh as shown in Fig. 6.

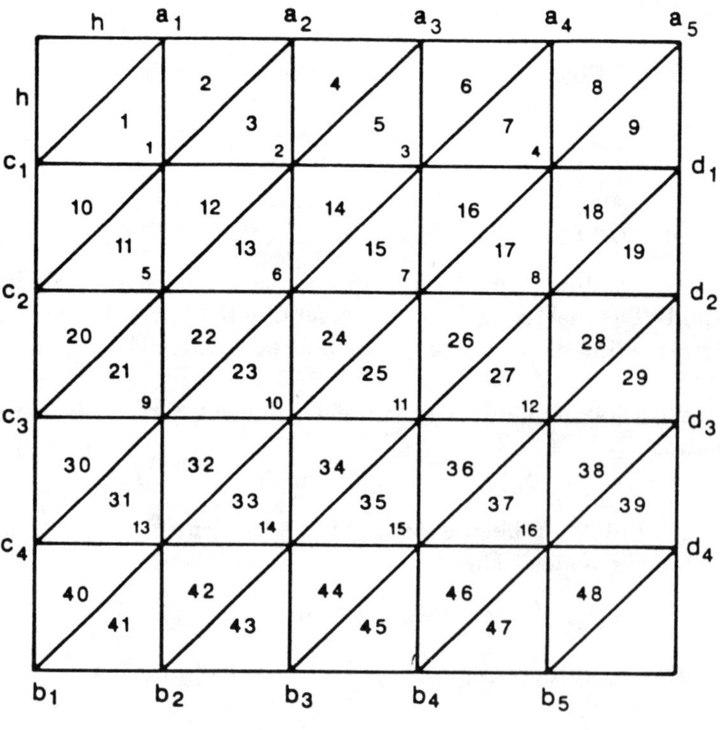

Figure 6. A simple rectangular finite element mesh.

Since we are interested only in the integrals in (15), we may use the center of mass of each element as the origin of the element. Therefore, there are only two types of elements in Fig. 6, namely, the even numbered type and the odd numbered type. The odd numbered elements have

$$N_{\text{odd}} = \begin{bmatrix} 1 & \frac{1}{3}h & \frac{2}{3}h \\ 1 & \frac{1}{3}h & -\frac{1}{3}h \\ 1 & -\frac{2}{3}h & -\frac{1}{3}h \end{bmatrix}, \quad N_{\text{odd}}^{-1} = \frac{1}{h^2}\begin{bmatrix} \frac{1}{3}h^2 & \frac{1}{3}h^2 & \frac{1}{3}h^2 \\ 0 & h & -h \\ h & -h & 0 \end{bmatrix} \tag{17}$$

and the even number elements have,

$$N_{\text{even}} = \begin{bmatrix} 1 & -\frac{1}{3}h & \frac{1}{3}h \\ 1 & \frac{2}{3}h & \frac{1}{3}h \\ 1 & -\frac{1}{3}h & -\frac{2}{3}h \end{bmatrix}, \quad N_{\text{even}} = \frac{1}{h^2}\begin{bmatrix} -\frac{1}{3}h^2 & \frac{1}{3}h^2 & \frac{1}{3}h^2 \\ -h & h & 0 \\ -h & 0 & h \end{bmatrix} \tag{18}$$

where the number sequence of X_i, Y_i are shown in Fig. 7.

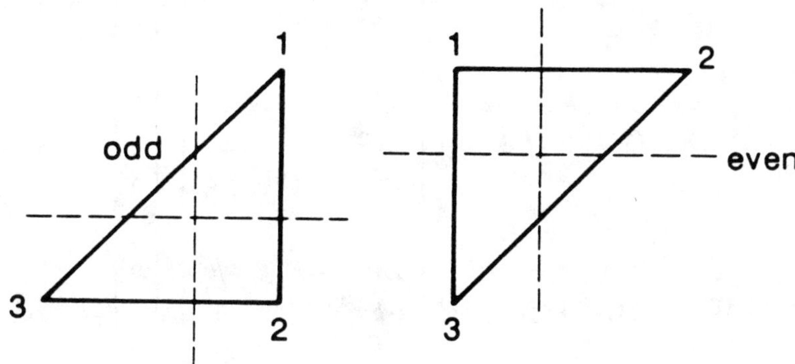

Figure 7. Even and odd numbered elements.

Considering the integration

$$\int_{\Delta s_1} \nabla\phi_1 \cdot \nabla\phi_1 \, ds = \int_{\Delta s_1} \phi_1^T N_1^{-1^T} \begin{bmatrix} 0 & 0 & 0 \\ 0 & 1 & 0 \\ 0 & 0 & 1 \end{bmatrix} N_1^{-1} \underline{\phi}_1 \, ds$$

$$= (1/h^2)[a_1, \phi_1, c_1] \begin{bmatrix} \frac{1}{3}h^2 & 0 & h \\ \frac{1}{3}h^2 & h & -h \\ \frac{1}{3}h^2 & -h & 0 \end{bmatrix} \begin{bmatrix} 0 & 0 & 0 \\ 0 & 1 & 0 \\ 0 & 0 & 1 \end{bmatrix} \begin{bmatrix} \frac{1}{3}h^2 & \frac{1}{3}h^2 & \frac{1}{3}h^2 \\ 0 & h & -h \\ h & -h & 0 \end{bmatrix} \begin{bmatrix} a \\ \phi_1 \\ c_1 \end{bmatrix} (1/h^2)$$

$$= (1/2h^2)[a_1, \phi_1, c_1] \begin{bmatrix} h^2 & -h^2 & 0 \\ -h^2 & 2h^2 & -h^2 \\ 0 & -h^2 & h^2 \end{bmatrix} \begin{bmatrix} a_1 \\ \phi_1 \\ c_1 \end{bmatrix} \tag{19}$$

The derivative w.r.t. ϕ_1, gives,

$$\frac{\partial}{\partial\phi_1} \int_{s_1} |\nabla\phi_1(\bar{r})|^2 ds = (1/h^2)[a_1, \phi_1, c_1] \begin{bmatrix} h^2 & -h^2 & 0 \\ -h^2 & 2h^2 & -h^2 \\ 0 & -h^2 & h^2 \end{bmatrix} \begin{bmatrix} 0 \\ 1 \\ 0 \end{bmatrix}$$

$$= (1/h^2)[a_1, \phi_1, c_1] \begin{bmatrix} -h^2 \\ 2h^2 \\ -h^2 \end{bmatrix} = (-a_1 + 2\phi_1 - C_1) \tag{20}$$

Performing a similar operation for element 2, we get

$$\int_s |\nabla\phi_2(\bar{r})|^2\, ds = (1/2h^2)[a_1, a_2, \phi_1] \begin{bmatrix} 2h^2 & -h^2 & -h^2 \\ -h^2 & h^2 & 0 \\ -h^2 & 0 & h^2 \end{bmatrix} \begin{bmatrix} a_1 \\ a_2 \\ \phi_1 \end{bmatrix}$$

and

$$\frac{\partial}{\partial\phi_2} \int_{s_2} |\nabla\phi_2(\bar{r})|^2\, ds = (1/h^2)[a_1, a_2, \phi_1] \begin{bmatrix} -h^2 \\ 0 \\ h^2 \end{bmatrix} = (-a_1 + \phi_1) \qquad (21)$$

the derivative w.r.t. ϕ_1 of the integrations of the rest of the elements that surrounds node 1, are:

$$\text{(Element 3)} \qquad (1/h^2)[a_2, \phi_2, \phi_1] \begin{bmatrix} h^2 & -h^2 & 0 \\ -h^2 & 2h^2 & -h^2 \\ 0 & -h^2 & h^2 \end{bmatrix} \begin{bmatrix} 0 \\ 0 \\ 1 \end{bmatrix} = (-\phi_2 + \phi_1) \qquad (22)$$

$$\text{(Element 12)} \qquad (1/h^2)[\phi_1, \phi_2, \phi_5] \begin{bmatrix} 2h^2 & -h^2 & -h^2 \\ -h^2 & h^2 & 0 \\ -h^2 & 0 & h^2 \end{bmatrix} \begin{bmatrix} 1 \\ 0 \\ 0 \end{bmatrix} = (2\phi_1 - \phi_2 - \phi_5) \qquad (23)$$

$$\text{(Element 11)} \qquad (1/h^2)[\phi_1, \phi_5, c_2] \begin{bmatrix} h^2 & -h^2 & 0 \\ -h^2 & 2h^2 & -h^2 \\ 0 & -h^2 & h^2 \end{bmatrix} \begin{bmatrix} 1 \\ 0 \\ 0 \end{bmatrix} = (\phi_1 - \phi_5) \qquad (24)$$

$$\text{(Element 10)} \qquad (1/h^2)[c_1, \phi_1, c_2] \begin{bmatrix} 2h^2 & -h^2 & -h^2 \\ -h^2 & h^2 & 0 \\ -h^2 & 0 & h^2 \end{bmatrix} \begin{bmatrix} 0 \\ 1 \\ 0 \end{bmatrix} = (-c_1 + \phi_1) \qquad (25)$$

Adding (20) through (25), we have

$$-2a_1 - 2c_1 + 8\phi_1 - 2\phi_2 - 2\phi_5 = 0$$

or

$$4\phi_1 - \phi_2 - \phi_5 = a_1 + c_1 \qquad (26)$$

Equation (26) is the finite element equation at node 1, which is a boundary node. An example of an interior node equation, say node 5, can be obtained by summing over elements number 13, 14, 15, 24, 23, and 22, and take derivative with respect to ϕ_6 which results in,

Element 13:	$-\phi_2$	$-\ \phi_5$	$+\ 2\phi_6$		
Element 14:	$-\phi_2$		$+\ \phi_6$		
Element 15:			ϕ_6	$-\ \phi_7$	
Element 24:			$2\phi_6$	$-\ \phi_7$	$-\ \phi_{10}$
Element 23:			ϕ_6		$-\ \phi_{10}$
Element 22:		ϕ_5	$+\ \phi_6$		
The sum is:	$-2\phi_2$	$-\ 2\phi_5$	$+\ 8\phi_6$	$-\ 2\phi_7$	$-\ 2\phi_{10} = 0$
or:	$-\phi_2$	$-\ \phi_5$	$+\ 4\phi_6$	$-\ \phi_7$	$-\ \phi_{10} = 0$ (27)

Which is identical to the 5-point finite difference equation. The reason we sum over the elements surrounding a particular node is that these are the only elements containing that node. After taking the derivative with respect to that nodal value, these are the only surviving matrix elements. The numbering system is such

that when we take the derivative with respect to nodal values ϕ_i we get the ith equation, e.g. the matrix element we get by summing over the elements around node 6, and their derivatives w.r.t. ϕ_6, all belong to the 6th equation.
Consider the integration over element 3, which gives

$$(1/h^2)[a_2, \phi_2, \phi_1] \begin{bmatrix} h^2 & -h^2 & 0 \\ -h^2 & 2h^2 & -h^2 \\ 0 & -h^2 & h^2 \end{bmatrix} \begin{bmatrix} a_2 \\ \phi_2 \\ \phi_1 \end{bmatrix} \tag{28}$$

Its derivative with respect to ϕ_1 is $(-\phi_2 + \phi_1)$ and with respect to ϕ_2 is $(-a_2 + 2\phi_2 - \phi_1)$, and they are part of the 1st equation and the 2nd equation of the matrix respectively. For the convenience of programming, we shall integrate the elements row by row, rather than integrating around each element. Differentiations will be performed with respect to each nodal value of the element integrated, and the coefficients are stored in the respective equations. Again, using the example of Fig. 5, the formation of the first 5 equations will be completed after we sweep the integration from element 1 to 19. Also we have formed part of the $5^{th} - 8^{th}$ equations. The $5^{th} - 8^{th}$ equations will be completed and the $9^{th} - 12^{th}$ equations partially formed after integrating up to element 29 etc. This integration strategy has some advantage in reducing computations and simplifying programming, especially in cases where the element number and node number are simply correlated.

To complete our discussion on integration of the wave equation, the second integration of (5) may be obtained as follows:

$$k^2 \int_{s_h} \phi_h(\bar{r})^2 ds =$$

$$k^2 \int_{s_h} [\phi_{(h)1}, \phi_{(h)2}, \phi_{(h)3}]\, N_{(h)}^{-1T} \begin{bmatrix} 1 \\ x \\ y \end{bmatrix} [1, \;\; x, \;\; y]\, N_{(h)}^{-1} \begin{bmatrix} \phi_{(h)1} \\ \phi_{(h)2} \\ \phi_{(h)3} \end{bmatrix} ds$$

$$= k^2 \int_{s_h} [\phi_{(h)1}, \phi_{(h)2}, \phi_{(h)3}]\, N_{(h)}^{-1T} \begin{bmatrix} 1 & x & y \\ x & x^2 & xy \\ y & xy & y^2 \end{bmatrix} N_{(h)}^{-1} \begin{bmatrix} \phi_{(h)1} \\ \phi_{(h)2} \\ \phi_{(h)3} \end{bmatrix} ds \tag{29}$$

The integration of the two variable polynomials over a triangle may be obtained using the center of gravity of the element as the origin of each element, i.e. $u = x - x_c, v = y - y_c$ where $x_c = \frac{1}{3}(x_{(h)1} + x_{(h)2} + x_{(h)3})$ and $y_c = \frac{1}{3}(y_{(h)1} + y_{(h)2} + y_{(h)3})$, The resulting integrations of the integrals

$$I_{ij} = \int_{s(h)} x^i y^j \, ds \tag{30}$$

are given in the following table,

$i+j$	I_{ij}
0	A (area of triangle)
1	0
2	$\frac{A}{12}\left(u_1^i v_1^j + u_2^i v_2^j + u_3^i v_3^j\right)$
3,4	$\frac{A}{30}\left(u_1^i v_1^j + u_2^i v_2^j + u_3^i v_3^j\right)$
5	$\frac{2A}{105}\left(u_1^i v_1^j + u_2^i v_2^j + u_3^i v_3^j\right)$

Table 1. Integrals of $\int x^i y^j \, ds$.

The derivatives of the integrations w.r.t. the nodal values can be obtained as in (20).

MATRIX INVERSION

It is obvious from the previous section that the finite element technique applied to the wave equation should result in a sparse matrix. More specifically the matrix will be banded if the node numbers are ordered in rows or columns. Using Fig. 6 as an example, the matrix configuration should have the following shape.

Let each row of Fig. 6 be a vector $\underline{\phi}_i$, the banded matrix of Fig. 8 is essentially a matrix 3-term recurrence formula of the type,

$$A_i \underline{\phi}_{i-1} + B_i \underline{\phi}_i + C_i \underline{\phi}_{i+1} = \underline{\alpha}_i \tag{31}$$

In Fig. 8, A_1 and C_4 are not zero matrices. They operate on the boundary values and are included in $\underline{\alpha}_1$ and $\underline{\alpha}_4$. The rest of A_i's and C_i's are unit matrices. In general, depending on the mesh geometry, A_i and C_i may not be unit matrices or diagonal. In fact, the vectors $\underline{\phi}_i$'s may not be of the same dimension, hence A_i and C_i are not necessarily square matrices and may not be invertible.

The 3-term recurrence formula of (31) may be solved by the following block by block elimination method. Assume each vector $\underline{\phi}_i$ and its neighbor are related by a constant matrix multiplier and a vector as

$$\underline{\phi}_{i-1} = R_i \underline{\phi}_i + \underline{s}_i \qquad (i = 1, 2,) \tag{32}$$

where R_i's are constant matrices and $\underline{s}_i$'s are constant vectors independent of the fields. Using (32), we may rewrite (31) as,

$$(A_i R_i + B_i)\underline{\phi}_i + C_i \underline{\phi}_{i+1} = \underline{\alpha}_i - A_i \underline{s}_i$$

or

$$\underline{\phi}_i = -(A_i R_i + B_i)^{-1} C_i \underline{\phi}_{i+1} + (A_i R_i + B_i)^{-1}(\underline{\alpha}_i - A_i \underline{s}_i) \tag{33}$$

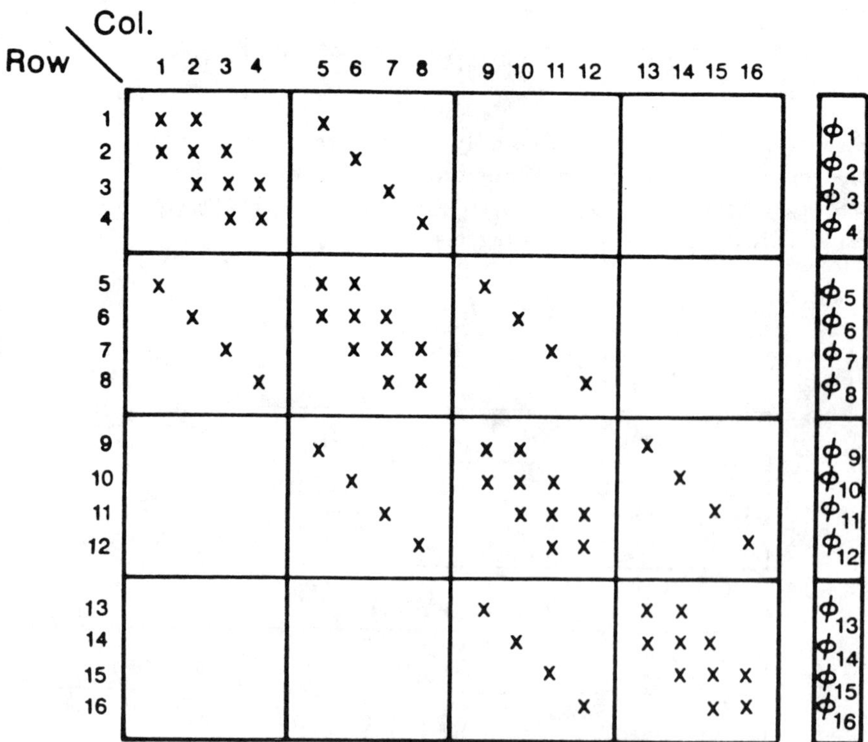

Figure 8. Matrix configuration of the mesh of Fig. 6.

Comparing (33) with (32) with $i = i + 1$, we get

$$R_{i+1} = -(A_i R_i + B_i)^{-1} C_i \tag{34}$$

and

$$\underline{S}_{i+1} = (A_i R_i + B_i)^{-1}(\underline{\alpha}_i - A_i \underline{s}_i) \tag{35}$$

The above recurrence formulas can be initiated with $i = 1$ in (31), where

$$B_1 \underline{\phi}_1 + C_1 \underline{\phi}_2 = \underline{a} + \begin{bmatrix} c_1 \\ 0 \\ 0 \\ d_1 \end{bmatrix} = \underline{\beta}_1 \tag{36}$$

and compare (36) with (32). Therefore

$$R_1 = -B_1^{-1} C_1$$

and

$$\underline{s}_1 = B_1^{-1} \underline{\beta}_1 \tag{37}$$

where

$$B_1 = \begin{bmatrix} 4 & -1 & 0 & 0 \\ -1 & 4 & -1 & 0 \\ 0 & -1 & 4 & -1 \\ 0 & 0 & -1 & 4 \end{bmatrix} \tag{38}$$

Comparing (37) with (34) and (35) it is clear that

$$R_o = 0 \quad \text{and} \quad \underline{s}_o = \underline{a} \tag{39}$$

It is necessary that $R_o = 0$ because $\underline{\phi}_o = \underline{a}$ is the boundary value, which cannot depend on the interior field. Using recurrence relation (34) we can obtain all R_i 's. After obtaining the R_i 's, we are able to obtain all $\underline{s}_i$'s from (35). Note the R_i 's are independent of the boundary values. All the information of the boundary values is contained in the vectors s_i, which are inexpensive to calculate when the boundary values are changed.

With the above described method we can get $\underline{\phi}_4$ of Fig. 6 by,

$$\underline{\phi}_4 = R_5 \underline{b} + \underline{s}_5 \tag{40}$$

Thus, we may get the field next to the boundary by operating on the boundary value by a matrix and adding a vector. If the interior fields are not required one may discard all the R_i 's except one so as to save memory space.

Now referring to Fig. 2, and for the convenience of illustration we shall construct the nodal vectors circumferentially. Assume the scatterer is a two dimensional dielectric body with a circular metal core. Start at the inner core and number the nodes circumferentially from $\theta = 0$ to $\theta = \pi$, and number the elements in the same manner. We load the matrices by sweeping through the elements as described earlier. After sweeping through the first two rows of elements we should obtain B_1, C_1, A_2 and part of B_2. From (37) we can already get R_1 and $\underline{s}_1$. Since B_1 and C_1 are no longer needed, we may use the space for C_2 and A_3. Sweeping through the next row of the elements, we should complete A_2, B_2, C_2, A_3 and part of B_3. At this point we may obtain R_2 and $\underline{s}_2$, and R_1 may be discarded. The process continues until we obtain R_N and $\underline{s}_N$.

It is noticed that the total memory space required to obtain R_N is six blocks and that, since the intermediate R_i 's are all discarded, all the $\underline{s}_N$'s associated with different boundary values must be computed simultaneously.

In the above discussion we have assumed the numbering to be circumferential. As a result we get a matrix R_N and a sequence of vectors $\underline{s}_N$. Each $\underline{s}_N$ is associated with a particular boundary value assigned to the exterior circle. We can obtain the nodal value next to the exterior circle by

$$\underline{\phi}_{N-1} = R_N \underline{\phi}_N + \underline{s}_N \tag{41}$$

where $\underline{\phi}_N$ is the boundary vector formed by the boundary values. It is noticed that in Figs. 2–4, the two layers of boundary nodes are all connected radially so as to facilitate the calculation of the normal derivatives at the boundary.

In most applications, however, the vectors $\underline{\phi}_i$ are formed radially, because there are fewer nodes for each radial line thus the matrix R_i will be smaller. This numbering system does not conveniently result in a matrix, such as R_N of (41), which immediately produces the normal derivative by operating on the boundary value. In this case, the method of generating a sequence of R_i and $\underline{s}_i$ is still similar to the previous case, except $\underline{\phi}_N$ will be the nodal values of ϕ on the top z-axis, and

$$\underline{\phi}_{N-1} = R_N \underline{\phi}_N + \underline{s}_N$$

represents the nodal values on the adjacent radial line. In general, when the nodal values on the ith radial line is known, we can obtain the next vector by

$$\begin{aligned}\underline{\phi}_{N-2} &= R_{N-1}\underline{\phi}_{N-1} + \underline{s}_{N-1} \\ &= R_{N-1}R_N\underline{\phi}_N + R_{N-1}\underline{s}_N + \underline{s}_{N-1}\end{aligned} \tag{43}$$

and

$$\begin{aligned}\underline{\phi}_{N-i} &= R_{N-i+1}R_{N-i+2}\dots R_N\underline{\phi}_N + R_{N-i+1}\underline{s}_N \\ &\quad + R_{N-i+1}\dots R_{N-2}\underline{s}_{N-1} + \dots + R_{N-i+1}\underline{s}_{N-i+2} \\ &\quad + \underline{s}_{N-i+1}\end{aligned} \tag{44}$$

It is evident that we may obtain any $\underline{\phi}_j$ from the boundary values along the top z-axis and all the R matrices and $\underline{s}$ vectors. It is noticed that we shall need many sequences of matrix products. Fortunately, we do not need the entire $\underline{\phi}_j$ on each radial line. We only need the nodal values next to the exterior and interior boundaries from which we obtain the normal derivatives.

After the continuous substitution of (44) we can represent the entire $\underline{\phi}_j$ vectors by Table 2.

In Table 2, $R_{1,j}$ means the matrix product $R_1R_2\dots R_j$. The subscript in each slot is determined by the row and columns of the slot as shown. The matrix product in each slot will multiply the $\underline{s}$ vector at the bottom of the column, and all the vectors in each row are summed to produce the $\underline{\phi}$ vector indicated on the right of the row. For example,

$$R_{N-2}R_{N-1}R_N\phi_N + R_{N-2}R_{N-1}\underline{s}_N + R_{N-2}\underline{s}_{N-1} + \underline{s}_{N-2} = \phi_{N-3} \tag{45}$$

Since, we only need the first and last elements of the vectors $\underline{\phi}_j$, we really need not compute and store all the matrix products. We only need, the first and last rows of the matrix products. We shall compute R_1 but retain only the first and last rows of R_1 and call it R'_1. After R_2 is computed, we shall compute R'_1R_2 which only has two rows. In effect the matrix products only involve the products of two row vectors and a matrix, which is not as time consuming as matrix products. Furthermore there is no need to store all the row vectors in Table 2. As soon as the row vectors are generated (they are generated from left to right in Table 2) their product with $\underline{s}$ vectors will be done and result in two scalar numbers, which are stored.

BOUNDARY MATCHING

In the previous sections, we have shown how to obtain the solutions of the wave equations of inhomogeneous media between two concentric circles given the boundary values on the circles. The sequence of solutions obtained by the finite element method will be called the base solutions. Note that the medium within the finite element region is in general inhomogeneous; it is difficult to separate the solutions

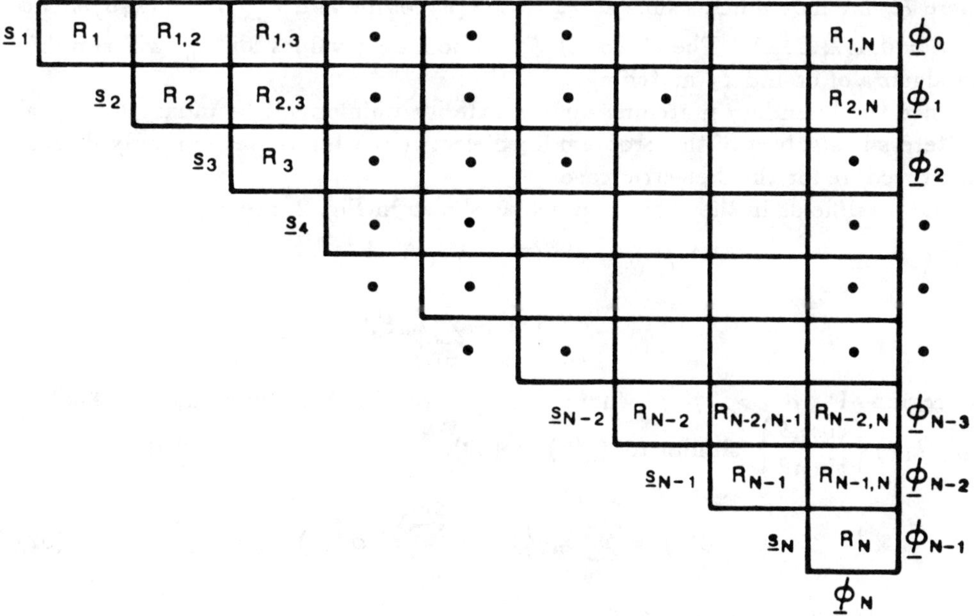

Table 2. Matrix products used to generate $\underline{\phi}_j$ from $\underline{\phi}_N$.

into "incident" and "scattered" components as we frequently do in free space. Therefore only the "total" field is computed in the finite element region. The boundary condition of the total field on a conducting surface will be either $\phi = 0$ or $\frac{\partial \phi}{\partial n} = 0$. For a dielectric scatterer, where the field within the center core is expressed analytically, we need the continuity conditions between the center and the finite element region.

In the unimoment method, we assume that the set of base functions we have generated is complete enough to represent the field inside the finite element region. For a conducting scatterer the base functions satisfy the following boundary conditions:

$$\text{(A)} \quad \phi^{\text{inc}} = E_z^{\text{inc}}, \qquad \begin{aligned} &\phi_n = 0 \quad \text{on scatterer} \\ &\phi_n = \psi_n(\theta) \quad \text{on } \rho = \rho_a \end{aligned} \tag{46}$$

where ρ_a is the radius of the exterior circumscribing circle shown in Fig. 2.

$$\text{(B)} \quad \phi^{\text{inc}} = H_z^{\text{inc}}, \qquad \begin{aligned} &\frac{\partial \phi_n}{\partial n} = 0 \quad \text{on scatterer} \\ &\phi_n = \psi_n(\theta) \quad \text{on } \rho = \rho_a \end{aligned} \tag{47}$$

where $\psi_n(\theta)$ is a set of linearly dependent functions. In this particular case of 2-dimensional scattering it is convenient to use the Fourier series,

$$\left.\begin{aligned} \psi_n(\theta) &= \cos n\theta \\ \psi_{N+n}(\theta) &= \sin n\theta \end{aligned}\right\} \qquad (n = 1, 2, \ldots, N) \tag{48}$$

For dielectric scatterers, there are two types of base functions, namely, type I and type II, which satisfy the following boundary conditions:

$$\phi^{\text{inc}} = E_z^{\text{inc}} \quad \text{or} \quad H_z^{\text{inc}}$$

$$\phi_n^{\text{I}} = \begin{cases} \psi_n(\theta), & \text{on } \rho = \rho_a \\ 0, & \text{on } \rho = \rho_d \end{cases} \tag{49}$$

$$\phi_n^{\text{II}} = \begin{cases} 0, & \text{on } \rho = \rho_a \\ \xi_n^i(\rho_d), & \text{on } \rho = \rho_d \end{cases} \tag{50}$$

where ξ_n^i are modal fields such as $\xi_n^i = J_n(k_1 d)\cos n\phi$ and $\xi_{n+N}^i = J_n(k_1 d)\sin n\phi$ for $n = 0, 1, 2, \ldots, N$. The choice of ξ_n^i as boundary values for ϕ_n^{II} will simplify the algebra of boundary matching.

Since the boundary matching on the exterior unimoment boundary for metal scatterers is a subset of the problem for dielectric scatterers we shall only discuss the procedure for the dielectric case.

The total fields in the three regions, as shown in Fig. 2, are

$$\begin{aligned}\phi(\bar{r}) &= \phi^{\text{inc}}(\bar{r}) + \phi^{\text{sc}}(\bar{r}) \\ &= \phi^{\text{inc}}(\bar{r}) + \sum_{n=0}^{2N} a_n \xi_n^0(\bar{r})\end{aligned} \tag{51}$$

in region I ($\rho > \rho_b$) where $\xi_n^0(\bar{r})$ are Hankel type modal fields $H_n^{(2)}(k_0 r)\begin{Bmatrix}\cos n\theta \\ \sin n\theta\end{Bmatrix}$ similar to $\xi_n^i(\bar{r})$ in (50).

$$\phi(\bar{r}) = \sum_{n=0}^{2N} b_n \phi_n^{\text{I}}(\bar{r}) + \sum_{n=0}^{2N} C_n \phi_n^{\text{II}}(\bar{r}) \tag{52}$$

in region II, ($\rho_d < \rho < \rho_a$), and

$$\phi(\bar{r}) = \sum_{n=0}^{2N} d_n \xi_n^i(\bar{r}) \tag{53}$$

in region III, ($\rho < \rho_c$).

The boundary conditions are:

$$\text{(a)} \quad \begin{Bmatrix} \phi(\bar{r}) \\ \dfrac{\partial \phi}{\partial \rho(\bar{r})} \end{Bmatrix} \quad \text{must be continuous at } \rho = \rho_a$$

$$\text{(b)} \quad \begin{Bmatrix} \phi(\bar{r}) \\ \dfrac{\partial \phi}{\partial \rho(\bar{r})} \end{Bmatrix} \quad \text{must be continuous at } \rho = \rho_d$$

Since the formulas in region I, (51), are all analytically represented, it would be tempting to find the radial derivatives analytically. But, that would be a bad choice because the derivatives in region II, (52) can only be found by the finite difference method, To equate analytical derivatives to finite difference derivatives of the same function will contribute extra errors, so it is advisable to use finite differences in (51) even though analytical derivatives are available. In doing so, the boundary conditions of continuity of (a) and (b) can be replaced by

$$\phi^{\text{inc}}(\bar{r})\Big|_{\rho=\rho_a} + \sum_{n=0}^{2N} a_n \xi_n^0(\bar{r})\Big|_{\rho=\rho_a} = \sum_{n=0}^{2N} b_n \psi_n^{\text{I}}(\bar{r})\Big|_{\rho=\rho_a} + \sum_{n=0}^{2N} c_n \psi_n^{\text{II}}(\bar{r})\Big|_{\rho=\rho_a} \tag{54}$$

and

$$\phi^{\text{inc}}(\bar{r})\Big|_{\rho=\rho_b} + \sum_{n=0}^{2N} a_n \xi_n^0(\bar{r})\Big|_{\rho=\rho_b} = \sum_{n=0}^{2N} b_n \psi_n^{\text{I}}(\bar{r})\Big|_{\rho=\rho_b} + \sum_{n=0}^{2N} c_n \psi_n^{\text{II}}(\bar{r})\Big|_{\rho=\rho_b} \tag{55}$$

for continuity at $\rho = \rho_a$, and similar formulas can be used for $\rho = \rho_d$. Equations (54) and (55) essentially equate the fields in region I and II on two adjacent circles to replace the continuity of the field and its normal derivatives. Equations (55) and (54) can be replaced by a set of weighting integrals, i.e. multiply both sides of (54) and (55) by $W_m(\theta)$ and integrate from $\theta = 0$ to $\theta = 2\pi$, ($m = 1, 2, \ldots, 2N$). An obvious and convenient selection of W_m will be $\psi_m(\theta)$ or (48). Each weighting integral of the equation resulted in a linear equation of a_n, b_n and c_n. Therefore, the following matrix equations result from the weighting integrals of (54) and (55)

$$\underline{\phi}_a^{\rm inc} + [A_a]\underline{a} = [B_a]\underline{b} + [C_a]\underline{c} \tag{56}$$

$$\underline{\phi}_b^{\rm inc} + [A_b]\underline{a} = [B_b]\underline{b} + [C_b]\underline{c} \tag{57}$$

where the matrix elements

$$A_a(m,n) = \int_0^{2\pi} \xi_n^0(\rho_a,\theta)W_m(\theta)\,d\theta \tag{58}$$

$$A_b(m,n) = \int_0^{2\pi} \xi_n^0(\rho_b,\theta)W_m(\theta)\,d\theta \tag{59}$$

and similar formulas apply to the matrices $[B_a]$, $[B_b]$, $[C_a]$, and $[C_b]$. Using the same procedures for the interior boundary, we get

$$[B_c]\underline{b} + [C_c]\underline{c} = [D_c]\underline{d} \tag{60}$$

$$[B_d]\underline{b} + [C_d]\underline{c} = [D_d]\underline{d} \tag{61}$$

which represent the weighted integrals equating (52) and (53). Notice that $\phi_n^{\rm I}(\rho_d) = 0$ as defined in (49), therefore $[B_d] = 0$. Since $\phi_n^{\rm II}(\rho_d) = \xi_n^i(\rho_d)$ we have $[C_d] = [D_d]$, hence $\underline{c} = \underline{d}$. From (60) we can easily solve for $\underline{b}$, i.e.

$$\underline{b} = [B_c]^{-1}\{[D_c] - [C_c]\}\underline{d} \tag{62}$$

Substituting (62) into (56) and (57), we get equations for $\underline{a}$ and $\underline{d}$.

It has now been demonstrated that with proper choice of boundary values of the numerically generated base functions, it is possible to introduce an interior analytical region without increasing the matrix size of the boundary matching process.

CONCLUSION

The basic steps of applying the Unimoment method have been presented. It should be understood that the methods, as presented, emphasize the pedagogical effects rather than sophistication. It is hoped that serious readers may succeed in developing their own Unimoment codes based on these simple methods, and move on to more sophisticated ones as they master the basic skills. The Unimoment method has already been applied to two dimensional scattering [3], scattering by bodies of revolution [4], scattering by buried objects [5], two-body scattering [6], and scattering by partially-buried objects [7]. Interested readers shall find the utility of the basic methods in many scattering problems which are much more complex than the ones discussed in this paper.

ACKNOWLEDGMENTS

Research sponsored by the U.S. Army Research Office Grant ARO DAAG29-84 and the U.S. Office of Naval Research Grant ONR N00014-84-K-0272.

The Editor thanks N. G. Alexópoulos, C. M. Butler, C. Lee, M. Morgan, D. R. Wilton and one anonymous Reviewer for reviewing the paper.

REFERENCES

1. Mei, K. K., "Unimoment method of solving antennas and scattering problems," *IEEE Trans. Antennas Propagat.*, Vol. AP–22, No. 6, 760–766, 1974.
2. Stovall, R. K., and K. K. Mei, "Application of a unimoment technique to a biconical antenna with inhomogeneous dielectric loading," *IEEE Trans. Antennas Propagat.*, Vol. AP–23, No. 3, 335–342, 1975.
3. Chang, S. K., and K. K. Mei, "Application of the unimoment method to electromagnetic scattering of dielectric cylinders," *IEEE Trans. Antennas Propagat.*, Vol. AP–24, No. 1, 35–42, 1976.
4. Morgan, M. A., and K. K. Mei, "Finite-element computation of scattering by inhomogeneous penetrable bodies of revolution," *IEEE Trans. Antennas Propagat.*, Vol. AP–27, No. 2, 202-214, 1979.
5. Chang, S. K., and K. K. Mei, "Multipole expansion technique for electromagnetic scattering by buried objects," *Electromagnetics*, Vol. 1, No. 1, 73–89, Jan.–March 1981.

6. Hunka, J. F., and K. K. Mei, "Electromagnetic scattering by two bodies of revolution," *Electromagnetics*, Vol. 1, No. 3, 329–347, July–Sept. 1981.
7. Chang, H., and K. K. Mei, "Scattering of EM waves by buried or partly buried body of revolution," *IEEE Antennas and Propagation International Symposium Digest*, 653–656, Los Angeles, June 1981.
8 Morgan, M. A., C. H. Chen, S. C. Hill and P. W. Barber, "Finite element-boundary integral formulation for electromagnetic scattering," *Wave Motion,* Vol. 6, 91–103, 1984.
9. Chang, H. and K. K. Mei, "Scattering of electromagnetic waves by buried and partly buried bodies of revolution," *IEE Trans. Geoscience Remote Sensing,* Vol. GE–3, 596–605, July 1985.
10. McDonald, B. H., and A. Wexler, "Finite-element solution of unbounded field problems," *IEEE Trans. Microwave Theory Tech.*, Vol. MTT–20, 841–847, Dec. 1972.
11. Sylvester, P., and M. S. Hsieh, "Finite-element solution of 2-dimensional exterior field problems," *Proc. IEE*, Vol. 118, 1743–1747, Dec. 1971.
12. Mei, K. K., R. Stovall, and D. Tremain, "On difference coupling methods of solving radiation problems," *URSI Fall Meeting Digest*, Sept. 1971.
13. Coffey, E. L. III, and W. L. Stutzman, "Hybrid computer solutions of electromagnetic scattering problems," *IEEE Antennas and Propagation International Symplosium Digest*, 82–85, Oct. 1976.
14. Mei, K. K., M. A. Morgan, and S. K. Chang, "Finite methods in electromagnetic scattering," *Electromagnetic Scattering*, edited by P. L. E. Uslenghi, 359–392, Academic Press, 1978.

Application of Hybrid Finite Element Method to Electromagnetic Scattering from Coated Cylinders

JIAN-MING JIN AND VALDIS V. LIEPA, MEMBER, IEEE

Abstract—**A hybrid finite element method is applied to solve the electromagnetic scattering (for both E- and H-polarizations) from coated cylinders. The coating material can be inhomogeneous and have ϵ and μ different from their free-space values. Three numerical examples are presented to show the validity and versatility of the method: 1) bistatic scattering from a coated circular cylinder, which verifies the formulation and the computer program, 2) backscattering from a coated rectangular cylinder, and 3) backscattering from an ogival cylinder with coated edges. Computations are compared with those obtained using the moment method codes and the exact solutions when applicable.**

Introduction

THE PROBLEM of electromagnetic scattering from cylinders coated with dielectric and/or magnetic materials has been studied intensively for many years by using exact (series), approximate (high frequency, low frequency) and numerical techniques. The exact analytical solutions are available only for few cases such as circular cylinder [1]. In nonspecular radar cross section study [2], the coated cylinder was mathematically modeled by a cylinder with an impedance boundary condition applied at the surface. The moment method was then used to solve the associated integral equations. Such an approach simplifies the formulation as well as makes the problem more practical to solve on a computer. However, the surface impedance boundary condition is an approximation which may cause errors when the coated body involves edges and small radii of curvature. Hence, a technique which solves the problem without using such an approximation is desirable, which then, for example, can be used to examine the validity of the impedance boundary condition for complex geometries, or just to compute the scattered fields.

Finite element method (FEM), which is usually based on variational principle, is a powerful numerical technique for solving complex geometries and media in bounded regions [3]. In the early 1970's, Silvester and Hsieh [4] and McDonald and Wexler [5] pointed out a way to solve unbounded problems which led to introduction of hybrid finite element methods (HFEM). In these methods exterior fields are expressed by an expansion of eigenfunctions or by integrals involving Green's functions, and such expressions are imposed in the functional as boundary constraints. These methods have been further used to solve two-dimensional antenna and static field problems [6], [7] and dynamic scattering problems [8], [9]. Also to solve unbounded problems, Mei *et al.* [10]–[12] developed the so called unimoment method in which the FEM was used to generate a set of trial functions inside a mathematical circle encompassing the scatterer. The fields outside were expanded in the Hankel functions and then coupled with interior problem on the circle. Morgan *et al.* [13] presented finite element-boundary integral formulation and applied to the problem of bodies of revolution. In that formulation FEM was used again to generate a set of basis functions which were used to compute the surface integrals contained in the extended boundary condition method.

In formulating the scattering problems circles are usually used as artificial boundaries to separate the interior and exterior regions. Although such approaches simplify the formulation, it is not an efficient technique when dealing with noncircular cylindrical problems, especially ones with thin geometries such as ogival cylinders. Using arbitrary boundaries which just enclose the scatterers can greatly reduce the region requiring FEM solution, and thus reduce the size of the resultant matrix. Here we discuss such a technique. The technique, in one way or another, has been used by numerous authors mentioned above and here we present a method based on a rather simple formulation and apply it to problems of current interest, such as scattering from cylinders coated with dielectric and/or magnetic materials for both E- and H-polarizations. In our formulation, two artificial boundaries Γ_A and Γ_B are used to separate the interior and exterior regions, with Γ_A enclosing Γ_B which in turn encloses all scatterers. The exterior field (outside Γ_B) is expressed by an integral over Γ_B involving free-space Green's function. This expression is imposed on the matrix equation derived by using FEM for the interior field (inside Γ_A). The use of the two boundaries, which was originally proposed by McDonald and Welxer [5], avoids the singularity problem typically encountered in Green's function integration and makes it easier to impose the continuity condition over the overlapping region between Γ_A and Γ_B.

Formulation

The problem considered is illustrated in Fig. 1 where a perfectly conducting cylinder is coated with dielectric and/or magnetic material. The constitutive parameters of the material

Manuscript received November 26, 1986; revised July 27, 1987.

The authors are with the Radiation Laboratory, Department of Electrical Engineering and Computer Science, The University of Michigan, Ann Arbor, MI 48109.

IEEE Log Number 8717993.

Reprinted from *IEEE Trans. Antennas Propagat.*, vol. 36, no. 1, pp. 50–54, Jan. 1988.

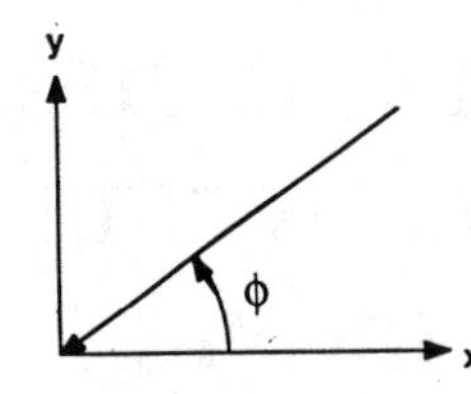

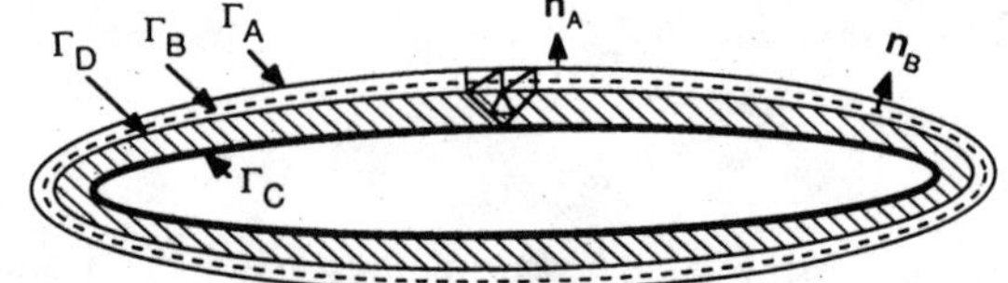

Fig. 1. A coated cylinder and artificial boundaries.

are denoted by complex permittivity ϵ and permeability μ. Assume an electromagnetic wave is incident in the direction normal to the axis of the cylinder. Both E- and H-polarized cases are considered and $e^{j\omega t}$ time convention is assumed. Let the boundary of the conducting cylinder be Γ_C and the outer boundary of the coating be Γ_D. The basic idea of the method is to draw one artificial boundary Γ_A enclosing Γ_D and another artificial boundary Γ_B lying between Γ_A and Γ_D. To minimize the region treated by FEM, Γ_A is drawn in such a way that only one layer of triangular elements is needed between Γ_A and Γ_D. We will use FEM in the region R_I defined by Γ_A and Γ_C, and use free-space Green's function integral formulation in the region outside Γ_B.

Finite Element Analysis

The field $\phi(= E_z$ for E-polarized case, and H_z for H-polarized case) in the inhomogeneous region R_I can be solved by minimizing the functional

$$F=\iint_{R_I}\left\{\frac{1}{\mu_r(\bar{r})}\nabla\phi\cdot\nabla\phi_a^*-k_0^2\epsilon_r(\bar{r})\phi\phi_a^*\right\}ds-\int_{\Gamma_A}\phi_a^*\nabla\phi\cdot\hat{n}_A\,dl \quad (1)$$

for E-polarized case, and

$$F=\iint_{R_I}\left\{\frac{1}{\epsilon_r(\bar{r})}\nabla\phi\cdot\nabla\phi_a^*-k_0^2\mu_r(\bar{r})\phi\phi_a^*\right\}ds-\int_{\Gamma_A}\phi_a^*\nabla\phi\cdot\hat{n}_A\,dl \quad (2)$$

for H-polarized case, where $k_0 = 2\pi/\lambda$ and λ is the free-space wavelength. The normal unit vector $\hat{n}_A$ is outward (Fig. 1) and the asterisk denotes complex conjugate. ϕ_a is the unknown solution of the adjoint problem to the original one that applies to ϕ. For lossless material, $\phi_a = \phi$. Subdividing R_I into M triangular elements and using linear interpolation for each element, one can express the field in R_I as

$$\phi(x, y)=\sum_{e=1}^{M}\sum_{i=1}^{3}N_i^e(x, y)\phi_i^e \quad (3)$$

where $N_i^e(x, y)$ $(i = 1, 2, 3)$ is the interpolating function for the eth element and ϕ_i^e is the field at the ith nodal point in the eth element. The same expansion is used for ϕ_a. Application of finite element analysis to (1) or (2) produces a matrix equation

$$K\phi=\psi \quad (4)$$

which in more detailed form can be written as

$$\begin{bmatrix} K_{AA} & K_{AD} & & \\ K_{DA} & K_{DD} & K_{DI} & K_{DC} \\ & K_{ID} & K_{II} & K_{IC} \\ & K_{CD} & K_{CI} & K_{CC}\end{bmatrix}\begin{bmatrix}\phi_A\\ \phi_D\\ \phi_I\\ \phi_C\end{bmatrix}=\begin{bmatrix}\psi_A\\ 0\\ 0\\ 0\end{bmatrix} \quad (5)$$

where the subscript A denotes the nodal points on Γ_A, D on Γ_D, I in the region between Γ_D and Γ_C, and C on Γ_C. Thus, ϕ_A represents the fields at nodal points on Γ_A, ϕ_D on Γ_D, ϕ_I in the material, and ϕ_C on Γ_C. ψ_A comes from the line integral on Γ_A in (1) or (2); however, it cannot be obtained without the complete knowledge about the boundary condition on Γ_A. This can be bypassed by discarding the first row of (5) since a simple relationship exists between ϕ_A and ϕ_D via the surface integral equation.

Surface Integral Formulation

In the region outside Γ_B, the solution to the Helmholtz wave equation, subject to the radiation condition at infinity, is the integral equation

$$\phi(\bar{r})=\phi^{\rm INC}(\bar{r})-\int_{\Gamma_B}\{G_0(\bar{r},\bar{r}_B)\nabla\phi(\bar{r}_B)\hat{n}_B-\phi(\bar{r}_B)\nabla G_0(\bar{r},\bar{r}_B)\cdot\hat{n}_B\}\,dl \quad (6)$$

where $\bar{r}_B$ is on Γ_B, $\hat{n}_B$ is an outward unit normal from Γ_B, and $G_0(\bar{r}, \bar{r}_B)$ is the two-dimensional free-space Green's function

$$G_0(\bar{r}, \bar{r}_B)=-\frac{j}{4}H_0^{(2)}(k_0|\bar{r}-\bar{r}_B|). \quad (7)$$

Applying (6) to calculate the field on Γ_A, and using the same field representation as (3) for $\phi(\bar{r}_B)$, one obtains the matrix equation

$$\phi_A=\phi_A^{\rm INC}+P_{AA}\phi_A+P_{AD}\phi_D \quad (8)$$

which gives

$$\phi_A=[1-P_{AA}]^{-1}(\phi_A^{\rm INC}+P_{AD}\phi_D). \quad (9)$$

Thus, we have converted the radiation condition at infinity into a boundary constraint on Γ_A.

Substituting (9) into (5) and defining

$$K'_{DD}=K_{DD}+K_{DA}[1-P_{AA}]^{-1}P_{AD}$$

$$\psi_D'^{\rm INC}=-K_{DA}[1-P_{AA}]^{-1}\phi_A^{\rm INC}$$

one obtains

$$\begin{bmatrix} K'_{DD} & K_{DI} & K_{DC} \\ K_{ID} & K_{II} & K_{IC} \\ K_{CD} & K_{CI} & K_{CC}\end{bmatrix}\begin{bmatrix}\phi_D\\ \phi_I\\ \phi_C\end{bmatrix}=\begin{bmatrix}\psi_D'^{\rm INC}\\ 0\\ 0\end{bmatrix}. \quad (10)$$

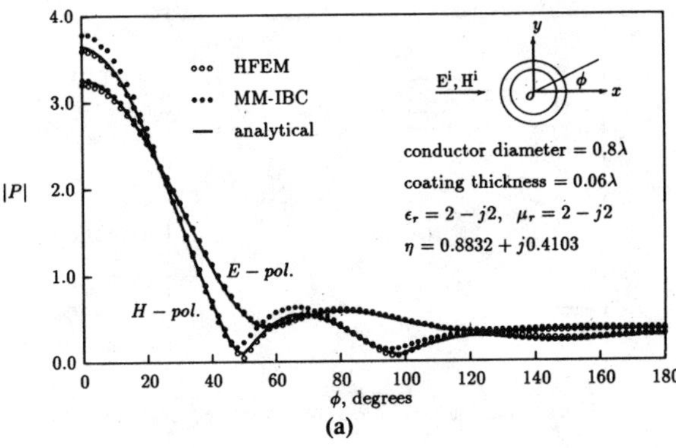

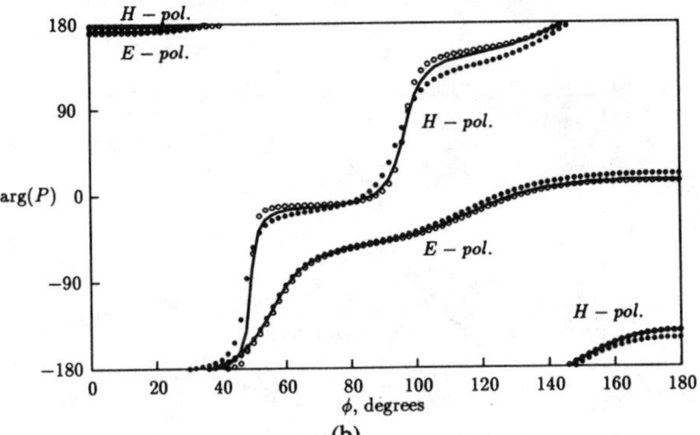

Fig. 2. Bistatic scattering of a coated circular cylinder. (a) Amplitude. (b) Phase.

The first row in (5) has been discarded here. It is noted that the continuity condition is enforced by using the same field representation (3) in the overlapping region enclosed by Γ_A and Γ_B for both the exterior and interior fields.

Numerically, for H-polarized case, the results are obtained by solving (10) since the second kind boundary condition is incorporated as natural boundary condition in the variational formulation. For E-polarized case, the matrix equation is smaller, since $\phi_C = 0$ on Γ_C. Once the nodal fields are found, the far field is calculated from (6) using large-argument approximation for the Hankel functions.

Numerical Results

Based on the formulation described above, a computer program was written to solve for bistatic and backscattered fields from coated cylinders with arbitrary cross sections for both E- and H-polarizations. The program has been used to compute numerous examples and three of them are presented here to show the validity and versatility of the technique. For the results shown here, the incident wave is assumed to be a plane wave and the coating homogeneous one. However, the program can handle the case of nonplane wave incidence and inhomogeneous materials as well.

A Coated Circular Cylinder

This is the case for which an exact solution is available [1] and hence is used to verify the program. Fig. 2 shows the results of numerical and exact (series) computations for both amplitude and phase of far-field factors [14] for E- and H-polarizations. The dimension of the cylinder, the coating thickness, and the coating electric properties are given in Fig. 2(a). For numerical computation the subdivision information is as follows: $k'h = 0.38$ in the direction normal to the surface and 1.14 along the circumference of the cylinder, where k' is the real part of the wavenumber $k(=\sqrt{\epsilon_r\mu_r}k_0)$ and h is the edge length of the triangular elements. The dimension of the resultant matrix equation is 64 for E-polarization and 96 for H-polarization. Also given are the results computed by using the moment method (MM) with impedance boundary condition (IBC) specified at the cylinder surface. The impedance value used is that appropriate for a coated conducting plane at normal incidence. In the figure the solid line shows the exact (analytical) data, and in general, both the HFEM and MM-IBC show a good agreement.

TABLE I
COMPARISON OF HFEM WITH MM IN COMPUTING THE SCATTERING FAR FIELD FROM A DIELECTRIC COATED CIRCULAR CYLINDER

Method	[matrix dim]	Forward field P: $\lvert P\rvert$	error	$\arg(P)$	err	Backward field P: $\lvert P\rvert$	error	$\arg(P)$	err
Analytical		3.334		161.1°		1.192		53.2°	
MM	[60×60]	3.374	1.20%	160.8°	0.3°	1.183	0.76%	53.8°	0.6°
MM	[72×72]	3.364	0.90%	160.9°	0.2°	1.191	0.08%	53.7°	0.5°
HFEM*	[48×48] [24×24]	3.303	0.93%	160.9°	0.2°	1.196	0.34%	52.8°	0.4°
HFEM*	[64×64] [32×32]	3.315	0.57%	161.0°	0.1°	1.195	0.25%	52.9°	0.3°

* In HFEM, two matrices are inverted: one is in (9) and the other in (10).

It might be of interest to compare the HFEM with volume MM solutions. Since the only volume MM program available to us is for E-polarized case and dielectric coating ($\mu_r = 1$), computations were made for the same size cylinder shown in Fig. 2, but for $\epsilon_r = 4 - j4$ and $\mu_r = 1$. The results are compared with HFEM and analytical data in forward and backward scattering directions and are presented in Table I.

Comparison shows that HFEM has essentially the same efficiency and accuracy as MM in this E-polarized case. However, using MM in the case of H-polarization, one would have to compute both normal and tangential equivalent electric current components and thus have a larger resultant matrix (almost twice as large) [15], [16]. Furthermore, when one considers materials with both ϵ and μ different from ϵ_0 and μ_0, the dimension of the matrix will become even larger (approximately three times) since there would be three current components to be considered in each cell within the material [16]. With HFEM technique the size of the matrix remains the same no matter what the material is: dielectric, magnetic or both.

A Coated Rectangular Cylinder

In this case, we will also examine results using HFEM and MM with IBC approximation. First, using HFEM the backscattering radar cross sections (σ) of bare (inner) and coated cylinders were computed and are shown in Fig. 3. Then the same were computed using MM with IBC. Again the normal incidence impedance value, η, for coated conducting plane

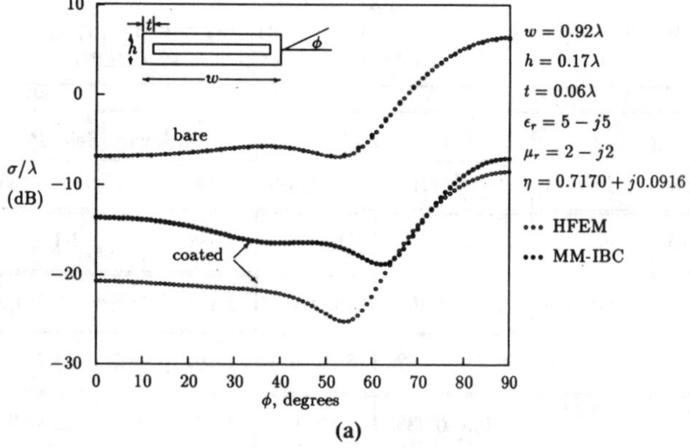

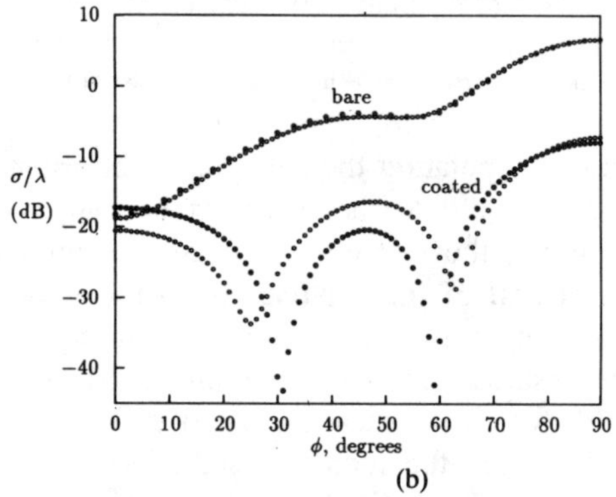

Fig. 3. Radar cross section for backscattering from coated and bare rectangular cylinders. (a) *E*-polarization. (b) *H*-polarization.

was used and the results are also given in Fig. 3. As expected, the MM for bare cylinder agrees very well with HFEM. However, for coated cylinders, the MM-IBC and HFEM solutions are quite different, especially around edge-on incidence. Recalling the better agreement between MM-IBC and the exact results for the coated circular cylinder case (see Fig. 2), we conclude that this discrepancy is caused by improper impedance value approximation at the edges. Techniques such as HFEM could be used to compute the fields and hence the η near such edges from which then better guideline to predict the η could be developed.

To reaffirm further the validity of HFEM solution on coated rectangular cylinder, we ran the same geometry case as in Fig. 3, but with $\epsilon_r = 4 - j4$ and $\mu_r = 1$ using HFEM, volume MM and MM with IBC, all for *E*-polarization. Results are shown in Fig. 4, and there we see an agreement within 0.5 dB between the HFEM and volume MM computations, but MM-IBC again differs by about 4 dB implying the shortcomings of IBC.

An Ogival Cylinder with Coated Edges

An ogival cylinder shape is often used to study the scattering behavior from bare and coated edges. Here we present a sample computation using HFEM for cases of bare, single-, and double-edge coated cylinders. In these cases the coating is homogeneous lossy dielectric, but of varying thickness as shown in a sketch in Fig. 5, where the computed backscattering is presented. The ogival cylinder is 1.0 wavelength wide and the maximum dielectric thickness at the edge is 0.1 (free-space) wavelength. From the results we see that the single and double-edge coating has very little effect on the return at

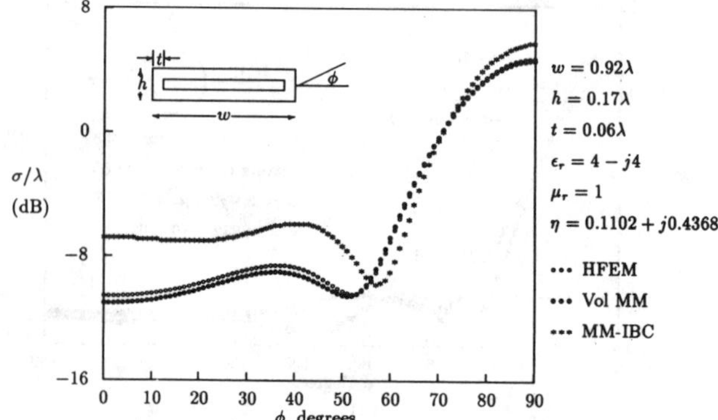

Fig. 4. Backscattering cross section of a dielectric coated rectangular cylinder.

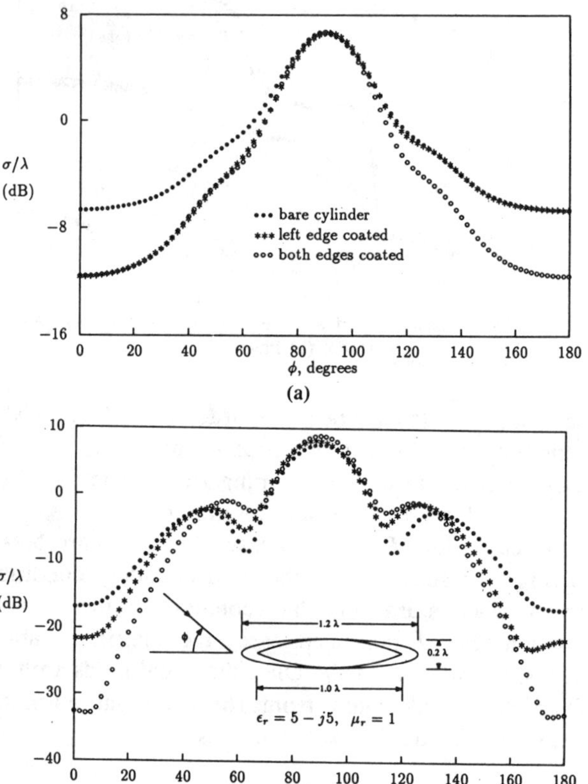

Fig. 5. Radar cross section for backscattering from ogival cylinders. (a) *E*-polarization. (b) *H*-polarization.

broadside incidence, but at edge-on the double-edge coating decreased the return by some 5 and 16 dB for *E*- and *H*-polarizations, respectively. By examining the return at edge-on from the single coated edge case it is evident that for *E*-polarization the leading edge is the dominant scatterer since coating the leading edge reduces the return ($\phi = 0°$) while by coating the rear edge the return ($\phi = 180°$) is unchanged. In case of *H*-polarization usually the dominant scatterer is the rear edge (due to traveling wave effects), but from this example such is inconclusive since the width of the cylinder is too small to launch and sustain appreciable traveling waves.

Conclusion

In this paper, the formulation of hybrid finite element method with application to the problem of scattering by dielectric/magnetic material coated cylinders is described. The method presented relaxes the restriction from the requirement that the artificial boundaries be circles to a condition that the boundries can follow the shape of the scatterer, and thus solves the problem numerically more efficiently. In addition to computations of scattered fields, validity of the impedance boundary conditions is briefly examined. By combining HFEM with physical optics formulations, a technique can be developed to solve electromagnetic scattering from electrically large geometries, such as coated wedges and half-planes.

References

[1] C. H. Tang, "Backscattering from dielectric-coated infinite cylindrical obstacles," *J. Appl. Phys.*, vol. 28, no. 5, pp. 628–633, May 1957.

[2] E. F. Knott and T. B. A. Senior, "Non-specular radar cross section study," Univ. Michigan Radiation Lab. Rep. 011764-1-T(AFAL-TR-73-422), Jan. 1974.

[3] Z. J. Csendes and P. Silvester, "Numerical solution of dielectric loaded waveguides: I-Finite element analysis," *IEEE Trans. Microwave Theory Tech.*, vol. MTT-18, pp. 1124–1131, Dec. 1970.

[4] P. Silvester and M. S. Hsieh, "Finite-element solution of 2-dimensional exterior field problems," *Proc. Inst. Elec. Eng.*, vol. 118, pp. 1743–1747, Dec. 1971.

[5] B. H. McDonald and A. Wexler, "Finite-element solution of unbounded field problems," *IEEE Trans. Microwave Theory Tech.*, vol. MTT-20, pp. 841–847, Dec. 1972.

[6] ——, *Finite Element in Electrical and Magnetic Field Problem*, M. V. K. Chari and P. P. Silvester, Eds. London, New York: Wiley, 1980, ch. 9.

[7] T. Orikasa, S. Washisu, T. Honma, and I. Fukai, "Finite element method for unbounded field problems and application to two-dimensional taper," *Int. J. Num. Meth. Eng.*, vol. 19, pp. 157–168, 1983.

[8] S. K. Jeng and C. H. Chen, "On variational electromagnetics: theory and application," *IEEE Trans. Antennas Propagat.*, vol. AP-32, pp. 902–907, Sept. 1984.

[9] Y. N. Guo, "Finite-element solution of electromagnetic scattering from inhomogeneous dielectric cylinder with arbitrary cross section," M.S. thesis, Nanjing Univ., Nanjing, China, 1985.

[10] K. K. Mei, "Unimoment method of solving antenna and scattering problems," *IEEE Trans. Antennas Propagat.*, vol. AP-22, pp. 760–766, Nov. 1974.

[11] S. K. Chang and K. K. Mei, "Application of the unimoment method to electromagnetic scattering of dielectric cylinders," *IEEE Trans. Antennas Propagat.*, vol. AP-24, pp. 35–42, Jan. 1976.

[12] M. A. Morgan and K. K. Mei, "Finite-element computation of scattering by inhomogeneous penetrable bodies of revolution," *IEEE Trans. Antennas Propagat.*, vol. AP-27, pp. 202–214, Mar. 1979.

[13] M. A. Morgan, C. H. Chen, S. C. Hill, and P. W. Barber, "Finite element-boundary integral formulation for electromagnetic scattering," *Wave Motion*, vol. 6, no. 1, pp. 91–103, Jan. 1984.

[14] J. J. Bowman, T. B. A. Senior, and P. L. E. Uslenghi, Eds., *Electromagnetic and Acoustic Scattering by Simple Shapes.* Amsterdam: North-Holland, 1969.

[15] J. H. Richmond, "TE-wave scattering by a dielectric cylinder of arbitrary cross-section shape," *IEEE Trans. Antennas Propagat.*, vol. AP-14, pp. 460–464, July 1966.

[16] R. F. Harrington, *Field Computation by Moment Methods.* New York: Macmillan, 1968.

A Note on Hybrid Finite Element Method for Solving Scattering Problems

J. M. JIN and V. V. LIEPA, MEMBER, IEEE

Abstract—**The hybrid finite element formulation is modified so that it results in a sparse banded symmetric matrix. This modification substantially improves the computational efficiency and enhances the capability of the method, which is demonstrated by numerical examples. A comparison with other numerical techniques is presented.**

I. INTRODUCTION

The hybrid finite element method (HFEM) was first introduced by Silvester and Hsieh [1] and McDonald and Wexler [2] in the early 1970's, and has been further improved and applied to solving various two-dimensional unbounded field problems, e.g., [3]–[9]. The method expresses the exterior fields using an eigenfunction series or an integral involving a Green's function. Such expressions are imposed on the variational equation or, more conveniently, on the matrix equation derived using the finite element method for the interior fields. As a result, HFEM produces a system matrix having a nonuniform block submatrix structure. Such a matrix, as pointed out by Mei and Morgan [10], [11], is not easily adaptable to banded matrix algorithms, and hence is numerically inefficient to solve using direct or iterative methods. As a consequence, HFEM in such a form is less efficient than the unimoment method [10]–[12] and its modification—the so-called hybrid finite element-extended boundary condition method (hybrid FEM–EBCM) [13], mainly because the unimoment method and the hybrid FEM–EBCM can take advantage of generating numerical basis functions by solving sparse or uniformly banded matrices.

In this communication, the HFEM formulation is modified in such a way that it also results in a sparse or uniformly banded matrix, rather than a partly full and partly sparse nonuniform matrix. The particular formulation used is that first proposed by McDonald and Wexler [2] and thereafter improved and used by others [5]–[9]. The modification is accomplished by changing the sequence of matrix substitutions, a procedure that has been discussed by McDonald and Wexler [6] for general HFEM formulations. The implementation is straightforward, but it substantially improves the computational efficiency and enchances the capability of the method. In the following we consider the technique applied to electromagnetic scattering, present sample computations, and compare with other methods.

II. FORMULATION

Consider a two-dimensional scattering problem illustrated in Fig. 1, where the contour Γ encloses the scatterer, and the contour Γ_A is an artificial boundary enclosing Γ. The details of the HFEM analysis of this problem can be found in [9]; however, for brevity, the notation used here is somewhat different. Application of the finite element analysis to the region enclosed by Γ_A results in the matrix equation

$$[K_{IA}]\{\phi_A\}+[K_{II}]\{\phi_I\}=0 \tag{1}$$

where the subscript A denotes the nodes on Γ_A, I the nodes interior to Γ_A, and $\{\phi\}$ is the discretized unknown field vector. There are no nodes between Γ_A and Γ. Equation (1) is referred to as the finite element equation.

To solve (1), another equation is needed to relate $\{\phi_A\}$ and $\{\phi_I\}$. Such an equation can be obtained by using a surface integral equation involving the free-space Green's function, which in discretized form is written as

$$[P_{AA}]\{\phi_A\}+[P_{AI}]\{\phi_I\}=\{\phi_A^{\text{inc}}\} \tag{2}$$

where $\{\phi_A^{\text{inc}}\}$ is the known incident field vector on boundary Γ_A. Equation (2) is referred to as the equivalent boundary constraint on Γ_A.

The solution for $\{\phi_A\}$ and $\{\phi_I\}$ can then be found by jointly solving (1) and (2). There are two approaches. The first, which is the one commonly used, imposes (2) on (1) and gives the final system equation

$$[K'_{II}]\{\phi_I\}=\{\psi_I\} \tag{3}$$

where

$$[K'_{II}]=[K_{II}]-[K_{IA}][P_{AA}]^{-1}[P_{AI}],\ \{\psi_I\}=-[K_{IA}][P_{AA}]^{-1}\{\phi_A^{\text{inc}}\}.$$

In this approach, two matrices have to be solved: one is complex and full matrix $[P_{AA}]$ having size of $N_A \times N_A$, and the other is a partly full and partly sparse complex matrix $[K'_{II}]$ having size of $N_I \times N_I$, where N_A is the total number of nodes on Γ_A and N_I is the total number of nodes interior to Γ_A. Usually, N_I is much larger than N_A, and hence the size of the scatterer to be treated by HFEM is mostly limited by the magnitude of N_I.

The second approach, which we present here, substitutes (1) into (2) and gives the equation

$$[P'_{AA}]\{\phi_A\}=\{\phi_A^{\text{inc}}\}$$

where

$$[P'_{AA}]=[P_{AA}]-[P_{AI}][K_{II}]^{-1}[K_{IA}].$$

Mathematically, this second approach is equivalent to the first one; however, computationally it is much more efficient. Here, one also needs to solve two matrices: one is a complex and full matrix $[P'_{AA}]$, but now the other is a symmetric and sparse matrix $[K_{II}]$, which becomes real-valued for lossless scatterers and can be narrowly banded if one numbers the nodes properly. A more obvious comparison is given in Table I. The difference between the first approach and the second approach is in the properties of the matrices $[K'_{II}]$ and $[K_{II}]$. Solving a symmetric, sparse or uniformly banded matrix $[K_{II}]$ is of course much easier and more efficient than solving a nonsymmetric matrix $[K'_{II}]$ with nonuniform block submatrix structure.

Three of the algorithms available for solving the matrix $[K_{II}]$ are briefly discussed here: the banded matrix algorithm, the frontal solution algorithm, and the Lanczos algorithm. The banded matrix algorithm decomposes the matrix into an upper triangular form, and then solves the equation by back substitution. In this process, neither storage nor operations are performed on zeros outside the band. This

Reprinted from *IEEE Trans. Antennas Propagat.*, vol. 36, no. 10, pp. 1486–1490, Oct. 1988.

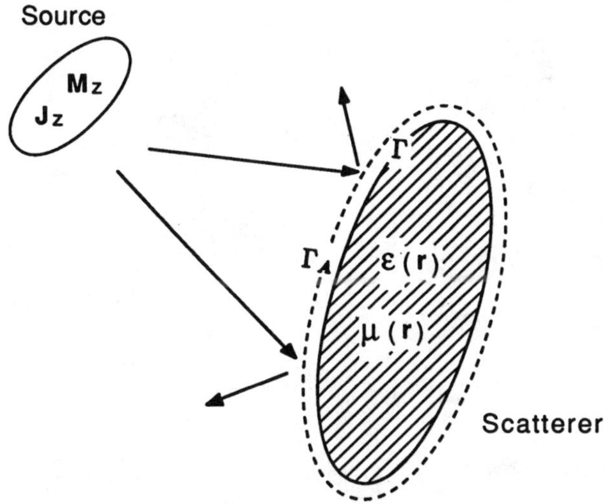

Fig. 1. Geometry of a general two-dimensional scattering problem.

TABLE I
COMPARISON OF THE TWO APPROACHES

Approach	Matrix to be solved	Matrix order	Properties of matrices
First approach	$[P_{AA}]$	N_A	Complex and full
	$[K'_{II}]$	N_I	Complex, partly full and partly sparse
Second approach	$[P'_{AA}]$	N_A	Complex and full
	$[K_{II}]$	N_I	Symmetric, sparse or banded. Real for lossless materials and complex for lossy materials

algorithm is used for computations presented in the next section. The frontal solution algorithm [14] applies Gaussian elimination during the assembly process of the matrices $[K_{II}]$ and $[K_{IA}]$ until all nodes have been processed. The unknowns are then solved by back substitution. This algorithm greatly reduces the storage requirement and avoids operations on zeros. The Lanczos algorithm, in contrast to the above two direct methods, solves the equation by iteration. Convergence is guaranteed if $[K_{II}]$ is a positive definite matrix. This algorithm performs computations such as $[K_{II}]\{p_I\}$ by a summation $\Sigma_{e=1}^{M} K_{II}^e\{p_I\}$, where K_{II}^e is the element matrix for the element e, M is the total number of elements, and $\{p_I\}$ denotes a vector used in the algorithm. Hence, like the frontal solution algorithm, this algorithm does not actually assemble the matrix K_{II}, thus minimizing the storage requirement and avoiding operations on zeros. A more detailed discussion of the three algorithms can be found in [15, pp. 54-72]. Of the three algorithms, the banded matrix one is most efficient for a matrix with dimension up to a thousand. However, it requires larger storage than do the other two. The frontal solution and the Lanczos algorithms are better for solving large scatterer problems having thousands of unknowns. Even though their programming is complicated, some well developed programs are available.

III. Numerical Results

To show the capability of the modified HFEM formulation, we present here four numerical examples, each of which contains about two thousand unknowns. The first example is a coated circular cylinder with the radius of the conducting cylinder of 1.0 λ (free-space wavelength) and the coating thickness of 0.5 λ (with $\epsilon_r = 2.0$, $\mu_r = 2.0$). The numerical results for bistatic scattering, along with the exact solutions, are shown in Fig. 2. Considering the subdivision information given below and the fact that the coating is lossless, this is good agreement. The results were obtained with $N_A = 188$ and $N_I = 2068$. The area of the coating was subdivided into 3760 triangular elements with the largest edge length one-tenth wavelength in the material.

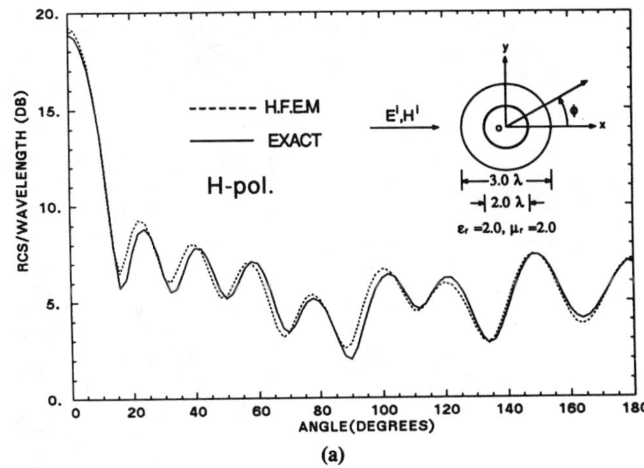

(a)

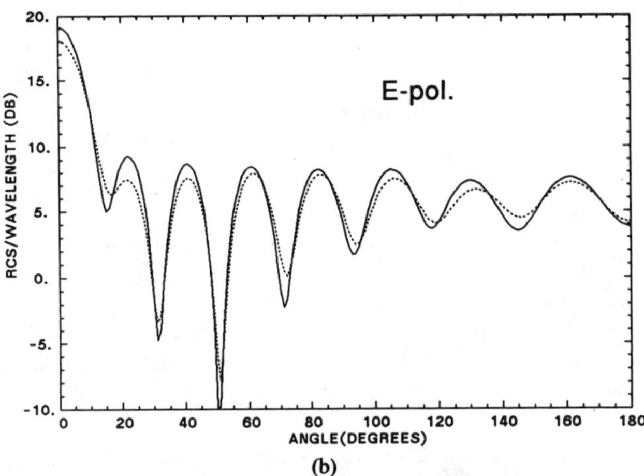

(b)

Fig. 2. Bistatic scattering pattern of a coated circular cylinder. (a) *H*-polarization. (b) *E*-polarization.

The second example is an inhomogeneous dielectric circular cylinder having radius 1.2 λ and the permittivity of the dielectric varying in the radial direction according to $\epsilon_r = (2.4\lambda - r)/r$. Since the permittivity tends to infinity at the center, a conducting cylinder of radius 0.2 λ is used to replace the dielectric there. The results for bistatic scattering are shown in Fig. 3. For computation, the cross section of the dielectric cylinder was subdivided into 3948 triangular elements, resulting in $N_A = 94$ and $N_I = 2068$.

The third example is an inhomogeneous rectangular cylinder of size 3.8 λ × 1.2 λ. The permittivity of the cylinder varies along the y-direction as $\epsilon_r = 1 + \cos(y\pi/a)$, with $a = 3.8$ λ being the height of the cylinder; the permeability is a constant, having $\mu_r = 1.5$. Both bistatic and backscattering results are given in Fig. 4. In this case, the number of triangular elements is 3648, resulting in $N_A = 200$, $N_I = 1925$.

The last example is an inhomogeneous triangular cylinder with the length of each side 2.77 λ. The permittivity and permeability vary

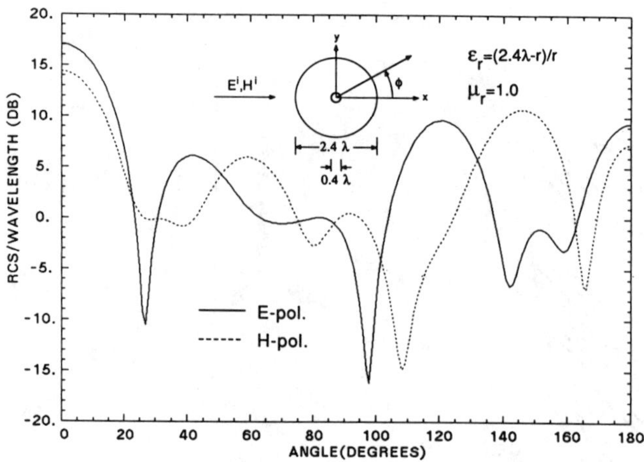

Fig. 3. Bistatic scattering pattern of an inhomogeneous dielectric circular cylinder.

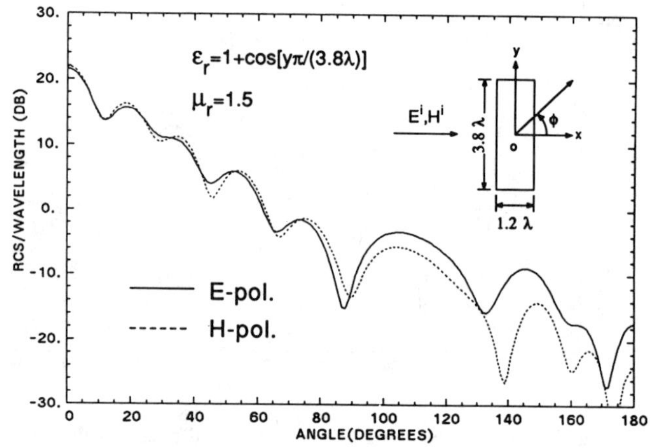

Fig. 4. Plane wave scattering by an inhomogeneous rectangular cylinder. (a) Bistatic scattering. (b) Backscattering.

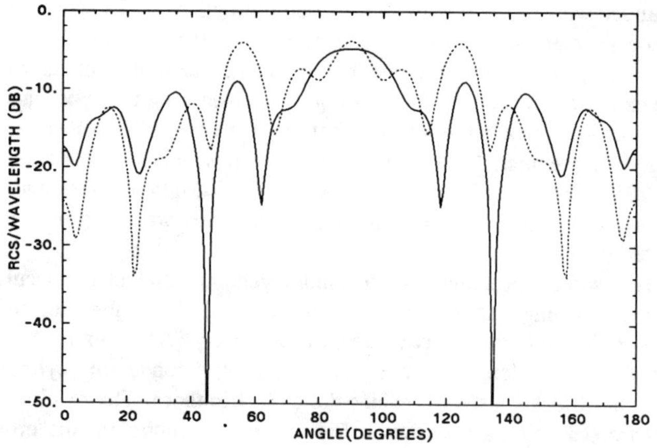

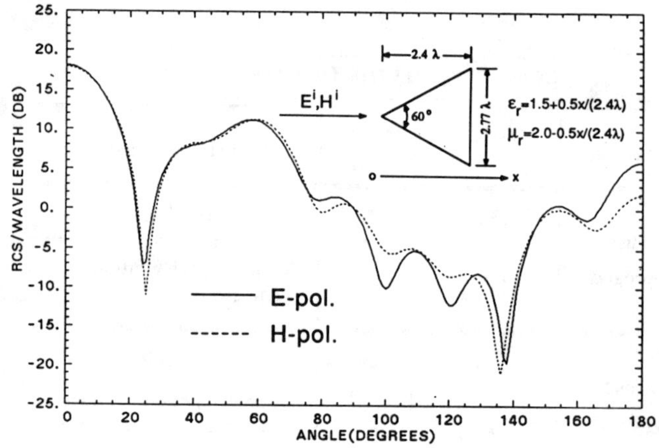

Fig. 5. Plane wave scattering by an inhomogeneous triangular cylinder. (a) Bistatic scattering. (b) Backscattering.

along the x-direction as $\epsilon_r = 1.5 + 0.5x/h$ and $\mu_r = 2.0 - 0.5x/h$, respectively, where h $(= 2.4\ \lambda)$ is the height of the triangle. The results are presented in Fig. 5, again for both bistatic and backscattering. In this computation, the number of triangular elements is 3844, resulting in $N_A = 186$, $N_I = 2016$.

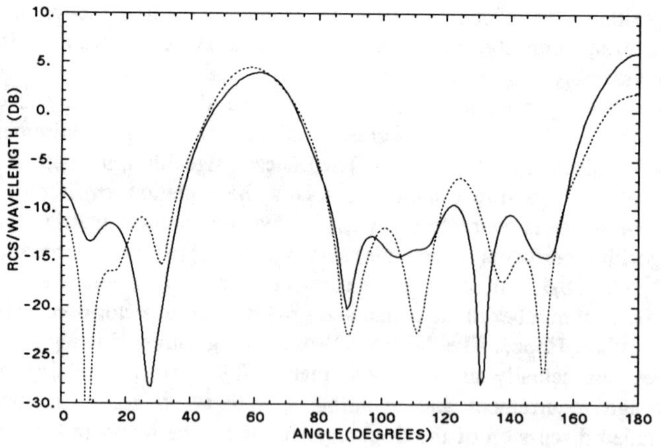

All above computations were performed on an Apollo Domain workstation (using Domain series 3000 and 4000 machines). The symmetry property of the first and second examples was not utilized, though it could have been easily adopted to reduce the number of unknowns.

TABLE II
COMPARISON OF THE METHODS

Method	Modified HFEM	Unimoment method	VIE method
Elements to generate	$4N_I+2N_A+2N_A^2$	$4N+2N_C+4Q^2$	$9N_I^2$
Matrices to solve	$N_I \times B$, $N_A \times N_A$	$N \times B'$, $2Q \times 2Q$	$3N_I \times 3N_I$

IV. Discussion

In order to give an objective evaluation of the present HFEM, we compare it with two other methods: the unimoment method [12] and the conventional volume integral equation (VIE) method [16], both of which are widely used to solve scattering by inhomogeneous cylinders having both ϵ and μ different from their free-space values.

In Table II we list the number of nonzero matrix elements that need be generated and the size of the matrices to be solved in each of the three methods. In the HFEM, $[K_{II}]$ has about $7N_I$ nonzero elements, since a node is usually connected to six adjacent nodes, and due to the symmetry of the matrix, only $4N_I$ elements need be generated. $2N_A$ is the number of nonzero elements in $[K_{IA}]$, and $2N_A^2$ the number of nonzero elements in $[P_{AA}]$ and $[P_{AI}]$. As previously shown, two matrices need to be solved: one is $[P'_{AA}]$ of size $N_A \times N_A$, and another one is the symmetric and sparse matrix $[K_{II}]$, which in banded form is of size $N_I \times B$ with B being the half-bandwidth (including the diagonal) of the matrix. In the unimoment method, as originally developed, an artificial circle is drawn to enclose the entire scatterer. Assume N is the total number of the interior nodes and N_C the total number of boundary nodes. One needs to generate $4N$ nonzero elements for the matrix $[Q]$ and $2N_C$ for the matrix $[T]$ (see [12, eq. (21)]. $4Q^2$ is the number of elements in a matrix of size $2Q \times 2Q$, resulting from the continuity conditions on the circle, where Q denotes the number of harmonics used to expand the field in the exterior region. In this method, also two matrices need to be solved: one is that with the size of $2Q \times 2Q$, and the other is the symmetric and sparse matrix $[Q]$, which in banded form has the size $N \times B'$, where B' is its half-bandwidth. In the VIE method, one needs to generate and solve a matrix of the size $3N_I \times 3N_I$, since at each node there are three unknown equivalent polarized current components.

In order to compare the present HFEM with the unimoment method we need to consider three cases. For a scatterer having its cross section close to a circle, the magnitudes of N_I and N are about the same, and so are the magnitudes of B and B'. Since the number $2Q$ can be smaller than N_A for far-field calculation, the unimoment method in this case is more efficient than the HFEM. However, for a slender scatterer whose cross section substantially deviates from a circle, N is much larger than N_I, and B' is also larger than B. Even though the number $2Q$ is possibly smaller than N_A, the HFEM in this case is expected to be competitive with the unimoment method. Finally, for the case of multiple scatterers, and especially when the scatterers are far apart, the HFEM may prove to be more efficient. This is because in the HFEM the artificial boundary Γ_A can be split into several contours, each enclosing one scatterer, while in the unimoment method a single circle is used to enclose all scatterers. However, a modification to the unimoment method, which uses a boundary conforming to the surface of the scatterer, would enhance its efficiency. One such modification—the hybrid FEM-EBCM was presented by Morgan *et al.* [13] for the case of dielectric bodies of revolution, and if applied to two-dimensional problems it would result in an efficient numerical technique.

Comparing the present HFEM with the conventional VIE method, we find that the HFEM, like the unimoment method and the hybrid FEM-EBCM, is far more efficient than the VIE method. For the problems of size comparable to those considered in Section III, using the VIE method, one would have to generate many more matrix elements, about four hundred times the number needed in the HFEM. The computational difference in solving the matrices is expected to be even larger, though no accurate number can be stated, since the computing time depends on the algorithm chosen for solving the matrix. We note that for the case of dielectric ($\mu_r = 1$) cylinders, the number of unknowns in the VIE method reduces to $2N_I$ for H-polarization and to N_I for E-polarization. We also note that the volume integral equation involving three equivalent polarized current components can be transformed to an integral equation having both volume and surface integrals but only involving one axial field component. Such an equation will result in a technique more efficient than the VIE method; however, it still would not be competitive with either the HFEM or the unimoment method.

V. Conclusion

In this paper, a modified formulation of HFEM is presented, which is shown superior to the previous HFEM in that it takes advantage of the important property of the finite element equation, i.e., its coefficient matrix is symmetric, sparse and banded. The modification improves the computational efficiency and enhances the capability of the method, as demonstrated by four numerical examples. The modified HFEM is objectively evaluated by comparing it with other methods. We conclude that, in addition to the unimoment method and the hybrid FEM-EBCM, the modified HFEM provides an efficient method to solve two-dimensional unbounded field problems involving inhomogeneous media.

Acknowledgment

The authors wish to thank the reviewer for his comments, which led to a comparison given in Section IV, and technical writer Mr. Rod Johnson for his help and suggestions.

References

[1] P. Silvester and M. S. Hsieh, "Finite-element solution of 2-dimensional exterior field problems," *Proc. Inst. Elec. Eng.*, vol. 118, pp. 1743–1747, Dec. 1971.

[2] B. H. McDonald and A. Wexler, "Finite-element solution of unbounded field problems," *IEEE Trans. Microwave Theory Tech.*, vol. MTT-20, pp. 841–847, Dec. 1972.

[3] S. P. Marin, "Computing scattering amplitudes for arbitrary cylinders under incident plane waves," *IEEE Trans. Antennas Propagat.*, vol. AP-30, pp. 1045–1049, Nov. 1982.

[4] S. K. Jeng and C. H. Chen, "On variational electromagnetics: Theory and application," *IEEE Trans. Antennas Propagat.*, vol. AP-32, pp. 902–907, Sept. 1984.

[5] S. Washisu, I. Fukai, and M. Suzuki, "Extension of finite-element method to unbounded field problems," *Electron. Lett.*, vol. 15, pp. 772–774, 1979.

[6] B. H. McDonald and A. Wexler, *Finite Element in Electrical and Magnetic Field Problem*, M. V. K. Chari and P. P. Silvester, Eds. London and New York, Wiley, 1980, ch. 9.

[7] T. Orikasa, S. Washisu, T. Honma, and I. Fukai, "Finite element method for unbounded field problems and application to two-dimensional taper," *Int. J. Num. Meth. Eng.*, vol. 19, pp. 157–168, 1983.

[8] Y. N. Guo, "Finite-element solution of electromagnetic scattering from inhomogeneous dielectric cylinder with arbitrary cross section," M.S. thesis, Nanjing Univ., 1985.

[9] J. M. Jin and V. V. Liepa, "Application of hybrid finite element method to electromagnetic scattering from coated cylinders," *IEEE Trans. Antennas Propagat.*, vol. 36, pp. 55–70, Jan. 1988.

[10] K. K. Mei, "Unimoment method of solving antenna and scattering problems," *IEEE Trans. Antennas Propagat.*, vol. AP-22, pp. 760–766, Nov. 1974.

[11] M. A. Morgan and K. K. Mei, "Finite-element computation of scattering by inhomogeneous penetrable bodies of revolution," *IEEE Trans. Antennas Propagat.*, vol. AP-27, pp. 202–214, Mar. 1979.

[12] S. K. Chang and K. K. Mei, "Application of the unimoment method to

electromagnetic scattering of dielectric cylinders," *IEEE Trans. Antennas Propagat.*, vol. AP-24, pp. 35–42, Jan. 1976.

[13] M. A. Morgan, C. H. Chen, S. C. Hill, and P. W. Barber, "Finite element-boundary integral formulation for electromagnetic scattering," *Wave Motion*, vol. 6, no. 1, pp. 91–103, Jan. 1984.

[14] B. M. Irons, "A frontal solution program for finite element analysis," *Int. J. Num. Meth. Eng.*, vol. 2, pp. 5–32, 1970.

[15] G. Sewell, *Analysis of Finite Element Method: PDE/PROTRAN.* New York: Springer-Verlag, 1985.

[16] R. F. Harrington, *Field Computation by Moment Methods.* New York: Macmillan, 1968.

The Bymoment Method for Two-Dimensional Electromagnetic Scattering

ANDREAS C. CANGELLARIS, MEMBER, IEEE, AND ROBERT LEE, STUDENT MEMBER, IEEE

Abstract—**The bymoment method is presented for the analysis of two dimensional electromagnetic wave scattering in an unbounded region. The method introduces a conforming surface to geometrically decouple the interior region containing the scatterer from the exterior region extending to infinity. The solution in the interior is generated using the standard finite-element solution of an interior Dirichlet boundary-value problem. The interior solution is then coupled to the exterior by an elegant use of Green's theorem in the unbounded exterior region. The validity of the method and the associated computer program is demonstrated by comparing our results to those of other methods for scattering by various cylindrical geometries.**

I. INTRODUCTION

THE APPLICATION of the finite-element method for the analysis of electromagnetic wave scattering by complex structures has been a major research topic over the past 15 years. This is mainly due to the fact that the numerical implementation of the finite element method results in very sparse matrices, in contrast to the full matrices generated by the more traditional integral equation formulation. Sparse matrices are computationally attractive because they allow significant reductions in computer memory, as well as computation time. Therefore, the finite-element method appears to be the perfect candidate for the analysis of electromagnetic wave scattering by very large, complex structures.

The major obstacle in the application of the finite-element method to problems involving unbounded regions is that differential equation-based methods are always formulated as boundary-value problems. From a computational point of view, this implies that an artificial boundary is required for the truncation of the computational domain. However, such a boundary should be transparent to the solution in the sense that an appropriate mechanism should be used to couple the solution inside the truncation boundary to the solution outside, ensuring the proper continuity of the fields along the boundary, as well as the radiation condition at infinity. Such a method was first introduced by Silvester and Hsieh [1], and McDonald and Wexler [2]. In their formulation, a surface integral equation for the fields in the exterior region was imposed as a boundary constraint on the finite-element solution in the interior region. This led to the so-called hybrid finite-element methods (HFEM). Since then several modifications and/or extensions of the method have appeared [3]–[5]. One of the disadvantages of the HFEM was that the resulting matrices have nonuniform block submatrix structures that are not easily amenable to banded matrix algorithms. Recently, however, an alternative formulation of the final matrix equations has been proposed so that it results in a sparse, banded, symmetric matrix, thus improving considerably the numerical efficiency of the HFEM [6].

Another approach, introduced by Mei [7], for coupling the interior and exterior solutions was the unimoment method. This method was later applied extensively to electromagnetic scattering problems involving two-dimensional inhomogeneous structures, as well as inhomogeneous penetrable bodies of revolution [8], [9]. A further generalization of the method, which removed the axisymmetric restriction of the inhomogeneous media, was presented recently [10]. The unimoment method uses a separable surface for the truncation boundary to decouple the solution in the interior region where the scatterer is present, from the solution in the exterior. Then the fields in the interior are found by employing the finite-element method to solve a standard Dirichlet boundary-value problem, while appropriate eigenfunction expansions for the specific coordinate system are used to represent the fields in the exterior region. Finally, the interior and exterior problems are coupled by imposing the continuity of the tangential electric and magnetic fields on the separable surface. At this point, we must also mention the recently developed transfinite element method, which uses the basic ideas of the unimoment method with a different functional that results in a computationally efficient solution [11]. The use of a separable surface seemed to be an inherent limitation of the unimoment method since the finite-element solution within a circular/spherical surface becomes rather inefficient for elongated scatterers that occupy only a small portion of the enclosed volume. A new method was then developed by Morgan, Chen, Hill and Barber [12], which combined the finite-element solution of the interior region with the surface integral equation used in the extended boundary condition method to circumvent the need for a separable boundary. However, this hybrid method was found to have convergence difficulties when dealing with very elongated objects.

Recently, the field feedback formulation was proposed in an attempt to overcome some of the disadvantages of the aforementioned methods [13]. This method employs a surface that conforms to the scatterer for decoupling the interior and exterior regions. The solution in the interior is then found by solving an interior Dirichlet boundary value problem with the

Manuscript received May 12, 1989; revised November 27, 1989. This work was supported by Sandia National Laboratories under Document 01-6833.

The authors are with the Electromagnetics Laboratory, Department of Electrical and Computer Engineering, The University of Arizona, Tucson, AZ 85721.

IEEE Log Number 9036706.

Reprinted from *IEEE Trans. Antennas Propagat.*, vol. 38, no. 9, pp. 1429–1437, Sept. 1990.

prescribed values for the tangential electric and/or magnetic fields on the enclosing surface. These boundary conditions are given in terms of appropriate expansions of vector basis functions whose coefficients are found by coupling the interior region solution to the unbounded region making use of the equivalence principle.

An alternative approach was proposed recently [14] for coupling the exterior and interior field solutions. In a manner similar to the field feedback formulation, this method, which we name the bymoment method, decouples the exterior and interior regions by using a surface that conforms to the scatterer. On this surface, the tangential electric and/or magnetic fields are expressed in terms of a set of appropriate basis functions whose coefficients are to be determined. The interior solution is then generated in exactly the same manner as the unimoment method and is expressed in terms of the unknown coefficients from the expansion of the tangential fields on the enclosing surface. Finally, an application of Green's theorem over a volume enclosed by a surface just inside the finite-element mesh and the surface at infinity allows us to couple the interior solution to the exterior region and solve for the unknown coefficients.

In what follows, we present the bymoment method as it applies to scattering by infinite cylindrical structures. As will become apparent from the discussion in the following section, the bymoment method tries to combine the attractive features of all the aforementioned methods.

II. Formulation of the Bymoment Method

Let us consider the case of an infinitely long cylinder in free space with an arbitrary cross section. The material properties of this cylinder can be either homogeneous or inhomogeneous. The coordinate system is chosen such that the z-axis is parallel to the axis of the cylinder. The field incident on the cylinder is assumed to have no variation in z. This allows us to decouple the problem into TM_z (E_z, H_x, H_y) and TE_z (H_z, E_x, E_y) modes. We can therefore solve either the TM_z or TE_z problem by using a single potential. For the TE_z case we can relate H_z to the magnetic potential ϕ_M [15] by

$$H_z = -k^2\phi_M. \tag{1}$$

For the TM_z case the electric potential ϕ_E is related to E_z by

$$E_z = -k^2\phi_E. \tag{2}$$

The wavenumber is given by $k = k_0\sqrt{\mu_r\epsilon_r^*}$ where the relative dielectric constant incorporates the conduction loss and is given by $\epsilon_r^* = \epsilon_r - i\sigma/\omega\epsilon_0$. The variable k_0 denotes the wavenumber for free space. The $e^{i\omega t}$ time variation is suppressed. For the purposes of discussion, we shall use the function ϕ to represent either potential. Since the material properties of the cylinder can be inhomogeneous, k is obviously a function of position. However, in applying the finite-element method, we assume that k is inhomogeneous in each element. Therefore, the potential ϕ satisfies the Helmholtz equation

$$(\nabla^2 + k^2)\phi(x, y) = 0 \tag{3}$$

within each of these elements, with the assumption that no sources are present. The operator ∇^2 is the Laplacian in x and y.

To determine the scattering from the cylinder we must solve (3) for both the interior and exterior regions of the cylinder due to an incident field whose potential is given by ϕ^{inc}. The interior problem is evaluated by using the finite-element method. We construct the FEM mesh such that the cylinder is wholly contained inside of it with the boundary elements of the mesh exterior to the cylinder. A pictorial representation is given in Fig. 1. The line ∂S defines the boundary of the mesh. The line $\partial S'$ is totally enclosed by ∂S and passes through the interior of all the elements on the boundary of the mesh. We denote the region inside of ∂S by S_1. The region exterior to the mesh is denoted by S_2. To solve the finite element problem, boundary conditions must be specified on the nodes located along ∂S. To obtain these values we couple the interior problem to the exterior problem by using the bymoment method. This coupling is accomplished by representing the potential on ∂S as a sum of known expansion functions multiplied by unknown coefficients. We then determine the coefficients by applying Green's theorem to the exterior region.

A. Solution Along ∂S

Since the problem is a two-dimensional one, ∂S represents a line enclosing the surface S_1. We define the potential on ∂S to be ϕ_s. Let us assume that ϕ_s is known. Because the y coordinate of a point on the line ∂S is dependent on its x coordinate, we may write ϕ_s as a function of a single variable t. We allow t to vary from 0 to d where d is the length of ∂S. If $\phi_s \in L_2(\partial S)$, we can represent this potential by a complete set of linearly independent functions $\Psi_n(t) \in L_2(\partial S)$ as follows:

$$\phi_s = \sum_{n=1}^{\infty} a_n\Psi_n(t), \qquad t \in [0, d] \tag{4}$$

where a_n are coefficients which are known since ϕ_s is given. Even though ϕ_s is not known at this moment, the representation in (4) will be useful for the FEM problem in S_1.

B. Finite-Element Solution

Let us again assume that ϕ_s is known. In that case, the potential in the interior region is found using a standard finite-element solution of an interior Dirichlet boundary value problem. We use the method of weighted residuals [16] to formulate the finite-element expression. This guarantees that the Helmholtz equation in (3) is true in the weak sense. We can write the expression as follows:

$$\iint [(\nabla^2 + k^2)\phi\psi_j]\, dS = 0 \tag{5}$$

where the weighting functions ψ_j $(j = 1, 2\cdots)$ constitute a set of first-order differentiable, scalar functions which we choose. By using Green's first identity, expression (5) becomes

$$\iint [\nabla\phi \cdot \nabla\psi_j - k^2\phi\psi_j]\, dS - \int [\psi_j\hat{n} \cdot \nabla\phi]\, dl = 0 \tag{6}$$

where $\hat{n}$ is the outward unit normal vector. The potential ϕ is expanded in a set of orthogonal basis functions. We then

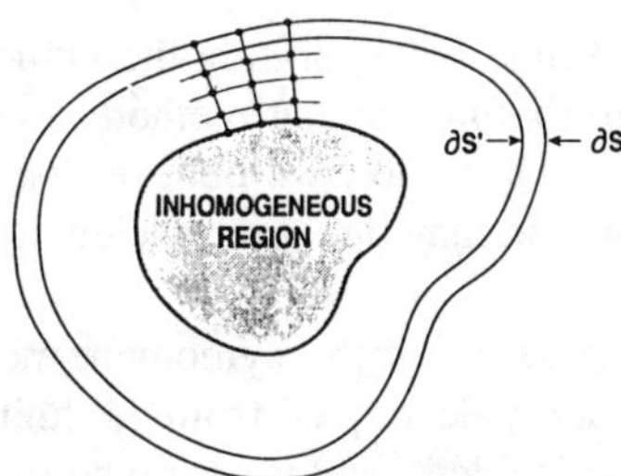

Fig. 1. Geometry: arbitrary cylinder in free space.

evaluate (6) over each element in the mesh. By applying the boundary conditions on the fields at the interfaces between adjacent elements, we can nullify the contributions from the line integral along these interelement boundaries. As an example let us consider the TM_z case. We choose the weighting functions ψ_j to be continuous between the elements. Unfortunately, $\hat{n} \cdot \nabla\phi$ is not necessarily continuous between the elements. It is given by

$$\hat{n} \cdot \nabla\phi = \frac{i\omega}{\epsilon_r^* k_0^2} H_t \tag{7}$$

where H_t represents the magnetic field component which is tangential to the interfaces separating the elements. Therefore, H_t is continuous between the elements. In order for the line integral in (6) to be continuous, we must multiply both sides of (5) by ϵ_r^*. This gives

$$\iint \epsilon_r^* [\nabla\phi \cdot \nabla\psi_j - k^2 \phi\psi_j] \, dS = 0. \tag{8}$$

Similarly, for the TE_z case, we obtain

$$\iint \mu_r [\nabla\phi \cdot \nabla\psi_j - k^2 \phi\psi_j] \, dS = 0. \tag{9}$$

This leaves only the line integral along the outer boundary ∂S. However, since we are solving the Dirichlet problem, the weighting functions ψ_j are chosen to be zero on ∂S; hence, the line integral along ∂S is zero and does not enter into the formulation. Therefore, the problem can be solved by evaluating a surface integral over each of the elements.

Let us now consider the case where ϕ_s is not known, as is actually the case. From the uniqueness theorem we know that, with the presence of any physical loss mechanism, the fields inside a given volume are determined uniquely by either the tangential electric or tangential magnetic fields on the surface enclosing the volume [15]. Since the potential is related to the tangential electric field for the TM_z case and the tangential magnetic field for the TE_z case, the potential in S_1 is uniquely determined by ϕ_s. Because of numerical considerations, we must truncate the sum in expression (4) to some finite value N. Therefore, we can approximately represent ϕ_s by N basis functions, each of which is multiplied by an unknown coefficient a_n. If we solve N finite-element problems with Ψ_n ($n = 1, 2, \cdots, N$) as the boundary condition for the nth finite-element problem, we obtain N numerical finite-element solutions which form a basis for the potential in S_1. We shall call each of the finite-element solutions a finite-element basis function. We denote the finite-element basis functions by $\Lambda_n(x, y)$ for $(x, y) \in S_1$. Because of linearity and the uniqueness theorem, we can write the potential as a superposition of the finite-element basis functions multiplied by the appropriate coefficient a_n. The expression is given by

$$\phi(x, y) = \sum_{n=1}^{N} a_n \Lambda_n(x, y), \qquad (x, y) \in S_1. \tag{10}$$

The choice for N is dependent upon convergence requirements and the choice of the basis functions Ψ_n. The coefficients a_n are determined by coupling the interior solution to the properties of the exterior region.

C. Determination of the Coefficients

To evaluate the coefficients, we must first define additional surfaces in our geometry. Let the surface enclosed by $\partial S'$ be denoted by S_1'. We also define S_2' to be the surface which encompasses the area of S_2 and the area between ∂S and $\partial S'$.

The potential in S_2' can be divided into a potential ϕ^{inc} due to the incident field and a potential ϕ^{sc} due to the scattered field. This is written as

$$\phi(x, y) = \phi^{\text{inc}}(x, y) + \phi^{\text{sc}}(x, y), \qquad (x, y) \in S_2'. \tag{11}$$

The potential ϕ^{sc} satisfies the Helmholtz equation

$$(\nabla^2 + k_0^2)\phi^{\text{sc}}(x, y) = 0, \qquad (x, y) \in S_2' \tag{12}$$

and the Sommerfeld radiation condition at infinity.

We now use (12) in applying Green's theorem to S_2'. This gives

$$\iint_{S_2'} [\Phi_j (\nabla^2 + k_0^2)\phi^{\text{sc}} - \phi^{\text{sc}} (\nabla^2 + k_0^2)\Phi_j] \, dS = \int_{\partial S'} \left[\Phi_j \frac{\partial \phi^{\text{sc}}}{\partial n} - \phi^{\text{sc}} \frac{\partial \Phi_j}{\partial n} \right] dl \tag{13}$$

where Φ_j ($j = 1, 2, \cdots$) are a set of linearly independent testing functions which are chosen to satisfy the Helmholtz equation over the surface S_2', i.e.,

$$(\nabla^2 + k_0^2)\Phi_j(x, y) = 0, \qquad (x, y) \in S_2' \tag{14}$$

and the radiation condition at infinity. Substituting (12) and (14) into (13), we see that the value of the surface integral is zero. This leaves only the line integral term,

$$\int_{\partial S'} \left[\Phi_j \frac{\partial \phi^{\text{sc}}}{\partial n} - \phi^{\text{sc}} \frac{\partial \Phi_j}{\partial n} \right] dl = 0. \tag{15}$$

By substituting (11) into (15) and then using (10) for ϕ, we may write the following equation:

$$\sum_{n=1}^{N} a_n \int_{\partial S'} \left[\Phi_j \frac{\partial \Lambda_n}{\partial n} - \Lambda_n \frac{\partial \Phi_j}{\partial n} \right] dl = \int_{\partial S'} \left[\Phi_j \frac{\partial \phi^{\text{inc}}}{\partial n} - \phi^{\text{inc}} \frac{\partial \Phi_j}{\partial n} \right] dl. \tag{16}$$

Analytical expressions for both ϕ^{inc} and Φ_j are given. We can therefore obtain analytical expressions for $\partial\phi^{\text{inc}}/\partial n$ and

$\partial \Phi_j/\partial n$. Also, the numerical values for both Λ_n and $\partial \Lambda_n/\partial n$ along $\partial S'$ can be obtained from the finite-element solution. We chose the line integral in (16) to be along $\partial S'$ rather than ∂S since we found that we obtained more accurate results when the integral is evaluated along the interior of the elements.

The only unknowns left in (16) are the coefficients a_n. By using the first N functions of Φ_j, we can form an $N \times N$ matrix equation to solve for the coefficients. We write the matrix equation as follows:

$$\begin{pmatrix} S_{11} & S_{12} & \cdots & S_{1N} \\ S_{21} & S_{22} & \cdots & S_{2N} \\ \vdots & \vdots & \ddots & \vdots \\ S_{N1} & S_{N2} & \cdots & S_{NN} \end{pmatrix} \begin{pmatrix} a_1 \\ a_2 \\ \vdots \\ a_N \end{pmatrix} = \begin{pmatrix} T_1 \\ T_2 \\ \vdots \\ T_N \end{pmatrix} \tag{17}$$

where S_{jn} and T_j are given by

$$S_{jn} = \int_{\partial S'} \left[\Phi_j \frac{\partial \Lambda_n}{\partial n} - \Lambda_n \frac{\partial \Phi_j}{\partial n} \right] dl, \tag{18}$$

$$T_j = \int_{\partial S'} \left[\Phi_j \frac{\partial \phi^{\text{inc}}}{\partial n} - \phi^{\text{inc}} \frac{\partial \Phi_j}{\partial n} \right] dl. \tag{19}$$

III. Numerical Considerations

One of the major questions regarding this method is its numerical efficiency. In the bymoment method, the finite-element problem must be solved N times for N different solutions on its boundary. This may seem numerically time consuming, but actually it is not. Since the only thing that changes in the N finite-element solutions is the forcing vector, the finite-element matrix must only be factored once. The only step which must be performed N times is the back-substitution. Since the computation time required for the back-substitution is small in comparison to the computation time required for the factorization of the matrix, the extra time needed to solve the N finite-element problems is not much greater than that needed to solve one problem. The bymoment method is superior to regular HFEM techniques since the finite-element matrix which results from HFEM is not banded. Therefore, let us compare the computational efficiency of the bymoment method to that of the modified HFEM which was developed by Jin and Liepa [6]. The modified HFEM manipulates the resulting matrix equations in such a manner that instead of solving a partly full and partly sparse complex matrix, it solves two matrices where one of the matrices is complex and full but of smaller dimension than the initial one, and the other is symmetric and sparse with the same dimension as the initial one. In [6], a comparison is given between the modified HFEM and the unimoment method showing that the unimoment method is more efficient for scatterers whose cross section is close to a circle and less efficient for slender scatterers whose cross section deviates substantially from that of a circle. This is due to the fact that the unimoment method requires a circular mesh, while the modified HFEM uses a mesh that conforms to the scatterer, resulting in a smaller finite-element matrix for slender scatterers. The bymoment method overcomes this disadvantage of the unimoment method while at the same time incorporating the numerically efficient algorithm used in the unimoment method to obtain the finite-element solutions. Therefore, we believe that in many cases, the bymoment method may be more efficient than the modified HFEM.

One further advantage of the bymoment method is that the incident field is totally decoupled from the finite-element solutions. Instead, the incident field is coupled through the surface integral in (16). Thus, the finite-element solutions need only be solved once for any number of incident fields. In what follows, we present numerical considerations which are specific to the bymoment method.

The choice of basis functions Ψ_n and the testing functions Φ_j have a major impact on the efficiency and accuracy of the method. For the basis functions, both entire-domain and subdomain functions can be used. Each choice has its advantages and disadvantages. The major advantage of using entire-domain basis functions is that, in most cases, we require a smaller N than in the subdomain case to obtain the same accuracy. Another advantage is the flexibility in choosing the number of basis functions. Because the exact solution requires an infinite number of basis functions, the number of basis functions N is determined by the criteria of convergence that the user specifies. In cases where accuracy is not as important as speed, a smaller value for N can easily be considered.

One disadvantage of using the entire-domain basis results directly from the flexibility in choosing N. The value of N for a given convergence criteria is not easily determined since it is dependent on the geometry of the mesh and the object inside the mesh. One method to circumvent this problem is to iterate the solution using different values of N. This is computationally efficient since the FEM solutions from the previous iteration can be used in the current iteration. Another method is to determine a relationship between N and the geometry characteristics. The major consideration in this relationship is the electrical size of the scatterer and the enclosing mesh. An attempt is made, later in this paper, to determine this relationship by considering a dielectric cylinder.

An additional restriction on the entire-domain basis functions Ψ_n is that they must be continuous on ∂S in order to obtain an accurate FEM solution. Since ∂S is a closed line, Ψ_n must be a function whose endpoints have the same value, i.e.,

$$\Psi_n(0) = \Psi_n(d). \tag{20}$$

The most readily adaptable functions are the periodic functions. Because of the oscillatory behavior of the periodic functions, the value of N is limited by how fine the mesh is on the boundary. For the sinusoidal case, Ψ_n is given by

$$\Psi_n = \begin{cases} \cos \dfrac{(n-1)}{2}\theta(t), & n \text{ odd} \\ \sin \dfrac{n}{2}\theta(t), & n \text{ even} \end{cases} \tag{21}$$

where $t \in [0, d]$ maps to $\theta(t) \in [0, 2\pi]$. Let the number of nodes on the boundary be given by M. If we assume that we need at least four elements along the boundary ∂S to approximate one period of the sine function, then we must have at least twice the number of nodes as basis functions, i.e., $N \leq M/2$.

The major advantage of the subdomain basis functions is the simplicity of their implementation into the finite element code. One very simple basis function is the triangle function. This function has support over two of the boundary elements with a maximum value of one at the boundary node common to the two elements and a minimum value of zero at the adjacent boundary nodes. The variation of the basis function from the common node to the two adjacent nodes is linear. Since the behavior of triangle function is the same as the basis functions used in the finite element problem, we can obtain the corresponding finite element basis function Λ_n by setting the value of one of the nodes on the boundary to one and setting the other nodes on the boundary to zero. Note that this requires that $N = M$. We are therefore restricted by our choice of basis functions to a fixed N. For meshes with a large number of nodes on the boundary the solution may be computationally intensive in both time and storage requirements, but since the basis functions Ψ_n are discretized in the same manner as the FEM problem, we expect the solution to be accurate. In many cases, the computational time required to obtain a solution using triangle basis functions becomes excessive. In these situations we can modify the triangle function so that it has support over four elements rather than two. This has the effect of halving the number of basis functions. The resulting loss of accuracy is dependent on the electrical spacing of the nodes on the boundary. Both of these triangle functions will be considered in the numerical results section. To differentiate between the two functions, we name the triangle function with support over two elements to be $T1$ and the triangle function with support over four elements to be $T2$.

There are numerous choices which can be made for the set of testing functions which satisfy (14). We can narrow the choices by requiring the testing functions to have closed-form expressions. One obvious choice is the set of cylindrical harmonic functions. These are

$$\Phi_j = H_j^{(2)}(k_0\rho)\left\{\begin{matrix}\cos j\phi \\ \sin j\phi\end{matrix}\right\}, \qquad j = 1, 2, \cdots. \tag{22}$$

The variables ρ and ϕ represent the location in cylindrical coordinates at which Φ_j is evaluated. For j small the evaluation of the Hankel function is fairly straightforward, but as j becomes large, problems arise in numerically evaluating the Hankel function. The large-order approximation for $H_\nu^{(2)}(x)$ is given by [17]

$$H_\nu^{(2)}(x) \sim \frac{1}{\sqrt{2\pi\nu}}\left(\frac{ex}{2\nu}\right)^\nu + i\sqrt{\frac{2}{\pi\nu}}\left(\frac{ex}{2\nu}\right)^{-\nu}. \tag{23}$$

As $\nu \to \infty$, the real part of the Hankel function goes to zero while the imaginary part of the Hankel function goes to infinity. Because both the real and imaginary parts of the Hankel function are important in the evaluation of (16), their values must be retained. Unfortunately, even using double precision, there is a limit to the value of j for which an accurate result in (16) can be obtained. Therefore, we must be careful in using this testing function when N is large.

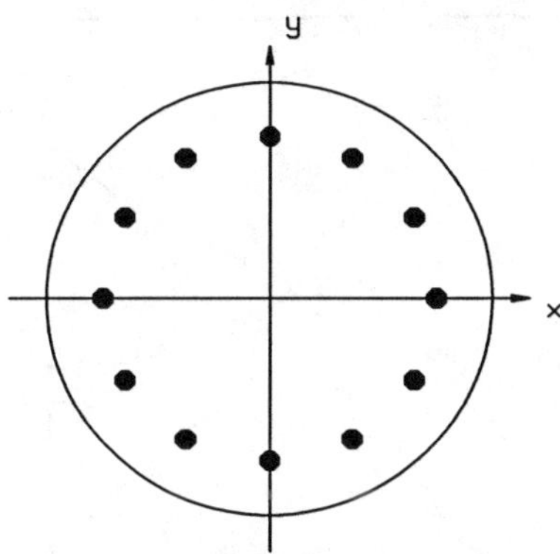

Fig. 2. Sample positioning of testing functions for a circular cylinder.

Another suitable testing function is the free-space scalar Green's function with the line source located in S_1'. The two-dimensional Green's function is given by

$$g(\bar{\rho}, \bar{\rho}') = H_0^{(2)}(k_0|\bar{\rho} - \bar{\rho}'|). \tag{24}$$

Let $\bar{\rho}_j'$, $j = 1, 2, \cdots, N$ be a set of points in S_1'. By substituting $\bar{\rho}_j'$ for $\bar{\rho}'$ in (24), we can form a set of linearly independent functions which we can use for Φ_j. The mathematical expression for Φ_j is

$$\Phi_j = H_0^{(2)}(k_0|\bar{\rho} - \bar{\rho}_j'|). \tag{25}$$

Therefore, the different testing functions represent solutions of (14) for different positions of the line source in S_1'. This is similar to the basis functions used in the generalized multipole technique [18]. Note that this testing function does not have the limitation on the size of N, but the proper placement of the line source is vital for obtaining an accurate solution. A random choice for $\bar{\rho}_j'$ may cause the matrix in (17) to be ill-conditioned. This results in gross inaccuracies in the solution. We found that we achieved accurate results by placing the line sources along contours which are concentric and conforming with ∂S. A typical placement of the line sources for a circular cylinder is shown in Fig. 2.

IV. Numerical Results

In this section, we present results using the bymoment method for several simple cylindrical geometries with known solutions. All of the results shown here were obtained from calculations done on a VAX 2000 workstation. Several of the subroutines in the computer program were taken from Akin's finite-element program [19]. The preprocessor FASTQ [20] was used for meshing our geometry and optimizing the nodal order in the mesh. The incident field that we use to obtain all of the following results is a plane wave propagating in the positive x direction. A mathematical description for ϕ^{inc} is

$$\phi^{inc} = -\frac{1}{k_0^2}e^{-ik_0x}. \tag{26}$$

The testing functions that we choose are the free-space Green's functions, as described in the previous section. Although the formulation is described in terms of potentials in

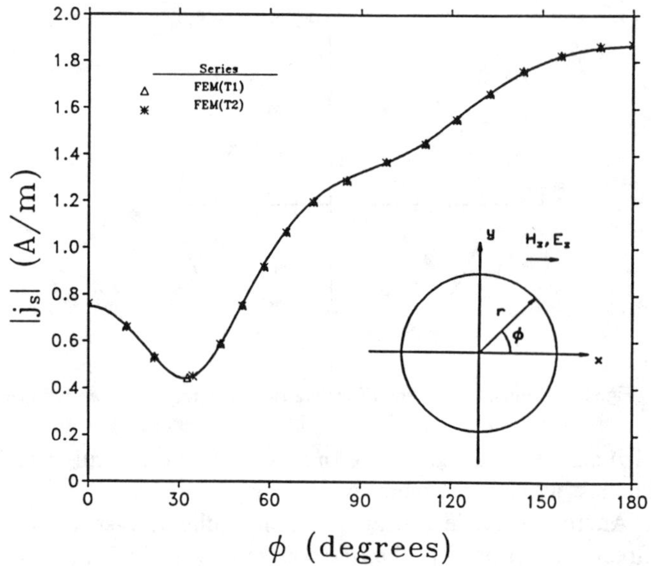

Fig. 3. Magnitude of the surface current density along the surface of a perfectly conducting cylinder ($r = 0.3\ \lambda$, TE_z case).

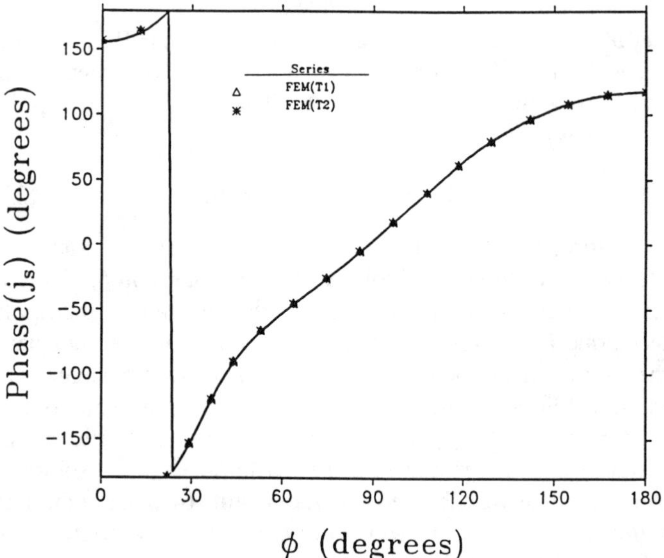

Fig. 4. Phase of the surface current density along the surface of a perfectly conducting cylinder ($r = 0.3\ \lambda$, TE_z case).

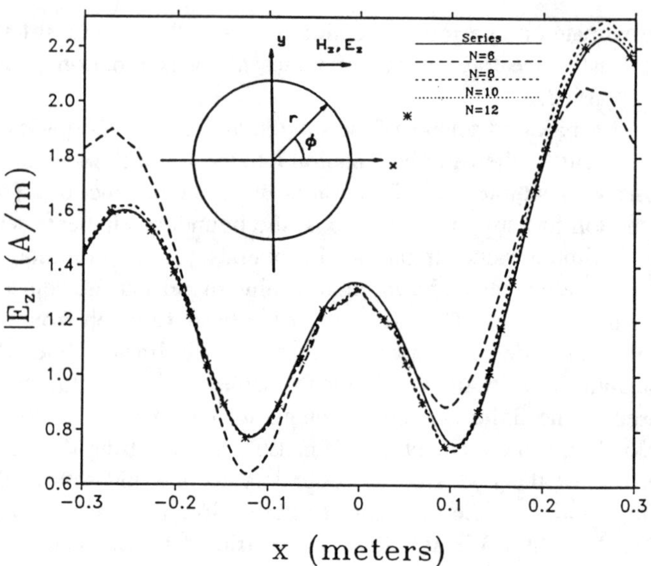

Fig. 5. Magnitude of the electric field E_z for a circular dielectric cylinder ($r = 0.3\ \lambda$, $\epsilon_r = 4$, TM_z case). The field is computed along x for $y = 0$. Comparison of entire function solutions to series solution.

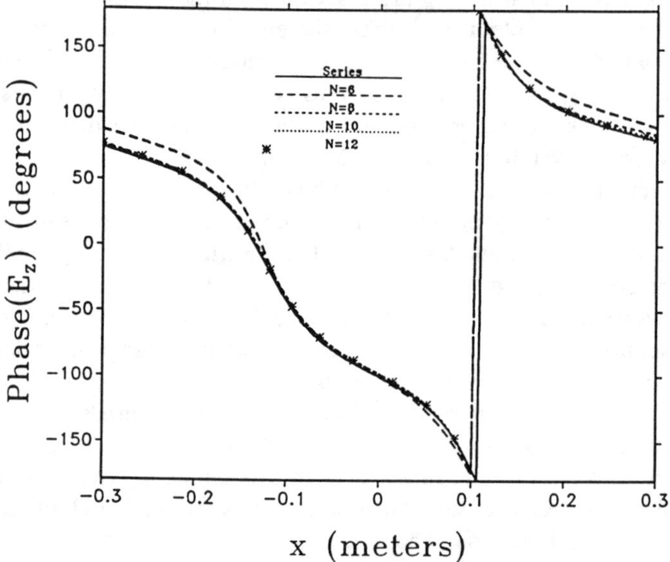

Fig. 6. Phase of the electric field E_z for a circular dielectric cylinder ($r = 0.3\ \lambda$, $\epsilon_r = 4$, TM_z case). The field is computed along x for $y = 0$. Comparison of entire function solutions to series solution.

order to generalize to both the TM and TE cases, the actual computer program uses field expressions to simplify the program.

The first geometry that we consider is a perfectly conducting circular cylinder of radius 0.3 λ in free space. Figs. 3 and 4 show the magnitude and phase, respectively, of the surface current density j_s on the cylinder under TE_z excitation. The results using both the $T1$ ($N = 40$) and $T2$ ($N = 20$) basis functions are compared against the series solution [15]. We see that FEM solutions are very accurate in both cases. Actually, the two finite-element solutions are practically indistinguishable, which implies that the use of the computationally less intensive $T2$ function is preferable for the given cylinder size and the specific discretization. In fact, if the geometry discretization is such that it corresponds to 20 nodes/wavelength, the use of the $T2$ function is expected to give sufficiently accurate results.

We next consider a dielectric circular cylinder ($\epsilon_r = 4$, $r = 0.3\ \lambda$) under TM_z excitation. In Figs. 5 and 6, we compare the series results [15] to the FEM results generated using the entire functions in (21) with different values of N. This gives us an indication of the convergence of the solution of a cylinder of this size using (21). The electric field is plotted as a function of x for $y = 0$. The apparent inaccuracy in the vicinity of $x = 0$ is not due to the finite-element solution. Instead, it is due to the linear interpolation used for the graphical representation of the solution between the nodes. We have very good agreement between the series solution and the FEM solution for $N = 12$. Even for $N = 8$, the error in the FEM solution is relatively small. There are no curves for $N > 12$ since we find no significant improvement in accuracy for larger values of N. For the sinusoidal basis functions we found that we can quantify the number of functions that are

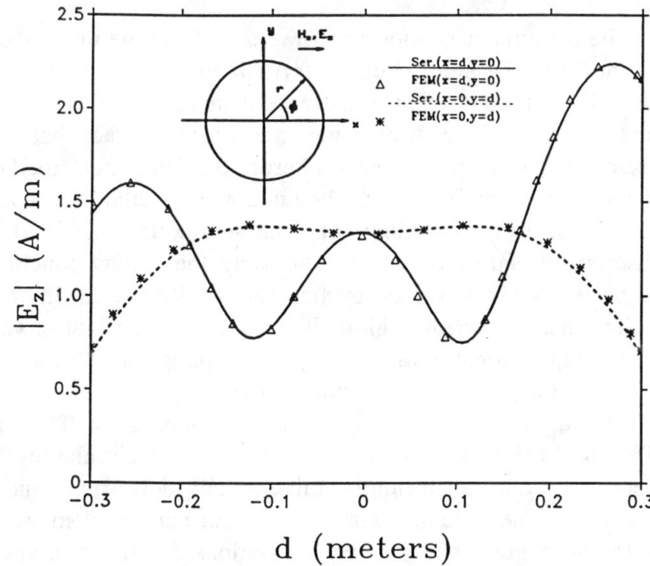

Fig. 7. Magnitude of the electric field E_z for a circular dielectric cylinder ($r = 0.3\ \lambda$, $\epsilon_r = 4$, TM_z case). The field is computed along x for $y = 0$ and along y for $x = 0$. Comparison of subdomain function solutions to series solution.

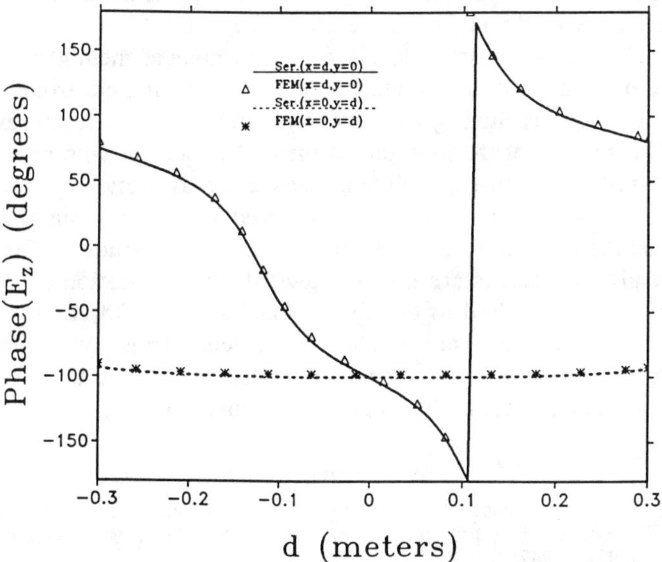

Fig. 8. Phase of the electric field E_z for a circular dielectric cylinder ($r = 0.3\ \lambda$, $\epsilon_r = 4$, TM_z case). The field is computed along x for $y = 0$ and along y for $x = 0$. Comparison of subdomain function solutions to series solution.

required for a given geometry size., If we specify the length of the perimeter of the scatterer to be P_{sc} and the length of the perimeter of the mesh to be P_{mesh}, then the number of basis functions that are required to obtain an accurate solution is given by

$$N \geq \max\left\{\frac{2P_{\text{sc}}}{\lambda_{\text{sc}}}, \frac{2P_{\text{mesh}}}{\lambda_0}\right\} \tag{27}$$

where for a homogeneous cylinder, λ_{sc} is the wavelength inside the scatterer and λ_0 is the wavelength in free space. For scatterers which are inhomogeneous λ_{sc} represents the smallest wavelength in the scatterer. For our dielectric cylinder we find that $N \geq 8$ which agrees with our numerical results. In Figs. 7 and 8 we consider the same geometry as before except we now compare the series solution to the FEM solution generated using the $T1$ basis functions. In addition to the results along the line $y = 0$, we present the results as a function of y for $x = 0$. Again we see that there is excellent agreement between the series and FEM solutions.

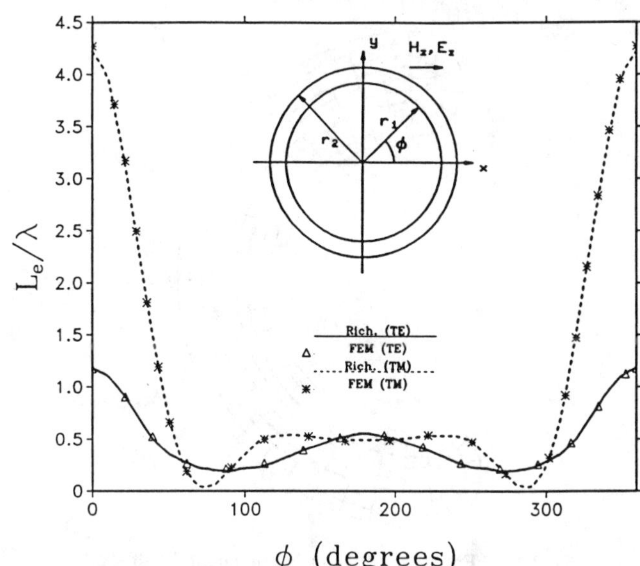

Fig. 9. Echo width for a dielectric circular shell ($r_1 = 0.25\ \lambda$, $r_2 = 0.3\ \lambda$, $\epsilon_r = 4$). Comparison of $T2$ basis function solution to Richmond's solution.

Richmond [21], [22] provides echo width results for several cylindrical geometries under both TE_z and TM_z excitations. For the TM_z case, the echo width L_e is defined by [15]

$$L_e = \lim_{\rho\to\infty}\left[2\pi\rho\frac{|E_z^{\text{sc}}|^2}{|E_z^{\text{inc}}|^2}\right]. \tag{28}$$

Similarly, for the TE_z case,

$$L_e = \lim_{\rho\to\infty}\left[2\pi\rho\frac{|H_z^{\text{sc}}|^2}{|H_z^{\text{inc}}|^2}\right]. \tag{29}$$

In this paper, we consider two of the geometries analyzed by Richmond. The first is a dielectric circular shell with an inner radius of 0.25 λ and an outer radius of 0.3 λ. The relative dielectric constant of the shell is four. Fig. 9 compares Richmond's solution to the FEM solution generated using the $T2$ basis functions. The two methods produce almost identical results for both the TE_z and TM_z cases. The second geometry is a dielectric circular half-shell ($r_1 = 0.25\ \lambda$, $r_2 = 0.3\ \lambda$, $\epsilon_r = 4$). This geometry is different from the others in that it does not have a circular boundary. We can therefore use this geometry to demonstrate the ability of the method to solve problems using grids that conform to the surface of the scatterer. The FEM mesh which we use for this problem is shown in Fig. 10. The three center rows of elements represent the dielectric shell. The resulting echo width solutions are shown in Fig. 11, where we compare them to Richmond's solutions. The sinusoidal basis functions ($N = 16$) are used for this example. Once again, the agreement is excellent.

V. Conclusion

The bymoment method has been presented for the analysis of two-dimensional electromagnetic scattering problems. A surface conforming to the shape of the scatterer is used

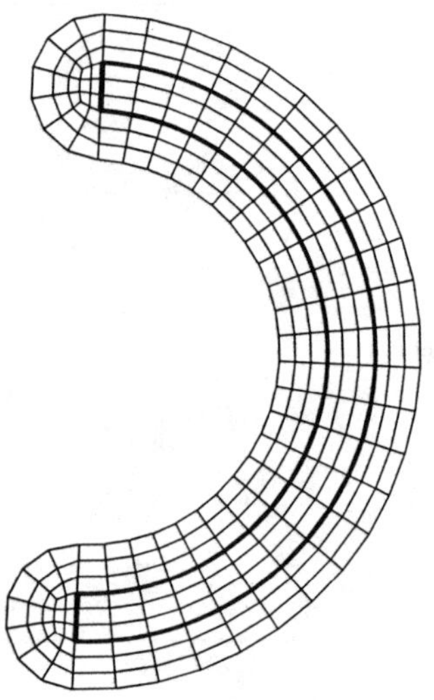

Fig. 10. FEM mesh of the dielectric circular half-shell.

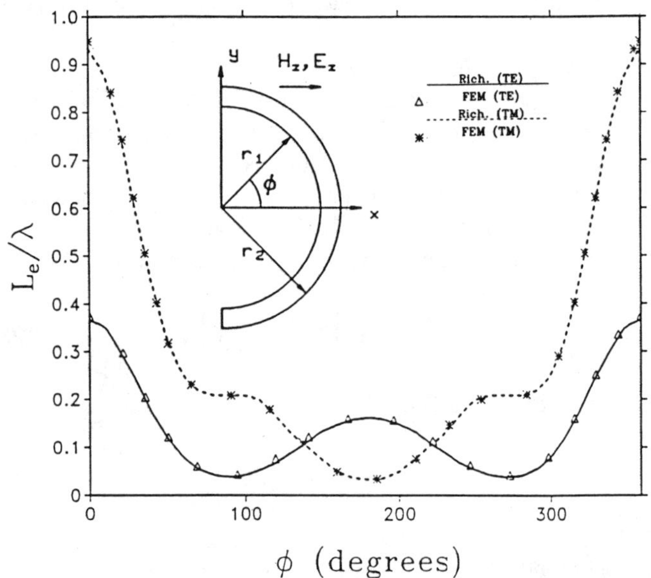

Fig. 11. Echo width for a dielectric circular half-shell ($r = 0.25\ \lambda$, $r_2 = 0.3\ \lambda$, $\epsilon_r = 4$). Comparison of sinusoidal basis function solution to Richmond's solution.

to divide the computational domain into an interior region where the scatterer exists, and an exterior unbounded region free from inhomogeneities. By using the tangential electric or magnetic field on this surface as the key unknown in the problem, the method allows the complete decoupling of the numerical solution in the interior from the solution in the unbounded region. This unknown tangential field is represented as a series expansion in terms of known basis functions with unknown coefficients. The solution in the interior is then expressed in terms of these unknown coefficients by superposing the finite-element solutions generated by solving the interior Dirichlet problems for each one of the known expansion functions. Finally, a linear system of equations for the unknown coefficients is generated by applying Green's theorem in the exterior unbounded region for the scattered field and a set of linearly independent functions which satisfies the Helmholtz equation and the Sommerfeld radiation condition in the exterior homogeneous region.

The bymoment method borrows ideas from the unimoment method and the field feedback formulation, and tries to combine them in an attempt to increase their numerical efficiency and versatility. Indeed, by using an enclosing surface that conforms to the shape of the scatterer, the bymoment method bypasses the requirement of the unimoment method for a separable surface, thus minimizing the volume that needs to be discretized. Furthermore, by choosing the testing functions to be the fields produced by line sources located at different points in the interior region, it introduces a straightforward and computationally simple way of coupling the interior numerical solution to the exterior region.

Several numerical examples were presented for TE_z and TM_z scattering by conducting and dielectric circular cylinders, as well as scattering by full and half-dielectric cylindrical shells. The excellent agreement of our numerical solutions with those generated by series solutions for the aforementioned geometries demonstrates the validity of the bymoment method. Furthermore, these numerical examples have allowed us to discuss the effects of the various choices of the basis functions used for the tangential fields representation on the accuracy of the numerical solutions.

One of the ultimate goals of the bymoment method is to result in a versatile and efficient tool in analyzing electromagnetic wave scattering by very large and complex structures. The success of the unimoment method in dealing with rather complex structures, including such cases as scattering from objects near the vicinity of planar media interfaces, suggests that the bymoment method will find similar applications. Currently, our efforts are directed towards the application of the bymoment method to two-dimensional scattering by multiple targets and scattering at oblique incidence from cylindrical structures. The extension of the bymoment method to three-dimensional geometries is also under investigation.

References

[1] P. Silvester and M. S. Hsieh, "Finite-element solution of 2-dimensional exterior field problems," *Proc. Inst. Elec. Eng.*, vol. 118, pp. 1743–1747, Dec. 1971.

[2] B. H. McDonald and A. Wexler, "Finite-element solution of unbounded field problems," *IEEE Trans. Microwave Theory Tech.*, vol. MTT-20, pp. 841–847, Dec. 1972.

[3] D. R. Lynch, K. D. Paulsen, and J. W. Strohbehn, "Hybrid finite element method for unbounded electromagnetic problems in hyperthermia," *Int. J. Num. Meth. Eng.*, vol. 23, pp. 1915–1937, 1986.

[4] K. D. Paulsen, D. R. Lynch, and J. W. Strohbehn, "Three-dimensional finite, boundary, and hybrid element solutions to the Maxwell equations for lossy dielectric media," *IEEE Trans. Microwave Theory Tech.*, vol. 36, pp. 682–693, Apr. 1988.

[5] J. M. Jin and V. V. Liepa, "Application of hybrid finite element method to electromagnetic scattering from coated cylinders," *IEEE Trans. Antennas Propagat.*, vol. 36, pp. 55–70, Jan. 1988.

[6] ——, "A note on hybrid finite element method for solving scattering problems," *IEEE Trans. Antennas Propagat.*, vol. 36, pp. 1486–1490, Oct. 1988.

[7] K. K. Mei, "Unimoment method of solving antenna and scattering problems," *IEEE Trans. Antennas Propagat.*, vol. AP-22, pp. 760–766, Nov. 1974.

[8] S.-K. Chang and K. K. Mei, "Application of the unimoment method to electromagnetic scattering of dielectric cylinders," *IEEE Trans. Antennas Propagat.*, vol. AP-24, pp. 35–42, Jan. 1976.

[9] M. A. Morgan and K. K. Mei, "Finite-element computation of scattering by inhomogeneous penetrable bodies of revolution," *IEEE Trans. Antennas Propagat.*, vol. AP-27, pp. 202–214, Mar. 1979.

[10] M. A. Morgan, "Generalized coupled azimuthal potentials for elec-

tromagnetic fields in inhomogeneous media," *IEEE Trans. Antennas Propagat.*, vol. 36, pp. 1735–1743, Dec. 1988.

[11] J. F. Lee and Z. J. Cendes, "Transfinite elements: A highly efficient procedure for modeling open field problems," *J. Appl. Phys.*, vol. 61, pp. 3913–3915, Apr. 1987.

[12] M. A. Morgan, C. H. Chen, S. C. Hill, and P. W. Barber, "Finite element-boundary integral formulation for electromagnetic scattering," *Wave Motion*, vol. 6, pp. 91–103, Jan. 1984.

[13] M. A. Morgan and B. E. Welch, "The field feedback formulation for electromagnetic scattering computations," *IEEE Trans. Antennas Propagat.*, vol. AP-34, pp. 1377–1382, Dec. 1986.

[14] A. C. Cangellaris and R. Lee, "A finite element method for solving electromagnetic scattering problems," presented at IEEE Antennas Propagat. Soc./URSI Conf., San Jose, CA, June 1989.

[15] R. F. Harrington, *Time-Harmonic Electromagnetic Fields.* New York: McGraw-Hill, 1961.

[16] K. H. Huebner, *The Finite Element Method for Engineers.* New York: Wiley, 1975.

[17] M. Abramowitz and I. A. Stegun, *Handbook of Mathematical Functions.* New York: Dover, 1972.

[18] C. Hafner, "MMP—A program Package Based on the Generalized Multipole Technique (GMT)," presented at IEEE Antennas Propagat. Soc./URSI Conf., San Jose, CA, June 1989.

[19] J. E. Akin, *Application and Implementation of Finite Element Methods.* Orlando, FL: Academic, 1982.

[20] T. D. Blacker, "FASTQ User Manual, Version 1.2," Sandia Rep., SAND88-1326, July 1988.

[21] J. H. Richmond, "Scattering by a dielectric cylinder of arbitrary cross section shape," *IEEE Trans. Antennas Propagat.*, vol. AP-13, pp. 334–341, May 1965.

[22] ——, "TE-wave scattering by a dielectric cylinder of arbitrary cross section shape," *IEEE Trans. Antennas Propagat.*, vol. AP-14, pp. 460–464, July 1966.

[23] E. B. Becker, G. F. Carey, and J. T. Oden, *Finite Elements: An Introduction.* Englewood Cliffs, NJ: Prentice-Hall, 1981.

A Combined Finite Element–Boundary Integral Formulation for Solution of Two-Dimensional Scattering Problems via CGFFT

JEFFERY D. COLLINS, STUDENT MEMBER, IEEE, JOHN L. VOLAKIS, SENIOR MEMBER, IEEE, AND JIAN-MING JIN, MEMBER, IEEE

Abstract—**A new technique is presented for computing the scattering by two-dimensional structures of arbitrary inhomogeneity. The proposed approach combines the usual finite element (FE) method with the boundary integral equation to formulate a discrete system. This is subsequently solved via the conjugate gradient (CG) algorithm. A particular characteristic of the method is the use of rectangular boundaries to enclose the scatterer. Several of the resulting boundary integrals are then convolutions and may be evaluated via the fast Fourier transform (FFT) in the implementation of the CG algorithm. The solution approach offers the principle advantage of having $O(N)$ memory demand and employs a one-dimensional FFT versus a two-dimensional FFT as required in a traditional implementation of the CG-FFT algorithm. The speed of the proposed solution method is compared with that of the traditional CG-FFT algorithm. Results are presented for several rectangular composite cylinders and one perfectly conducting cylinder. These are shown to be in excellent agreement with the moment method.**

I. Introduction

The demand for three-dimensional (3-D) methods has increased in recent years, and two-dimensional (2-D) methods often find their niche as testing grounds for 3-D applications. The step required to generalize a 2-D method to 3-D is almost always large in analytical and geometrical complexity. Most importantly, though, the demands in computation time and storage are often prohibitive for electrically large 3-D bodies. Vector and concurrent (i.e., hypercube, connection, etc.) computers are beginning to alleviate the first of these demands, but a minimization of the storage requirements is essential for treating large structures.

The traditional conjugate gradient fast Fourier transform (CGFFT) method [1]–[4] is one such frequency domain solution approach which requires $O(N)$ storage for the solution of N equations. This method involves the use of FFT's whose dimension equals that of the structure under consideration [5]–[7] and, therefore, demands excessive computation time when used in an iterative algorithm. Also, the standard CGFFT requires uniform rectangular gridding that unnecessarily includes the impenetrable portions of the scatterer. With these issues in mind, a new solution approach is proposed for solving scattering problems. The proposed method will be referred to as the finite element–conjugate gradient fast Fourier transform (FE–CGFFT) method and was inspired by Peters [8].

The FE–CGFFT method requires that the scatterer be surrounded by a double rectangular box. Inside the box boundaries, the Helmholtz equation is solved via the finite element method. The boundary constraint is obtained by an appropriate integral equation which implicitly satisfies the radiation condition. Along the parallel sides of the box, this integral becomes a convolution and is, therefore, amenable to evaluation via the FFT. The dimension of the required FFT in this hybrid method is one less than the dimensionality of the structure thus, making it attractive for 3-D simulations. Also, because it incorporates the finite element method, the FE-CGFFT formulation remains valid regardless of the structure's geometry and material composition.

The proposed method is similar to the moment method version developed by Jin [9]. Jin's method was in turn based on work published in the early 1970's by McDonald and Wexler [10] who introduced an approach to solve unbounded field problems. The proposed method is also similar to other methods, a few of which will be mentioned here. The unimoment method [11] uses finite elements inside a fictitious circular boundary and an eigenfunction expansion to represent the fields in the external region. The coefficients of the expansion are then determined by enforcing field continuity at the finite element mesh boundary. The coupled finite element–boundary element method [12] uses the finite element method within the boundary and the boundary element method to provide the additional constraint at the termination of the mesh. Unlike the proposed method, the solution in [12] was accomplished by direct matrix inversion (as in [9]), and the outer mesh boundary is not rectangular to take advantage of the FFT for the evaluation of the boundary integrals. Further, only one boundary was employed, and hence an analytical evaluation was required for the self-cell.

II. Analysis

Consider a cylindrical body of arbitrary cross section and composition (possibly coated) illuminated by the plane wave

$$\bar{\phi}^{\text{inc}}(\bar{\rho}) = \hat{z}\phi^{\text{inc}}(\bar{\rho}) = \hat{z}e^{jk_0\rho\cos(\theta-\theta_0)} \quad (1)$$

Manuscript received October 11, 1989; revised May 29, 1990. This work was supported in part by the Pacific Missile Test Center and the NASA Ames Research Center Grant NAG-2-541.

The authors are with the Radiation Laboratory, Department of Electrical Engineering and Computer Science, University of Michigan, Ann Arbor, MI 48109-2122.

IEEE Log Number 9038583.

Reprinted from *IEEE Trans. Antennas Propagat.*, vol. 38, no. 11, pp. 1852–1858, Nov. 1990.

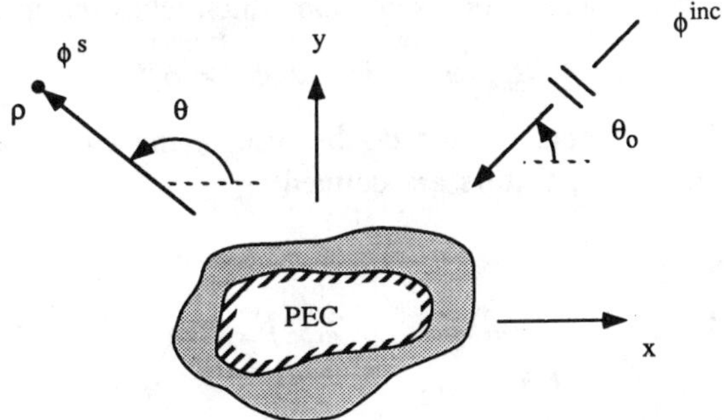

Fig. 1. Geometry of the problem.

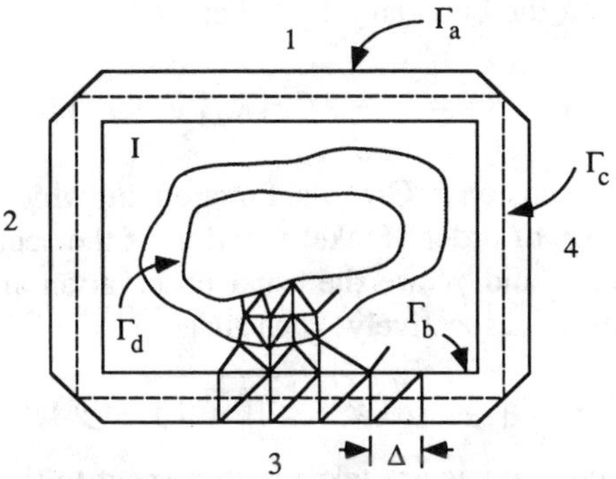

Fig. 2. Partially discretized body.

where $k_0 = 2\pi/\lambda$ denotes the free-space wavenumber and the other parameters are defined in Fig. 1 (A time dependence of $e^{j\omega t}$ has been assumed and suppressed.) To analyze the fields scattered from this object, two boundaries are placed tightly around the body as illustrated in Fig. 2. Inside the outer boundary, the finite element method is employed with the appropriate boundary condition enforced on the surface of the impenetrable boundary. The radiation condition is satisfied implicitly by invoking the boundary integral equation to relate the fields on Γ_a to those on Γ_b.

In Fig. 2, Γ_a $(= \sum_{i=1}^{4}\Gamma_{a_i}$ plus corner connections) is the field/observation contour, and Γ_c $(= \sum_{i=1}^{4}\Gamma_{c_i})$ is the integration contour placed midway between Γ_a and Γ_b $(= \sum_{i=1}^{4}\Gamma_{b_i})$. Also, Γ_d denotes the contour enclosing the impenetrable portion of the scatterer. Herewith, each side of Γ_a, Γ_b, and Γ_c will be numbered counterclockwise starting from the top side, as indicated in Fig. 2.

A. Derivation of the Finite Element Matrix

One of several approaches may be used to generate the finite element matrix, such as the variational approach or the method of weighted residuals. We will utilize the variational technique which involves the minimization of the functional [13]

$$F = \iint_{R_a}\left[\frac{1}{\mu_r(\bar{\rho})}\nabla\phi\cdot\nabla\phi_a^* - k_0^2\epsilon_r(\bar{\rho})\phi\phi_a^*\right]ds - \int_{\Gamma_a}\phi_a^*\nabla\phi\cdot\hat{n}_a\,dl \quad (2)$$

for E-polarization ($\phi = E_z$), and

$$F = \iint_{R_a}\left[\frac{1}{\epsilon_r(\bar{\rho})}\nabla\phi\cdot\nabla\phi_a^* - k_0^2\mu_r(\bar{\rho})\phi\phi_a^*\right]ds - \int_{\Gamma_a}\phi_a^*\nabla\phi\cdot\hat{n}_a\,dl \quad (3)$$

for H-polarization ($\phi = H_z$). The complex relative permittivity and permeability of the medium are denoted by ϵ_r and μ_r, respectively, and R_a denotes the region enclosed by Γ_a and Γ_d. Additionally, ϕ_a is the corresponding solution to the adjoint problem and the asterisk denotes complex conjugation.

To discretize (2) or (3) the region R_a is subdivided into N_e triangular finite elements. The field in R_a may then be represented as a summation

$$\phi(x, y) = \sum_{e=1}^{N_e}\phi^e(x, y) \quad (4)$$

where $\phi^e(x, y)$ represents the field within the eth element and is approximated by

$$\phi^e(x, y) = \sum_{j=1}^{3}N_j^e(x, y)\phi_j^e \quad (5)$$

and equals zero outside the eth element. In (5), N_j^e are the standard linear shape functions (see, for example, [14]) and ϕ_j^e denotes the field at the jth node of the eth element. Following the usual finite element procedure, the resulting system based upon the minimization of F can be written as

$$\begin{bmatrix} A_{aa} & A_{ab} & 0 & 0 \\ A_{ba} & A_{bb} & A_{bI} & A_{bd} \\ 0 & A_{Ib} & A_{II} & A_{Id} \\ 0 & A_{db} & A_{dI} & A_{dd}\end{bmatrix}\begin{bmatrix}\phi_a \\ \phi_b \\ \phi_I \\ \phi_d\end{bmatrix} = \begin{bmatrix} b_a \\ 0 \\ 0 \\ 0\end{bmatrix} \quad (6)$$

where ϕ_a, ϕ_b, and ϕ_d are vectors representing the unknown fields at the nodes on Γ_a, Γ_b, and Γ_d, respectively. The vector ϕ_I denotes the unknown fields at the nodes in region I, which is the portion of R_a that excludes the region between Γ_a and Γ_b. Also, the values of the elements in the submatrix A_{pq} are the contributions associated with the nodes in group p which are connected directly to the nodes in group q. Note that there are no nodes between Γ_a and Γ_b.

The vector b_a in (6) arises from the line integral contribution in (2) or (3) and without *a priori* knowledge of the relationship between $\hat{n}_a\cdot\nabla\phi$ and ϕ_a, it cannot be determined. We may, however, provide the appropriate constraint on this boundary by utilizing Kirchoff's integral equation (the 2-D version of the Stratton–Chu integral equations). This amounts to replacing the first block-row of the matrix (associated with ϕ_a on Γ_a) with a discrete form of this integral equation.

B. Evaluation of the Boundary Integral

To provide the appropriate boundary constraint on Γ_a, the integral equation

$$\phi(\bar{\rho}) = \phi^{\text{inc}}(\bar{\rho}) - \oint_{\Gamma_c}\left\{G(\bar{\rho}, \bar{\rho}')\left[\frac{\partial}{\partial n'}\phi(\bar{\rho}')\right] - \phi(\bar{\rho}')\left[\frac{\partial}{\partial n'}G(\bar{\rho}, \bar{\rho}')\right]\right\}dl' \quad (7)$$

is evaluated on the boundary Γ_a, where

$$G(\bar{\rho}, \bar{\rho}') = -\frac{j}{4} H_0^{(2)}(k_0 |\bar{\rho} - \bar{\rho}'|) \tag{8}$$

is the two-dimensional Green's function in which $H_0^{(2)}(\cdot)$ denotes the zeroth order Hankel function of the second kind. Furthermore, $\bar{\rho}$ and $\bar{\rho}'$ are the usual observation and source position vectors, respectively, such that

$$|\bar{\rho} - \bar{\rho}'| = \sqrt{(x - x')^2 + (y - y')^2}. \tag{9}$$

The normal derivatives are taken with respect to the outward normal vector on Γ_c.

The boundary integral in (7) may be written as a summation of four integrals, one for each side of the contour Γ_c to be evaluated on every side of Γ_a. By using the central difference formula to evaluate the normal derivatives of the unknown field ϕ, we obtain

$$\begin{aligned}
\phi(x, y_{a_1}) &= \phi^{\text{inc}}(x, y_{a_1}) \\
&\quad - \left\{ \int_{\Gamma_{c1}} [K_y^- G(x - x', y_{a_1}, y_{c_1}) \phi(x', y_{a_1}) \right. \\
&\quad - K_y^+ G(x - x', y_{a_1}, y_{c_1}) \phi(x', y_{b_1})] \, dx' \\
&\quad + \int_{\Gamma_{c2}} [K_x^+ G(x, x_{c_2}, y_{a_1}, y') \phi(x_{a_2}, y') \\
&\quad - K_x^- G(x, x_{c_2}, y_{a_1}, y') \phi(x_{b_2}, y')] \, dy' \\
&\quad + \int_{\Gamma_{c3}} [K_y^+ G(x - x', y_{a_1}, y_{c_3}) \phi(x', y_{a_3}) \\
&\quad - K_y^- G(x - x', y_{a_1}, y_{c_3}) \phi(x', y_{b_3})] \, dx' \\
&\quad + \int_{\Gamma_{c4}} [K_x^- G(x, x_{c_4}, y_{a_1}, y') \phi(x_{a_4}, y') \\
&\quad \left. - K_x^+ G(x, x_{c_4}, y_{a_1}, y') \phi(x_{b_4}, y')] \, dy' \right\}
\end{aligned} \tag{10}$$

for observation on Γ_{a_1}. Thus, either x_{pq} or y_{pq} is constant along Γ_{pq} for $p = a, b, c$ and $q = 1, 2, 3, 4$ (see Fig. 2). In (10) the operators $K_{x,y}^{\pm}$ are given by

$$K_x^{\pm} = \frac{1}{\Delta} \pm \frac{1}{2} \frac{\partial}{\partial x'} \tag{11}$$

$$K_y^{\pm} = \frac{1}{\Delta} \pm \frac{1}{2} \frac{\partial}{\partial y'} \tag{12}$$

where Δ is the displacement between Γ_a and Γ_b and is equal to the node spacing on each of these contours. It is seen that (10) contains four convolution integrals, and this trend prevails in the expressions associated with observation on the remaining sides of Γ_a.

To discretize (10), we use a pulse basis expansion for the nodal fields (centered at the nodal positions along contours Γ_a and Γ_b) and perform a midpoint integration to simplify the integrals in (10). Application of the point matching technique then yields a discrete representation which takes the form

$$(I + L_{aa})\phi_a - L_{ab} D_b \phi_b = \phi_a^{\text{inc}} \tag{13}$$

where D_b accounts for the double use of the endpoints on Γ_b. The matrix operators are defined as

$$I + L_{aa} = \begin{bmatrix} I + \mathscr{S}_{a_1b_1}^- & \mathscr{T}_{a_1b_2}^+ & \mathscr{S}_{a_1b_3}^+ R_x & \mathscr{T}_{a_1b_4}^+ \\ \mathscr{T}_{a_2b_1}^+ & I + \mathscr{S}_{a_2b_2}^- & \mathscr{T}_{a_2b_3}^+ & \mathscr{S}_{a_2b_4}^+ R_y \\ \mathscr{S}_{a_3b_1}^+ R_x & \mathscr{T}_{a_3b_2}^+ & I + \mathscr{S}_{a_3b_3}^- & \mathscr{T}_{a_3b_4}^+ \\ \mathscr{T}_{a_4b_1}^+ & \mathscr{S}_{a_4b_2}^+ R_y & \mathscr{T}_{a_4b_3}^+ & I + \mathscr{S}_{a_4b_4}^- \end{bmatrix} \tag{14}$$

$$L_{ab} = \begin{bmatrix} \mathscr{S}_{a_1b_1}^+ & \mathscr{T}_{a_1b_2}^- & \mathscr{S}_{a_1b_3}^- R_x & \mathscr{T}_{a_1b_4}^- \\ \mathscr{T}_{a_2b_1}^- & \mathscr{S}_{a_2b_2}^+ & \mathscr{T}_{a_2b_3}^- & \mathscr{S}_{a_2b_4}^- R_y \\ \mathscr{S}_{a_3b_1}^- R_x & \mathscr{T}_{a_3b_2}^- & \mathscr{S}_{a_3b_3}^+ & \mathscr{T}_{a_3b_4}^- \\ \mathscr{T}_{a_4b_1}^- & \mathscr{S}_{a_4b_2}^- R_y & \mathscr{T}_{a_4b_3}^- & \mathscr{S}_{a_4b_4}^+ \end{bmatrix} \tag{15}$$

in which I is the identity matrix and $R_{x,y}$ simply reverses the order of the unknown vector so that the convolutions may be performed properly. This is required solely because of the counterclockwise nodal numbering scheme used here. The operators $\mathscr{S}_{a_pb_q}^{\pm}$ imply convolutions and $\mathscr{T}_{a_pb_q}^{\pm}$ denote the cross-term submatrices. The first subscript in $\mathscr{S}$ or $\mathscr{T}$ is associated with the observation contour, while the second subscript corresponds to the integration contour. Clearly, the subscripts of $\mathscr{S}$ can only be associated with parallel contours if they are to maintain their convolutionality. This property permits a convenient and efficient computation of $\mathscr{S}_{a_pb_q}\phi$ as

$$\begin{aligned}
\mathscr{S}_{a_{\frac{1}{3}}b_q}^{\pm} \phi_{b_q} &= \text{DFT}^{-1}\Big\{ \text{DFT}\Big[G\big(x, y_{a_{\frac{1}{3}}}, y_{b_q}\big) \\
&\quad + G_y\big(x, y_{a_{\frac{1}{3}}}, y_{b_q}\big) \Big] \text{DFT}[\phi_{b_q}] \Big\}, \qquad q = 1, 3
\end{aligned} \tag{16}$$

$$\begin{aligned}
\mathscr{S}_{a_{\frac{2}{4}}b_q}^{\pm} \phi_{b_q} &= \text{DFT}^{-1}\Big\{ \text{DFT}\Big[G\big(x_{a_{\frac{2}{4}}}, x_{b_q}, y\big) \\
&\quad - G_x\big(x_{a_{\frac{2}{4}}}, x_{b_q}, y\big) \Big] \text{DFT}[\phi_{b_q}] \Big\}, \qquad q = 2, 4
\end{aligned} \tag{17}$$

where DFT denotes the discrete Fourier transform operator. The superscript plus sign corresponds to the upper subscript of a and likewise the superscript minus sign corresponds to the lower subscript of a. Also,

$$\begin{aligned}
&G_x\big(x_{a_{\frac{2}{4}}}, x_{b_{\frac{2}{4}}}, y\big) \\
&= -\frac{j}{4} \Delta k_0 \frac{H_1^{(2)}\left(k_0 \sqrt{\big(x_{a_{\frac{2}{4}}} - x_{b_{\frac{2}{4}}}\big)^2 + y^2}\right)}{\sqrt{\big(x_{a_{\frac{2}{4}}} - x_{b_{\frac{2}{4}}}\big)^2 + y^2}} \big| x_{a_{\frac{2}{4}}} - x_{b_{\frac{2}{4}}} \big|
\end{aligned} \tag{18}$$

$$G_y\left(x, y_{a_{\frac{1}{3}}}, y_{b_{\frac{1}{3}}}\right) = -\frac{j}{4}\Delta k_0 \frac{H_1^{(2)}\left(k_0\sqrt{x^2 + \left(y_{a_{\frac{1}{3}}} - y_{b_{\frac{1}{3}}}\right)^2}\right)}{\sqrt{x^2 + \left(y_{a_{\frac{1}{3}}} - y_{b_{\frac{1}{3}}}\right)^2}} \left| y_{a_{\frac{1}{3}}} - y_{b_{\frac{1}{3}}} \right| \tag{19}$$

where $H_1^{(2)}(\cdot)$ denotes the first-order Hankel function of the second kind. We clearly observe that a need to generate and store a square matrix to perform the operation $\mathscr{S}_{a_p b_q}\phi$ has been eliminated. The elements of the cross-term submatrices are given by

$$\left[\mathscr{T}^{\pm}_{a_p b_{\frac{2}{4}}}\right]_{ij} = G\left(x_i, x_{b_{\frac{2}{4}}}, y_{a_p}, y_j\right) + G_x\left(x_i, x_{b_{\frac{2}{4}}}, y_{a_p}, y_j\right), \qquad p = 1, 3 \tag{20}$$

$$\left[\mathscr{T}^{\pm}_{a_p b_{\frac{1}{3}}}\right]_{ij} = G\left(x_{a_p}, x_j, y_i, y_{b_{\frac{1}{3}}}\right) - G_y\left(x_{a_p}, x_j, y_i, y_{b_{\frac{1}{3}}}\right), \qquad p = 2, 4 \tag{21}$$

$$(I + L_{aa})^a = \begin{bmatrix} I + (\mathscr{S}_{a_1b_1})^a & (\mathscr{T}^{+}_{a_2b_1})^a & R_x^T(\mathscr{S}^{+}_{a_3b_1})^a & (\mathscr{T}^{+}_{a_4b_1})^a \\ (\mathscr{T}^{+}_{a_1b_2})^a & I + (\mathscr{S}_{a_2b_2})^a & (\mathscr{T}^{+}_{a_3b_2})^a & R_y^T(\mathscr{S}^{+}_{a_4b_2})^a \\ R_x^T(\mathscr{S}^{+}_{a_1b_3})^a & (\mathscr{T}^{+}_{a_2b_3})^a & I + (\mathscr{S}_{a_3b_3})^a & (\mathscr{T}^{+}_{a_4b_3})^a \\ (\mathscr{T}^{+}_{a_1b_4})^a & R_y^T(\mathscr{S}^{+}_{a_2b_4})^a & (\mathscr{T}^{+}_{a_3b_4})^a & I + (\mathscr{S}_{a_4b_4})^a \end{bmatrix} \tag{25}$$

$$(L_{ab}D_b)^a = \begin{bmatrix} D_b^T \end{bmatrix} \begin{bmatrix} (\mathscr{S}^{+}_{a_1b_1})^a & (\mathscr{T}^{-}_{a_2b_1})^a & R_x^T(\mathscr{S}^{-}_{a_3b_1})^a & (\mathscr{T}^{-}_{a_4b_1})^a \\ (\mathscr{T}^{-}_{a_1b_2})^a & (\mathscr{S}^{+}_{a_2b_2})^a & (\mathscr{T}^{-}_{a_3b_2})^a & R_y^T(\mathscr{S}^{-}_{a_4b_2})^a \\ R_x^T(\mathscr{S}^{-}_{a_1b_3})^a & (\mathscr{T}^{-}_{a_2b_3})^a & (\mathscr{S}^{+}_{a_3b_3})^a & (\mathscr{T}^{-}_{a_4b_3})^a \\ (\mathscr{T}^{-}_{a_1b_4})^a & R_y^T(\mathscr{S}^{-}_{a_2b_4})^a & (\mathscr{T}^{-}_{a_3b_4})^a & (\mathscr{S}^{+}_{a_4b_4})^a \end{bmatrix} \tag{26}$$

where

$$G_x\left(x_i, x_{b_{\frac{2}{4}}}, y_{a_{\frac{1}{3}}}, y_j\right) = \frac{-j}{8}k_0 \cdot \frac{H_1^{(2)}\left(k_0\sqrt{\left(x_i - x_{b_{\frac{2}{4}}}\right)^2 + \left(y_{a_{\frac{1}{3}}} - y_j\right)^2}\right)}{\sqrt{\left(x_i - x_{b_{\frac{2}{4}}}\right)^2 + \left(y_{a_{\frac{1}{3}}} - y_j\right)^2}} \left| x_i - x_{b_{\frac{2}{4}}} \right| \Delta \tag{22}$$

and

$$G_y\left(x_{a_{\frac{2}{4}}}, x_j, y_i, y_{b_{\frac{1}{3}}}\right) = \frac{-j}{8}k_0 \cdot \frac{H_1^{(2)}\left(k_0\sqrt{\left(x_{a_{\frac{2}{4}}} - x_j\right)^2 + \left(y_i - y_{b_{\frac{1}{3}}}\right)^2}\right)}{\sqrt{\left(x_{a_{\frac{2}{4}}} - x_j\right)^2 + \left(y_i - y_{b_{\frac{1}{3}}}\right)^2}} \left| y_i - y_{b_{\frac{1}{3}}} \right| \Delta. \tag{23}$$

As in the case of the $\mathscr{S}_{a_p b_q}$ operators, the superscript plus sign corresponds to the upper subscript of b and, similarly, the superscript minus sign corresponds to the lower subscript of b.

The $\mathscr{T}_{a_p b_q}$ must be stored explicitly. However, an interpolation table may be employed to ensure the $O(N)$ storage requirement independent of the structure's composition. This was not pursued in the implementation of the presented 2-D formulation, but would be necessary in a 3-D implementation of the method.

Upon replacing the first block row of (6) with (13), the complete system becomes

$$\begin{bmatrix} I + L_{aa} & -L_{ab}D_b & 0 & 0 \\ A_{ba} & A_{bb} & A_{bI} & A_{bd} \\ 0 & A_{Ib} & A_{II} & A_{Id} \\ 0 & A_{db} & A_{dI} & A_{dd} \end{bmatrix} \begin{bmatrix} \phi_a \\ \phi_b \\ \phi_I \\ \phi_d \end{bmatrix} = \begin{bmatrix} \phi_a^{\text{inc}} \\ 0 \\ 0 \\ 0 \end{bmatrix} \tag{24}$$

to be solved via the CG algorithm. Here, ϕ_a^{inc} is the vector containing the incident field evaluated on the nodes along Γ_a. The elements of the adjoint of the boundary integral operator required in the implementation of the CG algorithm are

where a denotes the adjoint operator and the superscript T denotes the matrix transpose.

In the above we outlined the major steps employed in the implementation of the proposed solution method. Of course, many details of importance to potential users were omitted, and the interested reader is referred to [15].

III. Computational Considerations

The fundamental advantage of this method is a reduction of storage requirements so that the scattering by electrically large bodies may be evaluated. To show that the low storage requirement of $O(N)$ is assured, we refer to Table I. Specifically, this table includes the memory requirements pertaining to the finite element matrix (FE), fast Fourier transforms (FT), boundary integral cross terms (cross) and conjugate gradient algorithm variables (CG). In the table, N_g is the total number of nodes and N_c is one less the number of elements connected to a particular node, and a typical value of five was assumed. Also, N_x and N_y are the number of nodes along Γ_a in the x and y directions, respectively.

TABLE I
SUMMARY OF MAJOR MEMORY CONSUMPTION

Major Memory Consumption ($N_x \geq N_y$)		
Item	Type	Count
FE	COMPLEX	$(\frac{N_c+1}{2})[N_g - 2(N_x + N_y)]$
FT	COMPLEX	$12N_x + 8N_y$
Cross	COMPLEX	$2N_x^2$
CG	COMPLEX	$4N_g$

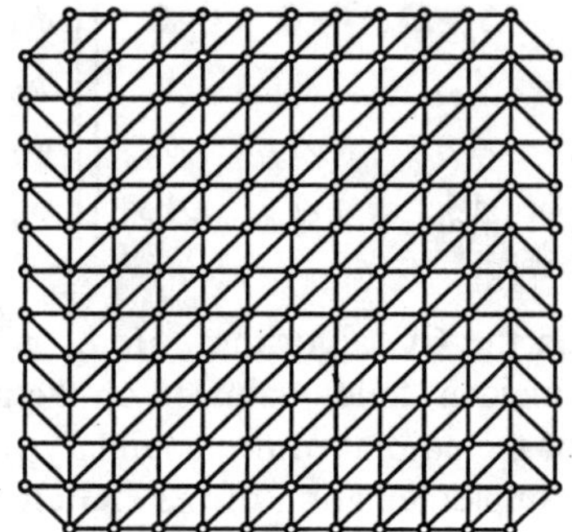

Fig. 3. Example of the mesh of a penetrable structure.

TABLE II
MEMORY CONSUMPTION FOR THE MESH ASSOCIATED WITH A PENETRABLE STRUCTURE

Item	Memory Consumption
FE	$(\frac{N_c+1}{2})(N_g - 4\sqrt{N_g})$
FT	$20\sqrt{N_g}$
Cross	$2N_g$
CG	$4N_g$
total	$\sim 9N_g$ $(N_c = 5)$

To put the quantities of Table I in terms of N_g, we consider the square mesh as shown in Fig. 3, corresponding to a penetrable body tightly enclosed by the "picture frame." As can be seen from the results in Table II, the $O(N_g)$ storage is guaranteed.

Since the primary importance of this method is storage reduction, a comparable level of efficiency with alternative methods is a bonus. A method with which a fair comparison is possible, is the standard CGFFT. This requires a 2-D FFT versus a few 1-D FFT's employed in the FE-CGFFT formulation. To perform the speed comparison we considered a $2 \times 2\lambda$ dielectric cylinder having $\epsilon_r = 4 - j.1$ and $\mu_r = 1$. It was found that the standard CGFFT algorithm (single incidence) required 170 s for each of the 33 iterations resulting in a total execution time of 5610 on an Apollo DN3500. The FE-CGFFT algorithm required 8.6 s for each of the 155 iterations for a total execution time of 1333 s. The proposed method is, thus, approximately four times faster, primarily because the FE–CGFFT method requires fewer FFT operations per iteration, combined with a fast finite element matrix-vector multiplication.

IV. RESULTS

To validate the accuracy of the method, the echowidth

$$\sigma = \lim_{\rho \to \infty} 2\pi\rho \frac{|\phi^s|^2}{|\phi^{\text{inc}}|^2} \tag{27}$$

was computed for several configurations. The far-zone scattered fields appearing in (27) can be computed from (see (7))

$$\phi^s(\bar{\rho}) = -\oint_{\Gamma_c} \left\{ G(\bar{\rho}, \bar{\rho}') \left[\frac{\partial}{\partial n'} \phi(\bar{\rho}') \right] - \phi(\bar{\rho}') \left[\frac{\partial}{\partial n'} G(\bar{\rho}, \bar{\rho}') \right] \right\} dl'. \tag{28}$$

This can be discretized by employing a pulse basis expansion with the pulses centered at the implied nodes on Γ_c in a manner similar to that used in (10). In doing so and employing the large argument approximation for the Hankel function, we obtain the far-zone field

$$\begin{aligned} \phi_{ff}^s(x, y) = -f_0(\rho) \cdot \Bigg\{ & \sum_{j=1}^{N_x} \left([j + k_0 \Delta \sin\theta] \{\phi_{a_1}\}_j - [j - k_0\Delta \sin\theta] \{\phi_{b_1}\}_j \right) \cdot f_1(\theta, y_{c_1}) e^{jk_0 x_j \cos\theta} \\ & + \sum_{j=1}^{N_y} \left([j - k_0\Delta \cos\theta] \{\phi_{a_2}\}_j - [j + k_0\Delta\cos\theta]\{\phi_{b_2}\}_j \right) \cdot f_2(\theta, x_{c_2}) e^{jk_0 y_j \sin\theta} \\ & + \sum_{j=1}^{N_x} \left([j - k_0\Delta\sin\theta]\{\phi_{a_3}\}_j - [j + k_0\Delta\sin\theta]\{\phi_{b_3}\}_j \right) \cdot f_1(\theta, y_{c_3}) e^{jk_0 x_j\cos\theta} \\ & + \sum_{j=1}^{N_y} \left([j + k_0\Delta\cos\theta]\{\phi_{a_4}\}_j - [j - k_0\Delta\cos\theta]\{\phi_{b_4}\}_j \right) \cdot f_2(\theta, x_{c_4}) e^{jk_0 y_j \sin\theta} \Bigg\} \end{aligned} \tag{29}$$

where

$$f_0(\rho) = \frac{1}{4}\sqrt{\frac{2j}{\pi k_0 \rho}} e^{-jk_0\rho} \tag{30}$$

$$f_1\left(\theta, y_{c_3^1}\right) = -e^{jk_0 y_{c_3^1} \sin\theta} \tag{31}$$

$$f_2\left(\theta, x_{c_4^2}\right) = -e^{jk_0 x_{c_4^2} \cos\theta}. \tag{32}$$

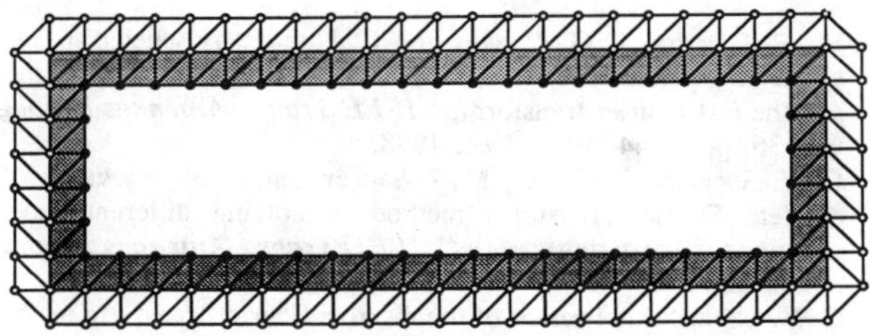

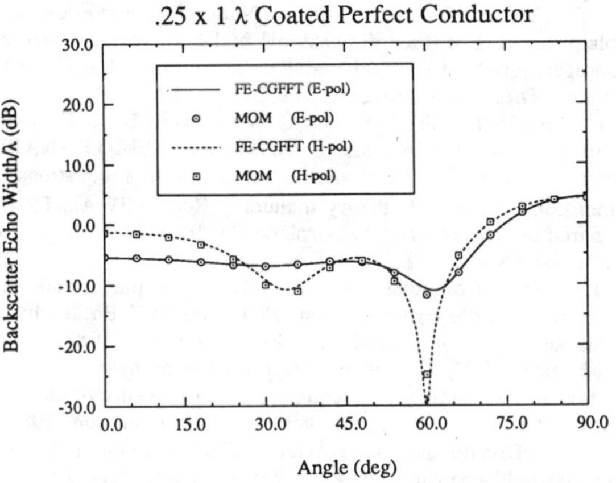

Fig. 4. E_z and H_z backscatter patterns for a 0.25 × 1 λ perfect conductor with a λ/20 thick material coating having $\epsilon_r = 5. - j.5$ and $\mu_r = 1.5 - j.5$.

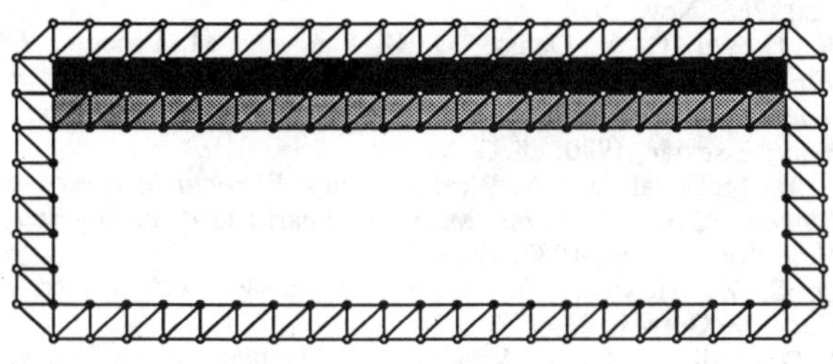

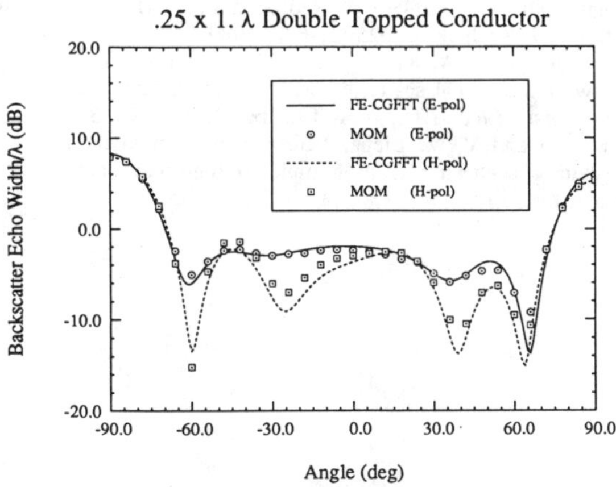

Fig. 5. E_z and H_z backscatter patterns for a .25 × 1 λ perfect conductor with two λ/20 thick material coatings on its top side. The lower layer has $\epsilon_r = 2. - j.5$, $\mu_r = 1.5 - j.2$ and the upper layer has $\epsilon_r = 4. - j.5$, $\mu_r = 1.5 - j.2$.

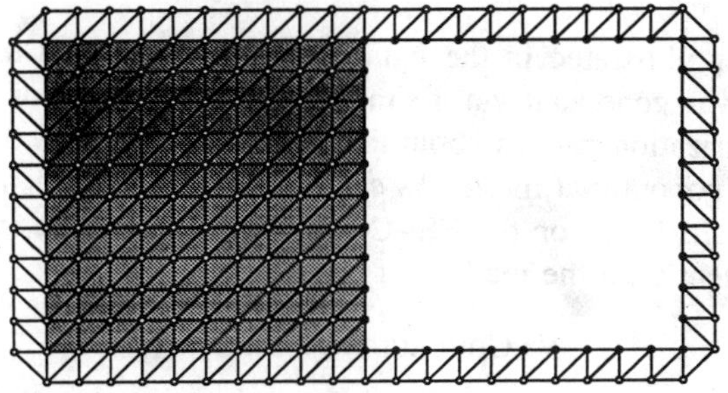

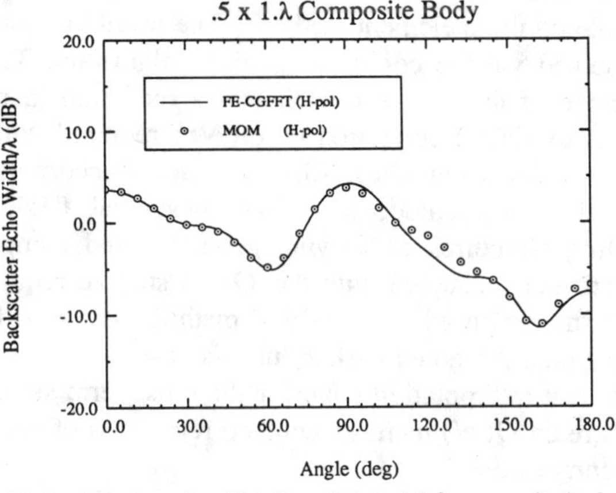

Fig. 6. H_z backscatter echowidth for a 0.5 × 1.0 λ composite body with a 0.5 × 0.5 λ dielectric having $\epsilon_r = 5. - j.5$, $\mu_r = 1.5 - j.5$ on the left and a 0.5 × 0.5 λ perfect conductor on the right.

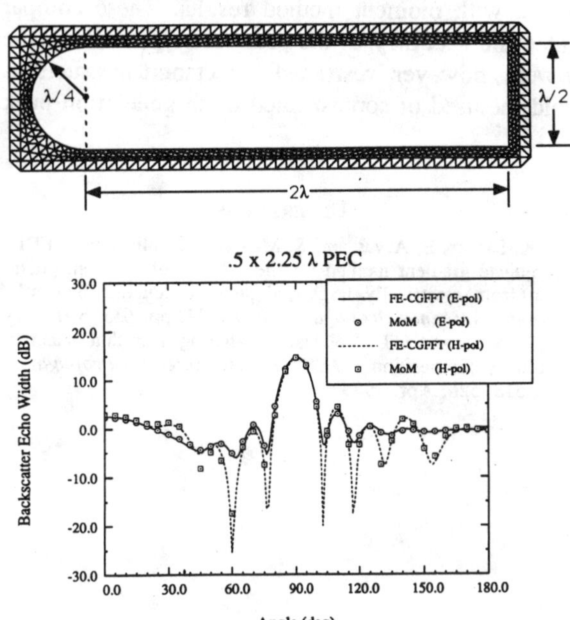

Fig. 7. E_z and H_z backscatter echowidth patterns for the illustrated perfectly conducting cylinder.

A number of echowidth patterns are presented in Figs. 4–6, which correspond to rectangular structures composed of perfect conductors and penetrable portions. In particular, Fig. 4 presents the E- and H-polarization patterns for a uniformly coated rectangular cylinder. The results in Fig. 5 correspond to a double coating (each 0.05 λ thick) on the top face of the rectangular cylinder. The pattern in Fig. 6 is the H-polarization echowidth by a 1.0 × 0.5λ rectangular cylinder whose left-half side is penetrable while the remaining portion is perfectly conducting. Deviations in these patterns from the comparison curves may be attributed to low sampling in the material media, since a λ/20 sampling was guaranteed only along the outer layer of elements (i.e., $\Delta = \lambda/20$) in all cases presented. Finally, Fig. 7 contains E- and H-polarization patterns for a nonrectangular conducting cylinder illustrated in the figure. It should be noted that the mesh was generated via a simple algorithm and error in the H-polarization pattern about the angle 45° may be attributed to this nonoptimal mesh. As can be seen from these figures, the results based on the FE–CGFFT formulation are in good agreement with the method of moments [16], [17].

V. Concluding Remarks

A procedure was developed for computing the scattering by 2-D structures. The procedure combines the boundary integral and finite element methods. The resulting system was then solved via the conjugate gradient algorithm. The main advantage of the new approach was a reduction in memory demand to $O(N)$ compared to $O(N^2)$ required with traditional solution techniques. Also, because it incorporates the finite element method, it allows substantial flexibility in modeling structures of varying geometry and composition. This attribute, coupled with the $O(N)$ storage requirement, makes the proposed FE–CGFFT method attractive for 3-D applications. As noted earlier, in this case one must employ an efficient interpolation table for the cross term submatrices to ensure an $O(N)$ memory demand regardless of the geometry's composition.

The computational efficiency of the method was examined and found favorable in comparison with competing methodologies. Several backscatter patterns were also presented and compared with moment method results. These comparisons validated the accuracy of the proposed FE–CGFFT method. They were, however, restricted to rectangular structures only to avoid the used of sophisticated mesh generation packages.

References

[1] T. K. Sarkar, E. Arvas, and S. M. Rao, "Application of FFT and the conjugate gradient method for the solution of electromagnetic radiation from electrically large and small conducting bodies," *IEEE Trans. Antennas Propagat.*, vol. AP-34, pp. 635–640, May 1986.

[2] T. J. Peters and J. L. Volakis, "Scattering from thin planar layers of arbitrary composition," *IEEE Trans. Antennas Propagat.*, AP-36, pp. 518–526, Apr. 1988.

[3] M. F. Cátedra, J. G. Cuevas and L. Nuño, "A scheme to analyze conducting plates of resonant size using the conjugate gradient method and the fast Fourier transform," *IEEE Trans. Antennas Propagat.*, vol. 36, pp. 1744–1752, Dec. 1988.

[4] C. Y. Shen, K. J. Gloner, M. I. Sancer, and A. D. Varvatsis, "The discrete Fourier transform method of solving differential-integral equations in scattering theory," *IEEE Trans. Antennas Propagat.*, to appear.

[5] C. Y. Shen, "Application of the discrete Fourier transform method to thin dielectric structures," *IEEE Trans. Antennas Propagat.*, vol. 37, pp. 1277–1283, Oct. 1989.

[6] M. F. Cátedra, E. Gago, and L. Nuño, "A numerical scheme to obtain the RCS of three-dimensional bodies of resonant size using the conjugate gradient method and the fast Fourier transform," *IEEE Trans. Antennas Propagat.*, vol. 37, pp. 528–537, May 1989.

[7] J. E. Bussoletti, F. T. Johnson, K. W. Sidwell, B. L. Everson, D. P. Young, R. H. Burkhart and S. S. Jamant, "EM-TRANAIR: Computer a program for the solution of Maxwell's equations in three dimensions: Volume I, theory manual," Rep. AFWAL-TR-87-3802, prepared by Boeing Military Airplane Co. for USAF Wright Aeronautical Labs., Sept. 1988.

[8] T. J. Peters, "Computation of the scattering by planar and non-planar plates using a conjugate gradient FFT method," Ph.D. dissertation, Radiation Lab., Univ. Michigan, Ann Arbor, MI, 1988.

[9] J. M. Jin and V. V. Liepa, "Application of hybrid finite element method to electromagnetic scattering from coated cylinders," *IEEE Trans. Antennas Propagat.*, vol. 36, pp. 50–54, Jan. 1988.

[10] B. H. McDonald and A. Wexler, "Finite-element solution of unbounded field problems," *IEEE Trans. Microwave Theory Tech.*, vol. MTT-20, pp. 841–847, Dec. 1972.

[11] K. K. Mei, "Unimoment method of solving antenna and scattering problems," *IEEE Trans. Antennas Propagat.*, vol. AP-22, pp. 760–766, Nov. 1974.

[12] K. L. Wu, G. Y. Delisle, D. G. Fang and M. Lecours, *PIER2: Finite Element and Finite Difference Difference Methods in Electromagnetic Scattering*. J. A. Kong and M. A. Morgan, Eds. New York: Elsevier, 1990, ch. 3.

[13] B. H. McDonald and A. Wexler, *Finite Element in Electrical and Magnetic Field Problems*, M. V. K. Chari and P. P. Silvester, Eds. New York: Wiley, 1980, ch. 9.

[14] O. C. Zienkiewicz, *The Finite Element Method*, 3rd ed. New York: McGraw-Hill, 1977.

[15] J. D. Collins and J. L. Volakis, "A Combined Finite Element and Boundary Integral Formulation for Solution via CGFFT of 2-Dimensional Scattering Problems," Radiation Lab., Univ. Michigan, Ann Arbor, MI, Tech. Rep. 025921-6-T, 1989.

[16] M. A. Ricoy, S. A. Kilberg, and J. L. Volakis, "Integral equations for two-dimensional scattering with further reduction in unknowns," *Proc. Inst. Elec. Eng.*, vol. 136, pt. H, pp. 298–304, Aug. 1989.

[17] J. M. Jin and V. V. Liepa, "Simple moment method program for computing scattering from complex cylindrical obstacles," *Proc. Inst. Elec. Eng.*, vol. 136, pt. H, no. 4, pp. 321–329, Aug. 1989.

Coupling of Finite Element and Moment Methods for Electromagnetic Scattering from Inhomogeneous Objects

XINGCHAO YUAN, MEMBER, IEEE, DANIEL R. LYNCH, AND JOHN W. STROHBEHN, SENIOR MEMBER, IEEE

Abstract—**A new hybrid formulation is proposed, which combines the method of moments (MM) with the finite element method (FEM) to solve electromagnetic scattering and/or absorption problems involving inhomogeneous media. The basic technique is to apply the equivalence principle and transform the original problem into interior and exterior problems, which are coupled on the exterior dielectric body surface through the continuities of the tangential electric field and magnetic field. The interior problem involving inhomogeneous medium is solved by the FEM and the exterior problem is solved by the MM. The coupling of the interior and exterior problems on their common surface results in a matrix equation for the equivalent current sources for the interior and exterior problems. By combining advantages of both methods, complicated inhomogeneous problems with arbitrary geometry are treated in a straightforward manner. The validity and accuracy of the formulation are checked by two-dimensional numerical results, which are compared with the exact eigenfunction solution, the unimoment solution, and Richmond's pure moment solution.**

INTRODUCTION

THE PROBLEM OF ELECTROMAGNETIC scattering and absorption due to inhomogeneous, lossy, and arbitrarily shaped dielectric bodies has been extensively dealt with in the literature because of its presence in many practical situations including microwave hyperthermia for cancer therapy, coupling to missiles with dielectric-filled apertures, and performance of communication antennas in the presence of dielectric and magnetic inhomogeneities. When the size of the dielectric object is neither large nor small compared to the wavelength of the excitation, asymptotic methods cannot be used and a rigorous solution of Maxwell's equations is required.

Many ways have been developed to treat such scattering problems. Among them are the pure method of moments solution using either surface integral [1] or volume integral equations [2]–[5], the finite difference time domain (FD-TD) method [6], [7], the unimoment method [8], [9], and hybrid methods combining the finite method (either finite element [10] or finite difference) with the integral equation method [11]–[16]. It is a well-known fact that integral equation methods such as the moment method treat unbounded problems very effectively but they become computationally intensive when complex inhomogeneities are present. In contrast, inhomogeneities are easily handled by finite methods. The finite element method requires less computer time and storage because it results in a sparse and banded matrix. The FD-TD method uses an iterative approach and is therefore not memory intensive. However, finite methods are most suitable for boundary value problems. Special steps need to be taken if an unbounded region is present. Clearly, any hybrid method that retains the most efficient characteristics of both finite methods and integral equation methods is computationally advantageous. Such hybrid methods developed include the hybrid moment method/FD-TD [11], the hybrid finite element method and extended boundary condition integral method (or FEBI) [12], and the hybrid finite element and the boundary element method (or BEM/FEM) [13]–[15]. A general hybrid formulation referred to as the field feedback formulation (or F^3) can be found in [16]. The basic techniques of these methods are to use a finite method to treat the bounded, inhomogeneous region and to use an integral equation method to treat the unbounded homogeneous region.

In this paper, we present a new hybrid formulation, which combines the method of moments (MM) [17] with the finite element method presented in [15]. This new MM/FEM hybrid has comparable advantages as the above hybrid FEBI and BEM/FEM methods, and in addition facilitates the use of established MM procedures and formulations in terms of equivalent current sources. The particular MM approach utilized here, adapted from [18], [19], is oriented toward scattering of incident waves originating from a distant source. The coupling is achieved explicitly through matching the boundary conditions on the dielectric body surface. Following a detailed formulation, we show numerical examples for two-dimensional problems under transverse magnetic (TM) and transverse electric (TE) plane wave excitations to assess the validity and accuracy of the method.

FORMULATION

An arbitrarily shaped, inhomogeneous, and lossy dielectric body under the illumination of an incident field $(\mathbf{E}^i, \mathbf{H}^i)$ is shown in Fig. 1. The dielectric body bounded by a surface S is characterized by $(\mu(\mathbf{r}), \epsilon(\mathbf{r}))$. The region inside the surface S is called region b and the unbounded region outside S is called region a, which is characterized by (μ_0, ϵ_0) and

Manuscript received August 25, 1988; revised March 15, 1989. This work was supported in part by NIH Grant R01 CA37245; and by NSF Grant CEE-835-2226.

The authors are with the Thayer School of Engineering, Dartmouth College, Hanover, NH 03755.

IEEE Log Number 8933573.

Reprinted from *IEEE Trans. Antennas Propagat.*, vol. 38, no. 3, pp. 386–393, 1990.

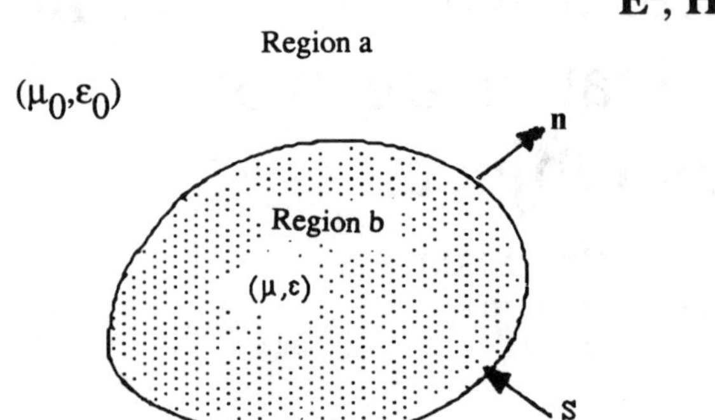

Fig. 1. Original problem: electromagnetic scattering due to an inhomogeneous dielectric body bounded by S.

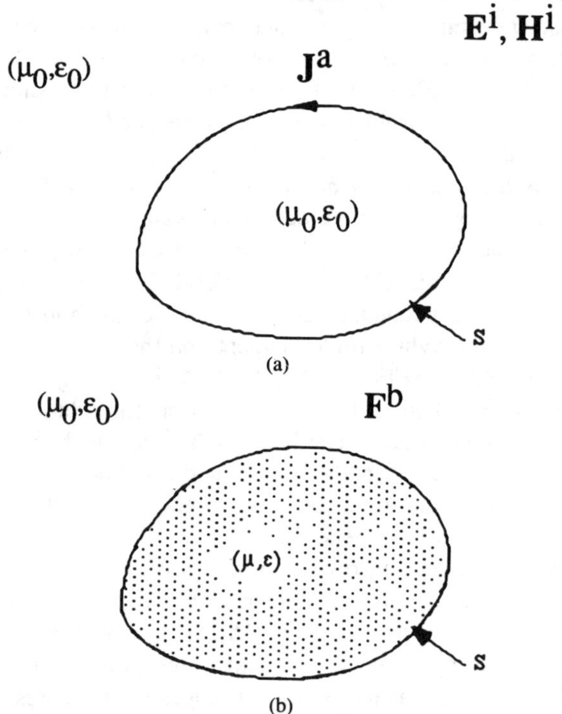

Fig. 2. The equivalence of the original problem. (a) Equivalent problem for region a, where $\mathbf{J}^a$ is the equivalent electric current on S. (b) Boundary value problem for region b, where $\mathbf{F}^b$ is the tangential magnetic field on S.

assumed to be vacuum without loss of generality. The objective is to determine the electromagnetic fields everywhere in space.

The strategy taken here is to use the equivalence principle [20, ch. 3] to transform the original problem into equivalent problems in regions a and b, which are coupled through the continuity conditions on the tangential components of $\mathbf{E}$ and $\mathbf{H}$ on the dielectric body surface S. A similar technique was used in [18] and [19] to solve the problem of electromagnetic scattering due to homogeneous dielectric cylinders partially covered by conductors.

The equivalent problem for region a is shown in Fig. 2(a). The electromagnetic fields $(\mathbf{E}^a, \mathbf{H}^a)$ in region a can be considered to be the superposition of the incident field $(\mathbf{E}^i, \mathbf{H}^i)$ and the scattered field $(\mathbf{E}^s, \mathbf{H}^s)$, where the scattered field is produced by an equivalent surface electric current $\mathbf{J}^a$ on S radiating in the infinite space filled with (μ_0, ϵ_0) everywhere. Specifically,

$$\mathbf{E}^a = \mathbf{E}^i + \mathbf{E}^s(\mathbf{J}^a) \tag{1}$$

$$\mathbf{H}^a = \mathbf{H}^i + \mathbf{H}^s(\mathbf{J}^a). \tag{2}$$

The equivalent problem for region b is shown in Fig. 2(b). The electromagnetic fields in region b are uniquely determined by the unknown tangential magnetic field on S, denoted by $\mathbf{F}^b$. That is,

$$\mathbf{E}^b = \mathbf{E}^b(\mathbf{F}^b) \tag{3}$$

$$\mathbf{H}^b = \mathbf{H}^b(\mathbf{F}^b) \tag{4}$$

where the superscript b on the field quantities indicates those actually present in the inhomogeneous dielectric body.

The equivalent electric current $\mathbf{J}^a$ and the tangential magnetic field $\mathbf{F}^b$ on S can be found by applying the boundary conditions on S. Namely, both the tangential electric field and the tangential magnetic field must be continuous across S, i.e.,

$$\mathbf{n} \times \mathbf{E}^a = \mathbf{n} \times \mathbf{E}^b, \qquad \text{on } S \tag{5}$$

$$\mathbf{n} \times \mathbf{H}^a = \mathbf{n} \times \mathbf{H}^b = \mathbf{F}^b, \qquad \text{on } S \tag{6}$$

where $\mathbf{n}$ is the unit normal on S pointing outward from region b, as indicated in Fig. 1. Substituting (1) and (2) into (5) and (6) and rearranging terms, we obtain

$$-\mathbf{n} \times \mathbf{E}^s(\mathbf{J}^a) + \mathbf{n} \times \mathbf{E}^b(\mathbf{F}^b) = \mathbf{n} \times \mathbf{E}^i, \qquad \text{on } S \tag{7}$$

$$-\mathbf{n} \times \mathbf{H}^s(\mathbf{J}^a) + \mathbf{F}^b = \mathbf{n} \times \mathbf{H}^i, \qquad \text{on } S. \tag{8}$$

Equations (7) and (8), together with the governing Maxwell's equations, are sufficient to determine the equivalent electric current $\mathbf{J}^a$ and the tangential magnetic field $\mathbf{F}^b$. The continuity of the normal components need not be addressed explicitly, since it follows from (7) and (8) and the Maxwell's equations.

Numerical Method

A. Moment Solution

Equations (7) and (8) can be reduced to weak-form matrix equations by the method of moments [17]. To apply the method of moments, we approximate $\mathbf{J}^a$ and $\mathbf{F}^b$ by

$$\mathbf{J}^a = \sum_{j=1}^{N} I_j^a \mathbf{J}_j \tag{9}$$

$$\mathbf{F}^b = \sum_{j=1}^{N} F_j^b \mathbf{J}_j \tag{10}$$

where $\{\mathbf{J}_j\}$ is a set of chosen expansion functions, N is the number of functions chosen, and $\{I_j^a\}$ and $\{F_j^b\}$ are two sets of coefficients to be determined. (Note that we employ the same expansion functions for the equivalent current in region a and the tangential magnetic field in region b. This is convenient but not necessary.) A symmetric product is defined by

$$\langle \mathbf{A}, \mathbf{B} \rangle = \int_S \mathbf{A} \cdot \mathbf{B} \, ds \tag{11}$$

where $\mathbf{A}$ and $\mathbf{B}$ are two vector functions defined on S. Assume $\{\mathbf{W}_i, \, i = 1, 2, \cdots, N\}$ to be a set of weighting functions defined on S. The weak-form discretization of (7) and (8)

$$\begin{bmatrix} [Z^a] & [Z^b] \\ [Y^a] & [Y^b] \end{bmatrix} \begin{bmatrix} I^a \\ F^b \end{bmatrix} = \begin{bmatrix} V \\ U \end{bmatrix} \tag{12}$$

is obtained by substituting (9) and (10) into (7) and (8) and taking the symmetric product (11) of both resulting equations with the weighting functions $\{\mathbf{W}_i\}$. In (12), I^a, F^b, V, and U are $N \times 1$ column vectors. The ith elements of I^a and F^b are simply I_i^a and F_i^b, respectively; and the ith elements of V and U are v_i and u_i given by

$$v_i = \langle \mathbf{W}_i, \mathbf{n} \times \mathbf{E}^i \rangle \tag{13}$$

$$u_i = \langle \mathbf{W}_i, \mathbf{n} \times \mathbf{H}^i \rangle. \tag{14}$$

$[Z^a]$, $[Z^b]$, $[Y^a]$, and $[Y^b]$ are $N \times N$ matrices and their ijth elements are given by

$$Z_{ij}^a = \langle \mathbf{W}_i, -\mathbf{n} \times \mathbf{E}^s(\mathbf{J}_j) \rangle \tag{15}$$

$$Z_{ij}^b = \langle \mathbf{W}_i, \mathbf{n} \times \mathbf{E}^b(\mathbf{J}_j) \rangle \tag{16}$$

$$Y_{ij}^a = \langle \mathbf{W}_i, -\mathbf{n} \times \mathbf{H}^s(\mathbf{J}_j) \rangle \tag{17}$$

$$\begin{aligned} Y_{ij}^b &= \langle \mathbf{W}_i, \mathbf{n} \times \mathbf{H}^b(\mathbf{J}_j) \rangle \\ &= \langle \mathbf{W}_i, \mathbf{J}_j \rangle. \end{aligned} \tag{18}$$

Equations (13)–(18) have to be evaluated before (12) can be solved to obtain I^a and F^b and in turn to assemble the equivalent sources $\mathbf{J}^a$ and $\mathbf{F}^b$ given in (9) and (10). Once the weighting functions $\{\mathbf{W}_i\}$ and the expansion functions $\{\mathbf{J}_j\}$ are chosen, (13)–(15) and (17) can be evaluated easily since the incident fields $(\mathbf{E}^i, \mathbf{H}^i)$ are given, and the electric field produced by $\mathbf{J}_j$ in the infinite space (μ_0, ϵ_0) is given by [20]

$$\mathbf{E}^s(\mathbf{J}_j) = \frac{-j\omega\mu_0}{4\pi} \int_S \mathbf{J}_j G(|\mathbf{r} - \mathbf{r}'|) \, ds' + \frac{\nabla}{4\pi j\omega\epsilon_0} \int_S \nabla_s' \cdot \mathbf{J}_j G(|\mathbf{r} - \mathbf{r}'|) \, ds' \tag{19}$$

where $G(\cdot)$ is the Green's function given by [20, eq. (5-88)] in two dimensions and by

$$G(|\mathbf{r} - \mathbf{r}'|) = \frac{e^{-jk_0|r-r'|}}{|r - r'|} \tag{20}$$

in three dimensions, where a harmonic time dependence $\exp(j\omega t)$ is assumed and suppressed, and $k_0 = \omega\sqrt{\mu_0\epsilon_0}$ is the wavenumber in region a. The magnetic field $\mathbf{H}^s(\mathbf{J}_j)$ is related to the electric field $\mathbf{E}^s(\mathbf{J}_j)$ by the Maxwell's equations. Specially, if the observation point $\mathbf{r}$ is not on an edge, the tangential magnetic field takes the following form [23, eq. (11.34)]

$$\begin{aligned} \mathbf{n} \times \mathbf{H}^s(\mathbf{J}_j) &= \lim_{r \to S} \frac{\mathbf{n} \times \nabla \times \mathbf{E}^s(\mathbf{J}_j)}{-j\omega\mu_0} \\ &= \mathbf{J}_j/2 + \mathbf{n} \times \frac{1}{4\pi} \int_S \mathbf{J}_j \times \nabla' G(|\mathbf{r} - \mathbf{r}'|) \, ds' \end{aligned} \tag{21}$$

where $\mathbf{r}$ approaches S from the region a side of S. Equations (19) and (21) may be substituted directly into (15) and (17) to evaluate Z_{ij}^a and Y_{ij}^a.

The evaluation of $\mathbf{n} \times \mathbf{H}^b(\mathbf{J}_j)$ in (18) is trivial since on S $\mathbf{n} \times \mathbf{H}^b(\mathbf{J}_j)$ is simply $\mathbf{J}_j$ itself the tangential magnetic field. Thus, (18) can be easily obtained once $\mathbf{W}_i$ and $\mathbf{J}_j$ are chosen. The evaluation of (16) requires knowledge of $\mathbf{E}^b(\mathbf{J}_j)$. We resort to the finite element method as follows.

B. The Finite Element Solution of $\mathbf{E}^b(\mathbf{J}_j)$

Since $\mathbf{F}^b$ is used to denote the equivalent tangential magnetic field on S and from (3), (4), and (10), $\mathbf{E}^b(\mathbf{J}_j)$ is the electric field due to the equivalent tangential magnetic field $\mathbf{J}_j$ defined everywhere on S. For piecewise approximation, $\mathbf{J}_j$ is usually zero except on a small portion of S (e.g. triangle patches of S). Hence, the problem of finding $\mathbf{E}^b(\mathbf{J}_j)$ in region b is a pure boundary value problem and its general solution is available by the finite element method. As described in [15], we first expand the unknown electric field in region b in terms of conventional real valued, scalar finite element functions ϕ_j defined in the volume enclosed by S:

$$\mathbf{E}^b = \sum_{j=1}^{N^t} \mathbf{E}_j^b \phi_j \tag{22}$$

where N^t is the total number of nodes including all interior nodes and those on the surface S, and $\{\mathbf{E}_j^b\}$ are the unknown electric field values at the nodes. The bases ϕ_j are typically locally supported simple polynomials, continuous with piecewise continuous first derivatives. We then obtain a weak form of the Maxwell's equations by requiring orthogonality with a set of weighting functions. In particular, we employ the Galerkin form [15][1]

$$\left\langle \left(\frac{1}{j\omega\mu} \nabla \times \mathbf{E}^b \right) \times \nabla\phi_i \right\rangle_v + \langle j\omega\epsilon \mathbf{E}^b \phi_i \rangle_v = \int_S \mathbf{n} \times \mathbf{H}^b \phi_i \, ds = \int_S \mathbf{F}^b \phi_i \, ds \tag{23}$$

where $\langle \cdot \rangle_v$ indicates integrating over the entire inhomogeneous volume bounded by S. Note that we do not require the expansion of $\mathbf{H}$ in the interior. Since $\mathbf{n} \times \mathbf{H}$ must be continuous across the internal dielectric interfaces, the boundary integral of (23) mutually cancels along such boundaries and the right-hand side of (23) survives only on the exterior boundary S. Hence, (23) expresses the electric field inside and on S in terms of the tangential magnetic field on S.

Substituting (22) into (23) and requiring the satisfaction of

[1] A harmonic time dependence $\exp(-j\omega t)$ was assumed in [15]. The sign is reversed for internal consistency.

(23) for $i = 1$ through N^t, we obtain the finite element matrix equation

$$[A][\mathbf{E}^b] = [\mathbf{F}^b] \tag{24}$$

where $[A]$ is a square, sparse, banded, and nonsingular $N^t \times N^t$ matrix. The detailed form of $[A]$ can be found in [14], [15]. $[\mathbf{E}^b]$ and $[\mathbf{F}^b]$ are $N^t \times 1$ column vectors containing the unknown electric field at nodes and the integral of the tangential magnetic field on the exterior boundary S times the weighting function ϕ_i. Therefore, $\mathbf{E}^b(\mathbf{J}_j)$ can be obtained by evaluating the right-hand side of (24) by replacing $\mathbf{F}^b$ in (23) with $\mathbf{J}_j$ and solving the matrix equation (24). N such solutions are required, but the matrix $[A]$ needs to be factored (e.g. LU decomposed) only once.

Finally, the obtained $\mathbf{E}^b(\mathbf{J}_j)$ on S is substituted into (16) to evaluate Z^b_{ij}. Equation (12) can then be solved to obtain I^a and F^b. Hence, the electromagnetic fields at any point in space can be obtained by using (1)–(4), (9), and (10); (19)–(21) for the exterior solution, and (24) for the interior solution. In the later case, retention of the LU decomposition of $[A]$ is efficient.

C. Dual Formulations

In the above formulation, we have used the equivalent sources $\mathbf{J}^a$ and $\mathbf{J}^b$ to produce the correct field in region a and b. Other equivalent sources are also possible. For instance, the combinations of $(\mathbf{M}^a, \mathbf{F}^b)$, $(\mathbf{J}^a, \mathbf{M}^b)$, and $(\mathbf{M}^a, \mathbf{M}^b)$ are all possible and valid equivalent sources, where $\mathbf{M}^a$ and $\mathbf{M}^b$ are equivalent magnetic current and tangential electric field on S, respectively. The determining factor in choosing which set of equivalent sources would be used is the computational simplicity and efficiency. In the two-dimensional TM case, $(\mathbf{J}^a, \mathbf{F}^b)$ is more convenient and $(\mathbf{M}^a, \mathbf{M}^b)$ is more convenient in the two dimensional TE case. For three-dimensional problems, it is not clear there is a preferred choice. In the case of using equivalent tangential electric field source $\mathbf{M}^b$ for the equivalence in region b, the following equation [15]

$$\left\langle \left(\frac{1}{j\omega\epsilon} \nabla \times \mathbf{H}^b \right) \times \nabla\phi_i \right\rangle_v + \langle j\omega\mu \mathbf{H}^b \phi_i \rangle_v = -\int \mathbf{n} \times \mathbf{E}^b \phi_i \, ds \tag{25}$$

should be used to find the magnetic field $\mathbf{H}$ inside S in terms of the tangential electric field $\mathbf{E}$ on S.

Two-Dimensional Scattering

The formulation given above can be applied to both two- and three-dimensional problems. However, only the two-dimensional results are given here. We wish to present the numerical results for three-dimensional problems in a future article.

In the two-dimensional case, the inhomogeneous dielectric body becomes an inhomogeneous infinitely long dielectric cylinder, which can be defined by its cross section on the xy plane assuming the cylinder axis coincides with the z-axis. The incident waves can be separated into transverse magnetic case and transverse electric case. If the incident field is a plane wave impinging upon the cylinder with an angle ϕ^i, then it can be defined by

$$\mathbf{E}^i = \mathbf{u}_z E_z = \mathbf{u}_z e^{jk_0(x \cos\phi^i + y \sin\phi^i)} \tag{26}$$

for the TM case, and

$$\mathbf{H}^i = \mathbf{u}_z H_z = \mathbf{u}_z e^{jk_0(x \cos\phi^i + y \sin\phi^i)} \tag{27}$$

for the TE case, where k_0 is a wavenumber in the free space. Note these two cases are completely dual to each other. That is, one can be obtained from the other by interchanging symbols according to [20, ch. 3]. In either TM or TE case, the equivalent electric or magnetic current sources will also be z-directed. Thus, a simple pulse function (as defined in [19]) can be used as the current expansion function in (9), (10) and also as the weighting function; and the impedance matrix $[Z^a]$ and the admittance matrix $[Y^a]$ may be expressed as in [19]. The admittance matrix $[Y^b]$ is a diagonal matrix since the electric (or magnetic) current expansion function resides only on a single segment of the contour. By a simple matrix manipulation of (12), one needs to solve only an $N \times N$ matrix equation for I^a instead of a $2N \times 2N$ matrix equation for both I^a and F^b simultaneously since the inversion of $[Y^b]$ is trivial. In the FEM solution, the same technique as in [15] is used. The components of $[\mathbf{F}^b]$ for a given $\mathbf{J}_j$ in (24) will, in this case, be nonzero on only two elements. Equation (24) is triangularized only once but the back substitutions are performed N times to obtain $\{\mathbf{E}^b(\mathbf{J}_j), j = 1, 2, \cdots, N\}$ in region b.

To show validity and accuracy of the hybrid MM/FEM method, we have compared our results with the exact eigensolutions and those appearing in the literature. Both far field and near field are examined. The echo width L_e is defined by [20, eq. (7-121)]

$$L_e = \lim_{\rho\to\infty} \left(2\pi\rho \frac{|E^s|^2}{|E^i|^2} \right) \tag{28}$$

for the TM case, and by

$$L_e = \lim_{\rho\to\infty} \left(2\pi\rho \frac{|H^s|^2}{|H^i|^2} \right) \tag{29}$$

for the TE case.

In the following numerical examples involving circular cylinders, the finite element mesh is generated by dividing the radius by N_r divisions and 2π by N_ϕ divisions. The total number of nodes N^t is $N_r \times N_\phi + 1$, where the last node resides at the center of the cylinder. The number N in (9) and (10) is equal to N_ϕ. The inner most circle is filled with triangles and the rest of the annular regions are filled with quadrilaterals. Hence, there are N_ϕ triangles and $(N_r - 1) \times N_\phi$ quadrilaterals in the finite element mesh.

Figs. 3 and 4 show the magnitude and phase of the electric field along the x axis (AA') and y axis (BB') inside a circular homogeneous dielectric cylinder, under a TM plane wave incidence ($\phi^i = 180°$). Fig. 5 shows the echo width of the same

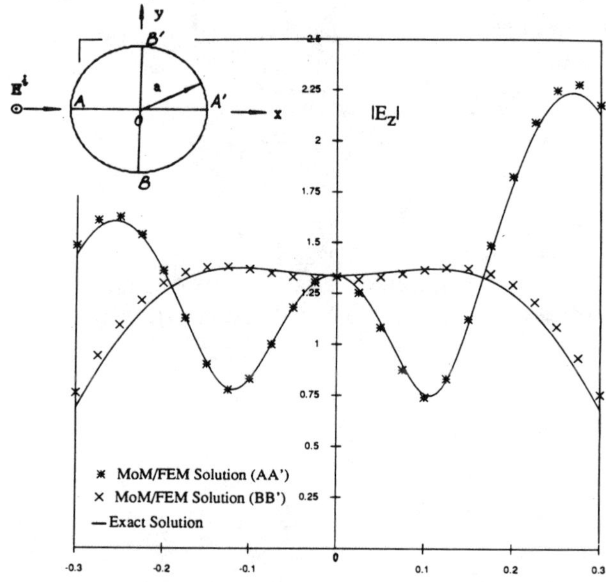

Fig. 3. Amplitude of the electric field along x axis (AA') and y axis (BB') inside a circular homogeneous dielectric cylinder ($a = 0.3\ \lambda$, $\sigma = 0.0$, $\epsilon_r = 4.0$, $f = 300$ MHz, $\phi^i = 180°$, and TM case). (MM/FEM solution compared to exact eigenfunction solution.)

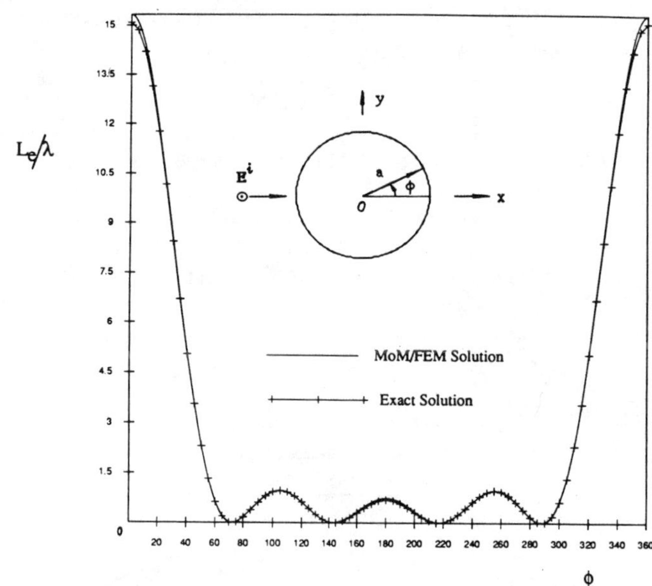

Fig. 5. Echo width of a circular homogeneous dielectric cylinder ($a = 0.3\ \lambda$, $\sigma = 0.0$, $\epsilon_r = 4.0$, $f = 300$ MHz, $\phi^i = 180°$, and TM case). (MM/FEM solution compared to exact eigenfunction solution.)

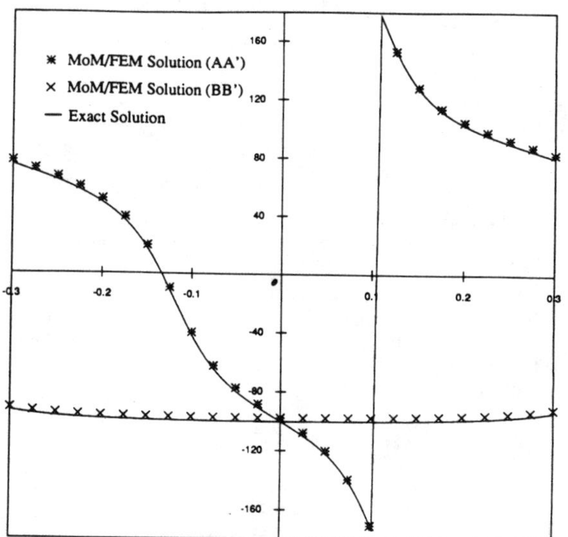

Fig. 4. Phase of the electric field along x axis (AA') and y axis (BB') inside a circular homogeneous dielectric cylinder ($a = 0.3\ \lambda$, $\sigma = 0.0$, $\epsilon_r = 4.0$, $f = 300$ MHz, $\phi^i = 180°$, and TM case). (MM/FEM solution compared to exact eigenfunction solution.)

cylinder under the TM plane wave incidence. In Figs. 3–5, the classical eigenfunction solution is obtained by summing the Fourier series up to 30 terms and the hybrid MM/FEM results are obtained by letting $N_r = 12$ and $N_\phi = 32$. The agreement of those two solutions is good with the maximum relative error within 4%. To show the convergence rate of the method, the numerical results obtained by varying N_r and N_ϕ are given in Table I. As N_ϕ increases from 22 to 42 (fixing N_r at 12), the relative error decreases from 6.74% to 2.068% (at point A). Note there are about 11 subsections per wavelength on the cylinder boundary when $N_\phi = 22$, and the numerical results are already in good agreement with the exact solutions. This is consistent with the general convergence rate of the moment method. When N_r increases from 12 to 20 (fixing N_ϕ at 32), little changes results, which shows the interior finite element solution is satisfactory with more than 20 subsections per wavelength. Finally, one may notice that slightly better results are obtained at the center of the cylinder (point O) than at the boundaries (points A and A'). Further study reveals that the elements span smaller area toward the center. We believe these results give very good evidence of the validity of the method.

To compare the MM/FEM method with other existing techniques, we compute the echo width of an off-centered circular dielectric cylinder under either TM or TE plane wave incidence, which is shown in Fig. 6 and was also obtained by the unimoment method [9]. The agreement is once again excellent. Figs. 7 and 8 show the echo widths, obtained by MM/FEM method and by Richmond [2], [22], of a circular and a semicircular dielectric shell under either TM or TE plane wave illumination. These two solutions generally agree very well except there are some discrepancies in the backscattering region for the semicircular shell under the TE plane wave incidence. The MM/FEM solution is shown to be converged by increasing N^t and N. It is not clear what caused the discrepancy. Nevertheless, Figs. 6–8 do demonstrate further the validity and accuracy of the MM/FEM method.

Since the FEM method treats the inhomogeneity in a very natural way, the hybrid MM/FEM method is capable of handling the scattering and/or absorption properties of any dielectric object with arbitrary shape and/or arbitrary inhomogeneity. Once a finite element mesh containing proper material properties is given, (for instance, generated by an automatic mesh generator), the electromagnetic fields at any point in space and other related quantities can be obtained by running the code without other changes required. Figs. 9 and 10 show the echo widths of a four layered circular dielectric cylinder under either TM or TE plane wave incidence. All dielectric layers are assumed to have the same conductivity σ. Also shown are the variations of the echo width with the change of the conductivity.

TABLE I
CONVERGENCE TEST ON CYLINDER IN FIG. 3

	Case 1 $N_r=12,\ N_\phi=22$		Case 2 $N_r=12,\ N_\phi=32$		Case 3 $N_r=12,\ N_\phi=42$		Case 4 $N_r=20,\ N_\phi=32$		
	MM/FEM	Error %	MM/FEM	Error %	MM/FEM	Error %	MM/FEM	Error %	Exact
$\|E_z\|$ at point A*	1.5276	6.74	1.4793	3.368	1.4607	2.068	1.4773	3.23	1.4311
$\|E_z\|$ at point O*	1.3160	1.68	1.3313	0.536	1.3369	0.114	1.3287	0.735	1.3385
$\|E_z\|$ at point A'	2.2295	4.611	2.1798	2.280	2.1614	1.415	2.1768	2.135	2.1312
L_e/λ at $\phi=0$	15.5116	2.842	15.3249	1.604	15.2464	1.06	15.3010	1.446	15.0830

* Points A, O, and A' are defined in Fig. 3.

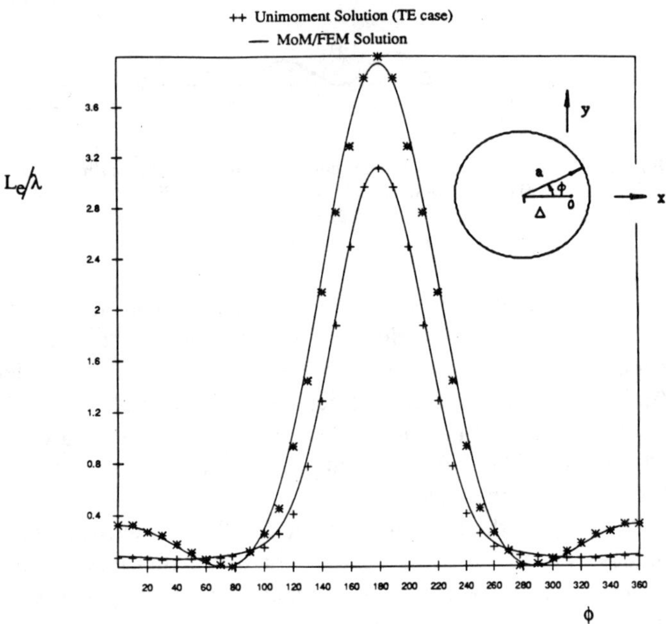

Fig. 6. Echo width of an off-centered circular homogeneous dielectric cylinder $a = 0.3\ \lambda$, $\Delta = 0.2\ \lambda$, $\sigma = 0.0$, $\epsilon_r = 2.0$, $f = 300$ MHz, $\phi^i = 0°$) under either TE or TM plane wave incidence. (MM/FEM solution compared to unimoment solution [9].)

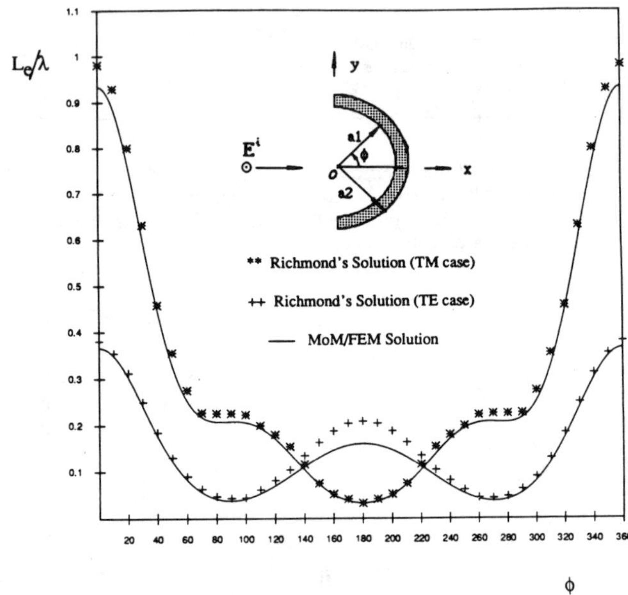

Fig. 8. Echo width of a semicircular dielectric cylindrical shell ($a_1 = 0.25\ \lambda$, $a_2 = 0.3\ \lambda$, $\sigma = 0.0$, $\epsilon_r = 4.0$, $f = 300$ MHz, $\phi^i = 180°$) under either TE or TM plane wave incidence. (MM/FEM solution compared to Richmond's solution [2], [20].)

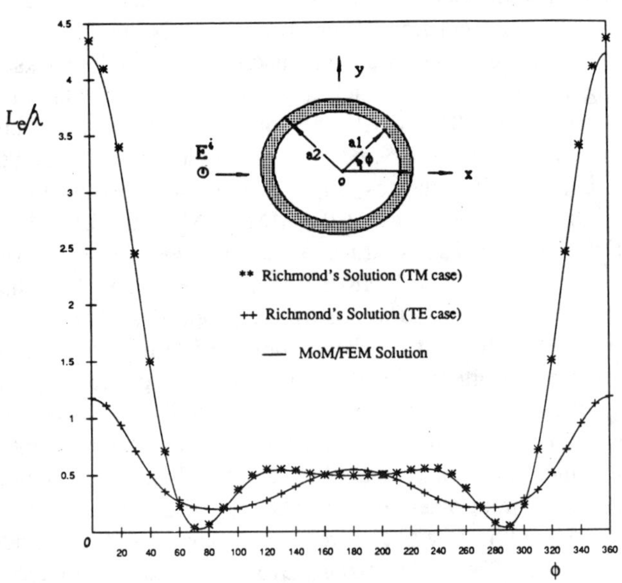

Fig. 7. Echo width of a circular dielectric cylindrical shell ($a_1 = 0.25\ \lambda$, $a_2 = 0.3\ \lambda$, $\sigma = 0.0$, $\epsilon_r = 4.0$, $f = 300$ MHz, $\phi^i = 180°$) under either TE or TM plane wave incidence. (MM/FEM solution compared to Richmond's solution [2], [20].)

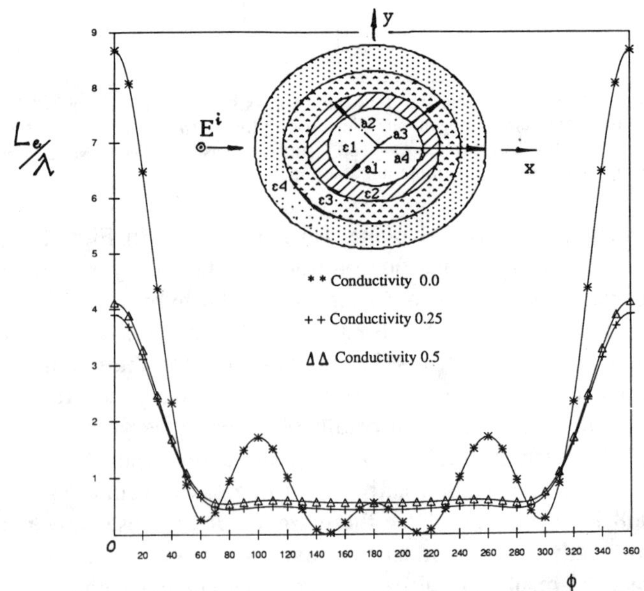

Fig. 9. Echo width of a four layered dielectric cylinder under TM plane wave incidence ($\phi^i = 180°$), where $a_1 = 0.15\ \lambda$, $a_2 = 0.20\ \lambda$, $a_3 = 0.25\ \lambda$, $a_4 = 0.3\ \lambda$, $\epsilon_{r1} = 8.0$, $\epsilon_{r2} = 6.0$, $\epsilon_{r3} = 4.0$, $\epsilon_{r4} = 2.0$, and all layers have the same conductivity σ.

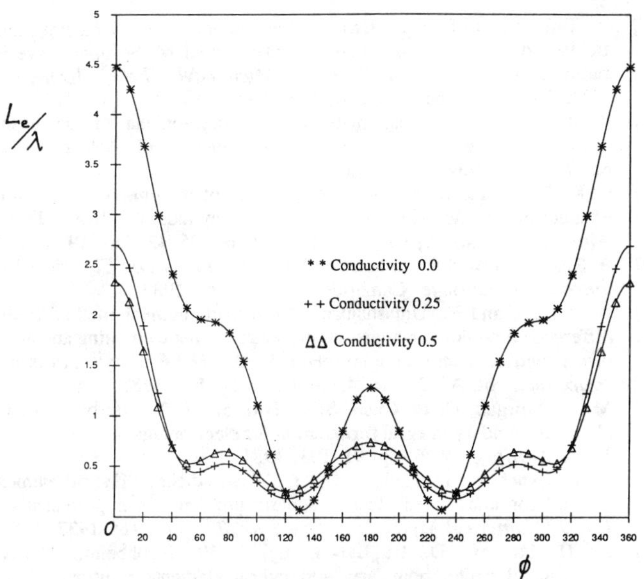

Fig. 10. Echo width of the same layered dielectric cylinder as in Fig. 9 under TE plane wave incidence ($\phi^i = 180°$).

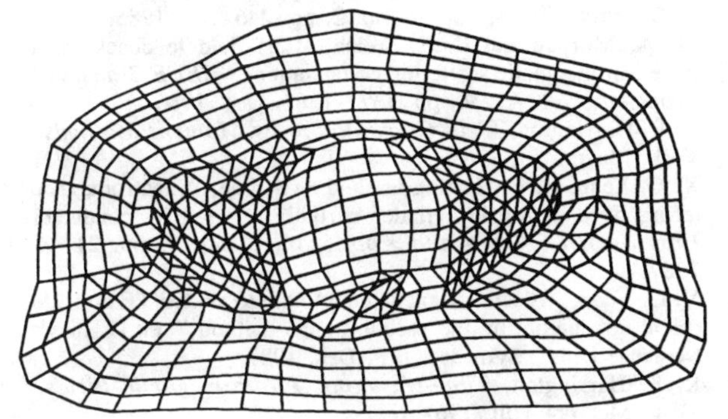

Fig. 11. The finite element mesh of a patient model (pelvic model B of [13]).

TABLE II
TISSUE PROPERTIES AT 100 MHz

	Bone	Fat	Muscle	Rectum	Tumor
ϵ_r	10	10.5	72	52	72
σ	0.02	0.22	0.89	0.61	0.89
μ_r	1.0	1.0	1.0	1.0	1.0

Finally, Fig. 11 shows the finite element mesh of a two-dimensional ([13, pelvic model B]). The tissue properties of this model at 100 MHz are given in Table II. Fig. 12 shows the distributions of the magnitude of the electric field and the absorbed power under the TM plane wave illumination ($\phi^i = 0°$ and $f = 100$ MHz). The scale is relative from 1 to 10.

CONCLUSION

The method of moments and the finite element method each is a very powerful numerical technique and has found many practical applications. Unfortunately, both have their limitations. The formulation proposed in this paper successfully combines these two methods in a very straightforward man-

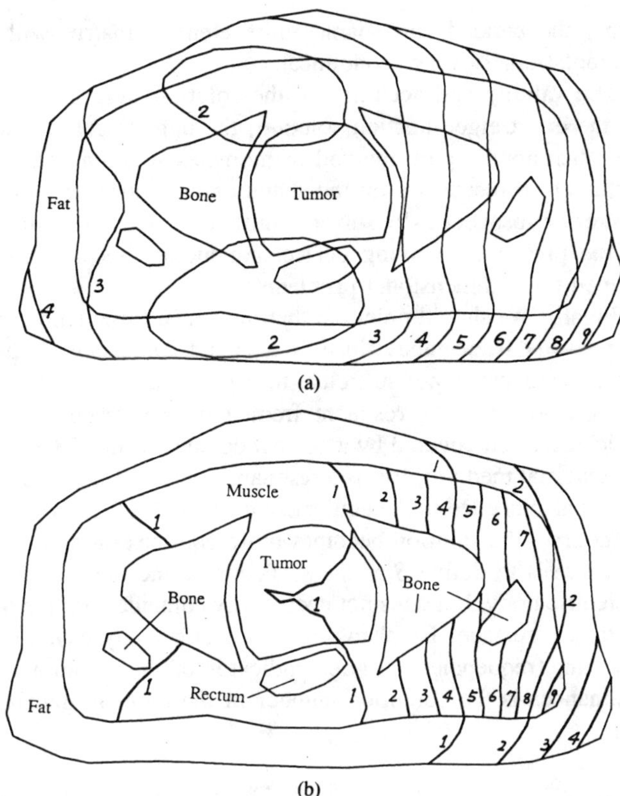

Fig. 12. The magnitude of the electric field and the absorbed power distributions inside the patients model defined in Fig. 12 under TM plane wave incidence ($\phi^i = 0°$). The scale is relative from 1 to 10. (a) Magnitude of the electric field distribution. (b) Absorbed power distribution.

ner, while utilizing both methods to their full capabilities. The method is especially suitable for problems involving inhomogeneous materials. Although there is no perfect conductor included here, the formulation can be easily extended to this limiting case, for instance, the dielectric coated with conductor or the conductor coated with dielectric layers. The formulation can even be applied to small aperture coupling problems since the interior solution does not involve the addition of the incident field and the scattered field. We have chosen the finite element method to solve the obtained boundary value problem. However, other numerical techniques, such as the finite difference method, may be used as well. To summarize, the hybrid MM/FEM method has the following advantages:

1) The formulation enables us to solve the interior and exterior problem separately. For the exterior electromagnetic fields, only the integration on the exterior surface is needed. The interior fields have been obtained naturally in the solution without requiring further efforts.

2) Since the interior and exterior problems are coupled only on the exterior surface, the finite element matrix needs to be computed and triangularized only once. Hence, little extra computations are needed if the incident waves change.

3) The versatility is developed in both method of moments and finite element method because on the one hand, the radiation properties are naturally included in the MM method (integral equation approach), and on the other hand the scattering problems involving inhomogeneous medium are easily and systematically handled by the FEM method (differential equation approach). Especially for three-dimensional prob-

lems, the banded and sparse finite element matrix will be computationally very economical.

The validity and accuracy of the solution were confirmed by the exact eigenfunction solution, the unimoment solution, and Richmond's pure method of moments solution. There is virtually no restriction on the shape or inhomogeneity of the scatter or absorber to be solved. Future efforts will be directed to the problem involving perfect conductors and arbitrarily shaped three-dimensional problems.

Finally, we should point out that the above formulation fails at resonant frequencies since at resonant frequencies a pure electric current is not sufficient to produce an arbitrary scattered field. Here, a resonant frequency is a frequency for which S, when covered by a perfect conductor and filled with the external medium, forms a resonant cavity. A magnetic current is also needed in the equivalent problem for region a [21]. Of course, the solution becomes more complex. A simple alternative is to define S as the surface that encloses the entire dielectric body, but does not necessarily coincide with the true dielectric surface. By changing S, we effectively change the resonant frequencies. A good indicator of the solution near resonances is the condition number of the moment matrix in (12).

References

[1] N. Morita, "Surface integral representation for electromagnetic scattering from dielectric cylinders," *IEEE Trans. Antennas Propagat.*, vol. AP-26, no. 2, pp. 261–266, Mar. 1978.

[2] J. H. Richmond, "Scattering by a dielectric cylinder of arbitrary cross section shape," *IEEE Trans. Antennas Propagat.*, vol. AP-13, no. 3, pp. 334–341, May 1965.

[3] D. E. Liversay and K. M. Chen, "Electromagnetic fields induced inside arbitrarily shaped biological bodies," *IEEE Trans. Microwave Theory Tech.*, vol. MTT-22, no. 12, pp. 1273–1280, Dec. 1974.

[4] M. J. Hagmann, O. P. Gandhi, and C. H. Durney, "Numerical calculation of electromagnetic energy deposition for a realistic model of man," *IEEE Trans. Microwave Theory Tech.*, vol. MTT-27, no. 12, pp. 804–809, Sept. 1979.

[5] D. H. Shaubert, D. R. Wilton, and A. W. Glisson, "A tetrahedral modeling method for electromagnetic scattering by arbitrarily shaped inhomogeneous dielectric bodies," *IEEE Trans. Antennas Propagat.*, vol. AP-32, no. 1, pp. 77–85, Jan. 1984.

[6] K. S. Yee, "Numerical solution of initial boundary value problems involving Maxwell's equations in isotropic media," *IEEE Trans. Antennas Propagat.*, vol. AP-14, pp. 302–307, May 1966.

[7] A. Taflove and M. E. Brodwin, "Computation of the electromagnetic fields and induced temperature within a model of the microwave irradiated human eye," *IEEE Trans. Microwave Theory Tech.*, vol. MTT-23, no. 12, pp. 888–896, Nov. 1975.

[8] K. K. Mei, "Unimoment method of solving antenna and scattering problems," *IEEE Trans. Antennas Propagat.*, vol. AP-22, no. 6, pp. 760–766, Nov. 1974.

[9] S. K. Chang and K. K. Mei, "Application of the unimoment method of electromagnetic scattering of dielectric cylinders," *IEEE Trans. Antennas Propagat.*, vol. AP-24, no. 1, pp. 35–42, Jan. 1976.

[10] P. P. Silvester and R. L. Ferrari, *Finite Elements For Electrical Engineers.* Cambridge, Cambridge Univ. Press, 1983.

[11] A. Taflove and K. Umashankar, "A hybrid moment method/finite-difference time-domain approach to electromagnetic coupling and aperture penetration into complex geometries," *IEEE Trans. Antennas Propagat.*, vol. AP-30, no. 4, pp. 617–627, July 1982.

[12] M. A. Morgan, C. H. Chen, S. C. Hill, and P. W. Barber, "Finite element-boundary integral formulation for electromagnetic scattering," *Wave Motion*, vol. 6, pp. 91–103, 1984.

[13] D. R. Lynch, K. D. Paulsen, and J. W. Strohbehn, "Hybrid element method for unbounded electromagnetic problems in hyperthermia," *Int. J. Numerical Methods in Eng.*, vol. 23, pp. 1915–1937, 1986.

[14] K. D. Paulsen, D. R. Lynch, and J. W. Strohbehn, "Three-dimensional finite, boundary, and hybrid elements solutions of the Maxwell equations for lossy dielectric media," *IEEE Trans. Microwave Theory Tech.*, vol. MTT-36, no. 4, pp. 682–693, April 1988.

[15] D. R. Lynch, K. D. Paulsen, and J. W. Strohbehn, "Finite element solution of Maxwell's equation for hyperthermia treatment planning," *J. Comput. Phys.*, vol. 58, no. 2, pp. 246–269, 1985.

[16] M. A. Morgan and B. E. Welch, "The field feedback formulation for electromagnetic scattering computations," *IEEE Trans. Antennas Propagat.*, vol. AP-34, pp. 1377–1382, Dec. 1986.

[17] R. F. Harrington, *Field Computation by Moment Methods.* New York, Macmillan, 1968.

[18] X. C. Yuan, R. F. Harrington, and S. S. Lee, "Electromagnetic scattering by a dielectric cylinder partially covered by conductors," *J. Electromagn. Waves and Appl.*, vol. 2, no. 1, pp. 21–44, Mar. 1987.

[19] X. C. Yuan, "Electromagnetic coupling into slotted TE and TM cylindrical conductors by the pseudo-image method," Ph.D. dissertation, Syracuse Univ., Syracuse, NY, Dec. 1987.

[20] R. F. Harrington, *Time-Harmonic Electromagnetic Fields.* New York, McGraw-Hill, 1961.

[21] J. R. Mautz and R. F. Harrington, "A combined-source solution for radiation and scattering from a perfectly conducting body," *IEEE Trans. Antennas Propagat.*, vol. AP-27, pp. 445–454, July 1979.

[22] J. H. Richmond, "TE-wave scattering by a dielectric cylinder of arbitrary cross section shape," *IEEE Trans. Antennas Propagat.*, vol. AP-14, no. 4, pp. 460–464, July 1966.

[23] J. Van Bladel, *Electromagnetic Fields.* New York: McGraw-Hill, 1964.

Three-Dimensional Electromagnetic Scattering from Inhomogeneous Objects by the Hybrid Moment and Finite Element Method

XINGCHAO YUAN, MEMBER, IEEE

Abstract —The hybrid moment and finite element method is used to obtain 3-D scattering and/or absorption from inhomogeneous, arbitrarily shaped objects. The surface of the object is approximated by triangles and the volume of the object is approximated by tetrahedrons. The electrical parameters are assumed constant in each tetrahedron. The Galerkin testing procedure is used. To avoid contaminations of spurious mode, a divergenceless vector basis function is used in finite elements. The calculated internal field and scattered field for a homogeneous sphere, a layered sphere, and a lossy prolate spheroid are compared with Mie series solutions and other numerical techniques. The accuracy and rate of convergence of the solution are discussed.

I. Introduction

Both the method of moments and the finite element method have been shown to be very powerful numerical techniques for solving electromagnetic problems [1], [2]. However, each of the methods has its own advantages and disadvantages when applied to different problems. The method of moments is an integral equation method which handles unbounded problems very effectively but becomes computationally intensive when complex inhomogeneities are present. In contrast, inhomogeneities are easily handled by the finite element method, which requires less computer time and storage because of its sparse and banded matrix. The matrix filling time is also negligible when simple basis functions are used. However, it is most suitable for boundary value problems. Therefore, the hybrid method that combines the method of moments and the finite element method (MOM/FEM), proposed in [3], is very advantageous for treating electromagnetic problems involving complex inhomogeneities such as in microwave hyperthermia applications. The formulation presented in [3] applies to both two- and three-dimensional problems, but only two-dimensional numerical results were given in [3]. It is the objective of this paper to demonstrate the validity and accuracy of the hybrid MOM/FEM method proposed in [3] for 3-D problems.

In the following, the hybrid moment and finite element method is briefly reviewed, followed by 3-D numerical examples. The accuracy and rate of convergence of the solution are then discussed.

II. The Hybrid MOM/FEM Method

Fig. 1 shows an arbitrarily shaped, inhomogeneous, lossy dielectric body under illumination of an incident field (E^i, H^i). The dielectric body bounded by surface S is characterized by $(\mu(r), \epsilon(r))$. The exterior region is characterized by (μ_0, ϵ_0). The objective is to determine the electromagnetic fields everywhere in space using the hybrid MOM/FEM method for 3-D problems.

A. The Hybrid MOM/FEM Method

A detailed formulation of the hybrid moment and finite element method is described in [3], where only 2-D numerical results were given. The basic technique is to apply the equivalence principle to transform the original problem into interior and exterior problems, which are coupled on the exterior dielectric body surface through the continuities of the tangential electric field and magnetic field. The interior problem involving the inhomogeneous medium is solved by the finite element method, and the exterior problem involving unbounded region is solved by the moment method.

There are two sets of unknowns on the exterior dielectric surface S. One is for the exterior region, which is an equivalent electric current source. The other is for the interior region, which is the tangential magnetic field on S. The application of boundary conditions on S results in two coupled operator equations, which are subsequently transformed into matrix equations using the moment method [3, eq. (12)]. If the Galerkin testing procedure is used, the matrix equation [3, eq. (12)] becomes

$$\begin{pmatrix} [Z^a] & [Z^b] \\ [Y^a] & [Y^b] \end{pmatrix} \begin{pmatrix} I^a \\ I^b \end{pmatrix} = \begin{pmatrix} V \\ U \end{pmatrix} \quad (1)$$

where I^a, I^b, V, and U are $N \times 1$ column vectors, assum-

Manuscript received December 8, 1989; revised March 30, 1990. This work was supported by the National Institutes of Health under Grant R01 CA37245 and the National Science Foundation under Grant CEE-835-2226.

The author is with the Thayer School of Engineering, Dartmouth College, Hanover, NH 03755.

IEEE Log Number 9036433.

Reprinted from *IEEE Trans. Antennas Propagat.*, vol. 38, no. 8, pp. 1053–1058, 1990.

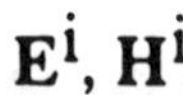

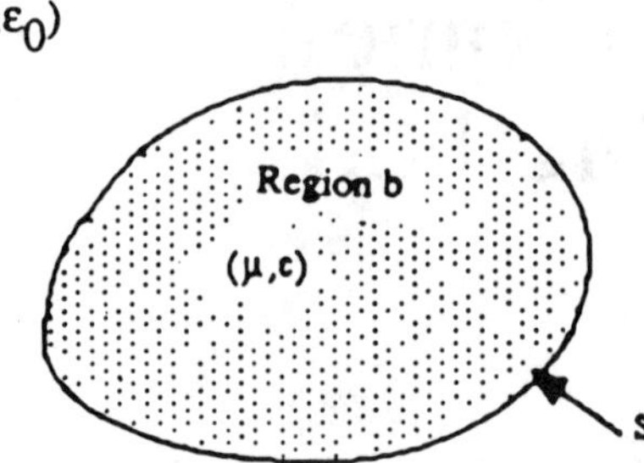

Fig. 1. Electromagnetic scattering and/or absorption due to an inhomogeneous dielectric body bounded by surface S.

ing there are N expansion functions. The ith elements of I^a and I^b are the complex coefficients of the two sets of unknowns for the exterior and interior regions [3, eqs. (9) and (10)], respectively. (Note that both unknowns are expanded in terms of the same basis functions for simplicity.) In (1), the first N rows and the second N rows enforce the continuities of the tangential electric field and magnetic field on S, respectively.

The ith elements of V and U are v_i and u_i, given by

$$v_i = \langle J_i, n \times E^i \rangle \tag{2}$$

$$u_i = \langle J_i, n \times H^i \rangle \tag{3}$$

where $\{J_i, i = 1,2,\cdots,N\}$ are the expansion functions used to approximate the surface equivalent currents. The quantity within the angle bracket $\langle \; \rangle$ denotes the symmetric product defined in [3, eq. (11)]. $[Z^a]$, $[Z^b]$, $[Y^a]$, and $[Y^b]$ are $N \times N$ matrices and their ijth elements are given by

$$Z_{ij}^a = \langle J_i, -n \times E^s(J_j) \rangle \tag{4}$$

$$Y_{ij}^a = \langle J_i, -n \times H^s(J_j) \rangle \tag{5}$$

$$Z_{ij}^b = \langle J_i, n \times E^b(J_j) \rangle \tag{6}$$

$$Y_{ij}^b = \langle J_i, J_j \rangle. \tag{7}$$

$[Z^a]$ and $[Y^a]$ are called impedance and admittance matrices and are related to the fields due to electric current sources radiating in free space [4]. They can be evaluated once J_i is chosen. $[Y^b]$ is simply the symmetric product of expansion functions. The jth column of $[Z^b]$ is related to the electric field in the interior region due to a tangential magnetic field J_j on the dielectric external surface, which is obtained by the finite element method as follows.

The electric field E in the interior region is expanded as

$$E = \sum_j^{N'} E_j W_j \tag{8}$$

where $\{E_j, j = 1,2,\cdots,N'\}$ is a set of unknown complex scalars and $\{W_j, j = 1,2,\cdots,N'\}$ is a set of vector basis functions defined within the volume bounded by surface S. The finite element matrix equation can be obtained using the following weak form of Maxwell's equation [5]:

$$\langle \nabla \times E, \nabla \times W_i \rangle_t - \langle k^2 E, W_i \rangle_t = j\omega\mu \oint_S W_i \cdot (n \times H)\, ds \tag{9}$$

where the surface integral is performed on the dielectric exterior surface S, $n \times H$ is the tangential magnetic field, and $k = \omega\sqrt{\mu\epsilon}$ is the wavenumber, which may be a function of location but is assumed constant within each element. The quantity $\langle \; \rangle_t$ denotes the symmetric product integrated over a volume rather than on a surface.

Substituting (8) into (9), we obtain the finite element matrix equation

$$[A][\bar{E}] = [\bar{F}] \tag{10}$$

with the ijth element of matrix $[A]$ given as

$$A_{ij} = \langle \nabla \times W_i, \nabla \times W_j \rangle_t - \langle k^2 W_i, W_j \rangle_t \tag{11}$$

and ith element of vector $[\bar{F}]$ given as

$$F_i = j\omega\mu \oint_S W_i \cdot (n \times H)\, ds. \tag{12}$$

The ith element of the vector $[\bar{E}]$ is E_i.

If $n \times H$ is replaced by J_j in (12), then $E^b(J_j)$ in the interior region can be obtained by solving (10). Those $E^b(J_j)$ on S can then be used to calculate (6). Note that the matrix $[A]$ in (10) needs to be factored only once. The back substitutions of $[A]$ are performed N times to obtain $[Z^b]$. The total interior electric field can be obtained by solving (10) using the $[\bar{F}]$ which is the summation of (12) multiplied by the ith element of I^b.

B. Basis Functions

The expansion functions $\{J_i, i = 1,2,\cdots,N\}$ for the unknowns on surface S are chosen to be those proposed in [4], which are given as

$$J_i = \begin{cases} \dfrac{l_i}{2A_i^+}(r - r_i^+) & r \text{ in } T_i^+ \\ \dfrac{-l_i}{2A_i^-}(r - r_i^-) & r \text{ in } T_i^- \\ 0 & \text{otherwise} \end{cases} \tag{13}$$

where l_i is the length of the ith edge which belongs to both triangles T_i^+ and T_i^-, r_i^+ are the position vectors of the node that are not related to the ith edge in triangle T_i^+, and A_i^+ is the area of triangle $T_i^\pm$. (Note the convention whereby the superscripts refer to triangle faces and the subscripts refer to the edges of triangles.) Clearly, J_i is associated with the ith edge, which occupies only two triangles sharing the same edge. This type of expansion function ensures the continuity of the normal component of the current across the common edge, but there are no normal components on the rest of the edges of the two triangles. Hence, it is free of line charges.

TABLE I
TETRAHEDRON EDGE DEFINITION

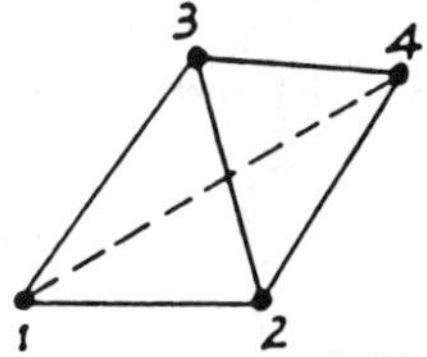

Edge #	i_1	i_2
1	1	2
2	1	3
3	1	4
4	2	3
5	4	2
6	3	4

The basis functions $\{W_i,\ i = 1, 2, \cdots, N'\}$ in the finite elements are chosen to be those proposed in [6]. They are defined within a tetrahedron and are associated with the six edges of the tetrahedron.

Assuming the four nodes of a tetrahedron are numbered such that it follows the left-hand rule and the edges are numbered according to Table I, then the vector basis functions associated with the $(7-i)$th edge of that tetrahedron are defined as [6]

$$W_{7-i} = \begin{cases} f_{7-i} + g_{7-i} \times r & r \text{ in the tetrahedron} \\ 0 & \text{otherwise} \end{cases} \tag{14}$$

with

$$f_{7-i} = \frac{b_{7-i}}{6V} r_{i_1} \times r_{i_2} \tag{15}$$

$$g_{7-i} = \frac{b_i b_{7-i} e_i}{6V} \tag{16}$$

where $i = 1, 2, \cdots, 6$; V is the volume of the tetrahedron; $e_i = (r_{i_2} - r_{i_1})/b_i$ is the unit edge vector of the ith edge; and $b_i = |r_{i_2} - r_{i_1}|$ is the length of the ith edge.

W_i has the following properties:

$$W_i \cdot e_j|_{r \text{ on } j\text{th edge}} = \delta_{ij} \tag{17}$$

$$\nabla \cdot W_i = 0 \tag{18}$$

$$\nabla \times W_i = 2 g_i \tag{19}$$

where $\delta_{ij} = 0$ if $i \neq j$ and $\delta_{ij} = 1$ if $i = j$. Equation (17) implies that the complex scalar E_j in (8) is the projection of the electric field onto the jth edge. Note that there are two numbering systems: one local and the other global. To ensure the continuity of $n \times E$ across all edges, a unique global edge direction is defined (e.g. always pointing from the smaller node number to the larger node number) so that (14) should be multiplied by (-1) if the local edge vector (as defined in Table I) does not have the same direction as the global edge direction. W_i forces no conditions on the normal component of E. However, with the above formulation one can show that the continuity of the electric flux can be satisfied within the degree of approximation [5]. Finally, since $\nabla \cdot W_i = 0$ the electric field obtained through (8) exactly satisfies $\nabla \cdot E = 0$, the divergence equation within the element. Therefore, the finite element solution is free of contamination of spurious solutions [5].

TABLE II
FINITE ELEMENT MODELS (TETRAHEDRA ONLY)

Model	NN	NE	NEQ	NBAND	NBEG	CPU Time (min.) for Computing $[Z^a]$, $[Y^a]$, and $[Y^b]$	$[A]^{-1}$	$[Z^b]$	TOTAL
(5,5,4)	169	800	1038	449	120	1.41	3.83	2.78	8.25
(5,5,8)	337	1760	2206	525	120	1.51	12.6	7.60	22.0
(8,8,4)	457	2240	2856	733	336	11.1	34.0	38.1	84.3
(10,10,4)	729	3600	4568	949	540	36.7	92.4	133.0	270.0

where NN is the number of nodes, NE is the number of tetrahedrons, NEQ is the number of independant edges of tetrahedrons, NBAND is the minimized bandwidth of [A], and NBEG is the number of boundary edges.

III. NUMERICAL RESULTS

To check the validity and accuracy of the hybrid MOM/FEM method, we let the incident wave be an x-directed plane wave:

$$E^i = u_x \exp\left(jk_0(\sin\theta^i \cos\phi^i x + \sin\theta^i \sin\phi^i y + \cos\theta^i z)\right) \tag{20}$$

$$H^i = \frac{\left(\cos\theta^i u_y - \sin\theta^i \sin\phi^i u_z\right)}{\eta_0} \cdot \exp\left(jk_0(\sin\theta^i \cos\phi^i x + \sin\theta^i \sin\phi^i y + \cos\theta^i z)\right) \tag{21}$$

where $k_0 = \omega\sqrt{\mu_0 \epsilon_0}$, $\eta_0 = \sqrt{\mu_0/\epsilon_0}$, and (ϕ^i, θ^i) are the incident angles.

In the following, the scattering from a dielectric sphere is studied since there exists an analytical (Mie series) solution for comparison. The finite element mesh for the sphere made of only tetrahedra is generated by dividing the radius with N_r divisions, the θ $(0 \leqslant \theta \leqslant \pi)$ with N_θ divisions, and the ϕ $(0 \leqslant \phi \leqslant 2\pi)$ with $2N_\phi$ divisions. (N_θ, N_ϕ, N_r) denotes the number of divisions used in each model. For example, (5,5,4) denotes that there are five divisions in the θ direction, ten divisions in the ϕ direction, and four divisions in the radial direction. Table II shows the number of nodes (NN), the number of tetrahedron elements (NE), the number of tetrahedron edges (NEQ), the full bandwidth of the finite element matrix $[A]$ (NBAND), and the number of boundary edges (NEG) used in various models.

Fig. 2 shows the magnitude of the electric field along the x axis inside a homogeneous sphere under the illumination of an x-directed plane wave, where $k_0 a = 1/2$, $\epsilon_r = 16$, $\phi^i = 0$, and $\theta^i = \pi$. The internal electric field is obtained by interpolating the finite element basis functions. For the model (8,8,4), there are approximately eight basis functions per wavelength (inside the sphere) in the θ and ϕ directions and 12 basis functions in the radial direction. For the model (10,10,4), there are approximately ten basis functions in the θ and ϕ directions and 12 basis functions in the radial direction. The solution seems to have converged. Further refinement of the mesh is prohibited owing to the available computer resources.

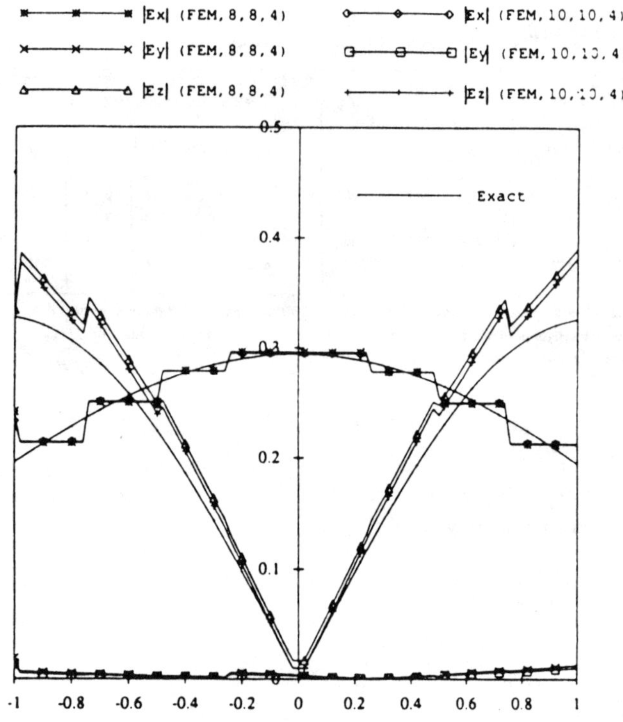

Fig. 2. The magnitude of the electric field inside a homogeneous dielectric sphere along the x axis where $k_0a = 1/2$, $\epsilon_r = 16$, $\theta' = \pi$, and $\phi' = 0$.

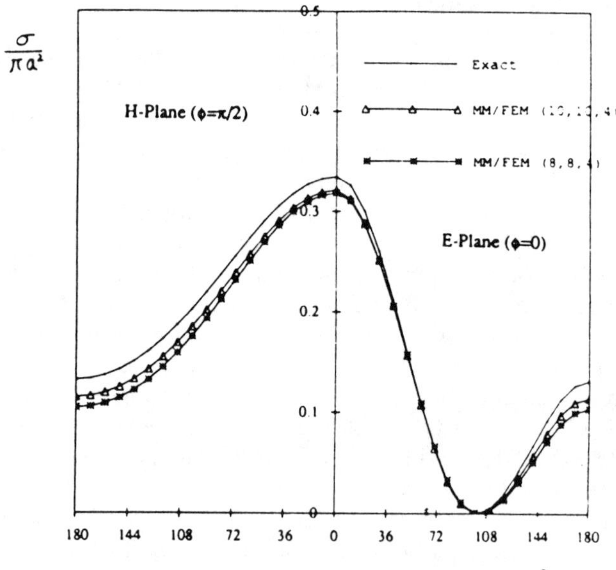

Fig. 3. The radar cross section of the same homogeneous dielectric sphere as in Fig. 2.

Nevertheless, the rate of convergence of the solution roughly falls into the general rule of thumb that at least ten basis functions per wavelength should be used. Fig. 3 shows the radar cross section of the same sphere. The numerical solution agrees well with the Mie series solution.

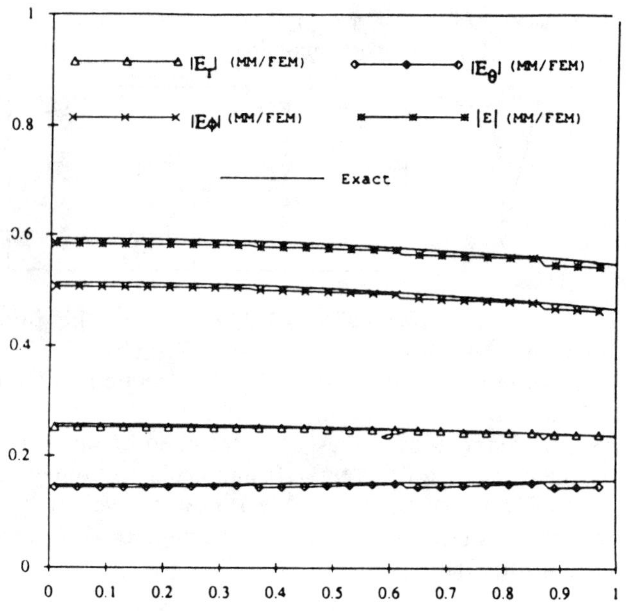

Fig. 4. The magnitude of the electric field inside a homogeneous dielectric sphere along a radial direction ($\theta = \phi = \pi/3$) where $k_0a = 1/2$, $\epsilon_r = 4$, $\theta' = \pi$, and $\phi' = 0$.

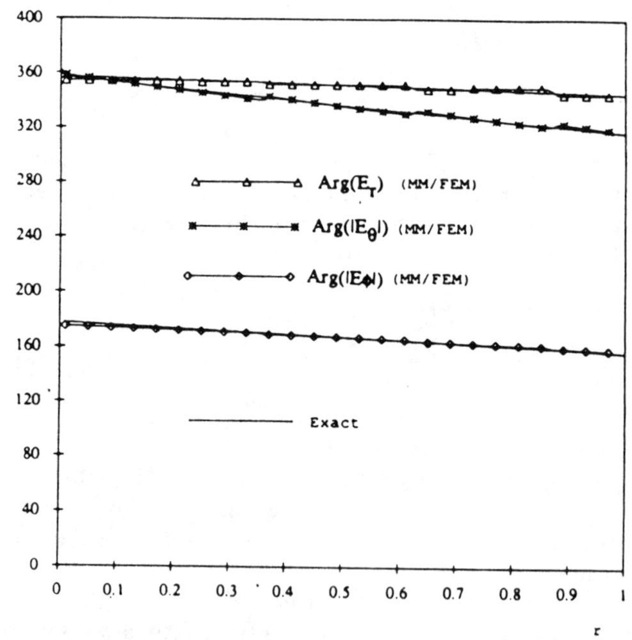

Fig. 5. The phase of the electric field inside the same homogeneous dielectric sphere as in Fig. 4 along a radial direction ($\theta = \phi = \pi/3$).

Figs. 4 and 5 show the magnitude and phase of the electric field along a radial direction ($\theta = \phi = \pi/3$), where $k_0a = 1/2$, $\epsilon_r = 4$, $\phi' = 0$, and $\theta' = \pi$. The numerical solutions agree very well with the Mie series (exact) solution. It is interesting to note the piecewise nature of the basis function appearing in the final solution. When the fields change rapidly, the discontinuities are more evident (see Figs. 2 and 4). In interpolating the solution, we took the

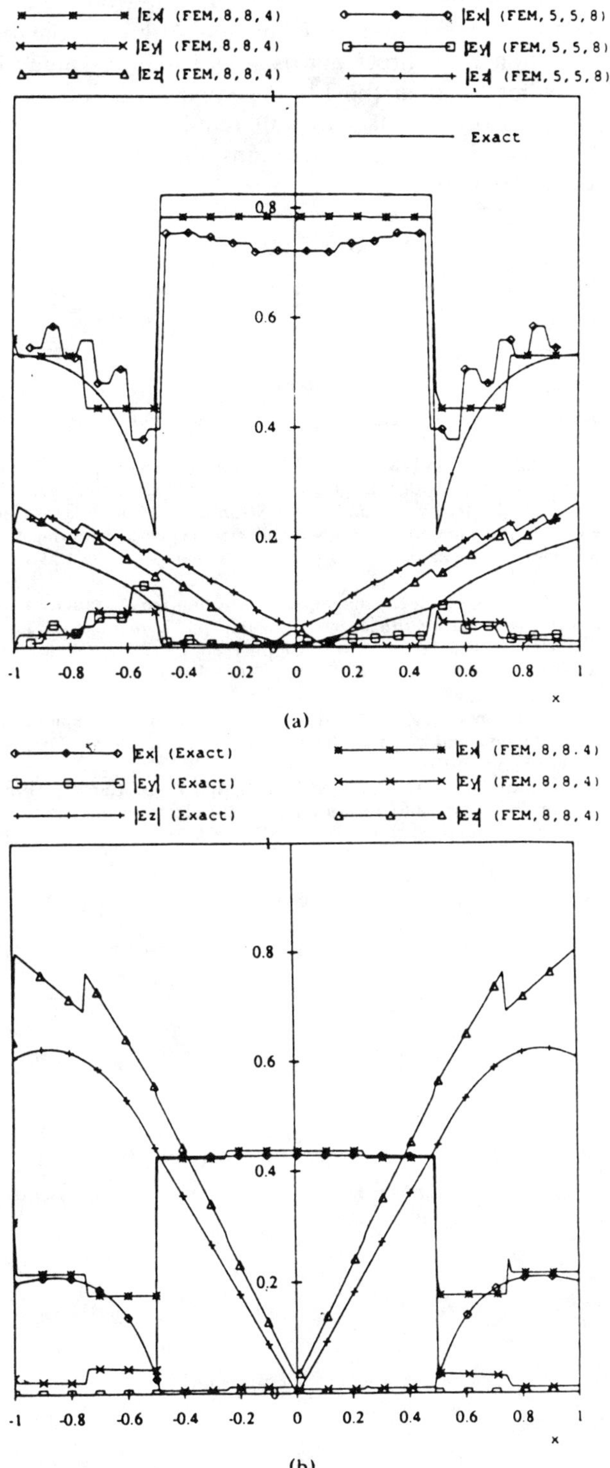

Fig. 6. The magnitude of the electric field inside a two-layered dielectric sphere along the x axis where $k_0a_1 = \pi/10$, $\epsilon_{r1} = 1$, $k_0a_2 = \pi/5$, $\theta' = \pi$, and $\phi' = 0$; (a) $\epsilon_{r2} = 4$ and (b) $\epsilon_{r2} = 16$.

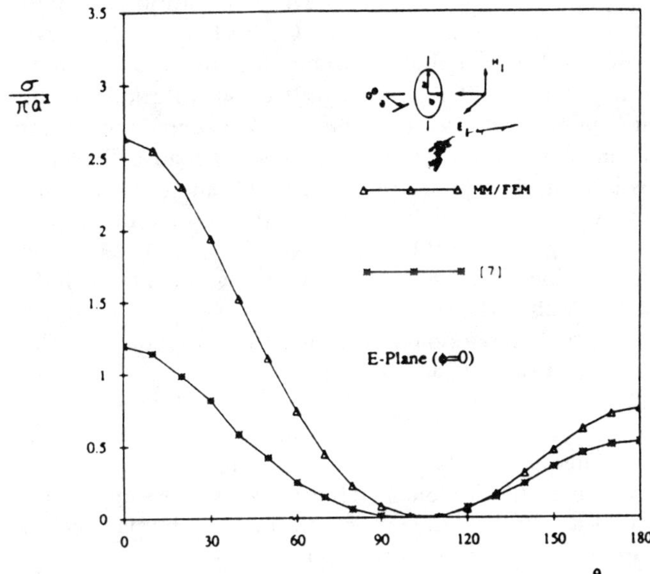

Fig. 7. The radar cross section of a homogeneous dielectric lossy prolate spheroid where $a/b = 2$, $k_0a = \pi/2$, $\epsilon_r = 4$, $\sigma = 0.00417234$, $\theta' = \pi$, and $\phi' = 0$.

approach that the first tetrahedron containing the observing point is used to compute the field at that point. Thus, it is arbitrary for the fields on tetrahedron faces since only the continuity of tangential components is satisfied exactly by the basis functions, whereas the continuity of electric flux is approximate. This effect can be seen in Figs. 2 and 6 (about the $x = -1$ region). However, the fields inside the tetrahedron are uniquely defined.

Fig. 6 shows the magnitude of the electric field on the x axis for a concentric sphere with inner sphere having radius $a/2$, $\epsilon_{r1} = 1.0$, and outer sphere having (a) $\epsilon_{r2} = 4.0$ and (b) $\epsilon_{r2} = 16.0$. Here $k_0a = \pi/5$, $\phi' = 0$, and $\theta' = \pi$. On the dielectric interface ($r = a/2$, $\theta = \pi/2$, $\phi = 0$), E_x is the normal component, which should be discontinuous by factors of 4 and 16. E_y and E_z are the tangential components, which should be continuous. Clearly, these are correctly obtained in the numerical solution. The model (5,5,8) has eight basis functions in the radial direction; hence the discontinuity in the normal component is better approximated than that obtained by model (8,8,4). However, (8,8,4) approximates the sphere better. Hence, the overall solution is smoother. To approximate the discontinuity in normal components well, a fine mesh near the dielectric interface is required. As another example, Fig. 7 shows the radar cross section of a prolate spheroid in comparison with that obtained in [7]. The agreement is reasonable.

It is worthwhile to discuss some points concerning the implementation of the method. When solving a 3-D problem, the CPU time and storage requirements are the determining factors in whether the problem can be solved. To reduce the amount of memory required for storing the moment matrix (eq. (1)) by half, some matrix algebra is performed so that I^a is first solved and then I^b, rather than I^a and I^b being solved simultaneously, as in (1). In particular, $[Y^b]$ is first LU decomposed and many back substitutions are performed to avoid computing the inverse of $[Y^b]$. Note that $[Y^b]$, as defined in (7), is a banded and sparse matrix for the expansion functions given by (13). Thus, additional computer time is saved.

Table II shows the CPU time used to obtain the solutions presented in this paper. The calculations are performed on a minisupercompter CONVEX C1. It is evident from Table II that the computations of $[Z^h]$, which are largely made of a number of back substitutions of the finite element matrix $[A]$ (eq. (10)), occupy the major portion of the total CPU time. This is because the back substitution of $[A]$ requires some CPU time (e.g., 16 s for model (8,8,4)) although it is very little compared with the LU decompositions of $[A]$ (5543 s for model (8,8,4)). The accumulation of such times can be significant, as indicated in Table II. Therefore, it is necessary to minimize the number of boundary triangles used to represent the object's exterior surface when possible.

The hybrid MOM/FEM method solves the problem of scattering and/or absorption from arbitrarily shaped inhomogeneous objects provided that a finite element mesh is available. A problem for future work is to apply this method to obtain the electric field inside a human model for microwave hyperthermia applications.

IV. Conclusion

The validity and accuracy of the hybrid MOM/FEM method for 3-D electromagnetic problems are examined. To avoid the contaminations of the spurious mode in the finite elements, divergenceless vector basis functions rather than usual scalar basis functions are used in finite elements. The numerical results are found to be in good agreement with the analytical solutions and other numerical techniques if at least ten basis functions per wavelength are used in the model. The piecewise-continuous nature of the basis functions prevents one from obtaining a smooth solution across elements. Although the back substitution in a direct matrix solver requires much less computer time than the LU decompositions, performing many such operations may still require significant computer time. Thus, such operations should be minimized whenever possible.

Acknowledgment

The author is grateful to Dr. D. R. Lynch for useful discussions.

References

[1] R. F. Harrington, *Field Computation by Moment Methods*. New York: Macmillan, 1968.

[2] P. P. Silvester and R. L. Ferrari, *Finite Elements for Electrical Engineers*. Cambridge, England: Cambridge University Press, 1983.

[3] X. Yuan, D. R. Lynch, and J. W. Strohbehn, "Coupling of finite element and moment methods for electromagnetic scattering from inhomogeneous objects," *IEEE Trans. Antennas Propagat.*, vol. 38, pp. 386–393, Mar. 1990.

[4] S. M. Rao, "Electromagnetic scattering and radiation of arbitrarily-shaped surface by triangular patch modeling," Ph.D. dissertation, University of Mississippi, 1980.

[5] X. Yuan, "On the use of divergenceless basis functions in finite elements," submitted to *Electron. Lett.*

[6] M. L. Barton and Z. J. Cendes, "New vector finite elements for three-dimensional magnetic field computation," *J. Appl. Phys.*, vol. 61, no. 8, pp. 3919–3921, Apr. 1987.

[7] M. A. Morgan, C. H. Chen, S. C. Hill, and P. W. Barber, "Finite element-boundary integral formulation for electromagnetic scattering," *Wave Motion*, vol. 6, pp. 91–103, 1984.

Finite Element Analysis of Electromagnetic Scattering from Inhomogeneous Cylinders at Oblique Incidence

Andreas C. Cangellaris, *Member, IEEE*, and Robert Lee, *Student Member, IEEE*

Abstract—**A finite element formulation of electromagnetic scattering from composite material cylinders at oblique incidence is presented. The cylinders are uniform along their axis which, without loss of generality, is assumed to be parallel to the z axis of a Cartesian coordinate system, while their cross section is of arbitrary shape and composition. The excitation can be either a plane wave of arbitrary polarization at oblique incidence with respect to the cylinder axis, or any traveling-wave field with z dependence of the form exp $(-i\beta z)$. The weighted-residual statement of the problem involves the z components of the electric and magnetic field, and is such that the continuity of the tangential electric and magnetic fields at media interfaces is directly enforced. The finite element solution is carried out using the bymoment method. Numerical results for echo widths and field distributions within cylinders of various shapes and compositions are presented. Comparisons of some of the results with those obtained from exact eigenfunction expansion and integral equation solutions demonstrate the validity of the method and the associated computer program.**

I. Introduction

Electromagnetic wave scattering from axially uniform cylinders with cross sections of arbitrary shapes and compositions has been extensively studied using both integral equation and finite element methods. However, most of the published results are for plane wave excitation at normal incidence, i.e., with the wave vector of the incident field parallel to a plane that contains the cross section of the cylindrical geometry. On the other hand, results for the general case of oblique incidence, where the wave vector of the incoming wave is at an angle other than perpendicular to the cylinder axis, are rather sparse, even though the first paper for plane wave scattering from homogeneous dielectric circular cylinders at oblique incidence was published by Wait as early as 1955 [1]. The problem has important practical applications in various areas, such as diffraction grating analysis and design and scattering by finite composite cylinders with cross-sectional dimensions much smaller than their length. Indeed, results for scattering from infinite cylinders may be usefully applied to the finite case as long as the angle between the incoming wave vector and the vector normal to the cylinder axis is small. This has been experimentally verified by Lind and Greenberg [2].

Wu and Chen [3] used the variational reaction theory [4] for a finite element solution of the problem for the general case of inhomogeneous, anisotropic cylinders with arbitrary shapes. An electric field integral equation formulation for the case of nonmagnetic cylinders was given by Su [5]. He used the conjugate gradient method and fast Fourier transform (FFT) to expedite the inversion of the resulting matrix equation. More recently, Rojas [6] presented a coupled integral equation formulation whose unknowns are the total electric and magnetic field vectors. He also accounted for the magnetic properties of the materials in the cylinder.

In this paper, the method of weighted residuals is used for the derivation of a convenient weak statement that is utilized for a finite element solution of the boundary value problem for the time-harmonic electromagnetic scattering from linear, isotropic, composite cylinders at oblique incidence. Assuming that the cylinder axis is parallel to the z axis of a Cartesian coordinate system, the excitation can be either a plane wave of arbitrary polarization incident at an angle other than perpendicular to the z axis, or any traveling-wave field with z dependence of the form exp $(-i\beta z)$, where β is constant. The cylinder is axially uniform while its cross section, in general, may be inhomogeneous, lossy, and of arbitrary shape.

First, the governing equations are presented along with the derivation of the weak statement of the boundary value problem. The weak form is given in terms of the axial components of the electric and magnetic field (E_z, H_z) and is derived using the method of weighted residuals. The continuity of the tangential electric and magnetic field components at different material interfaces is directly enforced and incorporated into the weak statement of the problem. Next, the weighted-residual statement is utilized for a finite element solution of the scattering problem by means of the bymoment method [10]. Numerical results are then presented from the application of the method to scattering from homogeneous and inhomogeneous material cylinders of various shapes, along with comparisons with results obtained using exact eigenfunction expansion and integral equation methods.

II. Weighted-Residual Statement Formulation

Consider a linear and isotropic infinite cylinder which is uniform along the z axis with its cross section being of

Manuscript received July 9, 1990; revised October 16, 1990. This work was supported by Sandia National Laboratories Document 01-6833.

A. C. Cangellaris is with the Electromagnetics Laboratory, Department of Electrical and Computer Engineering, University of Arizona, Tucson, AZ 85721.

R. Lee was with Electromagnetics Laboratory, Department of Electrical and Computer Engineering, University of Arizona, Tucson, AZ. He is now with the Department of Electrical Engineering, The Ohio State University, Columbus, OH, 43210-1272.

IEEE Log Number 9142924.

Reprinted from *IEEE Trans. Antennas Propagat.*, vol. 39, no. 5, pp. 645–650, May 1991.

arbitrary shape. The permittivity ϵ, the permeability μ, and the conductivity σ are assumed to be arbitrary functions of x and y but invariant with z. Without loss of generality, the medium exterior to the cylinder is taken to be free space. Let

$$(\vec{E}^i, \vec{H}^i) = (\vec{e}^i(x, y), \vec{h}^i(x, y))e^{-i\beta z} \tag{1}$$

be the time-harmonic incident field, where β is a constant and the conventional $e^{i\omega t}$ time dependence is suppressed. The incident field is defined to be the field everywhere in the absence of the cylinder, and its general representation in (1) includes various cases, such as a plane wave propagating at an angle θ with respect to the z-axis, in which case $\beta = \omega\sqrt{\mu_0\epsilon_0}\cos\theta$, or the field due to an infinitely thin straight wire of infinite length running parallel to the z axis and carrying a current $I(z) = I_0 \exp(-i\beta z)$.

The presence of the cylinder causes a scattered field $(\vec{E}^s, \vec{H}^s)$, which has the same space harmonic variation in z as the incident field due to phase matching

$$(\vec{E}^s, \vec{H}^s) = (\vec{e}^s(x, y), \vec{h}^s(x, y))e^{-i\beta z}. \tag{2}$$

The total field is the superposition of the incident and scattered fields

$$(\vec{E}, \vec{H}) = (\vec{e}(x, y), \vec{h}(x, y))e^{-i\beta z} = \left((\vec{e}^i + \vec{e}^s), (\vec{h}^i + \vec{h}^s)\right)e^{-i\beta z}. \tag{3}$$

It can be shown that for the assumed z dependence of the fields and for a homogeneous medium, the transverse electric and magnetic fields $\vec{e}_t = \hat{x}e_x + \hat{y}e_y$ and $\vec{h}_t = \hat{x}h_x + \hat{y}h_y$ can be expressed in terms of the transverse derivatives of the axial components e_z, h_z [7]

$$\vec{e}_t = \frac{1}{\beta^2 + \gamma^2}(i\beta\nabla_t e_z - i\omega\mu\hat{z} \times \nabla_t h_z) \tag{4}$$

$$\vec{h}_t = \frac{1}{\beta^2 + \gamma^2}(i\beta\nabla_t h_z - +(\sigma + i\omega\epsilon)\hat{z} \times \nabla_t e_z) \tag{5}$$

where $\nabla_t = \hat{x}\partial/\partial x + \hat{y}\partial/\partial y$, $\gamma^2 = i\omega\mu(\sigma + i\omega\epsilon)$, and the common z dependence $\exp(-i\beta z)$ has been suppressed for simplicity. Of course, the axial components satisfy the following Helmholtz equation

$$\nabla_t^2\begin{Bmatrix} e_z \\ h_z \end{Bmatrix} - (\gamma^2 + \beta^2)\begin{Bmatrix} e_z \\ h_z \end{Bmatrix} = 0. \tag{6}$$

As mentioned earlier, the material properties of the cylinder under consideration may vary with x and y. However, in applying the finite element method it is assumed that the material properties ϵ, μ, σ are constant in each element of the finite element grid. Therefore, (6) is satisfied within each element of the grid. Finally, from (4) and (5) it is apparent that the continuity of the tangential components of the transverse electric and magnetic fields at material interfaces couples e_z and h_z for $\beta \neq 0$ as first observed by Wait [1].

Using the method of weighted residuals [8] the weak forms of the two equations in (6) are

$$\langle(v^{-2}i\omega\mu\nabla_t^2 h_z)\phi_i\rangle_{\Omega_e} - \langle i\omega\mu h_z\phi_i\rangle_{\Omega_e} = 0 \tag{7}$$

$$\langle[v^{-2}(\sigma + i\omega\epsilon)\nabla_t^2 e_z]\phi_i\rangle_{\Omega_e} - \langle(\sigma + i\omega\epsilon)e_z\phi_i\rangle_{\Omega_e} = 0 \tag{8}$$

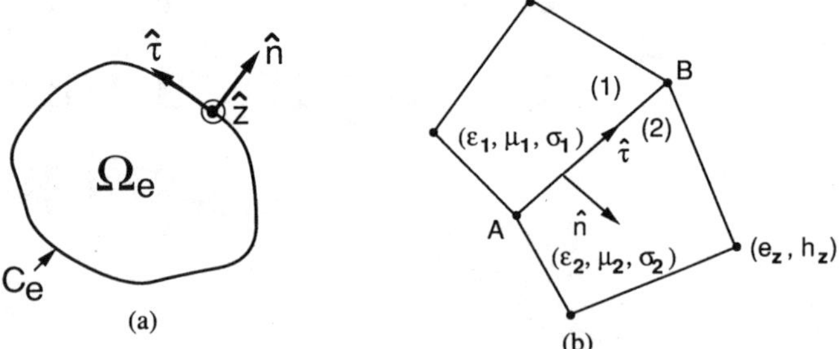

Fig. 1. (a) The localized coordinate system along the boundary C_e. (b) The localized coordinated system along the interelement boundary AB.

where $v^2 = \beta^2 + \gamma^2$, $\langle\cdot\rangle_{\Omega_e}$ indicates integration over the surface Ω_e on the xy plane, and ϕ_i, $i = 1, 2, \cdots$, are first-order differentiable functions of x and y. The first term in (7) and (8) may be integrated by parts and the above weak forms become

$$\langle(v^{-2}i\omega\mu\nabla_t h_z)\cdot\nabla_t\phi_i\rangle_{\Omega_e} + \langle i\omega\mu h_z\phi_i\rangle_{\Omega_e} = \oint_{C_e}(v^{-2}i\omega\mu\hat{n}\cdot\nabla_t h_z)\phi_i\, dl \tag{9}$$

$$\langle[v^{-2}(\sigma + i\omega\epsilon)\nabla_t e_z]\cdot\nabla_t\phi_i\rangle_{\Omega_e} + \langle(\sigma + i\omega\epsilon)e_z\phi_i\rangle_{\Omega_e} = \oint_{C_e}[v^{-2}(\sigma + i\omega\epsilon)\hat{n}\cdot\nabla_t e_z]\phi_i\, dl \tag{10}$$

where C_e is the curve enclosing Ω_e and $\hat{n}$ is the outward unit normal vector on C_e. Equations (9) and (10) constitute the weighted-residual statement of the boundary value problem under consideration. It is noted that the boundary integral terms in these equations involve the normal derivatives of h_z and e_z at the boundary. These terms are used for applying boundary conditions at material interfaces and for introducing the forcing terms at boundaries where h_z and e_z are not specified. In addition, they may be used for computing the normal derivatives of h_z and e_z on boundaries where h_z and e_z are specified.

An alternative form of caret (9) and (10) may be derived by expressing $\hat{n}\cdot\nabla_t h_z$ and $\hat{n}\cdot\nabla_t e_z$ in terms of the tangential derivatives of e_z and h_z along the boundary C_e and the components of $\vec{e}_t$ and $\vec{h}_t$ tangent to C_e. This is accomplished easily using (4), (5) and a local coordinate system $(\hat{n}, \hat{\tau}, \hat{z})$ along C_e as shown in Fig. 1(a), where $\hat{\tau}$ is the unit vector tangent to C_e. Dot multiplying (4) and (5) by $\hat{\tau}$ gives

$$v^{-2}i\omega\mu\hat{n}\cdot\nabla_t h_z = i\beta v^{-2}\hat{\tau}\cdot\nabla_t e_z - \hat{\tau}\cdot\vec{e}_t \tag{11}$$

$$v^{-2}(\sigma + i\omega\epsilon)\hat{n}\cdot\nabla_t e_z = \hat{\tau}\cdot\vec{h}_t - i\beta v^{-2}\hat{\tau}\cdot\nabla_t h_z. \tag{12}$$

In view of (11) and (12), (9) and (10) take the form

$$\langle(v^{-2}i\omega\mu\nabla_t h_z)\cdot\nabla_t\phi_i\rangle_{\Omega_e} + \langle i\omega\mu h_z\phi_i\rangle_{\Omega_e} = \oint_{C_e}(i\beta v^{-2}\hat{\tau}\cdot\nabla_t e_z - \hat{\tau}\cdot\vec{e}_t)\phi_i\, dl \tag{13}$$

$$\langle[v^{-2}(\sigma + i\omega\epsilon)\nabla_t e_z]\cdot\nabla_t\phi_i\rangle_{\Omega_e} + \langle(\sigma + i\omega\epsilon)e_z\phi_i\rangle_{\Omega_e} = \oint_{C_e}(\hat{\tau}\cdot\vec{h}_t - i\beta v^{-2}\hat{\tau}\cdot\nabla_t h_z)\phi_i\, dl. \tag{14}$$

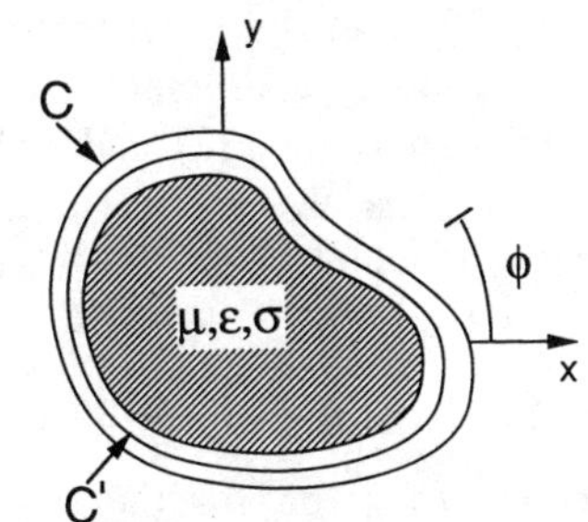

Fig. 2. Cross-sectional view of the cylinder.

This alternative form of the weak statement of the problem shows explicitly the coupling of e_z and h_z, and makes possible the direct enforcement of the continuity of the tangential components of the electric and magnetic field at material interfaces. This becomes clear in the following, where the Galerkin approximation and the finite element discretization procedure are described.

Following a standard Galerkin procedure, the unknowns h_z and e_z are expanded in terms of the set of N basis functions $\{\phi_1, \phi_2, \cdots, \phi_N\}$

$$\begin{Bmatrix} e_z \\ h_z \end{Bmatrix} = \sum_{j=1}^{N} \begin{Bmatrix} e_j \\ h_j \end{Bmatrix} \phi_j, \tag{15}$$

and the satisfaction of (13) and (14) is required for $i = 1$ through N over the surface Ω that encompasses the cross section of the cylinder as shown in Fig. 2. The $2N$ coefficients (e_j, h_j), $j = 1, 2, \cdots, N$, are the unknowns of the problem. This Galerkin approximation is then implemented on standard finite elements, where ϕ_i are typical finite element interpolation functions chosen such that they are continuous across interelement boundaries and their first-order derivatives are continuous within each element. The unknown coefficients (e_j, h_j), $j = 1, 2, \cdots, N$, are then the values of e_z and h_z at the N nodes of the finite element grid used for the discretization. As mentioned at the beginning of the section, material properties are assumed constant within each element of the finite element grid; hence, any boundaries where material properties change abruptly coincide with interelement boundaries. Since ϕ_j are continuous at interelement boundaries, the numerical approximations of e_z and h_z, as given by (15), are continuous by construction. Thus it only remains to enforce the continuity of the other two tangential components $\hat{\tau} \cdot \vec{e}_t$ and $\hat{\tau} \cdot \vec{h}_t$ at the interelement boundaries. However, in performing the usual element-by-element assembly of (13) and (14), i.e., evaluating (13) and (14) one element at a time and summing the contributions, the line integral contributions for a typical interelement boundary AB are (Fig. 1(b))

$$\int_{AB} \left[\left(i\beta v_1^{-2} \hat{\tau} \cdot \nabla_t e_z^{(1)} - i\beta v_2^{-2} \hat{\tau} \cdot \nabla_t e_z^{(2)} \right) - \left(\hat{\tau} \cdot \vec{e}_t^{(1)} - \hat{\tau} \cdot \vec{e}_t^{(2)} \right) \right] \phi_i \, dl, \tag{16}$$

$$\int_{AB} \left[\left(\hat{\tau} \cdot \vec{h}_t^{(1)} - \hat{\tau} \cdot \vec{h}_t^{(2)} \right) - \left(i\beta v_1^{-2} \hat{\tau} \cdot \nabla_t h_z^{(1)} - i\beta v_2^{-2} \hat{\tau} \cdot \nabla_t h_z^{(2)} \right) \right] \phi_i \, dl, \tag{17}$$

where the indices 1 and 2 refer to elements (1) and (2), respectively. It is then apparent that the enforcement of the continuity of $\hat{\tau} \cdot \vec{e}_t$ and $\hat{\tau} \cdot \vec{h}_t$ along the interface AB nullifies the second term in (16) and the first term in (17). This implies that the enforcement of the continuity of $\hat{\tau} \cdot \vec{e}_t$ and $\hat{\tau} \cdot \vec{h}_t$ at interelement boundaries can be accomplished automatically during the element-by-element assembly of the finite element equations.

III. Numerical Solution—The Bymoment Method

In this section the numerical solution of the scattering problem is carried out using the bymoment method. The bymoment method was first used for the finite element solution of the two-dimensional scalar wave equation in unbounded media [9], [10]. In what follows, the method is extended to the case of scattering from inhomogeneous cylinders at oblique incidence, where there are two unknown functions, $(e_z(x, y), h_z(x, y))$, to be determined from the numerical solution of two coupled scalar wave equations utilizing the weak statement in (13) and (14).

Consider the two-dimensional geometry shown in Fig. 2. The boundary curve C conforms to the cylindrical cross section, encloses it totally, and is taken to be at least one layer of grid elements away from it. Indeed, numerical experimentation has shown that for scatterers with smooth geometries the finite element mesh need only extend one element away from the boundary of the inhomogeneous region. The boundary curve C divides the unbounded region where the solution is sought into an interior region Ω containing the inhomogeneous scatterer and an exterior homogeneous region Ω_∞ which extends to infinity. From the uniqueness theorem it is known that, for the problem under consideration and for the chosen (e_z, h_z) formulation, the solution inside Ω can be found uniquely by specifying e_z and h_z everywhere on C [11]. Of course, e_z and h_z on C are not known *a priori*; hence, the solution inside Ω cannot be obtained directly. Instead, the following procedure is followed.

Let $e_z(t)$, $h_z(t)$ be the unknown functions that describe the restrictions of e_z and h_z on C, respectively. The variable t denotes position on C and varies from 0 to d, where d is the length of C. $e_z(t)$ and $h_z(t)$ are expanded in terms of a set of basis functions $\Psi_m(t)$ as follows:

$$\begin{Bmatrix} e_z(t) \\ h_z(t) \end{Bmatrix} = \sum_{m=1}^{M} \begin{Bmatrix} a_m \\ b_m \end{Bmatrix} \Psi_m(t), \qquad t \in [0, d] \tag{18}$$

where a_m, b_m are unknown coefficients and M is the number of expansion functions required to represent e_z and h_z. These expansion functions Ψ_m can be either subdomain or entire-domain functions [10]. Next, by specifying each one of the following $2M$ Dirichlet boundary conditions on C

$$\{e_z(t) = \Psi_m(t),\ h_z(t) = 0\}, \qquad m = 1, 2, \cdots, M \tag{19a}$$

$$\{e_z(t) = 0,\ h_z(t) = \Psi_m(t)\}, \qquad m = 1, 2, \cdots, M \tag{19b}$$

a set of $2M$ independent numerical field solutions are generated inside Ω using standard finite elements in conjunction with the weak statement in (13) and (14). Let $(\Lambda_m^{(e)}, \Xi_m^{(e)})$ be the numerical solution for (e_z, h_z) inside Ω corresponding to the mth boundary condition in (19a), and $(\Lambda_m^{(h)}, \Xi_m^{(h)})$ be the numerical solution for (e_z, h_z) corresponding to the mth boundary condition in (19b). Then, from linearity and the uniqueness theorem, e_z and h_z inside Ω can be written as a superposition of these finite element solutions multiplied by the appropriate coefficients a_m, b_m,

$$e_z(x, y) = \sum_{m=1}^{M} \left[a_m \Lambda_m^{(e)} + b_m \Lambda_m^{(h)} \right] \tag{20a}$$

$$h_z(x, y) = \sum_{m=1}^{M} \left[a_m \Xi_m^{(e)} + b_m \Xi_m^{(h)} \right]. \tag{20b}$$

From (20a) and (20b) it becomes apparent that in order to complete the solution the coefficients a_m and b_m need to be determined. This is done by coupling the interior solution to the exterior homogeneous region as explained next.

Considering the geometry of Fig. 2, let C' be a curve that passes through the interior of all the elements on the boundary of the grid. This curve lies entirely in the homogeneous region exterior to the scatterer and is totally enclosed by C. Let Ω'_∞ be the surface which extends from C' all the way to infinity. In Ω'_∞ the scattered field components e_z^s, h_z^s satisfy the source-free Helmholtz equation and the Sommerfeld radiation condition at infinity. Let Φ_j, $j = 1, 2, \cdots,$ be a set of linearly independent testing functions which are also chosen to satisfy the source-free Helmholtz equation inside Ω'_∞ and the radiation condition at infinity. The various choices for Φ_j have been discussed in detail in [10] and will not be repeated here. Application of the scalar Green's theorem for e_z^s and any of the Φ_j over Ω'_∞ gives

$$\int_{C'} \left[\Phi_j \frac{\partial e_z^s}{\partial n'} - e_z^s \frac{\partial \Phi_j}{\partial n'} \right] dl = 0, \tag{21}$$

where $\hat{n}'$ is the outward pointing unit normal vector on C'. Expressing the scattered field as the difference of the total and incident fields and using (20a), (21) becomes

$$\sum_{m=1}^{M} a_m \int_{C'} \left[\Phi_j \frac{\partial \Lambda_m^{(e)}}{\partial n'} - \Lambda_m^{(e)} \frac{\partial \Phi_j}{\partial n'} \right] dl + \sum_{m=1}^{M} b_m \int_{C'} \left[\Phi_j \frac{\partial \Lambda_m^{(h)}}{\partial n'} - \Lambda_m^{(h)} \frac{\partial \Phi_j}{\partial n'} \right] dl = \int_{C'} \left[\Phi_j \frac{\partial e_z^i}{\partial n'} - e_z^i \frac{\partial \Phi_j}{\partial n'} \right] dl. \tag{22}$$

A similar procedure for h_z^s gives

$$\sum_{m=1}^{M} a_m \int_{C'} \left[\Phi_j \frac{\partial \Xi_m^{(e)}}{\partial n'} - \Xi_m^{(e)} \frac{\partial \Phi_j}{\partial n'} \right] dl + \sum_{m=1}^{M} b_m \int_{C'} \left[\Phi_j \frac{\partial \Xi_m^{(h)}}{\partial n'} - \Xi_m^{(h)} \frac{\partial \Phi_j}{\partial n'} \right] dl = \int_{C'} \left[\Phi_j \frac{\partial h_z^i}{\partial n'} - h_z^i \frac{\partial \Phi_j}{\partial n'} \right] dl. \tag{23}$$

Equations (22) and (23) constitute the testing statements for calculating the unknown coefficients (a_m, b_m), $m = 1, 2, \cdots, M$. Indeed, applying (22) and (23) for M of the independent test functions Φ_j results in a system of $2M$ independent equations which can be solved for the $2M$ coefficients.

IV. Numerical Results

Numerical results are given for the case of plane wave scattering from material cylinders. The incident plane wave can be either TM_z ($h_z^i = 0$) or TE_z ($e_z^i = 0$), and is defined by

$$\begin{pmatrix} e_z^i \\ h_z^i \end{pmatrix} = \sin\theta\, e^{-ik_0 \sin\theta (x\cos\phi_i + y \sin\phi_i)} \tag{24}$$

where k_0 is the free space wavenumber, θ is the angle of incidence along the z-axis, and ϕ_i is the angle of incidence in the $x - y$ plane. In (24) e_z^i is used for the TM_z case, and h_z^i is used for the TE_z case.

The finite element solutions are generated from a mesh composed of four-node quadrilateral elements. The meshes themselves are created by a program called FASTQ [12]. The computational work was done on the VAX 2000 workstation and the VAX 8550. A modified Gauss–Choleski algorithm [13] is applied to solve the banded matrix equation. For all the geometries considered, entire-domain sinusoidal functions are used for the basis functions Ψ_m, and the testing functions Φ_j are normalized solutions to the scalar wave equation in free space due to a traveling-wave line source located at points (x_j, y_j), $j = 1, 2, \cdots, M$, in the region interior to C' as explained in [10]. They are given by

$$\Phi_j = H_0^{(2)}\left(k_0 \sin\theta \sqrt{(x - x_j)^2 + (y - y_j)^2} \right) \tag{25}$$

where the z-dependence is suppressed. For this testing function, (23) and (24) can be reduced to a much simpler form since

$$\int_{C'} \left[\Phi_j \frac{\partial e_z^i}{\partial n'} - e_z^i \frac{\partial \Phi_j}{\partial n'} \right] dl = -4ie_z^i(x_j, y_j), \tag{26a}$$

$$\int_{C'} \left[\Phi_j \frac{\partial h_z^i}{\partial n'} - h_z^i \frac{\partial \Phi_j}{\partial n'} \right] dl = -4ih_z^i(x_j, y_j). \tag{26b}$$

The first geometry considered is that of a homogeneous dielectric circular cylinder in free space. An eigenfunction solution for a TM_z incident wave is computed using the formulation from Wait [1] and then compared to the bymoment solution. The cylinder has a radius of $r/\lambda = 0.3$ and a relative permittivity $\epsilon_r = 4$. The incident wave is impinging the cylinder at an angle of $\theta = 45°$, $\phi_i = 180°$. The mesh contains 1321 nodes (2642 unknowns) resulting in a stiffness matrix with a maximum half-bandwidth of 94. The half-bandwidth is defined to be the number of matrix terms from and including the diagonal term to the last nonzero term in a given row. M is set to 14 and thus the full matrix which is inverted for the determination of the unknown coefficients (a_m, b_m) has dimensions 28×28. Results for the magnitude

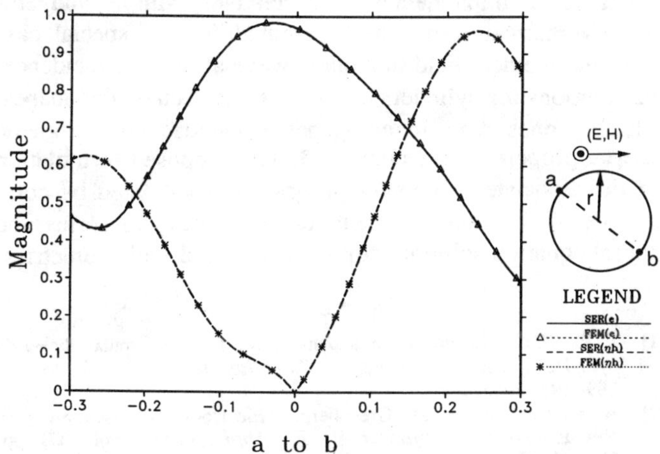

Fig. 3. Magnitude of the z-component of the fields along the interior line ab for a dielectric circular cylinder with $r/\lambda = 0.3$, $\epsilon_r = 4$, $\theta = 45°$, and $\phi_i = 180°$.

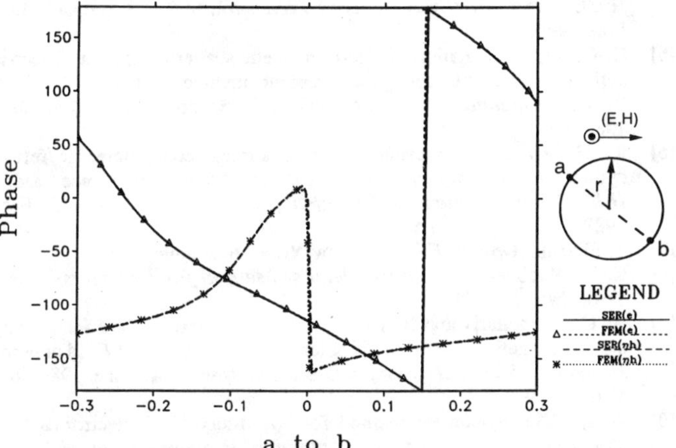

Fig. 4. Phase of the z-component of the fields along the interior line ab for a dielectric circular cylinder with $r/\lambda = 0.3$, $\epsilon_r = 4$, $\theta = 45°$, and $\phi_i = 180°$.

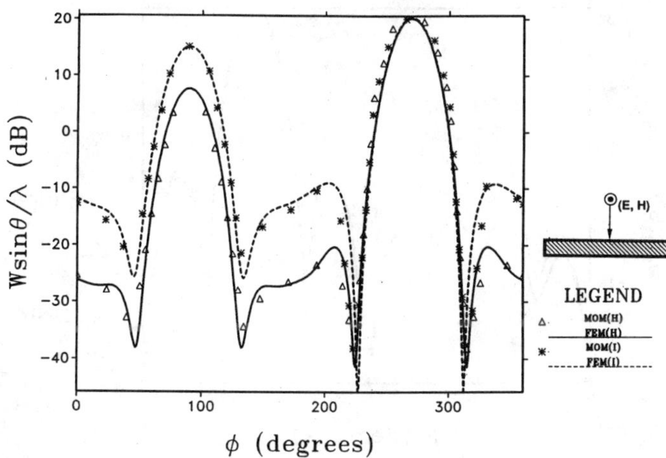

Fig. 5. Echo width of a 2 $\lambda \times 0.1\ \lambda$ rectangular cylinder for TE_z-polarized incident plane wave with $\theta = 45°$ and $\phi_i = -90°$.

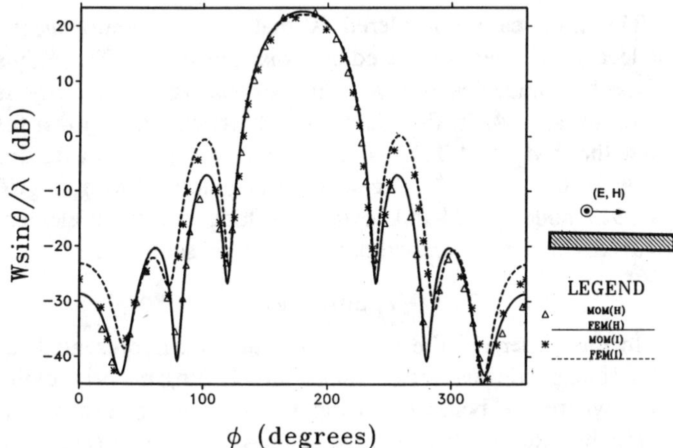

Fig. 6. Echo width of a 2 $\lambda \times 0.1\ \lambda$ rectangular cylinder for TE_z-polarized incident plane wave with $\theta = 45°$ and $\phi_i = 0°$.

and phase of e_z and ηh_z along the line labeled ab are given in Figs. 3 and 4 where η is the free space wave impedance. The line ab is oriented at a 45° angle from the horizontal. Wait's series solution is denoted by SER, and the bymoment solution is denoted by FEM. The agreement between the two methods is excellent. Next the case of a 2 $\lambda \times 0.1\ \lambda$ rectangular cylinder is considered with the longer length in the horizontal direction. Because of the large perimeter length, $M = 28$. The grid contains 394 nodes (788 unknowns) and the half-bandwidth of the finite element matrix is 30. Echo width results are obtained for this geometry and compared to the method of moments (MOM) results given by Rojas [6]. The expression for the echo width is

$$\frac{W}{\lambda} = \lim_{\rho \to \infty} k_0 \rho \frac{|E_z^s|^2 + |\eta H_z^s|^2}{|E_z^i|^2 + |H_z^i|^2}. \tag{27}$$

The direction of propagation of the incident TE_z plane wave is defined by $\theta = 45°$ and $\phi_i = 90°$. The results in Fig. 5 are for two separate cases. The first is the case where the cylinder is homogeneous with $\epsilon_r = 3 - j0.3$ and $\mu_r = 2 - j0.1$. The corresponding curves are labeled by MOM(H) and FEM(H). The results show good agreement. The second is the case where the cylinder is inhomogeneous. The curve denoted by MOM(I) has material properties of $\epsilon_r = 3 - j0.3 + 2\cos(\pi x/2\lambda)$ and $\mu_r = (2 - j0.1)|x|/2\lambda$ where the origin for x is taken to be the center of the cylinder. The material properties used for the curve marked FEM(I) are an approximation to that used in MOM(I). Instead of using a continuously varying function for ϵ_r and μ_r, the cylinder is divided into five homogeneous regions of equal area with the material properties of each region taking on the value of the continuously varying case evaluated at the center of the region. The material properties used for the five regions are $\epsilon_{r1} = \epsilon_{r5} = 3.62 - j0.3$, $\epsilon_{r2} = \epsilon_{r4} = 4.62 - j0.3$, $\epsilon_{r3} = 5 - j0.3$, $\mu_{r1} = \mu_{r5} = 1.8 - j0.04$, $\mu_{r2} = \mu_{r4} = 1.4 - j0.02$, and $\mu_{r3} = 1.0$. This was done in order to see whether one can approximate a continuously inhomogeneous cylinder with one that has piecewise homogeneous properties. In comparing the two curves, it can be seen that there is very little difference between the two solutions. The next set of cases (Fig. 6) uses the same geometry and parameters as those used in Fig. 5 except that ϕ_i is now set to 180°. For this angle of incidence, the differences between MOM(I) and FEM(I) are more evident than in the previous case. This is especially true for the forward scattering data.

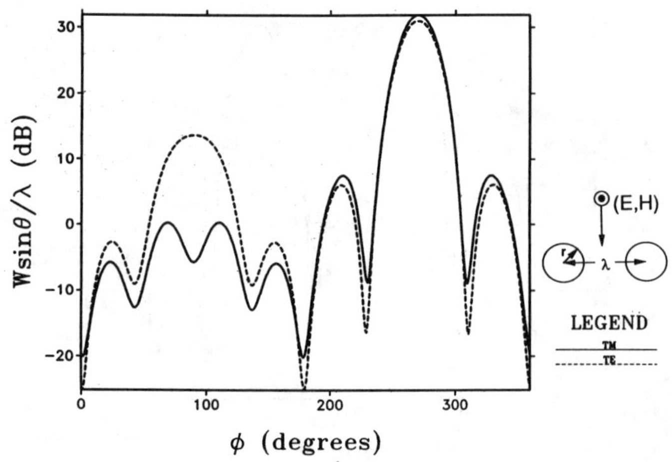

Fig. 7. Echo width of two dielectric circular cylinders of radius $r/\lambda = 0.3$ and relative permittivity $\epsilon_r = 4$ separated by a distance 1 $\lambda . \theta = 45°$ and $\phi_i = -90°$.

The final case considered is that of two homogeneous dielectric cylinders separated by a distance of 1 λ. The radius of each cylinder is 0.3 λ, and the relative permittivity is given by $\epsilon_r = 4$. In Fig. 7, echo width results are shown for both the TM_z and TE_z polarized incident plane waves for $\phi_i = -90°$, $\theta = 45°$. The solution is generated using a grid of 2317 nodes (4634 unknowns) resulting in a finite element matrix with a half-bandwidth of 166. M is set to 24.

V. Summary

In this paper a finite element formulation is presented for electromagnetic scattering from infinitely long material cylinders where the behavior of the fields in the z direction is given by exp $(-i\beta z)$. The weighted-residual expression incorporates the correct boundary conditions along material interfaces, resulting in the expected coupling of the **z** components of the electric and magnetic fields. The bymoment method is used to handle the coupling of the interior finite element solution to the exterior unbounded region. Because the bymoment method decouples E_z and H_z along the mesh boundary, its implementation is relatively simple and efficient. Numerical results are presented for the special case where the incident field is a plane wave at oblique incidence. The solutions for cylinders of circular and rectangular shapes, including ones with inhomogeneous electric and magnetic material properties, are computed. The proposed formulation and the associated computer program are validated by comparisons of the finite element method results to series and integral equation solutions for specific cylindrical geometries.

References

[1] J. R. Wait, "Scattering of a plane wave from a circular dielectric cylinder at oblique incidence," *Canadian J. Phys.*, vol. 33, pp. 189–195, 1955.

[2] A. C. Lind and J. M. Greenberg, "Electromagnetic scattering by obliquely oriented cylinders," *J. Appl. Phys.*, vol. 37, pp. 3195–3203, July 1966.

[3] R.-B. Wu and C. H. Chen, "Variational reaction formulation of scattering problem for anisotropic dielectric cylinders," *IEEE Trans. Antennas Propagat.*, vol. AP-35, pp. 640–645, May 1986.

[4] —, "On the variational reaction theory for dielectric waveguides," *IEEE Trans. Microwave Theory Tech.*, vol. MTT-33, pp. 477–483, June 1985.

[5] C.-C. Su, "Calculation of electromagnetic scattering from a dielectric cylinder using the conjugate gradient method and FFT," *IEEE Trans. Antennas Propagat.*, vol. AP-35, pp. 1418–1425, Dec. 1987.

[6] R. G. Rojas, "Scattering by an inhomogeneous dielectric/ferrite cylinder of arbitrary cross-section shape—Oblique incidence case," *IEEE Trans. Antennas Propagat.*, vol. 36, pp. 238–246, Feb. 1988.

[7] T. Okoshi, *Optical Fibers*. New York: Academic, 1982.

[8] K. H. Huebner, *The Finite Element Method for Engineers*. New York: Wiley, 1975.

[9] A. C. Cangellaris and R. Lee, "A finite element method for solving electromagnetic scattering problems," in *1989 IEEE Antennas Propagat. Soc. Int. Symp. Dig.*, San Jose, CA, June 1989, pp. 1116–1119.

[10] —, "The bymoment method for two-dimensional electromagnetic scattering," *IEEE Trans. Antennas Propagat.*, vol. 38, pp. 1429–1437, Sept. 1990.

[11] R. F. Harrington, *Time-Harmonic Electromagnetic Fields*. New York: McGraw-Hill, 1961.

[12] T. D. Blacker, "FASTQ User Manual, Version 1.2," Sandia Nat. Lab., Albuquerque, NM, *Rep. SAND88-1326*, July 1988.

[13] J. E. Akin, *Application and Implementation of Finite Element Methods*. San Diego, CA: Academic, 1982.

A Hybrid Finite Element Method for Scattering and Radiation by Microstrip Patch Antennas and Arrays Residing in a Cavity

Jian-Ming Jin, *Member, IEEE,* and John L. Volakis, *Senior Member, IEEE*

Abstract—**A hybrid numerical technique is presented for a characterization of the scattering and radiation properties of microstrip patch antennas and arrays residing in a cavity recessed in a ground plane. The technique combines the finite element and boundary integral methods to formulate a system for the solution of the fields at the aperture and those inside the cavity via the biconjugate gradient method in conjunction with the fast Fourier transform (FFT). By virtue of the finite element method, the proposed technique is applicable to patch antennas and arrays residing on or embedded in a layered dielectric substrate and is also capable of treating various feed configurations and impedance loads. Several numerical results are presented demonstrating the validity, efficiency and capability of the technique.**

I. Introduction

MICROSTRIP antennas and arrays are widely used on airborne vehicles because they are lightweight, conformal, and relatively easy and inexpensive to fabricate. Consequently, much literature exists on their operation, analysis, and design (see, for example, [1]–[3] and their references). In most cases, however, available analysis and design procedures are based on infinite substrate models, though in practice the substrate is finite. Moreover, existing methodologies have been directed at characterizing the radiation properties of the microstrip antennas and arrays, whereas the characterization of the radar scattering by such structures in deployed position has generally been neglected. The last is of importance for low radar profile applications, and recently some research has appeared on the subject [4]–[7]. In particular, Newman and Forrai [4], [5] and Pozar [6] have presented a scattering analysis of the microstrip patch based on an infinite substrate model and Sarkar *et al.* [7] employed an analysis based on a finite substrate model. Neither of these models is representative of the typical array periphery in deployment and moreover the associated methods of analysis have an $O(N^2)$ storage requirement, severely restricting them to small scale applications.

In this paper we present a hybrid numerical technique for a characterization of the scattering and radiation properties of microstrip antennas and arrays residing in a cavity recessed in a ground plane. The technique is an extension of that proposed in [8] and [9] for the analysis of three-dimensional cavity-backed apertures in a ground plane and slots in a thick conducting plane. It combines the finite element and boundary integral methods to formulate a system for the solution of the fields at the aperture and those within the substrate. Specifically, the finite element method is employed to formulate the fields within the cavity housing the substrate and the boundary integral equation is used to represent the fields above the ground plane. The last provides an additional relation between the electric and magnetic fields at the aperture and upon enforcing continuity of the tangential fields across the aperture, a coupled system of equations is obtained for the solution of the cavity and aperture fields.

To allow the analysis of large structures as is the case with microstrip arrays, the biconjugate gradient method is employed for the solution of the resulting system and in the process the fast Fourier transform (FFT) is used for an efficient evaluation of the convolutional boundary integrals over the aperture without a need to generate and store the matrix. Coupled with the sparseness of the finite element matrix, this reduces the storage requirement to $O(N)$ and, in addition, enhances the computational efficiency of the solution technique. Such a storage reduction is not shared by other methodologies considered in the past, making this technique attractive for simulating large array structures. Furthermore, by virtue of the finite element method, the proposed method of analysis is applicable to patch antennas and arrays residing on or embedded in an arbitrarily layered dielectric substrate. It also allows the modeling of various feed structures (either microstrip transmission lines or coaxial feeds) and impedance loads which may have a significant effect on the scattering of the antenna or array. The numerical results included in this paper demonstrate the method's validity, efficiency and capability. In addition, they illustrate

Manuscript received December 28, 1990; revised July 15, 1991.

The authors are with the Radiation Laboratory, Department of Electrical Engineering and Computer Science, University of Michigan, Ann Arbor, MI 48109-2122.

IEEE Log Number 9103351.

Reprinted from *IEEE Trans. Antennas Propagat.*, vol. 39, no. 11, pp. 1598–1604, Nov. 1991.

some of the scattering effects associated with the antenna's substructure, feed configuration and impedance loading.

II. Formulation

Consider the three-dimensional structure illustrated in Fig. 1 where a microstrip patch antenna or array is residing on or embedded in a substrate recessed in a ground plane. We are interested in a characterization of the scattering and radiation properties of this antenna, where in the first case the source is a plane wave illuminating the structure while in the case of radiation the source is the excitation generated by the antenna feed(s). In accordance with the finite element–boundary integral method discussed in [8] and [9], the problem can be formulated by first invoking the surface equivalence principle to divide the problem into two regions. One of the regions is the half-space above the antenna and the other comprises the cavity housing the substrate including the antenna elements and feed structures. Within the cavity the finite element method is employed to formulate the fields since this method is best suited for modeling inhomogeneities, including those due to the antenna elements and feeds. The boundary integral equation is employed to formulate the fields in the half-space and is enforced at the cavity's aperture leading to a relation between the aperture electric and magnetic fields. By enforcing continuity of the tangential fields at the aperture, this provides a boundary constraint necessary for the unique solution of the field within the cavity and at the aperture.

Let us first formulate the fields inside the cavity via the variational approach, although Galerkin's technique could be equivalently employed. In accordance with the variational approach, it is demanded that the cavity fields satisfy the variational equation:

$$\delta F(\mathbf{E}) = 0 \tag{1}$$

where the functional F is given by

$$F(\mathbf{E}) = \frac{1}{2}\iiint_V \left[\frac{1}{\mu_r}(\nabla\times\mathbf{E})\cdot(\nabla\times\mathbf{E}) - k_0^2\epsilon_r\mathbf{E}\cdot\mathbf{E}\right] dV + \iiint_V \left[jk_0Z_0\mathbf{J}^{\text{int}}\cdot\mathbf{E} - \frac{1}{\mu_r}\mathbf{M}^{\text{int}}\cdot(\nabla\times\mathbf{E})\right] dV + jk_0Z_0\iint_S (\mathbf{E}\times\mathbf{H})\cdot\hat{\mathbf{z}}\, dS. \tag{2}$$

In this, V represents the cavity volume, S is over the cavity aperture and (ϵ_r, μ_r) denote the relative permittivity and permeability of the material filling the cavity. Also, $k_0 = 2\pi/\lambda$ is the free space wavenumber and $Z_0 = 1/Y_0$ is the free space intrinsic impedance. We remark that (2) differs from the functionals given in [8] and [9] owing to the presence of the second volume integral involving the internal electric and magnetic sources $(\mathbf{J}^{\text{int}}, \mathbf{M}^{\text{int}})$ due to the antenna element feeds.

As seen, the surface integral in (2) involves the tangential electric and magnetic fields at the aperture and it is, therefore, necessary that we introduce an additional relation between these fields. This is provided by the boundary integral

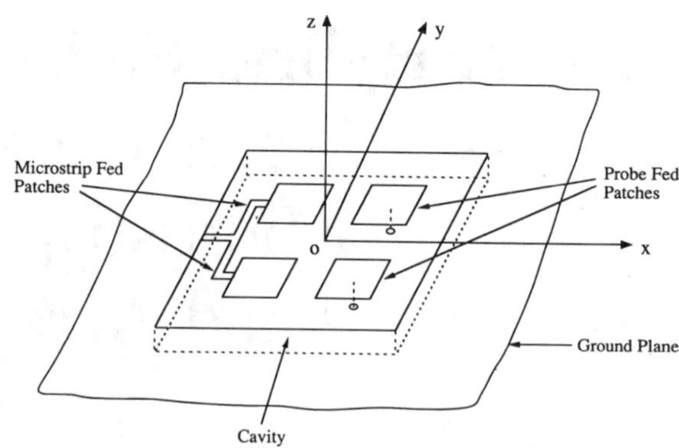

Fig. 1. Geometry of a microstrip patch array in a cavity.

equation for the fields in the upper half-space:

$$\mathbf{H}(\mathbf{r}) = \mathbf{H}^{\text{inc}}(\mathbf{r}) + \mathbf{H}^{\text{ref}}(\mathbf{r}) - 2jk_0Y_0\iint_S \overline{\overline{G}}_0(\mathbf{r},\mathbf{r}')\cdot[\mathbf{E}(\mathbf{r}')\times\hat{\mathbf{z}}]\, dS' \tag{3}$$

where $\mathbf{H}^{\text{inc}}$ denotes the incident field due to possible external sources $(\mathbf{J}^{\text{ext}}, \mathbf{M}^{\text{ext}})$, $\mathbf{H}^{\text{ref}}$ is that reflected by the ground plane without the aperture, and $\overline{\overline{G}}_0$ is the usual free-space dyadic Green's function. Since the tangential electric and magnetic fields must be continuous across the aperture, we may substitute (3) into (2) to find

$$F(\mathbf{E}) = \frac{1}{2}\iiint_V \left[\frac{1}{\mu_r}(\nabla\times\mathbf{E})\cdot(\nabla\times\mathbf{E}) - k_0^2\epsilon_r\mathbf{E}\cdot\mathbf{E}\right] dV + \iiint_V \left[jk_0Z_0\mathbf{J}^{\text{int}}\cdot\mathbf{E} - \frac{1}{\mu_r}M^{\text{int}}\cdot(\nabla\times\mathbf{E})\right] dV - 2k_0^2\iint_S [\mathbf{E}(\mathbf{r})\times\hat{\mathbf{z}}] \cdot\left[\iint_S \overline{\overline{G}}_0(\mathbf{r},\mathbf{r}')\cdot[\mathbf{E}(\mathbf{r}')\times\hat{\mathbf{z}}]\, dS'\right] dS - 2jk_0Z_0\iint_S [\mathbf{E}(\mathbf{r})\times\hat{\mathbf{z}}]\cdot\mathbf{H}^{\text{inc}}(\mathbf{r})\, dS \tag{4}$$

which involves only the electric field and can be discretized via the finite element method for a solution of the cavity and aperture fields.

To discretize (4), the cavity volume V is subdivided into small volume elements such as tetrahedra, triangular prisms, or rectangular bricks. Accordingly, by using the edge-based vector basis functions [10], the electric field $\mathbf{E}$ is expanded as

$$\mathbf{E} = \sum_{j=1}^{N} E_j\mathbf{W}_j \tag{5}$$

where N denotes the total number of element edges resulting from the subdivision, excluding those coinciding with conducting surfaces since the tangential components of the electric field associated with those edges are zero. Also, E_j denote the unknown expansion coefficients equal to element edge fields and $\mathbf{W}_j$ are the chosen vector basis functions. By substituting (5) into (4) and taking the partial derivative of F

with respect to E_i, we obtain

$$\frac{\partial F}{\partial E_i} = \sum_{j=1}^{N} E_j \iiint_V \left[\frac{1}{\mu_r} (\nabla \times \mathbf{W}_i) \cdot (\nabla \times \mathbf{W}_j) - k_0^2 \epsilon_r \mathbf{W}_i \cdot \mathbf{W}_j \right] dV + \iiint_V \left[jk_0 Z_0 \mathbf{J}^{\text{int}} \cdot \mathbf{W}_i - \frac{1}{\mu_r} \mathbf{M}^{\text{int}} \cdot (\nabla \times \mathbf{W}_i) \right] dV - 2k_0^2 \chi_i \iint_S [\mathbf{W}_i \times \hat{\mathbf{z}}] \cdot \sum_{j=1}^{N} \chi_j E_j \left[\iint_S \overline{\overline{G}}_0 \cdot [\mathbf{W}_j' \times \hat{\mathbf{z}}]\, dS' \right] dS - 2jk_0 Z_0 \chi_i \iint_S [\mathbf{W}_i \times \hat{\mathbf{z}}] \cdot \mathbf{H}^{\text{inc}}\, dS \tag{6}$$

where $\chi_i = 1$ when the ith edge lies on the aperture S and $\chi_i = 0$ elsewhere. We note that in deriving the third term in the right-hand side of (6) the differentiation was performed only with respect to E_i outside the square brackets whereas those electric field components resulting from the substitution of (3) into (2) were not differentiated. This is because $\mathbf{H}$ in the third term of (2) is viewed as a source of the electric field and is thus treated in the same manner as $(\mathbf{J}^{\text{int}}, \mathbf{M}^{\text{int}})$. The final system of equations is now obtained by imposing the stationarity condition (1), i.e., by letting $\partial F / \partial E_i = 0$ for $i = 1, 2, \ldots, N$. This yields

$$[A]\{E\} = \{B\} \tag{7}$$

where $\{E\} = [E_1, E_2, \ldots, E_N]^T$, $[A]$ is an $N \times N$ partly full and partly sparse matrix and $\{B\}$ is an $N \times 1$ column. Their elements are given by

$$A_{ij} = \iiint_V \left[\frac{1}{\mu_r} (\nabla \times \mathbf{W}_i) \cdot (\nabla \times \mathbf{W}_j) - k_0^2 \epsilon_r \mathbf{W}_i \cdot \mathbf{W}_j \right] dV - 2k_0^2 \chi_i \chi_j \iint_S [\mathbf{W}_i \times \hat{\mathbf{z}}] \cdot \left[\iint_S \overline{\overline{G}}_0 \cdot [\mathbf{W}_j' \times \hat{\mathbf{z}}]\, dS' \right] dS \tag{8}$$

$$B_i = -\iiint_V \left[jk_0 Z_0 \mathbf{J}^{\text{int}} \cdot \mathbf{W}_i - \frac{1}{\mu_r} \mathbf{M}^{\text{int}} \cdot (\nabla \times \mathbf{W}_i) \right] dV + 2jk_0 Z_0 \chi_i \iint_S [\mathbf{W}_i \times \hat{\mathbf{z}}] \cdot \mathbf{H}^{\text{inc}}\, dS \tag{9}$$

and we observe that $[A]$ is symmetric. We also remark that in evaluating the second term of A_{ij} it is convenient to transfer the derivatives from the Green's function to $\mathbf{W}_i \times \hat{\mathbf{z}}$ and $\mathbf{W}_j' \times \hat{\mathbf{z}}$ by invoking the divergence theorem as described in [8], thus reducing the singularity of the integrand.

It remains to choose the vector basis functions $\mathbf{W}_j$ for evaluating A_{ij} and B_i and this depends on the choice of volume elements. For tetrahedra and rectangular bricks they are given in [10] and [9], respectively, and for triangular prisms, they can be easily obtained from those associated with two-dimensional triangular elements. As usual, the basis functions are nonzero only within the associated volume elements and thus, the portion of the matrix $[A]$ contributed by the first term of (8) is very sparse. In contrast, the portion of the matrix contributed by the second term of (8) is fully populated, but is only associated with the fields at the aperture of the cavity.

Once the aperture and cavity fields are solved, the scattering and radiation patterns, and other design parameters such as the input impedance, mutual coupling, and radiation efficiency can be readily calculated.

III. Numerical Considerations

In this section we consider the modeling of the conducting patches, microstrip transmission lines, impedance loads, shorting pins and coaxial feeds in the context of the finite element–boundary integral formulation. We also discuss some computational aspects related to the implementation of the technique.

A. Conducting Patches and Microstrip Transmission Lines

The modeling of the conducting patches is straightforward and is carried out by setting the tangential electric field components to zero for the element sides coinciding with the patch. The modeling of microstrip transmission lines is treated in exactly the same manner. We remark, though, that owing to the use of the edge-based expansion functions, no special condition needs to be imposed on the transverse components of the electric field at the patch or stripline edges.

B. Impedance Loads and Shorting Pins

An impedance load of Z_L ohms can be modeled as a post of finite conductivity joining both the patch and the base of the cavity. Assuming a post of length l and cross section s, the required conductivity of the post to represent the load must be $\sigma = l/(Z_L s)$. Accordingly, its contribution to the matrix element A_{ij} is

$$\frac{jk_0 Z_0 l}{Z_L s} \iiint_v \mathbf{W}_i \cdot \mathbf{W}_j\, dv$$

where the integration is over the volume of the post. Provided the post is very thin, this reduces to $jk_0 Z_0 l^2 \mathbf{W}_i \cdot \mathbf{W}_j / Z_L$, where $\mathbf{W}_i \cdot \mathbf{W}_j$ takes its value at the position of the post. Further, if the thin post is coincident with the ith edge, then it only contributes to A_{ii} and this contribution is $jk_0 Z_0 l^2 / Z_L$. The short-circuit ($Z_L = 0$) condition is usually realized with a shorting pin and two approaches can be used in this case. One approach is to represent it with a post of high conductivity and another is to simply set the electric field along the post to zero. We have observed that the final solution is unaffected by the choice of the model.

C. Coaxial Feeds and Gap Generators

We consider two types of models for simulating a coaxial feed via this technique. For thin substrates the coaxial feed may be replaced by a current filament and in this case, the contribution to the matrix element B_i is simply $-jk_0 Z_0 l\mathbf{I} \cdot \mathbf{W}_i$ where $\mathbf{I}$ denotes the electric current of the filament and l is its length. For thicker substrates a magnetic frill current model may be employed. In this case, the inner conductor of the coaxial feed is modeled as a conducting post and an equivalent magnetic current is introduced over the aperture of the coaxial feed. This is similar to the treatment for the cavity aperture, except that the free space region is now replaced by the coaxial waveguide. Finally, one may resort to a simple gap generator for the excitation whose modeling amounts to specifying *a priori* the field across the gap.

D. Cavity Discretization and Memory Demand

As mentioned earlier, in a finite element discretization, the volume V is subdivided into small volume elements such as tetrahedra, triangular prisms, or rectangular bricks. For an arbitrarily shaped volume V, the most attractive elements are tetrahedra which, unfortunately, require more unknowns when the edge-based vector basis functions are employed. Therefore, wherever possible the triangular prisms or rectangular bricks are preferred, especially the latter because they yield the least number of unknowns. Another crucial characteristic of the rectangular bricks is that they permit the evaluation of the surface integrals via the FFT. Because of this, when an iterative solution such as the conjugate or biconjugate gradient method is employed, there is no need to generate and store the full submatrix, leading to an $O(N)$ storage requirement for solving the system (7). In particular, assuming that the number of surface elements at the cavity aperture is N_s, the implementation of the FFT for evaluating the surface integral requires a memory of $12N_s$ complex numbers and to this we must add $4N$ complex numbers for the implementation of the biconjugate gradient method, where N is the number of element edges for the discretization of the volume within the cavity. Thus, the total memory demand of the proposed method is approximately $4N + 12N_s$ (complex numbers).

IV. Numerical Results

In this section we present some numerical results for demonstrating the validity, capability and efficiency of the method as well as illustrating the effect of various substructures on the scattering cross section of the microstrip patch antenna.

To experimentally verify the proposed formulation for scattering computations, a low cross section body was designed and built to serve as the test platform. The backscatter radar cross section (RCS) of the test body by itself was measured to be less than -45 dBsm in the principal plane for incidences 10 degrees away from the broadside and grazing. The specific patch geometry which was measured was a 1.448 in $\times$ 1.083 in rectangular patch residing on a dielectric substrate having thickness $t = 0.057$ in and relative permittivity $\epsilon_r \approx 4.0$. The substrate was housed in a 2.89 in $\times$ 2.10 in $\times$ 0.057 in rectangular cavity recessed in a ground plane. The incident plane wave was polarized along the $\hat{\theta}$ direction and the backscatter RCS was measured as a function of incidence angle in the plane perpendicular to the patch and parallel to its long dimension. Fig. 2(a) shows the backscatter RCS at 9.2 GHz of the filled cavity structure with the patch placed on the surface of the substrate whereas Fig. 2(b) shows the backscatter RCS at the same frequency with the patch removed. As seen, the calculated pattern is in good agreement with the measured data. No comparison is shown near broadside and grazing due to the interference of the low cross-section body.

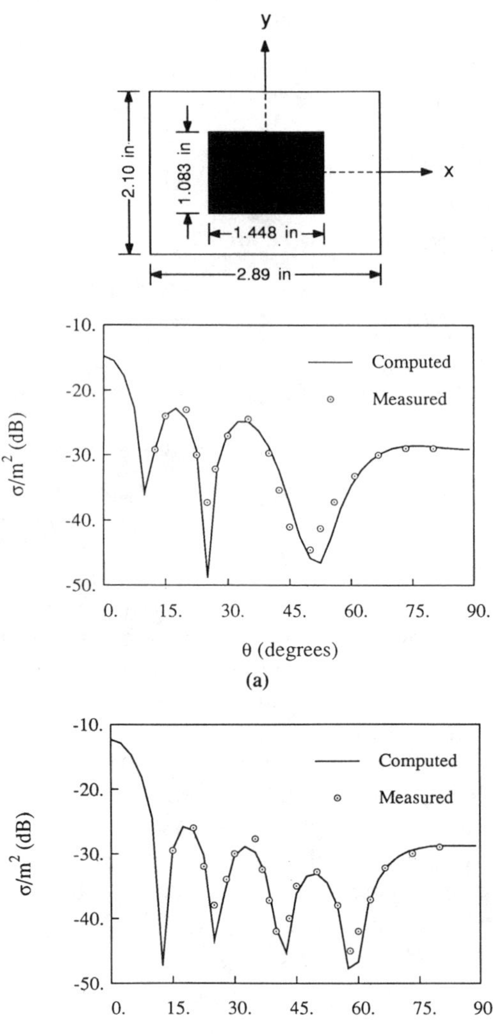

Fig. 2. Comparison of computed and measured θ-polarized xz-plane backscatter RCS of a cavity recessed in a ground plane at 9.2 GHz. (a) With the patch placed on the surface of the substrate. (b) With the patch removed.

To demonstrate the capability of the method as well as illustrate the effect of various substructures on the scattering cross section of the microstrip patch antenna, we now consider another antenna geometry consisting of a 3.66 cm $\times$ 2.60 cm rectangular conducting patch residing on a dielectric substrate having thickness $t = 0.158$ cm, relative permittivity $\epsilon_r = 2.17$ and a loss tangent of 0.001. The substrate is housed in a 7.32 cm $\times$ 5.20 cm $\times$ 0.158 cm rectangular cavity recessed in a ground plane. In the figures to follow, the incident plane wave is polarized along the $\hat{\theta}$ direction and an angle of incidence of $\theta^{inc} = 60°$ and $\phi^{inc} = 45°$ is assumed so that both $\hat{x}$ and $\hat{y}$ polarized modes are excited on the patch.

Fig. 3 shows the backscatter RCS spectrum for the structure. As expected, the introduction of the patch greatly increases the RCS of the structure due to the high field intensities at the patch periphery. The results compare favorably with the moment method solution and experimental data presented in [4]–[6] (for the patch residing on an infinite substrate) except for a slight shift of the resonant frequencies. As indicated in [4], the RCS response is characterized by a series of large peaks, each corresponding to a resonant mode on the patch.

Fig. 4 shows the RCS spectrum for the structure loaded with $Z_L = 50\ \Omega$ representing the presence of a 50 Ω coaxial feed. The results with the load located at ($x_L = -1.83$ cm, $y_L = -1.30$ cm) compare favorably with the moment method data given in [6] and those with the load located at ($x_L = 1.31$ cm, $y_L = 0.78$ cm) are also comparable with the moment method and measured data given in [5]. One observes that the impedance load can substantially affect the RCS of the patch structure; however, its position is crucial in that respect. This is primarily because at each location the load can influence a certain set of resonant modes but leave others unaffected.

In all of the above computations (except when the patch is removed), about 1280 unknowns were required for the simulation of the geometry. Notably, on the average only 100 iterations were required for the solution to converge to within 0.01 dB of the correct RCS level for each excitation with a corresponding CPU time of about 20 s on a Cray Y-MP832. When the patch is removed, 1585 unknowns were used and in this case the convergence was achieved in about 10 iterations (average) with a corresponding CPU time of 2 s per excitation. The faster convergence is obviously due to the smoother field distribution within the cavity in comparison with the case when the patch is present.

An example of the superstrate's effect on the RCS of the microstrip antenna is given in Fig. 5. The configuration is that depicted in Fig. 3 but is now covered with a dielectric layer of the same thickness and constitutive parameters as the substrate. The cavity's depth is accordingly increased to 0.316 cm to accommodate the whole structure. A comparison with the curves given in Fig. 3 shows that the inclusion of the superstrate lowers the resonant frequencies somewhat but the overall level of the RCS is not affected. About 2866 unknowns were required for the simulation of the geometry and an average of 140 iterations were needed to reach acceptable convergence for each excitation (the corresponding CPU time was about 35 s on a Cray Y-MP832). When the patch is removed, 3170 unknowns were used and an acceptable convergence was reached after 20 iterations with a corresponding CPU time of only 5 s.

To further demonstrate the capability of the method, a 3 × 3 microstrip patch array in an 18.3 cm × 13.0 cm cavity of depth of 0.158 cm was considered. The patches are of the same size as that shown in Fig. 3 and the substrate is of the same thickness and material constants. Further, the patch centers are 5.49 cm apart in the x-direction and 3.90 cm apart in the y-direction. Bistatic scattering patterns for this array are shown in Fig. 6 at $f = 2.62$ GHz and 3.55 GHz for normal incidence. Two sets of results are included for each frequency and these correspond to the cases where the patches are free standing ($Z_L = \infty$) or loaded with a 50 Ω resistor at their lower left corner. For this example, 6423 unknowns were used and for the open-circuit case the solution con-

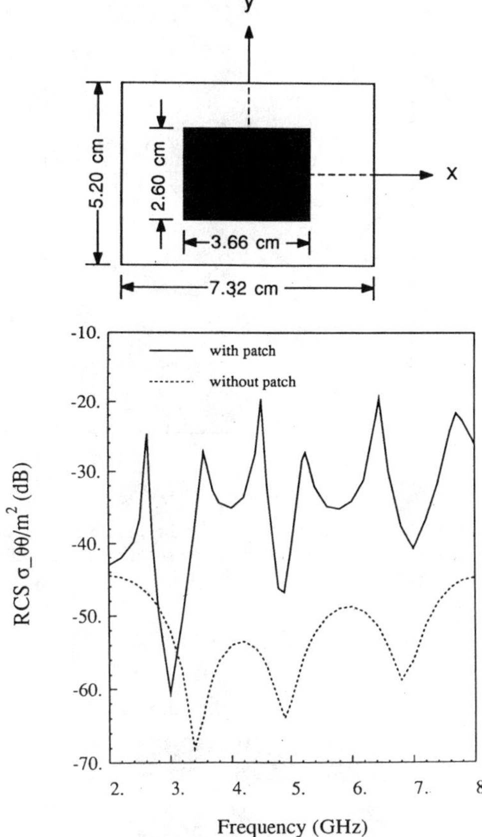

Fig. 3. Backscatter RCS versus frequency for a rectangular patch in a filled cavity. $\theta^{inc} = 60°$, $\phi^{inc} = 45°$, $\mathbf{E}^{inc} = \hat{\theta} E_0$.

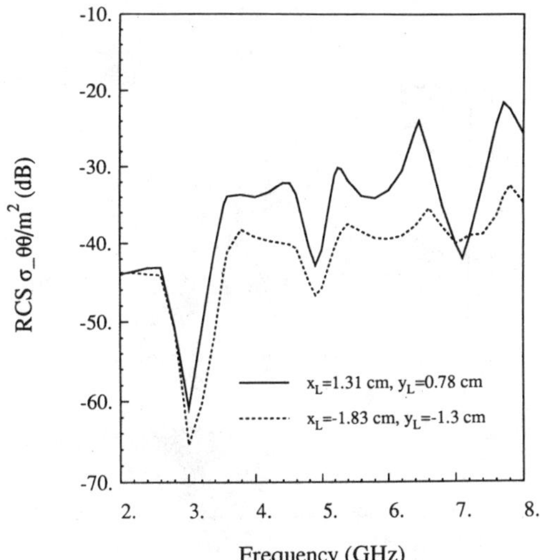

Fig. 4. Backscatter RCS versus frequency for the structure depicted in Fig. 3 loaded with $Z_L = 50\ \Omega$. $\theta^{inc} = 60°$, $\phi^{inc} = 45°$, $\mathbf{E}^{inc} = \hat{\theta} E_0$.

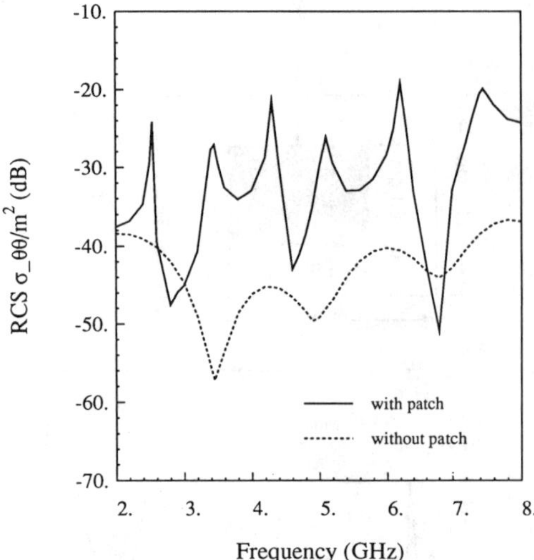

Fig. 5. Backscatter RCS versus frequency for the structure depicted in Fig. 3 but covered with a dielectric layer having the same thickness and material properties as the substrate. $\theta^{inc} = 60°$, $\phi^{inc} = 45°$, $\mathbf{E}^{inc} = \hat{\theta} E_0$.

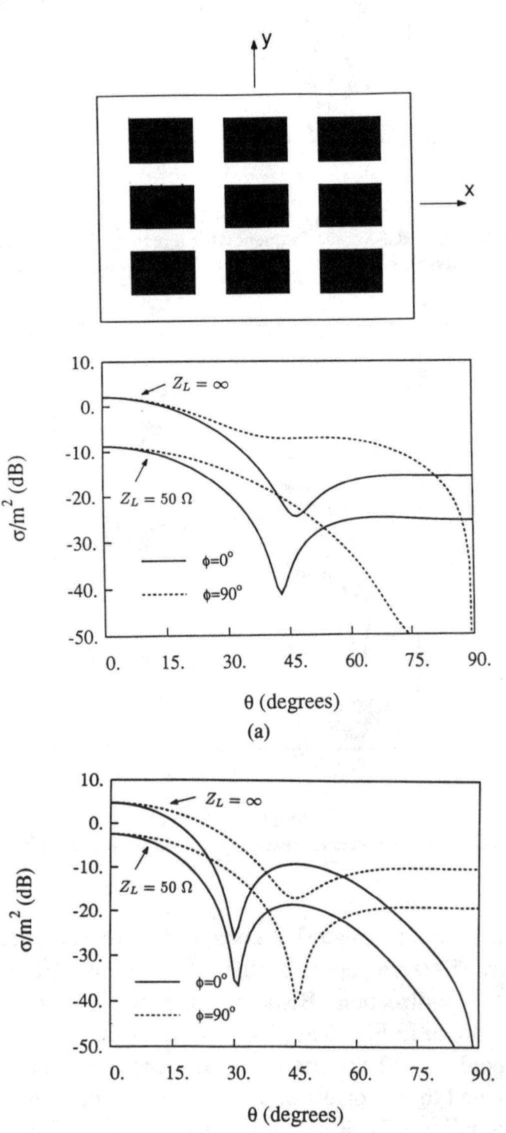

Fig. 6. Bistatic RCS for the 3 × 3 microstrip patch array at normal incidence. (a) $f = 2.62$ GHz, $\mathbf{E}^{inc} = \hat{x} E_0$. (b) $f = 3.55$ GHz, $\mathbf{E}^{inc} = \hat{y} E_0$.

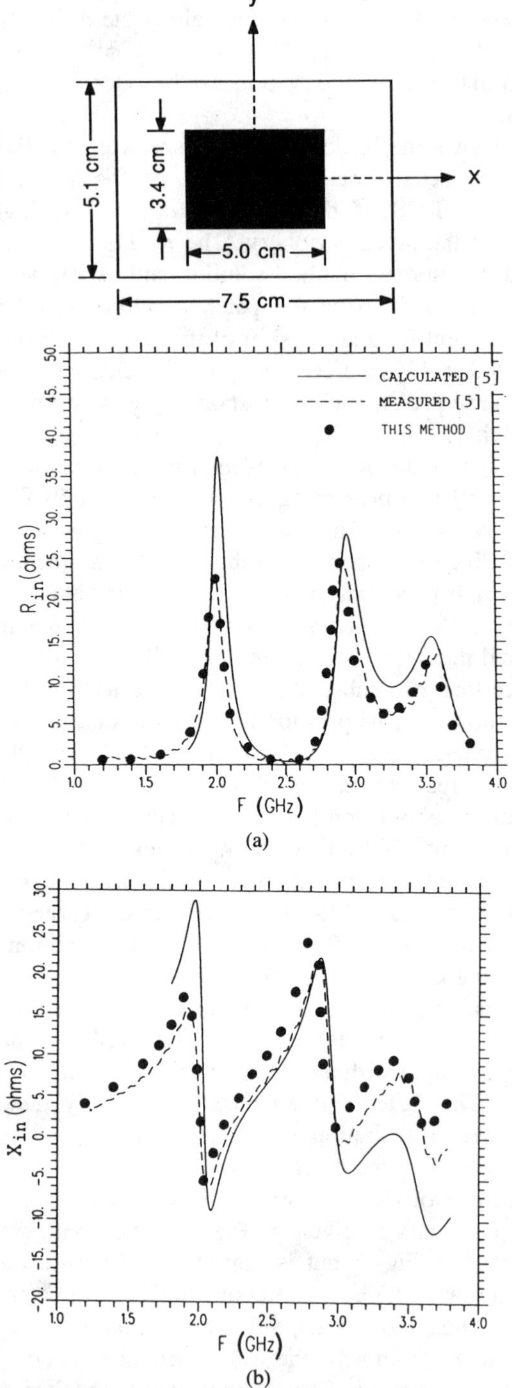

Fig. 7. Comparison of computed and measured input impedance of a loaded microstrip patch antenna as a function of frequency. (a) Real part. (b) Reactive part.

verged within 30 iterations (primarily because of normal incidence) and the corresponding CPU time was about 30 s for each excitation. For the loaded case, the average number of iterations increased to 60 and the corresponding CPU time was about 60 s for each excitation.

So far all of the presented computations pertained to scattering. However, the proposed technique is also suited for radiation pattern and input impedance computations. For the results to follow the input impedance was simply calculated as the voltage observed at the feed point due to a unit

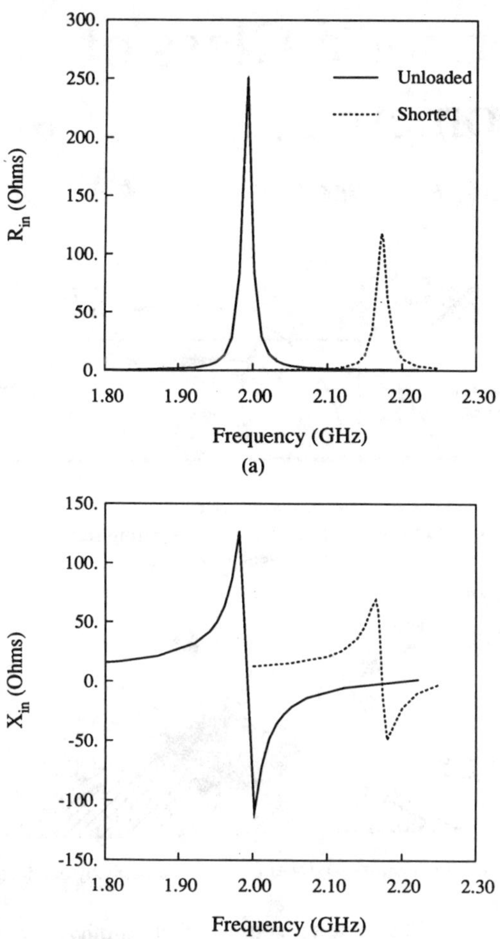

Fig. 8. Input impedance calculations for an unloaded and a shorted microstrip patch antenna. (a) Real part. (b) Reactive part.

electric current probe placed between the patch and the ground plane. Since the input impedance is more sensitive to possible numerical solution errors, here we present calculations for this quantity only, and to the best of our knowledge this is a first for the finite element method as applied to microstrip antennas. The specific antenna geometry is similar to that depicted in Fig. 3, but with different dimensions. In this case, the patch size is 5.0 cm × 3.4 cm and the cavity size is 7.5 cm × 5.1 cm. The substrate has thickness $t =$ 0.08779 cm, relative permittivity $\epsilon_r = 2.17$ with a loss tangent of 0.0015, and the patch is excited by a current probe applied at ($x_f = 1.22$ cm, $y_f = 0.85$ cm). Fig. 7 shows the input impedance for the antenna when a 50 Ω resistor is placed at ($x_L = -2.2$ cm, $y_L = -1.5$ cm) and our calculations are compared to the moment method solution [5] and data measured for a coaxially fed microstrip antenna. We note that the measured data and moment method solution [5] were based on the infinite substrate model, however, by increasing the cavity size our calculations showed that the employed cavity size has no effect on the input impedance. To further demonstrate the capability of the proposed hybrid finite element method, Fig. 8 shows the calculated input impedance when the 50 Ω resistor is replaced by a shorting pin and this result is compared to that generated without any impedance load.

V. Conclusion

A hybrid technique was presented which combines the finite element and boundary integral methods for a characterization of the scattering and radiation properties of microstrip patch antennas and arrays residing in a cavity. To allow the treatment of large structures, the biconjugate gradient method was employed in conjunction with the FFT for a solution with affordable memory requirement. It was found that the proposed solution method is computationally efficient and is capable of simulating rather complex geometries. By virtue of the finite element method, one can model multilayered superstrates and substrates, various feed structures or impedance loads, and the presented numerical results demonstrated the efficiency and capability of the technique. In addition, some included results illustrated the scattering effects associated with the antenna's substructure, feed configuration, and impedance loading. Comparisons with measured data were also presented demonstrating the accuracy of the scattering and input impedance calculations via the proposed method.

Acknowledgment

The authors wish to thank Dr. V. V. Liepa and Mr. C. Cheon for assisting the authors in performing the measurements given in Fig. 2.

References

[1] K. R. Carver and J. W. Mink, "Microstrip antenna technology," *IEEE Trans. Antennas Propagat.*, vol. AP-29, pp. 2-24, Jan. 1981.

[2] R. J. Mailloux, J. F. McIlvenna, and N. P. Kernweis, "Microstrip array technology," *IEEE Trans. Antennas Propagat.*, vol. AP-29, pp. 25-37, Jan. 1981.

[3] J. R. James and P. S. Hall, *Handbook for Microstrip Antennas*. London: Peregrinus, 1989.

[4] E. H. Newman and D. Forrai, "Scattering from a microstrip patch," *IEEE Trans. Antennas Propagat.*, vol. AP-35, pp. 245-251, Mar. 1987.

[5] D. P. Forrai and E. H. Newman, "Radiation and scattering from loaded microstrip antennas over a wide bandwidth," Ohio State University, Columbus, Tech. Rep. 719493-1, Sept. 1988.

[6] D. M. Pozar, "Radiation and scattering from a microstrip patch on a uniaxial substrate," *IEEE Trans. Antennas Propagat.*, vol. AP-35, pp. 613-621, June 1987.

[7] T. K. Sarkar, S. M. Rao, and A. R. Djordjević, "Electromagnetic scattering and radiation from finite microstrip structures," *IEEE Trans. Microwave Theory Tech.*, vol. 38, pp. 1568-1575, Nov. 1990.

[8] J. M. Jin and J. L. Volakis, "A finite element—boundary integral formulation for scattering by three-dimensional cavity-backed apertures," *IEEE Trans. Antennas Propagat.*, vol. 39, pp. 97-104, Jan. 1991.

[9] —, "Electromagnetic scattering by and transmission through a three-dimensional slot in a thick conducting plane," *IEEE Trans. Antennas Propagat.*, vol. 39, pp. 543-550, Apr. 1991.

[10] M. L. Barton and Z. J. Cendes, "New vector finite elements for three-dimensional magnetic field computation," *J. Appl. Phys.*, vol. 61, pp. 3919-3921, Apr. 1987.

Finite Element Solution for a Class of Unbounded Geometries

Omar M. Ramahi and Raj Mittra, *Fellow, IEEE*

Abstract—**An efficient and systematic procedure is described for the finite element solution of a class of electromagnetic radiation and scattering problems involving unbounded geometries. The numerical procedure is well suited for analyzing infinite metallic structures with cavity regions filled with inhomogeneous and anisotropic media. The formulation is based upon an approach that combines the finite element method (FEM) with the surface integral equation to truncate the mesh region. The efficiency of the proposed technique arises from the use of the Green's function of the first kind in the surface integral. Illustrative numerical representations that demonstrate the validity, versatility, and efficiency of the method are included.**

I. Introduction

SCATTERING problems involving infinite-sized geometries have traditionally been treated using the method of moments or the asymptotic techniques of physical optics and the geometrical theory of diffraction. However, these methods are not as well suited as in the finite element method (FEM) when locally inhomogeneous and anisotropic regions are present in an infinite metallic structure. The objective of this paper is to present a technique based on the combination of the finite element and integral equation methods for solving a class of scattering problems with unbounded geometries, such as cavities embedded in conducting walls.

In the past, finite element solutions of two-dimensional open-region scattering problems have mostly been confined to scatterers whose geometries are finite in size, i.e., those that can be confined within a closed region bounded by a circle or a sphere with a finite radius. To be numerically viable, the finite element method must be applied to a finite region by truncating the mesh to a finite size. Thus, for geometries with finite dimensions, the application of the finite element method poses two problems: first, the artificial boundary that truncates the region must be chosen such that the size of the solution region is manageable for computer implementation; second, the boundary condition that is enforced at the artificially boundary must model the proper behavior of the scattered field, which cannot be predicted *a priori*. For the purpose of illustrating these difficulties, we consider Fig. 1, which shows a scattering body next to an infinite perfectly conducting wall. Initially, we would be tempted to truncate the mesh region by the semicircular boundary $\partial\Omega_1$ or the rectangular body $\partial\Omega_2$, and then apply either the Engquist-Majda boundary operator [1], or the Bayliss-Turnkel operator [2]-[5] on $\partial\Omega_1$ on $\partial\Omega_2$, respectively. Our choice of either $\partial\Omega_1$ or $\partial\Omega_2$ (see Fig. 1) will be reasonable provided that the source of the incident field is located in the close proximity of the target, since these artificial outer boundaries will only experience outgoing waves impinging upon them. On the other hand, if the illuminating source is at infinity, we expect the scattered field to also have an incoming component due to the scattering by the portion of the conducting wall (see Fig. 1). The absorbing boundary conditions discussed in the literature [1]-[7] approximate the behavior of the outgoing waves only; hence, they are not particularly suited for unbounded scatterers. An additional difficulty with these unbounded scatterers is that the asymptotic behavior of their scattered fields cannot be predicted *a priori*. To circumvent these difficulties, we propose to combine the surface integral equation approach with the finite element method in a manner described in this paper. In this study, we consider the class of scattering geometries that comprise a metallic cavity of arbitrary shape and filling, embedded in a perfectly conducting wedge, defined in the cylindrical coordinate system, by the surface $\phi = \psi$ and $\phi = 2\pi - \psi$. A typical geometry of this type is illustrated in Fig. 2.

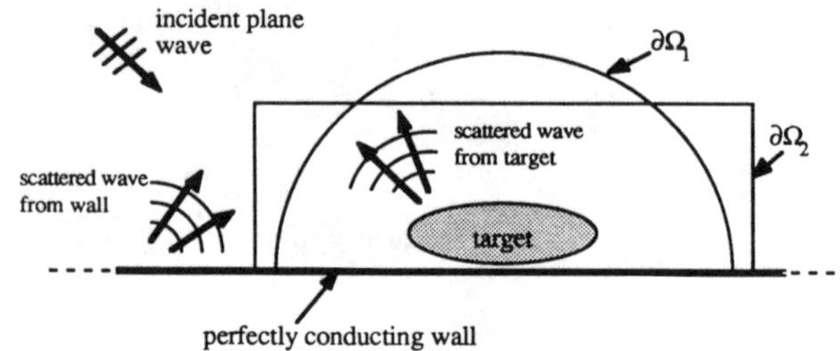

Fig. 1. Incoming and outgoing waves arising from scattering by unbounded geometries.

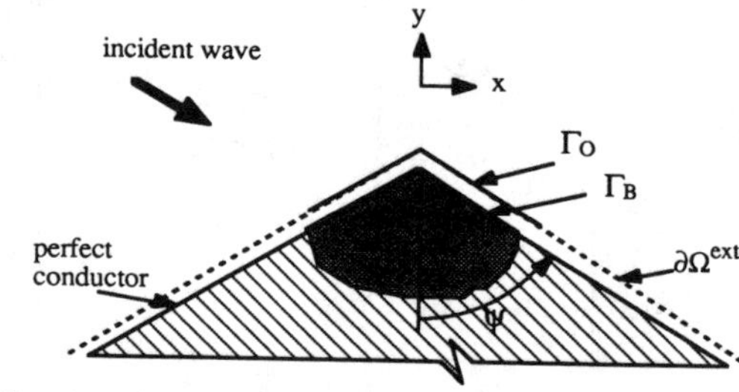

Fig. 2. An arbitrary cavity embedded in a perfectly conducting wedge.

The use of the surface integral as a boundary condition was first reported in the work of McDonald and Wexler [8], there it was applied to solve the Poisson equation. Others have used the method to solve acoustic and electromagnetic radiation problems [9]-[13]. In the previous work reported in the literature, the problems solved involved scatterers of finite extent. In contrast, as alluded to earlier, the purpose of this paper is to investigate unbounded scatterers. Previous approaches to combining FEM with surface integral equations have been found to produce system matrices that are partially dense. We show that by the use of the Green's function of the first kind in the surface integral, one can minimize the nonlocal behavior, which is typically the major drawback in the use of the surface integral equation as a boundary condition.

Manuscript received February 9, 1990; revised August 1, 1990.
The authors are with the Electromagnetic Communication Laboratory, Department of Electrical Engineering, University of Illinois, Urbana, IL 61801.
IEEE Log Number 9041262.

Reprinted from *IEEE Trans. Antennas Propagat.*, vol. 39, no. 2, pp. 244–250, Feb. 1991.

II. Surface Integral Equation

In this section we present the general formulation of the surface integral equation method as applied to a class of scattering geometries consisting of arbitrary cavities embedded in infinite metallic wedges, as shown in Fig. 2. Let Ω^{int} denote the interior region of the cavity including the layer trapped between Γ_O and Γ_B. Next, we use conventional finite element construction [14]–[16], in conjunction with the Green's theorem, to obtain the weak form of the Helmholtz equation

$$\begin{aligned}\int_\Omega \left\{\frac{1}{p(x)} \nabla\nu \cdot \nabla u - k^2 q(x)\nu u\right\} ds &= \int_{\partial\Omega} \frac{1}{p(x)} \nu \frac{\partial u}{\partial n} dl \\ &\quad + \int_\Omega \left\{\nabla\nu \cdot \left(\frac{1}{p(x)} - 1\right) \nabla u^{\text{inc}} + k^2(1 \right. \\ &\quad \left. - q(x))\nu u^{\text{inc}}\right\} ds \qquad (1)\end{aligned}$$

where u is the z-component of the unknown scattered field, u^{inc} is the z-component of the incident field and ν is the testing function. Depending on the field polarization, the medium parameters are defined as

$$\begin{aligned} p(\mathbf{x}) &= \mu(\mathbf{x}) \quad E\text{-polarized case} \\ q(x) &= \epsilon(x) \quad E\text{-polarized case} \\ p(\mathbf{x}) &= \epsilon(\mathbf{x}) \quad H\text{-polarized case} \\ q(\mathbf{x}) &= \mu(\mathbf{x}) \quad H\text{-polarized case.}\end{aligned}$$

Next, the solution region Ω^{int} (see Fig. 2) is divided into triangular elements, and the unknown field is approximated over each of the elements by a set of interpolating functions

$$u(\mathbf{x}) = \sum_{i=1}^{3} u_i \psi_i(\mathbf{x}). \qquad (2)$$

For the first-order linear triangular elements, the ψ_i's are linear interpolatory functions with a finite support covering a single triangular element, and u_i's are the nodal field values at the vertices of each element. By choosing the testing functions identical to those for the ψ_i's, and after substituting (2) into (1), we arrive at the typical finite element matrix system

$$\mathbf{MU} = \mathbf{B}$$

which is symbolically represented as

$$\begin{bmatrix} \mathbf{M}_{OO} & \mathbf{M}_{OB} & \mathbf{0} \\ \mathbf{M}_{BO} & \mathbf{M}_{BB} & \mathbf{M}_{BC} \\ \mathbf{0} & \mathbf{M}_{CB} & \mathbf{M}_{CC} \end{bmatrix} \begin{bmatrix} \mathbf{U}_O \\ \mathbf{U}_B \\ \mathbf{U}_C \end{bmatrix} = \begin{bmatrix} \mathbf{B}_O \\ \mathbf{B}_B \\ \mathbf{B}_C \end{bmatrix} \qquad (3)$$

where $\mathbf{M}$ is the typical stiffness matrix, and $\mathbf{U}_O$, $\mathbf{U}_B$, and $\mathbf{U}_C$ are column vectors corresponding to the nodal field values at the layer Γ_O, the layer Γ_B, and the interior region of the cavity, respectively. $\mathbf{B}$ is the right-hand column matrix that has the boundary condition information which is necessary for the well-posedness of the problem. Therefore, to fully characterize the matrix $\mathbf{B}$, particularly $\mathbf{B}_O$, a boundary condition must be enforced at Γ_B.

We now invoke the Green's function theorem to express the scattered field in Ω^{ext} in terms of the field values on the contour $\partial\Omega^{\text{ext}}$ via the surface integral equation [17]. This leads to the equation:

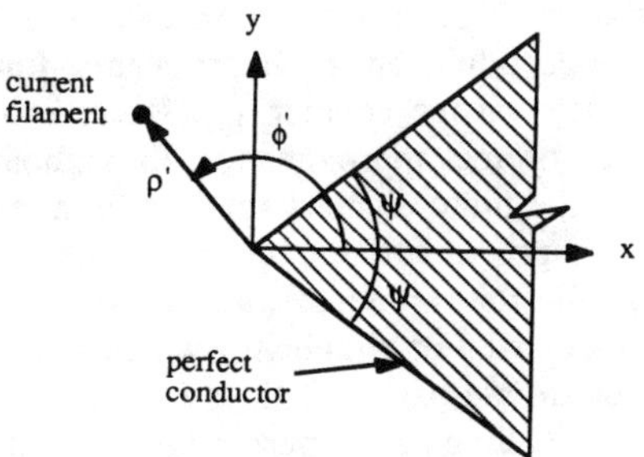

Fig. 3. Geometry for construction of the Green's function of the first kind.

$$u(\mathbf{x}) = \int_{\partial\Omega^{\text{ext}}} \{G(\mathbf{x}/\mathbf{x}')\, \nabla u(\mathbf{x}') \cdot \hat{\mathbf{n}}_B - u(\mathbf{x}')\, \nabla G(\mathbf{x}/\mathbf{x}') \cdot \hat{\mathbf{n}}_B\}\, d\zeta' \qquad (4)$$

where $\mathbf{x}'$ is the integration coordinate, $\hat{n}_B$ is the outward normal unit vector from $\partial\Omega^{\text{ext}}$, and G is the free-space Green's function.

Next, we enforce the surface integral equation (4) at Γ_O:

$$u(\mathbf{x}_O) = \int_{\partial\Omega^{\text{ext}}} \{G(\mathbf{x}_O/\mathbf{x}_B)\, \nabla u(\mathbf{x}_B) \cdot \hat{\mathbf{n}}_B - u(\mathbf{x}_B)\, \nabla G(\mathbf{x}_O/\mathbf{x}_B) \cdot \hat{\mathbf{n}}_B\}\, d\zeta'. \qquad (5)$$

The surface integral equation in (5) can be simplified by using a Green's function of the first kind, which is defined as the solution of

$$\nabla^2 G^{(1)}(\mathbf{x}/\mathbf{x}') + k^2 G^{(1)}(\mathbf{x}/\mathbf{x}') = -\delta(\mathbf{x} - \mathbf{x}')$$

subject to the boundary condition

$$G^{(1)} = 0 \text{ on } \partial\Omega^{\text{ext}}.$$

The geometry for the construction of $G^{(1)}$ is shown in Fig. 3. Because the contour $\partial\Omega^{\text{ext}}$ constitutes a complete coordinate surface, it is possible to construct the Green's function of the first kind in an analytic form. From [18], we have

$$\begin{aligned} G^{(1)}(\mathbf{x}/\mathbf{x}') &= \sum_\nu \frac{-\omega\mu\pi}{2(\pi - \psi)} H_\nu^{(2)}(k\rho') J_\nu(k\rho) \\ &\quad \cdot \sin\nu(\phi' - \psi) \sin\nu(\phi - \psi), \qquad \rho < \rho' \\ G^{(1)}(\mathbf{x}/\mathbf{x}') &= \sum_\nu \frac{-\omega\mu\pi}{2(\pi - \psi)} H_\nu^{(2)}(k\rho) J_\nu(k\rho') \\ &\quad \cdot \sin\nu(\phi' - \psi) \sin\nu(\phi - \psi), \qquad \rho > \rho' \end{aligned} \qquad (6)$$

where ν is given by

$$\nu = \frac{m\pi}{2(\pi - \psi)}, \qquad m = 1, 2, 3, \cdots.$$

Using the Green's function of the first kind (6), the surface integral equation reduces to

$$u(\mathbf{x}_0) = -\int_{\partial\Omega^{\text{ext}}} u(\mathbf{x}_B)\, \nabla G^{(1)}(\mathbf{x}_O/\mathbf{x}_B) \cdot \hat{\mathbf{n}}_B\, d\zeta'. \qquad (7)$$

The use of the surface integral equation as a boundary condition for the finite element region (region of the cavity) requires descritizing (7) into a matrix form to incorporate it into the solution procedure. This discretization can be accomplished in different ways depending on the interpolatory functions chosen to expand the field on the contour Γ_B. However, to maintain a consistent order of the approximation throughout the solution region vis-a-vis the finite element solution, and, furthermore, to assure the proper preconditioning of the matrix system such that the sparsity of the matrix is maximized, we employ piecewise linear basis functions. For simplicity, we choose basis functions that are identical to the basis functions used in the finite element region. Thus, expanding the scattering field over the contour Γ_B, in terms of linear basis functions, we have

$$u(\zeta) = \sum_{j=1}^{M} \sum_{i=1}^{2} u_i^{(j)} \psi_i^{(j)}(\zeta) \tag{8}$$

where $\psi_i(\zeta)$ is defined locally over the edge of each boundary element as

$$\psi_i(\zeta) = \begin{cases} 1 - \dfrac{\zeta}{h}, & i = 1 \\ \dfrac{\zeta}{h}, & i = 2. \end{cases}$$

The first summation in (8) is taken over all elements, numbering **M**, that are tangential to the contour Γ_O, and the u_i's are the nodal field values at the layer Γ_O. Substituting (8) into (7), we arrive at the discretized version of the surface integral equation, viz.,

$$u(\mathbf{x}_O) = -\int_{\partial\Omega^{\text{ext}}} \sum_{j=1}^{M} \sum_{i=1}^{2} u_i^{(j)} \psi_i^{(j)}(\mathbf{x}_B) \nabla G^{(1)}(\mathbf{x}_O/\mathbf{x}_B) \cdot \hat{\mathbf{n}}_B \, d\zeta'$$

which reduces to

$$u(\mathbf{x}_O) = -\sum_{j=1}^{M} \sum_{i=1}^{2} u_i^{(j)} \int_{\partial\Omega^{\text{ext}}} \psi_i^{(j)}(\mathbf{x}_B) \nabla G^{(1)}(\mathbf{x}_O/\mathbf{x}_B) \cdot \hat{\mathbf{n}}_B \, d\zeta'. \tag{9}$$

Finally, enforcing (9) at each of the Γ_O layer nodes gives the boundary matrix equation

$$\mathbf{U}_O = \mathbf{S}\mathbf{U}_B + \mathbf{T} \tag{10}$$

where **S** is an $M \times M$ matrix with elements defined as

$$\mathbf{s}_{jk} = \int_{\partial e^{(j)}} \psi_2 \nabla G^{(1)}(\mathbf{x}_j/\mathbf{x}_k) \cdot \hat{\mathbf{n}}_B \, d\zeta' + \int_{\partial e^{(j)}} \psi_1 \nabla G^{(1)}(\mathbf{x}_j/\mathbf{x}_k) \cdot \hat{\mathbf{n}}_B \, d\zeta'$$

and $\partial e^{(j)}$ is the edge of element j. The column matrix **T** in (10) has the contribution from the integration over the infinite metallic walls.

The above construction transformed the surface integral equation into an algebraic operation. Therefore, the matrix equation (10) can be imterpreted as an algebraic boundary operator for the system in (3). Adding (10) to (3), we eliminate the need to evaluate the column vector $\mathbf{B}_O$ by dispensing with the first row of the matrix **M** in (3) altogether, which results in

$$\begin{bmatrix} (\mathbf{M}_{BO}\mathbf{S} + \mathbf{M}_{BB}) & \mathbf{M}_{BC} \\ \mathbf{M}_{CB} & \mathbf{M}_{CC} \end{bmatrix} \begin{bmatrix} \mathbf{U}_B \\ \mathbf{U}_C \end{bmatrix} = \begin{bmatrix} \mathbf{B}_B - \mathbf{B}_{BO}\mathbf{T} \\ \mathbf{B}_C \end{bmatrix}. \tag{11}$$

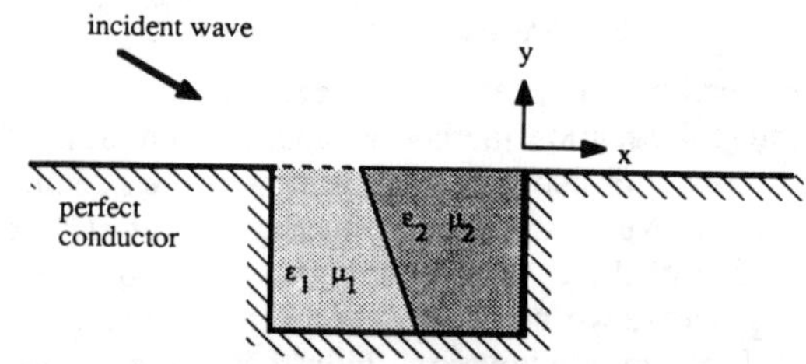

Fig. 4. A rectangular cavity with inhomogeneous filling embedded in a perfectly conducting wall.

The formulation developed above was based on the surface integral equation as it was applied to the scattered field. An alternative formulation involving the total field can, in certain applications, have some advantages over the scattered field formulation. When the total field is used as the primary unknown, the weak form of the Helmholtz equation has the form

$$\int_\Omega \frac{1}{p(\mathbf{x})} \nabla u^{\text{tot}} \cdot \nabla \nu - k^2 q(\mathbf{x}) \nu u^{\text{tot}} \, ds = \int_{\partial\Omega} \frac{1}{p(\mathbf{x})} \nu \frac{\partial u^{\text{tot}}}{\partial n} \, dl. \tag{12}$$

The surface integral equation that corresponds to (12) takes the form

$$u^{\text{tot}}(\mathbf{x}) = u^{\text{inc}}(\mathbf{x}) - \int_{\partial\Omega^{\text{ext}}} \{G(\mathbf{x}/\mathbf{x}') \nabla u^{\text{tot}}(\mathbf{x}') \cdot \hat{\mathbf{n}}_B - u^{\text{tot}}(\mathbf{x}') \nabla G(\mathbf{x}/\mathbf{x}') \cdot \hat{\mathbf{n}}_B\} \, d\zeta'. \tag{13}$$

If Green's function of the first kind is used, (13) reduces to

$$u^{\text{tot}}(\mathbf{x}) = u^{\text{inc}(1)}(\mathbf{x}) + \int_{\partial\Omega^{\text{ext}}} u^{\text{tot}}(\mathbf{x}') \nabla G^{(1)}(\mathbf{x}/\mathbf{x}') \cdot \hat{\mathbf{n}}_B \, d\zeta' \tag{14}$$

in which $u^{\text{inc}(1)}$ is given by

$$u^{\text{inc}(1)}(\mathbf{x}) = \mathbf{a} \int_\Omega \mathbf{J}(\mathbf{x}) G^{(1)}(\mathbf{x}/\mathbf{x}') \, d\Omega \tag{15}$$

where **a** is a complex constant. Equation (15) can be interpreted as the field solution due to the impressed current source distribution $\mathbf{J}(\mathbf{x})$ in the presence of a perfectly conducting wedge that coincides with the walls of the cavity. In the numerical examples presented in this paper the scattered field and the total field formulations were used; specifically, the total field formulation was used whenever the cavities contained inhomogeneous material.

III. Cavities Embedded in a Perfectly Conducting Plane Wall

To demonstrate the versatility and validity of the surface integral equation technique, we consider cavities of arbitrary geometry having an arbitrary inhomogeneity, embedded in a perfectly conducting wall. This class of cavities is a special case of a conducting wedge with $\psi = \pi/2$. This is shown in Fig. 4. For this geometry, the Green's function of the first kind simplifies to

$$G^{(1)}(\mathbf{x}/\mathbf{x}') = -\frac{jk}{4} H_0^{(2)}\left(k\sqrt{(x-x')^2 + (y-y')^2}\right) + \frac{jk}{4} H_0^{(2)}\left(k\sqrt{(x-x')^2 + (y+y')^2}\right).$$

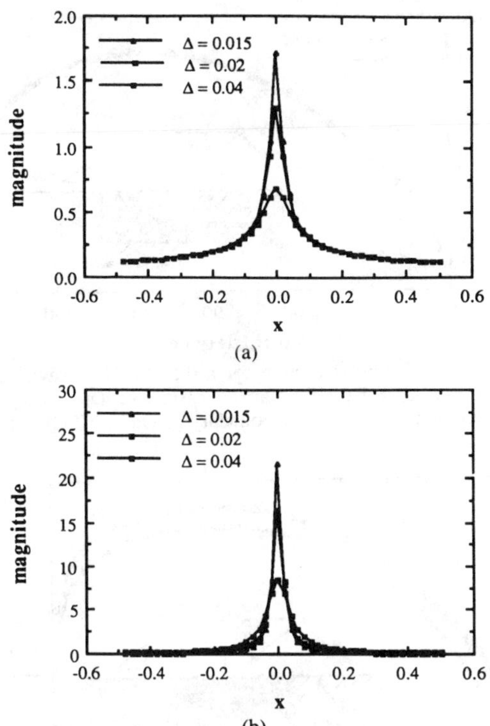

Fig. 5. (a) Effect of changing Δ on $G(x = 0/x')$. (b) Effect of changing Δ on the normal derivative of $G^{(1)}(x = 0/x')$.

Consequently, (5) reduces to

$$u(x_O) = -\int_{-\infty}^{x1} u(x') \frac{-jk\Delta}{2R} H_1^{(2)}(kR)\, dx' - \int_{x1}^{x2} u(x') \frac{-jk\Delta}{2R} H_1^{(2)}(kR)\, dx' - \int_{x2}^{\infty} u(x') \frac{-jk\Delta}{2R} H_1^{(2)}(kR)\, dx' \quad (16)$$

where $R = \sqrt{\{(x_O - x')^2 + \Delta^2\}}$, and $\Delta = y_0$.

The integration in (16) is observed to depend directly on Δ, which is the spacing between the layers Γ_O and Γ_B. We also observe the difficulty that arises from the integration over the infinite intervals: $(-\infty, x_1]$ and $[x_2, \infty)$. For the *E*-polarized case, the scattered field on the conducting wall is simply the negative of the incident field there; hence, the complete integration over $(-\infty, x_1]$ and $[x_2, \infty)$ is possible, at least theoretically. However, for the *H*-polarized case, *a priori* knowledge of the field on the infinite surface is not available; thus, the integration involves the nodes extending over the entire *x*-axis. This is not feasible because it calls for the solution of an infinite number of nodes.

One way of circumventing the problem of integration over the integrals $(-\infty, x_1]$ and $[x_2, \infty)$ is to use an educated guess of the solution at large distances from the cavity, based on the asymptotic behavior of the scattered field. Although a good guess may work well for some geometries, this scheme is not systematic, and it may not guarantee the accuracy of the solution for an arbitrary cavity. A systematic approximation, on the other hand, can be made if one takes a closer look at the behavior of the normal derivative of the Green's function appearing in (9). Figs. 5(a) and 5(b) show the behavior of the free-space Green's function G and the normal derivative of the Green's function of the first kind $\partial G^{(1)}/\partial n$, as a function of the separation Δ. We observe that the supports of G and $\partial G^{(1)}/\partial n$ would become smaller as the interval Δ becomes smaller. However, the important distinction between $\partial G^{(1)}/\partial n$ and G is that the former has a smaller support, which is due to the denominator factor arising from the normal differentiation. This distinctive feature of $\partial G^{(1)}/\partial n$, namely, its narrow support, in addition to simplifying the surface integral equation considerably, justifies neglecting the integration contribution over $(-\infty, x_1]$ and $[x_2, \infty)$ if Δ is chosen *small*. Another significant advantage offered by $\partial G^{(1)}/\partial n$ is that the integration over the entire Γ_B contour in (7) can be approximated by integrating over a much smaller interval that spans only the support of $\partial G^{(1)}/\partial n$. This, consequently, is expected to lead to a significant reduction in the nonlocal behavior of the surface integral, resulting in an increase in the sparsity of the final system matrix. It is important to realize that this distinctive advantage has been realized because the surface $\partial\Omega^{ext}$ (see Fig. 2) defines a separable geometry. When this construction is contrasted to the previous works where the surface integral equation has been used for solving open-region finite-size scatterers [8]–[13], one observes that in the works reported earlier, it was necessary to work with the free-space Green's function because the outer boundary that truncated the mesh region did not, in general, constitute a complete geometry.

For the purpose of illustrating the significance of using the Green's function of the first kind, we define the *bandwidth* as the number of nodes that are included in the surface integration over the Γ_B contour. In the next section we present numerical results for different representative cavities to demonstrate the validity of the above formulation. Also presented are results showing the effect of varying the *bandwidth* on the finite element solution.

IV. Numerical Results

Once the nodal field values on the opening of the cavity are obtained from the finite element solution, the radar cross section can be calculated using the surface integral equation for the total field (14). For the special case of a cavity embedded in a metallic wall (see Fig. 4), we use the equivalence theorem to replace the cavity and the metallic walls with a magnetic current radiating in free space. To this end, we postulate the zero field $\mathbf{E}_1$, to be inside the cavity ($y < 1$), and the exact total field $\mathbf{E}_2$, immediately outside the cavity ($y > 0$). The magnetic current that supports $\mathbf{E}_1$ and $\mathbf{E}_2$ will be the difference between the two fields, and is given by

$$\mathbf{M} = -\hat{\mathbf{n}} \times (\mathbf{E}_1 - \mathbf{E}_2) = -\hat{\mathbf{n}} \times \mathbf{E}_2.$$

Using the image theory, the magnetic current can be made to radiate in free space by eliminating the perfectly conducting wall of the cavity, which, in turn, has the effect of doubling the current $\mathbf{M}$.

The scattered field is then expressed as a convolution integral involving the free-space Green's function G:

$$\mathbf{E} = \int_{\text{aperture}} 2\,\nabla_{\mathbf{x}}\mathbf{x}\big[\mathbf{M}(\mathbf{x}')G(\mathbf{x}/\mathbf{x}')\big]\, d\mathbf{x}' \quad (17)$$

where ∇_x is the gradient operator with respect to the unprimed coordinates. G is expressed in the cylindrical coordinate system by

$$G = \frac{-j}{4} H_0^{(2)}(|\rho - \rho'|).$$

For the E-polarized case, (17) has the form

$$\mathbf{E} = -\hat{z}\int_{\text{aperture}} 2E_z|_{y'=0}\frac{\partial G}{\partial y}(\mathbf{x}/\mathbf{x}')\,d\mathbf{x}'$$

and for the H-polarized case, we have

$$\mathbf{E} = -\hat{x}\frac{1}{j\omega\epsilon}\int_{\text{aperture}} 2\frac{\partial H_z}{\partial y'}|_{y'=0}\frac{\partial G}{\partial y}(\mathbf{x}/\mathbf{x}')\,d\mathbf{x}' + \hat{y}\frac{1}{j\omega\epsilon}\int_{\text{aperture}} 2\frac{\partial H_z}{\partial y'}|_{y'=0}\frac{\partial G}{\partial x}(\mathbf{x}/\mathbf{x}')\,d\mathbf{x}'. \quad (18)$$

The far field can be calculated using a large argument asymptotic expansion for the Hankel function $H^{(2)}$:

$$\lim_{x\to\infty} H_\nu^{(2)}(x) = \sqrt{\frac{2j}{\pi x}}\, j^\nu e^{-jx}.$$

Using far-field approximations, the radar cross section (RCS) for the E-polarized case is given by

$$\text{RCS} = k\sin^2\phi\left[\int_{\text{aperture}} E_z|_{y'=0}\, e^{-jkx'\cos\phi}\,dx'\right]^2$$

where we have assumed $|E^{\text{inc}}| = 1$. For the H-polarized case, the RCS is given by

$$\text{RCS} = \left[\int_{\text{aperture}} \frac{\partial H_z}{\partial y'}|_{y'=0}\, e^{-jkx\cos\phi}\,dx'\right]^2$$

where $|H^{\text{inc}}| = 1$.

Based on the formulations discussed above, numerical codes were developed to solve for the field in the cavity. The bistatic as well as the monostatic radar cross sections were then computed from the solution of the total field on the aperture of the cavity. In all the numerical results presented, the scattered field formualtion was used for empty cavities, whereas the total field formulation was used for cavities having a dielectric filling. Three different representative rectangular cavities are discussed below for plane wave incidence. In all the cases studied, the mesh density was approximately 400 nodes per λ_c^2, where λ_c is the wavelength in the filling of the cavity.

In the first case, we considered a $0.5 \times 0.16\lambda$ rectangular cavity, where λ is the free-space wavelength. The cavity was filled with a dense lossy dielectric material. Different numerical experiments showed that good results are obtained when Δ, which is the separation between the layers Γ_O and Γ_B (see Fig. 2) is approximately 0.02 λ. Fig. 6 shows the bistatic radar cross section, E-polarized case. Also given are the results computed using the impedance boundary condition (IBC) [19]. In the IBC calculations [19], the cavity was considered as having an infinite depth, which was a reasonable approximation since the depth of the cavity is approximately twice the skin depth of the dielectric filling.

Fig. 7 shows the bistatic radar cross-section for near grazing incidence (10° off the $y = 0$ plane), for $\Delta = 0.02$, as the *bandwidth* is varied. This graph shows that excellent results are obtained for a *bandwidth* of four. Therefore, we conclude, for the E-polarized case under consideration here, that the system matrix can be made optimal in the sense that the sparsity will not be different from that resulting if the Neuman or the Dirichlet boundary condition were applied on the entire boundary of the cavity.

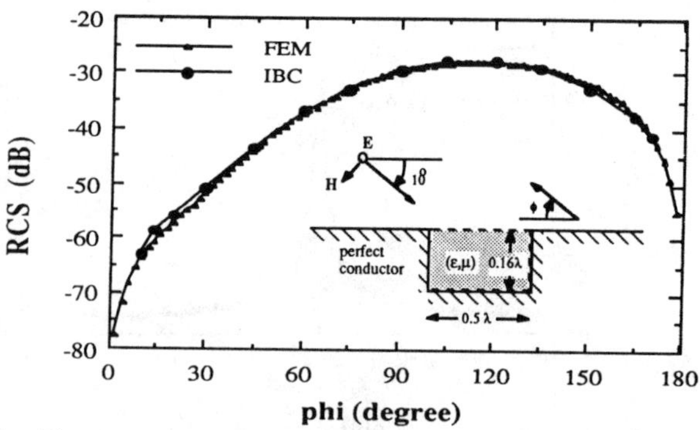

Fig. 6. Bistatic radar cross section for a $0.5 \times 0.16\ \lambda$ cavity with $\epsilon_r = 57.6 - j29.2$ and $\mu_r = 1.0$, E-polarized case, calculated using the finite element method and the impedance boundary condition (IBC) [19].

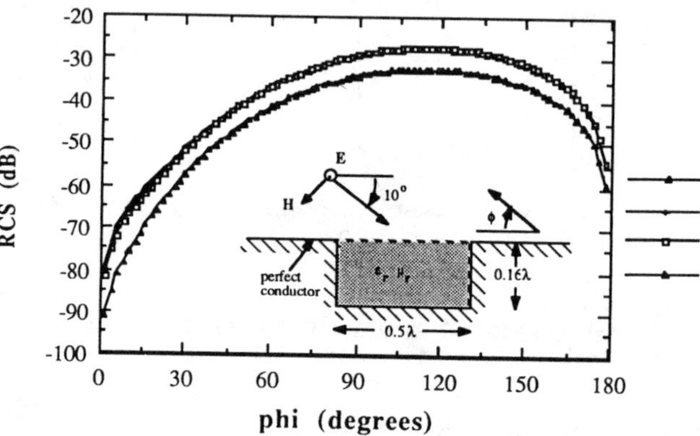

Fig. 7. Effect of varying the bandwidth (BW) on the bistatic radar cross section for a $0.5 \times 0.16\ \lambda$ cavity with $\epsilon_r = 57.6 - j29.2$ and $\mu_r = 1.0$, E-polarized case, calculated using the finite element method (FEM).

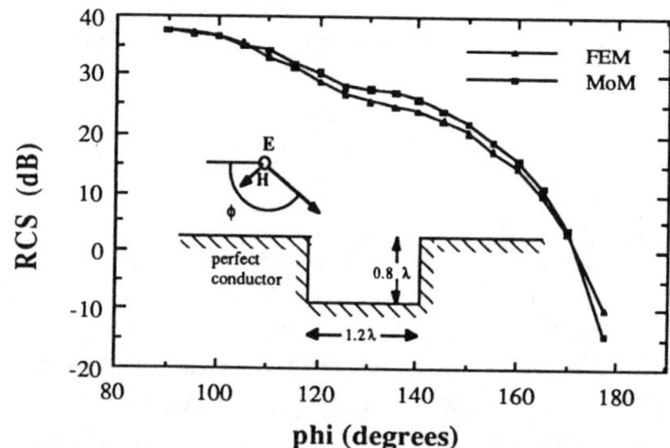

Fig. 8. Monostatic radar cross section for a $1.2 \times 0.8\ \lambda$ cavity, E-polarized case, calculated using the finite element method and the method of moments (MOM) [20].

In the second example, we considered a $1.2 \times 0.8\ \lambda$ rectangular cavity with no filling. Comparison of the results, in this case, was made with the method of moments. The monostatic radar cross sections were calculated for both the E-polarized and the H-polarized cases, and are given in Figs. 8 and 9. The solution for the H-polarized case is observed to be less accurate than that for the E-polarized case. This is believed to be directly related to the numerical differentiation, which is required for the evaluation of the total E-field on the aperture for the cavity (see (18)). Fig. 10 shows the bistatic radar cross section, H-polarized case, for normal incidence, and in Fig. 11, the effect of varying the bandwidth is also presented. For this H-polarized case, a bandwidth of 14 gives less than 1 dB error in the backscattering direction. The results presented in Fig. 11 show that a gradual decrease in the bandwidth gives gradual deterioration in the

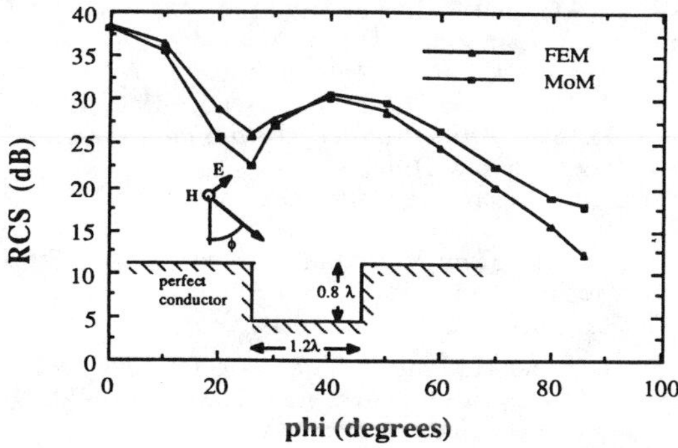

Fig. 9. Monostatic radar cross section for a 1.2 × 0.8 λ cavity, *H*-polarized case, calculated using the finite element method (FEM) and the method of moments (MOM) [20].

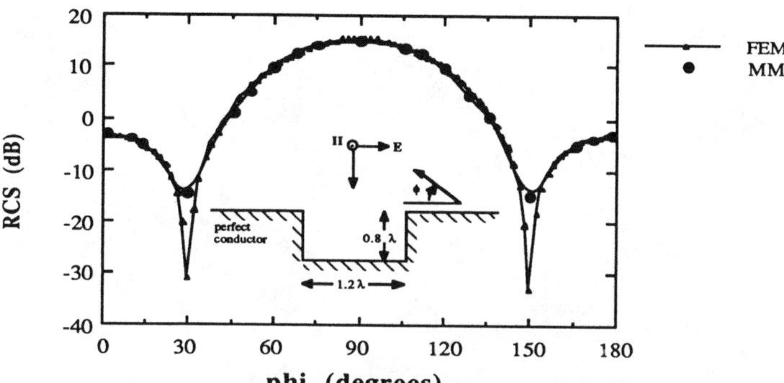

Fig. 10. Bistatic radar cross section for a 1.2 × 0.8 λ cavity, *H*-polarized case, calculated using the finite element method (FEM) and the mode matching technique (MM) [21].

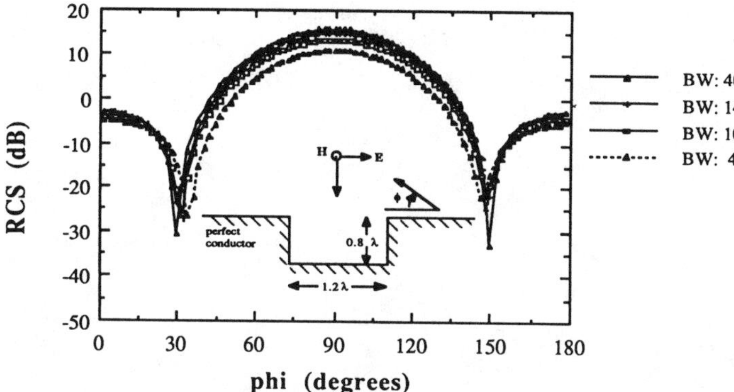

Fig. 11. Effect of varying the bandwidth (BW) on the bistatic radar cross section for a 1.2 × 0.8 λ cavity, *H*-polarized case, calculated using the finite element method (FEM).

accuracy. Although the bandwidth requirements are larger for this *H*-polarized case than for the *E*-polarized case discussed above a bandwidth of 14, nevertheless, gives significant improvement in the sparsity of the final finite element matrix. This can become a very important consideration when studying cavities having aperture widths of several free-space wavelengths.

Finally, we consider a 0.6 × 0.2 λ corrugated cavity filled with a dense lossy dielectric material of $\epsilon_r = 57.5969 - j29.1812$, and $\mu_r = 1.0$. The corrugation consisted of four metallic plates placed at equal distances from each other. The bistatic radar cross section is given in Fig. 12. For this case, no published results are available to compare with; however, our choice of this case was simply to demonstrate the versatility and adaptability of the surface integral equation technique when dealing with geometries that would be difficult to treat using alternative methods.

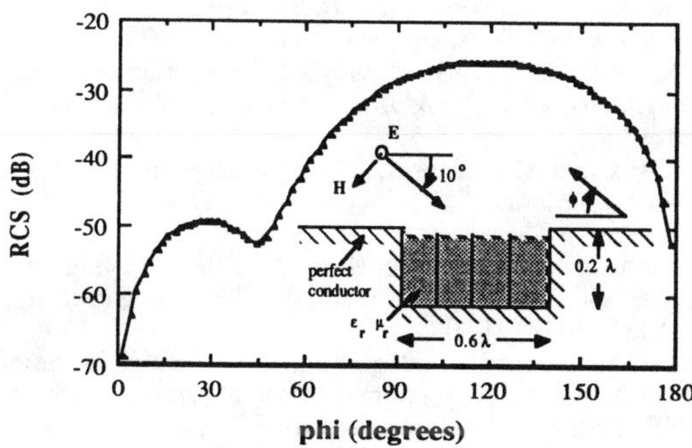

Fig. 12. Bistatic radar cross-section for a 0.6 × 0.2 λ corrugated cavity with $\epsilon_r = 57.6 - j29.2$ and $\mu_r = 1.0$, *E*-polarized case, calculated using the finite element method (FEM).

V. Conclusion

In this work, the finite element method was used to solve a class of unbounded geometries consisting of cavities in metallic wedges. To our knowledge, this topic has not been addressed before in a systematic way by using the finite element method. In this paper, a method for mesh truncation based on the use of the surface integral equation was employed. This method proved successful for solving problems with unbounded geometries, while largely preserving the sparsity of the finite element matrix. The optimization of the sparsity of the system matrix was achieved via the use of Green's function of the first kind.

The applications of the procedure described in this work can be extended to different geometries than those discussed above, particularly three-dimensional geometries comprising cavities in metallic cones. Because the Green's function of the first kind for the cone (the three-dimension extension of the wedge) can be derived analytically [22], it is expected that the minimization of the nonlocal behavior of the surface integral equations would be possible for this problem. A significant reduction of the mesh size can again be achieved in a manner similar to the case of the two-dimensional problem.

Acknowledgment

The authors wish to thank the reviewers for their comments and recommendations that enhanced the quality of this paper.

References

[1] B. Engquist and A. Majda, "Radiation boundary conditions for the numerical simulation of waves," *Math. Comput.*, vol. 31, no. 139, pp. 629–651, July 1977.

[2] A. Bayliss and E. Turkel, "Radiation boundary conditions for wave-like equations," *Comm. Pure Appl. Math.*, vol. XXXIII, pp. 707–725, 1980.

[3] A. Bayliss, M. Gunzburger, and E. Turkel, "Boundary conditions for the numerical solution of elliptic equations in exterior regions," *SIAM J. Appl. Math.*, vol. 42, no. 2, pp. 430–451, Apr. 1982.

[4] S. I. Hariharan, "Absorbing boundary conditions for exterior problems," NASA Contractor Rep. 177944, ICASE Rep. 85-33, July 1985.

[5] R. Mittra and O. M. Ramahi, "Absorbing boundary conditions for the direct solution of partial differential equations arising in electromagnetic scattering problems," in *Differential Methods in Electromagnetic Scattering*, vol. II, M. Morgan, Ed. New York: Elsevier, 1989.

[6] K. Mei, "Unimoment method of solving antenna and scattering problem," *IEEE Trans. Antennas Propagat.*, vol. AP-22, pp. 760–766, Nov. 1974.

[7] S. Chang and K. Mei, "Application of the unimoment method to electromagnetic scattering of dielectric cylinders," *IEEE Trans. Antennas Propagat.*, vol. AP-24, pp. 34–42, Jan. 1976.

[8] B. H. McDonald and A. Wexler, "Finite element solution of

unbounded field problems," *IEEE Trans. Microwave Theory Tech.*, vol. MTT-20, pp. 841-847, Dec. 1972.
[9] A. K. Aziz and R. B. Kellog, "Finite element analysis of a scattering problem," *Math. Comput.*, vol. 37, pp. 261-272, 1981.
[10] G. J. Fix and S. P. Marin, "Variational methods for underwater acoustic problems," *J. Comput. Phys.*, vol. 28, pp. 253-270, 1978.
[11] C. Johnson and J. C. Nedelec, "On the coupling of boundary integral and finite element methods," *Math. Comput.*, vol. 35, pp. 1063-1079, 1980.
[12] C.-C. Su, "A surface integral equation method for homogeneous optical fibers and coupled image lines of arbitrary cross sections," *IEEE Trans. Microwave Theory Tech.*, vol. MTT-33, pp. 1114-1119, Nov. 1985.
[13] J. Fin and V. Liepa, "Applicationof hybrid finite element method to electromagnetic scattering from coated cylinders," *IEEE Trans. Antennas Propagat.*, vol. 36, pp. 50-54, Jan. 1988.
[14] P. P. Silvester and R. L. Ferrari, *Finite Elements for Electrical Engineers.* Cambridge, U.K.: Cambridge Univ. Press, 1983.
[15] G. Strang and G. Fix, *An Analysis of the Finite Element Method.* Englewood Cliffs, NJ: Prentice-Hall, 1973.
[16] J. N. Reddy, *An Introduction to Finite-Element Method.* New York: McGraw-Hill, 1984.
[17] J. Stratton, *Electromagnetic Theory.* New York: McGraw-Hill, 1941.
[18] R. Harrington, *Time-Harmonic Electromagnetic Fields.* New York: McGraw-Hill, 1961.
[19] K. Whites, private communications.
[20] S. W. Lee and H. Ling, "Data book for cavity RCS," Univ. Illinois, Urbana, Electromagn. Lab. Tech. Rep. SWL89-1, 1989.
[21] S. Gedney, private communications.
[22] L. L. Balin and S. Silver, "Exterior electromagnetic boundary value problems for spheres and cones," *IRE Trans. Antennas Propagat.*, vol. AP-4, pp. 5-13, Jan. 1956.

Eliminating Interior Resonances in Finite Element-Boundary Integral Methods for Scattering

J. D. COLLINS, J. M. JIN, and J. L. VOLAKIS

***Abstract*—A modification in the implementation of the finite element–boundary integral method is proposed for eliminating the interior cavity resonances. It is demonstrated that by using the scattered rather than the total field as the system variable, and by introducing a small loss in the propagation constant associated with the boundary integral equation, internal resonance effects are eliminated. Results are presented for two-dimensional problems with the finite element mesh terminated on a circular enclosure.**

I. Introduction

The interior resonance corruption of boundary integral solutions for scattering computations is well known, and its treatment has been a subject of research for the last two decades. Methods based on the "complexification" of the wavenumber [1], the overspecification of the boundary conditions [2], [3], and the linear combinations of integral equations [4]–[6] have been proposed, while others have focused on the solution technique rather than the system formulation [7]. Not surprisingly, when the boundary integral equation is used to terminate the finite element mesh, the interior resonance corruption persists, and this has restricted the application of an otherwise promising method for large-scale computations of highly inhomogeneous structures.

To demonstrate the seriousness of the problem, Fig. 1(a) shows the backscatter echo width of a circular conducting cylinder of radius a_c for TM plane wave incidence. The mesh is terminated on a circle of radius $a_o = 1.01a_c$ on which the boundary integral equation is applied. The results, displayed as a function of ka_o, were obtained via the conventional finite element–boundary integral formulation (FE–BI) described in [8], [9]. The unknown in this implementation was the total field, and as seen, the solution is severely corrupted, a difficulty which persists for other scatterers as well. (For reference, the locations of the interior resonant frequencies are displayed by the vertical lines at the bottom of the figure.) The corruption is further evidenced in the near-zone scattered field plotted in Fig. 1(b), obtained by subtracting the incident from the total field generated via the aforementioned FE–BI method at $ka_o = 23.586$. To render the FE–BI method robust at all frequencies, it is essential to remove the problem associated with the interior resonances. However, traditional methods such as those described in [2]–[6] require either significant modifications to the original FE–BI formulation or substantial computing time, thus affecting the efficiency and performance of the method.

Further, for the "complexification" scheme proposed in [1] to be effective, we found that three different computations are required for each incidence angle when combined with the total field FE–BI formulation [8]. That is, the total field FE–BI solution must be repeated three times with different complex propagation constants (k), all slightly perturbed from the free-space wavenumber (k_0). An example of this behavior is seen in

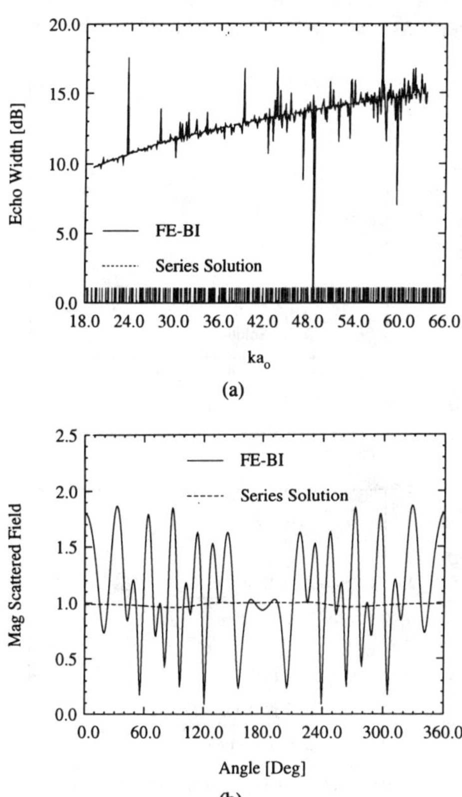

Fig. 1. Comparison of the far-zone and near-zone fields for TM plane wave incidence on a circular metallic cylinder as computed by the total field FE–BI method and the eigenfunction series. (a) Backscatter echo width versus ka_o—the lines over the horizontal axis correspond to the eigenvalues of a circular waveguide. (b) Magnitude of the TM scattered field on the enclosure at the resonant frequency $ka_o = 23.586$.

Fig. 2(a) for $k = k_0\alpha$. Clearly, the corruption in the far-field pattern has been significantly reduced for $\alpha = 1 - j0.001$, and even more so for $\alpha = 1 - j0.005$, but the patterns now deviate significantly from the reference solution. This behavior is again seen in the near-zone field in Fig. 2(b). (As a final step, we could also have chosen a third value of α, and quadratically extrapolated to a more accurate result corresponding to $\alpha = 1$ as in [1].) In the following section, we present a simple modification to the FE–BI formulation which renders it relatively immune to the resonance problem without the need to repeat the solution.

II. Method

The proposed approach for eliminating the failure of the FE–BI method at interior resonant frequencies is based on the observation that the specification of the tangential electric or

Reprinted from *IEEE Trans. Antennas Propagat.*, vol. 40, no. 12, pp. 1583–1585, Dec. 1992.

magnetic field excitation alone at the integration boundary is not sufficient to yield a unique solution [10]. Therefore, we move the excitation away from the integration boundary by employing the scattered field as the working variable. This implies that the finite element formulation functional is now of the form

$$F(\phi^s) = \frac{1}{2}\iint_\Omega \left[\frac{1}{\mu_r}\nabla\phi^s \cdot \nabla\phi^s - k_0^2\epsilon_r\phi^s\phi^s\right] ds - \int_\Gamma \phi^s\psi^s \, dl$$

$$+ \iint_\Omega \left[\frac{1}{\mu_r}\nabla\phi^i \cdot \nabla\phi^s - k_0^2\epsilon_r\phi^i\phi^s\right] ds - \int_\Gamma \phi^s\psi^i \, dl \quad (1)$$

where $\psi^s = \hat{n} \cdot \nabla\phi^s/\mu_r$ and $\psi^i = \hat{n} \cdot \nabla\phi^i/\mu_r$. Additionally, ϕ^s denotes the TM scattered field, and ϕ^i is the excitation field, whereas ϵ_r and μ_r denote the relative permittivity and permeability, respectively, of the region Ω enclosed by Γ. The sparse system of equations is generated by setting the first variation of F to zero, and this must be augmented by a discrete version of the boundary integral equation

$$\frac{1}{2}\phi^s(\bar{\rho}) = \oint_\Gamma \left\{\phi^s(\bar{\rho}')\left[\frac{\partial}{\partial n'}G(\bar{\rho}, \bar{\rho}')\right] - G(\bar{\rho}, \bar{\rho}')\left[\frac{\partial}{\partial n'}\phi^s(\bar{\rho}')\right]\right\} dl' \quad (2)$$

where $G(\bar{\rho}, \bar{\rho}') = H_0^{(2)}(k_0|\bar{\rho} - \bar{\rho}'|)/4j$ is the two-dimensional Green's function, $\bar{\rho}$ and $\bar{\rho}'$ are the position vectors of the observation and integration points, respectively, $k = 2\pi/\lambda$ is the wavenumber of free space, and $\partial/\partial n$ denotes differentiation with respect to the outward normal of Γ. In this form, the incident field is no longer part of the boundary integral equation, but is instead transferred to the finite element portion of the system.

The FE–BI system resulting from (1) and (2) may also be solved with the introduction of a small loss in the propagation constant appearing in (2), accomplished by replacing k with αk, where $\alpha = 1 - j\delta$. We found that this substantially improved the convergence of the employed biconjugate gradient solver. However, in contrast to the "complexification" scheme employed with the total field formulation, the solution based on (1) and (2) is relatively insensitive to δ (provided, of course, that δ is very small).

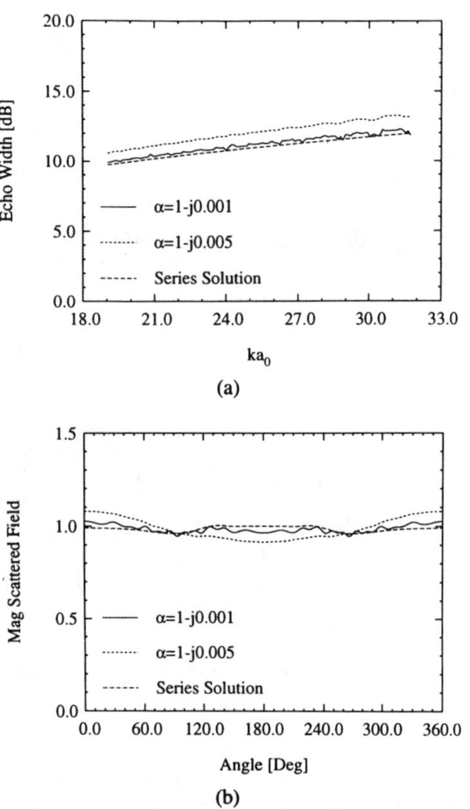

Fig. 2. Far- and near-zone fields for TM plane wave incidence on a circular metallic cylinder as computed by the total field FE–BI method with k replaced by $k\alpha$ in the BI equation. (a) Backscatter echo width versus ka_o. (b) Magnitude of the TM scattered field on the enclosure at the resonant frequency $ka_o = 23.586$.

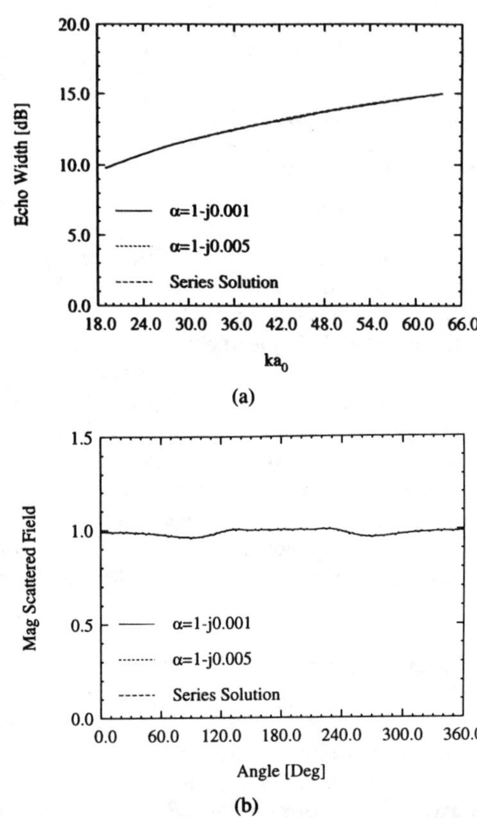

Fig. 3. Far- and near-zone fields for TM plane wave incidence on a circular metallic cylinder as computed by the scattered field FE–BI method with k replaced by $k\alpha$ in the BI equation. (a) Backscatter echo width versus ka_o. (b) Magnitude of the TM scattered field on the enclosure at the resonant frequency $ka_o = 23.586$.

III. Results

To demonstrate the effectiveness, efficiency, and accuracy of the proposed method, let us reconsider the problem of scattering by a circular cylinder via the scattered field FE–BI formulation. As seen in Fig. 3(a), the far field is no longer corrupted by the fictitious interior resonances, and the same holds for the near-zone field, as displayed in Fig. 3(b). The results, shown for $\alpha = 1 - j0.001$ and $\alpha = 1 - j0.005$, are also seen not to deviate from the series solution, although the convergence rate of the solver varied significantly as a function of α (although not shown, the results obtained for $\alpha = 1$ are virtually the same as these). For example, the unperturbed (no complexification, i.e., $\alpha = 1$) scattered field formulation converged in approximately $0.15N$ iterations over the frequency band considered in Fig. 2(a), where N denotes the unknown count. For $\alpha = 1 - j0.001$, the solution converged in $0.13N$ iterations, whereas for $\alpha = 1 -$

$j0.005$, convergence was achieved in approximately $0.08N$ iterations.

As another example, we consider the TM illumination of a $5.256\lambda \times 5.256\lambda$ metallic square cylinder. For the implementation of the scattered field FE–BI formulation, the boundary integral was enforced on a circle of radius $a_o = 3.754\lambda$ so that ka_o is the fifth zero of the Bessel function of order 6. With the incident field direction normal to one of the rectangle's faces, Fig. 4 depicts the corresponding bistatic scattered field obtained via the total and scattered field FE–BI formulation with $\alpha = 1 - j0.005$. Clearly, the pattern based on the scattered field FE–BI formulation agrees everywhere with the moment method data.

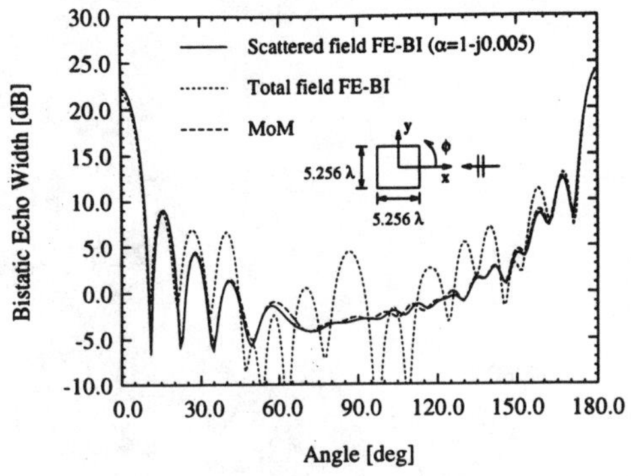

Fig. 4. Bistatic echo width for TM plane wave incidence on a $5.256\lambda \times 5.256\lambda$ metallic square cylinder with the boundary integral placed at a radius $a_o = 3.754\lambda$.

In contrast, the results based on the total field FE–BI formulation are substantially in error away from the back- and forward-scattering regions.

IV. Conclusion

We have presented a simple approach for an effective elimination of the interior resonance corruption associated with finite element–boundary integral solution methods. The approach involves use of the scattered field as the unknown variable in the implementation of the solution, and the introduction of a small loss in the wavenumber appearing in the boundary integral equation. Results were presented for TM incidence only, but the technique was found equally effective for TE incidence as well.

References

[1] W. D. Murphy, V. Rokhlin, and M. S. Vassiliou, "Solving electromagnetic scattering problems at resonance frequencies," *J. Appl. Phys.*, vol. 67, pp. 6061–6065, May 1990.

[2] A. D. Yaghjian, "Augmented electric- and magnetic-field equation," *Radio Sci.*, vol. 16, pp. 987–1001, Nov. 1981.

[3] R. Mittra and C. A. Klein, "Stability and convergence of moment method solutions," in *Numerical and Asymptotic Techniques in Electromagnetics*, R. Mittra, Ed. New York: Hemisphere, 1987 reprint.

[4] K. M. Mitzner, "Numerical solution of the exterior scattering problem at eigenfrequencies of the interior problem," in *Dig. 1968 URSI Radio Sci. Meet.*, Boston, MA, Sept. 1968, p. 75.

[5] M. B. Woodworth and A. D. Yaghjian, "Derivation, application and conjugate gradient solution of the dual-surface integral equations for three-dimensional, multi-wavelength perfect conductors," in *PIER 5: Progress in Electromagnetics Research*, T. K. Sarkar, Ed. New York: Elsevier, 1991, ch. 4.

[6] J. R. Mautz and R. F. Harrington, "H-field, E-field and combined-field solutions for conducting bodies of revolution," *AEÜ*, pp. 157–164, 1978.

[7] F. X. Canning, "Singular value decomposition of integral equations of EM and applications to the cavity resonance problem," *IEEE Trans. Antennas Propagat.*, vol. 37, Sept. 1989.

[8] J. D. Collins, J. M. Jin, and J. L. Volakis, "A combined finite element–boundary element formulation for solution of two-dimensional problems via CGFFT," *Electromagn.*, vol. 10, pp. 423–437, Nov.–Dec. 1990.

[9] J. D. Collins, J. L. Volakis, and J. M. Jin, "A combined finite element-boundary integral formulation for solution of two-dimensional scattering problems via CGFFT," *IEEE Trans. Antennas Propagat.*, vol. 36, Nov. 1990.

[10] A. F. Peterson, "The 'interior resonance' problem associated with surface integral equations of electromagnetics: Numerical consequences and a survey of remedies," *Electromagn.*, vol. 10, pp. 293–312, 1990.

Field Feedback Computation of Scattering by 2-D Penetrable Objects

M. A. MORGAN and T. B. WELCH

Abstract—**The field feedback formulation is applied to the solution of time-harmonic plane wave scattering by 2-D penetrable objects of arbitrary shape and composition. A conformal mesh, finite element algorithm is employed in the forward operator construction while a near-field Green's function integration is used in forming the feedback operator. Scattering validations for midresonance sized objects include a circular cylinder, a two-region bisected cylinder, a half-circular cylinder, a semicircular shell and a thin lossy planar strip.**

Introduction

Finite element and finite difference methods, to be denoted as *finite methods* [1], produce only local interactions between spatial node fields, which are explicitly related only to their nearest neighbors in space. This local connectivity produces highly sparse matrices in the time-harmonic case, thus allowing the use of numerically efficient algorithms for both memory allocation and linear system inversion.

In contrast, moment method approximated integral equations produce completely filled system matrices which embody interactions between each unknown, on a global basis [2]. Integral equations involving inhomogenous media require *volume* type formulations, wherein the full matrix generation severely constrains the limits on electrical size. For example, assuming an average node spacing of 0.1 λ within the material, the integral equation formulation for scattering by an inhomogenous dielectric cube having equal side dimensions of 5 λ_0 and average $\epsilon_r = 4$ would require filling and solving a $10^6 \times 10^6$ full matrix system.

Due to the localized interactions and sparse system matrices produced, finite methods have the potential to solve larger, and more complex, electromagnetic problems than can be handled by volume type integral equations. However, finite methods require supplemental conditioning on the outer mesh boundary to ensure correct far-zone behavior of solutions in unbounded regions, whereas integral equation formulations innately control radiation field behavior via the Green's function used. Several approaches to properly terminate the outer mesh boundary in finite methods have been developed over the past three decades. These include the infinite element technique [3], the unimoment method [4], [5], the boundary element method [6], [7], use of radiation boundary condition operators [8]–[10], and the field feedback formulation [11], [12].

Field Feedback Formulation

Time-harmonic ($e^{j\omega t}$-suppressed) plane wave scattering by 2-D penetrable objects is being considered, as illustrated in Fig. 1 for the special case of a semicircular shell. Both the structure and the fields are invariant to translations in z. Two orthogonal, linearly polarized incident plane wave vectors, $\overline{E}_i^{\text{inc}}$, are shown. Fields are denoted as transverse magnetic (TM) to z, and transverse electric (TE) to z, with respective subscripts $i = 1$ and 2. Three contours should be noted: the body surface, S_B, the outer geometrical boundary, S_0, and a Green's function integration contour, S_C.

A finite element mesh is used to span Ω while conforming to the body surface, as is depicted in Fig. 1. Complex values for $\epsilon_r(\bar{r})$ and $\mu_r(\bar{r})$ may be specified for each triangular element in the mesh to accommodate dissipative loss and spatial inhomogeneity. Nodal field values on S_0 are arranged as an M-dimensional column array. The finite element solution provides fields at nodes on, and inside of, S_B produced by specified nodal field values on S_0.

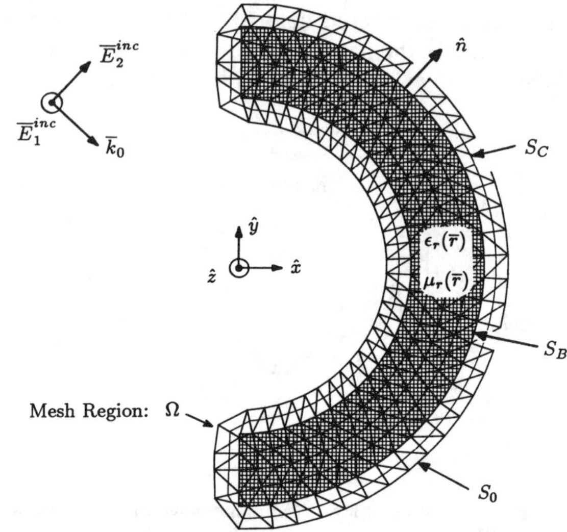

Fig. 1. Scattering problem (semicircular shell) with boundary contours and low density finite element mesh.

Three matrix operators are used in the F^3 scattering solution, as illustrated in Fig. 2. The *forward operator*, $\overline{\overline{\mathscr{A}}}_i$, transforms the S_0 boundary node column vector to yield an array of nodal tangential field values, and normal derivative, on the midpoint integration contour, S_C. This matrix is found by central difference interpolation of the nodal tangential fields on S_B (due to the finite element solution) and the boundary node tangential fields on S_0. The *feedback operator*, $\overline{\overline{\mathscr{B}}}$, is formed by numerically evaluating the scattered field at each node on S_0 through an appropriate Green's function integration over S_C of the tangential field and normal derivative. Scattered fields and radar cross section are found using the $\overline{\overline{\mathscr{C}}}$ matrix operator. This matrix is constructed from far-zone Green's function integration of the tangential field and normal derivative on S_C.

As shown in the F^3 system diagram of Fig. 2, the total field array for nodes on S_0 is $\overline{F}_0 = \overline{F}_0^{\text{inc}} + \overline{F}_0^{\text{sct}}$. The forward operator provides the total field array on S_C, which then produces the scattered field back on S_0, via the feedback operator. This system has an input–output relationship expressed by the geometric series

$$\overline{F}_C = \overline{\overline{\mathscr{A}}}_i \cdot \left[\overline{\overline{I}} + \overline{\overline{Z}}_i + \overline{\overline{Z}}_i \cdot \overline{\overline{Z}}_i + \overline{\overline{Z}}_i \cdot \overline{\overline{Z}}_i \cdot \overline{\overline{Z}}_i + \cdots \right] \cdot \overline{F}_0^{\text{inc}}$$
$$= \overline{\overline{T}}_i \cdot \overline{F}_0^{\text{inc}} \qquad (1)$$

Reprinted from *IEEE Trans. Antennas Propagat.*, vol. 40, no. 4, pp. 445–450, April 1992.

where

$$\bar{\bar{Z}}_i = \bar{\bar{\mathscr{B}}} \cdot \bar{\bar{\mathscr{A}}}_i \tag{2}$$

is the *closed-loop gain* operator of the F^3 system.

Stability of the feedback system is an important consideration. If the closed-loop gain operator has an array norm $\|\bar{\bar{Z}}_i\| \geq 1$, then the series in (1) will not converge, indicating an unstable system [13]. For a passive physical scattering system, exhibiting radiation damping and, perhaps, dissipation, it can be argued that $\|\bar{\bar{Z}}_i\| < 1$. High-Q scattering structures, such as reentrant cavities near resonant frequencies, will store energy via reactive power while relatively little energy loss occurs per oscillation cycle. This resonant behavior will correspond to $\|\bar{\bar{Z}}_i\| \to 1$, thus making convergence of (1) slow, if not impractical. Such difficulty in attaining accurate scattering solutions for high-Q structures near resonance is not confined to the F^3 approach, but spans the whole gamut of numerical methods. An advantage of the F^3 is that convergence and accuracy problems for such situations is well defined via closed-loop gain.

Assuming that $\|\bar{\bar{Z}}_i\| < 1$, the geometric series in (1) can be summed to give the feedback system transfer matrix

$$\bar{\bar{T}}_i = \bar{\bar{\mathscr{A}}}_i \cdot \left[\bar{\bar{I}} - \bar{\bar{Z}}_i\right]^{-1} \tag{3}$$

which is the approach employed here. As the system stability is reduced, so is the conditioning of the matrix inversion in (3), as is further discussed in [14].

Forward Operator

TM and TE vector fields in 2-D can be produced by the scalar Hertz potentials: $\psi_1(\bar{r}) = E_z(\bar{r})$ and $\psi_2(\bar{r}) = H_z(\bar{r})$, [15]. By using wavenumber coordinate normalization, $x = k_0 x'$ and $y = k_0 y'$, Maxwell's curl equations produce vector field generation from

$$\bar{H}(\bar{r}) = \frac{1}{\mu_r} \nabla\psi_1 \times \hat{z} \quad \text{TM case} \tag{4a}$$

$$\bar{E}(\bar{r}) = \frac{1}{\epsilon_r} \nabla\psi_2 \times \hat{z} \quad \text{TE case.} \tag{4b}$$

The Hertz potentials satisfy dual Helmholtz equations in inhomogeneous media,

$$\nabla \cdot \left[\alpha_i(\bar{r})\nabla\psi_i(\bar{r})\right] + \beta_i(\bar{r})\psi_i(\bar{r}) = 0, \quad \text{for } i = 1,2 \tag{5}$$

where $\nabla = \hat{x}\partial/\partial x + \hat{y}\partial/\partial y$, $\alpha_1 = 1/\mu_r(\bar{r})$, $\beta_1 = \epsilon_r(\bar{r})$, and $\alpha_2 = 1/\epsilon_r(\bar{r})$, $\beta_2 = \mu_r(\bar{r})$.

Solution of (5), with specified ψ_i on S_0, may be attained using the Euler–Lagrange variational approach, [16]. With specified Dirichlet boundary conditions on S_0, the $\psi_i(\bar{r})$ solving (5) also yields the stationary value of the complex functional

$$I_i = \int_\Omega\int \left[\alpha_i(\bar{r})\nabla\psi_i \cdot \nabla\psi_i - \beta_i(\bar{r})\psi_i^2\right] d\Omega. \tag{6}$$

The numerical solution of (6), as used here, employs triangular (simplex) elements for mesh generation inside of S_0, as shown in Fig. 1. Mesh construction begins by bisecting the object with a segmented curve and then connecting equispaced segments along this bisection to equispaced segments on the body perimeter, S_B. Outward normal vectors are then generated at each node on S_B to extend the mesh to the S_0 contour. Finally, triangular elements are inserted between each pair of segments connecting bisection nodes and S_0 nodes. A personal computer graphics CAD program is used to define the body contour, S_B, and the bisection contour. The mesh is then generated and displayed to the user prior to initiating the finite element solution [12].

Each node of the mesh is assigned an index, n, and the associated potential at the node is denoted by $\Psi_i(n)$. Global node ordering indexes the nonboundary (solution) nodes from 1 to N and orders the M boundary nodes on S_0 from $N+1$ to $N+M$. Unit amplitude, piecewise linear basis functions, $U_n(x, y)$, are then used to represent the solution within S_0

$$\psi_i(x, y) \doteq \sum_{n=1}^{N+M} \Psi_i(n) U_n(x, y). \tag{7}$$

Substituting (7) into (6) produces the quadratic form,

$$I_i \doteq \sum_{n=1}^{N+M} \sum_{m=1}^{N+M} \Psi_i(n)\Psi_i(m) V(n, m) \tag{8}$$

where the interaction between pairs of nodes is given by

$$V(n, m) = \int_\Omega\int \left[\alpha_i \nabla U_n(x, y) \cdot \nabla U_m(x, y) - \beta_i U_n(x, y) U_m(x, y)\right] d\Omega. \tag{9}$$

Integration in (9) is straightforward using *area coordinates* [17]. Element integrations are spliced together to form the self ($m = n$) and mutual ($m \neq n$) interaction terms between pairs of nodes.

The stationary value of (8) is found by differentiating with respect to each unknown nodal potential, and nulling the result, to produce a linear equation

$$\sum_{n=1}^{N} \Psi_i(n) V(n, m) = -\sum_{n=N+1}^{N+M} \Psi_i(n) V(n, m), \quad \text{for } m = 1, \cdots, N \tag{10}$$

where known boundary terms appear on the right-hand side. It is important to note that $V(n, m) = 0$, unless both the nth and mth nodes form vertices of a common triangular element. As a result, the system matrix defined by (10) will be quite sparse. Further, the system matrix can be made to have a triblock structure, thus allowing use of the Ricatti transform algorithm [5].

Recall that the forward matrix operator, $\bar{\bar{\mathscr{A}}}_i$, transforms the boundary values of $\psi(s_o)$ to obtain central difference interpolated values for $\psi_i(s_c)$ and $\partial\psi_i(s_c)/\partial n$. Let us denote the column array of M boundary nodal values on S_0 as

$$\bar{F}_0 = \left[\Psi(N+1), \Psi(N+2), \cdots, \Psi(N+M)\right]^T \tag{11}$$

where the indexing scheme in (10) is used and the superscript T indicates transpose. The M nodal solution values of $\Psi(n)$ along the S_B contour are, likewise, reordered to correspond to the nodes on the outer boundary. These are assembled into a column array, $\bar{F}_B$. The F_C column array is composed of M interpolated values of Ψ, followed by M central difference approximated values for the

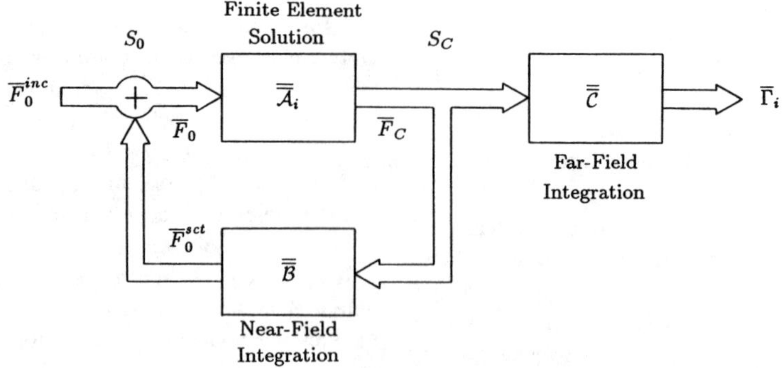

Fig. 2. Field feedback system.

normal derivative. Denoting the contour separation distance between S_B and S_0 as ΔR, the F_C array becomes

$$\bar{F}_C = \bar{\bar{\mathscr{A}}}_i \cdot \bar{F}_0 = \begin{bmatrix} \frac{1}{2}\{\bar{F}_0 + \bar{F}_B\} \\ \frac{1}{\Delta R}\{\bar{F}_0 - \bar{F}_B\} \end{bmatrix}. \tag{12}$$

The finite element method is used to find $\bar{F}_B$ for any $\bar{F}_0$ using

$$\bar{F}_B = \bar{\bar{\mathscr{U}}}_i \cdot \bar{F}_0 \tag{13}$$

where the $\bar{\bar{\mathscr{U}}}_i$-matrix is constructed from nodal solution values of $\psi(s_b)$, corresponding to individual unit nodal value boundary conditions on S_0. The forward operator may be found by substituting (13) in (12) and using the $M \otimes M$ identity matrix, $\bar{\bar{I}}$,

$$\bar{\bar{\mathscr{A}}} = \begin{bmatrix} \frac{1}{2}\{\bar{\bar{I}} + \bar{\bar{\mathscr{U}}}_i\} \\ \frac{1}{\Delta R}\{\bar{\bar{I}} - \bar{\bar{\mathscr{U}}}_i\} \end{bmatrix}. \tag{14}$$

Feedback Operator

The feedback matrix, $\bar{\bar{\mathscr{B}}}$, generates the scattered field portion of $\bar{F}_0$ from the nodal field terms contained in $\bar{F}_C$,

$$\bar{F}_0^{\text{sct}} = \bar{\bar{\mathscr{B}}} \cdot \bar{F}_C. \tag{15}$$

Near-field Green's function integrations of the field and its normal derivative over S_C are required to evaluate the scattered field on S_0,

$$\psi_i^s(s_o) = \int_{S_C} \left[G(s_o, s_c) \frac{\partial \psi_i(s_c)}{\partial n} - \psi_i(s_c) \frac{\partial G(s_o, s_c)}{\partial n} \right] ds_c \tag{16}$$

where $\hat{n}$ is the outward unit normal at s_c, [18]. The 2-D free-space Green's function,

$$G(s_o, s_c) = -\frac{j}{4} H_0^{(2)}(k_0 |\bar{r} - \bar{r}'|) \tag{17}$$

employs the second-kind Hankel function of order zero, where $R = |\bar{r} - \bar{r}'|$ is the distance between specified s_0 and s_c contour points.

Numerical integration of (16) uses a subsectional trapezoidal rule. The basis expansion in (7) and its normal derivative, when evaluated on S_C, are substituted into (16). This provides the scattered field at the nth nodal point on the S_0 contour,

$$\Psi_i^{\text{sct}}(n) \doteq \sum_{m=1}^{M} \left[\frac{\Psi_i^c(m)}{\partial n} \mathscr{B}(n, m+M) - \Psi_i^c(m) \mathscr{B}(n, m) \right]. \tag{18}$$

Feedback matrix elements are given by

$$\mathscr{B}(n, m) = \int_{S_C(m)} \left[\frac{\partial G(s_{o,n}, s_c)}{\partial n} u_m(s_c) \right] ds_c \tag{19a}$$

$$\mathscr{B}(n, m+M) = \int_{S_C(m)} \left[G(s_{o,n}, s_c) u_m(s_c) \right] ds_c \tag{19b}$$

where $u_m(s_c)$ represents $U_m(x, y)$ on S_C.

Because of the Green's function singularity, the integrations in (19) require careful consideration for closely spaced points on S_0 and S_C, (i.e., when $|n - m|$ is small). An adaptive integration algorithm was developed to maintain uniformly high integration accuracy, independent of source to evaluation point distance. When $n = m$, this algorithm typically subdivides the $S_C(m)$ source region into hundreds of segments. As the distance between points increases, the number of segments is reduced rapidly to a minimum of six. RMS errors of less than 1% were observed for a variety of tests involving circular contour integrations of plane waves and cylindrical harmonics.

Scattered Field Generation

The far-zone approximation to the integral in (16) gives the form [18],

$$\psi_i^s(\bar{r}) \doteq \sqrt{\frac{j}{8\pi k_0 r}} e^{-jk_0 r} \gamma_i(\theta) \tag{20}$$

where the pattern function,

$$\gamma_i(\theta) = \int_{S_C} \left[j \frac{\partial \psi_i(s_c)}{\partial n} + k_0 \hat{n} \cdot \hat{r} \psi_i(s_c) \right] e^{jk_0 \bar{r}' \cdot \hat{r}} ds_c \tag{21}$$

depends on direction, as specified by $\hat{r} = \cos(\theta)\,\hat{x} + \sin(\theta)\,\hat{y}$. The usual far-zone conditions required are: 1) $k_0 R = k_0 |\bar{r} - \bar{r}'| \gg 1$; 2) $r > D$, and; 3) $r > 2D^2/\lambda_0$, where D is the maximum 2-D object dimension.

The $\bar{\bar{\mathscr{C}}}$-matrix provides the pattern function in selected directions defined by θ_α, $\alpha = 1, 2, \cdots, K$,

$$\bar{\Gamma}_i = [\gamma_i(\theta_1), \gamma_i(\theta_2), \cdots, \gamma_i(\theta_K)] = \bar{\bar{\mathscr{C}}} \cdot \bar{F}_C. \tag{22}$$

Integration in (21) can be approximated by

$$\gamma_i(\theta_\alpha) \doteq \sum_{m=1}^{M} \left[\frac{\Psi_i^c(m)}{\partial n} \mathscr{C}(\alpha, m+M) - \Psi_i^c(m) \mathscr{C}(\alpha, m) \right] \tag{23}$$

where the elements of the $\bar{\bar{\mathscr{C}}}$-matrix use far-zone Green's function integrations,

$$\mathscr{C}(\alpha, m) = k_0 \int_{S_C(m)} u_m(s_c)(\hat{n} \cdot \hat{r}_\alpha) e^{jk_0 \bar{r}' \cdot \hat{r}_\alpha} ds_c \quad (24a)$$

$$\mathscr{C}(\alpha, m + M) = j \int_{S_C(m)} u_m(s_c) e^{jk_0 \bar{r}' \cdot \hat{r}_\alpha} ds_c. \quad (24b)$$

Bistatic radar cross section per unit length in 2-D, is given by

$$\sigma_i(\theta) = \lim_{r \to \infty} \frac{2\pi r |\psi_i^s|^2}{|\psi_i^{inc}|^2} = \frac{|\gamma_i(\theta)|^2}{4k_0}. \quad (25)$$

This is easily discretized for directions θ_α using the computed $\bar{\Gamma}_i$ array in (22).

Validations

Scattering computations made by the F^3 algorithm are compared to alternate numerical solutions for several 2-D penetrable objects.

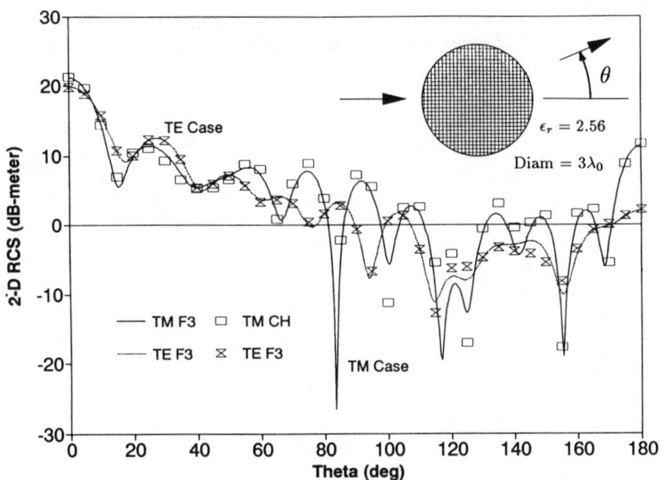

Fig. 3. Comparison of bistatic RCS for a circular dielectric cylinder having $3\lambda_0$ diameter and $\epsilon_r = 2.56$.

TM and TE incident polarizations at a frequency of 300 MHz were considered for each object. Comparisons are made for 2-D RCS in dB-meter (i.e., $10 \log_{10}(\sigma_i)$). Single-aspect bistatic scattering from $\theta = 0°$ (forward scatter) to $\theta = 180°$ (backscatter) is considered for all cases except the last, which compares multiple-aspect monostatic RCS. Since $\lambda_0 = 1$ m, RCS plots also represent normalized σ_i / λ_0 for frequency scaled calculations.

Calculations were performed on a 33 MHz, 80486-based personal computer using an extended DOS Fortran-77 compiler. Memory required for these computations was less than 4 MB. The finite element mesh was adjusted to give triangular elements having typical side dimensions of $\lambda/20$ within the material. For the semicircular shell shown in Fig. 1, the actual mesh used for the $2\lambda_0$ diameter example to be considered had about 16 times as many elements as is depicted. For the examples to be considered, computation times averaged about 2 min for the forward operator construction (finite element solution) while the feedback operator integrations required up to an hour. Reduction of the feedback operator integration bottleneck will be discussed in the Conclusion.

Bistatic RCS computations are shown in Fig. 3 for a dielectric circular cylinder with diameter of $3\lambda_0$ and $\epsilon_r = 2.56$. The F^3 calculation (denoted by F3) is compared to that found using a cylindrical harmonic (CH) series algorithm whose coefficients are evaluated by orthogonal Fourier mode matching of the tangential field components at the circular boundary [19].

A two-region bisected cylinder of $2\lambda_0$ diameter, where left-half $\epsilon_r = 2$ and right-half $\epsilon_r = 4$, is considered in Fig. 4. Comparison RCS was calculated using a specially developed CH program which employs separate left and right region cylindrical harmonic expansions. Full exterior domain expansions are used for the known incident field and the unknown scattered field. The special CH algorithm performs weighted integral (moment) matching of the tangential field components at the circular and bisection boundaries. Since the modes in the various regions are no longer fully orthogonal, as is the case for the homogeneous circular cylinder, a moment matrix system must be filled and inverted to evaluate the expansion coefficients.

Scattering computations for a half-circular dielectric cylinder having $2\lambda_0$ diameter and $\epsilon_r = 4$, is displayed in Fig. 5. Comparison computations were made by the same CH program used for the two-region cylinder in Fig. 4, but with the left-half $\epsilon_r = 1$. Strong convergence of the CH and F^3 algorithms, coupled with the good agreement shown in both Figs. 4 and 5, indicate that excellent accuracy is likely being achieved by both approaches.

Next, an example for scattering by a dielectric ($\epsilon_r = 4$) semicircular shell of outer radius 1.0 λ_0, and inner radius 0.75 λ_0, is shown in Fig. 6. Yet another special case CH program was developed to provide comparison RCS. This algorithm adds two additional vacuum filled regions to the shell to produce a circular cylindrical region of radius 1.0 λ_0. These added free-space regions are a left semicircular shell, which complements the right shell, and a circular core of radius 0.75 λ_0. CH expansions of the proper form are employed in each of the subregions, including that exterior to the outer radius. Moment matching is performed at the inner and outer circular boundaries, and at the vertical segments between the left and right semicircular shells. As with the bisected cylinder, a full moment matrix system results for the unknown expansion coefficients. Solution convergence was readily achieved for this CH program and superb agreement was observed in comparing to volume integral equation results from Richmond [20].

A final validation is shown in Fig. 7, for the case of a lossy ($\epsilon_r = 4.0 - j0.4$ and $\mu_r = 1.5 - j0.1$) thin planar strip of width $2\lambda_0$ and $0.05\lambda_0$ thickness. Comparisons are made to monostatic 2-D RCS provided by Baucke [21], using a volume integral equation (VIE) approach. Agreement is good for all aspects except near end-on for the TM case, where the RCS is about three orders of magnitude smaller than that for the TE case.

Conclusion

Theory and validations of the F3 have been presented for scattering by penetrable 2-D objects. RCS comparisons have been shown for a variety of structures having moderate electrical dimensions.

In applying the F^3, the forward and feedback matrix operators are computed independently and the scattering solution is then assembled by either iterative or direct procedures. This decoupling of the interior and exterior domains permits limited modifications to the shape and composition of the scattering object, when bounded by fixed mesh contours, while retaining the same feedback matrix. Field feedback system topology also clearly delineates the source of stability problems and ill-conditioning for high-Q scatterers near internal resonance frequencies.

Further enhancements of the F^3 are being pursued. As with integral equations, near-singular behavior of the integrand for terms on, and near, the diagonal of the feedback matrix requires careful and computationally tedious evaluation when direct numerical integration is used. Analytical evaluation of the singular portion of the integration is possible [22], thus reducing the segmentation required for the remaining integration. Going a step further, tests were run [12], which excluded integration contributions outside of a large neighborhood of the field point. These trials indicated that exclusion

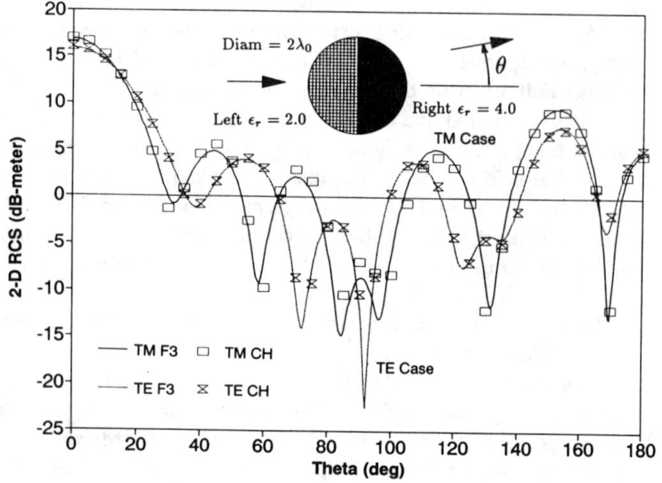

Fig. 4. Comparison of bistatic RCS for a two-region dielectric cylinder having $2\lambda_0$ diameter, left-half $\epsilon_r = 2$, and right-half $\epsilon_r = 4$.

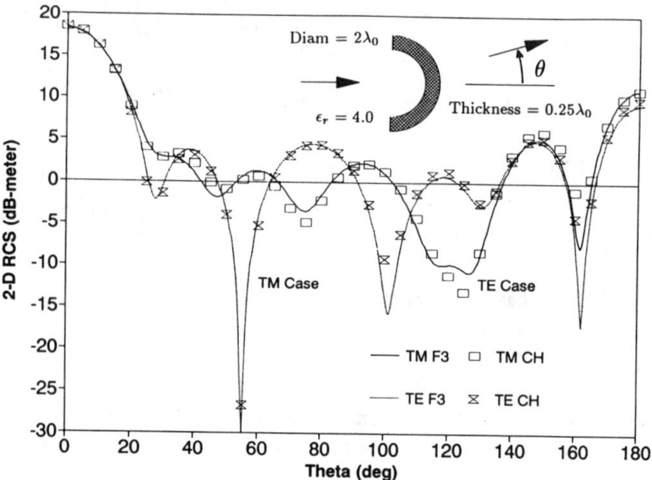

Fig. 6. Comparison of bistatic RCS for a semicircular dielectric shell with thickness = $0.25\lambda_0$, outer diameter = $2\lambda_0$ and $\epsilon_r = 4$.

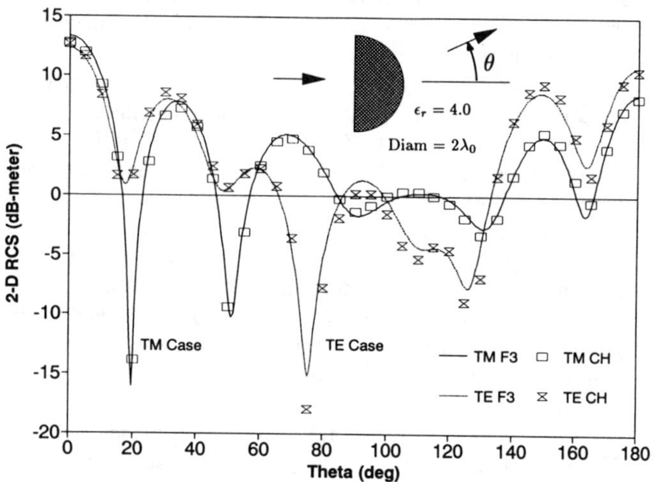

Fig. 5. Comparison of bistatic RCS for a half-circular dielectric cylinder having $2\lambda_0$ diameter and $\epsilon_r = 4$.

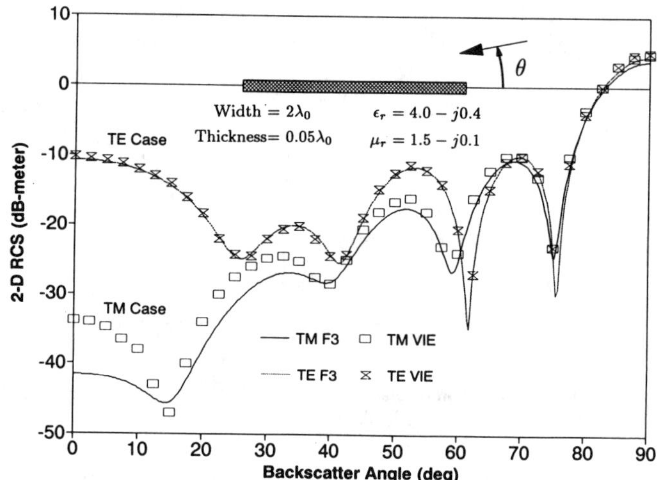

Fig. 7. Comparison of monostatic RCS for a thin lossy planar strip with width = $2\lambda_0$, thickness = $0.05\lambda_0$, $\epsilon_r = 4.0 - j0.4$, and $\mu_r = 1.5 - j0.1$.

beyond a distance of $1.0\ \lambda_0$ yields errors of between 1 and 2% in the integration. Such integration termination could yield significant computational savings when electrically large scatterers are considered.

An exciting area of progress in computational electromagnetics is in the development of radiation boundary conditions (RBC's) which are based upon local differential operator approximations to the behavior of outbound radiation fields. Since the F^3 uses *exact* analytical field relationships, it provides a starting point to develop new RBC's having innate flexibility and error control.

References

[1] K. K. Mei, M. A. Morgan, and S. K. Chang, "Finite methods in electromagnetic scattering," in *Electromagnetic Scattering,* P. L. E. Ushlenghi, Ed. New York: Academic, 1978, ch. 10.

[2] R. F. Harrington, *Field Computation by Moment Methods.* New York: Macmillan, 1968.

[3] P. Bettess, "Infinite elements," *Int. J. Num. Meth. Eng.,* vol. 11, pp. 53–64, 1977.

[4] K. K. Mei, "Unimoment method of solving antenna and scattering problems," *IEEE Trans. Antennas Propagat.,* vol. AP-22, pp. 760–766, Nov. 1974.

[5] M. A. Morgan and K. K. Mei, "Finite element computation of scattering by inhomogeneous penetrable bodies of revolution," *IEEE Trans. Antennas Propagat.,* vol. AP-27, pp. 202–214, May 1979.

[6] K. L. Wu, G. Y. Delisle, D. G. Fang and M. Lecours, "Coupled finite element and boundary element methods in electromagnetic scattering," in *Finite Element and Finite Difference Methods in Electromagnetic Scattering,* M. A. Morgan, Ed. New York: Elsevier Science, 1990, ch. 3.

[7] M. A. Morgan, C. H. Chen, S. C. Hill, and P. W. Barber, "Finite element-boundary integral formulation for electromagnetic scattering," *J. Wave Motion,* vol. 6, pp. 91–103, Jan. 1984.

[8] R. Mittra and O. Ramahi, "Absorbing boundary conditions for the direct solution of partial differential equations arising in electromagnetic scattering problems," in *Finite Element and Finite Difference Methods in Electromagnetic Scattering,* M. A. Morgan, Ed. New York: Elsevier Science, 1990, ch. 4.

[9] T. G. Moore, J. G. Blaschak, A. Taflove, and G. A. Kriegsmann, "Theory and application of radiation boundary operators," *IEEE Trans. Antennas Propagat.,* vol. 36, pp. 1797–1812, Dec. 1988.

[10] R. Janaswamy, "On the applicability of OSRC method to homogeneous scatterers," *IEEE Trans. Antennas Propagat.,* vol. 39, pp. 862–867, June 1991.

[11] M. A. Morgan and B. E. Welch, "The field feedback formulation for electromagnetic scattering problems," *IEEE Trans. Antennas Propagat.,* vol. AP-34, pp. 1377–1382, Dec. 1986.

[12] T. B. Welch, "Electromagnetic scattering from two-dimensional objects using the field feedback formulation," Electrical Engineer's Thesis, Elec. Comput. Eng. Dept., Naval Postgraduate School, Monterey, CA, Mar. 1989.

[13] R. Mittra and C. A. Klein, "Stability and convergence of moment method solutions," in *Numerical and Asymptotic Techniques in*

Electromagnetics, R. Mittra, Ed. New York: Springer-Verlag, 1975, ch. 5.

[14] M. A. Morgan, "Principles of finite methods in electromagnetic scattering," in *Finite Element and Finite Difference Methods in Electromagnetic Scattering,* M. A. Morgan, Ed. New York: Elsevier Science, 1990, ch. 1.

[15] J. A. Stratton, *Electromagnetic Theory.* New York: McGraw-Hill, 1941, pp. 349–354.

[16] P. M. Morse and H. Feshbach, *Methods of Theoretical Physics.* New York: McGraw-Hill, 1953, pp. 275–280.

[17] L. Lapidus and G. F. Pinder, *Numerical Solution of Partial Differential Equations in Science and Engineering.* New York: Wiley Interscience, 1982, ch. 2 and 3.

[18] J. Van Bladel, *Electromagnetic Fields,* New York: McGraw-Hill, 1964; (Reprinted) New York: Hemisphere, 1985, pp. 372–375.

[19] D. E. Barrick, "Cylinders," in *Radar Cross Section Handbook,* vol. 1, G. T. Ruck, Ed. New York: Plenum, 1970, ch. 4.

[20] J. H. Richmond, "TE-wave scattering by a dielectric cylinder of arbitrary cross-section shape," *IEEE Trans. Antennas Propagat.,* vol. AP-14, pp. 460–464, July 1966.

[21] R. C. Baucke, "Scattering by two-dimensional lossy, inhomogeneous dielectric and magnetic cylinders using linear pyramid basis functions and point matching," *IEEE Trans. Antennas Propagat.,* vol. 39, pp. 255–259, Feb. 1991.

[22] R. A. Rostant, "Efficient technique for calculating near-fields due to 2-D sources," Master's thesis, Elec. Comput. Eng. Dept., Naval Postgraduate School, Monterey, CA, Dec. 1990.

Inward-Looking and Outward-Looking Formulations for Scattering from Penetrable Objects

L. W. PEARSON, A. F. PETERSON, L. J. BAHRMASEL, and R. A. WHITAKER

Abstract—**This communication reviews a number of approaches for solving the problem of scattering from two-dimensional penetrable objects using the finite element method or other differential-equation based solution. Different ways of joining the numerical solution on a finite domain to a field representation that satisfies the Sommerfeld radiation condition in the infinite exterior domain are reviewed within the paradigm of the equivalence principle of electromagnetics. Depending on the region where the solution is deduced first in an inherently sequential process, these approaches can often be classified as either *inward-looking* or *outward-looking*. Although inward-looking formulations may exhibit computational advantages, outward-looking formulations appear preferable for electrically large domains because of internal resonance difficulties that may arise when solving the isolated interior problem first in an inward-looking formulation.**

I. Introduction

Recent research suggests that the direct numerical solution of the Helmholtz equation offers computational advantages over other frequency-domain procedures for modeling inhomogeneous scatterers in unbounded regions. These differential-equation-based numerical methods require the adaptation of a finite matrix method to an infinite solution domain. Either the computational domain must be truncated so it lies inside a closed boundary or the outermost elements on the finite element mesh must extend to infinity. In this communication, we consider the former possibility, and explore several ways in which the truncated field representation may be joined to an exterior field that satisfies the Sommerfeld radiation condition.

Most previous approaches can be placed into one of two broad categories, which we term *inward-looking formulations* and *outward-looking formulations*. Inward-looking formulations require that an explicit relationship between the tangential electric and magnetic fields on the boundary be computed from interior data as the first step of the solution process. This equation is then used to constrain unknown coefficients appearing within an exterior field representation. In outward-looking formulations, the first step of the solution process is to deduce a boundary relationship between the tangential electric and magnetic field that characterizes the infinite exterior region. This constraint is then used as a boundary condition for the differential equation describing the interior region. In this communication, we describe these two approaches and discuss their relative computational advantages. In addition, both formulations are susceptible to an "internal resonance" situation where the desired solution to the scattering problem is corrupted by a spurious cavity solution that also satisfies the equation. While outward-looking formulations can always be constructed in a manner that avoids the uniqueness difficulty, inward-looking approaches remain vulnerable and consequently may be less robust.

In either inward-looking or outward-looking procedures, the scattering problem must be partitioned into independent interior and exterior parts. In an attempt to unify the mathematical language of a number of previous papers, we employ the electromagnetics equivalence principle to provide a framework for the interior/exterior partitioning (Section II). Sections III and IV describe inward-looking and outward-looking formulations in detail. The discussion is limited to the case of a two-dimensional object excited by an electromagnetic field that is polarized with the electric field transverse (TE) to the axis of the scatterer. Through duality, all of the results can be translated to the transverse magnetic (TM) polarization. The results are also applicable to three-dimensional configurations, although the mathematical representations are substantially more complex.

II. Interior/Exterior Partitioning of a Scattering Problem Using the Equivalence Principle

Fig. 1 depicts the cross section of a scattering obstacle in space, illuminated by an electromagnetic field. The scatterer is infinite in extent and invariant in the direction perpendicular to the page, and is completely enclosed by a mathematical surface $\partial\Gamma$. We denote the field that would be present in the absence of the scattering obstacle as *incident* field. The total electric and magnetic fields, denoted, respectively, as $\bar{E}$ and $\bar{H}$, are the superposition of the fields associated with the incident wave and the *scattered* fields produced by sources induced within the obstacle. Without loss of generality, we consider the TE excitation where a z-component of $\bar{H}$ and transverse components of $\bar{E}$ are present.

Exterior to the surface $\partial\Gamma$, $\bar{E}$ and $\bar{H}$ are the same as fields produced by the incident field and the secondary electric and magnetic sources

$$J_t = \hat{t} \cdot \hat{n} \times (\hat{z}H_z) = -H_z \tag{1}$$

and

$$K_z = \hat{z} \cdot (\hat{t}E_t) \times \hat{n} = -E_t = -\frac{j\eta}{k}\frac{\partial H_z}{\partial n} \tag{2}$$

located on $\partial\Gamma$ and radiating in free space. The unit vector $\hat{n}$ is the outward normal to the surface, and $\hat{t}$ is the unit tangent vector defined so that $\hat{n} \times \hat{t} = \hat{z}$. The parameter $\eta = (\mu/\epsilon)^{1/2}$ denotes the

Reprinted from *IEEE Trans. Antennas Propagat.*, vol. 40, no. 6, pp. 714–720, June 1992.

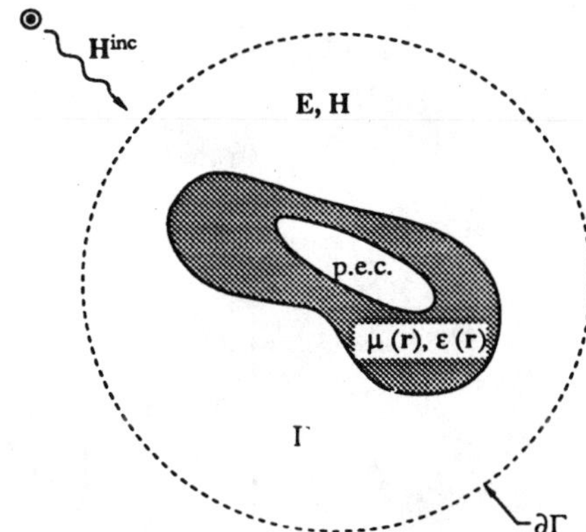

Fig. 1. Cross section of an infinite cylindrical scatterer containing inhomogeneous material. The scatterer is illuminated by an incident field that is TE to the cylinder axis.

intrinsic impedance of the exterior medium and $k = \omega(\mu\epsilon)^{1/2}$ is the wavenumber in the medium. The fields are time-harmonic with $e^{j\omega t}$ time dependence.

According to the equivalence principle of electromagnetics (c.f. [1, pp. 106–110]), these sources radiating in unbounded space add with the incident fields to produce the original fields $\bar{E}$ and $\bar{H}$ exterior to $\partial\Gamma$, and zero fields interior to $\partial\Gamma$, as shown in Fig. 2(a). Furthermore, the negatives of these two sources radiating in the presence of the obstacle act to produce the original fields $\bar{E}$ and $\bar{H}$ interior to $\partial\Gamma$ and zero fields exterior, as shown in Fig. 2(b).

For the equivalent exterior problem depicted in Fig. 2(a), we may write the magnetic field integral equation (MFIE) as

$$-H_z^{\text{inc}}(t) = J_t(t) - \lim_{\bar{\rho}\to\partial\Gamma} \int_{\partial\Gamma} \left\{ J_t(t')\frac{\partial G}{\partial n'} + \frac{jk}{\eta}K_z(t')G \right\} dt' \tag{3}$$

where

$$G = \frac{1}{4j} H_0^{(2)}(k\,|\,\bar{\rho} - \bar{\rho}'\,|) \tag{4}$$

and where the limit as $\rho \to \partial\Gamma$ is taken from the exterior. Equation (3) does not uniquely specify both secondary sources J_t and K_z. A second equation relating these sources may be obtained from the interior equivalent problem depicted in Fig. 2(b). The interior problem may be expressed formally as either a Dirichlet or Neumann boundary value problem involving the scalar Helmholtz equation for inhomogeneous media

$$\nabla \cdot \left(\frac{1}{\epsilon_r} \nabla H_z \right) + k^2 \mu_r H_z = 0. \tag{5}$$

Provided that either J_t or K_z is specified on the boundary $\partial\Gamma$, (5) can be solved for H_z throughout the interior. (If the obstacle contains impenetrable boundaries, then boundary conditions appropriate to those boundaries must be specified as well.) Furthermore, for any given J_t, (5) can be solved subject to the equivalent Dirichlet boundary condition to produce a corresponding K_z (which is proportional to the normal derivative of H_z on $\partial\Gamma$) and vice versa. Thus, the combination of (3) and (5) fully specifies the sources J_t and K_z. By imposing (3) and (5) simultaneously, we implicitly enforce continuity of the total fields across $\partial\Gamma$ for the two configurations in Fig. 2.

The partitioning outlined above produces a differential equation formulation defining the fields on a bounded region in infinite space.

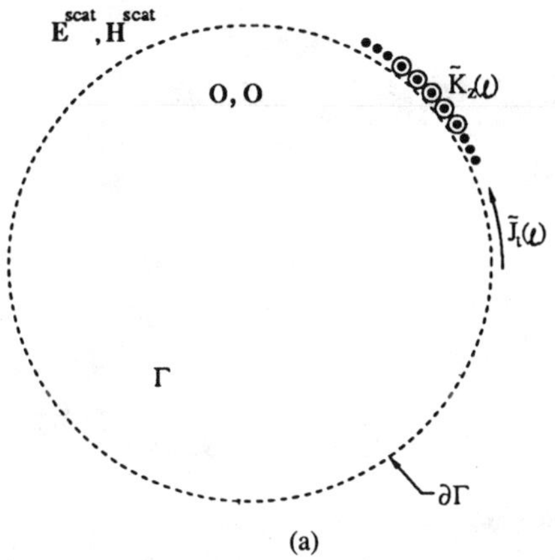

(a)

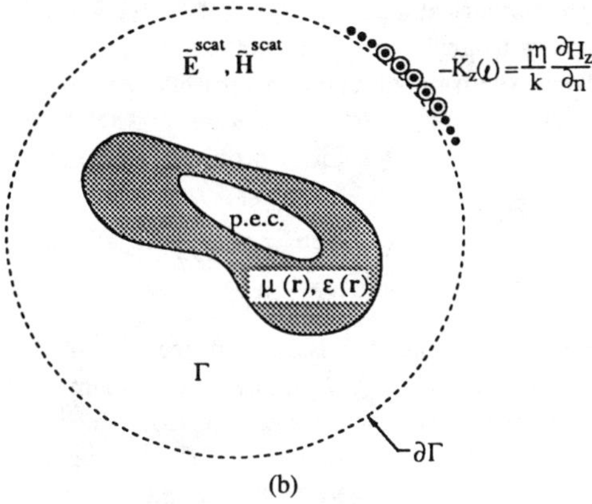

(b)

Fig. 2. Equivalent exterior and interior problems produced by an application of the equivalence theorem.

The Green's function incorporated into the MFIE satisfies the Sommerfeld radiation condition and ensures that the scattered fields propagate away from the inhomogeneity. However, there are theoretical difficulties associated with the uniqueness of solutions to both (3) and (5). The MFIE does not have a unique solution if $\partial\Gamma$ happens to coincide with a closed resonant cavity having perfect magnetic walls [2]–[5]. In a similar fashion, the scalar Helmholtz equation does not have a unique solution if the interior medium is lossless and happens to represent a resonant cavity appropriate to the type of boundary condition (Dirichlet or Neumann) being imposed on $\partial\Gamma$. The associated matrix operator may become very ill-conditioned at frequencies in the vicinity of these resonances, and gross errors in the numerical solution are possible. In practice, however, the difficulty with the MFIE is well understood and can be completely circumvented by employing the combined-field integral equation [3] or some equivalent remedy [2]–[5]. Unfortunately, there is no obvious way in which the scalar Helmholtz equation can be made robust while employing Dirichlet or Neumann boundary conditions.

III. Inward-Looking Formulations for Open-Region Scattering

In an inward-looking formulation, the equivalent surface current densities introduced in (1) and (2) are the primary unknowns to be determined. In the general context of the method of moments, we form the expansions

$$J_t(t) \cong \sum_{m=1}^{M} c_m f_m(t) \tag{6}$$

and

$$K_z(t) \cong \sum_{m=1}^{M} d_m g_m(t). \tag{7}$$

If substituted into (3), these produce

$$-H_z^{\text{inc}}(t) \cong \sum_{m=1}^{M} c_m \left\{ f_m - \lim_{\rho \to \partial\Gamma} \int_{\partial\Gamma} f_m(t') \frac{\partial G}{\partial n'} dt' \right\} - \frac{jk}{\eta} \sum_{m=1}^{M} d_m \int_{\partial\Gamma} g_m(t') G \, dt'. \tag{8}$$

Testing this equation with M linearly independent testing functions yields a system of M equations in $2M$ unknowns. The remaining information necessary to fully specify the $2M$ coefficients is obtained from the numerical solution of (5). To obtain this solution, it may be necessary to project the exterior expansion functions onto a set of boundary expansion functions compatible with the volumetric interior expansion functions used to discretize (5). If the interior problem is formulated as a Neumann problem, we form

$$g_m(t) \cong \sum_{n=1}^{N} \tilde{d}_n \tilde{g}_n(t) \tag{9}$$

where $\tilde{g}$ denotes the interior bases and the $\tilde{d}$ are the interior expansion coefficients. Each $\tilde{g}$ is applied as a Neumann boundary condition on the scalar Helmholtz equation (5), which is solved to produce a value for H_z on the boundary $\partial\Gamma$. If necessary, each of these solutions is then projected from the interior representation onto the basis $\{f\}$ according to

$$H_z(t) \cong \sum_{m=1}^{M} \gamma_{mn} f_m(t) \cong -J_t(t). \tag{10}$$

For each assumed $K_z(t)$, a corresponding $J_t(t)$ is found from (10). By carrying out this procedure for each of the basis functions $g_m(t)$, we obtain M equations constraining K_z and J_t through the interior field interactions. These augment the M equations representing the exterior problem previously obtained from the discretization of (3), to yield a $2M \times 2M$ matrix equation that can be solved for the boundary fields.

Perhaps the most widely used inward-looking method is the unimoment method proposed by Mei [6] and developed in [7]–[10]. The underlying characteristic of the unimoment method is the separation of the interior and exterior regions by a circular or spherical boundary, and the expansion of the exterior field in terms of cylindrical or spherical harmonics. The unimoment method can be deduced from (8) and (10) if we take $\partial\Gamma$ to be circular, with radius a. The surface currents can be expanded as

$$J_t(\phi) \cong \sum_{n=-M}^{M} j_n e^{jn\phi} \tag{11}$$

and

$$K_z(\phi) \cong \sum_{n=-M}^{M} k_n e^{jn\phi}. \tag{12}$$

Equation (3) can be enforced using testing functions $e^{-jn\phi}$, and through the use of the addition theorem for Hankel functions we obtain an $N \times 2N$ matrix equation (where $N = 2M + 1$) of the form

$$\begin{bmatrix} \alpha_{-M} & 0 & \cdots & 0 & \beta_{-M} & 0 & \cdots & 0 \\ 0 & \alpha_{1-M} & & 0 & 0 & \beta_{1-M} & & 0 \\ \vdots & \vdots & & \vdots & \vdots & \vdots & & \vdots \\ 0 & 0 & & \alpha_M & 0 & 0 & & \beta_M \end{bmatrix} \cdot \begin{bmatrix} j_{-M} \\ \vdots \\ j_M \\ k_{-M} \\ \vdots \\ k_M \end{bmatrix} = \begin{bmatrix} h_{-M} \\ \vdots \\ h_M \end{bmatrix} \tag{13}$$

where the diagonal elements of the submatrices are

$$\alpha_n = -\frac{j\pi ka}{2} J_n(ka) H_n^{(2)\prime}(ka) \tag{14}$$

and

$$\beta_n = \frac{\pi ka}{2\eta} J_n(ka) H_n^{(2)}(ka) \tag{15}$$

and the term on the right-hand side is given by

$$h_n = \frac{1}{2\pi} \int_0^{2\pi} H_z^{\text{inc}}(\phi) e^{-jn\phi} \, d\phi. \tag{16}$$

The prime in (14) denotes differentiation with respect to the argument of the Hankel function. (Because the boundary is separable, we can obtain (13) by forming the eigenfunction expansion of the exterior magnetic field and by joining these fields to the equivalent currents in Fig. 2 without recourse to the MFIE in (3). In this special case, the two procedures yield the identical system because the eigenfunctions of the integral operator were used to discretize the integral equation.)

To specify both equivalent sources J_t and K_z, the additional $N \times 2N$ system must be constructed based on the interior problem. The numerical solution proceeds by recasting the Helmholtz equation as a weak differential equation of the form

$$\iint_\Gamma \left\{ \frac{1}{\epsilon_r} \nabla T \cdot \nabla H_z - k^2 \mu_r T H_z \right\} dx\, dy = \int_{\partial\Gamma} \frac{1}{\epsilon_r} T \frac{\partial H_z}{\partial n} dt \tag{17}$$

where T is an arbitrary testing function which, with its first derivatives, is square integrable. Once the Neumann boundary condition

$$K_z = e^{jm\phi} \tag{18}$$

or, equivalently,

$$\frac{\partial H_z}{\partial \rho} = \frac{jk}{\eta} e^{jm\phi} \tag{19}$$

is substituted into the boundary integral on the right-hand side, (17) can be discretized using the finite element method and solved for the values of the total H_z field at each of the nodal points throughout the interior region. The fields of interest are those on the boundary $\partial\Gamma$, which can be decomposed into the basis functions $e^{jn\phi}$ to produce

$$H_z(\phi) \cong \sum_{n=-M}^{M} \gamma_{nm} e^{jn\phi} \cong -J_t(\phi). \quad (20)$$

This process must be repeated for each harmonic retained in the expansion of (11) and (12). By linearity, it follows that an arbitrary Neumann condition

$$K_z(\phi) = \sum_{m=-M}^{M} k_m e^{jm\phi} \quad (21)$$

would produce

$$J_t(\phi) = -\sum_{m=-M}^{M} k_m \left\{ \sum_{n=-M}^{M} \gamma_{nm} e^{jn\phi} \right\} = -\sum_{n=-M}^{M} \cdot \left\{ \sum_{m=-M}^{M} \gamma_{nm} k_m \right\} e^{jn\phi}. \quad (22)$$

By equating (11) and (22), we obtain the $N \times 2N$ system

$$\begin{bmatrix} 1 & 0 & \cdots & 0 & \gamma_{-M,-M} & \gamma_{-M,1-M} & \cdots & \gamma_{-M,M} \\ 0 & 1 & & 0 & \gamma_{1-M,-M} & \gamma_{1-M,1-M} & & \gamma_{1-M,M} \\ \vdots & & & \vdots & \vdots & & & \vdots \\ 0 & 0 & \cdots & 1 & \gamma_{M,-M} & \gamma_{M,1-M} & \cdots & \gamma_{M,M} \end{bmatrix} \cdot \begin{bmatrix} j_{-M} \\ \vdots \\ j_M \\ k_{-M} \\ \vdots \\ k_M \end{bmatrix} = \begin{bmatrix} 0 \\ \vdots \\ 0 \end{bmatrix}. \quad (23)$$

Equation (23) is a matrix relationship between the H_z-field and its normal derivative on the boundary $\partial\Gamma$, and takes into account interactions throughout the interior region Γ. Equation (13) provides an analogous relationship obtained from the exterior interaction of the field and its normal derivative. Together, (13) and (23) provide a $2N \times 2N$ matrix description of the scattering problem. Since three of the four blocks of the matrix are diagonal, it can be reduced to an $N \times N$ system for computational purposes.

There are two distinct situations in which the preceding inward-looking formulation may not have a unique solution. If the surface $\partial\Gamma$ represents a resonant cavity having perfect magnetic walls, an eigenvalue of the integral operator will vanish. (For the specific example of (13), this situation causes one row of the matrix to vanish—whenever $J_n(ka) = 0$ in (14) and (15).) This difficulty may be circumvented for the circular boundary case by simply avoiding radii with $J_n(ka) = 0$; for general boundary shapes, a robust remedy would be to employ the combined-field equation instead of the MFIE. However, if the interior region is lossless, the scalar Helmholtz equation does not have a unique solution for Neumann boundary conditions if the region bounded by $\partial\Gamma$ coincides with a resonant cavity having perfect electric walls. In this situation, the operator in (17) has at least one eigenvalue equal to zero, and it will be impossible to construct (23). There appears to be no robust remedy to this difficulty that can be applied *a priori* for general boundary shapes.

In addition to the unimoment method, a number of more general inward-looking formulations have appeared in the literature. Morgan *et al.* [11], Morgan and Welch [12], Jin and Liepa [13], Yuan *et al.* [14], and Cangellaris and Lee [15], [16] have presented two-dimensional schemes using surface integral equations on $\partial\Gamma$. A three-dimensional formulation employing a surface integral equation was recently proposed by Yuan [17]. These procedures permit the shape of the outer boundary to be arbitrary, but still suffer from the resonance problem described above.

IV. Outward-Looking Formulations for Open-Region Scattering

In an outward-looking formulation, the primary unknowns are the coefficients of the interior expansion of the H_z-field. The interior fields appear within the scalar Helmholtz equation, which must be augmented with a boundary condition on $\partial\Gamma$ in order to produce a solution that satisfies the Sommerfeld radiation condition at infinity. A two-dimensional formulation proposed by McDonald and Wexler in 1972 used a surface integral equation to truncate the computational domain [18]. Since then, a number of two-dimensional and three-dimensional outward-looking formulations have employed surface integral equations as radiation boundary conditions [19]–[26]. To illustrate this approach, consider the MFIE of (3), which can be discretized as discussed in (6)–(8) in order to produce an explicit relationship between the boundary values of H_z, the normal derivative of H_z, and H_z^{inc}. (To use our previous notation, K_z can be determined as an explicit function of J_t and H_z^{inc}.) Prior to discretization, this relationship can be expressed as

$$K_z(t) = \Lambda^{-1}\left\{ J_t(t) - \lim_{\rho \to \partial\Gamma} \int_{\partial\Gamma} J_t(t') \frac{\partial G}{\partial n'} dt' \right\} + \Lambda^{-1}\{H_z^{\text{inc}}\} \quad (24)$$

where Λ^{-1} denotes the inverse of the operator

$$\Lambda\{K_z\} = \int_{\partial\Gamma} \frac{jk}{\eta} K_z(t') G\, dt'. \quad (25)$$

After substitution into the right-hand side of (17), we obtain the formal representation

$$\iint_\Gamma \left\{ \frac{1}{\epsilon_r} \nabla T \cdot \nabla H_z - k^2 \mu_r T H_z \right\} dx\, dy + \frac{jk}{\eta} \int_{\partial\Gamma} T \Lambda^{-1} \left\{ H_z(t) - \lim_{\rho \to \partial\Gamma} \int_{\partial\Gamma} H_z(t') \frac{\partial G}{\partial n'} dt' \right\} dt = \frac{jk}{\eta} \int_{\partial\Gamma} T \Lambda^{-1}\{H_z^{\text{inc}}(t)\}\, dt \quad (26)$$

as the basic equation to be solved for the interior H_z-field. The integral equation radiation boundary condition (RBC) imposes no restrictions on the shape of the boundary. However, in order to implement this procedure numerically, the integral equation must be discretized as a first step to produce the inverse operator Λ^{-1} in matrix form (for instance, as an LU factorization).

In common with the inward-looking situation, some simplification is possible if $\partial\Gamma$ is circular. In that case, the integral equation can be made equivalent to an eigenfunction expansion of the exterior

H_z-field, following the discussion in Section III. Consequently, the required inversion can be performed analytically, to produce the exact RBC [27]–[29]

$$\left.\frac{\partial H_z}{\partial \rho}\right|_{\rho=a} = \int_0^{2\pi} H_z^{\text{inc}}(a, \phi') \cdot \left\{ \frac{j}{\pi^2 a} \sum_{n=-\infty}^{\infty} \frac{1}{J_n(ka) H_n^{(2)}(ka)} e^{jn(\phi-\phi')} \right\} d\phi' + \int_0^{2\pi} H_z(a, \phi') \cdot \left\{ \frac{k}{2\pi} \sum_{n=-\infty}^{\infty} \frac{H_n^{(2)\prime}(ka)}{H_n^{(2)}(ka)} e^{jn(\phi-\phi')} \right\} d\phi'. \quad (27)$$

Since integral operators give rise to full matrices, RBC's based on integral equations or eigenfunction expansions couple the fields around the entire boundary, and consequently reduce the sparsity of the associated finite element system. This behavior differs from the Sommerfeld type of radiation condition, which is entirely local, but appears a necessary trade-off in order to produce an "exact" radiation condition in the scatterer near zone. Approximate radiation conditions that attempt to restore the matrix sparsity have been proposed by Bayliss and Turkel [30], Engquist and Majda [31], and others [32], [33]. For illustration, the second-order Bayliss–Turkel RBC is given by

$$\frac{\partial H_z^s}{\partial \rho} = \left\{ \frac{-jk - \dfrac{3}{2\rho} + \dfrac{j3}{8k\rho^2}}{1 - \dfrac{j}{k\rho}} \right\} H_z^s + \left\{ \frac{-\dfrac{j}{2k\rho^2}}{1 - \dfrac{j}{k\rho}} \right\} \frac{\partial^2 H_z^s}{\partial \phi^2} \quad (28)$$

where $H_z^s = H_z - H_z^{\text{inc}}$. Although this RBC is restricted to a circular boundary, it only contains a second derivative along $\partial\Gamma$, and therefore produces no fill-in beyond that already arising from the Helmholtz operator. Local RBCs are approximate, and generally degrade in accuracy as $\partial\Gamma$ is brought closer to the scatterer surface. However, results using (28) for electromagnetics applications demonstrate that their accuracy can be quite good [34]–[36]. In fact, for many scatterer geometries there is no discernable difference between the numerical results obtained using (28) as an RBC and those found using (27) [29].

Incorporating a radiation boundary condition directly into the boundary integral on the right-hand side of the weak Helmholtz equation modifies the resulting matrix operator to one that no longer permits interior resonant cavity solutions. In other words, because of the radiation loss, uniqueness difficulties that arise for Dirichlet or Neumann boundary conditions no longer occur. If the RBC is obtained from a surface integral equation, difficulties that might arise in inverting the integral operator (such as (25), whose inverse does not exist when the surface $\partial\Gamma$ coincides with a resonant cavity) can be avoided by using the combined-field equation [3]. Formulations employing the exact RBC from (27) are completely robust, since the problematical factor $J_n(ka)$ is cancelled analytically. There have been a number of local RBC's developed to date, and these also appear to provide sufficient radiation loss to circumvent any difficulty. To illustrate this observation, the matrix condition number [37] obtained in the course of solving the scalar Helmholtz equation is displayed in Fig. 3, as a function of frequency, for two situations. In the first, a Neumann boundary condition is used on $\partial\Gamma$ to simulate the solution of the interior equation within an inward-looking formulation. In the second, the Bayliss–Turkel RBC from (28) is used as an example of an outward-looking formulation. The scalar Helmholtz equation is discretized with piecewise-linear interpolation functions, when modeling a circular region of free-space having radius a with a triangular-cell mesh having 85 nodes and 131 cells. The presence of spurious resonances is indicated by the sharp spikes in the figure, which arise for the case of Neumann boundary conditions but are not observed with the Bayliss–Turkel RBC. For this example, the problematical frequencies correspond to the TE resonant modes of an air-filled cavity with perfect electric walls.

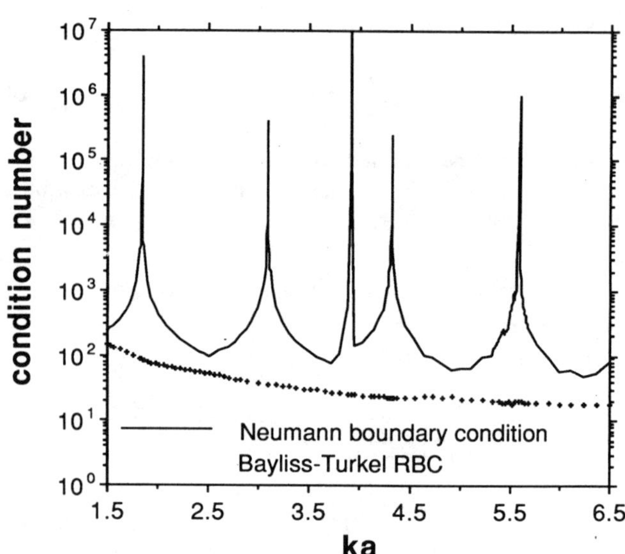

Fig. 3. Condition number estimate returned by LINPACK for the matrix operator representing the discretization of (17), when used to model a circular free-space region of circumference ka. The solid curve is obtained using a Neumann boundary condition, typical of an inward-looking formulation, while the broken curve was obtained using the Bayliss-Turkel radiation boundary condition from (28) within an outward-looking formulation. The numerical model employs piecewise-linear expansion and testing functions on a triangular-cell mesh having 85 nodes and 131 cells. Sharp spikes in the curve obtained using the Neumann condition indicate frequencies where the scalar Helmholtz equation admits spurious cavity solutions in addition to the desired result.

V. Comparison of Inward- and Outward-Looking Approaches

Inward-looking formulations offer several apparent computational advantages. The finite element system arising within an outward-looking procedure implemented with an "exact" RBC (surface integral equation or eigenfunction expansion) will not be as sparse as that of a comparable inward-looking scheme, because of the additional fill-in associated with the boundary interactions. Even if the Bayliss–Turkel RBC is employed, eliminating the additional fill-in, the surface $\partial\Gamma$ must usually be located farther away from the scatterer than with an inward-looking approach, in order to maintain the accuracy of the RBC. Furthermore, outward-looking formulations always produce complex-valued matrices, as a consequence of the radiation boundary condition. If the scatterer is lossless, the inward-looking procedure only requires a real-symmetric matrix. These observations suggest that inward-looking formulations will generally require less storage and computation for electrically large geometries. As with any sparse matrix procedure, the actual efficiency of a particular approach is highly dependent on the particular matrix solution algorithm employed.

The major disadvantage of inward-looking approaches is the fact that the interior problem is always being modeled as a closed region. Even if some loss is present, the system of equations may be nearly singular if the surface $\partial\Gamma$ happens to coincide with the surface of a resonant cavity. Although these resonances are typically quite narrow, and may be encountered infrequently, they are impossible to anticipate for electrically large domains with arbitrarily shaped boundaries. While partial remedies (such as modifying the boundary size or shape) can be imposed after the fact, the analyst has the additional burden of monitoring the matrix condition number or some comparable indicator in order to detect a problem, and subsequently modifying the finite element mesh in an attempt to move the boundary and obtain a valid solution. In the outward-looking formulations, however, radiation loss eliminates the resonances and ensures that the system of equations is nonsingular.

VI. Formulations Other Than Inward- and Outward-Looking

As an alternative formulation, a scattering problem may be posed as the simultaneous solution of the interior and exterior equations, without first solving one in order to substitute it into the other. For instance, the procedures described by Gong and Glisson [38] and Wu *et al.* [39] fall into this category. These formulations employed surface integral equations for the exterior problem. Lee and Cendes proposed a similar approach incorporating an exterior eigenfunction expansion [40], [41]. The principal drawback to these procedures is that they require the introduction of additional unknowns on $\partial\Gamma$, and therefore produce a larger order matrix equation than outward-looking formulations.

VII. Conclusion

Finite element computations for open-region electromagnetics problems must incorporate some type of radiation condition to ensure outward propagation as the distance from the scatterer becomes asymptotically large. Various approaches for accomplishing this can be viewed in a common framework using the electromagnetic equivalence theorem. Furthermore, most of the resulting formulations can be classified into two broad groups, which we term *inward-looking* and *outward-looking*.

Inward-looking formulations are generally more computationally efficient than outward-looking procedures, although the actual difference may be small depending on the particular RBC, the sparse matrix solution algorithm, etc. However, inward-looking formulations manifest numerical instabilities at selected frequencies when they are applied to electrically large solution domains. The instabilities are directly related to fictitious cavity resonances that are introduced by the need to model the interior problem as a closed region. Outward-looking formulations, on the other hand, prove stable because radiation loss at the exterior boundary is directly incorporated into the matrix formulation.

References

[1] R. F. Harrington, *Time-Harmonic Electromagnetic Fields.* New York: McGraw-Hill, 1961.

[2] R. Mittra and C. A. Klein, "Stability and convergence of moment method solutions," in *Numerical and Asymptotic Techniques in Electromagnetics*, R. Mittra, Ed. New York: Springer-Verlag, 1975.

[3] J. R. Mautz and R. F. Harrington, "H-field, E-field, and combined-field solutions for conducting bodies of revolution," *A. E. U.*, vol. 32, pp. 157–163, 1978.

[4] A. D. Yaghjian, "Augmented electric- and magnetic-field equations," *Radio Sci.*, vol. 16, pp. 987–1001, Nov. 1981.

[5] A. F. Peterson, "The interior resonance problem for surface integral equations of electromagnetics: Numerical consequences and a survey of remedies," *Electromagn.*, vol. 10, pp. 293–312, 1990.

[6] K. K. Mei, "Unimoment method for solving antenna and scattering problems," *IEEE Trans. Antennas Propagat.*, vol. AP-22, pp. 760–766, Nov. 1974.

[7] S.-K. Chang and K. K. Mei, "Application of the unimoment method to electromagnetic scattering by dielectric cylinders," *IEEE Trans. Antennas Propagat.*, vol. AP-24, pp. 35–42, Jan. 1976.

[8] K. K. Mei, M. A. Morgan, and S.-K. Chang, "Finite methods in electromagnetic scattering," in *Electromagnetic Scattering*, P. L. E. Uslenghi, Ed. New York: Academic, 1978.

[9] M. A. Morgan and K. K. Mei, "Finite element computation of scattering by inhomogeneous penetrable bodies of revolution," *IEEE Trans. Antennas Propagat.*, vol. AP-27, pp. 202–214, Feb. 1979.

[10] K. K. Mei, "Unimoment method for electromagnetic wave scattering," *J. Electromagn. Waves and Appl.*, vol. 1, pp. 201–222, 1987.

[11] M. A. Morgan, C. H. Chen, S. C. Hill, and P. W. Barber, "Finite element—boundary integral formulation for electromagnetic scattering," *Wave Motion*, vol. 6, pp. 91–103, Jan. 1984.

[12] M. A. Morgan and B. E. Welch, "The field feedback formulation for electromagnetic scattering computations," *IEEE Trans. Antennas Propagat.*, vol. AP-34, pp. 1377–1382, Dec. 1986.

[13] J.-M. Jin and V. V. Liepa, "A note on hybrid finite element method for solving scattering problems," *IEEE Trans. Antennas Propagat.*, vol. 36, pp. 1486–1490, Oct. 1988.

[14] X. Yuan, D. R. Lynch, and J. W. Strohbehn, "Coupling of finite element and moment methods for electromagnetic scattering from inhomogeneous objects," *IEEE Trans. Antennas Propagat.*, vol. 38, pp. 386–393, Mar. 1990.

[15] A. C. Cangellaris and R. Lee, "The bymoment method for two-dimensional electromagnetic scattering," *IEEE Trans. Antennas Propagat.*, vol. 38, pp. 1429–1437, Sept. 1990.

[16] A. C. Cangellaris and R. Lee, "Finite element analysis of electromagnetic scattering from inhomogeneous cylinders at oblique incidence," *IEEE Trans. Antennas Propagat.*, vol. 39, pp. 645–650, May 1991.

[17] X. Yuan, "Three-dimensional electromagnetic scattering from inhomogeneous objects by the hybrid moment and finite element method," *IEEE Trans. Microwave Theory Tech.*, vol. 38, pp. 1053–1058, Aug. 1990.

[18] B. H. McDonald and A. Wexler, "Finite-element solution of unbounded field problems," *IEEE Trans. Microwave Theory Tech.*, vol. MTT-20, pp. 841–847, Dec. 1972.

[19] R. C. MacCamy and S. P. Marin, "A finite element method for exterior interface problems," *Int. J. Math. and Math. Sci.*, vol. 3, pp. 311–350, 1980.

[20] S. P. Marin, "Computing scattering amplitudes for arbitrary cylinders under incident plane waves," *IEEE Trans. Antennas Propagat.*, vol. AP-30, pp. 1045–1049, Nov. 1982.

[21] J.-M. Jin and V. V. Liepa, "Application of hybrid finite element method to electromagnetic scattering from coated cylinders," *IEEE Trans. Antennas Propagat.*, vol. 36, pp. 50–54, Jan. 1988.

[22] J. D. Collins, J. L. Volakis, and J.-M. Jin, "A combined finite element-boundary integral formulation for solution of two-dimensional scattering problems via CG-FFT," *IEEE Trans. Antennas Propagat.*, vol. 38, pp. 1852–1858, Nov. 1990.

[23] O. M. Ramahi and R. Mittra, "Finite element solution for a class of unbounded geometries," *IEEE Trans. Antennas Propagat.*, vol. 39, pp. 244–250, Feb. 1991.

[24] K. D. Paulsen, D. R. Lynch, and J. W. Strohbehn, "Three-dimensional finite, boundary, and hybrid element solutions of the Maxwell equations for lossy dielectric media," *IEEE Trans. Microwave Theory Tech.*, vol. 36, pp. 682–693, April 1988.

[25] J.-M. Jin and J. L. Volakis, "A finite element—boundary integral formulation for scattering by three-dimensional cavity-backed apertures," *IEEE Trans. Antennas Propagat.*, vol. 39, pp. 97–104, Jan. 1991.

[26] —, "Electromagnetic scattering by and transmission through a three-dimensional slot in a thick conducting plane," *IEEE Trans. Antennas Propagat.*, vol. 39, pp. 543–550, Apr. 1991.

[27] S. K. Jeng and C. H. Chen, "On variational electromagnetics: Theory and application," *IEEE Trans. Antennas Propagat.*, vol. AP-32, pp. 902–907, Sept. 1984.

[28] R. B. Wu and C. H. Chen, "Variational reaction formulation of scattering problem for anisotropic dielectric cylinders," *IEEE Trans. Antennas Propagat.*, vol. AP-34, pp. 640–645, May 1986.

[29] L. W. Pearson, R. A. Whitaker, and L. J. Bahrmasel, "An exact radiation boundary condition for the finite element solution of electromagnetic scattering on an open domain," *IEEE Trans. Magnetics*, vol. 25, pp. 3046–3048, July 1989.

[30] A. Bayliss and E. Turkel, "Radiation boundary conditions for wave-like equations," *Comm. Pure Appl. Math.*, vol. 33, pp. 707–725, 1980.

[31] B. Engquist and A. Majda, "Absorbing boundary conditions for the numerical simulation of waves," *Math. Comp.*, vol. 31, pp. 629–651, 1977.

[32] A. F. Peterson, "Absorbing boundary conditions for the vector wave equation," *Microwave and Opt. Techn. Lett.*, vol. 1, pp. 62–64, Apr. 1988.

[33] W. C. Chew, *Waves and Fields in Inhomogeneous Media*. New York: Van Nostrand Reinhold, 1990, pp. 246–256.

[34] A. F. Peterson and S. P. Castillo, "A frequency-domain differential equation formulation for electromagnetic scattering from inhomogeneous cylinders," *IEEE Trans. Antennas Propagat.*, vol. 37, pp. 601–607, May 1989.

[35] B. J. McCartin, L. J. Bahrmasel, and G. Meltz, "Application of the control region approximation to two-dimensional electromagnetic scattering," in *Finite Element and Finite Difference Methods in Electromagnetic Scattering*, M. A. Morgan, Ed. New York: Elsevier, 1990.

[36] J. D'Angelo and I. D. Mayergoyz, "Finite element methods for the solution of RF radiation and scattering problems," *Electromagn.*, vol. 10, pp. 177–199, 1990.

[37] J. Dongarra, J. R. Bunch, C. B. Moler, and G. W. Stewart, *LINPACK User's Guide*. Philadelphia, PA: SIAM, 1979.

[38] Z. Gong and A. W. Glisson, "A hybrid equation approach for the solution of electromagnetic scattering problems involving two-dimensional inhomogeneous dielectric cylinders," *IEEE Trans. Antennas Propagat.*, vol. 38, pp. 60–68, Jan. 1990.

[39] K. L. Wu, G. Y. Delisle, D. G. Fang, and M. Lecours, "Coupled finite element and boundary element methods in electromagnetic scattering," in *Finite Element and Finite Difference Methods in Electromagnetic Scattering*, M. A. Morgan, Ed. New York: Elsevier, 1990.

[40] J. F. Lee, "New methods for predicting bistatic electromagnetic scattering from complicated and inhomogeneous shapes," M.S. dissertation, Carnegie-Mellon University, Pittsburgh, PA, 1986.

[41] J. F. Lee and Z. Cendes, "Transfinite elements: a highly efficient procedure for modeling open-field problems," *J. Appl. Phys.*, vol. 61, pp. 3913–3915, Apr. 1987.

A Finite Element Approach to the Electromagnetic Interaction with Rotating Penetrable Cylinders of Arbitrary Cross Section

Roberto D. Graglia, *Senior Member, IEEE,* Angelo Freni, *Member, IEEE,* and Giuseppe Pelosi, *Senior Member, IEEE*

***Abstract*— A method to evaluate the fields excited by an obliquely incident plane wave inside and outside a rotating cylinder of arbitrary cross section is presented. A differential formulation of the problem is developed with respect to the comoving reference frame of the scatterer on a closed circular domain including the scatterer cross section; the boundary conditions account for the far-field conditions through a series expansion of the total field on the domain contour. The results at oblique incidence are then particularized for normal incidence and a hybrid finite element numerical solution is presented and discussed for TM incidence. The present FEM approach and the related computer code are directly applicable to the study of rotating piecewise-homogeneous cylinders at normal incidence, as well as metallic cylinders coated by layers of penetrable materials.**

The model presented allows a comparison of the Galileian relativistic model with a heuristic model previously considered in [1]. Through a comparison at normal incidence between the quasi-stationary and the Galileian relativistic approach we derive a method to approximately reconstruct the Doppler frequency shift in the quasi-stationary method. Far-field and near-field numerical results for circular and arbitrary-shape (metallic or dielectric) cylinders are presented.

I. Introduction

THE study of the interaction of electromagnetic fields with moving bodies is one of the classical problems in the theory of electromagnetism. The electromagnetic analysis of moving bodies is relevant to many areas of electrical engineering where the electromagnetic effects induced by motion have to be modeled for device or system design. The area of interest ranges from radar engineering (e.g., in problems of remote sensing of velocity), electromagnetic compatibility applications, space science, and radio astronomy to power engineering, where the study and design of electrical machines are considered. An excellent survey of the theory of relativity applied to the electrodynamics of moving bodies and current engineering practice is provided by Van Bladel [2].

A very popular nonrelativistic approach to the motional problem is the *quasi-stationary* one, which consists in solving the scattering problem in the time domain by assuming the moving object is motionless at a given position at a given time. This approximate method, which has been proved to be inherently incorrect [3], has been used in conjunction with high-frequency techniques to study (large) rotating scatterers (see e.g., [4] and [5]).

Conversely, the relativistic study of the electromagnetic interaction with rotating bounded structures can be performed, in general, by following two different approaches:

1) In the first approach, the problem is formulated in the comoving frame of the scatterer, which is the same reference frame used by an observer rotating with the body. This procedure, which is best suited to geometrically complex structures, it is exact only for bodies in uniform motion and, when applied to study rotating bodies, it is often referred to as the jumping on the merry-go-round technique [1]. Application of this technique requires a transformation from the time–space coordinates of the laboratory frame to the coordinates of the object frame, where the so-called instantaneous rest-frame assumption (see e.g. [1] and [6]) is used to obtain the constitutive equations of the structure in the comoving frame. The Maxwell equations to be solved in the comoving frame are obtained by relativistic transformation of those valid in the laboratory frame; the solution obtained can then be back-transformed in the laboratory frame.
2) In the second approach, which is widely used to deal with structures invariant under rotation, one solves the Maxwell equations in the laboratory reference frame by using the Minkowski transformed constitutive equations to take into account the body motion. Approximate constitutive equations are obtainable under the *instantaneous velocity assumption* [3], which neglects relativistic acceleration effects. Typical applications of this second approach have considered the case of a circular cylinder rotating around its axis [1], [7], [8], of a rotating sphere [9]–[11] or, more generally, of a body with arbitrary revolution symmetry [12]. A discussion of the limits of this method and of its relationship with the first approach is provided in [13].

In this work we will study the electromagnetic interaction with an arbitrarily shaped cylindrical composite structure rotating, in free space, around an axis parallel to the cylinder axis, with constant angular velocity Ω. The problem will be formulated for an obliquely incident plane wave by working in the comoving frame of the scatterer and by use of the

Manuscript received April 10, 1992; revised September 29, 1992.

R. D. Graglia was with CESPA (CNR), Politecnico di Torino, Corso Duca degli Abruzzi 24, 10129, Torino, Italy. He is now with the Dipartimento di Elettronica, Politecnico di Torino, Italy.

A. Freni and G. Pelosi are with the Dip. Ingegneria Elettronica, Università di Firenze, Via di S. Marta 3, 50139 Firenze, Italy.

IEEE Log Number 9209229.

Reprinted from *IEEE Trans. Antennas Propagat.*, vol. 41, no. 5, pp. 635–650, May 1993.

instantaneous rest-frame assumption. In the comoving frame the total electromagnetic field is governed by a system of partial differential equations, to be solved numerically by a finite element algorithm [14] applied to a closed region having circular cross section and including the scatterer. The boundary conditions used on this external circular boundary account for the far-field conditions by expressing the total field on the contour in terms of a double series summation, one summation for each frequency component of the Doppler spectrum.

A similar approach in the comoving frame has already been used by Tanaka [15] to numerically study the scattering from a rotating perfectly conducting cylinder having arbitrary cross section. In [15], the boundary conditions were enforced on a finite set of points on the boundary of the cylinder (point-matching method). Thus, our technique can also be regarded as a finite element extension and improvement of the Tanaka approach, which is not easily applicable to penetrable cylinders and, moreover, relies on the appropriate location of the sampling points, a choice which may be delicate for particular shapes.

On the other hand, the rotating cylinder problem can also be considered numerically by working in the laboratory frame. In fact, apart from the approximate quasi-stationary approach, finite differences in the time domain (FD-TD) have been recently used to study a moving perfectly conducting mirror at normal and oblique incidence [16]. Although this FD-TD technique can be generalized to study arbitrarily shaped moving conductors, its generalization to deal with moving *penetrable* materials seems quite difficult since, in the laboratory frame, the constitutive equations of the penetrable material will be very complicated, and, in general, time dependent. Numerical procedures to study the electrodynamics of moving bodies have also been developed by other authors working in the area of the design of electrical machines (see e.g., [17] and [18]). Possible application of our technique to this field will be considered in future papers.

In Sections II and III we will formulate the differential problem and then we shall consider the particular case of normal incidence. Section II is devoted to deriving and discussing the partial differential equations, while in Section III we derive the related boundary conditions on the external circular boundary.

Through a comparison at normal incidence between our approach and the quasi-stationary one, in Section IV we first prove how it is possible to obtain the bistatic characteristics of a cylinder of arbitrary cross section at rest by solving only one problem in the comoving frame for vanishing rotation velocity; then, a method to approximately reconstruct the Doppler frequency shift in the quasi-stationary approach is provided.

The hybrid finite element algorithm for normal incidence is described in Section V. In Section VI we assess the accuracy of our method up to very high angular velocities by comparing the results of dielectric circular cylinders available in the literature with the ones obtained by our computer code. Results for rotating dielectric and metallic cylinders of different cross sections are also reported.[1]

[1] Part of this work has been presented in [19].

II. The Differential Formulation in the Comoving Frame

The laboratory $\mathcal{S} \equiv (\rho, \phi, z)$ and the comoving $\mathcal{R} \equiv (\rho, \theta, z)$ cylindrical reference frames are chosen with common origin and with the cylinder rotation axis as the common z axis. This choice, in the limit of Galileian relativity, gives the relation

$$\phi = \theta + \Omega t, \tag{1}$$

where Ω is the constant angular velocity of the rotating cylinder, measured counterclockwise with respect to the z axis. By introducing the dimensionless quantity $\beta = \Omega\rho/c$, the rotation velocity $\boldsymbol{v}_S$ at ρ is conveniently expressed as follows:

$$\boldsymbol{v}_S = \beta c \hat{\phi} = \frac{\beta Y_0}{\epsilon_0}\hat{\phi} = \frac{\beta Z_0}{\mu_0}\hat{\phi}, \tag{2}$$

where μ_0 and ϵ_0 are the free-space magnetic permeability and electric permittivity, respectively, $c = 1/\sqrt{\mu_0\epsilon_0}$ is the velocity of light in free space, $Z_0 = Y_0^{-1} = \sqrt{\mu_0/\epsilon_0}$ is the free-space impedance.

The cross-sectional view of a rotating cylinder in the reference frame $\mathcal{S}$ and $\mathcal{R}$ is shown in Fig. 1. With reference to Fig. 1(b), $\mathcal{R}$ has been divided into three regions: the cylindrical scatterer defines the inner region 1, delimited by the scatterer contour γ: the outer region 3 is external to the circle γ': $\rho = a$; while the shell region 2 is located between the two lines γ and γ'. Note that, in $\mathcal{S}$, the contour shape γ' is stationary and surrounded by free space, while the points of the circle γ' are moving with respect to $\mathcal{S}$ at speed $|\Omega|a = |\beta_a|c$.

The problem can now be subdivided into two parts: 1) an internal problem, relative to regions 1 and 2, which is modeled by partial differential equations and solved numerically in the comoving frame $\mathcal{R}$ by application of the finite element technique, and 2) an external problem, for region 3, where the solution in $\mathcal{S}$ can be represented in terms of series of special functions, which facilitate the derivation of the boundary conditions on γ' for the internal problem. In fact, our formulation requires only the transformation from observer to object frame of the free space fields on the contour γ' of region 3.

Application of the finite element method (FEM) requires the subdivision of the geometry of the internal problem into subdomains. This subdivision results in the approximations γ'_a and γ_a of the contour lines γ' and γ, respectively (Fig. 1(c)). Note that for a piecewise-homogeneous scatterer, similar geometrical approximations are obtained also for the boundaries between homogeneous subregions.

We shall denote with capital and small letters the field quantities measured in $\mathcal{S}$ and $\mathcal{R}$, respectively. The internal problem in the comoving frame is solved in the frequency domain, by considering each frequency component (ω_m) of the Doppler spectrum individually, and by omitting the factor $\exp(j\omega_m t)$.

In the following, we will avoid distinguishing between covariant and contravariant components and shall consider the (source-free) Maxwell equations in the rotating system in terms of the *ordinary* or *physical* components [1]. The

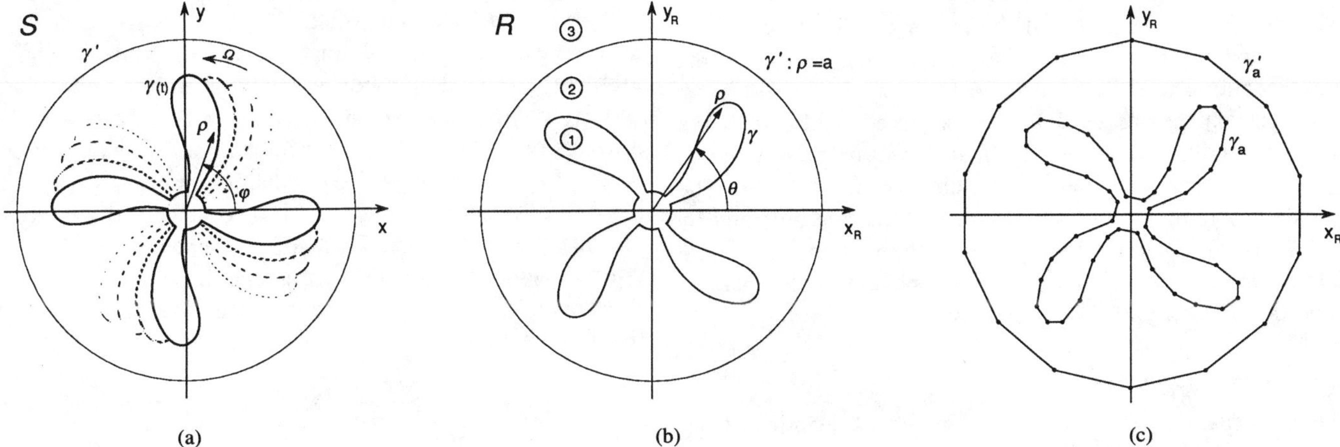

Fig. 1. Cross-sectional view of a rotating cylinder in free space. (a) In the laboratory frame $\mathcal{S} \equiv (\rho, \phi, z)$, the cylinder is rotating around the z axis at a constant angular velocity Ω. (b) In the comoving frame $\mathcal{R} \equiv (\rho, \theta, z)$, the inner region 1 is delimited by the scatterer contour γ, the outer region 3 is external to the circle γ', and the shell region 2 is located between γ and γ'. (c) In the comoving frame, application of FEM results in the approximations γ_a' and γ_a of the contour lines γ' and γ, respectively.

constitutive equations valid in the rotating axes take, by use of the instantaneous rest-frame assumption, the following vector form (cf. [1] and [2, sec. 9.2]):

$$\boldsymbol{d} = \frac{1}{\nu}\left(\epsilon \boldsymbol{e} - \frac{\beta}{c}\hat{\theta} \times \boldsymbol{h}\right) \qquad \boldsymbol{b} = \frac{1}{\nu}\left(\mu \boldsymbol{h} + \frac{\beta}{c}\hat{\theta} \times \boldsymbol{e}\right), \quad (3)$$

where $\nu = \sqrt{1-\beta^2}$, while μ and ϵ are the magnetic permeability and the electric permittivity at rest, respectively. When all the terms in β with order higher than the first are neglected, (3) yields

$$\begin{bmatrix} \boldsymbol{d} \\ \boldsymbol{b} \end{bmatrix} = \begin{bmatrix} \epsilon \boldsymbol{e} \\ \mu \boldsymbol{h} \end{bmatrix} - \delta_{\mathrm{vb}} \frac{\beta}{c}\hat{\theta} \times \begin{bmatrix} \boldsymbol{h} \\ -\boldsymbol{e} \end{bmatrix}, \quad (4)$$

with $\delta_{\mathrm{vb}} = 1$. In the limit $\beta = 0$ (i.e., for $\delta_{\mathrm{vb}} = 0$), (4) provides the Galileian constitutive equations in $\mathcal{R}$ which, of course, are identical to those valid in $\mathcal{S}$ when the material is at rest.

By relativistic transformation of the Maxwell equations valid in the observer frame and by use of the instantaneous rest-frame assumption, it is possible to derive partial differential equations for the comoving observer [1], [2, sec. 9.2]. However, these equations, which are nonlinear in β, are not valid at high $|\beta|$ values, since they neglect the coupling between gravitational and electromagnetic fields [1]. At relatively low angular velocities Ω (i.e., when $|\beta| \ll 1$), by considering only the first term in β one obtains (cf. [2, sec. 11.7, 11.9])

$$\nabla \times \begin{bmatrix} \boldsymbol{h} \\ \boldsymbol{e} \end{bmatrix} = j\omega_m \left(\begin{bmatrix} \epsilon \boldsymbol{e} \\ -\mu \boldsymbol{h} \end{bmatrix} - \delta_{\mathrm{vb}} \frac{\beta}{c}\hat{\theta} \times \begin{bmatrix} \boldsymbol{h} \\ \boldsymbol{e} \end{bmatrix} \right). \quad (5)$$

These are the Maxwell curl equations valid for an inertial observer at rest with respect to a medium whose *bianisotropic* constitutive equations are given by (4). In these equations, a unitary Lorentz contraction factor ν is understood and acceleration is neglected.

For $\delta_{\mathrm{vb}} = 1$, (4) and (5) define a linear model in β, which has been discussed by Van Bladel in [1] and used in conjunction with the Galileian transformation formula (1). In this heuristic model, use of (1) may be partially justified by noticing that, when second-order terms in β^2 are neglected, there is no need to distinguish between the time, t, as measured in the laboratory frame $\mathcal{S}$, and the time (t') measured by a comoving observer on γ', since both the time dilatation factor $(1/\nu)$ and the length-standard Lorentz contraction factor ν are second-order effects in β (cf, [2, ch. 7]). For the comoving observer, the only relativistic model consistent with (1) is provided by Galileian relativity (i.e., by choosing $\delta_{\mathrm{vb}} = 0$ in (4) and (5)).

The equations given by (5) simplify considerably when the constitutive material parameters of the scatterer and the fields are independent of z (i.e., $\partial/\partial z \equiv 0$); under these assumptions (e.g., in the simpler case of normal incidence on a longitudinally homogeneous cylinder), equations (5) yield the following two uncoupled partial differential equations:

$$\left(\nabla_t^2 + k_m^2 N^2 + 2j\delta_{\mathrm{vb}} k_m k_0 \frac{\Omega}{\omega_0}\frac{\partial}{\partial \theta}\right) \begin{bmatrix} e_z \\ h_z \end{bmatrix} = \begin{bmatrix} 0 \\ 0 \end{bmatrix}, \quad (6)$$

with

$$\delta_{\mathrm{vb}} = 0, 1 \qquad N = \frac{\sqrt{\mu\epsilon}}{\sqrt{\mu_0\epsilon_0}}$$
$$k_m = \omega_m\sqrt{\mu_0\epsilon_0} \qquad k_0 = \omega_0\sqrt{\mu_0\epsilon_0}, \quad (7)$$

and where N is the refractive index of the material at rest, while ω_0 is a reference angular frequency used to express the angular velocity Ω in terms of the dimensionless ratio Ω/ω_0 (see Section III).

It is easy to prove that the boundary conditions at the scatterer discontinuities are the classical relationships of the Galileian model (i.e., for $\beta = 0$ or $\delta_{\mathrm{vb}} = 0$) even for the β linear model (5) (cf. [1] and [2, sec. 9.8]). No special boundary conditions need to be specified on the circle γ' which delimits the internal region, since this circle is only a fictitious boundary. The field behavior on γ' is discussed in the following section.

III. Field Behavior and Boundary Conditions on γ'

In this section we derive the boundary conditions on γ' for the internal problem by transforming the expression of the free-space total field in $\mathcal{S}$ into a series expression valid in $\mathcal{R}$. In $\mathcal{S}$, the expansions used for the scattered field satisfy the far-field conditions and are given in terms of two unknown sets of scattering coefficients to be obtained numerically. The boundary conditions for the internal problem are obtained by evaluating, in $\mathcal{R}$, the series representation of the tangential field components on γ'. These components are easily obtained since the boundary γ' has a circular shape.

A. Field Transformations on γ'

In free space (region 3), to transform field quantities from $\mathcal{S}$ to $\mathcal{R}$ and vice versa one may use the results of general relativity, yielding (cf. [2, sec. 9.3] and [1])

$$\begin{bmatrix} \boldsymbol{e} \\ Z_0\boldsymbol{h} \end{bmatrix}_{(\rho,\,\theta,\,z)} = \left[\begin{array}{c|c} \underline{\underline{A}} & \underline{\underline{B}} \\ \hline -\underline{\underline{B}} & \underline{\underline{A}} \end{array}\right] \begin{bmatrix} \boldsymbol{E} \\ Z_0\boldsymbol{H} \end{bmatrix}_{(\rho,\,\phi,\,z)} \tag{8}$$

$$\begin{bmatrix} \boldsymbol{E} \\ Z_0\boldsymbol{H} \end{bmatrix}_{(\rho,\,\phi,\,z)} = \frac{1}{1-\beta^2} \left[\begin{array}{c|c} \underline{\underline{A}} & -\underline{\underline{B}} \\ \hline \underline{\underline{B}} & \underline{\underline{A}} \end{array}\right] \begin{bmatrix} \boldsymbol{e} \\ Z_0\boldsymbol{h} \end{bmatrix}_{(\rho,\,\theta,\,z)} \tag{9}$$

with:

$$\underline{\underline{A}} = \begin{bmatrix} 1 & 0 & 0 \\ 0 & \sqrt{1-\beta^2} & 0 \\ 0 & 0 & 1 \end{bmatrix}$$

$$\underline{\underline{B}} = \begin{bmatrix} 0 & 0 & \beta \\ 0 & 0 & 0 \\ -\beta & 0 & 0 \end{bmatrix}. \tag{10}$$

The field components on γ' are continuous in both $\mathcal{R}$ and $\mathcal{S}$, since γ' is in free space in $\mathcal{S}$, while the coefficients of the transformation matrix of (8) are continuous functions of the space variable ρ. By using (2) and neglecting second-order terms in β^2, (8) and (9) agree with Galileian relativity:

$$\boldsymbol{e} = \boldsymbol{E} + \boldsymbol{v}_s \times \boldsymbol{B} \qquad \boldsymbol{h} = \boldsymbol{H} - \boldsymbol{v}_s \times \boldsymbol{D} \tag{11}$$

$$\boldsymbol{d} = \boldsymbol{D} \qquad \boldsymbol{b} = \boldsymbol{B} \tag{12}$$

$$\boldsymbol{d} = \epsilon \boldsymbol{e} \qquad \boldsymbol{b} = \mu \boldsymbol{h}. \tag{13}$$

One can see from (13) and (4) that the only low-order-β model consistent with the instantaneous rest-frame assumption and with (1) is provided by Galileian ($\delta_{\text{vb}} = 0$) relativity.

B. Field on γ' in the Laboratory Frame

Let us consider in $\mathcal{S}$ one component of the Doppler spectrum, which can be represented by a monochromatic wave with angular frequency ω_l, z dependence factor $\exp(-jk_z z)$ and vector wave number $\boldsymbol{k}_l = \boldsymbol{k}_{tl} + k_z\hat{z}$, with

$$k_l = \omega_l\sqrt{\epsilon_0\mu_0} \qquad k_z^2 = k_l^2 - k_{tl}^2 \tag{14}$$

$$n_{tl} = \frac{k_{tl}}{k_l} = \cos\alpha_l \qquad n_{zl} = \frac{k_z}{k_l} = \sin\alpha_l. \tag{15}$$

For waves propagating along the z axis with $k_z^2 \neq k_l^2$ (i.e., with $k_{tl} \neq 0$), the transverse electromagnetic fields can be expressed in terms of the longitudinal components. Thus the problem can be simplified by studying in detail only the E_z and H_z components which, when $k_{tl} \neq 0$, may be expressed in terms of series of Bessel functions. For the sake of brevity, in the following we will exclude the case $k_{tl} = 0$, which has to be treated separately with a formalism different from the one presented here. Furthermore, when $k_z = 0$ (e.g., at normal incidence), one may formulate the cylindrical problem in terms of TE_z and TM_z modes, since in this case one has, for TM_z modes,

$$e_z = \left(1 - j\frac{\Omega}{\omega_l}\frac{\partial}{\partial\phi}\right)E_z \qquad h_z = H_z = 0$$
$$e_\rho = e_\theta = E_\rho = E_\phi = 0, \tag{16}$$

and for TE_z modes,

$$h_z = \left(1 - j\frac{\Omega}{\omega_l}\frac{\partial}{\partial\phi}\right)H_z, \qquad e_z = E_z = 0$$
$$h_\rho = h_\theta = H_\rho = H_\phi = 0. \tag{17}$$

In $\mathcal{S}$, we now assume the incident field $(\boldsymbol{E}^i, \boldsymbol{H}^i)$ to be a monochromatic plane wave of unit amplitude, propagating with angular frequency ω_0 in the direction of the unit vector $\hat{k}_0$. As shown in Fig. 2, the $\hat{x}$ axis is chosen to lie in the (k_0, z) plane, where α_0 is the incidence angle between the $\hat{x}$ axis and the unit vector $\hat{k}_0$, with $-\pi/2 < \alpha_0 < \pi/2$, and $\alpha_0 = 0$ at normal incidence. Hence:

$$\hat{k}_0 = \hat{x}\cos\alpha_0 + \hat{z}\sin\alpha_0$$
$$\boldsymbol{k}_{t0} = \hat{x}k_{t0} = \hat{x}k_0 n_{t0} = \hat{x}k_0\cos\alpha_0$$
$$\boldsymbol{k}_z = \hat{z}k_z = \hat{z}k_0 n_{z0} = \hat{z}k_0\sin\alpha_0 \tag{18}$$

$$\boldsymbol{E}^i = [\hat{y}\sin\delta + (\hat{z}\cos\alpha_0 - \hat{x}\sin\alpha_0)cos\,\delta] \cdot \exp(-jk_{t0}x)\exp[j(\omega_0 t - k_z z)],$$
$$Z_0\boldsymbol{H}^i = [-\hat{y}\cos\delta + (\hat{z}\cos\alpha_0 - \hat{x}\sin\alpha_0)\sin\delta] \cdot \exp(-jk_{t0}x)\exp[j(\omega_0 t - k_z z)]. \tag{19}$$

The axial components of the incident field (19) may be expressed in terms of series of Bessel functions J_m of the first kind and of integer order m as follows:

$$\begin{bmatrix} E_z^i \\ Z_0 H_z^i \end{bmatrix} = \exp[j(\omega_0 t - k_z z)]\cos\alpha_0 \begin{bmatrix} \cos\delta \\ \sin\delta \end{bmatrix} \cdot \sum_{m=-\infty}^{+\infty} (-j)^m J_m(k_{t0}\rho)\exp(jm\phi), \tag{20}$$

so that $\delta = 0$, $\pi/2$ (and $\delta = \pi$, $3\pi/2$) correspond to E_z (i.e., TM_z) and H_z (TE_z) incident polarization, respectively.

In region 3, while the incident field has been assumed to be monochromatic with angular frequency ω_0, for the laboratory observer the scattered field turns out to have an infinite, but discrete spectrum of frequencies, which includes the ω_0 component;[2] in some cases, a static component can

[2] One may note that the so far heuristic assumption of a scattered field with an infinite discrete spectrum of frequencies in presence of a monochromatic incident field is the same assumption that one can make when studying the scattering from a cylinder of nonlinear material.

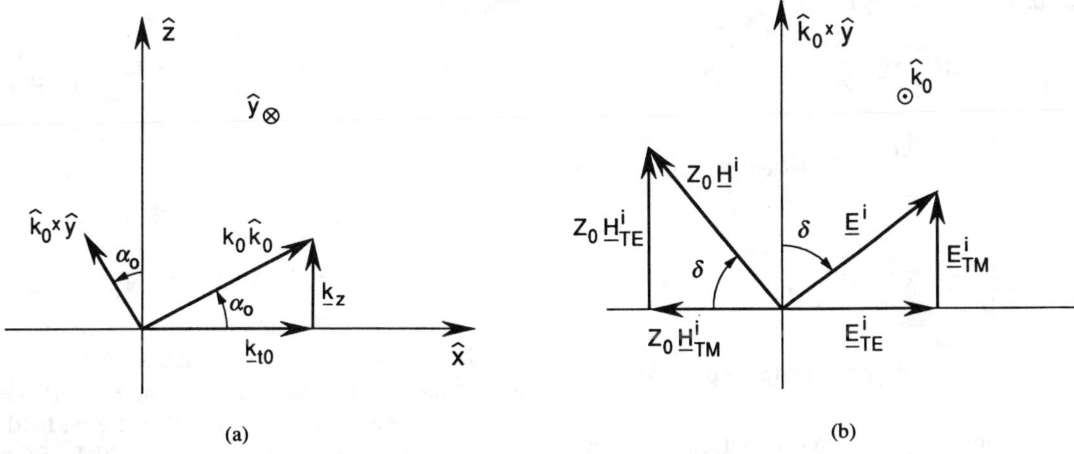

Fig. 2. (a) Incident plane wave propagating in $\mathcal{S}$ with angular frequency ω_0 in the direction of the unit vector $\hat{k}_0$. (b) TM_z and TE_z incident field components in the plane of the wave.

also be present. To account for the far-field conditions and for the "z-continuity condition" $\partial/\partial z \equiv -jk_z$, the longitudinal components of the scattered fields are conveniently written in terms of the two sets of unknown constants $\{a_{l,p}\}$, $\{c_{l,p}\}$ as follows ($\rho \geq a$):

$$\begin{bmatrix} E_z^s \\ Z_0 H_z^s \end{bmatrix} = \exp(-jk_z z) \cdot \sum_{l=-\infty}^{+\infty} \left\{ \delta_{\omega_l} \sum_{p=-\infty}^{+\infty} (-j)^p \begin{bmatrix} a_{l,p} \\ c_{l,p} \end{bmatrix} \mathcal{L}_p(\rho) \exp(jp\phi) + \frac{\omega_l}{\omega_0} \cos\alpha_l \exp(j\omega_l t) \sum_{p=-\infty}^{+\infty} (-j)^p \begin{bmatrix} a_{l,p} \\ c_{l,p} \end{bmatrix} \cdot \mathcal{C}_p(\rho) \exp(jp\phi) \right\}, \tag{21}$$

where:

$$\omega_l = \omega_0 + l\Omega, \qquad l;\, p \quad \text{integers} \tag{22}$$

$$\delta_{\omega_l} = \begin{cases} 1, & \text{for } \omega_l = 0, \\ 0, & \text{for } \omega_l \neq 0. \end{cases} \tag{23}$$

The longitudinal components of the "real" incident and scattered signals are obtained by adding to (20) and (21) their complex conjugates.

The eigenfunctions $\mathcal{L}_p(\rho)$ and $\mathcal{C}_p(\rho)$ of (21) are easily obtained via separation of variables (see e.g., [20]):

$$\mathcal{L}_p(\rho) = \begin{cases} 0 & \text{for } k_z = 0 \\ \operatorname{sign}(k_z) K_p(|k_z|\rho) & \text{for } k_z \neq 0 \end{cases} \tag{24}$$

$$\begin{aligned} \cos\alpha_l \mathcal{C}_p(\rho) &= (k_l\rho)^{-|p|} \qquad \text{for} \quad k_{tl} = 0 \\ \mathcal{C}_p(\rho) &= H_p^{(2)}(k_{tl}\rho) \qquad \text{for} \quad k_{tl} \neq 0 \end{aligned} \tag{25}$$

with

$$\begin{aligned} k_{tl} &= \sqrt{k_l^2 - k_z^2} \\ -\pi &< \angle k_{tl} \leq 0 \qquad \text{for} \quad \omega_l > 0 \\ -\pi &\leq \angle k_{tl} < 0 \qquad \text{for} \quad \omega_l < 0, \end{aligned} \tag{26}$$

and where K_p, $H_p^{(2)}$ are the modified Bessel function of the second kind and the Hankel function of the second type, respectively.

For $\omega_l < 0$ (and $k_{tl} \neq 0$), the functions $\mathcal{C}_p(\rho)$ are conveniently expressed in terms of the Hankel functions of the first type via analytic continuation:

$$\begin{aligned} H_p^{(2)}(k_{tl}\rho) &= -(-1)^p H_p^{(1)}(k_{tl}^{(1)}\rho) \\ k_{tl}^{(1)} &= -k_{tl}, \qquad 0 \leq \angle k_{tl}^{(1)} < \pi. \end{aligned} \tag{27}$$

Therefore, the eigenfunctions $\mathcal{C}_p(\rho)$ have been chosen so as to yield waves that, for real values of k_{tl}, originate from the z axis and propagate along $\hat{\rho}$. Similarly, the phase condition (26) is used to ensure that complex values of k_{tl} correspond to decaying waves.

In (21), the factor ω_l/ω_0 has been introduced for convenience; while the angular factor $\cos\alpha_l$ has been explicitly introduced on physical grounds, since, for oblique incidence (i.e., $\alpha_0 \neq 0$), the scattered propagating field is the sum of cylindrical waves whose vector wave numbers $\boldsymbol{k}_l$ are oriented according to the different angular directions α_l, because of the z-continuity condition (see the Appendix).

For $k_{tl} \neq 0$, the transverse dynamic fields are easily obtained from the longitudinal components, whereas for the static fields, separation of variables at $k_z = 0$ yields ($\rho \geq a$)

$$\begin{bmatrix} E_\rho^s \\ Z_0 H_\rho^s \end{bmatrix} = -\frac{1}{k_0\rho} \sum_{l=-\infty}^{+\infty} \sum_{p=-\infty}^{+\infty} \delta_{\omega_l} (-j)^p \cdot \begin{bmatrix} a_{l,p} \\ c_{l,p} \end{bmatrix} (k_0\rho)^{-|p|} \exp(jp\phi) \tag{28}$$

$$\begin{bmatrix} E_\phi^s \\ Z_0 H_\phi^s \end{bmatrix} = \frac{j}{k_0\rho} \sum_{l=-\infty}^{+\infty} \sum_{\substack{p=-\infty \\ p\neq 0}}^{+\infty} \delta_{\omega_l} (-j)^p \cdot \begin{bmatrix} a_{l,p} \\ c_{l,p} \end{bmatrix} \frac{p}{|p|} (k_0\rho)^{-|p|} \exp(jp\phi) \tag{29}$$

while, for $k_z \neq 0$, one has $(p \geq a)$

$$\begin{bmatrix} E_\rho^s \\ Z_0 H_\rho^s \end{bmatrix} = j\exp(-jk_z z) \sum_{l=-\infty}^{+\infty} \sum_{p=-\infty}^{+\infty} \delta_{\omega_l}(-j)^p \cdot \begin{bmatrix} a_{l,p} \\ c_{l,p} \end{bmatrix} K_p'(|k_z|\rho)\exp(jp\phi) \quad (30)$$

$$\begin{bmatrix} E_\phi^s \\ Z_0 H_\phi^s \end{bmatrix} = -\frac{\exp(-jk_z z)}{|k_z|\rho} \sum_{l=-\infty}^{+\infty} \sum_{p=-\infty}^{+\infty} \delta_{\omega_l}(-j)^p \cdot \begin{bmatrix} a_{l,p} \\ c_{l,p} \end{bmatrix} pK_p(|k_z|\rho)\exp(jp\phi). \quad (31)$$

Note that (21), (24), and (28)–(31) can model only scattered fields with static components given by the gradient of a scalar single-valued magnetic and/or electric potential.

C. Field on γ' in the Comoving Frame

The incident and scattered fields in $\mathcal{R}$ are obtained by using (1) and (11) and the previous results of this section. After some algebraic manipulations, the following results for k_z, $k_{tl} \neq 0$ are obtained $(\rho = a)$:

$$\begin{bmatrix} e_z^i \\ Z_0 h_z^i \end{bmatrix} = \sum_{m=-\infty}^{+\infty} \psi_m \left\{ \left(\frac{\omega_m}{\omega_0} - n_{z0}^2 \right) \frac{J_m(k_{t0}\rho)}{n_{t0}} \cdot \begin{bmatrix} \cos\delta \\ \sin\delta \end{bmatrix} + j \begin{bmatrix} \sin\delta \\ -\cos\delta \end{bmatrix} \cdot \frac{\Omega}{\omega_0} k_z \rho J_m'(k_{t0}\rho) \right\} \quad (32)$$

$$\begin{bmatrix} e_\theta^i \\ Z_0 h_\theta^i \end{bmatrix} = \sum_{m=-\infty}^{+\infty} \psi_m \left\{ \frac{m n_{z0}}{k_{t0}\rho} J_m(k_{t0}\rho) \begin{bmatrix} \cos\delta \\ \sin\delta \end{bmatrix} + j \begin{bmatrix} \sin\delta \\ -\cos\delta \end{bmatrix} J_m'(k_{t0}\rho) \right\} \quad (33)$$

$$\begin{bmatrix} e_z^s \\ Z_0 h_z^s \end{bmatrix} = \sum_{l=-\infty}^{+\infty} \sum_{p=-\infty}^{+\infty} \psi_p \Big\{ \delta_{\omega_l} \cdot \left[\boldsymbol{U}_{l,p}\operatorname{sign}(k_z) K_p(|k_z|\rho) - j\tilde{\boldsymbol{U}}_{l,p} \frac{\Omega}{\omega_0} k_0 \rho K_p'(|k_z|\rho) \right] + (1-\delta_{\omega_l}) \left[\boldsymbol{U}_{l,p} \left(\frac{\omega_m}{\omega_0} - \frac{\omega_0}{\omega_l} n_{z0}^2 \right) \frac{H_p^{(2)}(k_{tl}\rho)}{n_{tl}} + j\tilde{\boldsymbol{U}}_{l,p} \frac{\Omega}{\omega_0} k_z \rho H_p^{(2)'}(k_{tl}\rho) \right] \Big\} \quad (34)$$

$$\begin{bmatrix} e_\theta^s \\ Z_0 h_\theta^s \end{bmatrix} = \sum_{l=-\infty}^{+\infty} \sum_{p=-\infty}^{+\infty} \psi_p \Big\{ -\delta_{\omega_l} \boldsymbol{U}_{l,p} \frac{pK_p(|k_z|\rho)}{|k_z|\rho} + (1-\delta_{\omega_l}) \left[\boldsymbol{U}_{l,p} \frac{p n_{z0}}{k_{tl}\rho} H_p^{(2)}(k_{tl}\rho) + j\frac{\omega_l}{\omega_0} \tilde{\boldsymbol{U}}_{l,p} H_p^{(2)'}(k_{tl}\rho) \right] \Big\}, \quad (35)$$

where:

$$\boldsymbol{U}_{l,p} = \begin{bmatrix} a_{l,p} \\ c_{l,p} \end{bmatrix}, \qquad \tilde{\boldsymbol{U}}_{l,p} = \begin{bmatrix} c_{l,p} \\ -a_{l,p} \end{bmatrix}, \quad (36)$$

$$\omega_m = \omega_0 + m\Omega, \quad (37)$$

$$m = p + l, \quad (38)$$

$$\psi_q = (-j)^q \exp(j\omega_m t)\exp[j(q\theta - k_z z)]. \quad (39)$$

Equations (32)–(39) immediately prove the assumption on which (21) is based; that is, the scattered field in $\mathcal{S}$ has all the frequency components $\omega_l = \omega_0 + l\Omega$, $(l = -\infty, +\infty)$.

Thus, in the reference frame $\mathcal{R}$, the total field has an infinite discrete spectrum of frequencies which, in principle, may include a static component $(\omega_m = 0)$ only when the cylinder is rotating at stroboscopic velocities, for which the "stroboscopic ratio" ω_0/Ω is an integer. However, the condition $|\beta| \ll 1$ (on which (32)–(39) are based) implies that static components can hardly be observed in $\mathcal{R}$ and in $\mathcal{S}$. In fact, at stroboscopic velocities, $k_0 a$ can be written as follows:

$$k_0 a = \frac{\omega_0}{\Omega} \beta_a = -m_s \beta_a, \quad (40)$$

where $m_s = -\omega_0/\Omega$ is the value of the index m required by (37) to have $\omega_m = 0$, whereas β_a is small by assumption (at most, $|\beta_a|$ is of the order of 10^{-2}). When $k_{t0}a$ is fixed, the incident spectrum is limited by the amplitude factors $J_m(k_{t0}a)$ and $J_m'(k_{t0}a)$, which tend to zero for large values of m, provided that the condition $|m| > k_{t0}a$ is satisfied. As a rule of thumb, it is usually sufficient to consider terms up to order $m_{\max} \approx 2k_{t0}a \leq 2k_0 a$ to guarantee the accurate evaluation of the series (20), (32), and (33). This same condition, with $p+l = m$, applies to the series representations of the scattered fields (21), (34), and (35), since these series are also evaluated at $\rho = a$. Hence, for $2k_0 a \approx m_{\max} \geq 1$, (40) yields

$$|m_s| \simeq \frac{m_{\max}}{2|\beta_a|}, \qquad |m_s| > m_{\max} \text{ for } |\beta_a| < \frac{1}{2}. \quad (41)$$

Equations (41) and (38), where p and l may have opposite signs, prove that no significant static component is observed in $\mathcal{R}$ and in $\mathcal{S}$, provided that the condition $|\beta_a| < 1/10$ is satisfied. In fact, to obtain $\omega_l = 0$ at $l = l_s = m - p_s$, the indexes p_s and m have to satisfy the stroboscopic condition $(p_s - m) = \omega_0/\Omega$ or $p_s \simeq m + m_{\max}/(2\beta_a)$, which yields $|p_s|$, $|l_s| > 3m_{\max}$ for $|\beta_a| < 10^{-1}$; whereas $|m_s|$, $|p_s|$, $|l_s| > 48m_{\max}$ for $|\beta_a| < 10^{-2}$.

Obviously, the previous argument fails in the particular case when $|m_s| = m_{\max}$, which may occur only when dealing with $k_0 a < 1/2$ (where one may have to work with $m_{\max} = 1$, 2). In these cases, if $\omega_0 = \Omega$, 2Ω, one has to consider also the possibility of observing a static component both in $\mathcal{R}$ and in $\mathcal{S}$. In order to be able to exclude the presence of static components, in the following we will always consider $|\omega_0| \gg |\Omega|$, thereby restricting our attention to cases where the ratio $|\Omega/\omega_0|$ is small.

We may now observe that the conditions $|\beta| \ll 1$, $|\Omega/\omega_0| \ll 1$ imply that the $\omega > 0$ part of the spectrum of the longitudinal components of the real scattered signals (numerically) coincides with the expression reported in (21), since, under these conditions, all the coefficients of (21) related to $\omega_l \leq 0$ are (numerically) negligible.

D. Simplifications at Normal Incidence

At $k_z = 0$ and for $|\beta| \ll 1$, $|\Omega/\omega_0| \ll 1$, the longitudinal field components on γ' simplify as follows ($\rho = a$):

$$\begin{bmatrix} e_z^i \\ Z_0 h_z^i \end{bmatrix} = \begin{bmatrix} \cos\delta \\ \sin\delta \end{bmatrix} \sum_{m=-\infty}^{+\infty} (-j)^m \cdot \exp(j\omega_m t)\exp(jm\theta)\frac{\omega_m}{\omega_0} J_m(k_0\rho) \quad (42)$$

$$\begin{bmatrix} e_z^s \\ Z_0 h_z^s \end{bmatrix} = \sum_{l=-\infty}^{+\infty} \sum_{p=-\infty}^{+\infty} (-j)^p \cdot \exp(j\omega_m t)\exp(jp\theta)\frac{\omega_m}{\omega_0} \begin{bmatrix} a_{l,p} \\ c_{l,p} \end{bmatrix} H_p^{(2)}(k_l\rho), \quad (43)$$

where the conditions $\omega_m > 0$, $\omega_l > 0$ are understood. The azimuthal field components on γ' are not reported since, at normal incidence and in the absence of static components, one may deal with only the z components of the field in $\mathcal{R}$. The continuity of the azimuthal components of γ' is guaranteed by enforcing the continuity of the derivative of e_z and h_z with respect to ρ.

By separate consideration of each harmonic ω_m (i.e., by considering all the components of the scattered field with $\omega = \omega_m$), one notes that the angular factors in (42) and (43) have to be equal for a rotationally symmetric problem (e.g., a homogeneous scatterer with circular cross section rotating around its axis):

$$\left.\begin{aligned} \exp(jm\theta) &= \exp(jp\theta) \\ m &= p + l \end{aligned}\right\} \Rightarrow m = p \Rightarrow l = 0. \quad (44)$$

In such a case, one has only a single summation in (21), so that the scattered field in $\mathcal{S}$ is monochromatic with frequency ω_0.

IV. The Bistatic Characteristics of a Cylinder at Rest as Solution of One Galileian Relativistic Problem

Any approach to the motional problem, however poorly it may model the Doppler frequency shift phenomenon, agrees with the relativistic approach when the velocity of the (rigid) body goes to zero in the limit. Therefore, it should be possible to extract from the relativistic results the information required to use the quasi-stationary method, i.e., the bistatic characteristics of the cylindrical structure at rest. In fact, when the quasi-stationary approximation is used, one has to illuminate the scatterer from different directions of incidence, according to the positions that the rotating cylinder takes at the different observation times. For simplicity, we will restrict ourselves to normal incidence where duality makes it possible to discuss the TM_z-incidence case (see (6), (16), and (17)) by considering only the longitudinal component of the electric field (i.e., by setting $\delta = 0$ in (20) and (42)).

For an incident TM_z wave, normally propagating in $\mathcal{S}$ with a time dependence factor $\exp(j\omega_0 t)$ in the angular direction ϕ^i one has

$$E_z^i(\phi, \phi^i) = \exp(j\omega_0 t) \sum_{m=-\infty}^{+\infty} (-j)^m \cdot \exp[jm(\phi - \phi^i)] J_m(k_0\rho) \quad (45)$$

$$e_z^i(\theta, \phi^i) = \sum_{m=-\infty}^{+\infty} (-j)^m \exp(j\omega_m t) \cdot \exp[jm(\theta - \phi^i)]\frac{\omega_m}{\omega_0} J_m(k_0\rho), \quad (46)$$

which, for $\phi^i = 0$ (i.e., for a wave propagating in the positive x direction), yields

$$e_z^i(\theta, \phi^i = 0) = \sum_{m=-\infty}^{+\infty} (-j)^m \exp(j\omega_m t) \cdot \exp(jm\theta)\frac{\omega_m}{\omega_0} J_m(k_0\rho). \quad (47)$$

The previous formulas show that by phase-shifting each harmonic of the incident field in $\mathcal{R}$ one may simulate a variation from $\phi^i = 0$ to $\phi^i \neq 0$ of the incidence direction in $\mathcal{S}$. The phase delay or lead of each harmonic explicitly depends on the angular frequency ω_m and on the incidence angle $\phi^i (\neq 0)$, but not on the angular velocity Ω, which may assume any value, including $\Omega = 0$. Since the fundamental harmonic $\omega_m = \omega_0$ is never shifted and since no harmonic ought to be shifted for $\phi^i = 0$, the phase-shifting operation is the following:

$$\omega_m t|_{\phi^i = 0} \Rightarrow \omega_m t - m\phi^i. \quad (48)$$

For $\phi^i \neq 0$, the scattered field may be formally obtained by using (38), (43), and (48):

$$e_z^s(\theta, \phi^i) = \sum_{l=-\infty}^{+\infty} \exp(-jl\phi^i) \times \sum_{p=-\infty}^{+\infty} \exp(j\omega_m t)(-j)^p \exp[jp(\theta - \phi^i)] \cdot a_{l,p}\frac{\omega_m}{\omega_0} H_p^{(2)}(k_l\rho). \quad (49)$$

For any angular velocity Ω, the corresponding expression in $\mathcal{S}$ of the scattered field is then obtained by inspection of (21), (43), and (49):

$$E_z^s(\phi, \phi^i) = \sum_{l=-\infty}^{+\infty} \frac{\omega_l}{\omega_0} \exp(j\omega_l t)\exp(-jl\phi^i) \times \sum_{p=-\infty}^{+\infty} (-j)^p \exp[jp(\phi - \phi^i)] a_{l,p} H_p^{(2)}(k_l\rho). \quad (50)$$

As a matter of fact, (45), (50), and (46), (49) simply show how the rotating cylinder problem does not depend on the choice of ϕ^i made for the angular direction of incidence. In fact, the coefficients $a_{l,p}$ of (50) obtained at $\phi^i = 0$ allow one to express the scattered field in $\mathcal{S}$ for any given normal-incidence direction ϕ^i.

We consider now the evaluation of (50) in the limit as Ω goes to zero. By noticing that at $\Omega = 0$ one has $\omega_l = \omega_0$ and $k_l = k_0$, (50) yields

$$E^s_{z\,\Omega=0}(\phi, \phi^i) = \exp(j\omega_0 t) \sum_{p=-\infty}^{+\infty} (-j)^p \exp(jp\phi)\overline{a_p} H_p^{(2)}(k_0\rho) \tag{51}$$

$$\exp(jp\phi^i)\overline{a_p}(\phi^i) = \sum_{l=-\infty}^{+\infty} a_{l,p} \exp(-jl\phi^i), \tag{52}$$

where the coefficients $a_{l,p}$ in (52) are those valid for $\phi^i = 0$ and in the limit of $\Omega = 0$.

By noticing the classical (Galilelian) nature of the model (6) at $\Omega = 0$, the previous result shows how the bistatic problem for a cylinder at rest is a single problem, in the sense that it can be posed and solved at once for any ϕ-incidence direction. In fact, (51) and (52) prove that only a numerically finite double set of coefficients (i.e., the $\{a_{l,p}\}$ coefficient set at $\Omega = 0$) is required to express the bistatic cross sections of a cylinder at rest. The factor $\exp(+jp\phi^i)$ on the left-hand side of (52) reduces to baseband each coefficient $\overline{a_p}(\phi^i)$; then, for a given p, the coefficients $a_{l,p}$ are provided by the Fourier expansion (52).[3] Equations (51) and (52) permit the quasi-stationary method to be improved. In fact, to study at normal incidence a cylinder rotating at a low angular velocity Ω one may now use the quasi-stationary method and approximately reconstruct the Doppler frequency shift via (52), which provides the law to pass from the bistatic characteristics of the scatterer at rest to the $a_{l,p}$ coefficients, each one being associated with a different Doppler frequency component of the scattered field (cf. eq. (21)).

V. Numerical Algorithm for Normal TM$_z$ Incidence

The numerical algorithm we have implemented is a generalization of the one discussed in [21] and [22]. The internal problem (6) to find the total field $e_{z,m}$ of the mth harmonic is solved by application of the finite element method, by using linear elements defined on a conforming triangular mesh with n nodes [14]. The mesh covers the domain Σ bounded by the piecewise-linear approximation γ'_a of the circular contour γ' (see Fig. 1(c)). To each node of the triangular mesh corresponds one unknown and the numerical solution e_a is a piecewise-linear approximation of the true solution $e_{z,m}$:

$$e_{z,m} \approx e_a = \sum_{i=1}^{n} e_i f_i, \tag{53}$$

where $f_i (i = 1, n)$ are the well-known pyramidal base functions and e_i the corresponding complex expansion coefficients.

For each harmonic ω_m, we implemented with FEM a weak functional for the internal problem, since we used the Green theorem to reduce the Laplace operator appearing in (6):

$$\sum_{i=1}^{n} e_i \Big\{ \iint_\Sigma \nabla w_j \cdot \nabla f_i d\Sigma - k_m^2 \iint_\Sigma w_j N^2 f_i d\Sigma - 2j\delta_{\text{vb}} k_m k_0 \frac{\Omega}{\omega_0} \iint_\Sigma w_j \frac{\partial f_i}{\partial \theta} d\Sigma \Big\} = \sum_{i=1}^{n} e_i \int_{\gamma'_a} \frac{\partial f_i}{\partial n} w_j d\gamma'_a, \qquad (j = 1, n). \tag{54}$$

In (54), $N = 1$ in region 2 (i.e., in the shell region between γ_a and γ'_a, see Fig. 1(c)), w_j stands for the generic jth testing function, while $(\partial/\partial n)$ is the outward normal derivative taken on the boundary γ'_a of Σ. In the numerical implementation, a Galerkin testing procedure has been used and $w_j = f_j (j = 1, n)$.

The internal problem (54) is coupled to the external one by use of (43) and (47) to express the total field on γ', while the continuity on γ' of the $h_{\theta,m}$ component is obtained by enforcing the continuity of the derivative of $e_{z,m}$ with respect to ρ:

$$e_{z,m} = e^i_{z,m} + e^s_{z,m} \approx e_a \quad \text{on} \quad \gamma'$$
$$\frac{\partial e_{z,m}}{\partial \rho} = \frac{\partial e^i_{z,m}}{\partial \rho} + \frac{\partial e^s_{z,m}}{\partial \rho} \approx \frac{\partial e_a}{\partial \rho} \quad \text{on} \quad \gamma'. \tag{55}$$

Therefore, the source term (i.e., the one related to the incident field) is "active"' only on the contour γ' and it appears in the numerical solution of the partial differential equation only via the boundary conditions (55).

On the inner contour γ_a, the Neumann boundary condition for the continuity of the tangential component of the magnetic field can be enforced only when expansion functions of order higher than the first are used. Since we used linear elements, the Neumann condition on γ_a cannot be strictly enforced (cf. [23]). Then, in order to reduce the errors in the vicinity of the γ contour, we have to use a sufficiently dense mesh in the transition region where the refractive index N is discontinuous. The Dirichlet boundary condition on the inner contour γ_a are automatically satisfied because of the use of a conforming mesh. Since no special attention is required to deal with *internal* discontinuities of the refractive index, the implemented numerical code can deal with transversally piecewise-homogeneous cylindrical scatterers.

On the contour of perfectly conducting regions the total $e_{z,m}$ is zero and there is no need to locate nodes inside the perfectly conducting regions. For the nodes located on perfectly conducting boundaries, the finite element equation is simply substituted by the Dirichlet condition $e_{z,m} = 0$.

The Neumann condition on γ'_a are "explicitly" enforced in our numerical solution by using the series expansions (43) and (47) to rewrite the contour integral of (54) in the following form:

$$\sum_{i=1}^{n} e_i \int_{\gamma'_a} \frac{\partial f_i}{\partial n} w_j d\gamma'_a = \int_{\gamma'_a} \frac{\partial e_a}{\partial n} w_j d\gamma'_a \approx \int_{\gamma'_a} \frac{\partial (e^i_{z,m} + e^s_{z,m})}{\partial \rho} w_j d\gamma'_a. \tag{56}$$

[3] In other words, (51) and (52) can be regarded as the key expressions of a sampling procedure to approximate, from a discrete and finite set of samples of the scattered field (at different ϕ^i and ϕ), the expression of the scattered field at any point $\rho \geq a$ and relative to an arbitrary choice of the ϕ-incidence direction.

The Dirichlet condition on γ'_a are used to match the $(2M+1)$ unknown coefficients $a_{l,p}(p=-M,M;\ l=m-p)$ considered in (43) and in the contour integral (56). In fact, since we write n FEM equations (one for each unknown nodal coefficient e_i), we have to write also $(2M+1)$ Dirichlet equations on γ'_a in order to match for the remaining $(2M+1)$ unknowns. In order to allow for arbitrary M, the Dirichlet conditions at $\rho=a$ are imposed in an integral sense by evaluating, on the contour γ', $(2M+1)$ Fourier momenta of (43) plus (47) and equating them to those computed on γ'_a by using the FEM piecewise-linear approximation:[4]

$$\left[\frac{j^p}{2\pi a}\frac{\omega_o}{\omega_m}\sum_i e_i\int_{\gamma'_a} f_i\exp(-jp\theta)\,d\gamma'_a\right] - a_{l,p}H_p^{(2)}(k_la)\approx \delta_l J_m(k_oa)$$
$$l=m-p\ (p=-M,\,M:\,M>|m|),\qquad (57)$$

with:

$$\delta_l=\begin{cases}1 & \text{for } p=m,\\ 0 & \text{for } p\neq m.\end{cases}\qquad (58)$$

Equation (58) shows that the right-hand side of (57) is different from zero only for $p=m$. In order to have a good representation of the total field both in $\mathcal{R}$ and in $\mathcal{S}$ and to guarantee numerically accurate results, the summation indexes m and p and the Doppler summation index l must vary on symmetric intervals (e.g., $l=-l_{\max},\,l_{\max}$). Thus, by taking into account the condition $m=p+l$, the highest value M of the index p required in the numerical solution process is at least equal to $l_{\max}$ plus the value $m_{\max}$ of the maximum order index m considered. Furthermore, under the assumption of small $|\Omega/\omega_o|$ ratios, one can see how the condition $l_{\max}\simeq m_{\max}$ can be regarded as a safer operating condition.

To enforce on γ'_a the Neumann and Dirichlet conditions one has to numerically integrate functions with a fast varying phase factor $\exp\,[j\varphi(s)]$ of the following form:

$$\mathcal{I}=\int_{s_1}^{s_2}\mathcal{M}(s)\exp\,[j\varphi(s)]w(s)\,ds.\qquad (59)$$

These intergrals are evaluated by a preintegration process, based on a linear approximation of the functions $\mathcal{M},\varphi$, and w, and which yields the following matrix expression for the integrals $\mathcal{I}$:

$$\mathcal{I}\simeq\frac{s_2-s_1}{2}\exp\,(j\varphi_{\rm cm})[\mathcal{M}_1\quad\mathcal{M}_2]\,\underline{\underline{Q}}\begin{bmatrix}w_1\\ w_2\end{bmatrix}$$
$$\varphi_{\rm cm}=\frac{\varphi_2+\varphi_1}{2}\qquad \varphi_d=\frac{\varphi_2-\varphi_1}{2},\qquad (60)$$

where the subscript 1 (or 2) is used to denote the values of the functions $\mathcal{M}$, φ, and w at $s=s_1$ (or at $s=s_2$), while the quadrature matrix $\underline{\underline{Q}}$ is as follows:

$$\underline{\underline{Q}}=\begin{bmatrix}(\operatorname{sinc}\varphi_d-\text{aux}) & (\operatorname{sinc}\varphi_d+\text{aux})\\ \text{aux} & -\text{aux}\end{bmatrix} - j\varphi_d\,\text{aux}\begin{bmatrix}1 & 0\\ 0 & -1\end{bmatrix},\qquad (61)$$

with

$$\text{aux}=\frac{\operatorname{sinc}\varphi_d-\cos\varphi_d}{\varphi_d^2}.\qquad (62)$$

TABLE I
SPARSE MATRICES OF THE SYSTEM (63)

Matrix	Dimensions	Characteristics	Relevant Eqs.
$F_{\rm em}$	n^2	sparse	l.h.s. of (54)
$\mathcal{N}$	$n\times(2M+1)$	i–th node $\notin\gamma'_a$ ⇓	r.h.s. of (54) &
$\mathcal{K_N}$	$n\times 1$	i–th row $=0$	r.h.s. of (56)
D	$(2M+1)^2$	diagonal	l.h.s. of (57)
$D_{\gamma'_a}$	$(2M+1)\times n$	i–th node $\notin\gamma'_a$ ⇓ i–th column $=0$	
$\mathcal{K}_D$	$(2M+1)\times 1$	only the ℓ–th row $\neq 0$	r.h.s. of (57)

The linear system of order $(n+2M+1)$ relative to the mth harmonic is obtained by writing first the n FEM equations (54) according to the nodal order of the testing functions, then the $(2M+1)$ Dirichlet equations (57) for increasing l. Similarly, the expansion coefficients e_i are ordered into the vector $\boldsymbol{\mathcal{E}}$ according to the nodal order of the expansion functions, whereas the unknown scattering coefficients $a_{l,p}$ are ordered for increasing l into the vector $\boldsymbol{\mathcal{A}}$. The following *sparse*-matrix system is then obtained:

$$\left[\begin{array}{c|c}\underline{\underline{F}}_{\rm em} & \underline{\underline{\mathcal{N}}}\\ \hline \underline{\underline{D}}_{\gamma'_a} & \underline{\underline{D}}\end{array}\right]\left[\begin{array}{c}\mathcal{E}\\ \hline \mathcal{A}\end{array}\right]=\left[\begin{array}{c}\mathcal{K}_N\\ \hline \mathcal{K}_D\end{array}\right].\qquad (63)$$

The characteristics of the *sparse* submatrices appearing in (63) are summarized in Table I.

Since the dimension of the FEM square submatrix is in general rather high (i.e., $n\gg(2M+1)$), it is convenient to use a block solution of the whole system (63) based on the inversion of the $\underline{\underline{F}}_{\rm em}$ sparse submatrix only. To this purpose the numbering of the mesh nodes is optimized so as to obtain $\underline{\underline{F}}_{\rm em}$ with the smallest possible bandwidth; then, only a square submatrix of the inverse matrix $\underline{\underline{F}}_{\rm em}^{-1}$ is computed, since only a few rows (columns) of $\underline{\underline{\mathcal{N}}}(\underline{\underline{D}}_{\gamma'_a})$ are different from zero (see Table I). The scattering vector $\boldsymbol{\mathcal{A}}$ is obtained by solving the the following square matrix system:

$$[\underline{\underline{D}}-\underline{\underline{D}}_{\gamma'_a}\underline{\underline{F}}_{\rm em}^{-1}\underline{\underline{\mathcal{N}}}]\boldsymbol{\mathcal{A}}=\mathcal{K}_D-\underline{\underline{D}}_{\gamma'_a}\underline{\underline{F}}_{\rm em}^{-1}\mathcal{K}_{\mathcal{N}}.\qquad (64)$$

[4] More rigorously, from a numerical standpoint, this technique to enforce the Dirichlet condition may be regarded as the weighted-integrals minimization of the Dirichlet equation (i.e., the first of equations (55)) obtained by using as testing functions the exponential functions $\exp(-jp\theta)$ $(0\leq\theta<2\pi;$ $p=-M,\,M)$.

Then, the nodal vector $\mathcal{E}$ is computed by solving the first n equations of (63).

Conversely, since $\underline{\underline{D}}$ is diagonal (see (57)) and can be inverted at once, the following different numerical approach to first obtain the nodal vector $\mathcal{E}$ is possible:

$$\begin{cases} \underline{\underline{\mathcal{F}}}\mathcal{E} = \mathcal{K}_{\mathcal{N}} - \underline{\underline{\mathcal{N}}}\underline{\underline{D}}^{-1}\mathcal{K}_D \\ \underline{\underline{\mathcal{F}}} = \underline{\underline{F}}_{\mathrm{em}} - \underline{\underline{\mathcal{N}}}\underline{\underline{D}}^{-1}\underline{\underline{D}}_{\gamma_a'}, \end{cases} \tag{65}$$

where $\underline{\underline{\mathcal{F}}}$ is an $n \times n$ sparse matrix. As far as computational efficiency is concerned, we cannot make, at present, any comparison between these two numerical approaches, since we have not yet implemented the procedure of (65).

VI. Numerical Results and Validation

To study the normal incidence problem we developed a FORTRAN 77 package which has been tested on metallic and dielectric arbitrarily shaped cylinders. In the following, r is the maximum radial dimension of each of the cylinders considered, as measured from its rotation axis. In the numerical model, each cylinder is surrounded by a shell centered on the rotation axis and consisting only of triangles located in region 2 (i.e., in free space in $\mathcal{S}$). The external radius of this shell is equal to a, while the ratio a/r is equal to 1.1 for all the cases considered.

The results reported here are related to seven different cylindrical structures whose cross-sectional meshes are shown in Figs. 3 and 4. To generate these meshes, to prepare the relative data bases and to represent the results by colored-flooded contour plots we used the PATRAN preprocessor and postprocessor programs.[5]

Fig. 3 shows the mesh used to study a dielectric circular cylinder rotating around its axis, while the meshes reported in Fig. 4 are relative to both metallic and dielectric noncircular cylinders. The meshes used to study the metallic cylinders are reported in the lower part of Fig. 4, while the upper part corresponds to the dielectric cases. Since no triangle is required inside the metallic parts of a structure, the γ_a contours of the cylinders considered are directly represented in the lower part of Fig. 4. Note that the shell meshing of Fig. 4 for a dielectric case is equal to the corresponding metallic case. Table II summarizes the mesh data relative to Figs. 3 and 4.

The first case considered in Fig. 4 (i.e., case a_1, a_2) is an elliptic cylinder whose minor semiaxis is one half of the major one. Case (b) has been obtained by the union of two such ellipses: a first one vertically oriented as in case (a), plus a second one arranged horizontally. Finally, case (c) is a propeller-shaped cylinder where the radius of the circular hub is $0.18r$ and where each blade has been obtained by application of the Joukowski transformation.

From the data reported in Table II, one may observe that we were able to use a relatively large number of nodes in the (free-space) shell without loosing any accuracy in the solution.

Validation of our FEM code has been performed by studying with the mesh of Fig. 3 a circular dielectric cylinder of

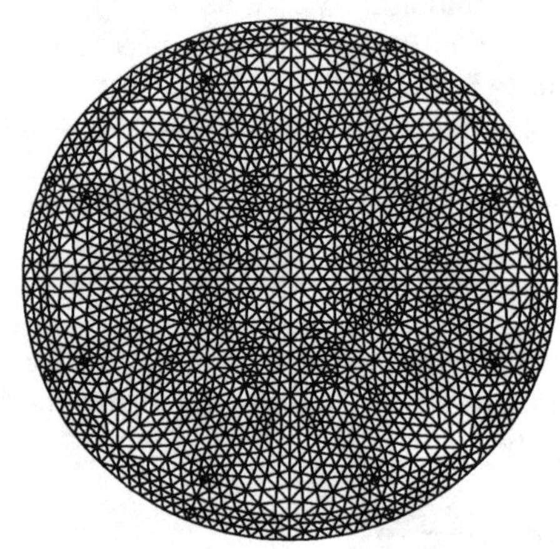

Fig. 3. Finite element mesh in the comoving frame for a dielectric circular cylinder. The mesh (external) radius is a, while the radius r of the cylinder is $r = a/1.1$.

radius r and relative electric permittivity ϵ_r, rotating around its center axis. Our FEM results have been compared with those obtained with the method described in [8]. Since this problem is rotationally symmetric, the scattered field in $\mathcal{S}$ has only the fundamental harmonic component ω_o. In fact, in this case and for $l \neq 0$, the coefficients $a_{l,p}$ obtained with our code have amplitudes below the numerical-noise level. In Fig. 5 we report the amplitude of the $a_{l,p}$ coefficients for $l = 0$, $p = m = -8, +8$, and relative to the three cylinders with $k_o r = 3$, $\epsilon_r = 3, 4$; $k_o r = 4$, $\epsilon_r = 4$. The solid-bar results refer to the nonrotating cylinders; the dashed-bar results refer to the rotating case $\beta_r = 0.03$, numerically studied by setting $\delta_{\mathrm{vb}} = 1$. From these results one may observe that rotation destroys the symmetry of the amplitude spectrum of the $a_{0,p}$ coefficients. By taking as reference the $\tilde{a}_{0,p}$ coefficients obtained from [8], in parts a_2, b_2, and c_2 of Fig. 5 we report the magnitude of the absolute errors $|a_{0,p} - \tilde{a}_{0,p}|$ for the three cases considered.

To better appreciate the smallness of the errors of the FEM solutions and the differences of the two models ($\delta_{\mathrm{vb}} = 0, 1$) considered, in Fig. 6 we report the laboratory-frame scattered field patterns of the three circular cylinders previously considered in Fig. 5. It should be noted that all the far-field patterns of this work report the amplitude of the coefficients $E^s_{z,l}(\phi) = (\omega_l/\omega_0)\sum_{p=-\infty}^{+\infty} a_{l,p}\exp(jp\phi)$ appearing in the far-field asymptotic expression of the scattered field $E^s_z(\rho, \phi)$:

$$E^s_z(\rho, \phi) \sim \sum_{l=-\infty}^{+\infty} \sqrt{\frac{2j}{\pi k_l \rho}} E^s_{z,l}(\phi) \exp\left[j(\omega_l t - k_l \rho)\right]. \tag{66}$$

In Fig. 6, the FEM results for $\delta_{\mathrm{vb}} = 1$ are reported by dashed lines, whereas dotted lines are used for the $\delta_{\mathrm{vb}} = 0$ results; the solid-line results refer to the method described in [8]. The results of Fig. 6(b) refer to $k_0 r = 3$, $\epsilon_r = 4$ and are in excellent agreement with those reported by Van Bladel in [2, fig. 10.13]. The effect of motion is more evident in the results of Fig. 6, where by comparison with the patterns relative to the stationary cylinders, one may see that rotation introduces a significant distortion of the scattering diagrams. However, one should keep in mind that this distortion is more evident in Fig.

[5]PATRAN is a software product of PDA Engineering, PATRAN Division, 2975 Redhill Avenue, Costa Mesa, CA 92626.

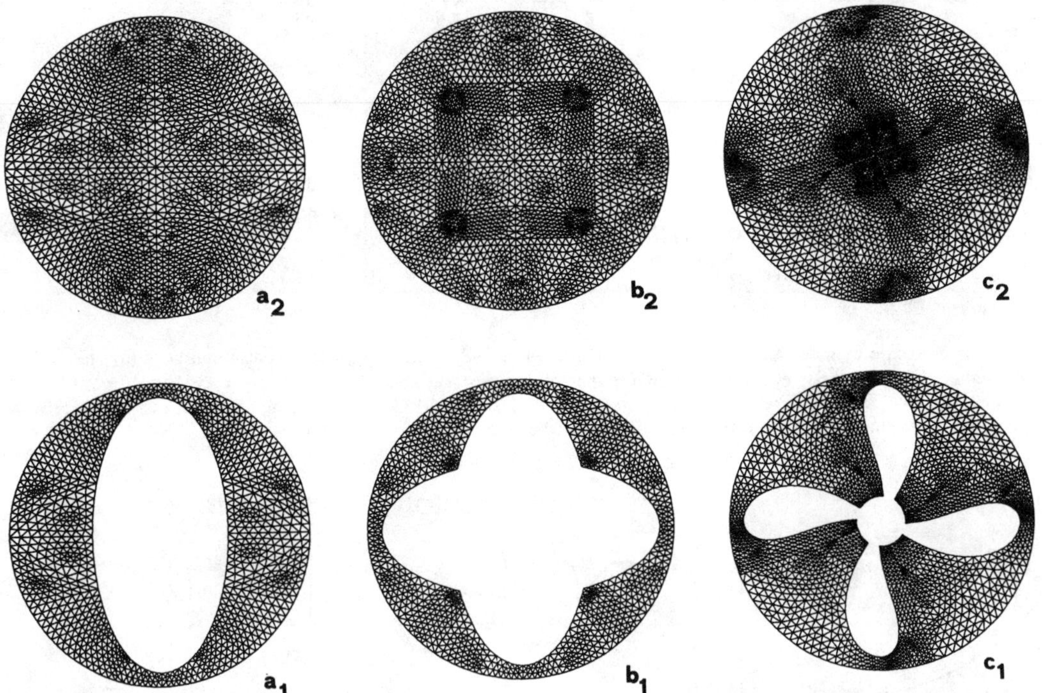

Fig. 4. Finite element mesh for dielectric (top) and metallic (bottom) noncircular cylinders. The shell meshing for the dielectric case is equal to the one of the corresponding metallic case: (a_1, a_2) elliptic cylinder; (b_1, b_2) crossed-ellipse cylinder; (c_1, c_2) propeller shaped cylinder.

TABLE II
MESH DATA RELATIVE TO THE MODELS OF FIGS. 3 AND 4

	Fig. 3	Fig. 4					
		a_1	a_2	b_1	b_2	c_1	c_2
B_w	70	26—56		23—70		45—85	
elements m_{el}	3744	1904—3448		1880—4640		3892—5644	
nodes $n = n_{ex} + n_{in}$	1937	1068—1789		1072—2389		2136—2893	
shell nodes n_{ex}	376	964		944		1896	
scatterer nodes n_{in}	1561	104—825		128—1445		240—997	
nodes on γ_a	112	104		128		240	
nodes on γ_a'	128	128		136		144	
$\approx m_{el}/n$ (%)	193	178—193		175—194		182—195	
$\approx n_{in}/n$ (%)	80.6	9.7—46.1		11.9—60.5		11.2—34.5	

B_w is the optimized semibandwidth of the F_{em} submatrix in (63), the total bandwidth B_{wt} being $B_{wt} = 2B_w - 1$.

6 because we have considered cylinders rotating at a very high speed (i.e., with $\beta_r = 0.03$); usually, the rotation velocities of interest are much smaller than this [1]. From the results of Fig. 6 one may also appreciate the importance of the $\partial/\partial\theta$ operation in the $\delta_{vb} = 1$ model (see eqs. (6)). In fact, in order to obtain with the β-linear model results significantly different from those of the Galileian model it is necessary to deal not only with a high β velocity parameter (e.g., of the order of 10^{-2}) but also with cases in which the variations of the total field with respect to θ are significant, i.e., for relatively large value of $k_0 r$ (as in Figs. 5(c_2) and 6(c_2)).

The results of Figs. 5 and 6 show that the errors of our FEM code increase for increasing electric permittivity and radius r, since the scatterer cell size in dielectric wavelength (i.e., C_s/λ_{d0}), for the mesh of Fig. 3, increases for increasing ϵ_r and r. Henceforth, λ_{d0} is the wavelength measured in the (motionless) material of the scatterer at the fundamental frequency ω_0. For example, the cell size in wavelength for the cylinder with $k_0 r = 4$, $\epsilon_r = 4$ is 1.54 times bigger than the one relative to the case $k_0 r = 3$, $\epsilon_r = 3$. Good results, as those of Fig. 6(b), were obtained with $C_s \approx \lambda_{d0}/16$, while the excellent results of Fig. 6(a) pertain to $C_s \approx \lambda_{d0}/19$. Thus, by using a mesh with cell size $\approx \lambda_{d0}/16$, one obtains very good numerical results. Note that, for a given accuracy, the cell size can be increased by using quadratic elements instead of linear ones [14].

The results of Figs. 7 and 8 are all relative to the noncircular cylinders of Fig. 4 with $k_0 r = 3$; for the dielectric cases we always considered $\epsilon_r = 3$, so that the biggest cell size inside the scatterers is of the order of $\lambda_{d0}/20$, to ensure numerical accuracy.

The results reported in Fig. 7 are relative to dielectric cylinders rotating with $\beta_r = 10^{-6}$ (i.e., $\Omega/\omega_0 = 1/3 \times 10^{-6}$),

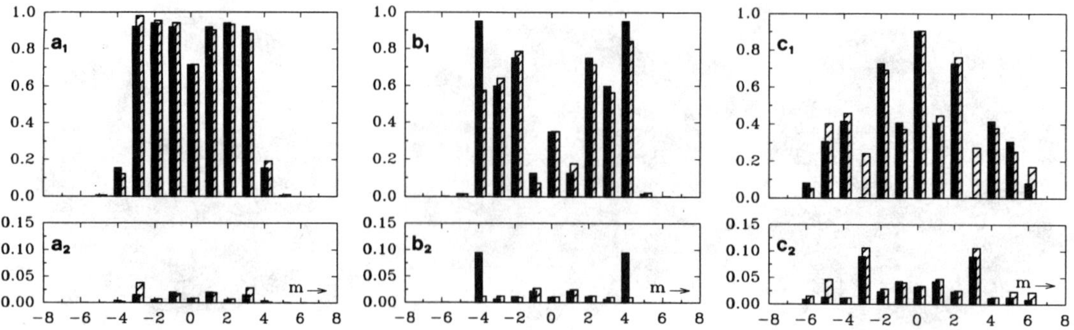

Fig. 5. Circular dielectric cylinder results for $l = 0$ and $p = m = -8, 8$. The $a_{l,p}$ coefficient amplitudes (a_1, b_1, c_1) and their respective absolute errors (a_2, b_2, c_2) are reported for the three cases: (a_1, a_2) $k_o r = 3$, $\epsilon_r = 3$, $\Omega/\omega_o = 10^{-2}$; ($b_1$, b_2) $k_o r = 3$, $\epsilon_r = 4$, $\Omega/\omega_o = 10^{-2}$; ($c_1$, c_2) $k_o r = 4$, $\epsilon_r = 4$, $\Omega/\omega_o = 0.75 \times 10^{-2}$. The solid-bar results are relative to the nonrotating case, while the dashed-bar results are relative to the rotating case $\beta_r = 0.03$.

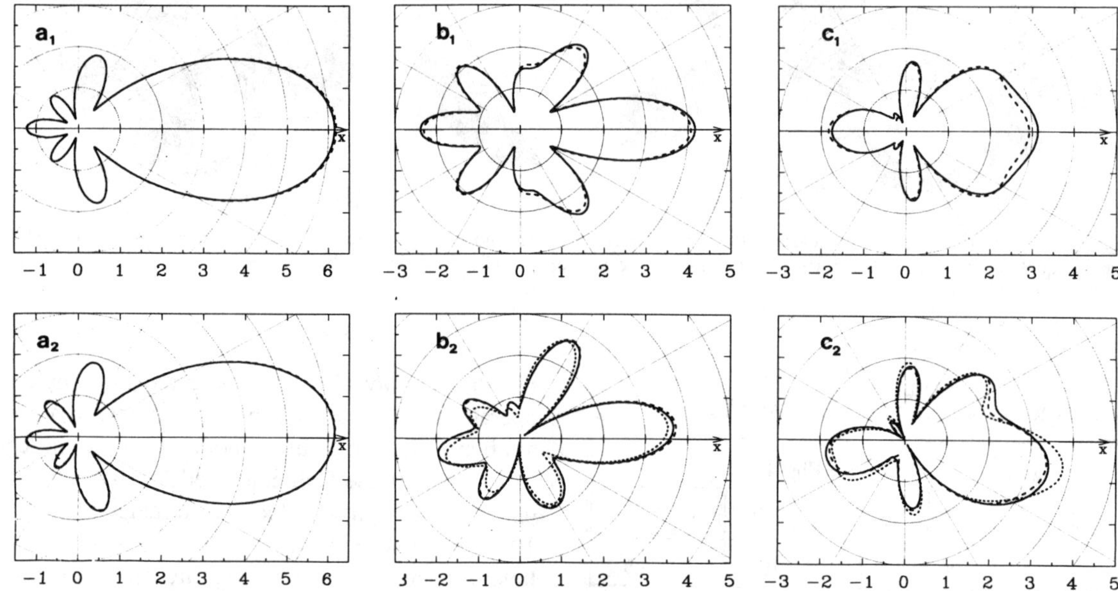

Fig. 6. Far-field patterns in the laboratory frame for circular dielectric cylinders with (a_1, a_2) $k_O r = 3$, $\epsilon_r = 3$; (b_1, b_2) $k_O r = 3$, $\epsilon_r = 4$; and (c_1, c_2) $k_O r = 4$, $\epsilon_r = 4$. The results a_1, b_1, and c_1 are relative to the nonrotating case while the results a_2, b_2, c_2 are relative to the rotating case $\beta_r = 0.03$. FEM results are reported by dashed ($\delta_{vb} = 1$) and dotted ($\delta_{vb} = 0$) lines; reference results by solid lines.

so that no difference between the $\delta_{vb} = 0$ and the $\delta_{vb} = 1$ model is detectable. The central part of this figure reports the laboratory-frame scattered field patterns at the fundamental frequency, ω_0, and at the Doppler frequencies ω_2 (i.e., for $l = 2$) and ω_4 ($l = 4$). From the results of Fig. 7 one may note how the excited Doppler harmonics $\omega_l = \omega_0 + l\Omega$ follow the index law $l = (2\pi/\Delta_\phi) \times q$ (with q an integer and $q = -\infty, +\infty$), where Δ_ϕ is the angular periodicity of the structure. In fact, for the elliptic cylinder ($\Delta_\phi = \pi$), the Doppler frequencies with odd values of l are not excited, while for the other two cylinders considered in Fig. 7 ($\Delta_\phi = \pi/2$) the excited Doppler frequencies correspond to values of l multiple of 4. It is also interesting to observe that the amplitude of the backscattered field at the fundamental harmonic, ω_o, may be equal to or lower than that relative to a Doppler harmonic (see the bottom panels in parts (a) and (c) of Fig. 7). Note also that the bar diagrams at the top of Fig. 7 reporting the amplitude of the $a_{l,p}$ coefficients are symmetric with respect to different values of the index p, when l varies. The amplitude of the $a_{l,p}$ coefficients relative to the nonexcited harmonics are numerically negligible, and therefore have not been reported.

The laboratory-frame scattered field patterns reported in Fig. 8 are relative to the nonrotating metallic and dielectric cylinders of Fig. 4, for the three different incidence directions $\phi^i = 0°$ (i.e., along the positive x axis), $\phi^i = 90°$, and $\phi^i = 225°$. To get these results we ran our FEM code with $\Omega = 0$ and for an incident plane wave propagating in the positive x direction; then, we used the computed $a_{l,p}$ coefficient set to reconstruct the bistatic characteristics of each cylinder according to the procedure described in Section IV. By comparing the patterns reported in Fig. 7 with those relative to the nonrotating dielectric cylinders given in Fig. 8 (top), one may appreciate the effects of rotation on the fundamental-harmonic pattern.

Our FEM computer code makes it possible to obtain in the scatterer comoving reference frame $\mathcal{R}$ the field behavior in the

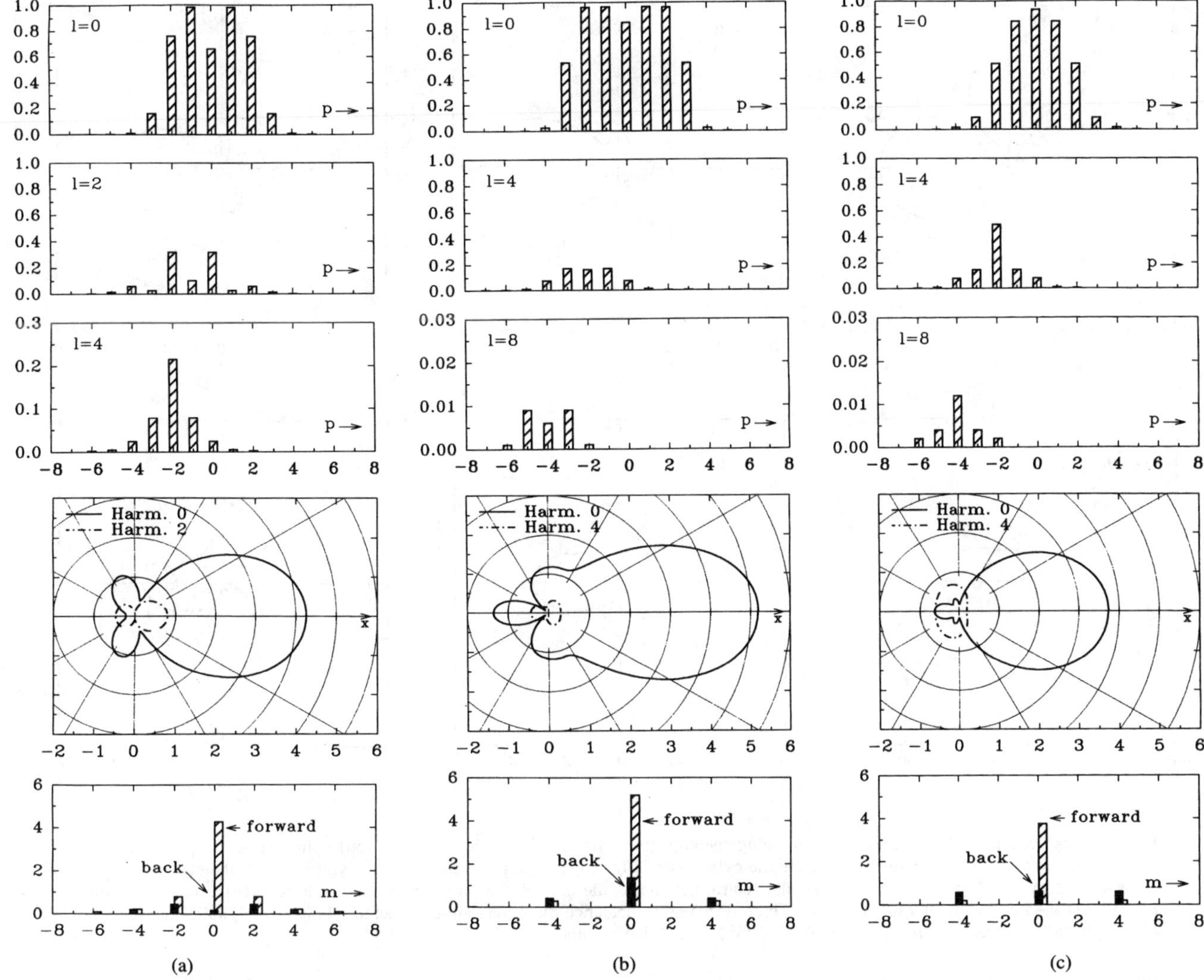

Fig. 7. Noncircular dielectric cylinders with $k_0 r = 3$ and $\epsilon_r = 3$ rotating at $\beta_r = 1 \times 10^{-6}$. The results for the elliptic cylinder of Fig. 4(a_2) are given in the left-hand column; the central column reports the results for the cylinder with a crossed ellipse cross section (Fig. 4(b_2); while the column on the right-hand side reports the results for the propeller-shaped cylinder of Fig. 4(c_2). The amplitude of the forward-scattering (dashed bars) and of the back-scattering (solid bars) $E^s_{z,l}$ coefficients versus the Doppler harmonic index l is represented at the bottom of the figure. The laboratory-frame scattered field patterns are given in the central part of the figure, while the top part gives the amplitude of the $a_{l,p}$ coefficients for $p = -8, 8$ and at three different values of the index l.

internal region. To this purpose, by coming back into the time domain, one may sum the fields $e_{z,m}$, computed by the FEM code and relative to the different harmonics, according to the following formula:

$$e_z(\rho, \theta, t) = \sum_{m=-\infty}^{+\infty} e_{z,m}(\rho, \theta) \exp(j\omega_m t)$$
$$= \exp(j\omega_0 t) \sum_{m=-\infty}^{+\infty} e_{z,m}(\rho, \theta) \exp(jm\Omega t). \quad (67)$$

The amplitude of the slowly varying field

$$\tilde{e}_z(\rho, \theta, t) = \sum_{m=-\infty}^{+\infty} e_{z,m}(\rho, \theta) \exp(jm\Omega t) \quad (68)$$

can then be graphically represented at different times.

In Fig. 9 we report, at six different evenly spaced instants, the colored-flooded contour plots of the $\tilde{e}_z$ amplitude in the propeller-shaped cylinder of Fig. 4(c_2). In this case the dielectric cylinder, with $\epsilon_r = 3$ and $k_0 r = 3$, is rotating clockwise with $\beta_r = -1 \times 10^{-6}$ (i.e., $\Omega/\omega_0 = -1/3 \times 10^{-6}$). In order to graphically represent the time variable, we drew our results in the laboratory-frame reference system by denoting with a star the north blade at $t = 0$; for the sake of clarity, the cross section of the rotating cylinder is also reported in the figure. In this connection, it is of importance to observe that one may assume the longitudinal component of the electric field in $\mathcal{S}$ to be approximately equal to the one in $\mathcal{R}$ only at very small rotation velocities since, in general, the longitudinal component of the electric field in $\mathcal{S}$ depends on the comoving field e_z as well as on $v_\mathcal{S} b_\rho$ (see (11) and (12)). The results of Fig. 9, obtained with $\delta_{\text{vb}} = 1$, coincide with those obtained with $\delta_{\text{vb}} = 0$. These results show that, under rotation, the

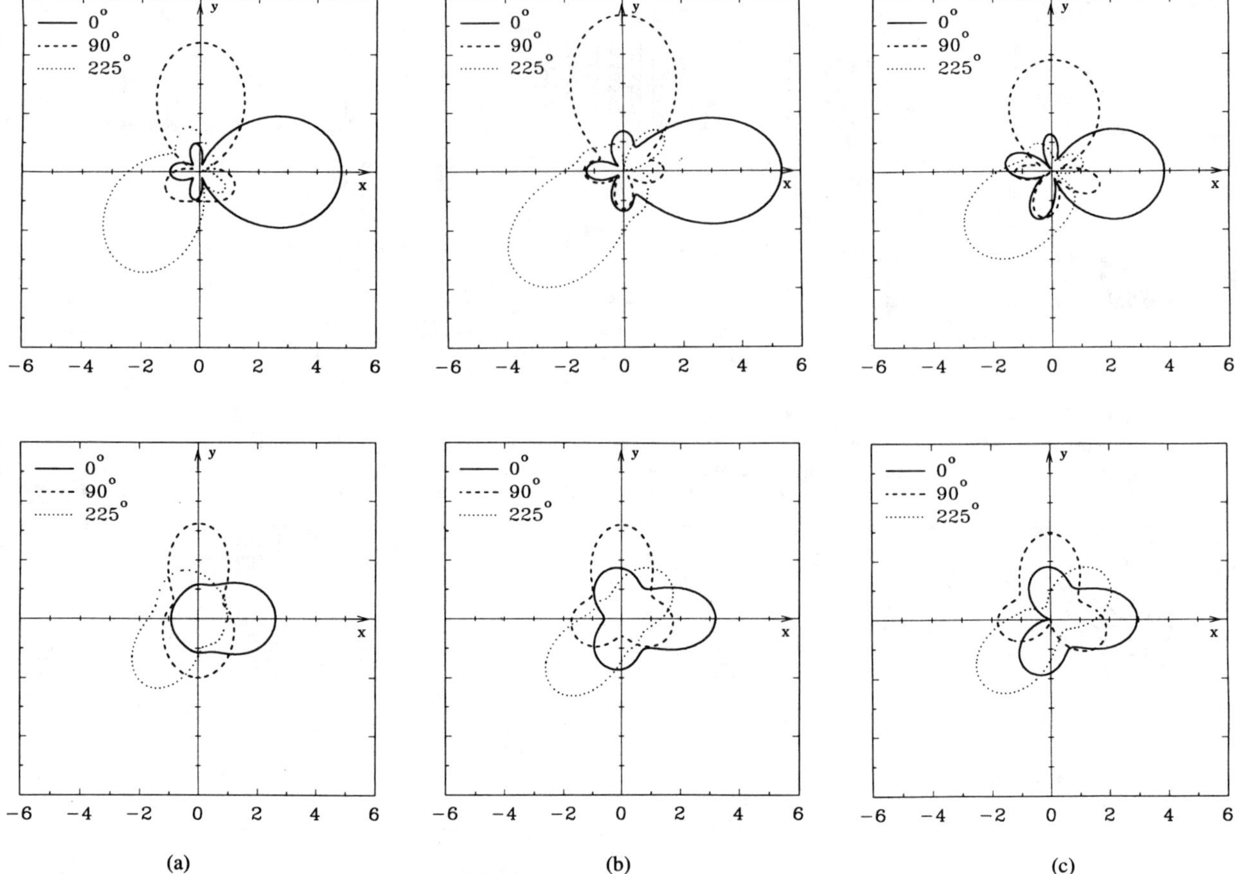

Fig. 8. Scattered far-field patterns for nonrotating metallic (bottom) and dielectric ($\epsilon_r = 3$, top) cylinders with $k_0 r = 3$. The left-hand column reports the results for the elliptic cylinders of Fig. 4 (a_1, a_2); the results for the cylinders with the crossed ellipse cross section of Fig. 4 (b_1, b_2) are given on the central column; while the column on the right-hand side reports the results for the propeller-shaped cylinders of Fig. 4(c_1, c_2). Three different ϕ-incidence directions are considered: (1) $\phi^i = 0°$, solid-line results; (2) $\phi^i = 90°$, dashed-line results; (3) $\phi^i = 225°$, dotted-line results.

maximum of the electric field amplitude in each blade is reached at the same point in $\mathcal{S}$ (*i.e.*, a given blade experiences the maximum field every $2\pi/\Omega$ s), whereas the field amplitude reaches a maximum in the hub of the propeller-shaped cylinder with a time periodicity equal to $\pi/(4\Omega)$.

The typical computation time required to run our code can be appreciated by considering that to obtain the results relative to the largest problem considered in this paper (e.g., the results of Fig. 9), 32 min. of CPU time were necessary on a VAX-9000.

VII. Conclusions

A differential approach to the electromagnetic problem of a rotating cylinder of arbitrary cross section illuminated by a plane wave at oblique incidence has been presented an discussed. The relevant partial differential equations, given in the comoving reference frame of the scatterer, allow the problem to be modeled on a limited, closed domain where the β velocity ratio is usually much smaller than unity. In our model, a flag variable (δ_{vb}) has been introduced to consider with the same formalism both the Galilean relativity model ($\delta_{vb} = 0$) and a heuristic linear model in β ($\delta_{vb} = 1$) that previously appeared in the literature. In the comoving frame, the boundary conditions were given on a circle γ' enclosing the cylindrical cross section of the scatterer in terms of a double series summation, one summation for each frequency component of the laboratory-frame Doppler spectrum. The conditions at which the scattered field may have a static field component have also been derived and discussed.

For normal incidence, a hybrid finite element computer code implementing our model has been presented and numerical comparisons between the Galileian and the β-linear model have been provided to show that Galileian relativity is suitable to model most practical applications.

Finally, a technique to reconstruct the Doppler frequency shift in the quasi-stationary method has been derived through a comparison at normal incidence between the quasi-stationary and the Galileian relativistic approach.

VIII. Appendix

This appendix is devoted to the study of the longitudinal components of the far-scattered fields in $\mathcal{S}$, thereby showing how the z-continuity condition implies different orientations for the vector wave numbers of the scattered cylindrical waves.

At oblique incidence, the scattered cylindrical wave propagating with angular frequency ω_0 is the only one related to

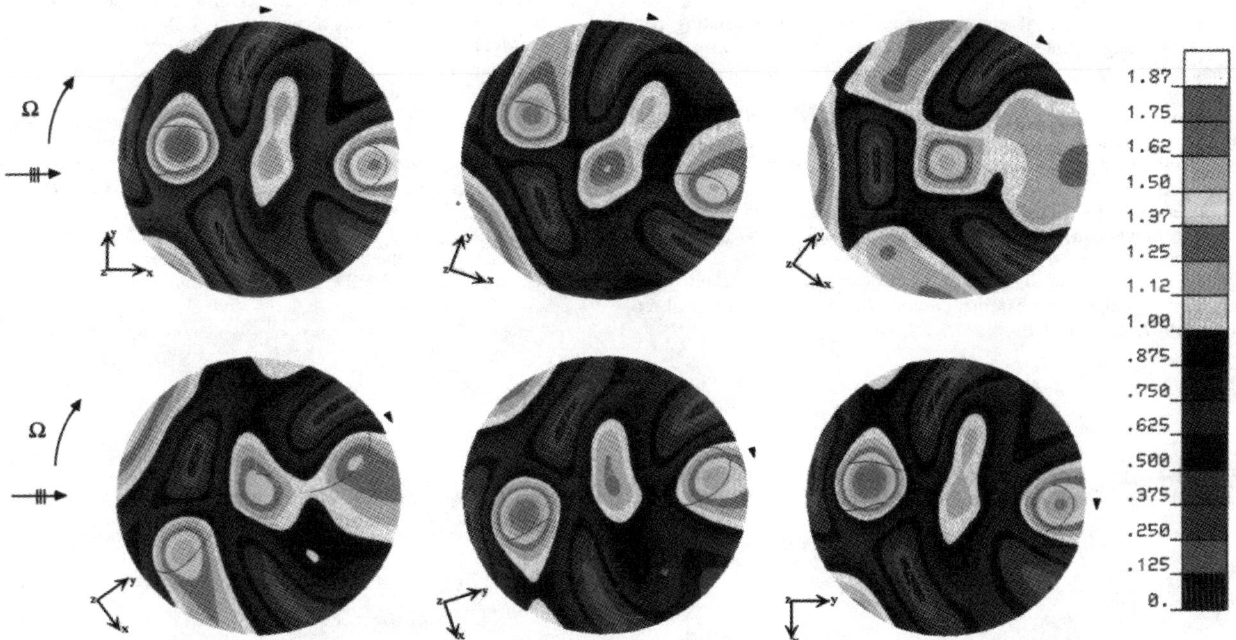

Fig. 9. Amplitude of the slowly varying total field $\tilde{e}_z$ in the comoving frame and in the internal region, at six different evenly spaced times, for the rotating propeller-shaped dielectric cylinder ($k_0 r = 3$, $\epsilon_r = 3$, $\beta_r = -1\times 10^{-6}$) of Fig. 4(c$_2$). To graphically represent the time variable, the results are drawn in the laboratory-frame reference system by denoting with a star the north blade at $t = 0$.

the incidence angle α_0. By retaining terms of the order $1/\sqrt{\rho}$, (15), (21), and (24)–(31) yield the following asymptotic form of the scattered field ($\rho \to \infty$):

$$\begin{bmatrix} \boldsymbol{E}^s \\ Z_0 \boldsymbol{H}^s \end{bmatrix} \sim (1+j) \sum_{l=-\infty}^{+\infty} \frac{\omega_l}{\omega_0} \frac{\exp(j\Psi_l)}{\sqrt{\pi k_{tl}\rho}} \cdot \left\{ (\hat{z}\cos\alpha_l - \hat{\rho}\sin\alpha_l) \begin{bmatrix} a_l \\ c_l \end{bmatrix} + \hat{\phi} \begin{bmatrix} c_l \\ -a_l \end{bmatrix} \right\}, \quad (69)$$

with

$$\Psi_l = \omega_l t - k_l(\rho\cos\alpha_l + z\sin\alpha_l) \quad (70)$$

$$\begin{bmatrix} a_l \\ c_l \end{bmatrix} = \sum_{p=-\infty}^{+\infty} \begin{bmatrix} a_{l,p} \\ c_{l,p} \end{bmatrix} \exp(jp\phi). \quad (71)$$

Thus, by considering (70), the following angular relationships are obtained:

$$-\frac{\pi}{2} < \alpha_o < \frac{\pi}{2} : \begin{cases} \omega_l \sin\alpha_l = \omega_o \sin\alpha_0 \\ |\omega_l|\cos\alpha_l = \omega_o \sqrt{\left(\frac{\omega_l}{\omega_o}\right)^2 - \sin^2\alpha_0} \end{cases} \quad (72)$$

$$\begin{cases} \alpha_l = \operatorname{sign}(\omega_l)\arctan\left[\frac{\sin\alpha_0}{\sqrt{(\omega_l/\omega_0)^2 - \sin^2\alpha_0}}\right] \\ \quad \text{for } \frac{|\omega_l|}{\omega_0} \geq |\sin\alpha_0| \\ \alpha_l = \operatorname{sign}(\alpha_0\omega_l)\left[\frac{\pi}{2} + j\operatorname{arccosh}\left(\frac{\omega_0|\sin\alpha_0|}{|\omega_l|}\right)\right] \\ \quad \text{for } \frac{|\omega_l|}{\omega_0} \leq |\sin\alpha_0|. \end{cases} \quad (73)$$

These relationships, together with (22), show that evanescent waves occur when the parameter Ωl is negative and $|\Omega l| > \omega_0(1 - |\sin\alpha_0|)$. In these instances, the relevant transverse wave number in free space is $k_{tl} = -j\sqrt{k_z^2 - k_l^2}$.

Acknowledgment

The authors would like to thank the various IEEE reviewers for their valuable comments. Among them, special thanks go to Prof. D. Censor, who suggested the comparison between the Galileian and the heuristic β-linear model. Helpful discussions with Prof. G. Ghione of the Politecnico di Torino are gratefully acknowledged.

References

[1] J. Van Bladel, "Electromagnetic fields in the presence of rotating bodies," *Proc. IEEE,* vol. 64, pp. 301–318, Mar. 1976.

[2] J. Van Bladel, *Relativity and Engineering.* Berlin: Springer-Verlag, 1984.

[3] D. Censor, "Theory of the Doppler effect: Fact, fiction and approximation," *Radio Sci.* vol. 19, pp. 1027–1040, July 1984.

[4] C. W. Chuang, "Backscatter of a large rotating conducting cylinder of arbitrary cross section," *IEEE Trans. Antennas Propagat.,* vol. 27, pp. 92–95, Jan 1979.

[5] I. J. Lahaie and D. L. Sengupta, "Scattering of electromagnetic waves by a slowly rotating rectangular metal plate," *IEEE Trans. Antennas Propagat.,* vol. 27, pp. 40–46, Jan. 1979.

[6] J. Van Bladel, "Relativistic theory of rotating disks," *Proc. IEEE,* vol. 61, pp. 260–267, Mar. 1973.

[7] D. De Zutter, "Scattering by a rotating circular cylinder with finite conductivity," *IEEE Trans. Antennas Propagat.,* vol. 31, pp. 167–169, 1983.

[8] G. Pelosi, E. Borchi, and A. Freni, "Scattering by rotating dielectric cylinder—Oblique incidence case," *J. Electromagnetic Wave and Appl.* vol. 4, no. 3, pp. 201–211, 1990.

[9] J. Van Bladel, "Rotating dielectric sphere in a low-frequency field," *Proc. IEEE,* vol. 67, pp. 1454–1455, 1979.

[10] D. De Zutter, "Scattering by a rotating dielectric sphere," *IEEE Trans. Antennas Propagat.,* vol. 28, pp. 643–651, 1980.

[11] D. De Zutter and D. Goethals, "Scattering by a rotating conducting sphere," *IEEE Trans. Antennas Propagat.,* vol. 32, pp. 95–98, 1984.

[12] R. D. Graglia and V. Daniele, "Integral equation formulation for moving media with applications," in *Proc. PIERS Con.* (Boston, MA, July 1989, p. 303.

[13] D. Censor and D. M. Le Vine, "The Doppler effect: Now you see it, now you don't," *J. Math. Phys.,* vol. 25, no. 2, pp. 309–316, Feb. 1984.

[14] O. C. Zienkiewicz, *The Finite Element Method,* 3rd ed. New York: McGraw-Hill, 1977.

[15] K. Tanaka, "Scattering of electromagnetic waves by a rotating perfectly conducting cylinder with arbitrary cross section: Point-matching method," *IEEE Trans. Antennas Propagat.,* vol. 28, pp. 796–803, Nov. 1980.

[16] F. Harfoush, A. Taflove, and G. A. Kriegsmann, "A numerical technique for analyzing electromagnetic wave scattering from moving surfaces in one and two dimensions," *IEEE Trans. Antennas Propagat.,* vol. 37. pp. 55–63, Jan 1989.

[17] D. Rodger, H. C. Lai, and P. J. Leonard, "Coupled elements for problems involving movement," *IEEE Trans Magn.,* vol. 26, pp. 548–550, Mar. 1990.

[18] F. Hecht and A. Marrocco, "A finite element simulation of an alternator connected to a non linear external circuit," *IEEE Trans. Magn.,* vol. 26, pp. 964–967, Mar. 1990.

[19] R. D. Graglia, G. Pelosi, A. Freni and R. Coccioli, "Electromagnetic interaction with a rotating cylinder of arbitrary cross section," in *Proc. IEEE AP-S Int. Symp.,* (Chicago, IL), vol. IV, July 1992, pp. 1853–1856.

[20] P. Moon and D. E. Spencer, *Field Theory Handbook.* New York: Springer, 1971.

[21] K. K. Mei, "Unimoment method of solving antenna and scattering problems," *IEEE Trans. Antennas Propagat.,* vol. 22, pp. 760–766, Nov. 1974.

[22] S. K. Chang and K. K. Mei, "Application of the unimoment method to electromagnetic scattering of dielectric cylinders," *IEEE Trans. Antennas Propagat.,* vol. 24, pp. 35–42, Jan. 1976.

[23] R. L. Ferrari and R. L. Naidu, "Finite-element modelling of high-frequency electromagnetic problems with material discontinuities," *Proc. Inst. Elec. Eng.* vol. 137, A, pp. 313–320, Nov. 1990.

Hybrid technique for analysing scattering from periodic structures

G. Pelosi
A. Freni
R. Coccioli

Indexing terms: Electromagnetic waves, periodic structures

Abstract: A hybrid technique which utilises the finite element method (FEM) and an expansion of the field in terms of the Floquet modes is presented for the analysis of the scattering from and through periodic structures, having arbitrary geometry, material properties and periodicity with respect to the wavelength. This formulation can play an important role in validating approximate methods and in the analysis of the interaction of an electromagnetic wave with complex periodic structures. The periodic boundary conditions which are imposed for applying the FEM procedure are discussed in detail. Numerical results for both TE_z and TM_z scattering are presented.

1 Introduction

The analysis of periodic structures is important in many fields such as optics, acoustics and electromagnetics. For example, it can be a useful tool in grating couplers [1], scattering from the sea surface [2, 3], leaky-wave antennas [4], properties of artificially soft and hard surfaces [5–7], etc. Furthermore, a periodic model, together with a Monte Carlo-based simulation, can be used to study the scattering of electromagnetic and acoustic waves by random, rough surfaces [8, 9].

Owing to the theoretical importance of this kind of structures, it has been studied extensively by means of analytical and numerical methods. The former rely on perturbational techniques and consequently allow the study of a limited number of issues [10]. The available numerical methods are based on integral equation formulations [11, 12], the point matching procedure [13], and mode matching methods [14]. In these latter methods, the field is expanded in terms of suitable propagating and evanescent modes in the various regions. Boundary conditions are then enforced to obtain a set of simultaneous equations whose solution provides the unknown amplitudes of the various modes. The drawback of this class of numerical methods is that their implementation is cumbersome when applied to complex structures, involving various materials which have an irregular composition profile. A different approach has been proposed by Lou *et al.* [8, 9] which combines the finite element method (FEM) with a Fourier modal expansion. However, it should be noted that this method is not general as it only allows one to analyse those directions of propagation of the incident plane wave that are related to the periodicity of the structure. In particular, only cases in which the intensity of the field at the extremes of the periodic cell assumes the same value can be considered. This restriction is not so important when the method is applied to the analysis of the scattering from rough surfaces, where the periodicity d is much greater than the incident field wavelength. The associated incidence angle discretisation is sufficiently fine in this case. However, this is not true for the analysis of generic periodic structures with a period d of the same order as the wavelength λ. Furthermore, the methodology employed in Reference 9 to match the field expanded in terms of the Fourier modes with the solution obtained via the FEM implies a numerical evaluation of the derivative of the field via a quadratic interpolation, and this can introduce numerical errors [15].

A hybrid FEM/MM (method of moments) technique for the analysis of the scattering by thick gratings has been proposed by Gedney and Mittra [16]. It should be noted that this method is not general since it only allows one to study gratings comprised of rectangular conductors filled with dielectric material of arbitrary profiles, this formulation is not suitable for analysing other types of periodic structures.

The hybrid formulation proposed here, still employs the FEM. Thus, using the Floquet modal expansion, instead of the Fourier modal expansion, combined with a new numerical formulation for imposing the periodicity of the field, allows one to eliminate the previous restrictions on the values of the incidence angle and periodicity, as well as on the properties and geometry of the grating. Hence the periodic structure can be completely arbitrary as to its form and material properties. In addition, explicitly enforcing the Neumann condition in the finite element formulation does not require any numerical evaluation of the derivatives.

2 Formulation

The geometry for the two-dimensional scattering by an infinite periodic structure is shown in Fig. 1. We assume

Reprinted with permission from *IEE Proceedings-H*, vol. 140, no. 2, pp. 65–70, April 1993.

that the incident plane wave is independent of the z-coordinate and has the electric field either parallel (TM_z, $H_z = 0$) or perpendicular (TE_z, $E_z = 0$) to the z-axis.

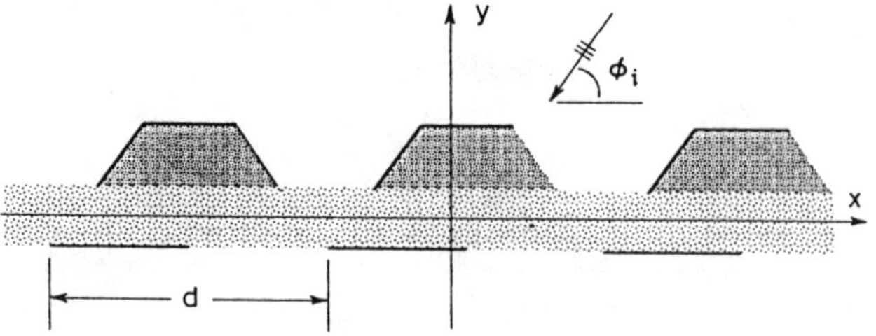

Fig. 1 *Geometry of problem*

The incident wave can be expressed as

$$u^i(x, y) = e^{jk(x \cos \phi_i + y \sin \phi_i)}\hat{z} = u^i(x, y)\hat{z} \quad (1)$$

where ϕ_i is the angle of incidence, k the propagation constant and $u^i = E^i_z$ or $u^i = H^i_z$ for TM_z or TE_z, respectively. A harmonic time dependence $\exp(j\omega t)$ is assumed and suppressed. Since the incident wave is a plane wave and the geometry of the scatterer is a periodic function of x with period d, the fields observed at two arbitrary points, separated by a distance d along a line parallel to the x-axis, are identical except for a constant phase shift $\exp(jkd \cos \phi_i)$. This means that it is necessary to determine the scattered field u^s only in a basic period $-d/2 \leqslant x \leqslant d/2$. In particular, it can be expanded in terms of Floquet modes as

$$u^s(x, y) = \sum_{m=-\infty}^{\infty} A^{U,D}_m \Psi_m(x) e^{-j\gamma_m |y|} \quad (2)$$

where $\{A^U_m\}$ and $\{A^D_m\}$ are the unknown coefficients in the upper and lower regions, respectively. In eqn. 2

$$\gamma_m = \sqrt{(k^2 - k^2_m)} \quad \text{for } k^2 > k^2_m \quad (3)$$

$$\gamma_m = -j\sqrt{(k^2_m - k^2)} \quad \text{for } k^2 < k^2_m \quad (4)$$

is the propagation constant along the y-axis of the mth harmonic. Moreover,

$$\Psi_m(x) = \exp(-jk_m x) \quad m = 0, \pm 1, \pm 2, \ldots \quad (5)$$

where

$$k_m = \left(\frac{2m\pi}{d}\right) - k \cos \phi_i \quad (6)$$

describe the transverse variation of the field.

The basic cell, between the boundaries Γ_L and Γ_R (Fig. 2), can be partitioned into three different regions. Region 1 is extended to the zone of complexity of the periodic structure and is delimited by the straight lines Γ_U and Γ_D. Regions 2 and 3 consist of two homogeneous media surrounding the periodic structure. The wavenumbers in the three regions are denoted by $k_1(x, y)$, $k_2 = k$ and k_3, respectively. It is now convenient to define an interior problem, relative to region 1, that is discretised by a FEM formulation, and an exterior problem for regions 2 and 3 where the scattered field is represented in term of a series as in eqn. 2.

In region 1, we need to solve the wave equation

$$[\nabla^2_t + k^2_1(x, y)]u^t(x, y) = 0 \quad (7)$$

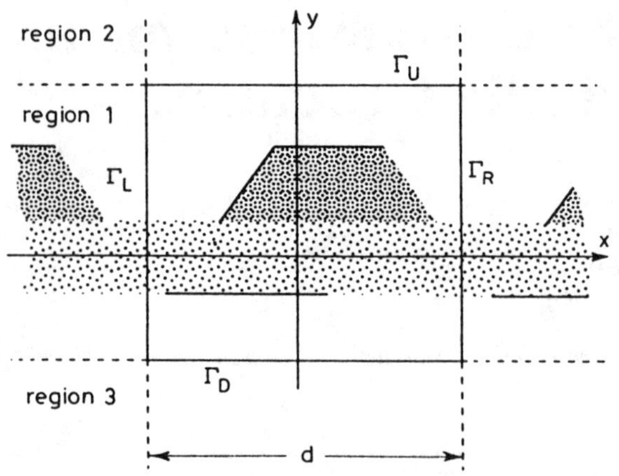

Fig. 2 *Basic cell of periodic structure*

where u^t denotes the total field, subject to appropriate boundary conditions. In particular, the continuity of the tangential component of the field and its normal derivative are imposed at the boundaries Γ_U and Γ_D, i.e.

$$u^t(x, y)\Big|_{\Gamma_U, \Gamma_D} = u^i(x, y)\Big|_{\Gamma_U, \Gamma_D} + u^s(x, y)\Big|_{\Gamma_U, \Gamma_D} \quad (8)$$

$$\frac{\partial u^t(x, y)}{\partial y}\Big|_{\Gamma_U, \Gamma_D} = \frac{\partial u^i(x, y)}{\partial y}\Big|_{\Gamma_U, \Gamma_D} + \frac{\partial u^s(x, y)}{\partial y}\Big|_{\Gamma_U, \Gamma_D} \quad (9)$$

Two additional conditions can be imposed at Γ_R and Γ_L that account for the periodicity of the field in the x direction, i.e.

$$u^t(x, y)\Big|_{\Gamma_R} = u^t(x, y)\Big|_{\Gamma_L} e^{jkd \cos \phi_i} \quad (10)$$

$$\frac{\partial u^t(x, y)}{\partial x}\Big|_{\Gamma_R} = \frac{\partial u^t(x, y)}{\partial x}\Big|_{\Gamma_L} e^{jkd \cos \phi_i} \quad (11)$$

To solve the wave equation, a FEM procedure is applied which uses weighted residuals [17] and the Green theorem to reduce the Laplace operator appearing in eqn. 7. This leads to the weak form of the Galerkin formulation

$$\iint_S (\nabla_t u^t \nabla_t \mathscr{W}_i - k^2_1 u^t \mathscr{W}_i)\, dS - \oint_\Gamma \frac{\partial u^t}{\partial n} \mathscr{W}_i\, d\gamma = 0 \quad (12)$$

In eqn. 12, $\Gamma = \Gamma_U + \Gamma_L + \Gamma_D + \Gamma_R$, S is the region delimited by Γ, $\mathscr{W}_i$ ($i = 1, 2, \ldots$) are suitable weighting first-order differentiable functions and $(\partial/\partial n)$ is the outwards normal derivative evaluated at the boundary Γ.

The contour integral appearing in eqn. 12 can be rewritten as

$$\oint_\Gamma \frac{\partial u^t}{\partial n} \mathscr{W}_i\, d\gamma = \int_{\Gamma_U} \frac{\partial u^t}{\partial y} \mathscr{W}_i\, dx - \int_{\Gamma_L} \frac{\partial u^t}{\partial x} \mathscr{W}_i\, dy - \int_{\Gamma_D} \frac{\partial u^t}{\partial y} \mathscr{W}_i\, dx + \int_{\Gamma_R} \frac{\partial u^t}{dx} \mathscr{W}_i\, dy \quad (13)$$

It is important to note that the boundary integral terms, relative to the upper and lower contours (Γ_U, Γ_D) involve the normal derivative of the field at the boundary, so that they can be used for applying the Neumann boundary condition expressed in eqn. 9 explicitly. This can be accomplished by substituting eqn. 9 in the first and third integrals of the second member of eqn. 13. It should be emphasised that the derivatives in the right member of eqn. 9 can be evaluated analytically.

The boundary condition stated in eqn. 11 can be accounted for as follows. Suppose that the weighting function $\mathscr{W}_l$ is centered at a node on the left boundary Γ_L. This function can be associated with the following integral equation

$$\iint_{S_l} (\nabla_t u^t \nabla_t \mathscr{W}_l - k_1^2 u^t \mathscr{W}_l)\, dS + \int_{\Gamma_L} \frac{\partial u^t}{\partial x} \mathscr{W}_l \, dy - \int_{\Gamma_U} \frac{\partial u^t}{\partial y} \mathscr{W}_l \, dx + \int_{\Gamma_D} \frac{\partial u^t}{\partial y} \mathscr{W}_l \, dx = 0 \quad (14)$$

in which S_l is the region where the weighting function $\mathscr{W}_l$ is defined. To simplify the formulation of the problem, a mesh is built with the nodes at the boundaries Γ_L and Γ_R having the same y-co-ordinate. Let us denote by $\mathscr{W}_r$ the weighting function centred in the node on the right boundary Γ_R having the same y-co-ordinate of the node in which $\mathscr{W}_l$ is centred. By applying the Floquet condition as expressed in eqn. 11, we obtain

$$\int_{\Gamma_L} \frac{\partial u^t}{\partial x} \mathscr{W}_l \, dy = e^{-jkd \cos \phi_i} \left\{ \iint_{S_r} (\nabla_t u^t \nabla_t \mathscr{W}_r - k_1^2 u^t \mathscr{W}_r)\, dS - \int_{\Gamma_U} \frac{\partial u^t}{\partial y} \mathscr{W}_r \, dx + \int_{\Gamma_D} \frac{\partial u^t}{\partial y} \mathscr{W}_r \right\} \quad (15)$$

in which S_r is the area where the weighting function $\mathscr{W}_r$ is defined. A similar expression can be found for the integral on the right boundary Γ_R. It should be noted that the last expression allows us to express the continuity of the derivative of the field through the left boundary Γ_L, in terms of the values of the field close to the right boundary Γ_R.

It is this feature that allows us to apply the method to arbitrary complex periodic structures, thus overcoming the requirement that the field is the same on the boundaries Γ_L and Γ_R as in References 9 and 16.

To guarantee the uniqueness of the wave equation solution, the continuity of the transverse field component u^t through the contour Γ must be imposed. This is done by enforcing expr. 10 at the nodes which lie on the contour Γ_R or Γ_L, in addition to expr. 8 at the nodes of the borders Γ_U and Γ_L. It is convenient to enforce this last condition via an integral formulation. Indeed, this allows us to choose a number of unknown coefficients $A_m^{U,D}$ less than the number of nodes on the contours Γ_U, Γ_D. By choosing $\Psi_m^*(x)$ as a weighting function, where * denotes the complex conjugate, we obtain the following integral equation:

$$\int_{\Gamma_U, \Gamma_D} u^t \Psi_m^*(x)\, dx = \int_{\Gamma_U, \Gamma_D} u^i \Psi_m^*(x)\, dx + \int_{\Gamma_U, \Gamma_D} u^s \Psi_m^*(x)\, dx \quad (16)$$

where $m = 0, \pm 1, \pm 2, \ldots, \pm M$ and M is the maximum number of modes used for representing the scattered field. Eqns. 1 and 2 can be used to express the incident and scattered fields, respectively, so that, by also taking into account the orthogonality of the functions $\Psi_m(x)$, we obtain

$$\int_{\Gamma_U, \Gamma_D} u^t \Psi_m^*(x)\, dx = d\delta_{0m} e^{jky_{U,D} \sin \phi_i} + dA_m^{U,D} e^{-j\gamma_m |y_{U,D}|} \quad (17)$$

where δ_{ij} is the Kronecker delta. The integral in eqn. 17 can be evaluated analytically once the base functions have been chosen.

3 Numerical solution and results

Region 1 is divided into N triangular elements so that, by using linear interpolation functions for each element, the corresponding field can be expressed as

$$u(x, y) = \sum_{i=1}^{N} \sum_{j=1}^{3} N_j^i(x, y) u_j^i \quad (18)$$

where $N_j^i(x, y)$, $j = 1, 2, 3$, are the interpolating first-order differentiable functions of the generic ith element and u_j^i is the field value at the jth nodal point, in the ith element. By applying the FEM to eqn. 7, together with the boundary conditions (eqns. 8–11), the following matrix equation is obtained

$$\begin{bmatrix} \underline{D}_a^{UD} & \underline{D}_u^{UD} \\ \underline{N}^{UD} & \underline{F} \end{bmatrix} \begin{bmatrix} \underline{a}^{UD} \\ \underline{u} \end{bmatrix} = \begin{bmatrix} \underline{I}^{UD} \\ \underline{I}_N^{UD} \end{bmatrix} \quad (19)$$

where subscripts U and D denote upper and lower boundaries, respectively. The matrices $\underline{D}_a^{UD}$, $\underline{D}_u^{UD}$ and the vector $\underline{I}^{UD}$ account for the continuity of the field on the boundary Γ_U and Γ_D, while $\underline{N}^{UD}$, $\underline{I}_N^{UD}$ guarantee the continuity of the derivative of the field at the same boundaries. The complexity of the structure is taken into account by the banded matrix $\underline{F}$.

The solution of eqn. 19 can be written as

$$\underline{a}^{UD} = [\underline{D}_a^{UD} - \underline{D}_u^{UD} \underline{F}^{-1} \underline{N}^{UD}]^{-1} \times [\underline{I}^{UD} - \underline{D}_u^{UD} \underline{F}^{-1} \underline{I}_N^{UD}] \quad (20)$$

$$\underline{u} = \underline{F}^{-1} [\underline{I}_N^{UD} - \underline{N}^{UD} \underline{a}^{UD}] \quad (21)$$

It can be shown that this form is most convenient for numerical evaluations [18]. It allows us to take the maximum advantage of the banded structure of the matrix $\underline{F}$, so that its inversion can be evaluated very efficiently. To this end, it must be noted that $\underline{F}$ has larger dimensions ($N \times N$) than the other matrices. An additional computational effort is necessary for the inversion of the matrix in the first brackets of eqn. 20. This is a full matrix having a number of columns and rows equal to the total number of the unknown coefficients by means of which the scattered and transmitted fields are represented. Furthermore, the columns of the matrix $\underline{D}_u^{UD}$ and the rows of the vector $\underline{I}_N^{UD}$ and of the matrix $\underline{N}^{UD}$ are different from zero only when they correspond to the nodal points which lie on the boundaries Γ_U and Γ_D. So, if we are only interested in analysing the scattering from or the transmission through the periodic structure, a knowledge of the columns of the $\underline{F}^{-1}$ matrix relevant to the above mentioned nodes is sufficient, allowing a considerable reduction in computational time.

One of the problems encountered in the implementation of the FEM concerns the node numbering of the mesh. The boundary condition (expr. 11) broadens the band of the FEM matrix unacceptably if no precautions are taken. This problem has been solved by optimising the node numbering procedure, taking into account the fact that the unknown coefficients of the field relevant to the nodes on the boundaries Γ_R and Γ_L at the same y-coordinate, appear in the same equations.

The accuracy of the proposed method has been verified by means of an energy conservation test. The percentage error in the conservation of energy is of the order of 10^{-3}%. Furthermore, the convergence of the method has been tested by analysing a self-complementary strip grating, loading a dielectric slab above a ground plane. The structure has a period $d = 0.75\lambda$, with the dielectric slab characterised by a thickness $h = d$ and a dielectric constant $\varepsilon_r = 3$. The analysis has been carried out for the two principal polarisations of the incident wave, impinging on the surface at $\phi_i = 30°$, by increasing the number of modes. Table 1, summarises both the number of modes and the relative normalised reflected power in the zeroth-order mode $|A_0|^2$ which are related to the analysis performed. They are compared with the results obtained by the method presented in Reference 14. It can be seen that our method converges with fewer harmonics. This is because the upper boundary of the basic cell lies in the homogeneous zone of the space above the dielectric slab, while in Reference 14 the mode matching is enforced on the discontinuity between the slab and the same homogeneous region. Of course, the wider area of the basic cell which is employed here, requires a higher number of unknowns in the FEM procedure. However, the banded nature of the resulting matrix limits the CPU time which is needed for the inversion of the matrix when the upper boundary in the homogeneous region is placed close to the strip grating.

In Fig. 3, the normalised reflected power in the zeroth-order mode plotted against the periodicity of the structure obtained with the present method against the results in Reference 19 is shown, for both TM_z and TE_z polarisations. Once again the agreement is good.

It is worth noting that, in some simpler cases, other techniques such as perturbational methods may be more attractive because of their simplicity and computational efficiency. However, the results obtained by this hybrid numerical method can be used as a benchmark for analytical and numerical techniques. It is important to note that the proposed procedure is accurate and efficient with respect to those available in the literature, particularly in the case of scattering from very complex periodic structures. In fact, this method combines the potentiality of the FEM in treating complex electromagnetic regions with the simplicity of the Floquet series, resulting in a very powerful and flexible technique for the analysis of periodic structures. For instance, Fig. 4 sketches a complex structure having three $0.2d$ width strips at $h_1 = d/2$, $h_2 = 3d/4$, $h_3 = d$, embedded in a dielectric slab with $\varepsilon_r = 3.0$, backed by a ground plane. This structure has been analysed with the proposed hybrid method and Fig. 5 shows the normalised reflected power in zeroth-order mode against its periodicity. In particular, a plane wave impinges on the grating with an angle $\phi_i = 30°$ and both TM_z and TE_z polarisations are considered.

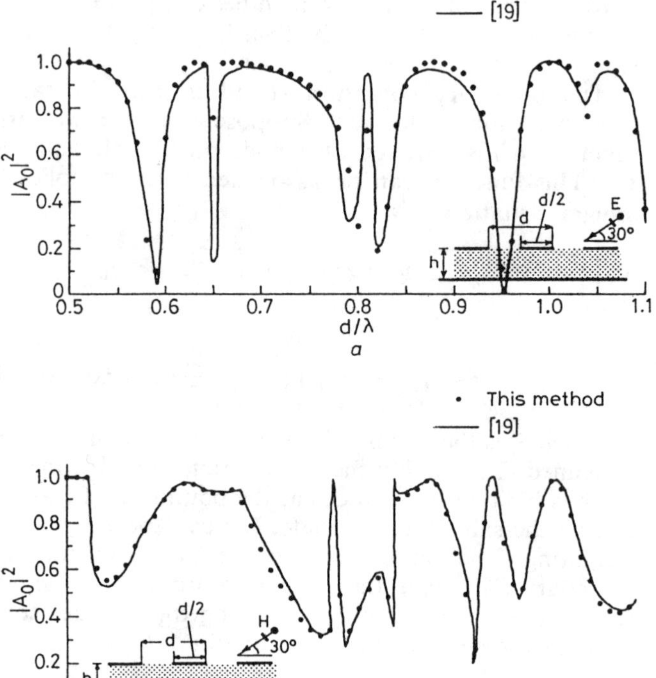

Fig. 3 *Normalised reflected power in zeroth-order mode against period for a self-complementary strip grating with $h = d$, $\varepsilon_r = 3$*

a $H_z = 0$ (TM_z case)
b $E_z = 0$ (TE_z case)
Incident wave impinges on surface at $\phi_i = 30°$
● This method
— Reference 19

Table 1: Normalised power reflected in zeroth-order mode for increasing number of modes for a self-complementary strip grating having $d = 0.75\lambda$, $h = d$ and $\varepsilon_r = 3$ with $\phi_i = 30$

Modes	$\lvert A_0\rvert^2$ for TM_z case ($H_z = 0$)			$\lvert A_0\rvert^2$ for TE_z case ($E_z = 0$)		
	Kalhor [14]	Present ($d_1 = 1.1d$)	Present ($d_1 = 1.3d$)	Kalhor [14]	Present ($d_1 = 1.1d$)	Present ($d_1 = 1.3d$)
5	—	0.898052	0.898525	—	0.292126	0.292071
7	0.8142	0.898055	0.898525	0.2968	0.292133	0.292071
11	0.8469	0.898055	0.898525	0.2925	0.292134	0.292071
15	0.8582	0.898055	0.898525	0.2955	0.292134	0.292071
19	0.8637	0.898055	0.898526	0.2981	0.292134	0.292071
23	0.8672	—	—	0.2999	—	—
35	0.8722	—	—	0.3033	—	—
39	0.8732	—	—	0.3039	—	—

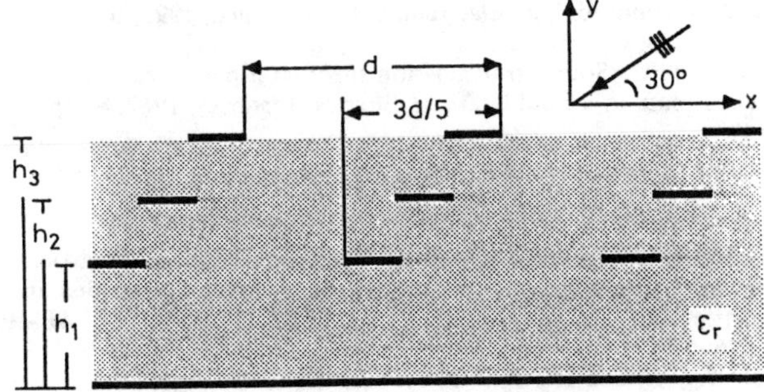

Fig. 4 *Complex structure analysed using this method*

$h_1 = d/2$, $h_2 = 3d/4$, $h_3 = d$, strip width = $0.2d$

Finally, it is worth pointing out that the method allows us, with a simple postprocessing, to also find the field in the complex electromagnetic region which is analysed via the FEM. Fig. 6 shows the field in region 1 for the aforementioned periodic structure with $d = 0.7\lambda$, $\phi_i = 30°$.

4 Conclusions

A hybrid technique which combines the potentiality of the FEM in treating complex electromagnetic regions with the simplicity of the Floquet series has been presented. This results in a very powerful and flexible method for the analysis of periodic structures which have arbitrary geometry and material properties. The method proposed has been validated by means of an energy conservation test and by comparison with some available results.

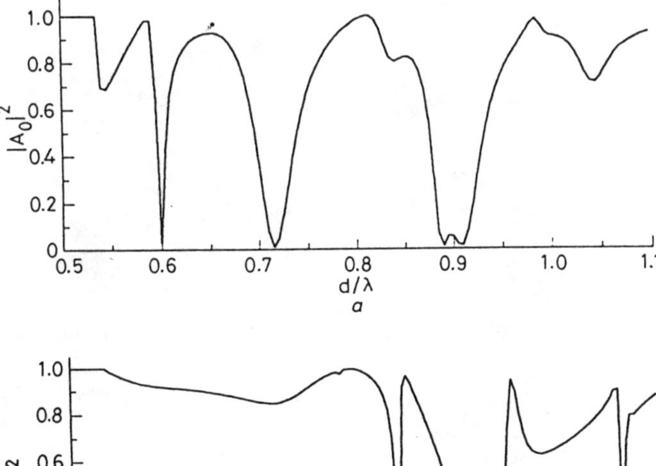

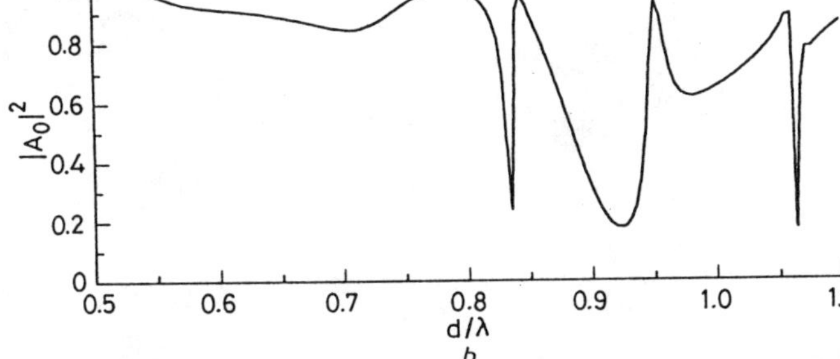

Fig. 5 *Normalised reflect power in zeroth-order mode against period for periodic structure of Fig. 4*

a $H_z = 0$ (TM_z case)
b $E_z = 0$ (TE_z case)
Incident wave impinges on surface at $\phi_i = 30°$

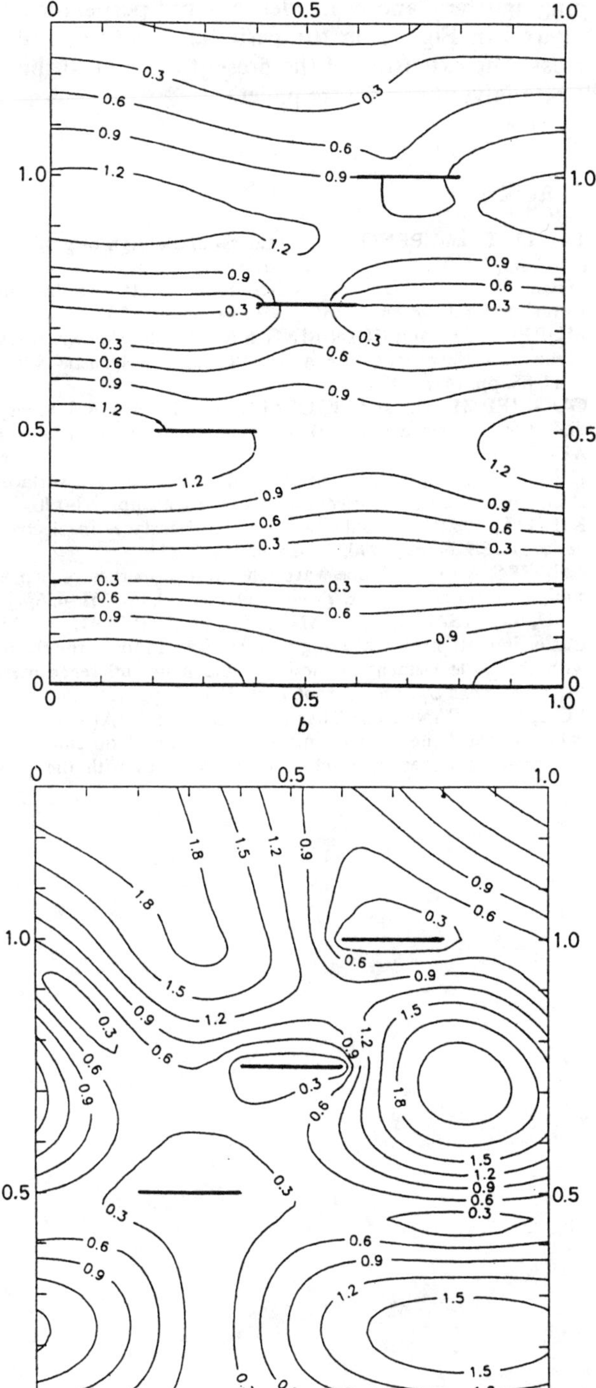

Fig. 6 *Field contours inside basic cell for same structure as in Fig. 4*

$d = 0.7\lambda$ $\varepsilon_r = 3$ $\phi_i = 30°$

a electric field for TM_z case ($H_z = 0$)
b magnetic field for TE_z case ($E_z = 0$)

Due to its flexibility and accuracy, the method appears to be a useful tool for the analysis of a large class of practical structures such as grating couplers or the scattering from artificially hard and soft surfaces, whose properties are exploited in many radiating systems. For this application, the case of a plane wave obliquely impinging (the plane of incidence is not perpendicular to the z-axis in Fig. 1) on the periodic structure is also of interest. The extension of the present method to this case will be analysed in a future paper.

5 References

1 TAMIR, T., and PENG, S.T.: 'Analysis and design of grating couplers', *Appl. Phys.*, 1977, **14**, pp. 235–254

2 MELLEN, R.H.: 'On underwater sound scattering by surface waves', *IEEE J. Oceanic Engng*, 1989, **14**, pp. 245–247

3 MURPHY, J.E., and CHIN-BING, S.A.: 'A finite-element model for ocean acoustic propagation and scattering', *J. Acoustic Soc. Am.*, 1989, **86**, pp. 1478–1483

4 GUGLIELMI, M., and BOCCALONE, G.: 'A novel theory for dielectric-inset waveguide leaky-wave antennas', *IEEE Trans.*, 1991, **AP-39**, pp. 497–504

5 KILDAL, P.S.: 'Definition of artificially soft and hard surfaces for electromagnetic waves', *Electron. Lett.*, 1988, **24**, pp. 168–170

6 KILDAL, P.S.: 'Artificially soft and hard surfaces in electromagnetics', *IEEE Trans.*, 1990, **AP-38**, pp. 1537–1544

7 ANDERS, AAS, J.: 'Plane-wave reflection properties of two artificially hard surfaces', *IEEE Trans.*, 1991, **AP-39**, pp. 651–656

8 LOU, S.H., TSANG, L., CHAN, C.H., and ISHIMARU, A.: 'Monte Carlo simulations of scattering of waves by a random rough surface with the finite element method and the finite difference method', *Microwave and Optical Technol. Lett.*, 1990, **3**, pp. 150–154

9 LOU, S.H., TSANG, L., CHAN, C.H., and ISHIMARU, A.: 'Application of the finite element method to Monte Carlo simulation of scattering of waves by random rough surfaces with the periodic boundary condition', *J. Electromag. Waves Appl.*, 1991, **5**, pp. 835–885

10 RYBAK, S.A.: 'Sound transmission through a periodically inhomogeneous plate in a liquid', *Soviet Physics-Acoustics*, 1962, **8**, pp. 83–87

11 KALHOR, H.A., and ARMAND, A.: 'Scattering of electromagnetic waves by gratings of conducting cylinders', *Proc. IEE*, 1975, **122**, pp. 245–248

12 KALHOR, H.A., and ILYAS, M.: 'Scattering of plane electromagnetic waves by a grating of conducting cylinders embedded in a dielectric slab over a ground plane', *IEEE Trans.*, 1982, **AP-30**, pp. 576–579

13 HINATA, T., and HOSONO, T.: 'On the scattering of electromagnetic waves by plane grating placed in homogeneous medium', *Trans. IECE Japan*, 1976, **59-B**, pp. 73–81

14 KALHOR, H.A.: 'Electromagnetic scattering by a dielectric slab loaded with a periodic array of strips over a ground plane', *IEEE Trans.*, 1988, **AP-36**, pp. 147–151

15 MEI, K.K.: 'Unimoment method for electromagnetic wave scattering', *J. Electromag. Waves Appl.*, 1987, **1**, pp. 201–222

16 GEDNEY, S.D., and MITTRA, R.: 'Analysis of the electromagnetic scattering by thick gratings using a combined FEM/MM solution', *IEEE Trans.*, 1991, **AP-39**, pp. 1605–1614

17 SILVESTER, P.P., and FERRARI, R.L.: 'Finite elements for electrical engineers' (Cambridge University Press, 1990).

18 JIN, J.M., and LIEPA, V.V.: 'A note on hybrid finite element method for solving scattering problems', *IEEE Trans.*, 1988, **AP-36**, 1486–1494

19 YAMASAKI, T., and HINATA, T.: 'Scattering of electromagnetic waves by plane grating with reflectors', *Trans. IECE Japan*, 1978, **61-B**, pp. 52–60

Radially Propagating Waves
II-Absorbing Boundary Conditions

Introduction

Exterior regions are treated in the finite element method by two fundamentally distinct techniques. One class, the so-called hybrid methods, seeks representations of the exterior region that may be thought equivalent to discretized Green's functions: they correctly model all exterior interactions and permit all possible electromagnetic interactions that may be admitted under the Maxwell equations. Absorbing boundary conditions belong to a second, and different class. Here an attempt is made to model absorption at the boundary, by constructing boundary conditions that admit only outgoing waves. The result is usually a locally defined boundary condition, with the attendant advantage of sparse matrices. However, this treatment is more restricted than the first class: it admits only a subset of the general field solutions, for it restricts attention to only those propagating fields that cross the absorbing boundary in an outward direction. This restriction is readily admissible in a large number of practical problems, so the absorbing boundary condition approach to wave problems has gained considerable prominence in recent years.

This chapter comprises the major recent works on absorbing boundary conditions and attempts to survey most of the available literature. It should be noted that many published papers in this area deal with scalar waves as encountered in acoustics. These have been included where they appear relevant, but no attempt has been made to survey the entire area of acoustics.

Absorbing Boundary Conditions

Solving an infinite-domain problem — such as a scattering or antenna problem — with the finite element method requires, as already stated, enclosing the bounded region in which the method is to be applied by means of a fictitious surface S. This enclosed finite region will be referred to as the "internal region" in the following. Appropriate boundary conditions must then be applied on the surface S so that the solution of Maxwell's equations in the domain delimited by S is a good approximation to the solution of the original unbounded domain problem. These boundary conditions must simulate the infiniteness of the original domain reasonably accurately.

All the methods that use non-local boundary conditions derived from integral equations or eigenfunction expansions of the scattered field have been reviewed and assessed in the previous section. This section of the survey is devoted to those techniques that make use of approximated local boundary conditions in conjunction with the finite element method. These techniques will be henceforth referred to as the "finite element method with absorbing boundary conditions."

The approximated local boundary conditions on the surface S must cancel out, in the internal region, the field due to the discontinuity created by the introduction of the surface S itself. Consequently, the surface S appears to be perfectly absorbing, which accounts for the name "absorbing boundary conditions."

Setting up absorbing boundary conditions is a problem analogous, from a certain point of view, to that of some radar applications in which boundary conditions on a target are sought to produce no scattered field (the black body problem). In a similar way, the boundary conditions on the surface S of a finite element model with absorbing boundary conditions must simulate the presence of a black body enclosing the inner domain, giving rise to no scattered radiation, as the free space does.

The basic difference between hybrid methods and the absorbing boundary condition method lies in the technique used to approximate the exact radiation boundary condition on S, as given by Green's integral. In hybrid techniques the Green's integral is numerically evaluated. In the absorbing boundary condition method, this integral is subject to an *a priori* approximation. For this reason, it should not be surprising that the accuracy of the finite element–absorbing boundary condition technique falls off when the hypotheses underlying the *a priori* approximation are no longer satisfied.

It is worth stressing that absorbing boundary conditions are local conditions, in contrast to hybrid formulations which express action at a distance. This means that their utilization in conjunction with finite numerical methods — finite difference, finite element method — preserves the sparsity and, where appropriate, band structure of the coefficient matrix of the associated system of linear equations. This characteristic makes absorbing boundary conditions particularly appealing for numerical implementations and a powerful tool for the solution of complex electromagnetic problems with large numbers of unknowns.

Combining the basic finite element method with absorbing boundary conditions still presents some problems. First of all, there are limitations on the shape of the bounding surface — usually circular, rectangular, or the corresponding three-dimensional versions of such contours — that can be used to enclose the domain of application of the finite element method. Second, some classes of absorbing boundary conditions make the coefficient matrix lose its symmetry. Most of the papers defining absorbing boundary conditions published after the fundamental works by Engquist and Majda [1977] and Bayliss and Turkel (1980) tend toward *a posteriori* checking of the validity of the absorbing boundary conditions themselves by means of numerical experiments. Consequently, a lot of work still remains to be done on the theory, as will be pointed out further below.

The theoretical development of absorbing boundary conditions began in the applied mathematics field at the end of the 1970s. Then their applications spread slowly in electromagnetics, where they have been used in solving infinite domain problems (antenna and scattering problems) in combination with finite methods. The first important theoretical research on absorbing boundary conditions for the numerical solution of the scalar wave equation in time domain is due to Engquist and Majda [1977]. They used the theory of pseudodifferential operators (Hitoshi 1974) to develop local scalar absorbing boundary conditions valid for two-dimensional problems. For example, for a surface S that corresponds to the $x = 0$ coordinate plane, the Engquist-Majda condition is expressed by the equation

$$L_{-}\Phi = 0 \qquad \text{at } x = 0\,, \tag{1}$$

where the operator L_{-} is defined as

$$L_{-} = D_x - \frac{1}{c} D_t \sqrt{1 - s^2}\,, \tag{2}$$

with

$$s^2 = c^2 \frac{D_y^2}{D_t^2}\,, \tag{3}$$

where c is the velocity of the light in the medium in which the electromagnetic wave propagates, and

$$D_x^2 = \frac{\partial^2}{\partial x^2}\,, \qquad D_y^2 = \frac{\partial^2}{\partial y^2}\,, \qquad D_t^2 = \frac{\partial^2}{\partial t^2}\,. \tag{4}$$

Equation (1), also known as one-wave equation, describes an absorbing boundary condition for a plane wave coming from the $x > 0$ half-plane and impinging on the $x = 0$ coordinate plane at an arbitrary angle of incidence. Indeed, it had already been used by Lindman (1975) as absorbing boundary condition for the finite difference approximation of the wave equation on a finite domain, but its theoretical justification is due to Engquist and Majda. They have proposed approximations of the equation $L_{-}\Phi = 0$ based on Taylor's series expansion, or Padé approximation, of $\sqrt{(1 - s^2)}$. The validity and accuracy of the absorbing boundary conditions that Engquist and Majda developed was verified by the authors themselves, who applied them in conjunction with a finite difference time domain formulation of the scalar wave equation. In this connection, it may be of interest to note that most of the numerical experiments to test the applicability of the absorbing boundary conditions, as well as many of the proposed improvements, have been implemented in the framework of a finite difference time domain scheme for the solution of the wave equation.

The work of Mur [1981] on absorbing boundary conditions is closely related to that of Engquist and Majda. Using a formulation analogous to theirs, Mur derived second-order scalar absorbing boundary conditions, still of a local nature and valid for the three-dimensional case. Their efficiency and limitations were tested by Mur by solving simple, two-dimensional, canonical problems. Mur's absorbing boundary conditions were later applied by Umashankar and Taflove (1982) to the finite difference analysis of electromagnetic scattering from complex objects.

Bayliss and Turkel (1980) also devised a technique to derive absorbing boundary conditions for wave-like equations, an alternative to the approach of Engquist and Majda. Subsequently, the application of their technique to the scalar Helmholtz equation was extensively analyzed by Bayliss, Gunzburger, and Turkel [1982]. The latter authors began from the asymptotic series for the solution of the scalar Helmholtz equation given by Karp (1961) and Wilcox (1956), for the two-dimensional and three-dimensional cases, respectively, and developed a sequence of differential operators of increasing order. When such operators are applied to the outgoing electromagnetic field, they give rise to absorbing boundary conditions of increasing accuracy for the solution of the problem in the region enclosed by the surface S. For instance, using a spherical coordinate system for the solution of a three-dimensional problem, the Wilcox expansion of an outgoing solution Φ of the scalar Helmholtz equation is

$$\Phi(\theta, \phi, r) = \frac{e^{-jkr}}{kr} \sum_{i=0}^{\infty} \frac{F_i(\theta, \phi)}{(kr)^i}, \tag{5}$$

and the sequence of differential operators can be expressed by the recursion relation

$$B_1 = \left(\frac{\partial}{\partial r} + jk + \frac{1}{r} \right), \tag{6}$$

$$B_m = \left(\frac{\partial}{\partial r} + jk + \frac{2m-1}{r} \right) B_{m-1} \quad \text{for } m > 1. \tag{7}$$

Since for any function having an expansion of the form in equation (5) it is true that

$$B_m \Phi = O\left(1/r^{2m+1}\right), \tag{8}$$

then imposing on a surface S the mth-order boundary condition $B_m \Phi = 0$ will force the solution Φ to match the first $2m$ terms of the series expansion (5). Although the Bayliss, Gunzburger, and Turkel absorbing boundary conditions can be easily imposed on circular contours or spherical surfaces, their use can be extended to arbitrary surfaces S (elliptic, conformal to the scatterer, etc.) as well. However, in such cases the application procedure becomes more involved.

It is worth underlining that the method introduced by Bayliss, Gunzburger, and Turkel allows appropriate boundary conditions to be determined for static and quasi-static problems also (e.g., for the Laplace equation); such conditions are called "asymptotic boundary conditions" (Khebir, Kouki, and Mittra 1990). The validity and accuracy of the Bayliss, Gunzburger, and Turkel absorbing boundary conditions have been tested by the authors themselves, who applied their conditions to the solution of the scalar Helmholtz equation and Laplace's equation in exterior regions, using the finite element method in the interior region.

Among the numerous works that have flowed from initial theoretical contributions already mentioned, attention should be drawn particularly to those by Bayliss, Goldstein, and Turkel (1985), Trefethen and Halpern [1986], and Blaschak and Kriegsmann (1988). Of these, the first analyzes the accuracy of the absorbing boundary conditions developed prior to 1985. Trefethen and Halpern [1986] proposed absorbing boundary conditions that also work well enough, as opposed to those suggested by Engquist and Majda, for waves impinging on the fictitious surface S with a direction consistently out of normal. Basically, the authors have extended the Engquist and Majda approach by improving the polynomial approximation of the radical $\sqrt{(1-s^2)}$, as had been also done by Lindman (1975) but using a different approximation scheme. This idea was further developed and generalized by Hingdon (1987). Finally, Blaschak and Kriegsmann (1988) furnished a comparative evaluation of the various absorbing boundary conditions proposed to date.

Until the end of the 1980s, publications relative to absorbing boundary conditions mainly had a theoretical character. For example, in the review paper by Moore, Blaschak, Taflove, and Kriegsmann [1988], the major portion of the bibliographic references are actually drawn from the literature of mathematics and mathematical physics, not engineering. Since 1988, there has been a remarkable increase in interest among practitioners of electromagnetic field computation in the application of absorbing boundary conditions combined with finite numerical methods (finite difference and finite element methods). This trend is clearly illustrated by the steadily increasing number of publications on absorbing boundary conditions for the analysis of scattering and antenna problems.

The first noteworthy application of absorbing boundary conditions in conjunction with the finite element method is that of Peterson and Castillo [1989], who solved a two-dimensional scattering problem. Since their work was published, scalar absorbing boundary conditions have been extensively analyzed and applied in solving antenna and scattering problems, both in two and three dimensions. The mechanisms through which errors are introduced in this class of problem by the finite element technique with absorbing boundary conditions have been pointed out by Mittra and Ramahi (1990), who have also proposed an algorithm for error compensation.

In contrast to the extensive work on scalar wave problems, there is still a relative shortage of published work about vectorial three-dimensional absorbing boundary conditions. Much research, both of the purely theoretical kind and in applications, is yet to be done to develop the technique fully.

The first part of this survey of publications on absorbing boundary conditions addresses studies devoted to their theoretical development. Its aim is to point out to the reader those major works and original ideas that have been at the basis of the development of absorbing boundary conditions. The second part concerns the applications, analysis, further theoretical evolution, and improvements introduced into absorbing boundary condition formulations within the framework of specific electromagnetics problems. The order of presentation follows a logical subdivision between two-dimensional, scalar three-dimensional, and vector three-dimensional absorbing boundary conditions, rather than the chronological order of their publication. A similar taxonomy of the field has been used by other authors. It seeks to bring into evidence the logical process followed in the development of each kind of boundary condition. Of course, research does not always follow such linear paths because interactions take place between new and old ideas. Hence, some of the published works involve more than one of the classes. In such cases, the works have been placed in the section to which they brought the greatest innovative contribution — admittedly a subjective judgment!

The last work included in this section is the review paper by Moore *et al.* [1988]. It should provide a helpful tool to readers who are interested in a brief description of the various kinds of absorbing boundary conditions proposed in the literature, but do not wish to delve too deeply into theoretical detail.

The classification used here is hardly unique. A useful alternative taxonomy, not followed here, could classify publications on absorbing boundary conditions on other bases, for example:

- the kind of separation surface S between the interior and exterior regions;
- the properties of the coefficient matrices they generate (symmetric, sparse, etc.).

Such possible classifications will be alluded to occasionally as individual topics and papers are examined more closely. The paragraphs to follow present a brief critical review of the publications included in the second part of the selection as well as of other interesting works on the subject. Special attention will be paid to the results obtained, both theoretical and practical.

Two-Dimensional Absorbing Boundary Conditions

Most of the papers included in the present collection concern absorbing boundary conditions for the numerical solution of the two-dimensional scalar wave equation. These boundary conditions have been extensively studied and applied, so there is a large amount of published material available on this subject. In the following, the most significant works dealing with the solution of two-dimensional scalar problems by means of the finite element technique with absorbing boundary conditions will be discussed. The papers have been subdivided into different categories on the basis of the geometric characteristics of the contour Γ (the two-dimensional version of the fictitious surface S) employed to separate the internal and external regions.

Rectangular or circular contour Γ

The first published work concerning the application of absorbing boundary conditions to an electromagnetic problem in the framework of a numerical formulation that uses the finite element method has already been mentioned: it is that of Peterson and Castillo [1989]. In their paper, finite elements are coupled to absorbing boundary conditions to analyze the frequency-domain electromagnetic scattering from cylinders of arbitrary cross-section and electrical properties. Peterson and Castillo used the total field as the problem unknown for both TM and TE polarization (with respect to the longitudinal axis of the cylindrical structure). They imposed second-order Bayliss and Turkel (1980) type absorbing boundary conditions on the circumference Γ that encloses the interior region. Their absorbing boundary conditions were modified, as suggested by Bayliss and Turkel (1980), to eliminate the higher-order derivative in n, where n is the coordinate along the direction of the outward normal $\mathbf{1}_n$ to Γ. This kind of formulation, which has been employed in most of the published works since, is highly suitable for numerical implementation because it is relatively stable and accurate (Bayliss and Turkel 1980). Peterson and Castillo also evaluated the accuracy of the method, varying both the scatterer dimension and the distance between the scatterer and the circumference Γ. Their results indicate that this technique provides accuracy comparable to that of the integral equation formulation for a given mesh density, if the contour Γ is located at a sufficient distance from the scatterer. However, this distance must be increased as the electrical dimension of the scatterer increases. Hence, hybrid techniques are to be preferred in the case of a large scatterer.

The next important work is that by Ramahi and Mittra [1989], in which the electromagnetic scattering from dielectric cylinders of arbitrary cross-section is analyzed. The absorbing boundary conditions imposed on the circular boundary Γ are still of second-order kind, but the authors have derived them by means of a procedure different from that proposed by Bayliss and Turkel in 1980. Details of this development were given by Mittra and Ramahi (1990). They regarded the scattered field as the unknown quantity in the finite element formulation, in contrast to Peterson and Castillo, who used the total field as the unknown. The results so obtained are more accurate than those provided by the total field formulation, and the error that affects the solution is inversely proportional to the magnitude of the scattered field.

In their review paper, Mittra, Ramahi, Khebir, Gordon, and Kouki (1989) showed two possible techniques for deriving absorbing boundary conditions. These agree with the conditions provided by Bayliss,

Gunzburger, and Turkel [1982], up to the $(k\rho)^{-3}$ terms, even though they are obtained through different procedures. The authors have also pointed out that absorbing boundary conditions are not suitable for simulating, with acceptable accuracy, the presence of either evanescent waves or waves traveling along the boundary Γ. Consequently, the results obtained by means of the finite element–absorbing boundary condition technique are likely to be inaccurate as long as the scattered field has a significant contribution from such kinds of waves at the truncation boundary Γ. Mittra and his collaborators (*op. cit.*) also proposed a method to improve the effectiveness of the absorbing boundary conditions in such cases. All the topics briefly developed in this paper were subsequently treated in more extensive detail by Mittra and Ramahi (1990).

The two-dimensional finite element technique has also been exploited for the analysis of planar devices and antenna problems, when the geometry of the structure allows considering the problem as two-dimensional. The paper by Webb [1990] deals with the first class of problems and is particularly valuable because it provides universal matrices, for polynomial orders one through four, to account for the additional terms arising from the imposition of second-order absorbing boundary conditions on the circular contour Γ. Webb has shown this method to be quite efficient in calculating the input impedance of a parallel-plate waveguide radiating into free space, and for finding the scattering parameters of various dielectric slab waveguide devices.

The papers by Povinelli and D'Angelo (1991) and Sumbar, Vermeulen, and Chute (1991) deal with two-dimensional antenna problems. Povinelli and D'Angelo used finite elements in conjunction with absorbing boundary conditions to evaluate the effect of both position and composition of radomes on the radiating performance of large-wavelength arrays considered as two-dimensional. They employed the second-order Bayliss and Turkel (1980) absorbing boundary conditions, imposed on a circular or rectangular contour Γ. Their work tends to show the potential of the technique itself rather than explain in detail the procedure of implementation.

A different approach to the analysis of antenna problems was proposed by Sumbar, Vermeulen, and Chute (1991) and applied to an axially symmetric structure. Their approach consists of imposing the Sommerfeld radiation boundary condition

$$\mathbf{H} = \frac{1}{\zeta}\mathbf{1}_r \times \mathbf{E} \tag{9}$$

directly on the boundary Γ. In equation (9), $\mathbf{1}_r$ denotes the direction of propagation, which points radially outward from the center of the antenna, while ζ is the characteristic impedance of the medium surrounding the antenna. Obviously, such a boundary condition is strictly valid only at great distances from the antenna, and then only if the medium is homogeneous. However, if the materials surrounding the antenna are electrically lossy, the magnitude of the outgoing wave, as well as the magnitude of the wave scattered from the boundary Γ into the problem domain, will be attenuated. Hence, it is possible to impose the condition expressed by equation (9) on a contour Γ not very distant from the antenna. This approach is very simple to implement and has the advantage of maintaining the sparsity and symmetry of the finite element matrices, with resultant savings in memory and computing time.

Elliptic contour Γ

In all the published works so far mentioned, the absorbing boundary conditions have been imposed on a rectangular or circular boundary Γ. The utilization of a rectangular domain is generally more advantageous, since it often provides a better fit to the antenna or scatterer shape and allows a relatively small geometric domain to be modeled by finite elements.

While imposing the Engquist and Majda absorbing boundary conditions on a rectangular boundary is straightforward and convenient, the use of the Bayliss and Turkel conditions on such a contour is cumbersome and moreover gives rise to a non-symmetric matrix (D'Angelo and Mayergoyz 1989). On the other hand, the Bayliss and Turkel boundary conditions, which are very simple to impose on a circular contour Γ or a spherical surface S, provide more accurate results than the Engquist and Majda conditions (D'Angelo and Mayergoyz 1989). For this reason, some effort has been devoted to devising procedures that allow applying the Bayliss and Turkel conditions on arbitrary contours (preferably conformal with the scatterer), so as to minimize the number of unknowns in the finite element formulation and maintain the symmetry of the matrix.

A valuable contribution to this field was made by Lee, Shin, Kong, and McCartin [1990]. These authors first demonstrated that by slightly modifying the Engquist and Majda approach, it is possible to obtain, on circular boundary, the same absorbing boundary conditions as can be derived by applying the annihilating procedure suggested by Bayliss and Turkel (1980) to the Wilcox expansion for the scattered field (Wilcox 1956). Furthermore, by following a procedure analogous to that of Engquist and Majda, Lee *et al.* developed scalar absorbing boundary conditions that can be imposed on elliptical boundaries. Enclosing the region treated via the finite element method with an elliptical boundary results in a reduction in the number of unknowns with respect to the case in which a circular boundary is used and the more slender the scatterer or the antenna shape, the greater the reduction in the number of unknowns.

Contour Γ conformal to the object (OSRC)
The differential operators B_m developed by Bayliss, Gunzburger, and Turkel have been found convenient not only in the framework of finite elements but also in the context of asymptotic high-frequency methods. This interesting application of such operators was proposed by Kriegsmann, Taflove, and Umashankar in 1987. They found that for convex two-dimensional conductive scatterers, a substantial simplification of the scattering problem can be achieved by imposing the absorbing boundary conditions directly on the surface of the object (for this reason, they called this approach "on-surface radiation boundary condition," or "OSRC"). To develop this technique, they approximated the contour Γ at the generic point $\boldsymbol{\rho}'$ by its osculating circle and locally defined the operators B_1 and B_2. Inserting either of these new operators into the integral expressing the scattered field

$$\Phi^s(\boldsymbol{\rho}) = \int_\Gamma \Bigg(G(\boldsymbol{\rho}, \boldsymbol{\rho}')\, \nabla\Phi^s(\boldsymbol{\rho}') \cdot \mathbf{1}_{n'} - \Phi^s(\boldsymbol{\rho}')\, \nabla G(\boldsymbol{\rho}, \boldsymbol{\rho}') \cdot \mathbf{1}_{n'} \Bigg)\, d\Gamma' , \tag{10}$$

where $G(\boldsymbol{\rho}, \boldsymbol{\rho}')$ is the free-space Green's function for the two-dimensional case, reduces the calculation of Φ^s to the evaluation of a line integral. Furthermore, the integrand is either known explicitly (TM case) or can be easily constructed by solving a differential equation along the contour Γ (TE case). Hence, this method can be easily implemented for two-dimensional, conducting cylinders of arbitrary cross-section and provides acceptable accuracy for engineering purposes. The authors pointed out both the similarities and differences between the on-surface radiation condition approach and the traditional physical optics approximation. Subsequently, such an analysis was developed by Jin, Volakis, and Liepa (1989), who compared the performance of the two approaches in dealing with conducting cylinders or cylinders with finite surface impedance.

The inability of the on-surface radiation boundary condition to model the edge-current singularity behavior has been addressed by Umashankar, Chun, and Taflove (1991), with good success. Using a concept analogous to the physical theory of diffraction, they have introduced in the expression of the scattered field a contribution due to the singular behavior of the induced current near edges. The results so obtained agreed well with those provided by the moment method also in the case of scatterers with sharp edges such as strips or triangular cylinders.

Recently, the on-surface radiation condition approach has been extended by Arendt, Umashankar, Taflove, and Kriegsmann (1990) to the case of scattering from convex homogeneous dielectric cylinders. Their procedure consists of expanding the field interior to the largest circle C completely inscribed into the scatterer in a series of Bessel functions. Then the analytic continuation method (Wilton and Mittra 1972) is applied to express the field at the point inside lying on the scatterer contour Γ. Eventually, imposing the absorbing boundary conditions at Γ, a linear equation system whose coefficient matrix is full is obtained. Each coefficient is constructed calculating a double, triple, or multiple series of Bessel functions according to the number of times that the continuation method has been applied to cover the whole scatterer cross-section.

So far, the on-surface radiation condition appears not to have been exploited in connection with the finite element method for the analysis of scattering from convex, inhomogeneous dielectric scatterers. This is probably due to the limitations of the on-surface radiation condition approach, which have been pointed out by Janaswamy (1991), who analyzed the effects of the permittivity and permeability values of the materials involved on result accuracy. This analysis indicated that the on-surface radiation boundary condition fails for certain values of the electric and magnetic constants (ϵ_r, μ_r) of the scatterer. The failure is primarily due to the inability of absorbing boundary conditions to deal with the reactive components of electromagnetic fields, which can be significant in the vicinity of dielectric scatterers.

Arbitrary contour Γ
Several authors have described techniques for imposing absorbing boundary conditions on a contour Γ conformable with the scatterer shape, which results in minimizing the size of the interior region that must be discretized by the finite element method. The first such work published appears to be that by Ma [1991], who presented an analytical approach to the problem. Other papers, notably those of Ramahi, Khebir, and Mittra [1991], and Jin, Volakis, and Liepa (1992), are based on numerical approaches.

Ma [1991] devised a set of absorbing boundary operators for the wave envelope function, which are expressed in the body-fitted coordinate system. Since it is possible to impose such operators on a boundary Γ

that exactly follows the shape of the scatterer cross section, this formulation minimizes the geometric domain for computation, thus saving memory and computing time. However, the procedure proposed by Ma is subject to a potentially serious restriction. It is valid for convex cylinders bounded by smooth curves (*i.e.*, having a radius of curvature that varies continuously along the contour of the scatterer). Moreover, smoothness only in the first derivative (with respect to distance along the curve) is not sufficient; the radius of curvature must be smooth to a higher order, depending on the order of the operator to be applied.

It is worth noting that the operators derived by Ma are equivalent to those developed by Bayliss and Turkel in the special case of a circular contour Γ. Moreover, the operators of finite order can be applied directly on the surface of the scatterer using the on-surface radiation condition approach, provided that the radius of curvature of the scatterer is not too large. For large radii, asymptotic expansions of the boundary operators break down, and an infinite-order boundary operator is to be applied.

Ramahi, Khebir, and Mittra [1991] have derived absorbing boundary conditions via numerical approximation of the analytically exact boundary condition. They expressed the outward normal derivative at a point n_0, lying on the contour Γ, as a linear combination of the field values at the nodes lying in the interior region and directly linked to n_0. This formulation is independent of the scatterer shape and it can yield more accurate results than the second-order Bayliss and Turkel absorbing boundary conditions do. Two difficulties, having different origins, arise in the application of this numerical technique. First, the accuracy of approximation depends on the nature and shape of the scattering body. The numerical absorbing boundary conditions have been derived by expressing the scattered field in terms of a series of outgoing cylindrical wave functions, a series uniformly convergent only outside a circular cylindrical region that completely encloses the scatterer; but it is subsequently employed everywhere on the arbitrary chosen contour Γ, parts of which must lie inside the circumscribing circle if Γ is not itself circular. The second difficulty is related to indeterminacy in the choice of terms in the series expansion used to derive the numerical approximation. Although the choice can be guided by an empirical criterion established by the authors through numerical and analytical studies (Mittra and Ramahi 1990), a certain degree of indeterminacy is still present and must be removed by trial and error.

Recently, Jin, Volakis, and Liepa (1992) have devised an interesting *ad hoc* approach to the open domain problem. Their proposed method consists in introducing a fictitious absorber (a loaded metallic surface) for enclosing the region discretized by the finite element method. The metallic surface employed to truncate the finite element mesh is loaded with three layers of different dielectric materials. The relative permittivities and permeabilities (ϵ_r, μ_r) of the dielectric are determined by optimizing the plane-wave absorption of a planar metallic infinite surface loaded by such a layered dielectric. Given a fixed thickness for each of the three dielectric layers, the optimization is carried out so as to minimize the reflection coefficient magnitude over the widest possible range of values of the plane wave incidence angle. Basically, this technique is similar to that proposed by Taflove and Brodwin (1975) in the framework of a finite difference formulation for the solution of electromagnetic scattering problems. The main difference between the two methods lies in the composition of the fictitious absorber. While Taflove and Brodwin employed a homogeneous but lossy dielectric material, Jin, Volakis, and Liepa have proposed a lossy stratified structure. This generalization has allowed the authors to reduce the required thickness of the fictitious absorber, thus minimizing the area of the interior region. The results presented by Jin, Volakis, and Liepa demonstrate that their method has advantages with respect to the traditional finite element method with absorbing boundary conditions. In fact, it provides the same degree of accuracy as the second-order Bayliss and Turkel absorbing boundary conditions even if the fictitious loaded metallic enclosure is placed near the scatterer.

The technique proposed by Jin, Volakis, and Liepa is based on well known concepts that have been extensively exploited in devising absorbing structures in the field of radar cross-section analysis. Consequently, it suffers from the same restrictions: since the optimization procedure is carried out for a fixed frequency, an infinitely-extending loaded metallic plane, and an incident plane wave, the technique is effective only over a narrow range of frequencies.

Three-Dimensional Scalar Absorbing Boundary Conditions

As already mentioned, the absorbing boundary conditions for three-dimensional scalar problems were obtained by Mur in 1981, following the procedure outlined by Engquist and Majda. However, Mur verified the validity of these expressions by analyzing only certain canonical two-dimensional problems. Subsequently, Umashankar and Taflove (1982) applied Mur's expressions to the analysis of electromagnetic scattering from complex objects in the framework of a finite difference time domain formulation.

Mittra, Ramahi, Khebir, Gordon, and Kouki (1989), still starting from the asymptotic expansion of the scattered field suggested by Wilcox, derived scalar three-dimensional absorbing boundary conditions similar to those introduced by Bayliss, Gunzburger, and Turkel. They employed a recursive technique that they had already presented for the two-dimensional case. However, they did not provide any numerical results to illustrate or verify their proposed absorbing boundary conditions.

The first application of scalar three-dimensional absorbing boundary conditions to practical problems in connection with a finite element formulation is due to D'Angelo and Mayergoyz (1990). They analyzed antenna and scattering problems using both the absorbing boundary conditions introduced by Bayliss, Gunzburger, and Turkel and those suggested by Engquist and Majda. It is worth noting that the use of a fictitious spherical surface S to truncate the finite element mesh is inefficient in the case of long and slender structures, while a rectangular surface is generally preferable. On the other hand, the imposition of the Bayliss, Gunzburger, and Turkel absorbing boundary conditions on planar or rectangular surfaces is cumbersome and does not yield accurate results (D'Angelo and Mayergoyz 1989). For this reason, D'Angelo and Mayergoyz (1990) found it preferable to impose the Engquist and Majda absorbing boundary conditions when the surface S is piecewise planar. The disadvantage of this formulation is that, although it preserves the banded nature of the global matrix, it destroys matrix symmetry. Consequently, a greater amount of main memory and computing time are needed to solve the electromagnetic problem.

Three-Dimensional Vector Absorbing Boundary Conditions

The first sequence of differential operators of increasing order suitable for imposing absorbing boundary conditions to vector fields was developed by Peterson in 1988 [included with this chapter]. He applied the procedure outlined by Bayliss and Turkel (1980) to the vectorial form of the Wilcox (1956) expansion for the outgoing solution of the Helmholtz equation. The mth-order operator in this sequence gives rise to a local absorbing boundary condition, much like its counterpart in the scalar case, which matches the true solution Φ to the first $2m$ terms of the Wilcox expansion.

Although the operators proposed by Peterson can be easily imposed on spherical surfaces, they have the drawback of yielding a nonsymmetric coefficient matrix. On the other hand, the application of the differential operators derived by Webb and Kanellopoulos [1989], which are equally suited to vector fields, can preserve the symmetry of the coefficient matrix. The latter set of boundary operators includes the Peterson operators as a special case and is still easily imposed on spherical surfaces. In a subsequent paper, Kanellopoulos and Webb [1991] demonstrated the superiority of their second-order differential operator over the first-order operator, both in accuracy — for a given distance of the fictitious surface S from the scatterer — and in speed of convergence as the surface S is placed at increasing distances from the scatterer. Since the computational effort of imposing the first- or second-order absorbing boundary conditions is practically the same, the second-order condition appears preferable for practical use.

A comparative evaluation of the two kind of second-order vector absorbing boundary conditions — those by Peterson [1988] and those by Webb and Kanellopoulos [1989] — was carried out by Peterson [1992]. On applying them to a set of spherical vector harmonics of increasing order (Stratton 1941) on a spherical surface, he found that the two absorbing boundary conditions act differently only on the $\mathrm{T}M_r$ part of the electromagnetic field (i.e., the part transverse with respect to r). Moreover, the Peterson conditions give rise to slightly smaller errors than those of Webb and Kanellopoulos. Peterson also pointed out that, much like the behavior observed for the two-dimensional Bayliss-Turkel conditions (Mittra and Ramahi 1990), the error in the three-dimensional vector absorbing boundary conditions increases with increases in the order of the vector spherical harmonic to which they are applied.

Recently, Chatterjee, Jin, and Volakis [1993] used the vector absorbing boundary conditions developed by Webb and Kanellopoulos [1989] to solve three-dimensional scattering problems. They chose first-order tetrahedral edge elements to implement the finite element method inside a fictitious spheric or cubic surface S, depending on the shape of the scatterer. Chatterjee, Jin, and Volakis used the scattered electric field as the unknown of the problem, and their results agree well with analytical ones or laboratory measurements even in the case of non-spherical fictitious surface placed at a small distance, only 0.3λ, from the scatterer edges. The encouraging results obtained by Chatterjee, Jin, and Volakis deserve further numerical investigations to determine which is the more accurate formulation — scattered or total electric field as the unknown — as well as to find the best kind of vector elements to use for solving such kind of problems — tetrahedral edge elements or covariant projection elements.

References

Arendt, S., K. R. Umashankar, A. Taflove, and G. Kriegsmann (1990): "Extension of on-surface radiation condition theory to scattering by two-dimensional homogeneous dielectric objects." *IEEE Transactions on Antennas and Propagation*, vol. **38**, pp. 1551–58, 1990.

Bayliss, A., C. I. Goldstein, and E. Turkel (1985): "On accuracy conditions for the numerical computation of waves." *Journal of Computational Physics*, vol. **59**, pp. 396–404, 1985.

Bayliss, A., M. Gunzburger, and E. Turkel [1982]: "Boundary conditions for the numerical solution of elliptic equations in exterior regions." *SIAM Journal of Applied Mathematics*, vol. **42**, pp. 430–51, 1982.

Bayliss, A. and E. Turkel (1980): "Radiation boundary conditions for wave-like equations." *Communications on Pure and Applied Mathematics*, vol. **33**, pp. 707–25, 1980.

Blaschak, J. G. and G. A. Kriegsmann (1988): "A comparative study of absorbing boundary conditions." *Journal of Computational Physics*, vol. **77**, pp. 109–39, 1988.

Chatterjee, A., J. M. Jin, and J. L. Volakis [1993]: "Edge-based finite elements and vector ABC's applied to 3-D scattering." *IEEE Transactions on Antennas and Propagation*, vol. **41**, pp. 221–26, 1993.

D'Angelo, J. and I. D. Mayergoyz (1989): "On the use of local absorbing boundary conditions for RF scattering problems." *IEEE Transactions on Magnetics*, vol. **25**, pp. 3040–42, 1989.

D'Angelo, J. and I. D. Mayergoyz (1990): "Finite element methods for the solution of RF radiation and scattering problems." *Electromagnetics*, vol. **10**, pp. 177–99, 1990.

Engquist, B. and A. Majda [1977]: "Absorbing boundary conditions for the numerical simulation of waves." *Mathematics of Computation,* vol. **31**, pp. 629–51, 1977.

Hingdon, R. L. (1987): "Numerical absorbing boundary conditions for the wave equation." *Mathematics of Computation*, vol. **49**, pp. 65–90, 1987.

Hitoshi, K. (1974): *Pseudo-Differential Operator.* Cambridge, MA: MIT Press, 1974.

Janaswamy, R. (1991): "On the applicability of OSRC method to homogeneous scatterers." *IEEE Transactions on Antennas and Propagation*, vol. **39**, pp. 862–67, 1991.

Jin, J. M., J. L. Volakis, and V. V. Liepa (1989): "A comparative study of the OSRC approach in electromagnetic scattering." *IEEE Transactions on Antennas and Propagation*, vol. **37**, pp. 118–24, 1989.

Jin, J. M., J. L. Volakis, and V. V. Liepa (1992): "Fictitious absorber for truncating finite element meshes in scattering." *IEE Proceedings–H*, vol. **139**, pp. 472–76, 1992.

Kanellopoulos, V. N. and J. P. Webb [1991]: "A numerical study of vector absorbing boundary conditions for the finite element solution of Maxwell's equations." *IEEE Microwave and Guided Wave Letters*, vol. **1**, pp. 325–27, 1991.

Karp, S. N. (1961): "A convergent 'farfield' expansion for two-dimensional radiation functions." *Communications on Pure and Applied Mathematics*, vol. **14**, pp. 427–34, 1961.

Khebir, A., A. B. Kouki, and R. Mittra (1990): "Higher order asymptotic boundary conditions for the finite element modeling of two-dimensional transmission line structures." *IEEE Transactions on Microwave Theory and Techniques*, vol. **38**, pp. 1433–38, 1990.

Kriegsmann, G. A., A. Taflove, and K. R. Umashankar (1987): "A new formulation of electromagnetic wave scattering using an on-surface radiation boundary condition approach." *IEEE Transactions on Antennas and Propagation*, vol. **AP-35**, pp. 153–61, 1987.

Lee, C. F., R. T. Shin, J. A. Kong, and B. J. McCartin [1990]: "Absorbing boundary conditions on circular and elliptical boundaries." *Journal of Electromagnetic Waves and Applications*, vol. **4**, pp. 945–62, 1990.

Lindman, E. L. (1975): "'Free-space' boundary conditions for the time dependent wave equation." *Journal of Computational Physics*, vol. **18**, pp. 66–78, 1975.

Ma, Y. C. [1991]: "A note on the radiation boundary conditions for the Helmholtz equation." *IEEE Transactions on Antennas and Propagation*, vol. **39**, pp. 1526–30, 1991.

Mittra, R. and O. Ramahi (1990): "Absorbing boundary conditions for the direct solution of partial differential equations arising in electromagnetic scattering problems." *Finite Element and Finite Difference Methods in Electromagnetic Scattering*, vol. II, M. Morgan, ed. New York: Elsevier, 1990.

Mittra, R., O. Ramahi, A. Khebir, R. Gordon, and A. Kouki (1989): "A review of absorbing boundary conditions for two and three-dimensional electromagnetic scattering problems." *IEEE Transactions on Magnetics*, vol. **25**, pp. 3034–39,

1989.

Moore, T. G., J. G. Blaschak, A. Taflove, and G. A. Kriegsmann [1988]: "Theory and application of radiation boundary operators." *IEEE Transactions on Antennas and Propagation*, vol. **36**, pp. 1797–812, 1988.

Mur, G. [1981]: "Absorbing boundary conditions for the finite-difference approximation of the time-domain electromagnetic-field equation." *IEEE Transactions on Electromagnetic Compatibility*, vol. **EMC–23**, pp. 377–82, 1981.

Peterson, A. F. [1988]: "Absorbing boundary conditions for the vector wave equation." *Microwave and Optical Technology Letters*, vol. **1**, pp. 62–4, 1988.

Peterson, A. F. [1992]: "Accuracy of 3-D radiation boundary conditions for use with the vector Helmholtz equation." *IEEE Transactions on Antennas and Propagation*, vol. **40**, pp. 351–55, 1992.

Peterson, A. F. and S. P. Castillo [1989]: "A frequency-domain differential equation formulation for electromagnetic scattering from inhomogeneous cylinders." *IEEE Transactions on Antennas and Propagation*, vol. **37**, pp. 601–7, 1989.

Povinelli, M. J. and J. D'Angelo (1991): "Finite element analysis of large wavelength antenna radome problems for leading edge and radar phased array." *IEEE Transactions on Magnetics*, vol. **27**, pp. 4299–302, 1991.

Ramahi, O., A. Khebir, and R. Mittra [1991]: "Numerically derived absorbing boundary conditions for the solution of open region scattering problems." *IEEE Transactions on Antennas and Propagation*, vol. **39**, pp. 350–53, 1991.

Ramahi, O. and R. Mittra [1989]: "Finite-element analysis of dielectric scatterers using the absorbing boundary conditions." *IEEE Transactions on Magnetics*, vol. **25**, pp. 3043–45, 1989.

Stratton, J. A. (1941): *Electromagnetic Theory.* New York: McGraw-Hill, 1941.

Sumbar, E., F. E. Vermeulen, and F. S. Chute (1991): "Implementation of radiation boundary conditions in the finite element analysis of electromagnetic wave propagation." *IEEE Transactions on Microwave Theory and Techniques*, vol. **39**, pp. 267–73, 1991.

Taflove, A. and M. Brodwin (1975): "Numerical solution of steady-state electromagnetic scattering problems using the time-dependent Maxwell's equations." *IEEE Transactions on Microwave Theory and Techniques*, vol. **MTT-23**, pp. 623–30, 1975.

Trefethen, L. N. and L. Halpern [1986]: "Well-posedness of one-way wave equations and absorbing boundary conditions." *Mathematics of Computation*, vol. **47**, pp. 421–35, 1986.

Umashankar, K., W. Chun, and A. Taflove (1991): "Simple analytical solution to electromagnetic scattering by two-dimensional conducting object with edges and corners: Part I: TM polarization." *IEEE Transactions on Antennas and Propagation*, vol. **39**, pp. 1665–71, 1991.

Umashankar, K. R. and A. Taflove (1982): "A novel method to analyze electromagnetic scattering of complex objects." *IEEE Transactions on Electromagnetic Compatibility*, vol. **EMC–24**, pp. 397–405, 1982.

Webb, J. P. [1990]: "Absorbing boundary conditions for the finite element analysis of planar devices." *IEEE Transactions on Microwave Theory and Techniques*, vol. **38**, pp. 1328–32, 1990.

Webb, J. P. and V. N. Kanellopoulos [1989]: "Absorbing boundary conditions for the finite element solution of the vector wave equation." *Microwave and Optical Technology Letters*, vol. **2**, pp. 370–72, 1989.

Wilcox, C. H. (1956): "An expansion theorem for electromagnetic fields." *Communications on Pure and Applied Mathematics*, vol. **9**, pp. 115–32, 1956.

Wilton, D. and R. Mittra (1972): "A new numerical approach to the calculation of electromagnetic scattering properties of two dimensional bodies of arbitrary cross-section." *IEEE Transactions on Antennas and Propagation,* vol. **AP-20**, pp. 310–17, 1972.

Absorbing Boundary Conditions for the Numerical Simulation of Waves

B. ENGQUIST and A. MAJDA

Abstract. In practical calculations, it is often essential to introduce artificial boundaries to limit the area of computation. Here we develop a systematic method for obtaining a hierarchy of local boundary conditions at these artificial boundaries. These boundary conditions not only guarantee stable difference approximations but also minimize the (unphysical) artificial reflections which occur at the boundaries.

Introduction. When calculating solutions to partial differential equations it is often essential to introduce artificial boundaries to limit the area of computation. Important areas of application which use artificial boundaries are local weather prediction (see [8] and [9]), geophysical calculations involving acoustic and elastic waves (see [6] and [7]), and a variety of other problems in fluid dynamics. One needs boundary conditions at these artificial boundaries in order to guarantee a unique and well-posed solution to the differential equation. In turn, this is a necessary condition to guarantee *stable* difference approximation. Of course, one hopes that these artificial boundaries and boundary conditions affect the solution in a manner such that it closely approximates the free space solution which exists in the absence of these boundaries. In particular, one would like to minimize the amplitudes of waves reflected from these artificial boundaries.

In this work we develop perfectly absorbing boundary conditions for general classes of wave equations by applying the recently developed theory for reflection of singularities for solutions of differential equations (see [2], [3], [4]). Unfortunately, these boundary conditions necessarily have to be nonlocal in both space and time and thus are not useful for practical calculations. Hence, in the main part of the paper, we derive a hierarchy of highly absorbing local boundary conditions which approximate the theoretical nonlocal boundary condition. Of course, it is also very important for applications that these boundary conditions generate well-posed mixed initial boundary value problems. (See [1] for a discussion of well-posedness.) The general approach is applied specifically to produce absorbing artificial boundaries for the acoustic wave equation in Cartesian and polar coordinates and for the linearized shallow water equations in two space dimensions.

We illustrate our results by the following hierarchy of absorbing boundary conditions at the wall $x = 0$ for the scalar wave equation,

$$w_{tt} = w_{xx} + w_{yy}, \qquad t, x \geqslant 0,$$

Reprinted from "Well-Posedness of One-Way Wave Equations and Absorbing Boundary Conditions," by L. N. Trefethen and L. Halpern, *Mathematics of Computation*, vol. 47, 1986, pp. 421–543, by permission of the American Mathematical Society.

(1) $$w_x - w_t|_{x=0} = 0,$$

(2) $$w_{xt} - w_{tt} + \tfrac{1}{2}w_{yy}|_{x=0} = 0,$$

(3) $$w_{xtt} - w_{ttt} - \tfrac{1}{4}w_{xyy} + \tfrac{3}{4}w_{tyy}|_{x=0} = 0.$$

Each of these bourdary conditions gives a well-posed mixed initial boundary value problem for the wave equation. We present an example in Section 1 where another obvious highly absorbing approximation generates a strongly ill-posed mixed problem. Thus, local absorbing approximations must be chosen carefully by using the normal mode analysis for mixed problems (see [1] for definitions) as a guideline. All of the above boundary conditions are perfectly absorbing at normal incidence. For a 45° angle of incidence, (1), (2), and (3) reflect waves with amplitudes, respectively, 17%, 3%, and .5% of the amplitude of the incident wave. Close to glancing, all reflection coefficients tend to unity. This is inherent in the problem; however, the corresponding reflected waves propagate slowly in the normal direction and hence do not affect the calculation in the interior.

In Section 1, we develop a theory of highly absorbing boundary conditions for second order wave equations. In Section 2, an analogous theory is developed for general symmetric hyperbolic systems. We illustrate this theory in Section 3 by explicitly calculating absorbing boundary conditions for the linearized shallow water equations; included in this section are some amusing observations from perturbation theory which have bearing here. Finally, in Section 4, mixed initial boundary problems with absorbing boundary conditions for the scalar wave equation and the linearized shallow water equations are approximated by difference schemes. The results of numerical experiments displayed here strongly indicate the value of our systematic approach as a practical tool.

The topic in this paper has been of considerable recent interest in the computational literature (see [9], [10], [11], [12], and [13]). In [9] and [11] the analogue of the first approximation (of Sommerfeld type) is implemented in nonlinear calculations while in [12] a special approximation is developed then analyzed in one space dimension. The wave equation in a half-space is studied in [10] and [13]. The starting point for [10] appears to be the same nonlocal perfectly absorbing condition developed in (1.6) below. The methods in [10] treat the nonlocal boundary condition directly and require storage of from three to six time levels on the boundary while in [13] solutions of the Neumann and Dirichlet problems are combined. The methods described in both of these papers are ingenious but rather special and appear to apply only to the constant coefficient scalar wave equation in a half-space. (In [13], some equations of elasticity are also analyzed.)

1. Highly Absorbing Boundary Conditions for Second Order Wave Equations.

In this section we shall begin with a special example which motivates the manner in which we construct nonlocal perfectly absorbing boundary conditions together with suitable highly absorbing local approximations.

1.A. *The Theoretical Nonlocal Boundary Condition.* We consider solutions of the wave equation

(1.1) $$\Box w = \frac{\partial^2 w}{\partial t^2} - \frac{\partial^2 w}{\partial x^2} - \frac{\partial^2 w}{\partial y^2} = 0$$

in the half-space, $x \geqslant 0$. We denote the dual variables to (y, t) by (ω, ξ). Special families of solutions to the wave equation representing waves travelling to the left are given by the plane waves

(1.2) $$w = e^{i(\sqrt{\xi^2-\omega^2}x+\xi t+\omega y)}$$

with $\xi^2 - \omega^2 > 0$, $\xi > 0$. If (ω, ξ) is held fixed, one first order differential boundary condition which annihilates w has the form,

(1.3) $$\left(\frac{d}{dx} - i\sqrt{\xi^2 - \omega^2}\right)w|_{x=0} = 0.$$

Thus, for all waves as in (1.2) with (ξ, ω) held fixed, the boundary condition in (1.3) produces no reflections. On the other hand, more general wave packets travelling to the left can be represented by

(1.4) $$w(x, y, t) = \iint e^{i(\sqrt{\xi^2-\omega^2}x+\xi t+wy)}\rho(\xi, \omega)\hat{w}(0, \xi, \omega)\,d\xi\,d\omega.$$

Here ^ denotes Fourier transform in the (y, t) variables, $w(0, \xi, \omega)$ is a special amplitude function with support for large (ξ, ω) in the cone $\xi^2 > \omega^2$, and $\rho(\xi, \omega)$ is a smooth function homogeneous of degree zero for $|\xi| + |\omega|$ large with support in $\xi^2 > \omega^2$ for (ξ, ω) large and identically one on a neighborhood of the support of $\hat{w}(0, \xi, \omega)$. We note that by Fourier's inversion formula, $w(0, y, t)$ is given by the inverse Fourier transform of the amplitude function, $\hat{w}(0, \xi, \omega)$. By superposition of the calculations for (1.3), the boundary condition which *exactly* annihilates singular wave packets travelling to the left of the form in (1.4) is given by

(1.5) $$\frac{dw}{dx} - \iint e^{i(\xi t+\omega y)} i\sqrt{\xi^2 - \omega^2}\rho\hat{w}(0, \xi, \omega)\,d\xi\,d\omega|_{x=0} = 0.$$

Now, we express the mapping

$$\iint e^{i(\xi t+\omega y)} i\sqrt{\xi^2 - \omega^2}\rho\hat{w}(0, \xi, \omega)\,d\xi\,d\omega$$

using the symbolic notation of *pseudo-differential operators* (see the article by Nirenberg in [3]). Suppose $b(y, t, \omega, \xi)$ is a formal asymptotic sum of terms with descending degrees of homogeneity in (ξ, ω) for (ξ, ω) large. Then a pseudo-differential operator is defined through the Fourier transform by the recipe,

$$b\left(y, t, \frac{\partial}{\partial y}, \frac{\partial}{\partial t}\right)w = \iint e^{i(\omega y+\xi t)}b(y, t, i\omega, i\xi)\hat{w}(\xi, \omega)\,d\xi\,d\omega.$$

Thus, the perfectly absorbing boundary condition in (1.5) can be written in the form

(1.6) $$\left(\frac{d}{dx} - \rho\left(\frac{\partial}{\partial y}, \frac{\partial}{\partial t}\right)\sqrt{\frac{\partial^2}{\partial t^2} - \frac{\partial^2}{\partial y^2}}\right)w|_{x=0} = 0.$$

Now, in general, it is not possible to design boundary conditions so that no reflections occur; nevertheless, it is possible to construct boundary conditions so that for an arbitrary *highly singular wave packet* (provided it contains no glancing directions) only low amplitude smooth reflections occur. (Of course, ρ can be adjusted to *exactly* annihilate the wave packets from (1.4).) In fact, by applying the construction of Lax and Nirenberg from [3] and the algebraic criterion for perfect reflection developed by Majda and Osher in [2], one verifies that necessarily the *only boundary conditions* with these properties are multiples of the boundary condition in (1.6) with ρ adjusted according to the largest angle of incidence of rays containing the singularities of ω.

1.B. *A Hierarchy of Highly Absorbing Local Boundary Conditions.* Unfortunately, the perfectly absorbing boundary condition developed above in (1.6) is necessarily nonlocal in *both space* and *time.* This boundary condition is impractical from a computational point of view since to advance one time level at a single point requires information from all previous times over the entire boundary. Thus, we shall develop highly absorbing local approximations to the perfectly absorbing boundary condition in (1.6). Necessarily, we want to build boundary conditions satisfying the following two additional criteria:

(1.7)
(1) These boundary conditions are local.
(2) The boundary conditions lead to a well-posed mixed boundary value problem for the wave equation.

We necessarily need (2) to be satisfied; otherwise, it would be impossible to construct *stable* difference approximations to the associated boundary value problem while, as remarked above, (1) is essential for reasonable control of the operation count.

We take the symbol of the boundary condition from (1.6) given by

$$d/dx - i\xi\sqrt{1 - \omega^2/\xi^2} \tag{1.8}$$

and approximate it at normal incidence ($\omega = 0$). Using the approximation $\sqrt{1 - \omega^2/\xi^2} = 1 + O(\omega^2/\xi^2)$ and recalling that $i\xi$ corresponds to $\partial/\partial t$, we obtain

1st approximation:

$$\left(\frac{\partial}{\partial x} - \frac{\partial}{\partial t}\right) w|_{x=0} = 0. \tag{1.9}$$

Using the next approximation (the first Taylor or Padé) to the square root, $\sqrt{1 - \omega^2/\xi^2} = 1 - \frac{1}{2}\omega^2/\xi^2 + O(\omega^4/\xi^4)$ in (1.8) and multiplying by $i\xi$, we obtain a symbol

$$i\xi\frac{\partial}{\partial x} + \xi^2 - \frac{1}{2}\omega^2;$$

and this yields the boundary condition,

2nd approximation:

$$\left(\frac{\partial^2}{\partial x \partial t} - \frac{\partial^2}{\partial t^2} + \frac{1}{2}\frac{\partial^2}{\partial y^2}\right) w|_{x=0} = 0. \tag{1.10}$$

Remark. The 1st approximation is a familiar one which works exactly for a single space dimension and could easily be derived from physical considerations. Numerical experiments (see Section 4 of this paper) indicate that the 2nd approximation provides a significant improvement over the first (see the discussion below for the theoretical reasons); the authors do not know of any derivation of this approximation from "ad hoc" physical reasoning.

When $\xi = 1$, ω has the interpretation of $\omega = \sin\theta$ where θ is the angle of incidence. Thus, if a wave of the form $a(\xi, \omega)e^{i\sqrt{\xi^2-\omega^2}x + i\omega y + i\xi t}$ strikes the boundary at the angle θ, the first approximation produces a reflected wave

$$be^{-i\sqrt{\xi^2-\omega^2}x + i\omega y + i\xi t}$$

with amplitude b given by

$$b = a\left(\frac{\cos\theta - 1}{\cos\theta + 1}\right),$$

while the second approximation produces an *even weaker* reflected wave

$$be^{-i\sqrt{\xi^2-\omega^2}x + i\omega y + i\xi t}$$

with amplitude

$$b = -a\left|\frac{\cos\theta - 1}{\cos\theta + 1}\right|^2.$$

We note that these conclusions are valid at both *low* and *high frequencies* (the numerical experiments in Section 4 confirm this).

We also claim that the 1st and 2nd approximation also satisfy the criterion in (2) above. Standard energy estimates prove that the 1st approximation is a maximal dissipative boundary condition and, therefore, is trivially well posed. The verification that the 2nd approximation is well posed is more subtle. We recall the general algebraic normal mode analysis for checking well-posedness (see the paper of Kreiss, [1]) specialized for the wave equation:

(1.11) A mixed initial boundary value problem, $\mathcal{B}$, for the variable coefficient wave equation is well posed if there are no solutions to the frozen coefficient half-space problems of the form

$$w(s, \omega) = e^{-\sqrt{s^2+\omega^2}x+i\xi t+i\omega y}$$

with $\operatorname{Re} s \geqslant 0$ and satisfying,

$$\Box w(s, \omega) = 0, \qquad \mathcal{B}w(s, \omega)|_{x=0} = 0;$$

furthermore, for $|s|^2 + |\omega|^2 = 1$ and $\operatorname{Re} s \geqslant 0$, $\mathcal{B}w(s, \omega)$ should satisfy $|\mathcal{B}w(s, \omega)| \geqslant c_0 > 0$.

Applying the criterion in (1.11) to the 2nd approximation, this approximation yields a well-posed problem unless

(1.12) $$-s\sqrt{s^2 + \omega^2} = (s^2 + \tfrac{1}{2}\omega^2) \quad \text{for some } s \text{ with } \operatorname{Re} s \geqslant 0.$$

Squaring the identity in (1.12) we obtain that necessarily $\omega = 0$ but when $\omega = 0$ the above identity yields $2s^2 = 0$, thus the problem is well posed.

It is natural to try higher order approximations to $\sqrt{1 - \omega^2/\xi^2}$ for even better absorption. The 2nd approximation involves approximating $\sqrt{1 - \omega^2/\xi^2} \cong 1 - \tfrac{1}{2}\omega^2/\xi^2$; this approximation coincides with both the first Taylor and first Padé approximations to $\sqrt{1 + x}$ at $x = 0$. Now, we investigate highly absorbing boundary conditions derived from the second Padé and Taylor approximations, respectively; these both reflect waves with amplitudes diminished by $|(\cos\theta - 1)/(\cos\theta + 1)|^3$ where θ is the angle of incidence.

We claim that the highly absorbing *boundary condition* derived from the *2nd Padé approximation is well posed.* To derive this boundary condition we use the second rational approximation

$$\sqrt{1 + x} = 1 + \frac{x}{2 + x/2} + O(|x|^3)$$

and use a similar analysis as in the above to arrive at the third order boundary condition

3rd approximation:

(1.13) $$\left(\frac{\partial^3}{\partial t^2 \partial x} - \frac{1}{4}\frac{\partial^3}{\partial x \partial y^2} - \frac{\partial^3}{\partial t^3} + \frac{3}{4}\frac{\partial^3}{\partial t \partial y^2}\right) w|_{x=0} = 0.$$

To check the well-posedness of this boundary value problem following (1.11) we need only show that there are no roots with $\operatorname{Re} s \geqslant 0$ for

(1.14) $$\left(s^2 + \frac{\omega^2}{4}\right)\sqrt{s^2 + \omega^2} + s^3 + \frac{3}{4}s\omega^2 = 0.$$

If we set $s/\omega = z$, we obtain

$$(z^2 + \tfrac{1}{4})\sqrt{1 + z^2} = -(z^3 + \tfrac{3}{4}z)$$

and squaring both sides and subtracting, we obtain $1/16 = 0$ thus there are no non-trivial solutions in (1.14) and the 3rd approximation is well posed.

We also claim that in contrast to the above situation, the *absorbing boundary condition corresponding to* the *second Taylor approximation* is *strongly ill posed.* Thus, this local approximation is *useless* for numerical purposes. To derive this boundary condition we use

$$\sqrt{1+x} = 1 + \frac{1}{2}x - \frac{1}{8}x^2 + O(x^3)$$

so that by following the above procedure we obtain the fourth order boundary condition,

4th approximation:

$$\left(\frac{\partial^4}{\partial x \partial t^3} - \frac{\partial^4}{\partial t^4} + \frac{1}{2}\frac{\partial^4}{\partial y^2 \partial t^2} + \frac{1}{8}\frac{\partial^4}{\partial y^4}\right) w\Big|_{x=0} = 0. \tag{1.15}$$

To check the well-posedness of the above boundary condition, we must check that there are no roots with $\operatorname{Re} s \geq 0$ to

$$-s\sqrt[3]{s^2 + \omega^2} - s^4 + \frac{1}{2}\omega^2 s^2 + \frac{1}{8}\omega^4 = 0. \tag{1.16}$$

Dividing by ω^4 and setting $z = s/\omega$, we obtain

$$-z\sqrt[3]{1 + z^2} - z^4 + \frac{1}{2}z^2 + \frac{1}{8} = 0. \tag{1.17}$$

Consider z with z real and $z \geq 0$. At $z = 0$, the above expression has the value $1/8$, while as z tends to ∞, the expression in (1.17) tends to $-\infty$. Thus a nontrivial *positive* real root exists and these boundary conditions are strongly ill posed.

Remark #1. These two examples indicate that one must be *careful* in choosing the *appropriate local approximation* to the nonlocal boundary condition *in order to guarantee stability.*

Remark #2. *In the situation where the initial data* or *inhomogeneous forcing functions have support up to the boundary,* one must be very careful in applying the above boundary conditions so that no additional discontinuities are introduced. (This difficulty exists for any mixed problem.) We illustrate this difficulty in the following example:

If we consider the free space solution of

$$w_{tt} - w_{xx} = 0, \qquad w|_{t=0} = 0, \qquad w_t|_{t=0} = 1,$$

then $w \equiv t$. On the other hand, if we straightforwardly compute using homogeneous absorbing boundary conditions,

$$\tilde{w}_{tt} - \tilde{w}_{xx} = 0 \quad \text{for } x \geq 0,$$

$$\tilde{w}_t - \tilde{w}_x|_{x=0} = 0, \qquad \tilde{w}|_{t=0} = 0, \qquad \tilde{w}_t|_{t=0} = 1,$$

we get a solution with discontinuity in the first derivative given by

$$\begin{aligned} \tilde{w} &= \tfrac{1}{2}(x + t), & t &\geq x, \\ &= t, & t &\leq x. \end{aligned}$$

To control this difficulty we must make our initial and boundary data compatible and solve an *inhomogeneous* absorbing boundary value problem

$$\approx{w}_{tt} - \approx{w}_{xx} = 0, \qquad x \geq 0,\ t \geq 0,$$

$$\approx{w}_t - \approx{w}_x|_{x=0} = g(t) \quad \text{for } t \geq 0,$$

$$\tilde{\tilde{w}}|_{t=0} = 0, \qquad \tilde{\tilde{w}}_t|_{t=0} = 1,$$

where $g(t)$ is a forcing function with $g(0) = \lim_{x\to 0^+} \partial\tilde{\tilde{w}}/\partial t - \partial\tilde{\tilde{w}}/\partial x = 1$ (the limits in the last equation are calculated using the Cauchy data). No additional singular waves are produced as long as $g(0)$ satisfies the above condition and with $g(t) \equiv 1$, we recover $\tilde{\tilde{w}} \equiv t \equiv w$. (General compatibility conditions for hyperbolic mixed problems are discussed in [5].)

1.C. *Modifications for Curved Boundaries and Variable Coefficient Wave Equations.* Here we sketch how more sophisticated versions of the above arguments yield nonlocal perfectly absorbing and highly absorbing local boundary conditions for variable coefficient wave equations. The arguments we give below work for general variable coefficient wave equations. For the application and simplicity in exposition we consider wave equations of the form

$$Lw \equiv \left(\left(\frac{\partial^2}{\partial x^2} + g(x, y)\frac{\partial^2}{\partial y^2} + h(x, y)\frac{\partial}{\partial x}\right) - \frac{\partial^2}{\partial t^2}\right) w = 0 \tag{1.18}$$

with $g(x, y) > c_0 > 0$ in the half-space $x > 0$; and we seek perfectly absorbing boundary conditions on the wall, $x = 0$. The key for the constant coefficient wave equation was that it could be factored (within a smooth error) by using pseudo-differential operators into

$$\square \cong \left(\frac{\partial}{\partial x} + \rho\sqrt{\frac{\partial^2}{\partial t^2} - \frac{\partial^2}{\partial y^2}}\right)\left(\frac{\partial}{\partial x} - \rho\sqrt{\frac{\partial^2}{\partial t^2} - \frac{\partial^2}{\partial y^2}}\right) \tag{1.19}$$

near the relevant rays. Here $\cong$ denotes "within a smooth error". The factor $\partial/\partial x - \rho\sqrt{\partial^2/\partial t^2 - \partial^2/\partial y^2}$ corresponds to propagation to the left, while $\partial/\partial x + \rho\sqrt{\partial^2/\partial t^2 - \partial^2/\partial y^2}$ corresponds to propagation of reflected waves travelling to the right. To minimize the amplitude of reflected waves travelling to the right, we chose the boundary condition in (1.6). For variable coefficient problems as in (1.18), the theory of pseudo-differential operators allows us to factorize general wave equations in the fashion as in (1.19). From a very special case of Lemma 1 on page 27 of the lectures by Nirenberg in [3], we conclude that the operator L in (1.18) can be factored in the form,

$$L \cong \left(\frac{\partial}{\partial x} + \lambda_+\left(x, y, \frac{\partial}{\partial y}, \frac{\partial}{\partial t}\right)\right)\left(\frac{\partial}{\partial x} - \lambda_-\left(x, y, \frac{\partial}{\partial y}, \frac{\partial}{\partial t}\right)\right),$$

where λ_- is a pseudo-differential operator with a symbol which has the formal asymptotic expansion,

$$\lambda_-(x, y, \omega, \xi) \cong \lambda_-^1(x, y, \omega, \xi) + \lambda_-^0(x, y, \omega, \xi) + \lambda_-^{-1}(x, y, \omega, \xi) ++,$$

where λ_-^j is a homogeneous function of degree j. By using Theorems 1 and 2 of the work of Majda and Osher in [2], we verify that the *perfectly absorbing boundary condition* is given by

$$\left(\frac{\partial}{\partial x} - \lambda_-\left(x, y, \frac{\partial}{\partial y}, \frac{\partial}{\partial x}\right)\right) w|_{x=0} = 0. \tag{1.20}$$

The coefficients, λ_-^j, are recursively determined by the composition formulas of pseudo-differential operators as in [3]. One calculates that the first coefficient has symbol given by

$$\lambda_-^1(x, y, \xi, \omega) = i\sqrt{\xi^2 - g(x, y)\omega^2}. \tag{1.21}$$

The second coefficient is determined recursively from the first to be

$$\lambda_-^0(x, y, \xi, \omega) = \frac{b}{2} + \frac{1}{4}\frac{\partial g\omega^2/\partial x}{(\xi^2 - g(x, y)\omega^2)} + \frac{ig(x, y)\partial g\omega^3/\partial y}{2(\xi^2 - g(x, y)\omega^2)^{3/2}}. \tag{1.22}$$

To make suitable local approximations, for simplicity we retain the first two terms $\lambda_- \cong \lambda_-^1 + \lambda_-^0$ in the asymptotic expansion of λ_- (retaining more terms clearly improves the method but is too cumbersome to display here). Thus, we must approximate the nonlocal boundary condition,

$$\begin{aligned}\frac{\partial}{\partial x} - \Bigg(\sqrt{\frac{\partial^2}{\partial t^2} - g(x, y)\frac{\partial^2}{\partial y^2}} + \frac{b}{2} - \frac{1}{4}\frac{(\partial g/\partial x)(\partial^2/\partial y^2)}{(\partial^2/\partial t^2 - g(x, y)\partial^2/\partial y^2)} \\ + \frac{g(x, y)\partial g/\partial y}{(\partial^2/\partial t^2 - g(x, y)\partial^2/\partial y^2)^{3/2}}\Bigg) w|_{x=0} = 0.\end{aligned} \tag{1.23}$$

If we expand the symbols at normal incidence suing $\xi^2 - g(x, y)\omega^2 \cong \xi^2$, we obtain the *first local approximation*

$$\left(\frac{\partial}{\partial x} - \frac{\partial}{\partial t} - \frac{b}{2}\right) w|_{x=0} = 0. \tag{1.24}$$

This approximation diminishes the amplitude of reflected waves by $O(|\omega/\xi|^2)$. Using $(1 - \omega^2/\xi^2)^{1/2} = 1 - \frac{1}{2}\omega^2/\xi^2$, etc., we obtain the *second local approximation* given symbolically by

$$\frac{\partial}{\partial x} - i\xi\left(1 - \frac{1}{2}g\frac{\omega^2}{\xi^2}\right) + \frac{b}{2} - \frac{1}{4}\frac{\partial g}{\partial x}\frac{\omega^2}{\xi^2}. \tag{1.25}$$

We note that the difference between (1.25) and the symbol of (1.23) is $O(|\omega/\xi|^3)$. Multiplying (1.25) by ξ^2, we obtain the second local approximation,

$$\left(\frac{\partial^3}{\partial x \partial t^2} - \frac{\partial^3}{\partial t^3} + \frac{1}{2}g\frac{\partial^3}{\partial t \partial y^2} + \frac{b}{2}\frac{\partial^2}{\partial t^2} + \frac{1}{4}\frac{\partial g}{\partial x}\frac{\partial^2}{\partial y^2}\right) w|_{x=0} = 0. \tag{1.26}$$

By freezing coefficients and applying the identical calculations from the 1st and 2nd approximation in (1.9), (1.10) above in the normal mode analysis from (1.11), we conclude that the absorbing approximations in (1.24) and (1.26) yield well-posed problems.

The Special Case of Polar Coordinates. We merely specialize the above formula to polar coordinates in two dimensions where we construct highly absorbing boundary conditions on the circle of radius *a*. Here L has the form

$$L = \frac{\partial^2}{\partial r^2} + \frac{1}{r^2}\frac{\partial^2}{\partial^2 \theta} + \frac{1}{r}\frac{\partial}{\partial r} - \frac{\partial^2}{\partial t^2},$$

where $\partial/\partial r$, $\partial/\partial \theta$ correspond to $\partial/\partial x$, $\partial/\partial y$ in the above formulas and $b = 1/r$, $g = 1/r^2$.

Using (1.24), the *first highly absorbing approximation* on $r = a$ for the region $r \leqslant a$ is given by

$$\left(\frac{\partial}{\partial r} + \frac{\partial}{\partial t} + \frac{1}{2a}\right) w|_{r=a} = 0. \tag{1.27}$$

Using (1.26), the *second highly absorbing* approximation on $r = a$ is given by

$$\left(\frac{\partial^3}{\partial r \partial t^2} + \frac{\partial^3}{\partial t^3} - \frac{1}{2a^2}\frac{\partial^3}{\partial t \partial^2 \theta} + \frac{1}{2a}\frac{\partial^2}{\partial t^2} + \frac{1}{2}\frac{1}{a^3}\frac{\partial^2}{\partial \theta^2}\right) w|_{r=a} = 0. \tag{1.28}$$

Remark. For $r = a$ with $r \longrightarrow \infty$, outgoing waves have the form,

$$\frac{f(r-t)}{r^{1/2}}\left(a(\theta) + O\left(\frac{1}{r}\right)\right).$$

We observe that the first approximation from (1.27) precisely annihilates $f(r-t)a(\theta)/r^{1/2}$ on $r = a$. The approximation from (1.28) also compensates for angular dependence, i.e., the wave is not quite a spherical wave.

2. Highly Absorbing Boundary Conditions for First Order Symmetric Systems. We consider general variable coefficient first order *strictly hyperbolic* systems of the form

$$\frac{\partial u}{\partial t} = \left(A_1(x, y)\frac{\partial}{\partial x} + A_2(x, y)\frac{\partial}{\partial y} + B(x, y)\right)u \tag{2.1}$$

in the region $x \geqslant 0$ where A_1, A_2 are *symmetric* $m \times m$ matrices (general symmetric systems in regions with curved boundary, at least locally, can be mapped to a half-space where they assume the form in (2.1)). Our objective is to design highly absorbing local boundary conditions for the system in (2.1). We assume that A_1 has the form

$$A_1 = \begin{pmatrix} A_1^1 & 0 \\ 0 & A_1^2 \end{pmatrix}, \tag{2.2}$$

where

$$A_1^1 = \begin{pmatrix} \lambda_1 & & 0 \\ & \ddots & \\ 0 & & \lambda_k \end{pmatrix}$$

with $\lambda_i(x, y) < 0$ and

$$A_1^2 = \begin{pmatrix} \lambda_{k+1} & & 0 \\ & \ddots & \\ 0 & & \lambda_n \end{pmatrix}$$

with $\lambda_{k+j} > 0$. As a consequence of strict hyperbolicity, $\lambda_i \neq \lambda_j$ for $i \neq j$.

2.A. *The Theoretical Nonlocal Perfectly Absorbing Boundary Condition.* We rewrite the equation in (2.1) as

$$\frac{\partial u}{\partial x} = A\frac{\partial u}{\partial t} + E\frac{\partial u}{\partial y} + \widetilde{B}u, \tag{2.3}$$

where

$$A = A_1^{-1} = \begin{pmatrix} \Lambda_1 & 0 \\ 0 & \Lambda_2 \end{pmatrix},$$

with $\Lambda_1 < 0$ and $\Lambda_2 > 0$, $E = -A_1^{-1}A_2$, $\widetilde{B} = -A_1^{-1}B$ and we define $M(\xi, \omega)$ by

$$M(\xi, \omega) \equiv Ai\xi + Ei\omega.$$

The theoretical construction to be used here parallels Nirenberg's factorization from part C of Section 1 and was developed by M. E. Taylor in [4] for general first order systems. More precisely, there exists a smooth matrix $V(\xi, \omega, x, y)$ invertible for all (ξ, ω) satisfying $|\omega|/|\xi| + |\omega| < c_0$ for some choice of c_0 and defining the symbol of a pseudo-differential operator so that if

$$w = V\left(\frac{\partial}{\partial t}, \frac{\partial}{\partial y}, x, y\right) u,$$

the equations in (2.3) assume the form

$$\frac{\partial w}{\partial x} = \begin{pmatrix} \Lambda_{11}\left(\frac{\partial}{\partial t}, \frac{\partial}{\partial y}, x, y\right) & 0 \\ \Lambda_{21}\left(\frac{\partial}{\partial t}, \frac{\partial}{\partial y}, x, y\right) & \Lambda_{22}\left(\frac{\partial}{\partial y}, \frac{\partial}{\partial t}, x, y\right) \end{pmatrix} w, \tag{2.4}$$

where $\Lambda_{11}(\partial/\partial t, \partial/\partial y, x, y)$ is a $k \times k$ pseudo-differential operator of order 1 with principal symbol, Λ_{11}, evaluated at $(0, 1, x, y)$, given by

$$\Lambda_{11}(0, 1, x, y) = \Lambda_1 = \begin{pmatrix} \lambda_1^{-1} & & 0 \\ & \ddots & \\ 0 & & \lambda_k^{-1} \end{pmatrix} \tag{2.5}$$

and satisfying $\Lambda_{11} < 0$ for (ξ, ω) with $|\omega|/|\xi| + |\omega| < c_0$. Here $\Lambda_{21}, \Lambda_{22}$ are $(m-k) \times k$, $m-k \times m-k$ matrix pseudo-differential operators of order 1, respectively.

Once the differential equation has the form in (2.4) it is easy to write down the *theoretical perfectly absorbing boundary condition* at least for all waves striking at angles sufficiently close to normal incidence. Recall that Λ_{11}^1 has all the negative real eigenvalues corresponding to light rays flowing inward into $x > 0$ for positive time (the doubting reader should consider the case of a single space dimension). Thus, the theoretical absorbing boundary condition annihilates these components for $x = 0$. Let $\pi_k(w_1, \ldots, w_k, w_{k+1}, \ldots, w) \equiv (w_1, \ldots, w_k)$; the *perfectly absorbing boundary condition* is given by

$$\pi_k w|_{x=0} = \pi_k V u|_{x=0} = 0. \tag{2.6}$$

To gain some insight into the nature of this nonlocal condition, we recall Taylor's construction of $V(\xi, \omega, x, y)$ (see Section 1 of [4]).

$$\begin{aligned} V(\xi, \omega, x, y) \cong\ & V_0(\xi, \omega, x, y) + \xi^{-1} V_{-1}(\xi, \omega, x, y) \\ & + \xi^{-2} V_{-2}(\xi, \omega, x, y) ++, \end{aligned} \tag{2.7}$$

where each V_{-j} is homogeneous of degree zero and $\cong$ is the asymptotic sense of pseudo-differential operators. One consequence of strict hyperbolicity is that $M(1, 0)$ is a diagonal matrix with distinct entries; therefore, there exists a constant c_0 so that for $|\omega|/|\xi| + |\omega| \leqslant c_0$, $M(\xi, \omega)$ also has distinct eigenvalues; and we choose $V(\xi, \omega, x, y)$ to be a smoothly varying matrix, homogeneous of degree zero, so that $M(\xi, \omega)$ becomes block lower triangular, i.e.

$$\begin{pmatrix} {}_1\Lambda_{11} & 0 \\ {}_1\Lambda_{21} & {}_1\Lambda_{22} \end{pmatrix} \equiv V_0 M V_0^{-1}. \tag{2.8}$$

It is obvious that $M(\xi, \omega)$ has distinct eigenvalues, in general, only in a conic neighborhood of normal incidence ($\omega = 0$) (since at a value of (ξ, ω) corresponding to glancing for one of the sound speeds, two real roots coalesce and then become complex); and the constraint $|\omega|/|\xi| + |\omega| \leqslant c_0$ restricts the value of ω to the region where these eigenvalues stay distinct. We set $w_0 = V_0 u$, then w_0 formally satisfies the equation in (2.4) up to error terms which at the symbol level are bounded in ξ. We write $w_1 = (1 + K_1)w_0$ with $K_1 V_0 = \xi^{-1} V_{-1}$ and attempt to write an equation for w_1 in the form of (2.4) at least up to terms of order ξ^{-1} at the symbol level. Using the composition formulas for pseudo-differential operators, we obtain that

$$(2.9)\quad \begin{aligned}&(1+K_1)V_0(M(\xi,\omega)+\widetilde{B})V_0^{-1}(1+K_1)^{-1}\\ &\quad = K_1V_0M(\xi,\omega)V_0^{-1}0V_0M(\xi,\omega)V_0^{-1}K_1 + V_0(\widetilde{B}-\partial_xV_0)V_0^{-1} + O(\xi^{-1}).\end{aligned}$$

We define $\widetilde{C}\equiv\widetilde{B}-\partial_xV_0$. From (2.9) the terms in the upper right-hand corner of (2.4) which are bounded in ξ will vanish if we choose K_1 to have the form

$$(2.10)\qquad K_1=\begin{pmatrix}0 & \widetilde{K}\\ 0 & 0\end{pmatrix},$$

where $\widetilde{K}$ is an $k\times m-k$ matrix satisfying

$$(2.11)\qquad \widetilde{K}_1\,{}_1\Lambda_{22}-{}_1\Lambda_{11}K_1+\widetilde{C}_{12}=0.$$

As was proved in Section 1 of [4], a unique choice of $\widetilde{K}_1$ satisfying the equations in (2.11) exists provided that

$$(2.12)\qquad {}_1\Lambda_{22}\equiv\Lambda_2 \quad\text{and}\quad {}_1\Lambda_{11}\equiv\Lambda_1$$

have disjoint spectra. The condition in (2.12) is satisfied here because the eigenvalues of Λ_2 are positive and those of Λ_1 are negative.

2.B. *Highly Absorbing Local Approximations.* To construct highly absorbing local approximations, we approximate the matrix $V(\xi,\omega,x,y)$ at normal incidence. Recall the asymptotic representation for V from (2.7). We take each $V_{-j}(\xi,\omega)$ and use Taylor's theorem to write

$$(2.13)\quad \begin{aligned}V_{-j}(\xi,\omega)&=V_{-j}\left(1,\frac{\omega}{\xi}\right)\\ &=V_{-j}(1,0)+\sum_{K=1}^{l}\left(\frac{\omega}{\xi}\right)^K\frac{1}{K!}\frac{\partial^K}{\partial\omega^K}V_{-j}(1,0)+O\left(\left|\frac{\omega}{\xi}\right|^{l+1}\right).\end{aligned}$$

Using (2.7) and (2.13) in the theoretical boundary condition in (2.6) and retaining only terms with an error $O(|\omega/\xi|+|1/\xi|)$, we obtain the *first approximation*,

1st approximation:

$$(2.14)\qquad \pi_kV_0u|_{x=0}=0.$$

Looking back at (2.3), we see that this is merely the projection on the inflowing characteristics for positive time for the one-dimensional equation derived from (2.1) by setting $A_2\equiv 0$. A simple energy estimate shows that this boundary condition is maximal dissipative and is *always well posed* for (2.1). We remark that if the 1-dimensional wave equation was written as a first order system using (u_t, u_x) as components, the boundary condition in (2.14) would coincide with the 1st approximation from Section 1.

To construct the *second approximation* with reflected error $O(|\omega/\xi|^2+1/|\xi|^2)$, we retain terms from V_0 up to order 1 and terms from V_{-1} up to order zero to get the symbol

$$\pi_K\left(V_0(1,0)+\frac{\omega}{\xi}\frac{\partial}{\partial\omega}V_0(1,0)+\frac{1}{\xi}V_{-}(1,0)\right).$$

Multiplying by ξ we arrive at the *second highly absorbing* approximation,

2nd approximation:

$$(2.15)\qquad \pi_k\left(\frac{d}{dt}V_0(1,0)+\frac{\partial}{\partial\omega}V_0(1,0)\frac{\partial}{\partial y}+V_{-1}(1,0)\right)u|_{x=0}=0,$$

where V_{-1} is determined by the lower order term B and the variable nature of the coefficients according to (2.9). We make the following conjecture:

The 2nd approximation always yields a well-posed boundary value problem for (2.1).

It is worthwhile mentioning that a highly absorbing approximation intermediate between the first and second approximations may be useful in situations where the effect of the lower order term B is rather large. We have with reflected error $O(|\omega/\xi| + 1/\xi^2)$, the 1½ *order approximation.*

$$\pi_K\left(\frac{d}{dt}V_0(1,0) + V_{-1}(1,0)\right)\mu|_{x=0} = 0. \tag{2.16}$$

This approximation is always well posed because its principal symbol is the same as the one in (2.14).

To complete the pattern, we retain the terms in (2.13) from V_{-j} up to order $l-j$ and then multiply the symbol by ξ^l to obtain the $l+1$*th highly absorbing approximation* with reflected error $O(|\omega/\xi|^{l+1} + 1/|\xi|^{l+1})$.

$l+1$*th approximation*:

$$\pi_K\left(\sum_{j=0}^{l}\sum_{K=0}^{l-j}\frac{\partial^{l-j-K}}{\partial t^{l-j-K}}\frac{1}{K!}\frac{\partial^K}{\partial\omega^K}V_{-j}(1,0)\frac{\partial^K}{\partial y^K}\right)u|_{x=0} = 0. \tag{2.17}$$

In view of the instability of the 4th approximation from Section 1 for the wave equation, it is *doubtful* that in general for $l > 1$ these boundary conditions would yield well-posed problems. Perhaps, using other approximations analogous to the Padé approximation from Section 1 would be useful here.

2.C. *Some Perturbation Calculations Applied to Highly Absorbing Approximations.* In order to use the 2*nd approximation* from (2.15) in explicit problems, we need to evaluate $\partial V_0(1,0)/\partial\omega$; as we shall see below, this reduces to a standard calculation in perturbation theory. Set $\omega/\xi = \delta$ and write

$$M\left(1,\frac{\omega}{\xi}\right) = \begin{pmatrix}\Lambda_1 & 0\\ 0 & \Lambda_2\end{pmatrix} + \delta\begin{pmatrix}E_{11} & E_{12}\\ E_{21} & E_{22}\end{pmatrix}. \tag{2.18}$$

Recall that V_0 is chosen to guarantee the relation in (2.8). We write V_0 as $V_0 = I + \delta\,{}_1V_0 + O(\delta^2)$, then it follows that

$$\frac{\partial V_0}{\partial\omega}(1,0) = {}_1V_0. \tag{2.19}$$

To calculate ${}_1V_0$ is a standard problem (identical to the one studied in [4] and (2.9), (2.10), (2.11) above). Substituting $V_0 = I + \delta\,{}_1V_0 + O(\delta^2)$ and requiring the upper right block of $V_0(A_1 + \delta E)V_0^{-1}$ to vanish to order $O(\delta^2)$, we obtain the equation

$$X\Lambda_2 - \Lambda_1 X + E_{12} = 0, \tag{2.20}$$

where we choose ${}_1V_0$ of the form

$${}_1V_0 = \begin{pmatrix}0 & X\\ 0 & 0\end{pmatrix}.$$

Following (2.9), (2.10), (2.11), X is uniquely determined by solving the linear equations in (2.20); thus, $\partial V_0(1,0)/\partial\omega$ is determined.

We shall now develop a special perturbation lemma under additional hypotheses.

These hypotheses concerning the $k \times (m-k)$ matrix, X, from (2.20) are that

$$k \geqslant m-k, \qquad X \text{ has rank } m-k. \tag{2.21}$$

In this situation, we prove that by choosing $V_0 = S + O(\delta^3)$ through the *linear change* of variables $S \equiv I + \delta R$ with R appropriately chosen, we have

$$S(A+\delta E)S^{-1} = \begin{pmatrix} \Lambda_{11} & O(\delta^3) \\ \Lambda_{21} & \Lambda_{22} \end{pmatrix}. \tag{2.22}$$

Thus, under these hypotheses, we have the *special* 2nd approximation,

*2nd approximation**:

$$\pi_k\left(\frac{d}{dt}V_0(1,0) + R\frac{\partial}{\partial y} + V_{-1}(1,0)\right)u|_{x=0} = 0. \tag{2.23}$$

If for simplicity A_1, A_2 have constant coefficients, the *special 2nd approximation** has the smaller reflected error $O(|\omega/\xi|^3 + 1/\xi^2)$ when compared to the *2nd approximation above*. (This is confirmed numerically.) We remark that the conditions in (2.21) are satisfied for the inflow case of the linearized shallow water equations (see the computations in Section 4 using 2nd approximation*) and often should be valid for the (hyperbolic) equations of fluid dynamics. We also remark here that the choice of ${}_1V_0(1,0)$ used in (2.22) is different from the one in (2.19), but V_0 is not unique anyway.

To verify the claim in (2.22) we choose R in the form

$$R = \begin{pmatrix} Y & X \\ 0 & 0 \end{pmatrix}. \tag{2.24}$$

When we compute $S(A+\delta E)S^{-1}$, a tedious calculation establishes that the terms of order δ, δ^2 appearing in the upper right-hand block of (2.22) and required to vanish have the form,

$$\begin{array}{ll} (\text{order } \delta) & -\Lambda_1 X + X\Lambda_2 + E_{12} = 0, \\ (\text{order } \delta^2) & -E_{11}X + Y(-\Lambda_1 X + E_{12}) + \Lambda_1 YX + XE_{22} = 0. \end{array} \tag{2.25}$$

If we choose X as in (2.20) above to satisfy the order δ equation and then substitute the order δ equation in the order δ^2 equation, we obtain with $Z = YX$

$$(\text{order } \delta^2) \qquad -Z\Lambda_2 + \Lambda_1 Z - E_{11}X + XE_{22} = 0. \tag{2.26}$$

Thus, Z is uniquely determined in exactly the same fashion as X above (since X is already known). To recover Y, we must determine Y from the relationship

$$Z = YX, \tag{2.27}$$

where Z, X are $k \times (m-k)$ matrices and Y is a $k \times k$ matrix. By the hypotheses in (2.21) the $m-k$ columns of X are independent and since $m-k \leqslant k$, we construct Y so that it maps the $(m-k)$ columns of X, respectively, onto the $m-k$ columns of Z (thus satisfying (2.27)) and on the orthogonal complement of the subspace of R^k spanned by the columns of X, choose $Y \equiv 0$.

Remark #1. The special second approximation yields reflected waves with amplitudes of the same order as the third approximation. The special second approximation utilizes only a first order boundary condition and from a numerical point of view, this approximation is much easier to discretize in a stable fashion than the second order boundary condition required by the third approximation.

Remark #2. Of course, the special second approximation can be derived whenever it is possible to solve the equation, $Z = YX$, even though the additional hypotheses from (2.21) are not satisfied.

3. Highly Absorbing Boundary Conditions for the Linearized Shallow Water Equations. We apply the theory developed in Section 2 to the linearized shallow water equations in the half-space $x \geqslant 0$. Thus, we consider a system of the form

$$\frac{\partial u}{\partial t} = \begin{pmatrix} a & 0 & c \\ 0 & a & 0 \\ c & 0 & a \end{pmatrix} \frac{\partial u}{\partial x} + \begin{pmatrix} b & 0 & 0 \\ 0 & b & c \\ 0 & c & b \end{pmatrix} \frac{\partial u}{\partial y} + \begin{pmatrix} 0 & f & 0 \\ -f & 0 & 0 \\ 0 & 0 & 0 \end{pmatrix} u \tag{3.1}$$

together with the physical restrictions, $0 < a^2 + b^2 < c^2, c > 0$. Here for simplicity we assume all matrices in (3.1) are constant. We diagonalize the normal matrix to $x = 0$ by the unitary map, U, where

$$U = \begin{pmatrix} \frac{1}{\sqrt{2}} & 0 & \frac{1}{\sqrt{2}} \\ 0 & 1 & 0 \\ \frac{-1}{\sqrt{2}} & 0 & \frac{1}{\sqrt{2}} \end{pmatrix}$$

and set $w = U^{-1}u$ to get from (3.1), a new equation for w of the form

$$\begin{aligned} \frac{\partial w}{\partial t} = {} & \begin{pmatrix} a-c & 0 & 0 \\ 0 & a & 0 \\ 0 & 0 & a+c \end{pmatrix} \frac{\partial w}{\partial x} + \begin{pmatrix} b & \frac{-c}{\sqrt{2}} & 0 \\ -\frac{c}{\sqrt{2}} & b & \frac{c}{\sqrt{2}} \\ 0 & \frac{c}{\sqrt{2}} & b \end{pmatrix} \frac{\partial w}{\partial y} \\ & + \begin{pmatrix} 0 & \frac{f}{\sqrt{2}} & 0 \\ -\frac{f}{\sqrt{2}} & 0 & -\frac{f}{\sqrt{2}} \\ 0 & \frac{f}{\sqrt{2}} & 0 \end{pmatrix} w. \end{aligned} \tag{3.2}$$

Writing this equation in the form of (2.3), (3.2) becomes

$$\begin{aligned} \frac{\partial w}{\partial x} = {} & \begin{pmatrix} \frac{1}{a-c} & 0 & 0 \\ 0 & \frac{1}{a} & 0 \\ 0 & 0 & \frac{1}{a+c} \end{pmatrix} \frac{\partial w}{\partial t} + \begin{pmatrix} \frac{-b}{a-c} & \frac{c}{(a-c)\sqrt{2}} & 0 \\ \frac{c}{a\sqrt{2}} & \frac{-b}{a} & -\frac{c}{a\sqrt{2}} \\ 0 & -\frac{c}{(a+c)\sqrt{2}} & -\frac{b}{(a+b)\sqrt{2}} \end{pmatrix} \frac{\partial w}{\partial y} \\ & + \begin{pmatrix} 0 & \frac{-f}{(a-c)\sqrt{2}} & 0 \\ \frac{f}{a\sqrt{2}} & 0 & \frac{f}{a\sqrt{2}} \\ 0 & \frac{-f}{(a+c)\sqrt{2}} & 0 \end{pmatrix} w. \end{aligned} \tag{3.3}$$

The boundary conditions we choose depend upon whether we have linearized about an *inflowing state* with $a < 0$ or about an *outflowing state* with $a > 0$.

3.A. *The Inflow Case.* Here we assume $a < 0$ so that the normal matrix to the boundary has the form

$$\begin{pmatrix} \Lambda_1 < 0 & 0 \\ 0 & \Lambda_2 > 0 \end{pmatrix},$$

where Λ_1, Λ_2 are given by

$$\Lambda_1 = \begin{pmatrix} \frac{1}{a-c} & 0 \\ 0 & \frac{1}{a} \end{pmatrix}, \qquad \Lambda_2 = \frac{1}{a+c}. \tag{3.4}$$

Following (2.14), we have the first absorbing approximation

1st approximation:

$$\begin{pmatrix} w_1 \\ w_2 \end{pmatrix}\Bigg|_{x=0} = 0. \tag{3.5}$$

Following the procedure at the end of Section 2, we choose X to satisfy (2.20) so that X is given by

$$X = \begin{pmatrix} 0 \\ -\frac{a+c}{\sqrt{2}} \end{pmatrix}. \tag{3.6}$$

Thus, the hypotheses in (2.21) for the *special 2nd approximation* are satisfied. We choose Z to satisfy (2.26) so that Z is given by

$$Z = \begin{pmatrix} -\frac{(a+c)^2}{4} \\ -\frac{b(a+c)}{2c}((1-\sqrt{2})a - \sqrt{2}c) \end{pmatrix}.$$

Thus, Y with $Z = YX$ is given by

$$Y = \begin{pmatrix} 0 & \frac{\sqrt{2}}{4}(a+c) \\ 0 & \frac{\sqrt{2}}{2}\frac{b}{c}((1-\sqrt{2})a - \sqrt{2}c) \end{pmatrix}. \tag{3.7}$$

Thus, R from (2.24) has been computed.

Finally, to calculate the effect to the lower order term corresponding to the Coriolis force on the special second approximation, we need to compute $K_1(1, 0)$ from the equation in (2.9)–(2.11). Thus, $K_1(1, 0)$ is given by

$$K_1(1, 0) = \begin{pmatrix} 0 \\ \frac{(a+c)f}{c\sqrt{2}} \end{pmatrix}. \tag{3.8}$$

Inserting the values for R in (2.24) obtained from (3.6) and (3.7) and the value of $V_{-1}(1, 0)$ determined by (3.8), we obtain the *special 2nd absorbing* approximation,

*2nd approximation**:

$$\left.\frac{\partial w_1}{\partial t} + \frac{\sqrt{2}}{4}(a+c)\frac{\partial w_2}{\partial y}\right|_{x=0} = 0, \tag{3.9}$$

$$\left.\frac{\partial w_2}{\partial t} + \frac{\sqrt{2}}{2}\frac{b}{c}(a - \sqrt{2}(a+c))\frac{\partial w_2}{\partial y} - \frac{a+c}{\sqrt{2}}\frac{\partial w_3}{\partial y} + \frac{(a+c)f}{c\sqrt{2}}w_3\right|_{x=0} = 0.$$

It is worthwhile mentioning that the 1½ *absorbing approximation* is given by

$$\left.\frac{\partial w_1}{\partial t}\right|_{x=0} = 0, \qquad \left.\frac{\partial w_2}{\partial t} + \frac{a+c}{c\sqrt{2}}fw_3\right|_{x=0} = 0. \tag{3.10}$$

3.B. *The Outflow Case.* Here we assume $a > 0$, so that

$$0 > \Lambda_1 = \frac{1}{a-c}, \qquad 0 < \Lambda_2 = \begin{pmatrix} \frac{1}{a} & 0 \\ 0 & \frac{1}{a+c} \end{pmatrix}.$$

In this case the *first absorbing approximation* is

$$w_1 |_{x=0} = 0. \tag{3.11}$$

Following the computations as in the inflow case, we calculate that

$$X = \left(\frac{a}{\sqrt{2}}, 0\right), \qquad K_1 = \left(-\frac{af}{c\sqrt{2}}, 0\right).$$

On the other hand, one calculates that $Z = (b_1, b_2)$ with $b_2 \neq 0$ so it is impossible to find a 1×1 matrix Y with $Z = YX$ so the special second approximation is not applicable. Thus, following (2.15) we have the *second absorbing* approximation

2nd approximation:

$$\left.\frac{\partial w_1}{\partial t} + \frac{a}{\sqrt{2}}\frac{\partial w_2}{\partial y} - \frac{a}{c\sqrt{2}}fw_2\right|_{x=0} = 0. \tag{3.12}$$

4. Difference Approximations for Absorbing Boundary Conditions–Numerical Experiments.

A. *The Scalar Wave Equation.* Here we discuss the difference approximation of the scalar wave equation and the highly absorbing boundary conditions from Section 1 and then present the results of numerical experiments which illustrate their effectiveness. We look at the scalar wave equation from (1.1) in the half-space, $x \geqslant 0$, and approximate the solution $w(t, x, y)$ on the grid $\{(t^n, x_j, y_k)\}$ by the mesh function $w^n_{j,k}$ ($t^n = n\Delta t$, $x_j = j\Delta x - \Delta x/2$, $y_k = k\Delta y$; $n = 0, 1, \ldots, j = 0, 1, \ldots, k = \ldots, -1, 0, 1, \ldots$). We use the notation D_+, D_-, and $D_0 = ½(D_+ + D_-)$ to denote the forward, backward, and centered divided differences, respectively. The equation in (1.1) is approximated by the simple explicit formula

$$(D^t_+D^t_- - D^x_+D^x_- - D^y_+D^y_-)w^n_{j,k} = 0, \tag{4.1}$$

$$n = 1, 2, \ldots, j = 1, 2, \ldots, k = \ldots, -1, 0, 1, \ldots .$$

The values of $w^0_{j,k}$ and $w^1_{j,k}$ are given initially.

We discretize the *first approximation* from (1.9) by

$$D^x_+(w^n_{0,k} + w^{n+1}_{0,k}) - D^t_+(w^n_{0,k} + w^n_{1,k}) = 0, \tag{4.2}$$

while we use the following difference approximation for the second highly absorbing boundary condition from (1.10):

$$(4.3)\quad D_0^t D_+^x w_{0,k}^n - \tfrac{1}{2} D_+^t D_-^t (w_{0,k}^n + w_{1,k}^n) + \tfrac{1}{4} D_+^y D_-^y (w_{0,k}^{n-1} + w_{1,k}^{n+1}) = 0.$$

For the numerical experiments presented below, we have discretized the Dirichlet and Neumann problems in the obvious fashion using $w_{0,k}^{n+1} = 0$ and $D_-^x w_{0,k}^{n+1} = 0$, respectively. It is easy to check that (4.1) coupled to (4.2) or (4.3) gives an explicit scheme with a local *truncation error of second order* in all variables. (This is also obvious for the approximations for the Dirichlet and Neumann boundary conditions which we have used.) Tests with initial data containing high frequencies gave numerical evidence of the stability of these approximations to the mixed initial boundary value problems. The boundary conditions can, of course, be coupled to any type of interior discretization.

In Figure 1 we display the results from a series of experiments for the scalar wave equation. The first Figure (1a) shows the initial values $w_{j,k}^0$. Each vertical trace corresponds to a vector $w_{j,k}^n$ with n and j fixed; different traces correspond to different j's. One hundred percent black is $w \equiv 1$ and 30% black is $w \equiv 0$ on these traces. The initial values $w_{j,k}^1$ are chosen using Taylor's theorem so that the circular wave form expands with time. For all experiments we have used the discrete Neumann boundary condition on the upper, lower, and right-hand walls. We will change our boundary condition only on the left wall corresponding to $x = 0$. In Figure (1b) we have the plot after 50 time steps using the perfectly reflecting Neumann boundary condition on $x = 0$. Figure (1c) displays the same solution after 100 time steps. (We used $\Delta x = \Delta y = 2.5\Delta t$, $0 \leqslant k \leqslant 50$, $0 \leqslant j \leqslant 100$.) Figures (1d), (1e), and (1f) all describe various solutions of (4.1) after 100 time steps. The discrete Dirichlet boundary condition was used on the wall, $x = 0$, in Figure (1d); waves are totally reflected. Figure (1e) represents the solution of (4.1) using the absorbing first approximation from (4.2); there is a weak reflected wave with amplitude in the L^2 norm about 5% of the incident amplitude. Figure (1f) represents the solution of (4.1) using the second absorbing approximation from (4.3); the weak negative reflection is not observed here. In fact, the reflected wave has amplitude in the L^2 norm smaller than 1% of the amplitude of the incident wave.

B. *The Shallow Water Equation.* As the last example, we will approximate the solution to the linearized shallow water equation from (3.1), (3.2) when $f = 0$. We have chosen to study the inflow case when $a < 0$, $b > 0$, and $c > 0$. The first and second (order) absorbing boundary conditions are then given by (3.5) and (3.9), respectively. It is outside of the scope of this paper to make a thorough study of possible difference approximations to that mixed initial boundary value problem. We will only report on some test runs performed in order to display the effect of the analytical absorbing conditions from (3.5) and (3.9). In the interior the differential equation was approximated by a standard Lax-Wendroff scheme. When approximating the boundary condition in (3.9), the Lax-Wendroff procedure, now in one space dimension and with added numerical dissipation, was used directly along the wall, $x = 0$. The difference equation needs one extra boundary condition. That was derived by extrapolating the w_3-component using $(D_+^x)^3 (w_3)_{0,j}^n = 0$.

Figure 2 shows the potential field or height, φ, with boundary of interest to the left at $x = 0$ (φ is the third vector component). At all other boundaries we used the first order absorbing condition from (3.5). The calculations were performed with $a =$

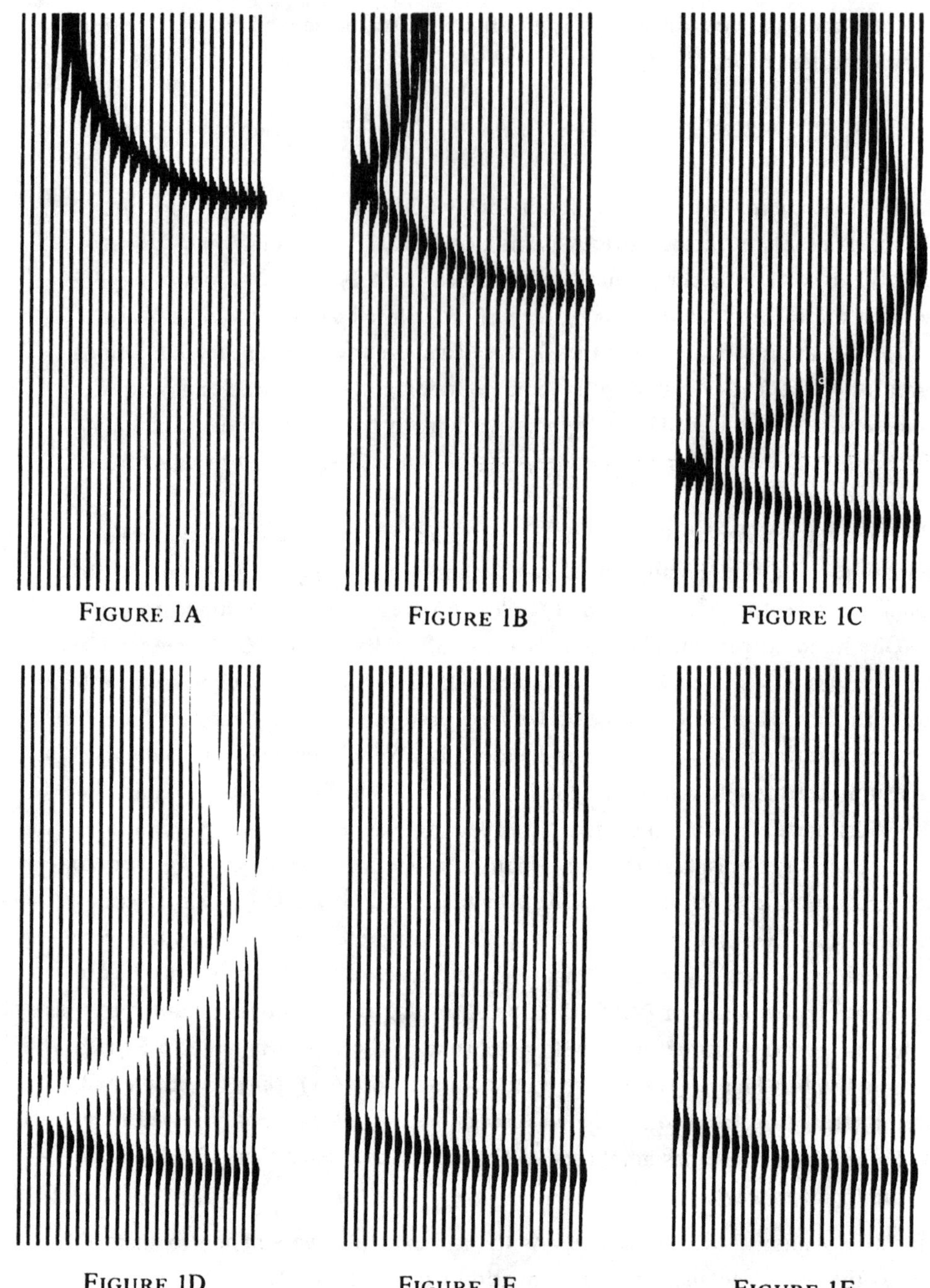

FIGURE 1A FIGURE 1B FIGURE 1C

FIGURE 1D FIGURE 1E FIGURE 1F

-1, $b = 2$, $c = 12$ and $25\Delta t = \Delta x = \Delta y$ on a 30×50 mesh. The initial φ is displayed in Figure (2a) where as in Figure 1, 100% black means the value 1 and 30% stands for zero. The first two components, the velocities u and v were taken as zero initially. In Figure (2b), φ is given after 30 time steps using the reflecting condition $u = v = 0$ at $x = 0$. Figures (2c) and (2d) show the corresponding picture for the absorbing conditions (3.5) and (3.9), respectively; but note that the scale in these two plots is multiplied by a factor of 3. In Figures (2e) and (2f) the reflected waves from (2c) and (2d) are displayed, respectively. They are computed by subtracting the solution of this mixed initial boundary value problem from the solution of the corresponding pure initial value problem. The scale is now changed by a factor of 30—that is, 100% black

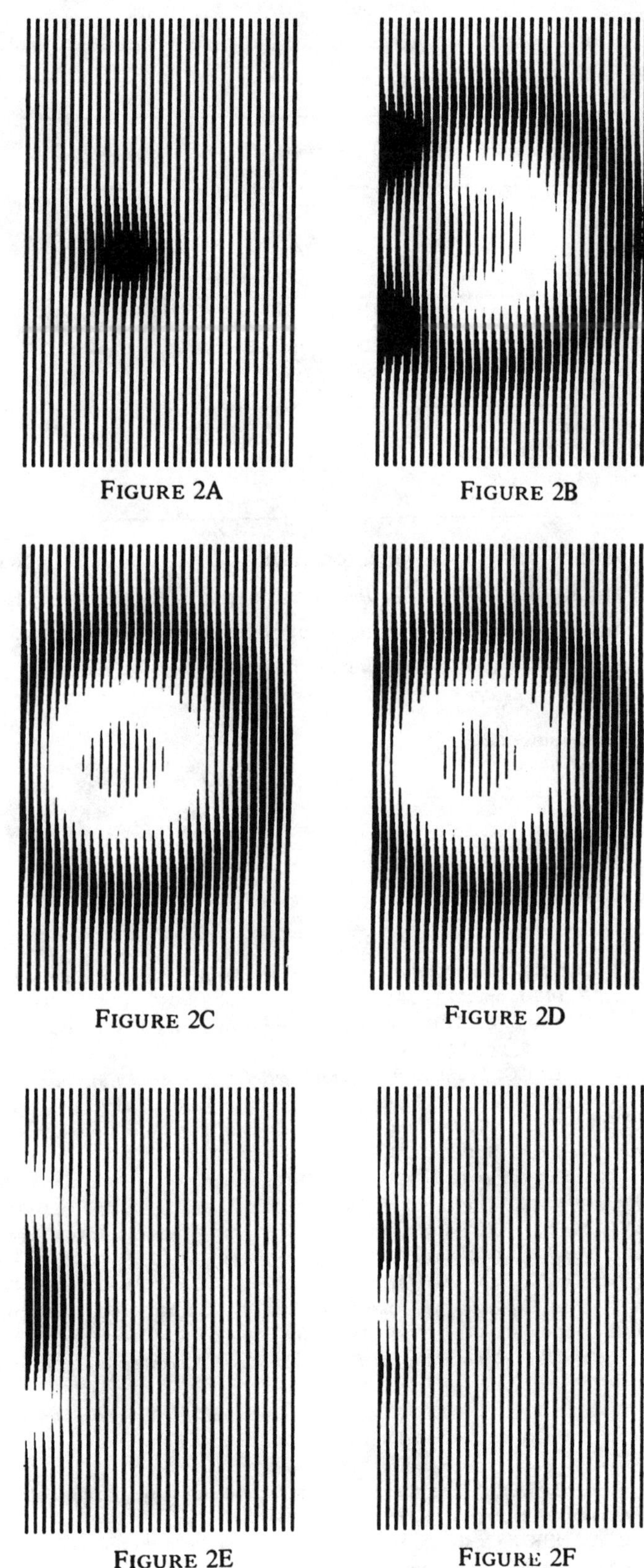

FIGURE 2A FIGURE 2B

FIGURE 2C FIGURE 2D

FIGURE 2E FIGURE 2F

stands for the value 1/30. The improvement in going from the first to the second absorbing condition is a factor of 3.5 in the L_2-norm for the reflected wave. Furthermore, the reflections seen in Figure (2f) using (3.9) are not penetrating as deeply into the region of computation as the reflections in Figure (2e) using (3.5). Thus, as expected the approximation from (3.9) produces weaker reflections. Below, we display the L_2-norm of the reflected waves for *all* dependent variables (after 30 time steps) for the example discussed above:

Boundary condition at $x = 0$	$\|u \text{ reflected}\|_2$	$\|v \text{ reflected}\|_2$	$\|\varphi \text{ reflected}\|_2$
perfect reflection $u = v = 0$	.621	.702	.695
First absorbing approx. from (3.5)	.103	.067	.053
Second absorbing approx. from (3.9)	.024	.024	.013

Thus, the first absorbing approximation from (3.5) produces reflected waves for u, v, and φ with amplitudes 17%, 10%, 7% of the incident waves, while the second absorbing approximation from (3.9) produces reflected waves for u, v, and φ with amplitudes 4%, 3.3%, and 2% of the incident waves, respectively.

Department of Computer Science
Uppsala University
Uppsala, Sweden

Department of Mathematics
University of California
Los Angeles, California 90024

1. H.-O. KREISS, "Initial boundary value problems for hyperbolic systems," *Comm. Pure Appl. Math.*, v. 23, 1970, pp. 277–298.

2. A. MAJDA & S. OSHER, "Reflection of singularities at the boundary," *Comm. Pure Appl. Math.*, v. 28, 1975, pp. 479–499.

3. L. NIRENBERG, *Lectures on Linear Partial Differential Equations*, C.B.M.S. Regional Conf. Ser. in Math., no. 17, Amer. Math. Soc., Providence, R. I., 1973.

4. M. E. TAYLOR, "Reflection of singularities of solutions to systems of differential equations," *Comm. Pure Appl. Math.*, v. 28, 1975, pp. 457–478.

5. F. J. MASSEY III & J. B. RAUCH, "Differentiability of solutions to hyperbolic initial-boundary value problems," *Trans. Amer. Math. Soc.*, v. 189, 1974, pp. 303–318. MR 49 #5582.

6. DAVID M. BOORE, "Finite difference methods for seismic wave propagation in heterogeneous materials," *Methods of Comp. Physics (Seismology)*, v. 11, 1972, pp. 1–37.

7. K. R. KELLY, R. M. ALFORD, S. TREITEL & R. W. WARD, *Applications of Finite Difference Methods to Exploration Seismology*, Proc. Roy. Irish Acad. Conf. on Numerical Analysis, Academic Press, London and New York, 1974, pp. 57–76.

8. P. J. ROACHE, *Computational Fluid Dynamics*, Hermosa Press, Albuquerque, N. M., 1972.

9. T. ELVIUS & A. SUNDSTRÖM, "Computationally efficient schemes and boundary conditions for a fine mesh barotropic model based on the shallow water equations," *Tellus*, v. 25, 1973, pp. 132–156.

10. E. L. LINDMAN, "Free space boundary conditions for the time dependent wave equation," *J. Computational Phys.*, v. 18, 1975, pp. 66–78.

11. I. ORLANSKI, "A simple boundary condition for unbounded hyperbolic flows," *J. Computational Phys.*, v. 21, 1976, pp. 251–269.

12. M. E. HANSON & A. G. PETSCHEK, "A boundary condition for sufficiently reducing boundary reflection with a Lagrangian mesh," *J. Computational Phys.*, v. 21, 1976, pp. 333–339.

13. W. D. SMITH, "A nonreflecting plane boundary for wave propagation problems," *J. Computational Phys.*, v. 15, 1974, pp. 492–503.

Absorbing Boundary Conditions for the Finite-Difference Approximation of the Time-Domain Electromagnetic-Field Equations

GERRIT MUR

***Abstract*—When time-domain electromagnetic-field equations are solved using finite-difference techniques in unbounded space, there must be a method limiting the domain in which the field is computed. This is achieved by truncating the mesh and using absorbing boundary conditions at its artificial boundaries to simulate the unbounded surroundings. This paper presents highly absorbing boundary conditions for electromagnetic-field equations that can be used for both two- and three-dimensional configurations. Numerical results are given that clearly exhibit the accuracy and limits of applicability of highly absorbing boundary conditions. A simplified, but equally accurate, absorbing condition is derived for two-dimensional time-domain electromagnetic-field problems.**

***Key Words*—Electromagnetic-field equations, time domain, finite-difference approximation, absorbing boundary conditions.**

I. INTRODUCTION

THE THREE-DIMENSIONAL finite-difference formulation [1] of time-domain electromagnetic-field problems is a convenient tool for solving scattering problems. The main advantages of such a method are that it can be easily applied to (infinitely) conducting obstacles and/or to dielectric and magnetic obstacles which can be either homogeneous or inhomogeneous. The obstacles can be of arbitrary shape. Furthermore, finite-difference techniques provide a very efficient way of solving Maxwell's equations [2]. In a finite-difference method, a space-time mesh is introduced and Maxwell's equations are replaced by a system of finite-difference equations on the mesh. The difficulty encountered when trying to solve field problems in this way arises from the fact that scattering problems are usually *open* problems, i.e., the domain in which the field has to be computed is unbounded. Since no computer can store an unlimited amount of data, a method has to be used for limiting the domain in which the field is computed. This is done by using a mesh of limited size, but one large enough to fully contain the obstacle, and by using a boundary condition on the outer surface of the mesh such that the unbounded surrounding is modeled as accurately as possible. Boundary conditions of this type are called absorbing boundary conditions. For finite-difference approximations of Maxwell's equations, absorbing boundary conditions have been described by Taylor *et al.* [3], who use a simple extrapolation method, and by Taflove and Brodwin [4], who simulate the outgoing waves and use an averaging process in an attempt to account for all possible angles of propagation of the outgoing waves. An alternative method, suggested by Taflove [5], is to introduce losses in the region that surrounds the structure to be modeled, thus absorbing both the outgoing waves and the waves that are reflected by the boundaries of the mesh. Numerical experiments, however, have shown that, in order to obtain accurate results, a relatively thick conducting layer is required which makes this way of absorbing the outgoing waves inefficient. Merewether [6] and Kunz and Lee [7] use the radiation condition at large distances from the center of the scatterer to obtain an absorbing boundary condition.

The absorbing boundary conditions mentioned above, although useful as first approximations, have the disadvantage of causing considerable reflections when the fields near the boundary of the mesh do not propagate in a specific direction (either the direction normal to the boundary of the mesh or the radial direction from the center of the obstacle). Furthermore, no general theory for improving upon these approximations is available. In this paper, a potentially superior method to those given above is derived for both the two- and three-dimensional electromagnetic-field equations; it is similar to the scalar derivation of Engquist and Majda [8]. The first approximation in the sense of this method appears to be comparable to those given above. The second and higher approximations are less subject to reflection problems, especially for fields grazing the outer boundary, making them highly absorbing boundary conditions. For two-dimensional fields, a simplified, but equally accurate, absorbing boundary condition will be derived. Numerical results will be presented to elucidate the usefulness of highly absorbing boundary conditions.

II. FINITE-DIFFERENCE APPROXIMATION OF MAXWELL'S EQUATIONS

Since, in this paper, we do not concern ourselves with obstacles or inhomogeneities but with the homogeneous region that surrounds them, we can, without loss of generality, confine our attention to Maxwell's equations for a vacuum region. In rectangular Cartesian coordinates we have

$$\mu_0 \partial_t \boldsymbol{H} = -\nabla \times \boldsymbol{E} \tag{1a}$$

$$\epsilon_0 \partial_t \boldsymbol{E} = \nabla \times \boldsymbol{H} \tag{1b}$$

where ϵ_0 and μ_0 denote the permittivity and the permeability of a vacuum, respectively. We now introduce a finite-difference approximation of (1) and, following Yee's notation [1], we denote a mesh point as

$$(i, j, k) = (i\delta, j\delta, k\delta)$$

Reprinted from *IEEE Trans. Elec. Comp.* vol. EMC-23, no. 4, pp. 377–382, Nov. 1981.

where $\delta = \delta x = \delta y = \delta z$ is the space increment, and any function of space and time as

$$F^n(i,j,k) = F(i\delta, j\delta, k\delta, n\delta t)$$

where δt is the time increment. By positioning the field components of $\boldsymbol{E}$ and $\boldsymbol{H}$ on the mesh in the way that is depicted in Fig. 1 and evaluating $\boldsymbol{E}$ and $\boldsymbol{H}$ at alternate half-time steps, Yee obtained finite-difference expressions that have a local trucation error of the second order in all increments. The finite-difference approximation for the x-component of (1a) and

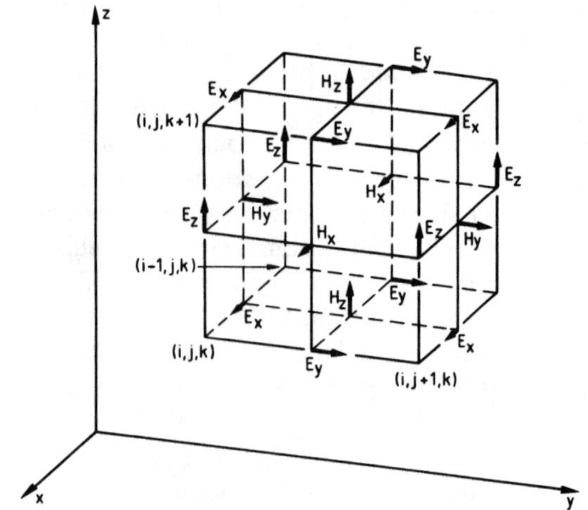

Fig. 1. The positions of the field components in Yee's mesh.

(1b) now reads

$$\begin{aligned} &H_x^{n+1/2}(i, j+1/2, k+1/2) \\ &\quad = H_x^{n-1/2}(i, j+1/2, k+1/2) - (\delta t/\mu_0\delta) \\ &\quad \cdot (E_z^n(i, j+1, k+1/2) - E_z^n(i, j, k+1/2) \\ &\quad - E_y^n(i, j+1/2, k+1) + E_y^n(i, j+1/2, k)) \end{aligned} \tag{2a}$$

$$\begin{aligned} &E_x^{n+1}(i+1/2, j, k) \\ &\quad = E_x^n(i+1/2, j, k) + (\delta t/\epsilon_0\delta) \\ &\quad \cdot (H_z^{n+1/2}(i+1/2, j+1/2, k) \\ &\quad - H_z^{n+1/2}(i+1/2, j-1/2, k) \\ &\quad - H_y^{n+1/2}(i+1/2, j, k+1/2) \\ &\quad + H_y^{n+1/2}(i+1/2, j, k-1/2)). \end{aligned} \tag{2b}$$

Similar finite-difference approximations for the y- and z-components of (1) can easily be derived. The condition for stability of (2) is

$$\delta t \leqslant \delta/(c_0\sqrt{3}) \tag{3}$$

where $c_0 = (\epsilon_0\mu_0)^{-1/2}$ denotes the speed of light in vacuo.

For a two-dimensional problem, we use only a single plane in the mesh and consequently (2) is considerably simplified. The condition for stability then reads

$$\delta t \leqslant \delta/(c_0\sqrt{2}). \tag{4}$$

In each of the coordinate directions, the mesh is truncated by enclosing it between two planes that are normal to the relevant coordinate axis and pass through mesh points (i,j,k), i, j, k integer. We then observe from Fig. 1 that all components of the electric-field vector $\boldsymbol{E}$ occurring in (2) applied to a particular point in the boundary of the mesh are tangential to this boundary while the relevant components of the magnetic-field vector $\boldsymbol{H}$ are normal to it. The latter field components can be evaluated by using the relevant finite-difference equation. The $\boldsymbol{E}$-field components, however, cannot be evaluated in this way since this would require $\boldsymbol{H}$-field components that are outside the mesh. In the next section, we shall describe absorbing boundary conditions that can be used for computing these $\boldsymbol{E}$-field components. For absorbing boundary conditions, it is assumed that the fields near the boundary are outgoing waves and, therefore, they can be applied to only scattered fields, i.e., those that find their origin somewhere near the center of the mesh. Consequently, incident fields, if present on the mesh, should be subtracted from the total field near the boundary of the mesh. Finally, we note that, upon elimination of $\boldsymbol{H}$ from (2), we obtain

$$(\partial_x^2 + \partial_y^2 + \partial_z^2 - c_0^{-2}\partial_t^2)\,\boldsymbol{E} = 0 \tag{5}$$

i.e., each component of the electric field independently satisfies the three-dimensional wave equation.

III. ABSORBING BOUNDARY CONDITIONS

In the previous section, we saw that absorbing boundary conditions for Maxwell's equations on the mesh depicted in Fig. 1 require absorbing boundary conditions for the three components of the electric field only. We also saw that each of these field components satisfies the three-dimensional scalar wave equation

$$(\partial_x^2 + \partial_y^2 + \partial_z^2 - c_0^{-2}\partial_t^2)W = 0. \tag{6}$$

In this section, we will present the necessary boundary conditions by using the method that is described in detail by Engquist and Majda [8]. We shall assume, without loss of generality, that the mesh is located in the region $0 \leqslant x$, and give boundary conditions for the plane $x = 0$.

A space-time plane-wave constituent traveling in the direction of decreasing x, with inverse velocity components s_x, s_y and s_z such that $s_x^2 + s_y^2 + s_z^2 = c_0^{-2}$, can be written as

$$\begin{aligned} W = \mathrm{Re}\,(\psi(t + (c_0^{-2} - s_y^2 - s_z^2)^{1/2}\,x \\ + s_y y + s_z z)) \end{aligned} \tag{7}$$

with $\mathrm{Re}(c_0^{-2} - s_y^2 - s_z^2)^{1/2} \geqslant 0$. For this outgoing wave, the first-order boundary condition

$$(\partial_x - c_0^{-1}(1 - (c_0 s_y)^2 - (c_0 s_z)^2)^{1/2}\partial_t)W|_{x=0} = 0 \tag{8}$$

would, for fixed values of s_y and s_z, determine a W on the outer surface that is consistent with an outgoing wave, i.e., it is absorbed. Since we do not know the angle of incidence of the wave approaching $x = 0$, an approximation in (8) is made. Writing

$$(1-(c_0s_y)^2-(c_0s_z)^2)^{1/2} = 1 + 0((c_0s_y)^2 + (c_0s_z)^2) \quad (9)$$

we obtain as a first approximation

$$(\partial_x - c_0{}^{-1}\partial_t)W|_{x=0} = 0. \quad (10)$$

Use of the next approximation to the square root

$$(1-(c_0s_y)^2-(c_0s_z)^2)^{1/2} = 1 - \tfrac{1}{2}((c_0s_y)^2 + (c_0s_z)^2) + 0(((c_0s_y)^2 + (c_0s_z)^2)^2) \quad (11)$$

yields the second approximation of the boundary condition

$$(c_0{}^{-1}\partial_{xt}{}^2 - c_0{}^{-2}\partial_t{}^2 + \tfrac{1}{2}(\partial_y{}^2 + \partial_z{}^2))W|_{x=0} = 0. \quad (12)$$

From (10) and (12), approximations of W can be determined on the outer boundary. Engquist and Majda, who arrived at the same boundary conditions using a different method, prove that these boundary conditions give well-posed initial-boundary-value problems. They also give a third approximation that they claim to be stable. Numerical experiments, in a two-dimensional configuration, with a difference approximation of the latter boundary condition, however, showed instabilities for values of $c_0\delta t/\delta$ near the maximum that is given by (4).

For two-dimensional electromagnetic-field problems, it is possible to simplify the second approximation. Assuming that the fields do not depend on z and are E-polarized, i.e., $E = E_z i_z$ and $H = H_x i_x + H_y i_y$, (12) applies to E_z only. Now, in this case, we have from (1a)

$$\mu_0\partial_t H_x = -\partial_y E_z. \quad (13)$$

Substituting (13) in (11), ($\partial_z \equiv 0$), with $W = E_z$, integrating with respect to t and using $E_z = 0$ for $t < 0$, we obtain

$$(\partial_x E_z - c_0{}^{-1}\partial_t E_z - (c_0\mu_0/2)\partial_y H_x)_{x=0} = 0 \quad (14)$$

which boundary condition is much simpler than (12) but still of the same order of approximation. We observe that the first two terms of (14) reduce to the first approximation. A similar boundary condition for the case of H-polarization can easily be derived.

IV. FINITE-DIFFERENCE APPROXIMATIONS OF ABSORBING BOUNDARY CONDITIONS

In this section, we present the finite-difference approximation of the absorbing boundary conditions from Section III. These approximations have a local truncation error of the second order in all increments. As we have seen above, we need absorbing boundary conditions for the E-field components that are tangential to the boundary of interest. Therefore, boundary conditions for the plane $x = 0$ are expressed in terms of E_y and E_z. We shall give the discretized form of the boundary condition for E_z at this boundary, the boundary conditions for E_y and the boundary conditions on the other planes following easily from the one for E_z. The finite-difference approximation of (10) was derived using centered differences in both the space and the time increments, it has a local truncation error of the second order in all increments. We prefer to present the actual formulas in a form that is directly manageable for the computer program. The first approximation (10) for E_z is discretized as follows:

$$E_z{}^{n+1}(0,j,k+1/2 = E_z{}^n(1,j,k+1/2) + \frac{c_0\delta t - \delta}{c_0\delta t + \delta}(E_z{}^{n+1}(1,j,k+1/2) - E_z{}^n(0,j,k+1/2)). \quad (15)$$

The second approximation (12) for E_z at the boundary $x = 0$ is discretized as

$$\begin{aligned} E_z{}^{n+1}&(0,j,k+1/2) \\ &= -E_z{}^{n-1}(1,j,k+1/2) + \frac{c_0\delta t-\delta}{c_0\delta t+\delta}(E_z{}^{n+1}(1,j,k+1/2) \\ &+ E_z{}^{n-1}(0,j,k+1/2)) + \frac{2\delta}{c_0\delta t+\delta}(E_z{}^n(0,j,k+1/2) \\ &+ E_z{}^n(1,j,k+1/2)) + \frac{(c_0\delta t)^2}{2\delta(c_0\delta t+\delta)} \\ &\cdot (E_z{}^n(0,j+1,k+1/2) - 2E_z{}^n(0,j,k+1/2) \\ &+ E_z{}^n(0,j-1,k+1/2) + E_z{}^n(1,j+1,k+1/2) \\ &- 2E_z{}^n(1,j,k+1/2) + E_z{}^n(1,j-1,k+1/2) \\ &+ E_z{}^n(0,j,k+3/2) - 2E_z{}^n(0,j,k+1/2) \\ &+ E_z{}^n(0,j,k-1/2) + E_z{}^n(1,j,k+3/2) \\ &- 2E_z{}^n(1,j,k+1/2) + E_z{}^n(1,j,k-1/2)). \end{aligned} \quad (16)$$

Finally, the second approximation (14) for the two-dimensional problem is discretized as

$$\begin{aligned} E_z{}^{n+1}(0,j) &= E_z{}^n(1,j) + \frac{c_0\delta t-\delta}{c_0\delta t+\delta}(E_z{}^{n+1}(1,j) \\ &- E_z{}^n(0,j)) - \frac{\mu_0 c_0}{2(c_0\delta t+\delta)} \\ &(H_x{}^{n+1/2}(0,j+1/2) - H_x{}^{n+1/2}(0,j-1/2) \\ &+ H_x{}^{n+1/2}(1,j+1/2) - H_x{}^{n+1/2}(1,j-1/2)) \end{aligned} \quad (17)$$

where we have deleted the z-dependence of the fields from our notation since the value of z is the same in all terms. Centered differences were used for deriving (16) and (17) and these finite-difference approximations also have a local truncation error of the second order in all increments. We note that the discretization of the first approximation (15) is identical to the discretization in [8]. The discretization of the second approximation (16), however, differs from the one in [8]. Ours has the advantage that it can be used closer to the vertices of the mesh that it requires less storage.

V. NUMERICAL RESULTS

In this section, we present numerical results that show the efficiency of highly absorbing boundary conditions. Since we need a relatively large-size mesh to exhibit clearly the properties of absorbing boundary conditions, we shall present results for two-dimensional configurations only; three-dimensional meshes of considerable size would give rise to storage requirements that were too large for the computer that was available to the author at the time this research was carried out. The insight gained by studying a two-dimensional configuration can easily be applied to three-dimensional configurations since the structure of (16) is the same for two- and three-dimensional problems. (The two-dimensional form of (16) is obtained by deleting from it the differentiations along the axis of cylindricity.) We present results for fields that do not depend on z and are E-polarized (i.e., $E = E_z i_z$). A square two-dimensional mesh is used ($0 \leqslant i \leqslant 34, 0 \leqslant j \leqslant 34$) with absorbing boundary conditions on all four sides. We have a monochromatic isotropic point source of wavelength λ that is switched on at $t = 0$ and we use $\delta = \delta_x = \delta_y = 2c_0 \delta t = 0.1\lambda$. This point source is modeled by adding a term representing a current

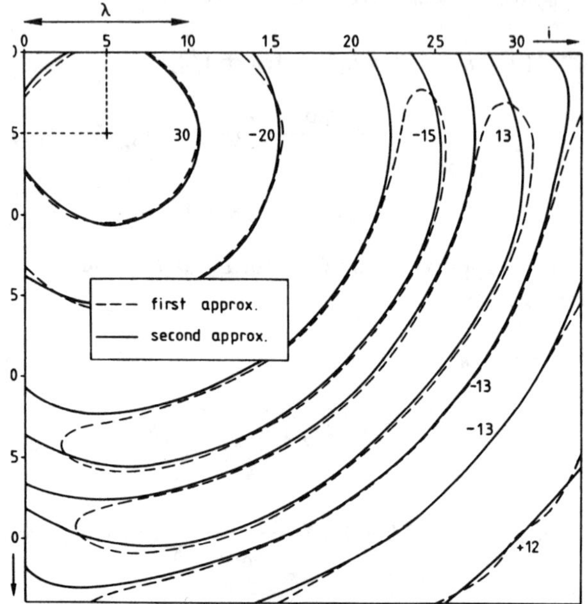

Fig. 2. Contour plot of the radiation pattern, after 141 time steps, of an isotropic source located at node (5, 5) of a 35*35-node mesh (arbitrary units).

$$I_z(t) = C \sin (2\pi c_0 t/\lambda)\epsilon(t) \tag{18}$$

at $r_s = (x_s, y_s) = (i\delta, j\delta)$ to the relevant finite-difference equation. In (18), $\epsilon(t)$ denotes the Heaviside unit-step function. The results are given after 141 steps.

An isotropic point source has a circular radiation pattern and we investigate how well this pattern is maintained on the truncated mesh. Fig. 2 gives a contour plot of the radiation pattern on the mesh for a point source that is located at node $(i, j) = (5, 5)$, which location was chosen to show, as clearly as possible, the efficiency of highly absorbing boundary conditions. The numbers in the contour plot are proportional to the local electric-field strength.

The contour plot is given for both the first and the second approximation. The first approximation turns out to yield relatively poor results, especially at large distances from the source. With the second approximation, however, an almost circular pattern is obtained with only slight deformations at points near the boundaries and far away from the source. These errors are caused by the fact that waves with grazing incidence on a boundary are not well absorbed, but partly reflected. For comparison, Fig. 3 gives the contour plot of the field of the same source that is now located closer to the boundary of the mesh, at node $(i, j) = (3, 3)$. A comparison with Fig. 2 is possible since the same contour lines are given. From Fig. 3, we observe that, near the boundary of the mesh, the second approximation is not accurate enough for sources this close to the boundary. It can be concluded that, on a mesh of the given dimensions, accurate results are obtained when the second approximation is used and when the source of the scattered field is located at about five or more nodes from the boundary of the mesh. With the first approximation, less accurate results are obtained and the source of the scattered field should be relatively far away from the boundary. Since (16) is easy to implement, it is advantageous to use the second approximation.

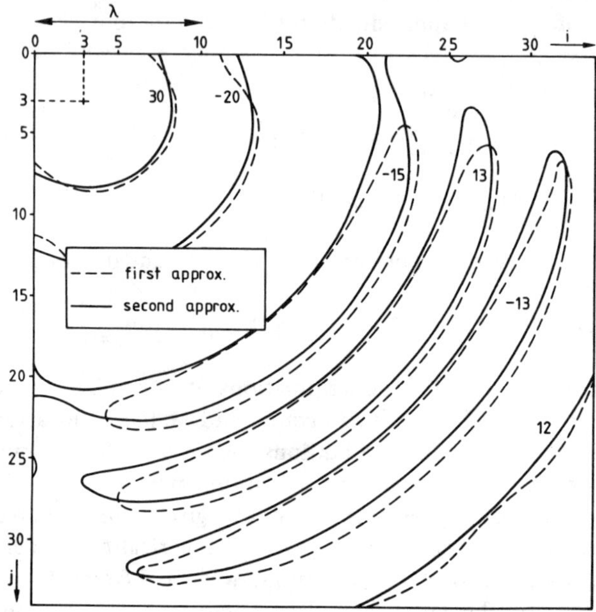

Fig. 3. Contour plot of the radiation pattern, after 141 time steps, of an isotropic source located at node (3, 3) of a 35*35-node mesh (arbitrary units).

As to the simplified second approximation for two-dimensional fields (17), we note that it gives exactly the same results as the more complicated one that follows from (16).

For the point source under consideration, the exact radiated field is the solution of

$$(\partial_x{}^2 + \partial_y{}^2 - c_0{}^{-2}\partial_t{}^2)E_z = \mu_0 \partial_t I_z(t)\delta(r - r_s) \tag{19}$$

that is rotationally symmetric around r_s and that consists of outgoing waves. This solution is easily obtained [10] as

$$E_z(r,t) = -(\mu_0/2\pi) \int_0^\infty \partial_t I_z(t - (|r - r_s|^2 + \zeta^2)^{1/2}/c_0) \cdot (|r - r_s|^2 + \zeta^2)^{-1/2}\, d\zeta. \tag{20}$$

In Fig. 4, this exact solution is plotted as a function of the distance from the point source and at the same instant in time as the results that are given in Fig. 2. For comparison we have,

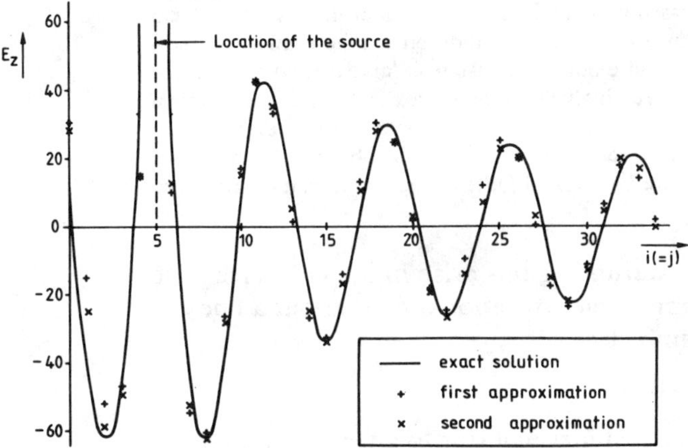

Fig. 4. A comparison of the results on the diagonal of the mesh of Fig. 2 with the exact solution.

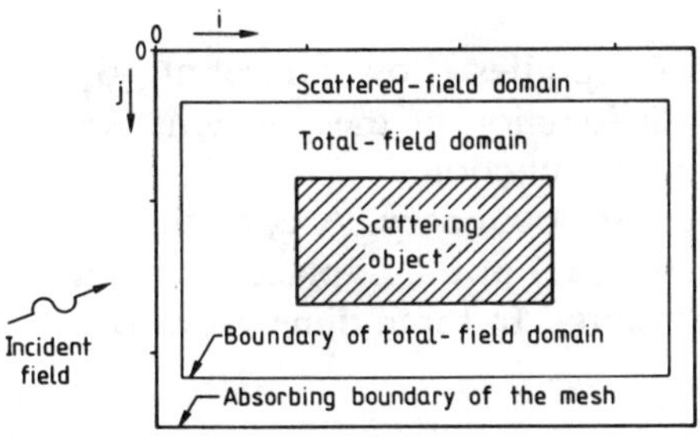

Fig. 5. The domain occupied by the mesh with the obstacle, the domain where the total field is computed, and the domain where the scattered field is computed (two-dimensional configuration).

in Fig. 4, also given the finite-difference results that are obtained on the diagonal of the mesh of Fig. 2 that passes through the point source. It turns out that, considering the fact that we have a relatively coarse mesh ($\delta = \lambda/10$), we have obtained accurate results.

VI. THE INCIDENT FIELD

In the present paper, we have used a point source to generate outgoing waves at the boundaries of the mesh. In practice, the outgoing waves are not generated by a source in the mesh but represent the scattered field E^s that is caused by the presence of an obstacle that scatters the incident field E^i (the incident field is assumed to be known.) If maximum accuracy is desired, it is advantageous to compute the total field $E = E^s + E^i$ rather than the scattered field [5], [9], especially in those regions where the scattered field almost cancels the incident field (behind or in the interior of scattering objects). Absorbing boundary conditions, however, cannot be applied to the total field.

To overcome this difficulty, a boundary can be introduced that is located close to the absorbing boundaries of the mesh (see Fig. 5.) Inside this boundary the total field is computed, outside it the scattered field is computed, thus allowing the application of absorbing boundary conditions. When (2) is evaluated for the total (scattered) field at a node inside (outside) this boundary, but adjacent to it, the total field will often be required from nodes that are located outside (inside) it. For the relevant nodes, (2) is modified. When computing the total field inside the boundary, the incident field is added to the scattered field that is obtained from nodes outside the boundary and, similarly, when computing the scattered field outside the boundary, the incident field is subtracted from the total field obtained from inside the boundary. Since the incident field is known exactly at any place or instant in time the above procedure does not involve any approximations, moreover, it does not require any storage nor does it cause any spurious wave reflection.

ACKNOWLEDGMENT

The author wants to thank A. T. de Hoop for his suggestions concerning the work presented in this paper.

REFERENCES

[1] K. S. Yee, "Numerical solution of initial boundary value problems involving Maxwell's equations in isotropic media," *IEEE Trans. Antennas Propagat.*, vol. AP-14, pp. 302–307, May 1966.

[2] C. L. Longmire, "State of the art in IEMP and SGEMP calculations," *IEEE Trans. Nucl. Sci.*, vol. NS-22, pp. 2340–2344, Dec. 1975.

[3] C. D. Taylor, D.-H. Lam, and T. H. Shumpert, "Electromagnetic pulse scattering in time-varying inhomogeneous media," *IEEE Trans. Antennas Propagat.*, vol. AP-17, pp. 585–589, Sept. 1969.

[4] A. Taflove and M. E. Brodwin, "Numerical solution of steady-state electromagnetic scattering problems using the time dependent Maxwell's equations," *IEEE Trans. Microwave Theory Tech.*, vol. MTT-23, pp. 623–630, Aug. 1975.

[5] A. Taflove, "Application of the finite-difference time-domain method to sinusoidal steady-state electromagnetic penetration problems," *IEEE Trans. Electromagn. Compat.*, vol. EMC-22, pp. 191–202, 1980.

[6] D. E. Merewether, "Transient currents induced on a metallic body of revolution by an electromagnetic pulse," *IEEE Trans. Electromagn. Compat.*, vol. EMC-13, pp. 41–44, May 1971.

[7] K. S. Kunz and K.-M. Lee, "A three-dimensional finite-difference solution of the external response of an aircraft to a complex transient EM environment: Part I—The method and its implementation," *IEEE Trans. Electromagn. Compat.*, vol. EMC-20, pp. 328–333, May 1978.

[8] B. Engquist and A. Majda, "Absorbing boundary conditions for the numerical simulation of waves," *Math. Comp.*, vol. 31, pp. 629–651, July 1977.

[9] R. Holland, L. Simpson, and K. S. Kunz, "Finite-difference analysis of EMP coupling to lossy dielectric structures," *IEEE Trans. Electromagn. Compat.*, vol. EMC-22, pp. 203–209, 1980.

[10] J. D. Achenbach, *Wave Propagation in Elastic Solids.* Amsterdam, The Netherlands: North-Holland, 1973, ch. 3, p. 94.

Boundary Conditions for the Numerical Solution of Elliptic Equations in Exterior Regions

A. BAYLISS, M. GUNZBURGER, and E. TURKEL

Abstract. Elliptic equations in exterior regions frequently require a boundary condition at infinity to ensure the well-posedness of the problem. Examples of practical applications include the Helmholtz equation and Laplace's equation. Computational procedures based on a direct discretization of the elliptic problem require the replacement of the condition at infinity by a boundary condition on a finite artificial surface. Direct imposition of the condition at infinity along the finite boundary results in large errors. A sequence of boundary conditions is developed which provides increasingly accurate approximations to the problem in the infinite domain. Estimates of the error due to the finite boundary are obtained for several cases. Computations are presented which demonstrate the increased accuracy that can be obtained by the use of the higher order boundary conditions. The examples are based on a finite element formulation but finite difference methods can also be used.

1. Introduction. Elliptic problems in exterior regions arise in many branches of physics. For example, the flow of an incompressible irrotational fluid about a body is described by the Laplace equation (e.g., Lamb [19])

$$\Delta u = 0.$$

The same equation arises in the study of electrostatics exterior to given surfaces (e.g., Stratton [30]). A different example is the exterior scattering problem for either acoustics or electromagnetism. In this case one wishes to solve the Helmholtz equation

$$\Delta u + k^2 u = 0,$$

with either Dirichlet or Neumann data specified on the bodies (Bowman, et al. [6], Müller [22]). For inhomogeneous media, k is a given function of the position. For some applications in plasma physics, k can be a nonlinear function.

In these cases infinity can be regarded as a separate boundary. A condition at infinity is required to make the exterior problem well posed. For the Laplace equation it is sufficient to impose a condition of regularity at infinity. In three dimensions this is

$$u = O\left(\frac{1}{r}\right), \qquad r \to \infty, \tag{1.1}$$

where r is the distance from a fixed (but arbitrary) origin (Kellogg [15]). For the Helmholtz equation, one can impose the Sommerfeld radiation condition (Sommerfeld [29])

$$\frac{\partial u}{\partial r} - iku = o\left(\frac{1}{r}\right), \qquad r \to \infty. \tag{1.2}$$

A more exact form of (1.2) is

$$\lim_{r \to \infty} \iint_{|y|=r} \left|\frac{\partial u}{\partial r} - iku\right|^2 dS = 0, \tag{1.3}$$

where the integral is over spherical shells centered at $r = 0$ (Rellich [27], Hellwig [13]). The radiation condition ((1.2)–(1.3)) states that the solution corresponds to outgoing waves (see Wilcox [35], [36], for more details).

In many instances one is interested in problems with variable coefficients that approach a constant state at infinity. An extension of the theory of radiation conditions to problems with variable coefficients was developed by Vainberg [33]. The techniques to be described are valid for the variable coefficient case provided the coefficients approach constants at infinity at a sufficiently rapid rate.

Reprinted from *SIAM Journal of Applied Mathematics*, vol. 42, no. 2, pp. 430–451, April 1982.

A numerical solution of an elliptic problem in an exterior region must be able to incorporate the radiation condition at infinity within the computational procedure. Solution techniques based on eigenfunction expansions or asymptotic methods automatically accomplish this by the proper choice of expansion functions. When the free space Green's function which satisfies the radiation condition is known (e.g., constant coefficients), the difficulty of imposing the radiation condition can be avoided by reformulating the problem as a Fredholm integral equation. Such formulations for the Helmholtz equation can be found in Chertock [9], Kleinman and Roach [16], and Burton and Miller [8]. Poggio and Miller [24] discuss a vector integral equation for the reduced Maxwell equations. Bayliss [4] shows that one can greatly increase the efficiency of the integral equation formulation by introducing an appropriate coordinate transform along the body. Schneck [28] has developed codes to solve the resultant integral equations for a range of applications.

The integral equation approach has several deficiencies. These methods are generally restricted to the constant coefficient case and require the inversion of a full matrix. This can result in storage difficulties, especially for three dimensional problems or problems with high frequencies. In addition, for many applications the matrix elements are expensive to compute [4]. Many mesh points are required to resolve the singularity in the kernel even for low frequencies.

For the Helmholtz equation, an additional difficulty with the integral equation formulation is the possibility of interior resonances. It is well known that for certain values of k the integral equation becomes singular [9], [16]. These resonances are connected with eigenvalues of associated interior problems. Various attempts have been made to overcome this difficulty, but they generally increase the complexity of the integral equation approach (Ursell [32]). It will be shown that the proper formulation of radiation conditions can eliminate the possibility of eigenvalues.

An alternative to the integral equation method is to couple an interior solution with a global functional of the solution on an artificial boundary. The global functional can be obtained by integral formulas using the free space Green's function or by using an expansion, typically obtained from separation of variables, to represent the solution exterior to the artificial boundary. Marin [21], Zienkiewicz et al. [37] and others have studied methods based on an integral relation over the artificial boundary, while Fix and Marin [10] have used a boundary condition based on separation of variables to solve problems in underwater acoustics.

These methods incorporate the exact radiation condition at the cost of a nonlocal boundary condition. A disadvantage of these methods is that the nonlocal coupling over the artificial boundary is equivalent to the full matrix that would be obtained from the integral equation. Furthermore, for the Helmholtz equation, spurious eigenvalues can also occur with this formulation. Goldstein [11] has suggested extending the boundary conditions of Engquist and Majda to the elliptic case. However, this method is restricted to only a range of frequencies depending on the expansion parameter. No calculations using this method have been carried out to date.

The method to be presented develops a sequence of local boundary conditions that are extensions of (1.1) and (1.2). These boundary conditions are then applied at a finite artificial boundary. As the order of accuracy of the boundary operator increases, the order of the highest derivative appearing in the boundary operator will also increase.

The artificial surface will generally be assumed, for the proofs, to be the sphere $r = r_1$. The resulting elliptic problem is then discretized and solved in the bounded region between the body and the artificial surface. The proposed boundary conditions are asymptotic in $1/r$. Hence, for a given accuracy one can bring the artificial boundary further in when using the higher order boundary conditions.

In many applications, the solution is required only in the vicinity of the body. The far field solution can be calculated by a quadrature formula, such as Green's formula, once u and $\partial u/\partial n$ are known along the body. For problems in potential flow, one frequently is only interested in the solution on the body. Because of this, we shall stress the accuracy of the boundary conditions as r_1 decreases. Furthermore, the error estimates will be for surface L^2-errors. These errors are due to the imposition of the generalized radiation conditions at a finite boundary.

These boundary conditions are related to a family of boundary conditions developed by Bayliss and Turkel [5] for time dependent problems. The boundary operators are differential relations which match the solution to an expansion in $1/r$ which is valid in a neighborhood of infinity. The boundary conditions generally involve derivatives of order greater than or equal to the order of the differential equation. Hence, these boundary conditions are different from the usual boundary conditions encountered in elliptic theory.

For the Laplace equation, the resultant discretization can be solved by fast iterative methods leading to a substantial improvement over the integral equation methods. For Helmholtz type equations, the error will have a dependence on the number of wave lengths between the body and the artificial surface. In several test cases it has been possible to constrict the computational region as k increases. All numerical results in this study were obtained with a finite element program based on a band Gaussian solver. This has large storage requirements which limited the investigation of high frequencies. This storage can be reduced by the use of iterative methods. An iterative method for the Helmholtz equation based on a decay law for a corresponding hyperbolic problem was developed by Kriegsman and Morawetz [17]. Kriegsman and Morawetz have also implemented time dependent boundary conditions similar to those proposed here (private communication). Brandt [7] and Nicolaides [23] have developed iterative methods based on the multigrid algorithm. These methods require Gaussian elimination on a coarse grid which provides some resolution of the solution. Hence, these methods have limited use for high frequencies. The scheme of Nicolaides can be combined with the finite element method described in the appendix. Extensions of the fast solvers based on capacitance methods are also feasible (Proskurowski and Widlund [25]).

In § 2 we develop the extension of the standard radiation conditions for both the Helmholtz and Laplace equations. Error estimates for Laplace's equation are given in § 3 and for the Helmholtz equation in § 4. These chapters can be skipped by those only interested in the computational procedure. In § 5 computational results are presented for both the Laplace equation and the Helmholtz equation. For the Helmholtz equation, both constant and variable indices of refraction are considered. The details of the finite element procedure are given in the appendix.

2. Construction of radiation boundary conditions. Let u be an outgoing solution to the three-dimensional Helmholtz equation

$$\Delta u + k^2 u = 0, \tag{2.1}$$

exterior to a sphere $r = r_0$. It is known (Atkinson [2], Wilcox [34]) that u can be represented by a convergent expansion

$$u = \frac{e^{ikr}}{kr} \sum_{j=0}^{\infty} \frac{F_j(\theta, \phi)}{(kr)^j}. \tag{2.2}$$

Here θ, ϕ denote the angular variables of an (r, θ, ϕ) spherical coordinate system. The series (2.2) is uniformly and absolutely convergent and can be differentiated term by term any number of times (Wilcox [34]). $F_j(\theta, \phi)$, $j \geqq 1$ can be obtained from the radiation pattern, F_0, by the formula

$$F_j(\theta, \phi) = \frac{1}{(2i)^j j!} \prod_{l=0}^{j} [l(l-1) + Q] F_0(\theta, \phi),$$

where Q is the Beltrami operator in the angular coordinates θ and ϕ,

$$Q = \frac{1}{\sin\theta} \frac{\partial}{\partial\theta}\left(\sin\theta \frac{\partial}{\partial\theta}\right) + \frac{1}{\sin^2\theta} \frac{\partial^2}{\partial\phi^2}.$$

The Sommerfeld radiation condition for any such solution is

$$-iku + u_r = o\left(\frac{1}{r}\right), \qquad r \to \infty.$$

In fact, it is clear from the expansion (2.2) that

(2.3) $$-iku + u_r = O\left(\frac{1}{r^2}\right), \qquad r \to \infty.$$

In numerical computations for which the exterior region is truncated at a finite value of r, say $r = r_1$, a possible boundary condition to impose is

(2.4) $$-iku + u_r\big|_{r=r_1} = 0.$$

However, this condition is very inaccurate, and in fact, it can easily be seen that it is not exact even for the first term in the expansion (2.2).

We will develop here a sequence of linear differential operators B_m which provide more accurate extensions of the condition (2.4) by annihilating the first m terms in the expansion (2.2). Thus, the condition that at $r = r_1$ the solution lies in the null space of the operator B_m can be considered as a procedure to match the solution to the first m terms in the expansion (2.2).

An example of such an operator is

$$B_1 u = \left(\frac{\partial}{\partial r} - ik + \frac{1}{r}\right) u.$$

It is easily verified that

$$B_1\left[\frac{e^{ikr}}{kr} F(\theta, \phi)\right] = 0,$$

for any function $F(\theta, \phi)$. Furthermore, it can be verified that for any function u having the expansion (2.2)

(2.5) $$B_1 u\big|_{r=r_1} = O\left(\frac{1}{r^3}\right).$$

To develop more accurate conditions we consider the sequence of operators

(2.6) $$B_m = \prod_{j=1}^{m} \left(\frac{\partial}{\partial r} - ik + \frac{2j-1}{r}\right) \equiv \left(\frac{\partial}{\partial r} - ik + \frac{2m-1}{r}\right) B_{m-1}.$$

A straightforward calculation verifies that

$$B_m p = 0,$$

for any function p of the form

$$p = \frac{e^{ikr} F(\theta, \phi)}{(kr)^l}, \qquad l = 1, \cdots, m.$$

Furthermore, if u is any function having an expansion of the form (2.2) then

(2.7) $$B_m u\big|_{r=r_1} \equiv h = O\left(\frac{1}{r_1^{2m+1}}\right).$$

It is clear that the leading order term in h will involve only the term of order $m+1$ in the expansion (2.2), and thus, the boundary condition $B_m u = 0$ will match the solution to the first m terms in the expansion (2.2). For any fixed k, the errors in the boundary condition $B_m u$ will decrease at a faster rate (in $(r_1)^{-1}$) as $r_1 \to \infty$. This is thus analogous to the use of higher order difference approximations where the errors decrease at a faster rate in a mesh size h. Thus one can expect a significant increase in efficiency by applying the higher order boundary conditions. We point out that the individual terms in (2.2) are not solutions to (2.1) (unless $k = 0$); however, the series (2.2) provides a description of the behavior of the solution as $r \to \infty$, which is generally not the case for expansions based on complete sets of solutions to (2.1) (see Aziz and Kellogg [3]). We further point out that the operators B_m in (2.6) can be written as

(2.8) $$\left(-ik + \frac{\partial}{\partial r}\right)^m + \text{lower order terms},$$

and thus they can be considered as generalizations of the Sommerfeld radiation condition.

In the study of the Helmholtz equation, we are interested in the behavior of the errors in different parameter ranges. Specifically,

1. The error for fixed k and m as the position of the artificial surface, i.e., r_1, varies.
2. The error for fixed k and r_1 as the order of the boundary operator m increases.
3. The error for fixed r_1 and m as the frequency k increases.

Case 2 is of greater importance than case 1 as it is generally more expensive to increase the size of the computational domain than to implement the higher order boundary conditions. Results to be established in §§ 3 and 4 demonstrate that the higher order boundary condition can lead to significantly increased accuracy when applied at a fixed r_1.

The constants involved in the order relations (2.5) and (2.7) will, in general, depend on k. This is because for arbitrary problems the angular functions $F_j(\theta, \phi)$ can be expected to grow with k. Thus, as k increases the errors in (2.5) and (2.7) will not, in general, be bounded uniformly in k. However, the quantity kr_1 is a natural non-dimensional quantity, which is, in fact, just the number of wave lengths to the artificial boundary $r = r_1$. The error obtained by taking only a fixed number of terms in the series (2.2) can be expected to depend significantly on this quantity. A consequence of this is that the radial resolution will not increase, or will increase slowly as k increases. Numerical results illustrating this will be presented in § 5.

We also consider the exterior Laplace equation. In this case it is well known that the analogue of (2.2) holds, i.e., solutions have the multipole expansion

$$u = \frac{1}{r} \sum_{j=0}^{\infty} \frac{F_j(\theta, \phi)}{r^j}. \tag{2.9}$$

It then follows that the differential operators obtained by setting $k = 0$ in (2.6)

$$B_m = \prod_{j=1}^{m} \left(\frac{\partial}{\partial r} + \frac{2j-1}{r} \right), \tag{2.10}$$

exactly annihilate the first m terms in (2.9). Similarly, for any u with the expansion (2.9),

$$B_m u \big|_{r=r_1} = O\left(\frac{1}{r^{2m+1}} \right). \tag{2.11}$$

Both (2.6) and (2.11) are valid for equations with variable coefficients provided the coefficients approach a constant state sufficiently rapidly at infinity. As indicated in the introduction, we will be mainly interested in computing the solution on some surface. The solution at far fields points can then be obtained by a quadrature based on Green's formula. In the next two sections, we prove theorems that demonstrate that in some cases the expected surface L^2-errors are achieved.

For computational ease one can replace all radial derivatives beyond the first by tangential (angular) derivatives. This is done by using the differential equation and is especially important for finite element applications. The exact forms of B_1 and B_2 are given by (5.2).

The previous discussion has concentrated on the three dimensional Helmholtz and Laplace equations. For the two-dimensional Helmholtz equation, the solution u has the convergent expansion (Karp [14])

$$u = H_0(kr) \sum_{j=0}^{\infty} \frac{F_j(\theta)}{r^j} + H_1(kr) \sum_{j=0}^{\infty} \frac{G_j(\theta)}{r^j}, \tag{2.12}$$

where H_0 and H_1 are the Hankel functions of the first kind of order 0 and 1.

As this expansion is difficult to work with, we use the asymptotic expansion [6]

$$u \sim \sqrt{\frac{2}{\pi kr}}\, e^{i(kr-\pi/2)} \sum_{j=0}^{\infty} \frac{f_j(\theta)}{r^j}. \tag{2.13}$$

The boundary conditions based on (2.13) are

$$B_m = \prod_{j=1}^{m} \left(\frac{\partial}{\partial r} + \frac{(2j-\frac{3}{2})}{r} - ik \right), \tag{2.14}$$

which are analogous to (2.6).

We have concentrated on the homogeneous Laplace and Helmholtz equations. The same boundary conditions can be used for the inhomogeneous equation

$$\Delta u + k^2 u = F. \tag{2.15}$$

The estimates obtained in the next two sections apply to (2.15), provided that F decays sufficiently rapidly for large r.

Boundary conditions based on the operators B_m are nonstandard because of the high order of the derivatives involved. The regularity of the solutions up to the boundaries, using $B_m u = 0$ at the artificial surface, is guaranteed by the Agmon, Dougalis, Nirenberg theory [1]. Using the results of Lopatinskii [20], we need only consider the half plane problem without any lower order terms. Thus, we have reduced the problem to

$$\begin{aligned} &\Delta u = 0, \qquad x \geqq 0, \quad -\infty < y, z < \infty, \\ &\frac{\partial^n u}{\partial x^n} = 0, \qquad x = 0. \end{aligned} \tag{2.16}$$

The only solution to (2.16) that decays as $x \to \infty$ is $u = 0$. Hence, the complimentary condition of [1] is satisfied and regularity follows (see also [31]).

When estimates for the error are obtained, it is important that the constants in the estimation do not grow too rapidly as the artificial boundary approaches infinity. It is also important, especially for the Helmholtz equation, to eliminate the possibility of spurious eigenvalues for the resulting interior problem. Indeed, we wish to obtain estimates that are uniform in k and do not have poles at discrete values of k. Such estimates will be obtained in the next two sections.

3. Error bounds for Laplace's equation. We wish to obtain estimates for the L^2 surface error that occurs when the regularity condition at infinity is replaced by the condition $B_m u = 0$ at a finite boundary. In this section we concentrate on Laplace's equation. As discussed in the introduction, the main interest is in the errors that occur in the solution and its derivative along the inner boundary.

For concreteness we shall consider the Neumann problem. All results are equally valid for the Dirichlet problem.

The problem that we wish to solve is

$$\Delta u = 0 \quad \text{in } \Omega, \qquad \frac{\partial u}{\partial n} = g \quad \text{on } \Gamma_1, \qquad u = O\left(\frac{1}{r}\right), \quad r \to \infty. \tag{3.1}$$

We replace this by the problem

$$\Delta v = 0 \quad \text{in } \Omega, \qquad \frac{\partial v}{\partial n} = g \quad \text{on } \Gamma_1, \qquad B_m v = 0 \quad \text{on } \Gamma_2, \tag{3.2}$$

(see Fig. 1).

Let w be the error, $w = u - v$. Then w satisfies

$$\Delta w = 0 \quad \text{in } \Omega, \qquad \frac{\partial w}{\partial n} = 0 \quad \text{on } \Gamma_1, \qquad B_m w = B_m u \equiv h \quad \text{on } \Gamma_2. \tag{3.3}$$

If Γ_2 is the sphere $r = r_1$, then by (2.6) $h = O(1/r_1^{2m+1})$.

In this section we will only consider the case that the artificial surface Γ_2 is the sphere $r = r_1$. We will consider two cases. In Theorem (3.1) we discuss the case that the body Γ_1 is also a sphere. In Theorems (3.2) and (3.3) we consider general bodies, but only treat the boundary conditions B_1 and B_2. This is not a major restriction since B_1 and B_2 are the boundary conditions of the greatest practical importance.

Since we are interested in surface errors, we introduce the notation

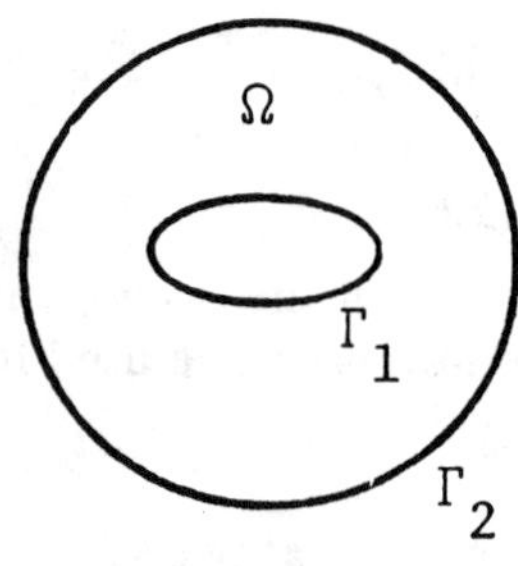

FIG. 1.

$$\|w\|_{(r)}^2 = \iint_{|y|=r} |w(y)|^2 \, dA_y. \tag{3.4}$$

If we introduce spherical coordinates (r, θ, ϕ), (3.4) becomes

$$\|w\|_{(r)}^2 = \int_0^{2\pi} \int_0^{\pi} |w(r, \theta, \phi)|^2 r^2 \sin\theta \, d\theta \, d\phi.$$

For simplicity we assume axial symmetry so that w is independent of ϕ. All the results are independent of this assumption.

For the first part, we assume that the body is also a sphere. The coordinates are scaled so that the surface is the sphere $r = 1$. We then have

THEOREM 3.1. *Let $g(\theta)$ be smooth and, hence, satisfy*

$$\int_0^{\pi} g^2(\theta) \sin\theta \, d\theta < \infty. \tag{3.5}$$

Let w be the solution to (3.3). *Then there exists a constant C, independent of m, such that*

$$\|w\|_{(r)} \leq \frac{Cr^{m+1}}{r_1^{2m+1} - 1}, \tag{3.6a}$$

$$\left\|\frac{\partial w}{\partial r}\right\|_{(r)} \leq \frac{Cr^m}{r_1^{2m+1} - 1}. \qquad 1 \leq r \leq r_1, \tag{3.6b}$$

Note. (1) The bounds on the derivatives are necessary for the Dirichlet problem.
(2) It follows from (3.6a) that

$$\|w\|_{(r_1)} = O\left(\frac{1}{r_1^{m+1}}\right), \qquad \|w\|_{(1)} = O\left(\frac{1}{r_1^{2m+1}}\right).$$

Hence, the smallest errors occur along the inner surface. This is analogous to the Saint-Venant's principle. Lax [19] has considered decay laws for volume norms for positive elliptic operators.

(3) Since C is independent of m, the estimate shows that we can improve the accuracy by fixing r_1 and increasing the order of the boundary operator.

Proof. We first consider the case that Dirichlet data is imposed, i.e., (3.1b) is replaced by $u = g$ on Γ_1. By assumption (3.5) and using axial symmetry, we have that

$$g(\theta) = \sum_{j=0}^{\infty} d_j P_j(\cos\theta), \tag{3.7}$$

where P_j are the Legendre polynomials. It then follows that

$$\int_0^{\pi} |g^2(\theta)| \sin\theta \, d\theta = \sum_{j=0}^{\infty} d_j^2.$$

If $g(\theta)$ is sufficiently smooth, it is also known that for any n (Gottlieb and Orszag [12])

$$|d_j| \leq \frac{C_n}{(j+1)^n}, \tag{3.8}$$

where C_n depends on n and g but not on j. Using separation of variables, the solution u to (3.1) is given by

$$u(r, \theta) = \sum_{j=0}^{\infty} \frac{d_j P_j(\cos\theta)}{r^{j+1}}, \tag{3.9a}$$

$$\frac{\partial u}{\partial r}(r, \theta) = -\sum_{j=0}^{\infty} \frac{(j+1) d_j P_j(\cos\theta)}{r^{j+2}}. \tag{3.9b}$$

It is easy to verify, by induction, that the boundary operators B_m given by (2.10) satisfy

$$B_m(r^k) = \prod_{l=1}^{m} (k+l) r^{k-m} \equiv A_{k,m} r^{k-m}. \tag{3.10}$$

Clearly, $A_{k,m} = 0$ for $-m \leqq k < 0$.

It follows from (3.2b), (3.9) and (3.10) that

$$B_m(w) = B_m(u) = \sum_{j=m}^{\infty} \frac{d_j P_j(\cos\theta) A_{-j,m}}{r^{j+m+1}}. \tag{3.11}$$

By separation of variables, the error w has the form

$$w(r, \theta) = \sum_{j=0}^{\infty} q_j P_j(\cos\theta)\left[r^j - \frac{1}{r^{j+1}}\right]. \tag{3.12}$$

The operator B_m is applied to (3.10) and the result is evaluated at $r = r_1$. Using the completeness of the Legendre polynomials, this series can be equated term by term with (3.11). We then have

$$q_j = \begin{cases} 0, & j = 0, \cdots, m-1, \\ d_j\left(\dfrac{A_{-(j+1),m}}{A_{j,m}} r_1^{2j} - 1\right)^{-1}, & j \geqq m. \end{cases} \tag{3.13}$$

From the definition of $A_{-j,m}$, it is easily verified that

$$\left|\frac{A_{-(j+1),m}}{A_{j,m}}\right| > 1,$$

and so

$$|q_j| \leqq \frac{|d_j|}{r_1^{2j-1} - 1}, \qquad j \geqq m \tag{3.14}$$

and

$$\left|q_j\left(r^j - \frac{1}{r^{j+1}}\right)\right| \leqq \frac{|d_j| r^j}{r_1^{2j-1} - 1} \leqq \frac{2|d_j| r^m}{r_1^{2m-1} - 1}, \qquad 1 \leqq r \leqq r_1, \quad j \geqq m.$$

Since, $\|w\|_{(r)}^2 = \sum_j |q_j(r^j - 1/r^{j+1})|^2 r^2$, the result (3.6a) follows. By differentiating the series for w (3.12), and repeating the process, (3.6b) follows.

The proof for the case of Neumann data is similar. In this case (3.9) and (3.12) are replaced by

$$u(r, \theta) = -\sum_{j=0}^{\infty} \frac{d_j}{(j+1)} \frac{P_j(\cos\theta)}{r^{j+1}},$$

and

$$w(r, \theta) = \sum_{j=0}^{\infty} q_j P_j(\cos\theta)\left[r^j \frac{j}{j+1} - \frac{1}{r^{j+1}}\right].$$

We next consider the case where (3.1) is to be solved exterior to an arbitrary domain Γ_1. We restrict the proofs by only considering $m = 1, 2$. The proof relies on the generalized maximum principle discussed by Protter and Weinberger [26]. To apply these methods, we assume that Γ_1 has the property that every $x \in \Gamma_1$ lies on the boundary of a ball contained in the exterior of Γ_1. We first need

LEMMA 3.1. *Let w satisfy*

(3.15) $\Delta w = 0$ *in* Ω, $\dfrac{\partial w}{\partial n} = 0$ *on* Γ_1, $B_j w = h$ *on* Γ_2, $j = 1, 2$.

Suppose there exist functions z_1, z_2 *such that*

(3.16) $\Delta z_1 \leqq 0$ *in* Ω, $\dfrac{\partial z_1}{\partial n} \leqq 0$ *on* Γ_1, $B_j z_1 \geqq h$ *on* Γ_2, $j = 1, 2$

and

$\Delta z_2 \geqq 0$ *in* Ω, $\dfrac{\partial z_2}{\partial n} \geqq 0$ *on* Γ_1, $B_j z_2 \leqq h$ *on* Γ_2, $j = 1, 2$.

Then

$$z_2 \leqq u \leqq z_1 \quad in\ \Omega.$$

We make several observations.

1. The normal derivative is taken in the direction away from the origin. Hence, for a sphere $\partial u/\partial n = \partial u/\partial r$.
2. Similar results hold when Dirichlet data is imposed on the inner body.
3. The results are valid for any uniformly elliptic equation in Ω.

Proof. For $j = 1$, the lemma is a restatement of [26, Thm. 12]. When B_2 is used, the result follows from the observation that the proof of Protter and Weinberger only requires that when w has a positive maximum on Γ_2, then $B_2 w$ is positive. This follows from the form of B_2.

Based on this lemma we have:

THEOREM 3.2. *For an arbitrary body* Γ_1 *with the boundary operator* B_1 *on* Γ_2, *the error satisfies*

$$|w| \leqq \frac{c}{r_1^2} \quad pointwise. \tag{3.18}$$

Furthermore, with the operator B_2 *on* Γ_2, *the error satisfies*

$$|w| \leqq \frac{C}{r_1^3} \quad pointwise. \tag{3.19}$$

Proof. The boundary condition $B_1 w$ can be expressed as

$$\left|\frac{\partial w}{\partial r} + \frac{w}{r}\right| = |h| \leqq \frac{C}{2r_1^3}.$$

We choose for the lemma $z_1 = C/r_1^2$ and $z_2 = -z_1$. Straightforward algebra shows that z_1 and z_2 satisfy (3.16) and (3.17), respectively. (3.18) then follows from the conclusion of the lemma. For the boundary condition B_2, we choose $z_1 = C/r_1^3$ and $z_2 = -z_1$, and again, (3.19) follows.

We note that the exact solution to the exterior problem decays at least as fast as $1/r$. Hence, the uniform error bounds given by (3.18) and (3.19) express a smaller relative error at the body Γ_1 than near the artificial surface $r = r_1$. For some applications one wishes error bounds for the normal derivative on Γ_1. The bounds can be obtained by deriving a Fredholm equation of the second kind for $\partial w/\partial n$ on Γ_1. A theorem valid for $k \geqq 0$ will be given in the next section.

4. Error bounds for the Helmholtz equation. The problem that we wish to solve is

$$\Delta u + k^2 u = 0 \qquad \text{exterior to } \Gamma_1, \tag{4.1a}$$

$$\frac{\partial u}{\partial n} = g \qquad \text{on } \Gamma_1, \tag{4.1b}$$

$$\frac{\partial u}{\partial r} - iku = o\left(\frac{1}{r}\right) \qquad \text{as } r \to \infty. \tag{4.1c}$$

As before, we replace condition (4.1c) by

$$B_m v = 0 \quad \text{on } \Gamma_2, \tag{4.1d}$$

where v also satisfies (4.1a, 4.1b). Let w be the error, $w = u - v$. Then w satisfies

$$\Delta w + k^2 w = 0 \qquad \text{in } \Omega, \tag{4.2a}$$

$$\frac{\partial w}{\partial n} = 0 \qquad \text{on } \Gamma_1, \tag{4.2b}$$

$$B_m w = B_m u \equiv h \quad \text{on } \Gamma_2, \tag{4.2c}$$

where B_m is given by (2.6) (see Fig. 1). For simplicity we only consider the case of axial symmetry, however, all results hold for the general three dimensional problem as well as for the Dirichlet problem. As in § 3, we restrict the proofs to the case that the artificial surface is the sphere $r = r_1$. We are interested in error estimates on surfaces and use the surface norm given by (3.4). We then have

THEOREM 4.1. *Given* (4.2) *and* $m = 1$, *the error* w *has the bound*

$$\|w\|_{(r_1)} \leqq \frac{\|h\|_{(r_1)}}{k}. \tag{4.3}$$

Proof. Let $\bar{w}$ denote the complex conjugate of w. Then from Green's theorem and (4.2) we have

$$\int_\Omega |\text{grad } w|^2 \, dV - k^2 \int_\Omega |w^2| \, dV = \int_{\Gamma_2} \bar{w} \frac{\partial w}{\partial r} \, dA. \tag{4.4}$$

Using the definition of B_1, we also have

$$\int_{\Gamma_2} \bar{w} \frac{\partial w}{\partial r} \, dA = ik \int_{\Gamma_2} |w^2| \, dA + \int_{\Gamma_2} wh \, dA. \tag{4.5}$$

We then substitute (4.5) into (4.4). Taking the imaginary part of the resulting equation yields

$$\|w\|_{(r_1)}^2 = \int_{\Gamma_2} |w|^2 \, dA = \frac{1}{k} \int_{\Gamma_2} wh \, dA \leqq \frac{\|w\|_{(r_1)} \|h\|_{(r_1)}}{k}. \tag{4.6}$$

Dividing both sides by $\|w\|_{(r_1)}$ gives the estimate (4.3).

Since $h = B_1 u$ we know that $|h| \leqq C/r^3$, where C depends only on g and k. Hence, (4.3) is equivalent to

$$\|w\|_{(r_1)} \leqq \frac{C}{k r_1^2}. \tag{4.7}$$

Using the definition of B_1, we also have

$$\left\| \frac{\partial w}{\partial r} \right\|_{(r_1)} \leqq \left(k + \frac{1}{r}\right) \|w\|_{(r_1)} + \|h\|_{(r_1)} \leqq \left(2 + \frac{1}{kr}\right) \|h\|_{(r_1)} \leqq \frac{C}{r^2}\left(2 + \frac{1}{kr}\right). \tag{4.8}$$

In Theorem 4.1 we considered the boundary operator B_1. We now consider the operator B_2, but restrict the body Γ_1 to be a sphere. We then have:

THEOREM 4.2. *Consider* (4.2), *with* $m = 2$, *and* Γ_1, *the sphere*, $r = 1$. *Then the solution* w *satisfies the bound*

$$\|w\|_{(r_1)} \leqq \left(1 + \frac{k r_1}{4}\right) \frac{\|h\|_{(r_1)}}{k^2}. \tag{4.9}$$

We note that by the construction of B_2 we have

$$|h| \leqq \frac{C}{r_1^5},$$

where C depends only on g and k. It follows that

$$\|h\|_{(r_1)} \leqq \frac{C}{r_1^4},$$

and (4.9) can be restated as

$$\|w\|_{(r_1)} \leqq \frac{C}{r_1^3}\left(\frac{1}{r_1 k^2}+\frac{1}{4k}\right). \tag{4.10}$$

A similar result holds for the Dirichlet problem.

Proof. By separation of variables (assuming axial symmetry), the solution to (4.2) has the expansion

$$w = \sum_{l=0}^{\infty} H_l(kr)P_l(\cos\theta), \tag{4.11}$$

where H_l are the spherical Bessel functions and P_l are the Legendre polynomials. Hence, H_l satisfies the ordinary differential equation

$$H_l'' + \frac{2}{r}H_l' + \left(k^2 - \frac{\lambda}{r^2}\right)H_l = 0, \qquad \lambda = l(l+1). \tag{4.12}$$

Similarly, by completeness, $B_m u \equiv h$ has an expansion

$$h = \sum_{l=0}^{\infty} h_l P_l(\cos\theta). \tag{4.13}$$

By orthogonality, it is sufficient to only consider one term in (4.11). We thus have reduced the problem to a one dimensional problem. Let $\phi = H_l(r)$, then

$$\phi'' + \frac{2}{r}\phi' + \left(k^2 - \frac{\lambda}{r^2}\right)\phi = 0, \qquad 1 \leqq r \leqq r_1, \tag{4.14a}$$

$$\phi'(1) = 0, \tag{4.14b}$$

$$B_2\phi = \phi'' + \frac{4}{r}\phi' + \left(\frac{2}{r} - k^2\right)\phi - 2ik\left(\phi' + \frac{2}{r}\phi\right) = h_l, \qquad r = r_1. \tag{4.14c}$$

We note that for the Dirichlet problem (4.14b) is replaced by $\phi(1) = 0$. For the Neumann problem, we define q by $q = \phi(1)$. We then introduce a normalized variable

$$\psi = \frac{\phi}{q}. \tag{4.15}$$

It follows from (4.14) that ψ is real. We then take the real and imaginary parts of (4.14c) and replace ϕ by ψ. Using (4.12) to eliminate the second derivatives, we obtain

$$\psi' + \frac{2}{r_1}\psi = -\operatorname{Im}\left(\frac{h_l}{2qk}\right) \equiv \tilde{h}_1 \tag{4.16a}$$

and

$$\frac{2}{r}\psi' + \left(\frac{2}{r_1^2} + \frac{\lambda}{r_1^2} - 2k^2\right)\psi = \operatorname{Re}\left(\frac{h_l}{q}\right) \equiv \tilde{h}_2, \tag{4.16b}$$

where both equations hold at $r = r_1$.

We need to consider two cases depending on which harmonic l was chosen.

Case I.

$$\frac{\lambda}{r_1^2} > k^2, \qquad \lambda = l(l+1).$$

In this case $\psi(1) = 1$, $\psi'(1) = 0$, together with (4.12), imply that ψ, ψ', ψ'' are all positive in a neighborhood of $r = 1$. This condition must persist at least until $r \geqq r_*$ where $\psi''(r_*) < 0$, $\psi'(r_*) = 0$ and $\psi(r_*) > 0$. However, combining these conditions with (4.12), we see that $r_* > r_1$. Since we are only interested in $1 \leqq r \leqq r_1$, we conclude that ψ, ψ' are positive in the whole interval. In particular, $\psi(r_1) > 0$, $\psi'(r_1) > 0$. Using (4.16) we have

$$\frac{2}{r}\psi(r_1) \leqq \psi' + \frac{2}{r_1}\psi \leqq |\tilde{h}_1| \leqq \frac{|h_l|}{2k|q|}.$$

Since $\phi = q\psi$, we have

$$|\phi(r_1)| \leqq \frac{kr_1}{4}\left|\frac{h_l}{k^2}\right|, \tag{4.17}$$

which is a stronger estimate than (4.9).

Case II.

$$\frac{\lambda}{r_1^2} \leqq k^2.$$

We now solve (4.16) for $\psi(r_1)$ and obtain

$$\psi(r_1)\left(\frac{\lambda}{r_1^2} - \frac{2}{r_1^2} - 2k^2\right) = \tilde{h}_2 - \frac{\tilde{h}_1}{r_1^k}. \tag{4.18}$$

Since $\lambda/r_1^2 \leqq k^2$, we have that

$$\left|\frac{\lambda}{r_1^2} - \frac{2}{r_1^2} - 2k^2\right| \geqq k^2 + \frac{2}{r_1^2}.$$

Hence, using (4.18) and $\phi = q\psi$, we have

$$k^2\left(1 + \frac{2}{r_1^2k^2}\right)\frac{|\psi(r_1)|}{|q|} \leqq \frac{2|h_l|}{|q|}\left(1 + \frac{1}{rk}\right)$$

or

$$|\phi(r_1)| \leqq \frac{|h_l|}{k^2}, \tag{4.19}$$

which again is a stronger estimate than (4.9). We finally point out that (4.9) and (4.16a) imply

$$\|\phi'(r_1)\| \leqq \frac{\|h\|}{k}\left[\frac{1}{2} + \frac{2}{r_1k}\left(1 + \frac{kr_1}{4}\right)\right] = O\left(\frac{1}{r_1^3}\right), \tag{4.20}$$

since $h = O(1/r_1^5)$.

The results of Theorems 4.1 and 4.2 can be stated as

$$\|w\|_{(r_1)}, \|w_r\|_{(r_1)} = O\left(\frac{1}{r_1^{m+1}}\right), \tag{4.21}$$

for $m = 1$ and 2. We now extend these results from the outer surface $r = r_1$ to the body Γ_1. Introducing the notation

$$\|w\|_{(\Gamma_1)} = \sqrt{\int_{\Gamma_1} |w|^2 \, dA}$$

for the surface L_2-norm, we then have:

THEOREM 4.2. *For the problem* (4.2), *we have*

$$\|w\|_{(\Gamma_1)} \leqq C[\|w\|_{(r_1)} + \|w_r\|_{(r_1)}], \tag{4.22}$$

where C depends only on k and Γ_1.

Note. (1) An entirely analogous theorem holds for $\|w_n\|_{(r_1)}$ when the Dirichlet problem is considered.

(2) C is continuous in k and the result holds for $k = 0$.

(3) Since $u = O(1/r)$ the relative error expressed in (4.22) is smaller than that expressed in (4.21).

Proof. Let $G(p, q)$ denote the free space Green's function

$$(4.23)\qquad G(p,q)=\frac{e^{ik|p-q|}}{4\pi|p-q|}.$$

Using Green's theorem applied to w in the region Ω (see Fig. 1) we have for $p\in\Omega$ (making use of (4.2b))

$$(4.24)\quad w(p)=\int_{|q|=r_1} G(p,q)w_n(q)-w(q)G_{n_q}(p,q)\,dA_q+\int_{q\in\Gamma_1} w(q)G_{n_q}(p,q)\,dA_q,$$

where the normal on Γ_1 points toward the exterior and the normal on $|q|=r_1$ is in the direction of increasing r. Upon letting p approach Γ_1 and using the standard jump relations of potential theory [15], we obtain

$$(4.25)\qquad I(w)=\int_{|q|=r_1}(G(p,q)w_n(q)-w(q)G_{n_q}(p,q))\,dA_q,$$

where I is the integral operator (see [9])

$$I(w)=\frac{w(p)}{2}-\int_{q\in\Gamma_1} G_{n_q}(p,q)w(q)\,dA_q.$$

It is clear that the right-hand side of (4.25) satisfies (4.22). It is shown in [9] that $I(w)$ is invertible except for a discrete set of values of k and this establishes (4.22) except for these interior resonances. At the interior resonances, we replace G by the modified Green's function of Ursell ([32]) and use the same proof.

5. Numerical results. We consider the three-dimensional Helmholtz equation in spherical coordinates with axial symmetry. Using r and θ as coordinates the equation becomes

$$(5.1)\qquad \frac{\partial^2 u}{\partial r^2}+\frac{2}{r}\frac{\partial u}{\partial r}+\frac{1}{r^2\sin\theta}\frac{\partial}{\partial\theta}\left(\sin\theta\frac{\partial u}{\partial\theta}\right)+k^2u=0.$$

The first two radiation boundary conditions are

$$(5.2a)\qquad B_1u=\frac{\partial u}{\partial r}-iku+\frac{u}{r}=0$$

and

$$(5.2b)\qquad B_2u=\frac{\partial^2 u}{\partial r^2}+\left(\frac{4}{r}-2ik\right)\frac{\partial u}{\partial r}+\left(\frac{2}{r}-4ik\right)\frac{u}{r}-k^2u=0.$$

To eliminate $\partial^2u/\partial r^2$, (5.2b) can be rewritten as

$$(5.2c)\qquad \frac{\partial u}{\partial r}=-\left(\frac{1}{r}-ik\right)u+\frac{\frac{\partial}{\partial\theta}\left(\frac{\partial u}{\partial\theta}\sin\theta\right)}{2r^2\left(\frac{1}{r}-ik\right)\sin\theta}.$$

We only consider the case that the body is a sphere, given by $r=r_0$. The radiation conditions are specified on the sphere $r=r_1$. On the body, Neumann data,

$$(5.3)\qquad \frac{\partial u}{\partial n}=h\quad\text{on }\Gamma_1,\qquad r=r_0$$

are given.

The Helmholtz equation is solved by a finite element code using linear elements. The radiation boundary condition is also written in weak form following an integration by parts. Details are given in the appendix. Richardson extrapolation in the theta direction is used for all results. This substantially increases the accuracy of the computed solutions, especially for the higher frequencies. The limiting factor in the method is the storage requirements of the banded matrix solver. The use of higher order elements or iterative techniques would reduce the storage requirements.

We first consider acoustic scattering by a point source at a fixed axial point q. We then have

$$\text{(5.4)}\quad \begin{aligned} &\text{(a)} && \Delta u + k^2 u = \delta(p-q), && r_0 < r < \infty, \\ &\text{(b)} && \frac{\partial u}{\partial r} = 0, && r = r_0, \\ &\text{(c)} && \frac{\partial u}{\partial r} - iku = o\left(\frac{1}{r}\right), && r \to \infty. \end{aligned}$$

Here p denotes the dependent variable, and q is the source point. (5.4) implies that the scatterer is hard, i.e., the normal acoustic velocity is zero on the body.

The singularity in (5.4) is eliminated by introducing v given by

$$v = u - \frac{e^{ik|p-q|}}{4\pi|p-q|}.$$

The equations for v are then

$$\text{(5.5)}\quad \begin{aligned} &\text{(a)} && \Delta v + k^2 v = 0, \\ &\text{(b)} && \frac{\partial v}{\partial r} = -\frac{\partial}{\partial r}\left(\frac{e^{ik|p-q|}}{4\pi|p-q|}\right)\bigg|_{r=r_0}, \\ &\text{(c)} && \frac{\partial v}{\partial r} - ikv = o\left(\frac{1}{r}\right), && r \to \infty. \end{aligned}$$

To solve this system according to the previously described procedure, we replace (5.5c) by

$$\text{(5.5d)}\qquad B_1 v = 0, \qquad r = r_1,$$

or

$$\text{(5.5e)}\qquad B_2 v = 0, \qquad r = r_1.$$

The "exact" solution was generated by using an integral equation code with a fine grid. This code had been previously checked by comparisons with analytic solutions presented in [6] to verify its accuracy.

In Table 1 we present the relative surface L^2-errors for various k. The data in this table was obtained with a fixed number of points ($N = 5$) in the radial direction.

TABLE 1

Relative surface L^2-errors for the Helmholtz equation

kr_0	r_1	m	error
11	.645	1	18.6
11	.645	2	5.4
23.6	.575	1	31.5
23.6	.575	2	6.5

For these computations the body has radius $r_0 = \frac{1}{2}$. The source is located on the axis at $r = .6$. Similar results have been obtained for a wide range of source positions.

The results given in Table 1 indicate that for the problem considered the error due to the artificial boundary decreases as k increases. Specifically, the grid in the normal direction can be chosen independent of k. One can also bring in the artificial surface so that it coincides with the inner boundary. In this case only an ordinary differential equation needs to be solved. However, this resulted in substantial errors and accurate solutions could not be obtained.

The results of Table 1 show that the first order boundary condition (5.2a) is not good enough for many computations. Except for the lowest of frequencies, 10% accurate solutions could not be obtained using (5.2a) because of storage difficulties that arise from using a large r_1.

We next consider the Helmholtz equation with a variable k. We choose k in (5.4a) as

$$k=\begin{cases}4\left(1-\frac{r-r_0}{.15}\right), & r_0\leqq r\leqq r_0+.15,\\ 3, & r_0+.15\leqq r.\end{cases} \tag{5.6}$$

We choose $r_0=\frac{1}{2}$ and consider a sequence of outer boundary positions r_1. The "exact" solution is generated by choosing the outer boundary sufficiently far away so that the solutions obtained by using the first or second order boundary conditions differed by less than 3%. The solution obtained by using B_2 at this r_1 was taken as the "exact" solution.

Since k is variable inside the region, the Green's function is not known. Hence, one cannot use Green's formula to calculate the far field solution given u and $\partial u/\partial n$ at $r=r_0$. Instead, u and $\partial u/\partial n$ on the outer boundary can be used to find the far field solution, since k is constant beyond the outer boundary. The value of $\partial u/\partial n$ on the outer boundary can be calculated from u and its tangential derivatives by using the boundary conditions (5.2a or 5.2c). The tangential derivatives can be eliminated by integration by parts in the Green's formula.

In Fig. 2 we plot the relative surface errors over the outer surface for different values of r_1. The improvement achieved by using B_2 (5.2c) rather than B_1 (5.2a) is evident.

As the final example, we consider Laplace's equation. The exact solutions are generated by imposing Neumann data corresponding to a monopole or dipole centered at an axial point p_s inside r_1 and displaced from the origin of coordinates. The solution is

$$u=\frac{1}{4\pi|p-p_s|} \tag{5.7a}$$

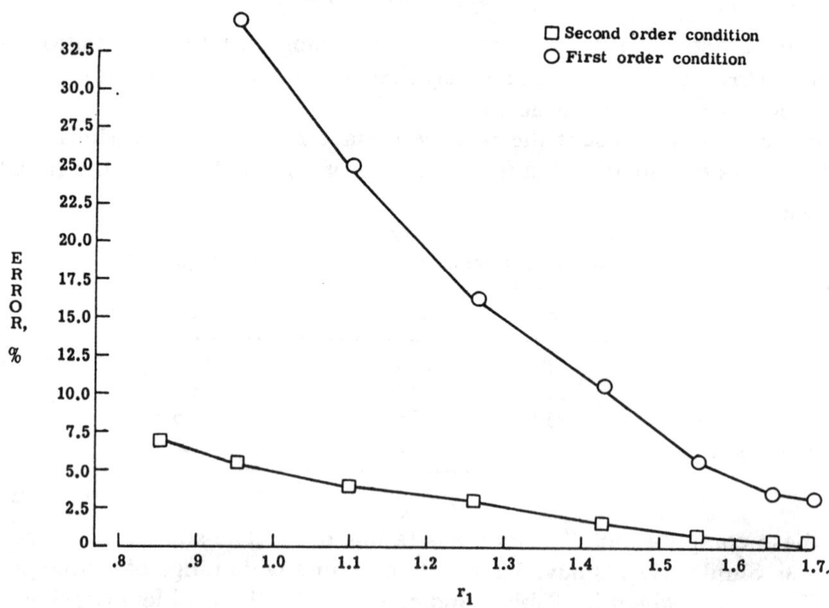

FIG. 2. *Errors for variable coefficient problem.*

and

$$u=\frac{\cos\theta'}{|p-p_s|^2}, \tag{5.7b}$$

for the monopole and dipole respectively. Here θ' denotes the polar angle in a polar coordinate system centered at p_s.

It is well known that general solutions to Laplace's equation in exterior domains can be expressed as a superposition of surface monopoles or surface dipoles [15]. Hence, the model problems (5.7) are relevant to realistic problems especially when p_s is chosen near the body surface.

In Table 2 we present the surface errors, over the inner surface ($r=\frac{1}{2}$), for several different cases. The solution types M and D denote the monopole (5.7a) or dipole (5.7b) solution, respectively.

The first order condition is exact for a monopole centered at the origin while the second order boundary condition is exact for both monopoles and dipoles centered at the origin. When the second order condition is used for the displaced dipole (5.7b) it is more accurate than the first order condition is for a displaced monopole (5.7a). For small p_s the dipole has no $1/r$ contribution and the first order condition is expected to be inaccurate. These results again confirm the substantial improvements that can be obtained by the use of (5.2c). The decay rate of the error as a function of r_1, as predicted in Theorem (3.1), has also been computationally verified.

Appendix. In this section we consider the problem of implementing the boundary conditions corresponding to B_1 and B_2. From (2.6), we see that these conditions have the general form

$$B_1 u = \frac{\partial u}{\partial r} + \gamma_1 u = 0 \tag{A.1}$$

TABLE 2
Relative surface L^2-errors for the Laplace equation

Solution type	p_s	r_1	m	error
M	.2	.57	1	14.7
M	.2	.57	2	1.1
M	.2	.64	1	9.4
M	.2	.64	2	0.6
D	.2	.57	1	65.0
D	.2	.57	2	8.9
D	.2	.64	1	38.3
D	.2	.64	2	4.9
M	.4	.60	1	25.4
M	.4	.60	2	4.1
M	.4	.65	1	17.7
M	.4	.65	2	2.7
D	.4	.60	1	35.4
D	.4	.60	2	10.5
D	.4	.65	1	21.6
D	.4	.65	2	7.2

and

$$B_2 u = \frac{\partial^2 u}{\partial r^2} + \beta_1 \frac{\partial u}{\partial r} + \gamma_1 u = 0. \tag{A.2}$$

Here α_1, β_1 and γ_1 are complex functions depending only on k and r and not on the angular variables which are given explicitly in (5.2a) and (5.2b).

For simplicity we will consider only axially symmetric problems, so that the Helmholtz equation

$$\Delta u + k^2 u = 0 \tag{A.3}$$

may be written as

$$\frac{\partial^2 u}{\partial r^2} + \frac{2}{r}\frac{\partial u}{\partial r} + \frac{1}{r^2 \sin\theta}\frac{\partial}{\partial \theta}\left(\sin\theta \frac{\partial u}{\partial \theta}\right) + k^2 u = 0. \tag{A.4}$$

The presence of the $\partial^2 u/\partial r^2$ term in (A.2) is nonstandard and difficult to implement directly. Since the boundary will be the sphere $r = r_1$, it is convenient to eliminate the $\partial^2 u/\partial r^2$ term in (A.2) by using (A.4), so that only tangential second derivatives appear in the boundary condition. This can also be done for arbitrary shapes of the artificial boundary. The result can be expressed in the form

$$B_2 u = \gamma_2 \frac{\partial u}{\partial r} + \frac{\beta_2}{\sin\theta} \frac{\partial}{\partial\theta}\left(\sin\theta \frac{\partial}{\partial\theta}\right) + \alpha_2 u = 0, \tag{A.5}$$

where $\alpha_2, \beta_2, \gamma_2$ depend only on k and r and not on θ and are given explicitly in (5.2c). The functional form of (A.5) also includes the boundary operator B_1 (with $\beta_2 = 0$).

The computational problem is to solve (A.3) in a region Ω exterior to an inner boundary Γ_1 and interior to the sphere $r = r_1$. On $r = r_1$, we impose (A.5), while on Γ_1 we may impose either Dirichlet or Neumann data. For concreteness we assume that Neumann data is specified, i.e.,

$$\frac{\partial u}{\partial n} = g(r, \theta) \quad \text{on } \Gamma_1. \tag{A.6}$$

The numerical method employed is a Galerkin finite element technique which we now describe.

Consider the following weak formulation of the problem (A.3), (A.5), (A.6). We seek a function u in a Hilbert space H_m such that

$$B(u, v) = F(v), \tag{A.7}$$

for all v in H_m, where

$$\begin{aligned} B(u, v) = &\iint_\Omega [\nabla u \cdot \nabla v - k^2 uv] r^2 \sin\theta \, d\theta \, dr \\ &+ \int_{r=r_1} \left[\frac{\beta_2}{\gamma_2} \frac{\partial u}{\partial\theta} \frac{\partial v}{\partial\theta} - \frac{\alpha_2}{\gamma_2} uv\right] r_1^2 \sin\theta \, d\theta \end{aligned} \tag{A.8}$$

and

$$F(v) = -\int_{\Gamma_1} g(r, \theta) v \, ds.$$

The boundary integral terms in (A.8) come from using (A.5) and integrating by parts in θ. We note that $r^2 \sin\theta$ comes from the three dimensional volume element. Our approximation problem is to choose a finite dimensional subspace $\mathcal{S}^h$ of H_m and then seek a $u^h \in \mathcal{S}^h$ such that

$$B(u^h, v^h) = F(v^h), \tag{A.9}$$

for all $v^h \in \mathcal{S}^h$.

It is easy to see that if u is a sufficiently smooth solution to (A.7), then u will be a solution to (A.3), (A.5) and (A.6). For $m = 1$ the above weak formulation almost leads to a standard finite element problem, once $\mathcal{S}^h$ is chosen to be some finite element space. Indeed, for $m = 1$, $\beta_2 = 0$, and we may choose $H_1 = H^1(r)$, i.e., the Sobolev space of functions with one distributional derivative. The only complication is the appearance of the term $(r^2 \sin\theta)$ in the integrals. For $m = 2$, we need more smoothness on the boundary $r = r_1$ due to the first term in the boundary integral in (A.8). The choice $H_2 = H^1(\Omega) \times H^1(r = r_1)$ suffices. For either $m = 1$ or $m = 2$, we may choose $\mathcal{S}^h$ to be the finite element space defined by subdividing Ω into triangular elements and then restricting the function in $\mathcal{S}^h$ to be continuous in Ω and linear in each triangle, i.e., a piecewise linear finite element space. Then we compute u^h in the standard manner, i.e., we choose a (local) basis for $\mathcal{S}^h$, expand u^h in terms of this basis, and let v^h in (A.9) range over the basis to obtain a system of algebraic equations for the nodal values of u^h.

Numerical experiments confirm that when both the θ and r grids are refined that h^2 convergence rates are obtained. These convergence rates were verified for both the first order $B_1u = 0$ and second order $B_2u = 0$ boundary conditions. Hence, the second order derivatives that appear in B_2 did not cause any deterioration in the rate of convergence. When only the θ mesh was refined and the errors were measured on the inner boundary, a convergence rate of order h was observed.

The computations presented in this paper were intended to exhibit the improvements due to increasing the order of the boundary operators. For this reason, it was desired to minimize discretization errors due to the finite element approximation. It has been verified that the higher order conditions did not affect the rate of convergence of the discretization scheme.

REFERENCES

[1] S. Agmon, A. Dougalis and L. Nirenberg, *Estimates near the boundary for solutions of elliptic partial differential equations satisfying general boundary conditions*, I., Comm. Pure Appl. Math., 12 (1959), pp. 623–727.

[2] F. V. Atkinson, *On Sommerfeld's "Radiative Condition"*, Philos. Mag., 40 (1949), pp. 645–651.

[3] A. K. Aziz and B. Kellogg, *A scattering problem in the Helmholtz equation*, in Advances in Computer Methods for Partial Differential Equations III, IMACS, 1979.

[4] A. Bayliss, *On the use of coordinate stretching in the numerical computation of high frequency scattering*, J. Sound Vibration, 60 (1978), pp. 543–553.

[5] A. Bayliss and E. Turkel, *Radiation boundary conditions for wave-like equations*, Comm. Pure Appl. Maths., 23 (1980), pp. 707–725.

[6] J. J. Bowman, T. B. A. Senior and L. E. Uslenghi, ed., *Electromagnetic and Acoustic Scattering by Simple Shapes*, North-Holland, Amsterdam, 1969.

[7] A. Brandt, *Multi-level adaptive solutions to boundary-value problems*, Math. Comput., 31 (1977), pp. 333–390.

[8] A. J. Burton and G. F. Miller, *The application of integral equation methods to the numerical solution of some exterior boundary-value problems*, Proc. Royal Society London A, 323 (1971), pp. 201–210.

[9] G. Chertock, *Integral equation methods on sound radiation and scattering from arbitrary surface*, Rep. 3938, David W. Taylor Naval Ship Research and Development Center, Bethesda, MD, June 1971.

[10] G. J. Fix and S. P. Marin, *Variational methods for underwater acoustic problems*, J. Comput. Phys., 28 (1978), pp. 253–270.

[11] C. I. Goldstein, *Numerical methods for Helmholtz type equations in unbounded regions*, in Mathematical Methods and Applications of Scattering Theory, Proceedings, Washington, DC, 1979.

[12] D. Gottlieb and S. Orszag, *Numerical Analysis of Spectral Methods: Theory and Applications*, CBMS Regional Conference Series in Applied Mathematics 26, Society for Industrial and Applied Mathematics, Philadelphia, 1977.

[13] G. Hellwig, *Partial Differential Equations*, Blaisdell Publishing Company, New York, 1964.

[14] S. N. Karp, *A convergent "far-field" expansion for two dimensional radiation functions*, Comm. Pure Appl. Math., 14 (1961), pp. 427–434.

[15] O. B. Kellogg, *Foundations of Potential Theory*, Dover, New York, 1953.

[16] R. E. Kleinman and G. F. Roach, *Boundary integral equations for the three-dimensional Helmholtz equation*, SIAM Rev., 16 (1974), pp. 214–236.

[17] G. Kriegsmann and C. S. Morawetz, *Numerical solutions of exterior problems with the reduced wave equation*, J. Comput. Phys., 28 (1978), pp. 181–197.

[18] H. Lamb, *Hydrodynamics*, Dover, New York, 1945.

[19] P. D. Lax, *A Phragmén-Lindelöf theorem in harmonic analysis and its application to some questions in the theory of elliptic equations*, Comm. Pure Appl. Math., 10 (1957), pp. 123–151.

[20] Ya. B. Lopatinskii, *On a method of reducing boundary problems for a system of differential equations of elliptic type to regular equations*, Ukrain. Mat. Z., 5 (1953), pp. 123–151.

[21] S. P. Marin, *A finite element method for problems involving the Helmholtz equation in two dimensional exterior regions*, Ph.D. Thesis, Carnegie-Mellon University, Pittsburgh, PA, March 1978.

[22] C. Müller, *Foundations of the Mathematical Theory of Electromagnetic Waves*, Springer-Verlag, New York, 1969.

[23] R. A. Nicolaides, *On some theoretical and practical aspects of multigrid methods*, Math. Comput., 33 (1979), pp. 933–952.

[24] A. J. Poggio and E. K. Miller, *Integral equation solutions of three-dimensional scattering problems*, in Computer Techniques for Electromagnetics, R. Mittra, ed., Pergamon Press, Oxford, 1973, pp. 159–265.

[25] W. Proskurowski and O. Widlund, *On the numerical solution of Helmholtz's equation by the capacitance matrix method*, Math. Comput., 30 (1976), pp. 433–468.

[26] M. H. Protter and F. Weinberger, *Maximum Principles in Differential Equations*, Prentice-Hall, Englewood Cliffs, NJ, 1967.

[27] F. Rellich, *Uber das asymptotishce verhatlen der Rosungen von $\Delta u + \lambda u = 0$ in unendlichen gebieten*, Jahresb. Deutsch. Math. Verein, 53 (1943), pp. 57–65.

[28] H. Schneck, *Improved integral formulation for acoustic radiation problems*, J. Acoust. Soc. Amer., 44 (1968), pp. 41–58.

[29] A. SOMMERFELD, *Partial Differential Equations in Physics*, Academic Press, New York, 1964.

[30] J. A. STRATTON, *Electromagnetic Theory*, McGraw-Hill, New York, 1941.

[31] M. TAYLOR, *Pseudo-Differential Operators*, Lecture Notes in Mathematics 416, Springer-Verlag, New York, 1974.

[32] F. URSELL, *On the exterior problems in acoustics*, Proc. Cambridge Philos. Soc., 74 (1973), pp. 117–125.

[33] B. VAINBERG, *Principles of radiation, limit absorption and limit amplitude in the general theory of partial differential equations*, Russian Math. Sur., 21 (1966), pp. 115–193.

[34] C. H. WILCOX, *A generalization of theorems of Rellich and Atkinson*, Proc. Amer. Math. Soc., (1955), pp. 271–276.

[35] ———, *Spherical means and radiation conditions*, Arch. Rat. Mech. Anal., 3 (1959), pp. 133–148.

[36] ———, *Scattering Theory for the d'Alembert Equation in Exterior Domains*, Lecture Notes in Mathematics 442, Springer-Verlag, New York, 1975.

[37] O. C. ZIENKIEWICZ, D. W. KELLY AND P. BETTESS, *Marriage a la mode—the best of both worlds: Finite elements and boundary integrals*, in Energy Methods in Finite Element Analysis, R. Glowinski, E. Y. Rodin, O. C. Zienkiewicz, eds., John Wiley, New York, 1978, pp. 81–108.

Well-Posedness of One-Way Wave Equations and Absorbing Boundary Conditions

L. N. TREFETHEN and L. HALPERN

Abstract. A one-way wave equation is a partial differential equation which, in some approximate sense, behaves like the wave equation in one direction but permits no propagation in the opposite one. Such equations are used in geophysics, in underwater acoustics, and as numerical "absorbing boundary conditions". Their construction can be reduced to the approximation of $\sqrt{1-s^2}$ on $[-1,1]$ by a rational function $r(s) = p_m(s)/q_n(s)$. This paper characterizes those rational functions r for which the corresponding one-way wave equation is well posed, both as a partial differential equation and as an absorbing boundary condition for the wave equation. We find that if $r(s)$ interpolates $\sqrt{1-s^2}$ at sufficiently many points in $(-1,1)$, then well-posedness is assured. It follows that absorbing boundary conditions based on Padé approximation are well posed if and only if (m,n) lies in one of two distinct diagonals in the Padé table, the two proposed by Engquist and Majda. Analogous results also hold for one-way wave equations derived by Chebyshev or least-squares approximation.

1. Introduction. The wave equation

$$u_{tt} = u_{xx} + u_{yy} \tag{1.1}$$

admits plane wave solutions

$$u(x,y,t) = e^{i(\tau t + \xi x + \eta y)} \tag{1.2}$$

for any ξ, η, $\tau \in \mathbf{R}$ that satisfy the dispersion relation

$$\tau^2 = \xi^2 + \eta^2. \tag{1.3}$$

In the (ξ,η)-plane, the curves of constant τ determined by this equation are concentric circles. The phase and group velocities of (1.2) are identical and equal to

$$\mathbf{v} = (v_x, v_y) = \left(-\frac{\xi}{\tau}, -\frac{\eta}{\tau}\right) = (\cos\theta, \sin\theta), \qquad \theta \in [0, 2\pi]. \tag{1.4}$$

Thus, for each frequency τ, (1.1) admits plane waves traveling with speed 1 in all directions.

In certain applications, what is ideally wanted is a *one-way wave equation* or *paraxial equation* that admits only half of these solutions: in this paper, all those with $v_x \leqslant 0$, i.e., $\theta \in [\pi/2, 3\pi/2]$. This idea has been introduced over the years in various fields, and some early references are given in the appendix to [27]. In the past decade and a half, one-way wave equations have come into heavy use in three areas. First, they were introduced by Claerbout in 1970 for **geophysical migration** of seismic waves, in which the aim is to construct images of geological formations underground by downward extrapolation of sound wave reflection data measured at the surface [7]. For subsequent developments see [4], [8], [26] and recent volumes of *Geophysics*. Second, they were introduced by Tappert and Hardin in 1973 for **underwater acoustics** calculations, where the simulation of waves in the ocean can be speeded up greatly by assuming one-way lateral propagation away from the source [27]. Further developments in this area are described in [22], [24], the latter of which contains many references, and in recent volumes of the *Journal of the Acoustical Society of America*. Finally, one-way wave equations were proposed again by Lindman in 1975 [23] and by Engquist and Majda in 1977 [10], [11] for application as approximate **absorbing boundary conditions** in numerical computations, where artificial boundaries must be introduced to limit a computational domain. For subsequent developments of this and related ideas, see [3], [9], [18] and recent volumes of the *Journal of Computational Physics*.

Reprinted from "Well-Posedness of One-Way Wave Equations and Absorbing Boundary Conditions," by L. N. Trefethen and L. Halpern, *Mathematics of Computation*, vol. 47, 1986, pp. 421–543, by permission of the American Mathematical Society.

Solving for ξ in (1.3), we get

$$\xi = \pm\tau\sqrt{1-\eta^2/\tau^2}, \tag{1.5}$$

where $\sqrt{\cdot}$ denotes the standard branch of the square root. By (1.4), the plus and minus signs correspond to leftgoing and rightgoing waves, respectively:

$$\xi = +\tau\sqrt{1-\eta^2/\tau^2} \Leftrightarrow v_x \leqslant 0, \tag{1.6a}$$

$$\xi = -\tau\sqrt{1-\eta^2/\tau^2} \Leftrightarrow v_x \geqslant 0. \tag{1.6b}$$

Consequently, the ideal one-way equation would be an equation with dispersion relation (1.6a), or equivalently

$$\boxed{\xi = \tau\sqrt{1-s^2}, \qquad s = \eta/\tau = -\sin\theta} \quad \text{IDEAL O.W.W.E.} \tag{1.7}$$

In the (ξ,η)-plane, the curves of constant τ determined by this equation are concentric semicircles. Since $\sqrt{1-s^2}$ is not rational, however, (1.7) is not the dispersion relation of a partial differential equation but of an equation containing a pseudodifferential operator. Such an operator is not local in space or time and cannot readily be implemented numerically, especially when the problem is generalized to allow variable coefficients.

To construct practical one-way wave equations, therefore, we will approximate (1.7) with the aid of a function

$$r(s) = p_m(s)/q_n(s),$$

where p_m and q_n are real polynomials of exact degrees $m \geqslant 0$ and $n \geqslant 0$ with no common zeros. We say that r is a rational function of *exact type* (m,n). The aim is for $r(s)$ to be a good approximation to $\sqrt{1-s^2}$ on $[-1,1]$. From any such approximation, we will derive a corresponding partial differential equation by replacing (1.7) by the dispersion relation

$$\boxed{\xi = \tau r(s), \qquad s = \eta/\tau = -\sin\theta} \quad \text{APPROXIMATE O.W.W.E.} \tag{1.8}$$

For example, suppose r is the Taylor approximant $r(s) = 1 - \frac{1}{2}s^2$. Substituting this for $\sqrt{1-s^2}$ in (1.7) gives

$$\xi = \tau\left(1 - \frac{1}{2}\frac{\eta^2}{\tau^2}\right),$$

or, equivalently,

$$\xi\tau = \tau^2 - \tfrac{1}{2}\eta^2,$$

which is the dispersion relation of

$$u_{xt} = u_{tt} - \tfrac{1}{2}u_{yy}.$$

This is the original one-way wave equation, now classical, and is called the *parabolic equation*. In the general case (1.8), we multiply both sides by $\tau^{\max\{m-1,n\}}q_n(\eta/\tau)$ to clear denominators, and get

$$\tau^{d-1}\xi q_n(\eta/\tau) = \tau^d p_m(\eta/\tau), \qquad d = \max\{m, n+1\}. \tag{1.9}$$

This equation has the form

$$P(\xi,\eta,\tau) = 0, \qquad \text{degree}(P) = d, \tag{1.10}$$

where P is a homogeneous polynomial with real coefficients in three variables. Thus it is the dispersion relation or symbol of a homogeneous partial differential equation of order d in x, y, t.

The most usual method for obtaining r is by Padé approximation [10], [11]. (Our example above is the Padé approximant of type $(2,0)$.) The disadvantage of Padé approximants is that although they are very accurate for $s \approx 0$, they are inaccurate near the singularities at $s = \pm 1$, and as a result, the corresponding one-way wave equations behave poorly for $\theta \approx \pi/2, 3\pi/2$. Therefore, it is tempting to consider alternative choices of r. Two reasonable candidates that we treat here are *Chebyshev* (L_∞) *approximation* and *least-squares* (L_2) *approximation*. (Certain least-squares approximations for one-way wave equations have been investigated previously [2], [23], [30].) There are many other possibilities too, such as *Chebyshev-Padé approximation* [12], and we hope to compare the practical merits of various approximation schemes in a later work. (See [33].)

The purpose of this paper is to investigate the well-posedness of one-way wave equations. In one-way wave calculations, as in many other areas of numerical computation, many seemingly reasonable approximations turn out to be ill-posed and hence useless in practice, and it is highly desirable to be able to identify these in advance. Depending on the application, various well-posedness questions arise, of which we will consider two:

Initial value problem (*IVP*). A one-way wave equation could be applied as a partial differential equation in the domain $t \geqslant 0$, $x, y \in \mathbf{R}$, subject to initial data at $t = 0$. Well-posedness refers to the existence of a unique solution whose norm at $t = t_0$ can be estimated in terms of the initial data.

Initial boundary value problem (*IBVP*). In absorbing boundary condition applications, the domain is $x, t \geqslant 0$, $y \in \mathbf{R}$, and the one-way wave equation is applied as a boundary condition along $x = 0$ for (1.1). Well-posedness is now the existence of a unique solution whose norm at $t = t_0$ and along $x = 0$ can be estimated in terms of the initial data.

Our main result is that for most methods of approximation of $\sqrt{1 - s^2}$, problems IVP and IBVP are well posed if and only if (m, n) lies in one of three diagonals of the table of approximants, or two in the case of even approximants. Our method consists of reducing each well-posedness question to algebraic criteria involving the function r, whose relationships to each other are studied systematically. Here is an intuitive explanation of why the result comes out so simply even for the relatively complicated problem IBVP: An absorbing boundary condition permits wave propagation in one direction, while ill-posedness amounts to the possibility of propagation in the other. Thus absorption and well-posedness are naturally related.

According to standard results for hyperbolic partial differential equations, our well-posedness criteria are also valid for one-way wave equations with smoothly varying coefficients [20].

Analogous three-diagonal results have been obtained previously for other problems. In the numerical solution of ordinary differential equations, the *Ehle conjecture*, proved by Wanner, Hairer, and Nørsett [31], asserts that a certain class of discrete approximations is stable in precisely three diagonals. More recently, Iserles and Strang have established a three-diagonal stability result for discrete approximations to hyperbolic partial differential equations [17]. The connections between these results and ours are at present not understood.

Well-posedness results for one-way wave equations have been proved previously in [1], [2], [9], [11], [30], [32], and elsewhere. In particular, Engquist and Majda showed in [11] that Padé approximation leads to a well-posed problem IBVP for all $m = n$ and $m = n + 2$, but an ill-posed one in case $(4, 0)$. Thus another way to summarize our results is as follows: We show that the boundary conditions proposed by Engquist and Majda are the only Padé absorbing boundary conditions that are well posed, and we extend an analogous conclusion to other classes of approximations and to initial-value problems.

Two topics must be mentioned that are not discussed in this paper. First, we do not consider the well-posedness of one-way wave equations as evolution equations in x rather than t, although this is their most common use in geophysics and underwater acoustics. Second, nothing is said here about discrete approximations to one-way wave equations, which may be unstable even when the differential equation is well posed. We hope to repair these omissions in the future.

2. Statement of Results and Outline of the Paper. Our results can be summarized as follows. Throughout, problems IVP and IBVP are based on the one-way wave equation derived by (1.8) from an arbitrary real rational function r of exact type (m,n) with $r(0) \neq 0, \infty$. Also, the integer χ_{mn} is defined by

$$\chi_{mn} = \begin{cases} 0 & \text{if } m+n \text{ is odd,} \\ 1 & \text{if } m+n \text{ is even.} \end{cases} \tag{2.1}$$

THEOREM 1. *Problem IVP is well posed if and only if the zeros and poles of $r(s)/s$ are real and simple and interlace along the real axis.*

THEOREM 2. *Problem IBVP is well posed if and only if the zeros and poles of $r(s)/s$ are real and simple and interlace along the real axis, and furthermore $r(s) > 0$ for $s \in [-1,1]$.*

These results imply:

COROLLARY 1. *Problems IVP and IBVP can be well posed only if $n \leqslant m \leqslant n+2$.*

COROLLARY 2. *IBVP well posed $\Rightarrow$ IVP well posed.*

Corollary 1 has the following converse:

THEOREM 3. *If $n \leqslant m \leqslant n+2$, and if $r(s)$ interpolates $\sqrt{1-s^2}$ at $m+n+1+\chi_{mn}$ points in $(-1,1)$, counted with multiplicity, then problems IVP and IBVP are well posed.*

The approximants one might consider in practice usually satisfy this interpolation condition. In particular, this is true of Padé, Chebyshev, and least-squares approximants. (The least-squares approximants in question are required to be even, for as we will explain, the situation becomes more complicated otherwise.) This implies finally:

THEOREM 4. *The Padé, Chebyshev, and least-squares families of one-way wave equation yield well-posed problems IVP and IBVP in precisely two distinct diagonals of the tables of approximants, namely $m = n$ and $m = n+2$.*

A remark should be made about odd values of m and n. Usually, as in Theorem 4, r is an even function, and the resulting one-way wave equation is symmetric about $\theta = \pi$. In this event m and n are even, so the case $m = n+1$ of Theorem 3 does not occur. But there are also applications in which one wants a one-way wave equation tuned asymmetrically to be accurate near some angle $\theta_0 \neq \pi$, and here, m or n may be odd. We have therefore considered it important to derive results for arbitrary m and n rather than assume these numbers are even.

Here is an outline of the remainder of the paper. Section 3 is devoted to investigating certain purely algebraic properties of r that constitute the heart of our argument. First, Lemma 1 establishes three equivalent statements to the interlace condition of Theorem 1, assuming $r(0) > 0$. Thus a fuller assertion than Theorem 1 would be that problem IVP is well posed if and only if r or $-r$ satisfies the conditions of Lemma 1. Next, Lemma 2 establishes three equivalent statements to the condition of Theorem 2; problem IBVP is well posed if and only if r satisfies the conditions of Lemma 2. One of these is an interpolation condition from which Theorem 3 will follow. Finally, Lemma 3 shows that a rational function r can interpolate $\sqrt{1-s^2}$ in at most $m+n+1+\chi_{mn}$ points in the plane. This conclusion has been used already in the proof of Lemma 2, and will be applied again later to establish nondegeneracy of Padé and Chebyshev approximants. Because of it, Theorem 3 would be unchanged if it read *at least* $m+n+1+\chi_{mn}$ points of interpolation.

Section 4 treats well-posedness of problem IVP by reducing it to one of the algebraic conditions of Lemma 1. This proves Theorem 1.

Section 5 treats well-posedness of problem IBVP by reducing it to one of the conditions of Lemma 2. This proves Theorems 2 and 3.

Finally, Section 6 shows that various families of even approximants to $\sqrt{1-s^2}$, including Padé, Chebyshev, and least-squares, satisfy the interpolation condition of Theorem 3. Together with Corollary 1, this proves Theorem 4.

3. Lemmas on the Rational Function r. Our first lemma identifies a class of functions that decrease monotonically along $\mathbf{R}$ except for simple poles. Parts of this result have appeared elsewhere, for example in [1], [2], [32].

LEMMA 1. *Let r be a real rational function of exact type (m, n) with $r(0) \neq 0, \infty$. The following conditions are equivalent:*

(a) *If s is finite with $\operatorname{Im} s > 0$, then $r(s)/s$ is finite with $\operatorname{Im} r(s)/s < 0$.*

(b) *$r(0) > 0$, and the zeros and poles of $r(s)/s$ in $\mathbf{C}$ are real and simple and interlace along the real axis.*

(c) *$r(s)/s$ can be written in the form*

$$\frac{r(s)}{s} = a - b_0 s + \sum_{k=1}^{n+1} \frac{b_k}{s - s_k} \tag{3.1}$$

for some $a, b_k, s_k \in \mathbf{R}$, where $s_1 < s_2 < \cdots < s_{n+1}$, $b_0 \geqslant 0$, and $b_k > 0$ for $k \geqslant 1$.

Proof. (a) $\Rightarrow$ (b). We will show that if any of the four assertions of condition (b) fails, then (a) cannot hold. First, if $r(0) < 0$, then $r(s)/s$ behaves like a positive multiple of $-1/s$ near $s = 0$, and necessarily maps $i\varepsilon$ into the upper half-plane for small enough ε. Second, if $r(s)/s$ has a zero or pole $s_0 \in \mathbf{C} - \mathbf{R}$, we can assume $\operatorname{Im} s_0 > 0$ by symmetry, and $r(s)/s$ must take values in both the upper and lower half-planes near s_0. Third, if $r(s)/s$ has a multiple zero or pole $s_0 \in \mathbf{R}$, then $r(s)/s$ must take values in both half-planes for s in a one-sided neighborhood of s_0 with $\operatorname{Im} s > 0$. Finally, suppose the zeros and poles are real and simple but do not interlace—there are two zeros $s_1, s_2 \in \mathbf{R}$ with no pole in-between, or the reverse. Then $r(s)/s$ has a local maximum or minimum σ at some point $s^* \in (s_1, s_2)$, so $r(s)/s - \sigma$ has a multiple zero at s^*, and this is inconsistent with (a) as before.

(b) $\Rightarrow$ (c). Condition (b) implies that $r(s)/s$ has exactly $n + 1$ simple poles and at most $n + 2$ simple zeros, hence order at most $O(s)$ at $s = \infty$. Therefore its partial fraction decomposition has the form (3.1) for some $a, b_k, s_k \in \mathbf{R}$ with $s_1 < \cdots < s_{n+1}$ and with $b_k \neq 0$ for $k \geqslant 1$, and if k_0 is the index with $s_{k_0} = 0$, then $b_{k_0} = r(0) > 0$. Now if $b_k < 0$ for some $k \geqslant 1$, there must be a pair b_k, b_{k+1} with $b_k b_{k+1} < 0$. But in this event there are an even number of zeros between s_k and s_{k+1}, contradicting the interlace condition. On the other hand, if $b_k > 0$ for all $k \geqslant 1$ but $b_0 < 0$, then $r(s)/s$ has constant sign and is therefore nonzero on $(-\infty, s_1)$ and (s_{n+1}, ∞). It follows that if the interlace condition holds, $r(s)/s$ has only n real zeros out of $n + 2$ zeros altogether, which implies the existence of a conjugate pair of zeros off the real axis, contradicting (b).

(c) $\Rightarrow$ (a). Trivial. □

The conditions of Lemma 1 are enough to ensure that $r(s)$ does not interpolate $-\sqrt{1-s^2}$ in the upper half-plane or at $s = 0$. For well-posedness of problem IBVP, this will have to be strengthened to exclude interpolation throughout $[-1, 1]$. In what follows, $\sqrt{1-s^2}$ denotes the branch of the square root defined by branch cuts $(-\infty, -1]$ and $[1, \infty)$ that takes the value 1 at $s = 0$.

LEMMA 2. *Let r be a real rational function of exact type (m, n) with $r(0) \neq 0, \infty$. The following conditions are equivalent:*

(d) *r satisfies the conditions of Lemma 1, and moreover $r(s) > 0$ for $s \in [-1, 1]$.*

(e) *$r(s) = -\sqrt{1-s^2}$ has no solutions in $\mathbf{C} - (-\infty, -1) - (1, \infty)$.*

(f) $n \leqslant m \leqslant n+2$, *and* $r(s) = +\sqrt{1-s^2}$ *has* $m+n+1+\chi_{mn}$ *solutions in* $\mathbf{C} - (-\infty, -1] - [1, \infty)$, *counted with multiplicity.*

Proof. (d) $\Rightarrow$ (e). For $\operatorname{Im} s > 0$, we have $\operatorname{Im}[-\sqrt{1-s^2}/s] > 0$, while by condition (a) of Lemma 1, $\operatorname{Im} r(s)/s < 0$. Therefore $r(s) = -\sqrt{1-s^2}$ has no solutions in the upper half-plane, or by symmetry, in the lower half-plane. On $[-1, 1]$ it cannot have any solutions either, since $-\sqrt{1-s^2} \leqslant 0 < r(s)$ by assumption.

(e) $\Rightarrow$ (f). Squaring $r(s) = \pm\sqrt{1-s^2}$ yields the polynomial equation

$$p_m^2(s) - (1-s^2)q_n^2(s) = 0, \tag{3.2}$$

which has exact degree $2d = 2\max\{m, n+1\}$, hence has $2d$ solutions counted with multiplicity, which can only lie in $\mathbf{C} - (-\infty, -1] - [1, \infty)$ since (e) prohibits solutions at $s = \pm 1$. Each of these is a solution of $r(s) = \sqrt{1-s^2}$ or of $r(s) = -\sqrt{1-s^2}$, but not both, so if (e) holds, there must be $2d$ solutions of the latter equation. Since $2\max\{m, n+1\} \geqslant m+n+1+\chi_{mn}$, with equality if and only if $n \leqslant m \leqslant n+2$, condition (f) now follows from Lemma 3 below.

(f) $\Rightarrow$ (d). Condition (f) can hold only if the estimates are sharp in the proof of Lemma 3 below, and this implies the interlace property (b) of Lemma 1; we omit the details. Moreover, $r(s)$ must be positive on $[-1, 1]$, for otherwise $r(s) = -\sqrt{1-s^2}$ would have a solution there, and for $n \leqslant m \leqslant n+2$, this would put the total number of solutions of (3.2) above the limit $2\max\{m, n+1\}$. □

Finally, Lemma 3, whose proof we have just appealed to, limits the number of times $r(s)$ can interpolate $\sqrt{1-s^2}$ in the plane. The following argument is based on ideas suggested to us by Peter Borwein of Dalhousie University. See also Lemma 4.1 of [17], where Rolle's theorem is used to limit the number of interpolation points on an interval. Arieh Iserles has further indicated to us that an alternative proof of Lemma 3 can be based on order stars [17], [31].

LEMMA 3. *Let r be a real rational function of exact type (m, n), and let L be the number of solutions of* $r(s) = \sqrt{1-s^2}$ *in* $\mathbf{C} - (-\infty, -1) - (1, \infty)$, *counted with multiplicity. Then*

$$L \leqslant m+n+1+\chi_{mn}. \tag{3.3}$$

Proof. If there is a solution at $s = \pm 1$, its multiplicity is defined to be 1, since $r(s)$ cannot match the infinite derivative there. But we can dispose of this possibility by considering a new rational function $r_\varepsilon(s) = r(s+\varepsilon)$ or $r_\varepsilon(s) = r(s+\varepsilon s)$, depending on the signs of $r(s)$ near the endpoints of $[-1, 1]$. For ε sufficiently small and of appropriate sign, $r_\varepsilon = \sqrt{1-s^2}$ will also have L solutions, all of them in the open region $\mathbf{C} - (-\infty, -1] - [1, \infty)$. So without loss of generality, we can assume $r(\pm 1) \neq 0$.

The number L is the number of zeros of

$$\phi(s) = p_m(s) - \sqrt{1-s^2}\, q_n(s)$$

in $\mathbf{C} - (-\infty, -1] - [1, \infty)$. For sufficiently large R, these must all lie in the region shown in Figure 1. Therefore ϕ has winding number L along the boundary $\alpha - \beta - \gamma - \bar{\gamma} - \bar{\beta} - \bar{\alpha}$. This implies that $\operatorname{Im}\phi(s)$ and $\operatorname{Re}\phi(s)$ must each have at least $2L$ zeros interlacing on this boundary. In what follows we assume $m \geqslant n+1$, and bound the number of zeros of $\operatorname{Im}\phi(s)$; for $m \leqslant n+1$, one proceeds analogously by bounding the number of zeros of $\operatorname{Re}\phi(s)$.

On β, $\phi(s)$ behaves approximately like $As^m - iBs^{n+1}$, where A and B are the leading coefficients of p and q, respectively. Therefore $\operatorname{Im}\phi(s)$ has at most $m+\chi_{mn}$ zeros on β, where the χ_{mn} results from the fact that if m and n are both even or both odd, the lower-order term iBs^{n+1} may bring about one extra crossing of the real axis. On $\bar{\beta}$ the count is the same. On α, γ, $\bar{\gamma}$, and $\bar{\alpha}$, we have $\operatorname{Re}\phi(s) = p_m(s)$ and $\operatorname{Im}\phi(s) = -i\sqrt{1-s^2}\, q_n(s)$, and therefore $\operatorname{Im}\phi(s)$ has at most $2n+2$ zeros on these contours, including those at $s = \pm 1$. Adding these bounds together gives

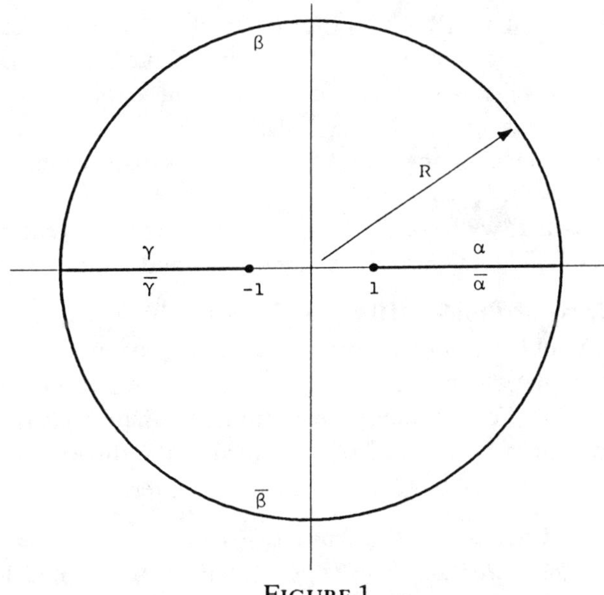

FIGURE 1
Contour for counting zeros of $r(s) - \sqrt{1 - s^2}$.

$$2L \leqslant 2m + 2\chi_{mn} + 2n + 2,$$

which reduces to (3.3). □

4. Well-Posedness of Problem IVP. Let r be a real rational function of exact type (m, n), and consider the corresponding homogeneous partial differential equation of degree $d = \max\{m, n + 1\}$ obtained from (1.8)–(1.10). Throughout this section, we assume $r(0) \neq 0, \infty$, which is no restriction in practice since $\sqrt{1 - s^2} = 1$ at $s = 0$. By (1.9), the condition $r(0) \neq 0$ amounts to the assumption that the coefficient of τ^d in $P(\xi, \eta, \tau)$ does not vanish, so that $t = 0$ is not a characteristic surface for the differential equation. Here is the standard result on well-posedness:

WELL-POSEDNESS CRITERION. *Problem IVP is well posed if and only if* $P(\xi, \eta, \tau) = 0$ *has no solutions with* $\xi, \eta \in \mathbf{R}$ *and* $\operatorname{Im} \tau < 0$.

A partial differential equation that satisfies this condition is said to be *hyperbolic*.

For a precise definition of well-posedness and a derivation of this criterion, which is due originally to Gårding, see Section 5.2(b) of [19] or Sections 12.3–12.5 of [15]. The essential idea is Fourier analysis. For suppose problem IVP admits as a solution a mode (1.2) with $\xi, \eta \in \mathbf{R}$ and $\operatorname{Im} \tau < 0$. By homogeneity, for any $\alpha > 0$, there is another solution with parameters $(\alpha\xi, \alpha\eta, \alpha\tau)$ that grows at the rate $e^{\alpha t |\operatorname{Im} \tau|}$, and as α can be arbitrarily large, the growth up to a fixed time t_0 cannot be bounded.

Here is the proof of Theorem 1.

THEOREM 1. *Problem IVP is well posed if and only if the zeros and poles of* $r(s)/s$ *are real and simple and interlace along the real axis.*

Proof. Since the zeros of $P(\xi, \eta, \cdot)$ are continuous functions of η, in the well-posedness criterion it is enough to look for normal modes with $\eta \neq 0$. By (1.9) and (1.10), since p_m and q_n have no common zeros, an equivalent well-posedness criterion is therefore that if $\xi \in \mathbf{R}$ and $\eta \in \mathbf{R} - \{0\}$, then

$$\frac{\tau}{\eta} \frac{p_m(\eta/\tau)}{q_n(\eta/\tau)} = \frac{\xi}{\eta}$$

has no solutions with $\operatorname{Im} \tau < 0$—or with $\operatorname{Im} \tau > 0$, since the solutions come in conjugate pairs. This is further equivalent to the statement that if $r(s)/s \in \mathbf{R}$ for some $s \in \mathbf{C} - \{0\}$, then $s \in \mathbf{R}$. Contrapositively, if $s \in \mathbf{C} - \mathbf{R}$, then $r(s)/s \in \mathbf{C} - \mathbf{R}$. Since $r(s)/s$ is a continuous function of s away from poles, this is equivalent to the statement that $r(s)/s$ maps the upper half-plane either into itself or into the lower half-plane. That is, either r or $-r$ satisfies condition (a) of Lemma 1. By that lemma, this is the same as saying that either r or $-r$ satisfies condition (b), which is the assertion of Theorem 1. □

As mentioned earlier, we have actually proved a stronger result: Problem IVP is well posed if and only if r or $-r$ satisfies any of the conditions of Lemma 1.

5. Well-Posedness of Problem IBVP. Again let r be a real rational function of exact type (m, n), and consider now the use of the corresponding one-way wave equation as a boundary condition at $x = 0$ for the wave equation (1.1) in the domain $x > 0$, $y \in \mathbf{R}$. For simplicity we continue to assume $r(0) \neq 0, \infty$, as in the last section. To determine well-posedness, we turn to the theory published by Kreiss in 1970 [20], specialized for the wave equation in the interior, and obtain:

WELL-POSEDNESS CRITERION. *Problem IBVP is well posed if and only if* $P(\xi, \eta, \tau) = 0$ *and* (1.3) *have no mutual solutions* $(\xi, \eta, \tau) \neq (0, 0, 0)$ *with* $\eta \in \mathbf{R}$, $\operatorname{Im} \tau < 0$, *and with* ξ *belonging to the branch* (1.6b).

Since τ and ξ are now complex, we have to be careful about what is meant by the branch (1.6b). For $\eta \in \mathbf{R}$ and $\operatorname{Im} \tau < 0$, it means $\xi = -\sqrt{\tau^2 - \eta^2}$, the analytic function of τ and η obtained by analytic continuation from the values $\xi = -\tau$ for $\eta = 0$. These values of ξ satisfy $\operatorname{Im} \xi > 0$. For $\tau \in \mathbf{R}$, ξ is defined by limits in the half-plane $\operatorname{Im} \tau < 0$, and satisfies $\operatorname{Im} \xi > 0$ if $|\tau| < |\eta|$, $\xi \in [-1, 1]$ if $|\tau| \geqslant |\eta|$.

For the definition of well-posedness and a derivation of this criterion, see [20] and [10], [11]. Again, however, the essential idea is to look for ill-posed normal modes. Suppose problem IBVP admits a solution (1.2) with $\operatorname{Im} \xi \geqslant 0$ and $\operatorname{Im} \tau \leqslant 0$. If $\operatorname{Im} \xi > 0$ (decay as $x \to \infty$) but $\operatorname{Im} \tau < 0$ (growth as $t \to \infty$), then by homogeneity, as in the last section, unbounded exponential growth as a function of t is possible. If $\operatorname{Im} \tau = \operatorname{Im} \xi = 0$ and (1.6b) holds, on the other hand, the solution is a plane wave with $v_x \geqslant 0$ that radiates energy from the boundary into the interior, generating weaker but still unbounded growth. See [13] for a presentation of the Kreiss theory from this physical point of view, and [28] for the analogous discussion of stability of finite difference models.

Here is the proof of Theorem 2.

THEOREM 2. *Problem IBVP is well posed if and only if the zeros and poles of* $r(s)/s$ *are real and simple and interlace along the real axis, and furthermore* $r(s) > 0$ *for* $s \in [-1, 1]$.

Proof. If there is a normal mode (1.2) with $\tau = 0$, then $\xi = \pm i\eta$, and by (1.9), the equation $P(\xi, \eta, \tau) = 0$ reduces to

$$\pm iA\tau^{d-1-n}\eta^{n+1} = B\tau^{d-m}\eta^{m}, \qquad d = \max\{m, n+1\}$$

for nonzero real constants A and B. Whether or not $m = n + 1$, this implies $\eta = 0$ and thus $\xi = 0$. Therefore, we can assume $\tau \neq 0$. Now for $\eta \in \mathbf{R}$ and $\operatorname{Im} \tau < 0$, the variable $s = \eta/\tau$ lies in $\mathbf{C} - (-\infty, 0) - (0, \infty)$. Taking limits $\operatorname{Im} \tau \to 0$ with $\tau \nrightarrow 0$ amounts to letting s range over all of $\mathbf{C}$, with points on the two sides of the cuts $(-\infty, 0)$ and $(0, \infty)$ viewed as distinct. By (1.6b) and (1.8)–(1.9), the well-posedness criterion now reduces to the condition that for s in this region, the equations $\xi = -\tau\sqrt{1 - s^2}$ and $\xi = \tau r(s)$ are never simultaneously satisfied. That is, the equation $r(s) = -\sqrt{1 - s^2}$ must have no solutions, where as in Section 3, the square root is defined by branch cuts $(-\infty, -1]$ and $[1, \infty)$ and the value 1 at $s = 0$.

On $(-\infty, -1)$ and $(1, \infty)$, $r(s) = -\sqrt{1 - s^2}$ cannot be satisfied, because the left-hand side is real or infinite while the right-hand side is imaginary, finite, and nonzero. Thus we have reduced the well-posedness criterion to condition (e) of

Lemma 2. By that lemma, an equivalent condition is (d), and by Lemma 1(b), this is equivalent to the assertion of Theorem 2. □

Again, we have actually proved a stronger result: Problem IBVP is well posed if and only if r satisfies any of the conditions of Lemma 2. In particular, by condition (e), it is well posed if and only if $r(s)$ interpolates $\sqrt{1-s^2}$ at $m+n+1+\chi_{mn}$ points in $\mathbf{C} - (-\infty, -1] - [1, \infty)$. Therefore, a sufficient condition for well-posedness is the existence of $m+n+1+\chi_{mn}$ points of interpolation in $(-1, 1)$, and together with Corollary 2, this proves Theorem 3.

If one or two points of interpolation lie at ± 1 instead of in $(-1, 1)$, problem IBVP is ill posed. This is a borderline case of weak ill-posedness, corresponding to a wave (1.2) that propagates tangentially to the boundary $x = 0$.

6. Families of One-Way Wave Equations. In this final section we will examine various families of approximants $r(s)$ to $\sqrt{1-s^2}$ on $[-1, 1]$, which correspond to various families of one-way wave equations. By Theorem 3, well-posedness is assured for $n \leqslant m \leqslant n+2$ if $r(s)$ interpolates $\sqrt{1-s^2}$ at $m+n+1+\chi_{mn}$ points of $(-1, 1)$, and conversely, the approximants one might consider in practice will almost always satisfy this condition. In fact, different methods of approximation of $\sqrt{1-s^2}$ can fruitfully be interpreted and compared as different strategies for the allocation of interpolation points in $(-1, 1)$—hence of angles in $(\pi/2, 3\pi/2)$ at which the ideal and approximate one-way dispersion relations coincide.

By a simple change of variables, an even approximation $r(s)$ to $\sqrt{1-s^2}$ on $[-1, 1]$ is equivalent to an approximation of the same type to the function $|s|$. The approximation of $|s|$ is a well-studied problem in approximation theory, which features the discovery by D. J. Newman in 1964 that whereas type $(n, 0)$ approximants have L^∞ error $O(1/n)$ as $n \to \infty$, type (n, n) approximants have much smaller errors $O(\text{const}^{-\sqrt{n}})$ [25]. Thus, fortunately, our results show that well-posedness correlates with accuracy as $n \to \infty$.

Until the end of the section, r is an even function. First, we will show that $\sqrt{1-s^2}$ can be interpolated in an arbitrary set of points symmetrically distributed in $(-1, 1)$. Then we will consider Chebyshev, Padé, and least-squares approximation, and show that each leads to the appropriate number of interpolation points, thereby proving Theorem 4. (These methods of approximation are illustrative; we make no claim about what method may be best in practice.) In Chebyshev and Padé approximation, uniqueness implies that $r(s)$ is automatically even, for otherwise, $r(-s)$ would be a distinct and equally good approximation. In least-squares approximation, uniqueness does not hold, so we have to impose evenness of r as an explicit constraint.

Symmetric interpolation in arbitrary points. The following construction was pointed out to us by Philip Roe, but was used earlier by Newman [25]. For any even $K \geqslant 2$, let $\pm s_1, \ldots, \pm s_{K/2}$ be a set of K points in $(-1, 1)$, counted with multiplicity, except that one of these pairs may be ± 1. Set $m = 2\lfloor K/4 \rfloor$ and $n = 2\lceil K/4 \rceil - 2$, so that $m = n$ or $m = n + 2$ and $K = m + n + 2$. Let p be a nonzero polynomial of degree $K/2$ that is zero at $\sqrt{1-s_k^2}$ for each k, and set

$$r(s) = \frac{p(t) + p(-t)}{(-p(t) + p(-t))/t}, \tag{6.1}$$

where $t = \sqrt{1-s^2}$. Since the numerator is even as a function of t, it is a polynomial in s of degree m, and since the denominator is even as a function of t, it is a polynomial in s of degree n. Also, since $|p(-t)| > |p(t)|$ for $t > 0$ and $p'(0) \neq 0$, $r(s)$ can have no poles or zeros in $[-1, 1]$ (we exclude the trivial case $K = 2$, $s_k = 1$, $r(s) \equiv 0$). Thus $r(s) = \sqrt{1-s^2}$ is equivalent to

$$p(t) + p(-t) = -p(t) + p(-t),$$

that is, $p(t) = 0$. In other words, (6.1) interpolates $\sqrt{1-s^2}$ at the points $\pm s_k$, and by Theorem 3, the corresponding one-way wave equation is well posed.

An alternative and more transparent presentation of this construction is implicit in Section 9 of a paper by Higdon [14]. If we write $\sqrt{1-s_k^2} = -\cos\theta_k$, then $p(\xi/\tau) = 0$ is equivalent to

$$\prod_{k=1}^{K/2} (\xi + \cos\theta_k \tau) = 0, \tag{6.2}$$

which corresponds to the partial differential equation

$$\prod_{k=1}^{K/2} (\partial_x + \cos\theta_k \partial_t) u = 0, \tag{6.3}$$

which is satisfied exactly by plane waves at the angles $\pm\theta_k$. Conversely, given a one-way wave equation in the form (6.3), let us use (1.3) to eliminate all powers of ξ higher than 1 in its dispersion relation (6.2). The result is a new dispersion relation that interpolates $\sqrt{1-s^2}$ in the points s_k and is equivalent to a rational expression (1.8). For example, Higdon points out that the Engquist-Majda absorbing boundary conditions can be generated in this fashion from $(\partial_x - \partial_t)^{K/2} u = 0$, an observation that highlights their close relation to the boundary conditions of Bayliss and Turkel (see Section 2 of [3]).

Now we turn to Chebyshev, Padé, and least-squares approximation. In what follows, let r denote a real rational function of type (M, N) but not necessarily *exact* type (M, N); that is, r has exact type (m, n) for some $m \leqslant M$, $n \leqslant N$. Assume further, until further notice, that M, N, m, and n are even integers and that r is an even function. The *defect* of r (as a function of type (M, N)) is the integer

$$\delta = \min\{M - m, N - n\} \geqslant 0. \tag{6.4}$$

Also, let K be the number of zeros of $\sqrt{1-s^2} - r(s)$ in $(-1, 1)$. By definition $K \leqslant L$, where L is the integer of Lemma 3, and so by that lemma, since $\chi_{mn} = 1$ when m and n are both even, we have

$$K \leqslant m + n + 2. \tag{6.5}$$

Chebyshev approximation. The Chebyshev approximant of type (M, N) to $\sqrt{1-s^2}$, denoted $r^*_{MN}(s)$, is defined by the condition

$$\left\| \sqrt{1-s^2} - r^*_{MN}(s) \right\|_\infty = \text{minimum},$$

where $\|\cdot\|_\infty$ is the supremum norm on $[-1, 1]$. It is well known that r^*_{MN} exists, is unique, and is characterized by the following "equioscillation condition" [5], [27]. Let r be a rational function of type (M, N), exact type (m, n), and defect δ. Then $r = r^*_{MN}$ *if and only if* $\sqrt{1-s^2} - r(s)$ *attains alternating values* $\pm\|\sqrt{1-s^2} - r(s)\|_\infty$ *at some sequence of points* $-1 < s_0 < \cdots < s_J \leqslant 1$ *with*

$$J \geqslant M + N + 1 - \delta. \tag{6.6}$$

Let $r = r^*_{MN}$ be the Chebyshev approximant to $\sqrt{1-s^2}$ of type (M, N). Between any two equioscillation points there must be a zero, so we have $K \geqslant J$, and by (6.6),

$$K \geqslant M + N + 1 - \delta. \tag{6.7}$$

Together with (6.5), this yields

$$M + N + 1 - \delta \leqslant m + n + 2,$$

or by (6.4),

$$M + N + 1 - \delta \leqslant (M - \delta) + (N - \delta) + 2.$$

Therefore $\delta \leqslant 1$. Since δ is even, however, we must have $\delta = 0$. Since K is even also, (6.5) and (6.7) now yield

$$m = M, \quad n = N, \quad K = m + n + 2. \tag{6.8}$$

(That is, the Chebyshev table for $\sqrt{1-s^2}$ is *normal* except for the even-odd degeneracy: It breaks into distinct 2×2 blocks of identical entries $r^*_{m,n} = r^*_{m+1,n} = r^*_{m,n+1} = r^*_{m+1,n+1}$.) This establishes the hypotheses of Theorem 3, proving Theorem 4 for Chebyshev approximation.

Padé approximation. The Padé approximant of type (M, N) to $\sqrt{1-s^2}$, denoted r^P_{MN}, is defined by the condition

$$\sqrt{1-s^2} - r^P_{MN}(s) = O(s^{\text{maximum}}) \quad \text{as } s \to 0.$$

Again one has existence and uniqueness (the former is trivial). For this particular case, r^P_{MN} is known explicitly and can be constructed from a continued fraction expansion [4], [11], [31], from hypergeometric function identities [16], or from (6.1). Alternatively, to emphasize the analogy to Chebyshev approximation, one can reason by means of the following equioscillation-type characterization [29]. Let r be a rational function of type (M, N), exact type (m, n), and defect δ. Then $r = r^P_{MN}$ *if and only if*

$$\sqrt{1-s^2} - r(s) = O(s^J) \quad \textit{as } s \to 0 \tag{6.9}$$

for some J satisfying (6.6). Since (6.9) asserts that $r(s)$ interpolates $\sqrt{1-s^2}$ at least J times at the origin, we again have $K \geqslant J$, and the argument leading to (6.8) goes through exactly as before. (That is, the Padé table for $\sqrt{1-s^2}$ is also normal except for even-odd degeneracy.) This proves Theorem 4 for Padé approximation.

Least-squares approximation. Rational least-squares approximation, unlike its polynomial counterpart, has certain troublesome properties—notably nonuniqueness and the possibility of local best approximations that are not global. To begin with, let us drop the assumption that M, N, m, and n are even. It is known that a least-squares approximant of type (M, N) to $\sqrt{1-s^2}$ exists, so let us denote such a function by $r^{(2)}_{MN}$. Here is the remarkable property that distinguishes this problem from Chebyshev and Padé approximation, proved by Cheney and Goldstein in 1967 [6]: $r^{(2)}_{MN}$ *always has* $\delta = 0$. As a corollary, first pointed out by Lamprecht [21], $r^{(2)}_{MN}$ cannot be even when M and N are odd, and so it cannot be unique. We do not know whether it is unique when M and N are even.

Nevertheless, certain conclusions about $r^{(2)}_{MN}$ can be reached despite its lack of uniqueness. Since $\delta = 0$, one can show that $\sqrt{1-s^2} - r^{(2)}_{MN}$ must be orthogonal on $[-1, 1]$ to all polynomials of degree $M + N + 1$ and, accordingly, must have $K \geqslant M + N + 1$ zeros in $(-1, 1)$. (Cheney and Goldstein point this out as their final corollary.) Since $K \leqslant m + n + 1 + \chi_{mn}$ by Lemma 3, we now reason as follows: If M and N are both even or both odd, then $m = M$ and $n = N$, but we are assured of only $m + n + 1$ interpolation points, not the $m + n + 2$ required by Lemma 3. On the other hand, if one of M and N is even and the other is odd, then the possibilities are $m = M$ and $n = N$, $m = M - 1$ and $n = N$, or $m = M$ and $n = N - 1$, and in each case, $m + n + 1 + \chi_{mn} = M + N + 1$, which guarantees a sufficient number of interpolation points.

We conclude that if one-way wave equations are derived by general least-squares approximation, problems IVP and IBVP are well posed if $M = N + 1$, but whether they are well posed for $M = N$ and $M = N + 2$ is uncertain.

Symmetric least-squares approximation. The situation returns to normal if, rather than permitting an arbitrary approximant of type (M, N), we require r to be even. (Formula (6.1) will now be a computational help, for we can take the interpolation points as independent variables and vary them systematically to minimize $\|\sqrt{1-s^2} - r(s)\|_2$.) Assume again that M, N, m, n, and K are even. The problem is equivalent to that of approximating $\sqrt{1-\sigma}$ on $[0, 1]$ by a rational function of type $(M/2, N/2)$ in the least-squares sense with weight function $\sigma^{-1/2}$. By an adaptation of the Cheney-Goldstein argument, we get $\delta = 0$ again, from which (6.8) follows. This completes the proof of Theorem 4.

Acknowledgments. Several of the ideas of Section 3 came to us from Peter Borwein of Dalhousie University, whose assistance is gratefully acknowledged. For additional advice we also thank Robert Almgren, Tony Chan, Robert Clayton, Robert Higdon, Arieh Iserles, and Philip Roe.

Note Added in Proof. Coefficients for several families of one-way wave equations are given in [33].

Department of Mathematics
Massachusetts Institute of Technology
Cambridge, Massachusetts 02139

Centre de Mathématiques Appliquées
Ecole Polytechnique
91128 Palaiseau Cedex, France

1. A. Bamberger et al., "The paraxial approximation for the wave equation: some new results," in *Advances in Computer Methods for Partial Differential Equations V* (R. Vichnevetsky and R. S. Stepleman, eds.), IMACS, 1984.

2. A. Bamberger et al., *Construction et Analyse d'Approximations Paraxiales en Milieu Hétérogène* I *and* II, Rapps. Internes Nos. 114 and 128, Centre de Mathématiques Appliquées, Ecole Polytechnique, 1984–85.

3. A. Bayliss & E. Turkel, "Radiation boundary conditions for wave-like equations," *Comm. Pure Appl. Math.*, v. 33, 1980, pp. 707–725.

4. A. J. Berkhout, "Wave field extrapolation techniques in seismic migration, a tutorial," *Geophysics*, v. 46, 1981, pp. 1638–1656.

5. E. W. Cheney, *Introduction to Approximation Theory*, McGraw-Hill, New York, 1966.

6. E.W. Cheney & A. A. Goldstein, "Mean-square approximation by generalized rational functions," *Math. Z.*, v. 95, 1967, pp. 232–241.

7. J. F. Claerbout, *Imaging the Earth's Interior*, Blackwell Publ. Co., 1985.

8. R. W. Clayton, *Wavefield Inversion Methods for Refraction and Reflection Data*, Ph.D. diss., Dept. of Geophysics, Stanford University, 1981.

9. R. W. Clayton & B. Engquist, "Absorbing boundary conditions for wave-equation migration," *Geophysics*, v. 45, 1980, pp. 895–904.

10. B. Engquist & A. Majda, "Absorbing boundary conditions for the numerical simulation of waves," *Math. Comp.*, v. 31, 1977, pp. 629–651.

11. B. Engquist & A. Majda, "Radiation boundary conditions for acoustic and elastic wave calculations," *Comm. Pure Appl. Math.*, v. 32, 1979, pp. 313–357.

12. W. B. Gragg & G. D. Johnson, "The Laurent-Padé table," in *Information Processing 74*, North-Holland, Amsterdam, 1974, pp. 632–637.

13. R. L. Higdon, "Initial-boundary value problems for linear hyperbolic systems," *SIAM Rev.*, v. 28, 1986, pp. 177–217.

14. R. L. Higdon, "Absorbing boundary conditions for difference approximations to the multi-dimensional wave equation," *Math. Comp.*, v. 47, 1986, pp. 437–459.

15. L. Hörmander, *The Analysis of Linear Partial Differential Operators.* II, Springer-Verlag, Berlin and New York, 1983.

16. A. Iserles, "A note on Padé approximation and generalized hypergeometric functions," *BIT*, v. 19, 1979, pp. 543–545.

17. A. Iserles & G. Strang, "The optimal accuracy of difference schemes," *Trans. Amer. Math. Soc.*, v. 277, 1983, pp. 779–803.

18. M. Israeli & S. A. Orszag, "Approximation of radiation boundary conditions," *J. Comput. Phys.*, v. 41, 1981, pp. 115–135.

19. F. John, *Partial Differential Equations*, 3rd ed., Springer-Verlag, Berlin and New York, 1978.

20. H.-O. Kreiss, "Initial boundary value problems for hyperbolic systems," *Comm. Pure Appl. Math.*, v. 23, 1970, pp. 277–298.

21. G. Lamprecht, "Zur Mehrdeutigkeit bei der Approximation in der L^p-Norm mit Hilfe rationaler Funktionen," *Computing*, v. 5, 1970, pp. 349–355.

22. D. Lee, *The State-of-the-Art Parabolic Equation Approximation as Applied to Underwater Acoustic Propagation with Discussions on Intensive Computations*, Tech. Doc. 7247, U.S. Naval Underwater Systems Center, New London, Conn., 1984.

23. E. L. Lindman, "'Free-space' boundary conditions for the time dependent wave equation," *J. Comput. Phys.*, v. 18, 1975, pp. 66–78.

24. S. T. McDaniel, "Parabolic approximations for underwater sound propagation," *J. Acoust. Soc. Amer.*, v. 58, 1975, pp. 1178–1185.

25. D. J. Newman, "Rational approximation to $|x|$," *Michigan Math. J.*, v. 11, 1964, pp. 11–14.

26. R. H. Stolt, "Migration by Fourier transform," *Geophysics*, v. 43, 1978, pp. 23–48.

27. F. D. TAPPERT, "The parabolic approximation method," in *Wave Propagation and Underwater Acoustics* (J. B. Keller and J. S. Papadakis, eds.), Lecture Notes in Phys., vol. 70, Springer-Verlag, Berlin and New York, 1977.

28. L. N. TREFETHEN, "Instability of difference models for hyperbolic initial boundary value problems," *Comm. Pure Appl. Math.*, v. 37, 1984, pp. 329–367.

29. L. N. TREFETHEN, "Square blocks and equioscillation in the Padé, Walsh, and CF tables," in *Rational Approximation and Interpolation* (P. R. Graves-Morris et al., eds.), Lecture Notes in Math., vol. 1105, Springer-Verlag, Berlin and New York, 1984.

30. L. WAGATHA, "Approximation of pseudodifferential operators in absorbing boundary conditions for hyperbolic equations," *Numer. Math.*, v. 42, 1983, pp. 51–64.

31. G. WANNER, E. HAIRER & S. P. NØRSETT, "Order stars and stability theorems," *BIT*, v. 18, 1978, pp. 478–489.

32. ZHANG GUAN-QUAN, "High-order approximation of one way wave equations," *J. Comput. Math.*, v. 3, 1985, pp. 90–97.

33. L. HALPERN & L. N. TREFETHEN, "Wide-angle one-way wave equations." (To appear.)

Theory and Application of Radiation Boundary Operators

THOMAS G. MOORE, STUDENT MEMBER, IEEE, JEFFREY G. BLASCHAK, STUDENT MEMBER, IEEE, ALLEN TAFLOVE, SENIOR MEMBER, IEEE, AND GREGORY A. KRIEGSMANN

Invited Review Paper

***Abstract*—A succinct unified review is provided of the theory of radiation boundary operators which has appeared principally in the applied mathematics and computational physics literature over the last ten years. With the recent introduction of the on-surface radiation condition (OSRC) method and the continued growth of finite-difference and finite-element techniques for modeling electromagnetic wave scattering problems, the understanding and use of radiation boundary operators has become increasingly important to the engineering community. In the OSRC method, specific radiation boundary operators are applied directly on the surface of an arbitrary convex target, substantially simplifying the usual integral equation for the scattered field. In the finite-difference and finite-element techniques, radiation boundary operators are used to truncate the computational domain near the target, while accurately simulating an infinite modeling space. Results are presented to illustrate the application of radiation boundary operators in both of these areas. Recent OSRC results include analysis of the scattering behavior of both electrically small and electrically large cylinders, a reactively loaded acoustic sphere, and a simple reentrant duct. New radiation boundary operator results include the demonstration of the effectiveness of higher order operators in truncating finite-difference time-domain grids.**

I. Introduction

WITH THE RECENT introduction of the on-surface radiation condition (OSRC) method [1] and the continued growth of finite-difference time-domain (FD-TD) [2] and finite-element [3] techniques for modeling electromagnetic wave scattering problems, the understanding and use of radiation boundary operators has become increasingly important to the engineering community. Radiation boundary operators have fundamentally different uses in the OSRC and finite-difference/finite-element methods. Finite techniques use radiation boundary operators in either the time domain or frequency domain to create a radiation boundary condition (RBC) which truncates a volumetric computational domain electrically close to a modeled target, and yet effectively simulates the extension of the computational domain to infinity. In contrast, the OSRC method uses the radiation boundary operator directly on the surface of the target to reduce the usual frequency-domain integral equation for the scattered field to either an integration of known quantities or a second-order ordinary differential equation. Each is simply implemented on the target surface. Although the OSRC and finite methods use radiation boundary operators in different manners, both techniques can be greatly enhanced by more effective radiation boundary operators.

The purpose of this paper is to provide a succinct unified review of key research that has been performed in the area of radiation boundary operators. Because much of this research has appeared in the applied mathematics and computational physics literature over the past ten years, its results and implications are generally not well known by the engineering electromagnetics community. This paper will also present some recent results from the application of these operators to engineering problems. In particular, we will examine two basic types of radiation boundary operators and give examples showing their use in both the FD-TD and OSRC methods. Specifically, in Section II we will discuss the theory behind radiation boundary operators. In Section III the radiation boundary operators will be used to construct new radiation boundary conditions for a two-dimensional FD-TD grid of higher order than those currently used; and the effectiveness of the new radiation boundary conditions will be tested. In Section IV, the radiation boundary operators will be used in the OSRC method to approximately solve the problem of scattering from a perfectly conducting cylinder. Section V concludes with a discussion of the research activities that are ongoing in the areas of radiation boundary operators and their applications.

II. Theory

There are two basic types of radiation boundary operators: mode annihilating and one-way wave equation approximations. Each of these radiation boundary operators possesses different characteristics and forms. In this section, the two different types of radiation boundary operators are examined in detail.

Manuscript received October 15, 1987; revised January 18, 1988. This work was supported in part by NASA Lewis Research Center Grant NAG 3-635 and in part by the National Science Foundation Grant MCS-8300578.

T. G. Moore and A. Taflove are with the Department of Electrical Engineering and Computer Science, Technological Institute, Northwestern University, Evanston, IL 60201.

J. G. Blaschak is with the Lincoln Laboratory, Massachusetts Institute of Technology, Lexington, MA 02173.

G. A. Kriegsmann is with the Department of Engineering Sciences and Applied Mathematics, Technological Institute, Northwestern University, Evanston, IL 60201.

IEEE Log Number 8823652.

Reprinted from *IEEE Trans. Antennas Propagat.*, vol. 36, no. 12, pp. 1797–1812, Dec. 1988.

A. Mode-Annihilating Operators

The first type of radiation boundary operator to be discussed is the mode-annihilating differential operator. This type of operator is based on the idea of killing the terms (herein referred to as "modes") of the far-field expansion of outward propagating solutions to the wave equation. One can view the idea of killing modes of the scattered fields as first being proposed by Sommerfeld in the form of the Sommerfeld radiation condition [4] which annihilates the first mode in the expansion. Later, researchers [5] extended the Sommerfeld theory and created an operator that annihilates the next mode in the expansion. Independently, other researchers created a general operator that kills an arbitrary number of modes in the expansion as derived and presented in [6]. It is the theory that appeared in [6] that will be reviewed in this section.

For this section we will proceed as follows. In Section II-A1) the scattered fields are written in terms of a far-field expansion, and the effect that the Sommerfeld radiation condition has on the expansion is presented. The operators derived in [6], are presented in Section II-A2) for the full three-dimensional case and are specialized to two dimensions in Section II-A3).

1) Far-Field Expansions and the Sommerfeld Radiation Condition: We consider here solutions $U(R, \theta, \phi, t)$ to the scalar wave equation

$$\nabla^2 U - U_{tt} = 0 \tag{1}$$

and the associated Helmholtz equation for time-harmonic waves

$$\nabla^2 U + k^2 U = 0 \tag{2}$$

where the wave speed c has been scaled to unity and the harmonic wave is assumed to have time dependency $e^{-j\omega t}$. The radiating solutions of the scalar wave equation (i.e., solutions propagating in directions which are outward from the origin of a spherical coordinate system) can be expanded in a convergent series of the form [7],

$$U(R, \theta, \phi, t) = \sum_{i=1}^{\infty} \frac{f_i(t-R, \theta, \phi)}{R^i}. \tag{3}$$

This result was extended to the time-harmonic case for both vector and scalar fields [8]. For the scalar Helmholtz equation, it is proved in [8] that

$$U(R, \theta, \phi) = \frac{e^{jkR}}{R} \sum_{i=0}^{\infty} \frac{f_i(\theta, \phi)}{R^i} \tag{4}$$

is a convergent expansion for scalar wave functions that satisfy the Sommerfeld radiation condition.

The Sommerfeld radiation condition [4], given by

$$\lim_{R\to\infty} R(U_R - jkU) = 0 \tag{5a}$$

where U_R denotes a derivative respect to R, is satisfied by the each term of (4). By using the correspondence $-jk = \partial/\partial t$, the Sommerfeld condition is extended to

$$\lim_{R\to\infty} R(U_R + U_t) = 0 \tag{5b}$$

which is satisfied by each term of the expansion in (3). The Sommerfeld radiation condition can be viewed as an operator on the far-field expansion of U giving the asymptotic result,

$$\left(\frac{\partial}{\partial R} - jk\right) U = O(R^{-2}) \tag{6}$$

in the limit $R \to \infty$. In other words, the Sommerfeld condition retains terms that are no greater than $O(R^{-2})$ in the expansion.

2) Higher Order Operators: With the goal of devising operators that annihilate terms up to any order in the far-field expansion of U, a sequence of operators B_n was proposed [6] for the expansion in (3). A similar sequence of operators was independently developed for the Helmholtz equation in two dimensions [5]. The former were extended [9] for the Helmholtz equation in both two and three dimensions. We restrict our review here to operators for the time-harmonic case [9], keeping in mind that results for waves of arbitrary time variation can be obtained by a simple substitution of $\partial/\partial t$ for the term $-jk$.

The derivation of B_n begins by multiplying a slightly rewritten version of (4) by R^n and then splitting the sum as shown:

$$R^n U(R, \theta, \phi) = \sum_{i=1}^{n} R^{n-i} e^{jkR} F_i(\theta, \phi) + \sum_{i=n+1}^{\infty} R^{n-i} e^{jkR} F_i(\theta, \phi). \tag{7}$$

Now define the intermediate operator,

$$L \equiv \frac{\partial}{\partial R} - jk \tag{8a}$$

and observe that applying L^n to both sides of (7) annihilates the first sum and makes the leading order term of the second sum be $O(R^{-n-1})$. We have

$$L^n(R^n U) = O(R^{-n-1}) \tag{8b}$$

which accomplishes the goal of annihilating the first n terms of the far-field expansion. A more useful way to express this result is as a single operator acting on U only. This is achieved by inductive arguments [9]. For $n = 1$,

$$L(RU) = \left(\frac{\partial}{\partial R} - jk\right) RU = O(R^{-2}) \tag{9a}$$

$$R\left(\frac{\partial}{\partial R} - jk\right) U + U = O(R^{-2}) \tag{9b}$$

which can be written as

$$\left(\frac{\partial}{\partial R} - jk + \frac{1}{R}\right) U = O(R^{-3}). \tag{9c}$$

The first operator in the sequence is then

$$B_1 = L + \frac{1}{R} \tag{10a}$$

which, when applied to both sides of (4), annihilates the first term of the expansion. Similarly,

$$B_2 = \left(L + \frac{3}{R}\right)\left(L + \frac{1}{R}\right) \tag{10b}$$

annihilates the first two terms. In general, the recursion relation,

$$B_n = \left(L + \frac{2n-1}{R}\right) B_{n-1} \tag{11}$$

produces an operator which annihilates the first n terms of the expansion in (4). The sequence of operators gives

$$B_n U = O(R^{-2n-1}) \tag{12}$$

for any function U satisfying the expansion in (4).

In the literature, B_n has been utilized as a boundary condition

$$B_n U = 0 \tag{13}$$

for the wave function U. This condition becomes more accurate, in powers of R^{-1}, as the order of the operator n increases. The original application of (13) was to truncate a computational domain while accurately modeling the outward propagation of waves to infinity. Further application of B_n, particularly B_2, is found in the OSRC method for computing scattering from two-dimensional, convex, conducting and homogenous dielectric bodies.

3) Operators for Two-Dimensional Wave Propagation: Extension of B_n for use with wave functions $U(r, \theta, t)$ in two space dimensions proceeds in the time harmonic case from an expansion presented in [10],

$$U(r, \theta) = H_0(kr) \sum_{i=0}^{\infty} \frac{F_i(\theta)}{r^i} + H_1(kr) \sum_{i=0}^{\infty} \frac{G_i(\theta)}{r^i} \tag{14}$$

which has the far-field result [9],

$$U(r, \theta) = \sqrt{\frac{2}{\pi kr}}\, e^{j(kr-(\pi/2))} \sum_{i=0}^{\infty} \frac{f_i(\theta)}{r^i} \tag{15}$$

that is analogous to (4). A sequence of boundary operators is defined [6] by the recursion relationship

$$B_n = \left(L + \frac{4n-3}{2r}\right) B_{n-1} \tag{16}$$

where

$$B_1 = L + \frac{1}{r} \tag{17a}$$

and

$$L \equiv \frac{\partial}{\partial r} - jk. \tag{17b}$$

The operator B_n annihilates the first n terms of the expansion (16) and yields

$$B_n U = O(r^{-2n-1/2}). \tag{18}$$

The utility of these operators will be demonstrated in Section IV in the OSRC calculation of electromagnetic scattering from a conducting circular cylinder illuminated by a TE polarized plane wave. The first two operators used there are

$$B_1 = \frac{\partial}{\partial r} + \frac{1}{2r} - jk \tag{19a}$$

and

$$B_2 = \frac{\partial}{\partial r} + \frac{1}{2r} - jk - \left(\frac{\partial^2}{\partial \theta^2} + \frac{1}{4}\right) \Big/ \left[2r^2 \left(\frac{1}{r} - jk\right)\right]. \tag{19b}$$

In the derivation of (19b) the recursion relation, (16) produces a second-order r derivative. It is conveniently eliminated by the substitution from the Helmholtz equation,

$$\frac{\partial^2}{\partial r^2} = -\frac{1}{r^2}\frac{\partial^2}{\partial \theta^2} - \frac{1}{r}\frac{\partial}{\partial r} - k^2. \tag{20}$$

B. One-Way Wave Equations

A partial differential equation which permits wave propagation only in certain directions is called a "one-way wave equation." Fig. 1 shows a finite two-dimensional Cartesian domain Ω on which the time-dependent wave equation is to be simulated. In the interior of Ω, a numerical scheme which models wave propagation in all directions is applied. On the outer boundary $\partial\Omega$, only numerical wave motion that is outward from Ω is permitted. The boundary must permit outward propagating numerical waves to exit Ω just as if the simulation were performed on a computational domain of infinite extent. A scheme which enacts a one-way wave equation on $\partial\Omega$ for this purpose is called a radiation boundary condition (RBC).

1) Derivation by Wave Equation Factoring[1]*:* The derivation of an RBC whose purpose is to absorb numerical waves incident upon the outer boundary of a finite-difference or finite-element grid can be explained in terms of operator factoring. Consider the two-dimensional wave equation in Cartesian coordinates,

$$U_{xx} + U_{yy} - U_{tt} = 0. \tag{21}$$

The partial differential operator here is

$$L \equiv D_x^2 + D_y^2 - D_t^2 \tag{22a}$$

which uses the notation,

$$D_x^2 \equiv \frac{\partial^2}{\partial x^2} \qquad D_y^2 \equiv \frac{\partial^2}{\partial y^2} \qquad D_t^2 \equiv \frac{\partial^2}{\partial t^2}. \tag{22b}$$

The wave equation is then compactly written as

$$LU = 0. \tag{23}$$

The wave operator L can be factored in the following manner:

$$LU = L^+ L^- U = 0 \tag{24a}$$

[1] It can be demonstrated that wave equation factoring can generate B_1 and B_2 described in Section II-A. However, this has not been shown for $n > 2$.

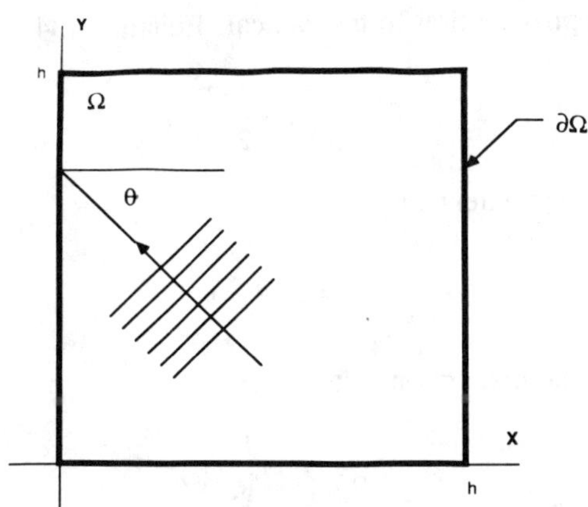

Fig. 1. Two-dimensional Cartesian computational domain.

where L^- is defined as

$$L^- \equiv D_x - D_t\sqrt{1-S^2} \tag{24b}$$

with

$$S = \frac{D_y}{D_t}. \tag{24c}$$

the operator L^+ is similarly defined except for a "+" sign before the radical.

In [11] it is shown that at a boundary, say at $x = 0$, the application of L^- to the wave function U will exactly absorb a plane wave incident at any angle and traveling in the $-x$ direction. Thus

$$L^- U = 0 \tag{25}$$

applied at $x = 0$ functions as an exact analytical RBC which absorbs wave motion from the interior of the spatial domain $\{\Omega = (x, y): 0 < x < h, 0 < y < h\}$. The operator L^+ performs the same function for waves traveling in the $+x$ direction that impact the other x boundary in Fig. 1 at $x = h$. The presence of the radical in (24b) classifies L^- as a pseudodifferential [11] operator that is nonlocal in both the space and time variables. This is an undesirable characteristic in that it prohibits the direct numerical implementation of (25) as an RBC.

Approximation of the radical in (24b) produce RBC's that can be implemented numerically and are useful in FD-TD simulations of the wave equation. The numerical implementation of an RBC is not exact in that a small amount of reflection does develop as numerical waves pass through the grid boundary. However, it is possible to design an RBC which minimizes the reflection as much as possible over a range of incident angles [12]. The RBC derived in [13] and applied in the simulation of electromagnetic scattering [14], uses a two-term Taylor series approximation to the radical in (24b).

$$\sqrt{1-S^2} \approx 1 - \frac{1}{2}S^2. \tag{26a}$$

This leads to the following approximate analytical RBC which can be numerically implemented at the $x = 0$ boundary:

$$U_{xt} - U_{tt} + \frac{1}{2}U_{yy} = 0. \tag{26b}$$

A generalization of (26) presented in [15] showed that the construction of numerically useful absorbing boundary conditions reduces to approximation of $\sqrt{1 - S^2}$ on the interval $[-1, 1]$ by the rational function,

$$r(s) = \frac{p_m(s)}{q_n(s)} \tag{27}$$

where p and q are polynomials of degree m and n, and $r(s)$ is said to be of type (m, n). By specifying $r(s)$ as a general type (2, 0) approximant, the radical is approximated by an interpolating polynomial of the form

$$\sqrt{1-S^2} \approx p_0 + p_2 S^2 \tag{28a}$$

resulting in the general second-order approximate analytical RBC,

$$U_{xt} - p_0 U_{tt} - p_2 U_{yy} = 0. \tag{28b}$$

The choice of the coefficients p_0 and p_2 is determined by the method of interpolation. Standard techniques such as Chebyshev, least-squares, or Padé approximation are applied with the goal of producing an approximate RBC whose performance is good over a wide range of incident wave angles. Expressions similar to (28) can be derived and applied at the other three boundaries of a two-dimensional FD-TD grid.

High order approximations to the radical in (24b) were proposed in [15] as a means to derive a more accurate approximate RBC. Use of the general type (2, 2) rational function,

$$\sqrt{1-S^2} \approx \frac{p_0 + p_2 S^2}{q_0 + q_2 S^2} \tag{29a}$$

gives the general third-order approximate analytical RBC,

$$q_0 U_{xtt} + q_2 U_{xyy} - p_0 U_{ttt} - p_2 U_{tyy} = 0. \tag{29b}$$

Appropriate selection of the p and q coefficients in (29) produces various families of RBC's, as suggested in [12] and [15]. For example, $q_0 = p_0 = 1$, $p_2 = -3/4$, and $q_2 = -1/4$ gives a Padé (2, 2) approximation in (29a) with the resulting RBC function better than (26b) for numerical waves impacting the grid boundary at near normal incidence. This results in the third-order RBC originally proposed in [11]. Other types of approximating polynomials "tune" the RBC to absorb numerical waves incident at specified angles other than normal, and are considered to be a means to improve wide-angle performance [12]. Results from a comparative study [16] of the performance of various families of RBC's are presented in Section III.

2) Derivation by Dispersion Relation: An alternate procedure for obtaining one-way wave equations is presented in the literature [12], [15]. We summarize the technique here for completeness. It is well known that if the dispersion relation for a linear constant-coefficient partial differential equation is known, then the equation itself is specified [17]. Thus if one can obtain the dispersion relation for a one-way wave equation, then an RBC appropriate for use on $\partial\Omega$ is obtained.

If a plane wave solution

$$U(x, y, t)=e^{j(\omega t+\xi x+\eta y)} \tag{30}$$

is substituted into (21), then

$$\omega^2=\xi^2+\eta^2 \tag{31}$$

is the dispersion relation for the wave equation which permits wave propagation in all directions of the x-y plane. The wave in (30) has velocity

$$\mathbf{v}=v_x\hat{\mathbf{x}}+v_y\hat{\mathbf{y}} \tag{32a}$$

where

$$v_x=-\frac{\xi}{\omega}=-\cos\theta \tag{32b}$$

$$v_y=-\frac{\eta}{\omega}=-\sin\theta \tag{32c}$$

and θ is the counterclockwise angle measured from the the negative x axis. By rewriting (31) as

$$\frac{\xi}{\omega}=\pm\sqrt{1-s^2} \tag{33a}$$

with

$$s=\frac{\eta}{\omega} \tag{33b}$$

a dispersion relation can be identified which corresponds to an equation that admits plane wave solutions propagating only in the $-x$ direction. This is obtained by choosing the positive branch of the square root in (33a) which corresponds to waves having velocity component v_z, in the $-x$ direction. Wave motion from the interior of Ω will be absorbed at the $x = 0$ grid boundary if an equation having the dispersion relation,

$$\xi=\omega\sqrt{1-s^2}. \tag{34}$$

is applied at that boundary.

Equation (34) is a dispersion relation for a pseudodifferential equation [11] and cannot be identified with a linear partial differential equation which can be implemented numerically on the $x = 0$ boundary. By approximating the radical in (34), it is possible to obtain a dispersion relation which can be identified with a partial differential equation that functions as an approximate analytical RBC. The same methods of approximation for the radical used in Section II-A1) can be applied here; however, s is now defined by (33b). Once a dispersion relation is obtained with approximates the exact relation in (34), the same RBC's are derived as in Section II-A1). We illustrate here application of the two-term Taylor series approximation to the radical. Equation (34) becomes

$$\xi=\omega\left(1-\frac{1}{2}\frac{\eta^2}{\omega^2}\right) \tag{35a}$$

which is equivalent to

$$\xi\omega=\omega^2-\frac{1}{2}\eta^2 \tag{35b}$$

which is the dispersion relation of

$$U_{xt}=U_{tt}-\frac{1}{2}U_{yy}. \tag{35c}$$

This is the same expression as the approximate analytical RBC given in (26b). Higher order RBC's follow directly.

III. Application of One-Way Wave Equations: FD-TD Radiation Boundary Conditions

In the simulation of electromagnetic wave scattering by finite techniques, one-way wave equations are used to truncate the computational domain in a manner which accurately models the propagation of scattered waves to infinity [13], [14]. This section summarizes recent results in applying the theory of one-way wave equations to the simulation of electromagnetic scattering by the FD-TD method. In particular, the promise of higher order RBC's is quantified by a reflection coefficient analysis and by numerical experiments. It is demonstrated that a reduction in grid boundary reflection is realized when a third-order RBC is applied on the boundary of a two-dimensional FD-TD grid.

A. Reflection Coefficient Analysis

Numerical radiation boundary conditions derived from approximate analytical one-way wave equations are not exact in that a small amount of reflection will be realized from numerical wave striking the grid boundary. For a numerical plane wave striking the $x = 0$ boundary in Fig. 1, the amount of reflection is dependent upon the angle of incidence θ. Now, scattered waves from a complex body can be viewed as a superposition of plane waves striking the computational boundaries over a wide range of incident angles. Therefore, the performance of a given RBC can be assessed by deriving a reflection coefficient R, which quantifies the amount of nonphysical reflection a plane wave produces as a function of θ when it interacts with the grid boundary. Clearly, a good RBC gives a small value of R over a wide range of θ. Such an RBC should perform well in the simulation of a realistic scattering situation because the grid boundaries would permit most of the scattered energy to exit the computational domain.

Consider the outgoing plane wave in Fig. 1. The wave has the form,

$$U_{\text{inc}}=e^{j(kt+kx\cos\theta-ky\sin\theta)}. \tag{36}$$

The total field at the boundary of the computational domain must satisfy the specific RBC in effect there. Postulating the existence of a reflected wave launched from the boundary, the

TABLE I
COEFFICIENTS FOR THIRD-ORDER RBC'S

Type of Approximation	p_0	p_2	q_2	Angles of Exact Absorption (°)
Padé	1.00000	−0.75000	−0.25000	0.00
L_a^∞	0.99973	−0.80864	−0.31657	11.7, 31.9, 43.5
Chebyshev points	0.99650	−0.91296	−0.47258	15.0, 45.0, 75.0
L^2	0.99250	−0.92233	−0.51084	18.4, 51.3, 76.6
C-P	0.99030	−0.94314	−0.5556	18.4, 53.1, 81.2
Newman points	1.00000	−1.00000	−0.66976	0.0, 60.5, 90.0
L^∞	0.95651	−0.94354	−0.70385	26.9, 66.6, 87.0

q_0 = 1.00000 for each technique.

TABLE II
COEFFICIENTS FOR SECOND-ORDER RBC'S

Type of Approximation	p_0	p_2	Angles of Exact Absorption (°)
Padé	1.00000	−0.50000	0.00
L_a^∞	1.00023	−0.51555	7.6, 18.7
Chebyshev points	1.03597	−0.76537	22.5, 67.5
L^2	1.03084	−0.73631	22.1, 64.4
C-P	1.06103	−0.84883	25.8, 73.9
Newman points	1.00000	−1.00000	0.0, 90.0
L^∞	1.12500	−1.00000	31.4, 81.6

total field at the $x = 0$ boundary has the form,

$$U = e^{j(kt + kx\cos\theta - ky\sin\theta)} + Re^{j(kt - kx\cos\theta - ky\sin\theta)} \tag{37}$$

where R can be determined by substituting U directly into the equation for the RBC used at the $x = 0$ boundary.

By substituting (37) into (28b) and (29b), reflection coefficient expressions as a function of incident angle are obtained for the general second- and third-order RBC's. They are, respectively,

$$R = \frac{\cos\theta - p_0 - p_2\sin^2\theta}{\cos\theta + p_0 + p_2\sin^2\theta} \tag{38}$$

and

$$R = \frac{q_0\cos\theta + q_2\cos\theta\sin^2\theta - p_0 - p_2\sin^2\theta}{q_0\cos\theta + q_2\cos\theta\sin^2\theta + p_0 + p_2\sin^2\theta} \tag{39}$$

where the coefficients p and q correspond to the approximating function used in the derivation of the RBC. Seven techniques of approximation are developed in [15] for this purpose. The techniques are: Padé; Chebyshev on a subinterval (L_a^∞); interpolation in Chebyshev points; least-squares (L^2); Chebyshev–Padé (or C–P); interpolation in Newman points; and Chebyshev (L^∞). Tables I and II show p and q coefficients for approximating functions of both second and third order. The mechanics of their derivation can be found in [15]. A type (2, 2) approximant produces a third-order RBC. Second-order RBC's are obtained from type (2, 0) approximants. Also shown in Tables I and II are angles of incidence at which the RBC's are designed to exactly absorb numerical plane waves. The Padé family concentrates absorption near $\theta = 0°$. The others distribute absorption angles through the range $[0, \pi/2]$ as a means to improve wide-angle performance [15]. A more general approach, which permits the design of boundary conditions for plane waves incident at arbitrary angles, is presented in [18] and [19].

Figs. 2 and 3 show the behavior of the reflection coefficient for the two best-performing RBC's as a function of incident angle on the range $[0, \pi/2]$. In all cases studied, the behavior of reflection coefficient for third-order RBC's is better than that of second-order RBC's. Fig. 2(a) shows R less than one percent for $0 < \theta < 45°$ for the third-order Padé RBC. Note that the Padé RBC's have a very low reflection coefficient for normal incidence. The distribution of exact absorption angles away from $\theta = 0°$ is illustrated in Fig. 2(b) for the L_a^∞ RBC. The nulls in the behavior of R are as predicted by the analysis presented in [15]. Fig. 3 compares the third-order Padé and the third-order L_a^∞ RBC's. By sacrificing performance near $\theta = 0°$, the L_a^∞ RBC extends the point at which R is less than one percent to about $\theta = 60°$.

B. Numerical Experiments

Numerical experiments are now reported which clearly measure the amount of nonphysical reflection a given RBC produces as a pulse propagates through a grid boundary. Fig. 4(a) shows two domains on which the two-dimensional FD-TD algorithm is computed simultaneously for the transverse magnetic (TM) case. On the boundary of the test domain Ω_T a test RBC is applied. Each point in Ω_T has a corresponding member in the substantially larger domain Ω_B. A line source is located at grid position (50, 25) in both domains. The source produces outward propagating, cylindrical waves which are spatially coincident in both domains up until time steps when the waves interact with the boundary of Ω_T. Any reflection from the boundary of Ω_T makes the solution at points within Ω_T differ from the solution at corresponding points within Ω_B. The wave solution at points within Ω_B represents the desired numerical modeling of free-space propagation up until time steps when reflections from its own boundary enter the region of Ω_B corresponding to Ω_T. By calculating the difference in the solutions in Ω_B and Ω_T at each point at each time step, a measure of the spurious reflection caused by the boundary of Ω_T is obtained.

We define at the nth time step

$$D(i, j) = E_z^T(i, j) - E_z^B(i, j) \tag{40a}$$

for all (i, j) within the test domain, where E_z^T is the solution

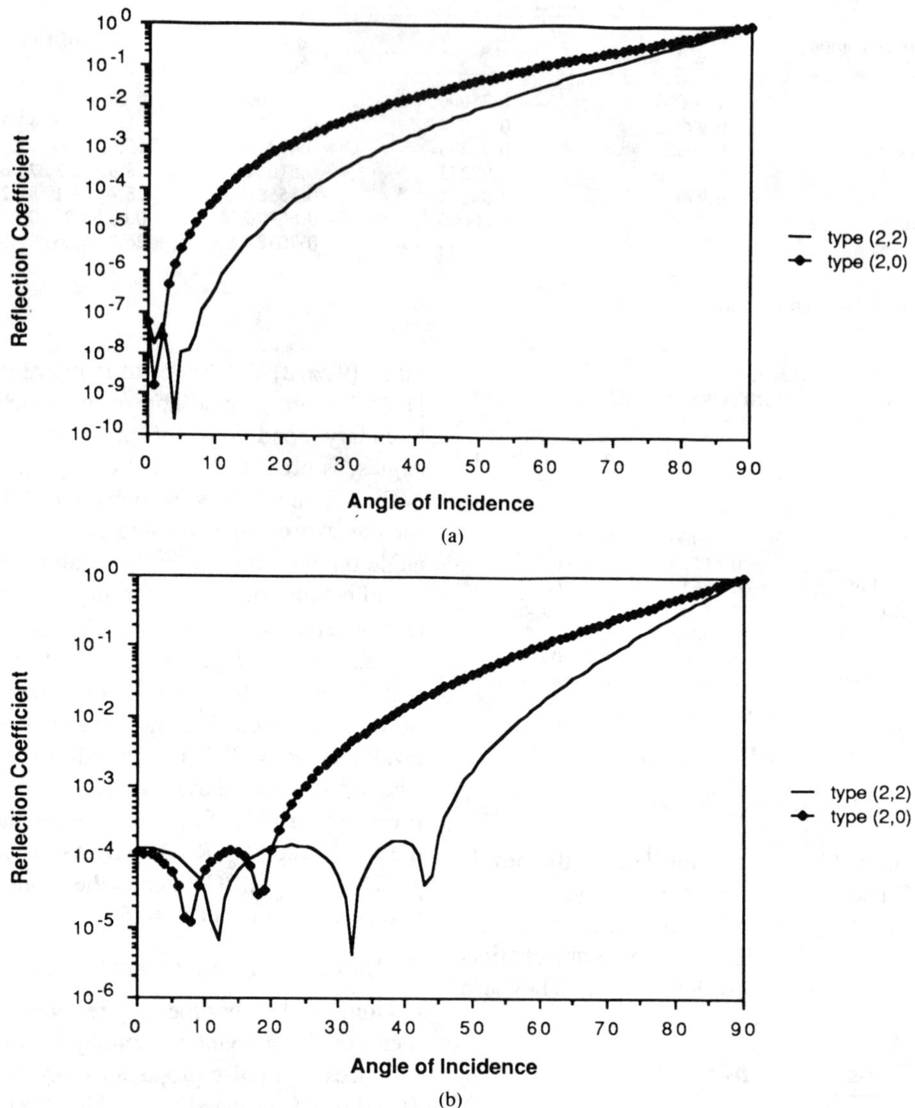

Fig. 2. Reflection coefficient versus wave angle of incidence. (a) Padé RBC. (b) Chebyshev on subinterval RBC.

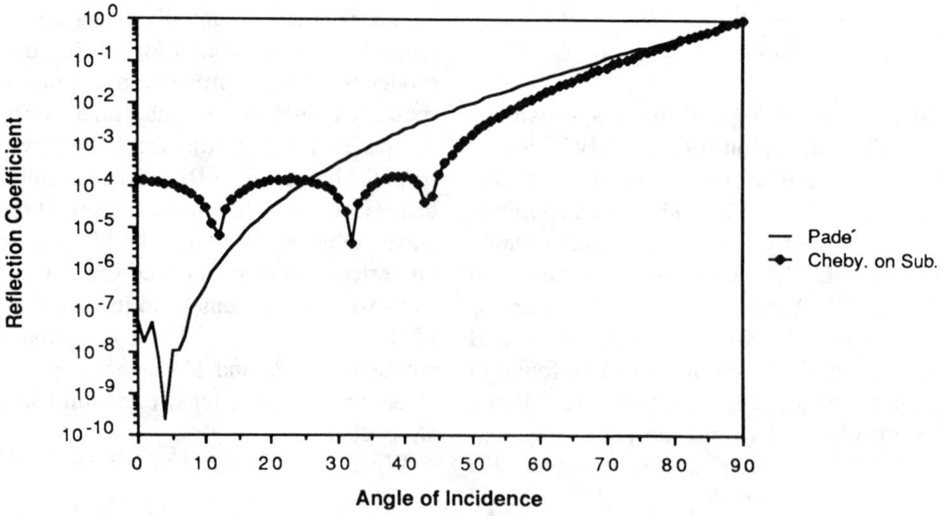

Fig. 3. Comparison of third-order RBC's.

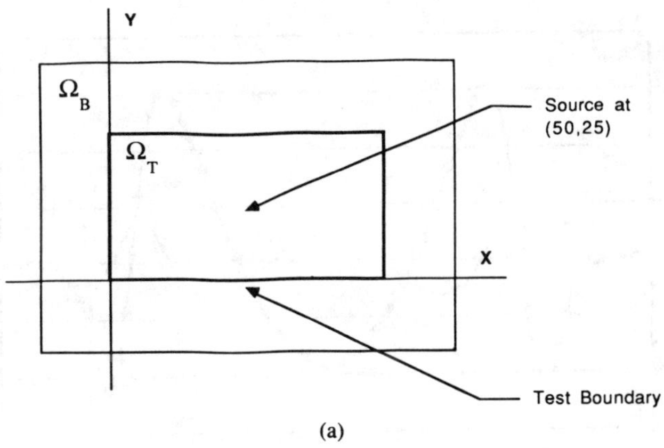

(a)

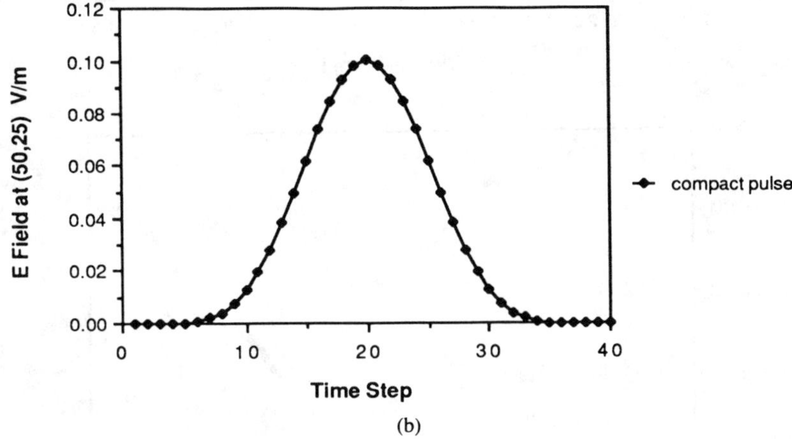

(b)

Fig. 4. Pulse studies. (a) Computational domains. (b) Wave source.

within Ω_T and E_z^B is the solution at points in Ω_B. $D(i, j)$ is the local error in the test domain caused by its grid boundary reflections. We also define a global reflected error measure,

$$E = \sum_i \sum_j D^2(i, j) \tag{40b}$$

for all (i, j) within Ω_T, which measures the total reflected error within the test grid at the nth time step.

The source used in the numerical experiments is the pulse obtained from [20] and is defined as follows:

$$E_z(50, 25, n) = \begin{cases} \alpha(10 - 15 \cos \omega_1 \xi + 6 \cos \omega_2 \xi - \cos \omega_3 \xi), & \xi \leq \tau \\ 0, & \xi > \tau \end{cases} \tag{41a}$$

where

$$\alpha = \frac{1}{320}$$

$$\omega_m = \frac{2\pi m}{\tau}, \qquad m = 1, 2, 3$$

$$\xi = n\delta t$$

$$\tau = 10^{-9} \tag{41b}$$

and δt is the time step used in the simulation. In all experiments we maintained $\delta t = 2.5 \times 10^{-11}$ sec and $\Delta = 2c\,\delta t$, where c is the speed of light in free-space and Δ is the space increment of the finite difference grid. The time profile of the pulse defined in (41a) is shown in Fig. 4(b). This pulse was selected because it has an extremely smooth transition to zero. As discussed in [20], the pulse has its first five derivatives vanish at $\xi = 0, \tau$ and is a good approximation to a smooth compact pulse.

This pulse has very little high-frequency content which is important because of the deleterious effects of grid dispersion (dependence of numerical wave phase velocity upon spatial wavenumber). Grid dispersion and its relation to RBC's is discussed in [21], and [22]. This problem is compounded in higher dimensions by anisotropies of the numerical wave phase velocity with wave vector angle in the grid [3], [23] and is a subject of current research aimed at further reduction of grid boundary reflection coefficients.

The source point in Fig. 4(a) is 25 cells from the boundary of Ω_T at $y = 0$. With the specification $\Delta = 2c\,\delta t$, disturbances at the source point require 50 time steps to propagate to the boundary at $y = 0$. At time step 70, the peak of the pulse just starts to pass through the boundary. We choose to observe the reflection at the first row of grid points away from the $y = 0$ boundary (along $J = 1$) at time step $n = 100$. This permits the bulk of the outgoing pulse to pass through the boundary and excite the largest observable reflection.

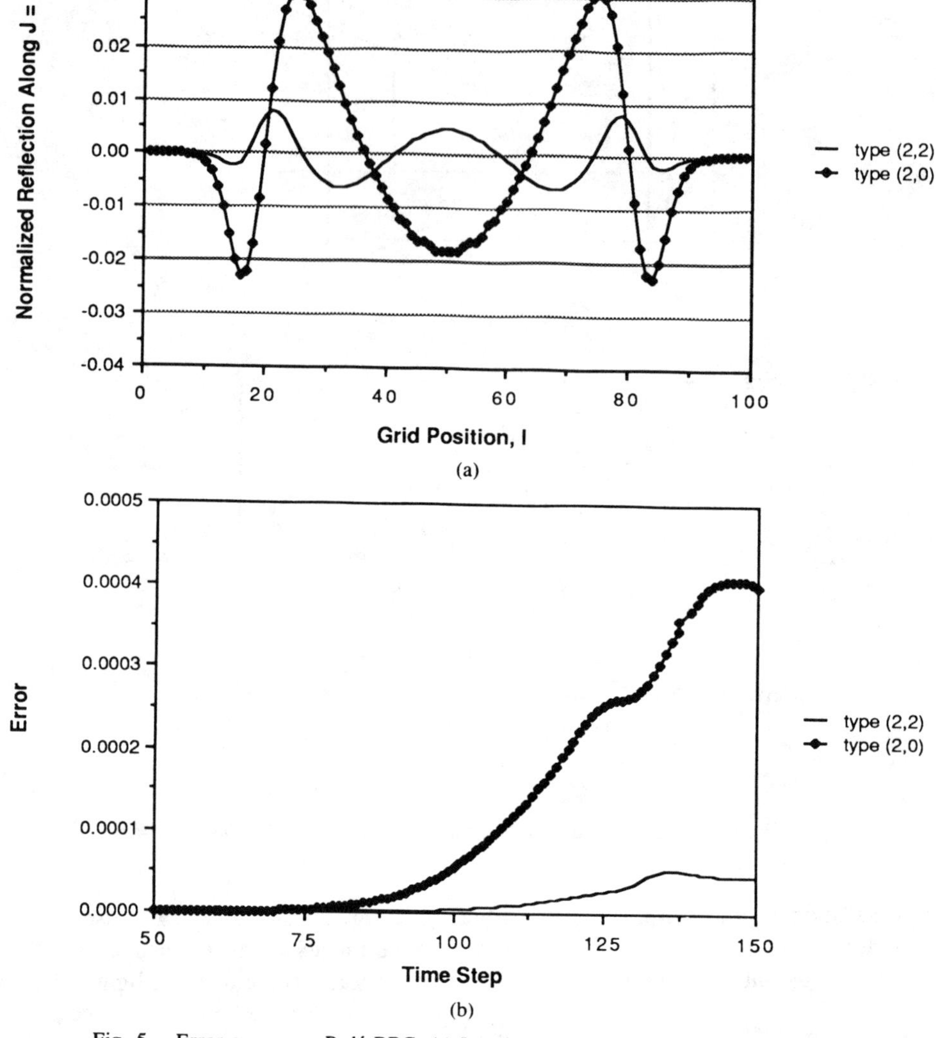

Fig. 5. Error measures, Padé RBC. (a) Local error at $n = 100$. (b) Global error.

Figs. 5 and 6 show the local and global reflection errors observed for the Padé and L_a^∞ RBC's. Fig. 7 compares the third-order conditions. In Figs. 5(a), 6(a), and 7(a), $D(i, 1)$ has been normalized with the peak value of the incident pulse which strikes the $y = 0$ boundary at time step $n = 70$ at grid position (50, 0). The pulse experiment results are in agreement with the reflection coefficient analysis by showing that higher order RBC's do perform better than lower order RBC's in actual simulations. However, comparison of the third-order L_a^∞ RBC to the third-order Padé RBC does not indicate any particular performance advantage. The improved wide-angle performance suggested in [15] is not evident in these experiments.

IV. Application of Mode-Annihilating Operators: OSRC

The on-surface radiation condition method [1] is a new analytical technique by which it is possible to construct accurate approximation of two- and three-dimensional scattering problems involving convex and simple reentrant targets. In this section, two areas of application will be examined. In the first application, the OSRC method will be applied to compute the scattering cross section of two canonical convex targets: 1) a circular cylinder illuminated by both a transverse electric (TE) and transverse magnetic polarized plane wave; and 2) an acoustic sphere with a constant surface impedance. In the second application, the OSRC method will be applied to the scattering of a plane wave by a canonical reentrant geometry: the open end of a semi-infinite flanged parallel-plate waveguide. Before either of the cases is examined, some background discussion on the OSRC method is necessary.

A. Background

The OSRC method is based upon the application of a radiation boundary operator, such as those discussed in Section II, directly on the surface of the target. The effect of this is to relate the surface currents to known field quantities through a simple expression; thus the problem reduces to solving an equation along the contour of the target. Only second-order operators will be considered here because they are the most widely used. The method used in this paper will be the same one used in [1]. For completeness, recently proposed variations will be reviewed as well.

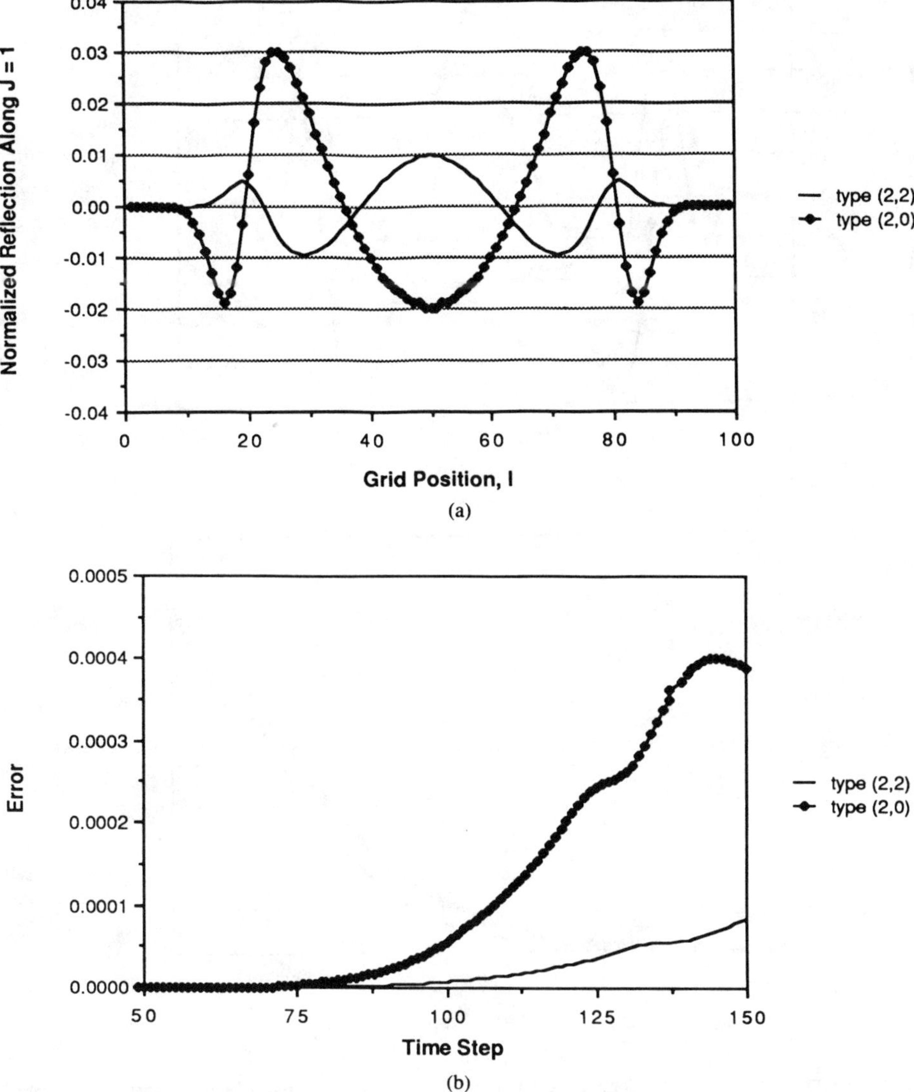

Fig. 6. Error measures, Chebyshev on subinterval RBC. (a) Local error at $n = 100$. (b) Global error.

The original method developed for two-dimensional electromagnetic targets is to apply a mode-annihilating radiation boundary operator locally at each point on the surface of the target [1]. This is conveniently done by noting that an osculating circle can approximate the target's surface locally at each point. The operator then works on the fields as if they were emanating from within the local osculating circle. In two dimensions, this is accomplished by making the following substitutions:

$$\frac{\partial}{\partial r} \rightarrow \frac{\partial}{\partial n} \tag{42a}$$

$$\frac{1}{r} \rightarrow \kappa(s) \tag{42b}$$

$$\frac{1}{r^2}\frac{\partial^2}{\partial \theta^2} \rightarrow \frac{\partial^2}{\partial s^2} \tag{42c}$$

where n is the outward normal, s is an arc length parameter, and $\kappa(s)$ is the curvature of the target at s. This produces the second-order two-dimensional surface boundary operator [1]

$$2(\kappa - jk)\frac{\partial}{\partial n} U = \frac{\partial^2}{\partial s^2} U + \left[2k^2 - \frac{3}{4}\kappa^2 + 3jk\kappa\right] U. \tag{42d}$$

A three-dimensional surface boundary operator developed for acoustic targets is presented in [24]. There, the following substitutions are made in the three-dimensional mode-annihilating radiation boundary operator:

$$\frac{\partial}{\partial R} \rightarrow \frac{\partial}{\partial n} \tag{43a}$$

$$\frac{1}{R} \rightarrow H(s) \tag{43b}$$

$$\frac{\nabla^2 s}{R^2} \rightarrow \nabla \cdot \nabla \tag{43c}$$

where $\nabla \cdot \nabla$ is the surface Laplacian and H is the mean curvature. This produces the second-order three-dimensional

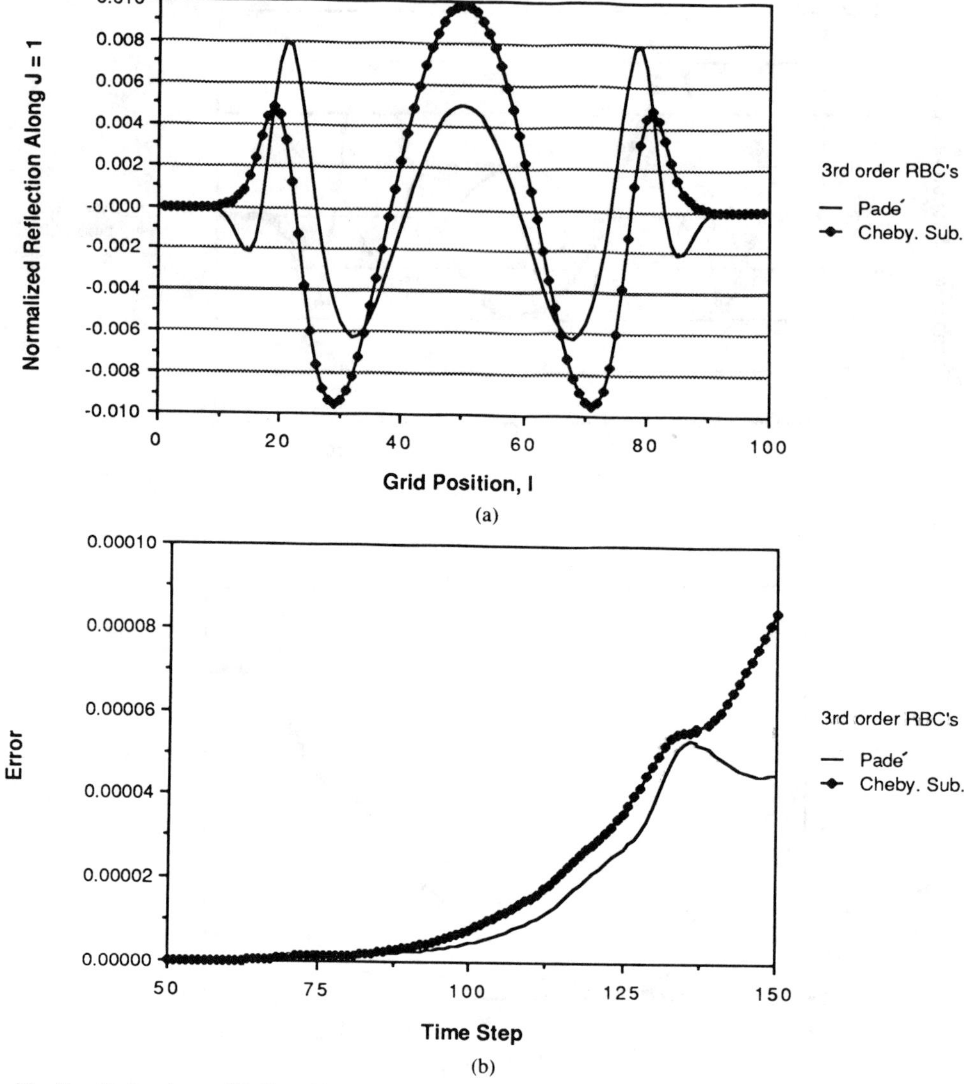

Fig. 7. Comparison of Padé to Chebyshev on subinterval. (a) Local error at $n = 100$. (b) Global error.

surface boundary operator,

$$-2jk\frac{\partial}{\partial n}U=\nabla\cdot\nabla U+2(-k^2-jkH)U. \tag{43d}$$

A variation of this surface boundary condition is presented in [25] in which a slightly different radiation boundary operator is used. It is

$$2(H-jk)\frac{\partial}{\partial n}U=\nabla\cdot\nabla U+2(H^2-k^2-2jkH)U. \tag{43e}$$

The fundamental difference between (43d) and (43e) is that the latter annihilates terms of order $(kR)^{-5}$ as compared to $(kR)^{-2}$ for the former.

Other methods have recently been presented for deriving a general surface boundary condition for OSRC. A sequence of surface boundary operators is derived in [26] by directly factoring the wave equation, as in Section II-B but in a general coordinate system based on the local properties of the target's surface. These results differ from (42d), (43d), and (43e) by a term proportional to the derivative of the curvature multiplied by k^{-2}. No published evidence exists at this time to indicate that this term has any beneficial effects. The most recent derivation of surface boundary conditions for OSRC was presented in [27], which demonstrates that the surface boundary condition can be derived directly from geometrical acoustics/optics by making the assumption that the surface of the target is a phase front. The resulting boundary condition in three dimensions is

$$2jk\frac{\partial}{\partial n}U=\nabla\cdot\nabla U+(2k^2+H^2-\kappa_G-jkH)U \tag{44a}$$

where κ_G is the Gaussian curvature. The corresponding surface boundary condition in two dimensions is

$$2(\kappa-jk)\frac{\partial}{\partial n}U=\frac{\partial^2}{\partial s^2}U+\left[2k^2-\frac{3}{4}\kappa^2+3jk\kappa\right]U-\frac{j}{4k}\frac{\partial^2\kappa}{\partial s^2}U-\frac{j}{k}\frac{\partial\kappa}{\partial s}\frac{\partial U}{\partial s}. \tag{44b}$$

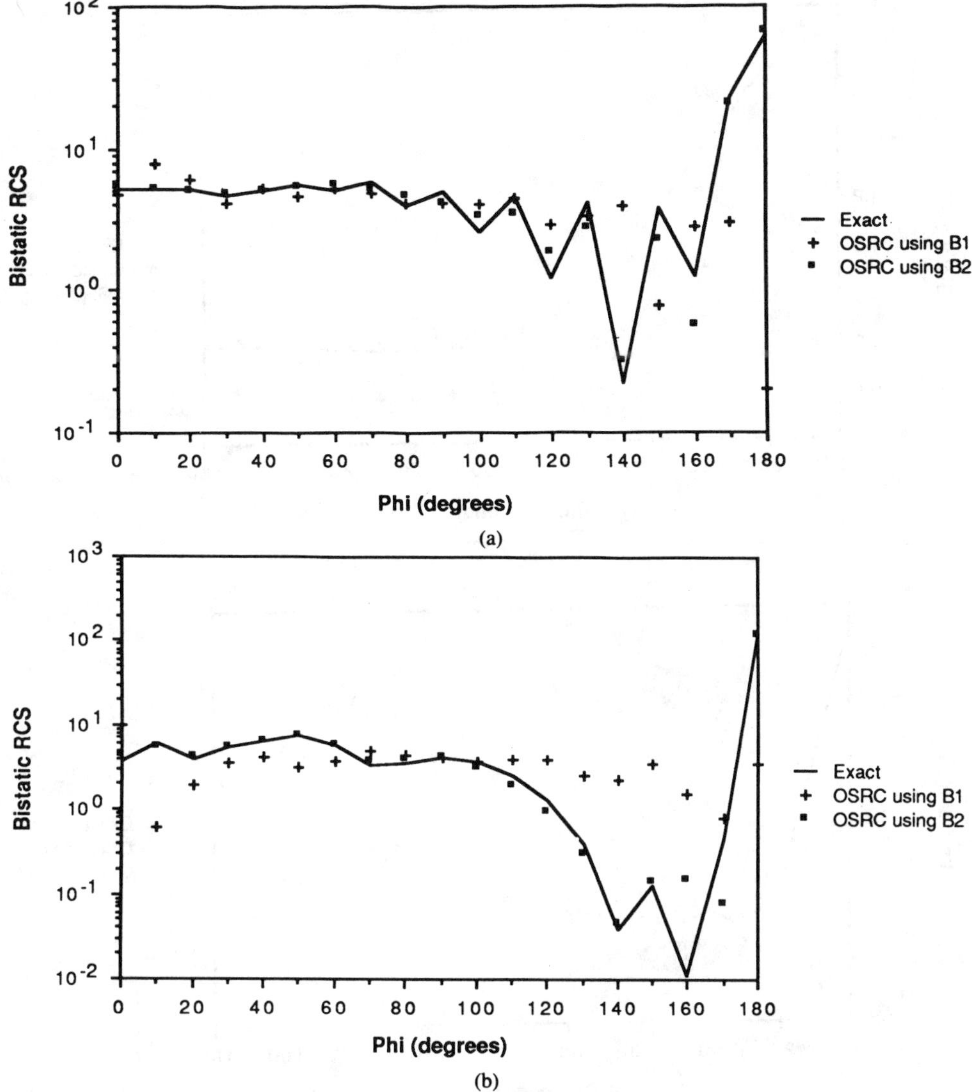

Fig. 8. OSRC computed bistatic RCS of conducting circular cylinder, TE case. (a) $ka = 10$. (b) $ka = 20$.

We note that (44b) also differs from (42d) in the terms proportional to the derivative of the target's curvature (which are formally $O(k^{-3})$ corrections), but, in turn, differs slightly from the operator derived in [26]. More importantly, [27] shows how this method is deriving surface boundary operators can be applied to the vector scattering case of OSRC. However, no validations have been published to date.

Now that the surface boundary conditions have been derived, the application of the OSRC method to scalar problems is straightforward. We set

$$B_n^s U^{scat} = 0 \tag{45}$$

on the surface of the target, where B_n^s is one of the surface boundary conditions described above. What results is an expression that relates the scattered field to its normal derivative at each point on the surface of the target. This is now combined with the usual relation between the incident and scattered field (or the normal derivative of the scattered field), as dictated by the problem.

B. Application to Scattering from Convex Targets

The use of OSRC is first illustrated by modeling scattering by a perfectly conducting circular cylinder of radius a. The cylinder is illuminated by either a transverse electric or transverse magnetic polarized plane wave. For either polarization, the second-order surface boundary condition (42d) is applied to the surface of the cylinder.

For TE polarization, with the magnetic field tangent to the target's surface, the OSRC method results in an ordinary differential equation (ODE) for the azimuthal surface current density J_ϕ:

$$C_2 \frac{d^2 J_\phi}{d\phi^2} - J_\phi = U_{\text{inc}}(1 - C_1 \cos \phi) - C_2 \frac{\partial^2 U_{\text{inc}}}{\partial \phi^2} \tag{46a}$$

where

$$C_1 = \frac{(8k^2 + j8k)}{-3 + 8k^2 + j12k} \qquad C_2 = \frac{-4}{-3 + 8k^2 + j12}. \tag{46b}$$

This is a very simple ODE since it has constant coefficients and thus may be solved analytically. The bistatic radar cross section patterns for a $ka = 10$ and $ka = 20$ cylinder, computed via a modal solution of (46), are plotted in Figs. 8(a) and (b). The OSRC results for the B_2 operator are seen to demonstrate excellent agreement with the exact solution over a

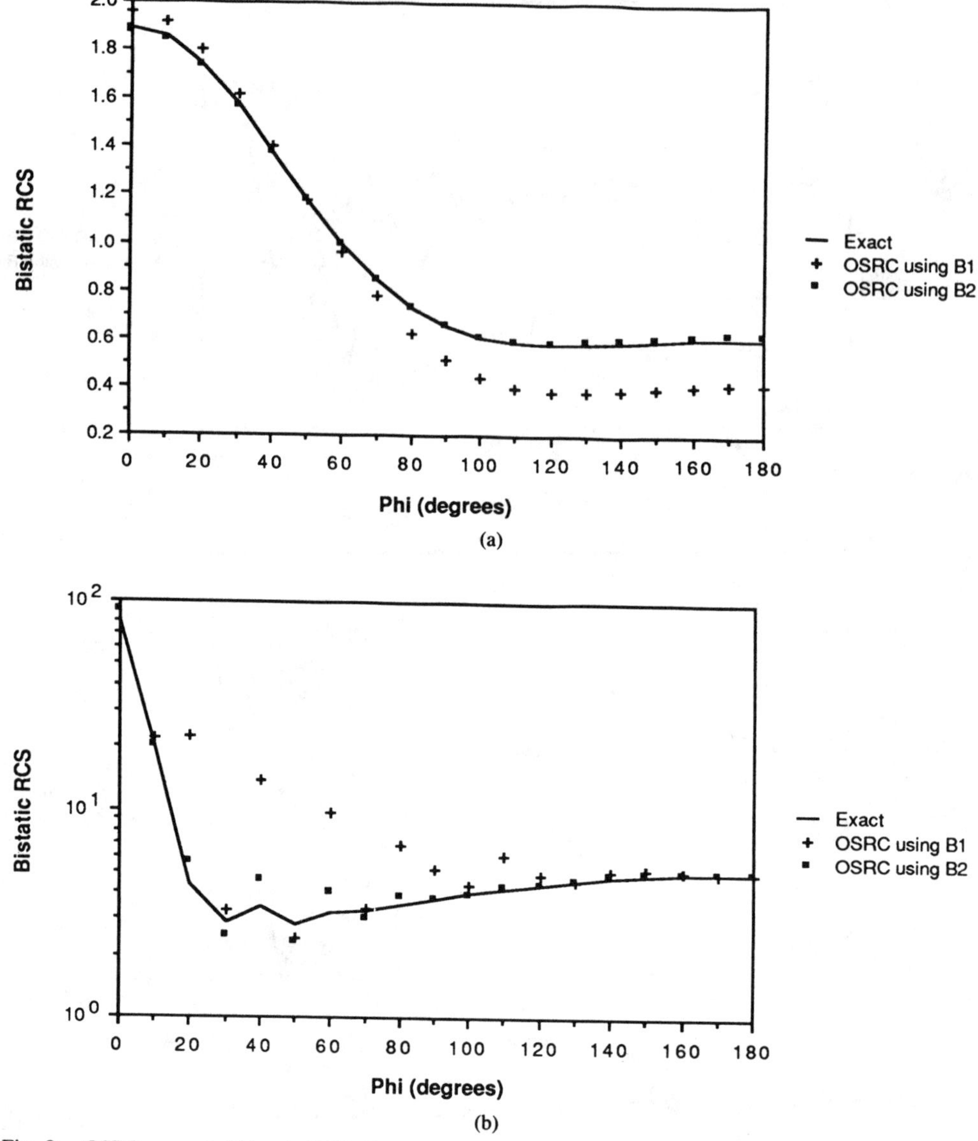

Fig. 9. OSRC computed bistatic RCS of conducting circular cylinder, TM case. (a) $ka = 1$. (b) $ka = 10$.

wide (30 dB) dynamic range. The solution to (46) can be accurately approximated by using the WKBJ method for $k \geq 5$ [28]. There an approximate formula for an arbitrary convex target is derived by the same technique.

For TM polarization, the electric field is tangent to the target's surface and thus the OSRC method gives a simple algebraic expression for the longitudinal surface current density J_z:

$$J_z = \frac{j}{\eta_0 k}\left[\frac{1}{2} + jk(\cos\phi - 1) - jk - \frac{j}{8k} - \frac{1}{2}\cos\phi + \frac{jk}{2}\sin^2\phi\right] e^{jk\cos\phi}. \quad (47)$$

The OSRC predicted bistatic radar cross section patterns for a $ka = 1$ and $ka = 10$ cylinder are plotted in Fig. 9(a) and (b). The OSRC results for the B_2 operator are again seen to demonstrate excellent agreement with the exact solution over a substantial dynamic range.

The application of OSRC to a basic three-dimensional convex target was first illustrated by modeling scattering by a soft acoustic sphere [24]. It was next applied to an acoustic sphere loaded with a constant impedance. Because this problem is solved in [25], only one example is presented here. In this example, an acoustic plane wave propagating in the $-z$ direction impinges upon a spherical target of radius a having a constant normalized surface impedance of $Z = 10$. Condition (43e) is applied to the surface of the sphere. After inserting (43e) into (45), a second-order partial differential equation results for the surface currents, whose solution is approximated by a simple two-term asymptotic expansion and is then used to determine the far fields. Fig. 10 shows the backscattered cross section versus k. Again, there is excellent agreement with the exact solution.

C. Application to Scattering from Reentrant Structures

The second area of application illustrates using OSRC to model scattering by simple reentrant structures. We consider the problem of a plane wave impinging on the open end of a semi-infinite flanged parallel-plate waveguide [29], shown in Fig. 11. A plane wave, at an angle α measured counterclock-

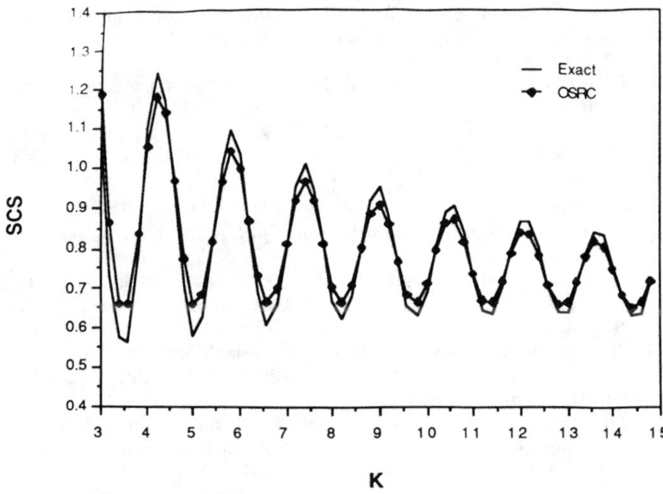

Fig. 10. Normalized back scattering cross section versus K for reactively loaded sphere ($Z = 10$). Exact solution is calculated from modal series representation. Dotted line is two-term OSRC solution.

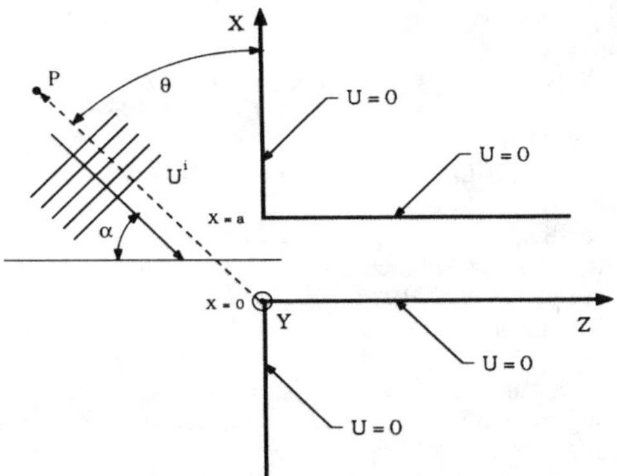

Fig. 11. Plane wave incident on open end of flanged, semi-infinite parallel-plate waveguide. $a = 1.5$ m, $f = 250$ MHz.

wise with reference to the $-z$ axis, illuminates the mouth of the waveguide. The flange and the walls of the guide are assumed to be perfectly conducting. The incident wave U^i is considered to be the y component E_y of the incident electric field vector. Thus, for the boundary conditions shown, TE modes are excited inside the guide. The operator (42d), with $\kappa = 0$, is applied to the field representations valid in the guide aperture and yields an expression for the coefficients in the modal representation of the waveguide fields. Knowledge of the modal coefficients then permits the derivation of a simple expression for the bistatic radar cross section of the field scattered by the aperture, and the fields penetrating into the waveguide. Results of calculations using this approach are presented in Figs. 12 and 13. Fig. 12 shows the bistatic cross section for a plane wave at $\alpha = 0°$. Fig. 13 show the magnitude and phase distribution of the field penetrating the guide at a distance of $z = 2$ m from the aperture. The OSRC results are compared to results obtained by FD-TD simulations. Excellent agreement is observed. The value of the OSRC solution is striking in its simple form and negligible computational requirements.

V. Future Research

Research on radiation boundary operators is presently directed at two basic goals. The first is the development of better truncation conditions for finite-difference and finite-element grids. This is aimed at reducing nonphysical reflections from the outer grid boundary which contribute to the numerical noise floor of the modeling procedure. Reducing the numerical noise floor will allow the simulation of scatterers with wider dynamic range. The second goal is the development of optimal mode-annihilating radiation boundary operators for the OSRC method.

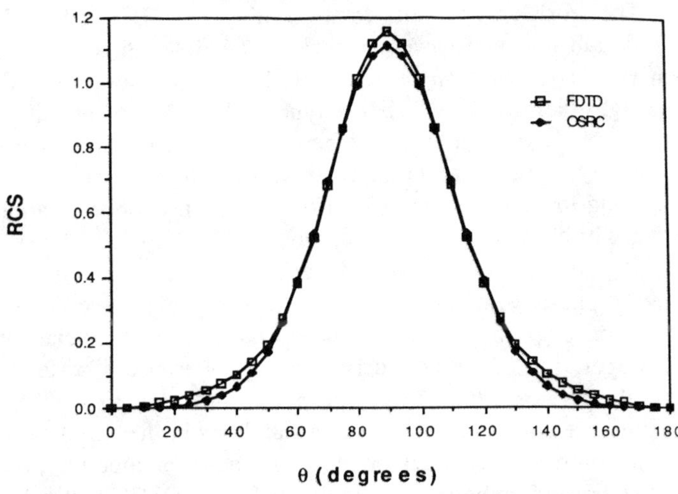

Fig. 12. Bistatic cross section for scattering from waveguide aperture due to plane wave at $\alpha = 0°$ and $f = 250$ MHz. Angle θ is as shown in Fig. 11.

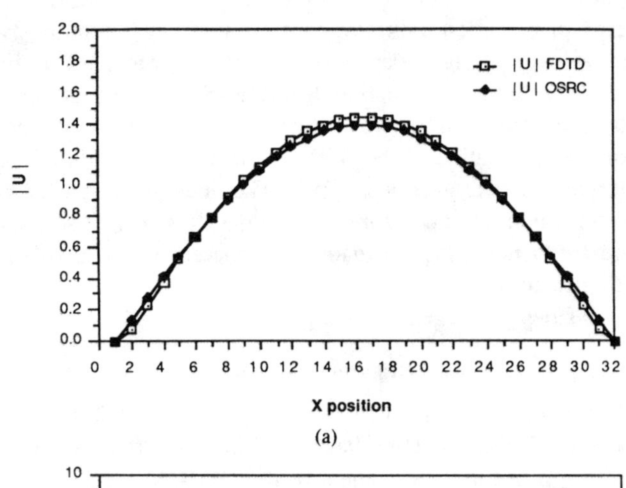

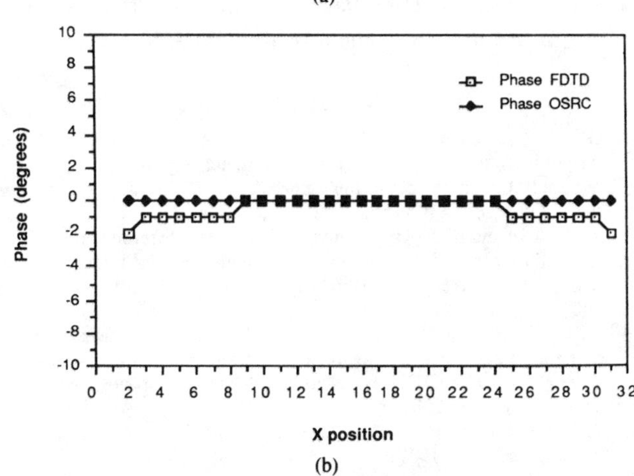

Fig. 13. Field penetration into waveguide at $z = 2$ m for plane wave at $\alpha = 0°$ and $f = 250$ MHz. (a) E-field magnitude. (a) E-field phase.

The work on operators for truncating FD-TD and FE-TD grids fall into two basic categories. The first category is the synthesis of wide-angle radiation boundary operators. Although there are two basic types of radiation boundary operators, the main thrust of the research is directed towards one-way wave equation approximations. Here, work is proceeding in deriving approximations to the dispersion relation that will absorb waves over a wide range of angles properly taking into account anisotropy and dispersion of numerical mode phase velocities. The second category concerns the difference approximation for the operator. This is crucial for the operation of the boundary condition, because a perfectly valid operator which is not properly converted to a finite-difference equation, may cause instabilities in the simulation.

In contrast to the work on the grid truncation operators, the work on surface boundary conditions for the OSRC method is based on the optimization of the mode-annihilating radiation boundary operators. A key goal of this research is to produce a surface boundary condition which can better predict tangential energy propagation along the target's surface. Presently, two methods are being investigated for this purpose. The first method postulates a special multiplicative operator that models tangential propagating waves. The second method depends upon the application of the OSRC in the time domain to obtain a target impulse response [30]. The idea is to adjust the coefficients in the operator to match the exact response over a wide frequency range. Finally, the operators are also being examined to assess their ability to predict the correct current singularities on targets with edges.

Acknowledgment

The authors wish to acknowledge the contributions of Professor Korada R. Umashankar of the University of Illinois at Chicago in the development of the OSRC theory.

References

[1] G. A. Kriegsmann, A. Taflove, and K. R. Umashankar, "A new formulation of electromagnetic wave scattering using an on-surface radiation boundary condition approach," *IEEE Trans. Antennas Propagat.*, vol. AP-35, pp. 152–161, Feb. 1987.

[2] A. Taflove and K. R. Umashankar, "The finite-difference time-domain (FD-TD) method for electromagnetic scattering and interaction problems," *J. Electromagn. Waves Appl.*, vol. 1, no. 4, pp. 363–387, 1987.

[3] A. C. Cangellaris, C. C. Lin, and K. K. Mei, "Point-matched time-domain finite-element methods for electromagnetic radiation and scattering," *IEEE Trans. Antennas Propagat.*, vol. AP-35, pp. 1160–1173, Oct. 1987.

[4] A. Sommerfeld, *Partial Differential Equations in Physics*. New York: Academic, 1949.

[5] G. A. Kriegsmann and C. S. Morawetz, "Numerical solutions of exterior problems with the reduced wave equation," *J. Comput. Phys.*, vol. 28, pp. 181–197, 1979.

[6] A. Bayliss and E. Turkel, "Radiation boundary conditions for wave-like equations," *Commun. Pure Appl. Math.*, vol. 23, pp. 707–725, 1980.

[7] F. G. Friedlander, "On the radiation field of pulse solutions of the wave equation," *Proc. Royal Soc. London Ser. A*, vol. 269, pp. 53–69, 1962.

[8] C. H. Wilcox, "An expansion theorem for electromagnetic fields," *Commun. Pure Appl. Math.*, vol. 9, pp. 115–132, 1956.

[9] A. Bayliss, M. Gunzburger, and E. Turkel, "Boundary conditions for the numerical solution of elliptic equations in exterior regions," *SIAM J. Appl. Math.*, vol. 42, pp. 430–451, Apr. 1982.

[10] S. N. Karp, "A convergent far-field expansion for two-dimensional radiation functions," *Commun. Pure Appl. Math.*, vol. 14, pp. 27–434, 1961.

[11] B. Engquist and A. Majda, "Absorbing boundary conditions for the numerical simulation of waves," *Math. Comput.*, vol. 31, pp. 629–651, July 1977.

[12] L. Halpern and L. N. Trefethen, "Wide-angle one-way wave equations," Dept. Math., Mass. Inst. Technol., Cambridge, Numerical Analysis Rep. 86-5, July 1986.

[13] G. Mur, "Absorbing boundary conditions for the finite-difference approximation of time-domain electromagnetic field equations," *IEEE Trans. Electromagn. Compat.*, vol. EMC-23, pp. 377–382, Nov. 1981.

[14] K. R. Umashankar and A. Taflove, "A novel method to analyze electromagnetic scattering of complex objects," *IEEE Trans. Electromagn. Compat.*, vol. EMC-24, pp. 392–405, Nov. 1982.

[15] L. N. Trefethen and L. Halpern, "Well-posedness of one-way wave equations and absorbing boundary conditions," *Math. Comput.*, vol. 47, pp. 421–435, Oct. 1986.

[16] J. G. Blaschak and G. A. Kriegsmann, "A comparative study of absorbing boundary conditions," *J. Comput. Phys.*, vol. 77, pp. 109–139, July 1988.

[17] G. B. Whitham, *Linear and Nonlinear Waves*. New York: Wiley, 1974.

[18] R. L. Higdon, "Absorbing boundary conditions for difference approximations to the multidimensional wave equation," *Math. Comput.*, vol. 47, pp. 437–459, 1986.

[19] ——, "Numerical absorbing boundary conditions for the wave equation," *Math. Comput.*, vol. 49, pp. 65–90, 1987.

[20] G. A. Kriegsmann, A. N. Norris, and E. L. Reiss, "Acoustic pulse scattering by baffled membranes," *J. Acoust. Soc. Amer.*, vol. 79, Jan. 1986.

[21] L. N. Trefethen, "Group velocity in finite difference schemes," *SIAM Rev.*, no. 24, pp. 114–136, 1982.

[22] ——, "Group velocity interpretation of stability theory of Gustafasson, Kreiss, and Sundstrom," *J. Comput. Phys.*, no. 49, pp. 199–217, 1983.

[23] A. Taflove and K. R. Umashankar, "Advanced numerical modeling of microwave penetration and coupling for complex structures—Final report," Lawrence Livermore National Lab., VCRL-15960, Contract 6599805, Sept. 1987.

[24] D. S. Jones, "An approximate boundary condition in acoustics," submitted to *J. Sound Vibration,* 1987.

[25] G. A. Kriegsmann and T. G. Moore, "An application of the on-surface radiation condition to the scattering of acoustic waves by a reactively loaded sphere," *Wave Motion*, vol. 10, pp. 277–284, 1988.

[26] W. F. Richards, D. R. Wilton, and J. X. Jiang, "Factorization of the three-dimensional Helmholtz equation with applications to the on surface radiation condition method," submitted to *IEEE Trans. Antennas Propagat.*, 1987.

[27] D. S. Jones, "Surface radiation conditions," submitted to *IMA J. Appl. Math.*, 1987.

[28] T. G. Moore, G. A. Kriegsmann, and A. Taflove, "An application of the WKBJ technique to the on-surface radiation condition," *IEEE Trans. Antennas Propagat.*, vol. 36, pp. 1329–1331, Sept. 1988.

[29] J. G. Blaschak, G. A. Kriegsmann, and A. Taflove, "A study of wave interactions with flanged waveguides and cavities using the on-surface radiation condition method," *Wave Motion*, in press.

[30] T. G. Moore, G. A. Kriegsmann, and A. Taflove, "An application of the OSRC method to time dependent problems," submitted to *IEEE Trans. Antennas Propagat.*, 1988.

Absorbing Boundary Conditions for the Vector Wave Equation

ANDREW F. PETERSON
Electromagnetic Communication Laboratory
University of Illinois
1406 West Green Street
Urbana, Illinois 51801

***Key Terms*—Finite element method, electromagnetic scattering, radiation condition**

***Abstract*—Approximate absorbing boundary conditions for the vector wave equation are developed. These permit a truncation of the computational domain required to accurately model an open-region electromagnetic scattering problem using the finite element method. Details are provided to illustrate the manner in which the boundary conditions can be coupled to a weak form of the vector wave equation.**

1. Introduction

Frequency domain integral formulations have been widely used for the numerical analysis of open-region electromagnetic scattering problems. Recent work suggests, however, that a significant improvement in computational efficiency can be realized through the direct numerical solution of differential equations [1–4]. Since the numerical treatment of differential equations requires the discretization of the entire domain of interest, some type of absorbing boundary condition is often applied in the near zone of the scatterer in order to truncate this region and limit the number of unknowns introduced into the problem. Several authors have proposed approximate absorbing boundary conditions for the scalar wave equation [5,6], and these appear to work well for a variety of two-dimensional electromagnetics problems [2,3]. Currently, there is considerable interest in the analysis of realistic three-dimensional structures. The purpose of the present communication is to present a class of approximate absorbing boundary conditions for the vector wave equation, applicable in the near zone of a three-dimensional scatterer.

2. Derivation of the Absorbing Boundary Conditions

In a homogeneous region of space exterior to some electromagnetic scatterer, consider the vector wave equation

$$\nabla \times \nabla \times \overline{H} - k^2\overline{H} = 0 \tag{1}$$

where $\overline{H}$ can represent the electric field, the magnetic field, or some vector potential function. For e^{jwr} time dependence, a solution to the wave equation is said to satisfy the radiation condition provided that

$$\lim_{r\to\infty} r\{\hat{r} \times (\nabla \times \overline{H}) - jk\overline{H}\} = 0 \tag{2}$$

The Wilcox expansion theorem states that in some region exterior to the scatterer, a solution to (1) and (2) may be written in the form [7]

$$\overline{H}(r,\theta,\phi) = \frac{e^{-jkr}}{4\pi r} \sum_{n=0}^{\infty} \frac{\overline{H}_n(\theta,\phi)}{r^n} \tag{3}$$

representing an outward-propagating field. Furthermore, the series can be differentiated term-by-term and remains uniformly convergent [7].

Equation (2) is one example of an absorbing boundary condition. However, if applied to the expansion of Eq. (3) at some finite value of r, the residual is

$$\hat{r} \times (\nabla \times \overline{H}) - jk\overline{H} = \frac{e^{-jkr}}{4\pi r} \sum_{n=0}^{\infty} \cdot \left\{ -jk \frac{\hat{r}(\hat{r} \cdot \overline{H}_n)}{r^n} + \frac{r\nabla(\hat{r} \cdot \overline{H}_n) + n\overline{H}_n^{\tan}}{r^{n+1}} \right\} \tag{4}$$

The vector identities

$$\hat{r} \times (r\nabla \times \overline{H}_n) = r\nabla(\hat{r} \cdot \overline{H}_n) - \overline{H}_n^{\tan} \qquad (5)$$

and

$$\hat{r} \times (\hat{r} \times \overline{H}_n) = -\overline{H}_n^{\tan} \qquad (6)$$

where

$$\overline{H}^{\tan} = \overline{H} - \hat{r}(\hat{r} \cdot \overline{H}) \qquad (7)$$

have been employed to simplify Eq. (4). Note that the operations $r\nabla(\hat{r} \cdot \overline{H}_n)$ and $r\nabla \times \overline{H}_n$ are actually independent of r. As a consequence of Eq. (1). The Wilcox expansion must satisfy

$$\nabla \cdot \overline{H} = 0 \qquad (8)$$

which produces

$$\hat{r} \cdot \overline{H}_0 = 0 \qquad (9)$$

it follows that the leading-order residual in Eq. (4) is $0(r^{-2})$.

For numerical implementation, it is necessary to enforce the absorbing boundary condition at an artificial surface enclosing the scatterer. To minimize the size of the computational domain and the number of unknowns to be determined, it is desirable to locate the artificial surface as close to the scatterer as accuracy permits. Since Eq. (4) is truly absorbing only in the far zone of the scatterer, a "higher-order" condition [i.e., one that will hold to higher powers of $1/r$ if applied to the Wilcox expansion of Eq. (3)] is necessary. We will derive such conditions based on the observation that the Wilcox expansion will exactly satisfy a true absorbing boundary condition, and can be used to systematically improve the degree of an approximate boundary condition. In principle, the greater degree to which the Wilcox expansion satisfies the higher-order condition, the closer the artificial boundary can be brought to the scatterer. As a general rule, we do not expect this type of boundary condition to remain valid if brought completely to the surface of the scatterer.

The form of the residual in Eq. (4) suggests the alternative operator

$$\hat{r} \times (\nabla \times \overline{H}) - jk\overline{H}^{\tan} = \frac{e^{-jkr}}{4\pi r} \sum_{n=o}^{\infty} \left\{ \frac{r\nabla(\hat{r} \cdot \overline{H}_n) + n\overline{H}_n^{\tan}}{r^{n+1}} \right\} \qquad (10)$$

The dominant behavior of Eq. (10) is

$$\frac{e^{-jkr}}{4\pi r} \left\{ \frac{r\nabla(\hat{r} \cdot \overline{H}_1) + \overline{H}_1^{\tan}}{r^2} \right\} = O(r^{-3}) \qquad (11)$$

To reduce the order of the residual error beyond that of Eq. (11), consider

$$\hat{r} \times [\nabla \times \{\hat{r} \times (\nabla \times \overline{H}) - jk\overline{H}^{\tan}\}]$$
$$= \frac{e^{-jkr}}{4\pi r} \sum_{n=1}^{\infty} \left\{ \left(jk + \left[\frac{n+1}{r} \right] \right) \frac{r\nabla(\hat{r} \cdot \overline{H}_n) + n\overline{H}_n^{\tan}}{r^{n+1}} \right\} \qquad (12)$$

The leading-order behavior of Eq. (12) is

$$\frac{e^{-jkr}}{4\pi r} \left\{ jk \frac{r\nabla(\hat{r} \cdot \overline{H}_1) + \overline{H}_1^{\tan}}{r^2} \right\} = O(r^{-3}) \qquad (13)$$

The form of Eqs. (11) and (13) suggests the dominant term in the residual can be eliminated by considering the higher-order operator

$$\{\hat{r} \times (\nabla \times \ ) - jk\} \{\hat{r} \times (\nabla \times \overline{H}) - jk\overline{H}^{\tan}\}$$
$$= \frac{e^{-jkr}}{4\pi r} \sum_{n=1}^{\infty} \left\{ (n+1) \frac{r\nabla(\hat{r} \cdot \overline{H}_n) + n\overline{H}_n^{\tan}}{r^{n+2}} \right\} \qquad (14)$$

Equation (14) has leading-order behavior

$$\frac{e^{-jkr}}{4\pi r} \left\{ 2 \frac{r\nabla(\hat{r} \cdot \overline{H}_1) + \overline{H}_1^{\tan}}{r^3} \right\} = O(r^{-4}) \qquad (15)$$

The form of Eqs. (15) and (11) suggests that an additional reduction in the residual can be obtained by the use of the modified operator

$$\left\{ \hat{r} \times (\nabla \times \ ) - jk - \frac{2}{r} \right\} \{\hat{r} \times (\nabla \times \overline{H}) - jk\overline{H}^{\tan}\}$$
$$= \frac{e^{-jkr}}{4\pi r} \sum_{n=1}^{\infty} \left\{ (n-1) \frac{r\nabla(\hat{r} \cdot \overline{H}_n) + n\overline{H}_n^{\tan}}{r^{n+2}} \right\} \qquad (16)$$

which has leading-order behavior

$$\frac{e^{-jkr}}{4\pi r} \left\{ \frac{r\nabla(\hat{r} \cdot \overline{H}_2) + 2\overline{H}_2^{\tan}}{r^4} \right\} = O(r^{-5}) \qquad (17)$$

It is apparent that the preceding procedure can be continued indefinitely, producing a family of absorbing boundary conditions of different order. If equated to zero, the left-hand side of Eq. (10) represents the "first-order" condition, which the Wilcox expansion satisfies to $O(r^{-3})$. The "second-order" condition is

$$\left\{ \hat{r} \times (\nabla \times \ ) - jk - \frac{2}{r} \right\} \{\hat{r} \times (\nabla \times \overline{H}) - jk\overline{H}^{\tan}\} = 0 \qquad (18)$$

Equation (3) satisfies this condition to $O(r^{-5})$. Higher-order conditions take the form

$$\left[\prod_{n=2}^{N} \left\{ \hat{r} \times (\nabla \times \ ) - jk - \frac{2(n-1)}{r} \right\} \right] \times \{\hat{r} \times (\nabla \times \overline{H}) - jk\overline{H}^{\tan}\} = 0 \qquad (19)$$

Equation (3) satisfies the Nth order condition to $O(r^{-2N-1})$. This family of absorbing boundary conditions appears to be analogous to the Bayliss-Turkel conditions developed for the scalar wave equation [6].

3. Error Due to the Absorbing Boundary Conditions

We turn our attention to the error introduced by the use of the approximate absorbing boundary conditions at a finite distance from the scatterer. Specifically, suppose we solve the wave equation exactly subject to the "second-order" condition from Eq. (18). The exact radiated or "scattered" field will have the form of the Wilcox expansion, i.e.,

$$\overline{H}^{\delta} = \frac{e^{-jkr}}{4\pi r}\left\{\overline{H}_0 + \frac{\overline{H}_1}{r} + \frac{\overline{H}_2}{r^2} + \frac{\overline{H}_3}{r^3} + \frac{\overline{H}_4}{r^4} + \frac{\overline{H}_5}{r^5} + \cdots\right\} \tag{20}$$

Since the error in the second-order boundary condition is $O(r^{-5})$, we can retain correct terms only up to $O(r^{-4})$. In the vicinity of the artificial boundary, the actual field produced by the procedure will be of the form

$$\overline{H}^{\delta}_{\text{app}} = \frac{e^{-jkr}}{4\pi r}\left\{\overline{H}_0 + \frac{\overline{H}_1}{r} + \frac{\overline{H}_2}{r^2} + \frac{\overline{H}_3}{r^3}\right\} + \frac{e^{-jkr}}{4\pi r}\left\{\frac{\overline{A}_4}{r^4} + \frac{\overline{A}_5}{r^5} + \cdots\right\} + \frac{e^{+jkr}}{4\pi r}\left\{\frac{\overline{B}_4}{r^4} + \frac{\overline{B}_5}{r^5} + \cdots\right\} \tag{21}$$

where the terms containing coefficients $\overline{B}_4$, $\overline{B}_5$, etc., represent a fictitious reflection produced by the artificial boundary. Note that the higher-order coefficients $\overline{A}_4$, $\overline{A}_5$, etc., in the outgoing terms are also incorrect.

For a solution to the wave equation subject to the absorbing boundary condition to be accurate, the boundary condition must be applied at a large enough value of r in order that the error terms are insignificant compared to the lower terms in Eq. (21). If the convergence of the series in Eq. (20) is rapid, this may be accomplished at a small value of r. Unfortunately, there is no guarantee that the convergence will be extremely favorable for electrically large scatterers.

4. Implementation Details

To illustrate the implementation of the absorbing boundary conditions, we focus on the second-order condition. Using Eq. (1) we rewrite Eq. (18) in the form

$$\hat{r} \times (\nabla \times \overline{H}) = \alpha(r)\overline{H}^{\tan} + \beta(r)\nabla \times [\hat{r}(\hat{r} \cdot \nabla \times \overline{H})] + \gamma(r)\nabla^{\tan}(\hat{r}\cdot \overline{H}) \tag{22}$$

where

$$\alpha(r) = jk \tag{23}$$

$$\beta(r) = \left[\frac{1}{2jk + 2/r}\right] \tag{24}$$

$$\gamma(r) = \left[\frac{jk}{2jk + 2/r}\right] \tag{25}$$

and $\nabla^{\tan}$ is defined according to Eq. (7).

If Eq. (1) is generalized to a medium where the relative permittivity ϵ_r and relative permeability μ_r are functions of position, the wave equation (now written specifically for the magnetic field $\overline{H}$) can be converted into the conventional "weak" equation [8,9]

$$\iiint_{\Gamma} \frac{1}{\epsilon_r}(\nabla \times \overline{T}) \cdot (\nabla \times \overline{H}) - \mu_r k^2 \overline{T} \cdot \overline{H} = \iint_{\partial\Gamma} \overline{T} \cdot \hat{r} \times (\nabla \times \overline{H}) \tag{26}$$

where Γ is the domain of interest and $\partial\Gamma$ is the spherical surface bounding that region. This surface is chosen so that all inhomogeneities are inside $\partial\Gamma$. $\overline{T}(x, y, z)$ is a vector testing function, typically chosen in the context of the finite element method to be identical to the expansion functions used to represent the $\overline{H}$ field. The absorbing boundary condition may be incorporated through the surface integral in Eq. (26). Note that the radiation condition only applies to the "scattered" or outward radiating $\overline{H}$ field, and depending on the specific formulation the fields may require separation into "incident" and "scattered" parts. Using Eq. (22) and appropriate vector identities, the part of the surface integral involving the scattered field can be written as

$$\iint_{\partial\Gamma} \overline{T} \cdot \hat{r} \times (\nabla \times \overline{H}^{3}) = \iint_{\partial\Gamma} [\alpha(r)\overline{T} \cdot \overline{H}^{5}_{\tan} + \beta(r)(\hat{r} \cdot \nabla \times \overline{T})(\hat{r} \cdot \nabla \times \overline{H}^{3}) + \gamma(r)\overline{T} \cdot \nabla^{\tan}(\hat{r} \cdot \overline{H}^{2})] \tag{27}$$

Since the incorporation of Eq. (22) only involves first-order tangential derivatives of the $\overline{H}$-field components, this condition appears especially appropriate for finite-element implementation using first-order linear interpolation functions, as discussed by Silvester and Ferrari [8]. Details of this type of implementation are available in the context of the scalar wave equation using the second-order Bayliss-Turkel boundary condition [2,6]. The second-order Bayliss-Turkel operator is somewhat analogous to Eq. (22) in that it only requires first-order tangential derivatives of the unknown function when implemented within a similar weak equation. It is worth noting that the second-order Bayliss-Turkel condition improves on the scalar Sommerfeld radiation condition by three powers of the variable r. Equation (22) is similar in that it improves on Eq. (4) by three powers of r.

References

[1] A. F. Peterson, "A Comparison of the Volume MFIE and Unimoment Methods for TE Scattering from Lossy, Inhomogeneous Dielectric Cylinders," *Digest of the 1987 IEEE Antennas and Propagation Society International Symposium*, Blacksburg, VA, June 1987, pp. 1118-1121.

[2] A. F. Peterson, "A Comparison of Integral, Differential and Hybrid Methods for TE-Wave Scattering from inhomogeneous Dielectric Cylinders," submitted to *J. Electromagnetic Waves Appl.*

[3] G. Meltz, B. J. McCartin, and L. J. Bahrmasel, "Application of the Control Region Approximation to Electromagnetic Scattering." *Abstracts of the 1987 URSI Radio Science Meeting*, Blacksburg, VA, June 1987, p. 185.

[4] R. T. Ling, "Numerical Solution for the Scattering of Sound Waves by a Circular Cylinder," *AIAA Journal,* Vol. 25, April 1987, pp. 560-566.

[5] B. Enquist and A. Majda, "Absorbing Boundary Conditions for the Numerical Simulation of Waves," *Math. Comp.*, Vol. 31, 1977, pp. 629-651.

[6] A. Bayliss and E. Turkel, "Radiation Boundary Conditions for Wave-like Equations," *Comm. Pure Appl. Math.*, Vol. 33, 1980, pp.707-725.

[7] C. H. Wilcox, "An Expansion Theorem for Electromagnetic Fields," *Comm. Pure Appl. Math.*, Vol. 9, 1956, pp. 115-134.

[8] P. P. Silvester and R. L. Ferrari, *Finite Elements for Electrical Engineers,* Cambridge University Press, 1983.

[9] J. N. Reddy, *An Introduction to the Finite Element Method,* McGraw-Hill, New York, 1984

A Frequency-Domain Differential Equation Formulation for Electromagnetic Scattering from Inhomogeneous Cylinders

ANDREW F. PETERSON, MEMBER, IEEE, AND STEVEN P. CASTILLO, MEMBER, IEEE

Abstract—**A procedure is described for calculating scattering cross-section data based on the numerical solution of differential equations in the frequency domain. The scatterers considered are two dimensional and consist of lossy, inhomogeneous dielectric, magnetic, and perfectly conducting material. The appropriate wave equation is combined with an approximate local absorbing boundary condition and discretized with the finite element method. Results for conducting and dielectric scatterers are presented to illustrate the accuracy and generality of the approach.**

I. Introduction

IN CONTRAST WITH the numerous formulations available in the literature for the analysis of perfect conducting or homogeneous electromagnetic scatterers, relatively few techniques have been developed for the treatment of highly inhomogeneous bodies. Several volume integral equation (VIE) formulations have been developed for the rigorous analysis of inhomogeneous dielectric bodies [1]–[6]. Scatterers containing magnetic materials can be analyzed using integral equations that are the electromagnetic dual to those employed for dielectric bodies. All integral equation methods require the solution of a fully populated matrix equation. Unfortunately, the necessary storage may be prohibitive for many problems of interest.

An alternative formulation involves the discretization of differential equations throughout the volume of the inhomogeneous scatterer [7]–[19]. These equations are convenient to employ within complex inhomogeneous regions, and can easily incorporate conducting boundaries. In addition, differential equations discretize to produce sparse matrix equations. Open-region scattering formulations involving frequency-domain differential equations fall into two general classes, depending on the global or local nature of the boundary condition used to truncate the computational domain. Some approaches are "hybrid" in nature, that is they combine a volume discretization of a differential equation (VDE) with a surface integral equation [7]–[11] or eigenfunction expansion [12]–[15]. The integral equation or eigenfunction expansion acts as a global boundary condition. In the alternate approach, a VDE is combined with an approximate local absorbing boundary condition applied directly in the near zone of the scatterer [16], [17]. A description of this technique is provided below. A recent variation on the latter approach involves transforming the Helmholtz wave equation into an alternate differential equation [18]. Such a procedure may simplify the discretization process as well as the selection of an accurate absorbing boundary condition.

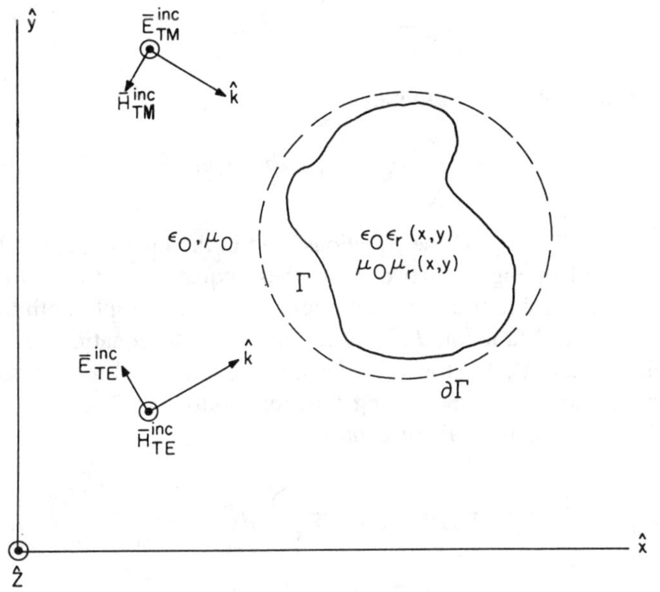

Fig. 1. Cross section of the cylindrical geometry under consideration.

This paper presents a VDE approach incorporating a local Bayliss–Turkel boundary condition for calculating the scattering cross section of two-dimensional structures. The method is similar to the "control region approximation" [19], but employs the finite element method instead of a finite difference approach to discretize the differential equation. Numerical examples are limited to cylinders containing isotropic dielectric and perfect conducting material, although the formulation is general enough to treat linear isotropic magnetic material. Extensions to the anisotropic case can be effected following an approach discussed in the literature [14].

II. Weak Forms of the Wave Equations for Scattering

A general two-dimensional scattering problem involves a source located somewhere in space, illuminating a cylindrical structure as depicted in Fig. 1. In the absence of the cylinder, this source produces the "incident" electromagnetic fields $\bar{E}^{inc}$ and $\bar{H}^{inc}$. The fields produced in the presence of the cylinder are the "total" fields $\bar{E}^{tot}$ and $\bar{H}^{tot}$. The "scattered"

Manuscript received March 12, 1988; revised October 26, 1988. This work was supported by a grant from the National Center for Supercomputing Applications at the University of Illinois.

A. F. Peterson is with the Electromagnetic Communication Laboratory, University of Illinois, Urbana, IL 61801.

S. P. Castillo is with the Department of Electrical and Computer Engineering, New Mexico State University, Las Cruces, NM 88003.

IEEE Log Number 8927011.

Reprinted from *IEEE Trans. Antennas Propagat.*, vol. 37, pp. 601–607, May 1989.

fields $\bar{E}^s$ and $\bar{H}^s$ are the difference between the "total" fields and the "incident" fields. Two-dimensional fields can be decomposed into the transverse electric (TE) polarization, where the $\bar{E}$-field has no $\hat{z}$-component, and the transverse magnetic (TM) polarization, where the $\bar{H}$-field has no $\hat{z}$-component. It suffices to consider these two polarizations separately.

In a medium where the relative permittivity and permeability vary with position, the scalar wave equations are

$$\nabla \cdot \left(\frac{1}{\mu_r} \nabla E_z^{\text{tot}} \right) + k^2 \epsilon_r E_z^{\text{tot}} = 0 \tag{1}$$

for the TM polarization and

$$\nabla \cdot \left(\frac{1}{\epsilon_r} \nabla H_z^{\text{tot}} \right) + k^2 \mu_r H_z^{\text{tot}} = 0 \tag{2}$$

for the TE case. These equations hold in the interior region Γ depicted in Fig. 1. To convert these equations into a form more amenable to direct numerical solution, multiply both by a "testing" function $T(x, y)$ and integrate the equations over the domain Γ. Vector identities may be employed to transfer differentiation to the testing function, following Reddy [20]. The result is the pair of equations

$$\iint_\Gamma \left\{ \frac{1}{\mu_r} \nabla T \cdot \nabla E_z^{\text{tot}} - k^2 \epsilon_r \, \text{T} E_z^{\text{tot}} \right\} dx\, dy = \int_{\partial\Gamma} T \frac{\partial E_z^{\text{tot}}}{\partial n} dl \tag{3}$$

$$\iint_\Gamma \left\{ \frac{1}{\epsilon_r} \nabla T \cdot \nabla H_z^{\text{tot}} - k^2 \mu_r \, \text{T} H_z^{\text{tot}} \right\} dx\, dy = \int_{\partial\Gamma} T \frac{\partial H_z^{\text{tot}}}{\partial n} dl. \tag{4}$$

We have assumed that all inhomogeneities are completely enclosed by the surface $\partial\Gamma$. Equations (3) and (4) are known as "weak" equations, since the order of differentiation is less than that of the original wave equations [20]. The surface integrals appearing on the right-hand side of these equations permit the convenient incorporation of boundary conditions on the surface $\partial\Gamma$.

III. Absorbing Boundary Conditions and the Finite Element Method

It is necessary to impose a boundary condition on $\partial\Gamma$ in order to extract a unique solution from the weak differential equations defined in (3) and (4). In the analytical treatment of an open-region scattering problem, the Sommerfeld radiation condition

$$\lim_{\rho \to \infty} \frac{\partial E_z^s}{\partial \rho} = -jkE_z^s \tag{5}$$

is enforced explicitly an infinite distance from the scatterer. For numerical analysis it is necessary to truncate the computational domain a finite distance from the scatterer, i.e., on the surface $\partial\Gamma$ in Fig. 1.

To derive an absorbing boundary condition that can be imposed in the near zone, note that an outward radiating field may be expressed in the asymptotic form

$$E_z^s(\rho, \phi) \approx \frac{e^{-jk\rho}}{\sqrt{\rho}} \left\{ E_0(\phi) + \frac{E_1(\phi)}{\rho} + \frac{E_2(\phi)}{\rho^2} + \frac{E_3(\phi)}{\rho^3} + \cdots \right\}. \tag{6}$$

Imposing the Sommerfeld radiation condition at finite ρ only forces the scattered field to agree with the first term in (6). Because of this, significant error will be introduced into the numerical solution if (5) is used directly in the near zone of a scatterer. Bayliss and Turkel have presented a higher order radiation condition [17]

$$\frac{\partial E_z^s}{\partial \rho} = \alpha(\rho) E_z^s + \beta(\rho) \frac{\partial^2 E_z^s}{\partial \phi^2} \tag{7}$$

where

$$\alpha(\rho) = \left\{ \frac{-jk - \dfrac{3}{2\rho} + \dfrac{j3}{8k\rho^2}}{1 - \dfrac{j}{k\rho}} \right\} \tag{8}$$

$$\beta(\rho) = \left\{ \frac{-\dfrac{j}{2k\rho^2}}{1 - \dfrac{j}{k\rho}} \right\}. \tag{9}$$

Equation (7) forces the scattered field to agree with the first four terms in (6), and can be imposed much closer to the scatterer than the Sommerfeld condition. Other absorbing boundary conditions have been developed and could be used as alternatives [16], [17], [21].

Consider TM-wave scattering from a cylinder containing dielectric and magnetic material. Assuming that the boundary $\partial\Gamma$ is circular, (7) can be substituted directly into the boundary integral in (3). Terms involving the incident field must be added to both sides of the resulting equation in order to preserve the total field as the primary unknown. Integration-by-parts is employed to eliminate the second-order derivative introduced by (7), yielding

$$\iint_\Gamma \left\{ \frac{1}{\mu_r} \nabla T \cdot \nabla E_z^{\text{tot}} - k^2 \epsilon_r \, \text{T} E_z^{\text{tot}} \right\} dx\, dy - \int_{\partial\Gamma} \left\{ \alpha \, \text{T} E_z^{\text{tot}} - \beta \frac{\partial T}{\partial \phi} \frac{\partial E_z^{\text{tot}}}{\partial \phi} \right\} \rho \, d\phi = \int_{\partial\Gamma} T \left\{ \frac{\partial E_z^{\text{inc}}}{\partial \rho} - \alpha E_z^{\text{inc}} - \beta \frac{\partial^2 E_z^{\text{inc}}}{\partial \phi^2} \right\} \rho \, d\phi. \tag{10}$$

For an arbitrary scatterer, analytical solutions to (10) are usually not available. Instead, we seek an approximate solution using the finite element method. Consider the division of the region Γ into triangular cells of constant permittivity and

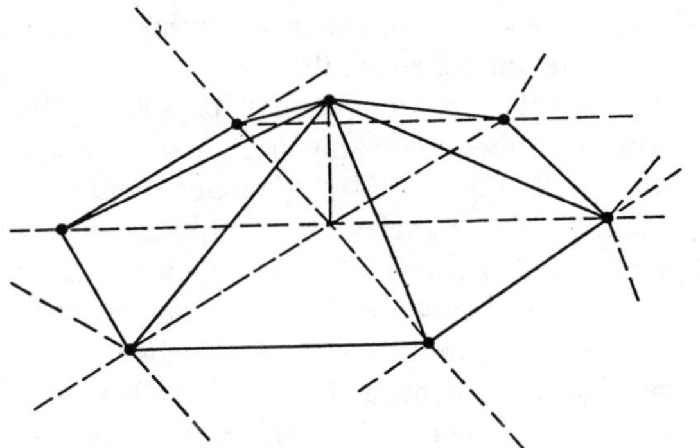

Fig. 2. Linear pyramid function employed with the first-order finite element method.

permeability. Throughout Γ, E_z^{tot} can be represented by linear interpolation functions. These function can be thought of as "pyramids" centered at each node of the mesh, such as depicted in Fig. 2. These interpolation functions vanish at all other nodes in the mesh. If the triangular-cell model of Γ contains N nodes, the representation of the field can be expressed

$$E_z^{\text{tot}}(x, y) \cong \sum_{n=1}^{N} e_n B_n(x, y) \tag{11}$$

where $B_n(x, y)$ denotes a "pyramid" centered at node n. If this expression is substituted into (10), N linearly independent equations can be generated using

$$T(x, y) = B_m(x, y), \qquad m = 1, 2, \cdots, N. \tag{12}$$

The resulting matrix equation has the form

$$\mathbf{Ae} = \mathbf{b} \tag{13}$$

where

$$A_{mn} = \iint_{\Gamma} \left\{ \frac{1}{\mu_r} \nabla B_m \cdot \nabla B_n - k^2 \epsilon_r B_m B_n \right\} dx\, dy - \int_{\partial\Gamma} \left\{ \alpha B_m B_n - \beta \frac{\partial B_m}{\partial \phi} \frac{\partial B_n}{\partial \phi} \right\} \rho\, d\phi \tag{14}$$

$$b_m = \int_{\partial\Gamma} B_m \left\{ \frac{\partial E_z^{\text{inc}}}{\partial \rho} - \alpha E_z^{\text{inc}} - \beta \frac{\partial^2 E_z^{\text{inc}}}{\partial \phi^2} \right\} \rho\, d\phi. \tag{15}$$

Because of the polynomial character of $B_n(x, y)$, the integrals appearing in (14) can be evaluated in closed form [20], [22].

Each row and column of the matrix represents a node in the model. Because of the linear expansion and testing functions employed, the only nonzero entries in a given column occur at rows corresponding to immediately adjacent nodes (i.e., those that share a common patch with the node number of that column). Thus, the matrix resulting from this discretization is sparse.

Higher order interpolation functions may be systematically employed as alternatives to the linear polynomials used above, without affecting the closed-form evaluation of the matrix elements [22]. However, the use of the higher order polynomials will increase the bandwidth of the sparse matrix and the complexity of the required scatterer model.

IV. Incorporation of Perfect Conducting Boundaries

The VDE formulation described above can easily treat perfectly conducting boundaries imbedded in the volume Γ of Fig. 1. If such a boundary is denoted $\partial\Gamma_c$ in order to distinguish it from the exterior surface $\partial\Gamma$, (3) can be modified to the form

$$\iint_{\Gamma} \left\{ \frac{1}{\mu_r} \nabla T \cdot \nabla E_z^{\text{tot}} - k^2 \epsilon_r \text{T} E_z^{\text{tot}} \right\} dx\, dy = \int_{\partial\Gamma} T \frac{\partial E_z^{\text{tot}}}{\partial n} dl + \int_{\partial\Gamma_c} \frac{1}{\mu_r} T \frac{\partial E_z^{\text{tot}}}{\partial n} dl \tag{16}$$

where Γ denotes the region between the two surfaces $\partial\Gamma$ and $\partial\Gamma_c$ and n denotes the surface normal direction out of Γ. The TM fields must satisfy the Dirichlet boundary condition

$$E_z^{\text{tot}} = 0 \qquad \text{on } \partial\Gamma_c. \tag{17}$$

Because the values of E_z^{tot} are prescribed on the surface of $\partial\Gamma_c$, all the necessary testing functions vanish on $\partial\Gamma_c$. The integral over $\partial\Gamma_c$ in (16) contributes nothing to the finite element system. Therefore, nodes on $\partial\Gamma_c$ (and in the conducting region) are excluded when constructing the matrix equation. Aside from this exclusion, the matrix elements given in (14) do not change.

For the TE polarization, the magnetic field at a perfect conducting surface $\partial\Gamma_c$ must satisfy the Neumann boundary condition

$$\frac{\partial H_z^{\text{tot}}}{\partial n} = 0 \qquad \text{on } \partial\Gamma_c. \tag{18}$$

Equation (4) can be modified to include the presence of the additional surface, yielding an equation for H_z^{tot} similar to (16). In this case, the surface integral over $\partial\Gamma_c$ vanishes because of (18). Values of H_z^{tot} on the surface $\partial\Gamma_c$ must be treated as unknowns, and mesh points on the surface $\partial\Gamma_c$ must be included when constructing the matrix equation. Aside from the fact that the mesh points within the conducting region are absent, the expression for the matrix elements remains unaltered from the TE equivalent of (14).

V. Scattering Cross-Section Calculation

Once the internal fields have been found from the solution of (13), the bistatic scattering cross section may be computed. In two-dimensional problems, the scattering cross section is also known as the *echo width* and is defined for the TM polarization as

$$\sigma_{\text{TM}}(\phi) = \lim_{\rho \to \infty} 2\pi\rho |E_z^s(\rho, \phi)|^2 \tag{19}$$

assuming that the field impinging on the scatterer has unity magnitude E_z^{inc}.

Although several alternative procedures for computing σ_{TM}

are possible, a simple approach employs an eigenfunction expansion of the boundary fields. In the TM case, the fields external to $\partial\Gamma$ have the form [23]

$$E_z^{\text{tot}}(\rho, \phi) = \sum_{n=-\infty}^{\infty} j^{-n}[J_n(k\rho)e^{-jn\theta} + \alpha_n H_n^{(2)}(k\rho)]e^{jn\phi} \quad (20)$$

where θ defines the polar angle into which the incident plane wave propagates. The coefficients can be found from the decomposition of the boundary E_z-field according to

$$\alpha_n = \frac{j^n \frac{1}{2\pi}\int_0^{2\pi} E_z^{\text{tot}}(a, \phi)e^{-jn\phi}\, d\phi - J_n(ka)e^{-jn\theta}}{H_n^{(2)}(ka)} \quad (21)$$

where a is the radius of the outer boundary. The scattering cross section is given by

$$\sigma_{\text{TM}}(\phi) = \frac{4}{k}\left|\sum_{n=-\infty}^{\infty} \alpha_n e^{jn\phi}\right|^2. \quad (22)$$

Similar expressions involving the H_z-field can be used to find the scattering cross section for the TE polarization.

VI. Computational Trade-Off Between VIE and VDE Approaches

A recent study has concluded that volume integral and differential equation formulations can produce similar accuracy for a given discretization level [24]. Therefore, the trade-off between VIE and VDE approaches depends primarily on two factors: the size of the region Γ that must be discretized in each case and the degree to which the sparsity arising within the VDE approach is exploited by the matrix solution algorithm. Since VDE formulations require the discretization of a region outside the scatterer itself, they can produce a much larger order system than that required by an integral equation formulation. Assuming that the VDE approach employs the approximate local boundary condition from (7), it may be necessary to locate the boundary $\partial\Gamma$ some distance from the scatterer (in order that the scattered field can be accurately expressed in terms of only the first four terms of (6)). If this condition is not satisfied at a boundary located a reasonable distance from the scatterer, VDE schemes incorporating global boundary conditions [9]–[15] will be more efficient. (Global boundary conditions can be imposed as close to the scatterer as desired, but decrease the sparsity of the finite element matrix.) In general, it appears that VIE approaches will be more efficient only for problems where the scatterer occupies a relatively small fraction of the region within a circumscribing circle. Otherwise, the additional unknowns introduced in a VDE formulation will not overwhelm the savings due to the sparse matrix. A more detailed discussion of these issues may be found in [24].

The relative efficiency of VDE formulations is a function of the extent to which the sparsity within the finite element system is exploited by the matrix solution algorithm. Direct methods for sparse systems include variations on Gaussian elimination incorporating banded or envelope storage schemes [25]. Although banded elimination algorithms require more storage than envelope schemes, they permit the incorporation of pivoting into the solver and may be more robust. An alternate direct method involving out-of-core storage is the frontal solution algorithm [26]. Although slower on conventional computers, the frontal approach appears attractive for certain parallel architectures. The efficiency of each of these direct methods can be enhanced by the use of row and column reordering algorithms to minimize the bandwidth, envelope, or frontwidth of the systems and reduce the fill-in that occurs during Gaussian elimination [25]. Although the system arising from this VDE approach involves a complex non-Hermitian matrix, several iterative methods are also available [27]. Iteration permits all of the sparsity to be exploited, but will likely be less efficient for problems involving multiple incident fields. Numerical results presented below were obtained using direct methods either employing envelope storage or a frontal algorithm.

VII. Numerical Results

The method described above has been applied to a variety of scatterer geometries, involving electrically small cylinders as well as cylinders whose cross sections are several wavelengths in maximum linear dimension. A variety of results are presented below to illustrate the VDE formulation. Provided that the Bayliss–Turkel condition is applied at a sufficient distance from the scatterer, the accuracy appears comparable to that of integral equation formulations for a given mesh density [24]. The cylinder models were created using an automated mesh generator, and contain a minimum density of 100 cells per square wavelength of cross-sectional area. This density appears appropriate for reasonable accuracy in the near fields and scattering cross section. Although we present no results for a continuously varying permittivity or permeability profile, the method could treat those problems using volume-averaged parameters in each cell to approximate the continuous profile with a piecewise-constant profile.

Consider a perfectly conducting circular cylinder of radius 0.25 λ (λ denotes wavelength) coated uniformly with a dielectric cladding of outer radius 0.30 λ and relative permittivity $\epsilon_r = 4$. Fig. 3(a) shows the electric field at radius 0.275 λ in response to a TM incident plane wave for several locations of the boundary $\partial\Gamma$. The solution is compared to the exact eigenfunction result. The accuracy improves as the boundary $\partial\Gamma$ on which the Bayliss–Turkel condition is applied is moved to greater ρ. For the case where the outer boundary is located at $\rho = 0.6\ \lambda$, the cylinder model required 304 nodes and 576 triangular cells. The array required to store the lower envelope of the sparse matrix was of size 4301. Fig. 3(b) shows the current induced on the conductor for the scatterer of Fig. 3(a) by a TE plane wave, with $\partial\Gamma$ located at $\rho = 0.37\ \lambda$. The cylinder model contained 142 nodes and 226 patches. The two-dimensional scattering cross section for the TE polarization is shown in Table I, compared to the exact eigenfunction result.

Fig. 4(a) shows a perfectly conducting wedge-shaped cylinder illuminated by a TM plane wave. For the boundary $\partial\Gamma$ located at $\rho = 2.0\ \lambda$, a plot of the bistatic scattering cross

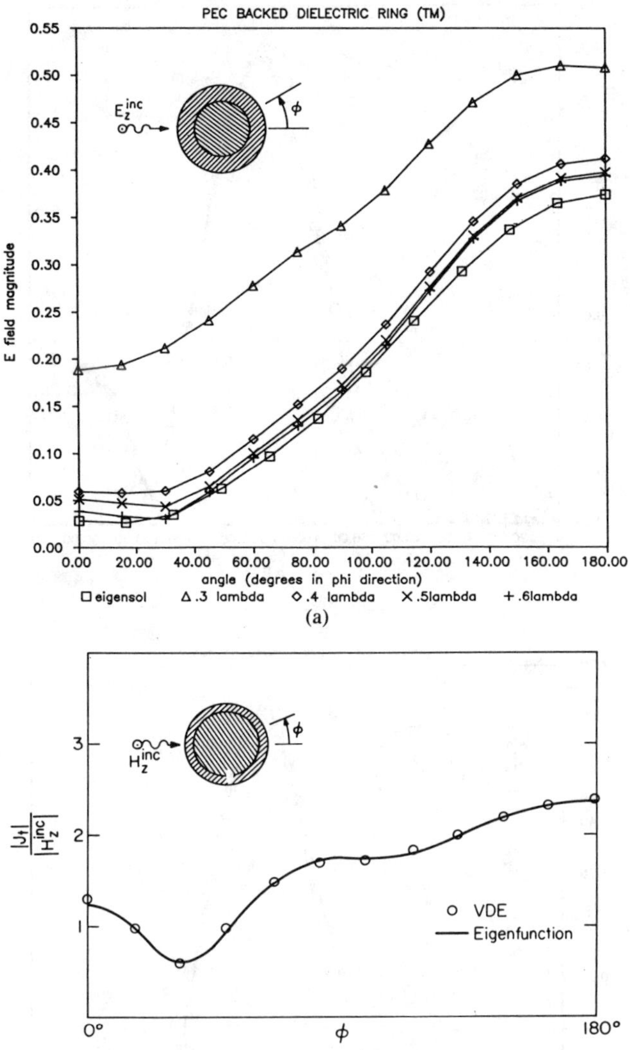

Fig. 3. (a) VDE solution for E_z^{tot} at $\rho = 0.275\ \lambda$ in a dielectric shell surrounding a circular perfectly conducting cylinder, as a function of the location of the outer boundary $\partial\Gamma$ (TM polarization). The conducting cylinder has radius 0.25 λ; the dielectric cladding has outer radius 0.3 λ. The dielectric cladding has $\epsilon_r = 4$ and $\mu_r = 1$. The exact eigenfunction solution is shown for comparison. (b) VDE solution for the surface current density induced by a plane-wave TE excitation on the dielectric-coated conducting cylinder of (a). The exact eigenfunction solution is shown for comparison.

TABLE I
BISTATIC SCATTERING CROSS SECTION OF THE DIELECTRIC-COATED CONDUCTING CYLINDER OF FIG. 3(a), WHEN ILLUMINATED BY A TE PLANE WAVE

ϕ	σ_{exact}	σ_{VDE}
0°	4.23	4.19
30	0.03	−0.11
60	−1.92	−1.44
90	1.03	1.10
120	−1.86	−2.09
150	−1.87	−1.77
180	−0.11	0.01

Conducting cylinder has radius 0.25 λ, and outer radius of dielectric cladding is 0.30 λ. Relative permittivity of cladding is $\epsilon_r = 4$. Exact solution is compared to the VDE result. Scattering cross section is presented in units of dB λ.

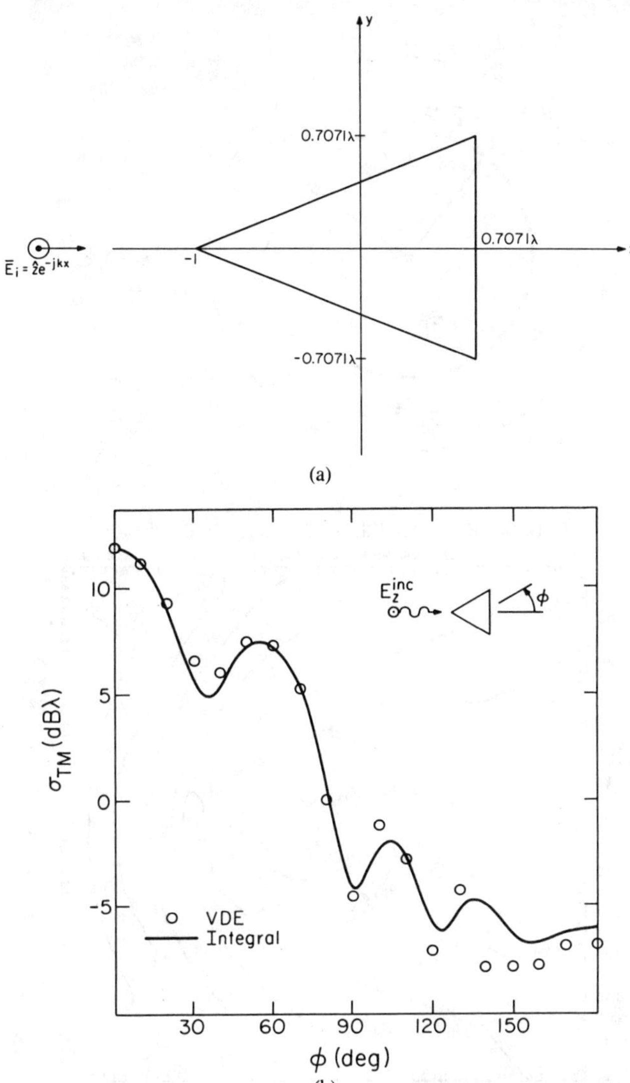

Fig. 4. (a) Geometry of a perfectly conducting triangular cylinder illuminated by a TM plane wave. (b) Bistatic scattering cross section of the cylinder shown in (a), compared with the result from an integral equation formulation. The VDE solution is obtained with the outer boundary $\partial\Gamma$ located at $\rho = 2.0\ \lambda$.

section σ_{TM} is presented in Fig. 4(b). A numerical solution of the electric-field integral equation using 156 pulse basis functions on the surface of the wedge is shown for comparison. The VDE result agrees reasonably well with the integral equation result. The cylinder model used with the VDE formulation consisted of 1599 nodes and 2952 cells.

Fig. 5(a) shows a perfect conducting circular cylinder of radius $a = 0.88\ \lambda$ containing a 90° slit. A contour plot of the total fields produced in and around the cylinder by an incident TM plane wave is presented in Fig. 5(b), based on a model containing 1585 nodes and 3052 elements. The outer boundary $\partial\Gamma$ is located at $\rho = 2.2\ \lambda$. The identical geometry has been analyzed by Ziolkowski and Grant [28]. Although certain features of our VDE solution differ somewhat from [28, fig. 8(a)], the peak magnitude of the electric field in the quasi-resonant cylinder agrees with their result. The lower envelope of the finite element matrix for this example was of size 51929, indicating that only about 4 percent of the storage associated with a full matrix of the same order was necessary to store the sparse system.

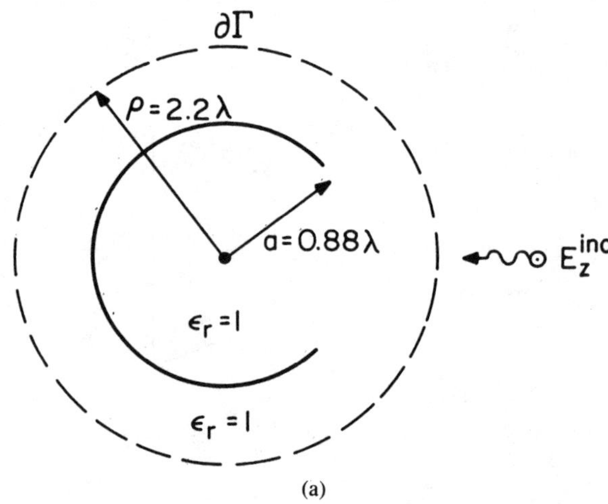

(a)

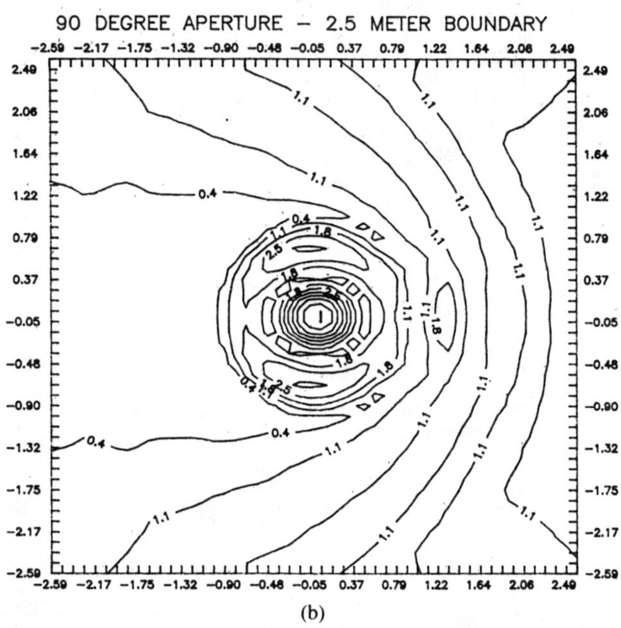

(b)

Fig. 5. (a) Geometry of a perfectly conducting cylinder containing a 90° slit. (b) A contour plot of the E_z-field in and around the cylinder of (a) produced by an incident plane wave (TM polarization).

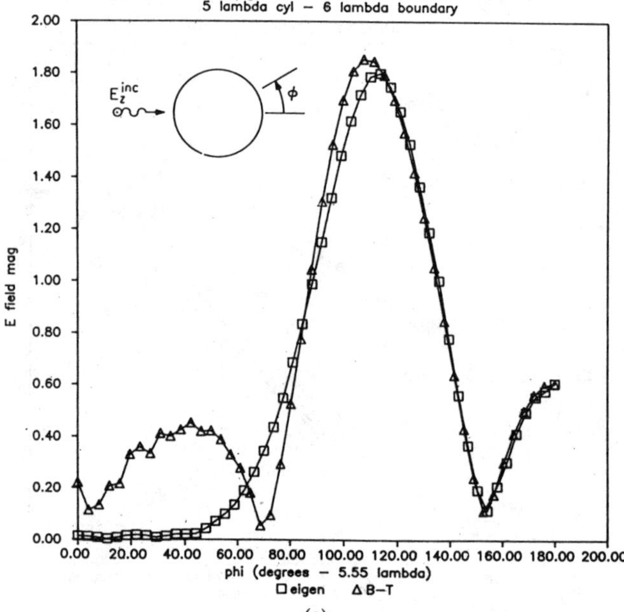

(a)

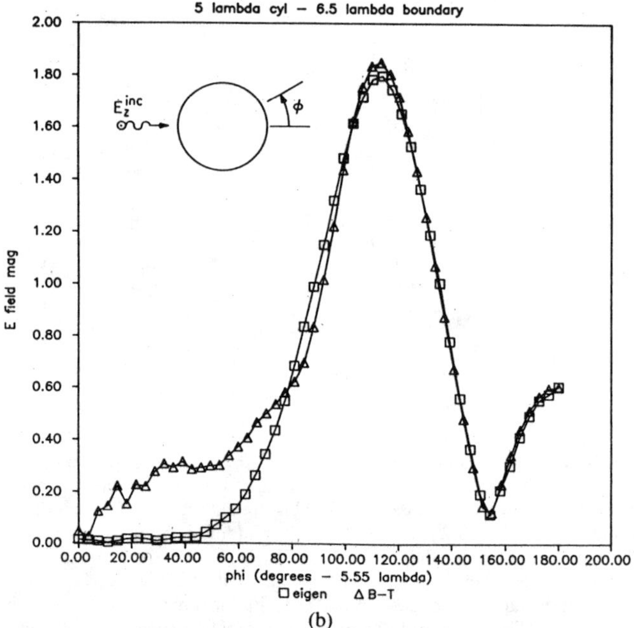

(b)

Fig. 6. (a) E_z^{tot} produced at $\rho = 5.55\ \lambda$ surrounding a perfectly conducting cylinder of radius 5.0 λ illuminated by a TM plane wave. The Bayliss–Turkel boundary condition of (7) is applied at $\rho = 6.0\ \lambda$. The exact eigenfunction solution is shown for comparison. (b) Same as (a), except with the absorbing boundary conditions applied at $\rho = 6.5\ \lambda$.

The data presented above illustrate that excellent accuracy can be obtained using the differential equation formulation. However, our results indicate that the validity of the Bayliss–Turkel absorbing boundary condition may be dependent on scatterer size. Figs. 6(a) and 6(b) show the E_z-field produced on a circle of radius 5 λ by a TM incident plane wave. Results are shown for the VDE formulation and the exact eigenfunction solution. The cylinder model for the case where ∂Γ is located at $\rho = 6.5\ \lambda$ contained 5712 nodes and 10680 cells. Observe the error in the VDE result for the fields located in the shadow region of the cylinder. As the scatterer size increases, it appears that the surface ∂Γ must be located a greater distance from the cylinder surface in order for the Bayliss–Turkel condition to be accurate. We have also studied these examples using the comparable absorbing boundary condition of Engquist and Majda [16], and the results appear virtually identical to those obtained using the Bayliss–Turkel condition.

VIII. Conclusion

A general numerical technique has been described for two-dimensional scattering cross-section calculations. The procedure involves the finite element solution of differential equations, and is well suited for the treatment of highly inhomogeneous and complex scatterers. Note that the approach is based on the incorporation of an absorbing boundary condition directly in the near zone of the scatterer, and therefore is dependent upon the accuracy of the radiation condition. For the examples presented, which involve electrically small or moderately sized scatterers, the method appears accurate. However, there are indications that the validity of the absorbing boundary condition may degrade for larger cylinders, forcing the surface ∂Γ to be located a relatively greater distance from the scatterer. If such is the case, alternative VDE methods incorporating a global boundary condition [9]–[15] may be more efficient for treating electrically large scatterers.

References

[1] J. H. Richmond, "Scattering from a dielectric cylinder of arbitrary cross section shape," *IEEE Trans. Antennas Propagat.*, vol. AP-13, pp. 334–341, May 1965.

[2] ——, "TE-wave scattering from a dielectric cylinder of arbitrary cross section shape," *IEEE Trans. Antennas Propagat.*, vol. AP-14, pp. 460–464, July 1966.

[3] D. E. Livesay and K. M. Chen, "Electromagnetic fields induced inside arbitrarily shaped biological bodies," *IEEE Trans. Microwave Theory Tech.*, vol. MTT-22, pp. 1273–1280, Dec. 1974.

[4] D. H. Schaubert, D. R. Wilton, and A. W. Glisson, "A tetrahedral modeling method for electromagnetic scattering by arbitrarily shaped inhomogeneous dielectric bodies," *IEEE Trans. Antennas Propagat.*, vol. AP-32, pp. 77–85, Jan. 1984.

[5] C. T. Tsai, H. Massoudi, C. H. Durney, and M. F. Iskander, "A procedure for calculating fields inside arbitrarily shaped, inhomogeneous dielectric bodies using linear basis functions with the moment method," *IEEE Trans. Microwave Theory Tech.*, vol. MTT-34, pp. 1131–1139, Nov. 1986.

[6] A. F. Peterson and P. W. Klock, "An improved MFIE formulation for TE wave scattering from lossy, inhomogeneous dielectric cylinders *IEEE Trans. Antennas Propagat.*, vol. 36, pp. 45–49, Jan. 1988.

[7] B. H. McDonald and A. Wexler, "Finite-element solution of unbounded field problems," *IEEE Trans. Microwave Theory Tech.*, vol. MTT-20, pp. 841–847, Dec. 1972.

[8] Z. J. Csendes, "A note on the finite-element solution of exterior-field problems," *IEEE Trans. Microwave Theory Tech.*, vol. MTT-24, pp. 468–473, July 1976.

[9] S. P. Marin, "Computing scattering amplitudes for arbitrary cylinders under incident plane waves," *IEEE Trans. Antennas Propagat.*, vol. AP-30, pp. 1045–1049, Nov. 1982.

[10] Z. Gong, "A hybrid equation approach for the solution of electromagnetic scattering problems involving arbitrarily shaped inhomogeneous dielectric bodies," Ph.D. dissertation, Univ. Mississippi, University, MS, 1985.

[11] M. A. Morgan and B. E. Welch, "The field feedback formulation for electromagnetic scattering computations," *IEEE Trans. Antennas Propagat.*, vol. AP-34, pp. 1377–1382, Dec. 1986.

[12] K. K. Mei, "Unimoment method for solving antenna and scattering problems," *IEEE Trans. Antennas Propagat.*, vol. AP-22, pp. 760–766, Nov. 1974.

[13] S. K. Chang and K. K. Mei, "Application of the unimoment method to electromagnetic scattering by dielectric cylinders," *IEEE Trans. Antennas Propagat.*, vol. AP-24, pp. 35–42, Jan. 1976.

[14] R. B. Wu and C. H. Chen, "Variational reaction formulation of scattering problem for anisotropic dielectric cylinders," *IEEE Trans. Antennas Propagat.*, vol. AP-34, pp. 640–645, May 1986.

[15] K. K. Mei, "Unimoment method for electromagnetic wave scattering," *J. Electromagn. Waves Appl.*, vol. 1, pp. 201–222, 1987.

[16] B. Engquist and A. Majda, "Absorbing boundary conditions for the numerical simulation of waves," *Math. Comp.*, vol. 31, pp. 629–651, 1977.

[17] A. Bayliss and E. Turkel, "Radiation boundary conditions for wave-like equations," *Comm. Pure Appl. Math.*, vol. 33, pp. 707–725, 1980.

[18] R. T. Ling, "Numerical solution for the scattering of sound waves by a circular cylinder," *AIAA J.*, vol. 25, pp. 560–566, Apr. 1987.

[19] G. Meltz, B. J. McCartin, and L. J. Bahrmasel, "Application of the control region approximation to electromagnetic scattering," *Abstracts 1987 URSI Radio Sci. Meet.*, Blacksburg, VA, June 1987, p. 185.

[20] J. N. Reddy, *An Introduction to the Finite Element Method.* New York: McGraw-Hill, 1984.

[21] A. F. Peterson, "Absorbing boundary conditions for the vector wave equation," *Microwave Opt. Technol. Lett.*, vol. 1, pp. 62–64, Apr. 1988.

[22] P. P. Silvester and R. L. Ferrari, *Finite Elements for Electrical Engineers.* Cambridge, U.K.: Cambridge Univ. Press, 1983.

[23] R. F. Harrington, *Time-Harmonic Electromagnetic Fields.* New York: McGraw-Hill, 1961.

[24] A. F. Peterson, "A comparison of integral, differential, and hybrid methods for TE-wave scattering from inhomogeneous dielectric cylinders," *J. Electromagn. Waves Appl.*, vol. 3, 1989 (to appear).

[25] A. George and J. W. Liu, *Computer Solution of Large Sparse Positive Definite Systems.* Englewood Cliffs, NJ: Prentice-Hall, 1981.

[26] B. M. Irons, "A frontal solution program for finite element analysis," *Int. J. Num. Methods Eng.*, vol. 2, pp. 5–32, 1970.

[27] A. F. Peterson and R. Mittra, "Iterative-based computational methods for electromagnetic scattering from individual or periodic structures," *IEEE J. Oceanic Engineering*, vol. OE-12, pp. 458–465, Apr. 1987.

[28] R. W. Ziolkowski and J. B. Grant, "Scattering from cavity-backed apertures: The generalized dual series solution of the concentrically loaded *E*-pol slit cylinder problem," *IEEE Trans. Antennas Propagat.*, vol. AP-35, pp. 504–528, May 1987.

Finite-Element Analysis of Dielectric Scatterers Using the Absorbing Boundary Condition

O. M. RAMAHI and R. MITTRA
ELECTROMAGNETIC COMMUNICATION LABORATORY
UNIVERSITY OF ILLINOIS, URBANA, IL 61801

Abstract

When solving open region scattering problems using the Finite Element Method (FEM), the infinite region exterior to the scatterer must be truncated by an artificial outer boundary in order to permit practical implementation on a digital computer. In this paper, it is shown how an absorbing boundary condition (ABC) can be enforced at this boundary to truncate the mesh at distances close to the surface of the scatterer, such that the ABC simulates not only the outward propagating nature of the scattered field but insures the well-posedness of the problem as well. It was found that the if the finite element problem is formulated using the scattered field as the primary unknown, the solution agrees well with the method of moments solution. On the other hand, the solution is found to have a significant errors if the problem is formulated using the total field.

Introduction

Open region scattering problems are frequently modeled using an integral equation formulation [1], and solved using the method of moments that leads to a full matrix system. For an inhomogeneous dielectric scatterer, the integral equation must be descretized over the entire volume of the scatterer, and, depending on the size of the scatterer, this might prove to be a costly operation in terms of storage and solution time. Alternatively, the dielectric scattering problem can be modeled using a partial differential equations technique such as the finite elements method (FEM) or the finite difference (FD) technique. Although theses techniques give rise to matrix systems that are considerably larger than those generated using the integral equation formulation, the FEM or FD matrices are highly sparse, and can thus be inverted efficiently using special software routines [2] that require the storage of only the non-zero elements of the matrix.

When using the PDE methods, the region surrounding the scatterer is truncated by using an artificial outer boundary in order to limit the number of unknowns. Next, an absorbing boundary condition (ABC) is imposed on the outer boundary so as to minimize the non-physical reflections from the boundary. In this work we used the type of asymptotic differential boundary operators which were reported in [3] and [4]. Mittra and Ramahi [4,5] have found that these differential operators, which are of the asymptotic type, work quite well for small scatterers, but are not as accurate for electrically large scatterers. In this study, only small dielectric scatterers are considered to demonstrate the application of the FEM formulation to the dielectric scattering problem.

For the problem of scattering by inhomogeneous dielectric bodies, two different approaches can be used to formulate the problem using partial differential equations techniques. In the first approach, the differential form of the wave equation is transformed into a variational form using the scattered field as the primary unknown, whereas this role is played by the total field in the second formulation. However, in both of these approaches, the absorbing boundary condition is applied only on the scattered field. Over each finite element, the variation in the material properties of the medium is represented in terms of piece-wise linear basis functions in the simplex coordinate system. The same basis function expansion is used to represent both the dielectric variation and the unknown using the two different approaches as applied to different dielectric scatterers. It is shown, that for a class of problems investigated in the paper, the scattered field formulation is more accurate than the corresponding total field approach. The error in the total field approach solution is pronounced in the forward scattering region.

Formulation

In this discussion we formulate the problem of an inhomogeneous dielectric scatterer by casting it into a form suitable for finite element implementation. Most of the details of the finite element formulation will be omitted, however, and the reader will be referred to the works cited in References [6-8] for further elaboration.

I. Formulation of the scattered field approach

Consider the arbitrary shape dielectric scatterer shown in Fig. 1. For a TM polarized incident plane wave, the scattered field, u^s, satisfies the wave equation

$$\nabla^2 u^s + k^2 u^s = F \tag{1}$$

where F is defined as

$$F = -(\nabla^2 u^i + k^2 u^i) \tag{2}$$

and where u^i is the incident field. For a plane wave incidence, u^i is given by

$$u^i = e^{-jk_0 x} \tag{3}$$

Substituting (3) in (2), we obtain

$$F = e^{-jk_0 x} [k_0^2 (1 - \varepsilon_r(x,y))] \tag{4}$$

where $\varepsilon_r(x,y)$ is the relative dielectric constant of the material. Now multiplying (1) by a testing function,v, and integrating over the domain Ω, we have,

$$\int_\Omega v(\nabla^2 u^s + k^2 u^s)\, ds = \int_\Omega v F \, dl \tag{5}$$

Using Green's identity, we arrive at the weak form of the wave equation

$$\int_\Omega (\nabla u^s \nabla v - k^2 u^s v)\, ds = \int_\Gamma v \frac{\partial u^s}{\partial n} dl + \int_\Omega v F \tag{6}$$

Reprinted from *IEEE Trans. Magn.*, vol. 25, no. 4, pp. 3043–3045, July 1989.

The excitation function, F, is expressed as a piecewise continuous linear function over each finite element, and is represented in terms of the same basis functions as used to represent the unknown scattered field, u^s. The derivation of the absorbing boundary condition appears in [3] and [4] and is repeated below for convenience. The boundary condition reads

$$\frac{\partial u^s}{\partial \rho} = \alpha(\rho)\, u^s + \beta(\rho)\, u^s_{\phi\phi} \qquad (7)$$

where $u^s_{\phi\phi}$ is the second order angular derivative of the scattered field, and α and β are give by

$$\alpha(\rho) = -jk - \frac{1}{2\rho} - \frac{j}{8k\rho^2} + \frac{1}{8k^2\rho^2} \qquad (8)$$

$$\beta(\rho) = -\frac{j}{2k\rho^2} + \frac{1}{2k^2\rho^3} \qquad (9)$$

Finally, substituting (7) in (6), and integrating by parts, we arrive at a form suitable for numerical implementation

$$\int_\Omega (\nabla u^s \nabla v - k^2 u^s v)\, ds = \int_\Gamma \alpha\, v\, u^s - \beta\, \rho^2\, v_l\, u^s_l \, dl + \int_\Omega v F \qquad (10)$$

where u^s_l and v_l are the tangential derivative along the contour Γ.

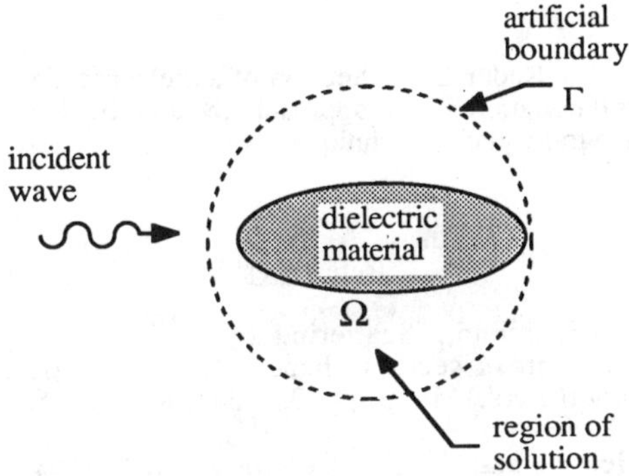

Fig. 1. Geometry for the partial differential equation method to the dielectric scatterer problem.

II. Formulation of the total field approach

The total field, u^t satisfies the Helmholz equation

$$\nabla^2 u^t + k^2 u^t = 0 \qquad (11)$$

Using a procedure similar to the one described in the last section, we arrive at the variational equation

$$\int_\Omega (\nabla u^t \nabla v - k^2 u^t v)\, ds = \int_\Gamma v \frac{\partial u^t}{\partial n}\, dl \qquad (12)$$

Expressing the total field in terms of the scattered and incident field, we have

$$\frac{\partial u^t}{\partial n} = \frac{\partial u^s}{\partial n} + \frac{\partial u^i}{\partial n} \qquad (13)$$

Using the absorbing boundary condition on the scattered field, we obtain

$$\frac{\partial u^s}{\partial \rho} = \alpha\, u^s + \beta\, u^s_{\phi\phi} = (\alpha\, u^t + \beta\, u^t_{\phi\phi}) - (\alpha\, u^i + \beta\, u^i_{\phi\phi}) \qquad (14)$$

Finally, substituting (14) in the contour integral of the variational equation, we have

$$\int_\Gamma v \frac{\partial u^t}{\partial \rho}\, dl = \int_\Gamma (v\, \alpha\, u^t + \beta\, \rho\, v_l\, u^t_l + v F)\, dl \qquad (15)$$

Where F is now given by

$$F = \frac{\partial u^i}{\partial \rho} - (\alpha\, u^i + \beta\, u^i_{\phi\phi}) \qquad (16)$$

For a plane wave incidence, F becomes

$$F = -e^{-jk_0\rho\cos\phi}\{jk_0\cos\phi + \alpha + \beta\,[(jk_0\rho\sin\phi)^2 + jk_0\rho\cos\phi]\} \qquad (17)$$

In the next section, we present some numerical results for three different dielectric scatterers, using the above two formulation procedures.

Numerical Results

To test the validity and accuracy of the two approaches discussed above, we considered the following three dielectric structures: a plane homogeneous dielectric slab, a circular cylindrical shell, and a semicircular cylindrical shell. All structures had a relative dielectric constant of 4. In Fig. 2, we show, as a representative example, how the solution region for the semicircular shell was subdivided into meshes for finite element implementation.

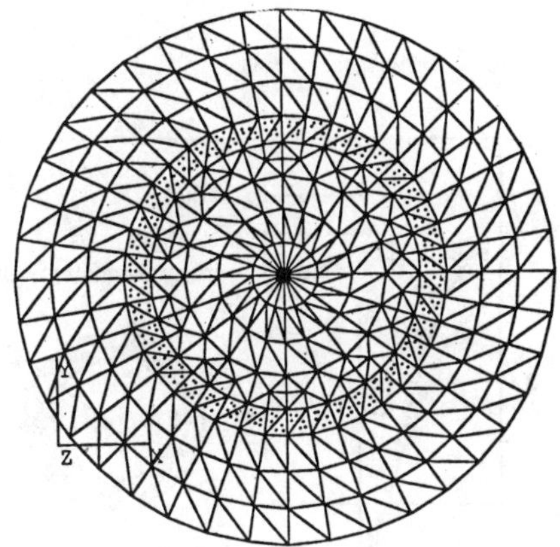

Fig. 2. Subdivision of the solution region into triangles for the circular shell.

Numerical results for the bistatic Radar Cross Section (RCS), computed by using the two approaches discussed in this paper, are given in Figures (3-5). These figures also exhibit the integral equation/method of moments solution (exact solution in the case of the circular shell) obtained by Richmond [1]. It is clear from these results that the scattered field approach gives more accurate solution than the total field formulation, especially in the forward scattering region. Further investigation that might help explain the cause of the difficulty with the total field formulation is currently underway.

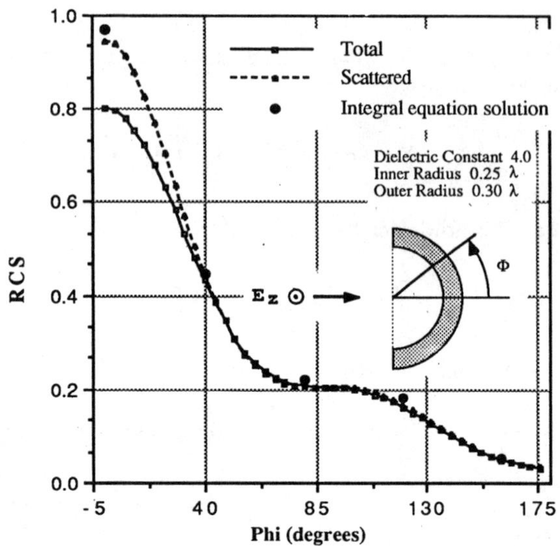

Fig. 3. Bistatic Radar Cross Section of a semicircular dielectric cylindrical shell calculated using the scattered field approach, the total field approach, and the integral equation technique.

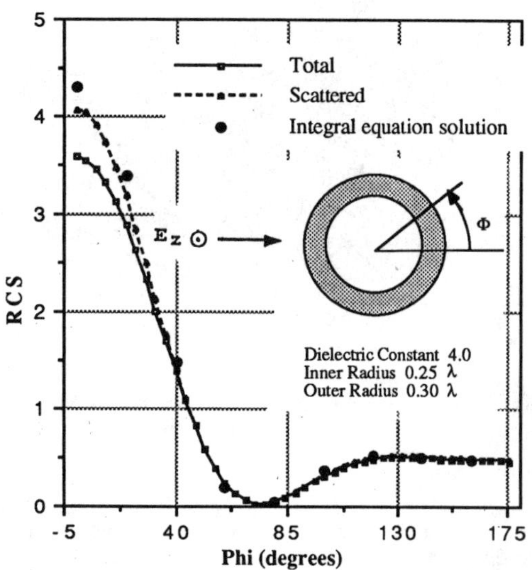

Fig. 4. Bistatic Radar Cross Section of a circular dielectric cylindrical shell calculated using the scattered field approach, the total field approach, and the integral equation technique.

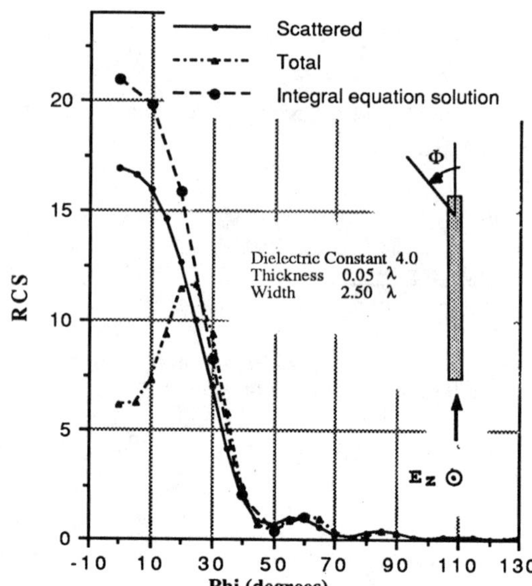

Fig. 5. Bistatic Radar Cross Section of a dielectric slab calculated using the scattered field approach, the total field approach, and the integral equation technique

References

[1] J. H. Richmond, "Scattering by a dielectric cylinder of arbitrary cross section shape," ***IEEE Trans. Antennas Propagat.***, vol. AP-13, pp. 334-341, May 1965.

[2] A. George and J. Liu, ***Computer Solutions of Large Sparse Positive definite Systems.*** Englewood Cliffs, N.J. : Prentice-Hall, Inc., 1981.

[3] A. Bayliss, M. Gunzburger and E. Turkel, "Boundary conditions for the numerical solution of elliptic equations in exterior regions," ***SIAM J. Appl. Math.***, vol. 42, pp. 430-451, April 1982.

[4] R. Mittra and O. M. Ramahi, "Absorbing Boundary Conditions for the Direct Solution of Partial Differential Equations Arising in Electromagnetic Scattering Problems," in ***Differential Methods in Electromagnetic Scattering,*** M. Morgan, ed., Vol II, New York, N. Y.: Elsevier Science, Inc., (to appear.)

[5] O. M. Ramahi and R. Mittra, "A systematic Approach to improving the absorbing boundary condition solution," URSI Radio Meeting Program and Abstracts, p.182, Syracuse, New York, June 1988.

[6] P.P. Silvester and R.L. Ferrari, ***Finite Elements for Electrical Engineers.*** Cambridge, U.K.: Cambridge University Press, 1983.

[7] G. Strang and G. Fix, ***An Analysis of the Finite Element Method.*** Englewood Cliffs, N.J. : Prentice-Hall, Inc., 1973.

[8] J. N. Reddy, ***An Introduction to Finite-Element Method.*** New York, N.Y. : McGraw-Hill, 1984

Absorbing Boundary Conditions for the Finite Element Solution of the Vector Wave Equation

J. P. WEBB and V. N. KANELLOPOULOS

KEY TERMS

Finite element method, electromagnetic scattering, absorbing boundary condition

ABSTRACT

Time-harmonic electromagnetic field problems in unbounded space may be solved by the finite element method applied inside a sphere which contains the region of interest, providing a suitable boundary condition is imposed on the surface of the sphere. The boundary condition must have the effect of absorbing out-going waves. A new second-order absorbing boundary condition (ABC) is proposed, similar to that introduced by Peterson [1], but capable of being incorporated in a variational principle and consequently leading to symmetric finite-element matrices.

1. INTRODUCTION

A time-harmonic electric or magnetic field is governed by the vector wave equation. The finite element method has been used to solve the vector wave equation numerically in a number of applications [2, 3]; it is especially useful when complicated dielectric materials are present. For unbounded problems, in which there are waves propagating outward to infinity, the finite element method can only be used once the infinite domain has been made finite. Various techniques have been proposed to do this. Among the most promising is the idea of imposing a special boundary condition on the surface of a sphere surrounding the region of interest (Figure 1). Such a boundary condition would ideally absorb the energy incident on the surface from within. Because absorbing boundary conditions do not affect the sparsity of the finite element matrices, their use does not significantly increase the cost of the finite element analysis.

Recently, Peterson [1] proposed first- and second-order absorbing boundary conditions of the Bayliss–Gunzburger–Turkel kind [4] for the vector wave equation. In this letter, we provide a different derivation leading to a more general result. The Peterson expressions can be recovered by suitable selection of an arbitrary parameter; on the other hand, another specific choice of the parameter gives a different second-order boundary condition which has the advantage of leading to symmetric finite element matrices.

J. P. Webb and V. N. Kanellopoulos
Computational Analysis and Design Laboratory
Department of Electrical Engineering
McGill University
3480 University Street
Montreal, H3A 2A7, Canada

2. ABSORBING BOUNDARY CONDITIONS

In the presence of inhomogeneous, isotropic, linear materials, the magnetic field **H** is governed by the vector wave equation:

$$\nabla \times \frac{1}{\epsilon_r} \nabla \times \mathbf{H} - k^2 \mu_r \mathbf{H} = 0 \tag{1}$$

where ϵ_r is the relative permittivity, μ_r is the relative permeability, and k is the wavenumber. The implicit time-dependence is e^{jkct}, where c is the velocity of light in vacuum. We shall consider cases in which sources may be represented by an inhomogeneous boundary condition on a finite surface S_d (Figure 1):

$$\mathbf{H} \times \mathbf{n} = \mathbf{H}_0 \times \mathbf{n} \quad \text{on } S_d \tag{2}$$

$\mathbf{H}_0$ is a known field and **n** is a unit normal to the surface. (In other cases, the source is a wave incident from infinity; the equivalence principle can be applied to transform these into problems where the sources are local, and where **H** represents just the scattered part of the field.)

A function satisfying (1) and representing outward-propagating waves at infinity is known as a vector radiation function. Wilcox [5] showed that, in the free space surrounding the materials and sources, a vector radiation function can be written as

$$\mathbf{H}(r, \theta, \varphi) = g \sum_{n=0}^{\infty} \frac{\mathbf{H}_n(\theta, \varphi)}{r^{n+1}} \tag{3}$$

where (r, θ, φ) are spherical coordinates and g is e^{-jkr}.

A similar expansion exists for scalar wave functions. Bayliss, Gunzburger, and Turkel [4] derived an expression for a general Nth-order differential operator that exactly annihilates the first N terms of the scalar expansion. We will define analogous operators B_N that annihilate the first N terms of (3). Such operators lead directly to absorbing boundary conditions.

The basic building block is the differential operator L_N, defined as

$$L_N(\mathbf{u}) = \hat{\mathbf{r}} \times \nabla \times \mathbf{u} - \left(jk + \frac{N}{r} \right) \mathbf{u}, \qquad N = 0, 1, 2, \ldots \tag{4}$$

where $\hat{\mathbf{r}}$ is a unit vector in the radial direction. It can be shown that for $N \geq 0$ and $n \geq 0$,

$$L_N\left(g\frac{\mathbf{H}_{nt}(\theta,\varphi)}{r^{n+1}}\right) = (n-N)g\frac{\mathbf{H}_{nt}(\theta,\varphi)}{r^{n+2}} \tag{5}$$

where subscript t denotes the $\theta - \varphi$ part of a vector. Also, for $N \geq 0$ and $n \geq 0$,

$$L_N\left(\nabla_t\left\{g\frac{\mathbf{H}_{nr}(\theta,\varphi)}{r^{n+1}}\right\}\right) = (n+1-N)\nabla_t\left\{g\frac{\mathbf{H}_{nr}(\theta,\varphi)}{r^{n+2}}\right\} \tag{6}$$

where subscript r denotes the radial part of a vector. In both cases, L_N has the effect of multiplying by i/r, but leaving the $\theta - \varphi$ dependence unchanged.

The operators B_N, $N = 1, 2, \ldots$, can now be defined:

$$B_N(\mathbf{u}) = (L_{N-1})^N(\mathbf{u}_t) + s(L_N)^{N-1}(\nabla_t u_r) \tag{7}$$

where s is an arbitrary number. The superscript N denotes that the operator L_{N-1} is applied N times; similarly for the superscript $N-1$. Applying Equations (5) and (6) repeatedly, it can be shown that

$$\begin{aligned} B_N&\left(g\frac{\mathbf{H}_n(\theta,\varphi)}{r^{n+1}}\right) \\ &= (n+1-N)(n+2-N)\cdots(n)g\frac{\mathbf{H}_{nt}(\theta,\varphi)}{r^{n+1+N}} \\ &\quad + s(n+1-N)(n+2-N)\cdots \\ &\quad (n-1)\nabla_t\left\{g\frac{\mathbf{H}_{nr}(\theta,\varphi)}{r^{n+N}}\right\} \end{aligned} \tag{8}$$

Noting that H_{0r} is zero [5], it can be seen that the right-hand side of (8) vanishes for $n = 0, 1, \ldots, N-1$, i.e., B_N annihilates the first N terms of (3). Further, for $n > N-1$ each of the terms on the right-hand side of (8) is proportional to $1/r^{n+N+1}$ (remember that in spherical coordinates, ∇_t includes a factor of $1/r$). Consequently, for any vector radiation function $\mathbf{H}$,

$$B_N(\mathbf{H}) = O\left(\frac{1}{r^{2N+1}}\right) \tag{9}$$

Thus, $B_N(\mathbf{H}) = 0$ is an approximate absorbing boundary condition on the surface S_a of a sphere of radius r. The approximation gets better as r increases: For $N = 1$, it improves at the rate $1/r^3$ and for $N = 2$ at the rate $1/r^5$.

After a certain amount of algebra, and making use of the fact that $\mathbf{H}$ satisfies the vector wave equation, the operators B_1 and B_2 can be written in a more usable form:

$$B_1(\mathbf{H}) = \hat{\mathbf{r}} \times \nabla \times \mathbf{H} - \alpha(r)\mathbf{H}_t + (s-1)\nabla_t H_r \tag{10}$$

$$\begin{aligned} B_2(\mathbf{H}) = &-\frac{1}{\beta(r)}\hat{\mathbf{r}} \times \nabla \times \mathbf{H} + \frac{\alpha(r)}{\beta(r)}\mathbf{H}_t \\ &+ \nabla \times [\hat{\mathbf{r}}(\nabla \times \mathbf{H})_r] + (s-1)\nabla_t(\nabla \cdot \mathbf{H}_t) \\ &+ (2-s)\alpha(r)\nabla_t H_r \end{aligned} \tag{11}$$

where $\alpha(r) = jk$ and $\beta(r) = 1/(2jk + 2/r)$.

3. IMPOSING ABSORBING BOUNDARY CONDITIONS WITH FINITE ELEMENTS

In order to impose the boundary condition $B_2(\mathbf{H}) = 0$ on S_a in a finite element analysis, it is first necessary to rewrite the condition as

$$P_2(\mathbf{H}) = \hat{\mathbf{r}} \times \nabla \times \mathbf{H} \quad \text{on } S_a \tag{12}$$

where P_2 is defined by

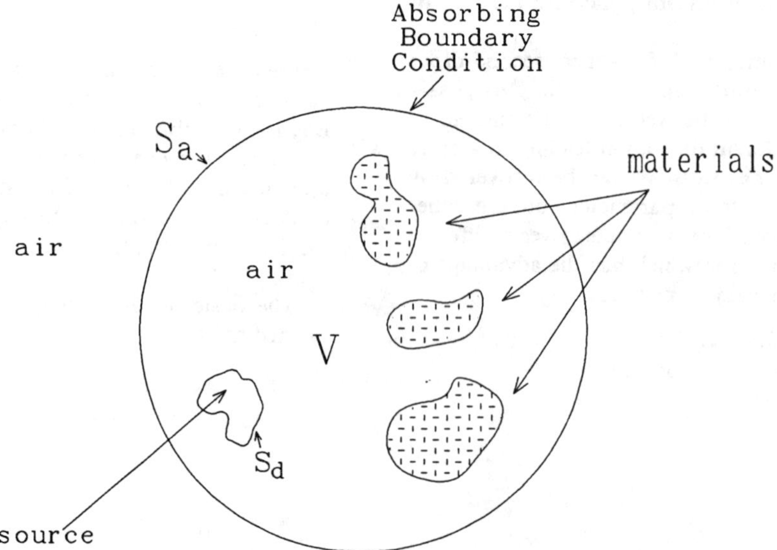

Figure 1 A typical time-harmonic electromagnetic field problem, showing the artificial spherical boundary S_a on which the absorbing boundary condition is imposed.

$$P_2(\mathbf{H}) = \alpha(r)\mathbf{H}_t + \beta(r)\nabla \times [\hat{\mathbf{r}}(\nabla \times \mathbf{H})_r] + (s-1)\beta(r)\nabla_t(\nabla \cdot \mathbf{H}_t) + (2-s)\alpha(r)\beta(r)\nabla_t H_r \tag{13}$$

The finite element method must solve the vector wave equation (1) inside a volume V, with inhomogeneous boundary condition (2) on S_d and absorbing boundary condition (12) on the spherical surface S_a. For bounded problems, when no absorbing boundary condition is necessary, a variational principle for the vector wave equation exists and can be used as the basis of a finite element method [2]. However, there seems to be no corresponding variational principle that also incorporates (12), for arbitrary s. A weighted-residual formulation is still possible, but this does not lead to a symmetric matrix problem. This is the approach taken by Peterson, whose boundary condition corresponds to the choice $s = 1$.

However, in the special case $s = 2$ the last term in (13) vanishes and a variational result is possible. It may be confirmed that, when $s = 2$, the field satisfying Equations (1), (2), and (12) is also the stationary point of the functional

$$F(\mathbf{H}) = \int_V \left[\frac{1}{\epsilon_r}(\nabla \times \mathbf{H})^2 - k^2\mu_r \mathbf{H}^2 \right] dV + \int_{S_a} \left[\alpha(r)\mathbf{H}_t^2 + \beta(r)(\nabla \times \mathbf{H})_r^2 - \beta(r)(\nabla \cdot \mathbf{H}_t)^2 \right] dS \tag{14}$$

Once trial functions are introduced for the field $\mathbf{H}$ and the essential boundary condition (2) is imposed, the problem of finding the stationary point of the functional F reduces to that of solving a matrix equation $Ax = b$, where A is a square, sparse, *symmetric*, matrix, and b is a column vector representing the known sources.

REFERENCES

1. A. F. Peterson, "Absorbing Boundary Conditions for the Vector Wave Equation," *Microwave and Optical Technology Letters*, Vol. 1, No. 2, April 1988, pp. 62–64.
2. J. P. Webb, G. L. Maile, and R. L. Ferrari, "Finite Element Solution of Three-Dimensional Electromagnetic Problems," *IEE Proc. H*, Vol. 130, No. 2, March 1983, pp. 153–159.
3. K. D. Paulsen, D. R. Lynch, and J. W. Strohbehn, "Three-Dimensional Finite, Boundary, and Hybrid Element Solutions of the Maxwell Equations for Lossy Dielectric Media," *IEEE Trans.*, Vol. MTT-36, No. 4, April 1988, pp. 682–693.
4. A. Bayliss, M. Gunzburger, and E. Turkel, "Boundary Conditions for the Numerical Solution of Elliptic Equations in Exterior Regions," *SIAM J. Appl. Math.*, Vol. 42, No. 2, April 1982, pp. 430–451.
5. C. H. Wilcox, "An Expansion Theorem for Electromagnetic Fields," *Comm. Pure Appl. Math.*, Vol. 9, 1956, pp. 115–134.

Absorbing Boundary Conditions on Circular and Elliptical Boundaries

C. F. LEE, R. T. SHIN, and J. A. KONG
DEPARTMENT OF ELECTRICAL ENGINEERING AND COMPUTER SCIENCE
AND RESEARCH LABORATORY OF ELECTRONICS
MASSACHUSETTS INSTITUTE OF TECHNOLOGY
CAMBRIDGE, MA 02139 USA

B. J. MCCARTIN
COMPUTER AND INFORMATION SCIENCE DEPARTMENT
THE HARTFORD GRADUATE CENTER
HARTFORD, CT 06120, USA

Abstract– Finite difference methods are becoming increasingly popular in the computational electromagnetics community. A major issue in applying these methods to electromagnetic wave scattering is to limit the computational domain to a finite size, which is accomplished by selecting an outer boundary and imposing absorbing boundary conditions to simulate free space. In this paper, the pseudo-differential operator approach is employed to derive absorbing boundary conditions for both circular and elliptical outer boundaries. The pseudo-differential operator approach employed by Engquist and Majda is modified to derive improved absorbing boundary conditions. In the case of the circular outer boundary, the modified pseudo-differential operator approach leads to a condition equivalent to that of Bayliss and Turkel's second-order condition. The modified pseudo-differential operator is then used to derive the second-order absorbing boundary condition for the elliptical outer boundary. The effectiveness of the second-order absorbing boundary condition on the elliptical outer boundary is illustrated by calculating scattered fields from various objects. It is shown that for elongated scatterers, the elliptical outer boundary can be used to reduce the size of the computational domain.

1. INTRODUCTION

Finite difference methods are becoming increasingly popular in the computational electromagnetics community. The flexibility of finite difference methods for inhomogeneous scatterers makes them more attractive than the method of moments. The sparse nature of the resulting matrices enables scientists to accurately analyze problems of electromagnetic scattering from large objects. For two-dimensional problems, typically only five to seven matrix elements per row are nonzero. Furthermore, the evaluation of the matrix elements requires only simple arithmetic, while, in the method of moments, singularities are frequently encountered and must be carefully treated. The difficulty with finite difference methods occurs when the radiation boundary conditions are considered. In the method of moments, the formulation is within or on the scatterers. The implication is that the number of unknowns is proportional to the volume or the surface area of the scatterers. On the other hand, finite difference methods are formulated in the entire space. An outer boundary must be used to achieve a finite computational domain. The boundary condition placed on the outer boundary should simulate unbounded space and only permit outgoing scattered waves.

The boundary conditions [1–4,8,11–16] used on the outer boundary may be exact or approximate. The exact boundary conditions, such as the modal expansion [1] or radiation integral approaches [2], require substantial memory and computation time [3,4]. The approximate boundary conditions [3,4] have been widely used, and are known to be efficient and offer sufficient accuracy. These are local operators and are able to preserve the sparsity of finite difference methods.

Engquist and Majda [3] used the pseudo-differential operator [5–7] approach to obtain boundary conditions for rectangular and circular boundaries. The rectangular boundary operator has been applied to many problems [3,8], but suffers a drawback in that the normal at the corner is not defined. The boundary conditions for a circular outer boundary do not have the problem of undefined normals and

Reprinted with permission from *Journal of Electromagnetic Waves and Applications*, vol. 4, no. 10, pp. 945–962, 1990.

have received wide attention in applications to both frequency and time domain problems.

For a circular boundary, Bayliss and Turkel [4] derived a boundary operator by assuming the Wilcox type expansion [9] for the scattered fields and then developing a series operator to eliminate the inverse power series. The most widely used operator is the second order operator, which has been demonstrated to be effective in bringing the outer boundary closer to the scatterer, thus reducing the size of the computational domain. Also, it has been demonstrated to have higher absorbability than the corresponding Engquist and Majda condition [10].

In this paper, we modify the pseudo-differential operator approach and obtain a circular boundary operator equivalent to Bayliss and Turkel's operator. Then, in an attempt to reduce the size of the computation domain for elongated scatterers, we employ the modified pseudo-differential operator to derive the elliptic boundary operator, which reduces to the circular operator when the boundary is circular. In Section 2, the pseudo-differential operator approach is modified to derive the improved circular boundary condition. The modified pseudo-differential operator approach is then used in Section 3 to derive the second-order absorbing boundary conditions for the elliptical outer boundary. The effectiveness of the elliptical boundary in reducing the size of the computational domain for elongated scatterers is illustrated in Section 4 by calculating bistatic and monostatic radar cross sections from a conducting ellipse, a strip, and a hypothetical "airfoil". In the following sections, the time dependence $e^{-i\omega t}$ is assumed.

2. RADIATION CONDITION ON CIRCULAR BOUNDARY

The pseudo-differential operator approach provides a systematic way to obtain absorbing boundary conditions in orthogonal coordinate systems. This technique has been employed by Engquist and Majda [3] to derive absorbing boundary conditions for both circular and rectangular boundaries. In this section, we apply this technique to circular boundaries. We begin with an outline of [3]. Then, following a simple observation, we apply a straightforward modification to obtain an absorbing boundary condition which is equivalent to [4]. The pseudo-differential operator approach is best described in the Fourier transform domain of the transverse coordinates, but the results are identical if we treat the transverse derivative operator as an algebraic quantity.

Outline of Engquist and Majda's Approach

Consider a two-dimensional (2D) scattering problem where the scattered field U_s is governed by the 2D Helmholtz equation

$$\left(\frac{\partial^2}{\partial r^2}+\frac{1}{r}\frac{\partial}{\partial r}+\frac{1}{r^2}\frac{\partial^2}{\partial \phi^2}+k^2\right)U_s=0 \tag{1}$$

We define

$$L=\frac{\partial^2}{\partial r^2}+\frac{1}{r}\frac{\partial}{\partial r}+\frac{1}{r^2}\frac{\partial^2}{\partial \phi^2}+k^2 \tag{2}$$

Using the factorization theorem [5–7], the L operator is factored into a product of incoming and outgoing wave operators as follows

$$L=\left(\frac{\partial}{\partial r}+\lambda_+\right)\left(\frac{\partial}{\partial r}-\lambda_-\right) \tag{3}$$

where λ_+ and λ_- are functions of k, r and $\partial/\partial\phi$. The operator within the first set of parentheses is the incoming wave operator while that within the second set is the outgoing wave operator which is an absorbing boundary operator

$$\left(\frac{\partial}{\partial r}-\lambda_-\right)U_s=0 \tag{4}$$

Hence we need to find λ_-. Following [3], we expand λ_- into the following series

$$\lambda_-=\lambda_-^{(1)}+\lambda_-^{(0)}+\ldots+\lambda_-^{(-n)} \tag{5}$$

in which $\lambda_-^{(j)}$ are recursively determined. To obtain the second order boundary operator, we need only the first two terms which are given by

$$\lambda_-^{(1)}=ik\sqrt{1+\frac{1}{k^2r^2}\frac{\partial^2}{\partial\phi^2}}\sim ik\left(1+\frac{1}{2k^2r^2}\frac{\partial^2}{\partial\phi^2}\right) \tag{6}$$

$$\lambda_{-}^{(0)} = -\frac{1}{2r} + \frac{1}{2r^3 k^2}\frac{\partial^2}{\partial\phi^2} \tag{7}$$

The second-order boundary operator, obtained by combining (6) and (7), is

$$\frac{\partial}{\partial r} = ik - \frac{1}{2r} + \frac{1}{2k^2 r^2}\left(ik + \frac{1}{r}\right)\frac{\partial^2}{\partial\phi^2} \tag{8}$$

These are the results obtained in [3]. By ignoring terms involving $1/r^2$ and higher-order terms from the second-order boundary operator, Sommerfeld's radiation condition can be obtained.

The first and second-order boundary operators derived by Bayliss and Turkel are given by [4]

$$\frac{\partial}{\partial r} = ik - \frac{1}{2r} \tag{9}$$

$$\frac{\partial}{\partial r} = \frac{1}{1+\dfrac{i}{kr}}\left(ik - \frac{3}{2r} - \frac{3i}{8kr^2} + \frac{i}{2kr^2}\frac{\partial^2}{\partial\phi^2}\right)$$

$$= ik - \frac{1}{2r} + \frac{ik}{2(kr)^2}\,\frac{\dfrac{\partial^2}{\partial\phi^2} + \dfrac{1}{4}}{1 - \dfrac{1}{ikr}} \tag{10}$$

Equations (8) and (10) have been applied to many problems; it has been demonstrated [10] that (10) provides higher absorbability than (8).

Modified Engquist and Majda Approach

In an attempt to improve the Engquist and Majda absorbing boundary condition, we recall Sommerfeld's radiation condition

$$\frac{\partial}{\partial r} = ik - \frac{1}{2r} \tag{11}$$

The above equation ensures that, in the far field, the scattered waves are cylindrical waves. We notice that the Sommerfeld radiation condition appears in the first-order condition of the Bayliss and Turkel's results. On the other hand, the ik term appears in (6) while $1/2r$ appears in (7). Thus, Sommerfeld's radiation condition cannot be obtained by using (6) alone, and this observation motivates us to modify the approach in [3] and to group the ik and $1/2r$ terms.

Considering an arbitrary function of r, we have the following operator relationship

$$\left(\frac{\partial}{\partial r} + \frac{1}{2r}\right)^2 + \frac{1}{4r^2} = \frac{\partial^2}{\partial r^2} + \frac{1}{r}\frac{\partial}{\partial r} \tag{12}$$

Substituting (12) into (2), we obtain

$$L = \left(\frac{\partial}{\partial r} + \frac{1}{2r}\right)^2 + k^2 + \frac{1}{r^2}\frac{\partial^2}{\partial\phi^2} + \frac{1}{4r^2} \tag{13}$$

Now we attempt to factor L into the following form

$$L = \left(\frac{\partial}{\partial r} + \frac{1}{2r} + \lambda_{+}\right)\left(\frac{\partial}{\partial r} + \frac{1}{2r} - \lambda_{-}\right) \tag{14}$$

In order to find λ_{-}, we multiply out (14) and obtain

$$L = \frac{\partial^2}{\partial r^2} + \frac{1}{r}\frac{\partial}{\partial r} - \frac{1}{4r^2} + \frac{1}{2r}(\lambda_{+} - \lambda_{-}) + (\lambda_{+} - \lambda_{-})\frac{\partial}{\partial r} - \frac{\partial\lambda_{-}}{\partial r} - \lambda_{+}\lambda_{-} \tag{15}$$

By comparing (15) with (1), we have

$$\frac{1}{r^2}\frac{\partial^2}{\partial\phi^2} + k^2 = -\frac{1}{4r^2} + \frac{1}{2r}(\lambda_{+} - \lambda_{-}) + (\lambda_{+} - \lambda_{-})\frac{\partial}{\partial r} - \frac{\partial\lambda_{-}}{\partial r} - \lambda_{+}\lambda_{-} \tag{16}$$

Since the left-hand side of the above equation does not involve $\partial/\partial r$, we assume

$$\lambda_{+} = \lambda_{-} = \lambda \tag{17}$$

and obtain

$$\frac{1}{r^2}\frac{\partial^2}{\partial\phi^2} + k^2 + \frac{1}{4r^2} = -\frac{\partial\lambda}{\partial r} - \lambda\lambda \tag{18}$$

To obtain the second order operator, we approximate λ by

$$\lambda \sim \lambda^{(1)} + \lambda^{(0)} \tag{19}$$

where the leading terms of $\lambda^{(1)}$ and $\lambda^{(0)}$ are $O(1)$ and $O(1/r^3)$, respectively. They are determined recursively by inserting (19) into (18). For $\lambda^{(1)}$, we equate the left-hand side of (18) to the terms for which the leading terms are $O(1)$. Assuming $\partial\lambda^{(1)}/\partial r$ is $O(1/r^3)$ [justified by (21)], we have

$$\frac{1}{r^2}\frac{\partial^2}{\partial\phi^2} + k^2 + \frac{1}{4r^2} = -\lambda^{(1)}\lambda^{(1)} \tag{20}$$

Therefore,

$$\begin{aligned}\lambda^{(1)} &= ik\sqrt{1 + \frac{1}{k^2 r^2}\frac{\partial^2}{\partial\phi^2} + \frac{1}{4k^2 r^2}} \\ &\sim ik\left(1 + \frac{1}{2k^2 r^2}\frac{\partial^2}{\partial\phi^2} + \frac{1}{8k^2 r^2}\right)\end{aligned} \tag{21}$$

For $\lambda^{(0)}$, we set the terms for which the leading terms are $O(1/r^3)$ to zero.

$$-\frac{\partial\lambda^{(1)}}{\partial r} - \lambda^{(1)}\lambda^{(0)} - \lambda^{(0)}\lambda^{(1)} = 0 \tag{22}$$

We cannot, in general, assume that $\lambda^{(1)}$ and $\lambda^{(0)}$ commute. But keeping in mind that the transverse derivatives are algebraic quantities, since we are in the Fourier transform domain, we have

$$\lambda^{(0)} = -\frac{1}{2\lambda^{(1)}}\frac{\partial\lambda^{(1)}}{\partial r} \tag{23}$$

Retaining the leading term for $\lambda^{(1)}$ in the denominator, we obtain

$$\lambda^{(0)} = \frac{1}{2k^2 r^3}\left(\frac{\partial^2}{\partial\phi^2} + \frac{1}{4}\right) \tag{24}$$

By adding (21) and (24), we obtain λ_-, and consequently

$$\frac{\partial}{\partial r} = ik - \frac{1}{2r} + \frac{ik}{2\,(kr)^2}\left(1 + \frac{1}{ikr}\right)\left(\frac{\partial^2}{\partial\phi^2} + \frac{1}{4}\right) \tag{25}$$

The above equation is the new absorbing boundary condition, which is similar to (10). By substituting (25) into the Helmholtz equation, it can be shown that the leading residual term is proportional to $1/r^4$. In fact the simple algebraic approximation for large kr

$$1 + \frac{1}{ikr} \sim \frac{1}{1 - \dfrac{1}{ikr}} \tag{26}$$

converts (25) into (10).

3. RADIATION CONDITION ON ELLIPTICAL BOUNDARY

In this section we apply the pseudo-differential operator approach to find an absorbing boundary condition for elliptical boundaries. For elongated scatterers, the elliptical outer boundary may be able to be used to reduce the computational domain.

The 2D Helmholtz equation in elliptical coordinate is given by

$$\frac{4}{d^2\left[\cosh^2 u - \cos^2 v\right]}\left(\frac{\partial^2}{\partial u^2} + \frac{\partial^2}{\partial v^2}\right)U_s + k^2\,U_s = 0 \tag{27}$$

where d is the interfocal distance and u, v are parameters of the elliptical coordinate system (Fig. 1). The parameters u and v are related to the rectangular coordinates by

$$x = \frac{d}{2}\cosh u\,\cos v, \qquad y = \frac{d}{2}\sinh u\,\sin v \tag{28}$$

in which $0 \le u < \infty$, $0 \le v < 2\pi$. Note that for a given interfocal distance d, the ellipse defined by constant u approaches a circle as u is increased. This is a desired property to have for the outer boundary, since the scattered field behaves like cylindrical wave as the distance is increased. Also, when $d = 0$, the ellipse reduces to the circle.

By examining both the cylindrical and elliptical coordinate systems, we find in the limit as $d \to 0$ or $u \to \infty$ that v becomes ϕ. In the cylindrical coordinate system, r is the distance along a constant ϕ line. Therefore, we introduce an additional parameter in the elliptical coordinate system n, which is defined to be the arc length along a constant v curve.

$$n = \int_0^u \frac{d}{2}\sqrt{\cosh^2 u' - \cos^2 v}\; du' \tag{29}$$

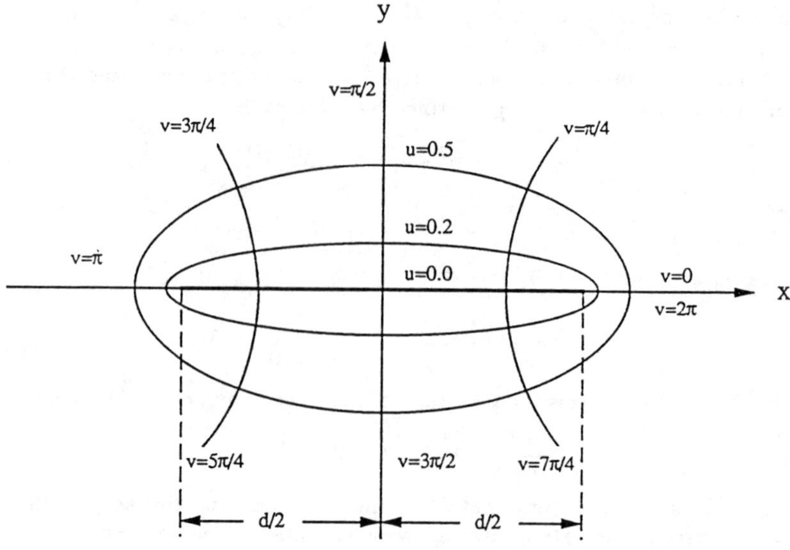

Figure 1. Elliptical coordinate system.

Now, n becomes the radius when the ellipse approaches a circle, and $\hat{n}$ is the unit normal vector. Partial derivatives with respect to n and u are related by

$$\frac{\partial}{\partial n} = \hat{n} \cdot \nabla = \frac{2}{d\,[\cosh^2 u - \cos^2 v]^{1/2}} \frac{\partial}{\partial u} \tag{30}$$

Using (29) and (30), we convert (27) to

$$\frac{\partial^2}{\partial n^2} + \frac{2 \sinh u \cosh u}{d[\cosh^2 u - \cos^2 v]^{3/2}} \frac{\partial}{\partial n} + \frac{4}{d^2[\cosh^2 u - \cos^2 v]} \frac{\partial^2}{\partial v^2} + k^2 = 0 \tag{31}$$

The above expression reduces to the cylindrical wave equation when the ellipse approaches a circle.

In order to derive the absorbing boundary conditions for the elliptical boundary, we complete the square, analogous to the circular case, and proceed to the factorization. By noticing that

$$\frac{\partial^2}{\partial n^2} + b \frac{\partial}{\partial n} = \left(\frac{\partial}{\partial n} + \frac{b}{2}\right)^2 - \frac{b^2}{4} - \frac{1}{2}\frac{\partial b}{\partial n} \tag{32}$$

we rewrite (31) as

$$\left(\frac{\partial}{\partial n} + \frac{\sinh u \cosh u}{d\,[\cosh^2 u - \cos^2 v]^{3/2}}\right)^2 + \frac{4}{d^2[\cosh^2 u - \cos^2 v]} \frac{\partial^2}{\partial v^2} + k^2 - A = 0 \tag{33}$$

where

$$A = \frac{2}{d^2[\cosh^2 u - \cos^2 v]^3} \left(\sin^2 v\,(\sinh^2 u + \cosh^2 u) - \sinh^2 u \,(\frac{1}{2} \cosh^2 u + 1)\right) \tag{34}$$

Equation (33) is then factored into the following form

$$\left(\frac{\partial}{\partial n} + \frac{\sinh u \cosh u}{d\,[\cosh^2 u - \cos^2 v]^{3/2}} + \lambda_+\right)\left(\frac{\partial}{\partial n} + \frac{\sinh u \cosh u}{d\,[\cosh^2 u - \cos^2 v]^{3/2}} - \lambda_-\right) = 0 \tag{35}$$

Then, following a similar approach as in the circular boundary case, we obtain

$$\begin{aligned} \lambda_+^{(1)} = \lambda_-^{(1)} = \lambda^{(1)} &= ik \sqrt{1 + \frac{4}{(kd)^2[\cosh^2 u - \cos^2 v]} \frac{\partial^2}{\partial v^2} - \frac{A}{k^2}} \\ &\simeq ik \left(1 + \frac{2}{(kd)^2[\cosh^2 u - \cos^2 v]} \frac{\partial^2}{\partial v^2} - \frac{A}{2k^2}\right) \end{aligned} \tag{36}$$

and

$$\begin{aligned} \lambda_+^{(0)} = \lambda_-^{(0)} = \lambda^{(0)} &= -\frac{1}{2\lambda^{(1)}} \frac{\partial \lambda^{(1)}}{\partial n} \\ &= -\frac{1}{d\lambda^{(1)}\,[\cosh^2 u - \cos^2 v]^{1/2}} \frac{\partial \lambda^{(1)}}{\partial u} \end{aligned} \tag{37}$$

Taking the leading order term from the $\lambda^{(1)}$ in the denominator, we obtain

$$\frac{\partial}{\partial n} = ik - \frac{\sinh u \cosh u}{d\,[\cosh^2 u\; - \;\cos^2 v]^{3/2}} + \frac{i2k}{(kd)^2[\cosh^2 u\; - \;\cos^2 v]} \cdot \Big(1 + \frac{2\sinh u \cosh u}{ikd\,[\cosh^2 u\; - \;\cos^2 v]^{3/2}}\Big)\Big(\frac{\sinh^4 u}{4\,[\cosh^2 u\; - \;\cos^2 v]^2} + \frac{\partial^2}{\partial v^2}\Big) \quad (38)$$

Then, making an approximation similar to the one used in (26) gives

$$\frac{\partial}{\partial n} = ik - \frac{\sinh u \cosh u}{d\,[\cosh^2 u\; - \;\cos^2 v]^{3/2}} + \frac{i2k}{(kd)^2[\cosh^2 u\; - \;\cos^2 v]} \cdot \frac{\dfrac{\sinh^4 u}{4[\cosh^2 u\; - \;\cos^2 v]^2} + \dfrac{\partial^2}{\partial v^2}}{1 - \dfrac{2\sinh u \cosh u}{ikd\,[\cosh^2 u\; - \;\cos^2 v]^{3/2}}} \quad (39)$$

The above equation is the second-order absorbing boundary condition on elliptical outer boundaries. In arriving at (39), the higher-order terms of A are ignored. The higher-order terms are determined by the power of the product of either $(d \cosh u)$ or $(d \sinh u)$, since they measure the size of the ellipse as does r in the case of a circle. We note that when $d = 0$, (39) reduces to (10).

At this point, it is worthwhile to examine the "on surface radiation condition (OSRC)" [11] interpretation of (10). According to [11], the r in (10) is replaced by the radius of curvature (R) and $(1/r^2)\partial^2/\partial\phi^2$ by $\partial^2/\partial s^2$ where s is the arc length. With these transformations, (10) becomes

$$\frac{\partial}{\partial n} = ik - \frac{1}{2R} + \frac{i}{2k}\,\frac{\dfrac{\partial^2}{\partial s^2} + \dfrac{1}{4R^2}}{1 - \dfrac{1}{ikR}} \quad (40)$$

For the elliptical outer boundary,

$$R = \frac{d\,[\cosh^2 u\; - \;\cos^2 v]^{3/2}}{2\,\cosh u\,\sinh u} \quad (41)$$

$$\frac{\partial^2}{\partial s^2} = \frac{4}{d^2\,[\cosh^2 u\; - \;\cos^2 v]}\,\frac{\partial^2}{\partial v^2} - \frac{2\sin 2v}{d^2[\cosh^2 u\; - \;\cos^2 v]^2}\,\frac{\partial}{\partial v} \quad (42)$$

Then (40) becomes

$$\frac{\partial}{\partial n} = ik - \frac{\sinh u \cosh u}{d\,[\cosh^2 u\; - \;\cos^2 v]^{3/2}} + \frac{i2k}{(kd)^2[\cosh^2 u\; - \;\cos^2 v]} \cdot \frac{\dfrac{\sinh^2 u \cosh^2 u}{4[\cosh^2 u\; - \;\cos^2 v]^2} + \dfrac{\partial^2}{\partial v^2}}{1 - \dfrac{2\sinh u \cosh u}{ikd\,[\cosh^2 u\; - \;\cos^2 v]^{3/2}}} \quad (43)$$

by ignoring the term involving $\partial/\partial v$. Comparing (39) with (43), we note that the OSRC interpretation gives similar results with the difference appearing only in the higher order terms.

4. NUMERICAL RESULTS AND DISCUSSIONS

The absorbing boundary conditions on circular outer boundaries have been studied and their effectiveness and accuracy have been demonstrated [3,4,10]. In this section, we focus on evaluating the absorbing boundary condition on the elliptical outer boundaries. The control region approximation [10] with triangular grids is used to discretize Maxwell's equations in the frequency domain (finite difference frequency domain, or FD-FD) and the absorbing boundary condition (39) is imposed on the outer boundary. The bistatic and monostatic two-dimensional radar cross sections are calculated and compared to the results obtained using the method of moments.

The three geometries considered are a conducting ellipse with major axis of 2.5λ, an aspect ratio of 0.1, and an interfocal distance of 4.97λ, as shown in Fig. 2, a conducting strip with a 5λ width and a conducting "airfoil". The FD-FD results will be compared with the method of moments (MoM) results. In the

MoM, the pulse basis and testing functions are employed with 240 divisions in those cases, resulting in a length on the order of 0.042λ for each division. The results will be presented in terms of two-dimensional RCS which is defined as

$$\sigma\,(\text{dBm}) = 10\log \lim_{r\to\infty} 2\pi r \left| \frac{E_s}{E_i} \right|^2$$

and expressed as dBm (in dB after normalized by 1 meter). The frequency is assumed to be 300 MHz in all cases.

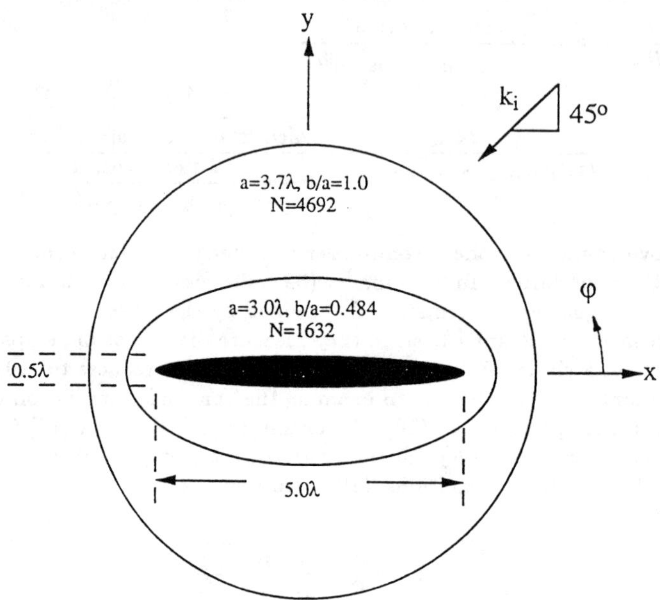

Figure 2. Circular and elliptical outer boundaries for a conducting ellipse: radius $r = 3.7\lambda$ for circular domain; $a = 3.0\lambda$ and $b/a = 0.484$ for elliptical domain.

The effectiveness of the elliptical outer boundary is first demonstrated by calculating the FD-FD results with circular and elliptical outer boundaries as shown in Fig. 2. The minimum separations from the elliptical and circular outer boundaries to the scatterer are 0.5λ and 1.2λ, respectively. Similar discretization is applied to both cases. A sample discretization for the elliptical boundary case is shown in Fig. 3. The number of unknowns for the elliptical and circular outer boundary cases are 1632 and 4692, respectively. The bistatic RCS as a function of scattering angle for an incidence angle of 45 degrees is shown in Figs. 4(a) and 4(b), for the electric field and magnetic field polarizations, respectively. As can be seen from the figures, the results obtained using the elliptical outer boundary seem to compare better with the MoM results than the circular boundary results. In order to simplify the comparison of results, an *rms* error is defined as follows

$$rms\,\text{dB} = \sqrt{\frac{\sum[\sigma\,(\text{dBm}) - \sigma_{MoM}\,(\text{dBm})\,]^2}{N}}$$

For the electric field polarization, the *rms* errors are found to 0.58 dB and 1.0 dB for the elliptical and circular boundaries. For the magnetic field polarization, the *rms* errors are found to be 1.5 dB and 2.1 dB for the elliptical and circular boundaries, respectively. This clearly demonstrates that the number of unknowns can be reduced significantly, with the same level of accuracy, for elongated scatterers by using the elliptical outer boundaries with absorbing boundary conditions given by (39).

In the case of circular outer boundary, the radius of the circle is the only parameter which needs to be determined. In the case of the elliptical outer boundary, there are two free parameters: the interfocal distance and ellipticity, or equivalently, the major and minor axes a and b. The absorbing boundary condition is derived based on the expansion of the scattered wave about the normal of the boundary. However, in general, it is difficult to know the optimum ellipticity due to the fact that one does not know the scattered wavefront prior to solving the

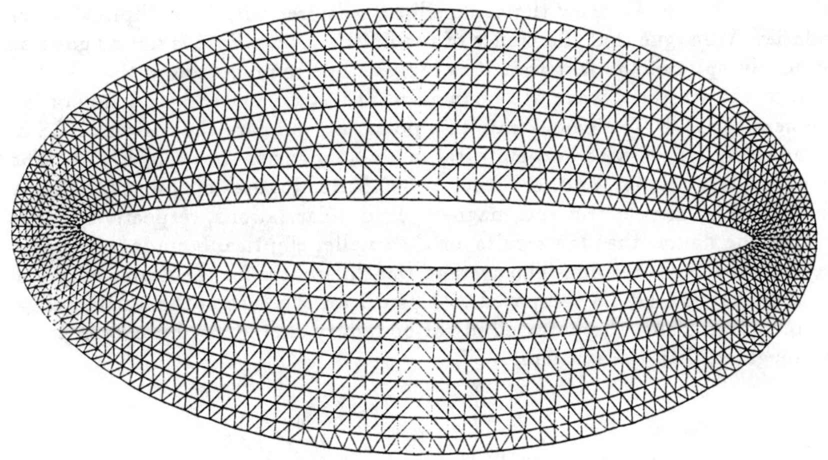

Figure 3. Triangular grid discretization of an elliptical computation domain with $a = 3.0\lambda$ and $b/a = 0.484$.

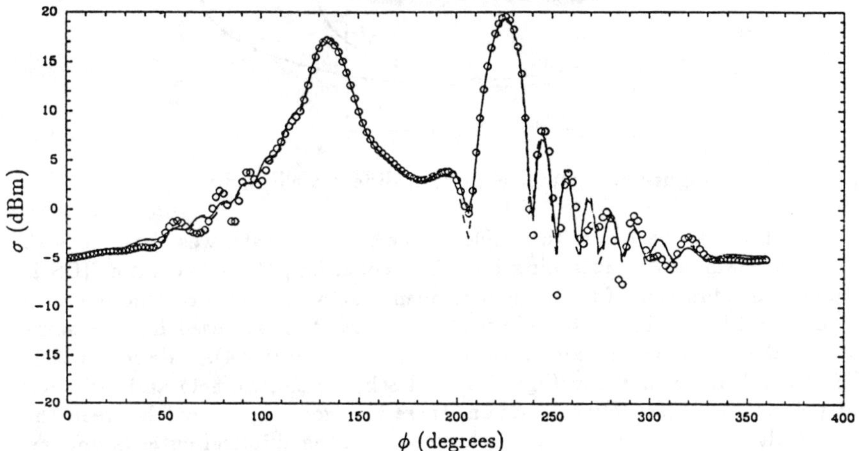

Figure 4(a). Bistatic RCS of a conducting ellipse with $a_o = 2.5\lambda$ and $b_o/a_o = 0.1$ at 300 MHz for electric field polarization for incidence angle of $\phi_i = 45°$. Solid curve: MoM; dashed curve: elliptical boundary with $a = 3.0\lambda$ and $b/a = 0.484$; circles: circular boundary with $r = 3.7\lambda$.

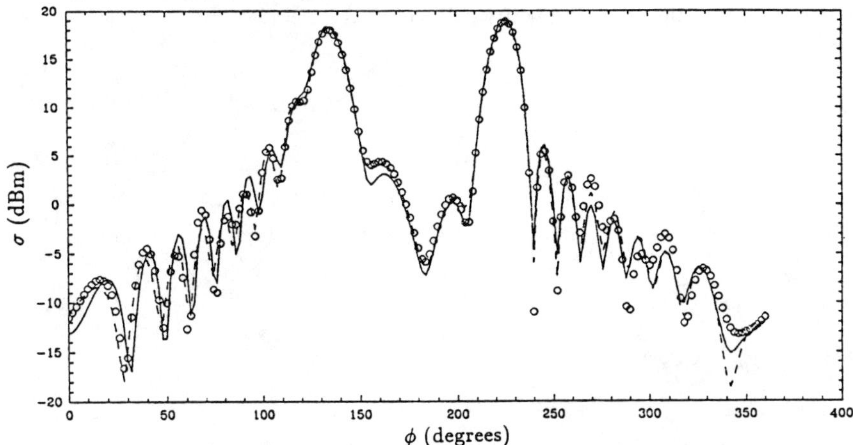

Figure 4(b). Bistatic RCS of a conducting ellipse with $a_o = 2.5\lambda$ and $b_o/a_o = 0.1$ at 300 MHz for magnetic field polarization for incidence angle of $\phi_i = 45°$. Solid curve: MoM; dashed curve: elliptical boundary with $a = 3.0\lambda$ and $b/a = 0.484$; circles: circular boundary with $r = 3.7\lambda$.

problems. In Fig. 5, we show the same ellipse scatterer with two elliptical outer boundaries. We argue that the choice of the flatter outer ellipse is not as good as the other, in spite of the fact that it is bigger. The main reason is that for the flat ellipse, the scattered wave will be incident at a steep angle to a portion of the ellipse near the end. To demonstrate this point, we compare the bistatic RCS of the conducting ellipse with two elliptical outer boundaries shown in Fig. 5. The angle of incidence is 45° and frequency is 300 MHz. Figures 6(a) and 6(b) show the bistatic RCS for electric and magnetic field polarizations, respectively. It is clear from the figures that the results for the smaller elliptical boundary compare better with the MoM results. The average *rms* differences for the flatter elliptical outer boundary is 1.90 dB compared to 1.04 dB for the case of the smaller elliptical boundary. Thus, the outer elliptical boundary should not be chosen to be more elongated than the scatterer.

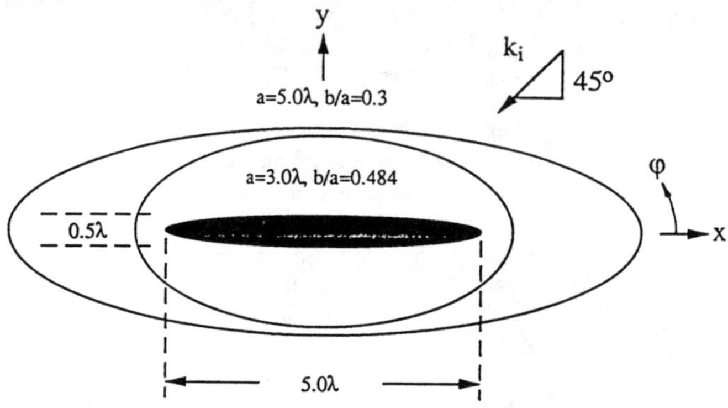

Figure 5. Outer ellipses of different ellipticity.

In all of the above cases, the angle of the incident wave was fixed at 45°. To further examine the absorbing boundary condition, the monostatic RCS is calculated as a function of the angle of incidence for a 5λ width conducting strip as shown in Fig. 7. The outer elliptical boundary that was used has the same shape as the one shown in Fig. 2 ($a = 3\lambda$ and $b/a = 0.484$). The monostatic RCS at 300 MHz is shown in Figs. 8(a) and 8(b) for electric field and magnetic field polarizations, respectively. Again, there is a good agreement between the MoM results and the FD-FD results obtained with the elliptical outer boundary. The average *rms* difference is 1.4 dB for the electric field and magnetic field polarizations.

The last geometry to be analyzed is a hypothetical "airfoil" shown in Fig. 9. This geometry contains a portion of the ellipse analyzed previously, a half circle, and two straight sections. The same outer elliptical boundary as shown in Fig. 7 is used. Figures 10(a) and 10(b) show the monostatic RCS at 300 MHz for electric field and magnetic field polarizations respectively. A good agreement between the MoM results and the FD-FD results is illustrated.

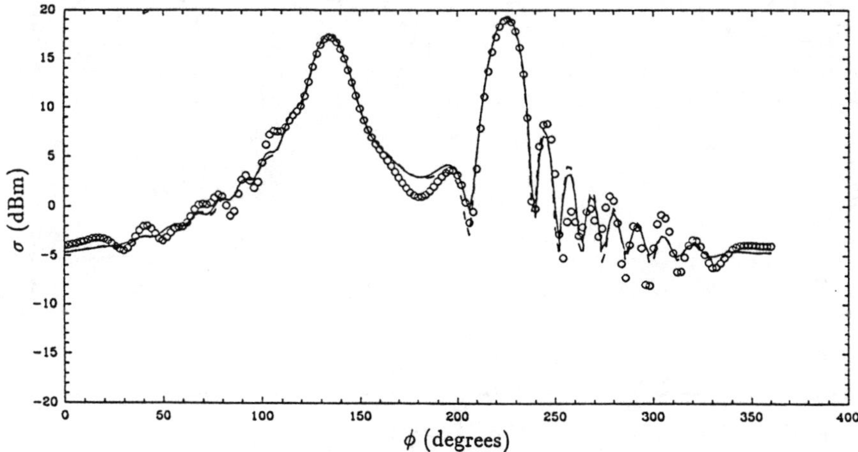

Figure 6(a). Bistatic RCS of a conducting ellipse with $a_o = 2.5\lambda$ and $b_o/a_o = 0.1$ at 300 MHz for electric field polarization for incidence angle of $\phi_i = 45°$. Solid curve: MoM; dashed curve: elliptical boundary with $a = 3.0\lambda$ and $b/a = 0.484$; circles: $a = 5.0\lambda$ and $b/a = 0.30$.

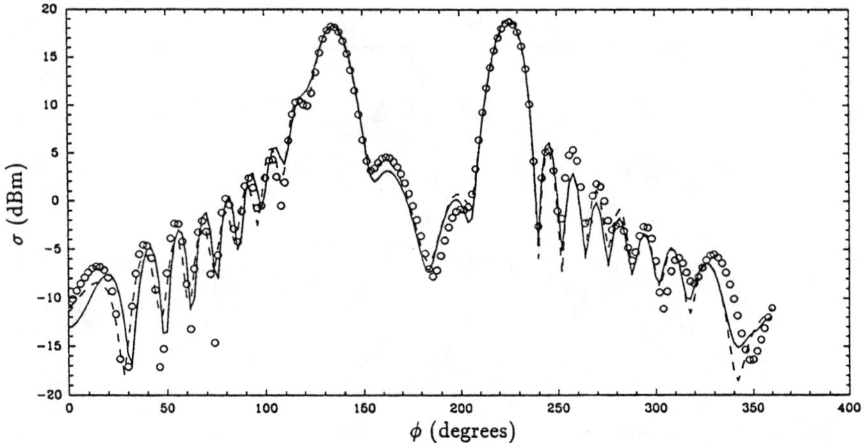

Figure 6(b). Bistatic RCS of a conducting ellipse with $a_o = 2.5\lambda$ and $b_o/a_o = 0.1$ at 300 MHz for magnetic field polarization for incidence angle of $\phi_i = 45°$. Solid curve: MoM; dashed curve: elliptical boundary with $a = 3.0\lambda$ and $b/a = 0.484$; circles: $a = 5.0\lambda$ and $b/a = 0.30$.

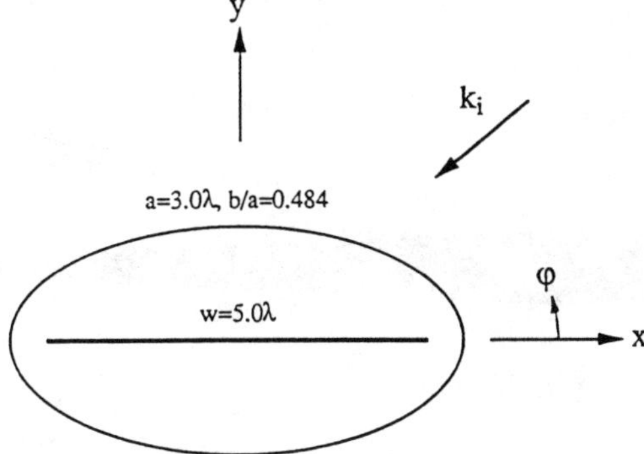

Figure 7. Geometrical description for FD-FD calculating of scattering by a 5λ conducting strip with an elliptical outer boundary ($a = 3.0\lambda$ and $b/a = 0.484$).

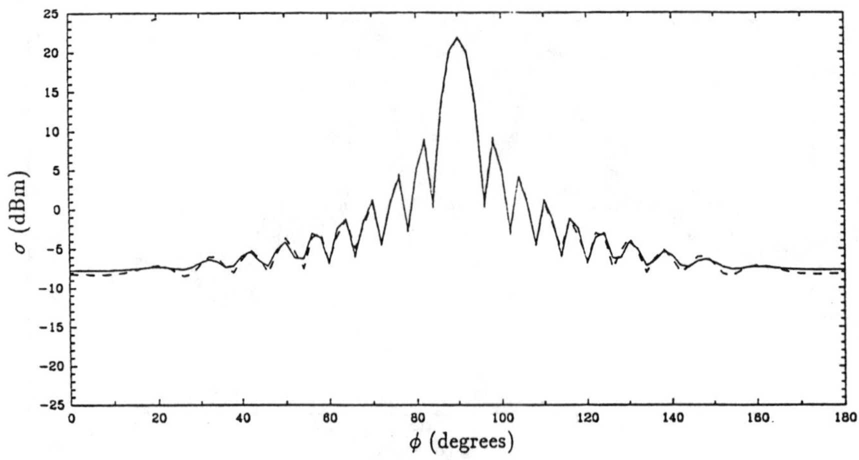

Figure 8(a). Monostatic RCS of 5λ conducting strip at 300 MHz for electrical field polarization. Solid curve: MoM; dashed curve: FD-FD with elliptical outer boundary.

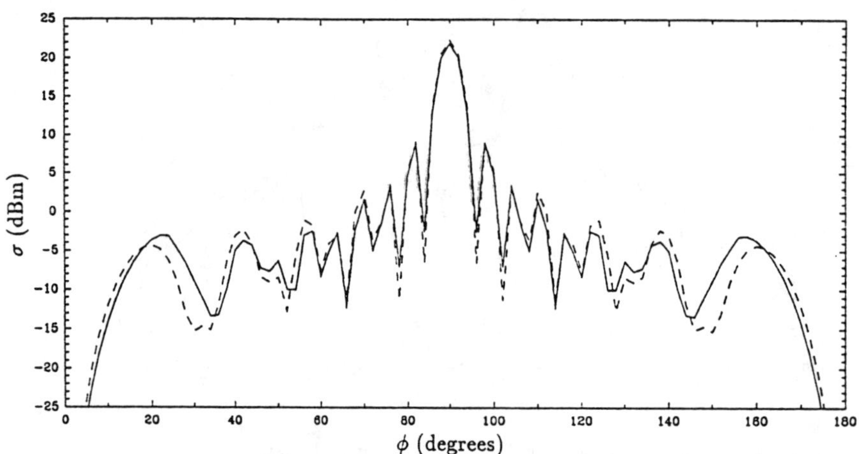

Figure 8(b). Monostatic RCS of 5λ conducting strip at 300 MHz for magnetic field polarization. Solid curve: MoM; dashed curve: FD-FD with elliptical outer boundary.

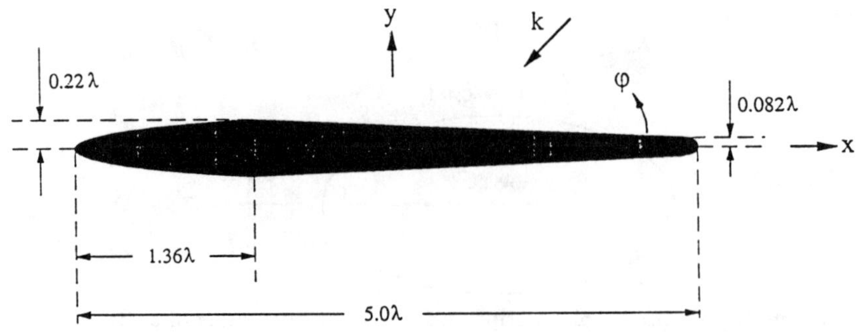

Figure 9. Geometrical description of a hypothetical "airfoil".

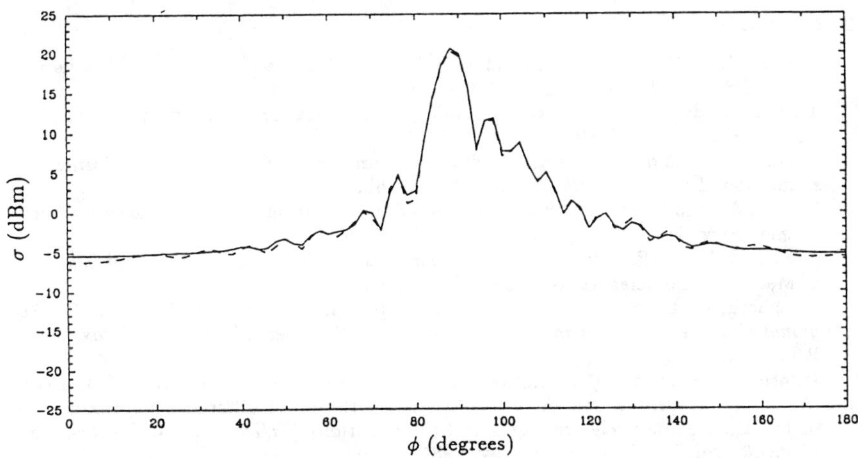

Figure 10(a). Monostatic RCS of a conducting "airfoil" at 300 MHz for electrical field polarization. Solid curve: MoM; dashed curve: FD-FD with elliptical outer boundary.

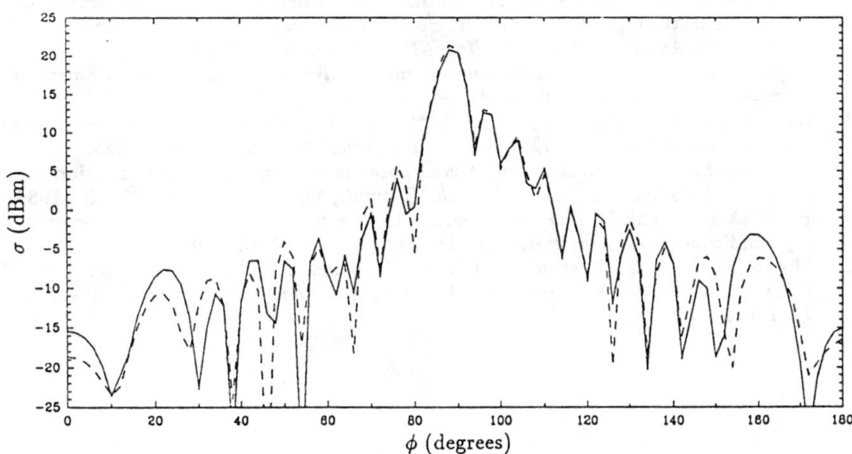

Figure 10(b). Monostatic RCS of 5λ conducting "airfoil" at 300 MHz for magnetic field polarization. Solid curve: MoM; dashed curve: FD-FD with elliptical outer boundary.

5. SUMMARY

We have demonstrated an effective method to modify the pseudo-differential operator approach used by Engquist and Majda to derive improved absorbing boundary conditions. The modified pseudo-differential operator approach is applied to both the circular and elliptic boundaries. The improved absorbing boundary condition for the circular boundaries is equivalent to the result obtained by Bayliss and Turkel, which has been shown to have higher absorbability. With a convenient change of variable and the modified pseudo-differential operator approach, we obtained the absorbing boundary conditions for the elliptic boundaries. This absorbing boundary condition is numerically demonstrated to be effective and can potentially offer substantial reduction in the requirements of computer resources for elongated scatterers.

ACKNOWLEDGMENTS

This work was supported by United Technologies Research Center, WaveTracer Contract, NSF Grant 8620029-ECS and ARO Contract DAAL03-88-J-0057.

The Editor thanks J. G. Blaschak, J. R. Lovell, T. G. Moore, and one anonymous Reviewer for reviewing the paper.

REFERENCES

1. Mei, K. K., "Unimoment method of solving antenna and scattering problems," *IEEE. Trans. Antennas Propagat.*, Vol. AP–22, No. 6, 760–766, 1974.
2. Ting, L., and M. Miksis, "Exact boundary conditions for scattering problems," *J. Acoust. Soc. Am.*, Vol. 80(6), 1986.
3. Engquist, B., and A. Majda, "Radiation boundary conditions for the numerical simulation of waves," *Math. Comp.*, Vol. 31, 629–651, 1977.
4. Bayliss, A., and E. Turkel, "Radiation boundary conditions for wave-like equations," *Comm. Pure Appl. Math.*, Vol. 33, 707–725, 1980.
5. Friedrichs, K. O., *Pseudo-Differential Operators, An Introduction,* Courant Institute of Mathematical Sciences, New York University.
6. Nirenberg, L., Lectures on *Linear Partial Differential Equations, C. B. M. S., Regional Conference Series in Mathematics, No. 17,* Amer. Math. Soc., Providence, R.I., 1973.
7. Hitoshi, K., *Pseudo-Differential Operators,* MIT Press, Massachusetts, USA, 1974.
8. Mur, G., "Absorbing boundary conditions for the finite difference approximation of the time domain electromagnetic field equations," *IEEE Trans. Electromagn. Compat.*, Vol. EMC–23, No. 4, Nov. 1981.
9. Wilcox, C., "An expansion theorem for electromagnetic fields," *Comm. on Pure and Applied Math.*, Vol. 9, 115–134, 1956
10. McCartin, B. J., G. Meltz, and R. Mittra, "Application of the control region approximation in conjunction with absorbing boundary conditions to the direct solution of the electromagnetic scattering problems," *Proc. 1988 URSI Radio Science Meeting,* Syracuse, New York, 1988.
11. Kriegsmann, G., A. Taflove, and K. Umashankar, "A new formulation of electromagnetic wave scattering using an on-surface radiation boundary condition approach," *IEEE Trans. Antennas Propagat.*, Vol. AP–35, No. 2, 153–161, 1987.
12. Majda, A., and S. Osher, "Reflection of singularities at the boundary," *Comm. on Pure and Applied Math.*, Vol. 28, 479–499, 1975.
13. **Bayliss, A., Goldstein, and E. Turkel, "On accuracy conditions for the numerical computation of waves,"** ***J. Computational Physics,*** **Vol. 59, 396–404, 1985.**
14. **Trefethen L. N., and L. Halpern, " Well-posedness of one-way wave equations and absorbing boundary conditions,"** ***Math. Comput.*,** **Vol. 47, No. 176, 421–435, 1986.**
15. **Blaschak J. G., and G. A. Kriegsmann, " A comparative study of absorbing boundary conditions,"** ***J. Computational Physics,*** **Vol. 77, 109–139, 1988.**
16. **Moore, T. G., J. G. Blaschak, and G. A. Kriegsmann, "Theory and application of radiation boundary operators,"** ***IEEE Trans. Antennas Propagat,*** **Vol. AP–36, No. 12, 1988.**

Absorbing Boundary Conditions for the Finite-Element Analysis of Planar Devices

J. P. WEBB, MEMBER, IEEE

Abstract —**Microwave and optical devices in which the fields are not confined to a finite region can nevertheless be analyzed with the finite-element method if special boundary conditions are used, to absorb outgoing radiation. The absorbing boundary conditions may be imposed by the addition of two terms to the usual functional for the scalar Helmholtz equation. New universal matrices are introduced to allow the additional terms to be easily assembled, for polynomial orders one through four. Results are given for the impedance of a parallel-plate waveguide radiating into free space and for the scattering parameters of three dielectric slab waveguide devices: a rectangular discontinuity, a feed structure, and a Y junction.**

I. Introduction

THE FINITE-ELEMENT method is an established technique for calculating the network characteristics of passive microwave components [1]–[3]. It is a technique which allows for arbitrary geometries and inhomogenous, lossy materials. In two dimensions, it can provide solutions relatively quickly on quite modest computers.

Some devices are enclosed by a conducting wall, so that there is no radiated energy. These are particularly well suited to finite-element analysis, because the fields are confined to a finite region. However, other devices—especially those designed for millimeter-wave and optical frequencies—are not enclosed. For example, the device may be made of dielectric with no metal present at all. Energy incident from the connecting waveguides radiates into the surrounding free space and is lost. The finite-element method cannot be directly applied in these cases, because the electromagnetic field exists out to infinity and an infinite region would have to be modeled. Truncating the region by enclosing the device with a virtual conducting wall some distance away is inadvisable: it will reflect waves back toward the device and may cause resonances.

As a result, finite-element methods proposed for open devices have been hybrid; that is, they couple the finite elements in the interior to another approach outside, e.g., an integral equation technique [4]–[6]. However, hybrid techniques are more complicated to implement than the finite-element method alone. Also, they reduce the sparsity of the finite-element matrices, which increases the computer time required, although this may be offset by the fact that they often reduce the size of the region that needs to be filled with elements. Perhaps because of their complexity, nobody seems to have used hybrid methods to analyze open devices which have more than two aligned waveguides connected.

Recently, the finite-element method has been applied to free-space scattering problems by surrounding the scattering obstacle with a sphere (or circle in two dimensions) and applying a special, absorbing boundary condition on the sphere [7], [8]. In this paper, the same approach is used to determine the scattering matrices of open devices, in two dimensions.

The next section reviews briefly the finite-element method as applied to enclosed devices [1]. It is followed by a discussion of absorbing boundary conditions and their application to open devices.

II. Admittance and Scattering Matrices

Consider P waveguides meeting a junction region Ω at port planes S_p, $p=1,\cdots,P$. All materials are linear, reciprocal, inhomogenous, and possibly lossy. This problem requires the solution of the vector wave equation in three dimensions, still a computationally expensive exercise. However, many devices can be modeled in two dimensions with reasonable accuracy, e.g., microstrip components, H-plane junctions of rectangular waveguide, and junctions of planar dielectric waveguide. Therefore, in this paper it will be assumed that there is no variation of the field in the y direction of a Cartesian coordinate system, and the analysis will take place in the $x-z$ plane. Under this assumption, and provided that the materials present are isotropic, Maxwell's equations decouple into two sets, involving E_y, H_x, H_z on the one hand and H_y, E_x, E_z on the other. Just the first set will be considered; solutions of the second set follow analogously. The y component of the electric field, denoted E, satisfies the scalar wave equation:

$$\nabla\cdot\frac{1}{\mu_r}\nabla E+k_0^2\epsilon_r E=0 \quad \text{in } \Omega \tag{1}$$

Manuscript received November 27, 1989; revised April 24, 1990. This work was supported by the Natural Sciences and Engineering Research Council of Canada.

The author is with the Department of Electrical Engineering, McGill University, 3480 University Street, Montreal, H3A 2A7, Canada.

IEEE Log Number 9036765.

Reprinted from *IEEE Trans. Microwave Theory Tech.*, vol. 38, no. 9, pp. 1328–1332, Sept. 1990.

where ∇ is the two-dimensional operator $\boldsymbol{a}_x\partial/\partial x+\boldsymbol{a}_z\partial/\partial z$; $\boldsymbol{a}_x$ and $\boldsymbol{a}_z$ are unit vectors in the x and z directions, respectively; μ_r and ϵ_r are the relative permeability and permittivity, in general functions of x and z; k_0 is the free-space wavenumber; and Ω is the junction region, an area of the $x-z$ plane.

Let e_{pk} be the y component of the electric field of the kth mode at port p of the device. The corresponding magnetic field will lie in the $x-z$ plane but will have components both normal and tangential to the port. Let $\boldsymbol{h}_{pk}$ be the tangential part. When the modal fields e_{pk}, $\boldsymbol{h}_{pk}$ are scaled correctly, they can be used to define the normalized voltage V_{pk} and current I_{pk} of the kth mode at port p [9, p. 147]:

$$E=\sum_{k=1}^{\infty} V_{pk}e_{pk}. \tag{2}$$

$$\boldsymbol{H}_t=\sum_{k=1}^{\infty} I_{pk}\boldsymbol{h}_{pk}. \tag{3}$$

E is the y component of the total electric field at port p, and $\boldsymbol{H}_t$ is the part of the total magnetic field that is tangential to port p. At any given frequency, only a finite number of modes at port p will be propagating; the rest will be evanescent. Let the number of propagating modes be D_p. Let $M=D_1+\cdots+D_P$. Then the response of the device to propagating waves at one frequency can be summarized by an M by M normalized admittance matrix, $[Y]$:

$$\{I\}=[Y]\{V\} \tag{4}$$

where $\{I\}$ and $\{V\}$ are M vectors of the currents and voltages at the ports. The M by M scattering matrix of the device can be obtained very simply once $[Y]$ is known [9, p. 173].

It is shown elsewhere [1] that the mnth entry of $[Y]$ is

$$Y_{mn}=-\frac{jb}{k_0\eta_0}\langle E^{(m)},E^{(n)}\rangle \tag{5}$$

where b is the thickness of the device in the y direction; η_0 is the intrinsic impedance of free space; $E^{(m)}$ is the electric field in the device when the mth entry of $\{V\}$ is set to unity and all the other entries are zero; and $\langle\cdot,\cdot\rangle$ is a symmetric, bilinear form, defined by

$$\langle f,g\rangle=\int_{\Omega}\left[\frac{1}{\mu_r}\nabla f\cdot\nabla g-k_0^2\epsilon_r fg\right]d\Omega. \tag{6}$$

Each of the fields $E^{(1)},\cdots,E^{(M)}$ corresponds to the excitation of just one mode at one port, with all the other ports shorted, and can be obtained with one finite-element analysis of the device [1]. The functional used for the finite-element analysis is just

$$F(E)=\langle E,E\rangle. \tag{7}$$

It should be noted that an alternative way of computing the scattering matrix has been suggested: the amplitudes of the outward-going waves at the ports can be used directly as unknowns in the finite-element solution [2], [3], [10]. If just part of the scattering matrix is required, this is a more efficient technique. However, to determine the full M by M scattering matrix, both techniques require M solutions in general (and a smaller number if the device has planes of symmetry). The use of the admittance expression (5), which is stationary with respect to perturbations in $E^{(m)}$ or $E^{(n)}$, may lead to greater accuracy for a given computational expense, although this has not been confirmed.

III. Absorbing Boundary Conditions

Suppose that in and around a device the electric field is governed by the scalar wave equation (1). Energy from the device flows outward into the surrounding free space, which is unbounded. Such a problem can be solved by enclosing the device with a circular boundary sufficiently far away and solving the wave equation inside, together with an absorbing boundary condition on the circle [7]. An infinite sequence of increasingly accurate absorbing boundary conditions has been developed. The second term of the sequence is relatively easy to impose. It is

$$\frac{\partial E}{\partial r}+a(R)E-b(R)\frac{1}{R^2}\frac{\partial^2 E}{\partial\theta^2}=0 \tag{8}$$

where (r,θ) are polar coordinates with respect to the center of the circle, radius R, and

$$a(R)=jk_0+\frac{1}{2R}-\frac{b(R)}{4R^2}\qquad b(R)=\frac{1}{2(jk_0+1/R)}. \tag{9}$$

For a junction of waveguides, it may not be possible to introduce a complete bounding circle, because of the presence of waveguides extending out to infinity. However, the unbounded region can be bounded by one or more circular arcs, on which condition (8) is imposed. Let there be Q arcs, $C_1,\cdots,C_Q$. In general the arcs may have different centers and radii, $R_1,\cdots,R_Q$.

To solve the wave equation with the absorbing boundary condition (8) on each arc, a modified bilinear form $\langle\cdot,\cdot\rangle$ must be used:

$$\langle f,g\rangle=\int_{\Omega}\left[\frac{1}{\mu_r}\nabla f\cdot\nabla g-k_0^2\epsilon_r fg\right]d\Omega+\sum_{q=1}^{Q}\int_{C_q}\left[a(R_q)fg+b(R_q)\frac{\partial f}{\partial s}\frac{\partial g}{\partial s}\right]ds \tag{10}$$

with s the distance along the arc. Provided each end of each arc is either a point at which $E=0$ or a point on a symmetry plane $\partial E/\partial n=0$, it can be shown that the stationary point of the functional $F(E)=\langle E,E\rangle$ satisfies the wave equation and the absorbing boundary condition on each arc, as required. The admittance matrix can still be computed from (5), using the new bilinear form (10).

In the finite-element method, the area Ω is broken into triangles. Each of the circular arcs is approximated by a polygon. Expression (10) can be written as a sum of contributions from all of the triangles. The contribution

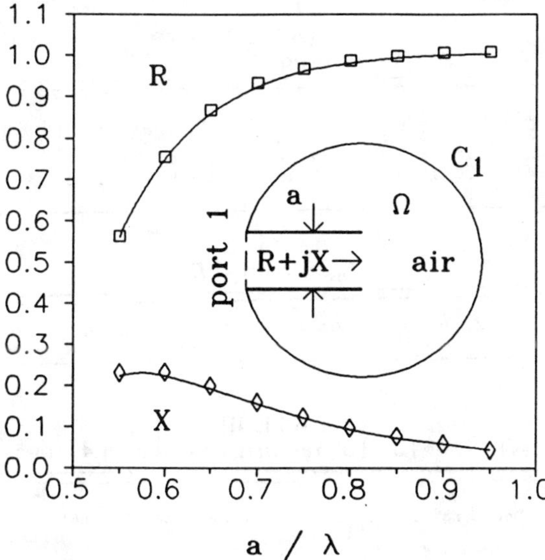

Fig. 1. Parallel-plate waveguide radiating into free space. The diagram is to scale. The curves show the normalized resistance (R) and reactance (X) of the TE_1 mode versus a/λ, where λ is the free-space wavelength. The squares and diamonds were obtained with the finite-element method, using one half of the region shown, and 201 fourth-order elements. The solid lines are from [12, sec. 4.8].

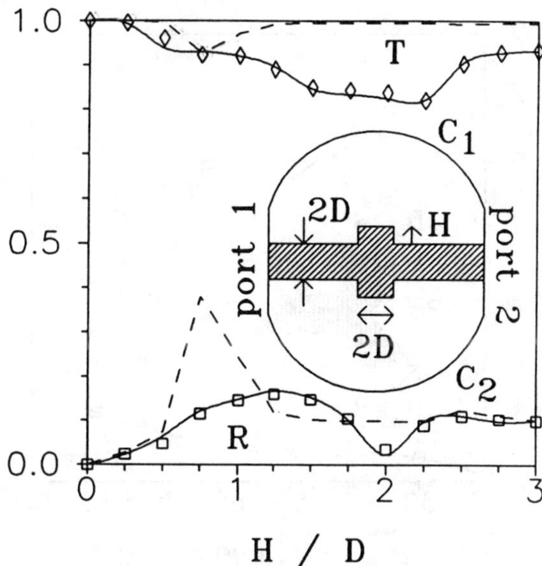

Fig. 2. A rectangular discontinuity in a dielectric slab waveguide. The diagram is to scale. The relative permittivity of the dielectric is 5. The curves show the reflection (R) and transmission (T) coefficients of the TE_0 mode when $k_0D = 1$. The squares and diamonds were obtained with the finite-element method, using one quarter of the region shown, and fourth-order elements. The solid lines are from [5]. The dashed lines show the results when the absorbing boundary condition on C_1 and C_2 is replaced by a perfect conductor.

from a single triangle reduces to

$$\{f\}^T[W]\{g\} \tag{11}$$

where $\{f\}$ and $\{g\}$ are N vectors containing the nodal values of the functions f and g at the N nodes of the triangle. N is 3, 6, 10, or 15 for first-, second-, third-, or fourth-order triangles respectively [11]. $[W]$ is an N by N symmetric matrix, given explicitly by the following:

$$[W] = \sum_{e=1}^{3}\left(\frac{1}{\mu_r}\cot\theta_e[Q^{(e)}] + a_e L_e[A^{(e)}] + \frac{b_e}{L_e}[B^{(e)}]\right) - \Delta k_0^2\epsilon_r[T] \tag{12}$$

where θ_e is the angle at vertex e of the triangle (numbered counterclockwise); $[Q^{(e)}]$ and $[T]$ are universal, precomputed matrices given in [11]; L_e is the length of edge e, which is the edge connecting vertex e to vertex $e+1$; $[A^{(e)}]$ and $[B^{(e)}]$ are new universal matrices (given in the Appendix); and Δ is the area of the triangle. If edge e lies on an arc, say on arc C_q, then a_e is $a(R_q)$ and b_e is $b(R_q)$; otherwise, a_e and b_e are zero.

IV. Results

The method was used to determine the impedance of the TE_1 mode in a parallel-plate waveguide that ends abruptly and radiates into free space (Fig. 1). The walls of the waveguide are assumed to be infinitely thin and perfectly conducting. This is a two-dimensional approximation to the problem of a rectangular waveguide radiating into free space when the dominant TE_{10} mode is incident.

Here, and with all the results to follow, the analysis is in the $x-z$ plane and the unknown is E_y. Fig. 1 shows the placement of the absorbing boundary condition, which was the same for all frequencies solved. The figure also shows the computed impedance, which agrees well with the analytic results given by Marcuvitz [12].

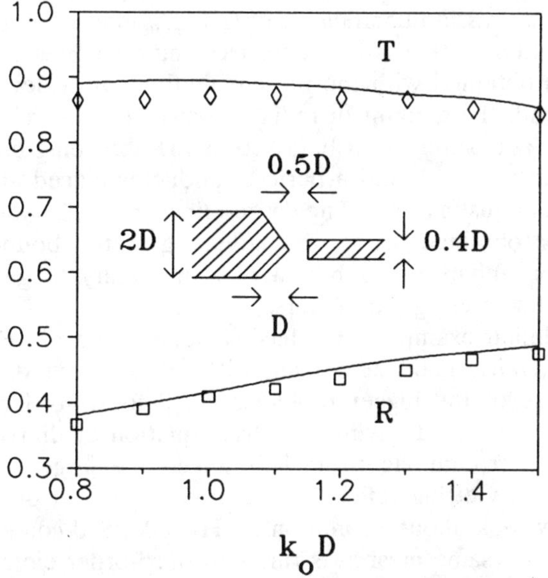

Fig. 3. Reflection (R) and transmission (T) coefficients of a slab waveguide feed structure for the TE_0 mode. The incident wave is from the left. The relative permittivity is 5. The squares and diamonds are from the finite-element method using 1010 fourth-order elements. The solid lines are from [5].

Fig. 2 illustrates the application of the method to a rectangular discontinuity in a slab waveguide. The modes of this waveguide are nonzero out to infinity; nevertheless, the fields decay exponentially with distance from the dielectric. In this example, each port was taken to be of

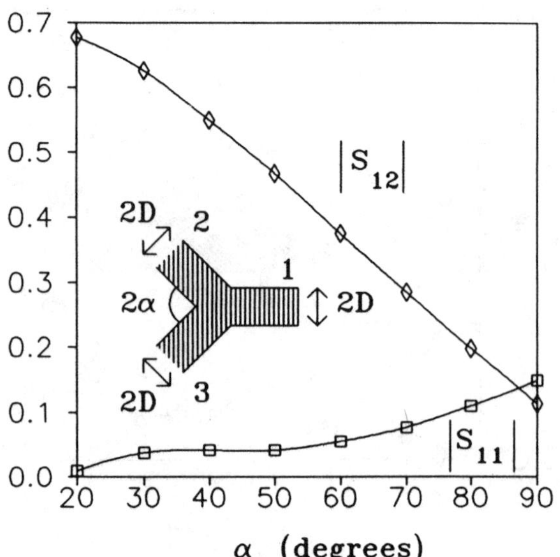

Fig. 4. Scattering parameters of a slab waveguide Y junction for the TE_0 mode. The dielectric constant is 2.25. Squares and diamonds are finite-element results, through which the solid line was passed.

width $4D$, i.e., twice the thickness of the dielectric. Fig. 2 shows the positions of the ports and also shows how absorbing boundary arcs are used to connect the ports together. The solid curves in the figure are the reflection and transmission coefficients of the TE_0 mode, computed by another method [5]. The squares and diamonds are the values obtained with the present method. In a device of this kind, the amount of radiated power is relatively low, and the question might be asked, is the absorbing boundary necessary? Would a perfect conductor placed at that radius do just as well? The dashed lines in Fig. 2 show the results obtained using a conducting outer boundary. Clearly, an absorbing boundary is necessary to get the correct scattering parameters.

A similar example is the feed structure shown in Fig. 3. Again, fairly good agreement with [5] is obtained. This was one of the bigger problems: 1010 fourth-order elements were used, giving a matrix equation of dimension 8277. A frontal algorithm [13] solved the large, sparse equation. Getting reflection and transmission at one frequency took about 11 min on an HP9000/840 computer. When the same mesh was run with third-order elements, the reflection and transmission coefficients changed by 0.001 at most, suggesting a high degree of convergence. The absorbing boundary arcs were at a radius of about $11D$, and increasing the radius had little effect on the results.

The method is not limited to two-port problems. Fig. 4 shows the scattering parameters of a Y junction. Unfortunately, no results are available for comparison. The difference between third- and fourth-order results was at most 0.001 for either scattering parameter.

V. Conclusions

The absorbing boundary condition described above is straightforward to impose and quite flexible. It provides a simple way to extend established finite-element techniques so that they may be applied to planar devices in which there is significant loss due to radiation.

TABLE I
Permutation Vectors: P

ORDER 1:	3	1	2												
ORDER 2:	6	3	5	1	2	4									
ORDER 3:	10	6	9	3	5	8	1	2	4	7					
ORDER 4:	15	10	14	6	9	13	3	5	8	12	1	2	4	7	11

TABLE II
Edge Vector: E

1	2	0	3	0	0	4	0	0	0	5	0	0	0	0

TABLE III
Matrices $[A']$ and $[B']$ ($[B']$ the Upper Triangle, $[A']$ the Lower)

ORDER 1. Denominator for [A'] = 6; denominator for [B'] = 1.

2		1	-1
1	2		1

ORDER 2. Denominator for [A'] = 30; denominator for [B'] = 3.

4		7	-8	1
2	16		16	-8
-1	2	4		7

ORDER 3. Denominator for [A'] = 1680; denominator for [B'] = 280.

128		1036	-1323	378	-91
99	648		3024	-2079	378
-36	-81	648		3024	-1323
19	-36	99	128		1036

ORDER 4. Denominator for [A'] = 5670; denominator for [B'] = 945.

292		4925	-6848	3048	-1472	347
296	1792		16640	-14208	5888	-1472
-174	-384	1872		22320	-14208	3048
56	256	-384	1792		16640	-6848
-29	56	-174	296	292		4925

Appendix
Matrices $[A^{(e)}]$ and $[B^{(e)}]$

To get entry i, j of $[A^{(e)}]$, first determine a new pair of indices, i', j', as follows:

$$
\begin{aligned}
i' &= E(i) && \text{if } e = 1 \\
&= E(P(i)) && \text{if } e = 2 \\
&= E(P(P(i))) && \text{if } e = 3
\end{aligned}
$$

and similarly for j'. The permutation vector P and the edge vector E are given in Tables I and II. (P is taken from the array IPTRI1 in [14].) If either i' or j' is zero, entry i, j of $[A^{(e)}]$ is zero. Otherwise, it is equal to entry i', j' of another symmetric matrix, called $[A']$. The lower triangle of each of the $[A']$ matrices can be obtained from Table III.

The same procedure can be used to obtain the i, j entry of $[B^{(e)}]$.

References

[1] J. P. Webb and S. Parihar, "Finite element analysis of *H*-plane rectangular waveguide problems," *Proc. Inst. Elec. Eng.*, p. H, vol. 133, no. 2, pp. 91–94, Apr. 1986.

[2] K. Ise and M. Koshiba, "Numerical analysis of *H*-plane waveguide junctions by combination of finite and boundary elements," *IEEE Trans. Microwave Theory Tech.*, vol. 36, pp. 1343–1351, Sept. 1988.

[3] J. F. Lee and Z. J. Cendes, "An adaptive spectral response modelling procedure for multiport microwave circuits," *IEEE Trans. Microwave Theory Tech.*, vol. MTT-35, pp. 1240–s1247, Dec. 1987.

[4] S. Washisu, I. Fukai, and M. Suzuki, "Extension of finite-element method to unbounded field problems," *Electron. Lett.*, vol. 15, no. 24, pp. 772–774, 22 Nov. 1979.

[5] S. J. Chung and C. H. Chen, "A partial variational approach for arbitrary discontinuities in planar dielectric waveguides," *IEEE Trans. Microwave Theory Tech.*, vol. 37, pp. 208–214, Jan. 1989.

[6] K. Hirayama and M. Koshiba, "Analysis of discontinuities in an open dielectric slab waveguide by combination of finite and boundary elements," *IEEE Trans. Microwave Theory Tech.*, vol. 37 pp. 761–768, Apr. 1989.

[7] A. Bayliss, M. Gunzburger, and E. Turkel, "Boundary conditions for the numerical solution of elliptic equations in exterior regions," *SIAM J. Appl. Math.*, vol. 42, no. 2, pp. 430–451, Apr. 1982.

[8] R. Mittra, O. Ramahi, A. Khebir, R. Gordon, and A. Kouki, "A review of absorbing boundary conditions for two and three-dimensional electromagnetic scattering problems," *IEEE Trans. Magn.*, vol. 25, pp. 3034–3039, July 1989.

[9] R. E. Collin, *Foundations of Microwave Engineering*. New York: McGraw-Hill, 1966.

[10] M. Koshiba and M. Suzuki, "Finite-element analysis of *H*-plane waveguide junction with arbitrarily shaped ferrite post," *IEEE Trans. Microwave Theory Tech.*, vol. MTT-34, pp. 103–109, Jan. 1986.

[11] P. Silvester, "A general high-order finite-element waveguide analysis program," *IEEE Trans. Microwave Theory Tech.*, vol. MTT-21, pp. 538–542, 1969.

[12] N. Marcuvitz, *Waveguide Handbook*. New York: McGraw-Hill, 1951.

[13] B. M. Irons, "A frontal solution program for finite element analysis," *Int. J. Numer. Meth. Eng.*, vol. 2, pp. 5–32, 1970.

[14] P. Silvester, "Permutation rules for simplex finite elements," *Int. J. Numer. Meth. Eng.*, vol. 18, pp. 1245–1259, Aug. 1982.

A Numerical Study of Vector Absorbing Boundary Conditions for the Finite-Element Solution of Maxwell's Equations

V. N. Kanellopoulos and J. P. Webb

Abstract—**In three-dimensional vector solutions to Maxwell's equations, boundary conditions of the Bayliss–Turkel kind can be used to absorb outgoing radiation. The boundary conditions were used with curvilinear brick finite elements to analyze spherical test problems for which the exact fields are known. Errors due to incomplete absorbtion decrease as the outer boundary is moved further away. The second-order boundary condition is appreciably more accurate than the first-order, at the same cost.**

I. Introduction

The finite element method is an effective way of solving Maxwell's equations numerically, particularly when dielectric and inhomogeneous materials are present. It is, for example, well-suited to the prediction of field intensities in living tissue, desirable in microwave hyperthermia treatment [1]. However, the method solves the equations in a finite volume. On the outer boundary of the volume, conditions must be imposed which represent the infinite free space beyond, into which electromagnetic waves may be flowing. One way of doing this is to couple to an integral equation on the boundary; this leads to a hybrid method that unfortunately lowers the sparsity of the finite-element matrices. An alternative is to use an absorbing boundary condition—a differential equation imposed on the boundary, and solved along with the equations within. Such absorbing conditions are only exactly correct when the boundary is at infinity. Nevertheless, they can be accurate enough at reasonable distances, and have the advantage of preserving sparsity.

The simplest absorbing boundary condition amounts to no more than the Sommerfeld radiation condition. However, higher order conditions exist that offer greater accuracy. The higher order forms were first developed for the scalar case [2]. Recently, three second-order vector conditions were presented [3]–[5]. Only one of these [4] is symmetric, i.e., preserves the symmetry of the finite-element matrix. To date, there has been very little numerical verification of any of these second-order conditions. In this letter, we give numerical results obtained using the symmetric form [4], including an examination of the reduction of error with the distance to the absorbing boundary.

Manuscript received June 5, 1991.

The authors are with the Computational Analysis and Design Laboratory, Department of Electrical Engineering, McGill University, 3480 University Street, Montreal, PQ, Canada H3A 2A7.

IEEE Log Number 9103568.

II. Theory

The solution to the problem is a stationary point of this functional:

$$F(H) = \int_V \left[\frac{1}{\epsilon_r} (\nabla \times H)^2 - k^2 \mu_r H^2 \right] dV + \int_{S_a} \left[jkH_t^2 + \beta(r)(\nabla \times H)_r^2 - \beta(r)(\nabla \cdot H_t)^2 \right] dS. \quad (1)$$

H is the unknown, complex, vector, magnetic field. V is the volume of interest and S_a is a spherical surface separating V from the infinite volume of free-space outside. k is the free-space wavenumber. Subscripts r and t denote the r and $\theta - \phi$ parts of a vector, respectively, where (r, θ, ϕ) are spherical coordinates based on an origin at the centre of $S_a \beta(r)$ is $1/(2jk + 2/r)\epsilon_r$ and μ_r are the relative permittivity and permeability, respectively. Each may be complex, to represent loss, and inhomogeneous.

The second integral causes a second-order absorbing boundary condition to hold on S_a [4]. If $\beta(r)$ is replaced by zero, the first-order boundary condition applies instead.

In order to find the stationary points of F, the volume V is divided into curvilinear bricks known as covariant-projection elements [6]. These bricks represent H as a mixed first-/second-order polynomial, whose tangential part is continuous from one brick to the next. They have been shown to provide good solutions, free from the effects of spurious modes, and make the imposition of vector boundary conditions as straightforward as it is in the scalar case. The unknowns are the covariant components of H at points throughout V, and requiring that F be stationary results in a linear matrix equation for the unknowns: $Ax = b$, where A is a large, square, sparse, symmetric matrix. Solving for x allows H to be determined at any point in V.

Covariant-projection elements do not impose the normal continuity of H, because it is not generally required: the normal continuity of the magnetic flux density arises naturally from the variational principle. However, because of the divergence term $\nabla \cdot H_t$ in the surface integral, it *is* neces-

Reprinted from *IEEE Microwave and Guided Wave Letters*, vol. 1, no. 11, pp. 325–327, Nov. 1991.

TABLE I

BOUNDARY CONDITIONS FOR THE CASE $(m, n) = (1, 1)$
(r, θ, ϕ): $0.03\lambda \times 22.5° \times 22.5°$

Boundary Surface	Boundary Condition
$r = 0.3\lambda$	Prescribed H_θ and H_ϕ of the (1, 1) function
$r = R$	Absorbing boundary condition
$\theta = 3°$	Prescribed H_r and H_ϕ of the (1,1) function
$\theta = 90°$	Electric wall
$\phi = 0°$	Electric wall
$\phi = 90°$	Magnetic wall

TABLE II

BOUNDARY CONDITIONS FOR THE CASE $(m, n) = (1, 2)$
(r, θ, ϕ): $0.03\lambda \times 15° \times 18°$

Boundary Surface	Boundary Condition
$r = 0.3\lambda$	Prescribed H_θ and H_ϕ of the (1, 2) function
$r = R$	Absorbing boundary condition
$\theta = 45°$	Electric wall
$\theta = 90°$	Magnetic wall
$\phi = 0°$	Electric wall
$\phi = 90°$	Magnetic wall

sary to impose normal continuity on $\boldsymbol{H}$ on the surface only. The second-order results shown next were obtained in this way.

III. RESULTS

Harrington [7] gives expressions for spherical TE wave functions, which are exact solutions of Maxwell's equations in spherical coordinates. Each function is characterized by two integers, (m, n); the ϕ variation increases with m and the θ variation increases with n. For each (m, n), there are two functions, one representing an outgoing wave, and one an incoming wave.

Now consider a boundary-value problem in which H_θ and H_ϕ on the surface of a sphere are constrained to be as they are for a spherical wave function (m, n), and outside the sphere is infinite free space. The solution is a single, outgoing, spherical wave function (m, n).

This boundary-value problem was solved using the finite element method, and the results compared with the exact answers. The cases $(m, n) = (1, 1)$ and $(m, n) = (1, 2)$ are considered here. Both cases are three-dimensional in the sense that the field varies with r, θ and ϕ. Because of symmetry, only sectors of the sphere were modeled; details are given in Tables I and II. The curvilinear nature of the elements made them easy to fit to the spherical inner and outer boundaries (Fig. 1). Note that in the case (1, 1), to exclude the difficult line $\theta = 0$, it was necessary to make the field satisfy the exact solution on $\theta = 3°$ as well as on $r = 0.3\lambda$.

Figs. 2 and 3 show how the error changes as the outer boundary $r = R$ is moved outwards, for first and second-order absorbing boundary conditions. The error shown is the largest value of

$$e = |H_{\text{FEM}} - H_{\text{exact}}|$$

over the volume modeled, expressed as a percentage of

$$|H_{\text{exact}}|$$

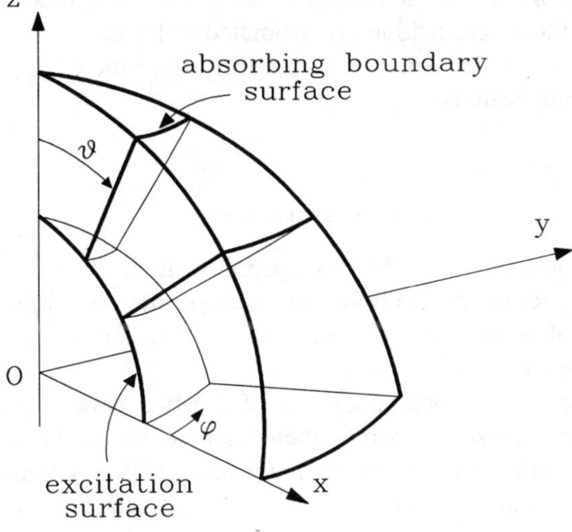

Fig. 1. Modeling a sector of a sphere with curvilinear bricks. Note the degenerate brick touching the z axis.

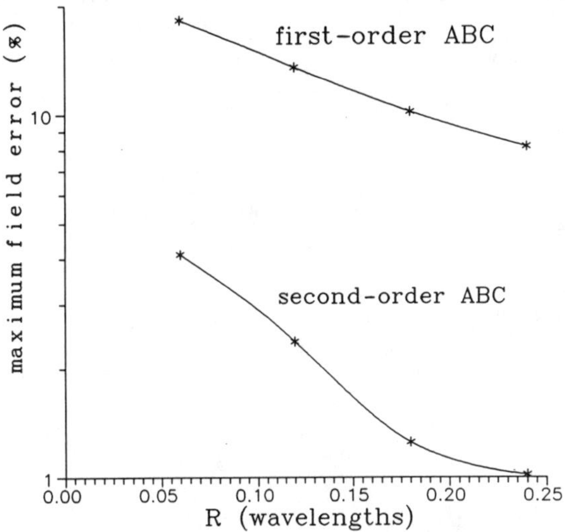

Fig. 2. Solution error versus R, the radius of the absorbing boundary, for the case $(m, n) = (1, 1)$.

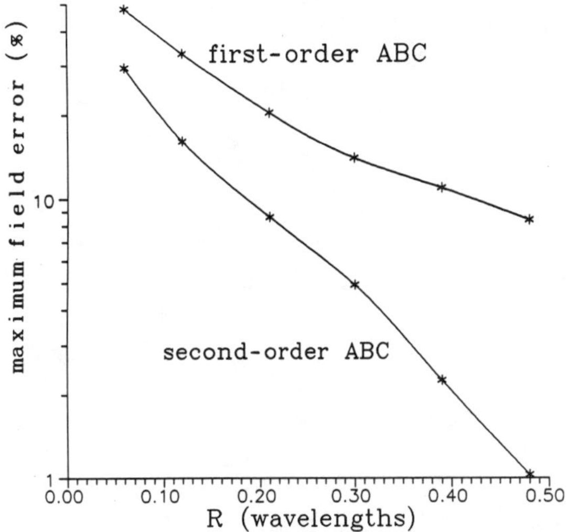

Fig. 3. Solution error versus R, the radius of the absorbing boundary, for the case $(m, n) = (1, 2)$.

at the point where the largest value of e occurs. Inaccuracies due to the discretization are estimated to be less than 1%, so the errors plotted arise almost entirely from the effects of the absorbing boundary.

Conclusion

The graphs show that, compared with the first-order absorbing boundary condition, the second-order condition gives errors that are smaller, and decrease faster as the outer boundary is moved away.

Since the computational cost of the two boundary conditions is almost the same, there can be no doubt that the second-order form is to be preferred. It allows reasonable accuracies to be obtained when the outer surface is still quite near.

Absorbing boundary conditions of order greater than two were not implemented. Such higher order conditions would introduce higher order derivatives, altering the sparsity pattern of the final matrix. Symmetric bilinear forms for orders greater than two have not been developed.

References

[1] K. D. Paulsen, D. R. Lynch, and J. W. Strohbehn, "Three dimensional finite, boundary and hybrid element solutions of the Maxwell equations for lossy dielectric media," *IEEE Trans. Microwave Theory Tech.*, vol. 36, pp. 682–693, Apr. 1988.

[2] A. Bayliss, M. Gunzburger, and E. Turkel, "Boundary conditions for the numerical solution of elliptic equations in exterior regions," *SIAM J. Appl. Math.*, vol. 42, no. 2, pp. 430–451, Apr. 1982.

[3] A. F. Peterson, "Absorbing boundary conditions for the vector wave equation," *Microwave and Optical Technol. Lett.*, vol. 1, no. 2, pp. 62–64, Apr. 1988.

[4] J. P. Webb and V. N. Kanellopoulos, "Absorbing boundary conditions for the finite element solution of the vector wave equation," *Microwave and Optical Technol. Lett.*, vol. 2, no. 10, pp. 370–372, Oct. 1989.

[5] J. D'Angelo and I. Mayergoyz, "Finite element methods for the solution of RF radiation and scattering problems," *Electromagn.*, vol. 10, pp. 177–199, 1990.

[6] C. W. Crowley, P. P. Silvester, and H. Hurwitz, "Covariant projection elements for 3-D vector field problems," *IEEE Trans. Magn.*, vol. 24, pp. 397–400, Jan. 1988.

[7] R. F. Harrington, *Time-Harmonic Electromagnetic Fields.* New York: McGraw-Hill, 1961, p. 269.

A Note on the Radiation Boundary Conditions for the Helmholtz Equation

Y.-C. MA

Abstract—**The radiation boundary conditions in two dimensions are derived in the body-fitted coordinate system using the method of successive approximations for the wave envelope function. Results are valid for a convex scatterer with continuous radius of curvature on the surface of the scatterer. In the special case when the computational boundary is circular, the boundary operators derived here are identical to the Bayliss operators. The on-surface radiation condition is examined. It will be shown that for a large conducting circular cylinder, a boundary operator of infinite order should be used due to the breakdown of the asymptotic expansions of the boundary operators on the surface of the scatterer. The leading order result based on the boundary operator of infinite order applied on the surface of the cylinder is the same as the result by the method of physical optics.**

I. Introduction

Solutions of the Helmholtz equation with radiation condition at infinity are of great practical interest. In the numerical computation of the solution of the Helmholtz equation, an exterior problem in an infinite domain is replaced by an approximate interior problem, resulting in a problem on a bounded domain. The approximate interior problem requires a suitable radiation boundary condition at some artificial boundary. Bayliss and Turkel [1] used the asymptotic behavior of the scattered field in cylindrical or spherical coordinate systems to establish a series of differential operators, known as the Bayliss operators, which when applied to the scattered field, annihilate the first n terms of the asymptotic series. Similar boundary conditions for the wave equation have also been studied [2], [3] and a succinct review can be found in [4].

The Bayliss operators of finite order are approximations to the exact radiation boundary condition. The exact boundary condition can be expressed in integral (nonlocal) form or can be expressed as differential (Bayliss) operators of infinite order. Numerical experiments have shown that the second-order Bayliss operator gives very accurate radar cross section (RCS) predictions for a class of scatterers when the artificial boundary is only a few wavelengths away from the scatterer [5].

For a two-dimensional problem, the Bayliss operators are defined on a circular artificial boundary. When the scatterers of interest are long and narrow, the use of Bayliss operators becomes numerically inefficient, and it is preferable to express the radiation boundary condition on a noncircular artificial boundary [5]. In this communication the radiation boundary conditions will be derived on a noncircular artificial boundary using the body-fitted coordinate. The artificial boundary is "parallel" to the shape of the scatterer and therefore the computational domain is minimized. The derivation given here is valid when the scatterer is convex and the radius of curvature of the scatterer is a continuous function of the arc length (measured from some reference point on the scatterer) along the surface of the scatterer. The higher order radiation boundary conditions also involve the derivatives of the radius of curvature of the scatterer with respect to the arc length. To apply higher order radiation boundary conditions the radius of curvature must be smooth to a certain order in the present approach.

The derivation given in this communication is based on the complex wave envelope function rather than the wave field itself [6]. The phase function is defined consistent with the radiation condition at infinity. The reason that the wave envelope function is used is due to the fact that the radiation boundary conditions are derived based on asymptotic expansions and method of successive approximations. In other words, the artificial boundary is at some large but finite distance from the scatterer in order for the asymptotic expansions to be valid. The wave envelope function is a slowly varying function of the radial position at large radial location and therefore the asymptotic expansion is simpler for the wave envelope function than for the wave field. The asymptotic nature of the wave envelope function can be easily understood in the present derivation. In the actual numerical computations one can use larger mesh size for grid points at a large distance from the scatterer if the wave envelope function is used. This is not possible if the wave field is used. On the other hand, for grid points close to the scatterer, one must use small mesh size for both the wave envelope function and the wave field. There is no advantage in using the wave envelope function in the near field.

In Section II the radiation boundary conditions will be derived for a two-dimensional convex scatterer with continuous (or smooth) radius of curvature on the surface of the scatterer. It is assumed that the computational boundary is at some large distance from the scatterer, and the derivation uses the asymptotic property of the wave envelope function on the computational boundary to generate an asymptotic expansion of the Helmholtz operator. The asymptotic expansion is similar to the Neumann series for solving integral equations, except that the integral operator in the governing equation is replaced by the differential operator. Unlike the Bayliss operators, the boundary operators derived in Section II do not involve normal derivatives higher than first order which are desirable. The boundary operators for the wave envelope function can be easily transformed to the boundary operators for the wave field. In the special case where the scatterer is a circular cylinder the computational boundary becomes a circle in the present approach. The boundary operators are identical to the Bayliss operators to the same order of accuracy.

Recently, a new formulation of electromagnetic wave scattering, known as the on-surface radiation condition (OSRC) approach, has been proposed [7]–[9]. This approach applies the second order Bayliss operator directly on the surface of a two-dimensional convex scatterer, instead of at some distance away from the scatterer. Significant agreements have been obtained by comparing solutions based on the OSRC approach and the exact solutions or solutions based on moment method [7]–[9].

In Section III the boundary operators obtained in Section II will be applied to the surface of a circular cylinder of large radius (large compared with the wavelength of EM waves) instead of at some distance from the cylinder. It will be shown that the asymptotic

Reprinted from *IEEE Trans. Antennas Propagat.*, vol. 39, no. 10, pp. 1526–1530, Oct. 1991.

expansions of the boundary operators break down. To estimate the surface current on the cylinder based on the OSRC approach, one need the boundary operator of infinite order. The zeroth-order effect of the boundary operator of infinite order on the surface of a cylinder of large radius can be summed up to a finite function. It will be shown the zeroth-order effect is the same as by the method physical optics. This suggests that the OSRC approach based on the second order Bayliss operator does not agree with the asymptotic result, i.e., physical optics result, for scattering from a body of very large radius of curvature.

II. Derivation of Boundary Operators in a Body-fitted Coordinate

As mentioned in the Introduction, it is preferable to express the boundary operators in a body-fitted coordinate in order to minimize the computational domain. The Helmholtz equation in the intrinsic coordinate (s, n) for a two-dimensional convex scatterer is given by [10];

$$\frac{\partial}{\partial s}\left[\frac{\partial\phi/\partial s}{1+n/a(s)}\right]+\frac{\partial}{\partial n}\left[\left(1+\frac{n}{a(s)}\right)\frac{\partial\phi}{\partial n}\right]+k^2\left(1+\frac{n}{a(s)}\right)\phi=0. \quad (1)$$

In (1) n is the normal distance from the surface (where $n = 0$) of the scatterer to the field point of interest and s is the arc length along the surface of the scatterer measured from certain reference point. ϕ is the wave field, k is the wavenumber, and $a(s)$ is the radius of curvature on the surface of the scatterer at s (see Fig. 1).

To analyze (1), the dimensionless coordinate will be used. Letting $n' = kn$, $a' = ka$, $s' = ks$ and omitting the primes, (1) becomes

$$\frac{\partial}{\partial s}\left[\frac{\partial\phi/\partial s}{1+n/a(s)}\right]+\frac{\partial}{\partial n}\left[\left(1+\frac{n}{a(s)}\right)\frac{\partial\phi}{\partial n}\right]+\left(1+\frac{n}{a(s)}\right)\phi=0. \quad (2)$$

The computational boundary for the boundary operators is taken to be at a large but constant value of n ($n \gg 1$). The (complex) wave envelope function is related to ϕ by the following relation;

$$\phi = e^{i(n+n_0)}f(s,n) \quad (3)$$

where n_0 is a constant.

Substituting (3) into (2) and rearranging various terms, the following equation can be obtained:

$$\left(2i+\frac{1}{2(n+a)}\right)\left(\frac{\partial f}{\partial n}+\frac{f}{2(n+a)}\right) = -\left[\frac{a}{n+a}\frac{\partial}{\partial s}\left(\frac{\partial f/\partial s}{1+n/a(s)}\right)+\frac{f}{4(n+a)^2}\right] - \frac{\partial}{\partial n}\left(\frac{\partial f}{\partial n}+\frac{f}{2(n+a)}\right). \quad (4)$$

The asymptotic property ($n \gg 1$) of the wave envelope function will be analyzed. Assuming $a^2\partial^2 f/\partial s^2$ is of same order as f for $n \gg 1$, the first term on the right-hand side of (4) (the term inside the square bracket) is a higher order term. The two independent asymptotic solutions of (4) are

$$f \sim (n+a)^{-\frac{1}{2}}f_1(s) \quad (5a)$$

$$f \sim (n+a)^{-\frac{1}{2}}e^{-2i(n+a)}f_2(s), \quad \text{for } n \gg 1. \quad (5b)$$

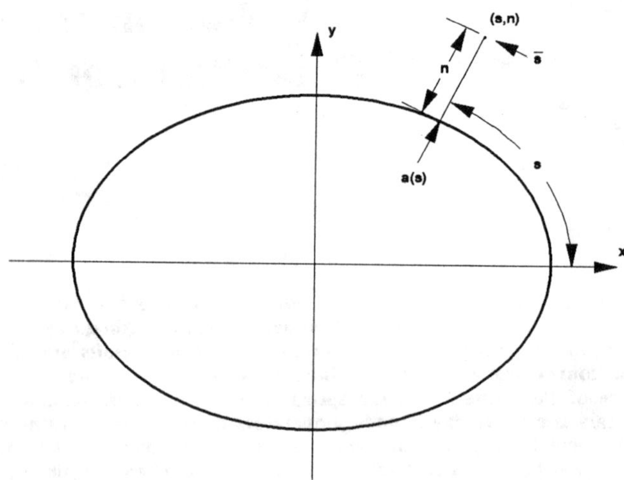

Fig. 1. Intrinsic coordinate for the Helmholtz equation in two dimensions.

The radiation condition of (4) requires that only outgoing waves are allowed at infinity. The solution given by (5b) corresponds to an incoming wave and must be discarded. The solution given by (5a) satisfies the radiation condition at infinity and is the solution of interest. For this solution the second term on the right-hand side of (4) $(\partial/\partial n(\partial f/\partial n + f/2(n+a)))$ is also a higher order term. The lowest order boundary operator is given by the left-hand side of (4);

$$\frac{\partial f}{\partial n}+\frac{f}{2(n+a)}=0. \quad (6)$$

The error incurred in the lowest order boundary operator is $0(n + a)^{-(5/2)}$.

The second boundary operator includes the first term of the right-hand side of (4);

$$\frac{\partial f}{\partial n}+\frac{f}{2(n+a(s))} = -\frac{1}{2i}\cdot\left[\frac{a(s)}{n+a(s)}\frac{\partial}{\partial s}\left(\frac{\partial f/\partial s}{1+n/a(s)}\right)+\frac{f}{4(n+a)^2}\right]. \quad (7)$$

The error in this boundary operator is of order $(n + a)^{-(7/2)}$. Sometimes it is preferable to express the tangential derivative along the computational boundary. Let $\bar{s}$ be the arc length measured along the computational boundary. $\bar{s}$ and s are related by the following relation:

$$d\bar{s} = \frac{n+a(s)}{a(s)}ds, \quad \text{or } \bar{s} = \int \frac{n+a(s)}{a(s)}\,ds. \quad (8)$$

With the relationship given by (8), the second boundary operator can also be expressed as

$$\frac{\partial f}{\partial n}+\frac{f}{2(n+a(s))} = -\frac{1}{2i}\left[\frac{\partial^2 f}{\partial \bar{s}^2}+\frac{f}{4(n+a)^2}\right]. \quad (9)$$

In applying (9), care must be taken in evaluating the partial derivatives. Since $\bar{s}$ is a function of s and n, the normal derivative $\partial/\partial n$ is measured such that the arc length (s) along the surface of the scatterer is fixed and the tangential derivative $\partial/\partial\bar{s}$ is measured along the computational boundary such that the normal coordinate (n) is fixed.

The second term on the right-hand side of (4) is of order $(n + a)^{-(7/2)}$ due to the radiation condition just mentioned. The third order boundary operator will therefore include this term. The third order boundary operator as well as higher order boundary operators can be obtained via iterations similar to the method of successive approximations. In essence the previous lower order operator is substituted into the second term on the right-hand side of (4) and then the normal derivative is performed. The third and fourth boundary operators are given by

$$\frac{\partial f}{\partial n}+\frac{f}{2(n+a)}=-\frac{1}{2i}\left(1-\frac{i}{n+a}\right)\frac{\partial^2 f}{\partial \bar{s}^2}+\frac{i}{8(n+a)^2}\left(1-\frac{i}{n+a}\right)f \quad (10)$$

$$\frac{\partial f}{\partial n}+\frac{f}{2(n+a)}=-\frac{1}{2i}\left(1-\frac{i}{n+a}-\frac{13}{8(n+a)^2}\right)\frac{\partial^2 f}{\partial \bar{s}^2}+\frac{i}{8(n+a)^2}\left(1-\frac{i}{n+a}-\frac{25}{16(n+a)^2}\right)f+\frac{1}{8i}\frac{\partial^4 f}{\partial \bar{s}^4}-\frac{aa'}{2(n+a)^3}\frac{\partial f}{\partial \bar{s}}-\frac{aa'^2+a^2a''}{8(n+a)^4}f, \quad (11)$$

where $a' = da/ds$ and $a'' = d^2a/ds^2$. The errors in (10) and (11) are $0(n+a)^{-(9/2)}$ and $0(n+a)^{-(11/2)}$, respectively.

In the special case where the scatterer is a circular cylinder, $n_0 = a =$ constant. The computational boundary becomes a circle of radius $R = n + a$. The arc length along the surface of the scatterer is $s = a\theta$ and the arc length along the computational boundary is $\bar{s} = R\theta$. The wave field is related to the wave envelope function by

$$\phi = e^{iR}f(R,\theta). \quad (12)$$

Equation (4) becomes

$$\left(2i+\frac{1}{2R}\right)\left(\frac{\partial f}{\partial R}+\frac{f}{2R}\right)=-\left(\frac{1}{R^2}\frac{\partial^2 f}{\partial \theta^2}+\frac{f}{4R^2}\right)-\frac{\partial}{\partial R}\left(\frac{\partial f}{\partial R}+\frac{f}{2R}\right). \quad (13)$$

The boundary operators given by (6), (9), (10), and (11) become

$$\frac{\partial f}{\partial R}+\frac{f}{2R}=0 \quad (14)$$

$$\frac{\partial f}{\partial R}+\frac{f}{2R}=-\frac{1}{2i}\left(\frac{1}{R^2}\frac{\partial^2 f}{\partial \theta^2}+\frac{f}{4R^2}\right). \quad (15)$$

$$\frac{\partial f}{\partial R}+\frac{f}{2R}=\frac{-1}{2iR^2}\left(1-\frac{i}{R}\right)\frac{\partial^2 f}{\partial \theta^2}+\frac{i}{8R^2}\left(1-\frac{i}{R}\right)f \quad (16)$$

$$\frac{\partial f}{\partial R}+\frac{f}{2R}=-\frac{1}{2iR^2}\left(1-\frac{i}{R}-\frac{13}{8R^2}\right)\frac{\partial^2 f}{\partial \theta^2}+\frac{i}{8R^2}\left(1-\frac{i}{R}-\frac{25}{16R^2}\right)f+\frac{1}{8iR^4}\frac{\partial^4 f}{\partial \theta^4}. \quad (17)$$

The boundary operators given by (14)–(17) defined on a circular computational boundary can be compared to the Bayliss operators. Since the Bayliss operators are defined for the wave field rather than for the wave envelope function, (13) must be used to transform the Bayliss operators in terms of the wave envelope function. The Bayliss operators also involve higher order radial derivatives but not the boundary operators derived here. The higher order radial derivatives can be replaced by tangential derivatives and lower order radial derivatives using the Helmholtz equation. The computations to relate the Bayliss operators to the boundary operators derived here are straightforward and will be omitted. To the same order of accuracy, the Bayliss operators are identical to the boundary operators given in this communication.

For a noncircular boundary, the first three boundary operators are identical to the boundary operators for a circular boundary, except that the local value of the radius of curvature should be used along the scatterer surface. Higher order (higher than the third) boundary operators also involve the rate of change and the higher derivatives of the radius of curvature along the scatterer surface which do not appear on a circular boundary.

III. Application of the Boundary Operators on the Surface of a Circular Cylinder of Large Radius

In this section the boundary operators derived previously will be applied on the surface of a two-dimensional perfectly conducting circular cylinder with large radius ($a \gg 1$ where a is the dimensionless radius normalized by the wavenumber of the incident wave) and examine the applicability of the OSRC approach.

Let us examine the OSRC approach based on the second boundary operator. From (15) one can compute the radial derivative of the wave envelope function on the surface of the cylinder and therefore one can compute the surface current and the radar cross section. On the other hand, the first term on the right-hand side of (15) ($(i/2)R^{-2}\partial^2 f/\partial\theta^2$) is actually of order one (except for $\theta \sim 0$ and $\theta \sim \pi$) on the surface of the cylinder instead of order a^{-2}. (This will become obvious.) In other words the radial derivative of the wave envelope function evaluated on the surface of the cylinder is of order one and the asymptotic expansions given in Section II break down. Likewise for the fourth-order boundary operator, (17), there are two terms on the right-hand side which are of order one when evaluated on the surface of the cylinder (terms proportional to $R^{-2}f_{\theta\theta}$ and $R^{-4}f_{\theta\theta\theta\theta}$). It suggests that a boundary operator of infinite order is needed if it is evaluated on the surface of a scatterer.

By examining the form of the boundary operators ((14)–(17)) one can deduce the general form of the boundary operator of infinite order;

$$\frac{\partial f}{\partial R}+\frac{f}{2R}=\frac{1}{2i}\sum_{n=1}^{\infty}\frac{a_n(R)}{R^{2n}}\frac{\partial^{2n} f}{\partial \theta^{2n}}+\frac{1}{R^2}\sum_{n=0}^{\infty}\frac{c_n}{R^n}f \quad (18)$$

with $a_n(R) = a_n^{(0)}(R) + (1/R)a_n^{(1)}(R) + \cdots$.

To examine the leading order effect on $\partial f/\partial R$, the second sum on the right-hand side of (18) can be ignored. For the first sum on the right-hand side of (18) only the leading order term ($a_n^{(0)}$) is important as $R = a \to \infty$. Each term in the sum is of same order (order one) and therefore the series cannot be truncated to a finite order. To leading order (18) can be simplified to

$$\frac{\partial f}{\partial R}+\frac{f}{2R}\sim\frac{1}{2i}\sum_{n=1}^{\infty}\frac{a_n^{(0)}}{R^{2n}}\frac{\partial^{2n} f}{\partial \theta^{2n}}. \quad (19)$$

Substituting (19) into (13) and evaluating the order one terms, it is straightforward to verify

$$a_1^{(0)} = -1. \quad (20)$$

The right-hand side of (19) can be used to evaluate $\partial^2 f/\partial R^2$ in (13). An recurrence relation for $a_n^{(0)}$ $(n > 1)$ can be obtained

$$a_n^{(0)} = \frac{1}{4}\sum_{l=1}^{n-1} a_l^{(0)}a_{n-l}^{(0)} \quad (n > 1). \quad (21)$$

Equation (21), together with (20), can determine $a_n^{(0)}$ for all values of n.

With (19)–(21) the surface current can be evaluated. For a transverse magnetic (TM) mode the quantity ϕ is the axial component of the scattered electric field. The boundary condition on the surface of the cylinder requires that the total axial electric field be zero, i.e.,

$$\phi(a,\theta) = -e^{-ia\cos\theta} \tag{22}$$

where the incident wave is in the negative x-direction. In terms of the wave envelope functioh, (22) becomes

$$f(a,\theta) = -e^{-ia(1+\cos\theta)}. \tag{23}$$

From (23) it is clear that $a^{-2n}\partial^{2n}f/\partial\theta^{2n}$ is of order one for any value of n.

Using (19) the value of $\partial f/\partial R$ to leading order is

$$\frac{\partial f}{\partial R}(a,\theta) \simeq \frac{1}{2i}\sum_{n=1}^{\infty}(-1)^n a_n^{(0)}\sin^{2n}\theta f(a,\theta) \tag{24}$$

where $f(a,\theta)$ is given by (23). The recurrence relation given by (23) can be used to prove the following:

$$1-\frac{1}{2}\sum_{n=1}^{\infty}(-1)^n a_n^{(0)}\sin^{2n}\theta = \cos\theta, \qquad \text{for } 0 \le \theta \le \frac{\pi}{2}, \tag{25a}$$

$$= -\cos\theta, \qquad \text{for } \pi/2 \le \theta \le \pi. \tag{25b}$$

To prove (25), take the square of the left-hand side of (25).

$$\begin{aligned}\left(1-\frac{1}{2}\sum_{n=1}^{\infty}(-1)^2 a_n^{(0)}\sin^{2n}\theta\right)^2 &= 1-\sum_{n=1}^{\infty}(-1)^n a_n^{(0)}\sin^{2n}\theta \\ &\quad +\frac{1}{4}\sum_{n_1=1}^{\infty}\sum_{n_2=1}^{\infty}(-1)^{n_1+n_2}a_{n_1}^{(0)}a_{n_2}^{(0)}\sin^{2(n_1+n_2)}\theta.\end{aligned} \tag{26}$$

The last term on the right-hand side of (26) can be simplified using (23);

$$\begin{aligned}\frac{1}{4}\sum_{n_1=1}^{\infty}\sum_{n_2=1}^{\infty}(-1)^{n_1+n_2}a_{n_1}^{(0)}a_{n_2}^{(0)}\sin^{2(n_1+n_2)\theta} &= \frac{1}{4}\sum_{M=2}^{\infty}\sum_{l=1}^{\infty}(-1)^M a_l^{(0)}a_{M-l}^{(0)}\sin^{2M}\theta \\ &= \sum_{M=2}^{\infty}(-1)^M a_M^{(0)}\sin^{2M}\theta.\end{aligned} \tag{27}$$

Substituting (27) into (26), we find

$$1-\frac{1}{2}\sum_{n=1}^{\infty}(-1)^n a_n^{(0)}\sin^{2n}\theta = \pm\cos\theta. \tag{28}$$

The sign in front of $\cos\theta$ in (28) can be determined by evaluating $\theta = 0$ and $\theta = \pi$. This completes the proof of (25).

The surface current in the z-direction to leading order can be estimated;

$$\begin{aligned}J_z = H_\theta &= \frac{j}{Z}\frac{\partial}{\partial R}(\phi + E_z^i) \\ &= \frac{1}{Z}\left(1-\frac{1}{2}\sum_{n=1}^{\infty}(-1)^n a_n^{(0)}\sin^{2n}\theta + \cos\theta\right)e^{-ia\cos\theta},\end{aligned} \tag{29}$$

where Z is the free space impedance.

With (25), the surface current becomes

$$J_z = \frac{2\cos\theta}{Z}e^{-ia\cos\theta}, \qquad \text{for } 0 \le \theta \le \pi/2, \tag{30a}$$

$$= 0, \qquad \text{for } \pi/2 \le \theta \le \pi. \tag{30b}$$

The surface current obtained by estimating the leading order effect of a boundary operator of infinite order is identical to the result obtained by the method of physical optics.

For a transverse electric (TE) mode the quantity ϕ is the axial component of the scattered magnetic field. The boundary condition on the surface of the cylinder is

$$\frac{\partial H_z^s}{\partial r} = \frac{\partial\phi}{\partial r} = -\frac{\partial H_z^i}{\partial r} = ik\cos\theta\, e^{ikr\cos\theta}. \tag{31}$$

In terms of dimensionless coordinates the wave envelope function satisfies

$$\frac{\partial f}{\partial R} + if = i\cos\theta\, e^{-ia(1+\cos\theta)}, \qquad \text{at } R = a. \tag{32}$$

Substituting (19) into (32) and keeping the leading order terms, it can be shown that

$$-\frac{1}{2}\sum_{n=1}^{\infty}\frac{a_n^{(0)}}{a^{2n}}\frac{\partial^{2n}f}{\partial\theta^{2n}} + f = \cos\theta\, e^{-ia(1+\cos\theta)}, \qquad \text{at } R = a. \tag{33}$$

The leading order solution of (33) can be expressed as

$$f = f_0 e^{-ia(1+\cos\theta)}, \qquad \text{at } R = a, \tag{34}$$

where

$$f_0 = \cos\theta \Big/ \left[1-\frac{1}{2}\sum_{n=1}^{\infty}(-1)^n a_n^{(0)}\sin^{2n}\theta\right].$$

Applying (25), $f_0 = 1$ for $0 \le \theta \le \pi/2$ and $f_0 = -1$, for $\pi/2 \le \theta \le \pi$. The surface current becomes

$$J_\theta = -H_z = -2e^{-ia\cos\theta}, \qquad \text{for } 0 \le \theta \le \pi/2, \tag{35a}$$

$$= 0, \qquad \text{for } \pi/2 \le \theta \le \pi. \tag{35b}$$

The results given by (35) are the same as the results by the method of physical optics.

The agreements of the above results with the method of physical optics for TE and TM modes are not surprising. The boundary operator of infinite order is an exact representation of the boundary condition. For a scatterer of large radius of curvature, the method of physical optics gives the leading order surface current. The leading order surface current based on the boundary operator of infinite order should therefore reduce to the same result as by the method of physical optics. It should be mentioned that the breakdown of the asymptotic expansions occurs on the surface of the scatterer. It is necessary to use the boundary operator of infinite order only if it is applied on the surface of a scatterer of large radius.

IV. Conclusion

In this note we have shown the derivation of a set of infinite boundary operators for the wave envelope function using the body-fitted coordinate. The derivation is the same as the method of successive approximations for the Neumann series of integral equations.

These operators involve first-order normal derivative and higher order tangential derivatives. In the special case when the computational boundary is circular, these operators are identical to the Bayliss operators to the same order of accuracy. The on-surface radiation condition has been examined by applying these operators on the surface of a conducting circular cylinder of large radius. A boundary operator of infinite order must be used to obtain the physical optics current.

References

[1] A. Bayliss and E. Turkel, "Radiation boundary conditions for wave-like equations," *Commun. Pure Appl. Math.*, vol. 33, pp. 705-725, 1980.

[2] B. Engquist and A. Majda, "Absorbing boundary conditions for the numerical simulation of waves," *Math. Comput.*, vol. 31, pp. 629-651, July 1977.

[3] G. Mur, "Absorbing boundary conditions for the finite-difference approximation of time-domain electromagnetic field equations," *IEEE Trans. Electromagn. Compat.*, vol. EMC-23, pp. 377-382, Nov. 1981.

[4] T. G. Moore, J. G. Blaschak, A. Taflove, and G. A. Kriegsmann, "Theory and application of radiation boundary operators," *IEEE Trans. Antennas Propagat.*, vol. 36, pp. 1797-1812, Dec. 1988.

[5] J. D'Angelo and I. D. Mayergoyz, "On the use of local absorbing boundary conditions for RF scattering problems," *IEEE Trans. Magn.*, vol. 25, pp. 3040-3042, 1989.

[6] R. T. Ling, "Numerical solution for the scattering of sound waves by a circular cylinder," *AIAA J.*, vol. 25, pp. 560-566, Apr. 1987.

[7] G. A. Kriegsmann, A. Taflove, and K. R. Umashankar, "A new formulation of electromagnetic wave scattering using a on-surface radiation boundary condition approach," *IEEE Trans. Antennas Propagat.*, vol. AP-35, pp. 153-161, Feb. 1987.

[8] T. G. Moore, G. A. Kriegsmann, and A. Taflove, "An application of the WKJB technique to the on-surface radiation condition," *IEEE Trans. Antennas Propagat.*, vol. 36, pp. 1329-1331, Sept. 1988.

[9] J.-M. Jin, J. L. Volakis, and V. V. Liepa, "A comparative study of the OSRC approach in electromagnetic scattering," *IEEE Trans. Antennas Propagat.*, vol. 37, pp. 118-124, Jan. 1989.

[10] G. T. Ruck, D. E. Barrick, W. D. Stuart, and C. K. Krichbaum, *Radar Cross Section Handbook*, vol. 1. New York: Plenum, 1970, p. 84.

Numerically Derived Absorbing Boundary Condition for the Solution of Open Region Scattering Problems

Omar M. Ramahi, Ahmed Khebir, and Raj Mittra, *Fellow, IEEE*

***Abstract*—A numerically derived absorbing boundary condition for the solution of open region scattering problems has been developed in this paper via the numerical approximation of the analytical behavior of the exact boundary condition. The important features of this boundary operator are that it is more accurate than other analytically derived differential operators having the same order, and that it can be applied to arbitrarily shaped scatterer geometries that can be handled most efficiently through the use of outer boundaries that conform to the body of the scatterer. Examples are given to demonstrate the improvement in accuracy and efficiency that can be achieved by the new numerical boundary condition.**

I. Introduction

When solving open region scattering problems, an artificial outer boundary must be introduced to limit the region surrounding the scatterer to a manageable size. To render the problem well-posed, a boundary condition must be imposed at the artificial outer boundary, such that it absorbs the waves scattered from the body, i.e., it minimizes the nonphysical reflections from the artificial boundary. In practice, however, it is desirable to bring the outer boundary as close to the scatterer as possible in order to reduce the size of the associated system matrix.

Several boundary conditions have been reported in the literature. These include the nonlocal-type conditions [1]–[5] which use surface integral equations as the boundary condition. The second class of boundary conditions, namely, the local-type, consists of boundary operators that attempt to absorb the waves reflected from the scattering body and, thus, simulate the condition in which the waves are entirely outgoing. These differential boundary operators, referred to henceforth in this paper as the ABC operators, are nonexact in the sense that they introduce nonphysical reflections in the solution region. A number of authors have investigated the ABC's, notably Bayliss *et al.* [6], who employed asymptotic analysis to derive a series of differential operators, henceforth referred to as the BGT operators, and Engquist and Majda [7], who used the pseudo-differential operator theory to derive differential operators. In contrast to the nonlocal type operators, the ABC's generate highly sparce matrices that can be inverted efficiently using a special software routines that require the storage of only the nonzero elements of the matrix [8]. Mittra and Ramahi [9]–[11] have found that the local differential operators reported in [6] and [7] work quite well for small scatterers, but are not as accurate for electrically large ones, even if higher orders of these operators were used.

Manuscript received January 26, 1990; revised September 5, 1990.

O. M. Ramahi and R. Mittra are with the Electromagnetic Communication Laboratory, Department of Electrical and Computer Engineering, University of Illinois, Urbana, IL 61801.

A. Khebir is with Corporate Research and Development, General Electric Co., Schenectady, NY 12301.

IEEE Log Number 9041790.

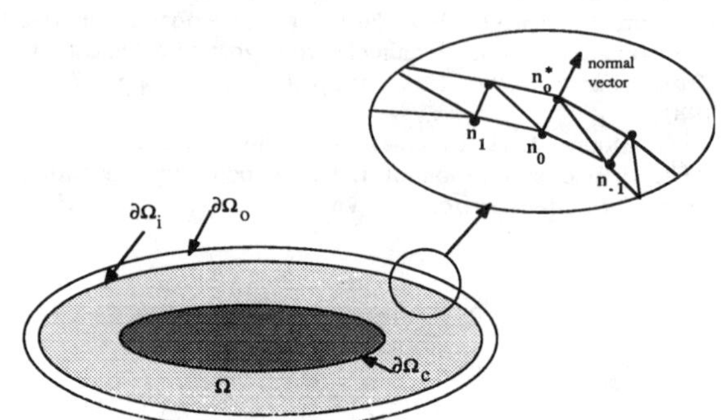

Fig. 1. Geometry for the formulation of the numerically derived boundary condition.

Furthermore, in [9]–[11], it was demonstrated that the absorbing boundary conditions derived from asymptotic analysis exhibit noticeable errors in the higher order harmonic content of the scattered field when the outer boundary is brought very close to the scatterer to reduce the number of mesh points, even when higher order forms of the ABC's are used. The purpose of this paper is to derive a more accurate ABC that circumvents the above-mentioned problem, by numerically approximating the exact analytical form of the boundary condition for a range of harmonic numbers n, which is determined by the size of the scatterer.

II. Formulation

In this section, we present the derivation of numerical absorbing boundary operator by approximating the analytical behavior of the propagating harmonics comprising the scattered field. The formulation presented is independent of the scatterer shape, and it applies to scatterers with dielectric and magnetic material coatings.

For the purpose of deriving a boundary operator applied to the scattered field in free-space, we consider the free-space wave equation:

$$(\nabla^2 + k^2)u = 0. \tag{1}$$

Fig. 1 shows the geometry of an arbitrary scatterer enclosed by an artificial outer boundary, $\partial\Omega_0$. The solution region is meshed such that the interior nodes closest to $\partial\Omega_0$ lie along the contour $\partial\Omega_i$. The nodes involved in the synthesis of the boundary operator are n_0^*, n_0, n_{-1}, and n_1 (see Fig. 1), with cylindrical coordinates (ρ_0^*, ϕ_0^*), (ρ_0, ϕ_0), (ρ_{-1}, ϕ_{-1}), and (ρ_1, ϕ_1), respectively. We then postulate a boundary condition that has a form that allows direct truncation of the mesh region. From past experience with differential boundary operators, the most direct way to truncate the mesh region is by establishing a relationship between the normal derivative of the field at the node n_0 and a weighted combination of the field values at the

Reprinted from *IEEE Trans. Antennas Propagat.*, vol. 39, no. 3, pp. 350–353, March 1991.

nodes lying directly to the interior, namely, the three nodes n_0, n_{-1}, and n_1. We have

$$\left.\frac{\partial u}{\partial v}\right|_{n0} = c_0 u_0 + c_{-1}u_{-1} + c_1 u_1 \tag{2}$$

where u_0, u_{-1}, and u_1 are the field values at the nodes n_0, n_{-1}, and n_1, respectively, and c_0, c_{-1}, and c_1 are complex coefficients yet to be determined. The next step is to express the solution to (1) in terms of the outgoing cylindrical wave functions

$$u(\rho,\phi) = \sum a_n H_n^{(2)}(k\rho)e^{jn\phi}. \tag{3}$$

Strictly speaking, the series in (3) is uniformly convergent only outside a cylindrical region that completely encloses the scatterer; however, one could invoke the Rayleigh hypothesis [12], [13] to justify using it, as an approximation, everywhere on the boundary $\partial\Omega_0$ and outside. The validity of this approximation will depend on the nature and shape of the scattering body [12], [13].

Returning now to (2), we seek the coefficients, c_{-1}, c_1, and c_0 that will give the most accurate approximation of the normal derivative of the unknown field, $\partial u/\partial v$, over the entire range of harmonics that significantly contribute to the series in (3). Substituting (3) into (2), we obtain

$$\begin{aligned}\frac{\partial}{\partial v}\left(H_n^{(2)}(k\rho_0^*)e^{jn\phi_0^*}\right) &= c_0\left(H_n^{(2)}(k\rho_0)e^{jn\phi_0}\right)\\ &+ c_{-1}\left(H_n^{(2)}(k\rho_{-1})e^{jn\phi_{-1}}\right)\\ &+ c_1\left(H_n^{(2)}(k\rho_1)e^{jn\phi_1}\right)\end{aligned} \tag{4}$$

where n ranges over all integer numbers. Next, we enforce (4) at three different harmonic numbers n_1, n_2, and n_3, that lie within the set of harmonics that are relevant to the series in (3), we obtain the matrix equation:

$$\begin{bmatrix} H_{n1}^{(2)}(k\rho_0)e^{jn_1\phi_0} & H_{n1}^{(2)}(k\rho_{-1})e^{jn_1\phi_{-1}} & H_{n1}^{(2)}(k\rho_1)e^{jn_1\phi_1} \\ H_{n2}^{(2)}(k\rho_0)e^{jn_2\phi_0} & H_{n2}^{(2)}(k\rho_{-1})e^{jn_2\phi_{-1}} & H_{n2}^{(2)}(k\rho_1)e^{jn_2\phi_1} \\ H_{n3}^{(2)}(k\rho_0)e^{jn_3\phi_0} & H_{n3}^{(2)}(k\rho_{-1})e^{jn_3\phi_{-1}} & H_{n3}^{(2)}(k\rho_1)e^{jn_3\phi_1} \end{bmatrix} \begin{bmatrix} c_0 \\ c_{-1} \\ c_1 \end{bmatrix} = \begin{bmatrix} \frac{\partial}{\partial v}\left(H_{n1}^{(2)}(k\rho_1^*)e^{jn_1\phi_0^*}\right) \\ \frac{\partial}{\partial v}\left(H_{n2}^{(2)}(k\rho_1^*)e^{jn_2\phi_0^*}\right) \\ \frac{\partial}{\partial v}\left(H_{n3}^{(2)}(k\rho_1^*)e^{jn_3\phi_0^*}\right) \end{bmatrix}. \tag{5}$$

The solution of the 3×3 matrix equation in (5) yields the complete description for the boundary condition in (2). Notice that we have not restricted the geometry of the artificial outer boundary $\partial\Omega_c$ to any particular shape, and, therefore, except for the special case of the circular outer boundary, the matrix system in (5) needs to be solved for each outer boundary node resulting in a *global* matrix system then can be interpreted as a boundary matrix operator.

The boundary condition in (2) employed three nodes on the $\partial\Omega_i$ boundary layer; hence, we call it the three-coefficient absorbing boundary condition (3-c ABC). Similar to the above formulation, more accurate higher order boundary conditions can be constructed (5-c ABC, 7-c ABC,$\cdots$, etc.) by including a higher number of the $\partial\Omega_i$ boundary nodes in the right-hand side of (2).

Up to this point we have not discussed how to select the three harmonic numbers m_1, m_2, and m_3, which are needed for the matrix in (5). If m_1, m_2, and m_3 were chosen to be 0, 1, and 2, then the numerical boundary condition will have an equivalent performance to that of the second-order operators reported in [6], [9], because it will be annihilating the perfectly propagating harmonics. The optimal choice, however, was revealed by extensive numerical experiments, which showed that when using the 3-c ABC, the best result is achieved when two of the harmonic numbers are selected from the propagation region and the third from the cutoff region. In the case of the 5-c ABC, the best performance is obtained when four of the five harmonic numbers are chosen in the lower range and the fifth one close to the cutoff. It is important to emphasize here that this empirical criteria is independent of the shape of the scattering body and that it only depends on its size.

III. Numerical Implementation

In the most conventional finite-difference implementation of open region scattering problems, a desirable boundary operator would be one that truncates the mesh region that surrounds the scattering body. The boundary condition given in (2) can be used for this purpose by eliminating the outer layer nodes through the use of the finite differencing to approximate the normal derivative u_n, at (ρ_0, ϕ_0) (see Fig. 1). Approximating the normal derivative of the field $\partial u/\partial v$ by finite differencing, we have

$$\left.\frac{\partial u}{\partial v}\right|_{n0} = \frac{u_0^* - u_0}{\Delta v} \tag{6}$$

where Δv is the distance along the normal vector between the nodes n_0^* and n_0. The boundary condition in (2) can be translated into a relationship between each of the nodes lying on $\partial\Omega_0$ and the three nodes of the $\partial\Omega_i$ layer, as shown in Fig. 1. We have

$$u_0^* = a_0 u_0 + a_{-1}u_{-1} + a_1 u_1$$

where $a_0 = c_0\Delta v + 1$, $a_{-1} = c_{-1}\Delta v$, $a_1 = c_1\Delta v$. By enforcing (6) at each of the $\partial\Omega_0$ nodes, we arrive at a matrix equation that defines the boundary relationship between the outer boundary nodes, $\mathbf{U}_0$ (those of $\partial\Omega_0$), and the adjacent nodes, $\mathbf{U}_i$, of the $\partial\Omega_i$ contour, which results in

$$\mathbf{U}_0 = \mathbf{B}\mathbf{U}_i. \tag{7}$$

The matrix $\mathbf{B}$ is interpreted as the boundary operator matrix.

The next step is to incorporate the boundary operator matrix into the finite element matrix. Recall the global finite element system matrix given by

$$\mathbf{MU} = \mathbf{F} \tag{8}$$

where $\mathbf{U}$ is a vector of all the unknown field values in the region $\Omega \cup \Omega'$, $\mathbf{M}$ is the matrix corresponding to the weak form integral of the wave equation, and $\mathbf{F}$ is a vector that depends mainly on

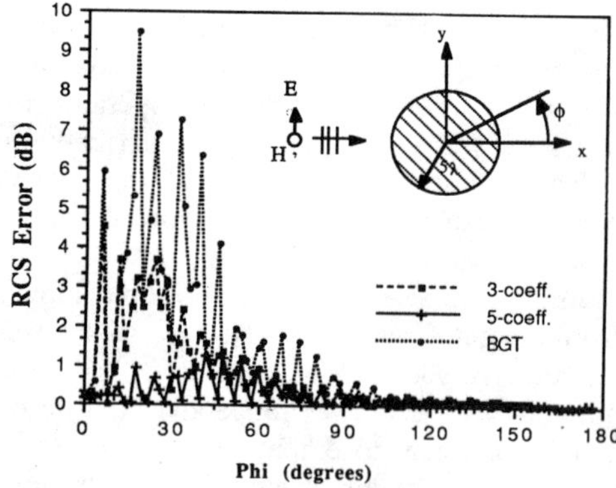

Fig. 2. Bistatic radar cross-section error for a 5λ peerfectly conducting circular cylinder, H-polarized case, using the 3-c ABC, the 5-c ABC, and the second-order BGT operator.

the line integral of the weak form which carries in it the information on the field behavior at the outer boundary that is needed to guarantee the well-posedness of the problem. To this end, we decompose the vector **U** into three vectors for the purpose of incorporating the boundary condition matrix in (7) into the system in (8): the first vector $\mathbf{U}_0$, containing the outer boundary nodes, the second vector $\mathbf{U}_i$, containing the $\partial\Omega_i$ contour nodes, and the third vector $\mathbf{U}_m$, containing the nodes interior to the $\partial\Omega_i$ contour. We have also divided **F** into three vectors, $\mathbf{F}_0$, $\mathbf{F}_i$, and $\mathbf{F}_m$, corresponding to the vectors, $\mathbf{U}_0$, $\mathbf{U}_i$, and $\mathbf{U}_m$, respectively, in order to isolate the contribution of the boundary line integral appearing in the weak form, namely, the $\mathbf{F}_0$ vector. Equation (8) can then be represented more explicitly as

$$\begin{bmatrix} \mathbf{M}_{00} & \mathbf{M}_{0i} & \mathbf{0} \\ \mathbf{M}_{i0} & \mathbf{M}_{ii} & \mathbf{M}_{im} \\ \mathbf{0} & \mathbf{M}_{mi} & \mathbf{M}_{mm} \end{bmatrix} \begin{bmatrix} \mathbf{U}_0 \\ \mathbf{U}_i \\ \mathbf{U}_m \end{bmatrix} = \begin{bmatrix} \mathbf{F}_0 \\ \mathbf{F}_i \\ \mathbf{F}_m \end{bmatrix}. \tag{9}$$

The matrix system in (9) remains ill-posed as long as $\mathbf{F}_i$ is not determined properly. In the conventional finite element solution of open-region scattering problems, a boundary operator that approximates the field behavior is introduced to replace the normal derivative in weak form, which leads to the evaluation of $\mathbf{F}_i$. Here, this procedure is avoided, and instead, the $\mathbf{F}_i$ vector is neglected by eliminating the first matrix row in (9). To this end, we substitute the boundary matrix operator in (7) into (9) to obtain

$$\begin{bmatrix} (\mathbf{M}_{i0}\mathbf{B} + \mathbf{M}_{ii}) & \mathbf{M}_{im} \\ \mathbf{M}_{mi} & \mathbf{M}_{mm} \end{bmatrix} \begin{bmatrix} \mathbf{U}_i \\ \mathbf{U}_m \end{bmatrix} = \begin{bmatrix} \mathbf{F}_i \\ \mathbf{F}_m \end{bmatrix}. \tag{10}$$

Numerical results will be presented in the following section to illustrate the versatility of the above-mentioned formulation.

IV. Numerical Results

The first geometry investigated is a perfectly conducting circular cylinder with a radius of 5 λ. The outer boundaries $\partial\Omega_0$ and $\partial\Omega_i$ were circular with radii of 5.15 λ and 5.151 λ, respectively. Because of the satisfactory performance of the analytic boundary operators when applied to the *E*-polarized case, as demonstrated in [9], we only show results for the *H*-polarized case. Fig. 2 shows the error in the bistatic radar cross section for the *H*-polarized case. For the 3-c ABC, the harmonics used were 1, 3, and 25; and the 5-c ABC, the harmonics used were 1, 2, 5, 7, and 25. The results presented show that when the three-coefficient condition was applied, the solution was more accurate than when the second-order BGT operator was used, even though the second-order operator and the 3-c ABC yield the same matrix bandwidth of 12. The 5-c ABC operator, on the other hand, increases the bandwidth of the matrix to 18, but yields an appreciable reduction in the error.

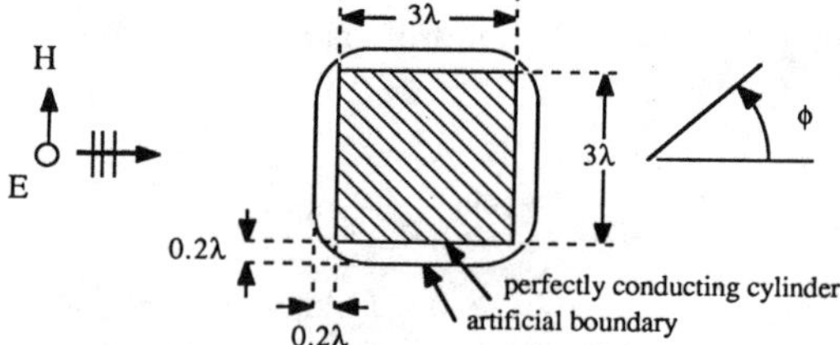

Fig. 3. Scattering problem geometry for a 3λ × 3λ perfectly conducting rectangular cylinder.

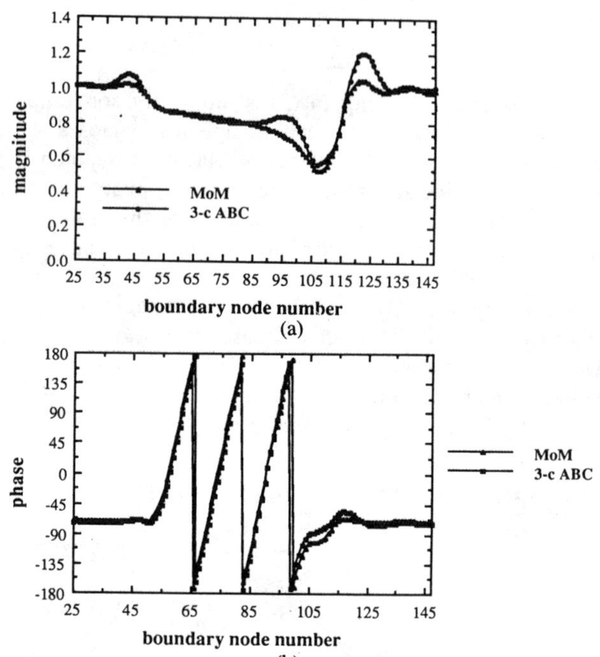

Fig. 4. Scattered near field on the outer boundary contour from a 3λ × 3λ perfectly conducting rectangular cylinder, starting at node 25 ($\phi = 0°$) and going counterclockwise to node 145 ($\phi = 180°$). (a) Magnitude. (b) Phase.

A nonseparable geometry was chosen for the second geometry to demonstrate the adaptability and potential efficiency of the method when arbitrary and conformable outer boundaries are used. The scatterer considered was a 3 λ × 3 λ perfect electric conductor rectangular cylinder. The artificial outer boundaries were chosen to be 0.2 λ and 0.22 λ from the edge of conductor (see Fig. 3). Fig. 4 shows the magnitude and phase of the near field for the *E*-polarized case as evaluated at the $\partial\Omega_0$ boundary layer.

V. Conclusion

By numerically approximating the analytical behavior of the exact boundary condition, numerical boundary operators were constructed that are simple in form and result in higher accuracy than other available local-type operators. The enhancement in accuracy is attributable to the inclusion of the evanescent harmonics behavior in the model. Another advantage in using numerically derived operators is that they can be applied to both separable and nonseparable outer boundaries. This feature allows the use of conformable outer boundaries that result in significant reduction in the mesh region surrounding the scatterer.

References

[1] B. H. McDonald and A. Wexler, ''Finite element solution of unbounded field problems,'' *IEEE Trans. Microwave Theory Tech.*, vol. MTT-20, pp. 841-847, Dec. 1972.

[2] R. MacCamy and S. Marin, ''A finite element method for exterior surface problems,'' *Int. J. Math. Math. Sci.*, vol. 3, no. 2, pp. 311-350, June 1980.

[3] S. I. Hariharan, ''Absorbing boundary conditions for exterior problems,'' NASA Contractor Rep. 177944, ICASE Rep. 85-33, July 1985.

[4] J. Jin and V. Liepa, ''Application of hybrid finite element method to electromagnetic scattering from coated cylinders,'' *IEEE Trans. Antennas Propagat.*, vol. 36, pp. 50-54, Jan. 1988.

[5] K. Mei, ''Unimoment method of solving antenna and scattering problem,'' *IEEE Trans. Antennas Propagat.*, vol. AP-22, pp. 760-766, Nov. 1974.

[6] A. Bayliss, M. Gunzburger, and E. Turkel, ''Boundary conditions for the numerical solution of elliptic equations in exterior regions,'' *SIAM J. Appl. Math.*, vol. 42, no. 2, pp. 430-451, Apr. 1982.

[7] B. Engquist and A. Majda, ''Radiation boundary conditions for the numerical simulation of waves,'' *Math. Comp.*, vol. 31, no. 139, pp. 629-651, July 1977.

[8] A. George and J. Liu, *Computer Solutions of Large Sparse Positive Definite Systems.* Englewood Cliffs, NJ: Prentice-Hall, 1981.

[9] R. Mittra and O. M. Ramahi, ''Absorbing boundary conditions for the direct solution of partial differential equations arising in electromagnetic scattering problems,'' in *Differential Methods in Electromagnetic Scattering*, vol. II, M. Morgan, Ed. New York: Elsevier, 1989.

[10] O. M. Ramahi and R. Mittra, ''A systematic approach to improving the absorbing boundary condition solution,'' *URSI Radio Meet. Program and Abstacts*, p. 182, Syracuse, NY, June 1988.

[11] R. Mittra, O. Ramahi, A. Khebir, R. Gordon and A. Kouki, ''A review of absorbing boundary conditions for two and three-dimensional electromagnetic scattering problems,'' *IEEE Trans. Magn.*, vol. 25, pp. 3034-3039, July 1989.

[12] M. L. Burrows, ''Equivalence of the Rayleigh solution and the extended-boundary-condition solution for scattering problems,'' *Electron. Lett.*, vol. 5, no. 12, pp. 277-278, June 1969.

[13] R. F. Miller, ''Rayleigh hypothesis in scattering problems,'' *Electron. Lett.*, vol. 5, no. 17, pp. 416-418, Sept. 1969.

Accuracy of 3-D Radiation Boundary Conditions for Use with the Vector Helmholtz Equation

A. F. PETERSON

Abstract—**Radiation boundary conditions (RBC's) developed for use with the vector Helmholtz equation in three dimensions are reviewed, including an alternate "second-order" radiation condition that preserves the symmetry of the matrix operator arising within a numerical implementation. The accuracy of these conditions is systematically tested as a function of the boundary location using outward-propagating spherical harmonics.**

I. Introduction

Numerical procedures for electromagnetic scattering from heterogeneous bodies often involve the discretization of differential equations into sparse matrix equations. These equations must be augmented with some form of radiation boundary condition (RBC) to ensure that the scattered fields are outgoing. By truncating the computational domain relatively close to the scatterer, the RBC also limits the number of unknowns. Unfortunately, exact near-zone radiation conditions appear in the form of surface integral equations or eigenfunction expansions that couple information around the entire scatterer boundary and create extensive fill-in within the sparse matrix operator. A significant improvement in efficiency can be realized through the use of approximate local radiation conditions, which maintain the same degree of matrix sparsity provided by the discretization of the Helmholtz equation. The second-order Bayliss–Turkel condition applied on a circular boundary is one example of a local RBC that can produce very accurate solutions for two-dimensional geometries [1], [2]. However, the scalar Bayliss–Turkel conditions are not directly compatible with the vector form of the Helmholtz equation required for 3-D electromagnetic analysis.

This communication reviews vector RBC's developed for the 3-D situation, including an alternate RBC that contains only even-order derivatives and therefore preserves the symmetry of the associated matrix operator. The accuracy of several radiation conditions is evaluated using outward-propagating spherical harmonics.

II. Vector Radiation Boundary Conditions

A family of vector radiation boundary conditions was introduced by Peterson [3]. These RBC's can be enforced on a spherical surface surrounding a 3-D scatterer. The general Nth-order condition

$$\left[\prod_{n=2}^{N}\left\{\hat{r}\times(\nabla\times)-jk-\frac{2(n-1)}{r}\right\}\right]\cdot\{\hat{r}\times(\nabla\times\bar{E}^s)-jk\bar{E}^s_{\tan}\}=0 \quad (1)$$

forces the scattered field $\bar{E}^s$ to agree with the first $2N$ terms of the Wilcox expansion [4]

$$\bar{E}^s(r,\theta,\phi)=\frac{e^{-jkr}}{4\pi r}\sum_{n=0}^{\infty}\frac{\bar{E}_n(\theta,\phi)}{r^n} \quad (2)$$

where (r,θ,ϕ) denote the usual spherical coordinates and $k=\omega(\mu\epsilon)^{1/2}$ is the wavenumber in the exterior medium. Of particular interest is the "first-order" condition

$$\hat{r}\times(\nabla\times\bar{E}^s)=jk\bar{E}^s_{\tan} \quad (3)$$

and the "second-order" condition

$$\hat{r}\times(\nabla\times\bar{E}^s)=\alpha(r)\bar{E}^s_{\tan}+\beta(r)\nabla\times[\hat{r}(\hat{r}\cdot\nabla\times\bar{E}^s)]+\gamma(r)\nabla^{\tan}(\hat{r}\cdot\bar{E}^s) \quad (4)$$

where

$$\alpha(r)=jk \quad (5)$$

$$\beta(r)=\frac{1}{2jk+\frac{2}{r}} \quad (6)$$

and

$$\gamma(r)=\frac{jk}{2jk+\frac{2}{r}}. \quad (7)$$

In (3) and (4), $\bar{E}^s_{\tan}$ denotes the tangential part of $\bar{E}^s$ on the spherical boundary and

$$\nabla^{\tan}\Psi=\hat{\theta}\frac{1}{r}\frac{\partial\Psi}{\partial\theta}+\hat{\phi}\frac{1}{r\sin\theta}\frac{\partial\Psi}{\partial\phi}. \quad (8)$$

The RBC's in (3) and (4) appear to be a natural extension of the scalar Bayliss–Turkel conditions to the vector case [3]. The second-order RBC involves derivatives of the same order as those appearing within the vector Helmholtz equation, and will produce comparable sparsity within a numerical implementation.

For three-dimensional analysis, the "curl–curl" form of the vector Helmholtz equation

$$\nabla\times\left(\frac{1}{\mu_r}\nabla\times\bar{E}^{\text{tot}}\right)-k^2\epsilon_r\bar{E}^{\text{tot}}=0 \quad (9)$$

can be multiplied with a vector testing function $\bar{T}$ and integrated over the computational domain Γ, to produce the "weak" equation

$$\iiint_{\Gamma}\frac{1}{\mu_r}\nabla\times\bar{T}\cdot\nabla\times\bar{E}^{\text{tot}}-k^2\epsilon_r\bar{T}\cdot\bar{E}^{\text{tot}}=-\iint_{\partial\Gamma}\bar{T}\cdot\hat{r}\times(\nabla\times\bar{E}^{\text{tot}}) \quad (10)$$

where the boundary $\partial\Gamma$ is spherical, and is assumed to enclose all inhomogeneities. The second-order RBC from (4) can be combined

Reprinted from *IEEE Trans. Antennas Propagat.*, vol. 40, no. 3, pp. 351–355, March 1992.

with the incident field and substituted into the right-hand side of (9) to produce

$$\iiint_{\Gamma} \frac{1}{\mu_r} \nabla \times \bar{T} \cdot \nabla \times \bar{E}^{tot} - k^2 \epsilon_r \bar{T} \cdot \bar{E}^{tot}$$
$$+ \iint_{\partial\Gamma} \alpha \bar{T} \cdot \bar{E}^{tot} + \beta(\hat{r} \cdot \nabla \times \bar{T})(\hat{r} \cdot \nabla \times \bar{E}^{tot})$$
$$+ \gamma \bar{T} \cdot \nabla^{tan}(\hat{r} \cdot \bar{E}^{tot})$$
$$= -\iint_{\partial\Gamma} \{\bar{T} \cdot \hat{r} \times (\nabla \times \bar{E}^{inc}) - \alpha \bar{T} \cdot \bar{E}^{inc}$$
$$- \beta \bar{T} \cdot \nabla \times [\hat{r}(\hat{r} \cdot \nabla \times \bar{E}^{inc})] - \gamma \bar{T} \cdot \nabla^{tan}(\hat{r} \cdot \bar{E}^{inc})\}. \tag{11}$$

Equation (11) provides a complete description of the scattering problem in terms of the electric field as the primary unknown.

The right-hand side of (4) contains a first-order derivative. When incorporated into (11), this term gives rise to an asymmetrical component of the resulting matrix operator. Although general-purpose sparse matrix solvers do not require a symmetric matrix, additional computation is necessary when treating a nonsymmetric system. In addition, procedures such as the biconjugate gradient algorithm can be significantly more efficient on symmetric systems [5]. Consequently, the first-order derivative is often viewed as an undesirable feature of this RBC.

III. Alternative Radiation Boundary Conditions

Using an idea originally proposed for the 2-D case by Parker *et al.* [6], we observe that

$$E_r = \frac{1}{k^2} \hat{r} \cdot \nabla \times \nabla \times \bar{E} = -\frac{1}{k^2} \nabla \cdot [\hat{r} \times (\nabla \times \bar{E})] \tag{12}$$

and consequently (4) and (12) can be combined to produce

$$E_r^s \cong -\frac{1}{k^2} \nabla \cdot \{\alpha(r) \bar{E}_{tan}^s + \beta(r) \nabla$$
$$\times [\hat{r}(\hat{r} \cdot \nabla \times \bar{E}^s)] + \gamma(r) \nabla^{tan}(\hat{r} \cdot \bar{E}^s)\}. \tag{13}$$

Provided that we ignore the higher order derivatives, a recursive substitution of this expression back into (4) yields

$$\hat{r} \times (\nabla \times \bar{E}^s) \cong \alpha(r) \bar{E}_{tan}^s + \beta(r) \nabla \times [\hat{r}(\hat{r} \cdot \nabla \times \bar{E}^s)]$$
$$+ \beta(r) \nabla^{tan}(\nabla \cdot \bar{E}_{tan}^s). \tag{14}$$

This RBC only involves second-order derivatives of tangential field components on the spherical boundary, and if used in conjunction with (10) preserves the symmetry of the matrix operator. Because terms containing higher order derivatives were neglected, this RBC may not be as accurate as (4).

Webb and Kanellopoulos recently generalized the approach of [3] to obtain a second-order RBC of the form [7]

$$\hat{r} \times (\nabla \times \bar{E}^s) \cong \alpha(r) \bar{E}_{tan}^s + \beta(r) \nabla \times [\hat{r}(\hat{r} \cdot \nabla \times \bar{E}^s)]$$
$$+ (s-1) \beta(r) \nabla^{tan}(\nabla \cdot \bar{E}_{tan}^s) + (2-s) \gamma(r) \nabla^{tan}(\hat{r} \cdot \bar{E}^s) \tag{15}$$

where α, β, and γ are defined in (5)–(7) and s is an arbitrary parameter. This RBC encompasses both (4) and (14) as special cases. By choosing $s = 1$, (15) reduces to (4); by selecting $s = 2$, (15) simplifies to (14). The choice $s = 2$ is recommended since it produces a symmetric matrix operator [7]. However, the accuracy of (15) as a function of s is also of particular interest. The following section considers the relative accuracy of each of the radiation conditions.

IV. Accuracy of the Radiation Boundary Conditions

An outward-propagating spherical harmonic will satisfy any "exact" RBC, and can be used to evaluate the relative accuracy of the approximate RBCs. The *mn*th spherical harmonic can be obtained using

$$\bar{E} = -j\frac{\eta}{k} \nabla \times \nabla \times \bar{A} - \nabla \times \bar{F} \tag{16}$$

where a TM-to-r electromagnetic field can be generated from [8]

$$\bar{A} = \hat{r} \frac{1}{\eta} b_{mn} \hat{H}_n^{(2)}(kr) P_n^{(m)}(\cos\theta) e^{jm\phi} \tag{17}$$

and a TE-to-r field is produced by

$$\bar{F} = \hat{r} c_{mn} \hat{H}_n^{(2)}(kr) P_n^{(m)}(\cos\theta) e^{jm\phi}. \tag{18}$$

In (16)–(18), $\eta = (\mu/\epsilon)^{1/2}$ denotes the intrinsic impedance of the exterior medium, and $\hat{H}_n^{(2)}$ and $P_n^{(m)}$ are the alternative spherical Hankel functions and associated Legendre polynomials, respectively, as defined by Harrington [8]. $P_n^{(m)} = 0$ for $m > n$. A general outgoing field can be represented by the superposition of harmonics for all values of m and n [8].

The components of the scattered electric field for the *mn*th harmonic are

$$E_\theta^s = b_{mn}\left(\frac{j}{r}\right) \hat{H}_n^{(2)\prime}(kr) \sin\theta \, P_n^{(m)\prime}(\cos\theta) e^{jm\phi}$$
$$+ c_{mn}\left(\frac{-jm}{r}\right) \hat{H}_n^{(2)}(kr) \frac{P_n^{(m)}(\cos\theta)}{\sin\theta} e^{jm\phi} \tag{19}$$

$$E_\phi^s = b_{mn}\left(\frac{m}{r}\right) \hat{H}_n^{(2)\prime}(kr) \frac{P_n^{(m)}(\cos\theta)}{\sin\theta} e^{jm\phi}$$
$$+ c_{mn}\left(\frac{-1}{r}\right) \hat{H}_n^{(2)\prime}(kr) \sin\theta \, P_n^{(m)\prime}(\cos\theta) e^{jm\phi} \tag{20}$$

and

$$E_r^s = b_{mn} \frac{n(n+1)}{jkr^2} \hat{H}_n^{(2)}(kr) P_n^{(m)}(\cos\theta) e^{jm\phi}. \tag{21}$$

If applied to this field, an exact RBC would produce a residual

$$\hat{r} \times \nabla \times \bar{E}^s$$
$$= \hat{\theta}\left\{ b_{mn}\left(\frac{jk}{r}\right) \hat{H}_n^{(2)}(kr) \sin\theta \, P_n^{(m)\prime}(\cos\theta) e^{jm\phi} \right.$$
$$\left. + c_{mn}\left(\frac{jmk}{r}\right) \hat{H}_n^{(2)\prime}(kr) \frac{P_n^{(m)}(\cos\theta)}{\sin\theta} e^{jm\phi} \right\}$$
$$+ \hat{\phi}\left\{ b_{mn}\left(\frac{mk}{r}\right) \hat{H}_n^{(2)}(kr) \frac{P_n^{(m)}(\cos\theta)}{\sin\theta} e^{jm\phi} \right.$$
$$\left. + c_{mn}\left(\frac{k}{r}\right) \hat{H}_n^{(2)\prime}(kr) \sin\theta \, P_n^{(m)\prime}(\cos\theta) e^{jm\phi} \right\}. \tag{22}$$

The first-boundary condition from (3) produces the residual

$$jk\bar{E}_{tan}^s = \hat{\theta}\left\{ b_{mn}\left(\frac{-k}{r}\right) \hat{H}_n^{(2)\prime}(kr) \sin\theta \, P_n^{(m)\prime}(\cos\theta) e^{jm\phi} \right.$$

$$+ c_{mn}\left(\frac{mk}{r}\right)\hat{H}_n^{(2)}(kr)\,\frac{P_n^{(m)}(\cos\theta)}{\sin\theta}\,e^{jm\phi}\Bigg\}$$

$$+\hat{\phi}\Bigg\{b_{mn}\left(\frac{jmk}{r}\right)\hat{H}_n^{(2)\prime}(kr)\,\frac{P_n^{(m)}(\cos\theta)}{\sin\theta}\,e^{jm\phi}$$

$$+c_{mn}\left(\frac{-jk}{r}\right)\hat{H}_n^{(2)}(kr)\sin\theta\, P_n^{(m)\prime}(\cos\theta)\,e^{jm\phi}\Bigg\}. \tag{23}$$

The error in the first-order RBC is evaluated by comparing (22) and (23). Regardless of whether we consider the θ or ϕ components of the residual, we obtain

$$\mathrm{err}_1^{\mathrm{TM}} = \left|\frac{\hat{H}_n^{(2)}(kr) - j\hat{H}_n^{(2)\prime}(kr)}{\hat{H}_n^{(2)}(kr)}\right| \tag{24}$$

for the TM part of the field, and

$$\mathrm{err}_1^{\mathrm{TE}} = \left|\frac{\hat{H}_n^{(2)\prime}(kr) + j\hat{H}_n^{(2)}(kr)}{\hat{H}_n^{(2)\prime}(kr)}\right| \tag{25}$$

for the TE part.

To obtain a similar measure of the error for the second-order boundary conditions from (4), (14), and (15) when applied to the mnth harmonic, observe that

$$\nabla\times\{\hat{r}\hat{r}\cdot\nabla\times\bar{E}^s\}$$

$$= \hat{\theta}c_{mn}\left(\frac{-jmn(n+1)}{r^3}\right)\hat{H}_n^{(2)}(kr)\,\frac{P_n^{(m)}(\cos\theta)}{\sin\theta}\,e^{jm\phi}$$

$$+\hat{\phi}c_{mn}\left(\frac{-n(n+1)}{r^3}\right)\hat{H}_n^{(2)}(kr)\sin\theta\, P_n^{(m)\prime}(\cos\theta)\,e^{jm\phi} \tag{26}$$

$$\nabla^{\tan}\{E_r^s\} = \hat{\theta}b_{mn}\left(\frac{jn(n+1)}{kr^3}\right)\hat{H}_n^{(2)}(kr)\sin\theta\, P_n^{(m)\prime}(\cos\theta)\,e^{jm\phi}$$

$$+\hat{\phi}b_{mn}\left(\frac{mn(n+1)}{kr^3}\right)\hat{H}_n^{(2)}(kr)\,\frac{P_n^{(m)}(\cos\theta)}{\sin\theta}\,e^{jm\phi} \tag{27}$$

and

$$\nabla^{\tan}\{\nabla\cdot\bar{E}_{\tan}^s\}$$

$$= \hat{\theta}b_{mn}\left(\frac{-jn(n+1)}{r^3}\right)\hat{H}_n^{(2)\prime}(kr)\sin\theta\, P_n^{(m)\prime}(\cos\theta)\,e^{jm\phi}$$

$$+\hat{\phi}b_{mn}\left(\frac{-mn(n+1)}{r^3}\right)\hat{H}_n^{(2)\prime}(kr)\,\frac{P_n^{(m)}(\cos\theta)}{\sin\theta}\,e^{jm\phi}. \tag{28}$$

Clearly, (26) involves only the TE component of the field while (27) and (28) involve only the TM part. The fact that the second-order RBC naturally separates the TE and TM fields in this manner suggests the roughly equal importance of the second and third terms on the right-hand side of (4), and also indicates that the alternative second-order conditions presented in (14) and (15) only differ from (4) in the TM part.

The error in the second-order RBC's can be determined by combining (23) with (26)–(28) according to (4), (14), or (15) and comparing the result with the "exact" representation obtained in (22). Identical expressions for the error are obtained from both the θ and ϕ components of the residual. For the TE part of the field, the error in any of the second-order RBC's is

$$\mathrm{err}_2^{\mathrm{TE}} = \left|\frac{\hat{H}_n^{(2)\prime}(kr) + \left\{j + \dfrac{n(n+1)}{2jk^2r^2 + 2kr}\right\}\hat{H}_n^{(2)}(kr)}{\hat{H}_n^{(2)\prime}(kr)}\right|. \tag{29}$$

The error in (15) for the TM part of the field is given by

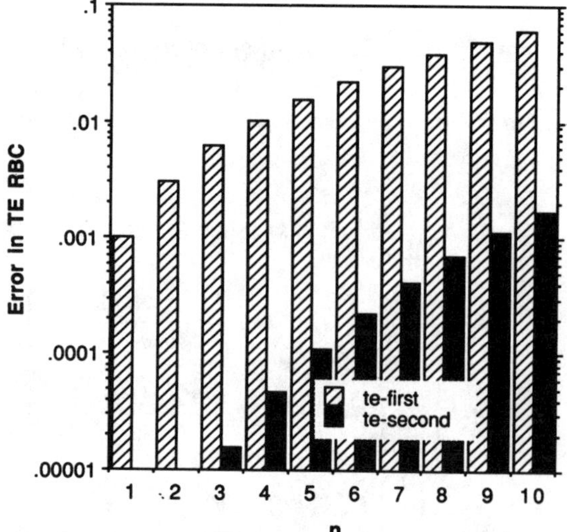

Fig. 1. TE error in the radiation boundary conditions applied on a spherical surface with $r = 5$ wavelengths as a function of harmonic index n. Both the first-order and the second-order RBC exhibit an error that increases with harmonic index n. For this radius, the second-order RBC from (4) is more than an order of magnitude more accurate than the first-order RBC from (3).

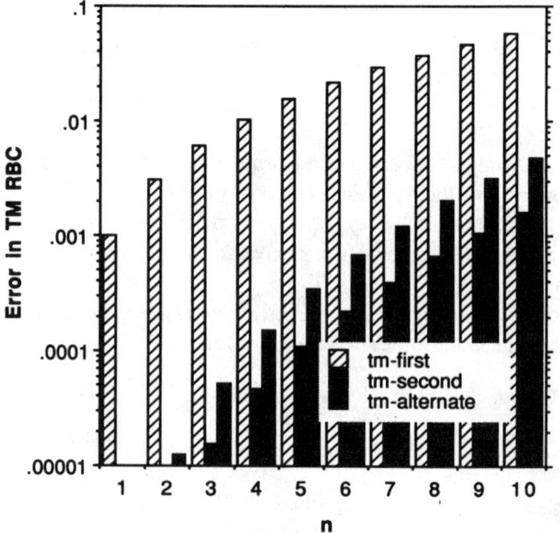

Fig. 2. TM error in the radiation boundary conditions applied on a spherical surface with $r = 5$ wavelengths as a function of harmonic index n. At this radius, the second-order RBC from (4) is the most accurate. The RBC from (4) and the alternate RBC from (14) both exhibit a significant improvement over the first-order RBC from (3).

$$\mathrm{err}_2^{\mathrm{TM}} = \left| \frac{\hat{H}_n^{(2)}(kr)\left\{1 - \dfrac{jn(n+1)(2-s)}{2jk^2r^2 + 2kr}\right\} - \hat{H}_n^{(2)\prime}(kr)\left\{j - \dfrac{n(n+1)(s-1)}{2jk^2r^2 + 2kr}\right\}}{\hat{H}_n^{(2)}(kr)} \right| \tag{30}$$

The error in the TM part of (4) can be obtained from (30) with $s = 1$; the error in the TM part of (14) can be found from (30) with $s = 2$. It is noteworthy that none of the previous error measures depend on the index m of the spherical harmonic.

Figs. 1 and 2 shows plots of the error in (3), (4), and (14) as a function of the index n for a spherical boundary fixed at $r = 5$ wavelengths. In common with behavior previously observed for the 2-D Bayliss-Turkel conditions [9], the error in the 3-D RBC's increases with increasing n. A plot of the error as a function of r for the $n = 15$ harmonic is presented in Fig. 3. The error associated with each RBC decreases as the boundary radius increases, suggesting that the conditions improve as they are applied at a greater distance from the scatterer. In general, the second-order RBC from (4) appears to be a significant improvement over the first-order condition from (3), and the alternate RBC of (14) is only slightly worse than (4).

Fig. 4 presents the TM error in the RBC of (15) as a function of s, for several choices of the harmonic index n and the boundary radius r. Each curve exhibits a well-defined minimum in the range $0 < s < 1$. These results suggest that, for greatest accuracy, the optimum value of s in (15) is neither 1 nor 2, but is in the vicinity of 0.5!

Although the primary purpose of this communication is to examine the relative accuracy of several approximately second-order RBC's, an indication of the absolute error can be gleaned from knowledge of the number of harmonics required for a given scatterer. For a scattering object that can be enclosed by a sphere of radius r, it is a general "rule-of-thumb" that the far-zone scattering can be described by approximately kr terms of a spherical harmonic expansion [10]. This guideline suggests the number of important harmonics for a given scatterer, and the number that must be accurately determined by a numerical solution. In fact, the error in the second-order RBC's is not small for $n = kr$, but rapidly decreases with larger values of r (for example, see Fig. 3). The actual harmonic content of a particular solution can be determined as a postprocessing step, and used to estimate the error for a given boundary location. Comparisons between the second-order vector RBC applied on a spherical boundary and the second-order Bayliss-Turkel RBC applied on a circular boundary of the same radius suggest that the error is roughly comparable for a given harmonic index n. This result is as expected, since the boundary conditions are developed from a similar expansion of the scattered field. Because the Bayliss-Turkel conditions are in widespread use, these results suggest that both the second-order RBC from (4) and the alternate proposed in (14) are likely to produce acceptable accuracy when used within a numerical implementation.

References

[1] A. F. Peterson and S. P. Castillo, "A frequency-domain differential equation formulation for electromagnetic scattering from inhomoge-

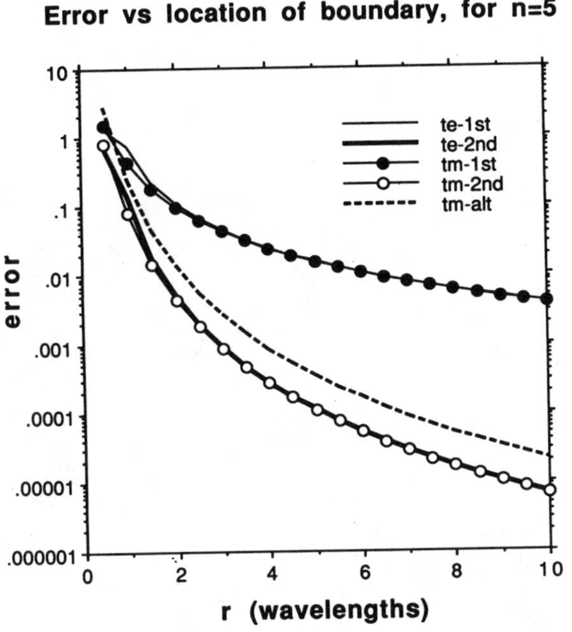

Fig. 3. TE and TM error in the first-order RBC of (3), the second-order RBC of (4), and the alternate second-order RBC of (14) plotted as a function of r for the $n = 5$ spherical harmonic.

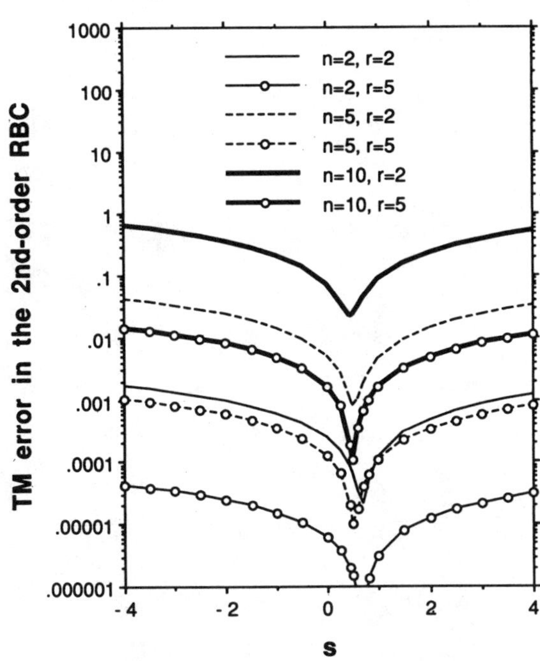

Fig. 4. TM error in the general second-order RBC of (15) as a function of s, for several choices of the harmonic index n and the boundary radius r in wavelengths.

neous cylinders," *IEEE Trans. Antennas Propagat.*, vol. 37, pp. 601-607, May 1989.

[2] A. Bayliss and E. Turkel, "Radiation boundary conditions for wave-like equations," *Comm. Pure Appl. Math.*, vol. 33, pp. 707-725, 1980.

[3] A. F. Peterson, "Absorbing boundary conditions for the vector wave equation," *Microwave Opt. Technol. Lett.*, vol. 1, pp. 62-64, Apr. 1988.

[4] C. H. Wilcox, "An expansion theorem for electromagnetic fields," *Comm. Pure Appl. Math.*, vol. 9, pp. 115-134, 1956.

[5] C. F. Smith, A. F. Peterson, and R. Mittra, "The biconjugate gradient method for electromagnetic scattering," *IEEE Trans. Antennas Propagat.*, vol. 38, pp. 938-940, June 1990.

[6] J. Parker, R. D. Ferraro, and P. C. Liewer, "A comparison of isoparametric edge elements and DnEt elements for 3-D electromagnetic scattering problems," in *Abstracts of URSI Radio Sci. Meet.*, Dallas, TX, May 1990, p. 377.

[7] J. P. Webb and V. N. Kanellopoulos, "Absorbing boundary conditions for the finite element solution of the vector wave equation," *Microwave Opt. Technol. Lett.*, vol. 2, pp. 370-372, Oct. 1989.

[8] R. F. Harrington, *Time-Harmonic Electromagnetic Fields*. New York: McGraw-Hill, 1961.

[9] R. Mittra and O. Ramahi, "Absorbing boundary conditions for the direct solution of partial differential equations arising in electromagnetic scattering problems," in *Finite Element and Finite Difference Methods in Electromagnetic Scattering*, M. A. Morgan, Ed. New York: Elsevier, 1990, pp. 133-173.

[10] W. V. T. Rusch, A. C. Ludwig, and W. C. Wong, "Analytical techniques for quasi-optical antennas," in *The Handbook of Antenna Design*, A. W. Rudge, K. Milne, A. D. Olver, and P. Knight, Eds. London, U.K.: Peregrinus, 1986, p. 116.

Edge-Based Finite Elements and Vector ABC's Applied to 3-D Scattering

A. Chatterjee, *Student Member, IEEE,* J. M. Jin, *Member, IEEE,* and J. L. Volakis, *Senior Member, IEEE*

Abstract—**An edge-based finite element formulation with vector absorbing boundary conditions is presented for scattering by composite structures having boundaries satisfying impedance and/or transition conditions. Remarkably accurate results are obtained by placing the mesh a small fraction of a wavelength away from the scatterer.**

I. Introduction

OF generic interest in electromagnetic scattering is the modeling of composite configurations comprising metallic and nonmetallic sections. In the case of man-made structures, abrupt material discontinuities and metallic corners are also encountered along with resistive sheets and thin ferrite coatings intended for controlling the scatterer's radar cross section (RCS). Differential equation methods, especially the finite element method (FEM), with its capability of handling arbitrary geometries and its versatility in modeling inhomogeneities and material discontinuities, have been a viable solution approach for bounded domain problems. However, for unbounded problems, as is the case with electromagnetic scattering, the solution is more involved since the finite element mesh needs to be truncated artificially at some distance from the object with a suitable boundary condition. These boundary conditions can be either global or local. Global boundary conditions are exact but lead to fully populated submatrices, thus spoiling the sparse, banded structure of the finite element system. In contrast, local conditions such as the absorbing boundary conditions (ABC's), are approximate but have the important advantage of retaining the sparsity of the matrix system. ABC's are essentially differential equations enforced at the mesh truncation boundary and are chosen to suppress nonphysical reflections from that boundary, thus ensuring the outgoing nature of the waves.

A variety of ABC's have been derived and widely employed in FEM solutions of open region two-dimensional scattering problems. However, the method's implementation and performance for scattering by three-dimensional geometries using edge-based finite elements has not received similar attention. The only three-dimensional implementation of the FEM for scattering has been a hybrid solution combined with the boundary element method (BEM) [1], [2] and those formulations combined with ABC's [3], [4]. The boundary element method, though exact, is equivalent to employing a global boundary condition for terminating the mesh and consequently leads to a full submatrix, restricting the method's utility to small geometries. For large-scale three-dimensional applications, it is necessary to employ an ABC for terminating the mesh to retain the $O(N)$ storage requirement, characteristic of the finite element method. However, the use of traditional node-based elements for solving electromagnetic problems complicates the enforcement of the required boundary conditions at material discontinuities and impedance boundaries or resistive transitions. Moreover, the inherent singularities of the electromagnetic fields at metallic corners limit their use altogether.

To avoid these difficulties, we consider an implementation of the FEM using vector basis functions whose degrees of freedom are associated with the fields along the six edges of a tetrahedron. In contrast to the traditional node-based elements, edge elements can treat geometries with sharp edges and are divergenceless. Moreover, they exhibit tangential continuity and normal discontinuity across interelement boundaries and material discontinuities, thus making them ideal for representing electromagnetic fields. They do, however, lead to a larger number of unknowns but this is balanced by the greater sparsity of the resulting finite element matrix. Thus the computation time required to solve such a system iteratively, with a given degree of accuracy, is less than the traditional node-based approach [5]. Our implementation is further coupled with a mesh termination scheme based on the vector ABC's derived in [6] and [7]. In contrast to the implementation proposed in [8], the one presented here preserves the symmetry of the finite element system, thus nearly halving the storage requirement and making it ideally suited for solution via a conjugate gradient type of algorithm. Further, the edge-based implementation discussed in [8] requires that the absorbing boundary be placed nearly a wavelength away from the scatterer, whereas in our implementation remarkably accurate results are obtained with the ABC's enforced a small fraction of a wavelength from the scattering body.

II. Formulation

Let us consider the problem of scattering by an inhomogeneous target associated with possible material discontinuities. To solve for the scattered fields via the FEM, it is necessary to enclose the scatterer—embedded inside the volume V—by an artificial surface S_0 on which the ABC is enforced (see Fig. 1). The ABC's to be considered in this paper are the Sommerfeld radiation condition, given by

$$\hat{\boldsymbol{n}} \times \nabla \times \boldsymbol{E}^s = -jk_0 \hat{\boldsymbol{n}} \times \hat{\boldsymbol{n}} \times \boldsymbol{E}^s, \qquad (1)$$

Manuscript received May 13, 1992; revised August 18, 1992.

The authors are with the Department of Electrical Engineering and Computer Science, University of Michigan, Ann Arbor, MI 48109-2122.

IEEE Log Number 9206179.

Reprinted from *IEEE Trans. Antennas Propagat.*, vol. 41, no. 2, pp. 221–226, Feb. 1993.

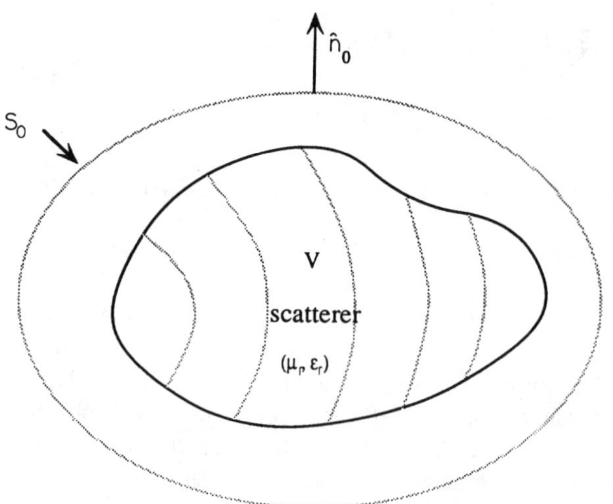

Fig. 1. Illustration of a scatterer enclosed by an artificial surface, S_0, on which the absorbing boundary condition is imposed.

and the second-order ABC [6], which can be written as

$$\hat{\boldsymbol{n}} \times \nabla \times \boldsymbol{E}^s = \alpha \boldsymbol{E}_t^s + \beta \nabla \times [\hat{\boldsymbol{n}}(\nabla \times \boldsymbol{E}^s)_n] + \beta \nabla_t (\nabla \cdot \boldsymbol{E}_t^s), \quad (2)$$

where $\alpha = jk$, $\beta = 1/(2jk + 2/r)$, $\boldsymbol{E}^s$ represents the scattered electric field, $\hat{\boldsymbol{n}}$ is the unit normal to the surface S_0, and the subscripts t and n denote the transverse and normal component to S_0, respectively. When these ABC's are employed on the artificial boundary S_0, they annihilate all field terms of $O(r^{-(2m+1)})$ and smaller, where m denotes the order of the ABC. The ABC's outlined above were derived for spherical surfaces [6] but in this work we have extended their application to S_0, which includes flat sections. In this case, the local curvature is used to replace $1/r$ in (2). For flat sections, the $1/r$ term, therefore, reduces to zero. This permits the construction of termination boundaries conformal to the scatterer, thus reducing the size of the computational domain.

The vector ABC's (1) and (2) can be combined and more conveniently written as

$$\hat{\boldsymbol{n}} \times \nabla \times \boldsymbol{E}^s = P(\boldsymbol{E}^s) \quad (3)$$

for the scattered field formulation, in which $\boldsymbol{E}^s$ is the working variable and

$$\hat{\boldsymbol{n}} \times \nabla \times \boldsymbol{E} = P(\boldsymbol{E}) + \boldsymbol{U}^{\text{inc}} \quad (4)$$

for the total field formulation, where the unknown is the total electric field. In (4),

$$\boldsymbol{U}^{\text{inc}} = \hat{\boldsymbol{n}} \times \nabla \times \boldsymbol{E}^{\text{inc}} - P(\boldsymbol{E}^{\text{inc}}) \quad (5)$$

where $\boldsymbol{E} = \boldsymbol{E}^s + \boldsymbol{E}^{\text{inc}}$ is the total field and $\boldsymbol{E}^{\text{inc}}$ is the incident electric field. Considering (4) to be the boundary condition employed at S_0, we can express the functional for the total electric field as

$$F(\boldsymbol{E}) = \int_V \left[\frac{1}{\mu_r} (\nabla \times \boldsymbol{E}) \cdot (\nabla \times \boldsymbol{E}) - k_0^2 \epsilon_r \boldsymbol{E} \cdot \boldsymbol{E} \right] dV + \int_{S_0} [\boldsymbol{E} \cdot P(\boldsymbol{E}) + 2\boldsymbol{E} \cdot \boldsymbol{U}^{\text{inc}}] \, dS, \quad (6)$$

where ϵ_r and μ_r are the relative permittivity and permeability, respectively.

The above functional can be generalized to account for the presence of impedance and resistive sheets or other discontinuous boundaries. In the case of a resistive card, the transition condition [9]

$$\hat{\boldsymbol{n}} \times (\hat{\boldsymbol{n}} \times \boldsymbol{E}) = -R\hat{\boldsymbol{n}} \times (\boldsymbol{H}^+ - \boldsymbol{H}^-) \quad (7)$$

must be enforced, where $\boldsymbol{H}^{\pm}$ denotes the total magnetic field above and below the sheet, R is the resistivity, in ohms per square, and $\hat{\boldsymbol{n}}$ is the unit normal to the sheet pointing in the upward direction (+side). For an impenetrable impedance surface, the appropriate boundary condition on that surface is

$$\hat{\boldsymbol{n}} \times (\hat{\boldsymbol{n}} \times \boldsymbol{E}) = -\eta \hat{\boldsymbol{n}} \times \boldsymbol{H}, \quad (8)$$

where $\hat{\boldsymbol{n}}$ is the unit normal to the surface and η is the surface impedance. Taking into consideration these boundary/transition conditions, the functional for the total electric field can be more explicitly written as

$$F(\boldsymbol{E}) = \int_V \left[\frac{1}{\mu_r} (\nabla \times \boldsymbol{E}) \cdot (\nabla \times \boldsymbol{E}) - k_0^2 \epsilon_r \boldsymbol{E} \cdot \boldsymbol{E} \right] dV + jk_0 Z_0 \int_{S_k} \frac{1}{K} (\hat{\boldsymbol{n}} \times \boldsymbol{E}) \cdot (\hat{\boldsymbol{n}} \times \boldsymbol{E}) \, dS + \int_{S_0} [\boldsymbol{E} \cdot P(\boldsymbol{E}) + 2\boldsymbol{E} \cdot \boldsymbol{U}^{\text{inc}}] \, dS, \quad (9)$$

where K is the surface resistivity (R) when integrating over a resistive card and equals the surface impedance (η) for an impedance sheet.

The formulation presented above is in terms of the total field but we can easily revert to a scattered field formulation by setting $\boldsymbol{E}^s = \boldsymbol{E} - \boldsymbol{E}^{\text{inc}}$ and noting that the scattered field satisfies the wave equation inside the domain of interest. The functional $F(\boldsymbol{E}^s)$ is given by

$$\begin{aligned} F(\boldsymbol{E}^s) = &\int_V \left[\frac{1}{\mu_r} (\nabla \times \boldsymbol{E}^s) \cdot (\nabla \times \boldsymbol{E}^s) - k_0^2 \epsilon_r \boldsymbol{E}^s \cdot \boldsymbol{E}^s \right] dV \\ &+ jk_0 Z_0 \int_{S_k} \frac{1}{K} (\hat{\boldsymbol{n}} \times \boldsymbol{E}^s) \cdot (\hat{\boldsymbol{n}} \times \boldsymbol{E}^s) \, dS \\ &+ \int_{S_0} \boldsymbol{E}^s \cdot P(\boldsymbol{E}^s) \, dS \\ &+ 2jk_0 Z_0 \int_{S_d} \frac{1}{\mu_r} \boldsymbol{E}^s \cdot (\hat{\boldsymbol{n}} \times \boldsymbol{H}^{\text{inc}}) \, dS \\ &+ 2 \int_{V_d} \left[\frac{1}{\mu_r} (\nabla \times \boldsymbol{E}^s) \cdot (\nabla \times \boldsymbol{E}^{\text{inc}}) - k_0^2 \epsilon_r \boldsymbol{E}^s \cdot \boldsymbol{E}^{\text{inc}} \right] dV \\ &+ 2jk_0 Z_0 \int_{S_k} \frac{1}{K} (\hat{\boldsymbol{n}} \times \boldsymbol{E}^s) \cdot (\hat{\boldsymbol{n}} \times \boldsymbol{E}^{\text{inc}}) \, dS \\ &+ f(\boldsymbol{E}^{\text{inc}}), \end{aligned} \quad (10)$$

where V_d is the volume occupied by the dielectric (portion of V where ϵ_r or μ_r is not unity), S_d encompasses all dielectric interface surfaces, and

$$\int_{S_0} \boldsymbol{E}^s \cdot P(\boldsymbol{E}^s) \, dS = \int_{S_0} [\alpha (\boldsymbol{E}_t^s)^2 + \beta (\nabla \times \boldsymbol{E}^s)_n^2 - \beta (\nabla \cdot \boldsymbol{E}_t^s)^2] \, dS$$

when the second-order ABC is employed. The function $f(\boldsymbol{E}^{\text{inc}})$ is solely in terms of the incident electric field and vanishes when we take the first variation of $F(\boldsymbol{E}^s)$. We remark that the scattered field formulation was implemented in our code.

III. Finite Element Discretization

To discretize the functional given in (10), the volume V is subdivided into a number of small tetrahedra, each occupying the volume $V^e(e = 1, 2, \cdots, M)$, where M denotes the total number of tetrahedral elements. Within each element, the scattered electric field is expressed as

$$\boldsymbol{E}^e = \sum_{j=1}^{m} E_j^e \boldsymbol{W}_j^e = \{\boldsymbol{W}^e\}^T\{E^e\} = \{E^e\}^T\{\boldsymbol{W}^e\}, \qquad (11)$$

where $\boldsymbol{W}_j^e$ are the edge-based vector basis functions [5], E_j^e denote the expansion coefficients of the basis and represent the field components tangential to the jth edge of the eth element, m is the number of edges making up the element, and the superscript stands for the element number. The basis functions used in our implementation have zero divergence and constant curl.

The system of equations to be solved for E_j^e is obtained by a Rayleigh–Ritz procedure which amounts to differentiating F with respect to each edge field and then setting it to zero. On substituting (11) into (10), taking the first variation in F, and assembling all M elements, we obtain the following augmented system of equations:

$$\left\{\frac{\partial F}{\partial E^e}\right\} = \sum_{e=1}^{M}[A^e]\{E^e\} + \sum_{s=1}^{M_s}[B^s]\{E^s\} + \sum_{p=1}^{M_p}\{C^p\} = 0. \qquad (12)$$

In this, M_s denotes the number of triangular surface elements on S_k and S_0, and M_p is equal to the sum of the surface elements on S_k and S_d and the volume elements in V_d. The elements of the matrices $[A^e]$, $[B^s]$, and $\{C^p\}$ are given by

$$A_{ij}^e = \int_{V^e}\left[\frac{1}{\mu_r}(\nabla\times\boldsymbol{W}_i^e)\cdot(\nabla\times\boldsymbol{W}_j^e) - k_0^2\epsilon_r\boldsymbol{W}_i^e\cdot\boldsymbol{W}_j^e\right]dV$$

$$B_{ij}^s = jk_0Z_0\left[\int_{S_k^s}\frac{1}{K}(\hat{\boldsymbol{n}}\times\boldsymbol{W}_i^s)\cdot(\hat{\boldsymbol{n}}\times\boldsymbol{W}_j^s)\,dS\right]$$
$$+\int_{S_0^s}[\alpha\boldsymbol{W}_{it}^s\cdot\boldsymbol{W}_{jt}^s + \beta(\nabla\times\boldsymbol{W}_i^s)_n\cdot(\nabla\times\boldsymbol{W}_j^s)_n$$
$$-\beta(\nabla\cdot\boldsymbol{W}_{it}^s)(\nabla\cdot\boldsymbol{W}_{jt}^s)]\,dS$$

$$C_i^p = 2jk_0Z_0\left[\int_{S_d^p}\frac{1}{\mu_r}\boldsymbol{W}_i^p\cdot(\hat{\boldsymbol{n}}\times\boldsymbol{H}^{\rm inc})\,dS\right.$$
$$\left.+\int_{S_k^p}\frac{1}{K}(\hat{\boldsymbol{n}}\times\boldsymbol{W}_i^p)\cdot(\hat{\boldsymbol{n}}\times\boldsymbol{E}^{\rm inc})\,dS\right]$$
$$+2\int_{V_d^p}\left[\frac{1}{\mu_r}(\nabla\times\boldsymbol{W}_i^p)\cdot(\nabla\times\boldsymbol{E}^{\rm inc}) - k_0^2\epsilon_r\boldsymbol{W}_i^p\cdot\boldsymbol{E}^{\rm inc}\right]dV,$$

where V_d^p is the volume of the pth tetrahedron inside the dielectric, S^s and S^p represent the surface areas of the sth and pth triangular surface elements, and the subscripts t and n denote the tangential and normal components of a vector, respectively. The boundary condition $\hat{\boldsymbol{n}}\times\boldsymbol{E}^s = -\hat{\boldsymbol{n}}\times\boldsymbol{E}^{\rm inc}$ must be imposed *a priori* on metallic boundaries; however, no special treatment is required at material discontinuities. Only the identification of the edges on material discontinuities or inhomogeneities is required to kick in the contribution from the surface integrals in (10).

The biconjugate gradient algorithm with diagonal preconditioning was used to solve the sparse, symmetric system of equations. The residual norm was usually set to less than 0.1% of the solution norm as a criterion for convergence since lower tolerances did not appear to offer significant improvement on the far-field values. The data structure was constructed such that only the nonzero elements of the upper triangular part of the symmetric, sparse matrix were stored in an $N_a \times k$ complex array. In our case, N_a was typically $1.1 \times N_u$, where N_u denotes the number of unknowns and k was equal to 12. The corresponding addresses were stored in a separate $N_a \times k$ integer array. The storage required in this scheme was about $25N_u$ and the number of distinct nonzero elements was typically $9N_u$.

IV. Results

A computer program was written for implementing the proposed FE-ABC formulation. This implementation was validated by computing the scattering for several configurations including metallic and dielectric bodies as well as structures satisfying resistive and impedance boundary conditions.

Fig. 2 compares the measured [10] bistatic cross section ($\theta^{\rm inc} = 180°$, $\phi^{\rm inc} = 90°$) of a metallic cube having an edge length of 0.755λ with the corresponding pattern computed by the three-dimensional FE-ABC code. The second-order vector ABC was employed on a spherical mesh truncation boundary which was placed only 0.1λ from the edge of the cube. About 33 000 unknowns were used for the discretization of the computational domain and the $[A]$ matrix contained a total of 264 000 distinct nonzero entries. The storage requirement of this matrix was consequently much smaller than that of the 1400 unknown moment method system (assuming the same sampling rate as the FEM of 14 points/λ), which had two million nonzero entries.

In Fig. 3, we plot the normal incidence backscatter RCS of a perfectly conducting cube as a function of its edge length. The meshes constructed for this experiment were terminated on conformal boundaries, i.e, on another cube placed a small distance (more than 0.15λ) from the scatterer. As seen, the agreement with measured data [10] is remarkably good over a 50 dB dynamic range.

Fig. 4 presents backscatter data for a cylinder of radius 0.3λ and height 0.6λ. The data from the three-dimensional finite element code again compare well with those obtained from a moment method-body of revolution code. The mesh was terminated on a spherical boundary at a distance of 0.3λ from the edge of the scatterer and the system consisted of nearly 33 000 unknowns. Convergence was achieved within about 350 iterations when the Sommerfeld radiation condition was imposed on the spherical mesh termination boundary. Each iteration took approximately 0.1 s on a Cray YMP after vectorization and on the average it was found that for $N >$

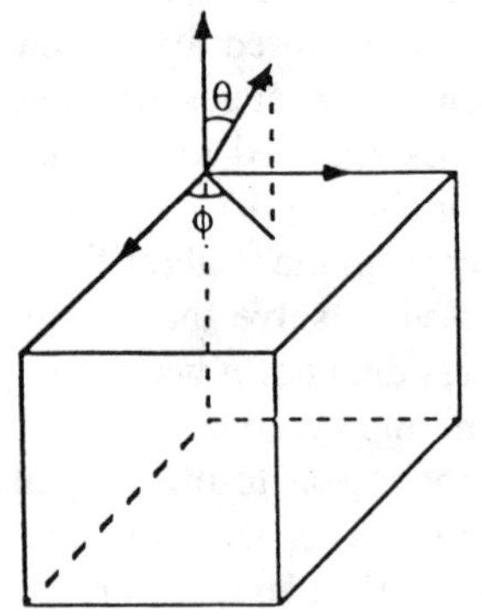

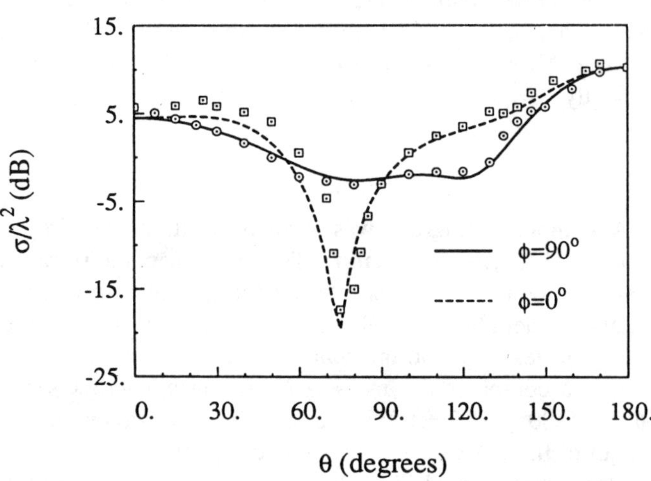

Fig. 2. Bistatic pattern of a metallic cube having an edge length of 0.755λ. The dotted circles and squares are measured data.

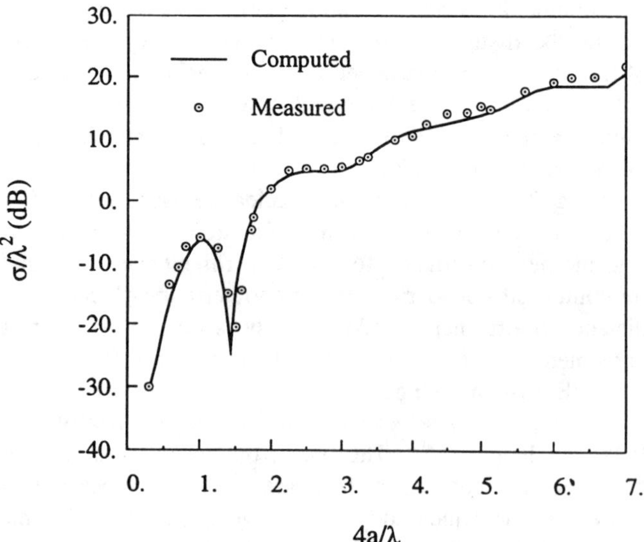

Fig. 3. Backscatter RCS at normal incidence for a metallic cube of side a.

25 000, the number of required iterations was approximately $N/100$. The agreement was quite good even on enclosing the metallic cylinder with a rectangular outer boundary placed 0.3λ from the edge of the scatterer.

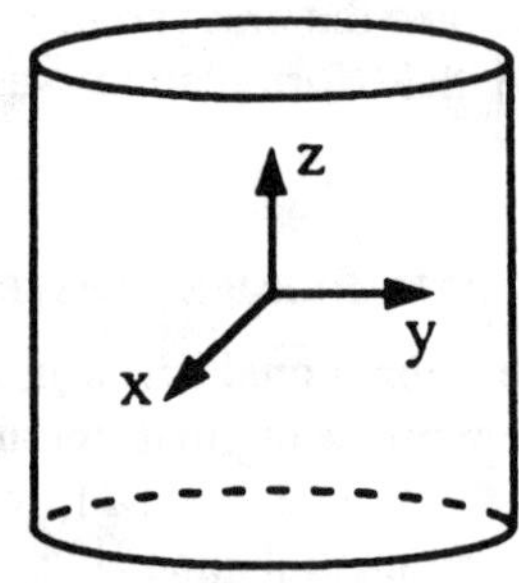

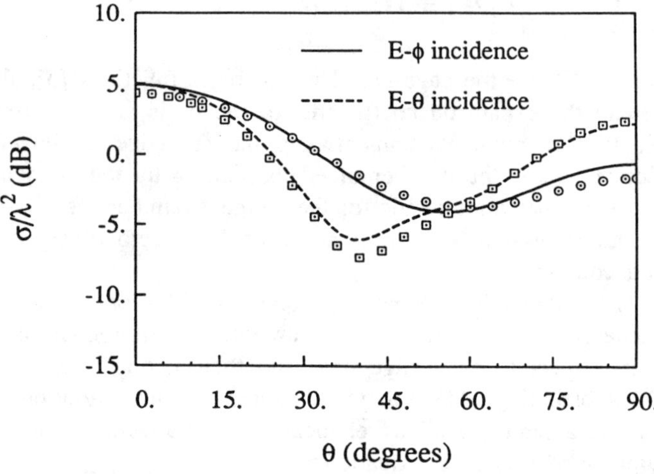

Fig. 4. Backscatter pattern of a perfectly conducting cylinder of radius 0.3λ and height 0.6λ. The axis of the cylinder coincides with the z axis of the Cartesian coordinate system. The dotted circles and squares are data obtained from a moment method code.

The results presented till now have been for perfectly conducting geometries. However, the real advantage of the FEM over integral equation techniques is the ease with which the former can handle material inhomogeneities and transition conditions. With this in mind, the remaining figures show backscatter and bistatic patterns for scatterers comprising resistive cards, dielectric material, and combinations of these. One of the test cases was a prolate spheroid, shown in Fig. 5, filled with lossy dielectric having $\epsilon_r = 4 - j1$, $k_0 a = \pi/2$, and $a/b = 2$, where a and b are the major and minor axes of the spheroid, respectively. The bistatic pattern ($\theta^{\rm inc} = 180°$; $\phi^{\rm inc} = 90°$) obtained from the FE-ABC solution agrees reasonably well with those obtained via the hybrid finite element–boundary integral method presented in [2]. However, the corresponding convergence rate for nonmetallic bodies and resistive/impedance sheets was found to be slower than that observed for metallic scatterers. A diagonal preconditioner was, therefore, used to accelerate the convergence of the biconjugate gradient algorithm with encouraging results.

For our last example, we compute the scattering from an inhomogeneous geometry with embedded resistive cards. In particular, the scatterer shown in Fig. 6 consists of an air-filled resistive card block $(0.5\lambda \times 0.5\lambda \times 0.25\lambda)$ joined to a metallic block $(0.5\lambda \times 0.5\lambda \times 0.25\lambda)$. In Fig. 7, we compare a principal plane backscatter pattern obtained from our 3D FE-ABC implementation with data computed using a traditional

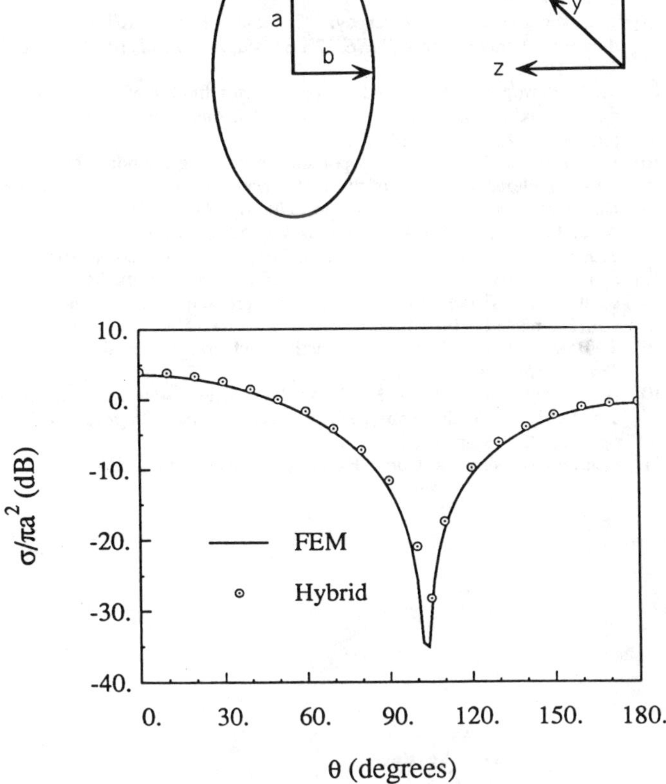

Fig. 5. Bistatic pattern of a lossy prolate spheroid ($\epsilon_r = 4 - j1$; $k_0 a = \pi/2$; $a/b = 2$), where a and b are the major and minor axes of the spheroid, respectively.

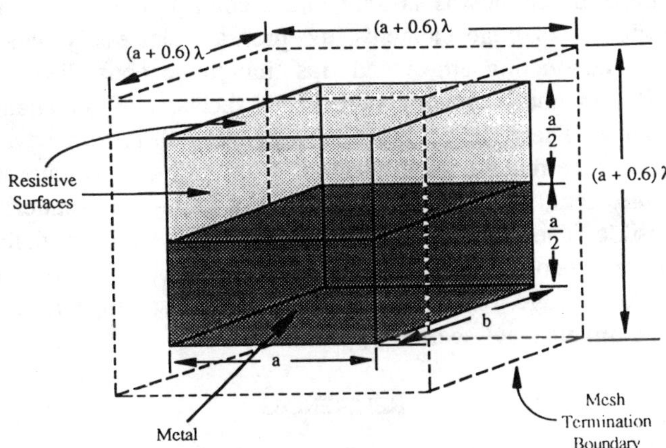

Fig. 6. Geometry of a cube ($a = b = 0.5\lambda$) consisting of a metallic section and an air-filled section, where the latter is bounded by a resistive surface having $R = Z_0$. The outer cube drawn with dashed lines represents the surface on which the ABC was enforced.

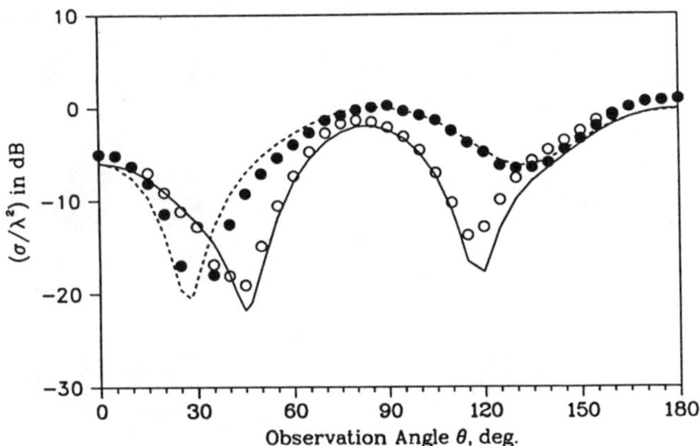

Fig. 7. RCS patterns in the x–z plane for the geometry in Fig. 6 with $H_z^{\mathrm{inc}} = 0$ and $E_z^{\mathrm{inc}} = 0$. The solid and dashed curves are the FE-ABC patterns and the circles and black dots are MOM data.

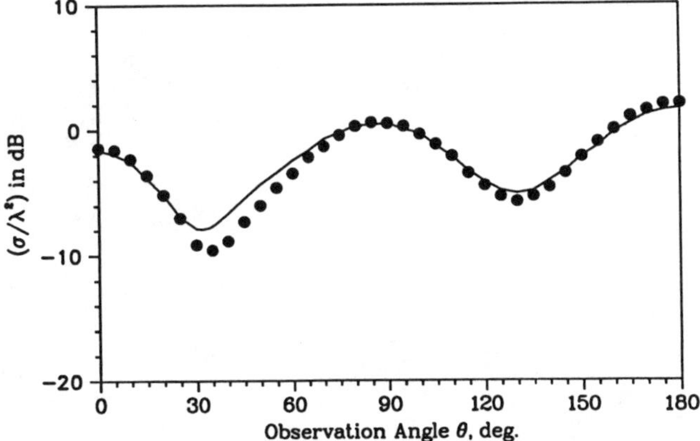

Fig. 8. RCS pattern in the x–z plane for the geometry in Fig. 6 with the air-filled section replaced with dielectric having $\epsilon_r = 2 - j2$. The solid curve is the FE-ABC pattern and the black dots are MoM data for the $E_z^{\mathrm{inc}} = 0$ polarization.

moment method code [11]. For the FE-ABC solution, the scatterer was enclosed within a cubical outer boundary placed only 0.3λ away from the scatterer. This resulted in a 30 000 unknown system which converged to the solution in about 400 iterations when using the Sommerfeld radiation condition and in 1600 iterations when the second-order ABC was used. For this geometry, the second-order ABC did not provide a significant improvement in accuracy (only about 0.1 dB) over the first-order condition. The same case was run with a higher discretization, resulting in a system of 50 000 unknowns; however, there was no significant difference in the far-field values with the earlier case. The geometry for the backscatter pattern shown in Fig. 8 is the same as in Fig. 6 with the air-filled section now occupied by a lossy dielectric having $\epsilon_r = 2 - j2$. The backscatter echo-area pattern for the $\phi\phi$ polarization as computed by our FE-ABC code is again seen to be in good agreement with corresponding moment method data [11].

V. Conclusions

In this paper, we have shown that the finite element technique with vector basis functions, when coupled with ABC's for mesh termination and the biconjugate gradient algorithm for the solution of the resulting system, is a viable procedure for computing the scattering by three-dimensional targets. We have found that these ABC's can be enforced only a small fraction of a wavelength from the scatterer's surface. This is probably due to the fast $(1/r)$ decay of the scattered fields. As a result, in addition to the sparsity of the matrix, the total number of unknowns is kept under control. Further, owing to the use of edge elements, the program can easily handle sharp conducting edges and tips, inhomogeneous dielectric and/or magnetic materials, resistive sheets, and impedance surfaces. These, in conjunction with the well-known advantages of the finite element method, result in low $O(N)$ storage requirement, making the computation of large-body scattering possible. These capabilities, along with the ease in modeling arbitrary geometries, make this formulation, to the best of our knowledge, one of the first suitable for solving practical three-dimensional scattering problems.

REFERENCES

[1] J. M. Jin and J. L. Volakis, "Electromagnetic scattering by and transmission through a three-dimensional slot in a thick conducting plane," *IEEE Trans. Antennas Propagat.*, vol. 39, pp. 543–550, Apr. 1991.

[2] X. Yuan, "Three-dimensional electromagnetic scattering from inhomogeneous objects by the hybrid moment and finite element method," *IEEE Trans. Microwave Theory Tech.*, vol. 38, pp. 1053–1059, Aug. 1990.

[3] I. D. Mayergoyz and J. D'Angelo, "New finite element formulation for 3-D scattering problems," *IEEE Trans. Magn.*, vol. 27, pp. 3967–3970, September 1991.

[4] J. D'Angelo and I. D. Mayergoyz, "Three-dimensional RF scattering by the finite element method," *IEEE Trans. Magn.*, vol. 27, pp. 3827–3832, Sept. 1991.

[5] M. L. Barton and Z. J. Cendes, "New vector finite elements for three-dimensional magnetic field computation," *J. Appl. Phys.*, vol. 61, no. 8, pp. 3919–3921, April 1987.

[6] J. P. Webb and V. N. Kanellopoulos, "Absorbing boundary conditions for finite element solution of the vector wave equation," *Microwave and Opt. Tech. Lett.*, vol. 2, no. 10, pp. 370–372, Oct. 1989.

[7] A. F. Peterson, "Absorbing boundary conditions for the vector wave equation," *Microwave and Opt. Tech. Lett.*, vol. 1, pp. 62–64, Apr. 1988.

[8] J. D'Angelo and I. D. Mayergoyz, "Finite element methods for the solution of RF radiation and scattering problems," *Electromagnetics*, vol. 10, pp. 177–199, 1990.

[9] T. B. A. Senior, "Combined resistive and conducting sheets," *IEEE Trans. Antennas Propagat.*, vol. 33, pp. 577–579, 1985.

[10] A. D. Yaghjian and R. V. McGahan, "Broadside radar cross-section of a perfectly conducting cube," *IEEE Trans. Antennas Propagat.*, vol. 33, pp. 321–329, Mar. 1985.

[11] Courtesy of Northrop Corp., B2 Division, Pico Rivera, CA.

Annotated Bibliography

Introduction

No anthology can be comprehensive, and no collection of reprints can form more than a sampling of the field. Many articles that might well have been included in this volume must therefore remain outside. To guide the reader to further material, the authors have commented extensively on the available literature in the main text of each chapter. This, however, is not sufficient in itself — the comments must perforce often be fleetingly brief, and anyway relate only to specific points of content. This chapter provides a more comprehensive guide to the literature, and does so in two ways. First, it includes a selection of review papers, which comment on the existing literature and include in themselves a large number of bibliographic references. Second, it contains a selected annotated bibliography. This bibliography includes most of the items referred to in the main text and a large number of others besides. Where the review papers comment on the same original material, opinions sometimes diverge — sometimes for reasons of emphasis, and occasionally because not all readers can agree.

The annotated bibliography of this chapter comprises some 250 bibliographic entries. Most of them refer to the same journals and conference proceedings as do the papers included in this volume. Most are in the English language, and most are published in Europe or America, so they should be easily accessible to readers of this book. They are organized in a loose grouping within the subject areas of the main chapters of this book. A very small number of articles have been listed twice, because they proved virtually impossible to classify uniquely.

This is an *annotated* bibliography, with an indication of the content given for each paper. The annotations are those of the authors; they are not abstracts, but often contain some indication of quality or probable value as well as content. They vary from purely factual ("this paper uses triangular elements") to the frankly evaluative ("this paper contains errors"). Some of the annotations amount to little more than brief abstracts, though author abstracts have rarely been used. Electrical engineers seem to be poor abstractors, and author abstracts are rarely quite so brief and informative as one might wish. In this age of ready access to automated bibliographic data bases, raw bibliographic data is easily obtained; the annotation is probably more valuable than the reference itself!

Review Papers

To stand tall, stand on the shoulders of giants. This ancient recipe serves the bibliographer well, for there are surely few fields where past giants have not labored mightily to make prior bibliographic work available. So it is for finite elements in wave propagation: while no comprehensive bibliography for this area exists yet, large portions of the field of finite elements are well served with review papers. In keeping with the spirit of an anthology, this bibliography too is in part an anthology, for it contains several review papers on specialist topics.

Basic to any finite element task is the geometric discretization of the problem region, usually known as "mesh generation." This is a topic of very broad scope, since much the same techniques apply anywhere finite element methods are used. The most recent general review of mesh generation methods was that of Ho-Le (1988), who attempted a general classification and demonstrated how the methods now used fit into it. This review is not far out of date, but additional bibliographic and evaluative material would be welcome, particularly in the area of adaptive meshing.

Finite element methods as applied to electromagnetics raise issues of mathematical physics and problem formulation that do not appear in many other applications. The coupled vector nature of electromagnetic fields does not appear to have a direct counterpart in applied mechanics, so some of the techniques well suited to the latter are not equally useful to the electrical engineer. On the other hand, fundamental properties — solenoidality and coupling of **E** with **H** at material interfaces, for example — apply to low and high frequency problems alike. Problems of eddy current flow and wave propagation, in particular, have so large a common mathematical foundation as to be almost indistinguishable. Much more has been done in the low-frequency area, however, so its literature is much richer in both original contributions and review articles. A good review of the literature, with applications ranging from resonant cavities to induction motors, was given by Konrad (1985). This review, now growing a bit aged, has since been augmented by two others, of much narrower scope but of mathematical value to the analyst concerned with wave problems. One, bearing an unlikely title but containing valuable comments as well as bibliographic references, was published by Tseng (1988), the other more recently by Tsukerman, Konrad, Bedrosian, and Chari (1993).

Guided-wave problems have been reviewed fairly regularly and in good detail. The excellent review paper of Rahman, Fernandez, and Davies [1991, see the chapter on rectilinear waves], augmented and updated by the later review by Davies (1993), is a good guide to the literature of this field, even for those not wholly in agreement with these reviewers' choice of working methods. The spurious mode problem, long a major hindrance to success in guided wave problems, was reviewed in detail by Koshiba, Hayata, and Suzuki (1987) and remains surprisingly current despite its date.

The infinite-domain problem has attracted a good deal of attention in recent years, and two extensive bibliographies are available. Emson's review [1988, with the chapter on rectilinear propagation] may well have been delayed in writing or publication; it is useful, but even a little less up to date than its date would suggest. The extensive, and more recent, work of Bettess and Bettess (1991) concentrates on mapped elements and dismisses several of the techniques useful to electromagnetics in a few lines. However, its bibliography is both sound and large.

Although the application of finite elements to problems of radial electromagnetic wave propagation is proceeding rapidly, no single comprehensive review paper seems to be available as yet. The more general reviews of Glisson (1989) and Peterson (1991), however, do devote sufficient space to this class of methods to be of substantial value and utility. In the narrower areas of modeling exterior regions, the bibliographic paper by Moore, Blaschak, Taflove, and Kriegsmann (1988) reviews the radiation boundary condition method, while Cooray and Costache (1991) give a good overview, unfortunately likely to date rapidly, of absorbing boundary conditions.

References

Bettess, P. and J. A. Bettess (1991): "Infinite elements for dynamic problems: Part 2." *Engineering Computations*, vol. **8**, pp. 125–51, 1991.

Cooray, F. R. and G. I. Costache (1991): "An overview of the absorbing boundary conditions." *Journal of Electromagnetic Waves and Applications*, vol. **5**, pp. 1041–54, 1991.

Davies, J. B. (1993): "Finite element analysis of waveguides and cavities — a review." *IEEE Transactions on Magnetics*, vol. **29**, pp. 1578–83, 1993.

Emson, C. R. I. [1988]: "Methods for the solution of open-boundary electromagnetic-field problems." *IEE Proceedings–A*, vol. **135**, pp. 151–58, 1988.

Glisson, A. W. (1989): "Recent advances in frequency domain techniques for electromagnetic scattering problems." *IEEE Transactions on Magnetics*, vol. **25**, pp. 2867–71, 1989.

Ho-Le, K. (1988): "Finite element mesh generation methods: A review and classification." *Computer Aided Design*, vol. **20**, pp. 27–38, 1988.

Konrad, A. (1985): "Electromagnetic devices and the application of computational techniques in their design." *IEEE Transactions on Magnetics*, vol. **MAG-21**, pp. 2382–87, 1985.

Koshiba, M., K. Hayata, and M. Suzuki (1987): "Finite-element method analysis of microwave and optical waveguides — trends in countermeasures to spurious solutions." *Electronics and Communications in Japan*, Part 2, vol. **70**, no. 9, pp. 96–107, 1987. (Translated from *Densi Tsushin Gakkai Ronbunshi*, vol. **69-C**, pp. 1477–86, 1986.)

Moore, T. G., J. G. Blaschak, A. Taflove, and G. A. Kriegsmann (1988): "Theory and application of radiation boundary operators." *IEEE Transactions on Antennas and Propagation*, vol. **36**, pp. 1797–812, 1988.

Peterson, A. F. (1991): "Analysis of heterogeneous electromagnetic scatterers: Research progress of the past decade." *Proceedings IEEE*, vol. **79**, pp. 1431–41, 1991.

Rahman, B. M. A., F. A. Fernandez, and J. B. Davies [1991]: "Review of finite element methods for microwave and optical waveguides." *Proceedings IEEE*, vol. **79**, pp. 1442–48, 1991.

Tseng, A. A. (1988): "Electroheating modeling in metal processing: A review." *Advances in Engineering Software*, vol. **10**, pp. 58–71, 1988.

Tsukerman, I. A., A. Konrad, G. Bedrosian, and M. V. K. Chari (1993): "A survey of numerical methods for transient eddy current problems." *IEEE Transactions on Magnetics*, vol. **29**, pp. 1711–16, 1993.

Bibliography

FINITE ELEMENTS AND WAVES

Arlett, P. L., A. K. Bahrani, and O. C. Zienkiewicz (1968): "Application of finite elements to the solution of Helmholtz's equation." *Proceedings IEE,* vol. **115**, pp. 1762–66, 1968.

Proposes first-order finite element methods for solving the Helmholtz equation and formulates element matrices for waveguides and resonators. Unfortunately, the authors fail to recognize the coupled nature of **E** and **H**, so their electromagnetics formulation is incorrect even though the finite elements are right.

Silvester, P. (1969): "High-order polynomial triangular finite elements for potential problems." *International Journal of Engineering Science,* vol. **7**, pp. 849–61, 1969.

High-accuracy triangular finite elements are derived for Laplace's, Poisson's, or Helmholtz's equations. General expressions for high-order interpolation polynomials are given, and element matrix tables are presented. (The tables for third and fourth order elements contain errors, corrected in the 1990 book with Ferrari.) The tabulated matrices are invariant with triangle size or shape; multiplication by cotangents of included angles and element area produces the element matrices.

Csendes, Z. J. and P. Silvester (1971): "Numerical solution of dielectric loaded waveguides. II — Modal approximation technique." *IEEE Transactions on Microwave Theory and Techniques,* vol. **MTT-19**, pp. 504–9, 1971.

The propagating modes of a dielectric loaded waveguide are expanded in the cutoff modes of the guide, which are obtained by finite element analysis. A theoretical comparison with the conventional finite element method is given; dispersion curves and contour plots are compared with analytic values and conventional finite element methods. The formulation does not eliminate spurious modes.

Benedek, P. and P. Silvester (1972): "Capacitance of parallel rectangular plates separated by a dielectric sheet." *IEEE Transactions on Microwave Theory and Techniques,* vol. **MTT-20**, pp. 504–10, 1972.

To determine the capacitance between two rectangular parallel plates separated by a dielectric sheet, the charge distribution on the plates is formulated in terms of a Fredholm integral equation of the first kind. This equation is solved numerically by a projective method using polynomial approximates. The resulting capacitance values are given in normalized graphical form, permitting capacitance determination for any practical values of dielectric constant and geometric parameters to within a few percent.

Silvester, P. and K. K. Chan (1972): "Bubnov-Galerkin solutions to wire-antenna problems." *Proceedings IEE,* vol. **119**, pp. 1095–99, 1972

Finite elements are developed for Hallén's equation using Lagrangian interpolation. For a single dipole and a 32-element Yagi-Uda array, good agreement with experiment is reported. Unlike many wire antenna solutions of the 1960s, this paper formulates a true finite element method, with current continuity the natural boundary condition rather than an enforced constraint.

Morgan, M. A., S.-K. Chang, and K. K. Mei (1977): "Coupled azimuthal potentials for electromagnetic field problems in inhomogeneous axially symmetric media." *IEEE Transactions on Antennas and Propagation,* vol. **AP-25**, pp. 413–17, 1977.

A potential formulation is described that is well suited to numerical computation of time-harmonic field problems in discontinuously and discretely inhomogeneous axially symmetric media. Typical numerical applications include solutions of scattering by arbitrarily shaped material bodies of revolution and radiation from inhomogeneously loaded rotationally symmetric antenna structures.

Konrad, A. (1977): "High-order triangular finite elements for electromagnetic waves in anisotropic media." *IEEE Transactions on Microwave Theory and Techniques,* vol. **MTT-25**, pp. 353–60, 1977.

Specialized functionals are introduced for traveling and circulating electromagnetic waves in planar and axisymmetric two-dimensional geometries. Singularities in the functional for axisymmetric geometries are eliminated by a transformation of the field components. The transformed fields are approximated by high-order interpolation polynomials over triangular regions in the $x - y$ and $r - z$ coordinate planes. A matrix expression assembled from constant element matrices and geometric factors relating to triangle shape, size, and position is obtained which is the discretized equivalent of the original functional. The necessary element matrices have been computed to sixth-order polynomial approximation.

Morishita, K. and N. Kumagai (1977): "Unified approach to the derivation of variational expression for electromagnetic fields." *IEEE Transactions on Microwave Theory and Techniques,* vol. **MTT-25**, pp. 34–40, 1977.

Variational expressions for a variety of electromagnetic parameters, such as a resonant frequency, a propagation constant, and an impedance matrix, can be formulated systematically from the principle of least action. It is pointed out that the Maxwell's equations themselves can also be derived from this principle. Computationally useful field formulations are deduced and listed.

Katz, I. N., A. G. Peano, and M. P. Rossow (1978): "Nodal variables for complete conforming finite elements of arbitrary polynomial order." *Computers & Mathematics with Applications,* vol. **4**, pp. 85–112, 1978.

Nodal interpolation functions are given for a hierarchal family of complete conforming triangular finite elements of arbitrary polynomial order, capable of first-derivative continuity. Results from a numerical example in plane stress analysis are presented.

Silvester, P. (1978): "Construction of triangular finite element universal matrices." *International Journal for Numerical Methods in Engineering,* vol. **12**, pp. 237–44, 1978.

The "universal" finite element matrices associated with various differential operators are all derivable by multiplication and addition of three fundamental matrices. The matrices are listed, and numerous matrices previously computed individually are rederived. The paper is analytically complete, but reference to earlier papers is essential if a full family of elementary matrices is required.

Rossow, M. P. and I. N. Katz (1978): "Hierarchical finite elements and precomputed arrays." *International Journal for Numerical Methods in Engineering,* vol. **12**, pp. 977–99, 1978.

Two-dimensional C0-continuous finite element approximating functions of arbitrary polynomial order are developed. These elements are hierarchical: that the nodal variables for polynomial order p are a subset of the nodal variables for order $p+1$. Elementary matrices, from which element matrices for a variety of operators may be composed, are given.

Chen, C. H. and C. D. Lien (1979): "A finite element solution of the wave propagation problem for an inhomogeneous dielectric slab." *IEEE Transactions on Antennas and Propagation,* vol. **AP-27**, pp. 877–80, 1979.

Variational equations equivalent to boundary value problems of electromagnetic fields are formulated and discretized on finite elements. Numerical results are given for various linear, parabolic, and sinusoidal profiles.

Chen, C. H. and Y. W. Kiang (1980): "A variational theory for wave propagation in a one-dimensional inhomogeneous medium." *IEEE Transactions on Antennas and Propagation,* vol. **AP-28**, pp. 762–69, 1980.

A compact variational theory permits one variational equation to handle permittivity and permeability profiles of various kinds, as well as both perpendicular and parallel polarizations.

Silvester, P. P. (1982): "Permutation rules for simplex finite elements." *International Journal for Numerical Methods in Engineering,* vol. **18**, pp. 1245–48, 1982.

Simplex finite element matrices possess a large number of symmetries that permit storage and manipulation of only a small part of the full matrices. The index permutation rules that apply to triangular and tetrahedral elements are described and tabulated.

Silvester, P. P. (1982): "Universal finite element matrices for tetrahedra." *International Journal for Numerical Methods in Engineering,* vol. **18**, pp. 1055–61, 1982.

Finite element matrices for arbitrary linear second-order differential operators are shown to be derivable from a few primitive matrices, by additions and multiplications. The primitives are derived, and their use is illustrated by examples.

Peng, J. P. and S. J. Salon (1984): "Hybrid finite element boundary element solutions using half-space Green's functions." *Journal of Applied Physics,* vol. **55, pt. 2B**, pp. 2198–200, 1984.

The hybrid finite element method, used with great success in electrostatics, magnetostatics, and eddy currents, is extended to problems whose exterior regions are composed of two linear materials: one embeds the finite element region, the other fills an infinite half-space. Both regions may contain sources. Examples are given and results compared to known solutions.

Hano, M. (1984): "Finite-element analysis of dielectric-loaded waveguides." *IEEE Transactions on*

Microwave Theory and Techniques, vol. **MTT-32**, pp. 1275–79, 1984.

A finite element analysis in which nonphysical spurious solutions do not appear is presented for the closed waveguide filled with various anisotropic media. The functional describing the behavior of the electromagnetic fields in anisotropic waveguides is stated, and trial functions exactly satisfying the boundary conditions required in the functional are derived. The finite element discretization is based on edge elements and uses a functional whose Euler equation is the three-component curl-curl equation derived from the Maxwell equations. Very close agreement was found with known solutions, and all nonzero eigenvalues correspond to true propagating modes of the waveguide.

Mur, G. and A. T. de Hoop (1985): "A finite-element method for computing three-dimensional electromagnetic fields in inhomogeneous media." *IEEE Transactions on Magnetics,* vol. **MAG-21**, pp. 2188–91, 1985.

Tetrahedral edge elements are developed. It is pointed out that they allow discontinuous jumps in the normal vector-field component and are therefore suitable for mixed-medium problems.

van Welij, J. S. (1985): "Calculation of eddy currents in terms of H on hexahedra." *IEEE Transactions on Magnetics,* vol. **MAG-21**, pp. 2239–41, 1985.

Isoparametric hexahedral elements (bricks) with edge-interpolated variables are formulated and applied to the solution of vector-field problems. The approximations are exactly solenoidal.

Kal, S. and N. B. Chakraborti (1985): "Finite-element analysis of helical phase shifters." *IEE Proceedings–H,* vol. **132**, pp. 231–36, 1985.

The finite element method is applied to a gyromagnetic medium with periodically varying spatial boundary conditions. Results are given for a normal, an inverted, and a ferrite-embedded helix ferrite phase shifter in circular cylindrical geometry.

Chang, H., S. Jeng, R. Wu, and C. H. Chen (1986): "Propagation of waves through magnetoplasma slab within a parallel-plate guide." *IEEE Transactions on Microwave Theory and Techniques,* vol. **MTT-34**, pp. 32–37, 1986.

A variational equation is established for wave propagation in a parallel-plate guide containing a magnetized inhomogeneous lossy plasma slab. This equation is solved by the finite element method to obtain the reflection coefficient and the field distribution in the slab. An anomalous numerical instability is found in this algorithm.

Cvetkovic, S. R. and J. B. Davies (1986): "Self-adjoint vector variational formulation for lossy anisotropic dielectric waveguide." *IEEE Transactions on Microwave Theory and Techniques,* vol. **MTT-34**, pp. 129–34, 1986.

A self-adjoint variational formula is given for the complex propagation constant, in terms of the magnetic field and real frequency, for an anisotropic lossy dielectric waveguide. Self-adjointness is achieved by using a real (power-related) inner product, so finite element discretization leads to an eigenvalue problem of the standard form.

Jeng, S. K., R. B. Wu, and C. H. Chen (1986): "Waves obliquely incident upon a stratified anisotropic slab: A variations reaction approach." *Radio Science,* vol. **21**, pp. 681–88, 1986.

A partial variational equation is derived from the variation of reactions and solved by the finite element method. Advantages of this approach include simplicity in formulation and numerical efficiency of solution. Results given include waves incident upon a stratified magnetoplasma.

Penman, J., M. D. Grieve, and A. F. Wilson (1986): "On the solution of open boundary field problems using the finite-element method." *IEE Proceedings–A,* vol. **133**, pp. 596–600, 1986.

Boundary conditions commonly used for approximating open boundary electromagnetic fields are examined, including the infinite elements and the effect of the boundary model on the interior potential distribution.

Cangellaris, A. C., C. C. Lin, and K. K. Mei (1987): "Point-matched time domain finite element methods for electromagnetic radiation and scattering." *IEEE Transactions on Antennas and Propagation,* vol. **AP-35**, pp. 1160–73, 1987.

The principal techniques used in time-domain computations and their supporting theory are reviewed in a tutorial fashion. Numerical results for two-dimensional scattering of Gaussian pulses are given.

Barton, M. L. and Z. J. Cendes (1987): "New vector finite elements for three-dimensional magnetic field computation." *Journal of Applied Physics,* vol. **61, pt. 2B**, pp. 3919–21, 1987.

Finite element vector approximation functions are defined, which interpolate to the tangential projection of the vector potential on each edge of a tetrahedral finite element. Continuity of the normal component of the vector potential is provided only approximately, as the natural interface condition inherent in variational procedures.

Crowley, C. W., P. P. Silvester, and H. Hurwitz, Jr. (1988): "Covariant projection elements for 3D vector field problems." *IEEE Transactions on Magnetics,* vol. **24**, pp. 397–400, 1988.

Covariant projection elements for vector finite elements facilitate imposing boundary and interface conditions on tangential components of the vector field. The elements' vector trial functions satisfy an inclusion condition, which guarantees that no spurious modes appear in the computed results. As a simple computational example, a quarter-section of a circular waveguide is modeled.

Pinchuk, A. R., C. W. Crowley, and P. P. Silvester (1988): "Spurious solutions to vector diffusion and wave field problems." *IEEE Transactions on Magnetics,* vol. **24**, pp. 158–61, 1988.

Deterministic problems involving the vector Helmholtz and diffusion equations often exhibit spurious solutions when they are solved using finite element methods. These nonphysical solutions arise from the same causes as the well-known spurious propagating modes encountered in high-frequency modal analyses. Using postprocessing techniques, these nonphysical solutions are shown to be responsible for inaccuracies in both modal and deterministic solutions throughout the frequency spectrum. Solution accuracy is recovered through use of mixed-order elements that are known to produce spectrally correct finite element solutions.

Sears, G. J. (1988): "Hierarchical quadrilateral elements in field analysis." *IEEE Transactions on Magnetics,* vol. **24**, pp. 404–7, 1988.

Hierarchical quadrilateral elements are discussed and evaluated. A hierarchy of linear and quadratic elements yields significant error reductions at the expense of additional variables.

Morgan, M. A. (1988): "Generalized coupled azimuthal potentials for electromagnetic fields in inhomogeneous media." *IEEE Transactions on Antennas and Propagation,* vol. **36**, pp. 1735–43, 1988.

The original coupled azimuthal potential formulation, which used two continuous potentials, is generalized, removing the restriction to axisymmetry of the inhomogeneous media while still representing the vector fields in terms of only two scalar potentials. Computational examples, using finite elements, are given for a simple three-dimensional boundary-value problem.

Inoue, K., K. Hayata, and M. Koshiba (1989): "Finite-element solution of three-dimensional periodic waveguide problems." *Electronics and Communications in Japan, Part 2 (Electronics),* vol. **72**, no. 8, pp. 68–77, 1989.

Tetrahedral elements are based on the variational expression in all three components of magnetic field. Spurious modes are encountered and fought with penalty methods.

Hano, M. (1989): "Efficient finite-element formulation for analyzing the propagation characteristics of waveguides." *Electronics and Communications in Japan, Part 2 (Electronics),* vol. **72**, no. 4, pp. 94–102, 1989.

A finite element representation for the wave number is given, with propagation constant as the parameter, as well as an alternative form with the propagation constant as eigenvalue and wave number as the parameter. No iterations or searchers are therefore required. The method is actually not very efficient, for the finite element matrix size is doubled, so the eigenvalue problems are rather costly to solve.

Tanner, D. R. and A. F. Peterson (1989): "Vector expansion functions for the numerical solution of Maxwell's equations." *Microwave and Optical Technology Letters,* vol. **2**, pp. 331–34, 1989.

Vector basis functions are developed for fields within triangles and tetrahedra, linear in the spatial variables and either solenoidal or irrotational. Resonance frequencies of some two-dimensional cavities are computed.

Kameari, A. (1990): "Calculation of transient 3D eddy current using edge-elements." *IEEE Transactions on Magnetics,* vol. **26**, pp. 466–69, 1990.

Quadratic 36-edge elements derived from 20-node isoparametric brick elements are formulated. They are used to solve the test problems of a square plate and a hollow sphere, with linear 12-edge and quadratic 36-edge elements. Results with these two element types agree well with each other and with the analytical solution. Current and field distributions are approximated better by quadratic elements with fewer degrees of freedom than the linear elements.

Ohtaka, M. and T. Kobayashi (1990): "A new vector variational expression for spurious-free finite-element analysis of waveguide eigenmodes." *Electronics and Communications in Japan, Part 2 (Electronics),* vol. **73**, no. 2, pp. 1–8, 1990.

A variational expression, with the propagation constant as the eigenvalue, is formulated in terms of the transverse electric and longitudinal magnetic field components of a waveguide. A mixture of vector elements (for the electric field) and scalars is used for discretization. The resulting formulations are free of spurious modes. Continuity of the normal component of the transverse electric field and the longitudinal magnetic field is required in this analysis. The superior features of this method are: (1) the propagation constant is the eigenvalue, (2) only three electromagnetic field components are used. The method it is applied to a dielectric loaded waveguide and a hollow circular waveguide.

Liu, Q. H. and W. C. Chew (1990): "Surface integral equation method for the analysis of an obliquely stratified half-space." *IEEE Transactions on Antennas and Propagation,* vol. **38**, pp. 653–63, 1990.

The radiation of a source in the presence of an obliquely stratified half-space is formulated in a surface integral equation. A generalized extinction theorem for the obliquely stratified half-space problem gives eight surface integral equations for the fields. These are solved by a one-dimensional finite element method.

Krauss-Varban, D. (1990): "Some aspects of finite element calculations applied to ULF magnetospheric cavity waves." *Computer Physics Communications,* vol. **59**, pp. 85–94, 1990.

Hydrodynamic pulsations of the cavity formed by the ionosphere and the magnetopause are analyzed to find cavity modes. Low-order elements with variable mesh size give good results, but when kinetic effects are considered, higher-order elements must be used.

Fernandez, F. A. and Y. Lu (1990): "Variational finite element analysis of dielectric waveguides with no spurious solutions." *Electronics Letters,* vol. **26**, pp. 2125–26, 1990.

A variational formulation for lossy anisotropic dielectric waveguides uses only transverse components of the magnetic field and avoids spurious solutions. It provides a direct solution for the propagation constant.

Ajose, S. O. and C. N. Ndujiuba (1990): "Wave propagation in a double-sided exponentially tapered rectangular waveguide." *IEEE Transactions on Microwave Theory and Techniques,* vol. **38**, pp. 1362–65, 1990.

Propagation in a rectangular waveguide with both sides exponentially tapered is discussed. The nonlinear wave equation in the guide is solved using the finite element approach.

Lawrence, B. G. and G. K. Cambrell (1990): "Finite-element field analysis for the scattering and immittance matrix elements of a general multiport microwave network." *Advances in Electrical Engineering Software,* P. P. Silvester, ed., pp. 297–306. Heidelberg and Southampton: Springer-Verlag and Computational Mechanics, 1990.

3D vectorial finite elements are applied to a general multiport microwave network to derive formulas for scattering, impedance, and admittance matrices. Edge elements are used to eliminate spurious modes, and dyadic impedance boundary conditions, especially those representing a matched port, are included.

Joseph, J., T. J. Sober, K. J. Gohn, and A. Konrad (1991): "Time domain analysis by the point-matched finite element method." *IEEE Transactions on Magnetics,* vol. **27**, pp. 3852–55, 1991.

The solution of Maxwell's equations using the point-matched finite element time-domain approach is explicit and does not require the solution of any matrix equations. Solution is carried out over a finite domain terminated using appropriate absorbing boundary conditions. Far-field quantities are obtained from the near-field solution by first computing the frequency-domain solutions from the time-domain solution using the fast Fourier transform and then using a near-field to far-field transformation algorithm.

Macnab, J. A. R., M. S. Towers, and A. McCowen (1991): "Finite and infinite elements applied to electromagnetic scattering from arbitrary objects in 2D." *International Conference on Computation in Electromagnetics* (IEE Conf. Publ. No. 350), pp. 164–66, 1991.

Mixed finite and infinite elements are used for time-independent problems. Far fields are determined directly from nodal values on infinite elements but can also be computed from surface or volume currents within the scatterer. Initial results for far fields are compared with analytic solutions.

Bogolyubov, A. N. and T. V. Edakina (1991): "Application of variational-difference methods to dielectric waveguide calculations." *Vestnik Moskovskogo Universiteta, Seriya 3 (Fizika i Astronomiya),* vol. **46**, no. 2, pp. 6–13, 1991.

Variational methods of analyzing dielectric waveguides and optical fibers are reviewed, with a view to solution by finite elements and finite differences of wave propagation in straight dielectric waveguides homogeneous along the guiding axis, of complex permittivity and with a complicated profile. Various types of variational formulation are described, including ways of bounding the region and taking into account the field behavior of the field at infinity. Some attention is devoted to spurious modes, but without fully outlining the state of that problem. The paper gives a good bibliography (47 references) of both Soviet and Western sources. Note: This paper is written in comprehensible English, not translated.

Lee, J. F., D. K. Sun, and Z. J. Cendes (1991): "Full-wave analysis of dielectric waveguides using tangential vector finite elements." *IEEE Transactions on Microwave Theory and Techniques,* vol. **39**, pp. 1262–71, 1991.

Tangential-component vector finite elements of second order are applied to dielectric waveguides, thus avoiding spurious modes. An iterative solver is combined with an adaptive mesh generator to provide automated solutions to specified accuracy. Numerical results are given for an image guide, microstrip, and a pedestal-supported stripline.

Webb, J. P. and R. Miniowitz (1991): "Analysis of 3-D microwave resonators using covariant-projection elements." *IEEE Transactions on Microwave Theory and Techniques,* vol. **39**, pp. 1895–99, 1991.

Resonators of arbitrary shape can be analyzed using covariant-projection elements, curvilinear hexahedra that impose only tangential field continuity. The method produces no spurious modes and works well even when sharp metal edges are present. The matrices involved are sparse; an appropriate sparse eigenvalue algorithm allows the method to run in modest amounts of memory.

Cangellaris, A. C. (1991): "Time-domain finite methods for electromagnetic wave propagation and scattering." *IEEE Transactions on Magnetics,* vol. **27**, pp. 3780–85, 1991.

Time-domain finite methods are widely used, but some concerns still remain about their accuracy and modeling versatility. Key characteristics of these finite methods are reviewed.

Peterson, A. F. and R. J. Baca (1991): "Error in the finite element discretization of the scalar Helmholtz equation over electrically large regions." *IEEE Microwave and Guided Wave Letters,* vol. **1**, pp. 219–22, 1991.

Discretization error in finite element solution of the scalar Helmholtz equation in open regions is studied for the simple case of scattering from dielectric slabs. The main kind of error is found to be phase error that increases progressively in a direction away from the boundary where the excitation is coupled into the computational domain. This error can be reduced by mesh refinement, increase in element order, or a scattered field formulation that couples the excitation into the equation in a different manner.

de Hoop, A. T. (1991): "Reciprocity, discretization, and the numerical solution of direct and inverse electromagnetic radiation and scattering problems." *Proceedings IEEE,* vol. **79**, pp. 1421–30, 1991.

A formulation based on Lorentz reciprocity unifies the finite element method and integral equation models, which are regarded as discretizations of weak formulations of differential or integral operator equations. In the complex frequency (Laplace transform) domain, the Lorentz reciprocity theorem encompasses all known weak formulations. Its discretization leads to discrete operator equations, including the finite element and approximated integral element models.

Cendes, Z. J. (1991): "Vector finite elements for electromagnetic field computation." *IEEE Transactions on Magnetics,* vol. **27**, pp. 3958–66, 1991.

Affine transformations are used to represent vectors and vector operations over triangular domains. Two-dimensional high-order vector elements are derived that are consistent with Whitney forms. One-form elements preserve the continuity of the tangential components of a vector field across element boundaries; Whitney two-form elements preserve continuity of the normal components. The one-form elements are supplemented with additional variables to achieve completeness in the range space of the curl operator.

Bardi, I. and O. Biro (1991): "An efficient finite-element formulation without spurious modes for anisotropic waveguides." *IEEE Transactions on Microwave Theory and Techniques,* vol. **39**, pp. 1133–39, 1991.

The electromagnetic field is described by the three components of a vector potential and a scalar potential, either electric or magnetic. Uniqueness of the potentials is ensured by the incorporation of the Coulomb gauge and by proper boundary conditions. No spurious modes appear, even for zero wave number. A variational finite element method is formulated in terms of the potentials so that a generalized eigenvalue problem with symmetric, sparse matrices results. Dielectric and ferrite loaded waveguides with closed and open boundaries are presented as examples.

Webb, J. P. and S. McFee (1991): "The use of hierarchical triangles in finite-element analysis of microwave and optical devices." *IEEE Transactions on Magnetics,* vol. **27**, pp. 4040–43, 1991.

Hierarchal triangular finite elements permit mixing of orders within the same mesh. They are particularly useful in microwave and optical problems where the underlying wave-like variation of the field is modeled better by high-order than low-order elements, but where the use of the same order or approximation everywhere is inefficient. Fields computed with hierarchical elements are shown for a range of example problems.

McFee, S. and J. P. Webb (1992): "Adaptive finite element analysis of microwave and optical devices using hierarchal triangles." *IEEE Transactions on Magnetics,* vol. **28**, pp. 1708–11, 1992.

Adaptive methods using hierarchal triangles can avoid repeated mesh refinement by varying the polynomial order of the elements in different parts of the mesh instead. A method useful for high-frequency finite element analysis is described.

Lee, R. and A. C. Cangellaris (1992): "A study of discretization error in the finite element approximation of wave solutions." *IEEE Transactions on Antennas and Propagation,* vol. **40**, pp. 542–49, 1992.

Errors caused by the spatial discretization in the finite element method for the two-dimensional scalar Helmholtz equation can be determined analytically for a uniform mesh of infinite extent. Numerical results show the effects on the error of the node density, the electrical size of the mesh, the direction of propagation of the incident wave, the type of element, and the type of boundary condition.

Boyse, W. E., D. R. Lynch, K. D. Paulsen, and G. N. Minerbo (1992): "Nodal-based finite-element modeling of Maxwell's equations." *IEEE Transactions on Antennas and Propagation,* vol. **40**, pp. 642–51, 1992.

Weak-form equivalents are given for Maxwell's equations, suited to conventional scalar finite elements, in terms of vector and scalar potentials for the electric field. Two specific gauges lead to Helmholtz weak forms free of spurious solutions; both have simple, physically meaningful boundary conditions. Jump discontinuities of normal components occur at dielectric interfaces.

Luo, Z. and N. A. Demerdash (1992): "A finite-element ballooning model for 2D eddy current open boundary problems for aerospace applications." *IEEE Transactions on Magnetics,* vol. **28**, pp. 2241–43, 1992.

A finite element ballooning method is described for field problems with open boundaries, in which eddy currents may exist everywhere, including the exterior ballooning region. Solutions are in agreement with analytical solutions for a simple eddy current example with an open boundary. The method is applied to the computation of fields surrounding two-wire parallel-pair transmission lines, one with a magnetic shielding and one with a conductive shielding.

Igarashi, H. and T. Honma (1992): "A vectorial finite element formulation for electromagnetic wave propagation in helical systems." *Journal of Applied Physics,* vol. **71**, pp. 5689–93, 1992.

The variational functional for the vector Helmholtz equation is written in twisted coordinates in which the problem becomes two-dimensional. Spurious solutions are removed using penalty functions. Finite element equations are derived for the covariant and contravariant components of the fields, with appropriate boundary conditions for each. A twisted waveguide with rectangular cross section is analyzed and good agreement with perturbational solutions is obtained for a small helical pitch.

Webb, J. P. (1993): "Edge elements and what they can do for you." *IEEE Transactions on Magnetics,* vol. **29**, pp. 1460–65, 1993.

Edge elements are described and contrasted with node-based elements. They provide freedom from spurious modes and permit Helmholtz theorem separation of vector fields. Currently available elements are briefly reviewed.

Rectilinearly Propagating Waves

Silvester, P. (1969): "Finite-element solution of homogeneous waveguide problems." *Alta Frequenza,* vol. **38**, pp. 313–17, 1969

First-order finite elements are formulated and

used for the solution of problems with known solutions. Presented at the URSI Symposium in May 1968, this is probably the first paper to describe a computer implementation of finite element methods applied to an electromagnetics problem.

Silvester, P. (1969): "A general high-order finite-element waveguide analysis program." *IEEE Transactions on Microwave Theory and Techniques,* vol. **MTT-17**, pp. 204–10, 1969.

A very general computer program for determining sets of propagating modes and cutoff frequencies of arbitrarily shaped waveguides uses high-order triangular finite elements. For convex polygonal guide shapes, the dominant eigenfrequencies are obtained to 4–5 significant figures. Use of the program is illustrated by calculating the propagating modes of a class of degenerate mode guides of current interest for which experimental data are available. Numerical studies of convergence rate and discretization error are described.

Csendes, Z. J. and P. Silvester (1970): "Numerical solution of dielectric loaded wave guides. I — Finite-element analysis." *IEEE Transactions on Microwave Theory and Techniques,* vol. **MTT-18**, pp. 1124–31, 1970.

A variational expression is derived for longitudinal **E** and **H** fields in a guide. It is discretized on high-order finite elements, and its use is illustrated by giving dispersion curves and mode structures of some dielectric-loaded guides.

Daly, P. (1971): "Hybrid-mode analysis of microstrip by finite-element methods." *IEEE Transactions on Microwave Theory and Techniques,* vol. **MTT-19**, pp. 19–25, 1971.

Hybrid-mode propagation in closed (shielded) microstrip is analyzed using longitudinal **E** and **H** components. Two modes have zero cutoff frequency: a quasi-TEM mode and a surface mode.

McAulay, A. D. and S. H. Charap (1974): "Numerical method for investigating the propagation of surface waves on dissipative guides." *Proceedings IEEE,* vol. **62**, pp. 402–3, 1974.

Surface waveguides of complicated shapes, with dissipative materials, are investigated, and the complex propagation constants for axially symmetric TM surface wave modes on a conducting wire and on a lossy dielectric rod are determined. The calculation extends beyond the limits of accuracy of perturbation methods.

Yeh, C., S. B. Dong, and W. Oliver (1975): "Arbitrarily shaped inhomogeneous optical fiber or integrated optical waveguides." *Journal of Applied Physics,* vol. **46**, pp. 2125–29, 1975.

Finite elements are applied to determining the characteristics of optical waves propagating along inhomogeneous waveguides of arbitrary cross-sectional shape. Good agreement with results arrived at by other methods is reported.

Vandenbulcke, P. and P. E. Lagasse (1976): "Eigenmode analysis of anisotropic optical fibres or integrated optical waveguides." *Electronics Letters,* vol. **12**, pp. 120–22, 4 March 1976.

The eigenmodes of an inhomogeneous, anisotropic optical waveguide with arbitrary cross-section are computed. Accurate results are obtained, even for guides formed by very small changes in the refractive index. The propagation characteristics of a channel guide embedded in $LiNbO_3$ are given as an example.

McAulay, A. D. (1977): "Variational finite-element solution for dissipative waveguides and transportation application." *IEEE Transactions on Microwave Theory and Techniques,* vol. **MTT-25**, pp. 382–92, 1977.

The complex propagation constants and fields are determined as a function of frequency for electromagnetic waves propagating along an inhomogeneous waveguide composed of dissipative materials, by a finite element method. Results are compared with analytical solutions for several simple dissipative waveguides for which analytical results are computed for comparison. A novel proposal is made for a rapid-transit collision-avoidance system.

McAulay, A. D. (1977): "The finite element solution of dissipative electromagnetic surface waveguides." *International Journal for Numerical Methods in Engineering,* vol. **11**, pp. 11–25, 1977.

A finite element method is developed for the propagation of electromagnetic waves along closed or surface waveguides of arbitrary shape containing dissipative materials. The method is applied to several structures for which analytic comparison results are available: a guide filled with dissipative material, a dissipative rod, and a dielectric coated wire.

Silvester, P. P., D. A. Lowther, C. J. Carpenter, and E. A. Wyatt (1977): "Exterior finite elements for 2-dimensional field problems with open boundaries." *Proceedings IEE,* vol. **124**, pp. 1267–70, 1977

Harmonic field problems of infinite extent are

treated by building a sequence of nested annular regions recursively. The size of the exterior region actually modeled grows very rapidly, and the computation is cheap.

Lowther, D. A., C. B. Rajanathan, and P. P. Silvester (1978): "A finite element technique for solving 2-D open boundary problems." *IEEE Transactions on Magnetics,* vol. **MAG-14**, pp. 467–69, 1978.

A recursion technique is described that models the exterior region by an annular "super-element" in which all the nodes except those on the inside edge have been eliminated. The annulus is generated and divided into elements automatically.

Yeh, C., K. Ha, S. B. Dong, and W. P. Brown (1979): "Single-mode optical waveguides." *Applied Optics,* vol. **18**, pp. 1490–504, 1979.

A modified finite element method is used to treat wave propagation along arbitrarily shaped single-mode dielectric waveguides with inhomogeneous index variations. Examples include triangular, elliptical, and rectangular fiber guides; the single material fiber; embossed integrated optics and diffused channel guides; and the optical stripline.

Mabaya, N., P. E. Lagasse, and P. Vandenbulcke (1981): "Finite element analysis of optical waveguides." *IEEE Transactions on Microwave Theory and Techniques,* vol. **MTT-29**, pp. 600–605, 1981.

Possible solutions to the spurious mode problem are discussed. For isotropic waveguides it is shown that all modes can be very accurately approximated by two different scalar finite element programs. A method for calculating the attenuation of leaky modes in a single-material integrated optic waveguide using this scalar finite element method is proposed.

Goldstein, C. I. (1982): "A finite element method for solving Helmholtz type equations in waveguides and other unbounded domains." *Mathematics of Computation,* vol. **39**, pp. 309–24, 1982.

This method replaces the boundary value problems in a region with infinite boundaries by a boundary value problem on a bounded domain, with the solution behavior near infinity incorporated in a nonlocal boundary condition. Detailed error theory is presented. The paper establishes much of the mathematical theory underlying this class of methods, but it may prove difficult reading for an engineering audience.

Oyamada, K. and T. Okoshi (1982): "Two-dimensional finite-element method calculation of propagation characteristics of axially nonsymmetrical optical fibers." *Radio Science,* vol. **17**, pp. 109–16, 1982.

A two-dimensional vector formulation is presented of finite elements for designing axially nonsymmetric single-mode optical fibers without instabilities resulting from polarization plane rotation of the propagated HE_{11} mode. The field outside the central core region is expanded in a series of modified Bessel functions to take account of the infinite exterior space.

Kagawa, Y. T. Murai, and S. Kitagami (1982): "On the compatibility of finite element-boundary element coupling in field problems." *COMPEL,* vol. **1**, pp. 197–217, 1982.

Combining finite elements and integral boundary elements, a hypothetical boundary is assumed in the unbounded domain. The usual finite elements are used in the inner region, boundary elements in the outer infinite region. On the coupling boundary, both potential and flux must be compatible. In the boundary element method, the same polynomial function is chosen for the potential and the flux, so compatibility cannot be assured without additional constraints. Several compatibility conditions are considered on the total flux or energy flow across the boundary.

Aubourg, M., J. P. Villotte, F. Godon, and Y. Garault (1983): "Finite element analysis of lossy waveguides-application to microstrip lines on semiconductor substrate." *IEEE Transactions on Microwave Theory and Techniques,* vol. **MTT-31**, pp. 326–31, 1983.

A finite element analysis of lossy shielded inhomogeneous waveguides or arbitrary shape is described, with applications to propagation on MIS or Schottky contact microstrip lines.

Rahman, B. M. A. and J. B. Davies (1984): "Finite-element analysis of optical and microwave waveguide problems." *IEEE Transactions on Microwave Theory and Techniques,* vol. **MTT-32**, pp. 20–28, 1984.

A vector H-field formulation is given for electromagnetic wave propagation for a wide range of guided-wave problems in arbitrarily anisotropic materials. Computed results are shown for several guides, including waveguides containing skew anisotropic dielectric. A paper of some review and bibliographic value (31 references).

Hurwitz, H., Jr. (1984): "Infinitesimal scaling — a new procedure for modeling exterior field problems." *IEEE Transactions on Magnetics,* vol. **MAG-20**, pp. 1918–23, 1984.

Treating unbounded regions as single elements requires basis functions that well approximate the solutions of the underlying partial differential equations. These can be obtained implicitly by infinitesimal scaling, which allows the coefficient matrix of an extended element to be determined without explicit knowledge of the basis functions.

Dasgupta, G. (1984): "Computation of exterior potential fields by infinite substructuring." *Computer Methods in Applied Mechanics and Engineering,* vol. **46**, pp. 295–305, 1984.

The exterior of an arbitrary convex region is conceived as an infinite assembly of geometrically similar finite element cells. A complete set of spatially discretized shape functions for the exterior, with the characteristic of outward decay, is evaluated by solving a quadratic eigenproblem whose matrices are furnished by the finite element system matrix of a typical cell. The virtual-work energies associated with pairs of these functions form a convergent geometrical series over a sequence of similarly shaped elements with increasing size. The exact infinite sum of the virtual work quantities yields the matrix representation of the unbounded region.

Koshiba, M., K. Hayata, and M. Suzuki (1984): "Approximate scalar finite-element analysis of anisotropic optical waveguides with off-diagonal elements in a permittivity tensor." *IEEE Transactions on Microwave Theory and Techniques,* vol. **MTT-32**, pp. 587–93, 1984.

A scalar finite element program avoids the spurious solutions that arise in vectorial axial-component formulations. Numerical examples are given for an anisotropic dielectric rectangular waveguide composed of a uniaxial medium and for other guides.

Rahman, B. M. A. and J. B. Davies (1984): "Finite-element solution of integrated optical waveguides." *Journal of Lightwave Technology,* vol. **LT-2**, pp. 682–88, 1984.

An H-vector variational functional is used to formulate finite elements for lossless optical waveguides with tensor permittivity. Infinite elements are used to model the guide exterior; spurious solutions are controlled by introducing a penalty for non-solenoidality. Results are given for a number of guide types and channel waveguide directional couplers.

Chiang, K. S. (1985): "Finite-element analysis of optical fibres with iterative treatment of the infinite 2-D space." *Optical and Quantum Electronics,* vol. **17**, pp. 381–91, 1985.

For weakly guiding fibers, an iterative process is used to eliminate the field variables in the (infinite) cladding so that the final algebraic eigenvalue equation only involves fields internal to the fiber. Numerical examples are given.

Wu, R. B. and C. H. Chen (1985): "A variational analysis of dielectric waveguides by the conformal mapping technique." *IEEE Transactions on Microwave Theory and Techniques,* vol. **MTT-33**, pp. 681–85, 1985.

Conformal mapping is used to model the region exterior to the guides. Numerical results are given for rectangular, strip, and channel waveguides.

Koshiba, M., K. Hayata, and M. Suzuki (1985): "Finite-element formulation in terms of the electric-field vector for electromagnetic waveguide problems." *IEEE Transactions on Microwave Theory and Techniques,* vol. **MTT-33**, pp. 900–905, 1985.

A vector finite element method for the analysis of general magnetically anisotropic waveguides is formulated in terms of all three components of the electric field. Spurious nonphysical solutions do not appear anywhere above the "air-line." To show the validity and usefulness of this formulation, examples are computed for dielectric-loaded waveguides and ferrite-loaded waveguides.

Wong, S. H. and I. R. Ciric (1985): "Method of conformal transformation for the finite-element solution of axisymmetric exterior-field problems." *COMPEL,* vol. **4**, pp. 123–35, 1985.

Axisymmetric exterior-field problems are solved by using standard finite elements in a bounded domain obtained by conformal mapping from the original unbounded one. The transformed functionals have very simple expressions, and the exact transforms of the original boundary conditions are used in the transformed domain. Consequently, no approximation is introduced. This method is particularly useful for material inhomogeneities and anisotropies within large regions.

Hayata, K., M. Koshiba, M. Eguchi, and M. Suzuki (1986): "Vectorial finite-element method without any spurious solutions for dielectric waveguiding problems using transverse magnetic-field component." *IEEE Transactions on Microwave Theory and Techniques,* vol. **MTT-34**, pp. 1120–24, 1986.

By formulating the finite element for dielectric waveguides in the transverse magnetic field components, **H** is guaranteed to be solenoidal, and spurious nonphysical solutions are completely eliminated. Numerical results for a rectangular metallic waveguide half-filled with dielectric are compared with exact and earlier finite element solutions. Dielectric rectangular waveguides are also analyzed for both isotropic and anisotropic cases.

Su, C. (1986): "A combined method for dielectric waveguides using the finite-element technique and the surface integral equations method." *IEEE Transactions on Microwave Theory and Techniques,* vol. **MTT-34**, pp. 1140–46, 1986.

A finite element formulation in terms of the transverse field components is combined with surface integrals to treat the propagation characteristics of inhomogeneous waveguides with single or multiple claddings. The combined method does not suffer from spurious modes, and it is able to find the cutoff frequencies of arbitrarily shaped, inhomogeneous dielectric waveguides with a single cladding. Numerical results are given for inhomogeneous elliptical waveguides, diffusion waveguides, and the corresponding directional couplers.

Hayata, K., M. Koshiba, M. Eguchi, and M. Suzuki (1986): "Novel finite-element formulation without any spurious solutions for dielectric waveguides." *Electronics Letters,* vol. **22**, pp. 295–96, 13 March 1986.

The dielectric waveguide problem is formulated variationally in terms of transverse magnetic field components. The field is made solenoidal implicitly so no spurious modes occur.

Pantic, Z. and R. Mittra (1986): "Quasi-TEM analysis of microwave transmission lines by the finite-element method." *IEEE Transactions on Microwave Theory and Techniques,* vol. **MTT-34**, pp. 1096–103, 1986.

Conventional finite elements, as well as singular and infinite elements, are used to solve for the potential and field distributions for computing the capacitance per unit length of a transmission line. A perturbation method is used to estimate losses.

Tzuang, C. K. and T. Itoh (1986): "Finite-element analysis of slow-wave Schottky contact printed lines." *IEEE Transactions on Microwave Theory and Techniques,* vol. **MTT-34**, pp. 1483–89, 1986.

Extensive high-order finite element analyses for printed-circuit slow-wave structures show good agreement with other theoretical results and with experiment. Schottky-contact microstrip and coplanar waveguide are compared to more general and practical slow-wave circuits.

Hayata, K., M. Koshiba, and M. Suzuki (1986): "Lateral mode analysis of buried heterostructure diode lasers by the finite-element method." *IEEE Journal of Quantum Electronics,* vol. **QE-22**, pp. 781–88, 1986.

Vectorial and scalar finite element methods are compared for the analysis of waveguides composed of active or lossy media. The gain difference between fundamental and higher modes are evaluated for GaInAsP/InP diode lasers. The scalar approximate finite element method appears to be useful for lateral mode analysis of diode lasers.

Matsuhara, M., T. Angkaew, and N. Kumagai (1987): "Finite element analysis of the waveguide eigenmodes — A novel method of employing transverse electromagnetic fields." *Electronics and Communications in Japan, Part 2 (Electronics),* vol. **70**, no. 2, pp. 85–91, 1987.

A finite element method is based on a variational expression for the propagation constant beta, using the transverse components of **E** and **H**. The true guided modes have real propagation constants, and all spurious modes have complex values of β.

Angkaew, T., M. Matsuhara, and N. Kumagai (1987): "Finite-element analysis of waveguide modes: A novel approach that eliminates spurious modes." *IEEE Transactions on Microwave Theory and Techniques,* vol. **MTT-35**, pp. 117–23, 1987.

A variational finite element method is formulated in terms of the transverse electric and magnetic field components. In this approach, all guided-mode solutions are real, while the spurious-mode solutions are not, allowing ready discrimination of spurious solutions. Two numerical examples are given for isotropic materials, one for a magnetically anisotropic case.

Baumeister, K. J. (1987): "Finite element modeling of electromagnetic propagation in composite structures." *1987 AP-S International Symposium Digest: Antennas and Propagation,* vol. **1**, pp. 272–75, 1987.

A finite element method is formulated for electromagnetic propagation in two-dimensional absorbing ducts. Reflection and transmission at entrance and exit boundaries are determined by coupling the finite element solutions to the eigenfunctions of an infinite uniform perfect conducting duct.

Hurwitz, H., Jr. (1988): "The surface projection approach to vector field calculations." *Mathematical and Computer Modelling,* vol. **11**, pp. 313–16, 1988.

A variational finite element method for the vector Helmholtz equation is derived. The trial functions satisfy the Helmholtz equation in each element and provide continuity in the surface projections of the field along the element boundaries, as expressed by nodal values of their surface projections. The element matrices may be found by a generalized form of infinitesimal scaling of greater complexity than in scalar problems. The method has been applied to open regions with an outgoing radiation condition at infinity.

Mustacich, R. V. (1988): "Scalar finite element analysis of electrooptic modulation in diffused channel waveguides and poled waveguides in polymer thin films." *Applied Optics,* vol. **27**, pp. 3732–37, 1988.

An approximate scalar finite element approach is used. The interaction of applied electric fields with optical fields is found over the 2-D finite element mesh and the guide eigenmodes provide the electrooptic changes in the propagation constants. Applications are shown for titanium-indiffused $LiNbO_3$ and poled waveguides in polymer thin films.

Pantic-Tanner, Z. and R. Mittra (1988): "Full-wave analysis of isolated and coupled microwave transmission lines using the finite-element method." *Archiv für Elektronik und Übertragungstechnik,* vol. **42**, pp. 370–74, 1988.

The hybrid (TE-TM) modes of isolated and coupled microwave transmission lines are analyzed using high-order nodal, singular, and infinite elements to solve for the modal propagation constant and the field distribution. The characteristic impedance is found using the power-current definition. A perturbational estimate is given of losses due to conductor and dielectric dissipation.

Hayata, K., M. Eguchi, and M. Koshiba (1988): "Self-consistent finite/infinite element scheme for unbounded guided wave problems." *IEEE Transactions on Microwave Theory and Techniques,* vol. **36**, pp. 614–16, 1988.

Unbounded guided wave problems are analyzed by decay-type infinite elements. To determine optimum decay parameters, two algorithms based on successive approximation are presented. The method is applied to an optical fiber problem.

Young, T. P. (1988): "Design of integrated optical circuits using finite elements." *IEE Proceedings–A,* vol. **135**, pp. 135–44, 1988.

Finite element methods are described for integrated optical devices. Their application to component design is illustrated by examples from electrostatics to the optics. The full H-vector formulation of the optical problem is outlined.

McDougall, M. J. and J. P. Webb (1989): "Infinite elements for the analysis of open dielectric waveguides." *IEEE Transactions on Microwave Theory and Techniques,* vol. **37**, pp. 1724–31, 1989.

Radially decaying infinite elements with multiple decaying exponential approximating functions are introduced. They are used in finite element analysis of unbounded, arbitrarily shaped, and arbitrarily inhomogeneous dielectric waveguides. Examples include slab, circular, and square dielectric waveguides.

Hayata, K., K. Miura, and M. Koshiba (1989): "Vectorial finite-element formalism for electromagnetic waveguiding structures with loss and/or gain." *Electronics and Communications in Japan, Part 2 (Electronics),* vol. **72**, no. 4, pp. 77–84, 1989.

A variational functional and a consequent finite element method are derived in terms of the transverse magnetic field components, with the propagation constant as the eigenvalue. The method is applicable to dispersive media with frequency-dependent properties and to structures containing conductive materials whose loss varies inversely with frequency.

Hayata, K., K. Miura, and M. Koshiba (1989): "Full vectorial finite element formalism for lossy anisotropic waveguides." *IEEE Transactions on Microwave Theory and Techniques,* vol. **37**, pp. 875–83, 1989.

A finite element method is developed for wave guiding structures containing lossy, anisotropic materials. The fields are formulated in terms of the transverse magnetic field components, and the eigenvalue of the final matrix equation corresponds to the propagation constant itself. To demonstrate this method, results are shown for a rectangular waveguide filled with lossy anisotropic dielectric; they agree with known results for both phase and attenuation.

Svedin, J. A. M. (1989): "A numerically efficient finite-element formulation for the general waveguide problem without spurious modes." *IEEE Transactions on Microwave Theory and Techniques,* vol. **37**, pp. 1708–15, 1989.

A finite element procedure without spurious modes is derived from the first-order Maxwell curl equations and comprises both the transversal and longitudinal components of the electric and magnetic fields, with all necessary boundary conditions on the tangential field components satisfied *a priori* by the trial functions. Spurious modes are absent.

Hayata, K., M. Eguchi, and M. Koshiba (1989): "Finite element formulation for guided-wave problems using transverse electric field component." *IEEE Transactions on Microwave Theory and Techniques,* vol. **37**, pp. 256–58, 1989.

Finite elements are formulated in transverse electric field components only. No spurious solutions can occur and matrix sizes are one-third smaller than with penalty-function methods. The formulation is tested by application to known waveguide problems.

Chew, W. C. and M. A. Nasir (1989): "A variational analysis of anisotropic, inhomogeneous dielectric waveguides." *IEEE Transactions on Microwave Theory and Techniques,* vol. **37**, pp. 661–68, 1989.

Variational formulations are given for anisotropic, dielectric waveguides using only the transverse components of E or H. These are completely equivalent because they are the transpose problems of each other. The theory is followed by a simple finite element implementation. No spurious modes occur in this formulation.

Gibson, A. A. P. and J. Helszajn (1989): "Finite element solution of longitudinally magnetized elliptical gyromagnetic waveguides." *IEEE Transactions on Microwave Theory and Techniques,* vol. **37**, pt. 1, pp. 999–1005, 1989.

The classic functional derivation (premultiplying the wave equation by the conjugate field and integrating over the problem region) is applied to define finite elements for the longitudinally magnetized gyromagnetic guide using the *z−directed* component wave equations. It is used to evaluate the propagation constants of an elliptical gyromagnetic waveguide with either an electric or magnetic wall.

Eswarappa, G. I. Costache, and W. J. R. Hoefer (1989): "Finlines in rectangular and circular waveguide housings including substrate mounting and bending effects — Finite element analysis." *IEEE Transactions on Microwave Theory and Techniques,* vol. **37**, pp. 299–306, 1989.

A finite element method is used to find the dispersion of dominant and higher modes in finlines, for various metallization thicknesses, substrate mounting groove depths, and bending of the substrate.

Imhoff, J. F., G. Meunier, X. Brunotte, and J. C. Sabonnadière (1990): "An original solution for unbounded electromagnetic 2D- and 3D-problems throughout the finite element method." *IEEE Transactions on Magnetics,* vol. **26**, pp. 1659–61, 1990.

An early version of the transformation techniques later worked out by Stochniol (*q.v.).*

Webb, J. P. (1990): "Finite element methods for junctions of microwave and optical waveguides." *IEEE Transactions on Magnetics,* vol. **26**, pp. 1754–58, 1990.

For calculating electromagnetic fields in microwave and optical components and finding scattering matrices, two approaches have been used: (1) exciting with given incident waves and calculating reflected waves, and (2) exciting with applied voltages and calculating currents. Optical components, usually unbounded, can be handled with absorbing boundary conditions. Advances in finite element technology have made this method competitive with integral-equation approaches, and distinctly better when there are several materials or complicated shapes.

Onuki, T. (1990): "Hybrid finite element and boundary element method applied to electromagnetic problems." *IEEE Transactions on Magnetics,* vol. **26**, pp. 582–87, 1990.

A novel discretization scheme in the hybrid method uses a mixed linear and constant boundary element. A boundary element formulation is given in terms of the vector potential, and another in the magnetic field intensity and the magnetic scalar potential. Besides two-dimensional, axisymmetric, and three-dimensional cases, the problem of two semi-infinite regions is investigated.

Svedin, J. A. M. (1990): "Propagation analysis of chirowaveguides using the finite-element method." *IEEE Transactions on Microwave Theory and Techniques,* vol. **38**, pp. 1488–96, 1990.

Waveguides including chiral media are analyzed using a full vector finite element formulation in terms of the electric and magnetic fields. This formulation admits no spurious modes. Both the necessary conditions on the tangential field components and the additional conditions on the normal field components are automatically satisfied by the trial functions, leading to comparatively small eigenvale problems. Extension to the novel class of chirowaveguides preserves these useful numerical properties. Finite element results for both metal and dielectric chirowaveguides are compared with exact solutions.

Hirayama, K. and M. Koshiba (1990): "Numerical analysis of arbitrarily shaped discontinuities between planar dielectric waveguides with different thicknesses." *IEEE Transactions on Microwave Theory and Techniques,* vol. **38**, pp. 260–64, 1990.

A combined finite element and boundary element method is applied to discontinuities between planar guides of different thicknesses. The waveguides connected to the discontinuity are dealt with analytically. Scattering characteristics and far-field patterns can then be calculated. The scattering from a step, a staircase transformer, and a tapered transformer are analyzed.

Feliziani, M. (1990): "A FEM approach to shielding effectiveness in braided-shield cables." *IEEE Transactions on Magnetics,* vol. **26**, pp. 929–32, 1990.

Magnetic field coupling between an external field and a braided-shielded cable is described by a homogeneous Helmholtz equation in the magnetic vector potential. Absorbing boundary conditions limit the area of finite element discretization. Results agree with the analytical solution for a test problem.

Wilkins, G. M., J. F. Lee, and R. Mittra (1991): "Numerical modeling of axisymmetric coaxial waveguide discontinuities." *IEEE Transactions on Microwave Theory and Techniques,* vol. **39**, pp. 1323–28, 1991.

Coaxial-to-coaxial discontinuities are solved by a finite element method. An absorbing boundary condition is applied at the input and output port boundaries. Numerical results are given for two particular geometries.

Fernandez, F. A. and Y. Lu (1991): "A variational finite element formulation for dielectric waveguides in terms of transverse magnetic fields." *IEEE Transactions on Magnetics,* vol. **27**, pp. 3864–67, 1991.

A variational finite element formulation free of spurious modes is proposed for anisotropic, inhomogeneous, lossy dielectric waveguides. This method permits direct solution for the propagation constant at a given frequency and preserves matrix sparsity.

Aubourg, M. and P. Guillon (1991): "A mixed finite element formulation for microwave devices problems. Application to MIS structure." *Journal of Electromagnetic Waves and Applications,* vol. **5**, pp. 371–86, 1991.

A mixed finite element formulation is presented for propagation characteristics of lossy or dissipative waveguides. A formulation in the transverse electric and magnetic field components is discretized using these mixed finite elements, leading to an absence of spurious modes. Propagation characteristics of an MIS microstrip line are calculated.

Ramahi, O. M. and R. Mittra (1991): "A surface integral equation method for the finite element solution of waveguide discontinuity problems." *IEEE Transactions on Microwave Theory and Techniques,* vol. **39**, pp. 604–8, 1991.

The exterior region is modeled by a surface integral representation, coupled to conventional finite elements in the interior. Illustrative numerical results are presented.

Farail, F., A. Roger, and M. Bouthinon (1991): "Finite element method for electromagnetic scattering in a lossy medium." *IEE Proceedings–H,* vol. **138**, pp. 201–6, 1991.

A two-dimensional H-field formulation for scattering by cylinders of arbitrary shape in an infinitely-extending lossy medium, and its implementation using a commercial finite element program, are presented. Because the medium is lossy, no exterior or infinite elements are used.

Stochniol, A. (1992): "An efficient open boundary finite element method for electromagnetic problems." *COMPEL,* vol. **11**, pp. 53–56, 1992.

Spatial mapping of infinite unbounded regions, exterior to the main region of interest, onto a finite region by means of an explicit coordinate transformation, allowing solution of the infinite problem by solving two coupled closed boundary problems.

Koshiba, M. and K. Inoue (1992): "Simple and efficient finite-element analysis of microwave and optical waveguides." *IEEE Transactions on Microwave Theory and Techniques,* vol. **40**, pp. 371–77, 1992.

A simple and efficient finite element method for microwave and optical waveguides is formulated using three components of the electric or magnetic field. Edge elements are used to eliminate spurious solutions. An eigenvalue problem is derived that involves only the edge variables in the transversal plane and can provide a direct solution for the propagation constant. Computed results are given for microstrip transmission lines and dielectric waveguides.

Stochniol, A. (1992): "A general transformation for open boundary finite element method for electromagnetic problems." *IEEE Transactions on Magnetics,* vol. **28**, pp. 1679–81, 1992.

A general transformation for the solution of open boundary field problems using the finite element method is presented. The infinite unbounded region is mapped onto a finite region, to which standard finite element techniques can be applied. This technique can be fully automated and allows material inhomogeneities as well as anisotropies.

Foo, S. L. and P. P. Silvester (1992): "Boundary-marching method for discontinuity analysis in waveguides of arbitrary cross section." *IEEE Transactions on Microwave Theory and Techniques,* vol. **40**, pp. 1889–93, 1992.

A recursive condensation algorithm previously used in diffusion problems and electrostatics is extended to waves. It is used to construct a matrix representation for an infinitely long waveguide of arbitrary cross-sectional shape. In numerical tests, scattering matrices for long guides converge to nearly full word-length in six to seven recursion steps. Discontinuity characteristics are within 1-2% of known results where they exist.

Feliziani, M. (1992): "Numerical solutions of low-frequency scattering problems." *IEEE Transactions on Magnetics,* vol. **28**, pp. 1224–77, 1992.

Two-dimensional finite elements are used to solve Helmholtz's equation in terms of scattered electric or magnetic fields, where the forcing terms are given in terms of incident fields. Wave impedance boundary conditions for the scattered fields are introduced and compared with tests.

Sachdev, S. and Z. J. Cendes (1993): "Combined tangential-normal vector elements for computing electric and magnetic fields." *IEEE Transactions on Magnetics,* vol. **29**, pp. 1456–59, 1993.

A new type of two-dimensional vector finite element possesses continuity of both normal flux components and tangential field components at element edges. Continuity is exact at element edge midpoints and approximate (exact in average) elsewhere.

Hybrid Methods

Silvester, P. P. and M. S. Hsieh (1971): "Finite-element solution of 2-dimensional exterior-field problems." *Proceedings IEE,* vol. **118**, pp. 1743–47, 1971.

This work is one of the first to deal with the finite element solution of the Laplace's equation on unbounded region. The infinitely extending region outside the arbitrary contour Γ, which encloses the region of major interest, is represented as a single finite element, and a variational approach is used to formulate the problem. The domain of integration of the appropriate functional is the whole unbounded space. For a solution to exist, the latter integral must exist. However, this requirement is met since it is known that finite energy is associated to any solution of Laplace's equation in exterior regions. The system of linear equations flowing from this formulation is augmented to include an integral relationship between the unknown field and its normal derivatives at Γ. The noteworthy aspect of this formulation is that the exterior region, treated as a finite element, fits precisely into the formalism of the interior problem. Details on computer implementation as well as time and memory requirements are discussed.

McDonald, B. H. and A. Wexler (1972): "Finite-element solution of unbounded field problems." *IEEE Transactions on Microwave Theory and Techniques*, vol. **MTT-20**, pp. 841–47, 1972.

This is one of the first works that combine the finite element method to an integral equation for the solution of unbounded field problems. The finite element method is applied to solve a partial differential equation on a region(s) of finite size, called picture-frame(s), delimited by a double contour. The integral equation, which represents the field in the remaining unbounded region, is formulated as a constraint upon the picture-frame(s) boundary. A noteworthy aspect of this formulation is that the boundary integral involves the free-space Green's function, thus special problem-dependent Green's functions need not be found. The introduction of a double contour permits avoidance of the problem of Green's function singularities in solving the integral equation, which is the major flaw in other formulations. The results presented, all relative to two-dimensional problems, include the static and harmonic fields of a parallel plate capacitor, a microstrip line on a dielectric substratum, and a radiating antenna with a dielectric obstacle.

Mei, K. K. (1974): "Unimoment method of solving antenna and scattering problems." *IEEE Transactions on Antennas and Propagation*, vol. **AP-22**, pp. 760–66, 1974.

The basic concept of the unimoment method consists in separating, by means of a constant coordinate contour Γ, the exterior boundary value problem of an antenna or scattering problem from an interior one. This is the major strength of this method, since it allows the internal problems to be solved in an extremely efficient manner employing the finite element method as applied to closed problems. In fact, the interior problem is treated as a Dirichlet boundary value problem and solved N times with different boundary conditions, so that N linearly independent solutions are generated. The scattered field outside Γ is represented by a harmonic expansion — with N harmonics — and then the continuity conditions on the field and its normal derivative are enforced at Γ between a linear combination of the N independent solutions of the internal problem and the sum of the incident and the scattered fields. Methods to solve the interior problem (techniques to solve sparse matrices) are discussed in this paper, since the successful application of the unimoment method depends on how fast the internal solutions can be generated. The efficiency of the unimoment is objectively evaluated by comparing its memory requirements and the computational effort needed to obtain the solution against those relative to the moment method.

Chang, S. K. and K. K. Mei (1976): "Application of the unimoment method to electromagnetic scattering of dielectric cylinders." *IEEE Transactions on Antennas and Propagation*, vol. **AP-24**, pp. 35–42, 1976.

The unimoment method is applied to calculate the scattered fields of dielectric cylinders of arbitrary cross-section or inhomogeneous material. The authors also introduced the method of "inhomogeneous element," which allows the use of finite elements comprising inhomogeneous material. Indeed, this method has not been widely used in applications of the finite element methods since it renders the finite element formulation more involved but not more efficient.

Bendali, A., D. Clair, and J. Tourneur (1982): "Finite-element approximation of electromagnetic diffraction by arbitrarily shaped surfaces." *Electronics Letters*, vol. **18**, pp. 641–42, 1982.

A polynomial finite element method is used for computing the surface current on a perfect conductor subject to an incident electromagnetic wave. Radiation patterns of two horns are compared with those obtained from experiment.

Bossavit, A. and J. C. Verite (1982): "A mixed FEM-BIEM method to solve 3-D eddy-current problems." *IEEE Transactions on Magnetics*, vol. **MAG-18**, pp. 431–35, 1982.

This paper introduces edge elements and solves eddy-current problems in three dimensions by blending finite elements with boundary integrals. It has been implemented in the linear harmonic case and used in studies of non-destructive testing.

Marin, S. P. (1982): "Computing scattering amplitudes for arbitrary cylinders under incident plane waves." *IEEE Transactions on Antennas and Propagation*, vol. **AP-30**, pp. 1045–49, 1982.

A non-local boundary condition is used to develop a variational formulation for the two-dimensional problem of electromagnetic scattering from cylinders with arbitrary cross-section. The main differences between this method and that suggested by McDonald and Wexler (1972) lie in using only one fictitious contour Γ to separate the interior and the exterior regions; and in introducing the boundary integral constraint on Γ directly inside the functional associated with the scattering problem. In particular, the first feature results in a shortcoming for the formulation, since it makes the method subject to non-physical resonance solutions that arise when the quantity $-k^2$ is close to an eigenvalue of the Laplacian in the interior region delimited by Γ.

Mayergoyz, I. D., M. V. K. Chari, and A. Konrad (1983): "Boundary Galerkin's method for three-dimensional finite element electromagnetic field computation." *IEEE Transactions on Magnetics,* vol. **MAG-19**, pp. 2333–36, 1983.

Electromagnetic field problems in unbounded regions are formulated in the Silvester-Hsieh manner by reducing the field equations in unbounded space to equivalent boundary Galerkin's criterion for regions occupied by conductors. A discretization based on the coupled boundary/volume Galerkin's criterion is presented.

Orikasa, T., T. Honma, I. Fukai, and S. Washisu (1983): "Finite element method for unbounded field problems and application to two-dimensional taper." *International Journal for Numerical Methods in Engineering,* vol. **19**, pp. 157–68, 1983.

The unbounded field treatment of McDonald

and Wexler, which excludes Green's function singularities, was extended and amended by the authors (*Electron. Letts.*, **15**, 772–74, 1979). The revised method is now applied to open-ended two-dimensional tapers, from which electromagnetic waves radiate into free space.

Jeng, S. K. and C. H. Chen (1984): "On variational electromagnetics: theory and application." *IEEE Transactions on Antennas and Propagation*, vol. **AP-32**, pp. 902–7, 1984.

A fundamental variational principle for electromagnetic problems is derived and subsequently tailored to a desired variational equation. The results obtained are applicable to unbounded domain problems. The deduced variational equation is numerically solved via a hybrid finite element method. This formulation is applied to solve the two-dimensional scattering from cylinders comprising isotropic, inhomogeneous dielectric material. In the TM_z case, the total E_z field outside a circular fictitious contour Γ is written in terms of an infinite sum of cylindrical waves. Such a series expansion is also used to derive an analytical expression for the outward normal derivative of the electric field at contour Γ, as a function of the the electric field itself. Substituting this relationship into the functional and minimizing the resulting expression via the finite element method, the problem can be solved. Obviously, the exact nature of the radiation boundary condition derived from the eigenfunction expansions of the exterior field spoils the sparse structure of the finite element matrix.

Morgan, M. A., C. H. Chen, S. C. Hill, and P. W. Barber (1984): "Finite element–boundary integral formulation for electromagnetic scattering." *Wave Motion*, vol. **6**, pp. 91–103, 1984.

The method presented in this paper is similar to the unimoment method in that it combines the finite element method, used to analyze the interior problem, to a spherical vector harmonic expansion, used to represent the scattered field in the exterior region. The interior and exterior solutions are then coupled through the transmission matrix method, which allows a fictitious contour Γ that conforms to the outer boundary of the scatterer to be used to separate the interior and exterior problems. Although this latter feature is a major strength of the finite elements with extended boundary conditions method, numerical results have shown that the method tends to have difficulties with attaining convergence of the solution if Γ becomes extremely elongated. This failure occurs because of the generally incomplete nature of the exterior region spherical harmonic expansion employed to represent the field over the surface of the scattering body.

Morgan, M. A. and B. E. Welch (1986): "The field feedback formulation for electromagnetic scattering computations." *IEEE Transactions on Antennas and Propagation*, vol. **AP-34**, pp. 1377–82, 1986.

A formulation is presented for the solution of three-dimensional electromagnetic scattering problems whose symbolic representation is a simple feedback system. Interior and exterior regions are decoupled by means of two arbitrary fictitious surfaces, one enclosing the other and both enclosing all discontinuities and sources. The field in the finite-size region delimited by the outer surface is described as a finite linear combination of numerically derived basis functions, which are obtained by solving interior Dirichlet boundary value problems. A surface integral equation involving the free-space Green's function enforces the continuity conditions between the interior and exterior problems. The method is similar to the unimoment method (Mei 1974) in that it separates the solution of the internal problem from the external one. Indeed, this is one of its major strengths together with its emploment of a fully arbitrary fictitious boundary to delimit the geometric domain for computation. On the contrary, its major flaw resides in the lack of standard procedures to derive the non-homogeneous Dirichlet boundary conditions necessary to start the solution technique. To illustrate the general formulation, a numerical example relative to the electromagnetic scattering from thin wire antennas is carried out in detail.

Wu, R. B. and C. H. Chen (1986): "Variational reaction formulation of scattering problem for anisotropic dielectric cylinders." *IEEE Transactions on Antennas and Propagation*, vol. **AP-34**, pp. 640–45, 1986.

The variational reaction theory is applied to obtain the (E_z, H_z) formulation for the scattering problem when a plane wave impinges obliquely upon an inhomogeneous, anisotropic, arbitrarily shaped dielectric cylinder. The radiation condition at infinity and the continuity condition over the boundary are automatically satisfied, for they constitute natural boundary conditions for the functional. This is obtained by choosing a circular fictitious contour to delimit the geometrical domain for computations and deriving an analytical relationship between the longitudinal component of the fields and their tangential component to the contour Γ. Such a relationship flows from the analytical expression of the scattered electromagnetic field exterior to Γ, which is given in terms of outgoing cylindrical wave functions.

Yamabuchi, T. and Y. Kagawa (1986): "Finite element approach to unbounded Poisson and Helmholtz problems using hybrid-type infinite element." *Electronics and Communications in Japan, Part 1 (Communications),* vol. **69**, no. 3, pp. 65–74, 1986.

Electric or sound fields with open boundaries, governed by the Poisson or Helmholtz equation, permit finite element analysis in conjunction with a prescribed-decay type of infinite element, without increasing the size and bandwidth of the system matrix beyond that of the finite region. Reasonable accuracy is obtained in the interior region, but the accuracy in the exterior is not satisfactory. The decay parameter can be set to minimize the finite element functional, and this condition promises reasonable results.

Lee, J. F. and Z. J. Cendes (1987*a*) : "Transfinite elements: A highly efficient procedure for modeling open field problems." *Journal of Applied Physics,* vol. **61**, pp. 3913–15, 1987.

The procedure called transfinite elements is presented in a form suitable to solve the Laplace's equation on unbounded domains. It employs a constant coordinate fictitious contour Γ to separate the interior and exterior problems and a finite linear combination of analytic basis functions to represent the solution outside Γ. A variational approach is used to formulate the problem. Although the domain of integration of the appropriate functional is the whole unbounded space, the integral surely exists, since it is known that finite energy is associated to any solution of Laplace's equation in exterior regions. The analytical solution of the exterior problem is substituted into the functional for the field, which is subsequently minimized via the finite element method. The system of linear equations flowing from this formulation is augmented to include the continuity condition between the interior and exterior solutions, which is imposed in a pointwise manner. The resulting system of linear equations is solved for the unknown potential.

Lee, J. F. and Z. J. Cendes (1987*b*) : "The transfinite element method for computing electromagnetic scattering from arbitrary lossy cylinders." *IEEE AP-S International Symposium Digest,* AP03–5, pp. 99–102, 1987.

This work briefly describes how to modify the transfinite element method to apply it to electromagnetic wave problems. Results relative to electromagnetic wave scattering from lossy cylinders are reported.

Mei, K. K. (1987): "Unimoment method for electromagnetic wave scattering." *Journal of Electromagnetic Waves and Applications,* vol. **1**, pp. 201–22, 1987.

A detailed and exhaustive description of the unimoment method is given in this paper. It covers the fundamentals of the unimoment method as well as all the steps required for its application to the two-dimensional problem of the electromagnetic scattering from a dielectric hollow cylinders of arbitrary shape: mesh formation, coefficient matrix construction and its efficient inversion, and eventually, boundary matching.

Jin, J. M. and V. V. Liepa (1988*a*) : "Application of hybrid finite element method to electromagnetic scattering from coated cylinders." *IEEE Transactions on Antennas and Propagation,* vol. **36**, pp. 50–54, 1988.

The hybrid finite element–boundary integral method developed by McDonald and Wexler (1972) is applied to the analysis of the electromagnetic scattering from coated cylinders. The choice of the two fictitious contours that delimit the picture-frame of interest is not subject to any restriction; hence, they are chosen so as to minimize the geometrical domain of computation. Details on the finite element mesh construction are given. The global system of linear equations obtained, whose coefficient matrix is partly full, is solved employing the same algorithm implemented by McDonald and Wexler, which inverts the coefficient matrix by blocks. Extensive numerical results are shown for both TE_z and TM_z polarization and compared with those provided by analytical method (when this is the case) or by a moment method–impedance boundary condition technique.

Jin, J. M. and V. V. Liepa (1988*b*) : "A note on hybrid finite element method for solving scattering problems." *IEEE Transactions on Antennas and Propagation,* vol. **36**, pp. 1486–90, 1988.

The solution algorithm widely used till then to solve the system of linear equations arising from the application of the finite element–boundary integral method to the analysis of scattering problems is modified. This results in a substantial improvement of its computational efficiency and capability, as demonstrated by numerical examples. However, this achievement harbors a shortcoming later pointed out by Pearson, Peterson, Bahrmasel, and Whitaker (1992): it renders the formulation subject to the "interior resonance problem." The modified hybrid finite element–boundary integral method is objectively evaluated by comparing its memory requirements with those relative to the unimoment method and the moment method.

Lee, J. F. and Z. J. Cendes (1988): "The transfinite element method for modeling MMIC devices." *IEEE Transactions on Microwave Theory and Techniques*, vol. **36**, pp. 1639–49, 1988.

The transfinite element method is applied to determine the fields and scattering parameters for a wide variety of stripline and microstrip devices. With minor modifications, the method can also be employed for analyzing waveguide junctions. Introducing special elements, the transfinite element method can also efficiently treat singular points in waveguide junctions.

Ling, R. T. (1988): "Application of computational fluid dynamics methods to a numerical study of electromagnetic wave scattering phenomena." *Journal of Applied Physics,* vol. **64**, pp. 3785–91, 1988.

Scattering of electromagnetic waves is analogous to the fluid dynamic problem of flow past an obstacle. The far-field radiation condition and interface conditions are analogous to free-stream and flow-tangency conditions. Numerical methods in computational fluid dynamics can therefore be carried over to scattering.

Moaveni, M. K., A. A. Rizvi, and B. A. Kamran (1988): "Plane-wave scattering by gratings of conducting cylinders embedded in an inhomogeneous and lossy dielectric." *Journal of the Optical Society of America, A (Optics and Image Science),* vol. **5**, pp. 834–42, 1988.

A combination of finite elements and boundary integrals is used. Numerical results are obtained for gratings with circularly cylindrical elements for either polarization.

Kuzuoglu, M. (1989): "Analysis of transmission of electromagnetic waves through apertures by a combination of the finite element and boundary element methods." *1989 URSI International Symposium on Electromagnetic Theory*, pp. 58–60, 1989.

Integral equations are used on appropriately chosen boundaries to describe the field exterior to the boundary, by differential equations in the interior. A weak formulation of the integral equations is coupled with a weak formulation of the interior boundary value problem. Although the integral equations involve singular kernels, only weakly singular integrals appear in the weak formulation.

Pearson, L. W., R. A. Whitaker, and L. J. Bahrmasel (1989): "An exact radiation boundary condition for the finite-element solution of electromagnetic scattering on an open domain." *IEEE Transactions on Magnetics*, vol. **25**, pp. 3046–48, 1989.

The exact radiation boundary condition valid for circular fictitious contour developed by Jeng and Chen (1984), and subsequently used by Wu and Chen (1986) in a variational formulation of electromagnetic problems, is introduced in a weighted residual statement of the two-dimensional scattering problem. The numerical results provided by this formulation are compared to those obtained via the finite element–absorbing boundary conditions (of Bayliss-Turkel type) method. In no case do results from the two methods differ by more than 2.5% in the root mean square average over all nodal values.

Barkeshli, S., H. A. Sabbagh, and D. J. Radecki (1990): "A novel boundary integral/infinite element method for the analysis of transient electromagnetic field coupling to a metallic enclosure." *1990 International Symposium Digest. Antennas and Propagation,* vol. **1**, pp. 122–25, 1990.

Transient electromagnetic field coupling into an enclosure (cavity) is modeled by three-dimensional finite elements coupled to the exterior region through the H-field integral equation.

Cangellaris, A. C. and R. Lee (1990): "The bymoment method for two-dimensional electromagnetic scattering." *IEEE Transactions on Antennas and Propagation*, vol. **38**, pp. 1429–37, 1990.

The interior region containing the scatterer is decoupled from the exterior unbounded region by introducing two arbitrary fictitious contours, one (Γ_o) enclosing the other (Γ_i), and both enclosing all discontinuities and sources. Much like in the unimoment method, the interior problem (region delimited by Γ_o) is solved many times so that N linearly independent solutions are generated. Subsequently, a linear combination of such interior solutions is coupled to the exterior region by means of the Green's theorem in the unbounded region external to Γ_i. Basically, the bymoment method accrues from the same concepts as the unimoment method: the major difference is the use of arbitrary fictitious contours instead of a constant-coordinate contour as employed in the unimoment method. This results in the necessity of increasing the number of the harmonics in the expression of the scattered field with respect to that generally employed in the implementation of the unimoment method.

Collins, J. D., J. M. Jin, and J. L. Volakis (1990): "A combined finite element–boundary element formulation for solution of two-dimensional scattering problems via CGFFT." *Electromagnetics*, vol. **10**, pp. 423–37, 1990.

The method developed in a previous work by the same authors (Collins, Volakis, and Jin 1990) is extended to two different types of boundary enclosure: circular and ogival. This extension is carried out with the main difference that only one fictitious contour Γ is introduced to separate the interior and exterior regions. While at a first glance this difference is insignificant, indeed it renders the method subject to non-physical solutions related to the "interior resonances problem." However, this shortcoming does not affect the computational efficiency and the memory demand of the CGFFT approach. It should be noted that while the circular enclosure leads to completely convolutional integrals that may be computed via the fast Fourier transform (FFT), ogival enclosure contours yield partially convolutional integrals, and the remaining must be efficiently evaluated to maintain the $O(N)$ memory demand.

Collins, J. D., J. L. Volakis, and J. M. Jin (1990): "A combined finite element–boundary integral formulation for solution of two-dimensional scattering problems via CGFFT." *IEEE Transactions on Antennas and Propagation*, vol. **38**, pp. 1852–58, 1990.

Using rectangular boundaries to enclose the scatterer, several of the boundary integrals arising in the application of the finite element–boundary integral method (McDonald and Wexler 1972) are convolutions and may be efficiently evaluated via the fast Fourier transform (FFT) in the implementation of the conjugate gradient algorithm to solve the global system of linear equations. The numerical results presented for both TE_z and TM_z polarizations relative to several rectangular composite cylinders show a good agreement with the moment method. The attractive features of this approach are its $O(N)$ memory demand and the use of a one-dimensional FFT versus a two-dimensional FFT as required in a traditional implementation of the CG-FFT algorithm.

Ise, K., K. Inoue, and M. Koshiba (1990): "Three-dimensional finite-element solution of dielectric scattering obstacles in a rectangular waveguide." *IEEE Transactions on Microwave Theory and Techniques*, vol. **38**, pp. 1352–59, 1990.

To save memory, substructuring is introduced in the finite element method. To combine the uniform waveguide region with the finite element region, both evanescent and propagating modes are considered in the analytical solution. Spurious fields like the spurious solutions generated in the finite element analysis of three-dimensional eigenvalue problems can appear; they are suppressed by a penalty function method.

Jeng, S. K. (1990): "Scattering from a cavity-backed slit in a ground plane — TE case." *IEEE Transactions on Antennas and Propagation,* vol. **38**, pp. 1523–29, 1990.

Two-dimensional transverse electric wave scattering from a cavity-backed slit in a ground plane is analyzed by the Harrington-Mautz generalized network formulation. The admittance matrix of the cavity of arbitrary shape and medium is obtained by the finite element method. The cavity admittance matrix is added to the radiation admittance matrix of the equivalent magnetic current on a ground plane, permitting solution for the equivalent magnetic current on the slit.

Jin, J. M. and J. L. Volakis (1990): "TM scattering by an inhomogeneously filled aperture in a thick conducting plane." *IEE Proceedings–H,* vol. **137**, pp. 153–59, 1990.

This analysis combines classical interior finite elements and boundary integrals. Numerical examples are presented that demonstrate the validity, versatility, and capability of the technique.

Jin, J. M. and J. L. Volakis (1990): "TE scattering by an inhomogeneously filled aperture in a thick conducting plane." *IEEE Transactions on Antennas and Propagation,* vol. **38**, pp. 1280–86, 1990.

The transmission and scattering properties of an aperture in a thick conducting plane, filled with an inhomogeneous composite material for transverse electric polarization, is analyzed using a combination of finite element and boundary integral methods. For large apertures, the conjugate gradient method and fast Fourier transform are used to accelerate solution of the resulting system of equations.

Yuan, X. (1990): "Three dimensional electromagnetic scattering from inhomogeneous objects by the hybrid moment and finite element method." *IEEE Transactions on Microwave Theory and Techniques*, vol. **38**, pp. 1053–58, 1990.

The finite element method employed in the internal region uses basis functions exactly solenoidal within each element (though not at element interfaces) to avoid contamination by spurious modes. The accuracy and rate of convergence of the solution are discussed.

Yuan, X., D. R. Lynch, and J. W. Strohbehn (1990): "Coupling of finite element and moment methods for electromagnetics scattering from inhomogeneous objects." *IEEE Transactions on Antennas and Propagation*, vol. **38**, pp. 386–93, 1990.

The equivalence principle is applied to transform the original problem into exterior and interior problems. The two problems are coupled at an arbitrary fictitious surface, enclosing all discontinuities and sources, through the boundary conditions on the tangential electric and magnetic fields. The fictitious surface may coincide with the dielectric body surface. The interior problem involving inhomogeneous material is solved by the finite element method, while the exterior one is solved by the moment method. This separation of the two problems is advantageous, especially for three-dimensional problems, because the banded and sparse finite element matrix for the solution of the internal problem will be computationally very economical and efficient. Indeed, this strength of the method also harbors its major flaw: the formulation is susceptible to non-physical solutions and fails near resonant frequencies. Here, a resonant frequency is a frequency for which S, when covered by a perfect conductor and filled with the external medium, form a resonant cavity. Only two-dimensional results are given and compared with those obtained by the exact eigenfunction solution, the unimoment solution, and moment method solution.

Boyse, W. E. and A. A. Seidl (1991): "A hybrid finite element method for near bodies of revolution." *IEEE Transactions on Magnetics,* vol. **27**, pp. 3833–36, 1991.

A hybrid finite element method for three-dimensional electromagnetic scattering uses an equivalent current approach. The major computational tasks involved in monostatic scattering calculations are analyzed, and various methods of implementing the near-field radiation condition are compared.

Cangellaris, A. C. and R. Lee (1991): "Finite element analysis of electromagnetic scattering from inhomogeneous cylinders at oblique incidence." *IEEE Transactions on Antennas and Propagation*, vol. **39**, pp. 645–50, 1991.

A finite element formulation is presented for electromagnetic scattering from cylinders where the behavior of the fields in the z direction is given by $\exp(-j\beta z)$. The weighted residual statement of the problem involves the $z -$ *components* of the electric and magnetic fields, and is such that the continuity of the electric and magnetic fields at media interfaces is directly enforced. Moreover, the correct boundary conditions along material interfaces results in the expected coupling of the $z -$ *components* of the electric and magnetic fields. The bymoment method is used to handle the coupling of the interior finite element solution to the exterior unbounded region. The numerical results for both TE_z and TM_z polarizations show good agreement with series and integral equation solutions.

Jeng S. K. and S. T. Tzeng (1991): "Scattering from a cavity-backed slit in a ground plane — TM case." *IEEE Transactions on Antennas and Propagation,* vol. **39**, pp. 661–63, 1991.

Transverse-magnetic wave scattering from a cavity-backed slit in a ground plane is analyzed by the Harrington-Mautz generalized network formulation. The admittance matrix of the cavity of arbitrary shape and medium is obtained by the finite element method. The computed admittance matrix is added to the radiation admittance matrix of the equivalent magnetic current on a ground plane, permitting solution for the equivalent magnetic current on the slit. Numerical results for coated staggered cavities are included.

Jin, J. M. and J. L. Volakis (1991*a*) : "A finite element–boundary integral formulation for scattering by three-dimensional cavity-backed apertures." *IEEE Transactions on Antennas and Propagation*, vol. **39**, pp. 97–104, 1991.

The equivalence principle is applied to decouple the interior cavity region, where the finite element method is employed to formulate the problem, and the exterior unbounded region above the ground plane, in which the field is represented by means of the boundary integral. Imposing opposite magnetic current densities at the two sides of the aperture surface guarantees the continuity of the tangential component of the electric field. The continuity of the tangential magnetic field is enforced directly into the expression of the functional pertaining to a closed cavity. This approach does not require the knowledge of the cavity's Green's function; therefore, it is applicable to arbitrary-shaped cavities and material fillings. Because nodal shape functions associated with rectangular brick elements are employed, a penalty term is added to the functional so as to mitigate the spurious solution problem.

Jin, J. M. and J. L. Volakis (1991*b*) : "Electromagnetic scattering by and transmission through a three-dimensional slot in a thick conducting plane." *IEEE Transactions on Antennas and Propagation*, vol. **39**, pp. 543–50, 1991.

The technique developed by the same authors in a previous work (Jin and Volakis 1991*a*) is applied to the problem of transmission through a three-dimensional slot in a thick conducting plane. The main difference between the two formulations lies in the implementation of the finite element method. While in the previous work nodal-based shape functions were used, in this work edge-based expansion functions associated with rectangular brick elements are employed.

Jin, J. M. and J. L. Volakis (1991*c*) : "A hybrid finite element method for scattering and radiation by microstrip patch antennas and arrays residing in a cavity." *IEEE Transactions on Antennas and Propagation*, vol. **39**, pp. 1598–604, 1991.

The finite element–boundary integral formulation previously developed by the same authors to deal with electromagnetic scattering from cavity-apertures on metallic plane (Jin and Volakis 1991*a*) is applied to the analysis of planar radiating systems residing in a cavity. This application of such formulation demonstrates all the potentiality and flexibility of the finite element method in analyzing very complex structures in a straightforward manner. The finite element implementation employs edge-based element, and the final system of linear equations is solved via the biconjugate gradient method in conjunction with the fast Fourier transform. This results in an $O(N)$ memory demand, which is a desirable feature, especially in three-dimensional problems.

Gedney, S. D. and R. Mittra (1991): "Analysis of the electromagnetic scattering by thick gratings using a combined FEM/MM solution." *IEEE Transactions on Antennas and Propagation*, vol. **39**, pp. 1605–14, 1991.

The algorithm proposed is based on the generalized network formulation and combines the finite element method to the moment method. The hybrid technique is obviously applied to a basic cell of the periodic structure. The equivalence principle is applied at a fictitious contour Γ to separately solve the interior and exterior problems, which are subsequently coupled through the continuity conditions at Γ. The finite element method is applied in the complex region of the basic cell, delimited by Γ, while the moment method, together with the Floquet's theorem, is employed to evaluate the electromagnetic field produced by the equivalent current densities in the unbounded region. The shortcoming of this implementation of the hybrid finite element and moment method resides in the lack of the periodicity condition in the finite element formulation of the interior problem. Consequently, the associated code can handle only periodic structures composed of rectangular conductors filled with dielectric materials of arbitrary profiles.

Kagami, Y., F. Nishimura, and Y. Kagawa (1991): "Numerical approach to unbounded axially symmetric electromagnetic radiation problems using hybrid-type infinite element with special attention to shielding effects." *Electrical Engineering in Japan,* vol. **111**, pp. 11–19, 1991.

An integral-equation type of external infinite element is obtained by dividing the unbounded space into a number of infinite elements, each of which is transformed to a boundary element derived from the associated integral equations. A three-dimensional axisymmetric heterogeneous field and the radiation field produced by a loop antenna are analyzed. The shielding effect of a ring-shaped shield is evaluated as an example of the method.

Ramahi, O. M. and R. Mittra (1991): "Finite element solution for a class of unbounded geometries." *IEEE Transactions on Antennas and Propagation*, vol. **39**, pp. 245–50, 1991.

An efficient finite element–boundary integral formulation procedure is described for the solution of electromagnetic scattering problems involving unbounded geometries, for example, metallic planes or wedges with localized inhomogeneities or anisotropies. This formulation differs from that developed by Jin and Volakis (1991*a*) in that it does not use the equivalence principle to decouple the exterior and interior problems. Indeed, the

Ramahi-Mittra approach enhances the efficiency of the standard finite element–boundary integral technique (McDonald and Wexler 1972) by using the Green's function of the first kind (i.e., the Green's function that satisfies the boundary conditions for the exterior problem). This results in a minimization of the non-local behavior of the boundary integral itself. The authors believe this can be exploited to reduce the bandwidth of the coefficient matrix of the system of linear equations, though the basis of their belief is not entirely clear.

Collins, J. D., J. M. Jin, and J. L. Volakis (1992): "Eliminating interior resonances in finite element–boundary integral methods for scattering." *IEEE Transactions on Antennas and Propagation*, vol. **40**, pp. 1583–85, 1992.

The method based on the "complexification" of the wave number is employed for eliminating the non-physical resonances of the interior problem. Furthermore, it is demonstrated that using the scattered rather than the total field as the system variable, the solution of the problem is relatively insensitive to the value of the attenuation factor α introduced in the propagation constant k (provided, of course, that α is very small).

Cwik, T. (1992): "Coupling finite element and integral equation solutions using decoupled boundary meshes." *IEEE Transactions on Antennas and Propagation*, vol. **40**, pp. 1496–504, 1992.

The problem of electromagnetic scattering by two-dimensional bodies is carried out by means of a hybrid finite element boundary integral method. Two independent meshes are used; one for the internal region, where the FEM is applied, and the other for the boundary, where the integral equation is discretized. The resulting system of linear equation is solved employing an inward-looking formulation, and a detailed discussion of the numerical stability, with particular attention to the interior resonance problem, is given.

Gakuru, M. K. and R. L. Ferrari (1992): "Finite-element analysis of high-frequency axisymmetric device problems." *IEEE Transactions on Magnetics,* vol. **28**, pp. 1635–38, 1992.

Finite element solution of wave propagation problems in devices with complete azimuthal symmetry in material properties and shapes. Open radiating boundaries are treated by simulated absorption by lossy materials or by a hybrid FEM (differential and integral) formulation based on the vector Green's function for the Helmholtz equation. Application is to an open-ended circular waveguide.

Gedney, S. D., J. F. Lee, and R. Mittra (1992): "A combined FEM/MoM approach to analyze the plane wave diffraction by arbitrary gratings." *IEEE Transactions on Microwave Theory and Techniques*, vol. **40**, pp. 363–70, 1992.

Essentially, this work constitutes an improvement of the formulation developed by Gedney and Mittra (1991) to analyze electromagnetic scattering by thick gratings. Boundary conditions that account for the periodicity of the structure are enforced as a natural boundary condition in the finite element formulation of the internal problem. This expedient permits to extend the application of the hybrid finite element and moment method to the analysis of electromagnetic scattering by complex periodic structures with arbitrary geometry and material composition.

Gedney, S. D. and R. Mittra (1992): "Electromagnetic transmission through inhomogeneously filled slots in a thick conducting plane — arbitrary incidence." *IEEE Transactions on Electromagnetic Compatibility,* vol. **34**, pp. 404–15, 1992.

A finite element discretization is combined with a generalized network formulation to compute the fields within the inhomogeneous interior cavity region. The method may preserve the sparsity of the finite element method matrix in some problems. Examples of inhomogeneously filled slots of various cross sections are presented.

Igarashi, H. and T. Honma (1992): "BE and FE solutions to open boundary potential problems with helical symmetry." *IEEE Transactions on Magnetics,* vol. **28**, pp. 1671–74, 1992.

Boundary element and finite element solutions to open boundary potential problems with helical symmetry are based on field equations in twisted coordinates in which helical potential fields are two-dimensional. Transfinite elements are used for modeling open boundaries. Magnetic fields around a helical solenoid are shown to be in good agreement with analytic solutions. Electric fields around a twisted-pair line are concentrated near the lines as the helical pitch is increased.

Jin, J.-M., J. L. Volakis, and V. V. Liepa (1992): "Fictitious absorber for truncating finite element meshes in scattering." *IEE Proceedings–H,* vol. **139**, pp. 472–76, 1992.

Fictitious absorbers are used to minimize

reflection from the artificial boundary enclosing the scatterer, thus simulating a free space environment. A three-layered fictitious absorber model is used to demonstrate the validity of this approach.

Morgan, M. A. and T. B. Welch (1992): "Field feedback computation of scattering by 2-D penetrable objects." *IEEE Transactions on Antennas and Propagation*, vol. **40**, pp. 445–50, 1992.

The field feedback formulation developed by Morgan and Welch (1986) is applied to the analysis of electromagnetic scattering from two-dimensional penetrable objects. An analysis of the stability of the numerical solution is given in terms of feedback system theory, and this approach clearly delineates the source of stability problems and ill-conditioning for high-Q scatterers near internal resonance frequencies. Numerical results for both TE_z and TM_z polarizations, and relevant to different scatterers, show good agreement with available results.

Pelosi, G., A. Freni, and R. Coccioli (1992): "Appropriate boundary conditions for applying the FEM to the analysis of scattering from periodic structures embedded in a stratified medium." *Microwave and Optical Technology Letters*, vol. **5**, pp. 513–16, 1992.

The problem of electromagnetic scattering from periodic structures in stratified media is solved by means of the transfinite element method. The boundary conditions suggested by the authors for this kind of problem allow the minimization of the size of the discretized region, which comprises the grating of the structures and only a portion of the stratified medium.

Fan, D. and Y. Wang (1993): "Global numerical boundary condition based PDE solution techniques for open-region electromagnetic field problems." *IEEE Transactions on Antennas and Propagation*, vol. **41**, pp. 253–60, 1993.

The underlying concept of the global numerical boundary condition technique is similar to those used, for example, by Pearson, Whitaker, and Bahrmasel (1989). The difference between the two methods resides in the relationship between the longitudinal component of the scattered field and its normal derivative at the fictitious contour Γ. While in the latter work an analytical relationship in a closed form is employed, here a numerical one is derived. Obviously, this harbors a loss in the accuracy, without bringing any particular advantage in the structure of the coefficient matrix, which remains partly full.

Graglia, R. D., A. Freni, and G. Pelosi (1993): "A finite element approach to the electromagnetic interaction with rotating penetrable cylinders of arbitrary cross section." *IEEE Transactions on Antennas and Propagation*, vol. **41**, pp. 635–50, 1993.

A differential formulation is developed with respect to the co-moving reference frame of the scatterer on a closed circular domain, including the scatterer cross-section. The numerical solution is obtained by applying the transfinite element method in a form that slightly differs with respect to that suggested by Lee and Cendes (1987*b*) both in enforcing the continuity conditions of the field and in solving the final system of equation.

Pelosi, G., A. Freni, and R. Coccioli (1993): "A hybrid technique for analyzing the scattering from periodic structures." *IEE Proceedings–H*, vol. **140**, pp. 65–70, 1993.

The finite element method is combined to a Floquet expansion of the scattered field. This technique gives rise to a very simple and efficient procedure to solve the problem of electromagnetic scattering from periodic structures having arbitrary geometry, material properties, and periodicity with respect to the wavelength. The periodic boundary conditions imposed for applying the finite element method to the solution of the internal problem are discussed in detail. Much like the work of Graglia, Freni, and Pelosi (1993), the transfinite element method is implemented in a form that slightly differs with respect to that suggested by Lee and Cendes (1987*b*) both in enforcing the continuity conditions of the field and in solving the final system of equation.

Absorbing Boundary Conditions

Wilcox, C. H. (1956): "An expansion theorem for electromagnetic fields." *Communications on Pure and Applied Mathematics*, vol. **9**, pp. 115–32, 1956.

Wilcox demonstrated that the outgoing solutions of the three-dimensional scalar and vector Helmholtz equations, provided that they satisfy appropriate regularity requirements, can be represented by means of a series expansion representation. The importance of this theorem lies in the fact that the Bayliss-Turkel technique for deriving a class of absorbing boundary conditions accrues from the Wilcox expansion of the scattered field.

Engquist, B. and A. Majda (1977): "Absorbing boundary conditions for the numerical simulation of waves." *Mathematics of Computation,* vol. **31**, pp. 629–51, 1977.

Starting from the acoustic wave equation in Cartesian and polar coordinates, Engquist and Majda derive the theoretical non-local absorbing boundary conditions by means of the theory of pseudo-differential operators. Such conditions are strictly valid for a plane wave impinging on a fictitious plane surface — when Cartesian coordinates are used — and for a cylindrical wave impinging on a fictitious cylindrical surface — polar coordinates. Increasing the order of approximation of such conditions gives rise to a hierarchy of local boundary conditions that minimize the artificial reflections that occur at the fictitious contour. The well-posedness of each approximated absorbing boundary condition is also discussed. The numerical results obtained by solving the scalar wave equation and the shallow water equation via a finite-difference time-domain scheme in conjunction with the absorbing boundary conditions developed illustrate the effectiveness of the approximated local boundary conditions derived.

Mur, G. (1981): "Absorbing boundary conditions for the finite-difference approximation of the time-domain electromagnetic-field equation." *IEEE Transactions on Electromagnetic Compatibility,* vol. **EMC-23**, pp. 377–82, 1981.

Finite-difference time-domain approximations of first- and second-order absorbing boundary conditions of Engquist-Majda type for two-dimensional and three-dimensional applications are presented. A simple two-dimensional antenna problem is analyzed to test the validity and accuracy of such boundary conditions. As expected, the results show that the second-order absorbing boundary condition provide better accuracy than the first-order one. Furthermore, efficiency and accuracy of such boundary conditions improve when they are imposed at increasing distances from the source.

Umashankar, K. R. and A. Taflove (1982): "A novel method to analyze electromagnetic scattering of complex objects." *IEEE Transactions on Electromagnetic Compatibility,* vol. **EMC-24**, pp. 392–405, 1982.

The electromagnetic scattering by arbitrarily shaped, extremely complex metal or dielectric objects is dealt with by means of a finite-difference time-domain method. The first step of the proposed technique consists of computing near total fields within a finite volume delimited by a fictitious rectangular surface that fully encloses the object. This is accomplished by employing the FD-TD method and second-order absorbing boundary conditions to truncate the geometrical domain for computation. The absorbing boundary conditions used coincide with those developed by Mur (1981). In the second step of the procedure, the equivalence principle is invoked at the fictitious rectangular boundary to transform the tangential near scattered fields to the far field.

Mason, J. L. and W. J. Anderson (1985): "Finite element solution for electromagnetic scattering from two-dimensional bodies." *International Journal for Numerical Methods in Engineering,* vol. **21**, pp. 909–28, 1985.

Scattering from bodies in free space is formulated as a boundary value problem. The finite element mesh extends into the far-field region, where a boundary condition is imposed using the asymptotic expression for the scattered field. Spurious (non-physical) standing waves appear in the results because the outer boundary condition does not adequately fulfill the radiation condition.

Bevensee, R. M. and S. T. Pennock (1987): "Radiation (absorbing) boundary conditions for electromagnetic fields." *AP-S 1987 International Symposium Digest: Antennas and Propagation,* vol. **1**, pp. 268–71, 1987.

To avoid spurious numerical reflection, the Panofsky-Phillips relations are used to extrapolate the vector field components parallel and perpendicular to a radial line drawn outward from the source distribution.

Kriegsmann, G. A., A. Taflove, and K. R. Umashankar (1987): "A new formulation of electromagnetic wave scattering using an on-surface radiation boundary condition approach." *IEEE Transactions on Antennas and Propagation,* vol. **AP-35**, pp. 153–61, 1987.

By applying a far-field expansion of the exact radiation condition directly on the surface of a convex, two-dimensional conductive body, the frequency-domain integral equation for the scattered field is reduced to just a line integral. For the transverse magnetic (TM_z) case, the integrand is known explicitly. For the transverse electric (TE_z) case, the integrand can be easily constructed by solving an ordinary differential equation around the scatterer surface contour.

Numerical results show that the method may present a useful alternative to integral equation and uniform high-frequency approaches for convex cylinders. The on-surface radiation boundary condition yields better accuracy than the physical optics approaches, for it models the presence of induced currents also in the shadow region. However, it is not able to account for the singular behavior of the currents at edges or corners.

Lin, C. C. and K. K. Mei (1988): "Time domain absorbing boundary condition in layered media." *International Journal of Numerical Modelling: Electronic Networks, Devices and Fields,* vol. **1**, pp. 115–28, 1988.

An absorbing boundary condition for simple layered media is obtained by a projection operator method that takes into account the presence of reflected waves at the medium interfaces. This projection operator is derived for one, two, and three dimensions. Examples of electromagnetic radiation and scattering in two dimensions are presented to illustrate the numerical implementation.

Moore, T. G., J. G. Blaschak, A. Taflove, and G. A. Kriegsmann (1988): "Theory and application of radiation boundary operators." *IEEE Transactions on Antennas and Propagation*, vol. **AP-36**, pp. 1797–812, 1988.

The paper provides a survey of the theory of radiation boundary operators covering the underlying concepts of both the mode-annihilating operators and one-way equations. Close attention is paid also to the applications. In particular, numerical experiments concerning the implementation of the one-way equations in the framework of the finite-difference time-domain solution scheme of two-dimensional electromagnetic scattering problems are presented. The applications of the mode-annihilating operators are carried out using the on-surface radiation boundary condition formulation. Numerical results attain with electromagnetic scattering from two-dimensional convex targets and reentrant structures. A three-dimensional acoustic problem is also analyzed.

Peterson, A. F. (1988): "Absorbing boundary conditions for the vector wave equation." *Microwave and Optical Technology Letters*, vol. **1**, pp. 62–64, 1988.

This is one of the earliest works providing a class of approximate absorbing boundary conditions for the vector wave equation. A class of increasing-order absorbing boundary operators is derived by applying a procedure similar to that suggested by Bayliss and Turkel (1980) to the Wilcox expansion of the scattered field. The weighted residual statement of the problem uses the total field as the system unknown. Implementation details of such boundary operators in the framework of such formulation are also given. The non-symmetric structure of the finite element matrix arising from this formulation constitutes the flaw of such class of boundary operators.

Petre, P. and L. Zombory (1988): "Infinite elements and base functions for rotationally symmetric electromagnetic waves." *IEEE Transactions on Antennas and Propagation,* vol. **36**, pp. 1490–91, 1988.

Infinite elements are described that are compatible with triangular or quadrilateral finite elements and satisfy the Sommerfeld radiation condition. Their application is illustrated by the radiation of a coaxial cable into a half-space.

D'Angelo, J. and I. D. Mayergoyz (1989): "On the use of local absorbing boundary conditions for RF scattering problems." *IEEE Transactions on Magnetics*, vol. **25**, pp. 3040–42, 1989.

The performances of second-order Engquist-Majda and Bayliss-Turkel absorbing boundary conditions are compared for different kinds of fictitious boundary geometry and distances from the scatterer. Numerical results show that the Bayliss-Turkel boundary condition is more accurate than the Engquist-Majda type for a given distance of the fictitious contour from the scatterer.

Jin, J. M., J. L. Volakis, and V. V. Liepa (1989): "A comparative study of the OSRC approach in electromagnetic scattering." *IEEE Transactions on Antennas and Propagation*, vol. **37**, pp. 118–24, 1989.

The on-surface radiation condition is applied to a variety of two-dimensional imperfectly conducting bodies, and the accuracy of the results obtained is compared with that provided by the physical optics (PO) approach and by the moment method (whose results serve as benchmark in this study). It is found that, for scatterers formed by smooth curved surfaces, the OSRC approach in general predicts more accurate results than the physical optics approximation, especially in the forward scatter region. On the other hand, the OSRC approach is not appreciably better when considering polygonal cylinders since, like the physical optics approximation, it cannot describe the correct current behavior near edges.

Mittra, R., O. Ramahi, A. Khebir, R. Gordon, and A. Kouki (1989): "A review of absorbing boundary conditions for two and three-dimensional electromagnetic scattering problems." *IEEE Transactions on Magnetics*, vol. **25**, pp. 3034–39, 1989.

This is a brief review of the possible mathematical techniques for deriving two-dimensional and three-dimensional absorbing boundary conditions. For the two-dimensional problem, the exact and approximate value of the outward normal derivative of the scattered field outside a fictitious circular contour are compared to investigate the error. On the basis of this investigation, a technique is proposed to improve the performance of two-dimensional absorbing boundary conditions, which strive to provide a better absorption of evanescent or near-evanescent waves. The question of using conformal (as opposite to cylindrical) outer boundary and derive absorbing boundary conditions for this geometry is also addressed.

Pearson, L. W., R. A. Whitaker, and L. J. Bahrmasel (1989): "An exact radiation boundary condition for the finite-element solution of electromagnetic scattering on an open domain." *IEEE Transactions on Magnetics*, vol. **25**, pp. 3046–48, 1989.

An exact radiation boundary condition is presented that embodies the Sommerfeld radiation condition. The price paid for exactness is that the boundary operator is global on the exterior boundary, so matrix sparsity suffers.

Peterson, A. F. (1989): "A comparison of integral, differential and hybrid methods for TE-wave scattering from inhomogeneous dielectric cylinders." *Journal of Electromagnetic Waves and Applications*, vol. **3**, pp. 87–106, 1989.

The work provides an objective comparative evaluation and computational trade-off between integral and differential equation formulations for electromagnetic scattering from inhomogeneous dielectric bodies. Three different frequency-domain formulations are considered: a volume integral equation, a volume differential equation combined with an approximate near-field absorbing boundary condition, and a hybrid method using both a surface integral and volume differential equation. Numerical results show that all three methods produce similar numerical accuracy for a given scatterer model. However, formulations based on volume discretization of integral equations involve computational and storage requirements far in excess of those needed for formulations incorporating differential equations. The hybrid approach generally requires less storage than the volume differential equation combined with an approximate near-field absorbing boundary condition, but may suffer from uniqueness problems for electrically large scatterers. The latter formulation can produce accurate results even if the absorbing boundary conditions are enforced at a fictitious surface very close to the scatterer surface, but seems to lose accuracy for a large scatterer.

Peterson, A. F. and S. P. Castillo (1989): "A frequency-domain differential equation formulation for electromagnetic scattering from inhomogeneous cylinders." *IEEE Transactions on Antennas and Propagation*, vol. **37**, pp. 601–7, 1989.

This work presents one of the first applications of the absorbing boundary conditions in the framework of a finite element formulation for the solution of electromagnetic scattering problems. The weighted residual statement of the problem uses the total field as the system unknown, and second-order absorbing boundary conditions of the Bayliss-Turkel type are applied at a circular boundary Γ to truncate the geometrical domain of computation. Provided that the Bayliss-Turkel condition is applied at a sufficient distance from the scatterer, the accuracy appears comparable to that of integral equation formulations for a given mesh density. Computational trade-offs between the volume integral equation and differential equation approaches are discussed considering the two following major factors: the size of the region that must be discretized in each case and the degree to which the sparsity arising within the differential equation approach is exploited by the matrix solution algorithm.

Ramahi, O. and R. Mittra (1989): "Finite-element analysis of dielectric scatterers using the absorbing boundary conditions." *IEEE Transactions on Magnetics*, vol. **25**, pp. 3043–45, 1989.

Second-order absorbing boundary conditions of the Bayliss-Turkel type are applied at a circular boundary Γ. It is found that the finite element–absorbing boundary condition formulation which uses the scattered field as primary unknown yields more accurate results than the finite element formulation with total field as the primary unknown.

Webb, J. P. and V. N. Kanellopoulos (1989): "Absorbing boundary conditions for the finite element solution of the vector wave equation." *Microwave and Optical Technology Letters*, vol. **2**, pp. 370–72, 1989.

A family of increasing-order absorbing boundary operators on a spherical surface for the vector wave equation is proposed. In particular, the second-order operator contains, as a special case, the vector operator previously developed by Peterson (1988). The expression of a functional having the second-order operator as a natural boundary condition is provided. This is the major strength of such a boundary operator, for it leads to symmetric finite element matrices, contrary to the Peterson second-order operator.

Arendt, S., K. R. Umashankar, A. Taflove, and G. Kriegsmann (1990): "Extension of on-surface radiation condition theory to scattering by two-dimensional homogeneous dielectric objects." *IEEE Transactions on Antennas and Propagation,* vol. **38**, pp. 1551–58, 1990.

By applying the outgoing radiation boundary condition on the surface of the convex homogeneous dielectric objects, the fields inside and outside the dielectric cylinders are decoupled, and a differential equation relationship is obtained between the external incident field excitation and the corresponding field distribution in the interior of a dielectric object. Either an analytical approach or a suitable numerical method may be used to solve the differential equation, thus obtaining the interior field.

Aydin, G. and A. Yildiz (1990): "The finite element solution of electromagnetic boundary value problems for open region with high order boundary radiation condition." *Journées Internationales de Nice sur les Antennes,* pp. 56–59, 1990.

Eight-noded isoparametric finite elements are combined with a local Bayliss-Turkel boundary condition for solving half-plane and wedge diffraction problems. Fine meshes are used near sharp corners.

Canning, F. X. (1990): "On the application of some radiation boundary conditions." *IEEE Transactions on Antennas and Propagation,* vol. **38**, pp. 740–45, 1990.

The Bayliss-Turkel first- and second-order boundary conditions, as well as the Sommerfeld and Mur (1981) boundary conditions, are examined to identify their underlying physical principle and inherent accuracy. The main purpose of this study is to aid in an appropriate choice of radiation boundary condition, as well as to calculate the actual errors in several of these conditions. It is found that the Bayliss-Turkel first- and second-order boundary conditions may be written in coordinate independent forms in which the adjustable parameter r corresponds to an estimate of the distance of a line source which approximately generates the field. Choosing too small a value for r could cause a large error, but choosing $r = \infty$ (which results in the Sommerfeld or Mur boundary conditions when substituted in the first-order or second-order Bayliss-Turkel boundary conditions, respectively), causes an error not large compared to that due to large angles of propagation with respect to the normal to the boundary. This work analytically proves, as already observed by other researchers, the decreasing accuracy provided by the absorbing boundary conditions when applied to scatterers of increasing dimensions.

D'Angelo, J. and I. D. Mayergoyz (1990): "Finite element methods for the solution of RF radiation and scattering problems." *Electromagnetics*, vol. **10**, pp. 177–99, 1990.

Two-dimensional and three-dimensional finite element–absorbing boundary condition formulations and computational results for electromagnetic scattering and radiation are presented. The two-dimensional formulations use standard finite elements with either the Engquist-Majda or the Bayliss-Turkel second-order absorbing boundary conditions. Their performances, when they are applied at rectangular or circular boundaries, are compared against the results given by a hybrid finite element–boundary integral formulation. The three-dimensional formulation employs vector basis functions and scalar second-order Engquist-Majda type absorbing boundary conditions enforced at a rectangular surface. Using Bayliss-Turkel absorbing boundary conditions at a spherical boundary is found impractical, since it generally produces a large number of unknowns. Problems related to the numerical solution of the system of linear equations arising from the different formulations are extensively covered.

Lee, C. F., R. T. Shin, J. A. Kong, and B. J. McCartin (1990): "Absorbing boundary conditions on circular and elliptical boundaries." *Journal of Electromagnetic Waves and Applications*, vol. **4**, pp. 945–62, 1990.

The pseudo-differential operator approach employed by Engquist and Majda is modified to derive improved absorbing boundary conditions. It is also shown that in the case of a circular fictitious boundary, the modified pseudo-differential approach leads to a condition equivalent to the second-order Bayliss-Turkel absorbing boundary conditions. The same approach is used to derive second-order absorbing boundary conditions for elliptical outer boundaries, which can be used to achieve a reduction in the size of the geometrical domain for the case of elongated scatterers. The validity and accuracy of such absorbing boundary conditions is tested by calculating bistatic and monostatic radar cross sections from a conducting ellipse, a strip, and another elongated object.

Peterson, A. F. (1990): "The 'interior resonance' problem associated with surface integral equations of electromagnetics: Numerical consequences and a survey of remedies." *Electromagnetics*, vol. **10**, pp. 293–312, 1990.

This paper illustrates the interior resonance problem and some of the proposed remedies, using the eigenvalue behavior of the equations when applied to two- dimensional circular cylinders. Although this approach lacks in generality, it lends significant insight into the problem and several of the remedies. It is found that the eigenvalue behavior alone is insufficient to explain the observed numerical results. In fact, in many cases, the incorrectness of the numerical solution is apparently not due to matrix ill-conditioning. It is primarily a result of the behavior of the excitation in the vicinity of a resonance. In fact, the tangential components of a single incident field at the scatterer surface lack sufficient information to uniquely determine the desired surface currents at a resonance frequency.

Webb, J. P. (1990): "Absorbing boundary conditions for the finite element analysis of planar devices." *IEEE Transactions on Microwave Theory and Techniques*, vol. **38**, pp. 1328–32, 1990.

The second-order Bayliss-Turkel absorbing boundary conditions are applied at a circular fictitious boundary enclosing the discontinuity of the two-dimensional planar device analyzed. The imposition of this kind of absorbing boundary condition gives rise to two additional terms in the usual functional for the scalar Helmholtz equation. New universal matrices are introduced to allow the additional terms to be easily assembled for polynomial orders one through four.

Wu, K. L., G. Y. Delisle, and D. G. Fang (1990): "EM scattering of an arbitrary multiple dielectric coated conducting cylinder by coupled finite boundary element method." *IEE Proceedings–H*, vol. **137**, pp. 1–4, 1990.

The field in the dielectric coating is formulated in terms of classical finite elements, the radiation condition is included by means of a boundary element treatment. Numerical results indicate that the method produces unique resonant frequencies.

Bahrmasel, L. J. and R. A. Whitaker (1991): "Convergence of the finite element method as applied to electromagnetic scattering problems in the presence of inhomogeneous media." *IEEE Transactions on Magnetics*, vol. **27**, pp. 3845–47, 1991.

Approximate radiation boundary conditions fit well with finite element or finite difference methods, especially for geometrically complex inhomogenous targets. They allow an exterior boundary value problem to be approximated by an interior problem. Second-order Bayliss-Turkel boundary conditions are discussed for two-dimensional inhomogeneous targets, and experimentally observed convergence rates are compared with predicted rates.

Chung, S. J. and C. C. Hsiung (1991): "A partial variational analysis of planar dielectric antennas." *IEEE Transactions on Antennas and Propagation*, vol. **39**, pp. 713–18, 1991.

A variational equation based on the partial variational principle is solved by the finite element method. The radiation and boundary conditions are incorporated by combining modal expansion with the Green's function approach for exterior field representation. Reflection coefficients, radiation patterns, and the directive and power gains of several antennas are given.

Cooray, F. R. and G. I. Costache (1991): "An overview of absorbing boundary conditions." *Journal of Electromagnetic Wave and Applications*, vol. **5**, pp. 1041–54, 1991.

The expressions of two-dimensional and three-dimensional, scalar, and vector absorbing boundary conditions developed to date are given, together with a brief description of the features of each of them.

D'Angelo, J. and I. D. Mayergoyz (1991): "Three dimensional RF scattering by the finite element method." *IEEE Transactions on Magnetics*, vol. **27**, pp. 3827–32, 1991.

Frequency-domain finite element formulations are coupled with absorbing boundary conditions using edge elements. An important paper at the time it was presented, it is of less long-term value as a survey.

Janaswamy, R. (1991): "On the applicability of OSRC method to homogeneous scatterers." *IEEE Transactions on Antennas and Propagation*, vol. **39**, pp. 862–67, 1991.

The performance of the second-order radiation boundary condition in two-dimensional scattering by homogeneous dielectric cylinders is evaluated. It is shown that the on-surface radiation condition formulation does not necessarily yield accurate results for all values of material constants. The cause of this erroneous results is directly related to the error in the reactive portion of the complex scattered power. Numerical studies with circular cylinders indicate that the outer boundary must be spaced at least a few tenths of a wavelength away from the surface in order to obtain uniformly acceptable results for all values of material constants.

Jin, J. M., J. L. Volakis, and J. D. Collins (1991): "A finite element–boundary integral method for scattering and radiation by two- and three-dimensional structures." *IEEE Transactions on Antennas and Propagation Magazine*, vol. **33**, pp. 22–32, 1991.

This work presents a review of a hybrid finite element–boundary integral formulation for scattering and radiation by two- and three-dimensional composite structures. First, the general formulation of the problem without reference to any specific geometry is presented. A number of two- and three-dimensional applications are then considered to demonstrate the accuracy, efficiency, and capability of the method. Of particular concern in these applications is the choice of the fictitious boundary enclosing the structure so that the resulting boundary integrals are convolutions. This feature allows such integrals to be evaluated via the fast Fourier transform, in conjunction with an iterative-solution approach, so as to maintain an $O(N)$ memory demand without a compromise in accuracy.

Kanellopoulos, V. N. and J. P. Webb (1991): "A numerical study of vector absorbing boundary conditions for the finite element solution of Maxwell's equations." *IEEE Microwave and Guided Wave Letters*, vol. **1**, pp. 325–27, 1991.

The accuracy of the Webb and Kanellopoulos (1989) vector absorbing boundary conditions is tested by considering canonical problem whose exact solution is known. A finite element formulation using curvilinear brick finite elements is implemented. The superiority of the second-order differential operator over the first-order one, in both accuracy and speed of convergence as the surface S is placed at increasing distances from the scatterer, is demonstrated.

Ma, Y. C. (1991): "A note on the radiation boundary conditions for the Helmholtz equation." *IEEE Transactions on Antennas and Propagation*, vol. **39**, pp. 1526–30, 1991.

The radiation boundary conditions in two dimensions are derived in the body-fitted coordinate system using the method of successive approximation for the wave envelope function. The derivation is valid in case of convex scatterers with radius of curvature which is a continuous function of the arc length along the contour of the scatterer. Higher-order radiation boundary conditions require the radius of curvature to be a smooth function of the arc length to a certain degree. In the special case when the fictitious boundary is circular, the boundary operators coincide with the Bayliss-Turkel ones. The application of the boundary conditions derived to the on-surface radiation condition is also examined.

Povinelli, M. J. and J. D'Angelo (1991): "Finite element analysis of large wavelength antenna radome problems for leading edge and radar phased array." *IEEE Transactions on Magnetics*, vol. **27**, pp. 4299–302, 1991.

The effect of both position and composition of radomes on the radiating performance of large-wavelength arrays considered as two-dimensional is evaluated by means of a frequency-domain finite element–absorbing boundary condition method. Bayliss-Turkel second-order absorbing boundary conditions are applied at a circular or rectangular fictitious boundary, depending on the geometry of the radiating system considered. This work tends to show the potentiality of the technique itself rather than explain in detail the procedure of implementation.

Ramahi, O., A. Khebir, and R. Mittra (1991): "Numerically derived absorbing boundary conditions for the solution of open region scattering problems." *IEEE Transactions on Antennas and Propagation*, vol. **AP-39**, pp. 350–53, 1991.

The scattered field is represented, much like in the unimoment method, in terms of outgoing cylindrical wave functions, while its outward normal derivative at a node lying on the fictitious arbitrary contour Γ is expressed as a linear combination of the value of the scattered field itself at adjacent nodes. The coefficients of such linear combination are determined by seeking those coefficients that will give the most accurate approximation of the normal derivative of the scattered field over the entire range of harmonics that significantly contribute to its series representation. In other words, an absorbing boundary condition is numerically derived by approximating the analytical behavior of the exact boundary condition. The operator obtained provides a greater accuracy than analytically derived boundary operators having the same order. Moreover, it can be applied at outer boundaries that conform to the body of the scatterer. Two major shortcomings affect the technique used to derive the operator. First, the series representation for the scattered field is uniformly convergent only outside a cylindrical region that completely encloses the scatterer. Second, a certain degree of indeterminateness attaining with the possible choices in the terms of the series expansion used to derive the numerical operator is present and must be removed by trial and error.

Sumbar, E., F. E. Vermeulen, and F. S. Chute (1991): "Implementation of radiation boundary conditions in the finite element analysis of electromagnetic wave propagation." *IEEE Transactions on Microwave Theory and Techniques*, vol. **39**, pp. 267–73, 1991.

An application of the finite element method to the analysis of azimuthally symmetric two-dimensional electromagnetic radiation problems is presented. The domain of interest is terminated by imposing the Sommerfeld radiation condition $\mathbf{H} = 1/\zeta \hat{\mathbf{r}} \times \mathbf{E}$ directly on the fictitious boundary Γ. Obviously, this approach is strictly valid only far away from the antenna and if the medium is homogeneous. However, if the medium surrounding the antenna is lossy, the amplitude of the wave scattered by the fictitious boundary will be attenuated, and as a result, the accuracy of the near-antenna field solution will be less sensitive to the shape and location of the fictitious contour. This kind of boundary condition has the advantage of maintaining the sparsity and symmetry of the finite element matrix, and it is very simple to implement. Numerical results approximate the electrical characteristics of bare and coated monopole antennas quite well.

Umashankar, K., W. Chun, and A. Taflove (1991): "Simple analytical solution to electromagnetic scattering by two-dimensional conducting object with edges and corners: Part I: TM polarization." *IEEE Transactions on Antennas and Propagation*, vol. **39**, pp. 1665–71, 1991.

A virtual boundary enclosing the scatterer cross-section is introduced, such that it is always parallel to the contour of the conductive cylinder's cross-section and is placed at a small distance from it. On such a virtual contour, the on-surface radiation boundary condition formulation is applied. This results in an integral expression for the scattered electric field which is simplified to the sum of two specific terms. The first consists of the contributions from the smooth segment of the scatterer contour, while the second consists of the contributions from the corners of the scatterer. The latter contributions can be viewed as due to localized electric current distributions which exist at the corners of the scatterer in the form of isolated line sources.

Chatterjee, A., J. M. Jin, and J. L. Volakis (1992): "Robust finite element solution for three-dimensional scattering." *Electronics Letters*, vol. **28**, pp. 966–67, 1992.

Edge-based finite elements with vector absorbing boundary conditions are used to solve for scattering by composite structures, with boundaries satisfying impedance conditions.

Jin, J. M., J. L. Volakis, and V. V. Liepa (1992): "Fictitious absorber for truncating finite element meshes in scattering." *IEE Proceedings–H*, vol. **139**, pp. 472–76, 1992.

A fictitious absorber comprising a metallic surface loaded by three layers of lossy material is introduced to truncate the finite element meshes. The thickness of the dielectric layers is fixed, while their dielectric and magnetic constants ϵ_r and μ_r are chosen so as to minimize the reflection of the scattered field from the artificial boundary, thus simulating a free-space environment. For the geometries considered, the solution based on the fictitious absorber truncation method has been found comparable in accuracy and efficiency to that employing the second-order absorbing boundary conditions. The attractive features of this technique reside in its capability of maintaining the sparsity and symmetry of the finite element method matrix and its capability of enabling existing commercial software, usually developed for bounded region problems, to treat open-region problems without any modification. Its major defect is due to the difficulties encountered in extending this procedure to the three-dimensional case.

Pearson, L. W., A. F. Peterson, L. J. Bahrmasel, and R. A. Whitaker (1992): "Inward-looking and outward-looking formulations for scattering from penetrable objects." *IEEE Transactions on Antennas and Propagation*, vol. **40**, pp. 714–20, 1992.

The inward-looking and outward-looking formulations concept, introduced by Peterson (1991) in a former paper, is further developed to justify the arising of the interior resonance problem in some finite element methods to treat unbounded domain problems. The inward-looking and outward-looking classification is based on the application of the equivalence principle of electromagnetics to separate the interior and exterior problem. Then, depending on the region where the solution is deduced first, the finite element methods for open-domain problems are classified as either inward-looking or outward-looking. Although inward-looking formulations may exhibit computational advantages, outward-looking formulations appear preferable for electrically large domains because of internal resonance difficulties that may arise when solving the isolated interior problem first.

Peterson, A. F. (1992): "Accuracy of 3-D radiation boundary conditions for use with the vector Helmholtz equation." *IEEE Transactions on Antennas and Propagation*, vol. **40**, pp. 351–55, 1992.

A comparative evaluation of the two kinds of second-order vector absorbing boundary conditions — those by Peterson (1988) and those by Webb and Kanellopoulos (1989) — is carried out. Tests are performed by applying them on a spherical surface to outgoing vector spherical harmonics of different orders. It is found that the two kinds of absorbing boundary conditions provide exactly the same accuracy on the TE_r part of the field, while on the TM_r part of the field the absorbing boundary conditions by Peterson give rise to slightly smaller errors than those by Webb and Kanellopoulos. The accuracy of both types of boundary conditions decreases with increasing the order of the vector spherical harmonic to which they are applied.

Rappaport, C. M. and L. J. Bahrmasel (1992): "An absorbing boundary condition based on anechoic absorber for EM scattering computation." *Journal of Electromagnetic Waves and Applications,* vol. **6**, pp. 1621–33, 1992.

A novel absorbing boundary condition is based on anechoic chamber absorber foam geometry, with complex permittivity and permeability. This absorbing boundary prevents reflections from much wider incident angles than lattice termination conditions. Since waves need not be normally incident, the boundary can be placed very close.

Chatterjee, A., J. M. Jin, and J. L. Volakis (1993): "Edge-based finite elements and vector ABC's applied to 3-D scattering." *IEEE Transactions on Antennas and Propagation*, vol. **41**, pp. 221–26, 1993.

The Webb and Kanellopoulos (1989) vector absorbing boundary conditions are applied at a non-spherical fictitious boundary to solve three-dimensional scattering problems. A variational formulation employing the scattered field as unknown is implemented. Numerical results agree well with analytical ones or laboratory measurements even in the case of a non-spherical fictitious surface placed at a small distance from the scatterer edges. This formulation has the very attractive feature of maintaining the sparse and symmetric structure of the finite element method as applied to the bounded problem; hence, it is one of the first suitable for solving practical three-dimensional scattering problems.

Zienkiewicz, O. C., J. P. de S. R. Gago, and D. W. Kelly (1983): "The hierarchical concept in finite element analysis." *Computers and Structures,* vol. **16**, pp. 53–65, 1983.

Review paper on hierarchical finite elements, useful but rather dated; 34 bibliographic references on this subject may have seemed a lot in 1983, but the literature has grown since.

Konrad, A. (1985): "Electromagnetic devices and the application of computational techniques in their design." *IEEE Transactions on Magnetics,* vol. **MAG-21**, pp. 2382–87, 1985.

A review of the finite element method in the design of electromagnetic devices, mainly in historical order, with the main emphasis on differential equation formulations. Numerous illustrative examples of both high and low frequency applications and an extensive bibliography (106 items).

Koshiba, M., K. Hayata, and M. Suzuki (1987): "Finite-element method analysis of microwave and optical waveguides — Trends in countermeasures to spurious solutions." *Electronics and Communications in Japan,* Part 2, vol. **70**, no. 9, pp. 96–107, 1987. (Translated from *Densi Tsushin*

Gakkai Ronbunshi, vol. **69-C**, pp. 1477–86, 1986.)
Thorough literature review, mainly concentrating on the English-language literature. Classifies and catalogues the methods available. 82 references.

Emson, C. R. I. (1988): "Methods for the solution of open-boundary electromagnetic-field problems." *IEE Proceedings–A,* vol. **135**, pp. 151–58, 1988.
Methods for the solution of open-boundary field problems are reviewed, emphasizing algorithms suited to use in conjunction with the finite element method. General requirements are outlined and the main characteristics of currently available methods are described, including their implicit assumptions and limitations. 74 references.

Ho-Le, K. (1988): "Finite element mesh generation methods: A review and classification." *Computer Aided Design,* vol. **20**, pp. 27–38, 1988.
A broad overview and review. Methods are classified into seven major categories by their sequence of node and element creation. Published methods are then reviewed. Only four methods are fully automatic. These are compared for element type and shape, mesh density control, and time efficiency. 73 references.

Tseng, A. A. (1988): "Electroheating modeling in metal processing: A review." *Advances in Engineering Software,* vol. **10**, pp. 58–71, 1988.
Brief introduction to electroheat problems, followed by formulation of electromagnetic, heat transfer, and fluid dynamic phenomena. Numerical methods used for electromagnetic modeling, with extensive review of finite elements and commercially available software packages for electromagnetic modeling. The review seeks to cover a wide range of computational methods, but finite elements occupy a major position. 64 references.

Glisson, A. W. (1989): "Recent advances in frequency domain techniques for electromagnetic scattering problems." *IEEE Transactions on Magnetics,* vol. **25**, pp. 2867–71, 1989.
Broad review of numerical techniques for scattering problems, all in the frequency domain, formulated in terms of an integral equation or a boundary value problem. 54 references.

Rahman, B. M. A., F. A. Fernandez, and J. B. Davies (1991): "Review of finite element methods for microwave and optical waveguides." *Proceedings IEEE,* vol. **79**, pp. 1442–48, 1991.
Significance and use of the different variational formulations, modeling of infinite domains for open-boundary waveguides, and avoidance of spurious solutions. Brief treatment of edge elements and some comments of matrix algebra relevant to the waveguide problem. 93 references.

Peterson, A. F. (1991): "Analysis of heterogeneous electromagnetic scatterers: Research progress of the past decade." *Proceedings IEEE,* vol. **79**, pp. 1431–41, 1991.
Reviews integral equation and finite element methods (including time domain) with emphasis on modeling penetrable bodies. Direct solution of frequency-domain partial differential equations may have a computational advantage over the integral equations. Time-domain solution of partial differential equations may be more efficient for electrically large structures. The development of these modeling procedures is summarized and unresolved questions identified. 102 bibliographic references cover a very broad area.

Cooray, F. R. and G. I. Costache (1991): "An overview of the absorbing boundary conditions." *Journal of Electromagnetic Waves and Applications,* vol. **5**, pp. 1041–54, 1991.
Overview of the two and three dimensional absorbing boundary conditions useful for open-region electromagnetic scattering problems. Their different types are listed, with brief description of problems solved to date. The review is interesting, but incomplete in this recently expanding field. 25 references.

Wen, X. Z. (1991): *Engineering Electromagnetism: Functional Methods.* Chichester: Ellis Horwood, 1991.
A graduate-level book that discusses various methods, including variational, weighted residuals, moments, boundary integrals, boundary elements, and finite elements, attempting to create a unified approach. Of bibliographic value (568 references).

Bettess, P. and J. A. Bettess (1991): "Infinite elements for dynamic problems: Part 2." *Engineering Computations,* vol. **8**, pp. 125–51, 1991.
A broad review of the literature is accompanied by a detailed exposé of mapped (transformed) elements of value to radially propagating evanescent waves (as encountered in guided wave problems). The bibliography refers in the main to Helmholtz equation problems. 169 references.

Davies, J. B. (1993): "Finite element analysis of waveguides and cavities — A review." *IEEE Transactions on Magnetics,* vol. **29**, pp. 1578–83, 1993.

Treats modal analysis (finding a few dominant modes) of uniform waveguides of optics, also classical microwaves. Two recent methods for avoiding the spurious mode problem are described. Complex modes and quasi-modes of optical guides containing nonlinear material are treated. 36 references.

Tsukerman, I. A., A. Konrad, G. Bedrosian, and M. V. K. Chari (1993): "A survey of numerical methods for transient eddy current problems." *IEEE Transactions on Magnetics,* vol. **29**, pp. 1711–16, 1993.

Time-stepping schemes and associated algebraic system solvers for transient eddy current computations are assessed. For 2-D problems, nested dissection and the quotient minimum degree method are preferred to the incomplete Cholesky conjugate-gradient method, provided enough memory is available. Multilevel preconditioning is highly efficient for magnetically homogeneous problems. The efficiency of multilevel methods for inhomogeneous problems, attractive theoretically, needs further practical investigation. 33 references.

Author Index

Subject Index

NOTE: Bold page numbers indicate figures, tables, and illustrations

N

O

P

Q

R

S

T

U

V

W

Editors' Biographies

PETER P. SILVESTER graduated from the Carnegie Institute of Technology (now Carnegie-Mellon University) in 1956. After a short period of industrial practice, he continued his studies at the University of Toronto, obtaining the M.A.Sc. in 1958, then at McGill University where he was awarded the Ph.D. in Electrical Engineering in 1964. He initially joined the Department of Electrical Engineering at McGill as Lecturer and now serves it at the rank of Professor.

Professor Silvester has devoted a large part of his career to numerical analysis of electromagnetic fields, with applications to magnetics, microwaves, geomagnetics, antennas, and bioelectricity. He has contributed extensively to the finite element art, both by developing new techniques and by applying finite elements in new areas. His current research topics include high-precision differentiation methods, as well as applied problems in magnetic separation, guided wave propagation, and plasma electrochemistry.

Dr. Silvester maintains research ties with colleagues in several other institutions, notably the University of Cambridge and the University of Florence. He has spent time as a General Electric Visiting Research Fellow in the GE corporate research laboratories, has been an SRC Senior Fellow and a Visiting Fellow Commoner of Trinity College, in addition to various shorter-term visiting fellowships and lectureships. He has also acted as consultant to a number of major corporations and government agencies. In addition to numerous articles and papers, Professor Silvester's publications include several textbooks and monographs on electromagnetics, numerical methods, and software engineering.

Dr. Silvester is a member of the major professional organizations in his field, and a Registered Engineer in Quebec. He is a Fellow of the IEEE.

GIUSEPPE PELOSI was born in Pisa, Italy, on December 25, 1952. He received the Laurea (Doctor) degree in Physics (*summa cum laude*) from the University of Florence in 1976. He first collaborated with the Department of Physics of the University of Florence, and then joined the Department of Electrical Engineering of the same University in 1979, where he is currently Associate Professor of Microwave Engineering.

Dr. Pelosi has been mainly involved in research in the field of numerical and asymptotic techniques for applied electromagnetics. His research interests included first radar systems and solid state physics and, more recently, extensions and applications of the geometrical theory of diffraction as well as methods for radar cross-section analysis of complex targets. His current research activity is mainly devoted to the development of numerical procedures based on the finite element method, with particular emphasis on radiation and scattering problems.

Professor Pelosi has maintained scientific collaboration with several international research centers: the Electroscience Laboratory of the Ohio State University; the Department of Electrical Engineering of McGill University; Le Centre Thomson d'Application Radar (LCTAR), France; ESA-ESTEC, The Netherlands. Dr. Pelosi is a Senior Member of the IEEE and a member of the Italian Electrical and Electronic Association, of the Applied Computational Electromagnetics Society, and the History of Sciences Society.